PHARMACOGNOSY

59th EDITION

FOR B. PHARM., M. PHARM. AND PHARM D. DEGREE COURSES IN PHARMACY
COVERS SYLLABI REVISED BY PHARMACY COUNCIL OF INDIA IN 2020

(THOROUGHLY REVISED AND ENRICHED)

Dr. C. K. Kokate
M. Pharm., Ph. D., F.G.A.E.S. (Germany)
Former Vice Chancellor,
KLE University,
JNMC Campus, Nehru Nagar,
Belgaum 590010

A. P. Purohit
M. Pharm., A. I. C.
Former Associate Professor and Head,
Department of Pharmacognosy,
Bharati Vidyapeeth's
Poona College of Pharmacy and Research Centre,
Pune 411038

S. B. Gokhale
M. Pharm., A. I. C.
Former Co-ordinator,
R. C. Patel Institute of
Pharmaceutical Education and Research
Shirpur 425405
Former Principal SVKM's College of Diploma in Pharmacy
Vile-Parle, Mumbai 400016

NIRALI PRAKASHAN
ADVANCEMENT OF KNOWLEDGE

N1275-A

ISBN 978-81-96396-15-2

PHARMACOGNOSY
Fifty Ninth Edition : May 2024
Price ₹ 1100/- : US $ 110.00

Published By :
Nirali Prakashan
Abhyudaya Pragati, 1312 Shivaji Nagar,
Off J.m. Road, Pune - 411 005. India
Ph. (020) 25512336 / 37
Email: Niralipune@pragationline.com

Printed By :
Yogiraj Printers And Binders
Survey No. 10/1a, Ghule Industrial Estate,
Nanded Village Road, Tal-Haveli, Dist. Pune,
Pune 411041. India

DISTRIBUTION CENTERS

Nirali Prakashan : **Dhayari Pune (Outstation):** Sur. No. 28/27, Dhayari-Katraj Road, Near Asian College, Dhayari, **Pune** 411 041. • Tel: (+91-020) 2469 0204
Email: bookorder@pragationline.com

Nirali Prakashan : 119, Budhwar Peth, Jogeshwari Mandir Lane, **Pune** 411 002.
Tel: (+91-020) 2445 2044, 6602 2708 • Email: bookorder@pragationline.com

Nirali Prakashan : Rasdhara Co-op. Hsg. Soc. Ltd. 'D' Wing Ground Floor, 385, S.V.P. Road, Girgaum, **Mumbai** 400004 • Tel: (022) 23856339/23869976
Email : niralimumbai@pragationline.com; mumbai@niralibooks.com

DISTRIBUTION BRANCHES

DELHI
Nirali Prakashan
Omkar Tower, Ground Floor, Office No. 2,
4575/15, Agarwal Lane, Darya Ganj,
New Delhi 110002 (Delhi)
Mobile : 9555778814/9818561840
Email : delhi@niralibooks.com

BENGALURU
Nirali Prakashan
Maitri Ground Floor, Jaya Apartments,
No. 99, 6th Cross, 6th Main, Malleswaram,
Bengaluru 560003 (Karnataka)
Mobile : 9686821074
Email : bengaluru@niralibooks.com

NAGPUR
Nirali Prakashan
Above Maratha Mandir, Shop No. 3,
First Floor, Rani Jhanshi Square, Sitabuld
Nagpur 440012 (Maharashtra)
Tel : (0712) 254 7129
Email : nagpur@niralibooks.com

KOLHAPUR
Nirali Prakashan
438/2, Bhosale Plaza, Ground Floor,
Khasbag, Opp.Balgopal Talim,
Kolhapur 416 012 (Maharashtra)
Mobile : 9850046155
Email : kolhapur@niralibooks.com

JALGAON
Nirali Prakashan
34, V. V. Golani Market, Navi Peth,
Jalgaon 425001 (Maharashtra)
Tel : (0257) 222 0395
Mobile : 94234 91860
Email : jalgaon@niralibooks.com

NAGPUR
Nirali Prakashan
Above Maratha Mandir, Shop No. 3,
First Floor, Rani Jhanshi Square, Sitabuld
Nagpur 440012 (Maharashtra)
Tel : (0712) 254 7129
Email : nagpur@niralibooks.com

Email: niralipune@pragationline.com Website: www.pragationline.com

Also find us on www.facebook.com/niralibooks @nirali.prakashan

Mission

It is our mission to disseminate authentic, scientific, authoritative, updated knowledge and information about all aspects of Natural Drugs and Phytopharmaceuticals to Pharmacy and Health Care Educational Institutes, Herbal Drug Industries, Regulatory Agencies, Research and Development Units, Complementary and Alternative Medicine Practitioners and Community Pharmacists.

EDITIONS

Earlier Editions

1st	September 1990
5th	July 1994
10th	June 1998
15th	July 2000
20th	February 2002
25th	December 2003
30th	February 2005
40th	January 2008
50th	September 2014
55th	November 2018
58th	June 2022
59th	May 2024

Pharmacognosy

Preface to the Fifty Ninth Edition

After the release or earlier edition, now we are pleased to release 59th edition, within ten months only.

Special features of this edition are:

- A complete chapter on 'Poisonous plants and poisonous animals.
- Four appendices such as country wise availability of natural drugs to locale their availability world wide.
- Disorders and their remedies with natural drugs.
- Addition of 12 natural drugs such as : Allspice, Avocado, Chaulinoogra, Coconut water, croton oil, Bitter orange oil Dhayali, Hamamalis, Kalijira, Pipper betal, raspberry and witch hazel.

We are open to recieve constructive criticism and timely suggestions from students and colleagues.

We are greatful to publisher Shri. D. K. Furia, Jignesh Furia and the staff of Nirali Prakashan Pune.

After the warm welcome of revised 58th edition by colleagues and students, we are pleased to present 59th edition in very short span.

Addition of two new chapters (Biologicals & Plant Based Vaccines) alongwith three natural drugs and four photograph to contents of book are the special features of Fifty Eighth Edition.

Several miscellaneous additions like Weed-barriers, Biostimulants are also made.

26th May 2024
Sankashti-Chaturthi

Kokate
Purohit
Gokhale

Preface to the First Edition

The discipline of pharmacognosy has undergone revolutionary changes in last four decades. Our understanding of systematic approach to quality control, therapeutic utility and clinical efficacy of herbal drugs has laid firm foundation of modern pharmacognosy. The advent of newer analytical techniques has made in roads in understanding the chemical complexity of tissues and their biogenetic behaviors. Even though, herbs were in use, since time immemorial in treatment of human ailments, it was only in recent times, they have been scientifically processed and incorporated as ingredients in different pharmaceutical dosage forms. The revival of phytopharmacy as reflected by its utility in modern drug therapy, is indeed, a significant contribution of all those engaged in research pertaining to this branch of pharmaceutical sciences.

It is our endeavour to provide comprehensive text of modern pharmacognosy to the students of degree classes in pharmacy and personnel engaged in pharmaceutical and plant-based industries. The efforts have been made to describe recent trends in cultivation technology, chemistry of bioconstituents, biogenesis and therapeutic utility of crude drugs. The progress in the commerce of biomedicinals and their production has been reviewed. The chemical classification of crude drugs has been adopted with an emphasis on providing their realistic pharmacognostic profiles.

The chapters on medicinal plant biotechnology including plant tissue culture, analytical pharmacognosy and biologicals are the important features of the text. Our aim remains to provide up-to-date relevant scientific information of natural drugs useful in understanding modern concepts of Pharmacognosy.

We are thankful to publisher Shri. Dineshbhai K. Furia of Nirali Prakashan for his co-operation in bringing out this book.

Ganesh Chaturthi
September 1990

Kokate
Purohit
Gokhale

Pharmacognosy

Contents

Pharmacognosy

Contents

Contents

Contents

Pharmacognosy

Contents

For colour photographs Scan QR Code given here and on cover page.

- Coloured Photographs of Medicinal Plants
- Coloured Photographs of Animals
- Habitats of Medicinal Plants in India
- Habitats of Medicinal Plants in World

SCAN ME for download pdf of Colour Pages

Pharmacognosy

LIST OF COLOURED MEDICINAL PLANT PHOTOGRAPHS APPENDED

Sr. No.	Name	Botanical name
1.	Acacia	*Acacia arabica*
2.	Aconite	*Aconitum napellus*
3.	Acorus	*Acorus calamus*
4.	Ajowan	*Trychyspermum ammi*
5.	Aloe	*Aloe barbadensis*
6.	Amla	*Emblica officinalis*
7.	Ammi	*Ammi majas*
8.	Amra	*Mangifera indica*
9.	Annato	*Bixa orellana*
10.	Arjuna	*Terminalia arjuna*
11.	Arnica	*Arnica montana*
12.	Ashoka	*Saraca indica*
13.	Ashwagandha	*Withania somnifera*
14.	Bael	*Aegle marmelose*
15.	Balsum of Tolu	*Myroxylon balsamum*
16.	Bearberry	*Arctostaphyllus uvaursi*
17.	Behra	*Terminalia belerica*
18.	Belladonna	*Atropa belladonna*
19.	Benafsha	*Viola odorata*
20.	Benzoin	*Styrax paralleloneurus*
21.	Bhuiamla	*Phyllanthus neruri*
22.	Bitter orange	*Citrus aurantium*
23.	Black catechu	*Acacia catechu*
24.	Black pepper	*Piper nigrum*
25.	Brahmi	*Bacopa monnieri*
26.	Camptotheca	*Camptotheca acuminata*
27.	Cannabis	*Cannabis sativa*
28.	Capsicum	*Capsicum annum*
29.	Cardamom	*Elettaria cardamomum*

contd...

Sr. No.	Name	Botanical name
30.	Castor	*Ricinus communis*
31.	Celery	*Apium graveolens*
32.	Chirata	*Swirtia chirata*
33.	Cinchona	*Cinchona succirubra*
34.	Cinnamon	*Cinnamomum zeylanicum*
35.	Citronella	*Cymbopogan nardus*
36.	Clove	*Eugenia caryophyllus*
37.	Coffee-herb	*Coffee arabica*
38.	Coriander	*Coriandrum sativum*
39.	Cotton	*Gossypium sp.*
40.	Crocus	*Crocus-sativa*
41.	Daruhalad	*Berberis aristata*
42.	Datura	*Datura metel*
43.	Davana	*Artemisia pallens*
44.	Derris	*Derris elliptica*
45.	Digitalis	*Digitalis purpurea*
46.	Dioscorea	*Dioscorea deltoida*
47.	Duboisia	*Duboisia myoporoidus*
48.	Echinacea	*Echinacea purpurea*
49.	Ephedra	*Ephedra gerardiana*
50.	Ergot of Rye	*Claviceps purpurea*
51.	Eucalyptus	*Eucalyptus globulus*
52.	Evening primrose	*Onethera biennis*
53.	Fennel	*Foeniculum vulgare*
54.	Garlic	*Allium sativum*
55.	Geranium	*Pelargonium graveolens*
56.	Giloe	*Tinospora corylifolia*
57.	Ginger	*Zingiber officinale*
58.	Gingko	*Gingko biloba*
59.	Ginseng	*Panax ginseng*
60.	Gloriosa	*Gloriosa-superba*
61.	Gokhru	*Tribulus terristris*

contd...

Sr. No.	Name	Botanical name
62.	Guar-herb	*Cyamopsis tetragonobobus*
63.	Gudmar	*Gymnema sylvestre*
64.	Guggul	*Commiphora weightii*
65.	Haridra	*Curcuma-longa*
66.	Henna	*Lawsonia inermis*
67.	Hypericum	*Hyperium perforatum*
68.	Hyoscyamus (Henbane)	*Hyoscyamus niger*
69.	Indian gentian	*Gentian lutea*
70.	Indian kino	*Pterocarpns marsupium*
71.	Ipecacunha	*Cephaelis ipecacunha*
72.	Isapgol	*Plantogo ovata*
73.	Jatamansi	*Nardostachis jatamansi*
74.	Jojoba	*Simmondsia chinensis*
75.	Kaladana	*Ipomoea hederaceae*
76.	Kalmegh	*Andrographis peniculata*
77.	Karaya	*Sterculia-urens*
78.	Kokum	*Garcinia indica*
79.	Lavender	*Lavendula officinalis*
80.	Lemon	*Citrus lemonis*
81.	Lemon grass	*Cymbopogan citratus*
82.	Linseed	*Linum usitatissimum*
83.	Lobelia	*Lobelia nicotianefolia*
84.	Long pepper	*Piper longum*
85.	Manjistha	*Rubia cordifolia*
86.	Milk thistle	*Silybum marianum*
87.	Momordica	*Momordica charantia*
88.	Myrrh	*Commiphora molmol*
89.	Neem	*Azadiracta indica*
90.	Nutmeg	*Myrestica fragrans*
91.	Nux-vomica	*Strychnous nux vomica*

contd...

Sr. No.	Name	Botanical name
92.	Olive herb	*Olea-europea*
93.	Papita	*Carica papaya*
94.	Peanut	*Arachis hypogea*
95.	Peppermint	*Mentha-piperata*
96.	Pineapple	*Ananus comosus*
97.	Poppy (opium)	*Papaver somniferum*
98.	Punarnava herb	*Boerrhaavia diffusa*
99.	Rauwolfia	*Rauwolfia serpentina*
100.	Safed musali	*Chlorophyllum borivillium*
101.	Safflower	*Carthamus tinctorius*
102.	Senna	*Cassia angustifolia*
103.	Shatawari	*Asparagus recemosus*
104.	Stevia	*Stevia rebaundiana*
105.	Taxus	*Taxus baccata*
106.	Tea-leaves	*Thea sinensis*
107.	Tobacco	*Nicotiana tobaccum*
108.	Tragacanth	*Astragallus gummifer*
109.	Tulsi	*Ocimum sanctum*
110.	Vasaka	*Adhatoda vasica*
111.	Vinca	*Catharanthus roseus*
112.	Yeshti	*Glycyrrhia glabra*

LIST OF COLOURED PHOTOGRAPHS OF ANIMAL DRUGS APPENDED

Sr. No.	Name	Zoological name
1.	Catharide	*Cantharis vesicatoria*
2.	Castoreum	*Castor canadensi*
3.	Civet Cat	*Viverra zibetha*
4.	Cobra	*Naja naja*
5.	Cochineal	*Dactylopius coccus*
6.	Cod Fish	*Gadus morrhea*
7.	Hog (Pig)	*Sus scrofa*
8.	Honey Bees	*Apis mellifera*
9.	Lac Insect	*Laccifer lacca*
10.	Leech	*Hirudo medicinalis*
11.	Musk Deer	*Moschus moschiferous*
12.	Mylabris	*Mylabris phalerita*
13.	Shark Fish	*Galeocerdo tigrinus*
14.	Sheep (Merino)	*Avis aries*
15.	Silk worm	*Bombyx mori*
16.	Sperm Whale Fish	*Physeter microcephalus*

Chapter 1...

GENERAL INTRODUCTION

[A] HISTORY, DEFINITION AND SCOPE OF PHARMACOGNOSY

NATURE always stands as a golden mark exemplify the outstanding phenomenon of symbiosis. The biotic and abiotic elements of nature are all interdependent. Plants are indispensable to man. The three important necessities of life namely food, clothing, shelter and most of other useful products are supplied to him by the plant kingdom. Nature has provided a complete storehouse of remedies to cure all ailments of mankind. The knowledge of drugs has accumulated over thousands of years as a result of man's inquisitive nature so that today we possess many effective means of ensuring healthcare.

The human being appears to be afflicted with more diseases than any other animal species. There can be little doubt then that he, very early, sought to alleviate his sufferings from injury and disease by taking advantage of plants growing around him. In the past, almost all the medicines used were from the plants, the plant being man's only chemist for ages. Today, a vast store of knowledge concerning therapeutic properties of different plants has accumulated. All phylla of plants viz. Thallophyta, Bryophyta, Pteridophyta and Spermatophyta (of which conservative estimates place the total number of known species at approximately 3,35,000) contain species that yield official and unofficial products of medicinal importance. By far, the greatest number of these are derived from plants and include three hundred or more recognised families of Spermatophyta.

The history of herbal medicines is as old as human civilization. The documents, many of which are of great antiquity, revealed that plants were used medicinally in China, India, Egypt and Greece long before the beginning of the Christian era. One of the most famous surviving remnants is ***Papyrus Ebers***, a scroll some 60 feet long and a foot wide, dating back to the sixteenth century before Christ. The text of document is dominated by more than 800 formulae and 700 different drugs. The drugs such as acacia, castor oil and fennel are mentioned alongwith apparent references to such compounds as iron oxide, sodium chloride, sodium carbonate and sulphur. Most of the medicinally active substances identified in the nineteenth and the twentieth centuries were used in the form of crude extract. In China, many medicinal plants had been in use since 5000 B.C. The oldest known herbal is *Pen-t'sao* written by emperor **Shen Nung** around 3000 B.C. It contains 365 drugs, one for each day of the year. Indians also, worked meticulously to examine and classify the herbs which they came across, into groups called Gunas. **Charaka** made fifty groups of ten herbs each of which, according to him, would suffice an ordinary physician's need. Similarly, **Sushrutha** arranged 760 herbs in 7 distinct sets based on some of their common properties. A large portion of the Indian population even today depends on the Indian System of Medicine - **Ayurveda**, **'An ancient science of life'.** The well known treatises in Ayurveda are ***Charaka Samhita*** and ***Sushrutha Samhita***.

Dhanvantari: It is believed that Dhanvantari emerged in the process of Samudra-Manthan (Churning of Sea).

Dhanvantari is the divine founder of Ayruveda (also known as father of Ayurveda) was physician of Gods. Lord Dhanvantari was born in Royal household of Kashi (Banaras) and became king, latter on.

Dhanvantari tought Ayurveda orally to one of his disciples **Sushruta**. Even at present birthday of Lord Dhanvantari is celebrated on **Dhanteras** two days prior to Diwali everywhere in India.

Fig. 1.1: The Great Contributors of profession of pharmacy

Following the contribution to medicine by **Hippocrates** *" Father of medicine"*, (460 - 360 B.C.), **Aristotle** (384 - 322 BC), **Dioscorides** (40 - 80 A D.) **Galen** (131 - 200 A. D.) and the early Arabian physicians, there was a period of approximately 1000 years during which little, if any, progress was made in the medical sciences. The first pharmacist, Galen, was known to have had a number of pain-relieving materials, including opium in his apothecary. For the next few hundred years, the formulations of the medicaments in use changed very little, although some bold attempts were made by chemical entrepreneurs such as **Paracelsus** (1493 -1541) to develop mineral salts which might have had the potential of being universal curative agents. However, throughout this period, the use of herbal extracts for medicinal purposes never lost its place.

(f) Dioscordies (040 - 080 AD)

(g) Galen (131 - 200 AD)

(h) Swede Linnaeus (1707 - 1778 AD)

(i) George Bentham (1800 - 1884)

(j) G.D. Hooker (1817 - 1911)

(k) Karl A.E. Prandtl (1849 - 1893)

Fig. 1.2 : The Great Contributors of profession of pharmacy

The importance of the extraction method and alcohol as an extractant was reported by **Le'mery** (1645-1715). **William Withering** in 1785 published an account of some of the medicinal properties of foxglove leaves based on ten years of experimentation. Percolation process was used for the crude drugs. Calumba, an alkaloidal drug, became official in 1788. In 1803, the French pharmacist, **Derosne** isolated narcotine from opium. In 1806, **Sertuerner** isolated morphine from opium and its role in alleviating pain was recognised. In the next few years, strychnine (1817), emetine (1817), brucine (1819), piperine (1819), quinine (1820) and colchicine (1820) were isolated. The French Pharmacist, **Pelletier** first reported the isolation of strychnine from ignatius beans and later from *nux vomica* seeds. A new extraction process for alkaloid was developed by **Stass** and **Otto** in 1852. Other important discoveries during this period were the isolation of nicotine from tobacco leaves (**Posselt** and **Reimann**, (1828), cocaine (**Neumann**, 1860), ouabain (**Hardy** and **Gallows**, 1877), pilocarpine (**Gerrard** and **Hardy**, 1875), ephedrine (**Nagai**, 1887) and podophyllotoxin (**Kuersten**, 1891). Isolation of ergometrine, digoxin, reserpine, theophylline and quinidine are the significant discoveries of the twentieth century.

Fig. 1.3: Seydler C.A.

In the nineteenth century, the term *'Materia Medica'* was used for the subject now known as "Pharmacognosy". While studying Sarsaparilla, it was **Seydler C.A.**, a German scientist, who coined the term "Pharmacognosy" in 1815 in the title of his work ***"Analecta Pharmacognostica"***. Pharmacognosy is derived from two Greek words viz. *Pharmakon* (a drug) and *Gignosco* (to acquire the knowledge of).

Pharmacognosy may be defined as a branch of bioscience which treats in detail medicinal and related products of crude or primary type obtained from plant, animal and mineral origins. In short, it is an objective study of crude drugs from natural sources treated scientifically and it encompasses the knowledge of the history, distribution, cultivation, collection, processing for market and preservation, the study of sensory, physical, chemical and structural characters and the uses of crude drugs. Pharmacognosy also includes study of other materials used in pharmacy such as suspending, disintegrating and flavouring agents, filtering aids, etc. and substances like antibiotics, allergens, hallucinogenic and poisonous plants, immunizing agents, pesticides, raw materials for the production of oral contraceptives, etc.

Swede Linnaeus (1707 - 1778) the great systematist classified the plants and introduced the system of naming the plants known as the binomial system which is still followed.

The progress achieved in botanical studies during 19th century had a direct influence on Pharmacognosy. Plant classification was further developed by **Bentham** and **Hooker** (1862-1863), **A.W. Eichler** (1883), **Engler** and **Prandtl** (1887-1898). In 1865, **G. Mendel's** important observations on plant hybrids were published. The introduction of the microscope, as an important analytical tool, was a landmark advancement in botanical research, especially due to the development of several techniques: clearing, mounting and staining of the preparations. The *Anatomical Atlas of Crude Drugs* was published by **Berg** in 1865. Later in the century, **Voehl**, **Tschirch** and others reported the anatomical characters of several powdered drugs which proved to be of great significance especially at a period when adulteration in both drugs and food articles was common. *'An Anatomical Atlas of Powdered Vegetable Drugs'* was compiled in 1904 by **Greenish** and **Collin**.

Pharmacognosy as an applied science has played a crucial role in the development of different disciplines of science. A pharmacognosist should possess a sound knowledge of the terms used to describe the vegetable and animal drugs as covered under botany and zoology, respectively. The knowledge of plant taxonomy, plant breeding, plant pathology and plant genetics is helpful in the development of cultivation technology for medicinal and aromatic plants. Plant - chemistry (phytochemistry) has undergone significant development in recent years as a distinct discipline. It is concerned with the enormous variety of substances that are synthesised and accumulated by plants and the structural elucidation of these substances. The technology involving extraction, purification and characterisation of pharmaceuticals from natural sources is a significant contribution to the advancement of natural and physical sciences. The knowledge of chemotaxonomy, biogenetic pathways for formation of medicinally active primary and secondary metabolites, plant tissue culture and other related fields is essential for complete understanding of Pharmacognosy. The basic knowledge of biochemistry and chemical engineering is essential for development of collection, processing and storage technology of crude drugs.

Pharmacognosy is an important link between Pharmacology and Medicinal Chemistry. As a result of rapid development of phytochemistry and pharmacological testing methods in recent years, new plant drugs are finding their way into medicine as purified phytochemicals, rather than in the form of traditional galenical preparations. The knowledge of pharmacology is essential for understanding action of drugs on animals and the human system. Pharmacognosy is the infrastructure on which depends evolution of novel medicines, as it is seen that several crude drugs are utilized for preparation of galenicals or as sources of therapeutically significant substances that cannot be synthesised economically. Further, the crude drugs also provide essential intermediates for final synthesis of active compounds. Phytopharmaceuticals or synthetic drugs derived from phytochemicals have to be ultimately incorporated in suitable dosage form which involves the knowledge of dispensing and preparative pharmacy, pharmaceutical technology and analysis.

In a nutshell, Pharmacognosy is an important bridge between the pharmaceutical and basic sciences. Pharmacognosy is a vital link between Ayurvedic and Allopathic systems of medicines. It provides a system wherein the active principles of crude drugs derived from natural origin can could be dispensed, formulated and manufactured in dosage forms acceptable to allopathic system of medicine.

WHO GUIDELINES ON GOOD AGRICULTURAL AND COLLECTION PRACTICES (GACP) FOR MEDICINAL PLANTS

At a WHO Informal Meeting on Methodologies for Quality Control of Finished Herbal Products, held in Ottawa, Canada from 20 to 21 July 2001, the entire process of production of herbal medicines, from raw materials to finished herbal products, was reviewed. It was recommended that WHO should give high priority to the development of globally applicable guidelines to promote the safety and quality of medicinal plant materials through the formulation of codes for good agricultural practices and good collection practices for medicinal plants.

Objectives:

The main objectives of these guidelines are to:

- Contribute to the quality assurance of medicinal plant materials used as the source for herbal medicines, which aims to improve the quality, safety and efficacy of finished herbal products.
- Guide the formulation of national and/or regional GACP guidelines and GACP monographs for medicinal plants and related standard operating procedures; and
- Encourage and support the sustainable cultivation and collection of medicinal plants of good quality in ways that respect and support the conservation of medicinal plants and the environment in general.

Terms relating to herbal medicines:

- **Contamination**: The undesired introduction of impurities of a chemical or microbiological nature or of foreign matter, into or on to a starting material or intermediate during production, sampling, packaging or repackaging, storage or transport.

- **Cross-contamination:** Contamination of a starting material, intermediate product or finished product by another starting material or product during production.
- **Herbal medicines:** It include herbs, herbal materials, herbal preparations and finished herbal products.
- **Herbs:** Herbs include crude plant material such as leaves, flowers, fruit, seed, stems, wood, bark, roots, rhizomes or other plant parts, which may be entire, fragmented or powdered.
- **Herbal materials:** Herbal materials include, in addition to herbs, fresh juices, gums, fixed oils, essential oils, resins and dry powders of herbs. In some countries, these materials may be processed by various local procedures, such as steaming, roasting, or stir-baking with honey, alcoholic beverages or other materials.
- **Herbal preparations:** Herbal preparations are the basis for finished herbal products and may include comminuted or powdered herbal materials, or extracts, tinctures and fatty oils of herbal materials. They are produced by extraction, fractionation, purification or concentration or by other physical or biological processes.
- **Finished herbal products:** Finished herbal products consist of herbal preparations made from one or more herbs. If more than one herb is used, the term mixed herbal product can also be used. Finished herbal products and mixed herbal products may contain excipients in addition to the active ingredients. In some countries, herbal medicines may contain, by tradition, natural organic or inorganic active ingredients that are not of plant origin (e.g. animal materials and mineral materials).

TERMS RELATED TO MEDICINAL PLANT CULTIVATION AND COLLECTION ACTIVITIES

The definitions below have been adapted from terms included in the glossary compiled by the Food and Agriculture Organization of the United Nations (FAO), available at the time of preparation of these guidelines.

Erosion

The process whereby water or wind moves soil from one location to another.

Types of erosion are :

(1) Sheet and rill: A general washing away of a thin uniform sheet of soil, or removal of soil in many small channels or incisions caused by rainfall or irrigation run-off;

(2) Gully: Channels or incisions cut by concentrated water run-off after heavy rains;

(3) Ephemeral: A water-worn, short-lived or seasonal incision, wider, deeper and longer than a rill, but shallower and smaller than a gully; and

(4) Wind: The carrying away of dust and sediment by wind in areas of high prevailing winds or low annual rainfall.

Integrated Pest Management (IPM)

The careful integration of a number of available pest control techniques that discourage pest population development and keep pesticides and other interventions to levels that are economically justified and safe for human health and the environment. IPM emphasizes the growth of a healthy crop with the least disruption to agro-ecosystems, thereby encouraging natural pest-control mechanisms.

SOURCES OF NATURAL DRUGS

India is one of the 17 mega biodiversity countries in the world which contributes about 7% of world biodiversity and also has 15 agro-climatic zones ranging from Himalayan to marine and desert to rain forest ecosystems. India has vast number of medicinal plants. According to database of NMPB (National Medicinal Plant Board), out of 17000-18000 flowering plants species in India, more than 7000 are medicinal.

Herbs can be obtained through various sources. Herbs can be collected from wild sources or from cultivated source such as field or green house. Cultivation of herb is becoming common practice all over world because it gives consistent and good quality raw material. Cultivated herbs are generally free from contamination.

Table 1.1: Various Sources of Natural Drugs

Sr. No.	Source	Significance	Examples
1.	Plant	Most important source of natural drugs, widely available, sustainable, renewable, cost effective source.	Cinchona, Ephedra, Opium, Senna, Aloe, Rauwolfia, Belladonna, Cinnamon, Tea, Corinder, Peppermint.
2.	Animal	Provides many important drugs, enzymes from ancient time.	Honey, Bees wax, Wool fat, Cod liver oil, Musk, Lakh, Lactose, Leech Cochineal, Lard, Shellac, Gelatin, Vaccines, Human blood.
3.	Marine	Huge source of many potential drugs, marine algae, plants, animals, fungi are not yet exploited or searched.	Agar, Algae, Toxins, Antibiotics.
4.	Mineral	Naturally occurring minerals, oxides are used as medicine from ancient time. They are stable and effective on various diseases.	Zinc oxide, Calamine, Shilajit, Sulphur, Talcum, Paraffines, Kaolin.
5.	Micro-organism	New drug discovered from micro-organisms.	Antibiotics, antimicrobials, enzymes, vitamins.
6.	Plant tissue culture	Biotechnological approaches.	Insulin, hormones, secondary-metabolites, antibiotics produced by fermentation.

1. Herbs can be collected from wild as well as cultivated source. But cultivation is becoming regular procedure throughout the world since it gives consistent and good quality of herbs. Cultivated drugs are free from adulteration or contamination.
2. White studying environment and climatic conditions biodiversity of India is of great importance. It has been recorded in the list of 17 mega-biodiversity countries. India has been divided into 15 Agro climatic zones covering that has seven percent of total world biodiversity.
   The climatic zones are marked as Himalayan to marine and from desert to rain forests.
3. According to database of NMPV (National Medicinal Plant Board) out of 18000 flowering plant species India has more than 7000 species of medicinal plants.

[B] TRADITIONAL AND ALTERNATIVE SYSTEMS OF MEDICINES

Traditional Chinese Medicine and Kampoh System

The Chinese system of medicine is still prevalent. This ancient system finds its references in the Yellow Emperor's classic of *Internal Medicine* (*Huang Di Nei Jing*) which is believed to be prepared between 200 BC and 100 AD. This herbal is based on the idea that all life is subject to natural laws. The hypothesis includes two quite different systems the *yin* and *yang* theory and the five elements (i.e. water, metal, earth, fire and wood). These two theories have been developed separately and differ in terms of diagnosis and treatment. The *yin* and *yang* theory says that everything in the universe consists of a dark (*yin*) and light side (*yang*). These are complimentary opposites like wet and dry, up and down or day and night. The five elements theory proposes that each element leads to the next in a continuous fashion like fire to metal, to wood, to earth, to water and so on. The elements are the five phases indicating the process of continuous movement of life. The elements play a dynamic role in the Chinese system of medicine like in making groups of herbal tastes and parts of body.

According to this system, diseased conditions are the expressions of imbalance in *yin* and *yang* like excess or deficiency of either of them. For example, shivering occurs due to excess of *yin* while excess of *yang* causes a fever. The treatment makes use of various herbs especially the formulations. The important herbs from this system are *Ephedra sinica, Rheum palmatum, Carthamus tinctorius, Clerodendron trichotomum, Panax ginseng, Schisandra, Chinensis, Schizonepeta tenuifolia, Agastache rugosa* etc.

The traditional chinese system of medicine has spread to Japan and Korea in a form called *Kampoh*, called as the traditional system of Japanese medicine. Although, it has developed its own characters, giving due importance to the Japanese style of simplicity and naturalness, still the basic ideas like *yin* and *yang* have a crucial role in *Kampoh* medicine. As compared to *Kampoh*, Korean system is very much similar to Chinese system and includes most of the herbs in it.

Ayurveda - Indian System of Medicine

Ayurveda-Ancient Science of Life is believed to be prevalent for last 5000 years in India. It is one of the most noted systems of medicine in the world. Ayurveda is based on the hypothesis that everything in the universe is composed of five basic elements viz. space, air, energy, liquid and solid. They exist in the human body in combined forms like *vata* (space and air), *pitta* (energy and liquid) and *kapha* (liquid and solid). *Vata, pitta* and *kapha* together are called *Tridosha* (three pillars of life). It is believed that they are in harmony with each other, but in every human being one of them is dominating which, in turn is called as the *prakruti* of that person. *Tridosha* exist in human body in seven forms called *Saptadhatu* viz. *Rasa* (lymph), *Rakta* (blood), *meda* (adipose tissue), *mamsa* (flesh), *majja* (nervine tissue), *Shukra* (reproductive tissue) and *asthi* (bones). These tissues are subject to wear and tear so that *mala* (excretory material) is formed from them. When *tridosha, saptadhatu* and *mala* are in balance with each other, it is called as healthy condition while imbalance causes a pathological condition. It is hypothesised that the five characters of the medicinal herbs viz. *rasa, guna, virya, vipak* and *prabhava* can be applied to treat various pathological conditions. Authentic information on Ayurveda has been compiled by ancient Indian Medicine practitioners in forms called *Samhita* and other similar books. Ayurvedic pharmacy (*Bhaishajya-Vigyan*) proposes five basic dosage forms like *swaras, kalka, kwath, hima* and *phant*. A number of other dosage forms like *churna, avaleha, ghrita, sandhana kalpa, bhasma* are prepared from them. Mostly, all of them are polyherbal formulations. Some important herbs in Ayurveda are *Rauwolfia sepentina, Asparagus racemosus, Cassia angustifolia, Sesamum indicum, Holarrhena antidysenterica, Withania, somnifera, Aconitum napellus, Piper longum, etc.*

Unani System of Medicine

The roots of this system go deep to the times of the well known Greek philosopher **Hippocrates** who is credited with it. **Aristotle Golen** (384 - 322 BC) Greek - Philosopher *"Father of natural history"* made valuable contributions to it. This system of Greek origin was further carried to Persia (Iran), where it has been improved by Arabian physicians.

Fig. 1.4: A Typical urban unani medical practitioner

This system is based on two theories viz. the Hippocratic theory of four humours and the Pythagorian theory of four proximate qualities. The four humours are blood, phelgm, yellow bile and black bile while the four qualities are the states of living human body like hot, cold, moist and dry.

They are represented as earth, water, fire and air. The Greek ideas were put by Arabian physicians as seven working principles (*Umur-e-Tabia*) and included elements, temperaments, humours, organs, life, spirit, energy and actions. They believed that these principles are responsible for the body constitution and its health, as well as, the diseased conditions.

The Unani system of medicine aims at treating the cause of disease and not its symptoms. For this purpose, thorough history of the patient is recorded in addition to his pulse, urine and stool examinations. The diseased condition is considered to be due to the imbalance between humours and accordingly, treatment is given. The drugs are polyherbal formulations and their collective effect is considered.

Unani system of medicine is called by various names in different parts of world such as Arab medicine, Greco-Arab medicine, Loniah medicine, Islamic medicine and also Oriental medicine.

Unani-medicines: Madar, Fufal, Gilo, Kabab chini, Karanj, Kulthi, Lodh, Qust, Sana, Tagar, Zeera, Siyah.

Homeopathic System of Medicine

Fig. 1.5: Samuel Hahnemann

In comparison to other traditional systems of medicine, Homeopathy is a newer one and has been developed in the eighteenth century by **Samuel Hahnemann** - a German physician and chemist. He proposed that the cause of disease itself can be used for its treatment. Hahnemann put forth the Law of Similars which says that like cures like (*Similae similibus curentur*). With this principle, he showed that cinchona can produce the symptoms of malaria. He succeeded in getting relevant results with a large number of extracts prepared from plants, animals and minerals. He compiled all these observations in what is called '*The Organon of Medicine.*'

In the homoeopathic system, the drug treatment is not specified, but the choice of drug depends on symptoms and the clinical condition of the patient. This is based on the concept of proving and prover. In a healthy person called prover, the symptoms created by different doses of drug extracts are noted which is called proving, and it specifically considers physical, mental and emotional changes of the prover. Consequently, these symptoms are compared with a patient with similar symptoms and accordingly, same type of extract is given for treatment. During the treatment, the drug extracts are extremely diluted, which is believed to cause potentiation and enhancement of curative effect. The drugs are extracted in the form of mother tincture, which is further diluted in terms of decimal or centesimal potencies. Various medicinal plants used in homoeopathy are:

Table 1.2: Sources of occurrence of Homeopathic medicines

Vegetable drugs	Animal drugs	Minerals and metals
Arnica, Belladonna, Marigold, Chamomile, Colchicum, Hemlock, Hyoscyamus, Hypericum, Ipecacunha, Lycopodium, Opium, Ergot, Thuja, Aconite, Nux vomica	Honeybees, Calcium carbonate, Cantharis	Arsenic oxide, Barium carbonate, Calcium phosphate, Kalashisa, Mercuric chloride, Antimony tartrate, Sulphur, Copper, Aluminium, Phosphorus, Platinum

Siddha System of Medicine

The term 'Siddha' means achievement and 'Siddhars' were saintly personalities, who attained proficiency in medicine through practice of Bhakti and Yoga. According to traditional belief Lord Shiva unfolded the knowledge of medicine to his wife Parvati which was then passed to Siddhars. This is the system of pre-vedic period identified with Dravidian culture and it is largely therapeutic in nature. Like Ayurveda, this system believes the role of three humors i.e. vatta, pitta and kapha and that all objects in universe are made up of five basic elements namely, earth, water, sky, fire and air. The identification of causative factors of diseases is done through pulse reading, the colour of body, the study of voice, urine examination, the status of digestive system and the examination of tongue. The literature of Siddha system is mostly in Tamil. A few natural drugs used in Siddha system of medicine are:

Abini (*Papaver somniferum*),	Alari (*Nerium indicum*),
Ethi (*Strychnos Nux-vomica*),	Gomathai (*Datura stramonium*),
Haikalli (*Euphoribia nerifolia*),	Ratha polam (*Aloe-barbadensis*).

Naturopathy and Yoga

Naturopathy is not merely a system of treatment, but also a way of life, which is based on laws of nature. The attention is particularly paid to eating and living habits, adoption of purificatory measures, use of hydrotherapy, mud packs, baths, massage, etc.

The system of Yoga is as old as Ayurveda. The eight components of Yoga are restraint, observance of austerity, physical postures, restraining of sense organs, breathing exercises, contemplation, meditation and Samadhi. Yoga exercises have potential in improvement of better circulation of oxygenated blood in the body, restraining the sense organs, improvement of social and personal behaviour and induction of tranquility and serenity in the mind.

Bach Flower Remedies

Bach flower remedies were discovered by **Edward Bach**, a physician in the early decades of the twentieth century. These include 38 remedies prepared from flowers of wild plants, bushes or trees. The remedies are prescribed as per the patient's state of mind as, depression, anger, fear, worry, etc. The prescription is meant for achieving vitality and a harmonious state of mind, the lack of which causes sickness. According to **Dr. Edward Bach**, the remedies enrich the body with vibrations of human's superior nature, rather than attacking a disease. It is believed here that change of outlook and peace of mind has a major role in healing of a disease. Some of the remedies prescribed are white chestnut, wild rose, *mimulus*, agrimony, chicory, gentian etc. Some states of mind for which they are prescribed are worry, vague, fear of unknown origin, despair, extreme mental anguish, extreme fright, lack of confidence, etc. For the purpose of medication, the mother tinctures are prepared and dispensed in a diluted form as in homeopathic potentiation.

Aromatherapy

It is one of the most ancient healing arts and traces its origin to 4500 BC, when Egyptians used aromatic substances in medicines. Greeks also used plant essences for aromatic baths and scented massage. In Ayurveda there is mention of scented baths (abhyanga). Prof. **Gantle Fosse**, a French cosmetic chemist coined the term *'Aromatherapy'* and described healing properties of essential oils. Many scientists at various universities are, now-a-days, investigating this method of healing. Different essential oils from various parts of plants are massaged into skin to treat a range of diseases to have an effect on the mind and emotions. They are massaged into the skin or inhaled or taken as bath. They have been shown to heal wounds, promote formation of scar tissue, treat acne and skin problems, pre-menstrual tension, rheumatism, poor circulation and also nervine disorders like headache, stress, insomnia, etc. Various essential oils used in aromatherapy are basil, bergamot, black pepper, calendula, caraway, eucalyptus, fennel, garlic, geranium, ginger, jasmine, juniper, lavender, rosemary and sandalwood.

Table 1.3: Approximate use of herbals by various systems of medicines

Sr. No.	System	No. of species
1.	Folk	4500 – 5000
2.	Ayurvedic	1750 – 2000
3.	Siddha	1000 – 1250
4.	Unani	600 – 800
5.	Homeopathy	350 – 500
6.	Tebetan (Amachi)	200 – 350
7.	Modern (Allopathic)	upto 200

[C] CLASSIFICATION OF CRUDE DRUGS

The term 'crude drug' generally applies to the products from plant and animal origin found in a raw form. However, the term is also applied to inclusion of pharmaceutical products from mineral kingdom in original form and not necessarily only of organic origin such as kaolin, bentonite, etc. The term 'crude drug' is referred in relation to the natural product that has not been advanced in value or improved in condition by any process or treatment beyond that which is essential for its proper packing and prevention from deterioration. Crude drugs are further grouped as **organized** (**cellular**) or **unorganized** (**acellular**) according to whether they contain a regular organized cellular structure or not. Organized drugs comprise those crude drug materials which represent a part of the plant and are, therefore, made up of cells. Unorganized drugs are a diverse group of solid and liquid materials which do not consist of parts of plants and are obtained from natural sources by a variety of extraction procedures.

In Pharmacognosy, the crude drugs may be classified according to

(i) their alphabetical status,

(ii) the taxonomy of plants and animals from which they are derived,

(iii) their morphology,

(iv) the chemical nature of their active constituents,

(v) chemotaxonomic status, and

(vi) their pharmacological actions and therapeutic applications.

It should be noted that, none of these systems give a total profile of natural drugs and have their own limitations.

1. Alphabetical Classification

The crude drugs are arranged according to the alphabetical order of their Latin and English names.

Some of the Pharmacopoeias and reference books which classify crude drugs according to this system are as follows:

(1) Indian Pharmacopoeia, (2) British Pharmacopoeia,

(3) British Herbal Pharmacopoeia, (4) United States Pharmacopoeia and National Formulary,

(5) British Pharmaceutical Codex, (6) European Pharmacopoeia (Latin titles),

(7) Encyclopaedia of Common Natural ingredients used in Drugs and Cosmetics.

e.g.: Acacia, benzoin, cinchona, dill, ergot, fennel, gentian *Hyoscyamus*, *ipecacuanha*, jalap, kurchi, liquorice, myrrh, *nux vomica*, opium, *Podophyllum*, quassia, *Rauwolfia*, senna, *Uncaria gambier*, vasaka, wool fat, yellow bees-wax, zedoary.

2. Taxonomical (Biological) Classification

The drugs are classified according to plants or animals from which they are obtained in phyla, orders, families, genera, species, subspecies, etc. This method of classification is based on the consideration of natural relationship or phylogeny among plants or animals. The crude drugs of plant origin are classified on the basis of one of the accepted systems of botanical classification. A large number of plant families have certain distinguishing characteristics that permit crude drugs from these families to be studied at one time. Thus, drugs obtained from plants having alternate leaves, cymose flowers and fruits that are berries or capsules (*hyoscyamus*, *Datura*, belladonna and stramonium) are considered with other members of Solanaceae. In case of animals, all arthropods are grouped, as are all mammals, fish and other phylogenetic types. This system of classification is criticized for its failure to recognise the organised and unorganised nature of the crude drug. At first sight, this classification looks appealing, but many drugs are not entire plants and represent parts of the plants that have been processed systematically. Further, the system fails to take into an account chemical nature of active constituents and therapeutic significance of crude drugs.

The taxonomical classification of few crude drugs derived from dicot plants is as follows.

(i)

Phylum	-	Spermatophyta
Division	-	Angiospermae
Class	-	Dicotyledons
Order	-	Rosales
Family	-	Leguminosae
Sub-family	-	Papilionaceae
Genus	-	*Glycyrrhiza, Astragalus, Myroxylon*
Species	-	*Glycyrrhiza glabra, Astragalus gummifer, Myroxylon balsamum*

(ii)

Phylum	-	Spermatophyta
Division	-	Angiospermae
Class	-	Dicotyledons
Sub-class	-	Sympetalae
Order	-	Tubiflorae
Family	-	Solanaceae
Genus	-	*Atropa, Hyoscyamus, Datura*
Species	-	*Hyoscyamus niger, Datura stramonium, Atropa belladonna.*

(iii)

Phylum	-	Thallophyta
Class	-	Ascomycetes
Order	-	Clavicipitales
Family	-	Clavicipitaceae
Genus	-	*Claviceps*
Species	-	*Claviceps purpurea*

3. Morphological Classification

The crude drugs are grouped according to the part of the plant or animal represented into organized and unorganized drugs. The organized drugs are divided into parts of plants like leaves, flowers, fruits, seeds, woods, barks, and subterranean parts like roots and rhizomes. The unorganized drugs are dried latex, gums, extracts, etc. Some of the examples of crude drugs under this type of classification are as follows.

Seeds	Nux-vomica, Strophanthus, Isabgol, Castor
Leaves	Senna, Digitalis, Vasaka, Eucalyptus
Barks	Cinchona, Kurchi, Cinnamon, Quaillaia
Woods	Quassia, Sandalwood, Sassafras, Red sanders
Roots	*Rauwolfia, Ipecacuanha, Aconite*, Jalap
Rhizomes	Turmeric, Ginger, Valerian, Podophyllum
Flowers	Clove, *Pyrethrum*, *Artemisia*, Saffron
Fruits	Coriander, Colocynth, Fennel, Bael
Entire drugs	Ephedra, Ergot, Cantharides, Belladonna
Dried latices	Opium, Gutta-percha, Papain

Resins and Resin combinations	Balsam of Tolu, Myrrh, Asafoetida, Benzoin
Dried Juices	Aloes, Kino, Red gum
Gums	*Acacia*, Tragacanth, Ghatti gum, Guar gum
Dried extracts	Gelatin, Catechu, Agar, Curare

This system of classification is more convenient for practical study especially when the chemical nature of the drug is not clearly understood.

4. Chemical Classification

The crude drugs are divided into different groups according to the chemical nature of their most important constituent. Since, the pharmacological activity and therapeutic significance of crude drugs are based on the nature of their chemical constituents, it would appear that chemical classification of crude drugs is the preferred method of study. The crude drugs containing alkaloids are grouped together, regardless of their morphology and taxonomical relationship.

Few examples of the drugs under chemical classification are as follows:

Glycosides	Digitalis, Senna, Cascara, Liquorice
Alkaloids	Nux vomica, Ergot, Cinchona, Datura
Tannins	Myrobalan, Pale catechu, Ashoka
Volatile oils	Peppermint, Clove, Eucalyptus, Garlic
Lipids	Castor oil, Bees wax, Lanolin, Cod liver oil, Kokum butter
Carbohydrates and Derived Products	Acacia, Agar, Guar gum, Pectin, Honey, Ispaghula
Resins and Resin Combinations	Colophony, Jalap, Balsam of Tolu
Vitamins and Hormones	Yeast, Shark liver oil, Oxytocin, Insulin
Proteins and Enzymes	Casein, Gelatin, Papain, Trypsin

The crude drugs belonging to different morphological or taxonomical categories may be brought together, provided there is some similarity in the chemical nature of active principles.

5. Pharmacological (Therapeutic) Classification

This system of classification involves the grouping of crude drugs according to the pharmacological action of their chief active constituent or their therapeutic uses. Regardless of morphology, taxonomical status or chemical relationship, the drugs are grouped together, provided they exhibit similar pharmacological action. Thus, Cascara, Castor oil, Senna, Jalap, Colocynth are grouped together as purgatives or laxatives because of their common pharmacological action. Similarly, Gentian, Cinchona, Nux vomica and Swertia are grouped as bitters. The drugs differing in mechanism of action, but with the same pharmacological effect is grouped together, e.g. bulk-purgatives, irritant purgatives, emollient purgatives, etc. Some of the drugs could be classified under two pharmacological headings, since they exhibit two different actions. For example, Cinchona is classified both as antimalarial and bitter tonic. An outline for pharmacological classification of crude drugs is as follows.

(i) Drugs acting on gastro-intestinal tract

Bitters	Gentian, Quassia, Cinchona
Carminatives	Dill, Mentha, Cardamom
Emetics	Ipecacuanha, Senega, Mustard
Anti-amoebics	Kurchi, Ipecacuanha
Bulk laxatives	Agar, Ispaghula, Banana
Purgatives	Senna, Castor oil
Peptic ulcer treatment	Derivatives of Glycyrrhetinic Acid (Liquorice), Raw Banana.

(ii) Drugs acting on respiratory system

Expectorants	Liquorice, Ipecacuanha, Vasaka
Antiexpectorants	Stramonium leaves (Atropine)
Antitussives	Opium (Codeine, Noscapine)
Bronchodilators	Ephedra, Tea (Theophylline)

(iii) Drugs acting on cardio-vascular system

Cardio tonics	Digitalis, Squill, Strophanthus
Cardiac depressants	Cinchona (quinidine), Veratrum
Vasco-constrictors	Ergot (ergotamine), Ephedra
Antihypertensives	Rauwolfia

(iv) Drugs acting on autonomic nervous system

Adrenergics	Ephedra
Cholinergics	Physostigma, Pilocarpus
Anticholinergics	Belladonna, Datura

(v) Drugs acting on central nervous system

Central analgesics	Opium (Morphine)
CNS Stimulants	Coffee (Caffeine), Tea, Cocoa
Analeptics	Nux-vomica, Lobelia, Camphor
CNS depressants	Hyoscyamus, Belladonna, Opium (Morphine, Codeine)
Hallucinogenics	Cannabis, Poppy Latex

(vi) Antispasmodics

Smooth muscle relaxants	Opium (Papaverine), Datura, Hyoscyamus
Skeletal muscle relaxants	Curare
(vii) Anticancer	Vinca, Podophyllum, Camptotheca, Taxus
(viii) Antirheumatics	Aconite, Colchicum, Guggul
(ix) Anthelmintics	Quassia, Male Fern, Vidang
(x) Immuno-modulatory agents	Ashwagandha, Tulsi, Ginseng, Asparagus, Picrorrhiza kurroa

They modulate immune responses and principally include immuno suppressive agents and also biological response modifiers which act by immuno stimulation.

(xi) Drugs acting on skin and mucous membrane	Olive oil, Wool fat, Bees wax, Arachis oil, sesame oil, Balsam of Tolu, Balsam of Peru.
(xii) Astringents	Myrobalan, Black Catechu
(xiii) Antimalarial	Cinchona, Artemisia
(xiv) Immunizing agents	Vaccines, Sera, Toxoids, Antitoxins
(xv) Drugs acting chemotherapeutically	Antibiotics
(xvi) Local anaesthetics	Coca

6. Chemotaxonomic Classification

The expanding knowledge of phytochemical screening has revealed the existence of close relationship between constituents of plants and their taxonomical status. The 'chemotaxonomy' has brought the plant chemist back to systematic botany in view of the fact that certain compounds have been found to characterize certain groupings. Chemotaxonomy establishes a relationship between position of the plant and attempts to utilize chemical facts for more exact understanding of the biological evolution and relationships. The characters more often studied in chemotaxonomy are secondary metabolites of pharmaceutical significance such as alkaloids, glycosides, flavonoids, etc. The knowledge of chemotaxonomy could serve as the basis for the classification of crude drugs. The location of berberine alkaloid in *Hydrastis, Berberis* and *Argemone,* distribution of rutin, ranunculaceous alkaloids and flavonoids in species of higher plants are of chemotaxonomic significance. DNA hybridization, amino acid sequencing in proteins and serotaxonomy are also gaining significance in this method of classification.

[D] SCHEME FOR PHARMACOGNOSTIC STUDIES OF A CRUDE DRUG

A systematic study of a crude drug under pharmacognostic scheme involves its description on the following lines.

1. Official title, synonyms, or vernacular names, if any
2. Biological source and family
3. Geographical source or habitat
4. History and introduction of crude drug
5. Cultivation, collection, processing for market and commerce in crude drugs
6. Morphological or macroscopical characters
7. Microscopic or histological studies
8. Chemical constituents and qualitative chemical tests
9. Pharmacological actions, therapeutic and other pharmaceutical uses and pharmaceutical preparations or formulations
10. Commercial varieties, substitutes and adulterants
11. Quality control of crude drugs and Phytopharmaceuticals derived from them.

A crude drug may have official or unofficial status. Unless otherwise indicated, the official title of a crude drug is the title by which it appears in the Pharmacopoeia or other official reference books. Official titles of the crude drugs are given both in English and Latin. Synonyms are the frequently used alternative terms, identical in sense to the official titles of the crude drugs. This includes common or vernacular names of the crude drugs.

The Latin name of the plant or animal with its family is called its botanical or zoological source respectively. Biological source of a crude drug also covers its nature, limits for the active constituents and any relevant information affecting the quality of the crude drug during processing or storage. The nature of the crude drug indicates whether it is in the fresh or dried form, e.g. digitalis consists of dried leaves of *Digitalis purpurea* (Scrophulariaceae) dried at a temperature below 60°C, immediately after collection and the dried leaves should not contain more than 5% moisture.

The indigenous source of a crude drug gives information regarding the place and the country where the crude drug is produced commercially on a large scale. *Cassia angustifolia* (Indian senna) is indigenous to Tinnevelley Tamil Nadu in India.

The history of a crude drug reveals information about its introduction to mankind. It gives us the information about the knowledge of a crude drug, its original place of growth and its introduction into modern medicine. History of some of the crude drugs like Rauwolfia roots, cinchona bark and opium is very interesting.

The cultivation technology for a crude drug has to be studied systematically with reference to the selection of proper strains of seeds or germplasm, types of soil systems, optimum climatic or ecological factors like light, temperature, rainfall, altitude and other factors like fertilizers, plant growth regulators, etc. The crude drugs are collected from the plants when they are rich in their active constituent content, which may be affected by the climatic conditions to a considerable extent. Drying of the plant material prior to packaging is necessary to minimize moisture content of a crude drug to protect it from microbial contamination or undesired chemical reaction during storage. The crude drugs can be dried either in sunlight or in shade or by use of artificial heat. In order to maintain high degree of quality in a crude drug, it is necessary to store and preserve it properly. Light adversely affects quality of crude drugs which are highly coloured, whereas excessive moisture favours enzymatic activity and facilitates fungal growth.

The macroscopical or morphological description of a crude drug includes size, shape, nature of outer and inner surfaces, type of fracture and organoleptic characteristics like colour, odour, taste, consistency, etc. The quality of several aromatic drugs like clove and cardamom can be determined by smell.

A microscopical study of an organized drug, either in entire or powdered form, is one of the important aspects of its histological evaluation. The arrangement of tissues in transverse and longitudinal sections and types of cells and cell contents are revealed by suitable histological study of a crude drug with the aid of a microscope. Certain microscopical characteristics like stomata, trichomes, calcium oxalate crystals, starch grains, stone cells, fibres, vessels, etc. are important anatomical characteristics of organized drugs. Microscopical evaluation of a crude drug also involves linear measurements, study of surface preparations of leaves, quantitative microscopical aspects and chemomicroscopy.

The utility of a crude drug is due to its therapeutically active constituent. Clove contains volatile oil of which eugenol is the chief constituent. Most of the times, the crude drug also contains several other constituents which are inert therapeutically. For example, caraway and many other umbelliferous fruits contain fixed oil and protein, in addition to the active principle which is volatile oil. Chemical tests are useful for the qualitative chemical evaluation of a crude drug. These tests are either general or specific and are usually performed for identification of the powdered drugs. Crude drugs like Senna, Rhubarb, Cascara, etc. give positive anthraquinone test. Vitali's test is performed for identification of tropane alkaloids of Solanaceous drugs.

The pharmacological actions of active constituents of crude drugs and their therapeutical or pharmaceutical uses form an integral part of a pharmacognostic scheme. Senna and Cascara are used as purgatives, whereas digitalis and strophanthus are cardiotonics. Rauwolfia is used in the treatment of high blood pressure and insanity, the activity being mainly because of the reserpine alkaloid. Some crude drugs and phytopharmaceuticals are used as pharmaceutical necessities for flavouring and imparting pleasant smell, Such agents and they are devoid of any significant pharmacological action. For example, acacia and tragacanth are used as emulsifying agents, honey as a sweetening agent, turmeric as a colouring agent and sandalwood oil for fragrance.

The study of different species and varieties of medicinal or aromatic plants is of paramount importance, especially when several closely related varieties are available commercially. In case of senna, *Cassia angustifolia* and *Cassia acutifolia* are the official drugs, even though several other species of Cassia such as C. *obovata, C. auriculata,* etc. are wildly grown. Several allied species of *Strophanthus kombe* are *S. hispidus, S. gratus, S. emini, S. sarmentosus,* etc. some of which are equally important. The allied drugs of Rauwolfia serpentina are *R. vomitoria, R. perakensis, R. canescens, R. micrantha and R. densiflora*. The word substitute itself is self-explanatory. Substitutes resemble the genuine drugs to a considerable extent in their morphology and type of chemical constituents. *Cinnamomum burmanii* and C. *loureirii* are substitutes for C. *zeylanicum*. *Digitalis thapsi* is used as a substitute for *D. purpurea*.

Adulteration is debasement of an article and the adulterant resembles the genuine drug with respect to its morphological appearance and organoleptic characteristics. However, it is devoid of the active constituents, which are present in the genuine drug. The adulterants of *S. nux-vomica* such as *S. nux-blanda* and *S. potatorum* do not contain strychnine. Cassia obovata or Dog senna is an adulterant of Indian senna.

Evaluation of a crude drug means its identification and determination of its purity and quality. Quality control of a crude drug and its pharmaceuticals can be attempted by different methods of evaluation depending upon the morphological and microscopical studies of the crude drugs or their physical, chemical and biological behaviour. Systematic identification of crude drugs and their quality assurance form an integral part of drug-description.

Chapter 2

INTRODUCTION TO PARTS OF MEDICINAL PLANT

THE CELL

In 1838 *Schleiden* and *Schwann* put forward the cell theory stating that plant or animal body is ultimately made up of minute cells and concluded that *the cell is the structural unit of life.*

The living organisms are of two types, either unicellular or multicellular.

In multicellular organisms the life activities are performed by co-ordination of several organs, these organs are made up of tissues, while the tissues in turn are aggregates of similar cells. Thus, a cell is a structural unit of life. While in unicellular organisms all activities are performed by the same cell.

The cell has a definite shape of its type with a few exceptions like amoeba, leucocytes, etc. Various shapes like oval, spherical, polyhedral, columnar, cylindrical, stellate and several others are found in plant as well as in animal cells. Normally, the size may vary from 0.5 to 20 μ and very exceptionally up to 200 – 300 mm as in the case of plant fibres and latex cells.

There are two types of cells. These are:

(i) Prokaryotic, (ii) Eukaryotic.

Prokaryotic cells are characterized by the absence of true nucleus, obviously the nucleolus, nuclear membrane are missing, DNA is without protein sheath and nuclear matter is in direct contact with the cytoplasm. The ribosomes are scattered in the matrix while other organelles and endoplasmic reticulum are missing in prokaryotic cells. Respiratory enzymes and photosynthetic pigments are present. Meiosis and mitosis are not observed in this type of cells.

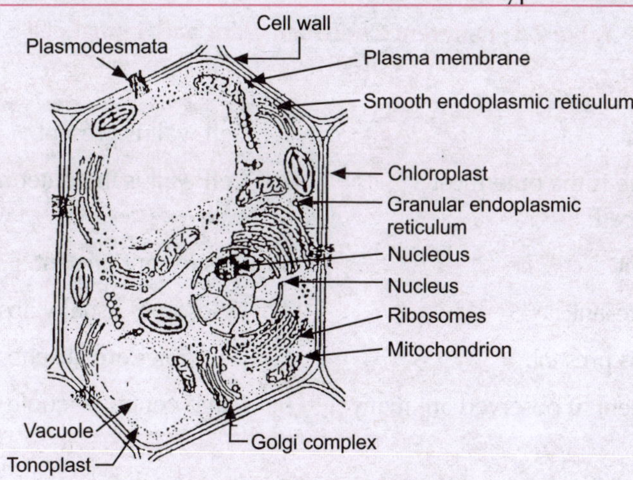

Fig. 2.1: Typical Plant Cell

Primitive type of organisms like blue-green algae, bacteria, represents this type of cells.

Eukaryotic cells possess well-marked true nucleus, with nuclear membrane, while DNA is covered with protein sheath. There is a distinct nucleolus in the nucleus, while plastids and mitochondria are represented in the cytoplasm. The other important and major component of the cytoplasm is endoplasmic reticulum associated with ribosomes. In eukaryotic cells, cell wall is made up of cellulose. Meiosis and mitosis are observed in these types of cells. All plant and animal cells belong to eukaryotic type of cells.

Difference between plant and animal cells:

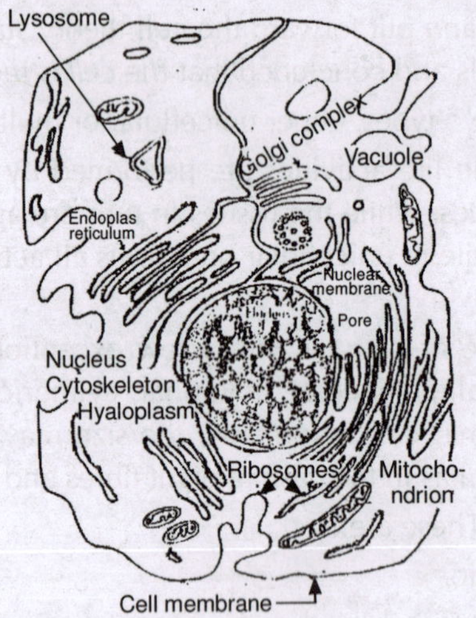

Fig. 2.2: Animal cell

Before knowing the structure and functions of various cell organelles, one must know the structural difference between a plant cell and an animal cell.

Table 2.1: Difference between plant and animal cell

Animal Cell	Plant Cell
1. Cell wall is absent.	1. Cell wall is present.
2. Plasma membrane is the outermost boundary of the cell.	2. Cell wall is the outermost boundary of the cell.
3. Plastids are absent.	3. Plastids are present.
4. Lysosomes are present.	4. Lysosomes usually absent.
5. Pair of centriole is present.	5. Centrioles are absent.
6. Vacuoles are absent, if observed are many and small.	6. Large central vacuole is present.
7. Phagocytosis or Pinocytosis is observed.	7. Phagocytosis or Pinocytosis is not observed.
8. Golgi apparatus is present with specific polarity.	8. Golgi apparatus is scattered in the cytoplasm.

The functions of various organelles of a plant cell are as under.

FUNCTIONS

1. **Cell wall:**
   (i) It offers rigid frame work and protection to protoplast.
   (ii) Thick and lignified cells of the plant provide mechanical support to the organ.
   (iii) Checks the rate of transpiration due to cuticular sheath.
   (iv) The developing wall pressure prevents the distention of protoplast. Due to various contents like cutin, lignin, wax etc. the cell wall results in permeability which is ultimately responsible for life of cell.
2. **Plasma membrane:**
   (i) Being selectively permeable controls transport of materials across it.
   (ii) Permits diffusion of water and fat soluble components. Fat insoluble components pass through the membrane by forming reversible compounds with membrane proteins.
3. **Endoplasmic reticulum:**
   (i) Due to ribosomes, it is involved in protein synthesis, also in glycogen and fat metabolism.
   (ii) Gives mechanical support to cytoplasm.
   (iii) Participate in exchange of materials by active and passive transport.
4. **Centriole:** A pair of granular bodies present in Eukaryotic-cell cytoplasm outside the nuclear membrane.
5. **Ribosomes:** Degradation and synthesis of proteins takes place in ribosomes.
6. **Centrosome:** Region of differentiated cytoplasm containing centriol.
7. **Golgi complex:**
   (i) Condensation of lipids, carbohydrates hormones takes place in golgi bodies.
   (ii) Participates in formation of Lysosomes.
8. **Mitochondria:**
   (i) Mainly responsible for transformation of chemical energy into biological energy in the form of ATP compounds.
   (ii) All enzymes involved in Krebs cycle are present in mitochondria. It is also responsible for transmission of hereditary characters (extra-nuclear).
9. **Plastids:** Play vital role in plant metabolism.
10. **Leucoplast:** Colourless plastid, storing starch.
11. **Chloroplasts:** Chloroplasts capture solar energy and convert it in chemical energy (photosynthesis).
12. **Chromoplast:** Coloured cell plastid-excluding chloroplast.
13. **Nucleus:** The nucleus controls all activities of the cell. Biogenesis of ribosomal proteins take place in nucleolus only, nucleolus takes part in cell division.
14. **Nucleolus:** Within the nucleus a single or more spherical bodies called nucleoli are present.
15. **Nucleic Acids:**

    Nucelotide: Phosphoric esters of nucleoside constituting the molecule of DNA or RNA.

    Nucleoside: Compound having a pentose sugar (Ribose or deoxyribose) and any one of the nitrogen bases like thymine, adenine, uracil or cytosine.

    DNA: Deoxynucleic acid covering genetic information.

    RNA: Rebose nucleic acid consisting of nucleotides.

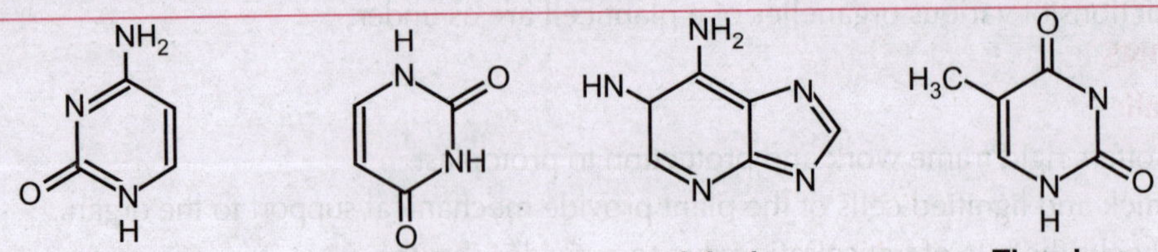

Cytosine Uracil Adenine Thymine

Thymine: One of the two pyrimidine bases of DNA.

Pyrimidine: Single ringed nitrogen bases present in Nucleic acids i.e. cytosine, thymine and uracil.

Adenine: Nitrogen base of purine type constituent of nucleic acids.

Cytosine: One of the pyrimidine base be present in Nucleic acid DNA.

Uracil: One of the three pyrimidine base constituent of nucleic acid. .

16. Chromosomes:

These play very important role in heredity, mutation and variation. Chromosomes have capacity of self-reproduction.

Golgi complex is present in all eukaryotic cells except mammalian red blood corpuscles. In plant cells as well as in invertebrate tissues, it is not present as a single body but there are many such organelles scattered throughout the cytoplasm.

Lysosomes are present in animal cells and not in the plant cells. Plastids are the organelles which are only present in plant cells. They are absent in bacteria and blue green algae.

Cell Inclusions (Ergastic Substances of Plants)

The non-living substances of plant metabolism are known as *ergastic substances*. They may be reserve foods, secretory and excretory or end products of metabolism.

(a) Reserve foods: The materials which occur as reserve food in a cell are the carbohydrates, proteins and lipids. They are present in insoluble forms. Their conversion to soluble form is covered by enzymatic processes.

(b) Excretory products: Animals are able to get rid of their excretory products in liquid or solid form. Plants excrete their waste in the form of insoluble products and are stored inside the cells only. Excretory products of the plants are named as tannins, resins, latex, volatile oils, chemicals like alkaloids, glycosides and mineral crystals, such as calcium oxalate, calcium carbonate, silica, etc.

(c) Secretory products: Various other products, not covered under any one of the above types, are formed in the plants and serve important functions in the plants and are known as the secretory products.

(a) Reserve Foods

(i) Carbohydrates: Many water soluble as also water insoluble carbohydrates are found in plants. Glucose, fructose and sucrose are all water-soluble. The water-insoluble contents are normally polysaccharides.

Inulin: It is the only polysaccharide soluble in water and is found in tubers of Dahlia and Dandelion. In alcohol, it forms sphere crystals.

Starch: It is another polysaccharide mostly found in tubers, foodgrains and seeds of the plant. It is found in the form like granules, concentric or eccentric. Starch gives blue colour with dilute solution of iodine.

Cellulose: A complex polysaccharide forming cell wall and is structural material of plants. *Mucilage* in senna, Isapgol, and linseed, forming gummy solution in epidermal cells is another example of polysaccharides.

(ii) Proteins: These are the nitrogenous substances either soluble or insoluble in water. They are of great importance in the structure and functioning of living cells. They are produced by and associated with living matter. They may be amorphous or crystalline. In maize, they are amorphous found in the form of layer as aleuronic layer or crystalline as aleuronic grain in castor seeds. They are converted to amino acids by proteolytic enzymes.

(iii) Lipids: These are reserves of food materials of plants. Those which are liquid at room temperature are called oils and solids are termed as fats. They are made up of fatty acids and glycerin. Oils are found in endosperm of the following seeds of Castor, peanut, and sesame.

(b) Excretory Products:

(i) Alkaloids: These are the basic nitrogenous secondary metabolites having marked physiological action, if taken internally. They occur in plants as salt of nicotinic, quinic, citric or oxalic acid. They are found in seeds, barks, leaves, roots, etc. Alkaloids are used as highly potent medicaments and possess curative properties. They are protective materials of plants discouraging animal or insect attacks. Alkaloids are detoxifying agents of plants and reservoirs for protein synthesis e.g. ergot, Vinca, Datura, etc.

(ii) Glycosides: Glycosides are the condensation products of sugar and glycol. Glycosides are secondary metabolites of pharmaceutical significance should in water and alcohol. Examples: digitalis, senna, rhubarb, etc.

(iii) Tannins: Tannins are present in cell sap, Tannins are soluble in water and alcohol. They give blue-black or green colour with iron compounds. Tannins have several medicinal properties; hence are used as drugs. Nutgalls, Myrobalan are the examples of tannin.

(iv) Resins: These are found in abundance in the resin ducts of the trunk of the trees or in rhizomes (ginger), fruits (capsicum) and other parts of the plants. They are insoluble in water, may be semi-solid or solid. Many a times, they are associated with gums or volatile oils.

(v) Latex: It is a white suspension wherein microscopically small particles or oil globules are suspended. They contain proteins, sugars, minerals and alkaloid salts in true solution, while gums, starch, resins are in suspended form. Latex is present in lateciferous tissue, characteristic to certain families like Euphorbiaceae, Caricaceae, Papaveraceae, etc. Banyan, madar, papaya, poppy, ficus are examples.

(vi) Volatile oils: These are the fragrant liquids found in plant. They are volatile, liquid and aromatic and are present in roots, leaves, barks, fruits, etc. They are insoluble in water and are soluble in alcohol. Normally volatile oils are carminative stimulants and antiseptic. Orange, lemon, coriander, cinnamon, ginger, eucalyptus and mentha are examples of volatile oil containing drugs.

(vii) Mineral crystals: These occur in cell wall or cell cavity. They are all insoluble in water. Common crystals are calcium oxalate, calcium carbonate and silica. They are found in roots, stems, leaves, fruits, etc. Various types of crystals of calcium oxalate are found like prisms, acicular, raphides, clusters, rosettes, etc.

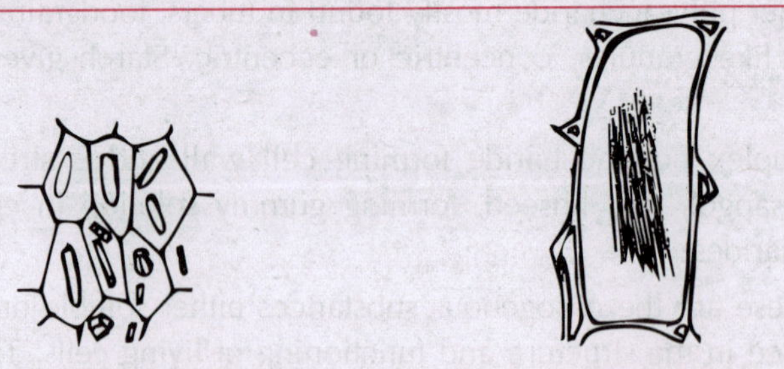

Prismatic crystals Acicular crystal (Raphides)

Fig. 2.3: Types of Calcium Oxalate Crystals

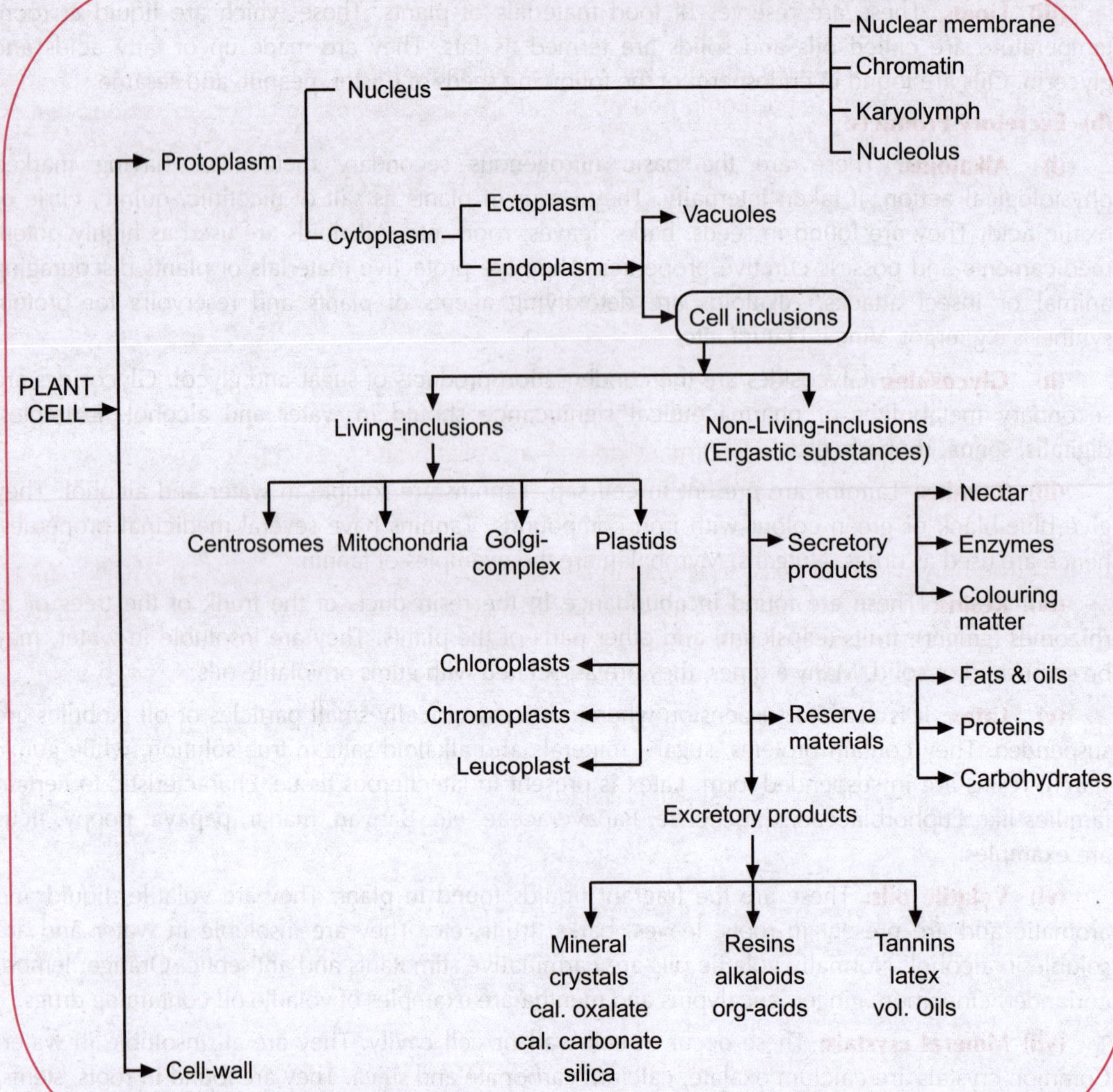

Various Organelles and Cell Inclusions of a plant cell

(c) Secretory products:

Various other products, not covered under anyone of the above types, are formed in the plants and serve important functions in the plants and are known as the secretory products, as follows.

Enzymes, colouring substances' nectar are the examples of this type.

Enzymes are nitrogenous water soluble compounds carrying out hydrolysis of carbohydrates and proteins. **Colouring substances** or pigments like chlorophyll are essential for photosynthesis, anthocyanin and several others. Flavonoid glycosides giving attractive colour to various flowers and leaves are the examples of this type.

Nectar is the sugary solution secreted by many flowers in special cells or glands to attract the insects for effective pollination.

PLANT TISSUES

An aggregation of similar cells or elements having same function is known as a tissue. The organs of plants or animals are composed of various tissues. Usually, in an organ various tissues perform interrelated functions.

Depending upon their origin, structure and physiology, the plant tissues can be classified as follows:

(I) Meristematic or embryonic tissues

(II) Permanent tissues

(I) Meristematic Tissues: They have the following characters.

1. The cells are small and isodiametric in nature.
2. The cells are very compact and without any intercellular spaces.
3. The cells are thin walled, normally cubical and with prominent nucleus.
4. Vacuoles in meristematic tissues are very small or even may be absent.
5. Their cells have capacity to multiply.
6. The newly formed cells due to enlargement and also due to morphological differentiation get transformed or changed into permanent tissue. Meristematic tissues in the plants are located at the growing points i.e. tips of the shoots and roots, between the bark and the wood of trees, in the bark themselves and also in any part of the plant wheresoever extensive growth of the plant takes place.

(II) Permanent Tissues: These exhibit the following characters:

1. The cells vary in size, shape and the nature of the protoplast.
2. The cells may be living or dead.
3. These are produced by meristematic cells.
4. They do not have any capacity to multiply.
5. They are not usually changed into other kind of tissues and normally retain their structural physiological characters.
6. They are composed of one or more type of cells.

Type of the cells, characters of the cell walls and the contents of the cells of the permanent tissues divide them in two categories namely, simple permanent tissue and complex permanent tissues.

(A) Simple permanent tissues

As described above, this consists of only one type of cells forming homogeneous structure. These are further sub-classified as:

(1) Epidermis, (2) Parenchyma, (3) Sclerenchyma, (4) Collenchyma, (5) Cork

1. Epidermis: It is a type of simple permanent tissue forming the outermost layer of plant structure and normally only one cell thick and is made up of parenchyma. It occurs as surface layer of leaves, flowers, fruits, seeds and younger parts of roots and stems.

Its function is to conserve the moisture supply to the inner tissues and offer protection to certain extent against mechanical infection.

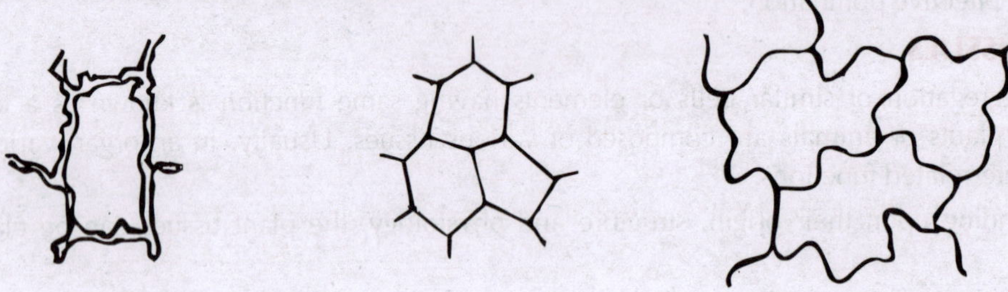

(1) (2) (3)

Fig. 2.4: (1) Beaded epidermal cells (2) Straight walled epidermal cells (3) Wavy epidermal cells

The outer walls of the epidermal cells are often thick and covered with a fatty substance cutin. The layer with cutin is called cuticle. Minute epidermal openings, i.e. stomata, epidermal hairs i.e. trichomes are the other characters of the epidermis. Epidermal cells are colourless except, when pigments are present.

Structure of epidermal cells may vary, i.e.

(a) Straight walled polygonal, e.g. senna leaf

(b) Thick walled beaded, e.g. Digitalis

(c) Wavy, e.g. Hyoscyamus

(d) Wavy and with striated cuticle, e.g. Belladonna

(e) Papillose, e.g. Stigma of pyrethrum flowers

2. Parenchyma: This is another type of permanent tissue of almost all organs of higher plants.

Parenchymatous cells are thin-walled polyhedral in shape and with large central vacuole in many cases. Aging results in the change in shape, wall thickening and the appearance of intercellular spaces.

The functions of parenchymatous cells are storage of food and water, food manufacture, excretion, secretion and also assimilation. Parenchyma is a living tissue.

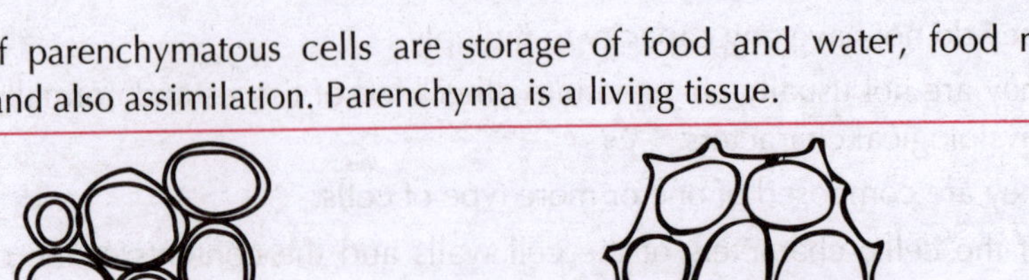

Fig. 2.5: (1) Parenchyma with intercellular spaces, (2) Simple parenchyma

The secondary thickening may give rise to (1) Simple parenchyma (2) Parenchyma with intercellular spaces (3) Lignified parenchyma and (4) Reticulate parenchyma.

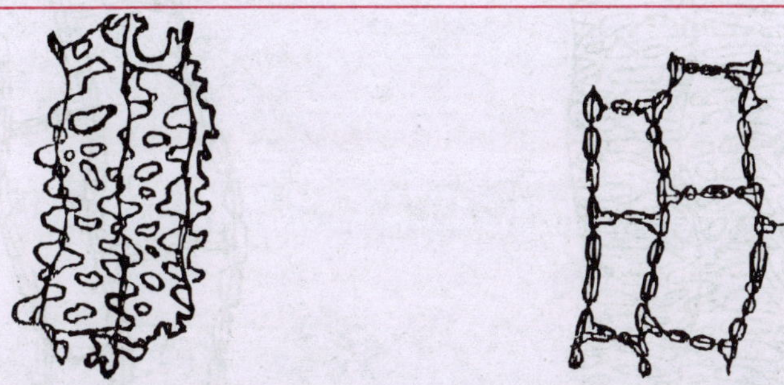

(1) Reticulate parenchyma (2) Pitted parenchyma

Fig. 2.6: Types of Parenchyma

3. Sclerenchyma: It is a supporting tissue and is purely mechanical in its function. It is found in stems, roots, leaves and other parts along with xylem. It is very hard and is responsible for strengthening of the organs in which it is present.

(1) Sclereid thickened (2) Sclereid Pitted (3) Stratified sclereid

Fig. 2.7: Types of sclerenchyma

It consists of two types of elements: (a) Stone cells (b) Fibres.

(a) Stone cells or Sclereids: These are more or less isodiametric, elongated or branched forms. Pitting or stratification may be also observed. They are found in groups, single or complete layer. Generally abundant in cortex, phloem of stems and roots, may be in seeds (cardamom) or flesh of fruits (pears).

(b) Fibres: Usually developed in bundles or in layers. Usually thick walled and have a narrow lumen and pointed ends. Walls of the fibres are usually lignified, but may also contain cellulose (ephedra contains non-lignified fibres flax-fibres are also non-lignified). Depending upon the seat of their occurrence they are said to be pericyclic fibres, xylem fibres or phloem fibres. A sheath of crystals is formed around sclerenchyma and is a salient character for identification as in senna-leaf.

4. Collenchyma: It is the typical supporting tissue found underneath the epidermis of many stems, leaf-stalks and leaf midribs. It is similar to parenchyma in structure except that primary cell wall is thickened to give mechanical strength to plant parts. Collenchyma may or may not have intercellular spaces, may contain chloroplasts. Thickening may be stratified or unevenly distributed around the lumen.

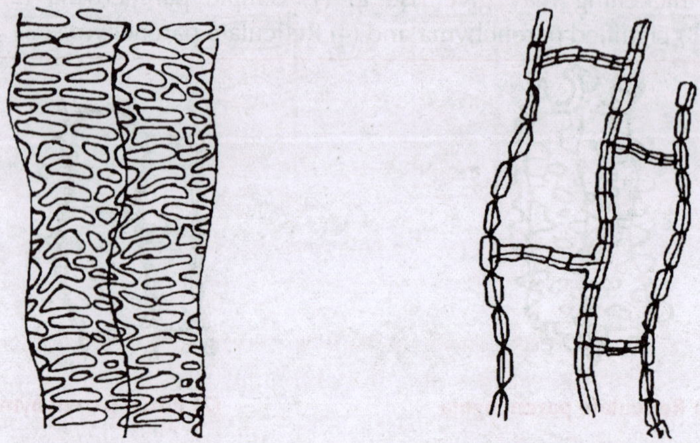

(1) Collenchyma with intercellular spaces (2) Irregularly thickened collenchyma

Fig. 2.8: Types of collenchyma

5. Cork: It is a simple permanent tissue made up of compactly arranged approximately prismatic cells found in radial rows.

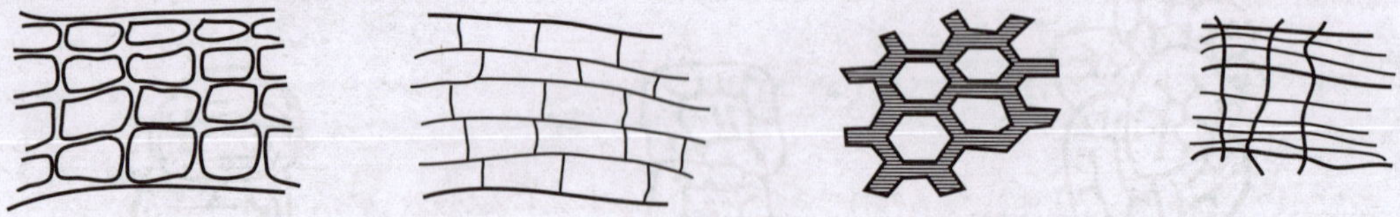

(A) Thick-walled (B) Thin-walled and Flattened outer surface (C) Thin-walled, Polygonal in surface view (D) Stratified

Fig. 2.9: Types of Cork Cells

No intercellular spaces are present between them. It forms the bark of stems or roots. It is protective in function and guards against mechanical injury to inner bark or cambium, and also prevents loss of water. Protoplasm in the cork cell dies and hence matured cork cells are dead.

(B) Complex Permanent tissues:

In this type of tissue several types of cells are present which usually engage in a group of closely related activities. Following are the various types of complex tissues found in plants.

(a) Xylem: It is the main water conducting tissue of the plant always associated with the phloem in the vascular tissues. The components of the xylem are parenchyma, fibres, and tracheary elements. Tracheary elements namely tracheids and vessel members are highly specialized cells elongated in structure and non-living in nature.

The thickening of xylem vessels is lignified. Tracheids are closed (imperforate) while the vessels are open at both the ends and form the tube. The thickening to the xylem elements may be annular, spiral, reticulate, pitted, border pitted and scalariform.

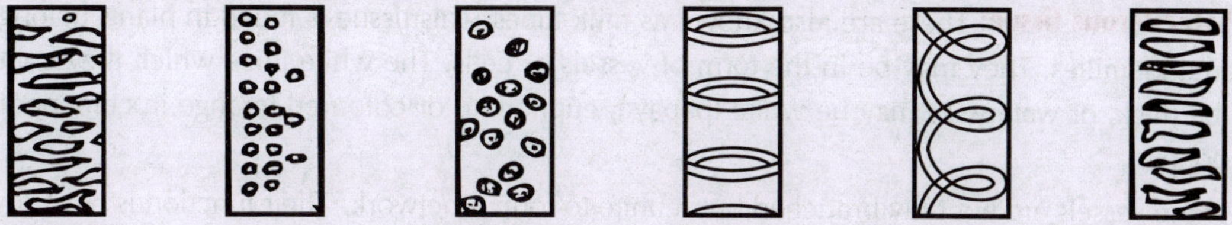

Fig. 2.10: Xylem Elements

(b) Phloem: It is mainly responsible for the transport of food manufactured in the leaves downward to stem and roots. Other types of cells constituting the phloem are, phloem parenchyma, sieve tubes, companion cells and phloem fibres. Sieve tubes contain living protoplasm. The end walls of sieve tubes contain circular perforations. There are perforations in between sieve tubes and companion cells. Phloem parenchyma, ray cells store the food while phloem fibres are meant for strength and support.

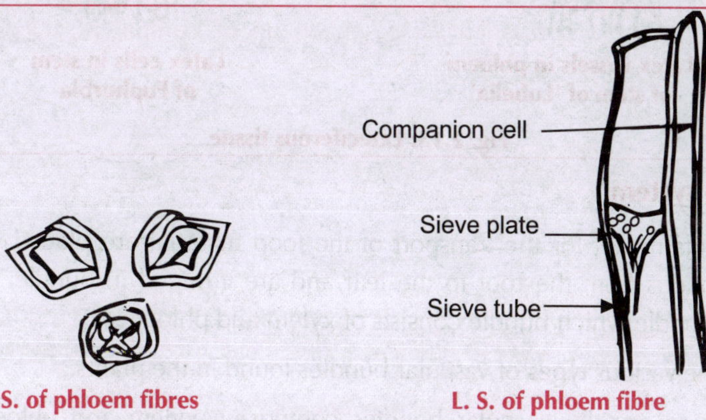

Fig. 2.11: Phloem-elements

(c) Secretory Structure: These are important features of certain plants containing secretions. They may be internal or external. These structures may be single celled or multicellular. Oil gland and resin cells are unicellular, while the ducts or glands are multicellular.

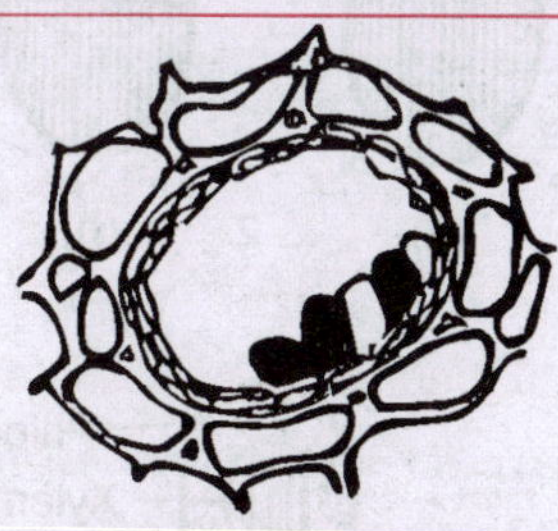

Fig. 2.12: Secretory-gland

They may get formed by splitting of cells i.e. *schizogenous* or by breakdown of group of cells i.e. *Lysigenous* or by combination of the former and the latter i.e. *schizolysigenous*. The secretions may be tannins, oils (volatile or fixed) mucilage and crystals.

Lateciferous tissue: These are also known as milk tubes. This tissue is found in plants belonging to certain families. They may be in the form of vessels or cells. The white juice which they contain may be thick, or watery, or may be white (papaya, euphorbia) or coloured (orange in chinopodium and poppy).

These vessels are not only branched but reunite to form a network. Their function is not known. Latex contains several chemicals like alkaloids, enzymes, proteins etc.

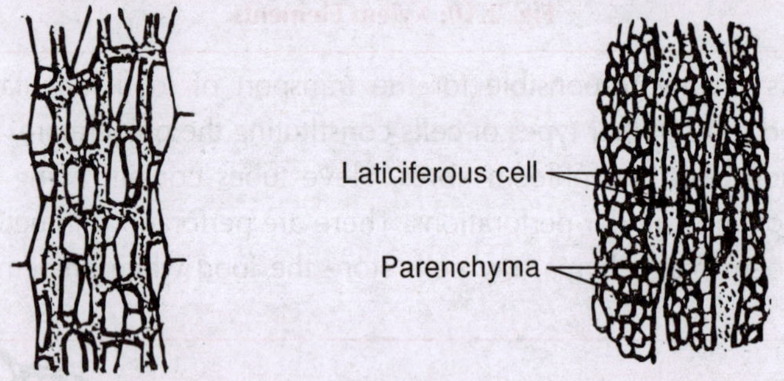

Latex vessels in phloem of stem of 'Lobelia'

Latex cells in stem of Euphorbia

Fig. 2.13: Lateciferous tissue

Conducting tissue system:

Vascular elements meant for the transport of the food material are placed vertically. Vascular elements are continuous from the root to the leaf and are grouped in bundle; such a structure is known as vascular bundle which bundle consists of xylem and phloem.

Following are the various types of vascular bundles found in the plants.

1. **Radial:** These are the vascular bundles comprising xylem and phloem only. They are arranged on alternate radii and hence known as radial. In roots the xylem is exarch with the proto-xylem pointing towards the periphery and metaxylem towards the centre.

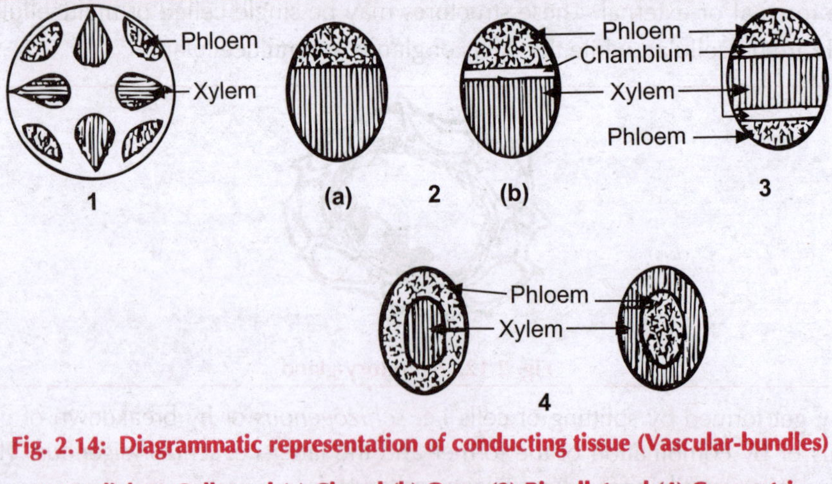

Fig. 2.14: Diagrammatic representation of conducting tissue (Vascular-bundles)

(1) Radial (2) Collateral (a) Closed (b) Open (3) Bi-collateral (4) Concentric

2. **Collateral:** In this case xylem and phloem are placed on the same radius xylem lies towards the centre and phloem towards periphery. There are two types of these vascular bundles.

(a) **Open collateral:** When cambium is placed in between xylem and phloem and occur in dicotyledonous plants.

(b) **Closed collateral:** When no cambium is present in between xylem and phloem this type is found in monocotyledonous plants.

3. **Bi-collateral:** In this case xylem lies in the centre with cambium and phloem on either side along the same radius. Bicollateral vascular bundles are found in *cucurbitaceae family*.

4. **Concentric:** Vascular bundles where in there is no cambium but (a) either the xylem lies in the centre surrounded by phloem (Centroxylic) or (b) Phloem in the centre surrounded by xylem (Centrophloic). The first type is reported in ferns while the second is reported in acorus (sweet flag).

MICROSCOPY OF PLANTS

Various types of the tissues studied so far are grouped in units called the tissue system of the plants. Normally it consists of

(a) Epidermal tissue system

(b) Ground tissue system

(c) Vascular tissue system

(a) Epidermal tissue system: The outermost layer of all parts of plants and its covering cuticle, outgrowth like trichomes, openings like stomata, water pores, etc.

Epidermis is meant for protection, aeration and control of loss of water. In case of roots, root hairs enable the absorption of water and minerals from the soil.

(b) Ground tissue system: The tissues other than epidermal and vascular bundles are included in this group. Actually it forms the bulk of the plant parts and consists of cortex, pericyclic medullary rays and pith in roots and stems. It also covers mesophyll of the leaves. Ground tissue is meant for, food formation, food storage, aeration and mechanical strength of the plants.

(c) Vascular tissue system: This tissue system consists of xylem, phloem with or without cambium. It is only meant for transport of raw or ready food material either raw or ready from one part to another part of the plant.

The detailed study of the histological characters of stem, root and leaves is beyond the scope of this book as it focuses only on the contents prescribed in the syllabus.

PARTS OF MEDICINAL PLANT

I. LEAVES

Leaves are flat, thin, and green appendages to the stem, containing, supporting and conducting strands in their structures. In pharmacognosy, the word leaf includes leaf, compound leaf and leaflets. Usually, the minimum percentage of active constituents is specified in leaves. The basic difference between the leaf and the leaflet is as follows.

Table 2.2: Difference between leaf and leaflet

Leaf	Leaflet
1. In case of leaves, bud or branch is present in the axil.	1. Bud or branch is absent in leaflets.
2. Leaves are arranged spirally and they are solitary in nature.	2. Leaflets are arranged in pairs.
3. Leaves lie in different planes.	3. Leaflets lie in the same plane.
4. Leaves are generally symmetrical at the base e.g. digitalis, belladonna, vasaka.	4. Leaflets are asymmetrical at the base e.g. senna, neem, rose.

Apart from the normal characteristics i.e. their arrangement on the stems, their apices, margin, petiole, presence or absence of stipules etc., leaves are characterized by certain diagnostic structures. Most of these diagnostic characters are microscopic one, such as stomata and trichomes.

Cellular Structure of Leaf:

Upper and lower epidermis is covered by waxy cuticle.

Epidermal cells are nonphotosynthetic except for guard cells that surround stomata.

Interior cells are called mesophyll cells and are photosynthetic. Palisade mesophyll are tightly packed under upper epidermis, and spongy mesophyll are composed of loose cells above lower epidermis.

Vascular bundles or veins project from the stem into the leaves with the xylem above the phloem.

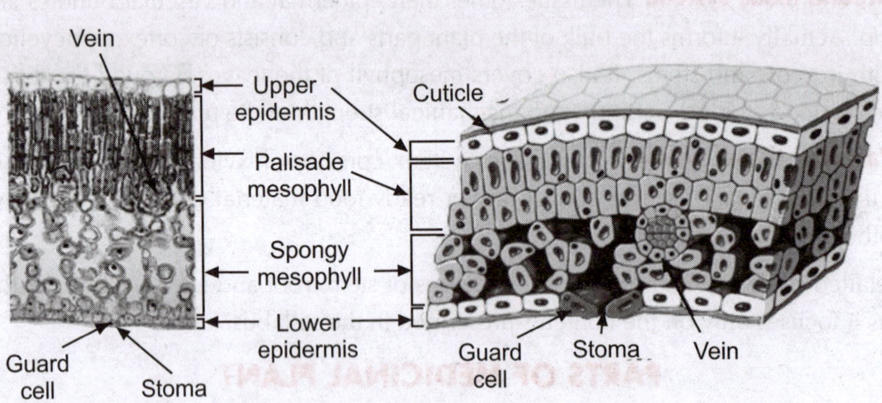

Fig. 2.15: Cellular structure of leaf

II. STEMS

The stem is an ascending axis of the plant developed from the plumule. It consists of nodes, internodes, buds and it gives rise to branches, leaves and flowers. The stem may be aerial, suberial and underground. Depending upon the presence of mechanical tissues, the stems may be weak, herbaceous or woody.

1. Weak stems: When the stems are thin and long, they are unable to stand erect, and hence may be one of the following types.

(a) Creepers or prostate stems: They grow flat on the ground without roots. e.g. grasses, gokharu, etc.

(b) Climbers: These are too weak to stand alone. They climb on the support with the help of tendrils, hooks, prickles or roots e.g. *Piper betel, Piper longum,* Gymnema.

(c) Twinners: These coil the support and grow further. They are thin and wiry e.g. Ipomoea and Phaseolus.

2. Herbaceous or woody stems: These are the normal stems and may be soft or hard and woody e.g. sunflower, sugarcane, ephedra, etc.

III. FLOWERS

The flower is a modified shoot meant for production of seeds. A typical flower consists of four different circles (whorls) arranged in a definite manner. A flower is built upon stem or pedicel with the enlarged end known as thalamus or receptacle. The four whorls of the flowers are as follows.

1. Calyx: It is the outermost whorl of flower and is generally green in colour. The individual member of calyx is called sepal.

2. Corolla: It is the second whorl of flower and is either white or bright coloured. Each member of corolla is known as petal. The number of petals varies with the type of flower.

3. Androecium: It is the third circle of flower and constitutes the male part. The individual component is called stamen which consists of filament, anther and connective.

4. Gynoecium: This is the fourth circle of the flower and constitutes the female part. Each component is known as carpel or pistil and is made up of stigma, style and ovary.

In pharmacognosy, the drugs to be studied as flowers are either entire flowers in botanical sense or the inflorescence or single part of the flower, used medicinally. The following are few examples of different parts of flowers.

(i) Inflorescence

(a) Raceme	Digitalis, mustard
(b) Panicle (compound raceme)	Gold mohair
(c) Capitula (Head)	Chamomile, arnica, Artemisia, sunflower, pyrethrum
(d) Umbel	Caraway, fennel
(e) Cymose	Jasmine
(f) Hypanthodium	Fig

(ii) Stigmas	Saffron
(iii) Corolla and stamens	Elder flowers
(iv) Petals	Rose, red poppy
(v) Flower buds	Cloves

IV. FRUITS

The ovules of the flowers, after fertilization, are converted into seeds, whereas the ovary wall develops further to form the protective covering over the seeds, which is known as fruit. In botany, this particular coating is also called *pericarp*.

Pericarp consists of three different layers.

1. **Epicarp:** It is the outermost coating of the pericarp; it may be thin, thick or woody.
2. **Mesocarp:** A layer in between epicarp and endocarp, usually pulpy or made up of spongy parenchymatous tissue.
3. **Endocarp:** The innermost layer of the pericarp may be thin, thick or even woody.

If the ovules do not fertilize, the seedless fruits are formed. Depending upon the number of carpels present in the flowers, the fruits fall into following categories:

1. Simple fruits,
2. Aggregate fruits, and
3. Compound fruits.

1. **Simple fruits:** These are formed from the single carpel or from syncarpous gynoecium. Depending upon the mesocarp, whether it is dry or fleshy, they are classified as dry fruits and fleshy fruits. Dry fruits are further classified into dehiscent and indehiscent fruits.
2. **Aggregate fruits:** These fruits are formed from many carpels or apocarpous gynoecium.
3. **Compound fruits:** In this particular case, many more flowers come together and form the fruits.

False fruits: Sometimes, apart from the ovary, the other floral parts like thalamus, receptacle or calyx grow and form the part of the fruit; such a fruit is known as false fruit or pseudocarp. Following are few examples of pseudocarp in which other parts of the flower forming important part of the fruits are shown in the bracket.

Strawberry (thalamus).

Cashew nut (Peduncle and thalamus).

Apple (thalamus).

Marking nut (peduncle).

Rose (thalamus),

Pharmaceutical fruits differ from botanical fruits because pharmaceutical fruits may or may not contain all the three layers. For example, Lemon and orange consist of only epicarp only tamarind and bael consist of mesocarp, while fennel and dill contain all the layers, enclosing seed.

FRUITS

- **Simple fruits**
  - **Dry Fruits**
    - **Dehiscent fruits**
      1. **Legume:** Gram, Pea, Senna, Tamarind
      2. **Capsule:** Cardamom, Cotton, Poppy, Datura
      3. **Follicle:** Anise, Calotropis, Rauwalfia
      4. **Siliqua:** Radish, Mustard, Cabbage
    - **Indehiscent fruits**
      1. **Achne:** Cannabis, Sunflower, Climatis, Rose
      2. **Caryopsis:** Rice, Wheat, Maize, Bomboo
      3. **Nut:** Cashew nut, Areca nut, Chest nut
      4. **Samara:** Dioscorea, Shorea
      5. **Schizocarps**
         - **Lomentum**: Acacia, Peanut
         - **Cremocarp**: Coriander, Caraway, Dill, Fennel
  - **Fleshy fruits (succulent fruits)**
    1. **Drupe:** Coconut, Mango, Plume, Olive
    2. **Berry:** Capsicum, Bael, Colocynth, Cucumber
    3. **Pome:** Apple, Pear
    4. **Hesperidium:** Orange, Lemon, Grape fruit
    5. **Balausta:** Pomegranate
- **Aggregate fruits**: Rasp berry, Star anise
- **Compound fruits**: Long pepper, Pineapple Mulberry, Hops

V. SEEDS

The seed is fertilized ovule. It represents a condensed form of life and it is a characteristic of phanerogams. Parenchymatous body of the ovule known as nucellus contains embryo sac surrounded by integuments (coatings). In the embryo-sac itself fertilization takes place giving rise to embryo. Thus, seeds are characterized by the presence of three parts namely embryo, endosperm and seed coat. Endosperm is the nutritive tissue nourishing the embryo. Endosperm may or may not be present in the seeds. Therefore, seeds are classified as follows:

1. Endospermic or albuminous seeds.
2. Non-endospermic or exalbuminous seeds.
3. Perispermic seeds.

1. Endospermic or albuminous seeds: A part of the endosperm remains until the germination of seed and is partly absorbed by embryo. It shows distinct presence of endosperm, e.g. colchicum, isapgol, linseed, nux vomica, strophanthus, etc.

2. Non-endospermic or exalbuminous seeds: During the development of seed, the endosperm is fully absorbed by embryo and endosperm is not represented in the natural seeds. e.g. sunflower, tamarind, cotton, soybean, etc.

3. Perispermic seeds: Herein, the nucleus develops to such an extent that it forms a big storage tissue and seeds are found to contain embryo, endosperm, perisperm, and seed coat. e.g. pepper, cardamom, nutmeg, etc.

Seeds are characterised by the following structures:

(a) Hilum: This is the point of attachment of seed to stalk.

(b) Micropyle: It is the minute opening of the tubular structure, wherefrom water is provided for the germination of seeds.

(c) Raphe: Raphe is described as longitudinal marking of adherent stalk of anatropous ovule.

Special Structures:

In some instances, apart from regular growth of seeds, additional growth is visible in the form of appendages.

1. **Aril** : Succulent growth from hilum covering the entire seeds, as observed in nutmeg (mace).
2. **Arillode** : Outgrowth originating from micropyle and covering the seeds, as seen in cardamom.
3. **Arista (awn)** : Stiff bristle like appendage with many flowering glumes of grasses, as found in strophanthus.
4. **Caruncle** : It is warty outgrowth from micropyle as seen in Castor, croton, viola etc.
5. **Strophiole** : Enlarged funicle e.g. *Datura fastuosa* and colchicum seed.
6. **Hairs** : *Gossypium* and Calotropis are the examples of this type of outgrowth.

These appendages are found to perform special functions sometimes e.g. hairs and awns of seeds help their dispersal.

VI. BARKS

The secondary external tissues lying outside the cambium in stem or root of dicotyledonous plants are known as the bark. Botanically, bark is also known as periderm. It consists of three layers namely cork (phellem); cork-cambium (phellogen) and secondary cortex (phelloderm). Commercially, barks consist of all the tissues outside the cambium. A young bark includes epidermis, cortex, pericycle and phloem. Barks are obtained from the plants by making longitudinal and transverse incisions through the outer layers followed by peeling. Barks may be obtained from stems or roots. Due to the excessive growth produced by the cambium and cork cambium, the external tissues get tangentially stretched or torn; hence, the epidermis is not found in the barks.

Methods of collecting Barks

Barks are collected in a season when they are maximum in their desired chemical constituents; i.e. cinnamon bark is collected in rainy season. Wild cherry bark is collected in antumn and cinchona bark is collected in rainy season. Various methods of collecting the barks are as under:

(1) Felling Method: This is very old method of collecting the barks in which tree is cut the base and bark is peeled out. This method is not used at present commercially, since if causes the destruction of trees.

(2) Uproozing method: As the word indicates in this method the roots of the plant are dug out of soil and bark to streeped off from the roots. This method is applied for collection of root bark of cinchona in Jawa at present.

(3) Coppicing: In this method plant is allowed to grow for the definite age and it is cut-off at specific distance from soil. The tree stamps which thus remain in the ground are allowed to send certain number of shoots which develop further independently to form the new plant. Again these new plants are cut off and bark is collected from their shorts. In this case cost of cultivation as well as the time is saved and hence this is the method of choice of collecting barks commercially at present. Cascarz, cinnamon and cinchona barks are collected by this method.

Characteristics of Barks

Barks exhibit several morphological and microscopical characters. The morphological characters need special attention, as they help in identification of the barks.

Shapes in Barks

The shape or form of the bark is dependent upon the method adopted for its preparation. It also depends on the type of incision made and the extent of any subsequent shrinkage of the tissues. When the bark is removed from the large trees and dried under pressure, the flats are produced, e.g. Quillaia and Arjuna. When the bark is removed from the small branches due to shrinkage of the soft tissues, it tends to curve forming a concavity on the inner side, yielding *curved* pieces; e.g. wild cherry, and cassia. If the concavity is on the outer side of the bark, it is described as *recurved*; e.g. kurchi.

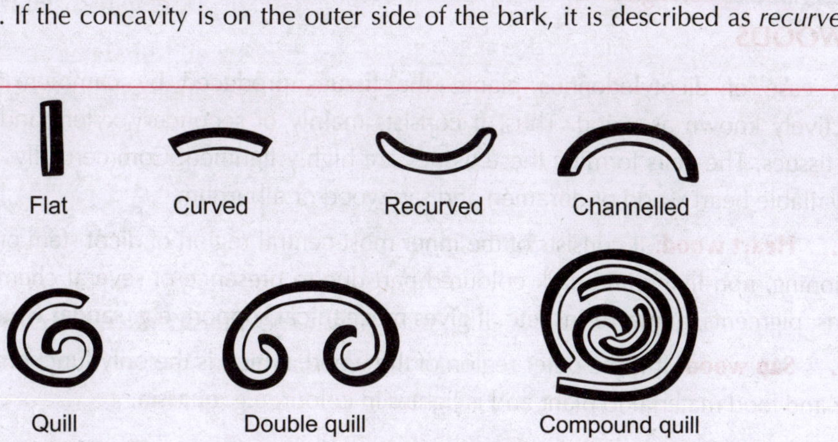

Fig. 2.16: Shapes in barks

When the shrinkage of the tissues is to a greater extent and it forms deep trough or channel, it is called a *channeled bark* e.g. ashoka, *Cinchona ledgeriana* and cassia. In some cases, one edge of a bark covers the other to form *quill* e.g. cascara and cinnamon. If both the edges of the bark roll independently forming quill, it is described as *double quill* e.g. Java cinnamon. In some cases, one quill of a bark is put inside other quill to form a *compound quill* e.g. cinnamon. Compound quill is a man-made shape of bark. It reduces the exposure of bark to atmospheric conditions and also saves the space in transport (Fig. 2.17).

Fractures in bark:

The appearance shown by the transversely broken surfaces of the bark is known as **fracture**. It is, sometimes, useful in identification of barks. The types of the fractures are as follows.

When fractured surface is smooth, it is described as a *short* fracture (cinnamon and kurchi). If the exposed surface exhibits small rounded appearance, it is described as a *granular* fracture (wild cherry and cassia). If the broken surface shows the presence of uneven projecting points, it is described as *splintery* fracture, as seen in cinnamon. The presence of numerous fibres on the transversely broken surface is described as a *fibrous* fracture (cinchona). If the exposed surface shows the arrangements of layers one over the other, it is described as a *laminated* fracture, as observed in Quillaia.

The various characters shown by the barks on the outer, as well as, inner surface are also diagnostically important. Amongst these, the colour, condition and presence of several growths like lichens, mosses etc., are characteristics to each bark. The presence of lenticels and development of cracks are additional characters of bark. Outer surface of the bark shows presence of cracks and fissures, which are due to lack of elasticity or due to increase in girth of the trees. **Fissures** are usually deep. **Wrinkles**, which are seen on outer surface of the bark, result due to shrinkage of inside soft tissues. **Furrows** are troughs between wrinkles. Inner surface of bark shows characteristics such as *striations*, which are longitudinal and parallel lines. Transverse wrinkles present on inner surface are described as **corrugations**.

VII. WOODS

In case of dicotyledonous plants, the tissues produced by cambium on inner side are collectively known as wood. Thus, it consists mainly of secondary xylem and smaller amount of other tissues. The cells forming these tissues are highly lignified. Commercially, two types of woods are available heart wood or duramen and sap wood or alburnum.

1. **Heart wood:** It consists of the inner most central region of dicot stem or root, which is non-functioning, non-living and dark coloured part due to presence of several chemical substances like tannins, pigments, gums, resins, etc. It gives mechanical support, e.g. **sandal wood**.

2. **Sap wood:** It is the outer region of the wood, which is the only functional wood conducting water and food material to plant and is lighter in colour, e.g. **quassia**.

VIII. UNDERGROUND DRUGS

Underground part of the plant may be either root or it may be a underground, aerial or subaerial modification of stem. The functions performed by modified stem or root are basically different, but taking into consideration their occurrence as underground parts of the plants, irrespective of their functions, they can be put together for the sake of convenience and study in the catagories as under:

Medicinally important Subterranean drugs

(a) **Underground aerial modifications** of stems are of the following types.

1. Rhizome
2. Tuber
3. Bulb
4. Corm.

1. Rhizome: The rhizomes grow horizontally under soil. They are thick, fleshy and characterized by the presence of nodes, internodes and scale leaves. They also possess bud in the axil of scale leaves. Adventitious roots may also be present on lower surface of rhizomes. The rhizomes may be branched and serve as storage organs. The examples of rhizomes are ginger, dioscorea, turmeric, rhubarb, male fern, etc.

Medicinally Important Subterranean Drugs:

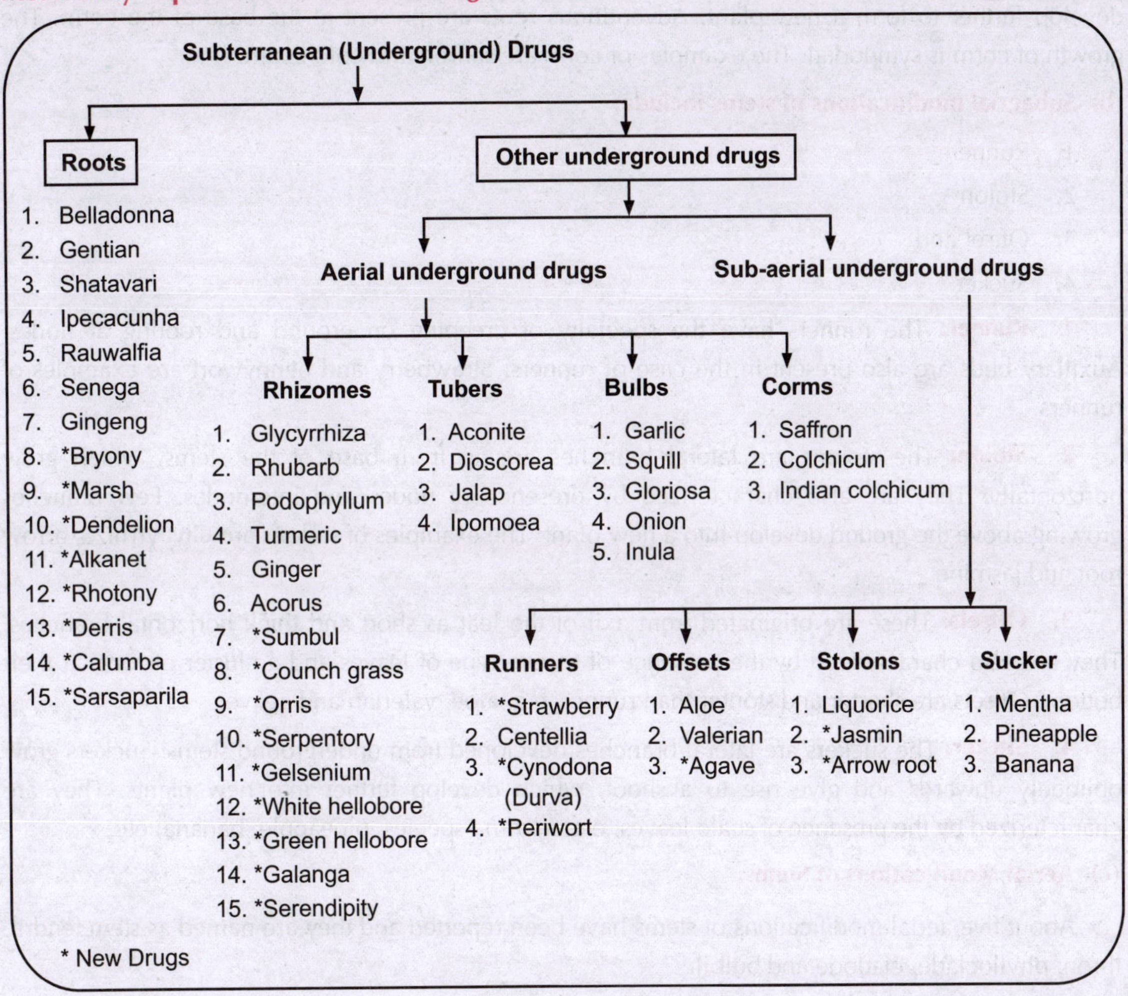

2. Tuber: The tubers are characterised by presence of 'eyes' from vegetative buds, which grow further and develop into a new plant. Adventitious roots are absent in tubers. Tubers are swollen underground structures of plant. Tubers are organs of storage, as well as, of propagation. The examples of tubers are potato, jalap, aconite, dioscorea, etc.

In India, roots and rhizomes cover about 33 percent of natural drugs used by pharmaceutical profession as a whole, for pharmaceutical formulations, Isolation or for export purposes.

3. Bulb: The bulb is a specialized underground shoot. The food material is stored in fleshy scales, which overlap the stem. The buds are present in axils of the scales and few of them develop in spring season at the expense of stored food material in the bulb. The reserve food material formed by leaves is stored at their bases and new bulbs are produced in subsequent year. The bulb is an organ of perennation and also vegetative propagation. The examples of bulb are garlic, onion, squill, Inula and *Gloriosa*.

4. Corm: The corms are also underground modifications of stems. Generally, they are stout and grow in vertical direction. They bear bud in the axil of the scale leaves and these buds then develop further to form a new plant. Adventitious roots are present at the base of the corm. The growth of corm is sympodial. The examples of corm are saffron and colchicum.

(b) Subaerial modifications of stems include:

1. Runner
2. Stolon
3. Offset and
4. Sucker

1. Runner: The runners have the specialty of creeping on ground and rooting at nodes. Auxillary buds are also present in the case of runners. Strawberry and pennywort are examples of runners.

2. Stolon: The stolons are lateral branches arising from base of the stems, which grow horizontally. They are also characterized by presence of nodes and internodes. Few branches growing above the ground develop into a new plant. The examples of stolons are Glycyrrhiza, arrow root and jasmine.

3. Offsets: These are originated from axil of the leaf as short and thick horizontal branches. They are also characterized by the presence of rosette type of leaves and a cluster of roots at their bottom. Offsets are shorter and stouter than runner, e.g. aloe, valerian and agave.

4. Sucker: The suckers are lateral branches developed from underground stems. Suckers grow obliquely upwards and give rise to a shoot, which develop further into new plants. They are characterized by the presence of scale leaves, e.g. *Mentha* species, pineapple, banana, etc.

(c) Aerial Modifications of Stems:

About five aerial modifications of stems have been reported and they are named as stem tendril, thron, phylloclade, cladode and bulbil.

Each of them is characterised as under.

(1) Stem tendrils: These are spirally curled branches by which these can attach themselves conviniently to the nearby objects and climb over them. Normally, they appear to be thin, wire-like leafless climbers.

Bud in the axil gets modified to tendril as in grape-vine (*vitis-vinifera*) Balloon vine.

(2) Thorns: These are described as strong, straight and hard pointed structures as in pomegranate, lemon, duranta. Thorns bear flowers, and even fruits. Prickles look like thorns but they are only outgrowth never bear flowers or fruits. Thorns are deep rooted while prickles are superficial. Typical example of prickles are rose, zizyphus, duranta.

(3) Cladodes: Some times flat or cylindrical branches arise from node of the stem which are short and green. Cladodes bear flowers.

Asparagus-racemosus and *Ruscus-aculeatus* are the common examples.

(4) Phylloclades: These are also flat, green branches of stems, containing nodes, internodes. They are characteristic to xerophytes (plants of deserts). Being green in colour they perform all functions of leaves including photosynthesis. Prickly pear, Opuntia dillenii and *Phyllocactus latifrons* are the common examples.

(5) Bulbils: These are produced by floral buds or vegetative buds only for the reproduction of plants. Bulbils get detached from the original plant i.e. mother plant and grows independently into a new plant. Onion, *diascorea, agave-americana* are the suitable examples.

IX. ROOTS

The roots are characterized by their downward growth into the soil. They do not have nodes and inter-nodes. The branching of roots arises from pericyclic tissues. The roots are covered by root caps or root heads.

Commercially, there is no clear demarcation between roots and rhizomes. Rhizomes studied in pharmacognosy may contain good proportion of root or even the roots may contain large amount of rhizomes.

Table 2.3: Utilisation of parts of medicinal plants by their use

Plant part	Percentage
1. Roots	29%
2. Whole plants	16%
3. Barks	14%
4. Fruits	10%
5. Seeds	07%
6. Leaves	06%
7. Stems	06%
8. Flowers	05%
9. Rhizomes	04%
10. Woods	03%

X. ENTIRE ORGANISMS

The term entire organism is used to denote the utility of major portion of a plant or animal as a drug. The plant drugs categorized as entire organisms are even called as herbs, which mean all the aerial parts of the plant, including leaves, flowers, seeds and smaller stems. This does not include the main axis of the stem. Thus, belladonna, cannabis, datura and lobelia are observed under this type, whereas, the insects like cantharides, cochineal and leech are covered under the animals. Ergot represents dried sclerotium of the fungus, whereas ephedra constitutes the stem portion of the plant. These two drugs are also classified as entire organisms.

Organised Crude Drugs

The term organised crude drugs itself is self-explanatory and refers to the fact they are either direct parts of plants or animals and have cellular structure. Organs are named as leaves, stems, flowers etc.

The unorganised drugs are not parts of plant or animals as such but are obtained from plant or animal parts by some extraction process such as oils, juices, lotions etc.

Unorganised Crude Drugs

Before dealing with the drugs in details, one should know how to distinguish clearly between organized (cellular) and unorganized (acellular) crude drugs.

Table 2.4: Difference between Organized and Unorganized drugs

Organised crude drugs	Unorganized crude drugs
1. As the term indicates these are 'organs' of plants or animals and are made up of cells or definite structure. These drugs are named as flowers, seeds, fruits, insects, etc.	1. These are derived from parts of plant or animal by some process of extraction and followed by purification, if necessary, e.g. juices, extracts, resins, etc.
2. These are solid in nature.	2. These are solid, semisolid or liquids in nature, e.g. oils, gums and balsams.
3. Botanical or zoological terminology can be used to describe these drugs.	3. Such terminology is inadequate to describe them, but one has to look for their physical characters, such as the solubility in various solvents, density, optical rotation, refractive index, etc. whichever is applicable.
4. Microscopic characters are one of the important criteria for the identification of organized drugs. **Examples:** Digitalis, cinchona, clove, fennel, jalap, ephedra, cochineal, etc.	4. Chemical tests and physical standards are confirmatory tests **Examples:** Aloe, agar, colophony, opium, castor oil, bees-wax, pepsin, etc.

Few unorganized crude drugs are commonly described by their physical characters only. The description for some of them is given below:

(i) Gums and Mucilages

Gums are translucent and amorphous substances, produced by plants. Gums are usually pathological products and are produced when the plant is growing under unfavourable conditions or is injured. Thus, they are the abnormal products of plant metabolism. The gums are produced by the process known as "*Gummosis*".

Gums are soluble or partly soluble in water. They are insoluble in alcohol and in most of the organic solvents. They form viscous adhesive solutions with water either by swelling or due to absorption. Aqueous solutions of gum are usually laevorotatory. Gums are plant hydrocolloids and may be anionic or non-ionic polysaccharides. On hydrolysis (acidic hydrolysis or prolonged boiling with water), gum yield sugar and uronic acids which form salts with calcium and magnesium. The uronic acids are glucoronic acid, galacturonic acid or aldobionic acid. Pharmaceutically, important gums are gum acacia, tragacanth, gum karaya, gum ghatti and guar gum.

Mucilages are also plant products similar to gums and are regarded to be the normal products of plant metabolism. They are produced inside the cells of the plant. Mucilages form slimy masses with water, but do not dissolve. Mucilages are esters of sulphuric acid, wherein ester group is a polysaccharide complex. Various examples along with method of formation of mucilages are given below -

1. From the cell wall of seed epidermis : Isabgol, linseed
2. From endodermis : Fenugreek
3. From leaf epidermis : Senna
4. From bark : Cinnamon, slipper-elm
5. From special secretion cells : Squill
6. From algae : Agar, chondrus

(ii) Resins and Resin Combinations

The resins are of two types:

1. Synthetic resins and
2. Natural resins

The scope of this book permits inclusion of only natural resins only as they alone pertain to pharmacognostic origin.

(a) Natural resins: The non-uniform physical and chemical nature of these substances make them difficult to define. They are obtained from plant, as well as, animal sources. The resin of animal source is shellac or lac, which finds number of applications in pharmaceutical industry.

Plant resins are natural or induced exudates. They are solids, liquids and semi-viscous materials from plants and are the end products of plant metabolism. Very often, they are clear, translucent yellow or brown materials insoluble in water and soluble in most of the organic solvents like alcohol, chloroform, carbon disulphide, benzene, etc. Upon heating, they soften, melt and finally burn with a sooty flame. On storage, some of them darken in colour due to chemical changes. Chemically, the natural resins are organic compounds without nitrogen. They represent various groups like acids, alcohols, esters and neutral substances called as resenes, e.g. podophyllum resin, jalap resin, colophony, sandarac, guaiacum, etc.

(b) Oleo-resins: When the natural plant resins are accompanied with the volatile oils in homogeneous form, they are known as oleo-resins. Canada balsam and copaiba are suitable examples of oleo-resins — do not compare these with **prepared oleo-resins**, which are prepared by percolating drugs containing volatile oil and resin together. The examples of this type are capsicum, black-pepper, turmeric, podophyllum and ginger oleo-resin.

(c) Balsams: Aromatic resinous substances of plant origin containing balsamic acids (i.e. benzoic and cinnamic acid) are known as balsams. Neither Canada balsam nor balsams of copaiba contain any balsamic acids and hence, they are not balsams in real sense. The examples of balsams are balsam of tolu, benzoin, storax and balsam of peru.

(d) Oleo-gum resins: These are the combinations of volatile oils, gum and resins. Sometimes, they also contain other substances like enzymes, e.g. myrrh and asafoetida.

The resins are found in different parts of plant or animal.

1. external glands: Indian hemp (cannabis)
2. in the heart wood of certain woods: guaiacum and red sanders

3. internal glands of plants: male ferns and
4. glands on surface of insects: shellac

Internal cavities producing resins in plant may be unicellular as in ginger, capsicum, and jalap or multicellular as in clove. The cavities producing resins may be schizogenous or schizolysigenous, according to the mode of their origin. Resins are usually formed by plants as their normal physiological products, while in some cases they are pathological products and are produced when plants are injured.

Glyricedia plant: No survival of human race under this plant and also to birds serpents, and are removed from hills in Pune area.

(iii) Dried Juices

The juices are obtained from fleshy leaves (aloes) or from stems of the trees (Kino). In all cases, incisions are made to respective part of the plant and juice coming out is collected and dried.

(iv) Latices

The latex is a product contained in special secretory tissues of certain plants. It is usually a white aqueous suspension, wherein microscopically small particles of oil globules are suspended. These natural suspensions of milky consistency may contain proteins, sugars, mineral and alkaloid salt in the true solution, whereas gums, starch and resins are in the suspended form. Examples of commercially important latices are rubber and chicle gum, while pharmaceutically important latices are opium, papain, ficin and gutta percha.

(v) Extracts

The extracts covered under crude drugs differ from galenical extracts. The extract of pharmacognostic origin consists of extracting the parts of the parts with water followed by concentration, while pharmaceutical preparations known as extracts are prepared by using alcoholic or hydro-alcoholic solutions and adjusting the product to a standard strength. Agar, sodium alginate, and catechu are few extracts of plant origin, while gelatin is the extract from animal source.

Table 2.5: Distribution of medicinal plants according to their habits

Plant type	Coverage
1. Trees	33%
2. Herbs	32%
3. Shrubs	20%
4. Climbers	12%
5. Others	03%

In colloquial language plants are described by various terms such as trees, herbs, shrubs, climbers and others but one must know how these terms differ from each other.

Tree: It is woody perennial plant with single thicken and branches that can gown to great height. For e.g. Neon mango.

Herbs: Plant with soft stem and may be annual, biannual or perennial, such as ginger, turmeric etc.

Shrubs: Plants with woody and hard stems, bromating profusely near the ground, larger than herbs i.e. Duronta, jasmine, croton etc.

Climbers: Plant with thin and long stems, grow with supports such as long pepper, calamas.

Poisonous plants and poisonous animals: Anything consumed in excess is troublesome and may be harmful, if not controlled. Consumption of natural drugs in excess is not exemption to that several plant drugs in excess even in small quantity are also harmful to human being such as. Aconite opium, Jamalgota and are known as poisonous plants which will be studied in detail separately.

Animals are also poisonous. List enclosed clarities suitably.

Morphological Nature of Natural Drugs:

1. Barks Arjuna Cascara segrada Cinchona Quillaia Wild cherry	**2. Leaves** Digitalis Senna Henbane Kalmegh Belladonna Pilocarpns	**3. Flowers** Ariica Artemisia Chammomile Clove Pyrethrum
4. Fruits Anise Cardamom Caraway Cassia pod Cumin Dill Lemon peels Orange peels	**5. Seeds** Cola seeds Isapgol Nutmeg Nux-vomica Physostigma Sweet almond	**6. Woods** Quassia Pterocarpus Sandal wood Rukta-chandan
7. Herbs Chaulmoogra Lobelia Parselay Tea-herb	**8. Roots** Chirrata Aconite Gentian Ipecacunha Jalap Senega Rauvolfia	**9. Rhizomes** Dioscorea Haridra Male fern Rhubarb
10. Latex Opium Papain	**11. Lipids** Pea-nut Halibut liver oil Cod-liver oil	**12. Vol oils** Peppermint Spearmint Lavender Camphor, Vetiver
13. Gums Acacia Tragacanth Locust bean Gum	**14. Animals** Bees wax Musk Shellac	

Harmful and Poisonous Animals:

I. Protozoa:

1.	*Entamoeba Hystolitica*	Microscopic	Amoebic dysentery
2.	*Plasmodium vivax*	Microscopic	Causes Malaria
3.	*Plasmodium falciferum*	Microscopic	Causes Malaria
4.	*Leishmonia*	Microscopic	Causes Kala-azar
5.	*Trypansoma Gambiense*	Microscopic	Causes Sleeping sickness
6.	*Balentidium Coli*	Microscopic	Balentidial dysentery

II. Platihelmenthes:

1.	*Fasciola-hepatica* (Liver-Fluke)	Persistant diarrhoea
2.	*Taena solium* (Pig tape-worm)	Body pain, and anaemic
3.	*Taenia saginata* (BEEF – Tapeworm)	Body pain, and anaemic
4.	*Diphyllo bothrium* (Fish-Tapeworm)	Body pain, and anaemic
5.	*Ascaris-lubricoides* (Rovnd-worm)	Colic pain, abdominal and nervous discomfort
6.	*Ancyclostoma-deodende* (Hork-worm)	Anaemia
7.	*Enterobias vernicularis* (Pin norm)	Sleeplessness

III. Arthropods:

8.	*Sarcoptes Seabie* (IT-MITE)	Seabies to human race
9.	*Pediculus hamanns* (Louse)	Head louse
10.	*Ticks and mites*	Blood suckers
11.	*Anopheles species* (Mosquito)	Maleria, Filaria
12.	*Musea-domestica* (House-fly)	Cholera, Typhoid, Diagghoea, Dysentry, Anemia
13.	*Pulex-irrilas* (Human-flea)	Plague

Chapter 3...

CULTIVATION, COLLECTION AND PROCESSING OF HERBAL DRUGS

(A) METHODS OF CULTIVATION

Though several countries in the world have a rich heritage of herbal drugs, very few can put claim for their procurement only from cultivated species. It is recently that some of these drugs have been subjected to systematic cultivation based on modern scientific information. Our reliance on wild sources of crude drugs and the lack of information of sound cultivation technology have resulted in gradual depletion of raw material from wild sources. Even though, cultivation of medicinal plants offers wide spectrum of advantages over their wild sources, it may be an uneconomical proposition for certain types of crude drugs which occur abundantly in their natural habitat, e.g. nux vomica, acacia, Myrobalan, etc. On the other hand, crude drugs like cardamom, clove, Indian hemp, poppy latex, tea, cinchona, ginger, linseed, isabgol, Ceylon cinnamon, saffron, peppermint, fennel, etc. are obtained from cultivated plants. The cultivation of vegetable drugs involves convergence of various agricultural and pharmaceutical factors, such as soil, climate, rainfall; irrigation, altitude, temperature; use of fertilizers, pesticides, genetic manipulation and biochemical aspects of natural drugs. When all such factors are precisely applied, the new approach to scientific cultivation technology emerges out.

The **advantages of cultivation** may be summarized as follows:

1. It ensures quality and purity of medicinal plants. Crude drugs derive their utility from chemical contents in them. If uniformity is maintained in all operations during the process of cultivation, drugs of highest quality can be obtained. Cultivation of rhizomes demands an adequate quantity of fertilizers and proper irrigation. Systematic cultivation results in raising a crop with maximum content of volatile oil and other constituents. The examples of ginger, turmeric and liquorice can be cited to illustrate this point. If the cultivated plants are kept free of weeds, the contamination of crude drugs can be conveniently avoided.

2. Collection of crude drugs from cultivated plants gives a better yield and therapeutic quality. However, it is a skilled operation and requires some professional excellence. If the collection of crude drugs for market is done from cultivated plants by skilled and well-experienced personnel, the high yield and therapeutic quality of drugs can be maintained. For example, collection of latex from poppy capsules and oleoresins from *Pinus* species, if done by experienced persons, can result in better yield of crude drugs. Preservation of green colour of senna leaves and minimizing the deterioration of cardiac glycosides in freshly collected leaves of digitalis can be achieved only by highly skilled labour.

3. Cultivation ensures regular supply of a crude drug. In other words, cultivation is a method of crop-planning. Planning a crop-cultivation regularizes its supply and as a result industries depending upon crude drugs do not face problem of shortage of raw material.

4. The cultivation of medicinal and aromatic plants also leads to industrialization to a greater extent. The cultivation of coffee and cocoa in Kerala has given rise to several cottage and small scale industries. The cultivation of cinchona in West Bengal has led to the establishment of the cinchona-alkaloid factory near Darjeeling. The government owned opium factory at Gazipur is an eloquent testimony to the significance of well planned cultivation of poppy.

5. Cultivation permits application of modern technological aspects such as mutation, polyploidy and hybridization.

Disadvantages of Cultivation

The high cost of cultivated drugs compared to collected from wild sources and the losses due to ecological imbalances such as storms, earthquakes, floods, droughts, etc. are major disadvantages of cultivation.

FACTORS AFFECTING CULTIVATION

Cultivation of medicinal and aromatic plants take cognizance of plant habitats and climatic requirements for their favourable growth.

The factors which are given special attention for cultivation are listed below:

1. Altitude, temperature and humidity
2. Rainfall or Irrigation
3. Soil and Soil fertilizers
4. Fertilizers and Manures
5. Pests and Pest Control

1. Altitude, Temperature and Humidity

Altitude is a very important factor in cultivation of medicinal plants. Tea, cinchona and eucalyptus are cultivated favourably at an altitude of 1000–2000 metres. Cinnamon and cardamom are grown at a height of 500–1000 metres, while senna can be cultivated at sea level. The following are the examples of medicinal and aromatic plants indicating the altitude for their successful cultivation. (Table 3.1)

Table 3.1: Altitude for Drug cultivation

Plant	Altitude (metres)
Tea	1000 - 1500
Cinchona	1000 - 2000
Camphor	1500 - 2000
Cinnamon	250 - 1000
Coffee	1000 - 2000
Clove	upto 900
Saffron	upto 1250
Cardamom	600 - 1600

Temperature is another factor affecting the growth of a plant. Excessive temperature, as well as, frost also affect the quality of medicinal plants adversely. The following are few examples of range conducive for luxuriant growth of certain medicinal plants.

Table 3.2: Optimum Temperature for Drug Cultivation

Plant	Optimum Temperature (°F)
Cinchona	60 - 75
Coffee	55 - 70
Tea	70 - 90
Cardamom	50 - 100

Camphor and coffee cannot withstand frost, whereas saffron needs only cold climate and pyrethrum requires dry weather for cultivation.

Range of minimum and maximum temperature available for cultivation of medicine plants in various countries of the world is mentioned against each of them

Table 3.3: Range of Temperature Favourable for Cultivation of Medicinal Plants

Country	Range of temperature in °C
1. Austrialia	10 – 30
2. France	01 – 20
3. Germany	– 05 – 25
4. Iran	05 – 25
5. Italy	10 – 20
6. Korea	05 – 25
7. Maritius	17 – 25
8. Malesia	20 – 30
9. Philipines	20 – 25
10. Russia	– 10 – 25
11. Shri-Lanka	22 – 30
12. South Africa	10 – 20
13. Stain	05 – 20
14. Sweeden	– 05 – 20
15. Switzerland	05 – 20
16. UAE	15.35
17. India (for various states)	
(a) Gujarat	20 – 30
(b) Jammu-Kashmir	05 – 20
(c) Maharasthtra	20 – 30
(d) Tamil Nadu	20 – 25
(e) Uttar-Pradesh	20 – 30
(f) West-Bangal	15 - 30

2. Rainfall or Irrigation

Hail fall and thunders are to be watched critically in the cultivation process of medicinal plants.

Except the xerophytes plants like aloe, acacia and few others, most of the plants need either proper arrangements for irrigation or sufficient rainfall for their favourable development. In a few cases, well distributed rainfall throughout the year is desired.

Irrigations: Now-a-days various new economical methods of irrigation are used even for cultivation of medicinal plants, replacing conventional dry-land methods. They are named as:

(i) Drip-irrigation method

(ii) Sprinkler irrigation method

(iii) Foliar-spray method (Fogging)

These methods are (a) Easy to operate, (b) convenient to handle (c) Labour and time saving, (d) Needs no technical or qualified workers with additional advantage of uniform distribution of water soluble fertilizers.

Fertigation: Providing water soluble fertilizers to crops by drip-irrigations is known as **Fertigation** which is cheapest and uniform method of providing fertilizers to plants, it also saves time.

3. Soil and Types of Soil

Soil is the most important natural resource as it supports growth of all plants. Soil provides mechanical anchorage as well as water and essential plant food elements for plant growth. The capacity of soil to supply plant nutrients in quantities and proportions required and to provide a suitable medium for plant growth is known as soil fertility. Soil makes chemical make-up and nutrients available to plants. Plant growth depends upon physical arrangement and nature of soil particles, organic matter content of soil and its living organisms.

The commonly known soil is the shallow upper layer and is the friable material in which plants find foothold and nourishment.

Clay is one of the highly weathered portions of the soil, consisting of finest particles. This provides the soil adhesive and cohesive properties and also holds plant nutrients with the result that nutrients are not lost through leaching.

Soil consists of mineral matter, air, water and organic matter. It is the mineral matter which makes a lot of difference in various forms of soil. Mineral matter may be coarse gravel, coarse sand or in the form of finest particles of clay and silt. Air and water give rise to pores, while purified and decayed plant and animal parts constitute the organic matter.

Depending upon the size of mineral matter, following names are given to the soil.

Table 3.4: Types of Soil

Types	Particle size (diameter)
Fine clay	Less than 0.002 mm
Coarse clay or silt	0.002 to 0.02 mm
Fine sand	0.02 to 0.2 mm
Coarse sand	0.2 to 2.00 mm

Depending upon the percentage covered by clay, soils are classified as given below.

Table 3.5: Classification of Soils based on Table 3.4

Sr. No.	Types of Soil	Percentage covered
1.	Clay soil	... More than 50% of clay
2.	Loamy soil	... 30 to 50% of clay
3.	Silt loam soil	... 20 to 30% of clay
4.	Sandy loam soil	... 10 to 20% of clay
5.	Sandy soil	... More than 70%
6.	Calcarious soil	... More than 20 % of lime

Any type of soil containing less than 0.5% of organic matter is considered as poor. If more than 1.5 to 5% of organic matter is present, it is described as rich soil. The soil with 0.5 to 1.5% of humus is termed as intermediate soil.

A soil good for plant growth should have half of the pore spaces filled with water and the rest with air, since good aeration is essential for root development.

The pH of soil decides favourable growth of plants and presence of micro-organisms. The maximum availability of plant nutrients is in between the pH range of 6.5 to 7.5. To bring the pH to neutral, acidic soils can be limed or alkaline soils can be reclaimed by application of gypsum.

Acidic soils are not suitable for leguminous plants due to poor development of nodular bacteria. Therefore, groundnut, sunflower seeds, cotton and rice grow better in alkaline soils only. Acidic pH is disadvantageous as it solubilizes more iron. Some of the plants like, tobacco, cinchona, tea, and potato grow well only in acidic soils. In alkaline soils, phosphorus is converted to insoluble forms of calcium phosphate and so it cannot be made available to aforementioned plants.

Soil Fertility:

It is the capacity of soil to provide nutrients in adequate amounts and in balanced proportion to plants. If cropping is done without fortification of soil with plant nutrients, soil fertility gets lost. It also diminishes through leaching and erosion. Soil fertility can be maintained by addition of animal manures, nitrogen-fixing bacteria or by application of chemical fertilizers. The latter is time-saving and surest of all above techniques.

Mulching:

Now-a-days there is lot of water shortage due to several reasons such as lack of rains, high temperature etc. Undesired use of water by society etc. and water saving has become a must.

This difficulty can be overcome by mulching. Mulching means to cover the irrigated land by suitable covering material made up of rexin or polythene sheets.

This saves not only the loss of water but evaporation of water also control improves quality of soil the undesired plants (weeds).

For effective, successful and profitable cultivation of medicinal plants, several compounds and chemicals are helpful and necessary. Cultivations of Medicinal plants is not so simple as cultivation of Jawar, Wheat, Bajra by usual agricultural practices.

Use of insecticides, pesticides is a must and unavoidable for protecting them against various pests and microbes.

Underlined are the few cultivation necessities for effective cultivation.

Cultivation Necessities

- Biostimulants
- Bio fertilizers
- Bio pesticides
- Micro organisms (Fungi)
- Pollen grains
- Plant growth regulators

4. Biostimulants

Biostimulants are the substances which in small concentration reduce the need for fertilizers and increase the plant growth, resistance to water and also abiotic stresses. Humic acid, algae-extract, plant growth regulators are the typical examples:

These are also efficient in favouring the good performance of plants vital processes and allow high yields and good quality product. Additionally when they are applied to plants enhance their nutrition efficiency abiotic stress tolerance and plant quality characteristics regardless of their nutrient contents.

Various chemicals are being used for formulation of biostimulant compositions. Such as humic acids, hormones algae extracts and plant growth promoting bacteria.

Biostimulants can be applied to seeds, plants and even soil.

Abiotic stress is defined as environmental condition whch reduces plant growth and yield below its optimal level.

Abiotic stresses may be drought salinity, cold and extreme temperatures. Biostimulants when applied correctly crops act directly on the physiological processes offering potential benefits for growth development, responses to water stress and toxic elements like aluminimum.

5. Fertilizers and Manures

Fiertilizers and their Types:

Plants also need food for their growth and development known as fertilizers. What plants need basically for their growth are the carbon dioxide, sunrays, water and mineral matter from soil. Thus, it is seen that with limited number of chemical elements, plants build up fruits, grains, fibres, etc. and synthesize fixed and volatile oils, glycosides, alkaloids, sugar and many more chemicals.

Fertilizers are chemicals when added properly to soil make the boil fertile i.e. enable it to produce crops more abundantly.

Chemical Fertilizers: As animals are in need of vitamins, plants are in need of 16 nutrient elements for synthesizing various compounds. Some of them are known as **primary nutrients** like nitrogen, phosphorous and potassium. Magnesium, calcium and sulphur are required in small quantities and hence, they are known as **secondary nutrients; trace elements** like copper, manganese, iron, boron, molybdenum, silica, zinc are also necessary for healthy plant growth. Carbon, oxygen, hydrogen and chlorine are provided from water and air, naturally.

Functions and Deficiency of Fertilizer Elements and Symptoms:

Every element has to perform some specific function in growth and development of plants. Its deficiency is also characterized by certain symptoms. Various parts of the plants are used in pharmaceutical industry for their active constituents, for example, leaves, rhizomes, roots etc. Some of the nutrients are responsible for growth of particular part of the plant body and hence their usefulness and deficiency needs to be described systematically.

Table 3.6: Fertilizers Elements, their Functions, Deficiency Symptoms and Products in which Available

Element of fertilizer	Function	Deficiency symptoms	Product in which available
1. Nitrogen (N)	1. Promotes vigorous vegetative growth 2. Builds up plant proteins and chlorophyll. 3. Improves the quality of leaf drugs.	Pale yellow appearance of the plant, shunted growth, less flowering, pre-mature shedding of leaves.	Compounds providing nitrogen to the plant (nitrogenous fertilizers) are namely ammonium sulphate, ammonium chloride, ammonium nitrate sulphate (ANS), calcium ammonium nitrate (CAN) and urea.
2. Phosphorus (P)	1. Being active part of the nucleus of the plant cell, it is essential for plants growth. 2. Stimulant of seed and fruit formation and also a stimulant of root development. 3. Brings early maturity of crops.	1. Shunted root growth. 2. Delays maturity and retards the plant growth in general.	Phosphorous containing fertilizers (i.e. Phosphatic fertilizers) are of two types - (i) Water soluble phosphate Single super phosphate, triple super phosphate (ii) Water insoluble phosphate: Di-calcium phosphate, biomineral rock phosphate.

Contd...

3. Potassium (K)	1. Mainly responsible for resistance of plants against diseases and adverse climatic conditions 2. Starch and sugar formation is enhanced and their movements in plant-parts are regularized.	1. Scorching and browning of tips of leaves. 2. Shrivelled seed, fruits, shunted growth.	Potassium chloride.
4. Sulphur (S)	Synthesis of various proteins and oils, formation of nodules and chlorophyll.	Leaves turn to yellowish-green colour.	Water wettable sulphur
5. Magnesium (Mg)	1. Essential constituent of green plant pigment chlorophyll. 2. Helps in carrying the phosphorus and other plant nutrients. 3. Essential for oil and fat formation in plants.	1. Discoloration of leaves and leaves tend to curve upwards. 2. Susceptible to fungal growth.	Magnesium sulphate.
6. Calcium (Ca)	1. Promotes root formation and responsible for hardness of plant tissue. 2. Detoxifying agents for organic acids in plants (by way of formation of calcium oxalate and calcium carbonate). 3. Translocation of carbohydrates necessary for mitosis.	1. Drying of plants, weaken the stem structure, falling of buds and blossoms. 2. Root growth retarded. 3. Hooking of the leaf-tips.	Calcium ammonium nitrate (CAN)
7. Zinc (Zn)	Formation of growth hormone like Indole acetic acid, essential part of most of enzymes, helps in utilization of phosphorus and nitrogen in plants.	1. Various symptoms in different plants. In general, plants are shunted. 2. Yellow coloured spots develop on leaves.	Zinc sulphate or Zinc oxide
8. Copper (Cu)	Activator for many plant activities such as development, reproduction and formation of vitamin 'A'.	Visible in certain plants. Brown spots on citrus fruits, yellowing of younger leaves in maize.	Copper sulphate.

Contd...

9. Manganese (Mn)	Involved in plant respiration process, catalyst for several enzymatic and physiological reactions in plants. Also in synthesis of chlorophyll.	Interveinal, chlorosis leaves turn to yellow colour.	Manganese sulphate
10. Iron (Fe)	Necessary for synthesis and maintenance of chlorophyll, as enzyme components.	Interveinal chlorosis of young leaves, yellowish green in colour of leaves.	Ferrous sulphate.
11. Boron (B)	Regulates potassium calcium ratio uptake and utilization of calcium in plants. Involved in lignin, protein synthesis.	Terminal buds turn to light green in colour, death of growing plants.	Borax.
12. Molybdenum (Mo)	For nitrogen utilization and nitrogen fixation.	Mottling of lower leaves, marginal necrosis, in-folding of leaves.	Sodium molybdate Ammonium molybdate Molybdenum trioxide
13. Silica (Si)	Increases the stiffness of cell walls and also strengthen them. Stiffness and hardness of leaves increases prevents fungal growth.	Susceptible to fungal infection.	Silicon-dioxide

ORGANIC FAMRING

This is method of agricultural production which avoids the use of synthetic products like fertilizers, pesticides, growth regulators and livestock feed additives. Successful organic farming depends on crop rotation use of manures (both of animal and vegetable origin) mineral rocks and all biological pest control methods to maintain the agricultural output, soil productivity and also tilt (to turn up with plough) to provide plant nutrients. Organic matter and water holding capacity plays very important role for the desired growth of medicinal plants.

Organic matter is formed from poultry manures, organic manures, farm yard manures, compost, oil cakes (of castor, karanj, neem) fishmeal and bonemeal.

In this connection it is to be stated that organic matter is the food of earth worms and micro-organisms of the soil. Micro-organisms decompose organic matter and convert it into humus.

Micro-organisms with suitable pH (i.e. 6.5 to 7.5) temperature (23 to 27°C) and optimum moisture, fix the atmospheric nitrogen in the soil and supply to the plants for their growth. The specific micro-organisms which are responsible for nitrogen fixation are rhizobium, azatobacter, azospirillium, bijericcia, azolla and blue-green algae species.

Humus is responsible mainly for water holding capacity of soil, more the organic matter, more the water holding capacity.

Organic farming works in harmony with nature which involves using techniques to achieve good crop yields without harming the natural environment or the people who live and work around them.

Organic farming began early in the twentieth century, primarily in Europe, and United States. The pioneers of the early organic movement were motivated by a desire to reverse the perennial problems of agriculture.

For example,

1. Erosion,
2. Soil depletion,
3. Decline of crop varieties,
4. Low quality food,
5. Livestock feed, and
6. Rural poverty.

They embraced a holistic notion that the health of a nation built on agriculture is dependent on the long-term vitality of its soil. The soil's health and vitality were believed to be embodied in its biology and in the organic soil fraction called *humus*.

Principles of Orgnaic Farming:

Organic farming is native to India. It is simply based on "feed the soil" which means feeding the soil food web. The soil food web is the living fraction of the soil, composed of bacteria, fungi, earthworms, insects, and a host of other organisms that digest organic matter and provide nutrition to growing crop plants. The concept of organic farming is mainly based on the principles such as:

- Nature is the best role model for farming, since it does not use any inputs nor demand unreasonable quantities of water.
- The entire system is based on intimate understanding of nature's ways.
- The soil in its system is a living entity.
- Living population of microbes and other organisms are significant contributors to soil, fertility on a sustained basis and must be protected and nurtured at all cost.
- The total environment of the soil, from soil structure to soil cover is most important.

Thus organic farming is defined **as a farming system which primarily aims at cultivating the land for raising crops in such a way, so as to keep the soil alive and in good health by use of organic wastes.**

Food and Agriculture Organisation (FAO) defined organic farming as *a unique production management system which promotes and enhances agro-ecosystem health, including biodiversity, biological cycles and soil biological activity.*

This is accomplished by using on farm agronomic, biological and mechanical methods in exclusion of all synthetic off-farm inputs".

Need of Organic Farming:

The relevance and need for an eco-friendly alternative farming system arose from the ill effects of the chemical farming practices adopted worldwide during the second half of the last century. The methods of farming evolved and adopted by our forefathers for centuries were less injurious to the environment. People began to think of various alternative farming systems based on the protection of environment which in turn would increase the welfare of the mankind by various ways like clean and healthy foods, low use of the non-renewable energy sources. Many systems of farming came out of the efforts of many experts and laymen. However, organic farming is considered to be the best among all of them because of its scientific approach and wider acceptance all over the world. Modern, intensive agriculture practices have caused many problems, related with soil, water environment, and health.

- Chemical fertilisers and herbicides are easily washed from the soil and pollute rivers, lakes and water courses.
- The prolonged use of chemical fertilisers results in soils with a low organic matter content which is easily eroded by wind and rain.
- Greater amount of chemicals are needed every year to produce the same yields of crops.
- Artificial pesticides can stay in the soil for a long period and enter the food chain where they begin to accumulate in animals and humans, causing health problems.
- Artificial chemicals destroy soil micro-organisms resulting in poor soil structure and aeration and decrease in the nutrient availability.
- Pests and diseases become more difficult to control as they become resistant to artificial pesticides. The numbers of natural enemies decrease because of pesticide use and habitat loss.

Organic farming provides long-term benefits to people and the environment as follows:

- Increase long-term soil fertility.
- Control pests and diseases without harming the environment.
- Ensure that water stays clean and safe.
- Use resources which the farmer already has, so the farmer needs less money to buy farm inputs.
- Produce nutritious food, feed for animals and high quality crops to sell at a good price.

Organic Fertilisers:

According to the Minnesota (USA) Department of Agriculture, a natural organic fertiliser has to be derived from either plant or animal materials containing one or more elements (other than carbon, hydrogen, and oxygen) that are essential for plant growth.

Thus, *organic fertilisers are defined as the materials used as fertiliser that occur regularly in nature, usually as a byproduct or end product of a naturally occurring processes.*

The types of organic fertilisers are as under

a) Compost

Compost fertiliser is organic matter (plant and animal residues) which has been degraded by the action of bacteria and other organisms, over a period of time. Materials such as leaves, fruit skins and animal manure can also be used to make compost. Compost fertiliser is cheap, easy to make and is a very effective material that can be added to the soil, to improve soil and crop quality. Compost has following advantages.

- Compost improves the structure of the soil. This allows more air into the soil, improves drainage and reduces erosion.
- Compost improves soil fertility by adding nutrients and by making it easier for plants accept up the nutrients present in the soil. This produces better yields.
- Compost improves the soil's ability to hold water. This stops the soil from drying out in times of drought.
- Compost can reduce pests and diseases in the soil and on the crop.

b) Green Manures

Green manures, often known as cover crops, and are plants which are grown to improve the structure, organic matter content and nutrient content of the soil.

They are a cheap alternative to artificial fertilisers and can be used to complement animal manures. Growing green manure is not the same as simply growing a legume crop, such as beans, in a rotation. Green manures are usually dug (put) into the soil when the plants are still young, before they produce any crop and often before they flower. They are grown for their green leafy material which is high in nutrients and provides soil cover. They can be grown together with crops or alone. Green manures have the following advantages.

- Increase and recycle plant nutrients and organic matter.
- Improve soil fertility and soil structure.
- Improve the ability of the soil to hold water.
- Control soil erosion and prevent weed growth.
- Stop nutrients being washed out of the soil, for example, when the ground is not used between main crops.

c) Blood Meal

Obtained from slaughter houses is an excellent source of nitrogen. It is used to side-dress leafy vegetables.

d) Bone Meal

It is one of the best sources of phosphorus. It is applied at the bottom of plants in the holes of bulbs, shrubs and trees.

e) Fish Meal (Liquid)

It is a good source of micronutrients. It is used specifically to feed container-grown plants after every two weeks.

f) Poultry Manure

The waste products from the poultry industry contain all three macronutrients i.e. N, P and K are used every where in agriculture floriculture and green houses.

Worm Composting:

Worm composting is the fastest and easiest way to recycle house hold food-waste. It is a natural method for recycling nutrient rich food scraps.

Worm composting is made in suitable containers made up of plastic or earthen pots or wood boxes. The containers or bags are filled with moistened soil bedding and **red worms**. The containers can also be prepared by using newspapers and leaves.

Fig. 3.1: Red worms

Process does not produce any unpleasant smell. Since the worms get desired food get reproduced plentifully. The faeces produced by worms are excellent soil conditioner for the crops. One can get rid of house garbage by producing ideal food for plants (crops). The process does not entertain fruit flies or *gnats* or insects which cause damage to the crop.

Advantages of Organic Farming:

- Organic fertilisers are made from naturally occurring sources.
- They lower the amount of greenhouse gases that are released into the atmosphere.
- Nutrients are released only when media is warm and moist.
- The slow-release nature of most organic fertilisers may slightly decrease the runoff of nutrients into local water systems when compared to some quick-release synthetic sources that release nutrients regardless of media conditions.
- Organic fertilisers not only provide essential nutrients to plants, but they also improve soil structure.
- Organic matter helps to break up heavy clay soil, improve air circulation and drainage, and it increases the capacity for sandy soils to retain moisture. It maintains good soil structure.
- Makes it easier for the roots to reach moisture and to absorb the nutrients in the soil.
- In terms of the end product, such as the quality of plant produced by a commercial greenhouse, organic fertilisers can be quite competitive with traditional synthetic sources for a detailed study comparing products of organic and inorganic fertilisers.
- When used over a long period of time during in-ground production, organic fertilisers may increase the quality of the soil, including an increase in the efficiency of nutrient utilisation.

Limitations of Organic Farming:

- Costs of organic fertilizers are significantly more than synthetic fertilisers.
- Organic certification requires documentation and regular inspections.
- Organic fertilisers, despite the advantages discussed above, still release nutrients into their surroundings; these nutrients can find their way into local streams, rivers, and estuaries just as nutrients from synthetic sources do.

- Motivations for using organic fertiliser vary from user to user. Users want to go organic in order to sell product, and members of society might be interested in organic fertiliser for home use because of environmental concerns. This would increase the cost of the final product.

Manures: Farm yard manure (FYM/compost), castor seed cake, poultry mannures, neem and karanj seed cakes vermi compost, etc. are mannures. Oil-cake and compost normally consist of 3 to 6 per cent of nitrogen, 2 per cent phosphates and potash 1 to 1.5 per cent potash. They are made easily available to plants. Bone meal, fishmeal, biogas slurry, blood meal and press mud are the other forms of organic fertilizers.

Cultivation Necessaties: Apart from soil, fertilisers, water, plant growth hormones many other compounds are necessary for effective, successful cultivation so get the maximum (Bumper) output of crops. These may be called as cultivation necessaries, which contribute to the maximum yield. Without these cultivation may not be economical or fruit ful in the ordinary conditions in the ordinary course of agricultural farming.

These may be enlisted as:

- Biostimulants
- Bio fertilisers
- No of pollen grains
- Fungal organisms like trichoderm

Biofertilizers: Inadequate supply, high costs and undesired effects if used successively are the demerits of fertilizers or manures and hence the cultivator has to opt for some other type of fertilizer. The biofertilizers are the most suitable forms that can be tried. These consist of different types of micro-organisms or lower organisms which fix the atmospheric nitrogen in soil for plants can use them for their day to day use. Thus, they are symbiotic.

The examples of microbial Biofertilizers are:

Rhizobium	*Azotobactor*	*Azosperillium*
Bijericcia	Blue-green algae	*Azolla* and

Phosphorus solubilizing organisms.

There are several advantages or benefits of using, Biofertilizers in agricultural practices. Therefore they are to be focused for their detailed information.

Biofertilizers are classified according their source of origin, in three classes:

1. Bacterial biofertilizers
2. Algal biofertilizers
3. Fungal biofertilizers.

Their details are as under.

1. Bacterial biofertilizers: these are further sub divided in three groups such as;

(i) Rhizobium

(ii) Azospirillum

(iii) Azotobactor

(iv) Phospate solubilizing bacteria.

2. Algal biofertilizers: These are separate group of biofertilizers with common occurance of bacteria and algae. These are cyanobacteria or blue-green algae. It is well known that algal plants are plants without true stems or leaves and are capable to carry photosynthesis and nitrogen fixation.

3. Fungal biofertilizers: Like other two types of biofertilizers fungal biofertilizers also boost the crop yield and enhance the soil fertility. In organic forming, eco-friendly farming, zero-budget farming and nature farming. They are equally important, and reduce the dependency on chemical fertilizers in agriculture.

Several elements of fertilizers like Nitrogen, Phosphorus, Silica, Potassium, sulphur, Magnesium, Zinc, Calcium, Manganese are required by plants for their proper growth. However these are present in complex forms, but biofertilizers make them soluble and make them available for crops.

Bio-mineral fertilizers: The combination of minerals and useful micro-organisms together are known as Biomineral fertilizers.

Rhizophore, the narrow zone of soil around the root shelters maximum no of bacteria as compared to other parts promotes plant growth. Mycorrhizal fungi (Mycorrhiza means fungus root) form symbiotic relationship with plant roots, about 90% of plants rely on mycorrhizal fungi for mineral nutrients i.e. phosphrous.

Utility and Importance of Biofertilizaers

1. Improve the soil structure
2. Improve sustainability by strengthening the soil profile.
3. Reduce the production cost of crop as compared to chemical fertilizers.
4. Yield of crops is increased.
5. Cyanobacteria improve the quality of soil as these bacteria produce growth hormones, proteins, vitamins and several minerals, also inhibit the spreading of soil born diseases.

Trichoderma Fungus

Normally it is felt that micro-organism are harmful or pathogenic, but there are several micro-organisms which are very useful and without which our life will not be safe and secured.

Micro-organisms are universal in their occurance and are found in cold climatic conditions, in forests, coastal salty area, in tundra* in air, in ocean, soil and all leaving animals including human being.

One of these millions is fungus TRICHODERMA. Trichoderma biologically is known as *Trichoderma-hersianum* or *Trichoderma vividi* Family: Hypocraceae.

Trichoderma is found on dead, decade or putrified matter, alongwith many natural or man-made chemicals and even on toxic compounds.

- It occurs on various food products.
- It is fast growing micro organism and resistant to environmental or climatic changes. It is also found abundantly on roots of several plants.
- It produces several pigments imparting vivid colours and shades such as yellow, red, green etc.

---

* Tundra means vast, flat, tree less arctic region with permanently, frozen soil, lying under the surface soil.

- It can produce fungicides, insecticides and even plant growth regulators while growing.
- When comes in contact with roots, the various plant growth regulators it can produce Indole acetic acid, Gibberlins, abscisic acid, salicyclic acid etc.
- Trichoderma also attacks other fungi causing diseases and produce various proteolytic enzymes and other enzymes like glucanuse, catinase, cellulase, and distroy their cell-wall and causes mico-parasitism.

Various metabolites and acids like Jasmonic acid:

- Due to proper utilization of leucoteroma fungus, heavy metal content, polluted water, soil contaminated with petroleum fractions can also be controlled.

Production of Chemicals:

- Lacton content produced by this fungal growth helps in proper development of root-system in plants.
- Trichoderma produces undermentioned chemical compounds of various applications such as:

Sr. No.	Antifungal chemical compounds produced by Trickoderma	Insectificidal chemical compounds produced by tricoderma
1.	6-pentyl pyron	Methyl pyron
2.	Peptaibols	p-cymin
3.	Herzianic acid	A pinine
4.	Viridin	8 methyl quinolin
5.	Tricodermin	Dec denoic acid methyl-ester
6.	MicoToxin	Di-hydro theophin

Uses: Apart from its above mentioned purposes it is unique in crop development, protection of plants from various harmful diseases. It encourages the growth of herbals in friendly association to boost the harvest.

METHODS OF PROPAGATION

Medicinal plants can be propagated by two usual methods as applicable to non-medicinal plants or crops. These methods are referred to as Sexual method and Asexual method. Each method has advantages and also disadvantages.

1. Sexual Method (Seed Propagation)

In the case of sexual method, the plants are raised from seeds and such plants are known as seedlings. Sexual method of propagation enjoys the following advantages.

1. Seedlings are long-lived (in case of perennial drugs) and bear more heavily (in case of fruits). Plants are sturdier.
2. Seedlings are comparatively cheaper and easy to raise.
3. Propagation from seed has been responsible for production of some chance-seedlings of highly superior merits which may be of great importance to specific products, such as orange, papaya, etc.

4. In case of plants where other vegetative methods cannot be utilized, propagation from seeds is the only method of choice.

Sexual method suffers from following limitations.

1. Generally, seedling-trees are not uniform in their growth and yielding capacity, as compared to grafted trees.
2. They require more time to bear, as compared to grafted plants.
3. The cost of harvesting, spraying of pesticides, etc. is more than the grafted trees.
4. It is not possible to avail of modifying influence of root stocks on scion, as in case of vegetatively propagated trees.

For propagation purpose, the seeds must be of good quality. They should be capable of high germination rate, free from diseases and insects and also free from other seeds, used seeds and extraneous material. The germination capacity of seeds is tested by rolled towel test, excised embryo test, etc. The seeds are preconditioned with the help of scarification to make them permeable to water and gases. If the seeds are not to be germinated in near future, they should be stored in cool and dry place to maintain their germinating power. Long storage of seeds should be avoided.

Before germination, sometimes a chemical treatment is given with stimulants like gibberellins, cytokinins, ethylene, thiourea, potassium nitrate or sodium hypochlorite. Gibberellic acid (GA_3) promotes germination of some type of dormant seeds and stimulates the seedling growth. Many freshly harvested dormant seeds germinate better after soaking in potassium nitrate solution. Thiourea is used for those seeds which do not germinate in dark or at high temperatures.

When enhance germination is required, special treatments to seeds are given, such as soaking the seeds in water for a day, e.g. castor seeds, and other slow germinating seeds. Sometimes, seeds are soaked in sulphuric acid, e.g. henbane seeds. Alternatively, testa is partially removed by grindstone or by pounding seeds with coarse sand, e.g. Indian senna.

Methods of Sowing the Seeds:

Numerous methods of sowing the seeds of the medicinal plants are in practice. Few of them using seeds for cultivation are described:

(a) Broadcasting: If the seeds are extremely small the sowing is done by broad-casting method. In this method, the seeds are scattered freely in well prepared soil for cultivation. The seeds only need raking. If they are deeply sown or covered by soil, they may not get germinated. Necessary thinning of the seedlings is done by keeping a specific distance, e.g. Isapgol, Linseed, Sesame.

(b) Dibbling: When the seeds of average size and weight are available, they are sown by placing in holes. Number of seeds to be put in holes vary from 3 to 5, depending upon the vitality, sex of the plants needed for the purpose and the size of the plant coming out of the seeds.

For example, in case of fennel* 4 to 5 fruits of fennel are put in a single hole keeping suitable distance in between two holes. In case of castor, only 2 or 3 seeds are put. In case of papaya, the plants are uni-sexual and only female plants are desired for medicinal purposes. Hence, 5 to 6 seeds are put together and after the sex of the plants is confirmed, healthy female plant is allowed to grow while male plants and others are removed.

---

* In case of fennel, seeds are located in the fruits only firmly and can not be separated, hence instead of seeds, fruits are used for sowing.

(c) Miscellaneous: Many a times the seeds are sown in nursery beds. The seedlings thus produced are transplanted to farms for further growth, such as cinchona, cardamom, clove, digitalis, capsicum, etc.

MAT-POT Technology for nursery

Now-a-days biologically degraded, pollution-free paper pots are used to present and store the seedlings. Commercially paper-pot machines are prepared in America, Italy and China on large scale. These paper-pots are biologically degraded in the soil by micro-organisms and do not cause any damage to the soil or seedlings. Paper-pots of 30 × 40 – 50 × 60 mm size and 3000 – 6000 pots per day are produced. Availability of raw-material a cheapness of the paper-pots are the other benefits of the process.

Special Treatment to Seeds

To enhance germination, special treatments to seeds may be given, such as soaking the seeds in water for a day. e.g. castor-seeds and other slow-germinating seeds. Sometimes, seeds are soaked in sulphuric acid e.g. hebane seeds. Alternatively, testa is partially removed by grind-stone or by pounding seeds with coarse sand. e.g. Indian senna. Several plant harmones like gibberllins, auxins are also used to cover the seeds before sweings.

2. Asexual Method

In case of asexual method of vegetative propagation, the vegetative part of a plant, such as stem or root, is placed in such an environment that it develops into a new plant.

Asexual propagation enjoys following **advantages**:

1. There is no variation between the plant grown and plant from which it is grown. As such, the plants are uniform in growth and yielding capacity. In case of fruit trees, uniformity in fruit quality makes harvesting and marketing easy.
2. Seedless varieties of fruits can only be propagated vegetatively e.g. grapes, pomegranates and lemon.
3. Plants start bearing earlier as compared to seedling trees.
4. Budding or grafting encourages disease-resistant varieties of plants.
5. Modifying influence of rootstocks on scion can be availed of.
6. Inferior or unsuitable varieties can be overlooked.

It suffers from following **disadvantages**:

1. In comparison to seedling trees, these are not vigorous in growth and are not long-lived.
2. No new varieties can be evolved by this method.

(a) Vegetative propagation: It is done by sowing various parts of the plants in well prepared soil. The following are the examples of vegetative propagation.

1. Bulbs ... Squill, garlic
2. Corms ... Colchicum, saffron
3. Tubers ... Jalap, aconite, potato
4. Rhizomes ... Ginger, turmeric
5. Runners ... Peppermint /spearmint

6. Suckers ... Mint, pineapple, chrysanthemum, banana
7. Offsets ... Aloe, valerian
8. Stolons ... Arrow-root, liquorice

(b) Aseptic methods of micropropagation: It is a novel method for propagation of medicinal plants. In micropropagation, the plants are developed in an artificial medium under aseptic conditions from fine pieces of plants like single cells, callus, seeds, embryos, root tips, shoot tips, pollen grains, etc. They are provided with nutritional, hormonal requirements.

This topic is discussed in detail under **Chapter 5 on Medicinal Plant Biotechnology.**

Transplanting:

Transplant: Method of removing one growing plant with roots to another plant or place. It saves time and expenses of cultivation: Capsicum, Rice.

Azolla

Azolla also known as mosquito fern, or duck-weed fern or water fern is one of the aquatic ferns of the family Salviniaceae.

It is extremely reduced in form and is specialised, it is like other typical ferns or mosses.

Azolla filienloides is one of the few species of fast growing ferns and gets doubled in growth in a week. It contains 25-30 per cent proteins, 10-15 per cent minerals and 07-10 per cent amino acids.

Uses:

In addition to its use in traditional cultivation as a bio-fertilizer for wet land paddy, due to its ability to fix nitrogen in soil azolla is finding increasing use for sustainable production of live-stock feed, since the azolla is rich in proteins, amino acids, vitamins and minerals. Feeding azolla to dairy cattles, pigs, ducks and chickens has reported in increased milk production and egg-laying capacity.

5. Pests and Pest Control

Population increases in arithmetic progression, whereas the agricultural produce is enhanced in geometric progression. What man has reaped must be protected. At the same time, future yields must be improved. But, unfortunately this is not so because, between the times a medicinal crop is harvested and consumed by man, considerable quantity of crude drug is wasted or destroyed by pests. Besides, there is loss in quality because these pests are allowed to grow on produce. The overall losses due to pest infestation tune to millions of rupees every year. The control of pests thus, assumes primary importance in the context of cultivation of medicinal and aromatic plants.

Pest is an undesired animal or plant species and pesticides are chemicals derived from synthetic and natural sources effective in small concentrations against pests.

CROP-TRAPS

To control the hazardous attack of insects and pests, in cultivation of medicinal plants special measures are taken into practice and most important and practicable of them is the use of **"CROP-TRAPS"**.

The use of crop-trap depends upon the type of medicinal plant under cultivation, since various types of crop-traps are available. To reduce (lessen) the effect of insects or pests on main crops another highly susceptible crop for the sake of insects/pests is also cultivated simultaneously side by side. So that the insects/pests will attack the another (Susceptible) crop first and will keep the main crop of medicinal plant free from insect/pest attack.

Various types of crop-traps available in the agricultural practice are as under.

(1) Funnel trap : For soyabean, grapes, peas, tomatoes etc.

(2) Fly 'T' trap : For mangoes, sapota, guava

(3) Bucket trap : For coconut, areca nut etc.

Table 3.7: Important crop-traps used in actual practice in agriculture

Main Crop	Crop Traps of Herbal Drugs
Cotton	Cow-pea or maize or proso-millet
Soyabean	Castor, sunflower
Ground nut	Sunflower
Tomato	Zendu (Tagetes Genda)
Cajanus	Jawar or Bajra
Fruit-crops	Zendu (Tagetes/Genda)

Now-a-days a new technique to control pests on the crops have been developed called 'Trap-crops' which consists of a crop which is more susceptible for the growth of pets is also cultivated alongwith the desired crops where by the pests will attack the trap-crop preferably and the main crop remains untouched. The infected trap crop is separately collected and destroyed.

Types of Pests

The different types of pests infesting medicinal plants are fungi, viruses, weeds, insects and non-insect pests including rodents.

(i) Fungi and Viruses: Different types of fungi are known to occur on medicinal plants.

Ascochyta atropae causes the formation of greyish-white irregular spots which further cause necrosis of leaves. The disease is called leaf necrosis. *Cercospora atropae* causes round to angular brown spots with chestnut coloured margins on both sides of leaves. It is called as leaf spot. *Phytophthora nicotianae* is a dreadful disease occurring on belladonna and other plants in which dropping of young leaves and branches, yellowing of older leaves and drying of whole apical portion occurs. This disease is called as phytophthora root rot. The association of *Fusarium solanii* and *Pythium butleri* causes damping off in young seedlings. The disease occurs on isolated branches of the roots of older plants. *Phytophthora erythrosceptica* causes damping off in young seedlings and wilt in matured plants. It causes black coloration of root. The disease is called phytophthora rot disease.

Several other pathogenic fungi infest the medicinal and aromatic plants and cause diseases, like *Pythium spinosum* (pythium rhizome rot); *Currularia prasadii* (leaf blight); and *Collectotrichmm fuscum* (anthracnose) on digitalis; *Septoria digitalis* and *Phyllosticta digitalis* (leaf spot); *Ascochyta kashmiriana* (leaf spot), *Peronospora hyoscyami* (mildew) and *Thiclavia basicola* (root rot) occurring on hyoscyamus *Fusarium oxysporum* (wilt), *Levellula taurica* (powdery mildew), *Cercospora rauuolfiae* (leaf spot) and *Alternaria tenuis* (leaf blight and bud rot) occurring on Rauwolfia; *Sclerotium rolfsfi* (wilt), *Corticium solanii* (root and root rot) and *Alternaria tennussima* (leaf spot) occurring on *Datura species*; *Fusarium solanii* (root rot); *Sclerotinia miner* (damping off), and *Ramularia bellunensis* (bud disease) occurring on pyrethrum; *Puccinia dioscorea* (bust) and *Cercospora dioscorea* (leaf spot) on *Dioscorea*. *Phytophthora species* (blight) on Vinca; *Erysiphe cinchoracearum* (powdery mildew), *Verticillium albotarum* (wilt), *Puccinia menthae* (rust), *Alternaria sp.* (leaf blight), *Curularia lunatus* (leaf spot) and *Sclerotium rolfsii* (sclerotium rot) on *Mentha species*; *Cercospora canescens* (leaf spot) on *Ocimum* species; *Cylindrocladium scoparium* on *Eucalyptus*; *Cercospora jasminicola* (leaf blight) and *Uremyces hobsonii* (rust) on *Jasminum* species and *Colectrotrichum graminicola* on lemongrass.

Many different viruses are responsible for some plant diseases. They are mosaic causing necrosis of leaves, petioles and stem of different Solanaceous plants. Tobacco mosaic virus, cucumber mosaic virus and tobacco ring spot virus are observed on digitalis and a strain of cucumber mosaic virus is detected on hyoscyamus. The viruses show disease symptoms on Rauwolfia, tobacco, datura, vinca and eucalyptus. Other viruses reported on medicinal plants are yellow vein mosaic, graft transmissible virus, distortion mosaic, rugose leaf curl, and *Ruga tabaci*.

(ii) Insects: It is found that the total number of insect species in the world is larger than the total number of species of all other forms of life put together. Throughout the world, about one million species of insects have been reported. These small creatures cause a drastic problem wherefrom the plants must be properly protected.

Various insect pests which attack medicinal plants are *Agrotis* species; *Heliothis armigera*, and *Odontotermes obesus*. Flea beetle, *Empoasca pteridis*, *Laphygma exigna* and *Odontotermes obesus* and *Phytomyza atricornis* are the insect pests which afflict the *Mentha* species.

Rauwolfia is attacked by *Diaphania nilgirica*, *Indomia cretaceus*, *Plantia viridicolis* and various beetles. Dill is affected by *Papilio machon* and *Hyadaphis coriandri*. Belladonna loses the leaves due to *Gonocephalum* species and *Agrotis flammatra*.

The other insect pests known to cause damage are caterpillar *Lepidopterus* larvae, cutworms, termites, weevil, Hessian fly, aphids, pyrilla, grasshoppers, locust, spiders, ticks, mites, etc.

All the insect pests belong to phylum Arthropoda of animal kingdom and they are further placed under two morphological groups, namely (a) biting and chewing, (b) piercing and sucking insects.

In most cases, the adult form in the metamorphosis of these insects is damaging in nature, but in certain instances the earlier stage like larvae is also destructive to plants.

(iii) Weeds: A weed is an undesired plant. Weeds are considered as dreadful pests because losses due to them are estimated to be more than those occurring due to other pests and diseases combined together. If the problem of control of weeds is not handled suitably, it leads to loss of nutrients, water, light and space, increase in cost of labour and equipment, low product quality and problems in marketeability besides, enhancing the rate of incidence of bacterial, fungal, viral and insects attack. Out of all these, losses of nutrients and moisture due to weeds are the problems of major concern to all plants yielding food, medicines, fibres and other economic raw materials.

Glyphoset is the only weedicide used at present to control the undesired growth of weeds.

Some weeds cause allergies, e.g. hay fever caused by ragweed, Medican tea, yellow dock, parthenium, etc. While the others, like corn cockle contains cyanophore glycoside and the seeds of this plant may have fatal effects. Dermatitis is caused by poison ivy, western poison oak, varnish tree, poison sumac, etc. Some plants growing as weeds may be poisonous like *Datura* and *Menispermus* species, etc.

Weed barriers:

Well developed countries now-a-days instead of using chemical weedicides are using weed barriers, for lawns public gardens. Terrous-gardans etc. This avoids the contamination of weedicides in main crops.

Weed-barriers are made from PVC films, papers or any other suitable coating material.

(iv) Non-insect pests: They are grouped into two categories:

1. Vertebrates like rats, monkeys, birds, rabbits and hares, squirrels, antelopes, deer, pigs, etc.
2. Invertebrates like nematodes, crabs, snails, mites, and symphylids.

The rodents have sharp and gnawing incisor teeth thereby causing considerable spoilage to stored crude drugs; besides, their faecal matter causes contamination of crude drugs.

METHODS OF PEST CONTROL

Different techniques followed for an integral approach to pest control are briefly summarized as under

(1) MECHANICAL PEST METHOD: It employs manual labour along with different devices for collection and destruction of pest.

The simple techniques used are hand-picking, pruning, burning and trapping of pests. A proper approach is made for collection and destruction of eggs, larvae, pupae and adults of insects.

The better way for protection from rodents like rats is construction of concrete warehouses. The warehouses with metal reinforcement corners on window frames. The rat and mouse traps are also used.

The method adopted for trapping flying insects is to emplace funnel shaped containers containing flavoured attractants comprising rose oil, anise oil, etc. mixed with saw dust. The insects can easily get an entry in the trap, but not an easy exit.

(2) AGRICULTURAL METHOD: It covers advanced plant breeding techniques capable of inducing genetic manipulations resulting in production of pest-resistant species. It has achieved much success in producing hybrid varieties, which are resistant to fungal and bacterial attack, as compared to limited success with insects. The systemic insecticides have been developed which are absorbed through the roots and reach to leaves by which all the foliage portion becomes distasteful for insects.

Another aspect in agricultural control is a deep plough to eradicate weeds, as well as, early stages of insects. If a plant is found out to be favoured by insects as major source of food, the land under cultivation of such plant should be subjected to crop rotation. Another method for checking supply of food to insects is by changing the environment which in many cases may lead to obstruction in their life cycle. Proper drainage serves this purpose to a great extent.

(3) BIOLOGICAL PEST CONTROL: This method is practiced by combating the pests, mostly the insects, with other living organisms. The latter is frequently the parasite form. If this method is properly designed, it may emerge as an effective, safe and economic method of pest control.

The chemical substances produced and released by some female insects are capable of eliciting a sexual response from the opposite sex, which could be properly exploited for biological control of pests. Such substances are called sex pheromones, e.g. 7, 8-epoxy-2 methyl octa decane from gypsy moth.

One of the experiments worth noting is about screw worms, which have been successfully carried out in U.S.A. The large numbers of laboratory reared male insects are sterilized by exposure to radioactivity with such a dose which shall not affect other physiological activities of that insect. Such irradiated insects are released in nature, in large enough number to oust the normal males existing in that area during competition for mating with virgin females. This leads to ineffective mating and sharp decline in progeny. There are several examples of biological control such as employing of Australian lady beetle 'lady bug' to feed on damaging insect called cottony cushion scale insect on citrus crop, rat terriers for rats, and various birds for insect pests. The larger harmful insects are many a time destroyed by hatching the eggs of certain types of flies and wasps.

(4) CHEMICAL PEST CONTROL: The control of pests is done with the use of chemical pesticides, which include insecticides, fungicides, herbicides and rodenticides.

Because of toxic effects of all, such chemicals they are used as pesticides, their use is regulated by the Insecticides Act in India; Federal Insecticide, Fungicide and Rodenticides Act in United States and the Agriculture (Poisonous substances) Regulations in the UK.

The chemical pesticides are further classified as rodenticides, insecticides, acaricides, fungicides, herbicides, predacides, ovicides, bactericides, arboricides, etc.

(i) **Rodenticides:** Warfarin, strychnine, arsenic trioxide, thallium sulphate, red squill, etc.

(ii) **Insecticides (Table 3.6):** DDT, gammaxine, methoxychlor, parathion, Malathion, sodium arsenate, pyrethroids, rotenoids, carbamates, etc.

DDT is the first insecticide reported in 1874 by Olhmer Zeiner. It was not known to him that it is powerful insectiside. In 1939 Swiss scientist PAUL Harmon Muller reported that it is a insecticide.

Ostrian scientist, Othmer Zether invented DDT in 1874 and Nobel Prize-winner, Swiss scientist Paul Herman Muller stated that it is a inseciside in 1933, DDT is banned in 2004 as insecticide.

(iii) **Acaricides (miticides):** Tetradifon, chlorobenzolate.

(iv) **Fungicides:** Bordeaux mixture, chlorophenols, antibiotics, quarternary ammonium compounds, etc.

(v) **Herbicides:** 2, 4-dichlorophenoxy acetic acid, calcium arsenate, sulphuric acid.

The insecticides are applied to vegetative parts for protective or eradicable activity in the form of spray, aerosol, solution, suspensions and fine dust. The stomach poisons are protective in action, while contact poisons act as eradicators. The fumigation keeps away the insect from the area under the influence of insect repellents. An ideal insecticide is required to satisfy following parameters.

(i) It should be non-toxic and non-injurious to medicinal plants and human beings.

(ii) It should be selective in action and highly toxic to insects in small concentrations.

(iii) The pesticides should be stable under ordinary conditions of storage, non-inflammable, non-corrosive and free from obnoxious odour.

(iv) It should be non-cumulative in soil and possess stability on treated surface (residual effect).

CHEMICAL STRUCTURE OF SYNTHETIC PESTICIDES

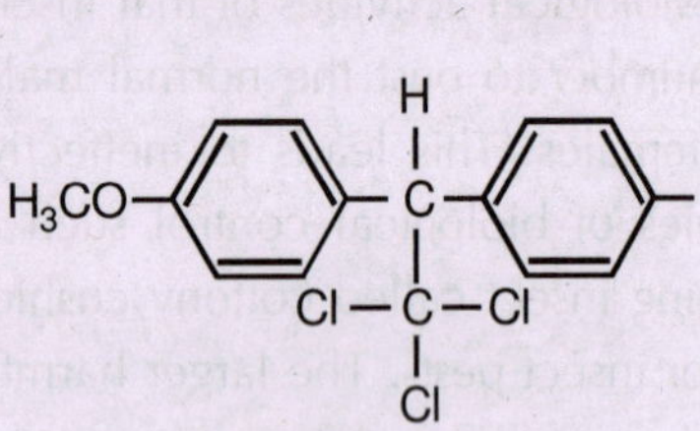

Methoxychlor

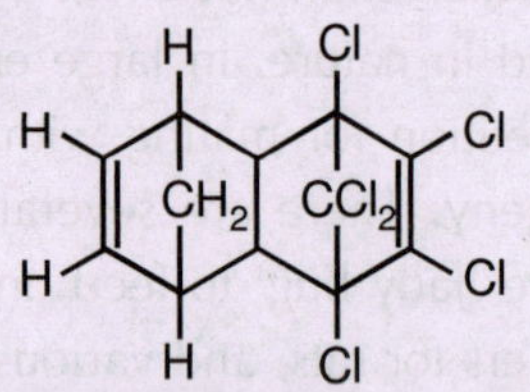

Aldrin

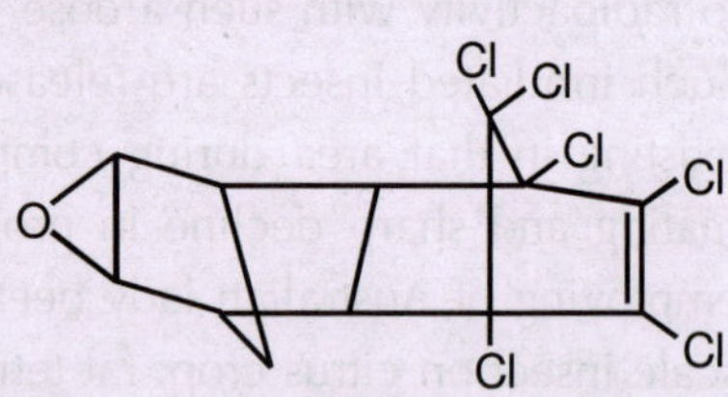

Dieldrin

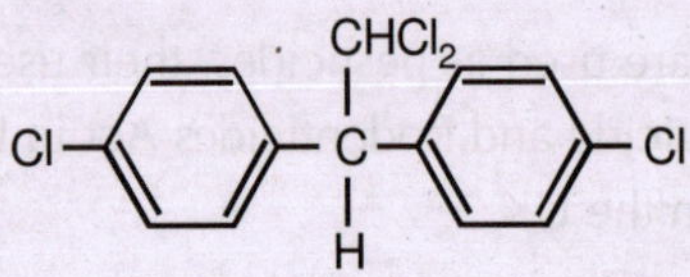

D.D.E.

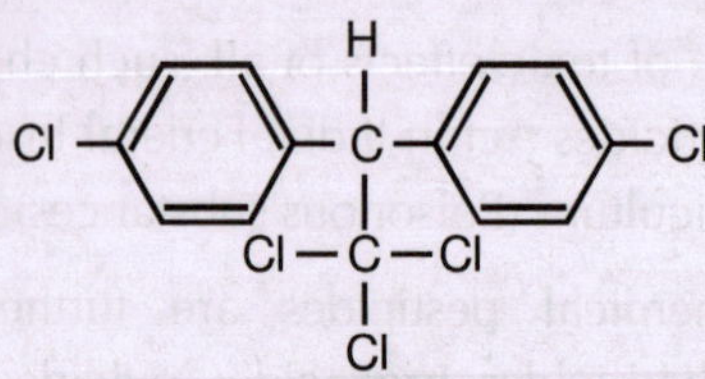

D.D.T.

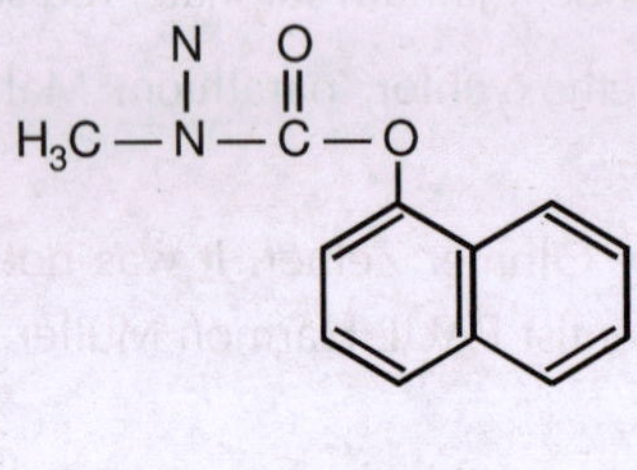

Carbaryl

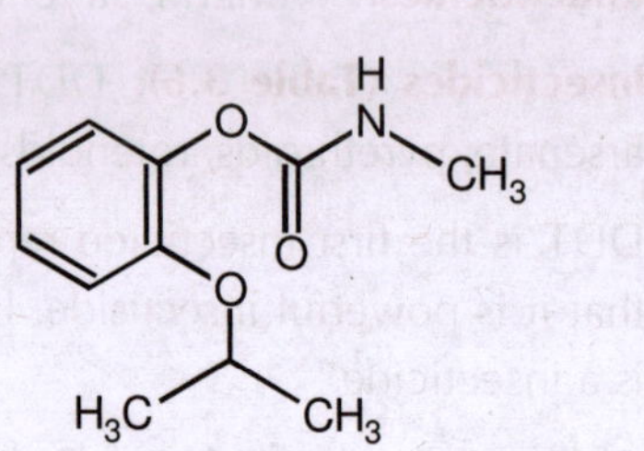

Baygon

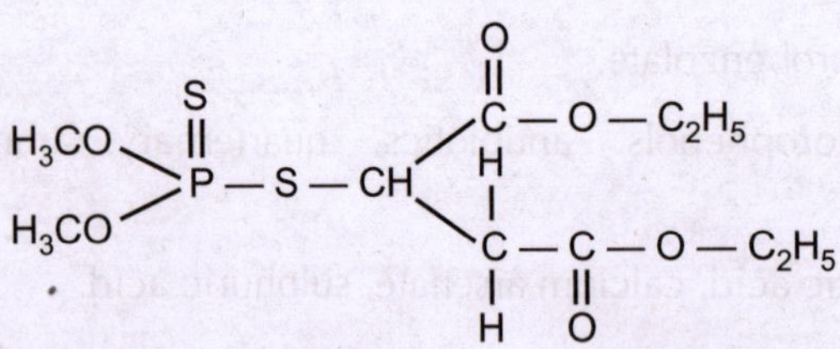

Malathion

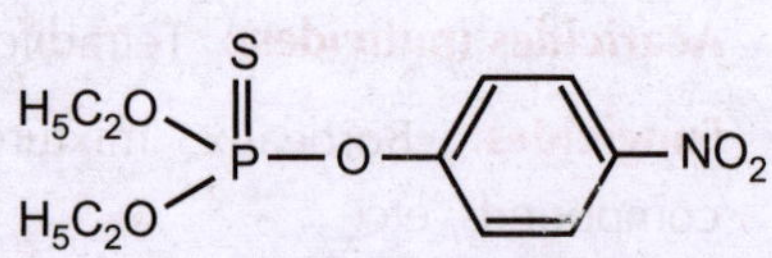

Parathion

Table 3.8: Types of Insecticides used in Cultivation and in Post-Harvest Technology

Types of Insecticides	Types of Insecticides
(a) Chlorinated hydrocarbons	**(b) Organophosphorus Compounds**
Aldrin	Carbophenothion
Benzene hexachloride (BHC, Gammaxine)	Demeton
DDT (Dichloro-diphenyl-Trichloroethane)	Dichlorvos
Dieldrin	Diaznon
Endrin	Dimethoate (Rogor)
Heptachlor	Malathion
Methoxychlor	Methyl parathion
Chlordane	Phorate
DDE (Dichloro-diphenyl ethylene)	Phosphamidon (Dimecron)
Trichlorophon	Schraden
	Tetraethyl pyro phosphate (TEPP)
	Thiometon
(c) Carbamates	**(d) Inorganic Compounds**
Carbaryl [Carbaril (Sevin)]	Calcium arsenate
	Lead arsenate
	Paris green
	Sodium fluosilicate
	Zinc phosphide
(e) Fumigants	**(f) Natural products**
Aluminium phosphide	Tobacco
Hydrogen cyanide	Pyrethrum
Ethylene dichloride	Derris
Methyl bromide	Neem
Phosphine	Sabadilla
	Ryania

The mode of action of different insecticides is given below (Table 3.9)

Table 3.9: Mode of Action of some Pesticides

Chemical Compound	Biological Effect
Effect on Animals	
1. Organophosphorus compounds and carbamates	Inhibition of acetyl cholinesterase
2. Chlorinated hydrocarbons and Pyrethroids	Neurotoxication
3. Nicotinoids	Inhibition of neuromuscular junction
Effect on Plants	
4. Carbamates, substituted urea, triazines	Inhibition of photosynthesis
5. Carbamates	Inhibition of oxidative phosphorylation
6. 2,4-D; 2,4,5-T.	Hormone analogues
7. Metals, sulphur	Unknown causes
8. 3 – amino – 1, 2, 4- triazole	Inhibition of chlorophyll synthesis
9. Chlorinated aliphatic hydrocarbons	Inhibition of pantothenate synthesis

BIOPESTICIDES FOR CROP PROTECTION:

Biopesticides (Biological pesticides) are pesticides derived from natural materials such as animals, plants, bacteria and certain minerals. Biopesticides also include naturally occurring substances that control pests and micro-organisms (microbial pesticides).

The grandfather of Charles Darwin, Erasmus Darwin in 1800 predicted the use of natural predators to minimize the pests in his book *Philosophy of Agriculture and Gardening*. But the importance of biopesticides early in the twentieth century was realized. After 1950 only the management of pest control by using biopesticides got the real impetus.

Chemical pesticides are hazardous and cause grave damage to health, ecosystem and water.

To boost up the food production, the use of chemical pesticides was favoured and to some extent it was useful.

But the uncontrolled and unscientific use of the chemical pesticides has lead to the disaster and has created several problems such as:

- Insecticide resistance
- Resurgence of secondary pests
- Environmental pollution and
- Residues in food, drinking water and even in the breast-milk.

Therefore agricultural scientists were after some alternatives and ultimate solution was biopesticides.

Table 3.10: Difference between Biopesticides and Chemical pesticides

Biopesticides	Chemical pesticides
1. These are completely non-toxic to the plants; hence increase chlorophyll and proteins in the plants.	1. These are poisonous hence destroy chlorophyll and proteins in the plants.
2. These are ecofriendly and do not cause any physiological damage to the organism.	2. Excess use of these chemicals cause environmental pollution thereby causing physiological damage and ecological imbalance.
3. These are used in very small quantity (4-5 sprays/crop/season). These prevent multiplication of insects.	3. These are required in large quantity (10-20 sprays/crop/season) and cause mutation in the insects hence new species of insects may appear.
4. These are biodegradable and do not retain any residue or toxins.	4. These retain toxic residue on the plants and are toxic to human beings.
5. No harmful effect on useful micro-organisms or pollinating insects or earthworms.	5. These cause harmful effect on useful micro-organisms or pollinating insects.
6. These are safe to handle, store, transport and spray.	6. These need careful handling, their transport and spraying should be done with caution.
7. Do not affect the soil fertility.	7. Adversely affect soil fertility.
8. Do not cause any chemical reaction and can be produced in large quantity and stored safely.	8. Sometimes cause toxic and harmful chemical reactions which are uncontrollable.
9. These are non-toxic to human beings, hence can be sprayed in habited areas.	9. These produce hazardous effects on human beings if sprayed in habited areas.
10. Biopesticides act as insecticides, fungicides, viricides, insect repellants and sometime act as growth nutrients.	10. Different products are necessary as insecticides, fungicides, viricides, and insect repellants and hence they are not economical.

As per the Gazette notification of Central Insecticide Board, Ministry of Agriculture New Delhi, Dated 26/3/1999 Biopesticides have been put under Insecticide Act 1968.

Following is the list of Biopesticides registered under Insecticides Act 1968 **added vide Notification No. GSR 224(E) dated 26/3/1999**

- Bacillus subtilis
- Gliocladium species
- Pseudomonae species

- Trichoderma species
- Beauveria bassiana
- Metarrhizium anisopliae var acridum
- Nomurea rileyi
- Verticillium lecanii
- Granulosis viruses
- Nuclear polyhedrosis viruses
- Neodiprion sertifer

Added vide Notification No. GSR 378(E) dated 24/6/2004

- Streptomyces griseoviridis
- Streptomyces lydicus
- Ampelomyes quisqualis
- Candida oleophila
- Furarium oxysporum
- Burkholderia cepacia
- Coniotyrium minitans
- Agrobacterium radiobacter strain-84
- Agrobacterium tumefaciens
- Pythium oligamdrum
- Erwinia amylovora
- Phlebia gigantean
- Penicillium islanidicum
- Placilomyces lilacinus
- Aspergillus niger strain AN27
- Vesicular arvescular mycorrhijal

Added vide Notification No. GSR 51 (E) dated 06/02/2006

- Mycothecirem verrucaria
- Serratia mercescens GP35

Polyhouse and Shade Housing

Polyhouse and shade housing facilitities be provided for healthy and insect free and pest free growth of medicinal plants. An area of 15 × 15 metre i.e. (675 sq. metres) is enough to keep the plants, under cultiation trials. Moisture content, temperature control, irrigation and even misting can be managed properly, as required by various crops.

Exposure to direct UV rays can also be controlled to arrest the damage of their active chemical constituents.

Fig. 3.2: Polyhouses (A1, A2) and Shadehouses (B1, B2)

PLANT GROWTH REGULATORS

Plant growth regulators are the organic compounds, other than nutrients which affect the morphological structure and/or physiological processes of plants in low concentrations. Phytohormones or plant hormones are naturally occurring growth regulators which in low concentrations control physiological processes in plants. More commonly, the term plant growth regulators is used, because it includes both the native (endogenous) and the synthetic (exogenous) substances; which modify the plant growth.

As the native plant growth regulators, five major kinds of substances are reported, viz. auxins, gibberellins, cytokinins, abscisic acid and ethylene. All of these, except ethylene and abscisic acid, are multiple forms of endogenous plant growth regulators. In general, the plant growth regulators or substances regulate cell enlargement, cell division, cell differentiation, organogenesis, senescence and dormancy. They are employed in seed treatment to achieve earlier growth and root development, quality improvement like protein level and amino acid balance, etc. Plant growth regulators have given a real boost to plant tissue culture techniques by which now it is possible to culture almost any part of the plant in-vitro. In the domain of pharmacognosy, **plant growth regulators** established their status specifically in enhancing the production of secondary metabolites used as drug.

(a) AUXINS

Auxin is a general term used to indicate substances that promote elongation of coleoptiles tissues. Indole acetic acid (IAA) is an auxin that occurs naturally in plants. They are either natural auxins which are produced by plants themselves or synthetic auxins, which have the same action as natural auxins. IAA is the principal auxin and other natural auxins are indole-3-acetonitrile (IAN) 4-chloroindole 3-acetic acid and phenyl acetic acid. The synthetic auxins are indole-3-butyric acid (IBA), 2-napthyloxyacetic acid (NOA), α-napthyl acetic acid (NAA), 1-napthyl acetamide (NAD), 2, 4-dichlorophenoxyacetic acid (2, 4-D), 2, 4, 5 trichlorophenoxy acetic acid and 5-carboxymethyl-N, N-dimethyl dithiocarbamate.

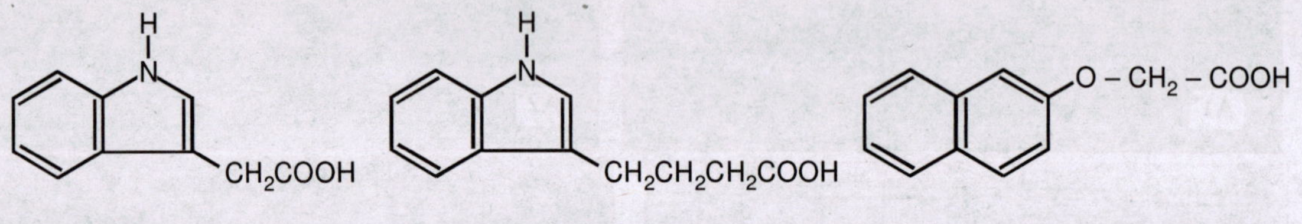

IAA **IBA** **NOA**

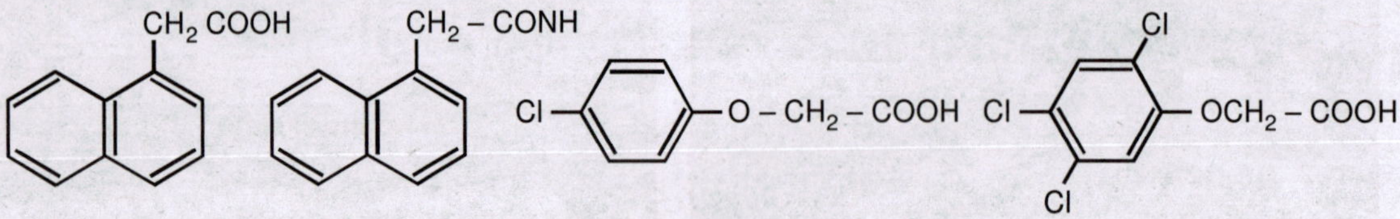

NAA **NAD** **2, 4 - D** **2, 4, 5 T**

Plant growth regulators (AUXINS)

Auxins are involved in different growth processes in plants like internodes' elongation, leaf growth, initiation of vascular tissues, cambial activity, fruit setting in absence of pollination, fruit growth, apical dominance, inhibition of root growth, influencing physical and chemical properties in leaf abscission and inhibition of lateral buds.

The proposed mechanism of action of IAA is its interaction with one or more components of biochemical systems involved in the synthesis of proteins. The other hypothesis suggested is the role of IAA to alter the active osmotic contents of cell vacuole during cell expansion or cell wall extension.

IBA and NAA in combination are used in rooting of cuttings. NOA is used as a fruit setting spray 2, 4-D and 2, 4, 5-T are used both as plant growth regulators and in higher concentrations, as selective weed killers, especially for dicot plants. IBA has shown promising results to induce rooting in cuttings for cinchona, *Pinus*, papaya and coffee. The addition of different auxins like IAA, NAA and 2, 4-D in tissue cultures of ergot has led to increase in Indole alkaloids. Treatment of derivatives of NAA given to seedlings and young plants of *Mentha piperita* has shown about 40 per cent increase in volatile oil content.

(b) GIBBERELLINS

They are a class of endogenous plant growth regulators and at present over 50 gibberellins are known. About 40 of these occur in green plants, while others are present in some fungi. They are present in different organs and tissues like roots, shoots, buds, leaves, floral apices, root nodules, fruits and callus tissues. The commercial formulations of gibberellins are used currently for promoting vegetative and fruit growth, breaking dormancy, flower initiation and induction of parthenocarpy.

Kurosawa, a Japanese physiologist, is credited for initiating the discovery of gibberellins from fungus *Gibberella fujikuroi* (previously known as *Fusarium heterospermum*) grown on rice. According to Paleg, gibberellins are compounds having gibbane skeleton and biological activity in stimulating cell division or cell elongation, or both.

The thorough research on gibberellins carried out in Japan, the United States of America and Great Britain has shown that gibberellins A isolated in 1938 is actually a mixture of at least 6 gibberellins referred to as GA_1, GA_2, GA_3, GA_4, GA_7 and GA_9. GA_3 is termed as Gibberellic acid.

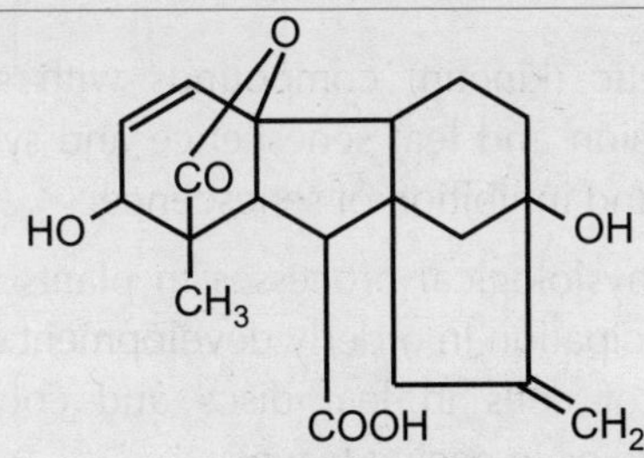

Gibberellic acid (Ga₃)

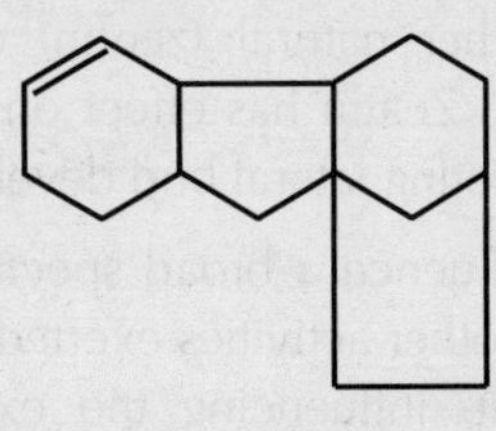

Gibbane skeleton

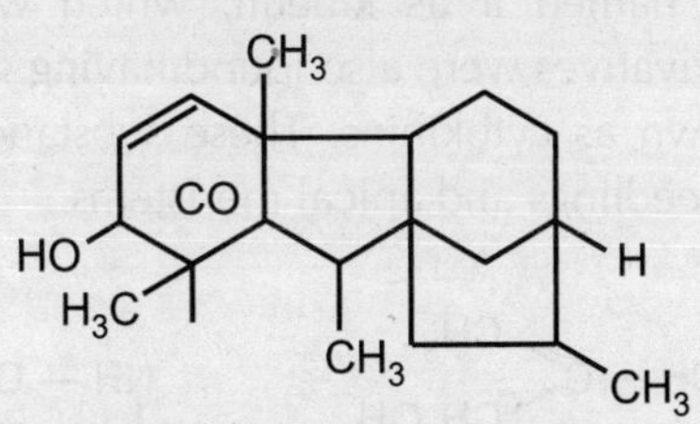

Gibberellane structure

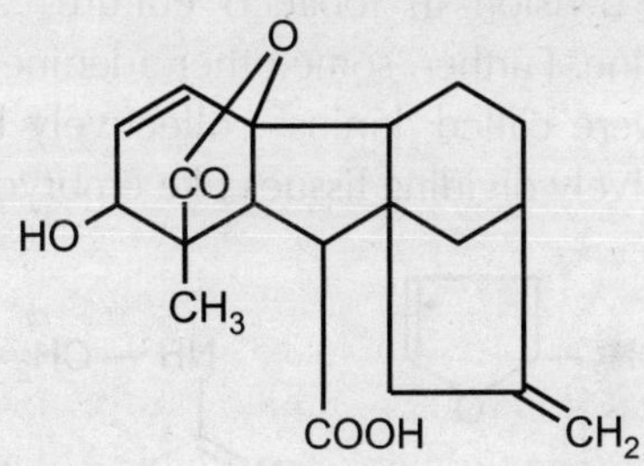

GA_7

All of these are the derivatives of gibbane ring skeleton. In addition to free gibberellins, they are also present in conjugated forms. GA has not yet been synthesized, but can be produced by large scale fermentation on commercial scale. The angiosperms, gymnosperms, ferns, algae, fungi and bacteria contain several forms of gibberellins, but no single plant contains all of them together. Many activities attributed to gibberellins are promotion of rapid expansion of plant cells, stimulation of seed germination, breaking dormancy of overwintering plants, induction of flowering under non-inductive conditions, marked increase in stem elongation, increase in the size of leaves and induction of parthenocarpy leading to seedless fruit sets.

The effect of gibberellins in cell division is an increase in cell size similar to the effect of auxins. It is observed that gibberellins are more effective in intact plants, while major auxin effects are on excised organs.

The applications of gibberellins are extended to various medicinal plants. The use of gibberellins in lower dose has shown increased yield of digitalis glycosides per shoot. The hormone tried with leaf and root culture of digitalis, showed higher production of digoxin. In case of Tinnevelley senna, GA shows a little positive effect on dry weight of shoot, but reduction in sennoside content of leaves. It is observed that GA treatment can cause an increase in height of castor plant up to five times, but does not show any change in fixed oil content. The treatment significantly causes reduction in alkaloid content of vinca, datura, hyoscyamus, etc.

The mechanism of action of Gibberellic acid appears mainly to induce activity of gluconeogenic enzymes during early stages of seed germination and this specificity ensures a rapid conversion of lipid to sucrose, which is further used in supporting growth and development of the embryonic axis to a competent root and shoot system. It is also found that gibberellins induce the synthesis of α-amylase and other hydrolytic enzymes during germination of monocot seeds. They are also involved in mobilizing seed storage reserves during germination and seedling emergence.

(c) CYTOKININS

These are either natural (zeatin) or synthetic (kinetin) compounds with significant growth regulating activity. Zeatin has effect on cell division and leaf senescence and synthetic cytokinins are useful in promoting lateral bud development and inhibition of senescence.

Cytokinins influence a broad spectrum of physiological processes in plants like promotion of cell division. The other activities exerted are participation in orderly development of embryos during seed development, influencing the expansion of cells in leaf discs and cotyledons, delaying breakdown of chlorophyll and degradation of proteins in ageing leaves.

Miller isolated the crystalline substance from autoclaved herring sperm DNA capable of inducing cell division in tobacco cultures and named it as kinetin, which was found to be 6-furfuryladenine. Further, some other adenine derivatives were also found having similar biological activity and were called 'kinins' collectively known as cytokinins. These substances are found in young and actively dividing tissues like embryos, seedlings and apical meristems.

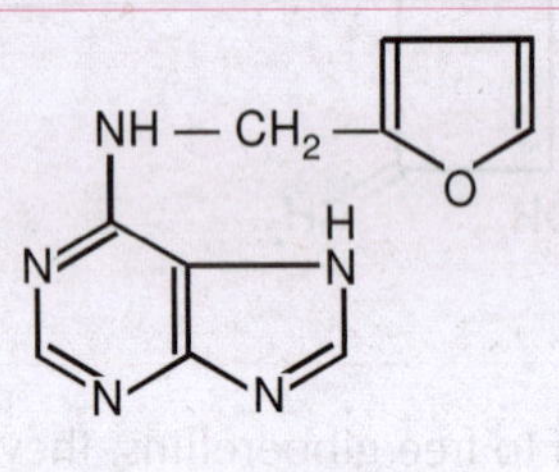

Kinetin

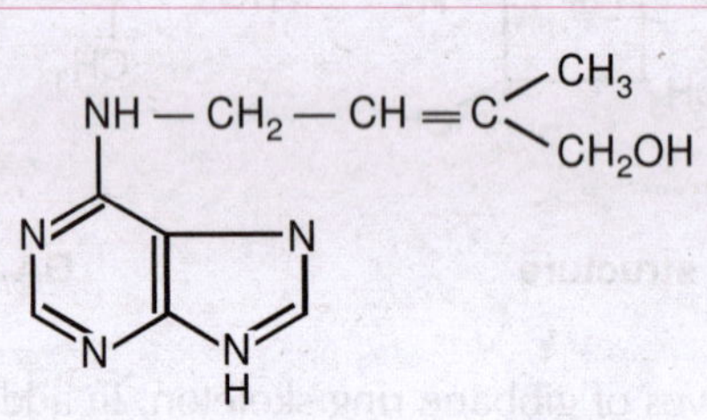

Zeatin

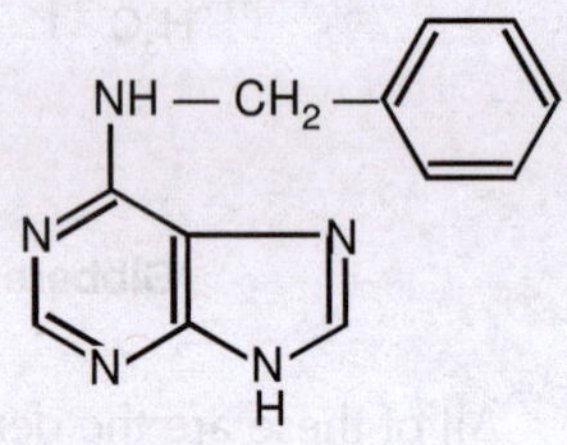

(6-Benzyl adenine)
(6-Benzyl aminopurine)

The naturally occurring cytokinins are zeatin, N^6 dimethylamino purine, and N^6-Δ^2-isopentenyl aminopurine. The synthetic cytokinins are kinetin, adenine, 6-benzyl adenine benzimidazole and N. N'-diphenyl urea.

Cytokinins are reported to increase marginally sennoside content in Tinnevelley senna leaves and also enhance the dry weight of shoots. In opium, they cause formation of elongated capsule and reduce alkaloid content. In *Duboisia* hybrids, the cytokinins activity present in extract of seaweed shows marked increase in both leaf content and also hyoscine content.

Kinetin is reported to play the role in nucleic acid metabolism and protein synthesis. In plant metabolism, it is proposed that some t-RNA contain cytokinins like activity. They have an action on some enzymes responsible for formation of certain amino acids.

(d) ETHYLENE

It is a simple organic molecule present in the form of volatile gas and shows profound physiological effects. It is present in ripening fruits, flowers, stems, roots, tubers and seeds. It is present in very small quantity in plant, say about 0.1 ppm — part per million — possibly, its quantity increases in local areas during the time of growth and development.

Ethylene is produced by incomplete burning of carbon rich substances like natural gas, coal and petroleum. Denny (1924) showed the yellowing of lemons due to stove gas. Also, the plant damage was noticed after the introduction of illuminating gas. Gane (1935) found that a gas evolved from ripe apples can also effect the ripening of green apples and showed that it was ethylene gas which had a role in ripening of other fruits.

Ethylene shows a broad array of growth responses in plants, which include fruit ripening, leaf abscission, stem swelling, leaf bending, flower petal discoloration, and inhibition of stem and root growth. It is commercially used for promotion of flowering and fruit ripening, induction of fruit abscission, breaking dormancy and stimulation of latex flow in rubber trees.

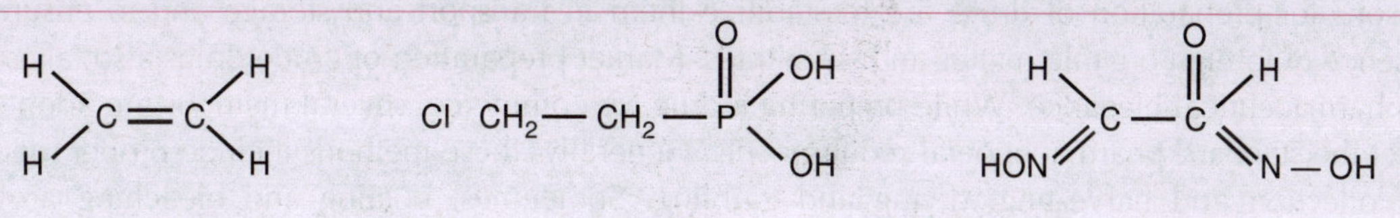

Ethylene **Ethephon** **Glyoxime**

(e) ABSCISIC ACID (ABA)

The physiological activities in plants like retaining or shedding of different organs such as leaves, stems, flowers and fruits have led to finding of natural growth inhibitor. A diffusible abscission - accelerating substance was found by Osborne (1955) in senescent leaves. Carns *et al.* isolated several abscission accelerating substances from cotton plants and named them as abscission I and abscission II.

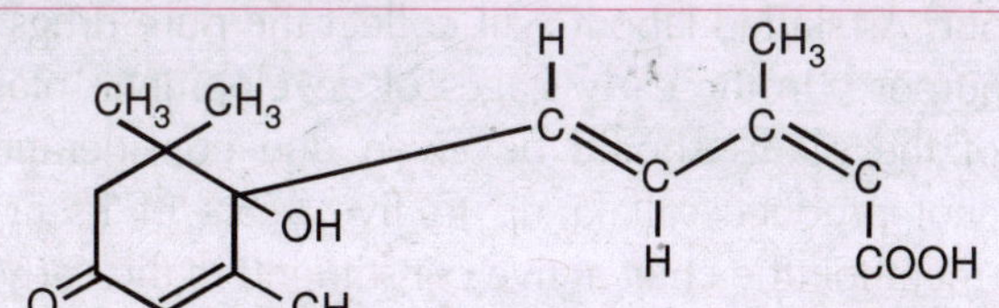

Abscisic acid (Abscisin II)

In an inhibitory way, ABA interacts with other plant growth substances. It inhibits the GA-induced synthesis of α-amylase and other hydrolytic enzymes. During maturation, ABA accumulates in many seeds and helps in seed dormancy. ABA concentrations are found to be enhanced in stress conditions, like mineral deficiency, injury, drought and flooding. ABA serves an important role as potential antitranspirant by closing the stomata, when applied to leaves.

A number of other synthetic growth inhibitors and retardants reported are maleic hydrazide, daminozide, glyphosine, chromequat chloride, S, S, S-tributyl phosphorotrithioate, ancymidol, chlorophonium chloride, piproctanyl bromide, etc. However, commercial use of these compounds is yet to be reported. A group of synthetic substances called morphactins is a potent inhibitor of auxin transport causing tropic responses, reduction of apical dominance and promoting lateral growth. Morphactins include chloroflurecol methyl, flurecol-butyl and TIBA (2, 3, 5-tri-iodobenzoic acid).

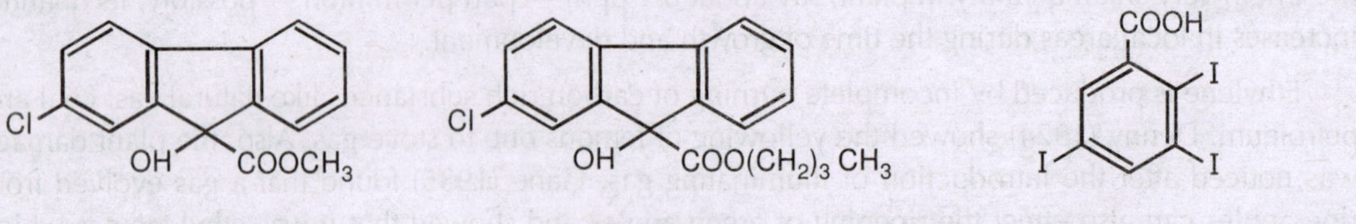

Chloroflurecol methyl **Flurecol-butyl** **(2, 3, 5-tri-iodobenzoic acid) TIBA**

(B) COLLECTION, PROCESSING FOR QUALITY ASSURANCE, PRESERVATION AND STORAGE OF HARBAL DRUGS

After collection of the crude drugs, they are required to be processed prior to marketing. The reasons for preparation of drugs are to stabilize them in transport and storage and to ensure the absence of foreign organic matter and substitutes. Market preparation of crude drugs also takes care of pharmaceutical elegance. While preparing a drug for commerce, several methods are adopted to meet the standard pharmacopoeial requirements. Generally, these methods include proper methods of collection and harvesting, drying and garbling. Sometimes, coating and bleaching are also necessary for converting the drug into suitable form for the market. While doing so, it should be taken care of that neither the action of the drug lowers nor does it change, due to the additives used in the process.

COLLECTION OF HERBAL DRUGS

Irrespective of the type of crude drug and area of collection, there cannot be two opinions that the drugs are collected suitably when they contain maximum concentration of active ingredients. The advantage of existing environmental conditions is also taken into consideration while collecting the crude drugs. While collecting the natural drugs on commercial scale, attention should be paid for the use of skilled labour. As skilled labour will collect the pure drugs without any admixture as they are trained for the purpose. In the early stages of development, plants do not produce active ingredients, hence age of the plants should be given due consideration for collection, say for instance *Boswellia* does not produce sallaki up to five years of its age; similarly mentha herb produces pulgone but no menthol (the chief active constituent) in the early stages.

As far as possible diseased plants or plants treated with pesticides are to be avoided for medicinal use and should be rejected during collection. At times the season of collecting the medicinal plants is also very specific i.e. Rhubarb rhizomes aconite roots contain their respective active constituents at the maximum, on the commencement of the warmer season only.

In general **rhizomes** and **underground parts of the plants** are to be collected when their aerial growth has ceased. The drugs which constitute leaf and the flowering tops of plants are collected just before they reach their flowering stage (maturity) e.g. senna, digitalis, vinca, belladonna, etc.

while the leaves of aloe are collected when they are sufficiently thick. **Flowers** need to be collected just before pollination or many a time, before their full expansion, e.g. saffron, clove buds, chamomile, arnica, etc. They are collected in dry weather and preferably during morning hours. **Barks** are generally collected in spring or early summer when cambium is active, as it is easy to detach them from the stem. Sometimes, they are collected in autumn (wild cherry) or in rainy season (cinnamon). Three different methods for collecting barks are (a) felling, (b) uprooting and (c) coppicing. In felling method, the tree is cut at base and bark is peeled off. In uprooting technique, the roots are dug out and bark is stripped off from roots and branches. In coppicing method, the plant is allowed to grow for a definite period and then it is cut off at specific distance from soil. The stumps which remain in ground are allowed to send shoots, which develop further independently yielding aerial parts. These new parts are cut off and bark is collected from shoots. As compared to other methods of collection of bark, this technique is more economical and less time-consuming. It is, therefore, the method of choice for collecting barks commercially. Cascara and cinnamon are collected by this method.

Seeds are collected just before ripening or the fruits in which they contained are fully ripe.

Saps and **exudates** are collected in winter or early in the summer.

Whole plants or **Entire drugs** are collected prior to any visual decline in any of the plant parts, or which the flowers are emerging.

No specific uniform method of collection of animal drugs is prescribed, since the animals as small as insects and animals of very large size like **cod-fish** from ocean are processed.

The **fruits** are collected depending upon the part of fruits used. They are collected either ripe or half ripe, but full-grown. For example, cardamom fruits are collected just before their dehiscence; bael and tamarind, after their maturity, while caraway, fennel and coriander are collected, when they are full ripe. The roots are collected in spring, before the vegetative process stops. Usually, the roots are sliced transversely or longitudinally to facilitate drying. Rhizomes are collected, when they have stored ample of reserve food material and contain maximum content of chemical constituents.

The **unorganized drugs** such as resins, gums, latices are collected, as soon as, they ooze out of the plant. Acacia gum is collected 2-3 weeks after making incisions on the bark of the tree and when it is sufficiently hard. Opium and papaya latices are collected after coagulation of latex. Turpentine oleoresin and balsam of peru are collected when the plant is about 8 - 10 years old.

PROCESSING FOR QUALITY ASSURANCE

Quality assurance of herbal drugs is of prime importance in every respect. Unless the quality is maintained the efficacy of herbal drugs is doubtful, as they are normally collected from natural sources by unskilled workers.

To maintain the quality many factors are to be considered which are directly related to their quality few of these are as under:

HARVESTING

Harvesting is an important operation in cultivation technology, as it reflects upon economic aspects of the crude drugs. An important point which needs attention over here is the type of drug to be harvested and the pharmacopoeial standards which it needs to achieve. Harvesting can be done efficiently in every respect by the skilled workers. Selectivity is of advantage in that the drugs other than genuine, but similar in appearance can be rejected at the site of collection. It is, however, a laborious job and may not be economical. In certain cases, it cannot be replaced by any mechanical means, e.g. digitalis, tea, vinca and senna leaves. The underground drugs like roots, rhizomes, tubers, etc. are harvested by mechanical devices, such as diggers or lifters. The tubers or roots are thoroughly washed in water to rid earthy-matter of. Drugs which constitute all aerial parts are harvested by binders for economic reasons. Many a time, flowers, seeds and small fruits are harvested by a special device known as seed stripper. The technique of beating plant with bamboos is used in case of cloves. The cochineal insects are collected from branches of cacti by brushing. The seaweeds producing agar are harvested by long handled forks. Peppermint and spearmint are harvested by normal method with mowers, whereas fennel, coriander and caraway plants are uprooted and dried. After drying, both they are thrashed or beaten and the fruits are separated by winnowing. Sometimes, reaping machines are also used for their harvesting.

DRYING

Before marketing a crude drug, it is necessary to process it properly to preserve it for a longer time and also to acquire better pharmaceutical elegance. This processing includes several operations or treatments, depending upon the source of the crude drug (animal or plant) and its chemical nature. Drying consists of removal of sufficient moisture content of crude drug to improve its quality and make it resistant to the growth of microorganisms. Drying inhibits partially enzymatic reactions. Drying also facilitates pulverizing or grinding of a crude drug. In certain drugs, some special methods are required to be followed to attain specific standards, e.g. fermentation in case of *Cinnamomum zeylanicum* bark and gentian roots. The slicing and cutting into smaller pieces is done to enhance drying, as in case of Glycyrrhiza, squill and Calumba. The flowers are dried in shade so as to retain their colour and volatile oil content. Depending upon the type of chemical constituents, a method of drying can be used for a crude drug; drying of flowers can be achieved by (1) Nautral (sun drying) and (2) Artificial means.

1. Natural Drying (Sun-drying)

In case of natural drying, it may be either direct sun-drying or in the shed. If the natural colour of the drug (digitalis, clove, senna) and the volatile principals of the drug (peppermint) are to be retained, drying in shed is preferred. If the contents of the drugs are quite stable to the temperature and sunlight, the drugs can be dried directly in sunshine (gum acacia, seeds and fruits).

2. Artificial Drying

Drying by artificial means includes drying the drugs in (a) an oven; i.e. tray-dryers; (b) vacuum dryers, and (c) spray dryers.

(a) Tray dryers: The drugs which do not contain volatile oils and are quite stable to heat or which need deactivation of enzymes are dried in tray dryers. In this process, hot air of the desired temperature is circulated through the dryers and this facilitates the removal of water content of the drugs (belladonna roots, cinchona bark, tea and raspberry leaves and gums are dried by this method).

(b) Vacuum dryers: The drugs which are sensitive to higher temperature are dried by this process. e.g. tannic acid and digitalis leaves.

(c) Spray dryers: Drugs which are highly sensitive to atmospheric conditions and also to temperature of vacuum drying are dried by spray drying method. The technique is followed for quick drying of economically important plant or animal constituents, rather than the crude drugs. Examples of spray drying are papaya latex, pectin, tannins, etc.

GARBLING (DRESSING)

The next step in preparation of crude drug for market after drying is garbling. This process is desired when sand, dirt and foreign organic parts of the same plant, not constituting drug are required to be removed.

This foreign organic matter (extraneous matter) is removed by several ways and means available and practicable at the site of the preparation of the drugs. If the extraneous matter is permitted in crude drugs, the quality of drug suffers and at times, it doesn't pass pharmacopoeial limits. **Excessive stems** in case of lobelia and stramonium need to be removed, while the **stalks** in case of cloves are to be deleted. Drugs constituting rhizomes need to be separated carefully from **roots, rootlets and also stem bases**. Pieces of iron must be removed with the magnet in case of castor seeds before crushing and by shifting, in case of vinca and senna leaves, the pieces of bark should be removed by peeling as in gum acacia.

Fig. 3.3: Garbling

PACKING

The morphological and chemical nature of drug, its ultimate use and effects of climatic conditions during transportation and storage should be taken into consideration while packing the drugs. Aloe is packed in goat skin. Colophony and balsam of tolu are packed in kerosene tins, while asafoetida is stored in well closed containers to prevent loss of volatile oil. Cod liver oil, being sensitive to sunlight, should be stored in such containers, which will not have effect of sunlight, whereas, the leaf drugs like senna, vinca and others are pressed and baled. The expensive drugs which are very sensitive to moisture need special attention, e.g. digitalis, ergot and squill. Squill becomes flexible ergot becomes susceptible to the microbial growth, while digitalis looses its potency due to decomposition of glycosides, if brought in contact with excess of moisture during storage. Hence, the chemicals which absorb excessive moisture (desiccating agents) from the drug are incorporated in the containers. Colophony needs to be packed in big masses to control auto-oxidation. Cinnamon bark, which is available in the form of quills, is packed one inside the other quill to facilitate transport and prevent volatilization of oil from the drug. The crude drugs like roots seeds and others do not need special attention and are packed in gunny bags while in some cases bags are coated with polythene internally. The weight of certain drugs in lots is also kept constant, e.g. Indian opium.

PRESERVATION OF HERBAL DRUGS

Preservation of crude drugs (herbals) need sound knowledge of their physical and chemical properties. Good quality of the drugs can be maintained if they are preserved well. The following illustrations shall help in understanding the factors to be considered while preserving the drug. All the drugs should be preserved in well closed and, possibly, in the filled containers. They should be stored in the premises which are waterproof, fireproof and rodent-proof. A number of drugs absorb moisture during their storage and become susceptible to the microbial growth. Some drugs absorb moisture to the extent of 25per cent of their weight. The moisture, not only increases the bulk of the drug, but also impairs the quality of crude drug. The excessive moisture facilitates enzymatic reactions resulting in decomposition of active constituents, e.g. digitalis leaves and wild cherry bark. Gentian and ergot receive mould infestation due to excessive moisture. Radiation due to direct sunlight also causes destruction of active chemical constituents as in the case of ergot, cod liver oil and digitalis. Form or shape of the drug also plays a very important role in preserving the crude drugs. Colophony in the entire form (big masses) is preserved nicely, but if stored in powdered form, it gets oxidized or loses its solubility in petroleum ether. Squill, when stored in powdered form becomes hygroscopic and forms rubbery mass on prolonged exposure to air. The fixed oil in the powdered ergot becomes rancid on storage. To maintain a good quality of powdered ergot, it is required that the drug should be defatted with lipid solvent prior to storage. Lard, the purified internal fat of the abdomen of the hog, is to be preserved against rancidity by adding Siam benzoin. Atmospheric oxygen is also destructive to several drugs; hence, such drugs are filled completely in well closed containers, or the air in the container is replaced by an inert gas like nitrogen, e.g. shark liver oil, papain, etc.

Apart from protection against adverse physical and chemical changes, the preservation against insect or mould attacks is also essential. Different types of insects, nematode, worms, moulds and mites infest the crude drugs during storage. Some of the main pests found in drugs are Coleoptera (*Stegobium paniceum* and *Calandrum granarium*), Lepidoptera (*Ephestia kuehniella* and *Tinea pellionella)*, and Arachnida or mites (*Tyroglyphus farinae* and *Glyophagus domesticus*.

STORAGE OF HERBAL DRUGS

Crude drugs are normally stored in glass containers or in gunny bags on commercial scale, however following measures will preserve the crude drugs suitably.

1. Floor area of store room should be tidy, dry, hygienic, without any cracks and easy to clean.
2. Air conditioning facility with dehumidifier is desired for the store room.
3. Crude drugs should be stored on shelves with sufficient distance from walls to permit free movement of trollies used for transport.
4. Protection against rodents, insects and line-stocks should be provided.
5. Arrangements to prevent pest-infestations mould formation should also be provided.
6. Well filled, well closed, air-tight containers are desired for storage of crude-drugs and should be kept away from direct sunlight and in cool place.
7. Arrangements to inspect the store room (premises) at regular intervals be made.
8. Packaging materials used to store should be non-polluting and should not absorb moisture.
9. Wooden boxes, paper bags, news papers should not be used for storage of crude drugs.
10. As a safety measure, fire-extinguishers should be available in store room.
11. Aluminium containers should not be used to store volatile oils.
12. Fibre glass or glass lined containers are most suitable for storage.

Chapter 4

CULTIVATION AND UTILIZATION OF MEDICINAL AND AROMATIC PLANTS IN INDIA

The wealth of India is stored in the enormous natural flora which has been gifted to her. Endowed with a wide diversity of agro-climatic conditions, **India is virtually the herbarium of the world**. India has all types of climatic conditions varying from temperate in the Himalayas to tropical in South India, dry in Central India to humid and wet in Assam and Kerala, thus providing conditions favourable for the growth of a variety of medicinal and aromatic plants provided suitable strains and sites are selected for their commercial cultivation.

The composition of soil differs from place to place in India. In the deltas around rivers in UP, Bihar and West Bengal alluvial soil is found. In parts of Maharashtra and North Karnataka, black soil is common. This is too heavy and sticky and needs to be lightened. Red soils are good for tree growth, but lack sufficient nutrients for smaller plants. Sandy soils do not retain enough water and need treatment. Because of vast areas and variety of agro-climates in India, a large number of medicinal and aromatic plants are found growing wildly. India, with her varied climatic conditions and topography has been considered as "Botanical Garden of the World" and this botanical wealth constitutes more than 2200 species of medicinal and essential oil containing plants.

The importance of medicinal and aromatic crops in the national economy and their potential for the rapid growth of phytopharmaceuticals, perfumery and allied industries in India has been emphasised from time to time. The experts feel that the drugs of natural origin with their wide range of action shall play an important role in health care particularly, in the rural areas of India. There exist today over 2500 registered manufacturers of these drugs in the country. The production of herbal formulations in 1998-99 was to the tune of ₹ 1800 crore approximately, inclusive of various ayurvedic products. In order to ensure this projection, it is necessary that more and more medicinal and aromatic crops be commercially cultivated as field crops. At present, bulk of the raw material is obtained from wild sources, whereas only a few are under systematic cultivation.

[A] CULTIVATION OF MEDICINAL AND AROMATIC PLANTS IN INDIA

Periwinkle *(Catharanthus roseus)* is cultivated in about 3000 hectares of land mainly in Tamil Nadu, Maharashtra, Karnataka and Andhra Pradesh. The yield of the roots per acre is approximately 250 to 500 kg depending on the nature of soil and amount of irrigation provided.

Cinchona plants were first cultivated in the Nilgiri Hills in 1860. Four species of cinchona have been commercially introduced as plantation crops in Nilgiri and Darjeeling. The total area under Cinchona plantation in India is approximately 4500 hectares. The Governments of West Bengal and Tamil Nadu own Cinchona plantations and factories in India. The West Bengal factory is located at Mungpoo and the Tamil Nadu factory is in Anamalai hills.

The plant, Zedoaria (*Curcuma zedoria*), occurs in the eastern Himalayas and in the deciduous forests of the coastal tract of Karnataka. It is widely cultivated in many parts of India and is propagated by tubers.

Dioscorea deltoidea is grown for its tubers throughout Himalayas at altitudes of 1000-3000 metres. It is fairly common in Jammu and Kashmir, Himachal Pradesh and parts of Uttar Pradesh. The depletion of natural forests has reflected in the high prices the drug companies have to pay since 1972 for the collection of tubers. In view of this, a number of drug companies like Cipla, Ciba, Searle and Wyeth have taken up the cultivation of other species of Dioscorea, especially *D. floribunda* and *D. composita* around Bengaluru and other places. A new composite clone of *D. floribunda* has been released for commercial cultivation by the Indian Horticultural Research Institute, (Bengaluru) and it is now grown in Karnataka, Goa and Tamil Nadu.

Isabgol (*Plantago ovata*) is an irrigated Rabi crop cultivated in light sandy-loam soil in approximately 35,000 hectares land mainly in Banaskantha and Mehsana districts of Gujarat and to lesser extent in Pali, Jalore and Sirohi districts of Rajasthan. Castor oil is procured from seeds of *Ricinus communis* grown in Gujarat, Andhra Pradesh and Tamil Nadu.

Despite its wide geographical distribution and edaphic tolerance, *Rauwolfia serpentina* has not lent itself to easy cultivation due to various factors. The cultivation of *R. serpentina* is restricted to small scattered locations in Maharashtra and Madhya Pradesh. At Dehra Dun, intensive cultivation is being done by one of the private pharmaceutical companies. The commercial supply of the roots is mostly from the wild source.

Castor crop is cultivated in about 0.6 million hectares of land while rapeseed oil is cultivated in about 3.5 million hectares of land; this accounts for 30 per cent of the total world production.

Solanum khasianum is cultivated for its berries on a commercial scale in Andhra Pradesh, Karnataka, Maharashtra and Gujarat. Extensive breeding and selection work on this crop have resulted in the development of several improved strains including the spineless variety. The commercial production of solasodine is undertaken near Yeola and Jalgaon in Khandesh region of Maharashtra.

Ipecacuanha (*Cephaelis ipecacuanha*) is mainly cultivated at Mungapoo in Darjeeling district in West Bengal. Currently, the cultivation of Ipecac is monopoly of the Government of West Bengal. The drug is also cultivated on a small scale in Sikkim and Assam.

Indian senna (*Cassia angustifolia*) is commercially cultivated in Tirunelveli, Madurai and Ramanathapuram districts of Tamil Nadu. The cultivation is also reported on small scale in Cudappa district of Andhra Pradesh. When grown for leaves alone under irrigated conditions, the yield obtained is 300 kg of dry leaves and 50 kg of dry pods per hectare.

The annual indigenous production of Neem and Mahua oils is 3 lakh tonnes and 50 thousand tones respectively.

Poppy plant (*Papaver somniferum*) is commercially cultivated in about 60,000 hectares in eleven districts of Madhya Pradesh, Uttar Pradesh and Rajasthan. The yield of opium per hectare is approximately 30 kg. Poppy cultivation, opium processing and marketing are controlled by the Narcotic Commissioner of India. The commercial cultivation of Indian hemp under Government vigilance is carried out in MP, UP, Bihar and Kashmir. The efforts to introduce the non-narcotic plant *Papaver bracteatum* have not been successful so far in our country.

One of the important items of export, covering approximately 70 per cent of world requirement, is a proteolytic enzyme — papain, mainly manufactured in Maharashtra from raw papaya fruits. The commercial production of pectin from the thalamus of Sun-flower is also carried out at Jalgaon in Maharashtra.

India is one of the few countries in the world where the essential oil industry was developed at a very early stage. The essential oils, perfumes and flavours have been associated with the Indian civilization for several thousand years. Because of availability of vast area and a variety of soil and climate, essential oil bearing plants of all types can be grown in one or the other area of the country.

Mentha arvensis subsp, *haplocalyx* var. *piperascens* (Japanese mint) is being cultivated commercially in the districts of Kashipur, Rampur and Moradabad in Uttar Pradesh. It has been introduced in Punjab, Haryana, Madhya Pradesh, Orissa and Maharashtra. *Mentha piperita* (peppermint) was first introduced by the Regional Research laboratory, Jammu in 1954. It has been found that the plant can be successfully cultivated in the hills of Kashmir, Himachal Pradesh and Kumaon (Uttar Pradesh). *Mentha citrata* (Bergamot mint) is commercially cultivated in the sub-mountainous terrains of Uttar Pradesh and Jammu and Kashmir.

As a result of agrotechnology developed by the Central Institute of Medicinal and Aromatic Plants, Lucknow and the Regional Research Laboratory, Jorhat, *Cymbopogon winterianus* (Java citronella) is now being commercially cultivated in the states of Assam, Uttar Pradesh, West Bengal; Karnataka, Tamil Nadu and Maharashtra where about 800 tonnes of oil is being produced annually from the last few years. The oil of lemon grass is obtained from two species of *Cymbopogon* namely *C. flexuosus* and *C. citratus*. The oil of palmarosa (ROSHA-OIL) is obtained from *Cymbopogon martini* in wild condition from the states of Madhya Pradesh, Maharashtra, Karnataka and Andhra Pradesh.

Pelargonium graveolens (geranium) is being cultivated in South India at Yercaud in the Shevaroy Hills, in Cinchona plantations at Nilgiri and in the Annamalai at an altitude of 975-2075 metres, covering an area of about 250 hectares and producing about 5 tones of volatile oil annually.

Eucalyptus citriodora, the source of eucalyptus oil, is commercially grown in the states of Tamil Nadu, Karnataka and Kerala. The linaloe oil is obtained by distillation of the husk of the berries of *Bursera deplechiana*. The plant is indigenous to Mexico and it was first introduced at Talaganj near Bengaluru in 1920 by Roerich and the linaloe estate was established. The plant of *Pogostemon cablin*, yielding oil of patchouli is a native of Malaysia and Indonesia. Recently, it has been commercially introduced in the tea gardens of Assam.

Cinnamomum zeylanicum is grown in the evergreen forests of the Western Ghats of Konkan south to Tirunelveli at low altitudes. The cultivation of the plant is confined to the lower elevations of Eastern Ghats in Cannanore, Calicut and Kottayam districts of Kerala and the lower hills of Nilgiri in Tamil Nadu.

The cultivated seed producing variety of *Vetivera zizanioides* is grown in Kerala and Tamil Nadu and the roots of this plant are used for the extraction of vetiver oil. *Coriandrum sativum* is grown commercially in the states of Madhya Pradesh, Tamil Nadu, Maharashtra, Andhra Pradesh and Rajasthan. A number of species of *Jasminum* are cultivated in India for their fragrant and ornamental flowers of which *Jasminum officinale* var. *grandiflorum* is the source of jasmine perfume. Aligarh, Kannauj, Jaunpur, Sikandarpur and Ghazipur are the main jasmine producing areas in the northern parts of India.

The sandalwood tree (*Santalum album*) is the source for commercial production of sandalwood oil which is called as the liquid gold of Karnataka. It is grown commercially in the states of Karnataka, Tamil Nadu and the boarders of Kerala and Andhra Pradesh. In Karnataka state, the bulk of sandal trees occur in open jungles. In Tamil Nadu, trees are naturally grown in districts of Nilgiri, Coimbatore, Salem and Vellore. The sandalwood trees are also found in the Kurnool and Cudappa districts of Andhra Pradesh.

The commercial cultivation of pepper, ginger, cardamom, cocoa, cinnamon, and areca nut is carried out in the state of Kerala. The head quarters of Hindustan Cocoa Products Limited are located at Chundale in the Wynad district of Kerala. There are separate boards established by the Ministry of Commerce, Government of India for spices, tea and coffee. The Government of India has recently announced establishment of independent Board for Medicinal plants. The cultivation of garlic in different parts of the country for its utilization in extraction of volatile oil has achieved commercial significance in recent years.

India also produces volatile oil from commercially grown plants. Appreciable quantities of volatile oils such as ajowan oil, calamus oil, cardamom oil, celery seed oil, cedar wood oil, davana oil, dill oil, lemon grass oil, ginger oil, lime oil and turpentine oil are obtained on commercial scale. The annual production of coriander is about 2,43,000 tonnes, which constitutes approximately 60 per cent of world demand. About 30 per cent of the global demand in cardamom and 10 per cent in saffron are met by India. The annual production of saffron in Kashmir is approximately 50 tones.

Table 4.1: Comparative study of Cultivation – Parameters of Natural Drugs

Sr. No.	Name of med. plant	Suitable Altitude	Temperature range	Required Rain fall in cms	Desired pH of soil	Propagation by using	Duration of crop	Yield in kgs per hectare
(1)	(2)	(3)	(4)	(5)	(6)	(7)	(8)	(9)
1.	Aconite	2400 to 3500 m	25 - 35°C	100-200 cm	4.5 - 4.5	seeds and root splits	3-4 years	1000-1200 kg
2.	Ashwagandha	up to 1500 m	30-35°C	65-75 cm	7.5-8.0	seeds	9 months	400-500 kg
3.	Daruhaldi	2000-3000 m	15-30°C	60-75 cm	6-8	seeds or stem cuttings	2 years	–
4.	Jatamansi	3300-5000 m	10-30°C	100-200 cm	1-5.5	seeds or root-splits	3 years	1300-1400 kg
5.	Liquorice	1200-1500 m	05-25°C	50-100 cm	6-8.5	stolons or tissue culture	3-5 years	5000-6000 kg
6.	Picrorrhiza	1800-2700 m	25-35°C	100-200 cm	neutral	seeds or rhizomes	2 years	500-600 kg
7.	Rauwolfia	1300-1400 m	10-38°C	100-200 cm	around 4.0	seeds or root stem cuttings	3-4 years	1000-1200 kg
8.	Safed musali	up to 1500 m	25-30°C	50-75 cm	6.5-7.0	seeds/tubers	9 months	800-900 kg
9.	Saussurea	2400-3500 m	25-30°C	60-70 cm	6-7	seeds/root cuttings	3 years	2000-3000 kg
10.	Shatavari	up to 1500 m	30-35°C	50-100 cm	7-8	seeds	2-3 years	6000-7000 kg
11.	Chandan	600-1100 m	10-45°C	80-135 cm	6-7	seeds	30-40 years	average 50 kg per tree
12.	Giloe	up to 500 m	25-40°C	average 50-100 cm	6-8	stem cuttings	perennial	900-1000 kg
13.	Ashoka	600-750 m	10-30°C	upto 100 cm	6-8	seeds	perennial	100 kg per tree
14.	Aloe	up to 1000 m	30-35°C	40-50 cm	8.5	suckers	2-3 years	15000-18000 kg
15.	Artemisia	3000-4000 m	250-30°C	60-75 cm	6-8	seeds	5-6 months	3000 kg
16.	Senna	600-900 m	30-35°C	25-40 cm	7-8	seeds	9 months	12000 kg

(1)	(2)	(3)	(4)	(5)	(6)	(7)	(8)	(9)
17.	Amla	1500-1700 m	25-30°C	60-80 cm	6.5-7.5	seeds	perennial	10500-17500 kg
18.	Vidang	900-1700 m	25-30°C	50-150 cm	6-8	seeds	2.5 years	1500-2000 kg
19.	Kokum	400-500 m	10-35°C	100-200 cm	6-8	soft wood graftings	perennial	850-900 kg
20.	Long pepper	sea-level to 1000 m	30-35°C	200-400 cm	6-8	seeds/suckers	perennial	1000 kg fruits and 500 kg roots
21.	Vanilla	from sea level up to 400 m	10-250°C	150-325 cm	6-8	stems/tissue-culture	perennial	50 kg
22.	Annato	300-1500 m	30-35°C	100-200 cm	6-8	seeds	perennial	1000-2000 kg
23.	Isapgol	1000-1800 m	30-35°C	40-60 cm	7.5 to 8.5	seeds	6 months	600-700 kg
24.	Jajoba	300-1600 m	30-40°C	15-50 cm	7.0 to 8.5	seeds	perennial	4000-4500 kg
25.	Brahmi	1300-1400 m	30-40°C	50-100 cm	6-8	seeds or plant cuttings	perennial	10000 kg
26.	Citronella	up to 1000 m	10-35°C	200-250 cm	6.0 to 7.5	seeds or plant cuttings	perennial	20000 kg
27.	Chiratta	1800-2100 m	20-35°C	50-100 cm	6-8	seeds	annual	2000-3000 kg
28.	Gymnema	up to 600 m	20-30°C	50-100 cm	6-8	seeds or stem-cuttings	perennial	400-500 kg
29.	Kalmegh	up to 1500 m	25-40°C	40-60 cm	6-8	seeds	4 months	2000-2500 kg
30.	Mentha	250-400 m	15-25°C	95-105 cm	neutral	suckers	perennial	20000-30000 kg
31.	Shankh pushpi	1200-1300 m	30-35°C	85-130 cm	5.5 to 7.0	seeds	annual	1000-2000 kg
32.	Tulsi	750-1500 m	30-40°C	40-60 cm	6 to 8	seeds	annual	15000-20000 kg

(1)	(2)	(3)	(4)	(5)	(6)	(7)	(8)	(9)
33.	Guggul	600-1100 m	30-40°C	10-40 cm	7.5 to 9.0	seeds or stem-cuttings	perennial	2000-3000 kg
34.	Phyllanthus	500-800 m	25-40°C	40-50 cm	6 to 8	seeds	annual	2000-3000 kg
35.	Davana	100-300 m	25-30°C	40-50 cm	6-8	seeds	annual	13000-14000 kg
36.	Calamus	up to 2000 m	35-30°C	100-200 cm	alkaline	rhizomes	perennial	2500-3000 kg
37.	Stevia	up to 200 to 250 m	10-30°C	100-150 cm	4 to 5.0	seeds or stem-cuttings	perennial	2000-4000 kg
38.	Black-pepper	up to 1200 m	10-30°C	200-300 cm	6 to 8	seeds or vine cuttings	perennial	300-400 kg

[B] UTILIZATION OF MEDICINAL AND AROMATIC PLANTS

The export of medicinal plants and herbs from India has been quite substantial in the last few years. India has been the major supplier of medicinal plants in the world market till 1976 when it was relegated to the second position by South Korea. With exports worth only ₹ 15 crore during 1978-79, the quantum of export has dropped to almost half of what it was in 1976-77, when India exported medicinal plants worth of ₹ 29 crore. During 1988-89, India exported crude drugs alone to the tune of about 62 crore. The items of export value are opium, psyllium husks and seeds, vinca rosea, kuth roots, nux-vomica, galanga and senna leaves and pods. India is the second largest producer of castor seeds in the world, producing about 1,25,000 tonnes per annum.

With development of phytochemical industry in India, domestic requirements for various medicinal plants grew considerably. Consequently, the Government of India has adopted restrictive export policy in respect of those crude drugs which were indiscriminately exploited in the forests. In accordance with the policy, the exports of Rauwolfia, podophyllum, Indian rhubarb, dioscorea, Saussurea, etc. from India were restricted. The exports of these drugs are, however, permitted by firms obtaining certificates from the Chief Conservator of Forests or officer authorized by him that the material is of plantation or nursery origin.

Medicinal and Aromatic Plants and Their Phytochemicals Exported from India

1. Seed husk and seed of Isabgol (*Plantago ovata*)
2. Opium alkaloids (*Papaver somniferum*)
3. Leaves, pods and total sennoside concentrate of Senna (*Cassia angustifolia*)
4. Root and root alkaloids of Periwinkle (*Catharanthus roseus*).

5. Quinine and quinidine alkaloids (*Cinchona officinalis* and allied species)
6. Ipecac root alkaloids (*Cephaelis ipecacuanha)*
7. Solasodine (berries of *Solanum viarum)*
8. Diosgenin/16 DPA (*Dioscorea floribunda* and allied species)
9. Menthol and mint oil (*Mentha arvensis*)
10. Celery seed (*Apium graveolens*)
11. Citral and ionones from oil of lemon grass (*Cymbopogon flexuosus*).
12. Gudmar herb (*Gymnema sylvestre*)
13. Kuth root and essential oil (*Saussurea lappa*)
14. Mehdi leaves (*Lawsonia Alba*)
15. Sindura fruits (*Bixa orelana*)
16. Papain (*Carica papaya*)
17. Podophyllum (*Podiphyllum nexandrum*)
18. Rauwolfia (*Rauwolfia serpentina*)
19. Valerian (*Valeriana wallichii*)
20. *Cassia tora.*
21. Guar gum (*Cyamopsis tetragonolobus*)

Following essential oils are also exported regularly:

1. Jasmine otto and absolute (*Jasminum grandiflorum*)
2. Agar wood oil (*Aquillaria agallocha*)
3. Sandal wood oil (*Santalum album*)

Medicinal Plants and their phytochemicals, (including adaptogens* and allied materials) imported in India

Very few medicinal plants are imported in the country. Further, except items 1, 2, 3 and 4 the amount involved is not very large.

1. Liquorice roots (*Glycyrrhiza glabra*)
2. Asafoetida (*Ferula asafoetida*)
3. Chirayata herb (*Swertia chirata*)
4. Gum-guggal (*Commiphora weightii*)
5. Gum-Arabic (*Acacia arabica*)
6. Pipali Root/unripe fruits (*Piper longum*)
7. Unnab fruits (*Zizyphus sativa*)

8. Akarkaara root (*Anacyclus pyrethrum*)
9. Asaroon herb (*Asarum europaeum*)
10. Ustakkoddus (*Lavendula stoeches*)
11. Red sage (*Salvia officinalis*)
12. Scamonia (*Convolvulus scamonia*)
13. Gaj pimpli (*Scindapus officinalis*)
14. Kankol (*Piper cubeba*)
15. Chopchini (*Smilax china*)
16. Maiphal (*Quircus infectoria*)
17. Clove (*Engenia caryophyllus*)
18. Nutmeg mace (*Myristica fragrance*)
19. Kapurchini (*Hedychium spicatum*)
20. Patang (*Caesalpinia sappan*)

Three essential oils are also imported regularly:

1. Oil of Geranium (*Pelargonium graveolens*)
2. Oil of Patchouli (*Pogostemon patchouli*)
3. Oil of Clove (*Eugenia caryophyllus*)

***Adaptogens:** These are biologically active substances which improve physical endurance for doing work even in adverse and difficult environmental conditions.

Adaptogens increase tolerance to change in environmental conditions and also resistance to noxious stimuli, like exposure to cold, heat, pain and infections organisms. Such health food supplements arrest aging processes and age induced deterioration in physical and mental performance, i.e. Ginseng Gingko, Withania etc.

India's export of essential oil during last few years has shown an erratic trend. The USA accounted for a major share of exports followed by Russia. India is also exporting volatile oils to France, Japan, Sudan, Germany and Switzerland. The principal items of imports into India are clove oil, geranium oil, lavender oil and patchouli oil.

Some of the pharmaceutical units engaged in extraction and utilization of Biomedicinals in India are Government Opium factories at Ghaziabad (UP) and Neemuch (MP); Amsar Pvt. Ltd., Indore (MP); Government factory for extraction of Cinchona alkaloids at Mungapoo (West Bengal); Maharashtra Fisheries Development Corporation Ltd. Mumbai for production of Shark liver oil; Jain Plastics and chemicals Ltd. at Jalgaon for production of papain, calcium caseinate and other spray dried products, Kothari Phytochemicals Ltd. at Madurai and Madras; INGA Laboratories Ltd., Mumbai; CIPLA at Bangalore and Mumbai; Cadila Laboratories Ltd., Ahmadabad; Zandu Pharmaceutical Works Ltd., Mumbai; Dabur (India) Ltd., Ghaziabad (UP); Hamdard Laboratories,

Delhi, Himalaya Drug House at Mumbai and Bangalore; Vicco Laboratories, Mumbai; Baidyanath Ltd., Nagpur; Franco-Indian Ltd., Mumbai; Wockhardt Ltd., Aurangabad; Glaxo (India) Ltd., Mumbai Organon Ltd., Calcutta; Sandoz (India) Ltd., Mumbai, Dhutpapeshwar at Panvel, Boeringher Knoll, Mumbai; Enzochem Laboratories at Yeola in Maharashtra; Hindustan Cocoa Products Ltd. at Chundale in Kerala; Cellulose Products of India, Ahmadabad; Ranbaxy Laboratories Delhi; Lupin, Mumbai, Anil Starch Products, Ahmadabad and Proteins Products of India, Cochin.

Fig. 4.1: Pharmaceutical export unit

The volatile oils, aromatic chemicals and related compounds are being manufactured by the Government Sandal Oil Factory, Mysore; S H Kelkar and Co. Pvt. Ltd., Mumbai; Meghalaya Phytochemicals in West Bengal; Camphor and Allied Products Ltd., Bareilly; Procter and Gamble (India) Ltd., Mumbai; Naarden Chemical Works Ltd., Mumbai; Ashok Thymol Factory, Indore; Royal Eucalyptus Oil Ltd., Nilgiri; Bhavana Chemicals Ltd., Vadodra; Jupiter Eucalyptus Oil Refinery, Nilgiri; Tata Oil Mills Co. Ltd., Tatapuram; Jammu Rosin and Turpentine Factory at Miransahib in Kashmir and Himachal Rosin and Turpentine factory at Nahan in H.P.

Some of the manufacturers of biological products are Hindustan Antibiotics Ltd., Pimpri; I.D.P.L., Rishikesh; Haffkine Biopharmaceutical Corporation Ltd., Mumbai; Central Research Institute, Kasauli; Serum Institute of India Ltd., Pune and Biological E. Ltd. Hyderabad.

The research and development in the field of medicinal and aromatic plants has acquired considerable importance in India. The Government organizations such as Council for Scientific and Industrial Research and Indian Council of Agricultural Research have already made significant progress in the cultivation technology of medicinal and aromatic plants. Significant contributions in the field of cultivation and/or utilization of natural medicinal agents have been made by Regional Research Laboratories at Jammu, Jorhat and Bhubaneswar; Central Institute for Medicinal and Aromatic Plants, Lucknow National Chemical Laboratories, Pune; National Bureau of Plant Genetic Resources and ICAR, New Delhi; Central Drug Research Institute, Lucknow; Forest Research Institute Dehradun; CHEMEXCIL, Mumbai, CIBA research centre, Goregaon Mumbai; Hoechst Research Centre Mumbai and number of other agencies in public and private sectors. The Indian research and development efforts are being directed to utilize new scientific discoveries as quickly as possible towards reducing the time lag between such discoveries and their applications in the field of medicinal and aromatic plants.

ENDANGERED MEDICINAL PLANTS

Many plants have been described in various authoritative texts of Indian system of medicine (Ayurveda) for their valuable medicinal activities. Over the years and centuries, they have been procured only from wild sources. Unfortunately, due to lack of application of scientific measures for their systematic cultivation, as well as unscientific over exploitation from nature, few plants are now-a-days on the verge of becoming extinct. Some of them have been duly identified by Govt. of India as endangered plants.

Unscientific exploitation and ignorance about their life cycle has consequently led to decline of such plants. About 40 plants have been declared as endangered.

Some of these plants are:

(i) Belladonna (*Atropa acuminata*)

(ii) Dhoop (*Jurenia macrocephala*)

(iii) Bhojpatra (*Betula utilis*)

(iv) Kalazeera (*Carum carvi*)

(v) Kapur Kachri (*Hedychium acuminatum*)

(vi) Tejbal (*Xanthoxylum alatum*)

(vii) Tejpatra (*Cinnamomum tamala*)

For preservation of endangered plants, action-plan has been prepared by Govt. of India, for which World Bank has indicated assistance.

The objectives of this plan include:

1. Harnessing the plant wealth in view of promoting the industries for production of phytochemicals, medicines, volatile oils and cosmetics.
2. Identification of commercially viable species in various agro-climatic zones.
3. Standardization of such agro-practices in view of their cultivation and, in turn, preservation.
4. Measures to boost up the research and development for the purpose of involving new herbal products and technology for the purpose of their manufacture.

MEDICINAL PLANTS PROHIBITED TO GET EXPORTED FROM INDIA

Ministry of commerce Govt. of India, New Delhi has prohibited the export of following medicinal plants, plant portions, their derivatives and extracts obtained from wild sources, since October 1998.

1. *Cycas beddomei (Beddomes cycad)*
2. *Vanda coerulea (Blue vanda)*
3. *Saussurea costus*

4. *Paphiopedilum species (Ladies slipper orchids)*
5. *Nepenthaes khasiana (Pitcher plant)*
6. *Renathera imscootiana (Red vanda)*
7. *Rauwolfia serpentina*
8. *Dioscorea deltoida*
9. *Podophyllum hexandrum*
10. *Euphorbia species*
11. *Orchidaceae species*
12. *Pterocarpus santalinus (Red - Sanders)*
13. *Taxus wallichiana (Yew)*
14. *Aconitum species*
15. *Gentiana kurroo*
16. *Picrorrhiza kurrooa*
17. *Swertia chirata*
18. *Gnetum species*
19. *Panax pseudo-ginseng*
20. *Aquilaria malaccensis*
21. *Cyatheaceae species*
22. *Frerea indica*
23. *Cycadaceae species*
24. *Dactylorhiza hatagirea*
25. *Coptis teeta*
26. *Coscinium fenestratum*
27. *Kamphergia galenga*
28. *Ceropegia species.*

Chapter 5

MEDICINAL PLANT BIOTECHNOLOGY

[A] GENETICS AS APPLIED TO MEDICINAL HERBS

George Johann Mendel in 1885 showed that certain hereditary factors operate in all biological species. The Danish biologist Wilhelm Johannesen called these factors as Genes. It is now known that the genes not only transmit hereditary characters, but also mastermind the entire process of life.

The genes are located in chromosomes which are situated in the nucleus of the cell.

It can be said that the genes form the riddle and the chromosomes represent the mystery. Much of the mystery surrounding the genes was cleared up with the discovery of the structure of the DNA announced by JD Watson and Francis Crick in April 1953. The structure of the DNA resembles a long rope ladder twisted around like a corkscrew. The two sides of the ladder are long chains of sugars and phosphates in repeated sequences. DNA contains the bases as follows:

Adenine — *A* Cytosine — *C*

Thymine — *T* and Guanine — *G*.

An A will form a bridge only with T and C with G. So the pairs, A-T, T-A, C-G and G-C form in a way a four letter alphabet with which messages can be spelt out. It makes up what is known as the genetic code which is not only complex but also extensive. In 1977, Fred Sangar pointed out that the DNA code of a virus, when decoded by computer came to a print out of 15 metres.

The genes control all functions of cell and body growth. The two main events in the life of most cells are multiplication (by division) and synthesis of proteins. Both these operations are carried out on the basis of blue prints coded in the genes.

Genetics has made valuable contributions to the improvement of plants having economic, medicinal and ornamental value, by applying various techniques like selective breeding. It has important role in mitosis and meiosis. Mitosis results in precise equal distribution of chromosomes from a parent nucleus to the daughter nuclei. Further, it maintains equilibrium in the amount of DNA contents of cells. The mitosis is associated with vegetative propagation of plants. The somatic number is the number of chromosomes in plant cell nucleus which in normal conditions remains constant, but varies with different species. Meiosis occurs in case of formation of reproductive cells (gametes or meiospores) in sexually reproducing species. Meiosis has also a genetic significance because of formation of four haploid nuclei from a single diploid one, in two successive divisions, thus balancing the doubling of chromosome number that results from fertilization. During the process of meiosis, the crossing over provides new combinations of genetical substance and hence, new combinations of characters in off springs. The process of segregation of chromosomes also occurs in meiosis.

The advances in techniques pertaining to genetic manipulations in plant cells have made much impact to improve yield of medicinal plants.

Mutation

Mutation is represented as variation in characters of the species. It is caused either due to environmental changes or changes in hereditary constitution. Normally, as a response to environmental changes, the variations are observed, but the original traits are restored when changes in environment are also withdrawn or disappeared. This type of change and restoring is not heritable and also not built into genotype. They are termed only as phenotypic variations and commonly called as modifications. It is evident that between organisms of similar genotype there are phenotypic variations. However, when a change occurs in the genome of an individual which is not caused due to environment, it may make a permanent evolutionary change. This is termed as mutation and represents a sudden change in genotype causing qualitative or quantitative alterations of genetic material. Mutations are of two types namely chromosomal mutations and point mutations. The former type of mutation is also called chromosomal aberration, which in many cases leads to changes in amount or position of genetic material. On the other hand, the changes with a gene or cistron of DNA molecule cause point mutations and it is permanent and heritable.

Mutation which occurs due to some unknown reason from nature is called as **spontaneous mutation**. This has been observed in some plants, bacteria, viruses, etc. Mutations can also be induced by artificial means with certain reagents called mutagens and are called induced mutations. The various mutagens used are exposure to UV rays, X-rays, ionizing radiations, certain chemicals, abnormal environment etc. The chemical mutagens used are nitrogen mustard, formaldehyde, nitrous acid, ethylethane sulphonate, 5-bromouracil, 2-aminopurine, manganese chloride etc.

The changes caused due to mutations include morphological and anatomical changes, as well as, changes in the chemical composition of the plants. This is significant for the medicinal plants. In some cases, favourable changes and yields in active constituents of plants have been achieved. Mutations may cause building the resistance of a medicinal plant towards certain diseases. But, in all these cases, the plant may become susceptible to climatic conditions, certain other diseases, retardation in growth, etc. These undesirable effects are to be eliminated by breeding and selection.

Polyploidy

Polyploidy has exhibited various useful effects on medicinal plants like digitalis, mentha species, poppy, plants containing tropane alkaloids, lobelia, etc. The specific number of chromosomes is a character of each species and is called genome which is observed in all types of organisms. The term euploidy is a type of ploidy in which genome contains whole set of chromosomes and euploidy includes monoploidy, diploidy and polyploidy. When the organism contains more than two genomes, it is called polyploidy. The polyploidy occurs in a multiple series of 3, 4, 5, 6, 7, 8, etc. of the basic chromosome or genome number and then accordingly, it is called triploidy, tetraploidy, pentaploidy, hexaploidy, heptaploidy and octaploidy respectively.

Polyploidy is caused through cell generation, physical agents like X-rays, centrifugation, temperature chocks and chemical agents, mainly colchicine, veratrine, sulphanilamide, hexachlorocyclohexane, and mercuric chloride. The chemical agents cause disturbance to mitotic spindle of dividing diploid cell and cause non-segregation of already duplicated chromosome and thus convert diploids into tetraploid cells. The phenomenon of polyploidy is of greater significance to medicinal plants. It may cause formation of **new species**, adaptability to various habitats and mainly accumulation of vitamins.

Perhaps, the best known chemical to cause polyploidy is colchicines, an alkaloid obtained from *Colchicum* species like *Colchicum autumnale*, *C. luteum* and *C. speciosum*. Colchicine prevents sister chromatids from separating into daughter nuclei at anaphase. These chromatids remain attached by their common centromere in C-metaphase. The chromatids eventually separate, but remain in the same nucleus. An interphase occurs, followed by a second C-metaphase, involving a doubled chromosome complement. Hence, the chromatid pairs are doubled in second C–metaphase. Likewise, the cell undergoes one, two or more than two rounds of DNA replication and causes polyploidy. The colchicines activity mentioned here is caused due to its interaction with disulphide bonds of spindle protein and by inhibition of conversion of globular proteins to fibrous proteins. The capacity of colchicines to induce polyploidy varies along with its different derivatives. Chemically induced mutations may lead to variations in biochemical composition of plant. Colchicines treatment given to medicinal plants has shown promising results in many cases. The observations in tropane alkaloids by way of polyploidy are more specific in stramonium, where yield of crop is enhanced by 60-150 per cent in 4 n form. The plants like lobelia, cinchona, belladonna; acorus, squill, cannabis, and poppy also show increased yield of respective compounds in 4 n form. Polyploidy may cause a reduction of total glycosides of *D. purpurea* and *D. lanata*, but in the later case raises slightly the contents of lanatosides A and B. The increase in chemical contents may not be coincidental with phenotype of medicinal plant. There may be reduction in size also along with enhancement in content of active constituents. Some plants do not show any change in chemical contents as a response to polyploidy.

Chemodemes (Chemical Races)

The knowledge about plant chemical races has surfaced, due to thorough chemical analysis of a huge number of different plants with the help of modern analytical instruments. **Chemodemes** are regarded as a group of plants of a species which have identical morphological characters, but differ in their chemical nature. Due to this, chemodemes are considered as chemically separate groups within species. The observation of chemodemes can be confirmed only by growing different plants of a species in identical conditions, preferably from the seeds and for many generations. By this way, it shows variations either in type or contents of certain constituents, like the secondary metabolites of medicinal importance. The chemical characters of chemodemes are hereditary.

Chemical races have been reported in *Digitalis purpurea* and *Digitalis lanata*. In the former, the races or strains identified are streptoside, digitoxin and digipurpurin. These chemical races in *D. purpurea* yield different proportions of glycosides obtained from digitoxin and gitoxin. Depending on content of lanatosides A and C the chemical races in *D. lanata* are *D. lanata* Ehrh, chemovarieties A and C. The chemodemes in *Rheum palmatum* are also known, which give differing values for rhein chrysophanol ratio. The different varieties of *Prunus communis* (almond) differ by the presence or absence of amygdalin. Chemical races have also been reported in *Claviceps purpurea*, *Duboisia myoporoides*, and *Duboisia leichharditii*. Three chemotypes are evident in *Withania somnifera*. Chemotype contains mainly, withaferin; chemotype II has similar compounds, and chemotype III contains withanolides (a mixture of steroidal lactones). Withaferin and withanolides have different medicinal actions.

The plants like *Cinnamomum zeylanicum, Ocimum menthaefolium, Cinnamomum camphora, Eucalyptus dives,* and *Ocimum sanctum,* all of which contain volatile oil, have chemical races with varied content of aromatic constituents.

The discovery of chemodemes has enabled the pharmacognosists to select high yielding chemical strains or to eliminate chemical strains which contain toxic principles. The chemodemes are induced by breeding in some species to manipulate active constituents, with a view to enhance the therapeutic efficacy.

ARTIFICIAL MUTATION

The artificial mutations are those which are induced artificially in the living organism by exposing them to abnormal environment, such as radiations, certain physical conditions like temperature and chemicals, all of which are called mutagens or mutagenic agents.

(1) Radiation mutations: The electromagnetic waves of short wavelength (ultra violet light, X-rays, gamma-rays, alpha and beta rays) are radiation mutagens. The X-rays and gamma rays are called ionizing radiations, and also include alpha particles, beta rays, thermal and fast neutrons. Due to ionizing radiations, in many cases, water molecule in a biological system releases one electron and becomes unstable and eventually splits into hydrogen ion and hydroxyl radical. Hydrogen ion reacts with O_2 and produces hydroperoxyl (HO_2) radical. Both these radicals viz. hydroperoxyl and hydroxyl are potent oxidizing agents. When chromosomes and their DNA are struck by such radicals, they react due to which sugar-phosphate part of DNA may be impaired leading to chromosomal mutations, like breaks, deletions, additions, inversions and translocations.

(2) Chemical mutations: Some chemical mutagens, like nitrogen mustard, formaldehydes, nitrous acid and ethyl ethane sulphonate alter chemical constitution of DNA bases and cause transitional substitution in DNA. But some compounds like 2-aminopurine, urethane, 5-bromouracil, caffeine, triazine, phenol and certain other carcinogens act as a base analogue and bring out copy error mutations in DNA. In general, the chemical mutagens have profound cellular effects like production of abnormal DNA (nitrogen mustard), inhibition of deoxyribonucleotide synthesis (deoxyadenosine), and inhibition of cytochrome oxidase with resultant peroxide formation (inorganic cyanides).

The artificial production of mutations in medicinal plants is an important milestone in the development of cultivation technology.

The higher solasodine content is achieved by applying radiation and chemical mutagens in *Solanum khasianum.* It is also reported that chemical mutagens have a successful use in increasing morphine content of *Papaver somniferum.* The tuber yield and diosgenin content of *Dioscorea bulbifera* is increased by radiation. The economically important characters of *Atropa belladonna* have been enhanced by radiations and chemical mutagens. The agronomic performance and harvest index of *Mentha arvensis* var. *piperascens* (Japanese mint) have been improved by exposure to gamma radiations. The effects of artificial mutations have been extensively studied on different species of mentha. The capsicum seeds (*Capsicum annuum*) treated with sodium azide and ethyl methane sulphonate have led to plants giving higher contents of capsaicin.

HYBRIDIZATION

The process through which hybrids are produced is called hybridization. A hybrid is an organism which results from crossing of two species or varieties differing at least in one set of characters. The resultant hybrids are monohybrids (one pair of different characters), dihybrids (two pairs of different characters) or polyhybrid (more than tow pairs of different characters). Hybridization helps in inducing in a single variety, the favourable characters of other varieties or species, and sometimes producing new and favourable characters which are not present in both the parents.

The hybridization of *Withania somnifera* Israeli chemotype II and *W. somnifera* South African chemotype has led to formation of a new hybrid which contains three new withanolides. The hybrids like *Digitalis purpurea* x *D. lanata* and *D. purpurea* x *D. lutea* contain principal glycoside as lanatoside A, along with lanatosides B and E, but devoid of lanatoside C or purpurea glycoside A. During the cross of *Solanum incanum* (1.8 per cent solasodine) and *S. melongena* (traces of solasodine), first generation bears more fruits (berries) with solasodine up to 0.5 per cent. Second generation has proved to be a high yielding source for solasodine.

A recent development in hybridization is through the medium of tissue culture. The protoplast cultures are employed for this purpose. Plant protoplasts are the cells without cell walls. Such protoplasts in cultures can be fused together (protoplast fusion or asexual hybridization). The fusion can be arranged in cells of same origin or between different species.

GENETIC ENGINEERING AND RECOMBINANT DNA TECHNOLOGY

In recent years, these techniques have taken greater strides in the field of drugs and pharmaceuticals. The method of artificial synthesis of new genes and their subsequent transplantation in the genome of an organism or the method of correcting defective genes of organism by molecular biological techniques form the discipline called genetic engineering.

Recombinant DNA technology involves gene splitting to change characters of plants and animals by implanting in them genes from other organisms and, in some cases, even from other species.

The genetically engineered products which already exist in market are human growth hormone, human insulin, *alpha*-interferon and hepatitis-B vaccine and the products coming in next few years are as follows

1. Interleukin-2, to fight certain types of cancer
2. Beta-endorphins, for treatment of pains
3. TPA-tissue plasminogen activator for treating heart attack
4. Somatostatin for treatment of pituitary diseases
5. Erythropoietin or EPO designed for treating anaemia in patients undergoing kidney dialysis.

The secondary metabolites from plants, which are expected to give promising results by genetic engineering include

1. Cardiac glycosides.
2. Morphine alkaloids.
3. Stigmasterol or diosgenin which give steroid hormones and corticosteroids.

In respect of such plants, the transfer of entire biosynthesis route to bacteria by gene-technology method is expected to yield economic processes.

Transgenic Plants

Application of genetic engineering techniques for the improvement of plants is fast developing area of research. The advent of modern technique of genetic modification has enabled to remove individual genes from one species and insert them into another, without the need for sexual compatibility.

Once the new gene has been inserted into the plant, offspring that will contain the copies of the new gene can be produced in the traditional manner.

Transgenic plants carry the stably integrated foreign genes. It is important that the new DNA remains stable and can pass the new gene to its offspring as well. The introduction of foreign genes is done by genetic transformation (gene transfer) or also called Transfection. It is carried out to get the same purpose, as viral or chemical transformation, by which a stable or transient change is obtained in the cells. But, it is accomplished with the help of DNA which is tailored as per the needs/aims.

Transgenic plants possess a wide variety of potential uses, among which crop improvement is most significant. It includes the goals like herbicide resistance, insect resistance, changes in oil and protein contents of the crops, enhanced quantities of products, etc. The overall transformation process includes the following steps:

- Gene transfer.
- Selection of transgenic plants.
- The recovery of transgenic whole plants.

Gene transfers or DNA delivery systems are of following types:

(1) Vector mediated:
  * Using *Agrobacterium tumefaciens*
  * Using viral vectors.

(2) Vector-free (Direct Gene Transfer)
  * Electrical methods (Electroporation)
  * Physical methods
  -- Microinjection, Silicon carbide whiskers, Microprojectile bombardment (Biolistic gene transfer), Sonication.
  * Chemical methods
  -- Using polyethylene glycol (PEG), Fusion of protoplasts with liposomes.

Gene Transfer

(a) By Using *Agrobacterium Tumefaciens*

This gram negative bacterium commonly occurs in soil, and causes 'crown gall disease' in higher plants. It enters the plant through the mechanically wounded tissue site. This infection causes local proliferation of undifferentiated tissue (tumorous outgrowth). *Agrobacterium tumefaciens* possesses a special ability to transfer some of its DNA into nuclear DNA of an infected host plant because of functions encoded on its "Ti" (tumour-inducing) plasmid. The transferred

DNA, or X-DNA, of this plasmid encodes substances which act as plant hormones, causing the proliferation of tissue seen in the infected plant. Ti-plasmid of this bacterium has been modified by means of genetic engineering to create "disarmed" plasmids (also called vectors) which can carry any DNA sequence into A. *tumefaciens*-infected plants, without tumorous growth on the host plant. These cloning vectors derived from Ti-plasmid have been used for replication in both *A. tumefaciens* and *Escherichia coli* bacterial strains. This makes possible the cloning of foreign DNA sequences into the vectors using *E. coli* host, as well as the further transfer of the completed vector to an *A. tumefaciens* host.

(b) By Electroporation

This technique is based on the use of short electrical pulses of high field strength to promote DNA uptake by protoplasts. The optimal electroporation conditions may vary in different plant species. Buffers greatly influence the gene transfer efficiency and also the protoplast survival rate.

(c) By Using Microinjection

Microcapillaries and microscopic devices are used to deliver DNA into protoplasts. It is considered as one of the most precise techniques. However, it is a very skilled process and expensive equipments are required.

(d) Biolistic Gene Transfer

The DNA is bombarded into the intact cell using a biolistic device such as particle gun. It can only be used for intact tissues. Because pressure is to be applied, it is not suitable for protoplasts. Like microinjection, this is also a very precise technique.

(e) Silicon Carbide Whiskers

Silicon carbide forms long, needle shaped crystals (whiskers) in solutions like in polyethylene glycol, 1-3 per cent. When cells are vortex mixed in the presence of whiskers and DNA, the DNA can be introduced into the cells following penetration by whiskers.

(f) Gene Transfer by Using Chemical Agents

Various chemical treatments have been used to stimulate DNA uptake by protoplasts. Among them polyethylene glycol is the most common. It acts by increasing the permeability of cell membranes. The optimal concentrations of PEG are 15-20 per cent.

Tobacco was the first plant to be genetically transformed in 1983 with cereals beginning in 1990. Recently, genetically modified Soya has reached the market.

Most cereal food crops in the Western countries have been made insect and pest resistant by incorporating suitable genes. Owing to this, percentage spoilage of crops per annum has declined sharply reduced. Consequently the use of these harmful chemicals has also been drastically reduced, proving to be beneficiary to the environment.

Transgenic Plants as a Source of Vaccines

It has been found that developing nations are still deprived of modern vaccines. This is mainly because of lack of refrigeration and other equipments needed for such programmes. It has been shown that the plants can be utilized for preparing vaccines given that they can be induced to synthesize appropriately folded bacterial and viral proteins that stimulate immune response.

Specific DNA sequence encoding for the expression of a surface antigen of a pathogen is isolated and ligated to a promoter (e.g. Cauliflower Mosaic virus 35 S promoters) which can regulate the production of the surface antigen in a transgenic plant. This gene is then transferred to plant cells using a procedure that results in its integration into the plant genome, such as through the use of *Agrobacterium tumefaciens* plasmid vector system.

Plasmid containing Hepatitis-B surface antigen gene containing a promoter, a marker gene, and flanking T-DNA sequences has been prepared and used in *A. tumefaciens* mediated transformation of tomato. Transgenic tomatos, producing the antigen were produced.

(B) PLANT TISSUE CULTURES AS SOURCES OF BIOMEDICINALS

The utility of plant tissue cultures in bioproduction of natural compounds under aseptic conditions by methods similar to those used to culture microorganisms has opened new approach in the field of biotechnology. The use of tissue culture technique for biosynthesis of secondary metabolites, particularly in plants of pharmaceutical significance, holds promise for the controlled production of plant constituents. Inspite of the recent developments on synthetic chemistry, higher plants are still an important source of medicinal compounds. In recent years, however, it has become difficult to maintain an ample supply of medicinal plants due to several factors, such as their ruthless exploitation, lack of conservation of the environment, increasing labour costs and economical or technical problems associated with the cultivation of medicinal plants. The technique of plant tissue cultures could afford possible solution to some of these problems. The major advantages expected from plant tissue culture systems over conventional cultivation technique may be summarized as follows.

1. The process offers the prospect of absolutely uniform biomass obtainable at all times and manageable under regulated and reproducible conditions, rarely possible in working with entire living potentials.

2. It is possible to use plant cell culture technique for synthesis of those medicinal compounds which are too difficult or impossible to synthesize chemically.

3. The useful natural compounds could be produced under controlled environmental conditions, independent of soil conditions and changes in climatic conditions.

4. The technique could be used to study biogenesis of secondary metabolites. It is possible to feed labelled precursors to cell cultures and deduce interpretations pertaining to metabolic pathways of desired compound.

5. It is possible to attempt biotransformation reactions in plant cell cultures. It is expected that specific modification of chemical structures of certain compounds may be achieved more easily in cultured plant cells rather than in micro-organisms or by chemical synthesis.

6. The cells of any plants, tropical or temperate, could be multiplied to yield specific metabolites produced by them.

7. The cultured cells could be maintained free from any microbial contamination and insect attack.

8. Another important application of plant tissue culture technique is immobilization of cells which could be used for various biotransformation or biochemical reactions. A particular strain of cells obtained from suspension cultures is immobilized by suspending it in sodium alginate solution, precipitating the alginate plus entrapped cells with calcium chloride solution, pelleting and allowing the product to harden.

The biotechnological potentials of plant tissue cultures have been reviewed by Reinhardt 1967, Tauscher 1973, Butcher 1977, Boehm 1977, Staba 1977, Tabata 1977, Zenk 1978 Kurz and Constabel 1979, Kokate 1979, Barz and Ellis 1981, Deus and Zenk 1982, Shargool 1982, Berlin 1984, Fowler 1984 and Misawa 1984.

Historical Development

The technique of plant tissue culture is now more than ninety years old, but is only in last two decades that implications of this technique have been realized and in particular its pharmaceutical potential appreciated. The hypothesis (1902) that isolated plant cells should be capable of cultivation on artificial medium was directed at the study of their capacity and characteristics as elementary organisms free from multicellular system of plant. Although, Haberblandt and coworkers were unable to demonstrate this hypothesis, other workers such as Robbins in the USA (1922) and Kotte in Germany (1922), successfully cultured excised plant roots, possibly inspired by the Haberblandt's hypothesis put forward by Haberblandt. In 1934, White reported successful continued culture of excised tomato root tips. In the same year, Gautheret published the first of his many classical papers on cambial tissue cultures and later in 1939 working independently of one another, White, Nobecourt and Gautheret reported the first unlimited culture of callus tissue. White (1939) reported continuous cultivation of a similar undifferentiated callus derived from procambial tissue of young stems of a hybrid *Nicotiana (N. glauca x N. langsdorffii)* cultured in the medium he had developed for excised tomato roots, but containing 0-5 per cent agar. Gautheret (1939) examined the behaviour of carrot explants on a medium compounded of inorganic salt mixture and dextrose containing aneurine, cystein hydrochloride and a low concentration of indole acetic acid. In 1954, Muir, Hildebrandt and Ricker reported growth of liquid cultures containing single cells and small clumps of cells of *Tagetes erecta* and *Nicotiana tabaccum.* In 1956, Reinert obtained evidence for occurrence of cell divisions in a similar suspension of single cells and cell groups from *Picea glauca.* Jones, Hildebrandt, Ricker and Wu in 1960 examined growth in hanging drop culture of separated cells from a callus of hybrid tobaccos (*N. tabaccum x N. glutinosa).*

Types of Cultures

Three main lines of technical development constituting methodology of plant tissue culture are concerned respectively with the culture of isolated plant organs (particularly of isolated roots but, to a lesser extent, of stem tips, leaf primordial, immature embryos and flower structures), the growth of callus masses on solidified media and the growth in liquid media of mixed suspensions of separated cells and small cell groups.

Plant tissue culture consists of a number of techniques for growing plant organ and cells. The organ cultures have been successfully obtained from roots, stem tips, leaves, flowers and fruits. Methods used to initiate such cultures are varied and depend upon nature of the organ acquired. Excised root cultures, for example, are initiated by placing sectioned or lengths of surface sterilized root tip in liquid medium containing essential macro and micro-nutrients and periodically subculturing or transferring to fresh medium. They demonstrate high growth rate and metabolic activity and a low level of variability therefore they are important tools in determining responses to varying conditions.

The tissue may be cultured on solid agar or in liquid nutrient medium. When grown on agar medium, the tissue forms a callus or a mass of unorganized cells. The technique of callus culture is convenient for starting and maintaining cell lines, as well as also for studies pertaining to organogenesis and meristems culture. The liquid suspension cultures consist of mixtures of cell aggregates, cell clusters and single cells. The growth rates of such cultures are generally much higher than on solidified medium and the technique provides for better control of the growth of biomass, because the cells are surrounded by the nutrient medium. For the same reason, the cell material should also be probably more uniform physiologically. Both callus and suspension cultures can be derived from tissues of most of the species, but the ease of starting the cultures varies with the type of plant and the tissue origin. Almost any part of a plant can be induced to produce a callus and a suspension culture. The necessary tissue can be obtained from roots, seedlings, pollen; stem and leaf portions and it usually grows as a mass of undifferentiated cells on enriched solidified medium.

CULTURE MEDIUM

Defined medium for the growth of cell cultures consists of inorganic salts, a carbon source, vitamins, growth regulators and some organic supplements. Since the pioneer studies of Gautheret (1939) and White (1939), it has been one of the main subjects of the researchers of plant tissue culture to devise a proper medium which supports sufficient growth of cell cultures. In consequence of these studies, we now have several well defined standard nutrient media which have been used most widely in basic or modified forms such as of Gautheret (1942), White (1943), Hilderbrandt et al (1946), Heller (1953), Nitsch and Nitsch (1956), Murashige and Skoog (1962), Eriksson (1965) and B_5 (1968).

An ideal nutrient medium for plant tissue cultures contains five classes of ingredients.

1. Inorganic Salts

The concentration of potassium and of nitrate should be at least 20-25 mM for each, whereas concentrations of 1-3 mM of phosphate, sulfate and magnesium appear to be adequate. Ammonium is essential although amounts in excess of 8 mM could be deleterious. The recommended micronutrients are iodide, boric acid and salts of zinc, manganese, molybdenum, copper, cobalt and iron. The later is usually incorporated in the chelated form.

2. Vitamins

Thiamine is the only vitamin which appears to be essential. Pyridoxine, myo-inositol and nicotinic acid are frequently added to improve the cell growth.

3. Carbon Source

Sucrose or glucose at a concentration of 2 - 4 per cent is the most suitable carbon source.

4. Growth Regulators

These substances are needed to induce cell division. The compounds most frequently used are naphthalene acetic acid (NAA) and 2, 4-dichlorophenoxy acetic acid (2, 4-D) in the molar concentrations of 10^{-7} to 5 10^{-5}. Both 2, 4-D and NAA are degraded very slowly by plant cells and are stable to autoclaving Cytokinins such as kinetin or benzyladenine (10^{-7} to 10^{-5} M) are sometimes required in conjunction with 2, 4-D or NAA to obtain good callus formation.

5. Organic Supplements

Protein hydrolyzates, yeast extract, malt extract and coconut milk (liquid endosperm) are used for enhancement in the growth rate of the cells in biomass.

The chemicals are dissolved in distilled water, the stock solutions of vitamins, micronutrients and growth hormones are added and the pH of the medium is adjusted to 5.5 to 5.7. The solution is made to volume and then 50 and 100 ml quantities are distributed into 250 ml Erlenmeyer flasks. The flasks are closed with stoppers with cotton wool plugs, and autoclaved at 120 C for 15 minutes. The flasks are removed for cooling, as early as possible. The agar medium is autoclaved in lots of 500 ml and subsequently poured into sterile containers. All media are stored at 10 C prior to use.

SURFACE STERILIZATION OF EXPLANTS

It is necessary to effect surface sterilization of the organ from which the tissue is to be aseptically excised or of the spore or seed whose germination shall yield the tissue explants. The commonly used surface sterilizing agents are sodium hypochlorite (1-2 per cent), bromine water (1-2 per cent), hydrogen peroxide (10-12 per cent) mercuric chloride (0.1-1 per cent) and silver nitrate (1 per cent).

The seeds are treated with 70 per cent ethanol for about 2 minutes, washed with sterile distilled water, treated with surface sterilizing agent for a specific period, once again rinsed with sterile distilled water and kept for germination under aseptic conditions. The seeds may be germinated aseptically by placing them on double layers of pre-sterilized filter paper in Petri dishes moistened sufficiently with sterile distilled water or on moist cotton plugs in Petri dishes or culture tubes. The seeds are germinated in dark at 26-28 C and small part of the seedling is utilized for the initiation of callus culture.

The aerial portion of plants such as bud, leaf and stem sections are sterilized by submerging for 2-3 minutes in 70 per cent ethanol followed by 2-3 rinses in sterile distilled water.

ESTABLISHMENT OF CULTURES

The surface sterilized plant material is aseptically transferred on solidified nutrient medium in flasks, glass jars or culture tubes and allowed to incubate at 26-28 C in dark. After 3-4 weeks, the callus should be about five times the size of the explants. Many tissue explants possess some degree of polarity with the result that callus is formed most easily at one surface. In stem segments, callus is formed particularly from that surface which in vivo is directed towards the root. The callus often develops more readily from the tissue not in contact with and particularly not immersed in the solidified culture medium. The maintenance of growth in callus tissue by subculturing requires the transfer on each occasion of a piece of healthy tissue every four weeks into the flask containing fresh solidified nutrient medium. Many cultures shall, however, remain healthy and continue a slow rate of growth for much longer periods without subculturing, if the standard incubation temperature of 26 C is lowered to 5-10 C. It has been observed that the growth of many cultures and particularly of those which form chlorophyll is stimulated by low-intensity illumination. Light either on a 12 hours cycle or continuous is, therefore, usually provided in the incubation chambers by fluorescent tubes.

The suspension cultures are generally initiated by transferring an established callus tissue to an agitated liquid nutrient medium in Erlenmeyer culture vessels (30-60 ml medium per 250 ml flask). The composition of the medium for establishment of suspension cultures should be same as defined for callus cultures except for addition of agar. The soft callus generally forms a suspension culture without much difficulty. The release of cells and tissue fragments from less friable callus masses and the maintenance of a good degree of cell separation may often be promoted by the presence in the liquid medium of a high auxin concentration; an appropriate balance between yeast extract and auxin or between auxin and kinetin. The suspension cultures are usually incubated at 25 C in darkness or low intensity fluorescent light. Continuous agitation of flask cultures is most commonly achieved by using a horizontal shaker which rotates at between 100 and 200 revolutions per minute. The culture flasks are sealed with double aluminum foils or parafilms to reduce evaporation during the process of culture growth. A cell suspension should be formed within 4 to 6 weeks. The cells grown in cultures are meristematic and generally undifferentiated and there is no significant evidence that cells of shoot or root origins are metabolically different. The suspension cultures are subcultured by the transfer at regular intervals of untreated or fractionated aliquots of the suspension to fresh medium.

There are several parameters for measuring growth of cultured cells such as measurements of cell number, packed cell volume, fresh and dry weight, total nitrogen, etc. None of these methods reflects growth in all its facets which include cell division, elongation and differentiation. The method of measuring growth of cultured cells by dry weight determination has the advantage of being a method which is simple, used quite commonly and gives an acceptable assessment of overall synthetic activity of cells.

Presently plant tissue culture is practised for two main purposes namely regeneration of plants and formation or production of useful secondary metabolites. Regeneration of whole, fertile plants from selected or biotechnologically engineered cells has been achieved in many cases. It is done mainly with the technique of organogenesis and in few cases with embryogenesis. Plant regeneration technology is applied for various purposes like production of virus-free plants, germplasm storage and studies on plant biology.

Organogenesis mainly includes root or shoot formation. The technique requires either the pre-existing meristematic primordials which are available through the medium of explants or induced meristematic primordials. The induction of meristems as well as the differentiations into shoot, root are achieved by providing the appropriate levels of plant hormones like auxins, cytokinins and also their relative proportions with each other. Certain observations and consequently guidelines in case of role of plant hormones are summarized here.

1. The balance between formation of root or shoot is governed by the ratio of auxin to cytokinins (auxin to cytokinins in 4:1 proportion causes shoot formation and auxin to cytokinin in 100: 1 proportion causes root formation). IAA favours shoot formation and NAA favours root formation.
2. High auxin concentrations cause the formation of meristems like cells.

By means of clonal propagation, technique of regeneration of plants has been commercially exploited for routine multiplication of agricultural and ornamental plant species.

For the development of virus free plants, meristems tip culture technique is used. Plant meristems are found to be virus free as many of plant viruses cannot enter or survive in meristematic tissue. The meristems tip culture is propagated from apical shoot tips and used as a source of sterile shoot cultures which are further regenerated into virus–free plants.

Followed by the success in developing the technique of isolating the protoplasts from cell walls, protoplast fusions have been achieved through the medium of protoplast culture. It is used for the purpose of somatic hybridization of plant cells. Cell wall can be efficiently hydrolyzed by using cell wall degrading enzymes like cellulose, hemicellulose and pectinase. Isolated protoplasts are mounted in proper medium containing mannitol/sorbitol for maintaining osmolarity. The protoplast fusion is done by chemical fusion or electrofusion. It may be a self-self fusion or binary fusion. The later is of only use for regeneration. Through protoplast fusion technique, it is possible to get desired plant characters like higher yields, disease resistance etc.

Besides regeneration of plants, plant tissue culture techniques have been much applied for development of secondary metabolites. The technique has been extended to large number of plants and a variety of secondary products have been reported to accumulate in cultures.

Along with cell suspension culture, other cultures have also been employed. Still cell suspension culture systems have been most widely used mainly due to potential for scale up to bioreactor levels. Through cell suspension culture, higher levels of secondary product yields have been reported for ginsenosides from *Panax ginseng,* shikonin from *Leptospermum erythrorhizon* serpentine from *Catharanthus roseus* etc.

The factors conventionally used for manipulation of yield of secondary metabolites in cell suspension cultures are light, temperature, osmotic stress, concentrations of nitrogen, carbohydrate and phosphate source, plant hormones, precursors and elicitors like fungal extracts.

For the accumulation of secondary metabolites in cultures, now-a-days hairy root cultures have been much used. Such cultures are developed by way of incorporating a segment of *"Ri-DNA"* from *Agrobacterium rhizogens* into the plant genome which brings out phenotypic characters called *'hairy roots.'* These are able to grow rapidly and indefinitely with profuse lateral roots. They grow on a medium without plant hormones and can accumulate levels of secondary metabolites similar to those occurring in natural roots. Many plants have been subjected for this purpose. However, this technique has shown success only in non-woody dicotyledonous plants.

PHYTOPHARMACEUTICALS IN PLANT TISSUE CULTURES

The exploration of biosynthetic capabilities of cell cultures derived from different higher plants has been systematically investigated in the last two decades by scientists all over the world. In many cases, however, the desired compounds are not produced, while in others only low amounts could be detected. The poor yields of secondary metabolites appear to be due to differences between young, dispersed and rapidly dividing cells, akin to meristems tissue, mature slow-growing cells in organized mass variation of product formation as a function of time in culture, repression or dormancy of desired biosynthetic pathway and non-excretion of products from culture cells. In order to avoid these problems, development of hairy root cultures, immobilized cell systems and techniques to encourage excretion of desired product into medium are being attempted. On the other hand there are examples which show that callus and suspension cultures are capable of synthesizing secondary metabolites with yields comparable to the intact plant. During the past some few years, promising findings have been reported for a variety of medicinally important substances, some of which could be produced on an industrial scale in the near future.

The productivity for economic considerations is defined as gram product/L culture/day. The yield of biomass, rate of growth, yield of desired product and the number of runs which can be carried out per year are the major factors affecting productivity on an annual basis of a bioreactor or process. In general, the secondary metabolites are synthesized after cell growth has ceased extending the bioreactor time and necessitating a change to production medium forming a process with an extended run time. The overall production of a secondary product in cultures depends upon the rate of accumulation within productive cells and proportion of such cells in culture.

It has been possible to establish large scale production of biomass containing useful secondary metabolites by defining nutritional requirements and ensuring proper environmental conditions for their growth. Only the cultures producing more than one gram of compound per litre with a value exceeding 500 to 1000 US dollars per kg are believed to be commercially exploitable. The total market size for interesting compounds should be large enough to warrant the capital expenditure needed to develop a tissue culture system. In view of these economical constraints, commercial production by tissue culture technique is limited to a few plant specific compounds. To date, over 30 classes of compounds have been produced in appreciable quantities by plant cell cultures. These include *Digitalis* glycosides, diosgenin-derived steroid hormone precursors, shikonin, rosmarinic acid, opium alkaloids (codeine and morphine), ginsenosides, ajmalicine and other indole alkaloids, including vinblastine and vincristine, and possibly complex mixture such as rose and jasmine oil. Mitsui Petrochemical Industries of Japan have been producing shikonin, a red coloured phenolic naphthoquinone compound from cell cultures of *Leptospermum* erythrorhizon, which is used as a dye and as an astringent. A German pharmaceutical company is in process of producing Digitalis glycosides by biotransformation in cell cultures. The research worldwide research in commercial production of anti-cancer alkaloids of *Catharanthus* shall be soon a practical reality. The factors that limit the success of plant culture include slow growth of plant cells and low accumulation of metabolite. The desired secondary product is often retained within the cells.

The selection of a specimen with a high content of desired pharmaceutical and use of specific precursor are prerequisite to the development of cell cultures with high yields. The introduction of enzyme immunoassay techniques for quantitative analysis of bioconstituents has revolutionized the selection of plants and cell clones with high yields. The techniques have been successfully applied in the bioproduction of vindoline, serpentine and ajmalicine. The presence of unusual ether lipids, 1(3), 2-diacylglycero 3 (1)-0-4′-(N, N, N-trimethyl) homoserines were isolated from algae cultures of *Chlorella fusca*.

The secondary products of pharmaceutical significance produced at high levels in plant cell cultures are indicated in Table 5.1.

Table 5.1: Secondary metabolites including Biomedicinals accumulated in High Levels by Plant Cell Cultures

Compound	Species	Yield (per cent dry wt)
Acetoside	Suringa vulgaris	5.0
Ajmalicine	Catharanthus roseus	1.8
Anthraquinone	Morinda citrifolia	18.0
Benzylisoquinolines	Coptis japonica	15.0
Berberine alkaloids	Berberis wilsonae	10.0
Diosgenin	Dioscorea deltoidea	7.8
Ginsenosides	Panax ginseng	17.0
Nicotine	Nicotiana tabaccum	5.0
Rosmarinic acid	Coleus blumei	21.4
Serpentine	Catharanthus roseus	2.2
Shikimic acid	Galium mollugo	10. 0
Shikonin	Leptospermum erythrorhizon	12.4

Bioproduction of Useful Metabolites in Hairy Root and Multiple Shoot Cultures

It may be premature to envisage the commercial impact of genetic manipulation on secondary metabolite production in tissue cultures. The bioproduction of scopolamine and hyoscyamine in hairy roots developed from *Scopolia japonica* has in some cases found to be lower or greater to that found in entire plant. The hairy root cultures of *Catharanthus roseus* capable of producing appreciable quantities of ajmalicine, serpentine and catharanthine have been established. The bioproduction of shikonin and betacynin in hairy root cultures of *Lithospermum erythrorhizon* and *Beta vulgaris* has been reported. The inherent ability of a large number of medicinal plants to generate multiple shoots in culture from axillary and shoot tip meristems has opened up new avenues for production of phytopharmaceuticals and preservation/propagation of elite genotype. The auxiliary meristems and shoot tip develop into shoots on a minimal inorganic basal MS medium both on agar surface and in liquid suspension. They can be propagated by routine periodic subcultures Pago or liquid medium and grown in multilitre fermentors of 2 L capacity flasks. The type of explants, hormonal combinations and growth conditions play an important role in maintenance of cultures. The bioactive shoot cultures have been established for *belladonna, solanum, dioscorea,* and *vinca.*

One can recognize the contribution that the plant tissue culture techniques have made in the study of fundamental problems of plant physiology and cytology, genetics, plant breeding and viral, bacterial and nematode plant relationships. The use of these techniques offers a unique avenue of approach to the analysis of the metabolic potentialities of higher plant cells, for studies of physiological interrelationships between cells, tissues and organs within the multicellular plant and for controlling the expression, in differentiation and morphogenesis, of the totipotency of such cells. In the application of *in-vitro* methods for improvements of genetic potential of plants for the production of primary and secondary compounds, the regeneration of plants from cell and tissue cultures acquires special significance. Some of the medicinal compounds localized in morphologically specialized organs or tissues have been produced in culture systems, not only by inducing specific organized structures in submerged cultures, but also by certain culture strains in the absence of organization.

Organogenesis refers to the process whereby tissue callus or cells may be induced to form shoots and complete plants and this process can be initiated by manipulations in the concentrations of growth hormones. Benzyladenine and other cytokinins alone or in combination with napthaleneacetic acid or indoleacetic acid and at times Gibberellic acid bring about differentiation and shoot formation. Meristem culture and stem tip vegetative propagation have allowed us to postulate clonal development of plant drugs and the intensely interesting field of somatic hybridization by isolated protoplasts could lead to exciting possibilities again in entire plant drugs. The interrelationship between plant tissue culture and different techniques of plant propagation can be briefly summarized as given in Table 5.2.

Table 5.2: Interrelationship Between Plant Tissue Culture and Propagation

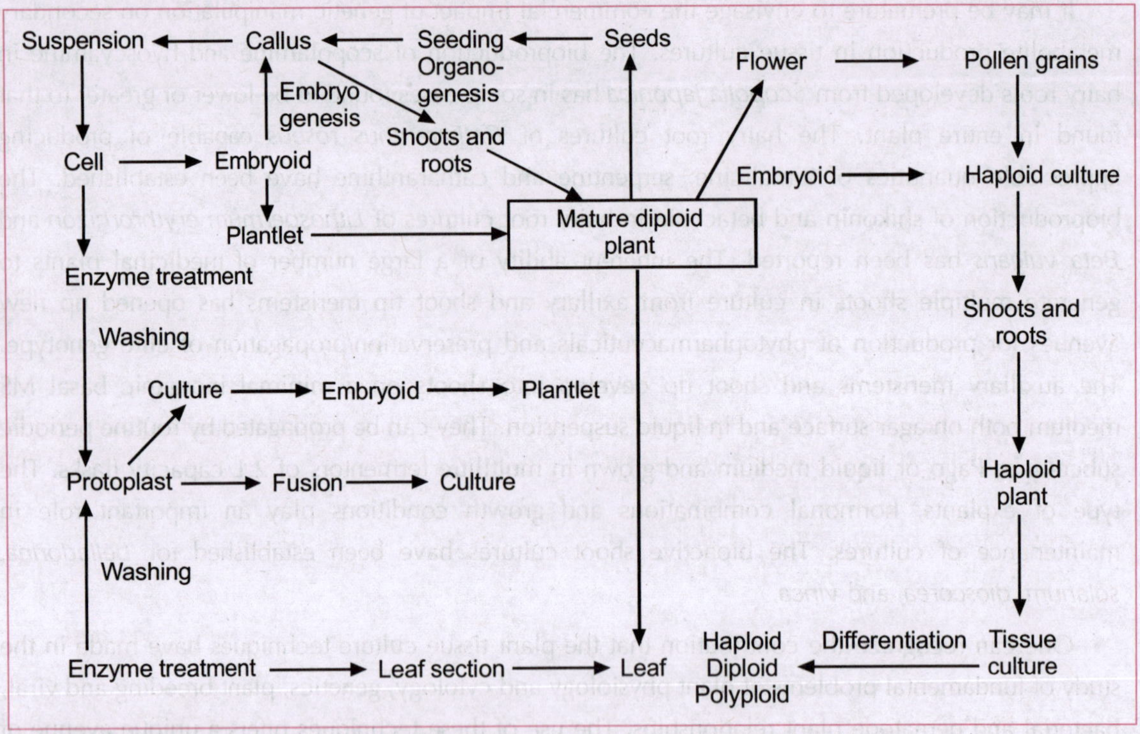

The technique of immobilizing plant cells or enzymes has increased the utility of plant cell biotechnology for production of pharmaceuticals. Immobilization by use of matrices such as calcium alginate, agarose, gelatin, carrageenan or polyacrylamide has number of advantages, including reuse of expensive biocatalysts, simplified control with automatic separation of the catalyst from the product and stabilization of the biocatalyst.

The recent advances in the field of production of medicinal substances by tissue cultures have strengthened our hope towards the realization of industrial production of phytopharmaceuticals in near future. The increasing awareness of pharmacists in general and pharmacognosists in particular of the biotechnological applications of plant tissue culture promises an ever increasing involvement in a branch of bioscience which could lead to production of new pharmaceutical products to be used in service of the mankind.

[C] AN INTRODUCTION TO BIOGENESIS OF PHYTOPHARMACEUTICALS

The living plant may be considered as a biosynthetic laboratory not only for the primary metabolites like sugars, amino acids and fatty acids that are utilized as food by man, but also for a multitude of secondary products of pharmaceutical significance such as glycosides, alkaloids, flavonoids, volatile oils, etc. A higher plant is a solar-powered biochemical factory which manufactures both primary and secondary metabolites from air, water, minerals and sunlight. Primary metabolites are substances that are widely distributed in nature, occurring in one form or another in virtually all organisms and are needed for general growth and physiological development, because of their basic cell metabolism. Secondary metabolites are biosynthetically derived from primary metabolites but are more limited in distribution, usually being restricted to a taxonomic group. They may represent chemical adaptations to environmental stresses, or they may serve as defensive, protective or offensive chemicals against microorganisms, insects and higher herbivorous predators. They are sometimes considered to be waste or secretory products of plant metabolism. In terms of cellular economy, secondary products are in general expensive to produce and accumulate, and are, therefore frequently present in plants in much smaller quantities than are primary metabolites.

The photosynthetic process in green plants is essential for all animal life on earth, since it initiates conversion of solar energy into organic carbon compounds which in turn are used to produce essential foods. The photosynthetic sequence in green plants is capture of light and use of its energy to liberate molecular oxygen from water to synthesize organic compounds from carbon dioxide, nitrate and sulphate. The major source of carbon is usually glucose, which is photosynthesized in green plants. The recent advances in the field of biochemistry have greatly clarified enzyme-catalyzed and interrelated reactions resulting in formation of primary metabolites and their role in the synthesis of secondary products, many of which are used by man. Since these secondary products do not occur in all plants, their biochemistry was ignored until several decades by those interested in chemical plant physiology. Nevertheless, many of them are extremely important as pharmaceutical agents.

The various biosynthetic reactions occurring in plant cells are enzyme-dependent, wherein enzymes act as catalysts of metabolism and it is through the control of enzymatic activity that plant metabolism is directed into specific biosynthetic pathways. The enzymatic reactions are reversible and in plants, many a time, the secondary metabolites are synthesized and hydrolyzed under the influence of more or less specific enzymes.

The elucidation of biosynthetic pathways in plants for the production of various plant metabolites has been extensively examined by means of isotopically labelled precursors. With the advancement of tracer technology, it is possible to incorporate isotopes into presumed precursors of plant metabolites and use as markers in biogenetic experiments. With the use of radioactive carbon (^{14}C) and hydrogen (^{3}H) and to a lesser extent sulphur (^{36}S) and phosphorous (^{32}P), it is possible to understand various biosynthetic pathways. A labelled nitrogen atom may give more specific information about the biosynthesis of alkaloids, proteins and amino acids. The most notable success by use of isotopically labelled precursors was achieved by Birch in the biosynthetic investigations of mould metabolites such as 6-methylsalicylic acid and griseofulvin from ^{14}C-labelled acetate.

The basic carbon reduction cycle by which carbon dioxide is converted to sugar phosphates is of primary importance, both as an energy yielding process and also as a source of various metabolic intermediates. Two biosynthetic pathways of special importance in breakdown of sugar are pentose phosphate cycle (direct pathway) and glycolysis. The hexose phosphate is oxidized first into carbon-dioxide and pentose phosphate in direct pathway. The pentose phosphate formed due to biosynthetic degradation may then be utilized as such or otherwise converted by a series of metabolic reactions into triose phosphate or by a reversal of glycolysis, into a hexose; however, in glycolysis, hexose phosphate is split hydrolytically to yield triose phosphate, which can then be oxidized. Although the total number of natural products for which biosynthetic investigations have been carried out are quite limited compared with the diversity and number of natural products, our accumulation of knowledge in biogenetic field makes it possible to predict the gross biogenetic origin of practically all plant products.

Table 5.3: The primary and secondary metabolites derived from carbon metabolism in plants

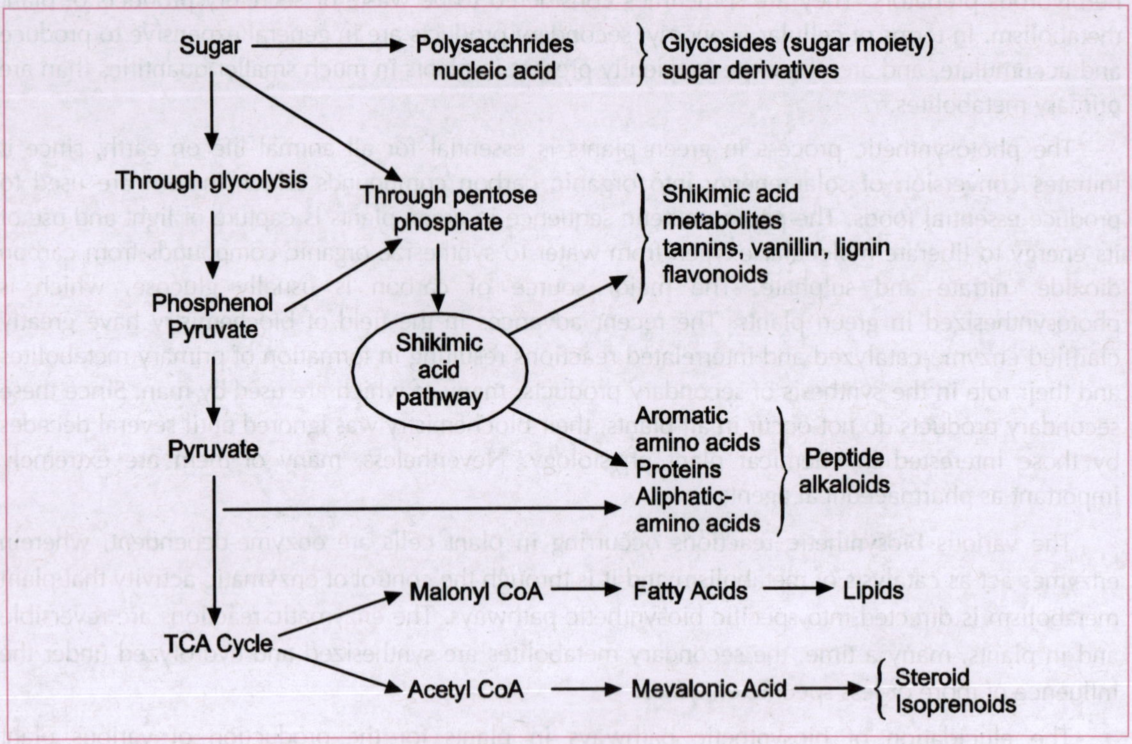

SHIKIMIC ACID PATHWAY

The Shikimic acid is a key intermediate from carbohydrate for the biosynthesis of $C_6 - C_3$ units (phenylpropane derivatives). Besides serving as precursor for the biosynthesis of amino acids, Shikimic acid is also an intermediate in production of tannins, flavones, coumarins and vanillin.

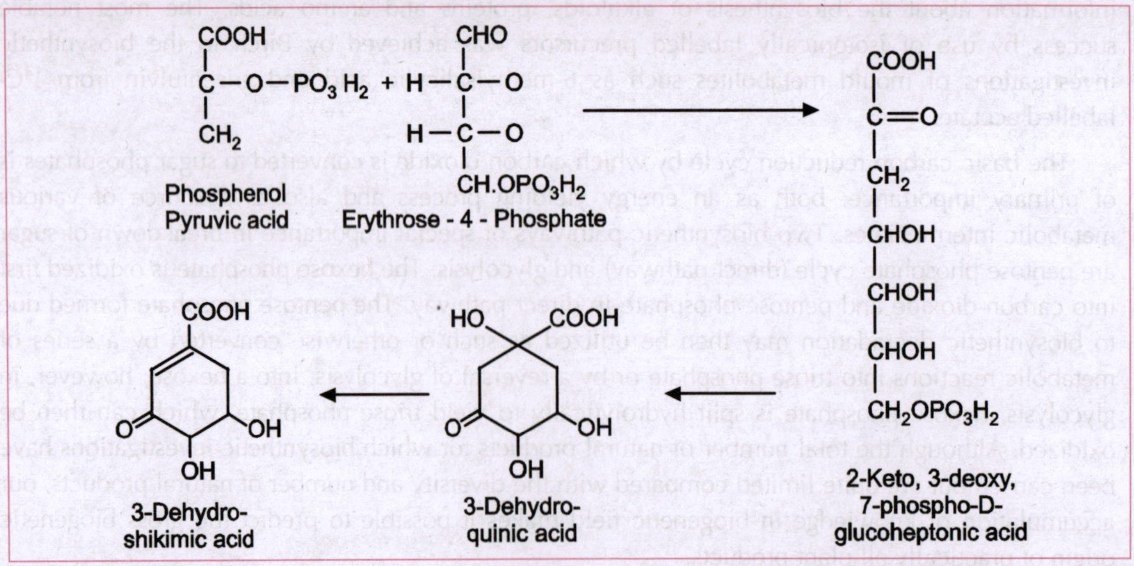

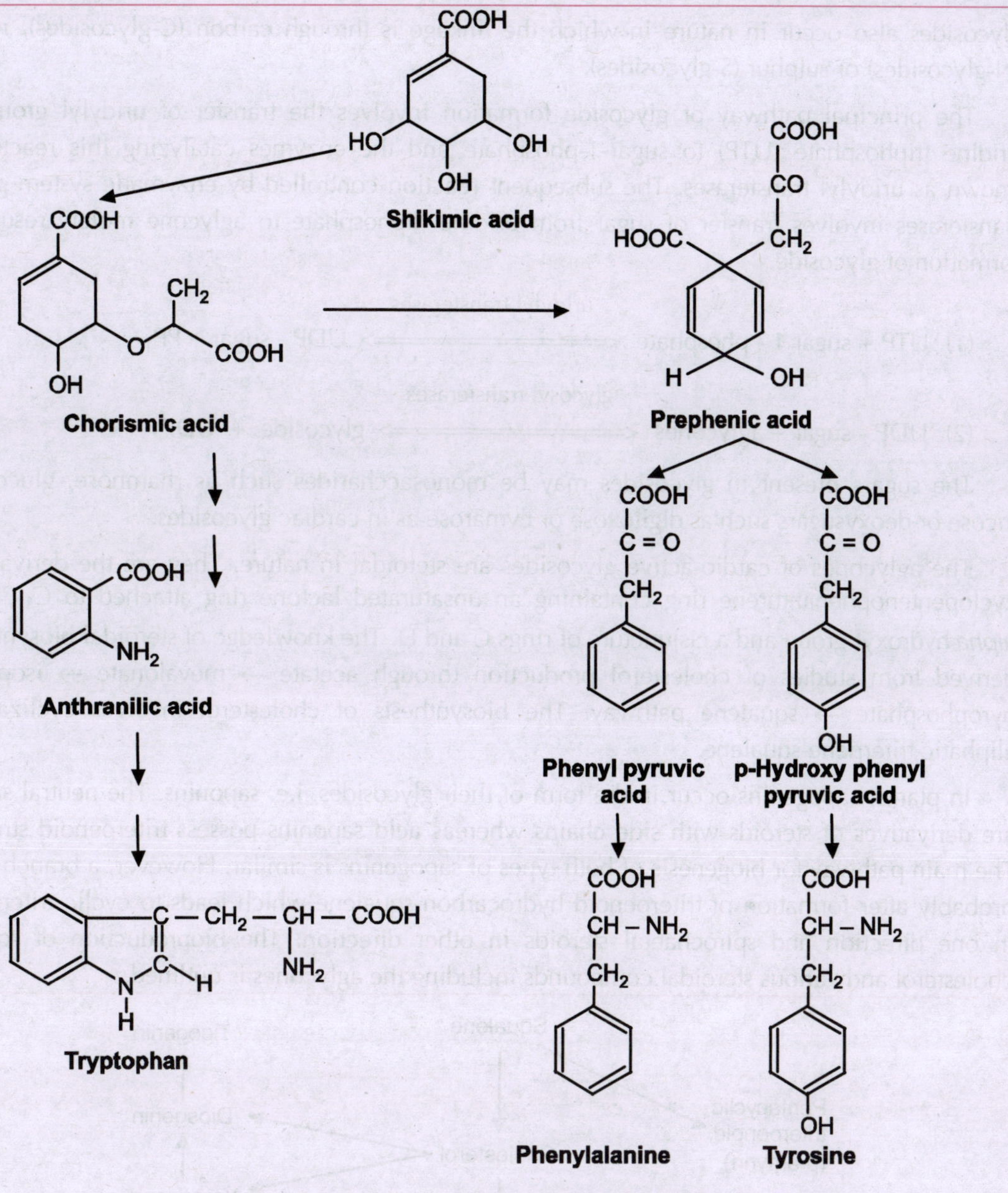

Production of Amino Acids by Shikimic Acid Pathway

BIOSYNTHESIS OF GLYCOSIDES

The metabolic process of glycoside formation essentially consists of two parts: The first part of biosynthesis is the reactions by means of which various type of aglycones are formed, where as the other part of biosynthesis process takes into account metabolic pathway involving coupling of aglycones with sugar moiety. The synthesis of glycosides in plant cells involves interaction of nucleotide glycoside such as UDP-glucose with alcoholic or phenolic group of second compound aglycones. Such glycosides, called as O-glycosides, are commonly found in plant. The other glycosides also occur in nature in which the linkage is through carbon (C-glycosides), nitrogen (N-glycosides) or sulphur (S-glycosides).

The principal pathway of glycoside formation involves the transfer of uridylyl group from uridine triphosphate (UTP) to sugar-1-phosphate and the enzymes catalyzing this reaction are known as uridylyl transferases. The subsequent reaction controlled by enzymatic system glycosyl transferases involves transfer of sugar from uridine diphosphate to aglycone moiety resulting in formation of glycoside.

(1) UTP + sugar 1 - phosphate $\xrightleftharpoons{\text{uridylyl transferases}}$ UDP - sugar + PP_1

(2) UDP - sugar + aglycones $\xrightleftharpoons{\text{glycosyl transferases}}$ glycoside + UDP

The sugars present in glycosides may be monosaccharides such as rhamnose, glucose and fucose or deoxysugars such as digitoxose or cymarose as in cardiac glycosides.

The aglycones of cardio-active glycosides are steroidal in nature. They are the derivatives of cyclopentenophenanthrene ring containing an unsaturated lactone ring attached to C_{17}, a 14 – *alpha* hydroxyl group and a cisjuncture of rings C and D. The knowledge of steroidal biosynthesis is derived from studies of cholesterol production through acetate → mevalonate → isopentenyl pyrophosphate → squalene pathway. The biosynthesis of cholesterol involves cyclization of aliphatic triterpene-squalene.

In plants, sapogenins occur in the form of their glycosides, i.e. saponins. The neutral saponins are derivatives of steroids with side chains whereas acid saponins possess triterpenoid structures. The main pathway for biogenesis of both types of sapogenins is similar. However, a branch occurs, probably after formation of triterpenoid hydrocarbon-squalene which leads to cyclic triterpenoids in one direction and spirochaetal steroids in other direction. The bioproduction of squalene, cholesterol and various steroidal compounds including the aglycones is outlined.

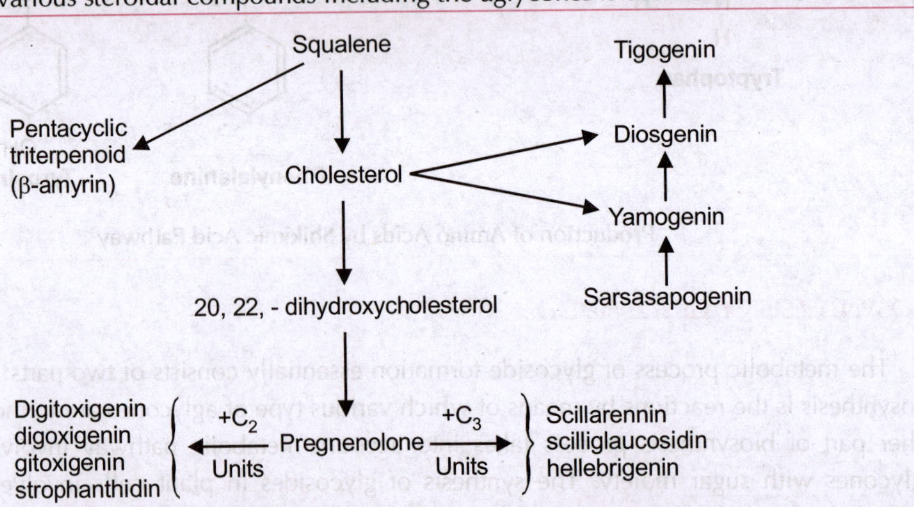

Some secondary products of cholesterol metabolism

The knowledge of biosynthesis of anthracene aglycones has been obtained from studies with micro-organisms, especially *Penicillium islanidicum*. An intermediate polyketomethylene acid is probably produced from 8 acetate units which on intramolecular condensation forms anthraquinone.

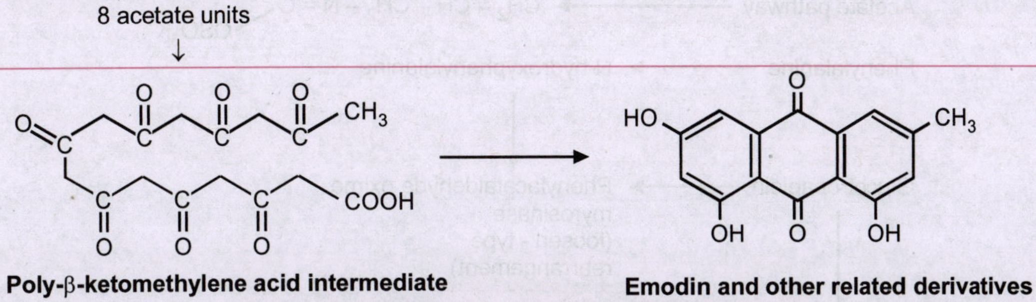

Biosynthesis of Emodin and Related Compounds

A second metabolic pathway for production of anthraquinone is through Shikimic acid as operative in some plants of Rubiaceae. The biosynthesis of alizarin reveals that ring A is derived from Shikimic acid, while mevalonic acid is incorporated into ring C.

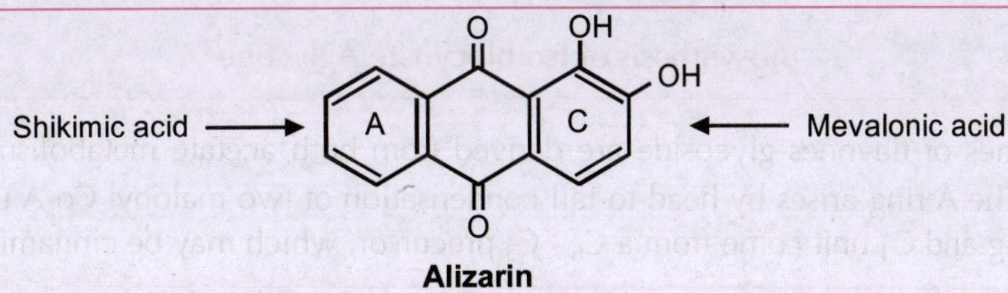

Biosynthesis of Alizarin from *Rubia tinctorum (Rubiaceae)*

The aglycones of pharmaceutically significant cyanogenetic glycosides are phenylpropanoid compounds derived from amino acids phenylalanine and tyrosine which are the products of Shikimic acid pathway. The aglycones of linamarin are derived from valine and that of lotaustralin from isoleucine.

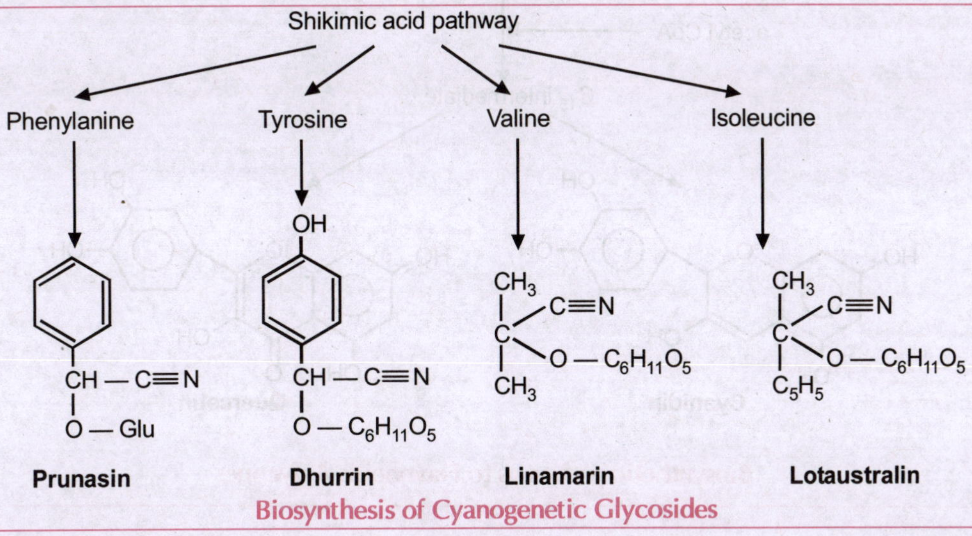

Biosynthesis of Cyanogenetic Glycosides

The aglycones of isothiocynate glycosides may consist of either aliphatic derivative biosynthesized via acetate pathway or aromatic derivatives produced biosynthetically via Shikimic acid route.

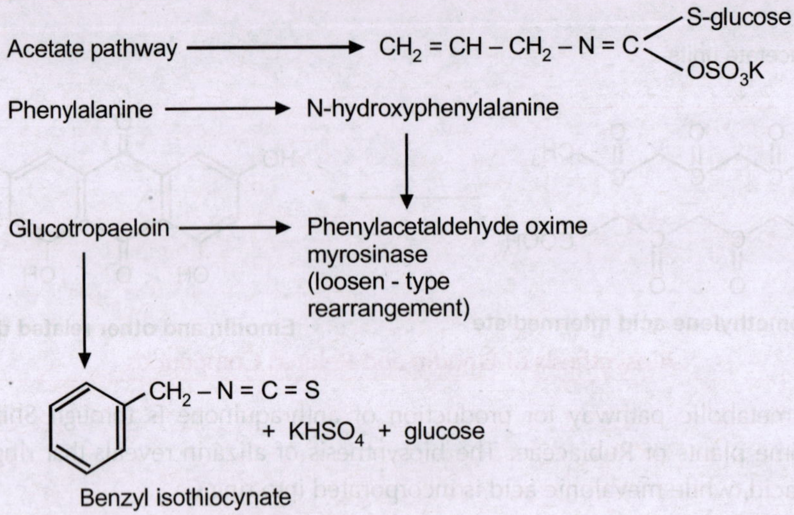

Biosynthesis of Isothiocynate Aglycone

The aglycones of flavones glycoside are derived from both acetate metabolism and Shikimic acid pathway. The A ring arises by head-to-tail condensation of two malonyl Co-A units and acetyl Co–A. The B ring and C_3 unit come from a C_6 - C_3 precursor, which may be cinnamic acid itself.

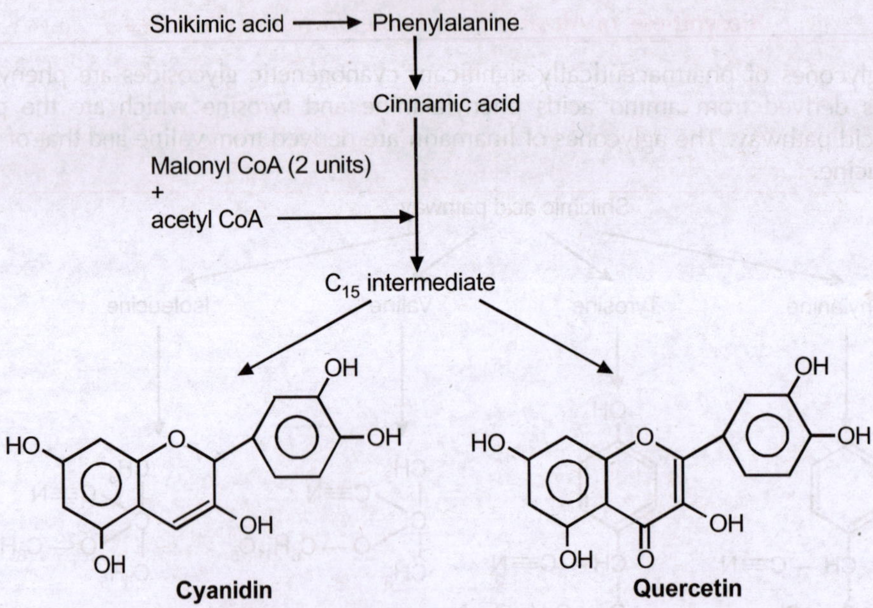

Biosynthetic pathways to Flavonoids Aglycones

The aromatic nuclei of alcohol, aldehyde, lactone and phenol glycosides are derived from C_6 - C_3 precursors formed via Shikimic acid pathway.

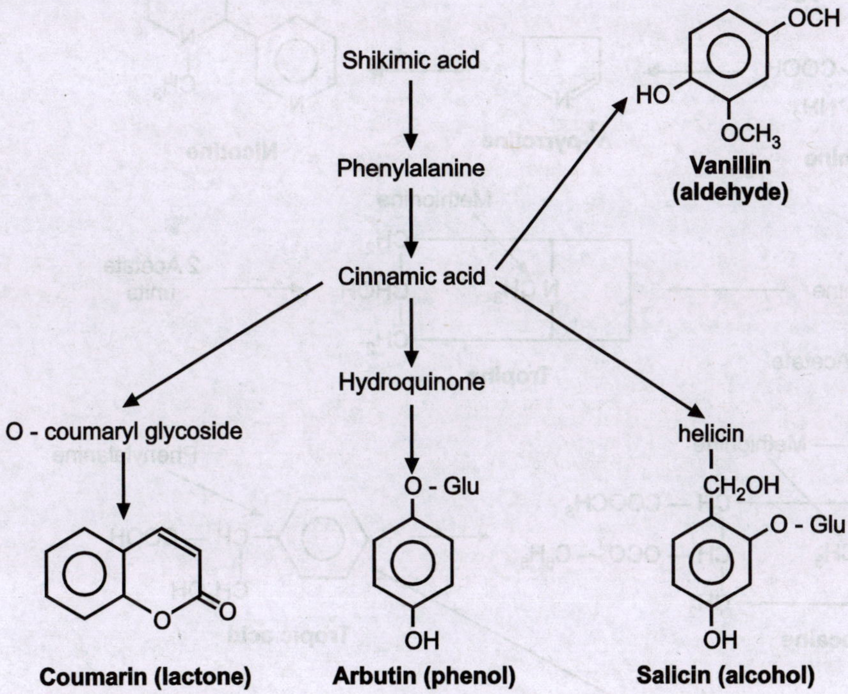

Biosynthesis of Lactone, Phenol, Alcohol and Aldehyde Glycosides

BIOSYNTHESIS OF ALKALOIDS

The biosynthesis of different groups of alkaloids has now been investigated to some extent using precursors labelled with radioactive atoms. Some work has, however, been published in the area of enzymology of alkaloid biosynthesis, some exceptions being in studies of *ergot* and *Amaryllidaceae* alkaloids. The biosynthetic pathways for pharmacognostically important alkaloids are given below.

1. Alkaloids Derived from Ornithine

The studies have revealed that ornithine is incorporated into both pyrrolidine and tropane alkaloids. Ornithine is incorporated stereospecifically and asymmetrically into pyrollidine ring of tropane nucleus, the alpha-carbon of ornithine becoming the C_1 of tropine nucleus. The remaining three carbon atoms are derived from acetate, thus completing piperidine moiety. Methionine serves as the methyl group donor whereas phenylalanine is the precursor of tropic acid. The different alkaloids derived from ornithine are grouped together.

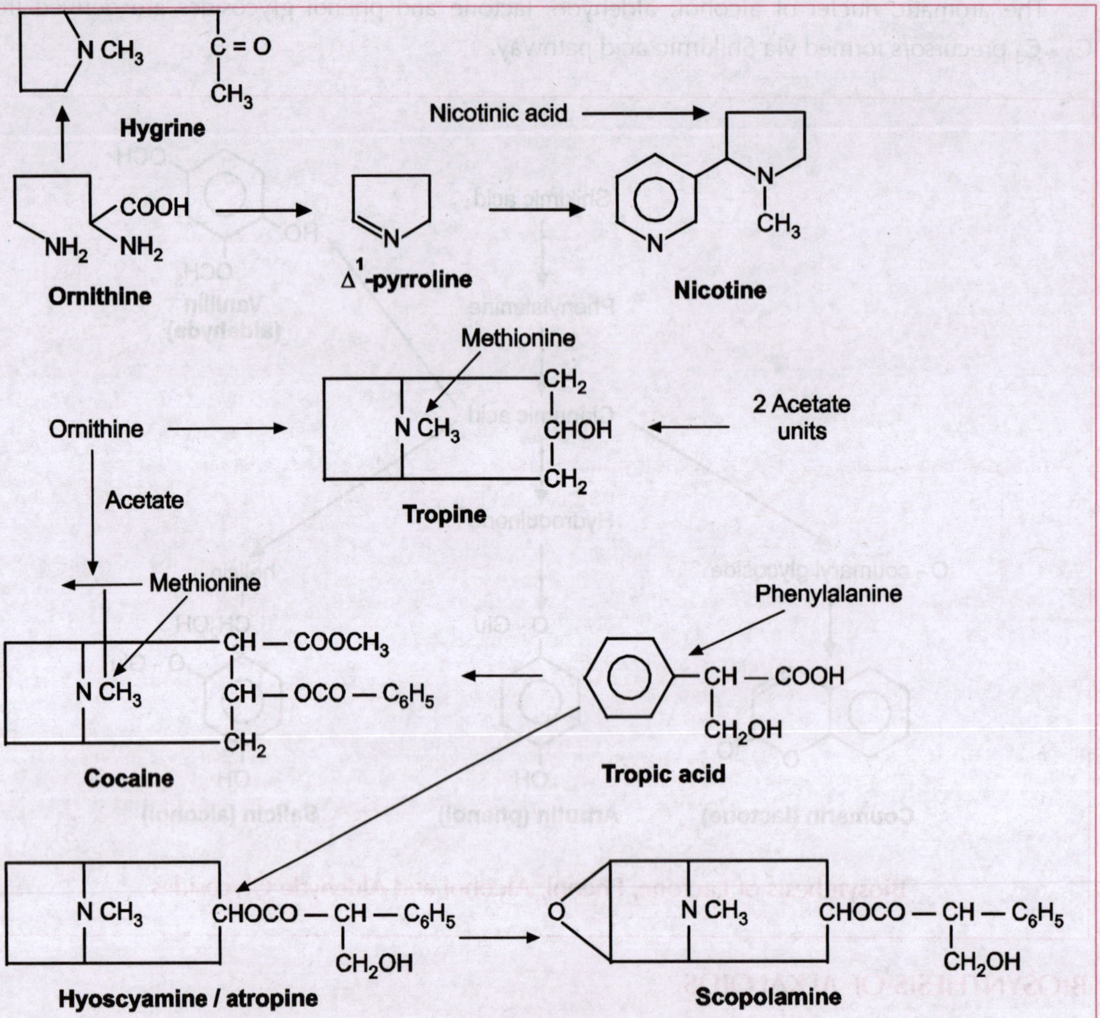

Alkaloids, Derived From Ornithine

2. Alkaloids derived from Lysine

Lysine and its associated compounds are responsible for the biogenesis of anabasine, lupinine, isopelletierine and other related alkaloids -

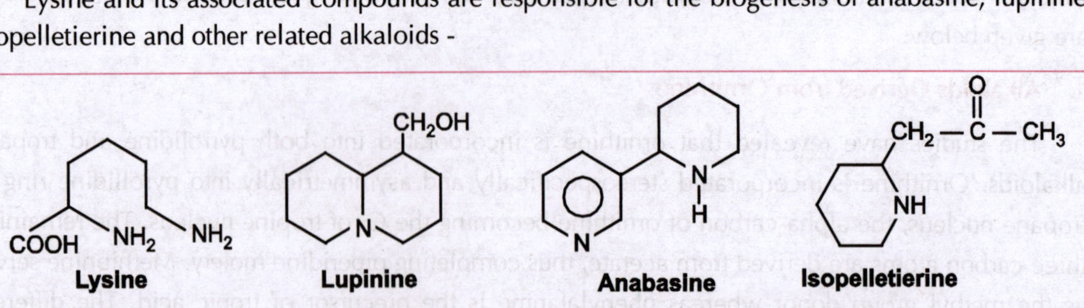

Alkaloids derived from Lysine

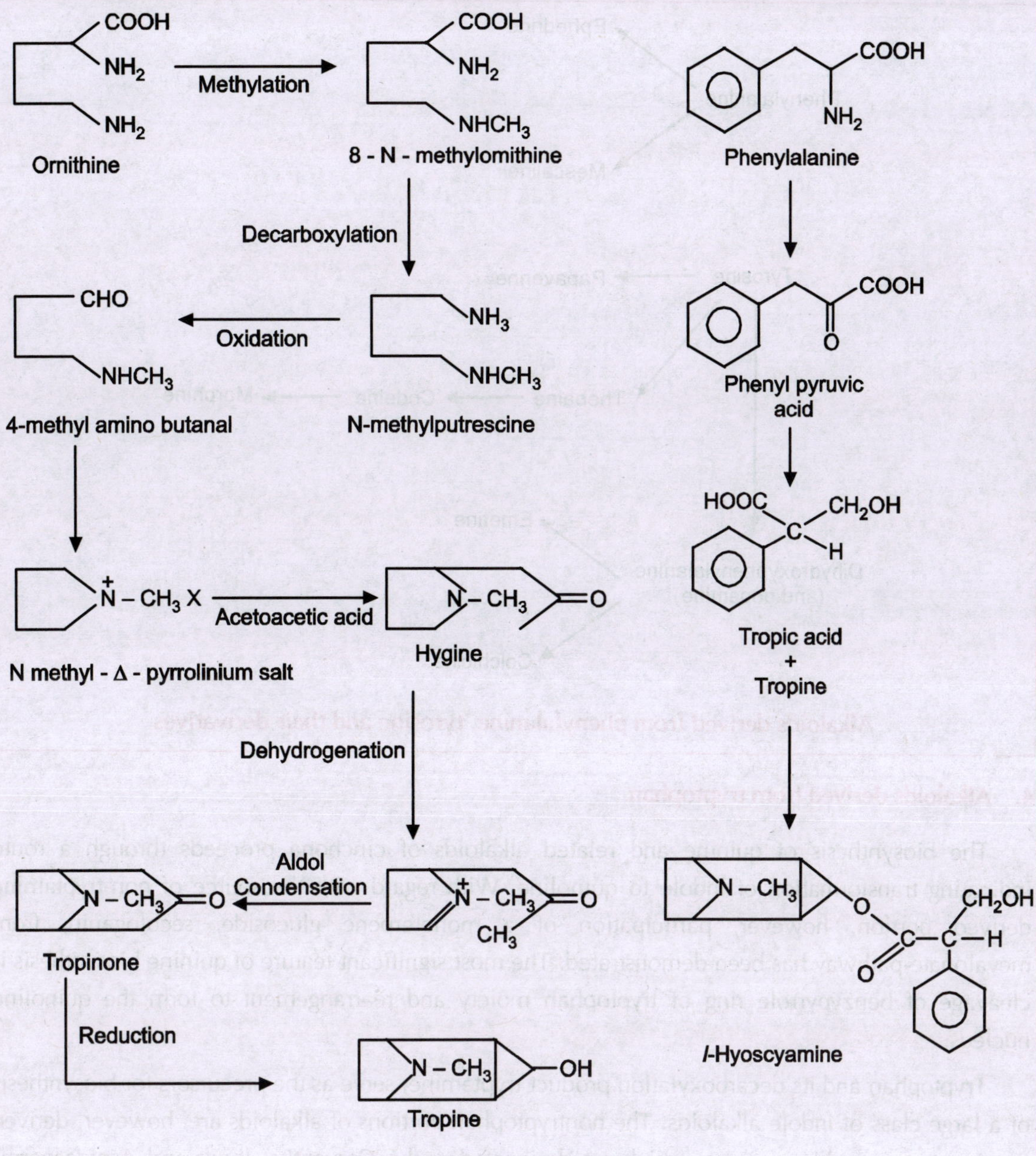

Biosynthesis of l-Hyoscyamine

3. Alkaloids derived from phenylalanine, tyrosine and related amino acids

These amino acids and their corresponding decarboxylation products serve as the precursors for a large number of alkaloids including ephedrine, colchicine and opium alkaloids. Earlier it was shown that tyrosine and dopamine could serve as precursors of morphine. It was also proved that the first of the opium alkaloids synthesized is thebaine, followed by codeine and morphine.

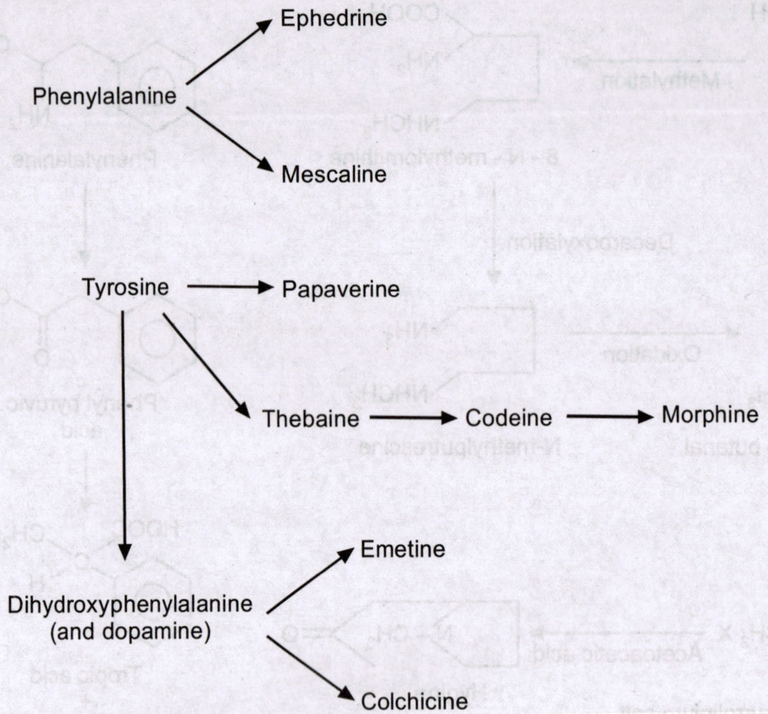

Alkaloids derived from phenylalanine, tyrosine and their derivatives

4. Alkaloids derived from tryptophan

The biosynthesis of quinine and related alkaloids of cinchona proceeds through a route indicating transformation of indole to quinoline. With regard with the source of non-tryptamine derived portion, however, participation of a monoterpene glucoside, secologanin, from mevalonate-pathway has been demonstrated. The most significant feature of quinine biosynthesis is cleavage of benzypyrrole ring of tryptophan moiety and rearrangement to form the quinoline nucleus.

Tryptophan and its decarboxylation product tryptamine, serve as the precursors for biosynthesis of a large class of indole alkaloids. The nontryptophan portions of alkaloids are, however, derived from monoterpenoid precursors which are designated as the *Coryanthe, Iboga* and *Aspidosperma* types. The reactive form of terpene involves an aldehyde group. The loss of one carbon atom during biogenesis to give C_9 unit appears to be largely common. It is believed that *Coryanthe* type monoterpenoid moiety is metabolically most primitive. Condensation of tryptamine (or tryptophan) with secologanin, a monoterpene glucoside, gives rise to a nitrogenous glucoside, vincoside, from which a great variety of indole alkaloids, including monomeric alkaloids in *Catharanthus roseus*, are formed. The Rauwolfia alkaloids such as reserpine, rescinnamine, serpentine, ajmaline, etc. are derived from a *Coryanthe* type monoterpenoid precursor.

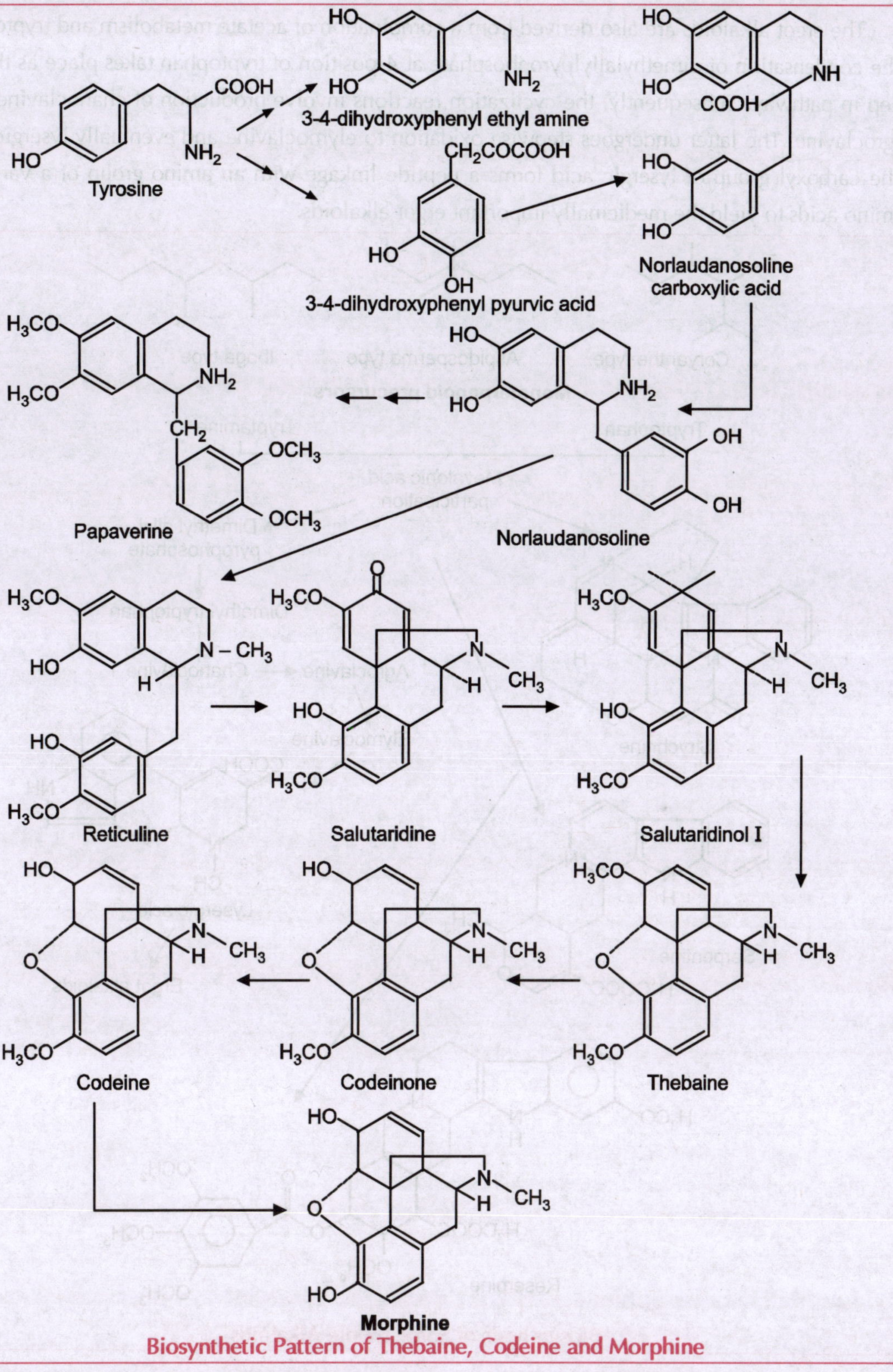

Biosynthetic Pattern of Thebaine, Codeine and Morphine

The ergot alkaloids are also derived from a combination of acetate metabolism and tryptophan. The condensation of dimethylallylpyrophosphate at 4-position of tryptophan takes place as the first step in pathway. Subsequently, the cyclization reactions involve production of chanoclavine I and agroclavine. The latter undergoes stepwise oxidation to elymoclavine and eventually lysergic acid. The carboxyl group of lysergic acid forms a peptide linkage with an amino group of a variety of amino acids to yield the medicinally important ergot alkaloids.

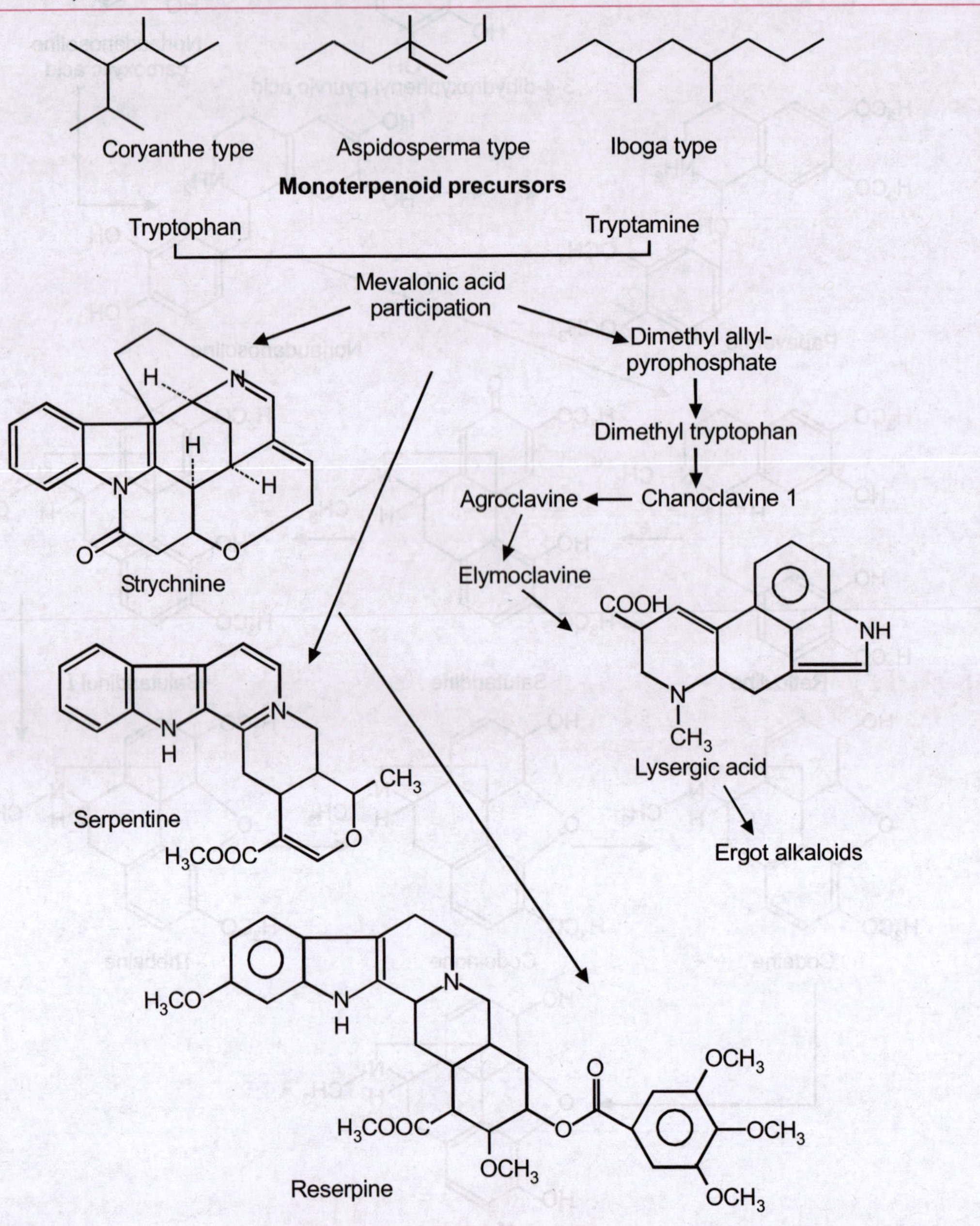

Biosynthesis of Some Indole Alkaloids

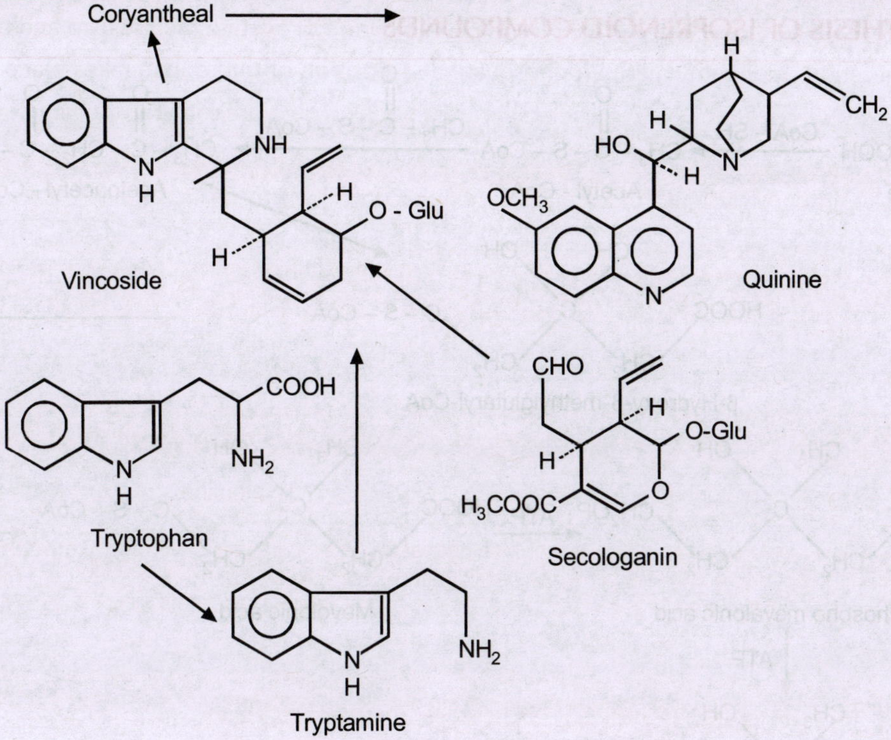

Biosynthesis of Some Quinoline Alkaloids

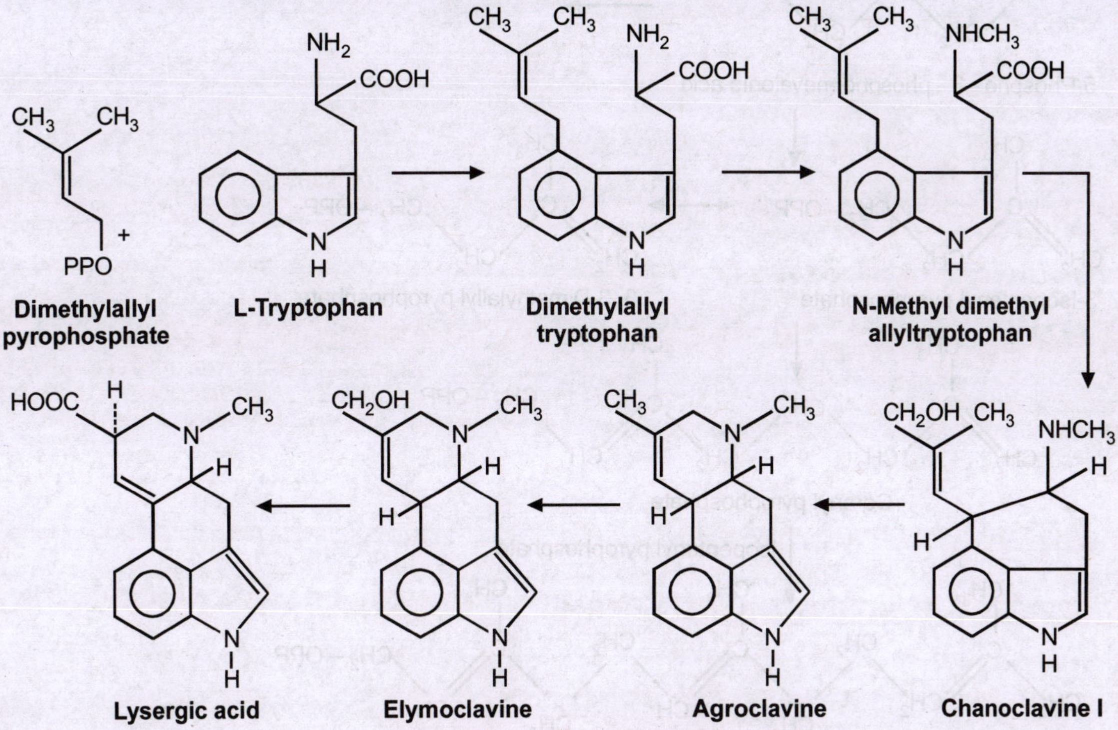

Biosynthesis of Lysergic acid

BIOSYNTHESIS OF ISOPRENOID COMPOUNDS

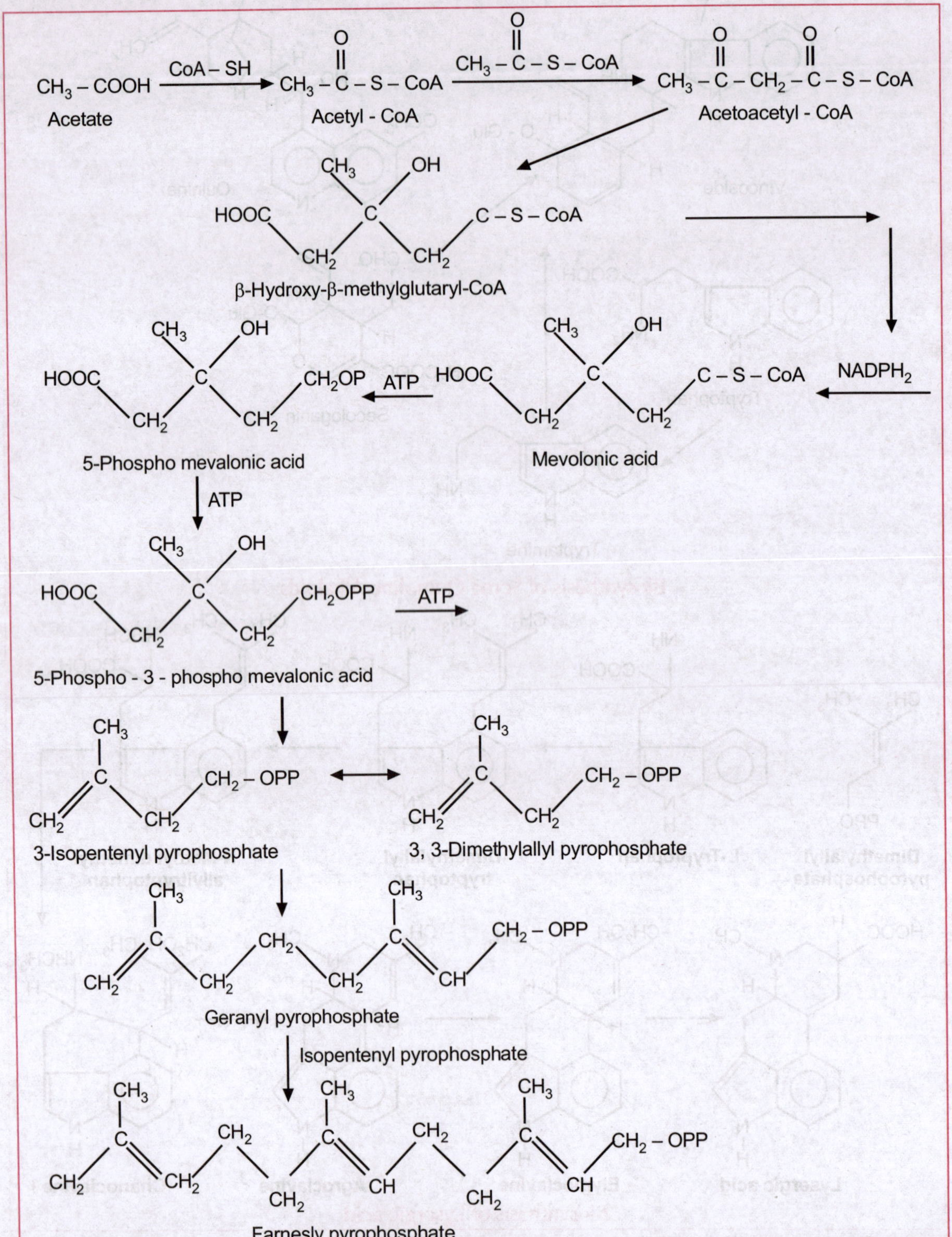

The "biogenetic isoprene rule" as published by Ruziicka in 1953 is the basis for formation of various isoprenoid compounds. The discovery of mevalonic acid (3, 5-dihydroxy-3-methylavaleric acid) in 1956 and demonstration of its incorporation by living tissues into these compounds, to which the isoprene rule applied, were milestones in understanding biogenesis of terpene derivatives. It is established by research involving tracer techniques, inhibitor studies and ionophoresis that C_5 compound-isopentenylpyrophosphate-is derived from mevalonic acid pyrophosphate by decarboxylation and dehydration. The C_{10} compound geranylpyrophosphate is formed by condensation of isopentenyl pyrophosphate with isomeric dimethylallylpyrophosphate. Further, C_6 units are added by participation of more isopentenylpyrophosphate units. From geranyl and farnesylpyrophosphates various isoprenoid structures are synthesized.

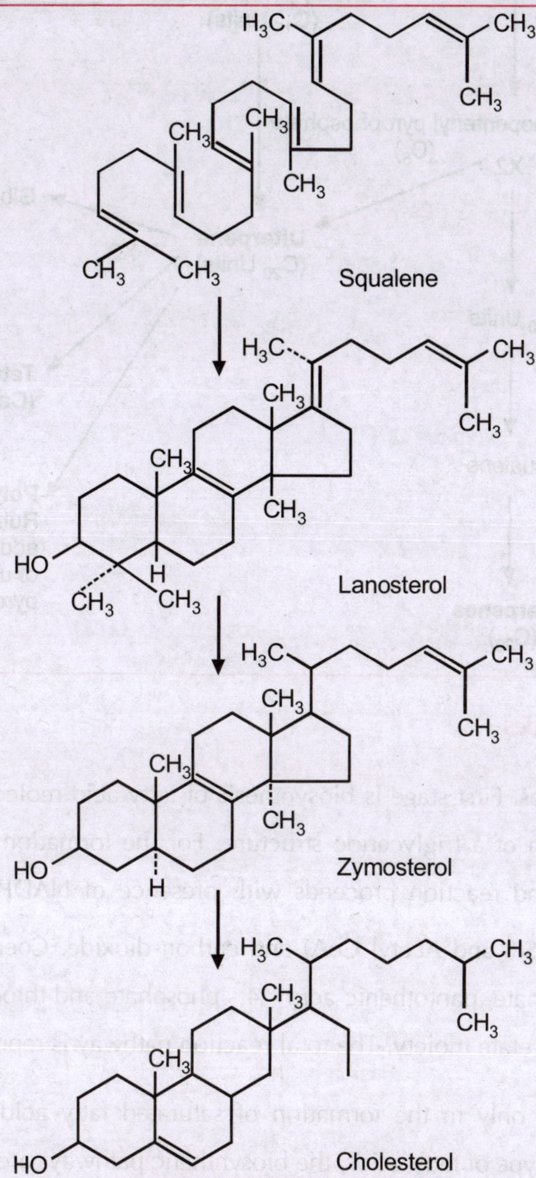

Biosynthesis of Cholesterol

Table 5.3: Biogenesis of Various Isoprenoid Compounds

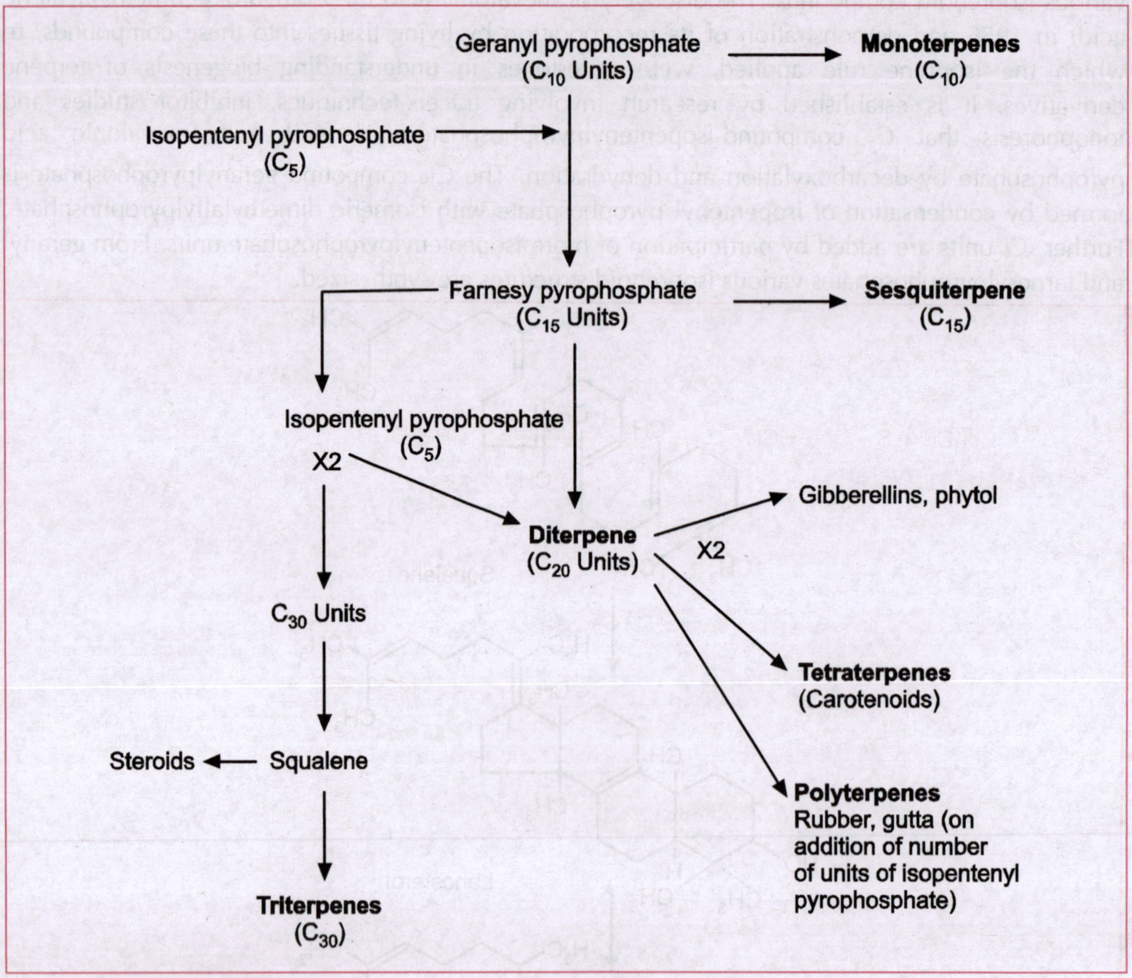

Biosynthesis of Triglycerides

It occurs in two stages. First stage is biosynthesis of fatty acid molecule while second stage is associated with formation of a triglyceride structure. For the formation of a fatty acid molecule, acetate is a precursor and reaction proceeds with presence of $NADPH_2$, Mn^{++}, ATP and two enzyme complexes (CoASH and Acetyl CoA) and carbon dioxide. Coenzyme A (CoA) comprises adenosine 3, 5 - diphosphate, pantothenic acid - 4 - phosphate and thioethanolamine. Acetyl CoA is the activated form of acetate moiety. The total reaction pathway is represented here.

This reaction occurs only in the formation of saturated fatty acids. In case of unsaturated, branched chain or other type of fatty acids, the biosynthetic pathways are not clearly known.

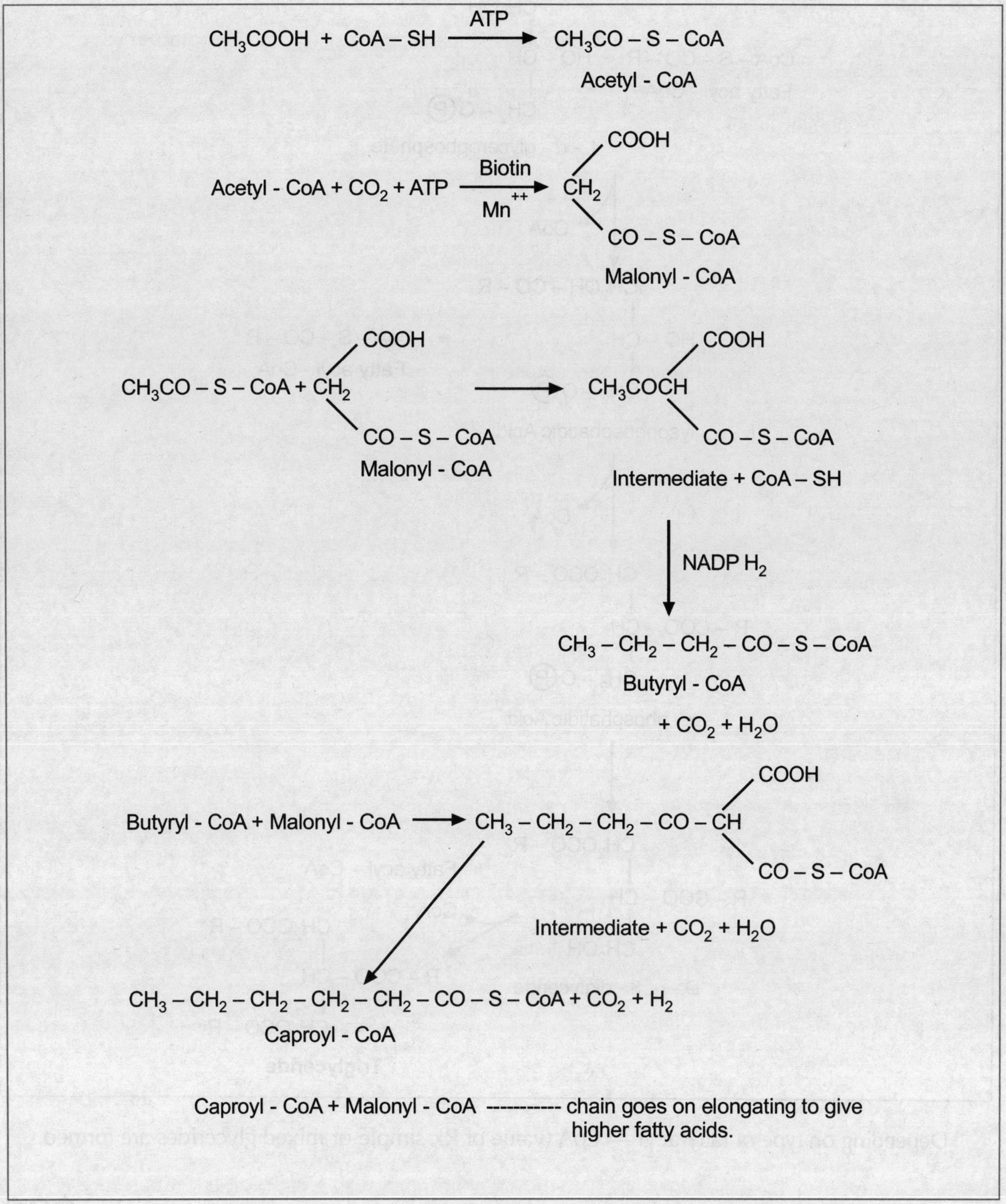

Second stage is the formation of triglyceride structure where required glycerol enters the reaction in the form of L – alpha – glycerophosphate while fatty acid comes as fatty acyl CoA.

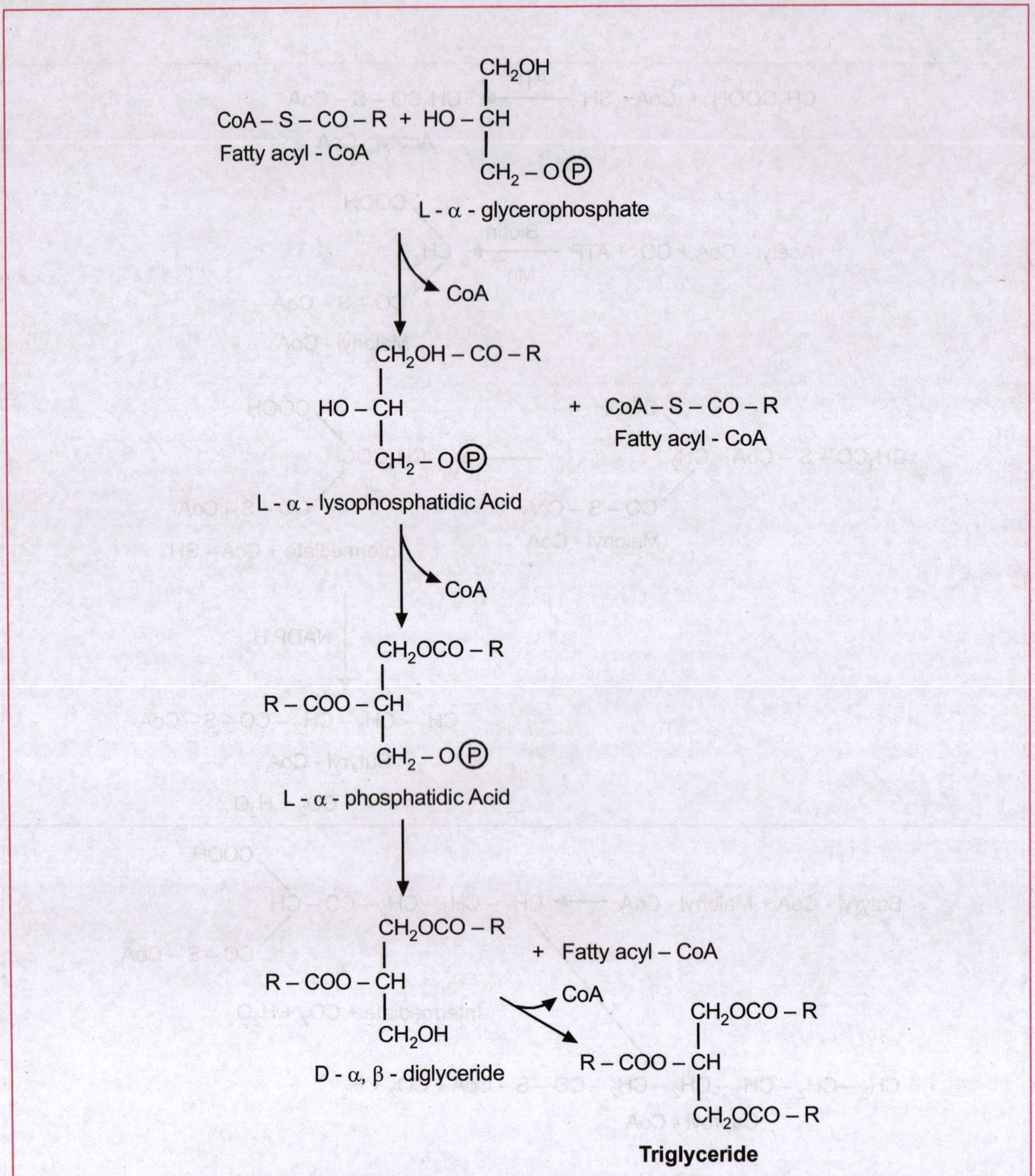

Depending on type of fatty acyl — CoA (value of R), simple or mixed glycerides are formed.

❖❖❖

Chapter 6...

PHYTOPHARMACEUTICALS RETROSPECT AND PROSPECTS

Since the dawn of civilization, the importance of medicinal plants in the treatment of a variety of human ailments has been immense. Since aeons, man has been dependent on the higher plants as a source of food and medicine. By experience and experimentation, he discovered over time that different plants possess different chemical composition. Till the beginning of the era of natural product chemistry i.e. mid-nineteenth century, these plants derived, medicinal agents remained structurally ill defined. In today's modern age of advancement in instrumentation technology, sophisticated techniques like Nuclear Magnetic Resonance (NMR) spectroscopy and mass spectrometry have markedly reduced the time and effort required for structural determination. Besides, these have also enabled structural productions to be made by analysis of even much smaller amounts of isolated plant constituents than were previously required.

In the last few decades, substantial work has been accomplished in the field of Pharmaceutically significant natural products. At no time in the history of mankind, progress has been so rapid and so meaningful as that achieved in the last quarter of the century in our understanding of plants and their constituents. The development of science of phytopharmaceuticals and the hopes for remedies in chronic diseases generated new enthusiasm in the research workers to develop herbal medicines. A considerable amount of work has been done to study the potentialities of herbal medicines and the modern science has accepted the potential of plant kingdom as a source of new biodynamic constituents. The laborious research work that has been carried out in many research institutes in different parts of the world has brought into limelight, the merits and qualities of various herbal medicines.

The natural plant products often serve as chemical models or templates for the design and total synthesis of new drug entities.

The concept of drug design of some of the synthetic molecules has emerged out of their Quantitative Structure Activity Relationship (QSAR) in terms of biodynamic constituents. For example, the belladonna alkaloids (atropine), quinine, physostigmine, cocaine, opiates (morphine and codeine) and Salicylic acid have served as models for design and synthesis of anti-cholinergics, antimalarials, anti-cholinesterases, benzocaine, procaine and other local anaesthetics and aspirin respectively. Eventhough several crude drugs derived from higher plants were reported to be useful in the treatment of cancer over 3000 years, it was only since 1959 that a concerted systematic effort has been made to screen the extracts of crude drugs for their inhibitory activity against animal tamour systems. During this period, about 50,000 plant extracts have been screened under the auspices, of National Institute of Health in the U.S.A. The increasing demand for traditional herbal drugs in China and South Korea, world-wide acceptability of Kempo medicines manufactured in Japan, and recent trends in utilization of herbal products in developed countries are milestones in revival of herbal therapy.

The growing concern in the recent past over the toxic effects of various synthetic drugs has forced the researchers, academicians and doctors to consider some steps for preventing the misuse of such drugs. The occurrence of side effects after a long term use of synthetic drugs is always feared during the treatment of chronic diseases. Such possibility is experienced to be of negligible extent in the case of herbal drugs and other medicines obtained from natural sources. These are some of the reasons which led the authorities in this field to reconsider the use of herbal medicines and to give enough attention to the development of science of phytopharmaceuticals. Considerable efforts have been made in last three decades for the development of technical know-how for extraction of phytochemicals on commercial scale by pharmaceutical and chemical industries and research institutions throughout the world. As a result of modern isolation techniques and pharmacological testing procedures, new plant drugs usually find their way into medicine as purified substances rather than in the form of older galenical preparations.

An examination of list of phytopharmaceuticals, as included in official publications, reveals that majority of them are derived from phylum Spermatophyta. Some useful oils, resins and the alkaloid ephedrine are obtained from the species of Gymnosperm, whereas the division Angiospermae, which is divided into monocotyledons and dicotyledons, provides a variety of useful drugs. The algae are a source of a limited number of drugs such as agar and alginic acid, but the full pharmacological significance of this group along with lichens and mosses is yet to be understood. Of the other divisions of plant kingdom, the fungi provide a number of life saving drugs, especially antibiotics.

Phytopharmaceuticals of Commercial Significance

The crude drugs are subjected to a suitable method of extraction and purification for the isolation of phytopharmaceuticals, which could be incorporated as active ingredients in the modern system of medicine.

The anthracene glycosides such as sennoside are extracted in the form of their stable calcium salts from the leaves and pods of senna. The indole alkaloids such as ergometrine and ergotamine are commercially extracted as their salts from the fungus of ergot, known as *Claviceps purpurea*. Ergometrine maleate is an oxytocic, whereas ergotamine Tartarate is used in combination with caffeine for the treatment of migraine. Caffeine is extracted from tea or tea waste, whereas nux vomica seeds serve as the raw material for the extraction of strychnine. The antihypertensive and tranquillising agent reserpine is commercially obtained from the roots and rhizomes of *Rauwolfia serpentina*. The miracle discovery of the recent years is the extraction of vinca alkaloids, especially vinblastine and vincristine sulfate from *Catharanthus roseus*.

Vinblastine sulphate is used for treatment of neoplasm and is recommended for generalised Hodgkin's disease and choriocarcinoma resistant to other therapy, whereas vincristine sulphate is recommended for the treatment of leukemia in children.

The commercial supply of tropane alkaloids, particularly scopolamine, is from different species of *Duboisia* such as *D. myoporoides*, and *D. leichhardtii*, whereas other tropane alkaloids l-hyoscyamine and atropine are obtained from *Atropa belladonna* and *Datura metel*. The drug,

8-methoxy psoralen, used in the treatment of leucoderma is extracted commercially from the inflorescence of *Ammi majus*. Pectin of pharmaceutical grade is only obtained from the vegetative source, especially the inner part of the rind of citrus fruits, apple pomace and sun flowers. Isoquinoline alkaloids such as morphine, papaverine, codeine and thebaine are extracted from opium, latex obtained from poppy capsules. The cardio-active glycoside, digoxin is commercially extracted from leaves of *Digitalis lanata,* whereas digitoxin, gitoxin and their aglycones are obtained from both *D. lanata* and *D. purpurea*. The ester pesticides obtained from pyrethrum, *Chrysanthemum cinerarifolium* are pyrethrin I and II. Several semi-synthetic analogs of these esters are being used as contact poisons for insects. In addition to pyrethrins, nicotine, sabadilla alkaloids and rotenoids are also used as contact insecticides and are derived from vegetative sources.

The alkaloids quinine and quinidine are commercially obtained from the bark of different *Cinchona* species. Quinine sulphate is an antimalarial drug. In recent years, it has regained considerable importance in the treatment of chloroquine-resistant *Falciparum malaria*. Quinidine sulphate occurs as fine needle like white crystals and is a cardiac depressant. It is used particularly to inhibit auricular fibrillation.

A steroidal sapogenin, hecogenin, is commercially extracted from *Agave veracuz* and *Agave cantata* (Amaryllidaceae). Hecogenin is used as the starting material for the manufacture of cortisones and sex hormones. Podophyllotoxin from *Podophyllum hexandrum* and anacardic acid from *Semecarpus anacardium* are two promising phytochemicals as anti-mitotic drugs. Asiaticoside is the glucoside useful in leprosy and extracted commercially in very small concentration from aerial parts of *Centella asiatica*.

The wonder drugs of plant origin reported in recent years are ginsenosides from *Panax ginseng* and have acquired commercial significance in view of its stimulative, alterative and general tonic properties. 18-β glycyrrhetinic acid from liquorice is the drug of choice in the treatment of peptic ulcers and hyperacidity.

Pilocarpine nitrate, a life saving drug useful in treatment of glaucoma, is extracted from leaves of *Pilocarpus jaborandi.*

Tannic acid of pharmaceutical grade is obtained from galls and myrobalans. The genus *Solanum* (Solanaceae) comprises widely distributed species, some of which contain steroidal glycoalkaloid solasodine. The berries of *Solanum khasianum* are rich in solasodine content, which is used as starting material for the production of sex hormones. Steroidal sapogenin, diosgenin, obtained from several Dioscorea species and *Costus speciosus* also serves as the starting material for the manufacture of corticosteroids.

A number of essential oils and their chemicals used as medicinal and flavouring agents are obtained from plant sources. Some of the examples of this category are the volatile oils of peppermint, spearmint, clove, eucalyptus, ginger, citronella, thyme and vetiver. The volatile products produce the characteristic odour of flowers, spices and perfumes. The most, exotic and expensive of the volatile oils are those used in perfumery. Geraniol, linalool and citronellal isolated from a number of volatile oils are widely used as components of perfume formulations. Citral from lemon grass oil is itself used in perfumery and soaps. It can also be converted to the ionones, one of which serves as an intermediate in commercial synthesis of vitamin A, while the other provides a substitute for the extremely expensive oil of violets.

The majority of spices, condiments, teas and other beverages such as coffee, kola and cocoa owe their unique and characteristic properties (aromas and flavours) to physiologically active secondary metabolites that they contain. Although today several of their purified principles (e.g. vanillin and caffeine) are produced by semi or total synthesis, high prices are still paid for compounds or mixtures extracted from their natural sources, especially if they are intended for use as flavouring agents or food additives. Spices are usually more diversified than the perfume principles containing not only terpenes, especially from dill, caraway, fennel, celery etc. but also aromatic aldehydes (cinnamic aldehyde from cinnamon), phenols (anethal from anise, eugenol from clove and safrole from sassafras) and sulphur compounds (Allium).

Plants continue to be important sources of new drugs, as evidenced by the recent approvals of several new plant derived drugs and synthetic drugs based on secondary plant metabolites. Etoposide is a new semi-synthetic antineoplastic agent derived from subterranean parts of *Podophyllum peltatum* reported to be useful in chemotherapeutic treatment of refractory testicular carcinomas, small cell lung carcinomas, non-Hodgkin's lymphomas and non-lymphocytic leukemias. The technical-know how for indigenous production of etoposide is claimed to have been developed by Indian Institute of Chemical Technology, Hyderabad. Cannabinoids are being developed for use in neurological disorders (e.g. epilepsy and dystonia) and as antihypertensives, antiasthmatics and potent analgetics. Atracurium besylate is a new synthetic skeletal muscle relaxant, which is structurally and pharmacologically related to curare alkaloids. Plant-derived drugs of clinical significance discovered in recent years include artemisinin, a rapidly acting antimalarial agent from chinese drug, *Artemisia annua*; forskohlin, a naturally occurring labdenone di-terpene with antihypertensive, positive inotropic and adenylcyclose - activating properties from *Coleus forskohlii* and sanguinarine an antiplaque alkaloid with both preventive and therapeutic effects on dental plaque. The other bioactive compounds of therapeutic significance include organosulphur compounds of garlic and onions as cardiovascular agents, guggulipids with antiarthrytic property from *Commiphora mukul*; ellagic acid as a prototype anti-mutagen and cancer preventing agent; vasicin as oxytocic agent from *Adhatoda vasica*. Intensely sweet plant metabolites such as stevioside, glycyrrhizin, hernandulcin are already used in Japan and Europe.

The technology involved in the extraction of phytochemicals of pharmaceutical significance in majority of the cases is the guarded secret of a pharmaceutical or chemical firm. An overall approach to the method of extraction of these phytopharmaceuticals represents the co-ordination of research work carried out by scientists of different disciplines. With the advancement in analytical techniques and instrumentation technology in last three decades, it has been possible to devise commercially feasible techniques for the extraction of several phytochemicals.

Current Trend of Market in Medicinal Plants

The trade in medicinal plants and their derivatives in pharmacy have declined in many industrialised countries owing to the volume of competitive synthetic products currently marketed. Overall, the trade in plant products has increased following their increased use in the food and cosmetic industries. A renewed interest in traditional medicine in Asia and in health foods in Europe and North America has provided new outlets for a number of plant products. From a data base compiled by the United Nations Statistical Office and International Trade Centre, the value of total imports of medicinal plants has increased from 355 million US dollars in 1976 to 551 million US dollars in 1980. American consumers paid about 8 billion US dollars in 1980 for prescriptions derived solely from higher plants.

At present, the United States is a major importer of medicinal plants amongst all the developed and developing countries. Japan is a major importer of medicinal plants among the Asian countries. The commerce in medicinal plants is prominent throughout Asia, with ginseng being an important individual item. The bulks of plant derivatives used in pharmaceutical preparations are extracted in the industrialized countries and then distributed throughout the world. The U.S.A. Federal Republic of Germany, China, Japan and Switzerland are the major suppliers of the finished products, although some developing countries such as India, Indonesia and Republic of Korea are marketing crude drugs and derivatives with increasing success.

The current unfavourable economic situation, which has caused a diminishing cash flow and increased interest rates, has distorted trade in plant products as many processors and traders have been forced to cut their inventories to a minimum. The commerce in ginseng and its products has shown dynamic growth in recent years, perhaps due to worldwide promotion by the health food industry. Some of the alkaloids and glycosides have maintained a steady market, with the exception of *Rauwolfia* alkaloids, which have fallen from favour in many industrialised countries. Cinchona bark and alkaloids, liquorice root and extract, papain, digitalis, ginseng and laxatives have large to medium sales volumes in industrialized countries. *Dioscorea* is not much in demand in international market and supplies of the crude drug shall have to be utilized in the producing countries. Any producer of diosgenin would normally have to compete with Chinese material, the price of which is unpredictable and not necessarily related to production costs. Papain is currently in short supply. Thailand seems to be the only country supplying *Rauwolfia serpentina* to the trade, since India no longer exports significant quantities, but retains it for domestic extraction. *Rauwolfia vomitoria* from Africa is available in sufficient quantities, but its demand is low. Supplies of senna from both India and Sudan are sufficient to satisfy the world demand.

The major exporter of psyllium and sandal wood oil and opium alkaloids is India. The Basic Chemicals Pharmaceuticals and Cosmetic Export Promotion Council of India claims that India provides 80 per cent of the psyllium available in the world market. World demand for psyllium is now about 15,000 tones annually with a growth rate equal to the growth rate for laxatives, which is about 3 per cent per annum.

The future of phytopharmaceuticals is because it serves as a cheap and steady source for varied range of therapeutic agents which are of great significance in the health care of mankind.

In last two decades, the scientific community has shown a fast increasing interest in pharmacognosy and natural products research. Some of the major reasons cited are search of lead molecules from natural sources for drug development; applying biotechnology for production of pharmaceuticals; therapeutic values of foods (i.e. nutraceuticals) and standardization of traditional medicines. This trend is well reflected through a number of new international scientific journals. In July, 1999 four major international societies in the field of pharmacognosy and natural products research held a joint meeting at Amsterdam; which attracted more than 1100 scientists.

In the new promising era of pharmacognosy, there are three major areas of studies which include: (a) Studies of new biologically active natural products, (b) Quality control of natural drugs and (c) Production of drugs from natural sources especially by biotechnological ways.

For the studies of new biologically active products recently high throughput screening (HTS) has been introduced. This technique uses the molecular targets by which a large number of drug samples (up to 1,00,000 in 24 hours) can be screened for a single activity. Thus a large number of extracts can be rapidly screened and hence HTS has been found to be suitable for bioassay-guided fractionation. New techniques for isolation and identification of new active compounds are being developed at a faster rate. For determination of biological active fractions, HPLC-on line techniques have been developed. Considering the advantages of plant cell cultures as novel sources of extracts for screening, biotechnology has emerged as a major area of interest in pharmacognosy. There is a lot of optimism associated for the possible role of plant cell culture in the production of pharmaceuticals. The promising area within biotechnology is genetic engineering at the advent of which it is also possible to introduce the production of therapeutically important products like vaccines in plants.

Along with the discovery of novel phytopharmaceuticals for combat of diseases, the subject of pharmacognosy has entered into a new era. This has been made possible, mainly due to various technologies made available for separation and isolation. There is a necessity for additional requirement in the knowledge of pharmacogosist for various phytochemicals techniques, advanced separation methods and also molecular biology and proteins. Molecular biology including DNA finger printing would be an essential technique for both quality control and characterization of medicinal plants. Polymerase chain reaction (PCR) could be very useful to find the vegetative contamination in herbal drugs at very low levels.

Taking a clue from traditional medicines and also with the advent of modern separation and isolation technologies, a number of phytopharmaceuticals have been screened out for their important roles in therapeutics. Some of such promising phytopharmaceuticals along with their source and activity have been summarised here

Table 6.1: Promising phytopharmaceuticals with their activity

No.	Phytopharmaceuticals	Natural Source	Activity
		ANTI-ULCER	
1.	Catechin	Leaves of *Artocarpus integra*	anti-ulcer
2.	Cyanidanol	Stereoisomer of catechin from seedcoat of *Anacardium occidentale*	anti-ulcer (inhibiting action on histidine decarboxylase)
3.	Isoliquiritin	Rhizomes of *Glycyrrhiza glabra*	anti-ulcer
4.	Sophoradin	*Sophora subprotostata*	anti-ulcer
		ANTI-PROTOZOAL	
5.	Quinine microcapsules in oral formulation	Bark of *Cinchona* species	anti-malarial
6.	Pycnamine	*Triclisi* species	anti-malarial
7.	Immuno toxins like pokeweed antiviral proteins	Seeds of *Phytolacca maricana*	anti-trypanosomal, anti-leishmanial
		CARDIOVASCULAR TREATMENTS	
8.	Colenol (forskolin)	*Coleus forskohlii*	Hypotensive
9.	Digoxin	Enzymatic conversion and further selective 12 β-hydroxylation of thevenerrin and nerrifolin from yellow oleander *Thevetia nerrifolia*	Cardiotonic
10.	Aescin	Seeds of *Aesculus hippocastanum*	anti-inflammatory
11.	Nimbidin	Seeds oil of *Azadirachta indica*	anti-inflammatory
12.	Curcumin	Rhizomes of *Curcuma longa*	anti-inflammatory

	ANTI-VIRAL AGENTS		
13.	Castarospermine	*Castarospermum australe*	anti-HIV
14.	Lyophilised infusion of hypericum	*Hypericum perforatum*	Influenza A and B
15.	Calanolide A	*Calophyllum lanigeum*	anti-HIV (reverse transcriptase inhibitor)
	IMMUNOMODULATORS AND ADAPTOGENS		
16.	Sitoindoside VII and VIII	Roots of *Withania somnifera*	antistress
17.	Polysaccharide fraction	*Echinacea angustifolia* and *E. purpurea*	immunomodulator
18.	Syringin and cordiol	*Tinospora cordifolia*	immunomodulator
	ANTI-CANCER		
19.	Bryostatin	Macrolide from bryozoans	anti-cancer (Partial agonist of protein kinase C)
20.	Rhizoxin	Fungal metabolite	anti-cancer (antimetabolite)
	ANTI-DIABETIC		
21.	Polypeptide P (P-Insulin) and charantin	*Momordica charantia*	anti-diabetic
22.	Gymnemic acids 1 - 4 and gurmarin	*Gymnema sylvestre*	anti-diabetic (both in insulin dependent and non-insulin dependent diabetis mellitus)

❖❖❖

Chapter 7...

ANALYTICAL PHARMACOGNOSY

The chemistry of plants is as divergent as the great variety of forms in which plants occur. The therapeutically important constituents are usually found to be associated with many inert substances such as colouring matters, cellulose, lignin, cutin, suberin, etc. The active principles are extracted from plant drugs and purified for therapeutic utility for their selective and regulated activity. The quality control of herbal crude drugs and their bioconstituents is of paramount importance in justifying their acceptability in modern system of medicine. One of the major problems faced by user industry is non-availability of rigid quality control profiles for herbal raw materials and their formulations. With the advent of new analytical tools and sophisticated instrumental technology, it is possible to suggest a practicable quality assurance profile for a crude drug or its bioactive constituent.

Given the medicinal properties attributed to a crude drug, it is necessary to maintain its quality and purity in commercial market. It is, however, observed that the drugs in commerce are frequently adulterated and do not comply with the standards prescribed for authentic drug.

[A] DRUG ADULTERATION

Adulteration is a practice of substituting original crude drug partially or wholly with other spurious substances but the later is either free from or inferior in chemical and therapeutic properties. Adulteration in simple terms is debasement of an article. The motives for intentional adulteration are normally commercial one and originate mainly with the intention of enhancement of profits. Some of the reasons that can be cited here are scarcity of drug and its high price prevailing in market. The adulteration is done deliberately, but it may occur accidentally in some cases. It is also very common with the contraband drugs. Adulteration means deterioration, admixture, sophistication, substitution, inferiority and spoilage. **Deterioration** is impairment in the quality of drug, while admixture is addition of one article to another due to ignorance or carelessness or by accident. **Sophistication** is the intentional or deliberate type of adulteration. **Substitution** occurs when some totally different substance is added in place of original drug. **Inferiority** refers to any substandard drug, and **spoilage** is due to the attack of microorganisms.

A variety of adulterants are found in natural drugs by which the methods employed for such practices can be enumerated. During the routine quality control, various tests are conducted for their detection. The present chapter deals with different techniques used in adulterating crude drugs and laboratory methods for their detection.

Types of Adulterants

Generally, the drugs are adulterated by substitution with substandard commercial varieties, inferior drugs or artificially manufactured commodities. The different types of adulterants found in market are given here.

(i) Substitution with substandard commercial varieties: The adulterants used here may resemble original crude drug by morphological, chemical or therapeutic characters, but are substandard in nature and hence cheaper in cost. This is rather the most common practice of adulteration. The examples are the presence of *Strychnous nux blanda* or *S. potatorum* in place of *S. Nux vomica* ; *Capsicum minimum* replaced by *C. annuum*; Indian senna substituted with Arabian senna and dog senna; gentian substituted by *kutki*; medicinal ginger replaced by its inferior varieties; African, Japanese and Cochin ginger.

(ii) Substitution with superficially similar inferior drugs: These inferior drugs used may or may not have any chemical or therapeutic value as that of original natural drug. Due to their morphological resemblance to authentic drug, they are marketed as adulterants. Belladonna leaves are substituted with *Ailanthus* leaves; saffron is admixed with dried flowers of *Carthamus tinctorius* ; scented bdellium is used for myrrh; mother cloves and clove stalks are mixed with clove; and beeswax is substituted by Japan wax.

(iii) Substitution with artificially manufactured substances: It has been also observed that substances artificially prepared to resemble original drug are used as substitutes. Generally, this practice is followed for much costlier drugs. Compressed chicory in place of coffee; paraffin wax made yellow coloured and substituted for beeswax properly cut and shaped basswood for nutmeg are some of the examples representing this type of adulteration.

(iv) Substitution with exhausted drugs: In this type, the same drug is admixed but is devoid of any medicinally active constituents as they are already extracted out. This practice is more common in case of volatile oil containing drugs like fennel, clove, coriander, caraway etc. Sometimes, natural characters of exhausted drugs like colour and taste are manipulated by adding other additives and then it is substituted, e.g. exhausted gentian made bitter with aloes, artificial colouring of exhausted saffron, etc.

(v) Besides these common practices, sometimes other methods are employed like use of synthetic chemicals to enhance the natural character as in the case of addition of benzyl benzoate to balsam of Peru, Citral to citrus oils like oil of lemon and orange oil, etc.

(vi) Presence of vegetative matter from the same plant: Sometimes, the other miniature plants growing along with medicinal plant are mixed with drug due to their resembling colour, odour and in some cases constituents. The lower plants like moss, liverworts and epiphytes growing on bark portion are mixed with cascara or cinchona. The stem portions are mixed along with leaf drugs like stramonium, lobelia and senna.

(vii) Harmful adulterants: Several times, the wastes from market are collected and admixed with authentic drugs. This is particularly noticed for liquids or unorganized drugs. The examples like pieces of amber coloured glass in colophony, limestone in asafoetida, lead shot in opium, white oil in coconut oils, cocoa butter mixed with stearin or paraffin indicate this type of adulteration practice. The addition of rodent faecal matter to cardamom seed is a very harmful adulterant.

(viii) Adulteration of powders: Besides the entire drugs, the powdered forms are frequently found to be adulterated. Some examples which can be cited here are dextrin in ipecacuanha, powdered liquorice or gentian admixed with powdered olive stones, exhausted ginger powder in powdered colocynth or ginger, red sanders wood in capsicum, etc. The powdered bark is frequently found to be adulterated with brick powder.

[B] DRUG EVALUATION

Evaluation of a drug means confirmation of its identity and determination of its quality and purity and detection of nature of adulteration. The evaluation of a crude drug is necessary because of three main reasons (a) biochemical variation in the drug, (b) deterioration due to treatment and storage, and (c) substitution and adulteration, as a result of carelessness, ignorance or fraud.

Over the years the nature and degree of evaluation of crude drugs has undergone a systematic change. Initially, the crude drugs were identified by comparison only with the standard description available. Due to advancement in the chemical knowledge of crude drugs, at present, evaluation also includes method of estimating active constituent present in the crude drug, in addition to its morphological and microscopic analysis. With the advent of separation techniques and instrumentation analysis, it is possible to perform physical evaluation of a crude drug which could be both qualitative and quantitative. The biological behaviour of crude drug extracts constitutes pharmacological evaluation.

The crude drugs can be identified on the basis of their morphological, histological, chemical, physical and biological studies. The different techniques involved in standardization of crude drugs are as follows.

1. MORPHOLOGICAL OR ORGANOLEPTIC EVALUATION

It refers to evaluation of drugs by colour, odour, taste, size, shape and special features, like touch, texture, etc. It is a technique of qualitative evaluation based on the study of morphological and sensory profiles of whole drugs. Organoleptic evaluation means conclusions drawn from impressions on organs of senses. The study of form of a crude drug is *Morphology* while description of the form is *Morphography*. The fractured surfaces in cinchona, quaillaia and cascara barks and quassia wood are important characteristics. Aromatic odour of umbelliferous fruits and sweet taste of liquorice are the examples of this type of evaluation. The ovoid tears of gum acacia, ribbon shaped characteristic of tragacanth, disc-shaped structure of nux vomica, conical shape of aconite, quills of cinnamon etc. are important diagnostic characters. The general appearance of the lot of a crude drug often indicates whether it is likely to comply with prescribed standards, such as percentage of seed in colocynth, stalk in clove, etc. Over drying, makes leaf drugs and flowers brittle and cause them to break in transit making the task of morphological evaluation difficult.

The wavy shape of Rauwolfia, pungent taste of capsicum and ginger, brown colour of cinnamon, odour and taste of spice-drugs like, asafoetida, black pepper, nutmeg, caraway, cummin, etc. are important diagnostic organoleptic characteristics.

2. MICROSCOPIC EVALUATION

This method allows more detailed examination of a drug and it can be used to identify the organized drugs by their known histological characters. It is mostly used for qualitative evaluation of organized crude drugs in entire and powdered forms.

Microscope, by virtue of its property to magnify, permits the minute structure understudy to be magnified to confirm the structural details of the drugs from plant origin. For the effective results, various reagents or stains can be used to distinguish cellular structure. Microscopic evaluation also covers study of the constituents by application of chemical methods to small quantities of drugs in powdered form or to histological sections of the drug (**Microchemistry** or **chemomicroscopy**). A drop of phoroglucinol and concentrated hydrochloric acid give red stain with lignin. Mucilage is stained pink with ruthenium red and also, when treated with corallin soda and few drops of sodium carbonate solution, cellulose swells and dissolves in cuoxam, while N/50 iodine solution stains blue starch and hemicelluloses.

Histological studies are made from very thin sections of drugs. The characteristics of cell walls, cell contents, starch grains, calcium oxalate crystals, trichomes, fibres, vessels, etc. can be studied in details, e.g. lignified trichomes in nux vomica, warty trichomes of senna, wavy medullary rays of cascara bark, glandular trichomes of mint etc. Some important types of Trichomes are shown below. Microscopic linear measurements and quantitative microscopy are also covered under this technique of evaluation. The powdered cloves do not contain sclereid or calcium oxalate crystals, but both of them are present in powdered clove stalks. Powdered clove fruits show presence of starch while it is absent in cloves. Presence of non-lignified vessels in powders of rhubarb and ginger indicate adulteration. The presence or absence of crystals of aloin indicates different varieties of aloes.

Other important histological aspect is the quantitative microscopy and linear measurements. The various parameters studied here are stomata number and index, palisade ratio, vein-islet number, size of starch grains, length of fibres, ete. Senna varieties are distinguished by differing stomata number and palisade ratio. The diameter of starch grains in *Cinnamomum cassia* is 10 microns; hence, useful for detecting adulterants. The number of sclerenchymatous cells per square mm of cardamom is useful for detecting different varieties of cardamom seed.

(a) Leaf Constants or Diagnostic Characters of Leaves

(i) Palisade ratio is defined as average number of palisade cells beneath each epidermal cell. It can be determined with powdered drugs.

Table 7.1: Palisade Ratios of various leaf-drugs

No.	Name of Plant	Range
1.	Adhatoda vasica	5.5 – 6.5
2.	Andrographis paniculata	03 – 4.5
3.	Atropa acuminata	05 – 12.4
4.	Atropa belladonna	05 – 7.0
5.	Azadirachta Indica	03 – 3.5
6.	Bacopa monniera	1.5 – 2.25
7.	Boerhaavia diffusa	3.1 – 4.5
8.	Cassia acutifolia	4.5 – 9.5 (upper) 3.5 – 7.0 (lower)
9.	Cassia angustifollia	5.5 – 10.0 (upper) 4.0 – 7.4 (lower)
10.	Centella asiatica	3.5 – 5.75
11.	Datura fastuosa	4.2 – 5.5
12.	Datura metel	05 – 6.5
13.	Datura stramonium	4.2 – 6.5
14.	Digilatis lanata	2.5 – 6.5
15.	Digitalls purpurea	3.7 – 4.2
16.	Digilalis thapsi	3.0 – 5.0
17.	Eucalyptus globules	5.5 – 6.5 (upper) 3.5 – 5 (lower)
18.	Hyoscyamus niger	03 – 06
19.	Lawsonia inermis	07 – 8.5 (upper) 4.5 – 5.7 (lower)
20.	Mellia azedarachta	7.2 – 8 (upper) 3.2 – 4 (lower)
21.	Mentha arvensis	3.5 – 05
22.	Mentha piperita	6.1 – 8.5
23.	Momordica charantia	2.5 – 3.8
24.	Nicotiana tobacum	3.5 – 04
25.	Ocimum sanctum	02 – 3.5
26.	Thymus vulgaris	1.7 – 3.2

(ii) Vein-islet number is defined as the number of vein-islets per sq. mm of the leaf surface midway between the midrib and the margin. Levin in 1929 determined vein-islet numbers of several dicot leaves.

The vein-islet numbers of various leaf drugs are given below:

Table 7.2: Vein-islet number of various leaf-drugs

No.	Name of drug	Range
1.	Adhatoda vasica	6 – 8
2.	Andrographis paniculata	9 – 12
3.	Atropa acuminata	8 – 11
4.	Azadirachta Indica	10 – 18
5.	Bacopa monniera	6 – 13
6.	Boerhaavia diffusa	9 – 13
7.	Cannabis sativa	20 – 30
8.	Cassia angustifollia	18 – 24
9.	Centella asiatica	3.5 – 6
10.	Datura fastuosa	18 – 24
11.	Datura metel	19 – 22
12.	Datura stramonium	12 – 16
13.	Digitalls purpurea	2.5 – 3
14.	Erythroxylon coca	8 – 14
15.	Eucalyptus globules	8 – 13.5
16.	Hyoscyamus muticus	16 – 24
17.	Hyoscyamus niger	20 – 28
18.	Lawsonia inermis	26 – 34
19.	Mellia azadarachta	10 – 18
20.	Mentha arvensis	9 – 11
21.	Mentha piperita	5 – 8
22.	Momordica charantia	7 – 12
23.	Nicotiana tobacum	12 – 20
24.	Ocimum sanctum	19 – 23

(iii) Vein-termination number is defined as the number of veinlet terminations per sq. mm of the leaf surface midway between midrib and margin.

(iv) Stomatal number is the number of stomata per sq. mm of epidermis of the leaf.

(v) Stomatal index is the percentage which the numbers of stomata form to the total number of epidermal cells; each stoma being counted as one cell.

It is calculated by using the following equation

$$SI = \frac{S}{E + S} \times 100$$

where, SI = Stomatal Index

S = Number of stomata per unit area

E = Number of epidermal cells in the same unit area

(b) Stomata

Epidermis of leaf shows different characteristics, e.g. cuticle Stomata, trichomes, water-pore, cell inclusions, etc.

A stoma is a minute epidermal opening present on aerial parts of the plants, with following characteristics. (a) A central pore (b) Two kidney shaped similar cells containing chloroplasts known as guard cells and varying number of subsidiary (epidermal) cells covering the guard cells.

The primary and most important function of stomata is gaseous exchange and the secondary function is transpiration. It is not essential that each plant must have stomata. The leaves of bryophytes and submerged leaves of aquatic parts do not contain stomata. Generally, stomata are present in green parts of the plant (mostly leaves), but absent in roots. Apart from the leaves, they are also present in the stems (ephedra), flowers (clove), and fruits (fennel). However, it is generally observed that stomata are abundantly present in dicot leaves. In some cases, they are present on the upper surface of leaves, while in others on lower surface only (coca and cherry). In some, the stomata are present on both surfaces of the leaves (senna, belladonna, datura, etc.). The distribution of stomata between upper and lower epidermis in dicot leaves shows great variation.

Types of stomata: Depending upon the type of the guard cells and arrangement of subsidiary cells, stomata are divided into four types:

1. Moss type
2. Gymnospermous type
3. Gramineous type
4. Dicotyledonous type

Out of these, fourth type of the stomata is of diagnostic significance. Dicotyledonous stomata are classified into following types depending upon the form and arrangement of subsidiary cells.

(i) Paracytic or rubiaceous or parallel-celled stomata: These type of stomata comprise two guard cells covered by two subsidiary cells, the long axes of which are parallel to that of stoma, e.g. coca and senna leaves (Fig. 7.1).

1. Parasitic 2. Diacytic

Fig. 7.1: Types of stomata

(ii) Diacytic or caryophyllaceous or cross-celled stomata: The guard cells are covered by two subsidiary cells, as in the case of parasitic stoma, but the arrangement of subsidiary cells on the guard cell is at right angle to that of stoma, e.g. peppermint, spearmint, and vasaka (Fig. 7.1).

(iii) Anisocytic or cruciferous or unequal-celled stomata: The number of guard cells is two, as in all other cases, but, the guard cells are covered by three subsidiary cells, of which one is markedly smaller than the other two, e.g. belladonna, datura and stramonium (Fig. 7.2).

(iv) Anomocytic or ranunculaceous or irregular-celled stomata: In this type, stoma is surrounded by varying number of subsidiary cells resembling other epidermal cells, e.g. buchu, digitalis and lobelia (Fig. 7.2).

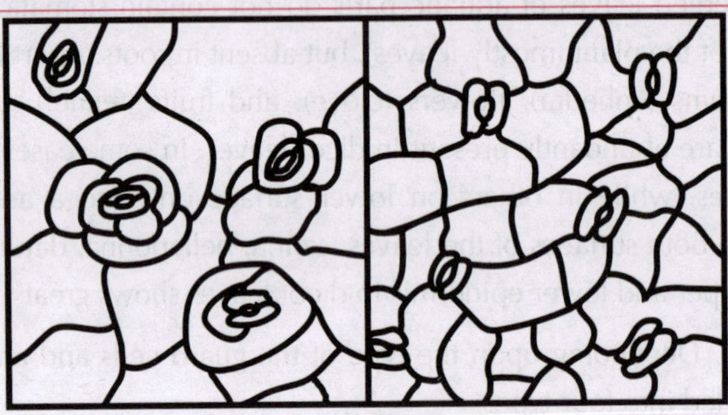

1. Anisocytic 2. Anomocytic

Fig. 7.2: Types of stomata

Indian Pharmacopoeia recognizes one more type of stomata known as Actinocytic stomata.

(v) Actinocytic or radiate-celled stomata: The two guard cells are surrounded by a circle of radiating subsidiary cells.

Table 7.3: Stomatal Number and Stomatal Index values

No.	Name of leaf-drug	Epidermis	Stomatal Number	Stomatal index
1	2	3	4	5
1.	Andrographis paniculata	Upper	65 – 128	12.0 – 17.8
		Lower	90 – 145	12.4 – 16.4
2.	Atropa acuminata	Upper	05 – 14	2.5 – 4.7
		Lower	78 – 95	15.5 – 17.9
3.	Atropa belladonna	Upper	07 – 10	–
		Lower	77 – 115	20.2 – 23.0
4.	Azadirachta Indica	Upper	–	–
		Lower	415 – 470	8 – 8.5
5.	Bacopa monniera	Upper	65 – 126	12.9 – 17.8
		Lower	90 – 145	12.4 – 16.4

1	2	3	4	5
6.	Boerhaavia diffusa	Upper	112 – 141	11.4 – 14.2
		Lower	118 – 146	9.7 – 14.4
7.	Cannabis sativa	Upper	–	–
		Lower	423 – 550	13.1 – 20.4
8.	Cassia angustifolia	Upper	220 – 260	17.0 – 21.4
		Lower	240 – 265	16.5 – 21
9.	Centella asiatica	Upper	66 – 82	9.2 – 12.7
		Lower	141 – 228	11 – 16.3
10.	Datura fastuosa	Upper	370 – 412	21.2 – 22.6
		Lower	365 – 388	22 – 25.6
11.	Datura innoxia	Upper	82 – 141	12.7 – 19.5
		Lower	105 – 163	21.2 – 24.0
12.	Datura metel	Upper	147 – 160	16.3 – 17.0
		Lower	200 – 209	16.8 – 22.8
13.	Datura stramonium	Upper	98 – 150	13.6 – 20.0
		Lower	200 – 207	14.4 – 15.4
14.	Digitalis purpurea	Upper	25 – 50	1.3 – 3.5
		Lower	–	–
15.	Erythroxylon coca	Upper	–	–
		Lower	230 – 510	11.5 – 15.6
16.	Eucalyptus globules	Upper	200 – 210	04 – 05
		Lower	220 – 235	3.5 – 4.2
17.	Hyoscyamus niger	Upper	165 – 195	18.6 – 20.0
		Lower	190 – 230	25.4 – 28.1
18.	Lawsonia inermis	Upper	100 – 142	7.5 – 10.6
		Lower	200 – 205	12.1 – 14.7
19.	Mellia azedarachta	Upper	–	–
		Lower	175 – 210	5.5 – 6.5
20.	Mentha arvensis	Upper	15 – 20	3.7 – 4.9
		Lower	160 – 188	18.6 – 20.6
21.	Mentha piperita	Upper	30 – 65	06 – 11.6
		Lower	325 – 375	19 – 21.6
22.	Momordica charantia	Upper	–	–
		Lower	450 – 475	14 – 15
23.	Nicotinana tabacum	Upper	80 – 100	21 – 24.0
		Lower	180 – 225	29 – 33.8
24.	Ocimum sanctum	Upper	64 – 72	08 – 8.5
		Lower	175 – 250	13.1 – 16.6
25.	Piper betel	Upper	–	–
		Lower	32 – 88	8.6 – 9.5

The techniques of determination of leaf constants can be used for microscopic evaluation of several leaf drugs such as senna, datura, digitalis, buchu, coca, belladonna, etc.

(c) Trichomes or Plant hairs

These are other important diagnostic characters helpful in the identification of drugs and detection of adulterants. Trichomes are the tubular elongated or glandular outgrowth of the epidermal cell. Trichomes are also known as plant hairs. Trichomes consist of two parts namely root (in the epidermis) and body (outside the epidermis). Trichomes are present in most of the parts of the plant such as leaves (senna and digitalis), seeds (nux vomica and strophanthus), fruits (*Helicteris isora* and Ladies finger), etc. Trichomes are as such functionless, but sometimes, perform secretory function. The trichomes excrete water and at times, volatile oil as in case of peppermint. Trichomes are present in most of the aerial parts of the plant, but are absent on roots. They are absent in coca, Sevin and hemlock and rarely present in buchu, and henna. Depending upon the structure and the number of cells present in trichomes, they are classified as follows:

1. Covering trichomes or non-globular trichomes or clothing trichomes;
2. Glandular trichomes;
3. Hydathodes or special type of trichomes.

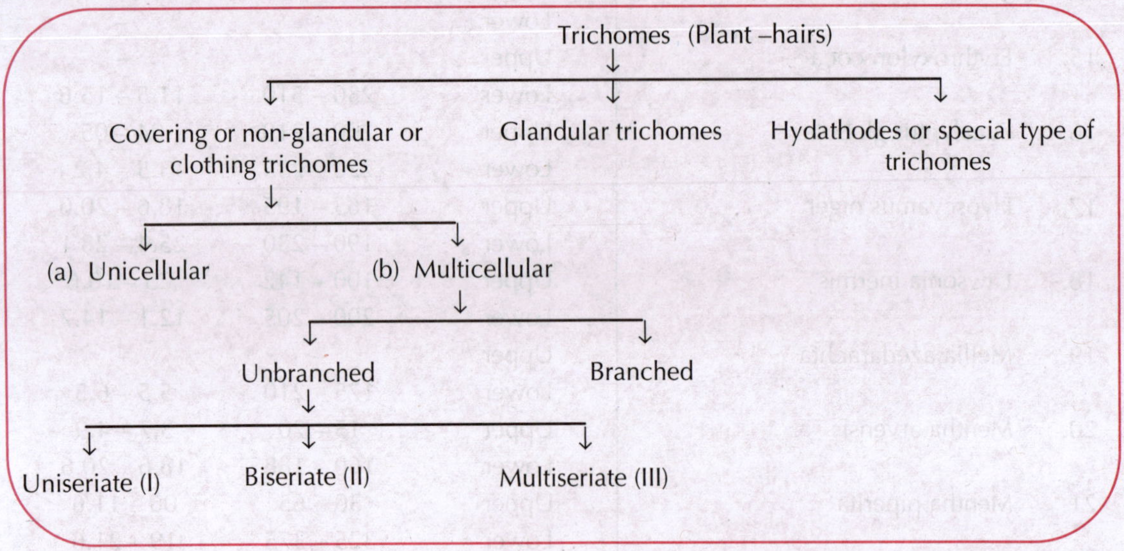

(I) COVERING TRICHOMES

(a) Unicellular

1. Lignified trichomes	Nux vomica, strophanthus
2. Short, sharply pointed, curved	Cannabis
3. Large, conical, strongly shrunken	Lobelia
4. Short, conical, unicellular	Tea, Buchu
5. Strongly waved, thick walled	Yerba, Santa

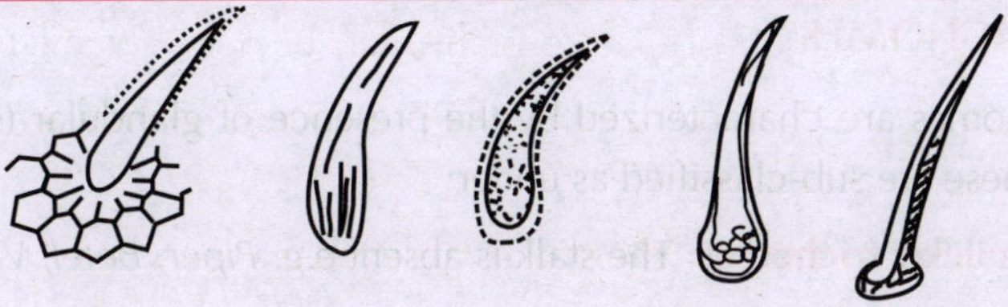

Unicellular, Uniseriate covering trichomes

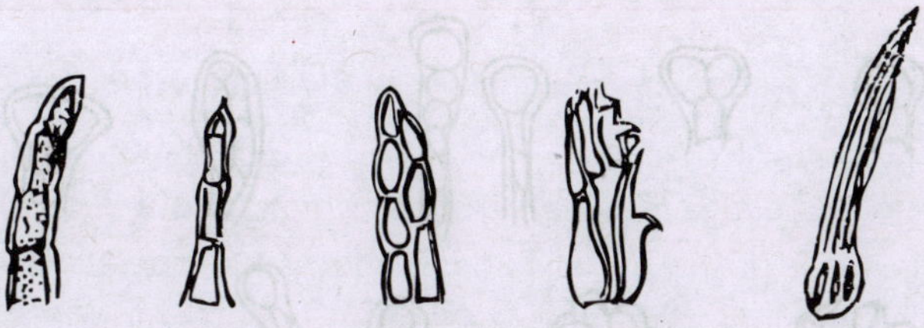

Multicellular Unbranched covering trichomes

Multicellular Branched trichomes

Fig. 7.3: Types of covering trichomes

(b) Multicellular — Unbranched trichomes

(i) Uniseriate

1.	Bi-cellular, conical	Datura
2.	Three celled long	Stramonium
3.	Three to four celled long	Digitalis
4.	Four to five celled long	Belladonna

(ii) Biseriate : *Calendula officinalis*

(iii) Multiseriate : Male fern

(c) Multicellular — branched trichomes

1.	Stellate	Hammamelis, *Helicteris isora*
2.	Peltate	(Plate like arrangement of surrounded cells) Humulus
3.	Candelabra	(Uniseriate branched axis) *Verbascum Thapsus*
4.	T shaped trichomes	Artemisia, pyrethrum

(II) GLANDULAR TRICHOMES

The glandular trichomes are characterized by the presence of glandular (spherical) cell all the top of the trichomes. These are sub-classified as under:

1. **Unicellular glandular trichomes:** The stalk is absent *e.g. Piper, Betel, Vasaka*.
2. **Multicellular glandular trichomes**

1. Trichomes with unicellular head and unicellular stalk, e.g. *Digitalis purpurea*.

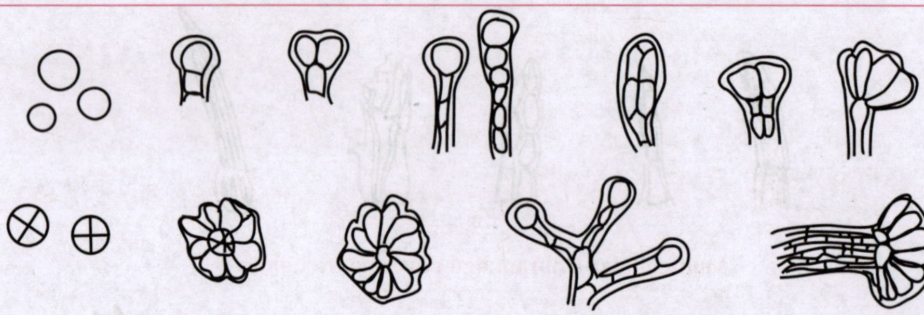

Fig. 7.4: Types of Glandular trichomes

2. Unicellular head and uniseriate multicellular stalk, e.g. *Digitalis thapsi,* belladonna, etc.
3. Multicellular head, multicellular, biseriate stalk, e.g. Santonica and plants of Compositae, such as sunflower, etc.
4. Unicellular stalk and biseriate head, e.g. *digitalis purpurea*.
5. Short stalk with secreting head formed of rosette or club shaped cells, e.g. *Mentha species*.
6. Trichomes with multicellular, multiseriate cylindrical stalk and a rosette of secretory cells, e.g. *Cannabis sativa*.
7. Multicellular multisteriate head and multicellular uniseriate stalk, e.g. Indian hemp and tobacco.

(III) HYDATHODES (SPECIAL TYPES OF TRICHOMES)

These are organs of absorption or secretion of water developed in certain plants e.g. *Piper betal,* London pride, etc.

(d) Calcium Oxalate Crystals

Several cell contents like aleurone grains, mucilage, tannin, fixed and volatile oil globules and several other inorganic components like calcium carbonate, calcium oxalate and silica are frequently found to occur naturally in the vegetable drugs. The inorganic crystalline compounds by virtue of their specific shapes can be utilized for the identification of herbal drugs. Due to this reason they are known as **diagnostic characters** of the plant.

Here a special reference is being given to different forms of calcium oxalate crystals in plant.

Calcium oxalate is a dimorphic salt of which both the types occur in the plant body. The crystals are either monoclinic or tetragonal in shape. Both the types differ in their characters such as water content, optical properties etc.

Monoclinic crystals contain only one molecule of water of crystallization i.e. $CaC_2O_4.H_2O$, while tetragonal crystals are $CaC_2O_4.3H_2O$.

There are six forms of calcium oxalate crystals in plants as follows .

(1) Cubical (Prisms): As the name indicates, these crystals are cubical in shape (i.e. their height, width and length is equal) and also have three equal axes, all at right angles to one another.

(2) Rhombic (Diamond Shaped): Here crystals have three axes, which axes are at right angles to each other and unequal in length.

(3) Tetragonal: This is characterized by presence of three axes, at right angles to one another, two axes equal which are lateral, while the third is either longer or shorter and is known as vertical or principal axis.

(4) Monoclinic: This form has three axes and all the three are unequal. The lateral two axes are at right angles to each other, while the third. i.e. Principal is at right angles to lateral axis. The monoclinic crystals shine more than the tetragonal system.

(5) Acicular: These are excessively long slender forms, with pointed ends and normally reported in bundles.

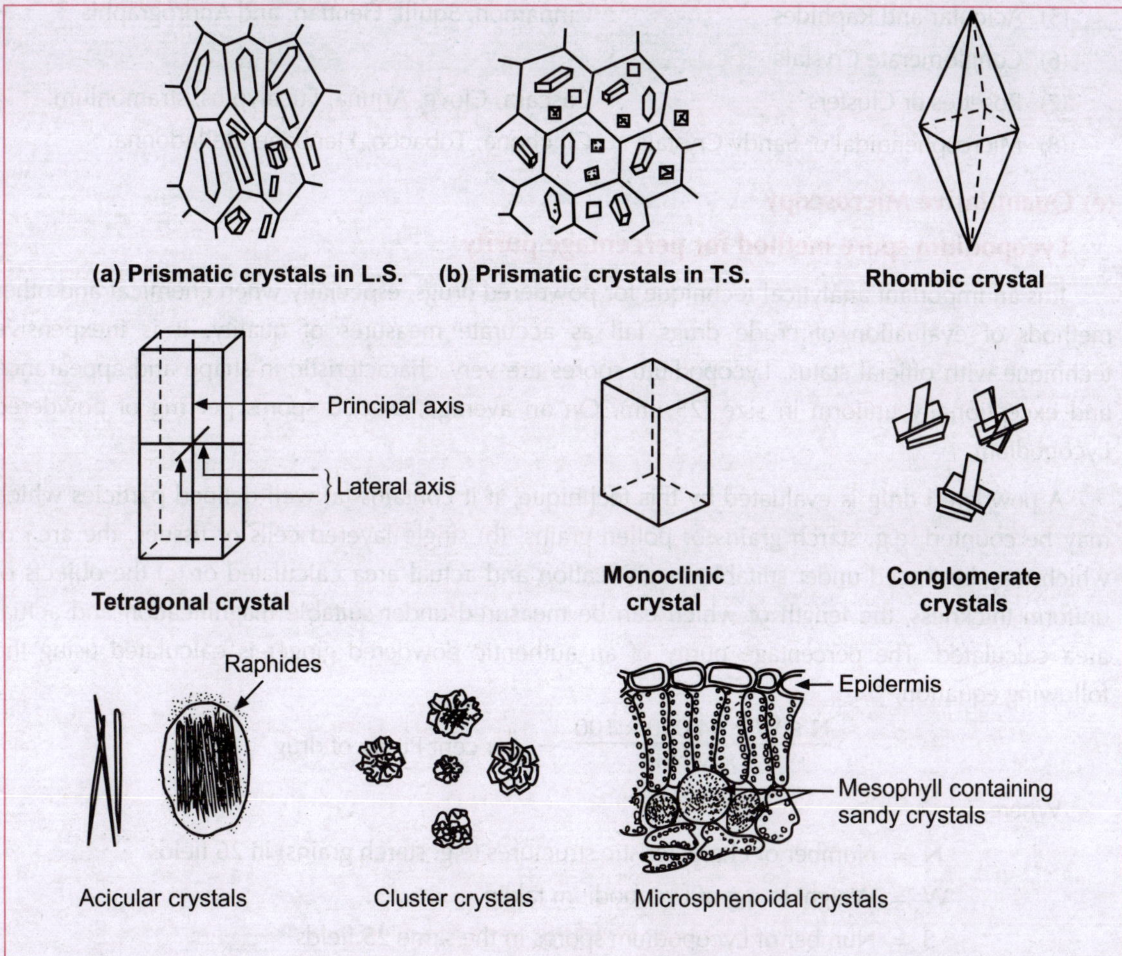

Fig. 7.5: Forms of Calcium Oxalate Crystals

(6) Conglomerate crystals: When the crystals get-together in specific mass by arranging in the form shown in Fig. 7.5 are named as conglomerate crystals.

(7) Rosettes or Clusters: These are also aggregate crystals. This type has shape similar to full expanded rose flower and hence the name rosettes.

Apart from above six types many times, very minute deltoid or arrow-shaped crystals occupying the full cells in which they occur, are also reported. They are known as **Microsphenoidal** or sandy crystals of calcium oxalate.

Following are a few examples of each of the forms described above.

(1) Cubic or Prisms	:	Quaillaia, Senna, Glycyrrhiza, Hyoscymus, Wild cherry.
(2) Rhombic or Diamond Shaped	:	Kurchi bark, Mimosa pudica
(3) Tetragonal	:	Hyoscyamus, Onion
(4) Monoclinic	:	Veratrum, Oak galls, Quaillaia.
(5) Acicular and Raphides	:	Cinnamon, Squill, Gentian, and Andrographis
(6) Conglomerate Crystals	:	
(7) Rosettes or Clusters	:	Cascara, Clove, Arjuna, Eucalyptus, stramonium.
(8) Microsphenoidal or Sandy Crystals	:	Cinchona, Tobacco, Henbane, Belladonna.

(e) Quantitative Microscopy

Lycopodium spore method for percentage purity

It is an important analytical technique for powdered drugs, especially when chemical and other methods of evaluation of crude drugs fail as accurate measures of quality. It is inexpensive technique with official status. Lycopodium spores are very characteristic in shape and appearance and exceptionally uniform in size (25 μm). On an average, 94,000 spores per mg of powdered Lycopodium.

A powdered drug is evaluated by this technique, if it contains (a) well defined particles which may be counted, e.g. starch grains or pollen grains, (b) single layered cells or tissues, the area of which may be traced under suitable magnification and actual area calculated or (c) the objects of uniform thickness, the length of which can be measured under suitable magnification and actual area calculated. The percentage purity of an authentic powdered ginger is calculated using the following equation,

$$\frac{N \times W \times 94,000 \times 100}{S \times M \times P} = \text{per cent Purity of drug}$$

Where,

N = Number of characteristic structures (e.g. starch grains) in 26 fields

W = Weight in mg of Lycopodium taken

S = Number of Lycopodium spores in the same 25 fields

M = Weight in mg of the sample, calculated on basis of sample dried at 105° C

P = 2,86,000 in case of ginger starch grains powder.

Lycopodium spore method can be used for evaluation of powdered clove, ginger, cardamom, nutmeg, umbelliferous fruits, etc.

Pollen Grains:

Flower is reproductive organ of the plant body and contains androcium (male) or gynoecium (female) or both or even may not contain any of these organs (Flowers less plants).

Androcium consists of three parts i.e. filament, anther and connective, Anther is filled with pollen-grains (also known as microspores).

Pollen grains are of very minute spherical particles from 10 – 20 mμ (micron) each pollen grain consists of single microscopic cell and has two coats namely exine and intine. Exine layer is tough and provided with reticulations of different types. Intine layer is very thin and delicate cellulose layer inside the exine.

Pollen grains are diagnostic characters and are useful in identification of plants though fertilization is their main object.

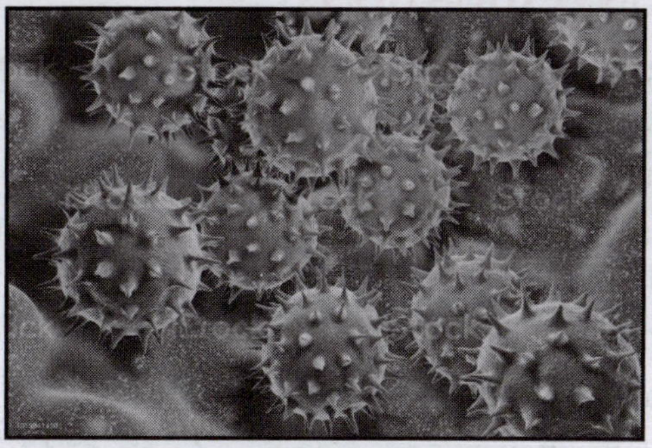

Fig. 7.6: Various pollen grains

3. CHEMICAL EVALUATION

It comprises different chemical tests and assays. The isolation, purification and identification of active constituents are chemical methods of evaluation. Quantitative chemical tests such as acid value, saponification value, etc. are also covered under this technique. Some of these tests are useful in evaluation of resins (sulphated ash, acid value); balsams (acid, saponification and ester values); volatile oils (acetyl and ester values); and gums (methoxy determination and volatile acidity). Preliminary phytochemical screening is a part of chemical evaluation. The qualitative chemical tests are useful in detection of adulteration. Colophony which is admixed as an adulterant for resins, balsams and waxes is detected by copper acetate. Halphen's test for cotton seed oil, Van Urk's reagent for ergot, Vitali's test for tropane alkaloids, murex idee test for purine bases, etc. are examples of specific tests.

The purity of crude drugs is ascertained by quantitative estimation of active chemical constituents present in them. The method may be useful in determining single active constituent or the group of related constituents present in the same drug.

The conventional titrimetric estimations as applicable to the estimation of alkaloids from crude drugs, ester and aldehyde contents of volatile oils, gravimetric methods etc. are the techniques of chemical assay. The alkaloidal drugs which are routinely analyzed for their total alkaloidal content and main alkaloid are opium (morphine), belladonna (atropine), ipecacuanha (emetine), nux vomica (strychnine), cinchona (quinine), Rauwolfia (reserpine), and ergot (ergometrine). Chemical assays are carried out for cineole in eucalyptus oil, aldehyde in lemon oil, carvone in caraway oil and dill oil. The chemical assays are also extended to anthraquinone and cardiac glycosides, content of balsamic acids like benzoic acid and cinnamic acid in balsam of tolu, balsam of peru and prepared storax. The application of physical methods involving instrumentation is a common complementary feature of chemical assays.

The chemical evaluation also covers phytochemicals screening carried out for establishing chemical profile of a crude drug.

(C) PHYTOCHEMICAL INVESTIGATIONS

The systematic investigations of plant material for its phytochemicals behaviour involve four different stages.

1. The procurement of raw material and quality control.
2. Extraction, purification and characterization of the constituents of pharmaceutical interest and in process quality control.
3. Investigations of biosynthetic pathways of particular compounds and
4. Quantitative evaluation.

The commonly employed technique for separation of active substance from crude drug is called **'Extraction'** which involves the use of different solvents. The plant material used for extraction should be properly authenticated or identified. The choice of the plant material for extraction depends on its nature and the components required being isolated. The dried powdered plant material is commonly used for extraction. The fresh plant parts when used are homogenized or macerated with a solvent such as alcohol. It is a general solvent for many potential constituents and as such may give problem in subsequent elimination of pigments, resins, etc. Water immiscible solvent, such as light petroleum is used for the extraction of fixed and essential oils, steroids and aglycones. Chloroform and ether are used for the separation of alkaloids and quinines. The extraction of organic bases like alkaloids usually necessitates basification of plant material if a water immiscible solvent is to be used while for aromatic acids and phenols, acidification may be required. The glycosides are soluble in water and alcohol, but insoluble in non-polar solvents. Tannins are phenolic matter soluble in water, alcohol and ethyl acetate. Extraction itself may be performed by repeated maceration with agitation, percolation or by continuous extraction using Soxhlet extractor.

Preliminary Phytochemical Screening

The plant is a biosynthetic laboratory, not only for chemical compounds such as carbohydrates, proteins and lipids that are utilized as food by man, but also for a multitude of compounds like glycosides, alkaloids, volatile oils, tannins, etc., that exert a physiological and therapeutic effect. The compounds that are responsible for medicinal property of the drug are usually secondary metabolites. A systematic study of a crude drug embraces, thorough consideration of primary and secondary metabolites derived as a result of plant metabolism. The plant material is subjected to preliminary phytochemicals screening for the detection of various plant constituents on following lines.

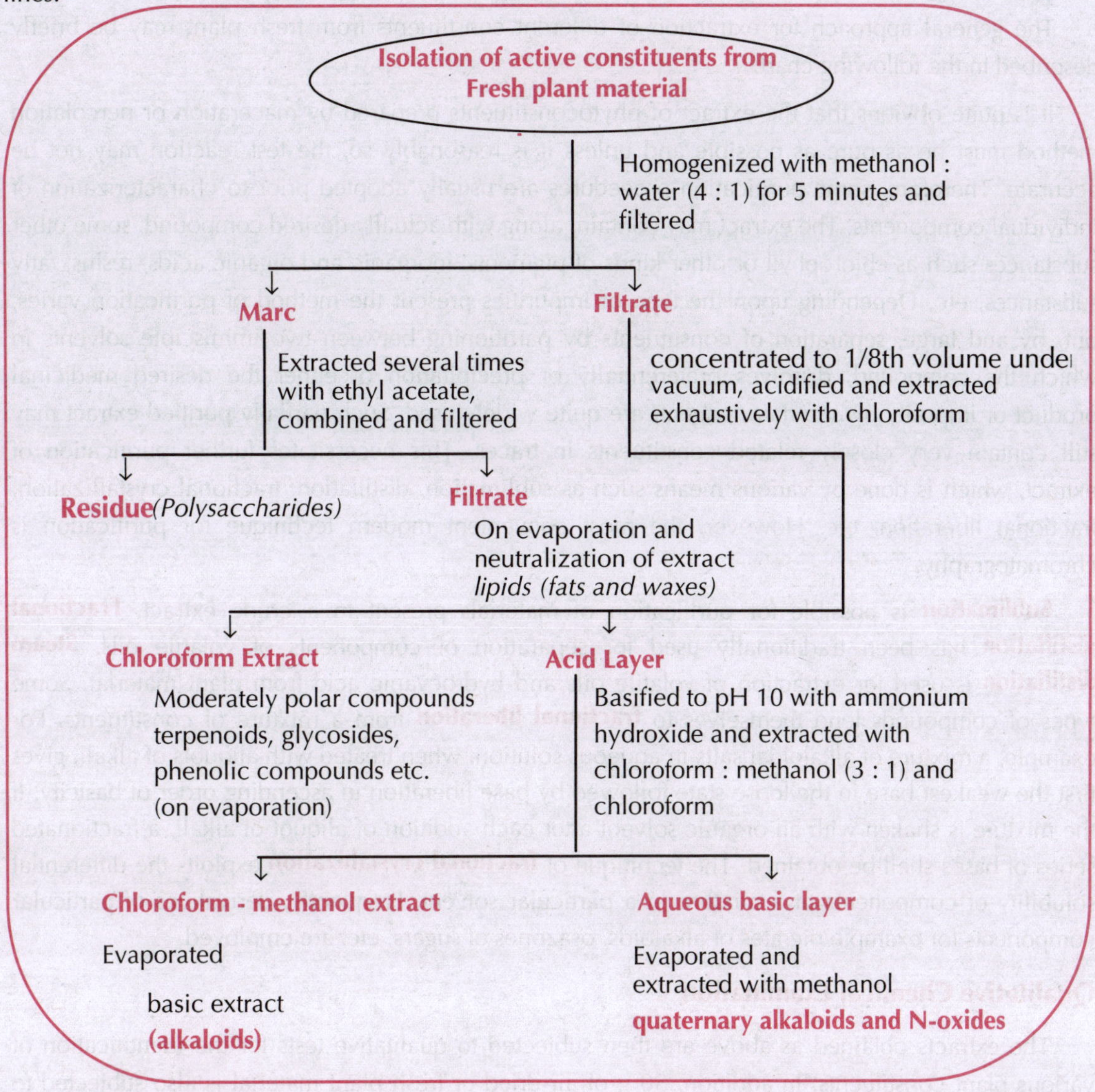

Successive solvent extraction: The air-dried powdered plant material is extracted in Soxhlet assembly successively with petroleum ether, benzene, solvent ether, chloroform, acetone, ethanol and methanol. Finally, the drug is macerated with chloroform water. Each time before extracting with the next solvent, the powdered material is dried in hot-air oven below 50° C.

Each extract is concentrated by distilling off the solvent and then evaporating to dryness on water-bath. The extract obtained with each solvent is weighed. Its percentage is calculated in terms of air-dried weight of plant material. The colour and consistency of the extract are noted.

The extracts with different solvents can also be prepared by successively macerating (cold extraction) the powdered drug in order of increasing polarity.

The general approach for extraction of different constituents from fresh plant may be briefly described in the following chart.

It is quite obvious that the extract of phytoconstituents prepared by maceration or percolation method must be as pure as possible and unless it is reasonably so, the test reaction may not be accurate. Therefore, some purification procedures are usually adopted prior to characterization of individual components. The extract may contain, along with actually desired compound, some other substances such as chlorophyll or other kinds of pigments, inorganic and organic acids, resins, fatty substances, etc. Depending upon the type of impurities present the method of purification varies, but, by and large, separation of constituents by partitioning between two immiscible solvents in which the compound, dissolves preferentially or precipitation of either the desired medicinal product or impurity by a certain reagent, are quite widely used. Such partially purified extract may still contain very closely related constituents in traces. This necessitates further purification of extract, which is done by various means such as sublimation, distillation, fractional crystallization, fractional liberation, etc. However, the most convenient modern technique for purification is chromatography.

Sublimation is possible for purification of materials present in a crude extract. **Fractional distillation** has been traditionally used for separation of components of volatile oils. **Steam distillation** is used for extraction of volatile oils and hydrocyanic acid from plant material. Some types of compounds lend themselves to **fractional liberation** from a mixture of constituents. For example, a mixture of alkaloidal salts in aqueous solution, when treated with aliquots of alkali, gives first the weakest base in the loose state followed by base liberation in ascending order of basicity. If the mixture is shaken with an organic solvent after each addition of aliquot of alkali, a fractionated series of bases shall be obtained. The technique of **fractional crystallization** exploits the differential solubility of components of a mixture in a particular solvent. Frequently, derivatives of particular components for example picrates of alkaloids, osazones of sugars, etc. are employed.

Qualitative Chemical Examination

The extracts obtained as above are then subjected to qualitative tests for the identification of various plant constituents. In addition, 50 g of air-dried or fresh plant material is also subjected to hydro-distillation to detect the presence of volatile oil.

1. Detection of alkaloids: The small portions of solvent free chloroform, alcoholic and water extracts are stirred separately with a few drops of dilute hydrochloric acid and filtered. The filtrate may be tested carefully with various alkaloidal reagents; Mayer's reagent (cream precipitate),

Dragendorff's reagent (orange brown precipitate), Hager's reagent (yellow precipitate) and Wagner's reagent (reddish-brown precipitate).

2. Detection of carbohydrates and glycosides:

(a) Small quantities (200 mg) of alcoholic and aqueous extracts are dissolved separately in 5 ml of distilled water and filtered. The filtrate may be subjected to Molisch's test to detect the presence of carbohydrates.

(b) Another small portion of extract is hydrolyzed with dilute hydrochloric acid for few hours in water-bath and is subjected to Liebermann-Burchard's, Legasl's and Borntrager's tests to detect presence of different glycosides.

(c) A small portion of extract is dissolved in water and treated with Fehling's, Barfoed's and Benedict's reagents to detect presence of different sugars.

3. Detection of phytosterols: The petroleum ether, acetone and alcoholic extracts are refluxed separately with solution of alcoholic potassium hydroxide till complete saponification takes place. The saponification mixture is diluted with distilled water and extracted with ether. The ethereal extract is evaporated and the residue (unsaponifiable matter) is subjected to Liebermann's and Burchard's tests.

4. Detection of fixed oils and fats: A small quantity of petroleum ether and benzene extracts is pressed separately between two filter papers. Oil stains on the paper indicate the presence of fixed oil.

A few drops of 0.5 N alcoholic potassium hydroxide is added to a small quantity of petroleum ether or benzene extract along with a drop of phenolphthalein. The mixture is heated on water bath for 1-2 hours. Formation of soap or partial neutralisation of alkali indicates the presence of fixed oils and fats.

5. Detection of saponins: About 1 ml of alcoholic and aqueous extracts is diluted separately with distilled water to 20 ml and shaken in a graduated cylinder for 15 minutes. One cm layer of foam indicates presence of saponins. The test solution may be subjected to test for haemolysis.

6. Detection of phenolic compounds and tannins: Small quantities of alcoholic and aqueous extracts in water are tested for the presence of phenolic compounds and tannins with dilute ferric chloride solution (5 per cent), 1 per cent solution of gelatin containing 10 per cent sodium chloride, 10% lead acetate and aqueous bromine solutions.

7. Detection of proteins and free amino acids: Small quantities of alcoholic and aqueous extracts are dissolved in a few ml of water and subjected to Millen's, Biuret and Ninhydrin tests.

8. Detection of gums and mucilages: About 10 ml of aqueous extract is added to 25 ml of absolute alcohol with constant stirring. The precipitate is dried in air. Then the precipitate is examined for its swelling properties and for the presence of carbohydrates.

9. Detection of volatile oil: About 50 g of powdered material is taken in a volatile oil estimation apparatus and subjected to hydro-distillation for the detection of volatile oil. The distillate is collected in the graduated tube of the assembly in which the aqueous portion is automatically separated from the volatile oil, if it is present in the drug, and returned back to the distillation flask.

4. PHYSICAL EVALUATION

Physical standards are to be determined for drugs, wherever possible. These are rarely constant for crude drugs, but may help in evaluation, specifically with reference to moisture content, specific gravity, density optical rotation, refractive index, melting point; viscosity, and solubility in different solvents. A few of them are described below.

(i) Moisture content: The percentage of active chemical constituents in crude drugs is mentioned on air-dried basis. Hence, the moisture content of a drug should be determined and should also be controlled. The moisture content of a drug should be minimized to prevent decomposition of crude drugs either due to chemical change or microbial contamination.

The moisture content is determined by heating a drug at 105°C in an oven to a constant weight (*Table 7.4*). For the drug containing volatile active constituents, the toluene distillation method is followed.

Table 7.4: Crude Drugs with limit for Moisture Content

Drugs	Moisture Content (per cent)	w/w
Aloes	not more than	10
Digitalis	not more than	05
Ergot	not more than	08
Acacia	not more than	15
Starch	not more than	15

(ii) Viscosity: Viscosity of a liquid is constant at a given temperature and is an index of its composition. Hence it can be used as a means of standardizing liquid drugs.

The following are the suitable examples.

Liquid paraffin — Kinematic viscosity not less than 64 centistokes at 37.8°.

Pyroxylin — Kinematic viscosity, 1100 - 2450 centistokes

(iii) Melting point: It is one of the parameters to judge the purity of crude drugs. In case of pure chemicals or phytochemicals, melting points are very sharp and constant. Since the crude drugs from animal or plant origin contain the mixed chemicals, they are described with certain range of melting point. The purity of the following crude drugs can be ascertained by determining their melting points in the range shown against each of them. (Table 7.5)

Table 7.5: Melting Point Range for a few Crude Drugs

Drugs	Melting Point (°C)
Colophony	75 - 85
Kokum butter	39 - 42
Cocoa butter	30 - 33
Bees wax	62 - 65
Wool fat	34 - 44

(iv) Solubility: The presence of adulterant in a drug could be indicated by solubility studies. Castor oil is soluble only in three volumes of 90 per cent alcohol, while the adulterated form may show good solubility in alcohol. Balsam of Peru is soluble in chloral hydrate solution and colophony is freely soluble in light petroleum. Asafoetida is soluble in carbon disulphide. Alkaloidal bases are soluble in chloroform, while alkaloidal salts are soluble in polar solvent. The glycosides are extractable with alcohol and water, while their aglycones moieties are soluble in non-polar solvents like benzene or solvent ether.

(v) Specific Gravity: It is also taken into consideration while evaluating the drugs. Variation prescribed value indicates that it is not up to the standard. If anyone of them do not pass the prescribed criteria it clearly indicate the drug is substanded or adulterated.

(vi) Optical rotation: Certain substances are found to have the property of rotating the plane of polarized light in the pure state or in the solution. Thus, they are described to be optically active and this property is known as optical rotation. Plane of polarized light may be rotated towards right (dextro rotatory) or towards left (laevo rotatory). Normally, the optical rotation is determined at 25°C using sodium lamp as the source of light.

Following is the list of few drugs of pharmacognostic origin with their optical rotation.

Table 7.6: Optical Rotation Values for a few Crude Drugs

Drugs	Angles of optical rotation		
Caraway oil	+ 75°	to	+ 80°
Castor oil	3.5°	to	+ 6.0°
Clove oil	0°	to	–1.5°
Honey	+3°	to	–15°
Eucalyptus oil	0°	to	+ 10°
Chinopodium oil	–30°	to	–8°

(vii) Refractive index: When a ray of light passes from one medium to another of different density, it is bent from original path. Thus, the ratio of the velocity of light in vacuum to its velocity in the substance is termed as refractive index of the second medium. Depending upon purity, it is constant for a liquid and can be considered as one of the criteria for its standardization. Refractive index of a compound varies with the wavelength of the incident light, temperature and pressure. Refractive indices of the following compounds are for sodium light and at a temperature of 25°C.

Table 7.7: Refractive Indices of Some Phytoconstituents

Drugs	Refractive Index
Arachis oil	1.4678 to 1.470
Caraway oil	1.4838 to 1.4858
Castor oil	1.4758 to 1.527
Clove oil	1.527 to 1.535

(viii) Ash Values and Extractives: As mentioned earlier, evaluation of drug basically needs its identification and can be done by morphological or microscopic characters. Many a time, even if the drug is identified it is of substandard quality due to either faulty collection or incorrect storage. Thus, to prove its acceptability as a drug, the following tests can be applied to it, wherever possible.

(a) Ash content

The residue remaining after incineration is the ash content of the drug, which simply represents inorganic salts, naturally occurring in drug or adhering to it or deliberately added to it, as a form of adulteration.

Physiological ash: Total ash of the drug is inclusive of 'Physiological ash' as well as non-physiological ash. Physiological ash is derived from the plant tissues119 while non-physiological ash consists of residue of the extraneous matter; sand, soil, etc. adhering to the herb itself. Many a time, the crude drugs are admixed with various mineral substances like sand, soil, calcium oxalate, chalk powder or other drugs with different inorganic contents. For determining ash, the powdered drug is incinerated to burn out all organic matter. Ash value is a criterion to judge the identity or purity of crude drugs (Table 7.8). Total ash usually consists of carbonates, oxides, phosphates, silicates and silica.

Table 7.8: Crude drugs with their Ash Contents

Drugs	Total Ash (% w/w)	Acid insoluble ash % (w/w)
Agar	–	1.00
Aloes	05.0	–
Ashoka	11.0	–
Bael	03.5	–

Contd...

Belladonna	–	3.00
Black catechu	06.0	–
Cannabis	15.0	5.00
Cardamom	06.0	3.50
Clove	07.0	0.75
Gelatin	03.6	–
Ginger	06.0	1.7 (Water soluble ash)
Valerian	12.0	–

Acid-insoluble ash, which is a part total ash insoluble in dilute hydrochloric acid, is also recommended for certain drugs. Adhering dirt and sand may be determined by acid-insoluble ash content.

(b) Extractives

The extracts obtained by exhausting crude drugs are indicative of approximate measures of their chemical constituents. Taking into consideration the diversity in chemical nature and properties of contents of drugs, various solvents are used for determination of extractives. The solvent used for extraction is in a position to dissolve appreciable quantities of substances desired.

1. Water-soluble extractives: This method is applied to drugs which contain water-soluble active constituents of crude drugs, such as tannins, sugars, plant acids, mucilage, glycosides, etc. (*Table 7.9*)

Table 7.9: Water-soluble Extractive Values of some Crude Drugs

Drugs	Water-soluble extractive (%w/w)	
Aloe	not less than	25.0
Glycyrrhiza	not less than	20.0
Linseed	not less than	15.0
Senna leaves	not less than	30.0
Senna pods	not less than	28.0
Ginger	not less than	10.0

2. Alcohol (ethanol)-soluble extractives: Alcohol is an ideal solvent for extraction of various chemicals like tannins, resins, etc. Therefore, this method is frequently employed to determine the approximate resin content of a drug. It is also used as an official method for assay in case of myrrh and asafoetida. Generally, 95 per cent ethyl alcohol is used for determination of alcohol-soluble extractive. In some cases, diluted alcohol may also be used, depending upon solubility of the constituents of crude drugs.

For following few drugs, the limit for alcohol soluble extractive is applicable.

Table 7.10: Alcohol-soluble Extractive Values of some Crude Drugs

Drugs	Alcohol-soluble extractive (%w/w)	
Aloe	not less than	10.0
Benzoin	not less than	90.0 (Siam benzoin)
		75.0 (Sumatra benzoin)
Asafoetida (90 % alcohol)	not less than	50.0
Ginger (90% alcohol)	not less than	04.5
Valerian (60% alcohol)	not less than	30.0
	Alcohol Insoluble Extractives	
Myrrh	not less than	70.0
Benzoin	not less than	24.0

3. Ether-soluble extractives: The types of ether-soluble extractive values determined for evaluation of crude drugs are volatile and non-volatile ether soluble extractives.

The volatile ether-soluble extractive represents volatile oil content of the drug while non-volatile ether-soluble extractives represent resin, fixed oils or colouring matter present in drugs. The limit for these extractives is applicable to following drugs.

Table 7.11: Non-volatile Ether soluble extractive values of some drugs

Drugs	Limit for non-volatile ether soluble extractives (% w/w)	
Capsicum	not less than	12.0
Male fern	not less than	01.5
Linseed	not less than	25.0
Nutmeg	not less than	25.0

(viii) Volatile Oil Content: Pharmaceutical significance of aromatic drugs is due to their odorous principles i.e. volatile oils. Such crude drugs are standardized on the basis of their volatile oil content. Some examples are quoted below.

Table 7.12: Volatile oil content of few crude drugs

Drugs	Volatile oil Content (%w/w)	
Caraway	not less than	2.5
Fresh lemon peel	not less than	2.5
Clove	not less than	15.0
Fennel	not less than	1.4
Dill	not less than	2.5
Cardamom seed	not less than	4.0

(ix) Foreign Organic Matter: The parts of the organ or organs other than those named in the definition and description of the drug are defined as foreign organic matter.

The maximum limit for the foreign organic matter is defined in the monograph of crude-drugs. If it exceeds the limits, quality of the drug deteriorates. The limit for foreign organic matter is specially mentioned for natural drugs of vegetable origin in their respective monographs.

(x) Test for microbial extermination is mandatory for all natural drugs.

PESTICIDAL RESIDUE:

Crude drugs (Herbal drugs) come in contact with various pesticides at least at three stages:

(1) During **Treatment of soil** which is done for disinfection of soil, before cultivation.

(2) **Treatment of pests,** during the normal growth of herbs (to get herbal drugs from them), to control various diseases which the herbs come across while growing.

(3) **Fumigation** during storage and preservation of herbal drugs.

Herbal drugs are used for treatment of various diseases for long period. Since the pesticides have cumulative effect, WHO (World Health Organisation) and FAO (Food and Agriculture Organisation) have laid down the limits for pesticideal residue in foods and animal feeds.

The same limits have also been recommended for herbal drugs. It s known that pesticides such as Chlorinated hydrocarbons and their related compounds as well as organo phosphorus compounds have long residual action as compared to other types of pesticides, therefore presence of organically bound chlorine and phosphorous is determined to assure their quality.

Chlorinated hydrocarbons are lipophilic in action and have affinity for fatty animal tissues, where as organophosphates effect nervous system adversely.

It is known that pesticides such as chlorinated hydrocarbons and their related compounds as well as organophosphorus compounds have long residual action as compared to other types of pesticides, therefore presence of organically bound chlorine and phosphrous is determined to assure their quality.

Following are the few examples of chlorinated hydrocarbon compounds and organophosphorus compounds respectively.

(a) Chlorinated hydrocarbon compounds:

DDT (Dichloro Diphenyl Trichloro ethane)

BHC (Benzene Hexachloride)

Alderin

Dialderin

Endrine

Endosulphan

Pentachloro phenol

Chlordane

(b) Organophosphorous Compounds

Malathion

Parathion

Methyl parathion

Fenitrothion

Carbamates

The physical or physico-chemical parameters listed above are useful in establishing quality profile of a crude drug, and constitute an important component of qualitative evaluation.

The utility of chromatographic techniques and instrumentation in establishing qualitative as well as quantitative profiles of crude drug constituents is well understood and systematically practiced in Quality Control and Quality Assurance laboratories of herbal drug based industries. The instrumental analysis of bio-constituents is reliable, quick, reproducible and more convincing.

The different chromatographic techniques and instrumental methods of analysis are also physical methods of evaluation. They are described in this chapter.

CHROMATOGRAPHIC TECHNIQUES

Among the various methods of separating plant constituents, the chromatographic procedure originated by Tswett is one of the most commonly used techniques of general application.

In 1944, Consden, Gordon and Martin introduced a method of partition chromatography using strip of filter paper for the analysis of amino acid mixture.

Chromatography represents a group of methods for separating molecular mixtures that depend on the differential affinities of the solute between two immiscible phases. One of the phases is a fixed bed of large surface area which is called a stationary phase while the other is a fluid or gas which moves through or over the surface is called the mobile phase. The stationary phase may be a porous or finely divided solid or liquid that has been coated as thin layer on an inert support material. The mobile phase may be a pure liquid or a mixture of solutions or it may be a gas or mixture of gases.

For the separation of some bioconstituents, it is necessary to use a two-dimensional chromatography using two different solvent systems. The resolved components of original mixture can be separately eluted from chromatogram by treating the cut-out spots with a suitable solvent and then determined quantitatively by some suitable instrumental method of analysis; fluorescence analysis, colorimetric, or ultra-violet absorption.

(a) Thin Layer Chromatography (TLC)

In 1958, Stahl demonstrated application of TLC in analysis, a method based on adsorption chromatography. It is at present an important analytical tool for qualitative and quantitative analysis of a number of natural products. The adsorbent such as silica gel G or C is coated to a thickness of 0.3 mm on clean TLC plates using commercial spreader; the plates are activated at 105° for 30 minutes and used. The selection of mobile phase depends upon type of constituents to be analyzed.

After the development of chromatogram by ascending technique, the resolved spots are revealed by spraying suitable detecting agent. TLC has certain advantages over paper chromatography. Fractions can be affected more rapidly with smaller quantities of mixture, spots are usually more compact and drastic reagent such as concentrated sulphuric acid can be used as spraying solutions. The TLC technique is useful in analysis of alkaloids, glycosides, isoprenoids, lipid components, sugars and derivatives and practically all bioconstituents. The Rf values (distance travelled by component / solvent front) may vary depending upon purity of solvent, nature of substance to be resolved, composition of solvent, presence of impurities, adsorbent used, polarity of the solvent, substance and absorbent etc.

TLC and HPTLC techniques, now a days, are important analytical tools for micro-analytical separation and determination of natural products. They have following advantages.

- simple to operate, economical, rapid (particularly HPTLC)
- always available for use (precolated plates are commercially available)
- method of detection does not place any restriction on choice of fluent, easily inspected.
- neutral, basic, acidic or purely aqueous eluent can be employed. The whole chromatographic system is flexible.

TLC can be exploited in the investigation and cultivation of medicinal plants. It is possible to run many samples of extracts from different chemical races simultaneously with authentic standards and high performance individuals can be recognized, selected and bred. This technique is also valuable for discovery of so-called chemical races.

(b) High Performance Thin Layer Chromatography (HPTLC)

It is very useful in qualitative and quantitative analysis of pharmaceuticals. It enjoys a practical application status, as it combines the art of chromatography with quickness at a moderate cost. HPTLC is a major advancement of TLC principle requiring shorter time and better resolution. The basic difference between conventional TLC and HPTLC is only in particle and pore size of the sorbents.

The plates are similar to conventional TLC plates. Layers of HPTLC are available in the form of precolates. Silica gel of very fine particle size is widely used as sorbent in HPTLC. The use of smaller particle size helps in greater resolution and sensitivity. About 3 - 6 cm solvent front migration is sufficient to effect proper separation. Whitman HPTLC plates are produced from 4 to 5 μm silica gel with an inert binder to form a 200 μm layer. About 7 cm development distance is achieved in about 4 minutes. Desaga also supplies HPTLC kit with 100 μl capillaries and plates.

Sample preparation in HPTLC needs a high concentrated solution, as very less amount of sample need be applied. The size of the sample spot must not exceed 1 mm in diameter. There are different techniques for the spotting of samples. One of them is self loading capillaries in which small volume of samples may be applied to the HPTLC plate surface using platinum-iridium tubing fused into the end of a length of glass tubing. Contact spotting, preadsorbent or concentrating zones, chemical focussing and programmed multiple developments are the other methods of spotting the samples.

The linear development method is most familiar technique in HPTLC. Here, the plate is placed vertically in solvent system in a suitable container. The solvent is usually fed by capillary action and chromatogram can be developed from both sides. Circular development, anti-circular device and multiple developments are some of the other methods which are used for development of plates.

HPTLC is rapidly gaining importance in biochemistry of natural products and in analysis of biofluids in the field of pharmacokinetics. The analytical profiles for cardenoloids, tropane, alkaloids, flavonoids, steroidal compounds, anthracene aglycones, lipids etc. have been developed using the technique.

TLC / HPTLC techniques are conveniently applied for compilation of profiles pertaining to varied range of bioconstituents such as berberine, quinine and related alkaloids, opium alkaloids, colchicines, chelidonine; sanguinarine serpentine, raubasine; asarone, elemicin, eugenol; thymol, coumarins, pulegon; sarsapogenin, lanatosides, digitoxigenin; purpurea glycoside A, flavones, diosgenin; silymarin, silybin, glycyrrhizin; robinin, aesculetin, polygallic acid; ginsenosides catechin, anthraquinone derivatives; valtrate, rotenone, aristolchic acids; gibberellins, antibiotics and number of other compounds of natural origin.

HPTLC is now-a-days applied to obtain *"Fingerprint"* patterns of herbal formulations, quantification of active ingredients and also detection of adulteration. The solvent systems (mobile phases) proposed for separation of certain herbal ingredients have been tabulated here.

Component	Mobile Phase
1. Amino acids, alkaloids	butanol - acetic acid - water (4: 5: 1)
2. Cardiac glycosides	Dichloromethane - methanol - formamide (8: 2: 1)
3. Flavonoids, coumarins glycosides	Ethyl acetate - methylethyl ketone - acetic acid - water (5: 3: 1: 1)
4. Saponins	Chloroform - methanol - water (7: 4: 1)
5. Terpene, essential oils, sterols	Hexane - acetone (9: 1)

HPTLC methods used for detection / estimation of some herbal constituents are outlined here.

1. 18 β - Glycyrrhetinic acid:

- Plant product : Liquorice formulation
- Stationary phase (S.P.) : Silica gel
- Mobile Phase (M.P.) : Ethylacetate - methanol - ammonia (10: 3: 1)
- Quantification : UV absorbance (in densitometry): 260 nm.

2. Panaxadiol and panaxatriol:

- Source : Marked formulation of ginseng.
- S.P. : Silica
- M.P. : Chloroform - ether (1: 1)

- Detection : Spraying with 10 per cent sulphuric acid in methanol, heating at 105°C for 10 minutes
- Quantification : UV absorbance (in densitometry): 544 nm and 52 nm.

3. **Flavonol glycosides:**
   - Source : Gingko biloba leaf extract.
   - S.P. : Silica gel
   - M.P. : Chloroform - benzene - ethanol - acetic acid - water (11: 4: 2: 1: 2)
   - Detection : Spraying with 8 per cent $AlCl_3$ in ethanol.
   - Quantification : UV absorbance 370 nm in densitometry.

4. **Carvone:**
   - Source : Extract of *Cuminum cyminum*.
   - S.P. : Silica gel
   - M.P. : Chloroform - acetone (100: 2)
   - Detection : By dipping in anisaldehyde sulphuric acid reagent, heating at 80°C for about 10 minutes.
   - Quantification : UV absorbance in densitometry at 410 nm.

5. **Cholesterol:**
   - Source : Bear gall-bladder powder.
   - S.P. : Silica gel
   - M.P. : Ethyl acetate - acetone - pet. ether (2: 1: 11)
   - Detection : Spraying with 10 per cent sulphuric acid in alcohol, heating at 100°C for 5 minutes.
   - Quantification : UV absorbance in densitometry at 400 nm.

6. **Aloin:**
   - Source : Tincture of *Aloe vera*.
   - S.P. : Silica gel
   - M.P. : Ethyl acetate - formic acid - water (17: 2: 3)
   - Quantification : UV absorbance in densitometry at 350 nm.

(c) Gas Liquid Chromatography (GLC)

It was introduced by James and Martin in 1952, based on the suggestion of Martin and Synge. Gas chromatography separates volatile substances by percolating a gas stream over a stationary phase. The basis of separation in GLC is the partitioning of the sample in and out of the film of liquid spread over an inert solid. GLC is the most selective and versatile form of gas chromatography since there is a wide range of liquid phases usable up to 450°C. The choice of the carrier gas shall determine the efficiency of the GLC system. A compromise among inertness, efficiency and operating cost makes nitrogen or helium the most common carrier gas.

In pharmaceutical analysis, GLC is applied to three types of problems; (a) Assay of the starting materials and drug substances, (b) Quantification of drug in formulations, and (c) Assay of impurities and / or solvents in the raw material or in drug substances. The important applications of GLC include examination of many volatile oils, plant acids, alkaloids of opium, tobacco, conium and belladonna, the resins of cannabis; steroidal compounds, cardioactive glycosides, aglycones, sugars and amino acids, etc.

Many drugs which contain carboxylic acid or primary amine functional groups give badly tailed peaks due to interactions of the functional groups with the stationary phase. To overcome the problem faced by these compounds, special techniques such as pyrolysis GC, photolysis GC and derivatisation have been developed. In pyrolysis GC, the sample for analysis is heated to a very high temperature in an extremely short time and the fragments of the thermally destroyed samples are swept into the gas chromatograph.

Modification of the sample structure may be accomplished by derivatisation and the most widely applied general derivatisation is silylation. The compounds containing labile functional groups namely alcohols, amines, carboxylic acids and thiols are converted into corresponding ethers, silylamines, esters and thioesters. The common derivatising agents used are trimethylchlorosilane, hexamethyldisilazone and N. O-bis-trimethylacetamide.

(d) High Performance Liquid Chromatography (HPLC)

The term liquid chromatography is used to refer to those methods in which the separation takes place with a packed column. The packing material is the stationary phase and it may be solid or a liquid coated over an inert material. A liquid mobile phase is used as the eluent. In HPLC, the mobile phase is forced through the column under high pressure either with autocratic elution or gradient elution.

It has become the most versatile, safest, dependable, fastest and sensitive chromatographic technique for the quality control of drug components. The drugs like morphine, papaverine, codeine, emetine, antibiotics, steroids, ergot alkaloids, cardiac glycosides, sennoside, capsaicin, vitamins, and rhubarb constituents etc. are analyzed by HPLC. Digoxin and gitoxin is analyzed using RP column, 65 per cent methanol.

HPLC is beginning to make inroads into the officially accepted procedures. HPLC is being used in the laboratories of many pharmaceutical companies for routine assay of new drugs, as well as substitutes for older, more troublesome assays for marketed drugs. It is also very widely used in the studies of drug levels in biological fluids.

Derivatisation in HPLC is undertaken primarily to increase the sensitivity of detection for a given compound. A compound can be derivatised before or after the column separation. Precolumn derivatisation is most widely used because it allows the greatest flexibility in the choice of the reactions to be used. The most popular derivatives are those which introduce a UV or visible chromospheres into the molecule.

A number of commercial HPLC instruments are available with various modifications and advancements. In microbore chromatography, the column used is very narrow (1 mm or less). Flow rate of the mobile phase is also very less (100 μl min). This method is useful for regular analysis of drugs in the pharmaceutical industry and it saves the amount of solvents used as mobile phase. In supercritical fluid chromatography, the mobile phase is held at conditions of temperature and pressure such that it has its viscosity and diffusibility midway between gas and liquid form. The advantage of supercritical fluid chromatography is high column efficiency and short time taken for the analysis.

(e) Column Chromatography

Basically, it is a liquid chromatography in which mobile phase in the form of liquid passes over the stationary phase packed in a column. The column is either a glass or metallic column. The column adsorption chromatography is the oldest one and still practiced currently for extraction purposes. Column chromatography has been derivatised into other forms like gel permeation, ion exchange, affinity and column partition.

In the column adsorption chromatography, fairly large number of adsorbents is used like starch, calcium carbonate, magnesia; lime, silica gel, alumina; charcoal and fueller's earth. To optimize the resolution, various mobile phases used, either singly or in combination, are petroleum ether, cyclohexane, chloroform; acetone, water, pyridine and organic acids. For the better separation, it is essential to consider the polarity of the sample, adsorbent and mobile phase.

(f) Gel Permeation Chromatography (Gel Filtration, Molecular Sieves)

It is also called as Size Exclusion Chromatography. The separation occurs not on the basis of adsorption or partition, but on the effective size of solutes present in the solution. For the separation purpose, the stationary phases used are cross-linked polymers which give an open network with a large number of pores of fairly uniform size. During the flow of mobile phase through such stationary phase, the very large sized molecules cannot enter the pores and hence get "excluded" and continue to travel further along with mobile phase. But, the other molecules with low molecular size enter freely into different pores and hence find a longer path through the column. During the elution, the largest molecules in the mobile phase elute first, followed by the molecules with decreasing order in sizes. This phenomenon is observed because the extent, to which the molecules are retained, depends on the size of the included molecules relative to the size of the pores.

The various types of gels used as stationary phases are either soft gels or semi-rigid or rigid gels. Commercially available soft gels are generally cross-linked carbohydrate materials like agarose (sepharose), dextran (sephadex) and polyacrylamide (Bio-Gel). The semi-rigid or rigid gels available are alkylated dextran, cross-linked polystyrene and controlled-porosity glass beads. Soft gels are used with primarily aqueous mobile phases and the process is called as gel filtration. The other gels mentioned above are used along with non-aqueous mobile phases like acetone, pyridine, chloroform, etc., and here the process is called as gel permeation.

Gel filtration is employed for the separation of large bimolecular like proteins and polypeptides. It is also used for desalting protein solutions, determination of molecular size, studies of plasma-binding of drugs, etc.

(g) Affinity Chromatography

This technique is mainly used for separation of proteins, peptides, enzymes; antigens and antibodies, nucleic acids, ribosomes, etc. The adsorbent used is one of the biological substance (which may be termed as receptor) having a specific affinity for other substance. These two substances are the biologically interacting pairs. Such adsorbent is attached to a porous stationary phase and placed in a column. When a mixture containing the other complement of the adsorbent (interacting pair) is passed through the stationary phase, selective separation occurs. It is designed as a reversible adsorption process and hence during elution, the complementary part adsorbed is collected in pure form by dissociating the interacting pair with the help of changing the ionic strength or pH of the column buffer. Affinity chromatography is efficiently applied for separation of enzymes, utilizing the technique of enzyme substrate interaction. In the place of substrate, the ligand which is a competitive inhibitor for enzyme is used. Such ligand is attached to matrix by interlacing a long chained hydrocarbon in between them. The hydrocarbon termed as "arm" enhances the access of inhibitor towards enzymes. The arms used are aminophenyl, sulfhydryl, carboxylic acid or aminoalkyl substituted compounds.

This technique is applied for the separation of m-RNA, purification of enzymes, isolation of antibodies, etc.

SPECTROPHOTOMETRIC METHODS

(i) Ultra-violet and Visible Spectrophotometry

Ultraviolet (UV) and visible absorption techniques encompass analytical methods based upon measurement of light absorption by substances in the wavelength region from 190 - 900 nm. The region from 190 to 380 nm is known as the UV region and from 380 - 900 nm, the visible region of the spectrum. Absorption in the UV-visible region arises from electronic transitions within the molecule.

A variety of instruments are available for measuring light absorption in the UV-visible region of the spectrum. In a single beam instrument, light passes through a monochromator and then through the sample and into the detector. A dual beam instrument measures ratio of the intensity of beam coming through sample cell to a second beam which does not pass through the sample. A rapid scanning spectrophotometer employs multichannel detectors. Differential spectroscopy is a method of compensating for the presence of extraneous materials in a sample, which would otherwise interfere with the spectrum of the drug being determined. In dual wavelength Spectrophotometry, two monochromatic beams at different wavelengths are passed through the same sample. One wavelength is generally characteristic of the drug, while the other is carefully selected so that the absorbance is equivalent to the level of absorptive interference anticipated at the analytical wavelength. Thus, the second radiation beam is analogous to the reference cell employed in conventional difference Spectrophotometry. In derivative spectroscopy, the absorbance of a sample is differentiated with respect to wavelength to generate the first, second or higher order derivatives. This method has found significant application in forensic and bio-medical analysis.

The use of a separation step followed by measurement of absorption in UV-visible region is the most widely used analytical procedure in pharmaceutical analysis today. Often, the UV spectrum serves as a confirmatory evidence of identity in support of other analytical data and is also used as detector in HPLC.

A variety of natural products of pharmaceutical significance are analyzed by spectroscopic technique, e.g. lobeline (249 nm), reserpine (268 nm), morphine (286 nm), cardioactive glycosides (217 nm), colchicines (360 nm), vanillin (301 nm), etc. The phytoconstituents analyzed in visible range are alkaloids of ergot (550 nm using p-dimethylamino benzaldehyde), morphine (442 nm) by nitroso-reaction), anthraquinone (505 nm after treatment with alkali), cardioactive glycosides (590 nm by Keller-Kiliani reaction), cyanogenetic glycosides (630 nm by pyridine pyrazolone colour reaction), etc.

Sample of dry powdered drug when treated with 1 N normal sodium hydroxide in methanol, (a), 1 N HCl (b) and Nitric acid diluted equal vol. of water (c) and exposed to ultra-violet radiation develops colour as indicated in the columns a, b and c, against each of the following crude drugs, respectively.

Table 7.13: UV Radiation Tests for Natural Drugs

		Testing solutions		
No.	**Crude-Drugs**	**(a)**	**(b)**	**(c)**
(1)	Aconite	Dull blue green	Dull green	Very dark green
(2)	Almond	Dull blue green	Blue green	No colour
(3)	Anise	Light bright blue	Reddish brown	Dull greenish brown
(4)	Aspidium	Dark greenish brown	Brown	No colour
(5)	Berberis	Bright greenish yellow	Bright greenish yellow	Bright greenish yellow
(6)	Black-pepper	Blue	Brownish green	Colourless
(7)	Cantharis	Greenish brown	Greenish blue	Greenish blue
(8)	Capsicum	Yellowish orange	Brownish green	Blue green
(9)	Caraway	Brownish green	Reddish brown	Very dark brown
(10)	Cardamom	Blue	Greenish blue	Bluish green
(11)	Chondrus	Light reddish brown	Light greenish brown	Greenish brown
(12)	Cochineal	Purple green	Bluish red	Dark brown
(13)	Coriander	Bluish green	Reddish blue	Dark yellow

(14) Fennel	Bluish green	Brownish purple	Colourless
(15) Ginger	Bright yellowish green	Yellowish brown	Colourless
(16) Jalap	Light blue	Bluish green	Dark green
(17) Linseed	Greenish blue	Dull greenish blue	Dark greenish yellow
(18) Myristica	Greenish brown	Blue	Colourless
(19) Physostigma	Blue	Light bluish green	Dark green
(20) Pyrethrum	Greenish brown	Brownish blue	Dark brownish green
(21) Senega	Bright green	Purple	Dark greenish brown
(22) Spearmint	Orange brown	Dull purple red	Colourless
(23) Thyme	Brownish green	Reddish purple	Colourless
(24) Ura-ursi	Brown	Reddish purple	Reddish purple
(25) Valerian	Light bluish green	Dull bluish green	Dark green

(ii) Infra Red Spectroscopy

Infra Red (IR) spectroscopy is the study of the reflected, absorbed or transmitted radiant energy in region of electromagnetic spectrum ranging from wavelength 0.8 - 500 nm. A more commonly used measurement is the frequency and is expressed in wave number. The IR spectrum is usually divided into three regions namely - near IR (12500 - 4000 cm^{-1}), mid IR (4000 to 400 cm^{-1}) and far IR (400 - 20 cm^{-1}). Only the mid IR region is usually referred to simply as infra red and is widely used in the analysis of drugs and pharmaceuticals.

Infrared spectrophotometers can be single or double beam instruments. Fourier transform spectrophotometer is the recent advancement in the field of IR spectroscopy, which has a number of advantages over dispersive instruments. They can scan quickly and therefore be used along with gas or liquid chromatography for recording spectra, as compounds are eluted.

The widespread use of IR spectroscopy for the identification of drugs, polymorphic modifications, excipients and raw materials used in pharmaceutical manufacturing is due to its sensitivity and the ease with which spectra can be obtained on any type of sample including insoluble solids, polymers, solutions and gases. Identical IR spectra of two samples (super-imposable spectra) is a very good evidence that the two samples have the same chemical structure. IR spectrometric analysis is a very useful tool in the detection of functional groups of bio-molecules, thus aiding in their structural elucidation. The quantitative analysis of antibiotics, alkaloids, quinine and strychnine (6.2 and 6.06 μ), steroidal sapogenins (11.11 μ) etc. is also possible.

(iii) Fluorescence Analysis

The organic molecules absorb light usually over a specific range of wavelength, and many of them re-emit such radiations. This phenomenon is called as luminescence. When the re-emission of absorbed light lasts only whilst the substance is receiving the exciting rays, the phenomenon is defined as fluorescence.

The drugs like belladonna leaf and root, wild cherry bark, gambier catechu, aloes, jalap, etc. show fluorescence in visible range. The ultraviolet light obtained from mercury vapour lamp or tungsten produce fluorescence in many natural products, which do not visibly fluoresce in day light. Fluorescence lamps are fitted with a suitable filter which eliminates visible radiation from the lamp and transmits ultraviolet radiation of definite wavelength. Under fluorescent light, cinchona bark shows a number of luminous yellow and a few light blue patches. The other drugs evaluated qualitatively are Calumba (yellow, whereas phloem and cambium show dark green fluorescence); Hydrastis rhizome (golden yellow); *Rheum emodi* (brown); *Rheum rhaponticum* (violet), ergot powder (red); olive oil (deep golden yellow); Arachis and sesame oils (blue fluorescence); Berberis (yellow); gentian (light blue); aconitine (light blue); emetine (orange); precipitated chalk (greyish-violet) etc.

The location of resolved components on the developed TLC or paper chromatogram is also possible by observing it under UV lamp detection of menthol by Guruvich method on TLC plates is possible. Quantitative fluorescence analysis using the fluorescence produced by a compound in UV light for quantitative evaluation is possible with the help of an instrument fluorimeter or spectro-fluorimeter. Sennoside, extracts of rhubarb and Berberis, as well as, cinchona alkaloids may be evaluated by this technique.

Table 7.14: Fluorescence characters under UV light

Herbal drugs	Nature of fluorescence
1. Cinchona	Purple blue
2. Gentian root	Whitish blue
3. Ipomoea	Deep purple-violet
4. Quassia	Whitish blue
5. Rhubarb	Violet
6. Wild cherry	Mauve

(iv) Nuclear Magnetic Resonance Spectroscopy (NMR)

NMR is the branch of spectroscopy dealing with the absorption of radio frequency radiation by substances held in a magnetic field. Absorption results from interaction of radiation with magnetic moment of nuclei in the sample and it occurs at different frequencies for nuclei with chemically different environments within a molecule. NMR spectra are normally prepared using minimal resonance signals. Many solvents are available like water, chloroform, acetone, etc. as their deuterated analogues.

NMR has been an extremely important tool for elucidation of molecular structure, especially the stereochemistry and configuration. The technique reveals position of protons in a complex molecule.

It is an excellent choice for identification test in pharmaceutical analysis. In quantitative analysis, it is an absolute technique, in the sense that a reference standard of the substance being analyzed, is not required.

NMR has found many applications in the determination of impurities and minor components in mixtures because of ease, speed and specificity of the analysis. NMR can also be coupled with mass spectrometer.

(v) Mass Spectrometry

Mass spectroscopy is concerned with the electron ionization, subsequent fragmentation of molecules, determination of the mass to charge ratio (m/e) and relative abundances of ions which are produced. Knowing the fragmentation pattern, a possible structure of the original molecule can be suggested.

The most important application of mass spectrometry is in determination of molecular weight of compounds. Mass spectrometry, either alone or in combination with a chromatographic separation, is probably the most effective method for the identification of drug constituents. The complete mass spectra for about 100 drugs, metabolites and other compounds of pharmaceutical interest are reproduced in *Pharmaceutical Mass spectra*, London, published by Pharmaceutical Press London, 1985. Mass spectrometer has been successfully coupled to both gas chromatograph and HPLC.

(vi) X-Ray Difractions

Many compounds are capable of crystallizing in more than one type of crystal lattice i.e. calcium oxalate, sulphur and mica. At a particular temperature and pressure, only one crystalline form is thermodynamically stable. Since, the rate of phase transformation of a metastable polymorph to the stable one can be quite slow, it is common to find several polymorphs of crystalline pharmaceutical compounds existing under normal handling conditions. Solubility, melting point, density, hardness, crystal shape, optical and electrical properties vapour pressure, etc. vary with polymorphic forms. Hence, polymorphism plays an important role in the field of pharmaceutical sciences.

(vii) Radioimmunoassays (RIA)

The popularity of radioimmunoassay is increasing because of high sensitivity, lack of necessity of a preliminary extraction stage and applicability to the analysis of large numbers of samples. This technique uses an antibody specific for the drug being assayed and a labelled form of the same drug. The label may be a particular radioisotope, an active enzyme or a C^{14}, and iodine- (I^{125}) is commonly used isotopes in radioimmunoassay.

Radioimmunoassay is the method of choice for cardiac glycosides, insulin, lysergide, cannabinoids and opiates, where blood concentrations and sample volumes are low. Other applications include screening of drug abuse, investigation of poisoning cases and therapeutic drug monitoring.

It is sufficient to state that even after utilizing classical chemical methods of degradation, phytochemists and pharmacognosists are coming to rely more and more on the use of physical techniques to establish structures of new bioconstituents and to identify known phytoconstituents. Ultraviolet, infrared, mass and nuclear magnetic resonance spectroscopy together with X-ray crystallographic and optical rotatory dispersion methods have all played significant role in the development of natural product chemistry and drug analysis. Many problems of structural elucidation, which 40 years ago were incapable of investigation, either due to paucity of material or because of lack of suitable chemical methods, can now be readily solved in a research laboratory with the aid of sophisticated instrumental techniques.

The compound purified by the chromatographic technique is recrystallized from suitable solvent, the sharp melting point, confirming the purity of the active constituent, is determined and subjected to identification by co-chromatography with the authentic sample, if available. The elemental analysis is carried out and suitable molecular formula is assigned to unknown compound. The isolated compound is subjected to infrared spectroscopic analysis which gives an idea about the functional groups present in the complex molecule. Mass spectrometric analysis gives the fragmentation pattern and molecular weight of the compound whereas nuclear magnetic resonance spectroscopic analysis reveals the position of protons in a complex molecule.

The recent advances in the field of electronics and the flexibility in the use of computers and microprocessors have increased the use of modern instruments in the field of pharmaceutical sciences. The principle remaining the same, the modern techniques have a number of ancillary units attached to the main instrument, which determine the selectivity and accuracy of analysis. Interfacing of two or more instruments on line has also become common and is frequently used to isolate, identify and determine the purity of compound. With the advancement in the field of medicine and a variety of pharmaceutical formulations available today, the modern analytical techniques play a pivotal role in the field of pharmaceutical sciences.

5. BIOLOGICAL EVALUATION

When the estimation of potency of crude drug or its preparation is done by means of its effect on living organisms like bacteria, fungal growth or animal tissue or entire animal, it is known as *bioassay*. This method is generally called for, when standardization is not adequately done by chemical or physical means and also for conformity of therapeutic activity of raw material and finished product.

In other words, bioassay is the measure of sample being tested capable of producing biological effect as that of the standard preparation. Such activity is represented in units known as International Unit (IU). The specific biological activity contained in each IU of the few drugs is mentioned as under:

Digitalis	- 1 IU is contained in 76 mg of standard preparation
Vit. A	- 1 IU is present in 0.344 micrograms of standard preparation
Vit. D	- 1 IU is contained in 0.025 micrograms of standard preparation
Heparin	- 1 IU is present in 7.7 micrograms of standard preparation

Biological assay methods are mainly of three types, (a) toxic, (b) symptomatic, and (c) tissue methods.

In toxic and symptomatic techniques, the animals are used, whereas in tissue method, the effect of a drug is observed on isolated organ or tissue. Among the drugs that are subjected to bioassay are cardiac glycosides, natural pesticides and antibiotics.

(D) BIOLOGICAL TESTING OF HERBAL DRUGS

In standardization/evaluation of herbal drugs, assessment of biological efficacy is found to be most assuming method. Hence, protocols followed for assessment of some types of biological activities have been given here. In all these methods, requirements are a suitable animal model for testing and control, methodology for experiment and assessment of results.

(a) Hepatoprotective Activity

For this purpose, male/female albino rats are used. A number of chemicals, drugs, industrial pollutants or hepatitis virus, ethyl alcohol and immunogenic reactions are known to cause hepatitis.

Most commonly used liver damaging agents are either carbon tetrachloride which causes acute toxicity or alcohol, paracetamol, rifampicin may be used which cause subacute or chronic toxicity. The parameters used to assess liver toxicity are

(a) Physiological (hexobarbital hypnosis.)

(b) Biochemical serum estimation of enzymes like SGPT, SGOT, super oxide dismutase and other factors like blood proteins, cholesterol, triglyceride levels etc.

(c) Histopathological — (liver tissue necrosis and fatty degeneration).

After ensuring the induction of toxicity, drug treatment is given for 10 - 15 days and recovery from damage/toxicity is verified by testing the same parameters.

For testing Hepatoprotective activity, cultured hepatocytes are used in *in-vitro* method. For this purpose, hepatic cells are isolated using collagenase and their viability is determined. Their cultures are prepared and then a liver damaging agent and herbal drugs are added. The hepatocytes culture with later additions is maintained for a suitable period after which enzyme activity e.g. transaminase levels are found out to assess liver functions.

(b) Hypoglycaemic Activity

In traditional systems, a number of plant extracts have been used for their hypoglycaemic activity like *Karela (Momordica charantia), Jambul (Syzigium cumini),* fenugreek (*Trigonella foenum-gracecum) Gudmar (Gymnema sylvestre).* They are used in treatment of diabetes mellitus. For the testing purpose, experimental diabetes is induced in animals generally by alloxan or streptozocin. Due to alloxan, a urea derivative, selective necrosis of pancreatic islet — β cells occurs. By using a suitable dose of alloxan, moderate diabetes (Fasting Blood sugar levels 180 - 250 mg/ml) is induced and in such cases herbal drug extracts are tested.

Fermentation of *Streptomyces achromogenes* produces a cytotoxic nitrourcido glucopyranose derivative called streptozocin, which induces diabetes. Many a time, normal animals are also used to test hypoglycaemic activity of plant extracts. This is a non-diabetic animal model. Before treatment with plant extract, glucose loading is given or animals are fasted and then glucose tolerance test is carried out.

Animals used for these tests are rabbits, rats or mice. Streptozocin in a single oral dose (rats — 80 mg/kg and mice — 150 mg/kg) develops diabetes in 4 - 7 days. For alloxan, a single injectable dose for all types of animals (140 - 180 mg/kg) is given in marginal ear vein of rabbit or intraperitoneally in rats or mice. Alloxan causes diabetes in two days in rats/mice and in seven days in rabbits.

In all the cases, blood glucose levels are measured by two classical methods using glucose oxidase or ortho-toluidine. Glucose autoanalysers are also used. In glucose tolerance tests, sometimes, insulin levels are measured to find out whether glucose itself or plant extracts are stimulating insulin secretion, to cause hypoglycaemic effect. Radio-immunoassay (RIA) or enzyme linked immunosorbate assay (ELISA) are done for measurement of insulin levels.

(c) Anti-fertility Testing

Although a number of herbal drugs have been used for abortificient, antifertility activity, few of them show scientific evidence like embelin from *Embelia ribes* as abortifacient or gossypol from *Gossypium* species as male antifertility agent. Due to overgrowing population in developing countries, and severe side effects of synthetic antifertility drugs, plant drugs are gaining increasing significance in this context. Such activities of these drugs are known to occur in following ways.

(a) In females

 (i) Destruction of zygotes (early abortion)

 (ii) Prevention of ovulation, fertilization or implantation.

(b) In males

 (i) Spermicidal (ii) anti-androgenic

Antifertility drugs include contraceptives and abortifacients. Contraceptive drugs are used to prevent pregnancy while abortifacients are used to terminate pregnancy. In females, the estimation of antifertility activity measures the pregnancy rate and hence, includes anti-ovulation and anti-implantation drugs. Inhibition of spermatogenesis or sperm motility is considered in males.

(I) Protocols for Anti-fertility activity in Female Rats: Female rats are placed singly and eventually acclimatized for about 8 - 10 days. Normal oestrus cycle is assured by taking vaginal smears after 10-14 days. The animals are then given daily doses of herbal extracts to see the changes like oestrogenicity; increase in weight of uterus, cornification of vagina. Further the animals are paired with male rats which are proven breeders. Meanwhile, daily dose of herbal extract is continued. Mating is detected with the help of vaginal smears. Absence of mating may indicate possibly the lack of oestrogen and/or progesterone. The reason may be antigonadotrophic activity or peripheral antagonism. In the event of mating, one group of rats is continued with daily dose of drug for next seven days. After this their uteri, ovaries and oviducts are examined. The absence of ovulation is possibly due to contraceptive activity. In the case of fertilization, the ova are examined to see transportation and implantation and luteolytic effects. Anti-implementation activity may be interpreted if any changes are observed. This is also called as post-coital contraceptive effect. In the second group of animals, dose is continued up to day 19. If there is no weight gain, it may be contraception or anti-implantation. Bleeding or the loss of gained weight may be due to abortifacient effect.

(II) Antispermatogenic activity in male rats: After acclimatization, adult male rats are given a daily dose of herbal drug extract for 60 days. Between days 12 and 15, each male rat is mated with a female rat (undergoing proestrus or estrus cycle). Subsequent mating is arranged between days 56 and 60. After this, testes of each male rat are morphologically studied and weight is noted. Sperms are examined under microscope. Besides, Histopathological studies of testes are also carried out. If there is no fertilization due to mating between days 12 and 15, it is considered as functional sterility; while that between days 56 and 60 will be due to antispermatogenic activity of herbal drug extract.

(III) Spermicidal activity: It is a simple method and can be carried out *in vitro* directly on human semen. The fresh sample of human semen taken on slide is added with two drops of herbal drug extract in Sorensen's phosphate buffer, mixed well and observed under microscope for motility of sperms. Immobility of sperms occurs, in case of spermicidal herbal drug extract.

An occlusion in vas deferens in male rat can occur due to granuloma type tissue formation there, leading to male sterility. It can be assessed by carrying *in vitro* testing in male rats.

(d) Anti-inflammatory Activity

A number of drugs from plant sources are known to cause anti-inflammatory effects and used in conditions like rheumatoid arthritis, gout, dysmenorrhoea etc. There are many sites in body where these types of drugs can act. Inflammation may occur due to mechanical causes, infections or auto-immune diseases. In the damaged tissues in such cases, lysosomal enzymes and other products are released by polymorphonuclear leukocytes. This leads to number of complex inter-related pathways, finally release of kinins, prostaglandins and histamine causing inflammation as vascular effects.

The principle underlying the testing of anti-inflammatory activity is the reduction of local oedema induced in rat paw by injecting irritant, inflammatory substances. Most common inducer is carrageenan which is a mucopolysaccharide isolated from Irish Sea moss *Chondrus cripus*, given in the dose of 0.1 ml, 1 % w/v in saline. Before the test, the antagonist (herbal drug extract) is given orally 1 hour or 30 minutes intraperitoneally before the test. Agonist (carrageenan) is injected into plantar surface of the right hind paw and volume of paw is measured just after injection, and then every after every hour subsequently for five times. Volume of paw is measured by volume displacement method e.g. using plethysmometer.

Another method employs albino adult mice, wherein erythema and odema is produced in pinna of ear using agonist as croton oil (1 mg/ear). Test plant extracts are applied to the pinna of ear, and 15 minutes after croton oil is applied over the same area. Measurement of anti-inflammatory effects is done by simple method like expressing as O-no activity, + → slight and ++ → pronounced. Odema is measured by vernier callipers.

(e) Neuropharmacological Activity

Plant Kingdom is rich of various medicinal species having their effects on nervous system. They are used or abused for medicinal or narcotic purposes. Some of them like Opium, *Pilocarpus,* and *Brahmi* have potential therapeutic uses also. Now a days the pharmacological spectrum and biological efficiency of such herbal drugs can be suitably established due to development of various Neuropharmacological testing protocols. They are outlined here as:

(i) Testing of herbal drugs with effects on central nervous system (CNS).

(ii) Testing of herbal drugs with effects on autonomic nervous system.

CNS acting drugs are classified as stimulants — cocaine (*Erythroxylon coca)*, and tetrahydrocannabinol (*Cannabis sativa)*, depressants — morphine (*Papaver somniferum)*, tranquilisers (*Rauwolfia* spp.), hallucinogens — psilocin (*Psilocybe mexicana)*, and antidepressants. CNS acting drugs are tested by following methods in rodents.

(i) Locomotor activity in mice.

(ii) Locomotor co-ordination.

(iii) Penetrazole convulsions in mice.

(iv) Barbiturate at sleeping time.

Locomotor activity is measured on mice by using an instrument called activity cage in which the movement of animals across a light beam is measured and taken as locomotion count. The test represents the CNS stimulant or depressant profile. Locomotion co-ordination is assessed by means of rotating bar on which mice are placed.

Activity of anti-convulsant herbal extracts is assessed in mice, where the convulsions are induced by Penetrazole (80 - 120 mg/kg. intraperitoneally). The factors measured are time of onset of convulsion after drug treatment, number of convulsive periods during observation period, number of convulsing animals and mortality rate.

Barbiturate sleeping time is used to assess stimulant or depressant effects on CNS. Sleeping time is measured in terms of righting reflex. To induce sleep, barbiturates like pentobarbitone (55 mg/kg) or hexobarbitone (35 mg/kg) are used. Prolongation or reduction of sleeping time may indicate depressant or stimulant activity of herbal drug.

For testing the effects of herbal drug extracts on **autonomic nervous system,** following types of in vitro preparations are used with their possible purposes as mentioned along with:

1. Guinea pig ileum (for non-specific anti-spasmodic activity).
2. Isolated Rabbit Jejunum (Adrenergic activity and mechanism).
3. Rat Phrenic Nerve - Diaphragm (Muscle relaxant activity).
4. Frog Rectus activity on skeletal muscles.

1. Guinea - Pig Ileum preparation

It has a number of uses in testing of ANS drugs because of its specific characters. Guinea pig ileum contains a network of intrinsic parasympathetic nerves (Auerbach and Meissuer's plexuses) present between circular and longitudinal muscle layers. It also contains non-adrenergic non-cholinergic nerves. Hence, guinea pig ileum is useful for drugs causing nerve mediated and directly stimulated contractions. Guinea pig ileum does not give large spontaneous contractions. Ganglion blocking or stimulating activity of drugs can be studied because Auerbach plexus in guinea pig ileum has parasympathetic ganglion bodies and they can be stimulated by nicotinic receptor agonists.

For the purpose of setting up the guinea pig ileum preparation, food is withdrawn for overnight, and then the animal is killed. Suitable length of ileum (except first 5 cm. next to catecum) is separated from mesenteric attachment and put in physiological salt solution. Ileum is emptied off its contents. A piece measuring about 1.5 - 2 cm length is mounted in physiological salt solution in jacketed organ bath at constant temperature (32 - 37°C) and properly aerated. Guinea pig ileum shows a response to parasympathomimetics (acting either as anticholinesterases or ACH receptor agonist) and parasympatholytic drugs (acting either to prevent action of ACH or its release). During these experiments, graded doses of ACH are given for ileum stimulation, after which herbal drug is given.

Due to drug treatment, if ACH type contractions are produced, and persist even after giving a dose of histamine H_1 - receptor antagonist, then it may be indicated that it is a cholinergic drug. This can be verified by blocking effect with atropine. This does not distinguish between muscarinic and nicotinic receptor activity. If the contractions caused by herbal drug are blocked by a ganglion blocking agent such as hexamethonium or coniine, then it may be indicated that herbal drug has a nicotinic receptor mechanism. Besides, cholinergic activity, the contractions occurred due to drug may be because of its activity at other receptors like histamine.

If the herbal drug produces inhibition of contractions, it may have anticholinergic (parasympatholytic) activity. For verification of muscarinic receptor blockade, dose response curves are found out which may show competitive or non-competitive antagonism. Nicotinic receptor type blockade may be indicated using hexamethonium.

Besides muscarinic and nicotinic receptor blockade, parasympatholytic activity is observed due to morphine also which prevents release of ACH from cholinergic nerve endings. Such type of activity in drugs can be found by mounting longitudinal muscle strip of guinea pig ileum in between two platinum electrodes.

2. Isolated Rabbit Jejunum Preparation

This type of preparation is used to find out nerve mediated and direct muscle stimulated adrenergic effects. For mounting purpose, rabbit jejunum is isolated along with mesenteric vessels as the vessels are associated with sympathetic nerves. The mesentery vessels are carefully cut off and then isolated rabbit jejunum with periarterial nerves is mounted in physiological salt solution in jacketed organ bath at a constant temperature (32 - 37 °C). The nerves are connected through an electrode and stimulated. If these are relaxed by plant drug, it may have either noradrenaline like effect or through other mechanism like calcium channel blockade or opening potassium channels. If the action of drug is blocked by adrenergic receptor antagonists, it may be indicated that drug has an effect on adrenoreceptors.

A drug may show blocking effect on periarterial nerve stimulation (adrenergic neuron blocking property) and hence, indirectly cause relaxation of jejunum.

3. Rat Phrenic Nerve — Diaphragm Preparation

Phrenic nerve-diaphragm is isolated from frontal part of right thoracic wall. It is mounted on a special phrenic nerve electrode. Here the broad end of diaphragm (cut into triangular form) is tied at two points while muscle of diaphragm is allowed to remain across the length of platinum electrode (at one end). The preparation is joined to recording device by tendinous end of triangle.

Inhibiting indirectly, the stimulation of diaphragm muscle by the drugs is considered important because this effect is mediated through neuro-muscular junction.

(f) Testing for Anti-Ulcer Activity:

Various factors are responsible for increase in acidity, diet, alcohol, mental and physical stress, besides certain drugs especially non steroidal anti-inflammatory molecules. Gastric and duodenal ulcers range from simple gastritis to ulcers observable by X-rays. Plant drugs mainly used as anti-ulcerogenic include liquorice, atropine, hyoscine and to a less extent gefarnate from white cabbage juice.

For testing of anti-ulcer activity of an herbal drug, the ulcer is induced by either of the following agents.

(1) Chemicals: Alcohol (1 ml/kg body weight, orally), aspirin (200 mg/kg body weight orally) endomethacin, serotonin (5 - HT), prednisolone.

(2) Stress: The stimuli like immobilization and cold. (Animal is immobilized in a cylindrical cage and maintained at 3 - 5°C). Animal model used is rat (male, wistar); sometimes guinea - pigs are also used. The animals are grouped as (a) Those treated with normal saline only (b) Those treated with ulcerogen in saline or ulcerogenic stress procedure (c) Test groups: Along with ulcerogen (chemical/stress), plant extracts in a suitable dose range (d) Standard reference group: Along with ulcerogen (chemical/stress), a known anti-ulcer drug like liquorice or ranitidine.

During the procedure, the ulcers are induced by administering ulcerogen (chemical/stress); After an hour, the animals are sacrificed, and the stomach or duodenum (for histamine-induced ulcers) are given a slit along the curvature. These organs are opened for assessment of damage due to ulcer. In some experiments, the gastric acid is measured. For this purpose, before slitting along curvature, the stomach is ligated and the contents are drained and collected.

The severity of ulcer damage, due to ulcerogen and also the protection offered by herbal anti-ulcer agents, is assessed by 'Ulcer Index'. For assessing ulcer index, the number and size of ulcerative lesions is considered. Ulcer index is expressed on simpler scale as

0 = No damage

1 = Redness of mucosa

2 = Erosion of mucosa

3 = Ulceration

In elaborate ways, it is expressed in following way.

0 = Absence of ulcer

1 = Slightly dispersed and haemorrhagic ulcers (less than 2 mm length)

2 = One ulcer up to 5 mm in length

3 = More than one ulcer

4 = One ulcer above 5 mm in length

10 = Total ulceration and haemorrhage.

For the further testing of ulcer healing activity, the experiments carried out are:

1. Estimation of sialic acid content — for the assessment of mucus formation which indicates cytoprotective activity.
2. DNA determination — It gives an idea about stimulation of cell proliferation.

Anti-insect Activities

The procedures recommended by World Health Organization have been briefed here. To find out either of the activities, the batches of test insects are exposed to a series of concentrations of herbal extracts and the mortality / repellency of insects or larvae at each level of concentration are determined. The period of exposure depends on type of insects.

Larvicidal activity: For this purpose, larvae of *Aedes aegypti* (yellow fever mosquito) are introduced into 100 ml. beaker filled half its capacity with water. Acetone extract of herbal drug (in a known dose) is added to water and larvae are provided with animal food and yeast powder. Percentage mortality is found out by recording the number of larvae dying in 24 hours.

Adulticidal Activity: (Contract toxicity test): The film of herbal extract in acetone of a known concentration is prepared in a petridish (10 cm diameter) and dried. Adults of red flour beetle (*Triboleum castaneum*) are put into the petridish, covered and maintained at 27°C and 60 per cent relative humidity. Mortality count is taken after 24 hours.

Repulsion Test for Stored product pest

It is also known as Arena ring test. It may be carried out on red flour beetle as test insects. A filter paper disc is cut in two halves (test and control). The test disc is treated with herbal extract of known dose and dried. The control disc is treated with solvent, dried and both discs (halves) are joined. About 30 adult insects are released at centre of disc and covered with glass ring. The distribution of insects in the two halves is observed every 10 minutes for three and half hours.

Mosquito repellance test: Mosquito repellence activity is judged on the basis of protection period offered by herbal extract against mosquito bite. The protection period is measured in terms of time until the first bite. The test is carried out against 3-6 days old, blood starved sucrose fed yellow fever mosquito (*Aedes aegypti).* Human hand covered with snugly fitting polythene bag is introduced in a cage containing about a thousand hungry mosquitoes. The insects are allowed to bite on back of hand through muslin cloth screen stuck (pretreated herbal extract) over a small window prepared in the muslin cloth screen, rather than direct skin exposure. Number of bites received in five minutes is counted. If there are no bites in initial five minute exposure, the test hand is exposed repeatedly after every consecutive 30 minutes for 5 minutes, till the time a confirmed bite is received. Number of hours before the confirmed bite represent the protection period offered by herbal extract under trial.

Microbiological Assays

Drug substances that either suppress or influence the growth of micro-organisms are generally analyzed by microbial method. Antibiotics and a few vitamins fall into this category. The procedures employed in microbial assay of antibiotics may be divided into two broad groups:

(a) Cylinder plate method (cup): Assay of antibiotics is based on measurement of the diameter of microbial growth inhibition surrounding the cylinders containing various dilutions of test compound which are placed on the surface of a solid nutrient medium, previously inoculated with a culture of a suitable organism. Inhibition produced by the test compound is compared with that produced by known concentration of a reference standard. Numerous assay designs have been employed in many laboratories depending on the facilities available.

(b) Turbidimetric method: Based on inhibition of microbial growth as indicated by measurement of turbidity (transmittance) of suspensions of a suitable micro-organism in a fluid medium, to which have been added graded amounts of the test compound. Changes in the transmittance produced by the tested compound are compared with those produced by known concentrations of standard.

Microbiological evaluation technique is employed for antibiotic principles of plant extracts. The screening for antibacterial and antifungal activities of bioconstituents is done by this technique.

Microbiological limits (Absence of micro-organisms)

To control the contamination in crude drugs either by natural source or even during their processings, several official books, in their method of standardisation have prescribed the microbial limit tests. Indian pharmacopoeia also has laid down the following norms for a few of the crude drugs as mentioned below.

Gum acacia powder	:	1 gram sample of acacia powder is free from *E. coli*
Guar gum	:	1 gram is free from *E. coli and salmonella,* and total bacterial count should not be more than 5000 per gram.
Starch	:	1 gram is free from *E. coli* and *Salmonella.*

Tragacanth powder : 1 gram is free from *E. coli* and *Salmonella*.

Dried Yeast : 1 gram meets the requirement of test for absence of *Salmonella* and *E. coli*.

Gelatin : One gram sample should confirm the absence of *E. coli*,10 gram is free of *Salmonella* and total bacterial count is not more 1000 per gram.

QUALITY CONTROL OF HERBAL DRUGS

Owing to long standing and time proven uses of herbal drugs along with higher safety margins, World Health Organization (WHO) has taken necessary steps to ensure quality control with modern techniques and application of suitable standards for this purpose. The pharmacopoeias of different countries include monographs indicating quality parameters and standards for various herbal drugs, and also for some of their products. The individual parameters included in these monographs have been discussed herinbefore.

Monograph guidelines: For the purpose of quality control of herbal drugs, WHO has prepared accordingly the guidelines. The objectives put forth are provisions for recommended general test methods and also the general limits for contaminants for herbal drugs. A typical monograph for herbal drugs as per WHO guidelines is as follows.

MONOGRAPH TITLE

I. Botanical

- Sensory Evaluation : Visual Macroscopy / Touch / Odour / Taste
- Foreign Matter : Foreign plants, foreign animals, foreign mineral, etc.
- Microscopy : Histological observation, Histochemical detection, measurements, etc.

II. Physicochemical

- TLC : Ion, measurements, etc.
- Ash : Total, Acid insoluble, Water soluble.
- Extractable Matter : In hot water, cold water and ethanol
- Water Content and Volatile Matter : LOD, Azeotropic
- Volatile oils : By steam distillation.

III. Pharmacological

- Bitterness Value : Units equal to bitterness of solution of quinine hydrochloride
- Haemolytic Activity : On ox blood by comparison with reference — saponins
- Astringency : Fraction (Tannins) that binds to standard — hides powder
- Swelling Index : In water
- Foaming Index : Foam height produced by 1 gm material under specified conditions.

IV. Toxicological

- Pesticide Residues : Total organic chloride and Total organic phosphorus
- Arsenic : Strain produced on $HgBr_2$ paper in comparison to standard stain.
- Heavy Metals : Cadmium and Lead

V. Microbial Contamination:

Total viable aerobic count

Pathogens: Enterobacteriaceae, *E. coli,*

Salmonella, P. aerogenosa, S. aureus

Aflatoxins: By TLC using standard Aflatoxins (B_1, B_2, G_1, G_2) mixture.

VI. Radioactive Contamination:

Among the various pharmacopoeias, the monograph for herbal drugs in **British Herbal Pharmacopoeia** has the following parameters.

MONOGRAPH TITLE

Definition

Introductory Passage

Constituents

Therapeutics

Regulatory Status

References

Relevant excerpts from regulator guidelines of certain EC countries

Therapeutic section

BHC has mentioned that, where the research in depth has not been carried out, the actions and uses described are largely based on unpublished observation or common experience. Therefore sections on therapeutics are based on the assessment of highly experienced practitioners of Phototherapy and supported by references wherever possible.

W.H.O. Guidelines for General Limits (Contaminants in Herbal Drugs)

Following limits have been proposed for contaminants:

1. Harmful organic matter : Totally free from it
2. Innoccuous foreign matter : Free from it (to maximum possible extent)
3. Pesticidal residues, arsenic and heavy metals : Maximum residue limits $= \dfrac{\text{BW} \times \text{ADI} \times \text{Extraction factor}}{\text{Safety factor } 100 \times \text{MDI}}$

Where, ADI = Average daily intake

BW = Body weight

MDI = Mean daily intake of drug

4. Microbial load (Contamination)

Following are the WHO suggestions for upper limit for number of micro-organisms per gram of material.

Type of Micro-organisms	Crude drug for Processing	Ready for Internal Use	Ready for Topical Use
(i) Salmonellae	–	Nil	Nil
(ii) Enterobacteriaceae	–	10^3	10^4
(iii) Total aerobic	–	10^5	10^7
(iv) E. coli	10^4	10	10^2

5. Aflatoxins : Totally free from it

6. Radioactive

Contaminants : As per recommendations of International Atomic Energy Agency (IAEA), Vienna, Austria. (Wherever applicable only)

❖❖❖

Chapter 8

CARBOHYDRATES AND DERIVED PRODUCTS

Formerly, carbohydrates were defined as a group of compounds composed of carbon, hydrogen and oxygen in which the later two elements are in the same proportion as in water and were expressed by a formula $(CH_2O)n$, i.e. hydrates of carbon. The word carbohydrate can be traced back to Germans, who called them *'kohlenhydrates'*. It was then termed carbohydrates in English. The definition did not remain valid as it was misleading with a few compounds like (1) acetic acid expressed as CH_3COOH (i.e. $C_2H_4O_2$) and lactic acid as $CH_3CHOHCOOH$ (i.e. $C_3H_6O_3$), which are not carbohydrates and (2) sugars like rhamnose and fucose represented by the formula $C_6H_{12}O_5$. The carbohydrates are defined as *polyhydroxy aldehydes* or *polyhydroxy ketones* or *compounds that on hydrolysis produce either of the above*. They are substances of universal occurrence and are much abundant in plants, rather than in animals.

Carbohydrates are grouped into two major classes: simple sugars (saccharides) and polysaccharides. Low molecular weight carbohydrates are crystalline, soluble in water and sweet in taste; e.g. glucose, fructose, and sucrose. The high molecular weight carbohydrates (polymers) are amorphous, tasteless and relatively less soluble in water e.g. starch, cellulose, gums, pectins, inulin, etc. Depending upon the chemical structure, saccharides are subdivided as monosaccharides, disaccharides and trisaccharides.

[A] MONOSACCHARIDES

Monosaccharides are sugars, which cannot be further hydrolysed to simple sugars. However, they are classified according to the number of carbon atoms in sugar molecules.

1. **Bioses:** They contain two carbon atoms. They do not occur free in nature.

2. **Trioses ($C_3H_6O_3$):** They contain three carbon atoms, but in the form of phospheric esters, e.g. glyceraldehyde.

3. **Tetroses ($C_4H_8O_4$):** They contain four carbon atoms, e.g. erythrose and threose.

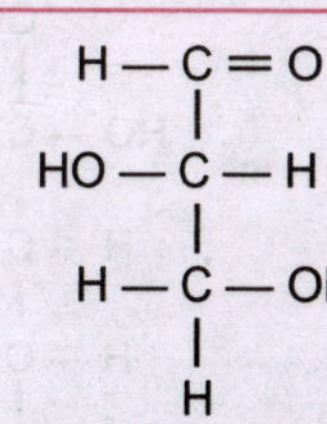

Glyceraldehyde

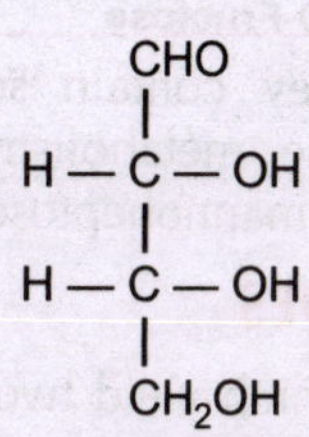

D-erythrose

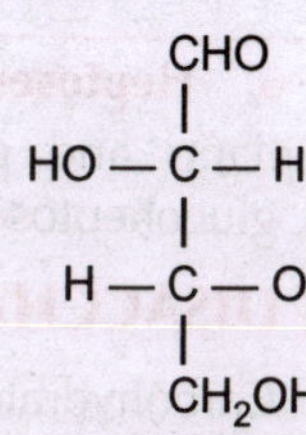

D-threose

4. Pentoses ($C_5H_{10}O_5$): They are very common in plants and are the products of hydrolysis of polysaccharides like hemcelluloses, mucilage and gums, e.g. arabinose, ribose and xylose.

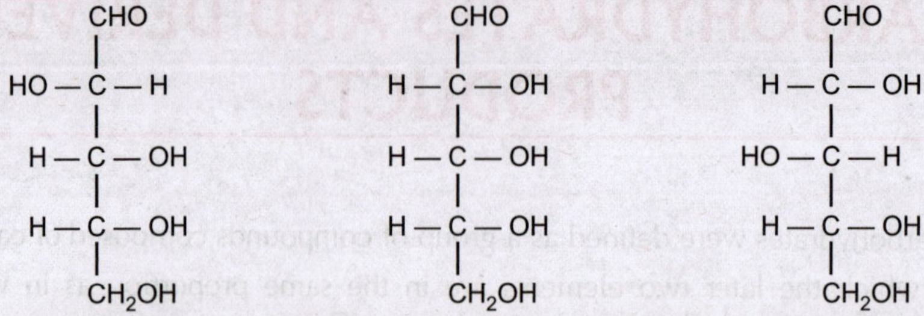

Arabinose **D-ribose** **D-xylose**

5. Hexoses: They are the monosaccharides containing six carbon atoms and are abundantly available carbohydrates of plant kingdom. They are further divided into two types - aldoses and ketoses, obtained by the hydrolysis of polysaccharides like starch, inulin, etc.

Aldoses: Glucose, mannose, galactose

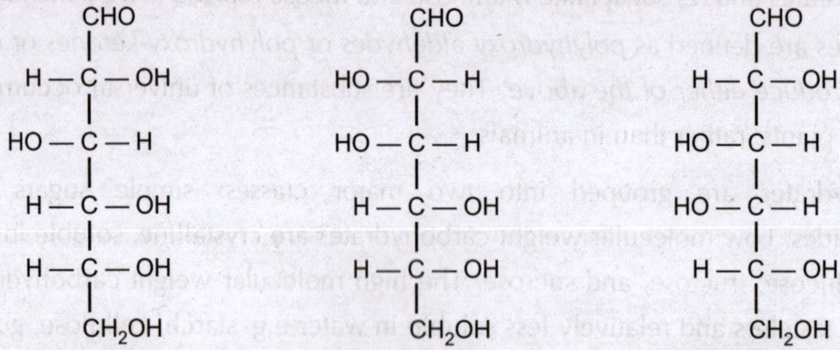

D-glucose **D-mannose** **D-galactose**

Ketoses: Fructose, D-sorbose and rhamnose

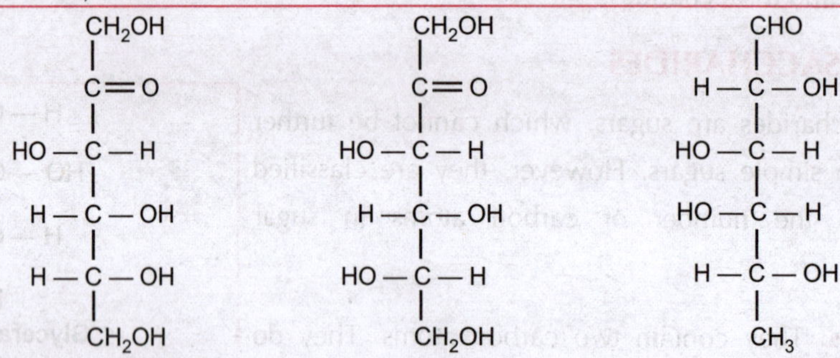

D-Fructose **D-Sarbose** **D-Rhamnose**

6. Heptoses: They contain seven carbon atoms, vitally important in the photosynthesis of plants and glucose metabolism of animals and are rarely found accumulated in plants, e.g. glucoheptose and mannoheptose.

[B] DISACCHARIDES

Carbohydrates, which yield two molecules of monosaccharides upon hydrolysis, are called as disaccharides.

$$\text{Sucrose (Cane-sugar)} \xrightarrow{\text{Hydrolysis}} \text{glucose + fructose}$$

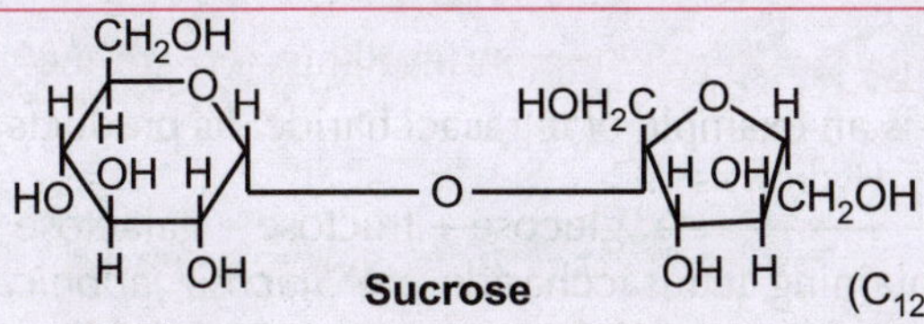

Sucrose ($C_{12}H_{22}O_{11}$)

Maltose (Malt-sugar) $\xrightarrow{\text{Hydrolysis}}$ glucose + glucose

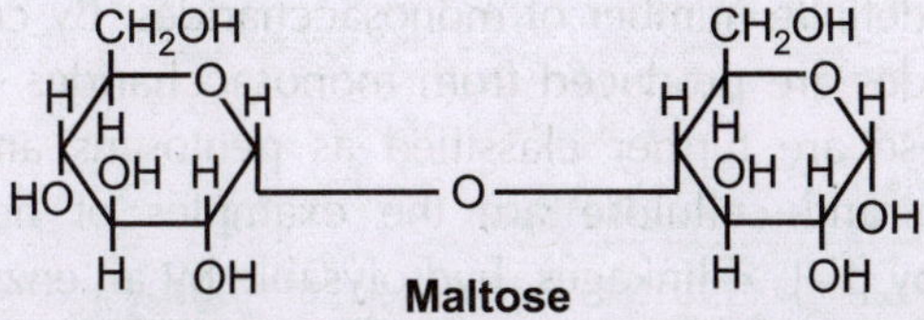

Maltose ($C_{12}H_{22}O_{11}$)

Lactose (Milk-sugar) $\xrightarrow{\text{Hydrolysis}}$ glucose + galactose

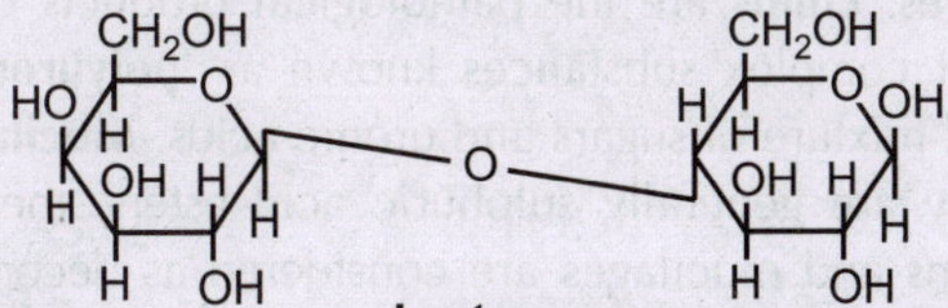

Lactose ($C_{12}H_{22}O_{11}$)

[C] TRISACCHARIDES

As the name indicates, these liberate three molecules of monosaccharides on hydrolysis.

Raffinose $\xrightarrow{\text{Hydrolysis}}$ glucose + fructose + galactose (in beet and manna)

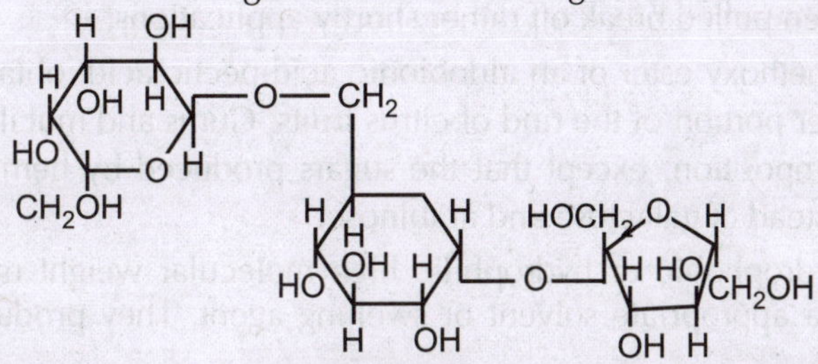

Raffinose ($C_{18}H_{32}O_{16}$)

Gentianose $\xrightarrow{\text{Hydrolysis}}$ glucose + glucose + fructose (gentian roots)

Scillatriose (squill), mannotriose (manna) and phanteose (psyllium) are the other examples of trisaccharides.

Following are the examples of trisaccharides along with their products of hydrolysis.

Sr. No.	Name	Source	Products of Hydrolysis
1.	Gentianose	*Gentiana* sp.	Glucose + glucose + fructose
2.	Planteose	Isapgol, Psyllum seeds	Glucose + fructose + gelactose
3.	Rhamniose	Rhubarb (*Rhamnus infectoria*)	Rham + rhamnose + gelactose
4.	Scillatriose	Squill (*Urgenia maritima*)	Glucose + glucose + rhamnose
5.	Manneotriose	Manna (*Fraxinus* – sp.)	Glucose + galactose + galactose

[D] TETRASACCHARIDES

Stachyose or manneotetrose is an example of tetrasaccharide. Its products of hydrolysis are:

$$\text{Stachyose (manneotetrose)} \xrightarrow{\text{Hydrolysis}} \text{glucose + fructose + galactose + galactose}$$

The examples of plants containing tetrasaccharide are *Stachys japonica* and manna *Fraxinus urnus*.

[E] POLYSACCHARIDES

On hydrolysis, they give an indefinite number of monosaccharides. By condensation, with the elimination of water, polysaccharides are produced from monosaccharides. Depending upon the type of product of hydrolysis these are further classified as pentosans and hexosan. Xylan is pentosans, whereas starch, inulin and cellulose are the examples of hexosans. Cellulose is composed of glucose units joined by β -1, 4 linkages, hydrolysable by an enzyme cellulose present in animals, whereas starch contains glucose units connected with α-1, 4; α-1, 6 units.

Polyuronides i.e. alginic acid, pectin), gums and mucilages are the other pharmaceutically important polysaccharide derivatives. **Gums** are the pathological products consisting of calcium, potassium and magnesium salts of complex substances known as **'polyuronides'**. On prolonged boiling with dilute acids, they yield mixture of sugars and uronic acids. **Mucilages** are physiological products related to gums and they are generally sulphuric acid esters, the ester group being a complex polysaccharide. Both gums and mucilages are considered as decomposition products of cellulose.

Cellulose:

It is a polusaccharide linked will glucose units and is chief constellates of fiber of plant cells. It is tremendously used in textile and paper industry. Cellulost denuatives such as methyl cellulose, ethyl cellulose, pyroxylone cellulose, carboxy methyl cellulose (CMC) has large industrial and pharmaceutical uses when pulled break off rather shortly applications.

Pectin is a neutral methoxy ester of an aldobionic acid-pectic acid, obtained as a water soluble compound from the inner portion of the rind of citrus fruits. Gums and mucilages are closely related to hemicelluloses in composition, except that the sugars produced by hemicelluloses are glucose, mannose and xylose, instead of galactose and arabinose.

Gums are either hydrophobic or hydrophilic high molecular weight molecules, usually, with colloidal properties, in a appropriate solvent or swelling agent. They produce gels, highly viscous suspensions or solutions.

Depending upon types of occurrence, gums can be classified as follows.

Gums

(A) Natural gums

1. Seaweed gum kelp : agar, algin, carrageenan, laminaran.
2. Plant exudates : acacia, tragacanth, karaya, ghatti.
3. Seed gums : guar gum, isabgol, tamarind, locust bean gum.
4. Plant extracts : pectin, larch, arabinogalactan.
5. Animal extract : chitin.

(B) Prepared gums

1. Biosynthetic gums xanthan, scleroglucan, dextrans.
2. Starch and its derivatives, dextrins.
3. Cellulose derivatives.

Chemical Tests for Carbohydrates

1. **Molisch's test:** The test is positive with soluble as well as insoluble carbohydrate. It consists of treating the compounds with α-naphthol and concentrated sulphuric acid which gives purple colour.

2. **Reduction of Fehling's solution:** To the solution of carbohydrate, equal quantity of Fehling's solutions A and B is added. After heating, brick red precipitate is obtained.

The summary of carbohydrate and related drugs is given in Table.

Table 8.1: Carbohydrates and Derived products

Sr. No.	Name of drug & Synonym	Biological Source	Constituents	Uses
1.	Acacia (Indian gum)	*Acacia arabica* (Leguminosae)	arabin, enzyme oxidase	Emulsifying and suspending agent.
2.	Agar (Agar agar)	*Gelidium amansii* (Gelidaceae)	agarose, agaropectin	Laxative, bacteriological cultures.
3.	Alfalfa	*Medicago sativa Leguiminosae*	vitamins specially vit. K	Nutritive with vitamins
4.	Algin	*Macrocystis pyrifera, Laminaria hyperborsea (phaeophyceae)*	alginic acid	Thickening and suspending agent.
5.	Bael	*Aegle marmelos (Rutaceae)*	marmelosin and furocoumarin	Digestive, antidiarrhoeal
6.	Caramel	by heating sucrose	burnt sugar	Colouring agent.
7.	Carrageenan (Irish moss)	*Chondrus cryspus* Rhodophyceae	kappa-carrageenan lambda carrageenan.	Demulcent, antidiarrhoeal, pharmaceutical aid.
8.	Dextran	Enzymic fermentation of sucrose	α - D-glucopyranosyl units	Binding, suspending and stabilizing agent, blood plasma extender.
9.	Dextrin	Controlled hydrolysis of starch	polysaccharide	Adhesive in paper, textile and food industries.
10.	Echinacea	*Echinacea purpurea (Compositae)*	Arabinogalactan	Immunostimulant
11.	Fenugreek	*Trigonella foenum graecum Fabaceae*	mucilage 3% Fatty oil 5% Trigonellin, choline Saponin 0.5 – 1-5%	Anti-inflammatory
12.	Guar gum	*Cyamopsis tetragonolobus* (Leguminosae)	guaran	Binding agent, emulgent, disintegrating agent.

Contd...

13.	Gum ghatti	*Anogeissus latifolia* (Combretaceae)	Two distinct polysaccharides, oxidase	Emulgent, pharmaceutical aid.
14.	Gum karaya	*Sterculia urens* (Sterculiaceae)	A polysaccharide containing 8 % acetyl and 37 % uronic acid residues and no methoxyl group.	Thickening agent, emulgent.
15.	Inulin	*Inula helenium* (Compositae)	fructo-furanose unit	source of fructose in foods and drinks.
16.	Isapgol	Plantago ovata (Plantaginaceae)	pentosans, mucilage, aldobionic acid	Demulcent, laxative, pharmaceutical aid.
17.	Liquid glucose	Partial hydrolysis of starch	dextrose, maltose	Sweetening agent, non-crystallizing agent.
18.	Locust bean gum	*Ceratonia siliqua* (Leguminosae)	88 per cent D-galacto-D-manoglycon	thickener, pharmaceutical aid.
19.	Manna	*Fraxinus ornus* (Oleaceae)	mannitol, mannotriose	laxative.
20.	Pectin	*Citrus limonis, C. aurantium* (Rutaceae)	D-galacturonic acid pectric acid	Adsorbent, thickening agent.
21.	Starch	*Zae mays, Oryza sativa* (Gramineae)	amylose, and amylopectin	disintegrating agent, demulcent, nutritive.
22.	Tragacanth	*Astragalus gummifer* (Leguminosae)	tragacanthin, tragacanthic acid, bassorin	Thickening agent, demulcent.
23.	Xanthan gum	fermentation of glucose	mannose, glucose, glucuronic acid	Emulsifying agent, stabilizer.
24.	Chitin and Chitosan	Shells of Lobster crab, cell walls of lower plants	2-acetamide-2-deoxy-cellulose	Wound healing, adhesive to glass and plastics, sizing of rayon.
25.	Cyclodextrins	Products of bacterial digestion of celluloce	Cyclic olego sacchorides	Oral and topical applications
26.	Honey	*Apis species* (Apidae)	natural invert sugar,	Demulcent, sweetening agent.
27.	Lactose	Disaccharide from mammalian milk	*Disaccharide containing galactose and glucose*	Diluent in tablets and pharmaceuticals.

ACACIA

Synonyms

Gum acacia, Gum arabica, Indian gum

Biological Source

Indian gum is the dried gummy exudation obtained from the stem and branches of wild *Acacia arabica* belonging to family Leguminosae.

Geographical Source

The plant is found in India, Sri Lanka, Sudan, Morocco and Africa. In India, it occurs in Punjab, Rajasthan and Western Ghats. About 85 per cent of world supply of gum acacia is from Sudan.

Cultivation and Collection

It is a common member of dry monsoon forests of India. It is an evergreen tree with short trunk. It is not cultivated on commercial scale. Gum is collected from wild grown plants, made free of bark and foreign organic matter, dried in sun, which also results in partial bleaching of gum. (*Fig. 8.1*)

Fig. 8.1 (a): Acacia tree

Description

Colour	:	Tears are cream-brown to red in colour, while powder is light brown in colour.
Odour	:	Odourless
Taste	:	Bland and mucilaginous
Size and Shape	:	Irregular brown tears of varying size

Extra Features

The tears are glossy and marked with minute fissures and are brittle in nature. The pieces of broken tears are with angular fragments and glistering surfaces, breaking with difficulty, and with conchoidal fracture.

Solubility

It is soluble in water; the watery solution is viscous and acidic. It is insoluble in alcohol.

Fig. 8.1 (b): Acacia tears

Chemical Constituents

It consists principally of arabin, which is a complex mixture of calcium, magnesium and potassium salts of Arabic acid. Arabic acid on hydrolysis gives L-arabinose, L-rhamnose, D-galactose and D-glucuronic acid. It also contains an enzyme oxidase and peroxidase.

Identification

1. Solution of lead sub acetate gelatinizes the aqueous solution of Indian gum.
2. It does not produce a pink colour with the solution of ruthenium red.
3. On addition of solutions of hydrogen peroxide and benzidine in alcohol to aqueous solution of gum, it turns blue in due to oxidase enzyme.

Standards of Quality

Moisture: Not more than 15%

Sulphated ash: Not more than 5.0%.

Acid insoluble ash: Not more than 1.0%

Indian gum should not contain tannin, starch and dextrin.

Uses

Acacia is a demulcent. It is also administered intravenously in haemolysis. In the form of mucilage, it is used as a suspending agent, specifically in mixtures with resinous substance. Acacia is a-good emulsifying agent for fixed oils, volatile oils and also for liquid paraffin. It is a good binding agent and is used in the preparation of lozenges, pastilles and compressed tablets. It is a gum of choice, as it is compatible with other plant hydrocolloids, as well as, starches and carbohydrates. In combination with gelatin, it is used to form coacervates for microencapsulation of drugs.

Decolourization of Acacia can be done by ethyl alcohol after drying at 60ºC which will also inactiate oxidase enzyme suitably.

Tests for Iso-propyl Purity of Indian Acacia

1. Dilute 1 ml of the solution of gum with 10 ml of water and keep for a few hours. No sedimentation should take place.
2. To 1 ml of solution, add 4 ml. of water, boil, cool and add 2 drops of N/10 iodine. Brown colour indicates presence of dextrin, whereas blue colour is due to starch. This test should be negative with the authentic drug.
3. To the gum acacia solution, add a drop of hydrogen peroxide and tincture of guaiacum - colour changes to blue.
4. With a few drops of 0.1 per cent ferric chloride to 1 ml. of the solution, there is change in colour to blue or black (due to tannins).

Substitutes and Adulterants

The B.P. variety consists of gum obtained from wild *Acacia senegal* Wild, (Leguminosae), a plant of African origin and grown in Africa. The tears are rounded or ovoid and about 5 - 40 mm in diameter. Tears are yellowish white in colour. It can be used as a substitute to Indian gum.

Indian gum is adulterated with gum ghatti, obtained from *Anogeissus latifolia* (Combretaceae), which is distinguished from the genuine drug by the following characters. Its outer surface is dull and without fissures. It shows very slight precipitate with lead sub-acetate solution and its aqueous solution is highly viscous. Starch, tragacanth, dextrin and Sterculia gum are the other adulterants of acacia.

Storage

Acacia or powdered acacia should be stored in cool dry place in air-tight containers.

AGAR

Synonyms

Agar has several synonyms: Agar-agar, Japanese-Isinglass, Vegetable gelatin, Chinaighas.

Biological Source

It is the dried gelatinous substance obtained from *Gelidium amansii* (Gelidaceae) and several other species of red algae like, *Gracilaria lichenoides* (Gracilariaceae) and *Pterocladia* (Gelidaceae).

Geographical Source

Agar is produced commercially in Japan, Australia, New Zealand, the USA, and India. In India, it is produced in the coastal regions of Bay of Bengal.

Preparation

In Japan, the red-algae are grown on the bamboos spread in the ocean. The collection of the material is done between May and October. The seaweeds are scrapped from the bamboos, dried and shaken. This is necessary to bleach the product to some extent and even to remove and foreign material like shells, sand, etc. The entire material is then taken to the high altitudes where it is washed and bleached by exposing to sun. It is boiled for 5 - 6 hours with large quantity of dilute acidified water (about 1 part of algae with 60 parts of water). The extract is then strained while hot through the cloth and transferred to wooden troughs. On cooling, jelly is produced. These rectangular solid pieces of jelly are then passed through the netting under pressure. Narrow strips, thus formed, are allowed to melt during the daytime in the sun, which removes the excess of water. This operation is continued for several days to remove the excess of water. The manufacture of agar takes place in the winter, and moisture is removed by successively freezing, thawing and drying at about 35°C. Japan is taking the advantage of natural climatic conditions since centuries for the preparation of agar. In America, the modern method of deep-freezing is being utilized for the same purpose.

Description

Colour	:	Depending upon the shape and the form, it is yellowish-grey or white to nearly colourless.
Odour	:	Odourless
Taste	:	Mucilaginous
Shape	:	It is found in various forms like strips, sheets, flakes or coarse powder.
Size	:	Sheets are 45 - 60 cm long and 10 - 15 cm wide. Bands are about 4 cm. wide, while strips are 4 mm in width. Strips are translucent, lustrous and slender while the flakes are greyish-white in colour.

In India, the raw material being used is known as *Galidiella accrosa* and indigenous production of agar is estimated to be about 200 tones. In India, the process has been developed for the extraction of agar from the species of *Hypnea*.

It is insoluble in cold water, but forms a gelatinous mass after cooling hot solution. It is soluble in boiling water and insoluble in organic solvents.

Chemical Constituents

Agar consists of two different polysaccharides named as agarose and agaropectin. Agarose is responsible for **gel strength** of agar and is composed of D-galactose and 3.6 uni anhydro L-galactose units. It contains about 3.5% cellulose and 6% of nitrogen containing substance. Agaropectin is responsible for the **viscosity** of agar solutions. It is believed to be a sulphonated polysaccharide in which galactose and uronic acid units are partly esterfied with sulphuric acid.

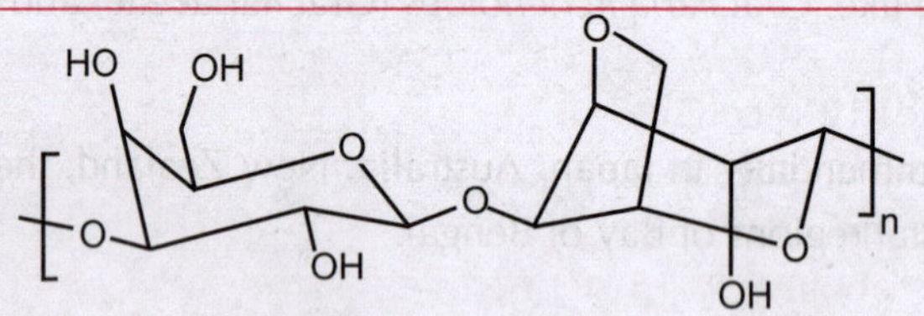

Agarose

Identification

1. Boil about 1.5 g agar with 100 ml. water. Cool the solution to room temperature. It forms a stiff jelly.
2. When mounted in the solution of ruthenium red and examined under microscope, the mounted particles acquire pink colour.
3. To 0.2 per cent solution of agar in water, add solution of tannic acid; no precipitate is produced.

Standards of Quality

Acid-insoluble ash	:	Not more than 1.0 per cent
Sulphated ash	:	Not more than 5.0 per cent
Foreign organic matter	:	Not more than 1.0 per cent
Loss on drying	:	Not more than 18.0 per cent
Starch	:	Negative with iodine solution

Uses

Agar is used as an emulsifying agent and bulk laxative. It is used in the preparation of jellies, confectionery items and in microbiology, it is employed in preparation of bacteriological culture medium.

ALFALFA

Synonyms

Lucerne, Medicago, Purple medick.

Biological Source

It consists of dried leaves, sprouts and seeds of the plant *Medicago sativa,* family: Fubaceae (Leguminosae).

Geographical Source

It is cultivated in many countries such as USA, California, South Dakota, Wisconsin are the leading states to cultivate. It is also cultivated throughout the world.

Cultivation and Collection

It is cultivated on the large scale as livestock forage (fodder). Arabs fed alfalfa to their horses claiming their animals more strong. The plants lives for 4 to 8 years and can survive more than 20 years. It can be cut 3 to 4 times in a year. The 6.8 - 7.5 pH of soil is favourable for growth. It is sown in spring or fall. Normally 13 to 20 kg seeds per hectare are required for cultivation. Alfa alfa fields are rotated as it shows **autotoxicty** and hence it is difficult to grow alfalfa seeds, in existing stands of alfalfa and rotated with corn or wheat, before reseeding. Total yields 7 to 8 tonnes per hectare per year. Potassium and phosphorus are required to grow alfa-alfa favourably.

Description

Plant is known since 12th century. It is perennial herb with trifoliate dentage leaves and underground stems. Stems are woody and upto 1 metre in length. It sends about 5 to 15 stems. After maturity the leaflets get elongated and has cluster of small blue to purple flowers. Fruits are spiraled with 2 to 3 turns, containing approximately 10-20 seeds. It flowers from July to September.

Fig. 8.1(c): Transport of Alfalfa

Macroscopic Characters

Colour : Leaves are Green

Odour : Slight

Taste : Bitter

Size : 0.5 – 2 cm

Shape : Leaflets are trifoliate.

Margin dentate, acute apex glabraus, lower surface covered with whitish hairs.

Leaflets often whole, folded with serrate margin acute apex, upper surface is glabrous, lower surface has scattered long hairs.

Stem pieces are hollow, ridged upto 5 mm diameter, flower fragments are pink.

Microscopic Characters

Epidermis of leaf is wavy. Walled cells with anomocytic stomata on both the surfaces, covering trichomes on under surface, which are thick walled with warty cuticle glandular trichomes on undersurface which are uniseriate normally with four cells and multicellular head of 2-4 cells, fibres and calcium oxalate crystal sheaths are also present pollen grains spherical, oil globulaes and aleuron grains.

Chemical Constituents

It is rich in vitamins specifically vitamin K, leaves are commercial source of chlorophyll and carotene. Leaves are rich in soluble dietary fibres and proteins. It also contain vit. A, B, B6, D, E but no vitamin C.

Standards of Quality

Total Ash	:	Less than 10.0 per cent
Acid insoluble ash	:	Less than 2.0 per cent
Water soluble extractives	:	Not more than 20.0 per cent

Uses

Ayurveda (Indian system of medicine prescribes alfalfa seeds and sprouts for improving blood cells while leaves and stems for proteins and minerals. It is also used as source of chlorophyll and carotene. It is used as cattle feed in England.

It may be used for lowering cholesterol and treating menopausal symptams. Herb can be taken in the dose of 5 – 10 gm three times a day.

ALGIN

Synonyms

Soldium alginate, Sodium polymannuronate

Biological Source

Sodium alginate is the sodium salt of alginic acid. Alginic acid is a polyuronic acid composed of reduced mannuronic and glucoronic acids, which are obtained from the algal growth of the species of family Phaeophyceae. The common species are *Macrocystis pyrifera, Laminaria hyperborea, Laminaria digitata and Ascohyllum nodosum*. It is purified carbohydrate extracted from brown sea-weed (algae) by treatment of dilute alkali.

Geographical Source

Seaweeds are found in Atlantic and Pacific oceans, particularly in coastal lines of Japan, the USA, Canada, Australia, and Scotland. In India, it is found near the coast of Saurashtra. The USA and the UK are largest production centres of Algin.

History

It was discovered by **Stanford** in 1880. The commercial production of Algin first began in 1929 in USA. Since then it is produced in the UK, France, Norway and Japan. The present total Algin production is estimated to be more than 15,000 tonnes per annum.

Preparation of Sodium Alginate

The brown coloured algae are used for extraction of alginic acid. The colour is due to carotenoid pigment present in it.

M. pyrifera, the principal source for global supply, is a perennial plant that lives from 8 - 12 years, and grows, as much as, 30 cm per day. This giant kelp is found mainly in Pacific Ocean. It grows on stands from 15 metres to 1.5 km in width and several km in length. The mechanical harvesting is done about four times a year.

Alginic acid is present in the cell wall. The seaweeds are harvested, dried, milled and extracted with dilute sodium carbonate solution which results in a pasty mass. It is then diluted to separate insolouble matter. Soft water is only used for extraction purposes, so as to avoid in compatibilities. It is treated with calcium chloride or sulphuric acid for conversion into either calcium alginate or insoluble alginic acid, which is collected and purified by thorough washing. If calcium is used, it is treated with hydrochloric acid. Alginic acid so collected, is treated with sodium carbonate for neutralisation and conversion into sodium salt. The alginic acid content on dry solid basis varies from 22 to 35 per cent in all the varieties of brown algae.

Description

Colour : White to buff coloured powder

Odour : Odourless

Taste : Tasteless

It is available either as a coarse or fine powder. It loses about 20 per cent of its weight on drying. Sodium alginate is incompatible with calcium salts, phenyl mercuric acetate and nitrate, crystal violet, alcohol in the concentrations above 5 per cent and heavy metals. It is also precipitated below pH 3.0.

Sodium alginate is readily soluble in water forming viscous colloidal solution and insoluble in alcohol, ether, chloroform and strong acids. 1 per cent solution of gum at 20°C may have a viscosity in the range of 20 - 400 centipoises.

Chemistry of Algin

Alginic acid is a linear co-polymer. It contains D-mannopyranosyluronic acid linked with L-glucopyranosyluronic acid units. Alginic acid can absorb 300-400 times its weight of water.

Standards of Quality

Ash : 18 – 24 per cent

pH : 7.2 (1.0% w/v aqueous solution)

LOD : Not more than 15.0 per cent

Identification

1. The aqueous solution of sodium alginate forms copious precipitate with calcium chloride solution.
2. 1 per cent solution in water forms heavy gelatinous precipitate with dilute sulphuric acid.

Uses

High and medium viscosity grades of sodium alginate are used in the preparation of paste, creams and for thickening and stabilizing emulsions. It is a good suspending and thickening agent, but a poor emulsifying agent. It is used as binding and disintegrating agent in tablets and lozenges. In food industry, it is used for the preparation of jellies, ice cream, etc. It is also used in textile industry. For pharmaceutical purposes, when desired, it is sterilized by heating in an autoclave. The solution of sodium alginate should not be stored in metal containers. It is preserved by the addition of 0.1 per cent of chloroxylenol, chlorocresol, benzoic acid or parabenes. Potassium, aluminium and calcium alginates are also used medicinally.

BAEL

Synonyms

Bael Fruits, Bel, Indian Bael, Bengal Quince.

Biological Source

Bael consists of unripe or half ripe fruits of the plant known as *Aegle marmelos* Corr., family Rutaceae.

Geographical Source

It is indigenous to India and found in Myanmar and Sri Lanka.

Macroscopic Characters

A deciduous tree about 25 feet in height with straight sharp thorns and tripholiate aromatic leaves.

Fruits are globase, 5-10 cm in diameter, grey with smooth hard aromatic rind and numerous, oblong, compressed seeds. Pulp is red in colour with mucilaginous and astringent taste.

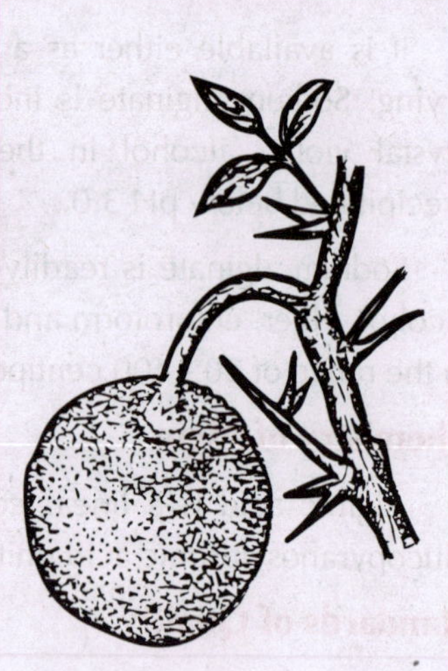

Fig. 8.2: A branch of *Aegle marmelos*

Description

Colour	:	Fruits are grey. Pulp is pale yellowish red.
Odour	:	None for fruits but pulp is aromatic.
Taste	:	Muokginons and astring etc.
Size	:	5 – 10 cm
Shape	:	Globose, smooth

Extra features: The fruit has very hard rind. Fruits contain numerous compressed seeds.

Chemical Constituents

The chief constituent of the fruit is marmelosin (0.5 per cent), which is a furocoumarin. Other coumarins are marmesin, psoralin, umbelliferone. The drug also contains carbohydrates (11 - 17 per cent), protein, volatile oil and tannins. The pulp also contains good amount of vitamin C and vitamin A. Two alkaloids O-methylhalfordinol and isopentylhalfordinol have been isolated from the fruits. Aegeline can be isolated from leaves.

		R_1	R_2
Marmelosin A	:	H	H
Marmelosin B	:	H	CH_3
Marmelosin C	:	CH_3	H

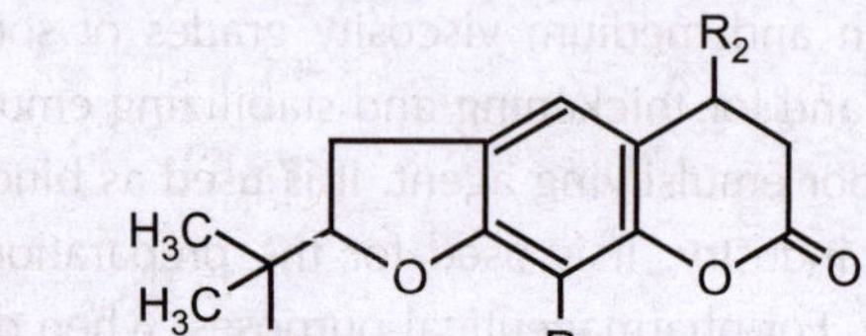

Aegle marmalosine as orange viscous oil is isolated from fruits. Yellow dye is obtained from rind of the fruit and is used in calico printing.

Standards of Quality

Ash	:	Not more than 3.5 per cent
Acid insoluble ash	:	Not more than 0.25 per cent
Alcohol (990%) soluble extractives	:	Not less than 40.0 per cent.
Water soluble extractives	:	Not less than 30.0 per cent.

Uses

It is used as digestive, appetizer and also used in the treatment of diarrhoea and dysentery. It is also a tonic. Bael is also anthelmintic, antispasmodic and has astringent properties.

CARAMEL

Synonym

Burnt sugar colouring.

Method of Preparation

Caramel is prepared by heating a concentrated aqueous solution of sucrose or glucose until the sweet taste is destroyed and uniform dark brown viscous liquid results by adding small quantities of sodium carbonate or mineral acid or alkali during heating.

Description

Colour	:	Caramel is a thick dark brown liquid or dark brown deliquescent powder.
Odour	:	The odour is characteristic of burnt sugar.
Taste	:	Caramel has a pleasant bitter taste.
		When incinerated, it swells and forms charcoal. When spread in thin layer on glass slide there appears a homogeneous reddish brown and transparent colour.
Solubility	:	Caramel is soluble in water and in dilute alcohol, but insoluble in organic solvents.

Standards of Quality

Ash	:	Not more than 8.0 per cent
Specific gravity	:	Not less than 1.30 per cent
Sulphated ash	:	Not more than 2.0 per cent
Residue on Ignition	:	Not more than 8.0 per cent
Colour	:	One part dissolved in 1000 parts of water gives a clear, distinct yellowish orange colour and not changed or precipitated by exposing to sun-light for six hours. 5 per cent aqueous solution gives no precipitate with 0.5 ml. of phosphoric acid.

Chemical Tests

(1) **Marsh test:** To 6 ml of solution of caramel add 6 ml of alcohol and mix. Add 15 ml of March reagent (which consists of 100 ml of the pure amyl alcohol 3 ml of syrupy phosphoric acid and 5 ml of water) shake the mixture gently for 2 minutes avoiding forming of emulsion. Allow the layers to separate completely. A coloured aqueous layer indicates presence of caramel.

(2) 5.0 per cent aqueous solution of caramel gives no precipitate with 0.5 ml of phosphoric acid.

Uses

Caramel is used as colourant for food, confectionery, vinegars; liquours, malt beverages; tobacco flavouring, and in elixirs, syrups and other oral liquid products in medicine.

CARRAGEENAN

Synonyms

Carrageenan, Chondrus extract, Irish moss extract

Biological Source

It is sulphated polysaccharide extract of the sea weed called carrageen; or Irish moss, the red-algae obtained from *Chondrus crispus* (Rhodophyceae).

Geographical Source

These red algae grow on the North Atlantic coast of Europe and North America.

France, Spain, Denmark and the USA are the major producers of carrageenan in the world market.

Method of Preparation

Carrageenan is found in the intercellular matrix and cell wall of the algae and constitutes about 60 - 80 per cent of dry weight.

The dried seaweed is first cleaned with cold water and mechanical devices so as to remove salt and other extraneous matter. It is further extracted with hot water containing sodium hydroxide or calcium hydroxide and by adjusting the pH to slightly alkaline range. The extract is obtained by filtration and the purity of the product is attained by precipitation with isopropyl alcohol or ethyl alcohol.

Description

It is a water soluble colloid, dried to form free flowing powder. The molecular weight is 10,000 - 50,000 and it is stable at pH 9.0.

Depending upon the method of preparation, two forms of drugs are available.

One which is extracted with cold water gives viscous solution in water, and the other extracted with hot water forms gel on cooling.

Colour : Off coloured powder. Leaves are light blue in colour.

Odour : Characteristic of marin-powder.

Taste : Bland

Chemical Constituents

Chemically carrageenan have two forms. Depending upon the position of sulphate and presence or absence of anhydrogalactose, two types of Carrageenan are known i.e. Kappa (k) and Lambda (λ).

k-form has ability to form thermo-reversible gel. It is soluble in polar solvents, in hot milk, and hot concentrated sugar solution above 60°C. The sodium salt of this form is soluble in cold water while calcium and potassium salts are insoluble in cold water.

k - carrageenan contains D-galactose, 3, 6 anhydrous-D-galactose and ester sulphate group.

λ carrageenan contains D-galactose and its mono and bisulphate esters.

k - carrageenan is a good gelling agent, while λ - carrageenan is non-gelling and is a good thickener. *i*-carrageenan, μ - carrageenan and ν carrageenan are other types of k - carrageenan only.

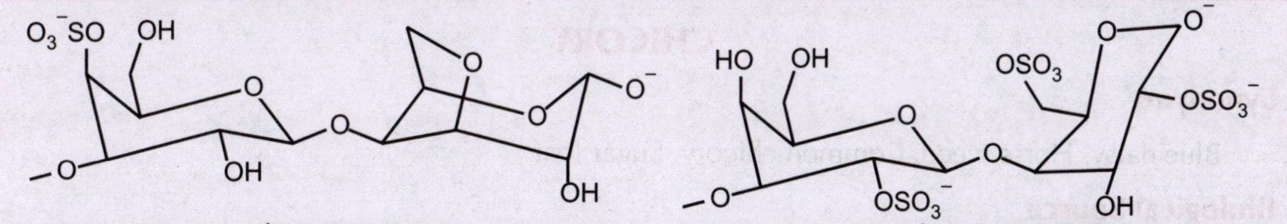

k carrageenan λ carrageenan

Carrageenan differ from agar which contains sulphated esters but carrageenan also contains 20 - 40 per cent sulphite on dry weight basis.

Standards of Quality

Loss on drying	:	2.0 – 2.5 per cent
Total ash	:	2.0 – 5.0 per cent

Uses

Carrageenan is used as emulsifying agent, stabilizing agent, gelling, and viscosity builder in food products. Tooth pastes, creams, lotions and other cosmetic products are prepared by using carrageenan.

In food industry, it is utilised in milk products, ice creams, chocolate, puddings, jams, gels in the concentration of 0.5 - 1.0 per cent.

CELLULOSE

Cellulost is a polysaccharide linked with glucose units and it is chief constituent of fibers of plant cells.

It is tremendously used in paper and textile industry and for several cellulose derivatives such as methyl cellulose, ethyl cellulose, pyroxyline, cellulose acetate, carboxy methyl cellulose (CMC). Cellulose has large industrial and pharmaceutical applications.

Description

It is white in colour, harsh to touch.

Insoluble in water and organic solvents. When pulled it breaks off rather shortly.

It gives no reaction of lignin.

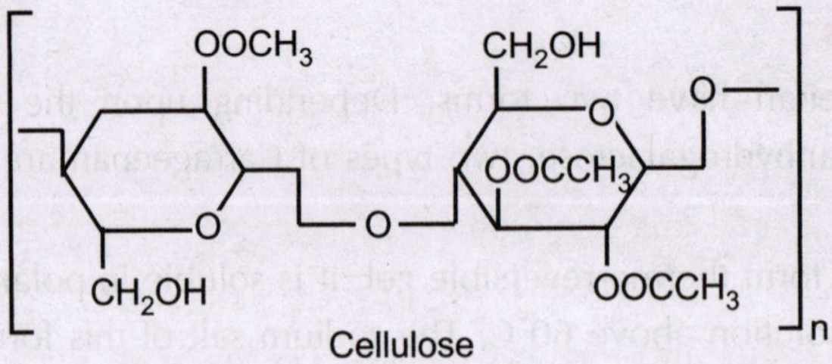

Cellulose

Uses

Used as raw material for several pharmaceutical derivatives such as methyl cellulose, carboxy methyl cellulose (CMC) etc.

Basic material for textile and paper industry. Nitrocellulose is used in explosives, collodiones and laquars.

CHICORY

Synonyms

Blue daisy, Horse weed, Common chicory, Sugar loaf.

Biological Source

Chicory consists of dried leaves and roots of the plant *Cichorium intybus* family: Asteraceae.

Chicory leaves

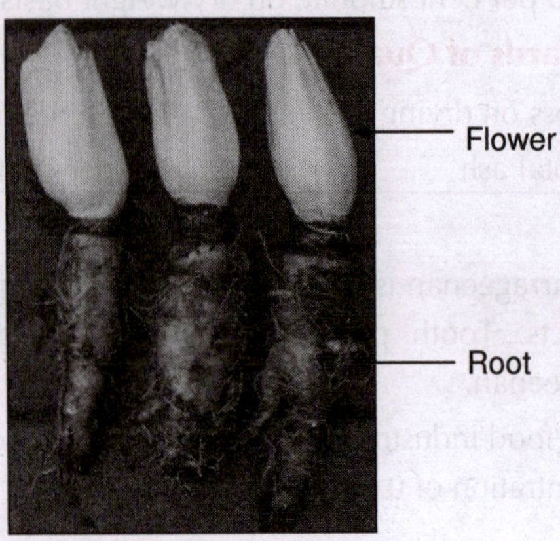

Fig. 8.2 (a): Chicory herb leaves, flower and roots

Geographical Source

It is native to Europe and cultivated in North America, China, New Zealand, Australia and India. In Europe, it is cultivated in Spain, France, Greece, Syria, Palestine, Lebanon, Belgium and Italy.

Cultivation and Collection

Chicory is cultivated indoors in absence of sunlight, to prevent the leaves from turning green.

Description

The healthy chicory leaves are oblong light blue, rarely white which are very delicate and slightly bitter and spicy in taste. Whiter the leaves less bitter is the taste. It is odourless.

Description

Colour	:	Leaves are light blue in colour.
Odour	:	Odourless
Shape	:	Oblong, curly stalked, unlobed.
Size	:	Leaves are 2.0 to 5.0 cm.
Taste	:	Delicate bitter and spicy.

Extra Features

Chicory plants have tough groomed hairy stems. Plants are 30 to 100 cm tall. It flowers from July to October. Leaves are curly stalked, lanceolate, unlobed, smooth and creamy white to green light blue leaves are served by wrapping in blue paper. Leaves are collected in autumn.

Chemical Constituents

Chicory mainly contains Inulin about 20 per cent. Inulin is a polysaccharide carbohydrate.

It also contains flavonoids, sequiterpene lactones, tannins, vitamins, minerals and volatile oil.

It is source of soluble dietary fibers, Lactucin and lactucopicrin are the two sesquiterpene lactones which are responsible for bitter taste. Cichorin, aeseulin, aesuletin and dihydrocoumarin are the other contents of chicory.

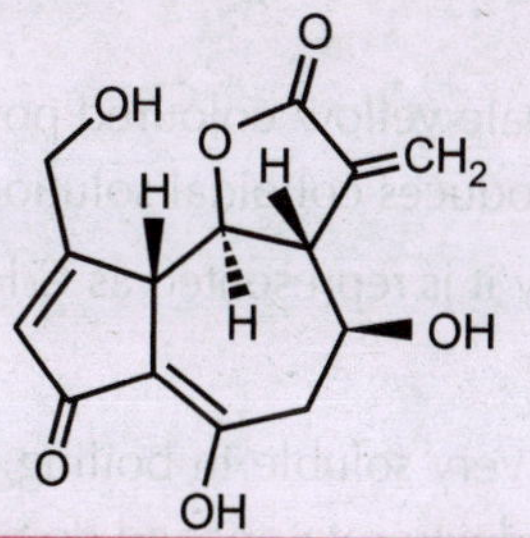

Lactucin

Uses

It is mainly used as dietary supplement or as food additive. Chicory is mainly cultivated as coffee - substitute. Chicory leaves are used in skin-inflammation and swellings also for rheumatism, gall stones and gout.

DEXTRAN $(C_6H_{10}O_5)_n$

It is a polysaccharide produced by growing bacteria on the substrate sucrose, containing a backbone of D-glucose units linked predominantly α-D (1-6).

Commercially *Leuconostoc mesenteroides and Leuconostoc dextranicum* organisms are used.

Chemically can be represented as $(C_6H_{10}O_5)_n$.

Various grades of dextrans are produced depending upon the method used for preparation.

Dextrans are water soluble and insoluble in alcohol. Molecular weight of dextrans varies from 40000 - 75000. All dextrans exclusively contain α-D-glucopyranosyl units.

Uses

Dextran is used as plasma volume expander. Dextran 40 is also used as blood flow adjuvant.

DEXTRIN

Synonyms

Starch gum and British gum.

Biological Source

Dextrin is produced by incomplete hydrolysis of starch with dilute acids or by heating dry starch.

Preparation of Dextrin

There are two types of Dextrin available commercially. All types of starches are used for preparation.

(a) White Dextrin: In this type large amount of acid is reacted with starch at low temperature and for a very short period of time. It is white in colour and gives bluish colour with iodine.

(b) Yellow Dextrin: It is produced by heating starch at high temperature for longer period of time in presence of small amount of acid; the product is yellowish in colour.

Description

White or pale yellow coloured powder, with slight characteristic odour. It is tasteless, light, free flowing and produces colloidal solution with water.

Chemically it is represented as $C_6H_{10}O_5 . XH_2O$

Solubility

Dextrin is very soluble in boiling water, slowly soluble in cold water, and insoluble in alcohol and ether. It is dextrorotatory and do not reduce Fehling's solution.

Standards of Quality

pH of 10 per cent solution	:	3.5 – 4.5
Ash	:	10 per cent (Max.)
Sulphated ash	:	0.5 per cent (Max.)
Moisture content	:	10.0 per cent (max)
Free acidity	:	1.0 ml of 0.1N NaOH (max.)
Proteins	:	0.5 per cent (max.)
Ethanol-soluble substances	:	1.0 per cent (max.)

Identification

(a) Microscopic: Dextrin granules exhibit the same histological appearance of the starch from which the dextrin has been prepared and can be seen under microscope i.e. concentric striations in case of maize starch. No distinct striations in case of potato.

(b) Prepare 2.0 per cent solution of dextrin in water by boiling, cool it, to the suspension and add a drop of Iodine solution. Purple colour is produced.

(c) It does not reduce **Fehling's** solution, gives reddish brown colour with iodine solution.

Uses

Dextrin is a pharmaceutical necessity and mainly used as tablet excipients, in preparation as mucilage and in match sticks and explosives.

DEXTROSE or D-GLUCOSE ($C_6H_{12}O_6$)

Synonyms

Grape sugar

It is naturally occurs in grapes and many other fruits. It can be synthesized for further use.

Preparations

Normally it is made available by controlled enzymatic hydrolysis of corn-starch on commercial base.

Description

Dextrose is white, crystalline solid, odourless and sweet in taste. M.P. – 419. It is optically active and is dextro-rotatory. It contains aldehydic group.

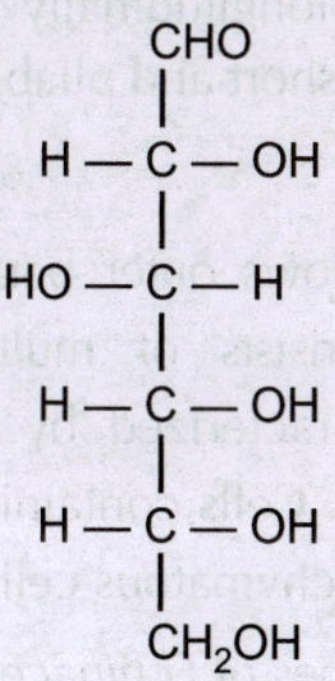

D-glucose

Uses

It is nutritive as it is instant source of energy, also used as sweetening agents and preservative.

It can be given by mouth, by enema or by subcutaneous or intravenous injection. It is present in anticoagulant citrate-dextrose solution. In the form of liquid glucose it it used in carbonated beverages, ice-creams bakery product and in canning industry.

ECHINACEA

Synonyms

Black Sampson, Cone flower purple, Cone flower Brauneria.

Biological Source

Echinaceae consists of dried underground parts roots/rhizomes of *Echinacea purpurea or E. augustifolia* or *E. pallida* collected when the plant is in full bloom family Asteraceae (Compositae).

Geographical Source

It is a perennial herb of South Western and Central parts of the United States of America especially Kansas, Arkansas, Missouri and Oklahoma. It is also found in Canada.

Macroscopic Characters

Fig. 8.3: Echinacea herb

Colour	:	Pale brown to yellowish brown
Taste	:	Right in the beginning sweet followed by bitterness with a tingling effect on the tongue.
Odour	:	Aromatic (mild)
Size	:	Rhizomes 15 mm in diameter Roots 5 - 10 mm in diameter
Shape	:	Roots are cylindrical slightly tapering spirally twisted.
Extra features	:	Rhizomes are crowned with the attachment of aerial stems, show annulations at times. While roots are wrinkled longitudinally and deeply furrowed, fracture is short and pliable.

Microscopic Characters

The transverse section of the root shows outer bark, a distinct cambium, broad xylem and central pith in the rhizome. Cork consists of multilayered thin-walled cells followed by parenchymatous cortex. Rhizome is characterized by lignified phloem-fibres in the pericycle; medullary rays are non-lignified and wide. Cells containing brown pigments and inulin in the form of spherocrystals are also seen in the parenchymatous cells.

The distinguishing characters of rhizomes of *Echinacea augustifolia* and that of *Echinacea pallida* are as under.

In *E. augustifolia* oleoresin canals, which are 80 - 150 μm in diameter containing yellowish orange pleo-resin, are present outside the central cylinder while lignified fibres which are 500 - 800 μm in length are scattered and in groups and are surrounded by phytomelanin deposition.

E. pallida oleo-resinous canals are present inside as well as outside the cylinder whereas lignified fibres have isolated presence in the periphery of the cortex and contain no phytomelanin.

Microscopy of roots of both varieties is similar.

Chemical Constituents

Echinacea species contain various types of phytochemicals, among which the polysaccharides (present in cell wall) namely Arabinogalactan and two fucogalactoxyloglucans have been found to be mainly involved in immunostimulant activities. The other constituents are inulin / echinacoside (caffeic acid glycoside), high molecular weight polysaccharide called echinacin, polyacetylene, tussilagine and isotussilagine (both pyrrolizidine alkaloids).

Standards of Quality

Total ash	>	9.0 per cent
Acid-insoluble ash	>	3.0 per cent
Water soluble extractives	<	15.0 per cent
Moisture	>	10.0 per cent

Uses

Echinacea is used as immunostimulant, mainly for prevention or in the treatment of viral infections causing colds and flu. It also acts as antiseptic and peripheral vasodilator.

The polysaccharides causing immunostimulant activity are found to stimulate macrophages, by which signal molecules namely IL - 6 (interferon), IL - 1 and TNF - α are produced. This triggers the immune system causing migration of neutrophils from bone marrow to blood stream. Stimulated macrophages also cause Phagocytosis. Although like infecting bacteria, these polysaccharides also stimulate immune system, still they are nontoxic. Echinacea has been found to be highly promising immunostimulant from nature.

FENUGREEK

Synonym

Methi, Trigonella, Greek hay.

Biological Source

It consists dried leaflets and ripe dried seeds of cultivated plants *Trigonella foenum - graecum* Linn; Family : Fabaceae. It should contain not less than 0.1% w/w of trigonelline on dry basis.

This is annual herb raised from seeds growing to the maximum height of 30 centimeters.

Geographical Source

Found wild in Punjab and Kashmir. But now cultivated all over India.

Description of Fenugreek Herb

Leaves are compound and about 3-4 cm in length. Entire leaflets normally three, lancolate or obovate.

Flowers are axillary 1-2 sessile **yellow**, pods 2-3 cm long, thin, pointed with 10-20 seeds containing a peculiar **aroma.**

Seeds are 4-5 × 2-3 mm in size and having **oblique furrow** along the length. They are irregularly rhomboidal or square in outline, odourless and yellow in colour or olive green to dark brown.

Taste is bitter and mucilaginous.

Seeds are official in BPC 1949 and British Herbal Pharmacopoeia 1983.

Organoleptic Characters

Colour	:	Seeds are brown to reddish brown in colour.
Odour	:	Characteristic, strong, aromatic.
Taste	:	Bitter
Size	:	3 - 5 × 2 - 3 mm and 1.5 to 2.0 mm thick.
Shape	:	Rhomboidal with rounded edges.
Extra Features	:	A prominent groove on wide surface dividing the seed in two parts. Radical is present in smaller part, while cotyledons are present in larger part. Seeds are hard and flattened.

(a) Methi herb

(b) Methi seeds

Fig. 8.4: Fenugreek herb and seeds

Microscopic Characters

T. S. of methi exhibits thick cuticle covering epidermal cells. Epidermal cells are polygonal, small and thickened. Hypodermic consists of large cells towards periphery and narrower cells towards centre.

They are circular in shape, thickened and with beaded walls. Parenchyma of testa contains elongated, rectangular cells and beaded walls. Endosperm is irregularly thickened of elongated cells. Endospermic cells contain mucilage. Calcium oxalate crystals, sclereids and starch grains are absent.

Chemical Constituents

Methi seeds contain about 3.0% of mucilage 5.0% fixed oil, steroidal saponins 0.5 – 1.5% and 20% of proteins. Two alkaloids trigonelline 0.13% and choline are also present in the seeds. Steroidal saponin Trigofoenoside-A on hydrolysis release diosgenin, yamogenin trigogenin, smilagenin and sarsapogenin. Flavonoids, vitexin, saponaritine, triterpenoid saponin betulin and lupeol are also reported.

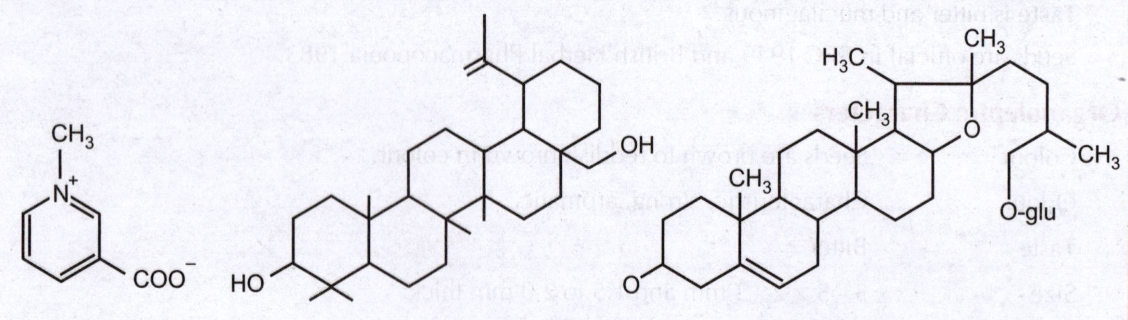

Trigonelline Betulin Trigofoenoside-A

Standards of Quality

Ash	:	Not more than 5.0 per cent
Acid insoluble ash	:	Not more than 2.0 per cent
Loss on drying	:	Not more than 12.0 per cent
Water soluble extractives	:	Not less than 30 per cent

Uses

Anti-inflammatory, anti-cholesterolemic, hypoglyceamic and galactogogue. It is used as demulcent laxative and expectorant.

GUAR GUM

Synonyms

Guar flour, Jaguar gum.

Biological Source

Guar gum is the powder of the endosperm of the seeds of *Cyamopsis tetragonolobus* Linn, belonging to family Leguminosae.

Geographical Source

It has been grown for centuries in India and Pakistan, and was introduced in 1900 in the USA. Its commercial production was started in 1953. It is also available in Sudan, China, Australia and south Africa. 80 per cent of world production is produced in India only. In India it is mainly produced in Rajsthan only.

Isolation of Guar Gum

Guar gum is industrially manufactured from the white, well developed seeds that are freed from foreign matter. The seeds are put into a grinder to get bifurcated guar seeds. The seeds are separated into husk and cotyledons containing embryo. The husk contains fibrous matter and the gum is located into endosperms of embryo. Guar seeds contain 14 - 17 per cent of hull, 35 - 40 per cent of endosperm and 45 - 50 per cent of germ. Cotyledons separated from the endosperm by winnowing and sifting fetch a very high price in the market as a cattle feed. The endosperms i.e. crude guar gum known as Guar meal is pulverised by means of micropulverizer and grinding for 15 minutes.

The endosperm being harder is not affected by micropulverizer. The portion of cotyledons adhering to the endosperm is soft and is converted into fine powder which is separated by sifting. The crude guar gum is now free of cotyledons, the main impurity in the gum. The crude guar gum thus separated is put into pulverizer and grinding is continued for 3 - 4 hours followed by sifting. This process is repeated about 5 to 6 times for several hours to give white coloured guar gum. Finally, it is sifted through sieves of 40 - 60 mesh to give granular and powdered gum. In India, about 15 units are producing guar gum on industrial scale.

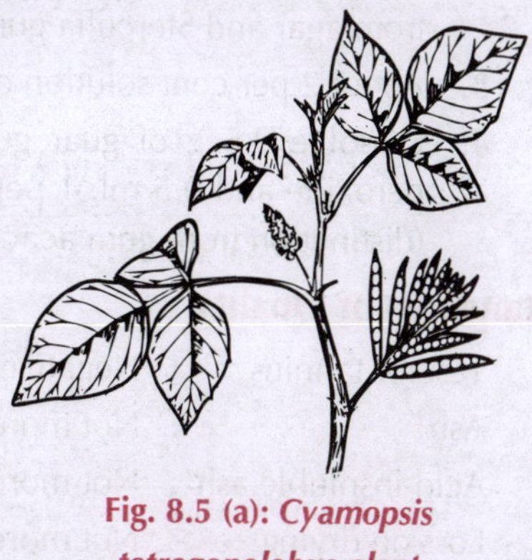

Fig. 8.5 (a): *Cyamopsis tetragonolobus* plant

Description

Colour	:	Pale yellowish to colourless
Odour	:	Odourless
Taste	:	Gummy
Size	:	Fine powder
Extra features	:	It swells rapidly in water, with translucent suspension

Fig. 8.5 (b): Guar seeds

Chemical Constituents

Guar meal contains moisture 8.0 %, Fats 5.3 %, Ash 5.2%, and Fibre 5.9 %.

The contents of guar gum are divided into water-soluble and water insoluble parts. The water soluble fraction constituting about 85 per cent of the gum is known as guaran, which is a high molecular weight hydro-colloidal polysaccharide. Guaran on hydrolysis yields 65 per cent galactose and 35 per cent of mannose, which is combined through glycosidic linkage. Guar gum also contains 5-7 per cent of proteins. The proposed structure of guaran is given here. It also contains 1.13 % calcium, 2.45 % magnesium, 0.3% potassium, phosphorus 3.1%, Iron 40 ppm, zinc 36 ppm and manganese 16 ppm.

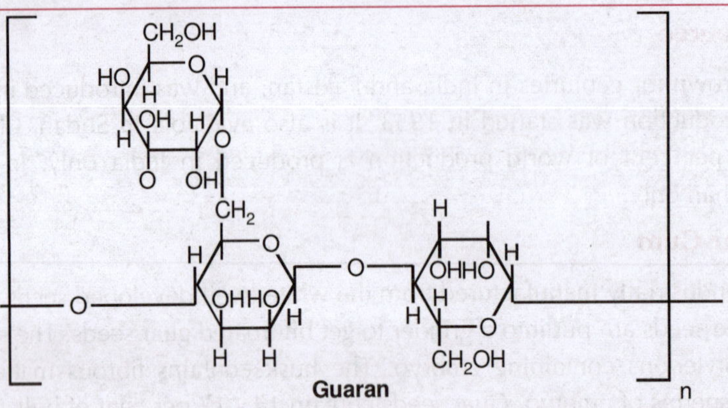

Guaran

Identification

1. It does not acquire olive green colour with weak solution of iodine.
2. With solution of ruthenium red, the gummy solution does not acquire pink colour (distinct from agar and Sterculia gum).
3. About 2 per cent solution of lead acetate gives precipitate with the solution of guar gum.
4. Dissolve 0.5 g of guar gum in 20 ml of water by shaking. To it, add 0.5 ml hydrogen peroxide and 0.5 ml, 1 per cent solution of benzidine in alcohol. No blue colour is formed (distinction from gum acacia).

Standards of Quality

Test for tannins	:	Negative
Ash	:	Not more than 1 per cent
Acid-insoluble ash	:	Not more than 0.5 per cent
Loss on drying	:	Not more than 8 per cent

pH	:	5.5 – 7.5 (1.0 per cent w/w solution)
Test for proteins	:	Not more than 5.0 per cent.

Uses

About 1 per cent mucilage of guar gum possesses similar viscosity to that of mucilage of acacia and a 3 per cent mucilage is similar to mucilage of tragacanth. It has 5 - 8 times thickening power than starch. It is used as a protective colloid, a binding and disintegrating agent, bulk laxative, appetite depressant and in peptic ulcer therapy. Guar gum is a good emulsifying agent. Industrially, this is used in paper manufacturing, printing, polishing, textiles and also in food and cosmetic industries. It is one of the important poultry feed.

It is extensively used as flocculent in ore-dressing and treatment of water.

GUM GHATTI

Synonym

Button tree, Akle wood.

Biological Source

It is a gummy exudate obtained from the stem bark of plant, *Anogeissus latifolia* Wallich, family Combretaceae.

Geographical Source

It is a quite large tree (Fig. 8.6) found in the dry deciduous forests of India and also to some extent in Sri Lanka. The plant has grey bark and red leaves during the dry season.

Collection and Purification

Artificial incisions to stem bark increase the yield. The best crops are obtained in absence of rains and the gum is picked up in dry season especially in April. After collection, the grading of gum is done taking into consideration the colour of the product. Light coloured gum is supposed to be superior. Gum is dried in the sun for several days and then pulverised. It undergoes the process of sifting, aspiration and also density-table separations to remove most of the impurities.

It is exported from India in the crude form to USA the processing of the gum is done in United States.

Fig. 8.6: *Anogeissus latifolia* twig

Description

Gum ghatti occurs in the form of amorphous translucent rounded tears or large vermiform masses. It has light to dark brown colour. The light brown coloured variety is supposed to be best. Tears are about 1 cm. in diameter, without any cracks on the surface. Fracture is uniform and glassy. Gum ghatti has a bland taste and no odour.

Gum ghatti forms glairy, ropy and colourless viscous mucilage with water. A 5 per cent solution in water has 50 - 400 cps viscosity. It is insoluble in ethyl alcohol.

Now-a-days, spray-dried gum ghatti containing no insoluble matter is available in the market.

Chemical Constituents

Similar to gum acacia, gum ghatti is a calcium-magnesium salt. It consists of L-arabinose, D-galactose, D-mannose, D-xylose and D-glucuronic acid. Gum-ghatti also contains oxidase enzyme.

Identification Tests

1. Aqueous solution of gum ghatti (5.0 per cent) gives fine precipitate with Millon's reagent.
2. Aqueous solution gives slight precipitate with lead subacetate (while aqueous solution of gum acacia gives copious precipitate with lead subacetate).
3. 2.0 per cent gelatin solution in water gives precipitate with aqueous solution of gum.

Standards of Quality

Acid-insoluble ash	:	0.05 to 0.5 per cent
Moisture contents	:	12 - 15 per cent
Ash	:	1.5 - 4.0 per cent

Uses

It is mainly used as an excellent emulsifying agent. It can emulsify. It is extensively used in sizing of-paper and in calco printing. It is also used internally. It is more difficult systems as compared to acacia. Aging of gum ghatti solutions increases their viscosity.

Due to its high L-arabinose content, it is used for commercial arabinose isolation.

It is also employed to emulsify petroleum and non-petroleum waxes.

GUM KARAYA

Synonyms

Indian tragacanth; Sterculia gum; Karaya gum.

Biological Source

Gum karaya is a dried gummy exudate obtained from the stems of tree *Sterculia urens* (Roxburgh); *Sterculia villosa* (Roxburgh), *Sterculia tragacantha* (Lindley) or other species of *Sterculia* (Sterculiaceae)

Geographical Source

Sterculia urens is wide spread in India, especially in north central parts.

Collection and Purification

The trees which have a girth of 1 metre are selected and blazes are made. The number of blazes per tree should not exceed two. Immediately after tapping, the exudation oozes out of which exudation is maximum during the first 24 hours. The large irregular tears which weigh in pounds are picked and sent to the collecting centres. Tapping is done during March-April up to June or till the commencement of monsoon.

During rainy season, the yield of gum is reduced. The plants are tapped again in September. The yield per tree is about 1 - 5 kg per year, and on an average the tree is tapped about five times during its lifetime. The large tears are broken into small pieces, which also enhance drying process.

The pieces of bark, sand particles and foreign organic matter are removed. The size reduction and air flotation of loose bark ensure purification to certain extent. Sand particles are removed by gravity. Granulated or crystal gum karaya has particle size between 6 and 30 mesh. Fine powdered variety usually passes through 150 mesh screen.

Fig. 8.7: Karaya twig

Description

Colour	:	White to brown coloured tears.
Odour	:	Slight acetous odour.
Taste	:	Bland, mucilageous.
Size	:	Tears 1 - 5.0 cm in diameter.
Shape	:	Irregular or vermiform pieces.
Solubility	:	Insoluble in water form colloidal solution.
Extra features	:	Swells 50 – 100 times in water.

Chemical Constituents

Karaya gum contains about 8.0 per cent of acetyl group and more than 37% of uronic acid residues. On acid hydrolysis, it gives D-galactose, L-rhamnose, D-galacturonic acid, aldobiuronic acid and an acid trisaccharides. Gum karaya gives pink colour with solution of ruthenium red (not observed in tragacanth).

Standards of Quality

Foreign organic matter	:	Not more than 3.0 per cent
Acid insoluble ash	:	Not more than 1.0 per cent

Uses

It swells about 60-100 times in water. It is neither digested nor absorbed by the body and hence, it is a good bulk laxative. It is also used as denture adhesive in dental treatment and in pharmaceuticals as emulsifier, thickener and stabiliser. Karaya gum is listed in the food chemical codex and hence, used on large scale in foods such as ice pops, cheese spread, sherbets, and ground meat products. It is also used in paper and textile industries.

During 1995-96 and 1996-97, India has exported Karaya gum of worth ₹ 675 lac and ₹ 894 lac respectively.

INULA

Synonyms

Hydrous inulin, Alant starch, Dehlia, Elecampane root.

Biological Source

Inula consists of roots (bulbs) of Dehlia, *Inula helenium* (Compositae) mainly, roots of Dendelion, *Taraxacum officinale*, (Compositae). Burdock root, *Saussurea lappa* (Compositae) or chicory roots, *Chichorium intybus* (Compositae), *chicary chiroricum inlybus* and *echincea* purparea

Inula is the chief source and it contains 40 - 50 per cent of Inulin.

Geographical Source

Inula is cultivated in Germany, France, and Belgium; Chicory and Dandelion in North America; while Saussurea in India.

Description

Inula has following characteristics

Colour	:	Colurless to white,
Odour	:	Odourless
Taste	:	Tasteless
Extra features	:	Soluble in Hot water and insoluble in organic solvents.

It is not hydrolysed by mammalian enzymes and is excreted unchanged thus, it is useful for testing the renal efficiency.

Fig. 8.8 (a) : Inula recemosa herb

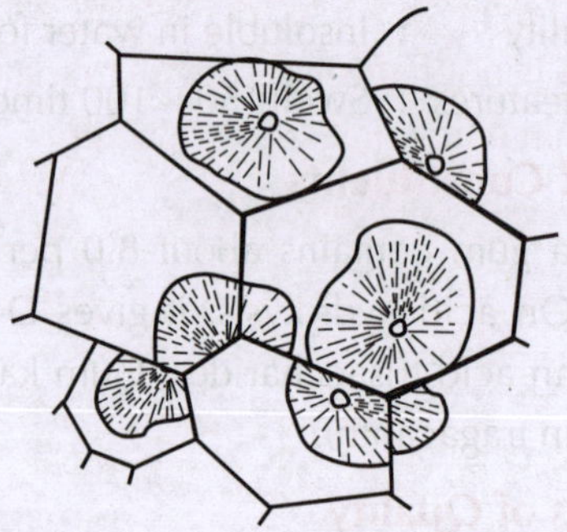

Fig. 8.8 (b) : Sphaerites of Inulin

It crystallises in the form of sphaerites (spherical crystals) in organic solvents.

Isolation

Commercially it is obtained from bulbs of the Dahlia plant. The bulbs are washed thoroughly to remove the earthy matter and vegetable debris and crushed. The pulp so obtained is heated to dissolve the inulin in large volume of purified water. The solution is filtered while hot to remove the impurities and then precipitated and purified by using water soluble organic solvent. Washed several times in organic solvent and dried.

Chemical Constituents

Inula contains mainly polysaccharide inulin. Inulin is a polymer consisting of 35-50, 1-2 linked fructo-furanose units, terminated with one glucose molecule and represented by following formulae.

Polysaccharide of Compositae family and is represented by formula $(C_6H_{10}O_5)_n$ = 3 or 4.

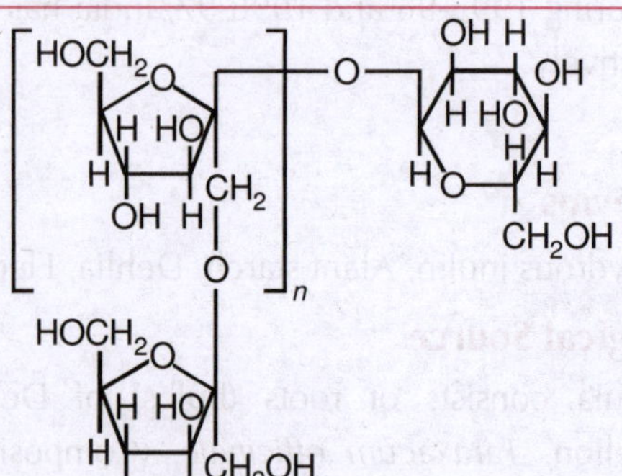

Inulin

Standards of Quality

Ash	:	Not more than 5.0 per cent.
Acid insoluble ash	:	Not more than 2.0 per cent.
Water insoluble extractives	:	Not less than 30.0 per cent.
Specific gravity	:	1.35
Optical rotation	:	–36.5 to 40.5
Sulphated ash	:	Not more than 0.1 per cent

Chemical Tests

To the test solution add solution of a-naphthol and sulphuric acid, brownish red colour is formed.

Uses

Inulin is not hydrolysed by mammalian enzymes and hence excreted unchanged, this useful for testing renal efficiency.

1. Used as diagnostic agent.
2. In preparation of culture media as fermentative identifying agent for certain bacteria.
3. It manufacture of fructose by using an enzyme inulase.
4. It is a ingredient of diabetic bread.
5. It can be sterilized by anticlaving.

ISAPGOL

Synonyms

Ispaghula, Isabgul, Indian Psyllium, Isabgol.

The origin of the word isapgol is form the Persian words ISAP (the horse) and GHOL (the ear). Thus, the literal meaning of word isapgol is the ear of the horse. The seeds and the husk of the seeds are used in medicine since eighteenth-century. About 10 species of the drugs are available in India. Seeds are very small in size. One thousand seeds weigh about 01.5 g. Isapgol has high export potential.

Biological Source

Isapgol consists of dried seeds of the plant known as *Plantago ovata* (Syn: *Plantago indica*; *Plantago afra*) Forskal, family Plantaginaceae. In the pharmaceutical field, the seeds and the dried seed coats, known as isapgol husk, are used.

Geographical Source

The plant is cultivated largely in Gujarat, Punjab and South Rajasthan. The factory for preparation of husk is located at Sidhpur in North Gujarat. In Maharashtra, it is found to be grown successfully near Pune. About 50 thousand hectares of area is said to be under cultivation for the drug in India.

Cultivation and Collection

Isapgol (Fig. 8.8) is a Rabi crop and needs well-drained loamy soil with a pH 7.5 to 8.5 and cool and dry weather. Heavy rains and cloudy weather at its maturity affect the yield adversely. The drug is cultivated by broadcasting method, in the month of November. About 6-12 kg seeds are needed per hectare. RI-87, RI-89, GI-1, GI-2, HI-1, HI-2, HI-5, MIB-121 are some of the species preferred for cultivation in Gujarat state. Irrigation is done 7-8 times at an interval of 8-10 days.

Ammonium sulphate is found to be satisfactory fertilizer for the plant. 25 kg of nitrogenous fertilizers and 25 kg per hectare are most suitable spray of 5.0% aldrin on the soil controls root-eaters. Aerofungin is used to control mid-dew. The crop is harvested in March/April and the average yield of the seeds per hectare is 7.4 quintals. It is collected by cutting the plant just above the ground, converting it to sheaves and drying. Thrashing is done and the thrashed material is winnowed and sieved to maintain quality.

Fig. 8.9: Isapgol Plant

India is producing 48000 tones of seeds annually at present.

The world demand for psyllium and isapgol seeds and husk is increasing (approx. 50,000 tonnes) and the main markets are in U.S.A., France, West Germany, and U.K. The export of isapgol husk and seeds together during 1995-96 was ₹ 155 crores and during 1996-97 it was ₹ 137 crores.

Macroscopic Characters

Colour	:	Pinkish-grey to brown
Odour	:	None
Size	:	10 - 35 mm in length and 1 - 1.75 mm in width
Shape	:	It is ovate cymbiform

Seeds are hard, transparent and smooth with grey or reddish brown oval spot in the centre of the convex surface. Concave surface contains the hilum covered with thin membrane having two perforations. 1 gm accommodates 500 - 600 seeds.

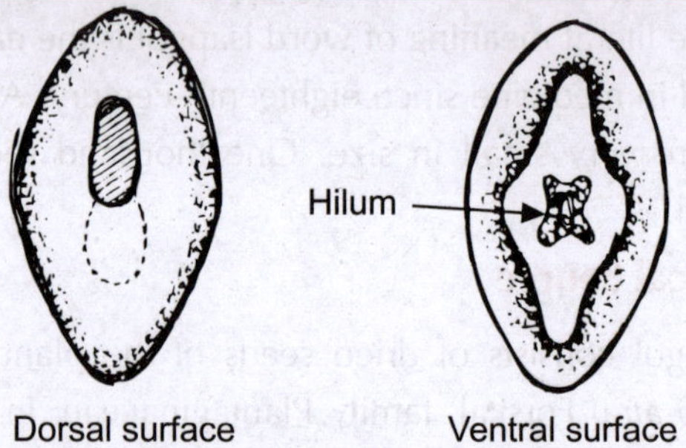

Fig. 8.10: Macroscopic Characters of Isapgol Seed

Isapgol Husk

Isapgol husk consists of epidermis and collapsed adjacent layers removed from dried ripe seeds of *Plantago ovata*.

Isapgol seeds are processed to take out seed coats, commercially known as husk. Husk makes up about 25 - 27 per cent of the seed. The seeds are thoroughly dried and sieved to get rid of foreign organic matter. Seeds are crushed in flat stone grinders by passing several times through them, so as to effect complete removal of the coating. The crushed material is then winnowed to separate the kernels and husk. Husk is sieved to get different grades and sizes. Thus, it consists only of the epidermis and its adjacent layers removed from the dried seeds of *Plantago ovata*.

Macroscopical Characters

It is in the form of pale buff ovate flakes with more or less lanceolate shape. The pieces are 1 - 2 mm in size; flakes are odourless, smooth and free flowing.

Indian psyllium or Isapgol seeds are preferred in the world market for the following reasons.

1. Indian psyllium seeds are available at a lower price.
2. The mucilage content of this species is more.
3. It yields practically colourless mucilage.
4. The husk of the Indian psyllium cracks off under slight mechanical pressure and even it can be easily separated from the seed.

Chemical Constituents of Seed and Husk

Isapgol husk and seeds contain mucilage which is present in the epidermis of the seeds. Chemically, it consists of pentosans and aldobionic acid. The products of hydrolysis are xylose, arabinose, galacturonic acid and rhamnose. Fixed oil and proteins are other constituents of the isapgol seeds.

Swelling factor is the criterion for purity of Isapgol. Swelling factor of the drug is a quantitative swelling due to mucilage present in the drug.

Standards of Quality for Seeds

Total ash	:	Not more than 4.5 per cent
Acid insoluble ash	:	Not more than 0.45 per cent
LOD	:	Not more than 12 per cent
Swelling factor	:	12 – 14
Swelling factor for husk	:	90

Chemical Tests

Swelling factor is determined by putting 1 g of the drug in the measuring cylinder (25 ml capacity) in 20 ml water with occasional shaking. The volume occupied by the seeds after 24 hours of wetting is measured.

Being the mucilage chemically, isapgol gives pink colour with the solution of ruthenium red.

Uses

The seeds, as well as, husk are used as demulcent, laxative, emollient in the treatment of chronic constipation, amoebic and bacillary dysentery. Isapgol husk is preferred to the seeds as the husk contains more mucilage and seeds are said to be irritant as compared to the husk. Mucilage of the isapgol is used in the preparation of tablets and also as a stabilizer in the ice cream industry. The crushed seeds are, many a time, used in the form of poultice for rheumatic pain. Gujarat Drug Chemical Ltd. at Mehsana is engaged in manufacturing of isapgol formulations.

The product formed by removing cations from the mucilage by treatment with cation exchange resins followed by spray drying is an acid form of polysaccharide. This finds special pharmaceutical applications as enteric coating material, tablet disintegrator and also used in the sustained release drug formulations. The mucilage of isapgol has a property of glairiness or stringiness, which is desired in certain cosmetic formulations.

PSYLLUM SEED

Synonym

Flaeaseed, Spanish psyllum

Biological Source

This consists of dried seeds of the plant *Plantago psyllum* Fam.: Plantagenaceae.

Geographical Source

Psyllum seeds are maily cultivated in France and also in Spain, Cuba and to some extent in India.

Description

Colour	:	Dark brown with shining
Odour	:	Odourless
Taste	:	Tasteless, (Mucilagenous)
Size	:	2 – 3 mm × 0.75 to 1 mm
Shape	:	Rounded, oblong
Extra features	:	Transparent, slight constriction across the centre with white membranous patch in the centre of brown furrow.
Weight of 100 seeds	:	90 to 100 mg.
Swelling factor	:	12.75

Chemical Constituents

Epidermal cells of seeds are full of mucilage. Mucilage of these seeds yield of hydrolysis D-xylose, L arabinuse, D galactose and D galactoronic acid.

Uses

In diarrhoe and dysentry used as poultice and in as mechanical laxative.

Substitutes

Several other species of *Plantago* have been investigated of which *Plantago rhodosperma* distributed in Missouri and Louisiana and *Plantago wrightiana* are important. The earlier variety contains 17.5 per cent of mucilage, while the latter contains about 23.0 per cent. In all respects, these two varieties favourably compare with the official drugs.

Plantago purshii and *Plantago aristata* are substitutes of official drug with good mucilage forming capacity. *Plantago asiatica*, found in Andhra Pradesh and Tamil Nadu is also used as a substitute for isapgol seeds.

Adulterants

Isapgol seeds are adulterated with the seeds of Plantago lanceolata. These seeds are oblong, elliptical in shape and yellowish - brown in colour. The swelling factor of these seeds is about five.

LIQUID GLUCOSE

Synonyms

Corn syrup, Starch syrup

Source

It is purified concentrated aqueous solution of nutritive saccharides.

Method of Preparation

Liquid glucose is obtained by partial hydrolysis of edible starch or starchy substances by action of Hydrochloric acid. It is purified and concentrated aqueous solution of nutritive saccharides.

The degree of hydrolysis of starch is measured as Dextrose equivalent, which is percent of Dextrose calculated on dry basis.

Description

It is highly viscous transparent liquid yellow coloured non-crystallizing liquid with slight characteristic smell and sweet taste. It is less sweet than sucrose.

The viscosity of liquid glucose depends upon density, temperature and its Dextrose equivalent. It is soluble in water, glycerine and slightly soluble in alcohol. Liquid glucose has following additional characteristics.

Chemical Constituents

Liquid glucose mainly contains 40 per cent D (+) glucose (Dextrose), maltose, dextrin and about 20 per cent water.

Identification

Liquid glucose differs from sucrose in the respect that it readily reduces Fehling's solution producing a red precipitate of cuprous oxide.

Standards of Quality

Moisture content	:	13 - 18 per cent
Total solids	:	82 - 87 per cent

Sulphated ash	:	0.5 per cent max.
pH (50% solution)	:	4.5 - 5.5
Sulphur dioxide	:	70 - 400 ppm.
Arsenic and copper	:	Not more than 1.0 ppm of (each) content
Dextrose equivalent	:	4.5 - 5.5

Uses

Its hygroscopic and non-crystalizing property is very important tool to improve the quality of liquid preparations in pharmaceutical industry. Liquid glucose should not be used instead of glucose for parenteral purposes. It is also used as sweetening agent. Its high dextrose equivalent exhibits moderate osmotic pressure and also inhibits microbial spoilage.

Industrially, it is used in jams, jellies, creams, ice creams, tobacco and confectionery products.

SUGAR-CANE

Biological Source

It is the plant *Saccharum-officinarum* fam. Graminae, important source of sugar.

Geographical Source

Sugar canes are commercially cultivated in Cuba, Lousiana, Moriceous, Phillipines, Hawai, Indonesia and India. Total world production of sucrose is equally divided between cane-sugar and beet-sugar.

Production of Cane Sugar

India is leading in production of cane sugar (sucrose) and producing highest quantity of sugar in world.

Sugarcanes (stems) are crushed between series of heavy iron-rollers. The juce produced is boiled with lime to neutralise the plant acids, which would otherwise change the sucrose to invert sugar (i.e. mixture of glucose and fructose) and also to coagulate albumines and which are then removed. Juice is filtered and is bleached with sulphur di-oxide. Concentrated further and allowed to crystallise sucross. Residue produced is called mangoes, which is also very important raw material for many industrial chemical like ethyl alcohol. Yield of sucrose is approximately 11 to 12 per cent.

Chemical Nature:

Cane sugar or sucrose is odorless, white, crystaline solid colourless with sweet taste.

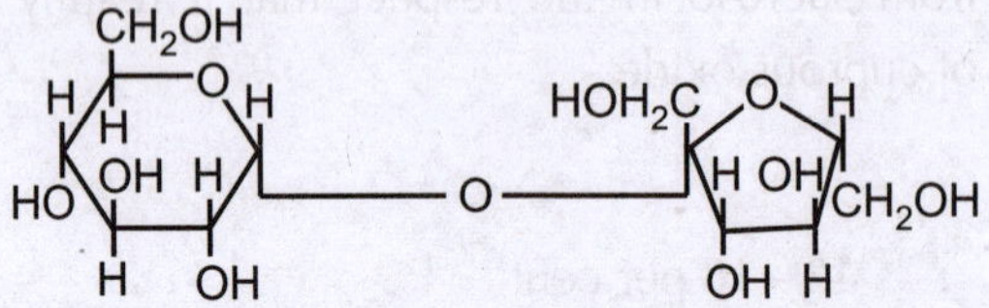

Sucrose $C_{12}H_{22}O_{11}$

Sucrose is soluble in water, insoluble in alcohol. It is optically active and its specific rotation is +66.5. Starting material for fermentative production of ethenol, glycerol, citric acid. Granuting agents and for coating of tablets.

Uses

Sugar is used as sweetening agent, demulant and nutrient in pharmaceutical and food industires.

SUGAR-BEET

Synonym

Beet root

Biological Source

Sugar-beets are the rhizomes of *Beta-vulgaris* fam. Chenopodiceae.

Geographical Source

Sugar beets are produced in Germany, Austria, Russia, France and USA. In India it is cultivated as vegetable and not source of sucrose.

Fourty five countries in the world are cultivating sugar beets for manufacture of cane-sugar (Sucrose) on commercial scale and are producing 30.0 per cent of sucrose of the world production.

Chemical Constituents

Sugar beets contains 18.0 per sent sucrose.

Benefits of production of sucrose from sugar beets are self explanatory and being.

(1) Short duration of sugar-beet crop compared to sugar cane.

(2) Less expensive – For cultivation.

(3) Leaves and wastage produced during processing of beets are used as forage (Animal feed).

(4) Overflow of water, during food season (In June/July) does not cause damage to the bet-crops.

(5) No specific requirement of soil and even waste-land, can also be used for cultivation of beet.

(6) Harvesting of sugar beets for production of sucrose is done after the season of collection of sugar-canes by industries.

(7) Season for collection of cane-sugar can be increased. So as to encourage the employment of workers in sugar-processing units for longer time.

(8) Sugar-beets crop normally do not suffer from any microbial damage (pests), as it is a under-ground crop.

Uses

Major source for industrial production of sugar.

LOCUST BEAN GUM

Synonyms

Arobon, Carob gum, Ceratonia, Carat.

History

The cultivation of locust bean trees was known before Christian era. **Dioscorides** referred its curative properties in the first century AD. In Sicily, the carob trees were probably planted in 16th century. Arabs used carob seed as a unit of weight and it was labelled as **carat**, which in turn has become unit of weight for precious stones.

Biological Source

This gum consists of endosperms of the seeds of *Ceratonia siliqua* Linn, family Leguminosae.

Geographical Source

The plant is found in Cyprus, Sicily and Egypt. It is cultivated easily but very sensitive to low temperature. It is produced in Spain, Greece, Algeria, Morocco, Israel, Italy and Portugal.

Fig. 8.11 (a): Ceratonia siliqua (Locust bean)

Fig. 8.11 (b): Strophan thus Kombe

Cultivation and Collection

Carob is an evergreen tree growing to a height of about 10 metres and has luxuriant perennial foliage. The tree grows on rocky soil and has very long roots that penetrate up to 18 - 25 m and survive in an area where there is very little rain fall. It starts bearing fruits only after three years.

The plants start flowering in January to March every year, and the fully matured fruits are ready by October-November. The fruits (pods) are harvested by shaking the twigs and picked up from the earth by hand and sent to market.

The locust bean fruits or carob are dark chocolate coloured, 10 - 20 cm long, 2.0 - 5 cm wide and 0.5 - 1.0 cm thick.

The seeds are ovoid dark brown, very hard and weigh 0.21 g or 3.2 grains (i.e. one **CARAT**). The locust bean tree can be planted or even grafted. **Seeds are uniform in weight**.

Preparation and Purification of Gum

Locust bean pods consist of 9096 of pulp and 8 per cent of kernels. The pods are fed to kibbling machine. The kernels separated consist of 30 - 33 per cent of husk, germ (about 25 per cent) and endosperm (42 - 46 per cent). Preparation of high quality gum consists of separation of endosperm from embryo. Successful removal of outer dark coloured husk decides the quality of the gum.

Decorticated seeds (dehisced seeds) split lengthwise and are separated from the embryo. The presence of yellow germ (i.e., embryo) increases the rate of fermentation of the gum solutions in further products. Hence, it must be thoroughly removed from the endosperm. It is then pulverised and graded according to the mesh size. The normal mesh sizes available in Italian market are 150, 175 and 200.

Description

Colour	:	Colurless
Odour	:	Odourless
Taste	:	Mucilageneous
Solubility	:	Incompletely disperses in water, Insoluble in alcohol.
Extra features	:	Translcent, hard, horny difficult to break, Natural, Polysaccharidc.
Viscosity	:	160 centipoise of 2.0 percent dispersion at 25°C in water.
		Gum solution is pseudoplastic.

Chemical Constituents

Locust bean gum contains about 88 per cent of D-galacto-D-mannoglycan, 4 per cent pentan, 6 per cent protein, 1 per cent cellulose, and 1.0 per cent ash. The ratio of D-galactose to D-mannose is approximately 20 : 80.

Commercial locust bean gum contains no specks and gum particles should not exceed 8-10 per cent. The natural moisture content of gum is about 14.0 per cent.

Identification

1. Locust bean gum mucilage when boiled with 5.0 per cent potassium hydroxide solution becomes clear and shows no yellow colour as in agar and tragacanth or brown colour as in Sterculia gum.

Uses

It is useful as a stabiliser, thickener and binder in cosmetics; adsorbent and demulcent therapeutically.

It is used as sizing and finishing agent in textiles and also as drilling mud additive. In food industry, it is used as substitute for starch.

XANTHAN GUM

Synonym

Keltrol.

Biological Source

Xantham gum is a polysaccharide produced by certain species of bacteria. It is produced by pure fermentation of glucose using the bacterium named as *Xanthomonas compestris,* which is non-toxic and non pathogenic to human beings.

The high molecular weight (> 10^6) polysaccharide produced in culture broth is isolated by using isopropyl alcohol, which is dried and milled to give commercial Xanthan gum.

Description

Colour : Cream coloured powder

Odour : Odourless

Solubility : Soluble in cold and hot water giving highly viscous solution, this is stable towards change in pH and also to heat.

The aqueous solution of Xanthan gum forms films on evaporation. Aqueous solutions are pseudoplastic.

Chemical Constituents

Xanthan gum consists of chain composed of D-glucose, D-mannose and D-glucuronic acid with short side chains. Additionally, pyruvic acid is present as side chain to the extent of 2.5 - 4.8 per cent of the molecule. Xanthan gum is available in the market as its sodium or potassium salts.

Standards of Quality

Loss on drying	:	Not more than 15.0 per cent
Total ash	:	6.5 – 15.0 per cent
Specific gravity	:	1.600 at 25ºC
Refractive index	:	1.333 (1.0% aqueous solution)
pH (1.0 per cent solution)	:	6.0 – 8.0

Uses

Xanthan gum is a very good emulsifying agent, stabilizer and thickener. It is used in pharmaceutical and cosmetic industries including toothpaste. It is extensively used in food industry for dairy products, salad dressings and canned products.

MANNA

Biological Source

It is saccharine exudation obtained from the stem of *Fraxinus ornus* Linn. Oleaceae.

Geographical Source

It is small tree widely found in Mediterranean basin and Southern Europe. It is also reported in Spain and commercially cultivated in Sicily.

Macroscopical Characters

Manna is available in the form of flakes.

Colour : Yellowish-white

Odour : Agreeable

Taste : Sweet

Size : Pieces of 10 - 15 to 2 - 3 cm

Shape : Three sided, flakes

Extra features : It is very much friable showing crystalline structure and concentric layering. It is soluble in water and insoluble in organic solvents.

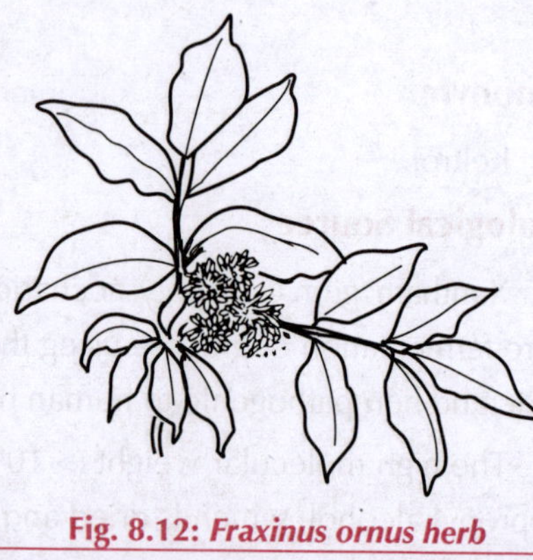

Fig. 8.12: *Fraxinus ornus* herb

Chemical Constituents

Manna contains 40 - 60 per cent mannitol, along with 5 - 15 per cent mannotriose, 10 - 15 per cent manneotetrose. While dextrose, mucilage and small quantity of a fluorescent substance fraxin, are the other contents of manna.

Use

It is a laxative.

MANNITOL

Synonyms

Manna sugar, D-mannitol.

It is a hexahydric alcohol obtained by isolation from manna or sea weeds. It can also be obtained chemically by reduction of mannose.

It is a white, crystalline, odourless non hygroscopic and sweet powder. The crystals are orthorhombic prisms. The melting point is 166-168°C. It is freely soluble in water and insoluble in alcohol. Its specific gravity is 1.52, optically inactive or slightly laevorotatory.

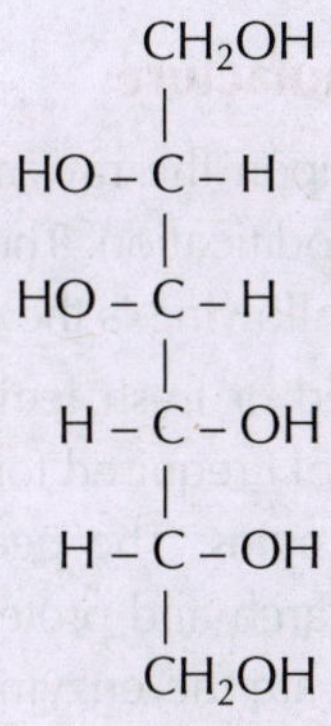

Mannitol

Uses

Mannitol is used as excipient for tablets, diluent for liquids, as stabiliser, thickener and nutritive sweetner, texturizing agent, nutrient and diagnostic agent. Mannitol does not get absorbed from gastro-intestinal tract. In the form of injection, it is not metabolised and is eliminated by glomerular filtration and hence, used as diagnostic aid and also as an osmotic diuretic.

PECTIN

Pectin is a complex carbohydrate component found in nature in the middle lamella of plant cells. Chemically, pectin is a neutral methoxy ester of pectic acid. Pectins are polyuronides and consist of mixture of pectic substances like protopectin, pectin, pectinic acid and calcium pectate. They are obtained from the inner portion of the rind of citrus fruits or other vegetative matter, such as sun- flower, papaya etc. Pectin is a reversible colloid i.e. it may be dissolved in water, precipitated dried and redissolved, without altering its physical properties. Pectin, by addition of water, forms the lumps which on heating convert into solution. The solution is clear to transmitted light and cloudy to reflected light. Under certain conditions, in presence of sugar and acid, it forms jelly like mass.

Pectin is available from number of plants, belonging to different families. Following are few important sources of pectin with content

Lemon peel	:	10 - 15 per cent
Orange peel	:	10 - 12 per cent
Apple pomace	:	10 - 15 per cent
Carrots	:	10.0 per cent
Sunflower-heads	:	05.0 per cent

Papaya, guavas and mangoes are also rich sources of pectin. Pectin has several industrial and pharmaceutical applications. Most of the requirement is met by import. In India, very a few units are manufacturing pectin on commercial scale. The USA, Switzerland and other European countries are producing pectin either from citrus peels or from apple pomace. Pectin is standardised on 'jelly grade' i.e. it's 'setting power' by the addition of sugar (100, 150 and 200 jelly grades are supposed to be standard grades for food and medicinal use).

Biological Source

Pectin is the purified carbohydrate product obtained by acid hydrolysis from inner portion of the rind of citrus peels i.e. *Citrus Simon* or *Citrus aurantium*, family Rutaceae.

Method of Manufacture

Depending upon the raw material from which it is to be isolated the process of manufacture needs suitable modification. The type of pectin (i.e. low or high methoxy group) needed should also be considered. Following is the general process for isolation of pectin from citrus peels.

The preserved or fresh lemon peels are heated with 20 times its weight of water at 90°C for 30 minutes. The pH required for maximum extraction is 3.5 - 4.0. The pH can be adjusted by lactic citric or tartaric acids. The peels after boiling are pressed and solution is cleared by settling or centrifugation. Starch and proteins are removed by suitable enzymic hydrolysis and the solution is heated to deactivate the enzymes and decolorized with active carbon or any other suitable agent. Pure pectin is precipitated out by using water soluble organic solvent, washed again and dried in vacuum. Pectin is incompatible with calcium, so necessary precaution should be taken to keep it away from its metallic salts throughout the process of extraction. It is packed in the containers or polyethylene bags for marketing.

Description

Colour	:	Cream or yellowish-coloured powder
Odour	:	Odourless
Taste	:	Mucilaginous

Extra Features

It is coarse or fine light powder and hygroscopic in nature.

It is soluble in 20 parts of water, aqueous solution being viscous opalescent, colloidal and mobile. It is insoluble in alcohol and other organic solvents.

Chemical Constituents

Total hydrolysis of pectin yields D-galacturonic acid, methyl alcohol, small amount of galactose and arabinose.

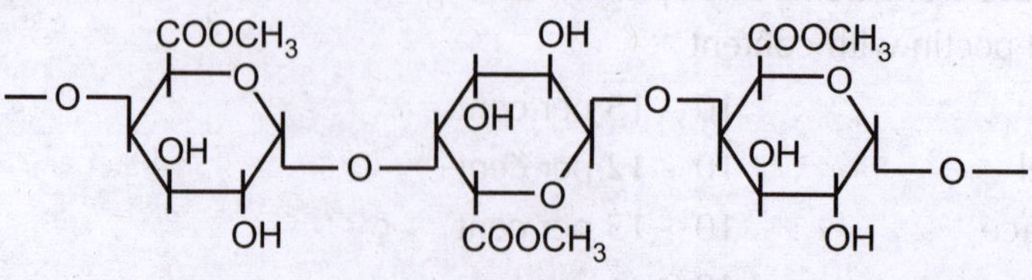

Pectin

Pectin should not contain less than 7.0 per cent of methoxy groups and 78.0 per cent of galacturonic acid, calculated with reference to the ash-free and dried substance.

Identification Tests

1. 10 per cent aqueous solution forms stiff gel on cooling.
2. To 5 ml (1 per cent) solution add 1 ml (2 per cent) solution of potassium hydroxide and set aside at room temperature for 15 minutes. A transparent gel or semi-gel forms (distinction from tragacanth). Acidify gel with dilute hydrochloric acid and shake well. A voluminous, colourless, gelatinous precipitate forms which when boiled becomes white and flocculent.

Standards of Quality

Ash	:	Not more than 4.0 per cent
Acid insoluble ash	:	Not more than 0.4 per cent
Loss on drying	:	Not more than 10.0 per cent
Microbial limit	:	1 g is free from *Salmonellae*

Uses

Pectin is used as an adsorbent in the treatment of diarrhoea and as a haemostatic for internal or external haemorrhage. It is used as an emulsifying agent, a gelling agent in acid medium, and as a plasma substitute. In food industry, it is used as a thickening agent for sauces, jams, ketchups, etc. It is extensively used in cosmetic preparations.

Pectin is more stable in acidic medium. Pectin in combination with gelatin has been suggested for use as an encapsulating agent in pharmaceutical formulations to promote sustained release.

STARCH

Synonym

Amylum

Biological Source

Starch consists of polysaccharide granules obtained from the grains of maize (*Zea mays* Linn.); rice (*Oryza sativa* Linn.); or wheat (*Triticum aestivum* Linn.); belonging to family Gramineae or from the tubers of potato (*Solanum tuberosum* Linn.), family Solanaceae.

Geographical Source

Most of tropical, as well as, subtropical countries prepare starch commercially.

Preparation of Starch

Depending upon the raw material to be used for processing or type of the starch to be produced, different processes are used for the commercial manufacture of starch.

Potato Starch

The potatoes are washed to remove the earthy matter. They are crushed or cut and converted into a slurry. Slurry is filtered to remove the cellular matter. As potatoes do not contain gluten, they are very easy to process further. After filtration, the milky slurry containing starch is purified by centrifugation and washing. Then, it is dried and sent to the market.

Rice Starch

The broken pieces of rice sequel polishing are used for processing. The pieces of rice are soaked in water with dilute sodium hydroxide solution (0.5 per cent), which causes softening and dissolution of the gluten. After this, the soaked rice pieces are crushed and starch prepared as described under potato starch.

Maize Starch (Corn Starch)

Maize grains are washed thoroughly with water to remove the adhered organic matter after which they are softened by keeping in warm water for 2 - 3 days. Sufficient sulphur dioxide is passed to the medium to prevent fermentation. The swollen kernels are passed through attrition mill to break the grains, so as to separate the endosperm and outermost coating of the grains. At this point, special attention is given to separate the germ (embryo). This is effected by addition of water, wherein germs float and are separated. The water which is used to soften the grains dissolves most of the minerals, soluble proteins and carbohydrates from the grains. The water, being rich in all these contents, is used as a culture medium for the production of antibiotics like penicillin (*corn steep liquor*). The separated germs are used to prepare the germ oil by expression method and are known as corn oil. The oil contains fatty acids like inoleic and linolenic acids and vitamin E. It is used commercially, for the preparation of soap. The starchy material contains gluten; most of this is removed by simple sieving and then by washing. Starch being heavier, settles at the bottom and is followed by gluten. Several treatments with cold water wash the starch effectively, which is then centrifuged or filter-pressed and finally, dried in flash dryers on a moving-belt dryer.

Wheat Starch

Wheat being the major article of food is restrictedly used for preparation of starch. In this process, the wheat flour is converted into dough and kept for a while. The gluten in the dough swells and the masses are taken to grooved rollers, wherein water is poured over them with constant shaking The starchy liquid coming out of the rollers is processed conveniently to take out the starch, which is then dried and packed suitably.

Description

Starch occurs as fine powder or irregular, angular masses readily reducible to powder.

Colour	:	Rice and starch grains are white, while wheat is cream coloured and potato is slightly yellowish.
Odour	:	Odourless
Taste	:	Mucilaginous
Size and Shape	:	Starch grains vary depending upon the types which can be described as under.

Microscopic Characters

1. **Rice Starch:** The granules are simple or compound. Simple granules are polyhedral, 2 - 12 micron in diameter. Compound granules are ovoid and 12 - 30 μ × 7 - 12 μ in size. They may contain 2 - 150 components.

2. **Wheat Starch:** Simple lenticular granules which are circular or oval in shape and 5 - 50 μ in diameter. Granules contain hilum at the centre and concentric faintly marked striations. Rarely, compound granules with 2 - 4 components are also observed.

3. **Maize Starch:** Granules are polyhedral or rounded, 5 - 31 in diameter, with distinct cavity in the centre or two to five rays cleft.

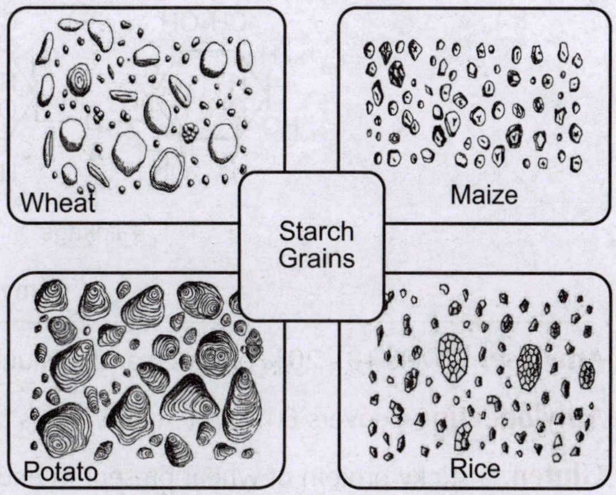

Fig. 8.13: Commercial Starches

4. **Potato Starch:** Generally, found in the form of simple granules, which are sub-spherical, somewhat flattened irregularly ovoid in shape. Their sizes vary from 30 - 100μ. Hilum is present near the narrower end with well-marked concentric striations.

Starch is insoluble in cold water and also alcohol.

Chemical Constituents

Starch contains chemically two different polysaccharides; Amylose (β-amylose) and amylopectin (α-amylose), in the proportion of 1 : 2. Amylose is water soluble and amylopectin is water-insoluble, but swells in water and is responsible for the gelatinising property of the starch. Amylose gives blue colour with iodine, while amylopectin yields bluish black colouration. When starch contains gluten protein compound and is sticky in nature. It is absent in rice, maize, potato starch.

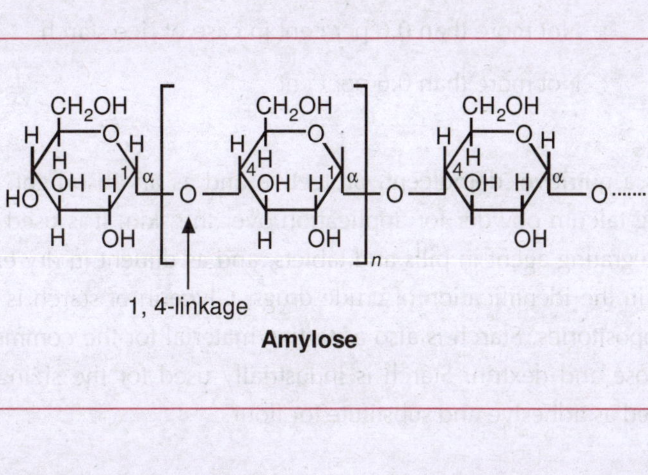

Amylose

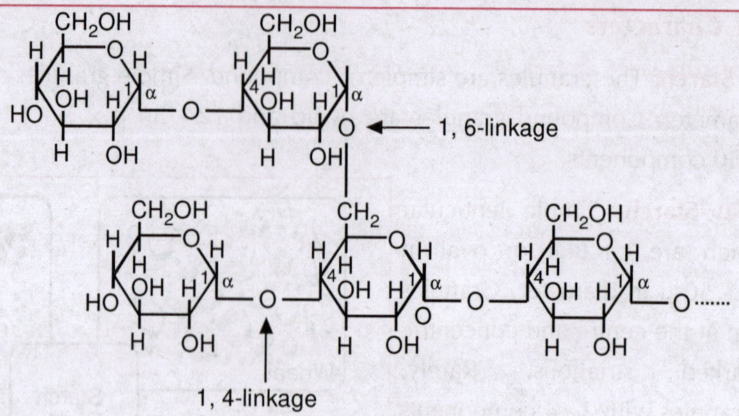

Amylopectin

Amylose: Covers 10 - 20% of starch and it is unbranched polymer.

Amylopectin: It covers 60 - 80% starch and is branched polymer.

Gluten, a sticky protein of wheat present in endosperm. It retains the carbon-dioxide produced during fermentation and makes the dough spongy and porus.

Identification

1. Boil 1 g of starch with 15 ml of water and cool. The translucent viscous jelly is produced.
2. The above jelly turns deep blue by the addition of solution of iodine. The blue colour disappears on warming and reappears on cooling.

Standards of Quality

For pharmaceutical purpose, starch should have the following standards

Loss on drying	:	Not more than 15 per cent for rice, wheat and maize starches
	:	Not more than 20 per cent for potato starch
Ash	:	Not more than 0.3 per cent for potato, wheat and maize starches
	:	Not more than 0.6 per cent in case of rice starch
Sulphated ash	:	Not more than 0.6 per cent

Uses

Starch is used as a nutritive, demulcent, protective and as an absorbent. Starch is used in the preparation of dusting talcum powder for application over the skin. It is used as antidote in iodine poisoning, as a disintegrating agent in pills and tablets, and as diluent in dry extracts of crude drug. It is a diagnostic aid in the identification of crude drugs. Glycerin of starch is used as an emollient and as a base for suppositories. Starch is also a starting material for the commercial manufacture of liquid glucose, dextrose and dextrin. Starch is industrially used for the sizing of paper and cloth. Gluten of starch is used as adhesive and substitute for flour.

Substitutes and Adulterants

Tapioca starch or Cassava or Brazilian arrowroot. This starch is obtained from *Manihot esculenta* (Euphorbiaceae) and is used as substitute for starch.

Soluble Starch

It is prepared from the potato or maize starch by chemical process, such as treatment with dilute hydrochloric acids so as to destroy the gelatinising property of the starch. Microscopically, it shows no change in its characters. It is soluble in water and forms a transparent mobile liquid. It may be sterilised by autoclaving.

Mucilage of starch is used as an emollient for skin and is also a base for some enemas.

Arrow Root Starch

There are several starches in the market known as arrow-root starch, their botanical identity is difficult. Very common in the field are cassava starch (*Manihot utilissima*) sweet potato starch (*Ipomoea batatas*), Convolvulaceae and canna starch *Canna edulis*, Cannaceae

In the pharmaceutical industry West Indian arrowroot and East Indian arrow-root varieties are well known.

ARROW ROOT STARCH (WEST INDIAN)

Synonyms

Maranta starch, Natal arrow root, Arrow root

Biological Source

It is obtained from rhizomes of *Maranta arundinaceae* Linn. family Marantaceae

Description

Shape	–	Ovoid to ellipsoid
Size and aggregation	–	7 - 30 - 45 - 75 μm, simple
Hilum	–	Cleft at broad end
Striations	–	Faint and concentric
Appearance between crossed polarisers	–	Well defined cross

Uses

West Indian arrow root starch is used as a pharmaceutical aid.

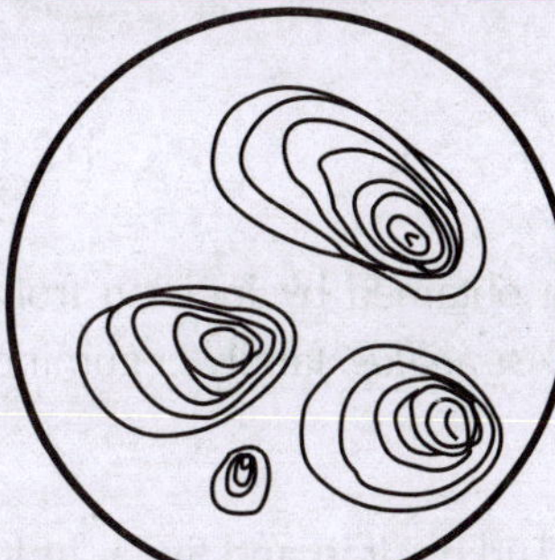

Fig. 8.14: Arrow-root starch (West-Indian)

ARROW ROOT STARCH (EAST INDIAN)

Synonym

Curcuma starch.

Biological Source

It is obtained from rhizomes of *Curcuma angustifolia* Roxburg or *C. leucorhiza* Roxburgh fam. Zingiberaceae

Description

Colour	:	White
Shape	:	Flattened avoid with slight terminal projection
Size and aggregation	:	15 - 30 - 60 μm; simple
Hilum	:	Point in terminal projection
Striations	:	Faint, numerous and transverse
Appearance between cross polarisers	:	Well-marked cross

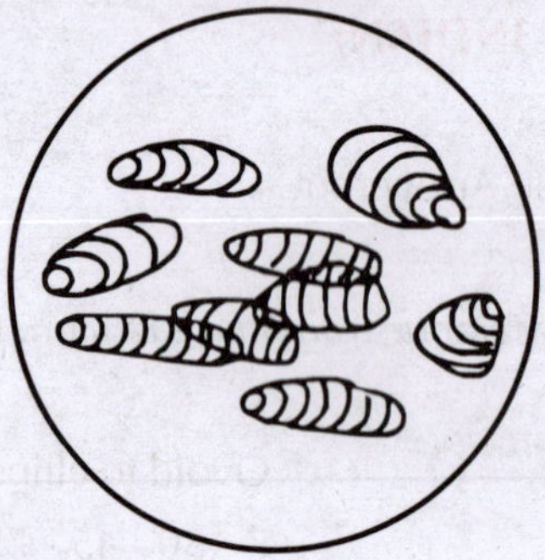

Fig. 8.15: Arrow-root starch (East-Indian)

Use

East Indian arrow root is used as pharmaceutical aid.

TRAGACANTH

Synonyms

Gum Tragacanth, Tragacantha.

Biological Source

It is the dried gummy exudation obtained by incision from stems and branches of *Astragalus gummifer* Labill and other species of *Astragalus*, family Leguminosae.

Geographical Source

It is indigenous to Iran, Greece, Turkey, Iraq and Syria. In India, Garhwal, Kumaon and Central Punjab are the areas where few species of tragacanth are found. North Syria and Iran supply Persian tragacanth. Smyrna tragacanth is exported from Smyrna port in Asiatic Turkey.

Fig. 8.16: Tragacanth herb

Collection of Tragacanth

Most of the shrubs from which tragacanth is collected grow at an altitude of 1000 - 3000 m. The shrubs are thorny. The mode of formation of tragacanth is entirely different from that of acacia, the gum exuding out immediately after an injury. It is collected from wild grown plants.

Most of the drug comes from Persian source only. Tragacanth gum is formed as a result of transformation of the cells of pith and medullary rays into gummy substance. Incisions are more on various parts of the stem and fluid which oozes out is collected after drying. Tragacanth is found in irregular flattened flakes with ribbon like appearance depending upon the incisions made on the plant. It is collected from April to November every year.

Description

Colour : The flakes are white or pale yellowish-white.

Odour : Odourless

Taste : Mucilaginous

Shape : Tragacanth occurs in the form of thin, flattened ribbon like flakes, more or less curved

Size : Flakes are approximately 25 × 12 × 2 mm in size.

Fig. 8.16 (a): Tears of Tragacanth

The gum is horny, translucent with transverse and longitudinal ridges. Fracture of the drug is short.

It is partly soluble in water wherein it swells to a homogeneous, adhesive and gelatinous mass. It is insoluble in alcohol.

Chemical Constituents

Tragacanth contains two fractions of which one is soluble in water. The water-soluble portion of the tragacanth is known as tragacanthin, constituting about 8 - 10 per cent of the gum. Water-insoluble portion is known as bassorin (60 to 70%). Tragacanth contains about 15 per cent of methoxy group which swells in water. It is the constituent of the gum responsible for its high viscosity. Normally, 1.0 per cent solution of tragacanth has a viscosity of not less than 250 centipoises The products of hydrolysis of tragacanth are tragacanthic acid, galacturonic acid, D-galactopyranose, L-arabino-rhamnose and D-xylopyranose. Galacturonic acid is produced by hydrolysis of tragacanthic acid. It also contain 3.8% of methoxyl group.

Standards of Quality

Foreign organic matter	:	Not more than 1.0 per cent
Sulphated ash	:	Not more than 4.0 per cent
Moisture content	:	Not more than 15.0 per cent
Specific gravity	:	1.250 to 1.385
Ash contents	:	Not more than 3.00 per cent
Methoxyl group	:	Not less than 3.25 per cent

Identification Tests

1. When solution of tragacanth is boiled with few drops of 10 per cent aqueous ferric chloride solution, deep yellow precipitate is formed.
2. A stringy precipitate is formed by dissolving tragacanth and precipitated copper oxide in concentrated ammonium hydroxide.
3. When it is warmed with sodium hydroxide solution, canary yellow colour is developed. With strong iodine solution, it gives green colour.

Uses

It is used as a demulcent and as an emollient in cosmetics Tragacanth is used as a thickening, suspending and as an emulsifying agent. It is used along with acacia as a suspending agent. Mucilage of tragacanth is used as a binding agent in the tablets and also as excipients in the pills. Tragacanth powder is used as an adhesive. It is also used in lotions for external use and also in spermicidal jellies, as a stabiliser for ice-cream in 0.2 - 0.3 per cent concentration and also in sauces.

CARBOXY METHYL CELLULOSE

Synonyms

CMC, Cellolaxepolycell.

Polymer sodium carboxy-methyl cellolose is a sodium salt of poly carboxy methyl-ether of cellulose, and is one of several cellulose derivatives used in medical and pharmaceutical profession.

Other derivatives of cellulose are ethyl cellulose, hydroxy ethyl cellulose, hydroxy propyl-methyl cellulose and cellulose acetate phthalate.

Its one percent solution has 2000 CPS viscosity.

Uses

As thickening agent, suspending agent, as tablet excipient and bulk laxative and also in preparation of pyroxylin.

Pyroxylin is used in preperation of collodion and flexible colladion which are topical protectives.

It is highly hygroscopic white powder, soluble in water.

Available in various viscosities of 1.0 per cent solution. Solutions are stable between pH2 to pH10, used as food stabiliser.

CHITIN AND CHITOSAN

CHITIN:

It is the polysaccharide derivative containing amino and acetyl groups and is the most abundant organic constituent in the skeletal material of the invertebrates. It is found in mollusks, annelids, arthropods and also, as a constituent of the mycelia and spores of many fungi.

During the canning of lobster, crab and shrimp, substantial quantities of crustacean waste material is formed. The waste mainly consists of shells and heads. The hard crustacean shell contain 15 - 20 per cent chitin while soft shells of shrimp are reported to have 15 - 30 per cent of chitin, Mycelia of *Penicillium* species contain about 20 per cent of chitin. Chitin is amorphous solid, practially insoluble in water, dil acids, dilute and concentrated alkalies, alcohol, and other organic solvents, but soluble in concentrated hydrochloric acid sulphuric acid, phosphoric acid and unhydrous acetic acid. Chitin obtained from different sources have different physical standards.

Deacetylated chitin is known as **chitosan**.

CHITOSAN:

Preparation of Chitosan

The shells are pulverized to fine powder and treated with 5 per cent hydrochloric acid for about 24 hours to remove all calcium and other impurities of the shell (shells contain 60 - 75 per cent of calcium carbonate). The proteins of the shells are removed by subjecting the above extract with proteolytic enzymes like pepsin or trypsin. The product, thus obtained, is pink in colour. It is bleached by acidified hydrogen peroxide for 5 - 6 hours at ambient temperature. Bleached product obtained by this method is once again processed for deacetylation. Deacetylation is done at 120°C with two parts of potassium hydroxide, one part of ethyl alcohol and one part of ethylene glycol. It is tested to ascertain the acetyl content. When the acetyl content is low, the reaction is stopped. The deacetylated product is known as chitosan.

Description

It is an amorphous solid, practically insoluble in water, dilute acids and concentrated alkalis, alcohol and organic solvents; It is soluble in concentrated hydrochloric and sulphuric acids.

Properties

Good quality of chitosan with a concentration of 1.25 per cent in dilute acetic acid has very high viscosity, i.e. 120 cps. Its molecular weight is 1,43,000 - 2,10,000. It is a cationic polysaccharide. It contains approximately 6.5 per cent of nitrogen.

Chemical Constituents

Chemically, chitin is a 2-acet amido-2-deoxycellulose, because it is composed of N - acetyl - D -glucosaminyl units linked together by β - D (1 - 4) linkages.

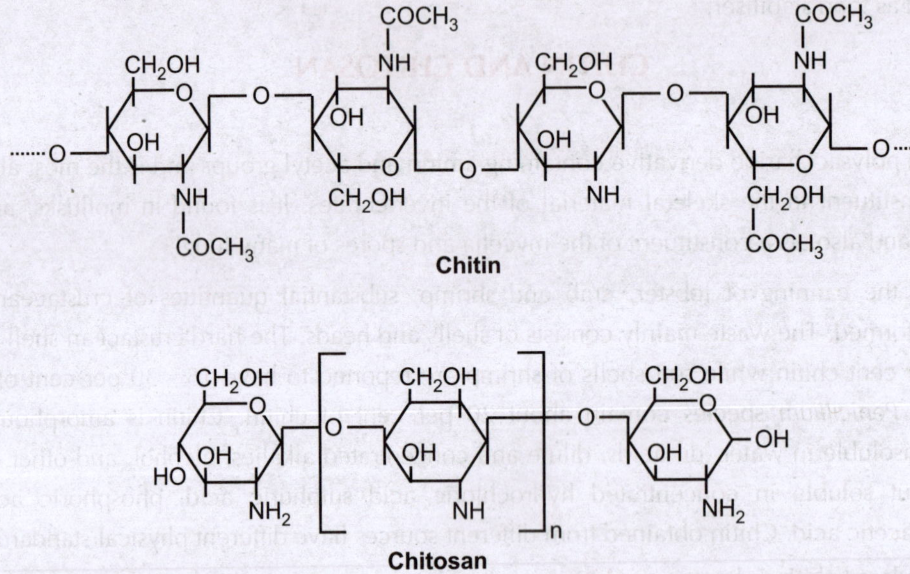

Identification Tests

1. Soak chitosan in iodine solution and to it add 10 per cent sulphuric acid; a deep violet colour is developed.
2. Dissolve chitosan in 50 per cent nitric acid and allow it to crystallise the sphero-crystals of chitosan-nitrate are formed. If the crystals are examined by polarised light, using crossed nicols, a distinct cross is observed.

Standards of Quality

pH	:	4.0 – 6.0 (1.0 per cent w/w aqueous solution)
Density	:	1.35 – 1.45
Mol. wt.	:	1,43,000 – 2,10,000

Uses

Due to cross linkages between glucose molecules in chitin, it is indigestible to humans like cellulose. Therapeutically, it is used in wound healing preparations. Chitin is used as a sizing agent for rayon, cotton, wool and even for synthetic fibres. It has great adhesiveness with glass and plastics. Industrially chitin and chitosan are used in the process of water treatment.

TAMARIND

Synonym

West Indian Tamarind, Indian datepalm, Tamarind pulp.

Biological Source

Tarmarind consists of dried pulp of the ripen fruits of *Tamarindus indica* Linn. free of its brittle epicarp. It belongs to family Leguminosae.

Geographical Source

It is indigenous to Tropical Africa and cultivated in India and West Indies. It is found wild in Madhya-Pradesh, Andhra Pradesh, Karnataka and Maharashtra.

Cultivation and Collection

It is a large tree and bears racemes of flowers, followed by indehiscent legumes of varying size. Cultivation of tamarind is done by sowing the seeds. Plantation is done at a distance of 10 × 10 metres. The plant starts bearing after 10 years. Shivrai, Yogeshwari no. 263, PKM-1, AKT-10 are the species, specially used for cultivation. The fruits are collected from the plants when they are fully ripe and about to fall on the ground. The fruits are collected by knocking with the bamboos or by beating in April and May. The fruits are made free of its woody and brittle epicarp. Plant survives for 60 – 80 years. The mesocarp which is fleshy and reddish brown in colour forms the drug. Before using it is processed to remove hard black seeds which are 10 to 12 mm in size and 4 to 5 in number in every legume. For keeping quality, salt and preservatives are generally added to the pulp. Each plant produces about 100 to 200 kg fruits per year.

Macroscopical Characters

Colour : Reddish brown

Odour : Pleasant and fruity.

Taste : Sweet and acidic.

Size : The entire legumes are about 5 to 10 cm in length and 2 cm in width.

Shape : It is found as moist sticky mass with yellowish brown fibres and rarely seeds.

Fig. 8.17: Tamarind fruits

Chemical Constituents

Tamarind contains 30 to 40% of invert sugars; 10% tartaric acid, about 8%, of sodium potassium tartarate; and pectin. Seeds are full of starch and are used for sizing in textile industry. Dry bark of tamarind contains 15 – 25% Tannin.

Uses

(i) It is used as a mild laxative and pleasant acid refrigerant.

(ii) Bark is used for dyeing alongwith leaves and fruits of acacia. It is also used in the preparation of Confection of Senna.

(iii) Seeds are used for isolation of starch.

CYCLODEXTRINS

Cyclodextrins are a group of structurally related natural products formed during bacterial digestion of cellulose. These cyclic oligosaccharides consist of (α-1,4)-linked α-D-glucopyranose units and contain a lipophilic central cavity and a hydrophilic outer surface.

In the year 1891, French scientist **A. Villiers** first discovered cyclodextrins (CD) as degradation products of starch and named this dextrin "cellulosine". In 1903, the Austrian microbiologist **Franz Schardinger**, who is supposed to be founder of cyclodextrin chemistry, identified two crystalline compounds (A, B) similar to cellulosine from bacterial digest of potato starch and replaced the name "cellulosine" by α-dextrin and β-dextrin. These were later known as α-cyclodextrin and β-cyclodextrin. In 1935 **Freudenberg** and **Cramer** suggested the existence of larger cyclodextrin known as γ-cyclodextrin.

In the pharmaceutical industry they are used as complexing agents to increase the aqueous solubility of poorly soluble drugs and to increase their bioavailability and stability. In addition, cyclodextrins can be used to reduce gastrointestinal drug irritation, convert liquid drugs into microcrystalline or amorphous powder, and prevent drug–drug and drug–excipient interactions etc. Thus, the CDs are well known to form inclusion complexes with a variety of organic compounds, among them.

Properties of Cyclodextrins

Thus three well characterized and commercially available members of this family are α-, β-, and γ-CDs (with 6, 7 or 8 glucose units respectively). The α-cyclodextrin, also known as Schardinger's α-dextrin, α-CD, ACD, and C6A; β-cyclodextrin, also known as Schardinger's β-dextrin, β-CD, BCD, and C7A; γ-cyclodextrin, also known as Schardinger's γ-dextrin, γ-CD, GCD, and C8A. Due to the chair formation of the glucopyranose units, cyclodextrin molecules are shaped like cones with secondary hydroxyl groups extending from the wider edge and the primary groups from the narrow edge. This gives cyclodextrin molecules a hydrophilic outer surface, whereas the lipophilicity of their central cavity is comparable to an aqueous ethanolic solution. They have limited aqueous solubility due to the strong intermolecular hydrogen bonding in the crystal state. The natural α-CD and β-CD, unlike γ-CD cannot be hydrolysed by human salivary and pancreatic amylases.

Importance of cyclodextrins (CDs):

1. ***Enhancement of solubility:*** CDs increase the aqueous solubility of many poorly soluble drugs by forming inclusion complexes with their apolar molecules or functional groups.
2. ***Enhancement of bioavailability:*** When poor bioavailability is due to low solubility, CDs play an important role to improve solubility. When drug is complexed with CD, dissolution rate and, consequently, absorption are enhanced.
3. ***Improvement of stability:*** CD complexation improves the chemical, physical and thermal stability of drugs. When a molecule is entrapped within the CD cavity, it is difficult for the reactants to diffuse into the cavity and react with the protected drug molecule.
4. ***Reduction of irritation***: Drug substances that irritate the stomach, skin or eye can be encapsulated within a CD cavity to reduce their irritancy. As the complex gradually

dissociates and the free drug is released, it gets absorbed into the body and its local free concentration always remains below the levels that might be irritating to the mucosa.

5. ***Odour and taste masking***: Unpleasant odour and bitter taste of drugs can be masked by complexation with CDs.
6. ***Prevention of incompatibility***: Drugs are often incompatible with each other or with other inactive ingredients present in a formulation. Encapsulation of the incompatible ingredients within a CD molecule stabilises the formulation and prevent drug-drug or drug additive interaction.

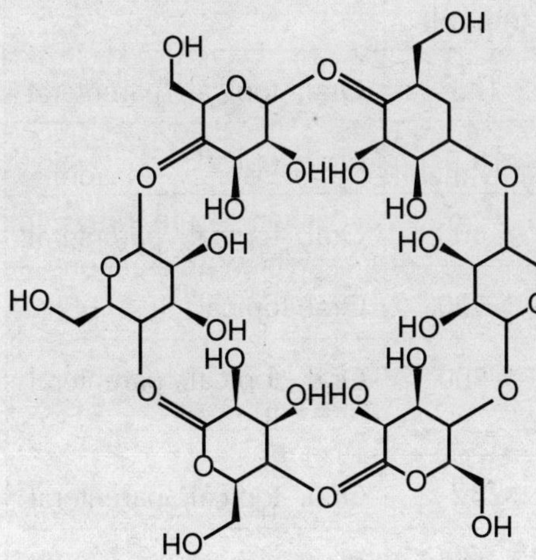

(a) α-cyclocextrin (α-DC)

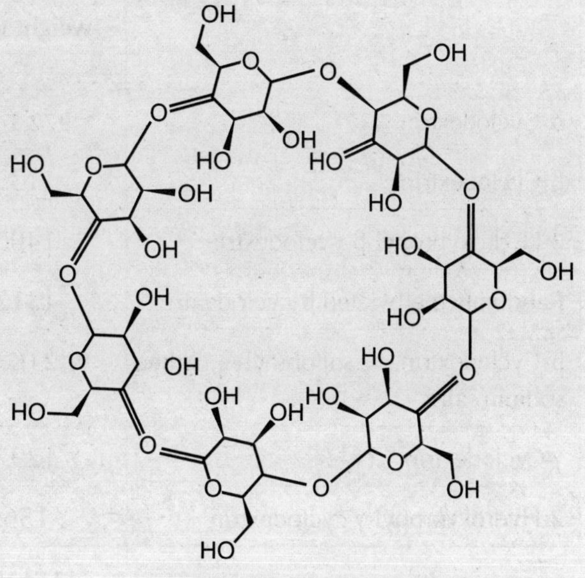

(b) β-cyclodextrin (β-CD)

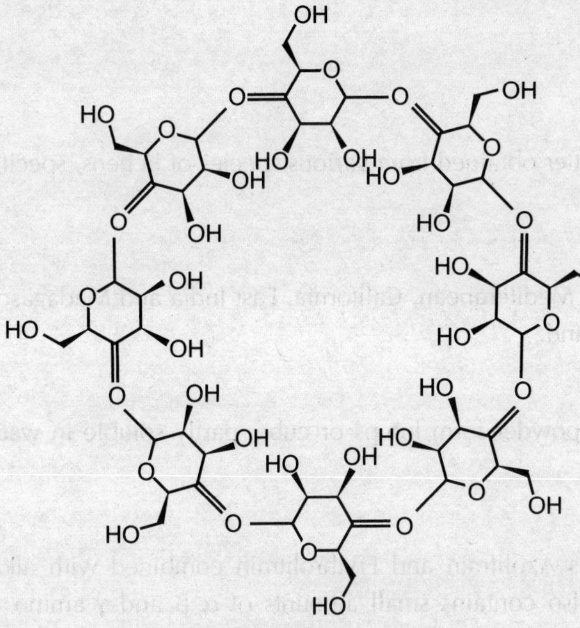

(c) γ-cyclodextrin (γ-CD)

Uses:

Hydrophilic cyclodextrins are considered non-toxic at low to moderate oral dosages. The natural cyclodextrin and its derivatives are used in topical and oral formulations, but only α-cyclodextrin and the hydrophilic derivatives of β- and γ-cyclodextrin can be used in parenteral formulations.

Table 8.2: Cyclodextrins that are found in marketed pharmaceutical products

Cyclodextrin	Molecular weight (Da)	Solubility in water (mg/ml)	Applications
α-Cyclodextrin	972 145	145	Oral, topical, parenteral
β-Cyclodextrin	1135	18.5	Oral, topical
2-Hydroxypropyl-β-cyclodextrin	1400	> 600	Oral, topical, parenteral
Randomly methylated β-cyclodextrin	1312	> 500	Oral, topical
β-Cyclodextrin sulfobutyl ether sodium salt	2163	> 500	Oral, topical, parenteral
γ-Cyclodextrin	1297	232	Oral, topical, parenteral
2-Hydroxypropyl-γ-cyclodextrin	156	> 500	Oral, topical, parenteral

LITMUS

Synonyms

Lacmus, Tumsole.

Biological Source

It is blue colouring matter obtained from various species of lichens, specifically from *Variolaria, Lecanora* and *Roccella*.

Geographical Source

Scandinavia, shores of Mediteranean, California, East India and Madagascar. It exclusively gets manufactured only in Holland.

Morphology

It is available in blue powder form lumps or cube, partly soluble in water and ethyl alcohol. Mol. Weight = 3300.

Chemical Constituents

Litmus chiefly contains Azolitmin and Erythrolitmin combined with alkalies. Licanoric acid, orcein and erythrolin. It also contains small amounts of α, β and γ amino and hydroxy, orecin. Orecin is a biological stain.

Uses

Litmus is used as acid-base indicator and has red colour at pH 4 and blue colour at pH 8.3 used in preparation of litmus paper. In microscopy to colour culture media. For diagnostic purpose and colouring beverages.

HONEY

Synonyms

Madhu, Honey Purified, Mel

Biological Source

Honey is a sugar secretion deposited in honey comb by the bees, *Apis mellifera, Apis dorsata,* and other species of *Apis*, belonging to family Apidae, order Hymenoptera.

Geographical Source

Honey is produced in Africa, Australia, New Zealand; California and India.

Preparation of honey for the Market

The nectar of the flowers is a watery solution containing 25 per cent sucrose and 75 per cent water. The worker bee sucks this nectar through its hollow tube of mouth (proboscis) and deposits in honey sac located in the abdomen. The enzyme invertase present in saliva of the bee converts nectar into invert sugar, which is partially utilised by the bee and the remaining is deposited into honeycomb.

Collection of Honey

About 70-80 villages in the vicinity of Mahabaleshwar are engaged in collection of honey for about eight months of the year.

Government of Maharashtra offers 90.0% of expenses to collect the honey. Honey collection centres of their societies are offered Rs. 400/- per kg rate for honey and Rs. 400 – 500/- per kg for Bees wax.

Honey Directorate through KHADI and GROMMODHYOG maha mandal of Agriculture department has been entrusted the work of collection of honey and bees wax, by Maharashtra government. Honey collection boxes and other amenities are also provided to collecting societies. The societies collect about 30000 to 40000 kg of honey per year. MANGHAR village of Mahabaleshwar Taluka is known as Madhache goan. i.e. Honey village.

Honeycomb is smoked to remove the bees and honey is obtained by applying the pressure to it or allowing it to drain naturally. The honey of commerce is heated to 80°C and allowed to stand. The impurities which float over the surface are skimmed off and liquid diluted with water to produce honey of 1.35 densities. Natural honey has the density of 1.47. Many a time, honey is extracted from the comb by centrifugation. It must be free from foreign substances. Honey is liable to fermentation, unless it is suitably processed.

Fig. 8.18: Collection of Honey comb for Isolation of honey

Honey is heated to 80°C before it is sent to the market, so as to avoid fermentation. It should be cooled rapidly or else it darkens in colour on keeping. If necessary (and if not prepared by centrifugation method), honey is required to be filtered through wet cloth or flannel.

Description

Colour : Pale yellow to yellowish-brown liquid

Odour : Characteristic, pleasant

Taste : Sweet and faintly acidic

Chemical Constituents

Honey is an aqueous solution of glucose 35 per cent (± 3 per cent), fructose 45 per cent (± 5 per cent) and sucrose about 2 per cent. The proportion of sugars may vary depending upon the source of nectar and the enzymatic activity responsible for converting nectar into the honey. The other constituents of honey are maltose, gum, traces of succinic acid, acetic acid, dextrin, formic acid, colouring matters, enzymes (invertase, diastase, and inulase) and traces of vitamins. Proteins and pollen grains from various flowers are also found in honey.

Since, honey is a saturated solution of sugar, on keeping, it starts crystallising. A product which contains crystallised dextrose is called as **Granulated honey**. Heating of honey serves the purpose of minimising the granulation.

Artificial invert sugar, an adulterant of honey contains furfural which is detected by **Fiehe's** test. It gives instant red colour with resorcinol in hydrochloric acid.

Chemical Test for Honey:

1. To an aqueous solution of honey (2 ml) Fehlings solution 1 and 2 are added and the reaction mixture is heated on steam bath for 5 – 10 minutes, a brick red colour is produced due to presence of reducing sugars.
2. **Fiehe's test:** Honey is shaken with petroleum ether (5 ml) for 5 – 10 minutes. The upper ethereal layer is separated and evaporated in a china dish. On addition of 1% of resorcinol in HCl (1 ml) a transient red colour is formed in natural honey while in adulterated honey, the colour persists for some time.

Standards of Quality

Weight per ml : 1.35 - 1.36 g

Specific rotation : +3° - –10°

Total ash : 0.1 - 0.8 per cent

It is soluble in water, but insoluble in alcohol.

It has to pass limit tests for chloride and sulphate

It is syrupy thick liquid, translucent when fresh and on keeping it becomes opaque and granular due to the crystallisation of glucose.

Uses

Honey is used as a demulcent and sweetening agent. It is readily assimilated and hence is a good nutrient to infants and patients. It is antiseptic and applied to burns and wounds. It is a common ingredient of several cough mixture, cough drops and vehicle for ayurvedic formulations. Currently, it is used in preparation of creams, lotions, soft drinks and candies also.

India has only exploited 10 per cent of its honey potential. India is producing 11000 tones of honey per annum. Per capital consumption of honey in India is only 8.0 gms while in Germany is 1800 gms.

According to Khadi and Village Industries Commission, absence of modern technology inadequate marketing and infrastructure for manufacture of honey are the reasons for poor development.

LACTOSE

Synonyms

Milk sugar, Lactose monohydrate, Lactin *Saccharum lactis*.

Biological Source

Lactose is a natural disaccharide consisting of galactose and glucose and obtained from milk of most mammals. Being nutritive it is manufactured by most of the developed countries commercially.

Description

Lactose occurs as white or almost white, crystalline powder with odourless and faintly sweet taste. It is stable in air but readily absorbs odours.

Several varieties of lactose are available

(a) Anhydrous α lactose

(b) α-lactose monohydrate

(c) Anhydrous β-lactose.

Chemistry of Lactose

Lactose monohydrate is the monohydrate of O - β - D galactopyranosyl - (1 - 4) – α - D glucopyranose. Its elemental formula is $C_{12}H_{22}O_{11}$. It is hydrolysed to glucose and gelatose in gastro-intestinal tract by lactase enzyme in the body.

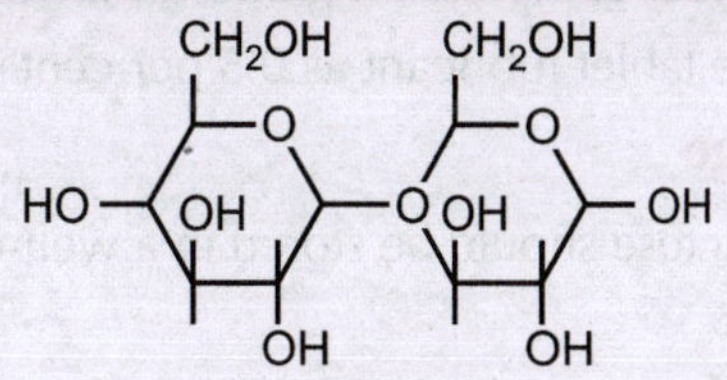

Lactose

Standards of Quality

Bulk density	:	0.62 g/cm^3
Specific rotation	:	+ 54.4 to + 55.9°
Loss on drying	:	Not more than 5.0 per cent
Ash content	:	Not more than 1.0 per cent
Sulphated ash	:	Not more than 0.1 per cent
Hygroscopically	:	It absorbs insignificant amount of moisture at 25° C (at relative humidity upto about 90 per cent)
Solubility	:	Lactose is freely soluble in water while practically insoluble in chloroform, ethanol and ether.

Standards for Different Varieties

Sr. No.	Varieties	Density	True Density	m.p.	Moisture content
1.	Anhydrous α-lactose	---	---	223°C	7 per cent w/w
2.	Anhydrous β-lactose	1.589	1.552	252.2°C	1 per cent w/w
3.	α-lactose monohydrate	1.540	1.552	201.202°C	5 per cent w/w

Chemical Tests

1. On warming a mixture of 5 ml of sodium hydroxide and 5 ml of a hot saturated solution of lactose the liquid becomes yellow and finally brownish red; adding several drops of cupric sulphate forms a red precipitate of cuprous oxide forms.
2. Dissolve 0.25 g of lactose in 5 ml of water and then add 5 ml of 10 M ammonia and than heat in a water bath at 80ºC for 10 minute, a red colour develops.

Uses

Lactose is widely used as a filler or diluent in tablets, capsules, and infant feed formulae besides in dry powder inhalations.

It is used in combination with sucrose to prepare sugar coating solutions (Approx.1: 3).

Various lactose grades are available usually fine grades of lactose are used in the preparation of tablet by wet granulation method.

Direct compression grades of lactose are more compressible and consist of spray dried lactose, which contains pure α lactose monohydrate along with small amount of amphorous lactose, while other specially produced direct compression grades of lactose do not contain amphorous material but may contain glossy areas to improve compressibility.

Direct compression grades of lactose combined with microcrystalline cellulose or starch usually require tablet lubricant as 0.5 per cent w/w magnesium stearate.

Storage

Lactose should be stored in a well-closed container in a cool dry place.

Chapter 9...

DRUGS CONTAINING GLYCOSIDES

A number of medicinal plants contain complex organic molecules which are in conjugation with sugar moieties, mostly monosaccharides. The number of these sugar molecules may be one or more. Such compounds are called as glycosides and exert therapeutically significant effects on humans and animals. Many glycosides are used in traditional and modern medicines because of their cardiotonic, purgative, analgesic, anti-rheumatic, demulcent and other useful actions.

Glycosides may be defined, in general, as the organic compounds from plants or animal sources which on enzymatic or acid hydrolysis give one or more sugar moieties along with nonsugar moiety. The former is called as glycone and the later as aglycones or genin. Chemically, they are the acetayls or sugar ethers, formed by interaction of hydroxyl group each of non-sugar and sugar moieties, with a loss of water molecule. The hydroxyl group of aglycones may be alcoholic or phenolic and in some cases from amines also. The sugars involved in glycosides are of different types, but most commonly, it is β - D - glucose. The other sugars found are galactose, mannose, rhamnose, digitoxose, cymarose, etc.

Stereochemically, these are alpha and beta glycosides which may be considered as a theoretical aspect, because the plants contain only beta glycosides. The linkage between glycone and aglycones is called glycosidic linkage and on the basis of this linkage, alpha and beta stereoisomers are assigned. In a simplest form, the glycoside with these two isomers can be synthesized from union of methyl alcohol and glucose.

The classification of glycosides is based either on the chemical nature of aglycone part or therapeutic activity exhibited by the same. Another mode of classification is based on the type of linkage existing between the glycone and aglycones part. A brief description of all these methods of classifications is given below.

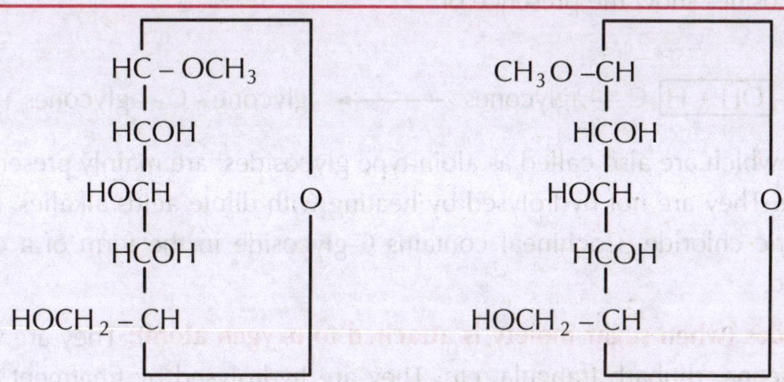

α - methyl glycoside   β - methyl glycoside

According to chemical nature of aglycones moiety, they are grouped into:

1. Anthraquinone or anthracene glycosides,
2. Sterols or cardiac glycosides,
3. Saponin glycosides,
4. Cyanogenetic or cyanophoric glycosides,
5. Isothiocynate glycosides,
6. Flavonoids Flavonol glycosides,
7. Coumarins and Furanocoumarin glycosides,
8. Aldehyde glycosides,
9. Phenol glycosides,
10. Steroidal glycolkaloids,
11. Glycosidal bitters or miscellaneous glycosides.

The glycosides may be classified, in relation to their therapeutic activity into different groups like cathartics, cardiotonics, analgesics, anti-rheumatics, anti-ulcer etc.

Sometimes Glycosides are also classified on the basis of type of sugar or the glycone part of their structure. Accordingly, there are glycosides with glucose, rhamnosides with rhamnose, pentosides with pentose like ribose, etc. The sugars involved in the structure are normally the common type of sugars. Only in few cases, sugars like digitoxose, cymarose are present.

Another approach for their classification is by considering the linkage across glycone and aglycones part. Basically, all types of glycosidal linkages are occured by interaction of – OH group of glycone and hydrogen coming through any of the radicals like CH, – OH, – SH and – NH present on aglycones part. Hence, by elimination of one water molecule, linkage or a bridge is formed and the type of glycoside formed is named by putting the element as prefix like C-glycoside, N-glycoside, O-glycoside or S-glycoside.

To illustrate the individual pattern, the following brief account of such glycosides is given.

(1) C-glycosides (when sugar moiety is linked to carbon atom): Some of the anthraquinone glycosides such as cascarosides from cascara and aloin from aloe, as well as, some members of flavone type of glycosides show the presence of
C-glycosides.

$$\text{Glycone} — \boxed{\text{OH + H}}\ \text{C} — \text{aglycones} \longrightarrow \text{glycone - C - aglycones} + H_2O$$

C - glycosides, which are also called as aloin-type glycosides, are mainly present in members of Liliaceae, e.g. Aloe. They are not hydrolysed by heating with dilute acids/alkalies, but by oxidative hydrolysis with ferric chloride. Cochineal contains C-glycoside in the form of a colouring matter called carminic acid.

(2) O - glycosides (when sugar moiety is attached to oxygen atom): They are very common in higher plants e.g. senna, rhubarb frangula, etc. They are hydrolysed by treatment of acid or alkali into aglycones and sugar, i.e. glucoranillin, amygdaling.

$$\text{Glycone - O}\boxed{\text{H + HO}}\ — \text{aglycones} \longrightarrow \text{glycone O - aglycones} + H_2O.$$

(3) S - glycosides (when sugar is linked to sulphur atom): Their occurrence is restricted to isothiocynate glycosides like sinigrin from black mustard.

$$C_3H_5 - \underset{\underset{\displaystyle N-O-SO_3K}{||}}{C} - S - C_6H_{11}O_5$$

They are formed by interaction of sulphydryl group of aglycones and hydroxyl group of glycone.

$$\text{Glycone - }\boxed{\text{OH + H}}\text{ S - aglycones} \longrightarrow \text{glycone - S - aglycones} + H_2O$$

(4) N - glycosides: The most typical representative example of N - glycosides is nucleosides, where the amino group of base reacts with - OH group of ribose/deoxyribose and ultimately gives N - glycosidic form.

$$\text{Glycone - }\boxed{\text{OH + H}}\text{ N - Glycone} \longrightarrow \text{Glycone - N - aglycones} + H_2O$$

The glycosides contain sugars but still the physical, chemical and therapeutic properties are dictated by aglycones part. The sugar moiety facilitates absorption of glycosides and helps in transportation of aglycones portion at the site of action.

Glycosides are crystalline or amorphous substances which are soluble in water and dilute alcohol with an exception of resin glycosides, but insoluble in organic solvent like chloroform or ether. The aglycones moiety is soluble in non-polar solvents like benzene or ether. Glycosides are easily hydrolysed by water, mineral acids and enzymes. They show optical activity, normally with laevo rotatory effects. Glycosides do not reduce Fehling's solution until they are hydrolysed. They are believed to participate in growth regulation and protection of plant.

METHOD OF ISOLATION OF GLYCOSIDES:

Glycosides are usually accompanied by specific enzymes (usually emulsion and maltase) capable of causing their hydrolysis, emulsion causes hydrolysis of β-Glycosides, while maltase causes hydrolysis of α-glycosides and therefore enzymatic hydrolysis is one of the criteria to ascertain the nature of linkages in the glycosides.

Sta-Otto Method

The general method of extraction of glycosides is outlined here. The drug containing glycoside is finely powdered and the powder is extracted by continuous hot percolation using Soxhlet apparatus with alcohol as solvent. During this process, various enzymes present in the plant parts are also deactivated due to heating. The thermolabile glycosides, however, should be extracted at temperature preferably below 45°C. The extract is treated with lead acetate to precipitate tannins and thus eliminate non-glycosidal impurities. The excess of lead acetate is precipitated as lead sulphide by passing hydrogen sulphide gas through solution. The extract is filtered, concentrated to get crude glycosides. From the crude extract, the glycosides are obtained in pure form by making use of processes like fractional solubility, fractional crystallization and chromatographic techniques such as preparative thin layer and column chromatography. The characterization of isolated purified compounds is done by IR, UV, visible, NMR and mass spectrometry and elemental analysis.

The synopsis of glycoside containing crude drugs with their chief constituents and therapeutic uses is given in Table 9.1.

Table 9.1: Summary of Glycoside Containing Crude Drugs

I. Anthracene Glycosides

	Name of Drug and synonym	Biological Source	Active constituents	Uses
1.	Aloes (Kumari)	Dried juice of leaves of *Aloe vera, Aloe barbadensis, Aloe ferox,* Liliaceae	Barbaloin, aloe-Emodin	Purgative
2.	Cascara (Sacred bark)	Dried bark of *Rhamnus purshiana*, Rhamnaceae	Cascarosides A, B, C and D	Mild purgative
3.	Hypericum (Goat-weed)	Dried aerial parts of Hypericum perforatum, Hypericaceae	Hypericin, Hyperforin	Anti-depressant
4.	Rhubarb (Rheum)	Dried rhizome of *Rheum palmatum*, Polygonaceae	Rhein, aloe emodin	Purgative, bitter stomachic
5.	Alexandrian (Senna)	Dried leaflets of *Cassia acutifolia*, Leguminosae	Sennoside A and B	Purgative
6.	Indian senna (Senna leaf)	Dried leaflets of *Cassia angustifolia*, Leguminosae	Sennoside A and B	Purgative
7.	Senna pods (Senna-fruit)	Dried nearby ripe fruits of *Cassia acutifolia* or *C. angustifolia*, Leguminosae	Sennoside A and B	Purgative

II. Steroidal Glycosides or Cardiac Glycosides

	Name of Drug and synonym	Biological Source	Active constituents	Uses
1.	Digitalis (Fox glove leaves)	Dried leaves of *Digitalis purpurea, Scrophulariaceae,* and dried leaves of *Digitalis lanata*, Scrophulariaceae.	Purpurea glycoside A B, digitoxin lanatosides A, B and C, digoxin	Cardiotonic
2.	European squill (Scilla)	Dried sliced bulbs of *Urginea maritima,* Liliaceae	Scillaren A and B	Cardiotonic
3.	Indian squill (Urginea)	Dried bulbs of *Urginea indica*	Scillaren A and B	Cardiotonic
4.	Strophanthus (Arrow-poison)	Dried ripe seeds of *Strophanthus kombe*, Apocynaceae	Strophanthidin	Cardiotonic
5.	Thevetia (Lucky-nut-tree)	Dried seeds of *Thevetia nerifolia*, Apocynaceae	Thevetin	Cardiotonic

III. Saponin Glycosides

1.	Brahmi (Bacopa)	Leaves and stems *Bacopa moniera* Scrophulariaceae	Bacosides A and B	Nervine tonic
2.	Dioscorea (yam)	Dried Rhizomes of *Dioscorea deltoidea,* Dioscoreaceae	Diosgenin (steroidal sapogenin)	Synthesis of medicinal steroids
3.	Ginseng (Panax)	Dried root of *Panax ginseng,* Araliaceae	Ginsenosides & panaxosides (triterpenoid saponins)	Adaptogen, tonic and stimulant
4.	Gokhru (Tribulus)	Dried fruits *Tribulus terrestris* Zygophyllaceae	Steroidal sapogenins	Diuretic
5.	Jalbrahmi (Centella)	Dried herb of Centella asiatica umbelliferae	Asiaticoside	Nervine tonic
6.	Momordica (Karela)	Dried fruits of *Momordica charantia* Cucurbitaceae	Charantin, Momordicin	Hypoglycemic
7.	Quillaia (Soap bark)	Dried inner bark of *Quillaia saponaria,* Rosaceae	Quillaia sapotoxin (triterpenoid saponin)	Reflex expectorant
8.	Safed musali	Peeled tuberous roots of *Chlorophytum borovillianum* Liliaceae	Hecogenin	General tonic
9.	Senega	Dried roots of *Polygala senega,* Polygalaceae	Senegin, polygallic acid (triterpenoid saponin)	Stimulant expectorant
10.	Shatavari (Shutmuli)	Dried roots and leaves *Asparagus racemosus* Liliaceae	Shatavarin I, II	Galactogogue
11.	Liquorice (Glycyrrhiza)	Dried roots and stolon of *Glycyrrhiza glabra,* Leguminosae	Glycyrrhizin (triterpenoid saponin), 18 - β - glycyrrhetinic acid	Expectorant, treatment of peptic ulcer

IV. Cyanogenetic or Cyanophoric Glycosides

1.	Bitter almond	Dried ripe seeds of *Prunus amygdalus,* Rosaceae	Amygdalin	Demulcent, sedative
2.	Wild cherry (Wild black cherry)	Dried bark of *Prunus serotina,* Rosaceae	Prunasin	Mild sedative, flavouring agent

V. Isothiocynate Glycosides

1.	Mustard seeds (Black mustard)	Dried ripe seeds of *Brassica nigra,* Cruciferae	Sinigrin	Counter irritant, rubefacient externally and emetic internally

VI. Flavonoid Glycosides

1.	Buck wheat	Dried fruits of *Fagopyrum esculentum,* Polygonaceae	Rutin (also from *Ruta graveolens)*	Treatment of capillary bleeding
2.	Ginkgo	*Dried leaves of Ginkgo biloba* Ginkgoaceae	Ginkgolide A, B, C	Vascular disorders
3.	Milk-Thistle (Silymarin)	*Dried seeds of Silybus marianum* Asteraceae	Silybin Silycrystin	Liver disorders

VII. Coumarin and Furanocoumarin Glycosides

1.	Ammi	Dried fruits of *Ammi majus,* Umbelliferae	Xanthotoxin	Treatment of vitiligo
2.	Psoralea fruit (Bavchi)	Dried ripe fruits of *Psoralea corylifolia,* Leguminosae	Psoralen, corylifolin	Treatment of leucoderma
5.	Tonka bean	Dried seeds of *Dipteryx-odorata, leguminosae*	Coumarin	Flavouring agent
4.	Visnaga (Khella)	Dried ripe fruits of *Ammi visnaga,* Umbelliferae	Khellin, visnagin	Smooth muscle relaxant, coronary vasodilator
5.	Cantharides (Spanish fly)	Dried beetles *Cantharis vasicatoria,* Meloidae	Cantharidin	Counter, irritant, Rubefacient

Contd...

6.	Cochineal (Red scale insect)	Dried full grown female insects enclosing young larve of Coccus cacti (coccidae)	Carminic acid	Colouring agent
7.	Mylabris	Dried female insects of *Mylabris pustulata* and other species	Cantharidin	Counter irritant, rubefacient

VIII. Aldehyde Glycosides

1.	Anantmul (Sariva)	Dried roots of *Hemidesmis Indicus* Asclepiadaceae	Iso-vanillin	Anti-inflammatory, flavouring agent
2.	Vanilla (Vanilla pods)	Unripe fruits of *Vanilla planifolia,* Orchidaceae	Gluco-vanillin	Flavouring agent

IX. Phenol Glycosides

	Bearberry (Uva ursi)	Dried leaves of *Arctostaphylos uvaursi,* Ericaceae	Arbutin	Diuretic in urethritis

X. Steroidal Glyco-alkaloids

	Solanum	Dried fruits of *Solanum khasianum* Solanaceae	Solasodine	For steroidal synthesis

XI. Glycosidal Bitters and Miscellaneous Glycosides

1.	Chirata (chirayata)	Dried plant of *Swertia chirata,* Gentianaceae	Gentiopicrin from sweroside),	Stomachic, antipyretic
2.	Garcinia	Dried de-seeded fruits *Garcinia combogia* Family: Guttiferae	Hydroxy citric acid, tartaric acid	Anti-rhumatic condiment
3.	Gentian	Dried roots of *Gentiana lutea,* Gentianaceae	Gentiopicrin, amaro gentin	Bitter stomachic, tonic
4.	Gudmar (Gymnema)	Dried leaves of *Gymnema* sylvestre Asclepiadaceae	Gymnemic acid	Anti-diabetic
5.	Guduchi	Dried leaves and stems of *Tinospora cordifolia,* Family: Minispermaceae	Tinosporoside cardiofoliside	Bitter tonic and general debility
6.	Henna	*Lawsonia inermis* Lythraceae	Lawson	Anti-fungal

Contd...

7. Kalmegh (Andrographis)	Dried leaves and tender shoots of *Andrographis paniculata*, Acanthaceae	Andrographolide	Bitter tonic, anthelmentic, hepato-protective
8. Manjishta	Dired stems of *Rubia cordifolia* Family: Rubiaceae	Rubiandin, purpurin (anthraquiuones)	In leucoderma and in cosmetic products to gain glow and lustre.
9. Picrorrhiza (Indian gentian)	Dried rhizomes of *Picrorrhiza kurroa*, Scrophulariaceae	Picroside I, kutkoside amarogentin	Bitter tonic, hepatopro-tective bitter
10. Quassia (Bitter wood)	Dried stem wood of *Picrasma excelsa*, Simarubaceae	Picrasmin, quassin	Bitter stomachic, anthelmentic
11. Stevia	Dried roots and leaves of *stevia reboundians* Asteraceae	Sterioside Rebandioside	Antidiabetic Sweetening agent
12. Vajradanti (Yellow nail dye plant)	*Barleria prionitis* Acanthaceace	Saponin glycosides prioniside A, Prioniside B, Barlerinoside.	In treatment of stiffness of limbs, sciatica, to reduce obesity, joint pains, promotes hair growth, urinary infections.
13. Kava-Kava (Kawa)	Dried rhizomes of *Piper methysticum* Piperaceae	Methysticin, Kawain	Anti inflammatory and Anxylolytic
14. Black-Cohosh (Black-snake root)	Dried roots of *Cimicifuga, recemopse* *Ranunculaceae*	Fucinolic acid, Actein	For menopausal symptoms
15. Oleander (Nerium)	Dried roots and leaves of *Nerium-oleander* fam.: Apocyanaceae	Glyrasides neriodern, Neuoaerin, oleandrin	Cardiotonic and diuretic

I. ANTHRACENE GLYCOSIDES

These constitute a major class of glycosides. They are mainly found in the dicot plant and families like Euphorbiaceae, Ericaceae, Lythraceae, Polygonaceae, Rhamnaceae, Rubiaceae, Leguminosae, Verbenaceae, etc. The family Liliaceae from monocots also shows the presence of C-glycosides. Some of the fungi and lichens also contain anthraquinone glycosides. But, it is observed that lower plants like Bryophytes, Pteridophytes and Gymnosperms are devoid of such glycosides. It is postulated that the aglycone part of these glycosides is formed by head-to-tail condensation of acetate units. This group of glycosides comprises of different aglycone moieties like anthraquinone, anthrone, anthranol, dianthranol, oxanthrone and dianthrone.

In different drugs like aloe, senna, rhubarb, cascara, aglycones are present in their derivate forms. The parent molecule for all these aglycones i.e. anthraquinone is present in different forms along with methyl, hydroxymethyl, carboxyl, dihydroxy phenol, trihydroxy phenol or free carboxylic acid groups. In a reduced form, anthraquinone is present as anthranol or anthrone which are isomeric with each other. Anthrone is pale yellow substance and without any solubility in alkali, while anthranol is brownish-yellow and soluble in alkali. Anthranol shows strong fluorescence in alkali, but anthrone is non-fluorescent by nature.

Some plants contain oxanthrone which is the intermediate substance from anthraquinone to anthranol. In some plants, the anthrone molecule orients in bimeric form called dianthrone which is therapeutically more important.

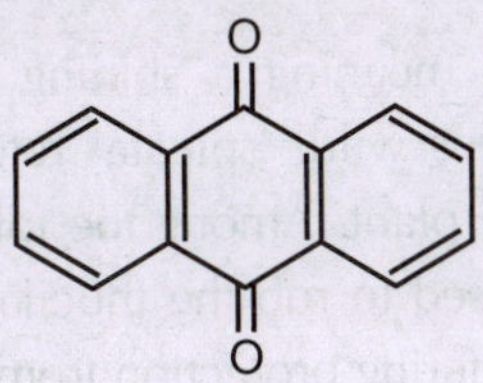

Anthraquinone

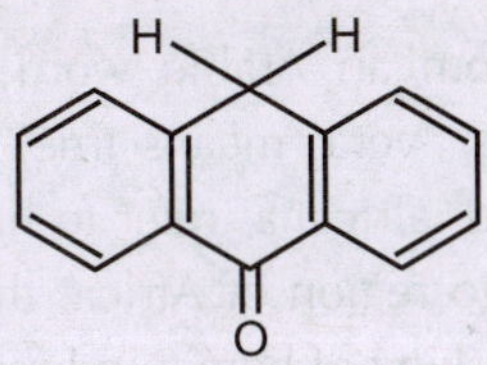

Anthrone

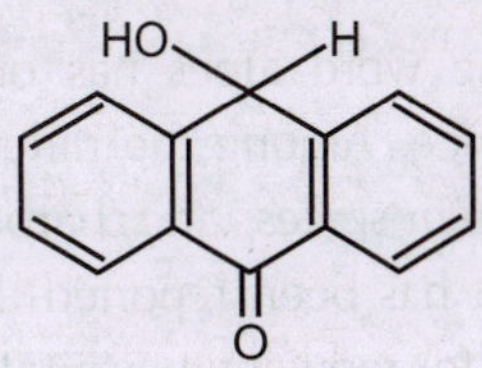

Oxanthrone

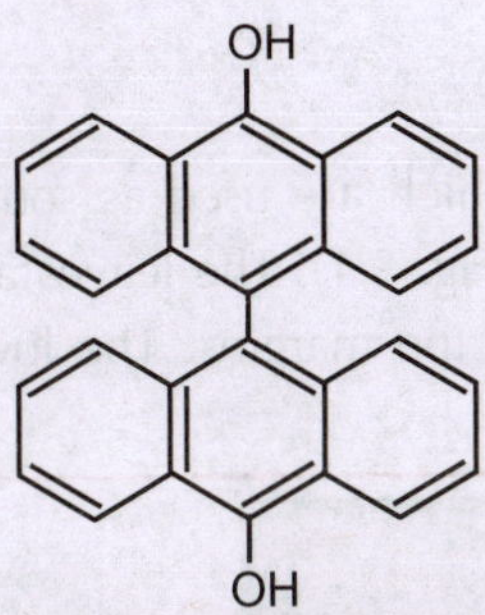

Dianthranol

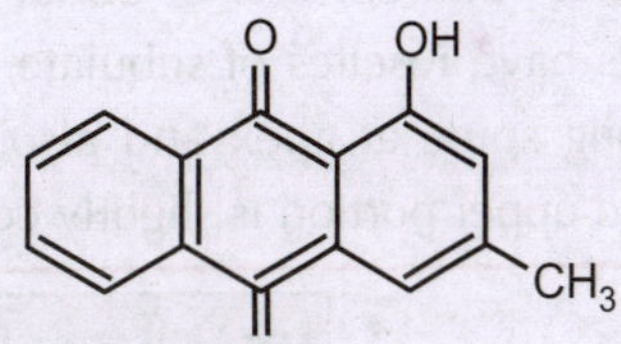

Chrysophanic acid

The reduced anthraquinone are biologically more active. In a fresh drug, these aglycones are present in reduced form, but are hydrolysed and oxidized during their storage. They are present along with different sugars like glucose, rhamnose, arabinose and primeverose.

Borntrager's Test

Anthraquinone derivatives are generally detected by Borntrager's test. In this test, the drug is powdered and further extracted with ether or any water immiscible organic solvent. The filtered ethereal extract is made alkaline either with caustic soda or ammonia, by which the aqueous layer shows, after shaking, pink, red or violet colour. Borntrager's test is negative in case of anthranol (reduced forms). Anthrones are detected with their fluorescence tests.

ALOES

Synonyms

Aloe, Musabbar, Kumari.

Biological Source

Aloes is the dried juice of the leaves of *Aloe barbadensis* Miller, known as Curacao aloes; or of *Aloe perryi* Baker, known as Socotrine aloes; or of *Aloe ferox* Miller and hybrids of this species with *Aloe africana* Miller and *Aloe spicata* Baker, known as Cape aloes, belonging to family Liliaceae.

Geographical Source

Aloes is indigenous to eastern and southern Africa and grown in Cape colony, Zanzibar and islands of Socotra. It is also cultivated in Caribbean islands, Europe and many parts of India, including North West Himalayan region.

History

The word aloes has originated from an Arabic word viz. alloch meaning a shining bitter substance. Among the different species, 'vera' means true, 'ferox' means wild, 'spicata' refers to flowers in spikes, and 'barbadensis' and 'africana' refer to habitat of the plant. Among the folklore uses, it has been reported that in Congo region of Africa, the natives used to rub the mucilage of leaves for reducing perspiration and masking of human odour, thereby offering protection from wild animals. It is also known that *Aloe barbadensis* was used along with burnt alum for healing sore eyes.

Cultivation, Collection and Prepartaion of Aloe

The genus *Aloe* consists of about 200 species, some of which are used as sources of Aloes. These plants have rosettes of subulate, succulent large leaves (Fig. 9.1). The leaves are sessile and have a strong spine at apex and also number of spines along the margins. The lower portion is rounded and upper portion is slightly concave.

Fig. 9.1: Cultivated Aloe Plants

For the **cultivation**, root suckers are used for propagation. The plants grow even in poor grades of soils and in dry climatic conditions. The root suckers are planted in the rows about 50 cm apart.

Water logging near the plant must be prevented. The roots do not penetrate much in the soil. For the purpose of manure, a mixture of nitrogen, potassium and phosphorus is used. The leaves are cut for the first instance in second year of cultivation and the drug is obtained from leaves for twelve years. After twelve years, the plants are completely harvested by uprooting and once again the land is worked for replantation. During the collection of leaves, a cut is given to leaves near their bases, by which the juice located in parenchymatous cells of pericycle exudes out, due to the pressure exerted by mucilage cells. A single incision is sufficient for drawing out all the juice from entire system of pericyclic cells. The preparation of various types of aloes is outlined here.

Preparation of Aloe

(i) Barbados or Curacao aloes: It is prepared in islands of Aruba and Bonaire in West Indies. After giving the transverse cuts near the bases of the fleshy leaves of *Aloe barbadensis,* the cut leaves are placed along the sides of V-shaped wooden troughs. Because of the spines on the leaves, they are put into kerosene tins immediately after cutting and then brought to wooden troughs and kept in tilted positions, to drain out all the juice. The juice collected by this way is allowed to boil in large copper pans. During boiling, the latex evaporates and juice is further thicknened. The thick juice is then poured into gourds or metal containers where it hardens. It is brought in market under the name Barbados or Curacao aloe.

(ii) Cape aloes: It is obtained from *Aloe ferox* and its hybrids in South Africa. The transversely cut leaves are arranged in a circular manner in the basin shaped depression, dug in the ground which is lined either with goat skin or canvas. The leaves are placed in such a way so as to overlap the cut ends. They are kept in this position for 5 - 6 hours, till all the juice exudes out and is collected in goat skin.

The collected juice is transferred to large iron kettle where it is boiled and stirred continuously with the help of a wooden paddle type instrument. When the boiling juice attains a desired concentration, it is poured into wooden cases. It becomes a solid mass which is brought into commerce under the name of Cape aloe.

(iii) Socotrine aloes: It is prepared from *Aloe perryi* occurring in island of Socotra and mainland of East Africa. The juice of the leaves is collected in goat skin and allowed to become semisolid in nature. It is exported in a pasty condition under the name of Socotrine aloe.

(iv) Zanzibar aloes: The origin of this variety is not correctly known and sometimes,it is also regarded simply as a variety of Socotrine aloe. The juice is poured into skins of some small carnivorous animals, where it solidifies and as such packed in wooden boxes. Zanzibar aloe is also sometimes, called as monkey-skin aloe, although the skin is not that of monkeys.

Description

(i) Curacao aloes: It occurs as "Cape" Barbados and livery variety. Both varieties have strong odour, somewhat resembling to iodoform and an intensely bitter taste. It occurs as brownish-black, opaque mass. It is waxy and somewhat resinous. It shows an uneven fractured surface. Livery curacao aloe and 'Capey' Barbados aloes are distinguished by their transparency.

(ii) Cape aloes: It occurs as dark brown or greenish-brown to olive brown masses. It has a nauseating and bitter taste. It breaks with a glassy fracture. It has a sour but distinct odour.

(iii) Socotrine aloes: It occurs as brownish-yellow, opaque mass with a pasty or semisolid consistency. The fractured surface is conchoidal. Socotrine aloes has an extremely bitter, nauseous taste with unpleasant odour. It is totally soluble in 60 per cent alcohol.

(iv) Zanzibar aloes: This variety show a liver brown colour with a dull, waxy, smooth and even fracture. It has a characteristic but not much disagreeable odour and a bitter taste.

Microscopic Characteristics of Aloe Leaf

T. S. of the leaf exhibits (Fig. 9.2) outermost cuticle followed by epidermis, palisade tissue and mucilaginous parenchymatous mesophyll.

Mesophyll encloses vascular bundles which are covered with pericyclic layer. Inside the pericycle few large elongated thin walled aloeitic cells are located. These aloeitic cells contain highly viscous, yellow juice (aloe gel). The leaves are sessile and when they are cut at the base, the juice flows down and which is to be collected. Few calcium oxalate crystals are present in the parenchyma.

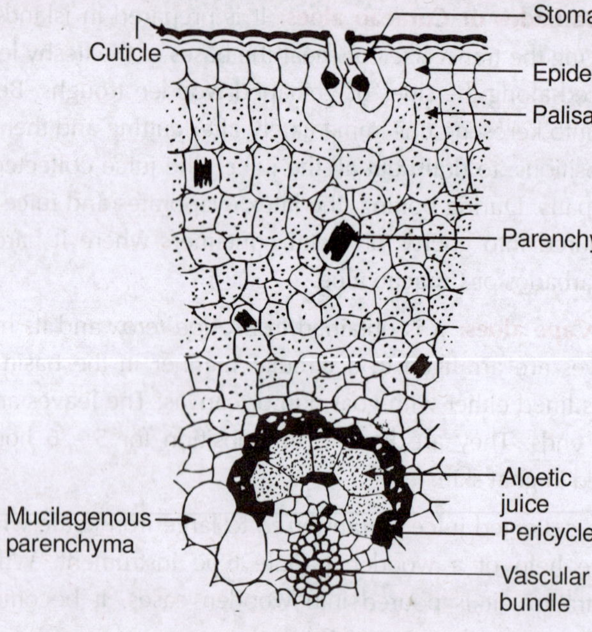

Fig. 9.2: T. S. of Aloe-leaf

Microscopic Characters of Aloe Powder

The microscopy of the different varieties of aloe is more significant for identification of their powdered forms. The microscopic characters are studied in lactophenol, because in this reagent the particles are gradually solubilized so that crystals are clearly and quickly observed.

(i) Curacao aloes: It shows fragments consisting of a large number of very small needles or slender prisms.

(ii) Cape aloes: This variety appears as transparent, brown, angular or irregular fragments.

(iii) Socotrine aloes: It is characterised by fragments consisting of quite large prisms either present in group or in dispersed form.

(iv) Zanzibar aloes: It exhibits irregular lumps in which modular masses are embedded.

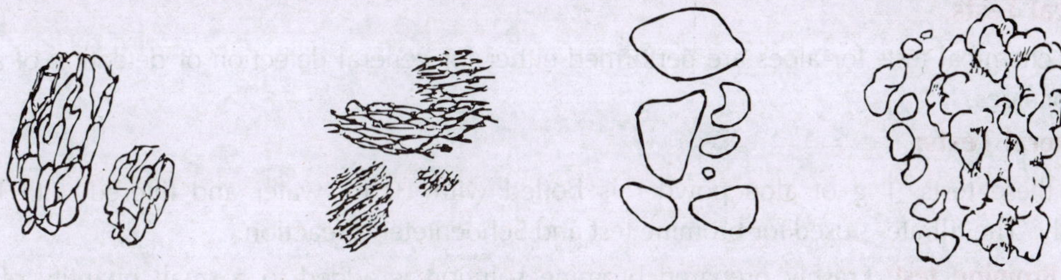

Curacao Cape Socotrine Zanzibar

Fig. 9.3: Microscopic characters of Aloe powder

In commerce, aloes is available both as crystalline and amorphous substance. It generally appears that amorphous form is cape aloe and crystalline one is curacao aloe, depending on their response to the tests. If both the varieties are mounted in cresol and viewed under polarised light, crystalline aloe, insoluble in cresol shows a shining bright colour. Amorphous aloe, as it gets quickly dissolved, remains invisible.

Chemical Constituents

All the varieties of aloe are the major sources of anthraquinone glycosides 10 – 130 per cent. The principal active composition of aloe is aloin, which is a mixture of glucosides, among which barbaloin is the chief constituent. It is chemically aloe-emodin anthrone C - 10 glucoside and it is water soluble.

Barbaloin is a C-glycoside and it is not hydrolysed by heating with dilute acids or alkalis. Ferric chloride decomposes barbaloin by oxidative hydrolysis into aloe-emodin-anthrone, little aloe-emodin and glucose.

Along with barbaloin, aloes also contains isobarbaloin, β–barbaloin, aloe-emodin and resins.

The drug also contains aloetic acid, homonataloin, aloesone, chrysophanic acid, chrysamminic acid, galactouronic acid, choline, choline salicylate, saponins, mucopolysaccharides, glucosamines, hexuronic acid, coniferyl alcohol, etc.

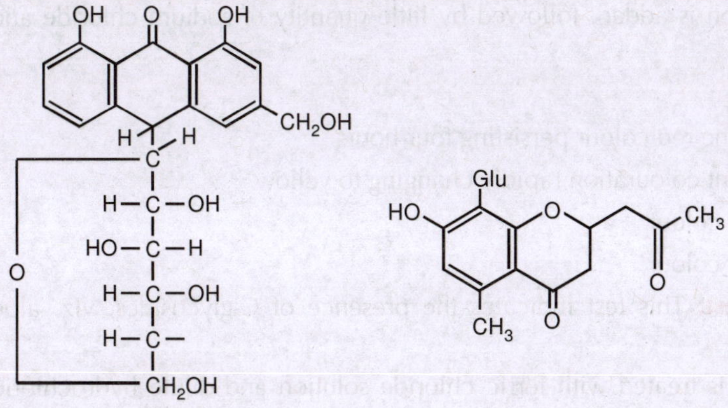

Barbaloin **Aloesin**

The amount of barbaloin in different commercial varieties varies to a large extent. Curacao aloes contains about 22 per cent of barbaloin. Indian variety, generally Aloe vera, contains very less quantity (3.5 - 4 per cent). Curacao aloes contains two and half times quantity of aloe-emodin, as compared to Cape-aloe-emodin.

The resin of aloe principally contains Aloesin. It is a type of a C-glucosyl chromome.

Aloesin is also responsible for purgative action of aloes.

Chemical Tests

The chemical tests for aloes are performed either for general detection or detection of specific variety of aloes.

(a) General Tests:

For these tests, 1 g of aloe powder is boiled with 10 ml water and filtered with help of kieselguhr. The filtrate is used for bromine test and Schoenteten's reaction.

(1) Bromine test: Freshly prepared bromine solution is added to a small quantity of above filtrate. The test gives a pale yellow precipitate of tetrabromalin.

(2) Schoenteten's reaction (Borax test): Little quantity of above filtrate is treated with borax and shaken well till the borax dissolves. When few drops of this solution are added to a test tube nearly filled with water, a green fluorescence appears.

(b) Special Tests:

These tests are meant for distinguishing different varieties of aloe vera.

(1) Nitrous acid test: Crystals of sodium nitrite along with small quantity of acetic acid are added to aqueous solution of aloes. The observations are as follows:

(i) Curacao aloes - sharp pink to carmine colour

(ii) Cape aloes - faint pink colour

(iii) Socotrine and Zanzibar aloes - very less change in colour

This test is due to isobarbaloin.

(2) Nitric acid test: This test is carried out either by directly applying nitric acid to drug or to its aqueous solution. The observations are as follows.

(i) Curacao aloes - deep brownish-red colour

(ii) Cape aloes - brownish colour changing to green

(iii) Socotrine aloes - pale brownish - yellow colour

(iv) Zanzibar aloes - yellowish brown colour

(3) Cupraloin test (Klunge's isobarbaloin test): To very dilute aqueous solution of aloes, a drop of saturated copper sulphate solution is added, followed by little quantity of sodium chloride and excess of 90 per cent alcohol.

Following results are observed:

(i) Curacao aloes - wine red colour persisting four hours

(ii) Cape aloes - faint colouration rapidly changing to yellow

(iii) Socotrine aloes - no colour

(iv) Zanzibar aloes - no colour

(4) Modified anthraquinone test: This test indicates the presence of C-glycosides, viz. aloe emodin.

The aqueous solution of aloes is treated with ferric chloride solution and dilute hydrochloric acid to bring out the oxidative hydrolysis of aloe-emodin. The hydrolysis sets free anthraquinones which are collected in organic solvent like carbon tetrachloride or ether. The organic layer is separated and shaken with dilute ammonia. The ammoniacal layer shows rose-pink to cherry red colour, indicating the presence of C-glycosides, viz. aloe emodin.

Standards of Quality

Loss on drying	:	Not more than 10.0 per cent
Water soluble extractives	:	Not less than 50.0 per cent
Ash	:	Not more than 5.0 per cent

Uses

Aloes is used as a purgative. Its effect is mainly on colon. It has a stronger purgative action in the series of all crude drugs with anthracene glycosidal content. To counter effect the gripping action, it is given with carminatives.

Aloin is preferred now-a-days to aloes, both of which are official. Besides purgative property, aloes enjoys many other uses. It is an ingredient of compound tincture of benzoin (Friar's balsam).

Aloe gel: It is formed in inner parenchymal cells of the leaf, is a slightly viscous and clear liquid. During collection, it should not get contaminated with aloe juice. Such gel is used in topical therapeutic applications and also in many cosmetic products, but the therapeutic value, if taken orally, is questionable. The gel possesses good moisturizing properties and also has formulation role for oil in water (**approved by U.S.F.D.A.**) preparation. It shows anti-inflammatory properties due to the chemical contents like salicylates, carboxypeptidases (inactivating bradykinen) and magnesium lactate (interfering with the conversion of histidine to histamine in the mast cells). The polysaccharide and sugar content have the role for hydrocolloid dressing and also osmotic bactericides. Aloe gel also increases the removal of dead tissue due to its Aloctine - A content which stimulates macrophage production. It is believed that only fresh gel probably has a role in treatment of burns and wounds.

It is also used in the treatment of pains and itchings and also to slow down ulceration and keratosis, Aloe gel is used in skin cosmetics as a protective due to its antiwrinkle properties. Aloe is also used externally for painful inflammation.

Following table distinguishes both the products suitably.

Table 9.2: Differentiation between Aloe gel and Aloes

	Aloe gel	Aloes
Part used	Mucilaginous tissue located in leaf-parenchyma of *Aloe* species. Family Liliaceae.	Dried leaf juice of *Aloe barbadensis, A. ferox* and hybrids, *A. africana, A. spicata,* Family Liliaceae.
Constituents	Mono and polysaccharides, tannins, sterols, cycloxygenase, saponins, vitamins and minerals. Polysaccharides contain xylose, arabinose and galactose.	Anthraquinones: 30 per cent mainly C glucosides known as aloin. Mixture contains barbaloin, isobarbaloin, emodin and aloe-emodin.
	Lipids contain cholesterol, gamolenic acid and arachidonic acid.	Resins: Aloesin, aloesone.
Use	Used externally in the form of ointments, creams to assist healing of wounds, burns, eczema, and also in psoriasis.	In the treatment of constipation.

Other Preparations

(1) Aloin: It is a mixture of crystalline principles obtained from aloes. It varies both in chemical composition and physicochemical properties, depending upon the source of aloes from which it is prepared. Generally, it is pale yellow to dark yellow and microcrystalline powder. It is odourless with an intense taste. It is soluble in water and 90 per cent alcohol. It is to be stored in tight light-resistant container. Its effective dose is 15 - 60 mg for purgation.

(2) Aloes powder: Its characters vary as per the source of aloes from which it is prepared. The source is identified with the help of microscopic observations.

(3) Compound benzoin tincture (Friar's balsam): It consists of benzoin, prepared storax, balsam of tolu, aloes and 90 per cent alcohol.

Adulterants and Substitutes

(1) Natal Aloes: It resembles cape aloes in microscopic characters, and hence used as a substitute. Natal aloes contains natalion, homonatalion and resin with nataloresinotannol. It is a weak purgative.

(2) Mocha Aloes: This is a brittle, black and glassy aloes with strong odour.

(3) Aloe gets adulterated with black catechu, pieces of iron and stones. Alcoholic extract of aloe under UV light gives deep brown colour, while black catechu gives black colour.

CASCARA

Synonyms

Cascara bark, cascara sagrada, sacred bark, chittem bark, cortex *Rhamni purshianae*.

Biological Source

Cascara sagrada is the dried bark of *Rhamnus purshiana* de Candolle, belonging to family Rhamnaceae. It is collected at least one year before use.

Geographical Source

It is obtained both from wild and cultivated shrubs and small trees in North Carolina, Washington, Oregen regions of United States. It is also cultivated in Western Canada and Kenya.

Description of Herb

This tree grows to approximately 20 feet tall. It is characterized by reddish-brown bark and thin leaves that range from 3-5 inches long with finely serrated edges.

This tree yields berries that are red when young and ripen to purple or black. Each fruit contains three seeds.

Macroscopical Characters

Colour : It is purplish-brown externally and reddish brown internally.

Odour : Characteristic nauseous.

Taste : Persistently bitter.

Size : About 1 to 4 mm in thickness with varying size.

Shape : Usually, in the broken pieces which are small and flat. It occurs in quills or channels.

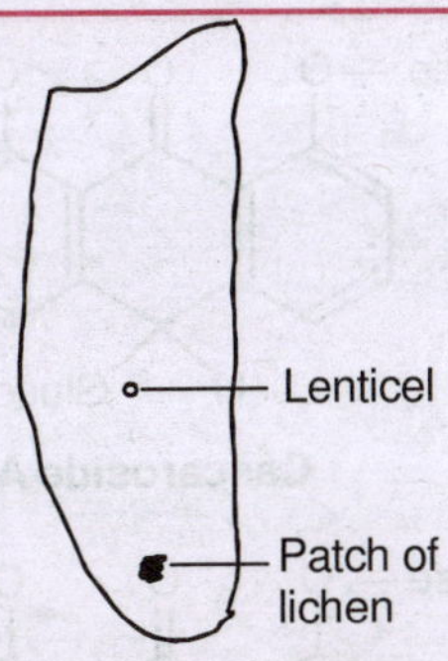

Fig. 9.4 (a): Cascara sagrada Bark

Fig. 9.4 (b): Cascara sagrada Plant

Extra Features

Outer surface of the bark is very smooth and shows the presence of scattered lenticels, lichens and cork. Externally, liverworts and sometimes, insects are also present on the bark. Bark shows internally, longitudinal striations.

Microscopic Characters (Fig. 9.5)

The transverse section shows cork, cortex, sclereid, primary and secondary phloem. Cork has several layers of cells which contain yellowish-brown substance showing purple colour with alkali. Cortex is composed of collenchyma, externally and cellulosic parenchyma on inner side. The latter contains a large number of calcium oxalate crystals. The cortex region also contains small and large groups of sclereid. The major portion of bark is occupied by secondary phloem, in which phloem fibres are lignified and arranged in tangential bands, alternating with sieve tubes. Many cells of phloem parenchyma contain prisms of calcium oxalate and thus, form a crystal sheath to each group of phloem fibres.

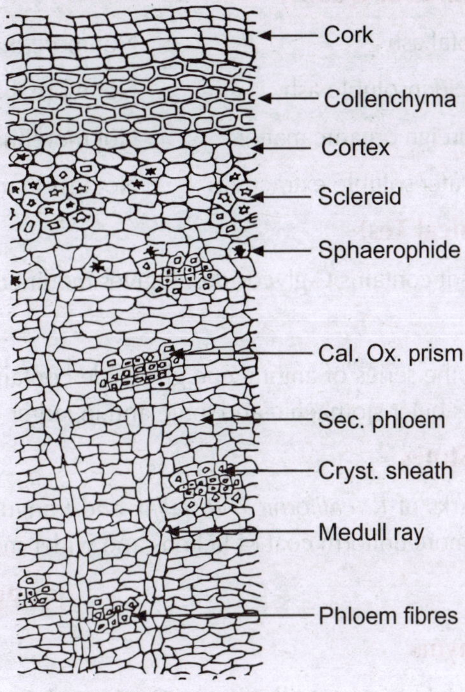

Fig. 9.5: T. S. of Cascara Bark

Chemical Constituents

The freshly collected cascara bark contains anthranol derivatives which cause gripping and emetic effects. The drug becomes acceptable by storing for at least one year, during which anthranol portion is oxidised to anthraquinones.

Cascara contains both O-glycosides and C-glycosides. It contains four primary glycosides which are called as cascarosides A, B, C and D.

Cascaroside	Configuration	R
A	10 β	OH
B	10 α	OH
C	10 β	H
D	10 α	H

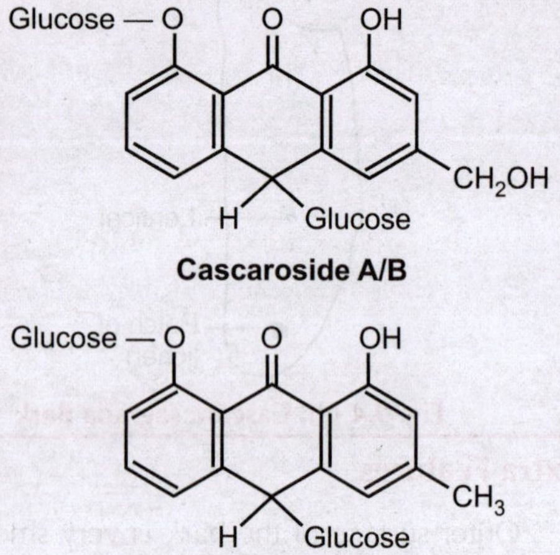

Cascaroside A/B

Cascaroside C/D

The bark also contains barbaloin and chrysaloin, various dianthrone of emodin, aloe-emodin, chrysophanol; heterodianthrones like palmidins A, B and C; free aloe-emodin, emodin chrysophanol and O–glycosides derived from them.

A lactone present in the drug makes it bitter in taste.

Standards of Quality

Total ash	:	Not more than 8.0 per cent
Acid insoluble ash	:	Not more than 2.0 per cent
Foreign organic matter	:	Not more than 1.0 per cent
Water soluble extractives	:	Not less than 15.0 per cent

Chemical Test

As it contains C-glycosides, it gives modified Borntrager's test.

Uses

In the series of anthracene glycoside containing drugs, cascara is mild drug. In small doses, it is used as bitter stomachic and tonic and in larger doses as purgative.

Substitutes

Barks of *R. californica, R. fallax* and frangula bark are substitutes for cascara. The former bark has a more uniform coat of lichens and wider medullary rays than the original species.

HYPERICUM

Synonyms

St. John's wort, Millepertuis, **Goat weed**, Bassant.

History

St. John's wort is a perennial flowering plant and has been long used in folk medicine. World-wide, there are more than 370 Hypericum species out of which 25 species are found in North America. *Hypericum perforatum* is native to Europe and grows widely in New found land west of British Columbia and also in N. Carolina and California. It is named Saint John's Wort, because it blooms around June 24, Saint John, the Baptist's day. It is also known by other names like Klamath weed and goat weed. The species is named as perforatum because the translucent glands present on its leaves look like perforations.

Biological Source

This consists of dried aerial parts of the plant known as *Hypericum perforatum* Linn. Fam. Hypericaceae collected during flowering season or shortly before flowering.

Geographical Source

It is indigenous to England and found in Europe and Australia. In India it is found distributed in Western Himalaya upto an altitude of 1000 - 3000 metres.

Macroscopic Characters

Hypericum is a rhizomatous perennial herb, about one metre in height.

- Colour : Leaves green coloured marked with black spots,
- Odour : Distinct, balsamic
- Taste : Astringent and bitter
- Shape : Leaves are oblong, entire.
- Flowers : Yellow corolla 2.5 cm in diameter in corymrose cymes. Flowers are small five petaled and marked by tiny dots of reddish brown pigments.
- Capsules : Ovoid, 1 cm long
- Seeds : Very small
- Stem : 2 – edged

Fig. 9.6: Hypericum plant

Chemical Constituents

Hypericum contains naphthodianthrones (anthraquinone derivatives) hypericin and pseudo-hypericin, hyperoside and flavonoids. Hypericins (0.06 to 0.75 per cent) include hypericin red fluorescent pigment, pseudohypericin and hyperforin. Hypericin is 4, 5, 7, 4', 5', 7' hexahydroxy - 2, 2' dimethyl-meso-naphthodianthrones.

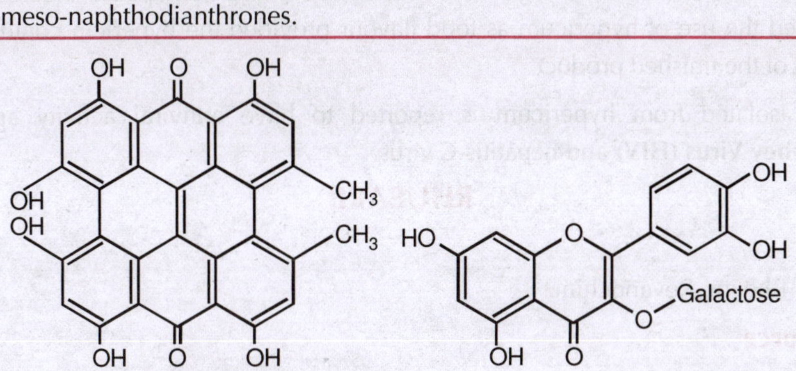

Hypericin **Hyperoside**

They are characteristic of St. John's Wort. The other constituents, ubiquitous to plant kingdom are flavonoids (2 to 4 per cent) which include biflovonoids (e.g. amentoflavone). It also contains xanthones (0.0004 per cent) hyperforin (2.8 per cent), procyanidines (8 per cent), Kaempferol, rutin and volatile oil (0.1 to 1 per cent) Hypericum preparations are presently standardised by their hypericin content. The extract used in most of the clinical trials has been standardised to contains not less than 0.3 per cent hypericin. Herb also contain tannins, resinous substance, rutin, alkaloids, fixed oil, vit. C and hyperin. Fixed oil contains phytosterol and hydrocarbons.

Standards of Quality

Foreign organic matter	:	Not more than 3.0 per cent
Alcohol soluble extractives	:	Not less than 13.0 per cent
Water soluble extractives	:	Not less than 24.0 per cent
Loss of dryig, at 105ºC	:	Not more than 10.0 per cent
Total ash	:	Not more than 7.0 per cent
Acid insoluble ash	:	Not more than 0.8 per cent

Hypericin is soluble in aqueous alkaline solution giving red colour up to pH 11.5, while green with red fluorescence above pH 11.5.

Uses

Hypericum is described as an anti-depressant. It is prescribed for treatment of anxiety, moderate depression and seasonal affective disorders. It has been licensed in Germany for these treatments. Based on the reports, presently it is believed that it is either hypericin or hyperforin or both in combination show the biological activity. The reports also indicate that, contrary to synthetic antidepressants, St. John's Wort shows this activity by several possible mechanisms. These mechanisms include (i) Monoamine oxidase (MAO) inhibition, (ii) Inhibition of serotonin reuptake (iii) Inhibition of gamma - aminobutyric acid (GABA) and (iv) inhibition of interleukin - 6 (IL - 6).

Several groups of compounds of various chemical nature act synergistically to produce the enhanced mood. Hypericin has shown irreversible monoamine oxidase inhibitor activity.

Hypericum is indicated in neuralgia fibrocitis, sciatica, and menopausal neurosis. Council of Europe has listed the use of hypericum as food flavour provided the hypericin content is not more than 0.1 mg/kg of the finished product.

Hypericin isolated from hypericum is reported to have antiviral activity against **Human Immunodeficiency Virus (HIV)** and hepatitis-C virus.

RHUBARB

Synonyms

Radix rhei, Rheum, Revandchini.

Biological Source

Rhubarb consists of the dried rhizome of *Rheum emodi* Wall (Indian rhubarb), *Rheum palmatum* Linn and *Rheum webbianum* Royal (Chinese-rhubarb), family Polygonaceae. It is collected from 6 - 7 year old plants just before flowering season and marketed with cortex intact or partially decorticated.

Geographical Source

It is obtained both from wild and cultivated species grown in regions ranging from south east China to Tibet and Korea. It is also obtained from Kashmir and Sikkim in India; and West Germany and other European countries.

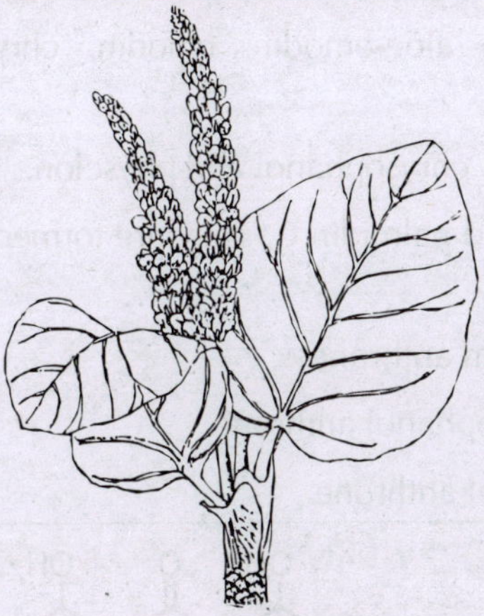

Fig. 9.7: Rheum - palmatum twig

Fig. 9.8: Rhubarb *Rheum emodi* twig

Collection and Preparation

The drug is mostly obtained from wild plants. High altitude (more than 3,000 metres) is a pre-requisite for growth of rhubarb. It is a drought resistant plant. It is propagated by seeds or from rhizomes. The drug is obtained from 6-10 year old plants. The rhizome portion is dug up in spring or autumn. Aerial portion dies in winter and rhizomes remaining in ground regenerate in next spring. The collected rhizomes are cleaned, decorticated and dried. The larger rhizomes are cut transversely or longitudinally. The pieces are threaded by piercing a hole in them and such threaded pieces are dried under shade. Some of the pieces are artificially dried. The drug is packed in wooden cases and exported from Shanghai or Tientsin.

In India, it is collected from Kashmir, Kulu and Manali regions.

Macroscopic Characters

Colour	:	Rhubarb rhizomes are brown to yellow in colour.
Odour	:	Fragrant and sharp
Taste	:	Bitter and astringent
Size	:	2 to 20 × 1.3 to 2.0 cm
Shape	:	Sub-cylendrical barrel-shaped, irregulat longitudinal wrinkles compact and firm.

Extra Features : Round pices are prepared from small rhizomes while flat pieces from large rhizomes. Pieces of rhizomes show holes through which they are threaded. They break with an angular and granular fracture.

Chinesei rhubarb is available in three forms viz. Shensi, Canton and high dried rhubarb, each with flats or rounds.

Chemical Constituents

Rhubarb contains anthraquinone glycosides and astringent principles. The anthraquinone glycosides range from 2 to 4.6 per cent and are categorised into four groups.

(1) Anthraquinones with a carboxyl group like rhein and glucorhein.

(2) Anthraquinones without a carboxyl group like aloe-emodin, emodin, chrysophanol, physcion and also their glycosidal forms.

(3) Anthrones and dianthrone of aloe-emodin, emodin, chrysophanol and physcion.

(4) Heterodianthrones like palmidin A, palmidin B, and palmidin C. They are formed from two different anthrone molecules.

Palmidin A $\longrightarrow$ aloe - emodin anthrone + emodin anthrone

Palmidin B $\longrightarrow$ aloe - emodin-anthrone + chrysophanol anthrone

Palmidin C $\longrightarrow$ emodin-anthrone + chrysophanol anthrone

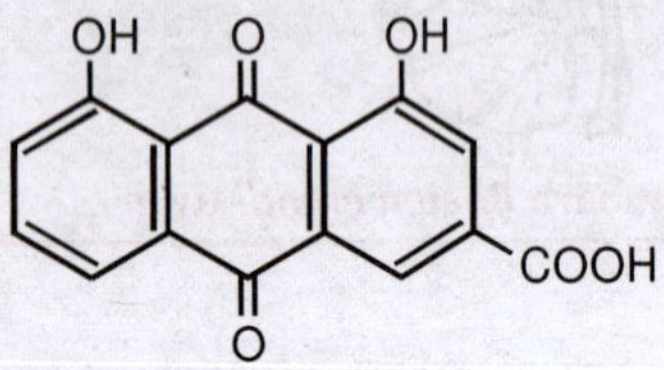

Rhein

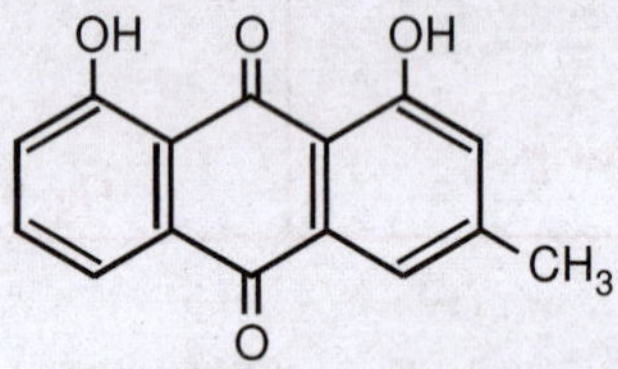

Chrysophanol

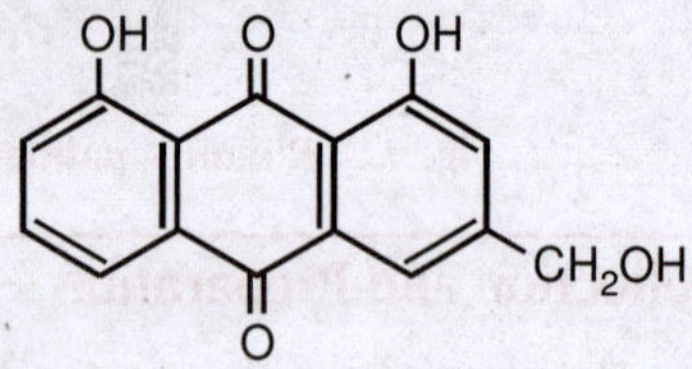

Aloe-emodin

The astringent part mainly consists of gallic acid as glucogallin, along with tannin, catechin and epicatechin. The drug also contains rheinolic acid, pectin, starch, fat and calcium oxalate. The amount of calcium oxalate ranges over 3 to 40 per cent and hence the total ash also varies.

Chemical Tests

(1) Rhubarb gives positive result for modified Borntrager's test.

(2) Rhubarb shows red colour with addition of alkalies due to presence of anthraquinone glycosides.

Standards of Quality

Total ash	:	Not more than 12.0 per cent
Acid insoluble ash	:	Not more than 1 per cent
Foreign organic matter	:	Not more than 1 per cent
Dilute alcohol soluble extractive	:	Not less than 30.0 per cent
Loss on drying	:	Not more than 12.p per cent.

Uses

Rhubarb is used as a bitter stomachic, in the treatment of diarrhoea and as a purgative.

Indian Rhubarb (Himalayan Rhubarb, Revandchini)

It consists of dried rhizomes and roots of *R. webbianum* Royle, *R. emodi* Wallich or other related *Rheum* species found in India and Nepal. It occurs as unpeeled or partly peeled pieces. They resemble to the Chinese variety, but are light in weight. Indian Rhubarb contains anthraquinone glycosides. It shows a strong deep violet fluorescence in ultra violet light. Indian Rhubarb contains hydrolysable tannins. It does not contain rhaponticin.

Rhapontic Rhubarb

It is obtained from rhizomes of *R. rhaponticum*. It is found in southern Siberia and the surrounding regions of Volga river. It is pinkish in colour and shrunken in nature. Rhapontic rhubarb lacks the presence of rhein, emodin or aloe-emodin, but it contains rhaponticin, which is a crystalline glycoside and is a derivative of diphenyl ethylene.

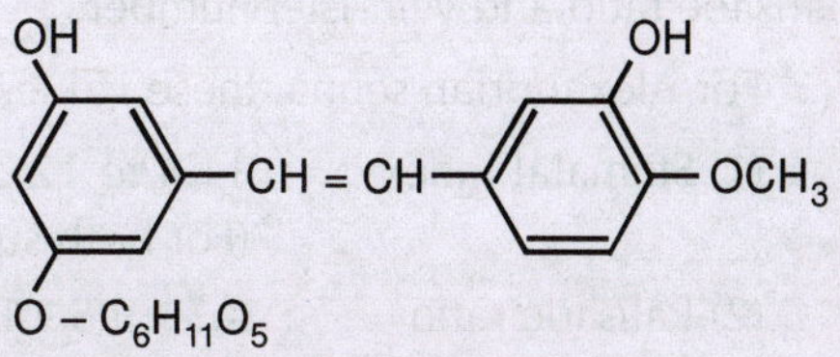

Rhaponticin

This glycoside has an estrogenic activity. It shows a blue fluorescence in ultra violet light.

ALEXANDRIAN SENNA

Synonyms

Folia sennae Alexandrina, Cassia senna, Egyptian senna.

Biological Source

It consists of dried leaflets of *Cassia acutifolia* Delile, belonging to family Leguminosae.

Geographical Source

Middle and upper Nile territories, tropical Africa and specially Sudan.

Collection and Preparation

Alexandrian senna is obtained mainly from wild sources growing in Kordofon region of white Nile Province and also in Khartoum area. The collection of leaflets along with branches is done when plants are bearing unripe fruits and dried under sunlight. The leaves are cleaned and graded into entire leaves, entire and broken leaves, siftings and pods.

The process of separation is carried out by sifting whereby first the large stalks and pods are separated from which pods are manually plucked out. The material which passes through sieves contains entire leaves, broken leaves, heavier stalks and sand. This is mechanically tossed by which entire and broken leaves are cleared out from other part. Such leaves are once again sifted, so that it separates out mixed leaves, small fragments and very small leaflets. Entire leaves are hand picked from mixed leaves and packed in mats or bales without pressing.

Macroscopic Characters

Colour	:	Pale greyish-green colour
Odour	:	Odourless
Taste	:	Mucilagenous and bitter
Size	:	2-4 cm × 7-12 mm
Shape	:	Ovate-lanceolate

Extra features : Margin is entire and curled.

Leaves are brittle and then, base is more asymmetrical, more pubescent.

Microscopic Characters

Both Alexandrian senna and Tinnevelley senna are very much identical in their histological characters. The main criterion applied for the differentiation among two species is the various leaf contents viz. stomatal index, palisade ratio and vein-islet number.

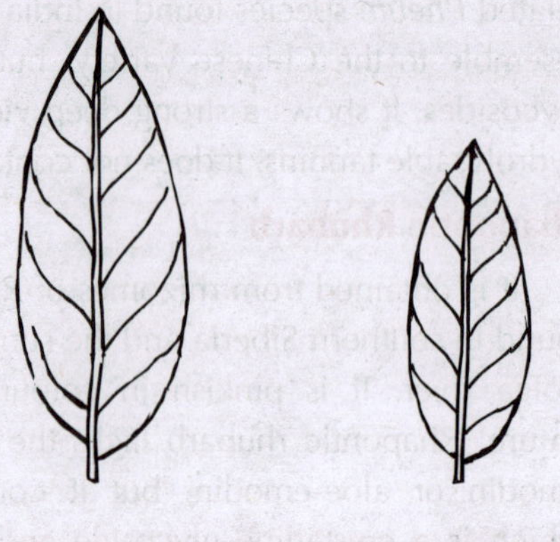

Fig. 9.9: Alexandrian senna leaf lets

For Alexandrian senna, these values are

(1) Stomatal index : 11.4 to 12.2 to 13.3 (For both surfaces)

(2) Palisade ratio : 4.5 - 9.5 - 19 (upper surface)
3.5 - 7.0 - 14.5 (lower surface)

(3) Vein-islet number : 25 - 29.5

Chemical Constituents

All the chemical constituents of Alexandrian senna are same as that of Tinnevelley senna. The percentage of sennosides in leaves and pods is more (3 - 3.6 per cent) as compared to Indian senna.

Uses

Alexandrian senna leaves are used as taxative.

Adulterants and Substitutes

It is found adulterated with Dog senna, Palthe senna, Bombay, Mecca or Arabian senna.

INDIAN SENNA

Synonyms

Senna leaf, Sennae folium, Senai-ki-patti, Tinnevelley senna, Cassia senna.

History

Senna was known to physicians from very old days and was included in Unani medicine. About 26 species of genus Cassia have been reported to contain anthracene derivatives either in free form or glycosides. Out of them, *Cassia angustifolia* (Indian Senna) and *Cassia acutifolia* (Alexandrian senna) are official in different pharmacopoeias, because of cathartic activity and also because they are available in large quantities. The other species with known cathartic activity are *C. alata, C. fistula, C. obovata, C. podocarpa, C. sieberiana, C. sofora, C. dentata, C. javanica etc.*

Biological Source

It consists of dried leaflets of *Cassia angustifolia or Cassia senna Vahl* belonging to family Leguminosae. It contains not less than 1.0 per cent w/w of sennosides A and B calculated on dried basis.

Geographical Source

Indian Senna is cultivated and collected in India. Its cultivation is mainly done in Tinnevelley, Madurai and Ramanathapuram districts of Tamil Nadu. Cultivation is attempted in Cudappa district of Andhra Pradesh, and to some extent it is collected from Kutch in Gujarat State and Rajasthan.

Cultivation of Senna herb

In India 25000 hectares of land is under cultivation of senna in India, producing about 22500 tones of leaves and 7500 tones of fruits per annum.

Senna is mainly cultivated in Tinnevelley, Ramanathapuram districts and their adjacent areas in Tamil Nadu State and is exported by ships from Tuticorin port.

Senna is a small shrub of 1.5 m, having compound paripinnate leaves and each leaf has got 3 - 5 - 7 pairs of leaflets.

For the cultivation, generally red loamy or coarse gravelly soil is selected, but it also grows on alluvial loamy soil. Before sowing, the land is ploughed twice, harrowed and levelled. First sowing is done in February-March, while second sowing is carried out in October to November i.e. after rains. Normally, the sowing is done by broadcasting method. For earlier germination, the seed surface is abraded or triturated with sand. About 15 - 25 kg of seeds are required per hectare of land and seeds are sown thinly. The plants are grown as a semi-irrigated or light-irrigated crop. Heavy irrigation, water logging, etc. are avoided. Also, the plant cannot survive at very low temperatures. The plants are allowed to grow for about five months till the growth of flower stalks occurs. Due to the cutting given to flower stalks, the further branching occurs. Though, it is a leguminous plant, still the application of nitrogenous fertilizers is beneficial because of lack of root nodules.

Collection of leaves

After 2 - 3 months of sowing, the harvesting of leaves is done in three stages. The first plucking is done when the leaflets are thick, full-grown and greenish in colour. After one month, second plucking is done and last plucking is carried out after 4 - 6 weeks. Normally, the plants are uprooted after third plucking of leaves. It is observed that the content of sennosides in leaves is maximum when they are fully grown and goes on decreasing along with maturation of pods. Senna crop which is taken after the rice crop on the same land in same season gives best yields and senna grown on dry land without proper irrigation gives poor results.

Preparation of leaves for market

After collection, the leaflets are dried under shade or indoor by spreading them in thin layers. Drying takes place within 7 - 10 days and leaves show yellowish-green colour. The leaves are tossed to separate pods. They are then packed into bales under hydraulic pressure.

Pressing the leaves under hydraulic press produces. Transverse lines on its leaves. Which are absent in Alexendrian senna leaves. About 1,200 kg of leaves and pods are obtained in one hectare of irrigated land and 700 kg of drug from rain fed land.

Fig. 9.10: Indian senna twig

Yield of leaves α-pods

During storage, the drug should be protected from light. But, generally it does not lose the potency atleast upto 5 years.

Macroscopic Characters

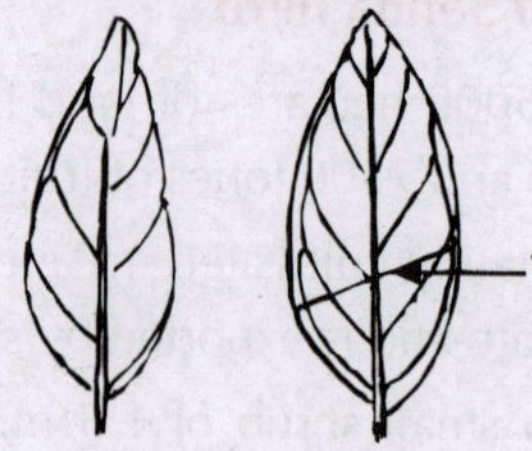

Fig. 9.11: Indian senna leaflets

Colour	:	Yellowish-green
Odour	:	Slight
Taste	:	Mucilaginous, bitter and characteristic
Size	:	7 to 8 mm in width and 25 to 60 mm in length
Shape	:	Leaves are lanceolate, entire, apex is acute with spine at the top. Bases of the leaflets are asymmetrical with transverse lines, more prominent on lower surface, while the trichomes are present on both the surfaces.

Microscopic Characters

Senna represents the typical histological characters of isobilateral leaf which include the following. The epidermis shows presence of unicellular, conical, thick-walled warty trichomes. Trichomes are slightly curved at their bases and are present on both the surfaces. Rubiaceous or parasitic stomata are present on epidermal surfaces. Palisade tissue is present on both the sides, consisting of rectangular cells, enclosing cluster-crystals of calcium oxalate. The spongy mesophyll and the conducting tissues are represented by the leaves as shown in the Fig. 9.12. A patch of sclerenchyma towards upper epidermis and above xylem (also known as pericyclic fibres) is present. The presence of cluster sheath and collenchyma are characteristics of the senna leaves.

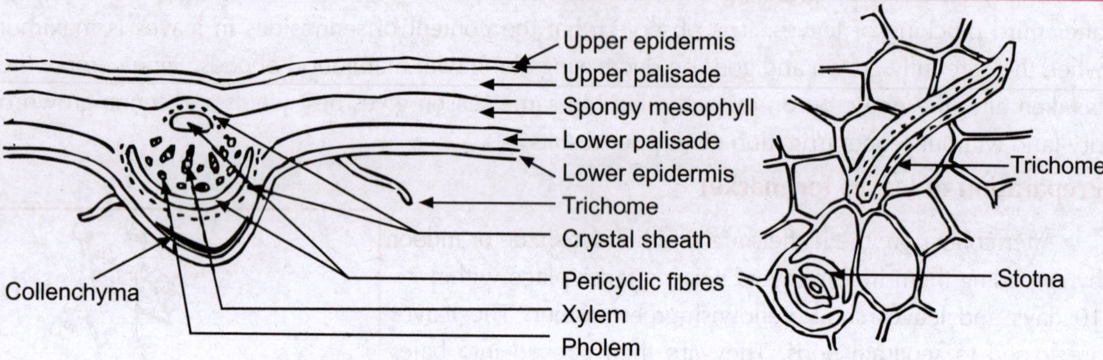

(A)T. S. of Senna leaf-let (B) Lower epidermis of Senna leaf

Fig. 9.12: Microscopic characters of senna

The following are the quantitative microscopic constants of Indian and Alexandrian senna.

		Indian senna	Alexandrian Senna
1. Vein-islet no.	-	19.5 to 22.5	25 - 29.5
2. Stomatal index	-	17 to 20	11.4 - 13.3
3. Palisade ratio	-	7.5 (upper epidermis)	9.5
	-	5.1 (lower epidermis)	7.0

Chemical Constituents

Senna contains mainly two anthraquinone glycosides called as sennoside A and sennoside B (not less than 2.5 per cent) which account for its purgative property. Tutin in 1913 isolated rhein and aloe-emodin. Stoll and his colleagues in 1941 reported isolation of crystalline sennosides A and B.

Sennosides A and B are stereoisomers of each other. They are dimeric glycosides with rhein dianthrone as aglycone. i.e. 10,10' bis (9, 10 - dihydro - 1 - 8 - dihydroxy - 9 - oxoanthracane - 3 carboxylic acid). In sennoside B the aglycone is in meso form, and in sennoside A, it is dextro-rotatory. They also differ by the way of linkage of glucose to the aglycone fraction. The purgative activity of sennosides A and B accounts upto 40 - 60 per cent activity of crude drug. It is also reported that, sugar part of these glycosides has transporting function for the aglycone upto large intestine and also protective action so that oxidation of aglycone to other very less active anthraquinone is prevented.

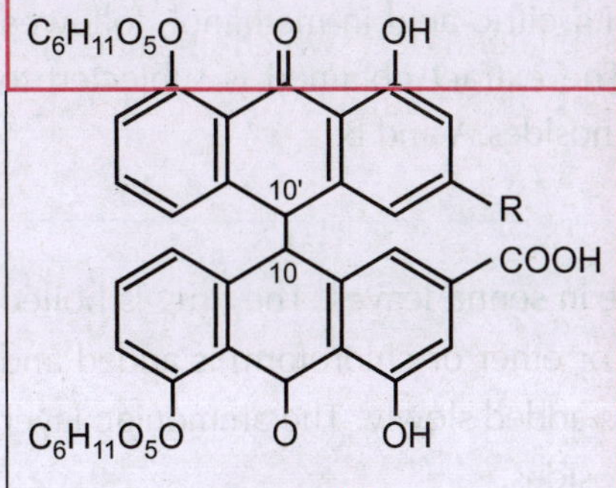

Sennosides

R	10 – 10'	Glycoside
COOH	*trans*	sennoside A
COOH	*meso*	sennoside B
CH_2OH	*trans*	sennoside C
CH_2OH	*meso*	sennoside D

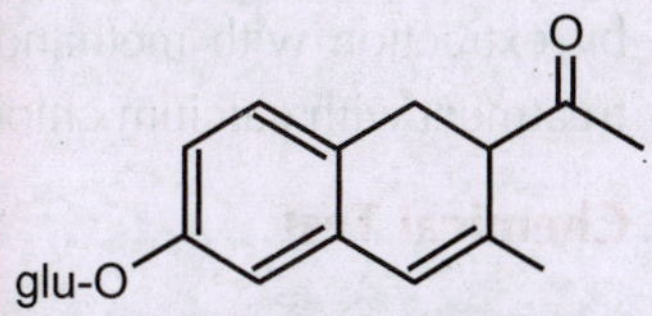

Tinevellin glucoside

Senna leaf also contains other anthraquinone glycosides in small amounts. They are sennosides C and D, rhein 8 - glucoside, rhein 8-diglucoside, aloe-emodin, 8-glucoside, aloe-emodin, anthrone diglucoside, rhein, kaempferol, aloe-emodin and isorhamnetin. It also contains phytosterol, mucilage, resin, myricyl alcohol, salicylic acid, chrysophanic acid and calcium oxalate.

It is also reported that besides anthraquinone glycosides, senna contains two naphalene glycosides called Tinnevelley glycoside and 6 - hydroxy musizin glycoside.

The contributions of Fairbairn, Lemli, Friedrich and Friedman pertaining to pharmacological work on senna are significant.

Chemical Constituents of Indian senna and Alexandrion senna are identical.

Standards of Quality

Foreign Organic Matter	Not more than 02.0 per cent
Total ash	Not more than 12 per cent
Acid insoluble ash	Not more than 2.0 per cent
Moisture	Not more than 10.0 per cent
Water soluble extractives	Not more than 30 per cent
Heavy metals:	
Lead	Not more than 10 ppm
Cadmium	Not more than 0.3 ppm

Isolation of Sennosides

Many different methods have been suggested for extraction of sennosides from senna. For commercial purposes, calcium sennosides in different strengths are extracted for purpose of stability of compounds.

The drug powder is either treated with 90 per cent methanol or 80 per cent acetone for 6 hours and then for 3 hours with cold water which leads to an extract containing about 17-18 per cent sennosides. This method enables to extract about 62 per cent of sennosides.

By the use of solutions of non-ionic surfactants and polyethylene glycols in 70 per cent ethanol, sennosides and other anthracene derivatives can be extracted.

For the isolation of sennosides, the use of non-polar synthetic resins with porous structure has been also proposed.

In another method, the senna leaf powder is macerated with citric acid in methanol, followed by extraction with methanol-toluene mixture and ammonia. The extract obtained is subjected to treatment with calcium chloride so as to get calcium salts of sennosides A and B.

Chemical Test

Borntrager's test is employed for presence of anthraquinone in senna leaves. The drug is boiled with dilute sulphuric acid, filtered and to the filtrate, benzene, or ether or chloroform is added and shaken well. The organic layer is separated to which ammonia is added slowly. The ammoniac layer shows pink to red colour due to presence of anthraquinone glycosides.

Uses

Indian senna and its preparations are used as purgative in perpetual constipation. The anthraquinone glycosides of senna are absorbed first in intestinal tract after which the aglycone part is separated and excreted in colon. These excreted anthraquinone irritate and stimulate the colon thereby movements are increased due to local action. The increase in peristalsis also causes reduction in the water absorption. This results in soft and bulky faeces.

It is believed that the griping effect caused by senna is due to its resin or emodin content. To counteract the gripping effect, some times the drug is dispensed along with carminatives. The drug given by parenteral mode is secreted in colon and thereby, causes the therapeutic action.

Adulterants and Substitutes

Tinnevelley senna is found to be adulterated with Dog senna (*C. obovata*), Palthe senna (*C. auriculata*) and Mumbai, Mecca and Arabian senna mild variety of *C. angustifolia* from Arabia.

In Dog senna, the leaves are obovate in shape and with tapering apex. They show papillose cells in lower epidermis. They contain about 1 per cent of anthraquinone glycosides.

Palthe senna is characterized by the absence of anthraquinone glycosides. The leaflets show long hair and when the leaves are boiled with chloral hydrate solution, it shows crimson colour.

The leaflets of Bombay, Mecca and Arabian senna are brownish-green in colour and more elongated and narrower in shape.

SENNA PODS

Synonyms

Senna legumes, Sennae fructus, Senna fruit.

Biological Source

Senna pods are the dried nearly ripe fruits of *Cassia acutifolia* Delile (Alexandrian senna pods), and of *C. angustifolia* Vahl (Indian senna pods),' family-Leguminosae and should contain not less than 1.0 per cent of sennoside A and B on dried basis.

Geographical Source

It is same as in case of Indian and Alexandrian senna leaflets.

Collection and Preparation for Market

The pods are collected from the same plants from which leaves are collected. Hence, the cultivation procedure does not need any special description. During harvesting both pods and leaves are collected and dried. During sifting, the pods are separated and manually segregated into different qualities. The superior quality contains only entire pods without any foreign organic matter, while lower qualities are admixed with broken pods and leaves, pieces of stems and foreign organic matter. The inferior qualities of drug are normally used for preparation of galenicals.

Macroscopical Characters

Colour	:	Senna pods are pale green to greenish-brown.
Odour	:	None
Taste	:	Slight
Size	:	4 to 5 cm long and 2 to 2.5 cm in width
Shape	:	The pods are flat, thin and broadly oblong. Apex is rounded with a slightly projecting point formed by the base of the style.

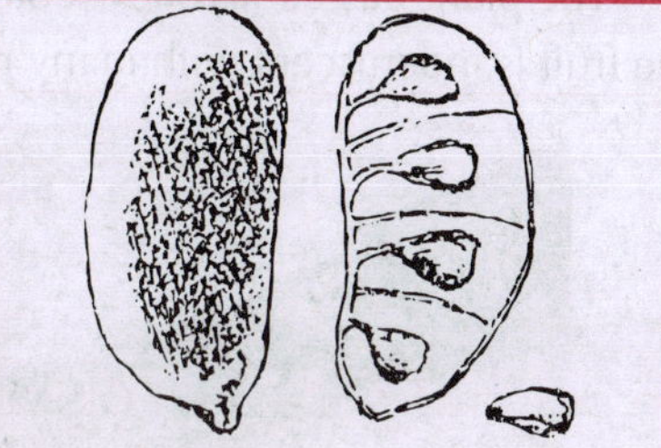

Fig. 9.13(a): Senna Pods with Seeds

Extra Features

Seeds which are about six in number and obovate in shape are with bluntly pointed projection at hilum end.

Chemical Constituents

Senna fruits contain about 2.5 to 4.5 per cent of sennosides A and B (Not less than 2.2 per cent). Indian senna pods are inferior in glycoside content (1.5 - 3 per cent). The drug responds to Borntrager's test. The pods are superior to leaves, as they contain more percentage of glycosides. The glycosides are present in pericarp of pods. The seeds contain a little quantity of sennosides. The pods also contain sennoside A_1 containing (–) sennidin 8, 8' diglucoside.

Standards of Quality

FOM	:	Not more than 1.0 per cent
Total ash	:	Not more than 6.0 per cent

Acid insoluble ash	:	Not more than 2.0 per cent
Water soluble extractives	:	Not less than 25.0 per cent
Moisture	:	Not more than 12.0 per cent

Uses

Senna pods are used as a purgative due to anthraquinone derivatives. Pods do not contain mucilage and are easy to handle for the commercial extraction of anthraquinone glycosides. Sennosides are extracted as their calcium salts. The pods are preferred to leaves, as they have less gripping action.

BAHAVA

Synonyms

Pudding pipe, Purging Cassia, Drum-stick, Cassia pod.

Biological Source

This consists of dried ripe fruits of (PODS) *Cassia-fistula*. family-*Leguminosae*.

Geographical Source

It is native to India, found in Konkan forests, in Khandesh, also planted as ornamental plant, as roadside tree in Shree Lanka, Myanmar and Egypt.

Description

The plant has dark yellow coloured fragrant flower and develops long cylindrical fruit. Typically the fruit is indeniscent with many more seeds.

Fig. 9.13: Cassia-pod opened and cassia-pod bundle

Macroscopic Characters

Colour	:	Pods are brown or chocolate brown.
Odour	:	Odour less
Taste	:	Pulp of pod is sweetish.
Size	:	40-70 × 2.0 – 2.5 cm
Shape	:	Cylindrical and elongated.

Fruits are straight, smooth, finely striated distal end is rounded. Internally pod is divided into buff-coloured segments containing one seed each.

Number of seeds is 25 – 50, which are oval, flat and 8 × 10 × 2.5 mm in size.

Chemical Constituents

Pulp of the bahava fruit is only used medicinally and contains about 50 per cent of sugar. It contains 1.0 per cent of oxymethyl anthraquinones responsible for taxative action. Pulp also contain cetyl alcohol, Kaepherol, Rhein and Fistulin. Alcoholic extract of the pulp also contains fistulic acid. Bark of the fistula contains 10 – 12 per cent tannin.

Standards of Quality

Water soluble extractives.

Pulp of Bahava fruit contains:

(i) Not less than 30.0% of water soluble extractives.

(ii) Sugars: Approximately 60.0% on dried basis.

Uses

Pulp is used as laxative and is safe for children and pregnant women.

A light red dye is obtained from bark with alum as a **mordant**.

II. STEROIDAL GLYCOSIDES OR CARDIAC GLYCOSIDES

The occurrence, chemistry and pharmacology of natural compounds with cardiac actions have been widely studied. Though, only 3 - 4 plants are used in therapeutics their presence in varied concentrations has been reported in number of plants.

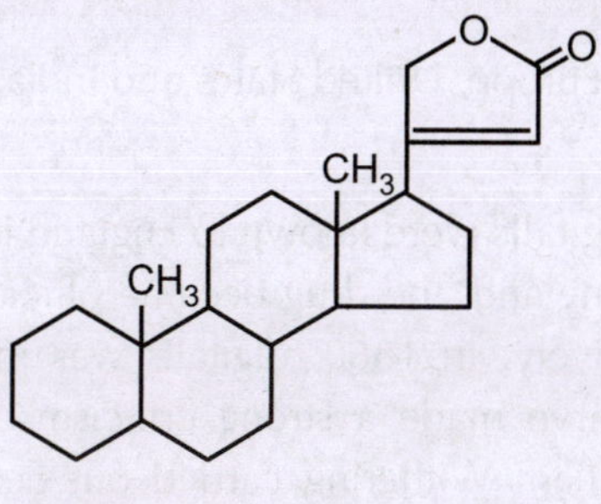

Cardenolide

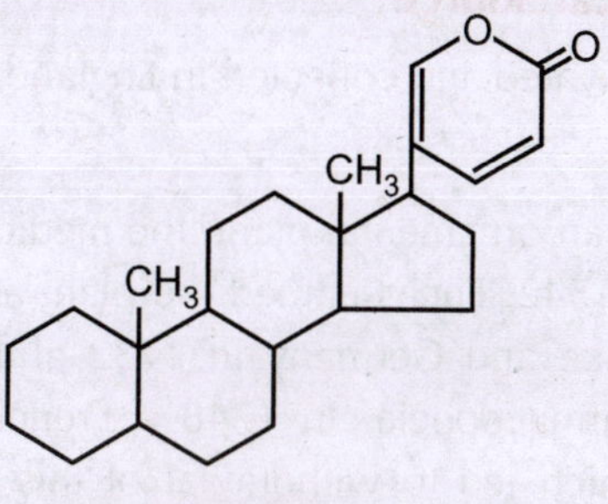

Bufadienolide

Chemically, the aglycone part of cardiac glycosides is a steroidal moiety. They are either C_{23} or C_{24} steroids, because of either five members or six members lactone ring respectively.

Those with five membered lactone ring are called as cardenolides, while with six members lactone ring are termed as bufadienolides. The lactone of cardenolide contains only one double bond and is attached to steroidal nucleus through C - 17 position. In bufadienolide, the lactone ring contains 2 double bonds and is attached to steroidal nucleus through 17 β position. The presences and attachment through a specific position to steroidal nucleus of the lactone is essential for proper cardiac activity. The term bufadienolide has been derived from bufalin, obtained from skin of toads.

The sugar part is attached through C - 3, β - linkage. Various types of sugars have been reported to be present in cardiac glycosides, such as glucose, fucose, rhamnose, digitoxose, digitalose, cymarose, sarmentose, thevatose, etc. Though, the sugars do not potentiate the medicinal activity of aglycone part, they are useful in solubilization of aglycone thereby, beneficial in absorption and distribution in the body. At a time upto three sugar molecules are attached to aglycone. It has been also shown that increase in number of OH groups on aglycone leads to quicker onset of action and enhanced metabolism.

The occurrence of cardenolides from nature is quite high as compared to those of bufadienolides.

Though, the numbers of plants contain cardiac glycosides, they are broadly restricted to angiosperms. Leguminosae, Sterculiaceae, Cruciferae, Scrophulariaceae, Euphorbiaceae etc. show presence of cardenolides. Very few families like Liliaceae and Ranunculaceae contain bufadienolides. The toad poison present on its dorsal part is a bufadienolide.

Congestive Cardiac Failure (CCF), which is a consequence of different diseases, is treated by cardiac glycosides. They act as cardiotonic by increasing the force of systolic contractions. The cardiac glycosides have a significant position in modern therapeutics, because of their utility and also because they have not yet been synthesized satisfactorily.

DIGITALIS

Synonyms

Digitalis leaves, Foxglove leaves.

Biological Source

Digitalis consists of dried leaves of *Digitalis purpurea*, family Scrophulariaceae, dried at a temperature below 60°C, immediately after collecting the leaves. The leaves should contain not more than 5 per cent of moisture.

Geographical Source

It is cultivated and collected in England, other parts of Europe, United States and India.

History

Besides an ornamental plant, the medicinal uses of Digitalis were known to England in the 10th century. In 1542, Fuchs named the plant as digitalis. In England, the drug became official in 1650 and in France and Germany in 1732 and 1771 respectively. In 1860, digitalis was included in Russian Pharmacopoeia. In 1748 a French Scientist Salarve made a strong criticism on uses of digitalis, which led to withdrawal of this drug. But, William Withering carried out detailed and elaborate studies on digitalis, due to which this drug acquired its importance from 1785 onwards.

Cultivation of Digitalis herb

It is a biennial or perennial herb of about 1 to 2 meters. In India, it is cultivated in Kashmir (Tannary and Kishtawar) and also in Mungapoo and Nilgiri hills. It is propagated by seeds of the selected strain, containing high glycosidal content. It needs calcareous, acidic sandy, light soil with traces of manganese. Soil is sterilized by steam before sowing. It grows suitably in shady situation, luxuriantly at an altitude of 1600 to 3000 m. Favourable temperature range 20 - 30°C and rainfall 30-40 cm per annum. The seeds of digitalis are very small in size i.e. 100 seeds weigh 40 to 70 mg. They are mixed with fine sand and sown in the nursery beds in March/April. About 2.5 kg seeds are needed per hectare. The young seedlings are transplanted in September and November. The crop is manured properly and kept free of weeds. The plantation is done twice a year. In the first year, the plant bears rosette leaves and in the second year sessile leaves, which are collected.

From the centre of rosette leaves in the second year a single, erect flowering axis arises with sessile leaves terminating in one-sided receme of flowers about 3-8 cm long.

The plant flowers in the month of April and is followed by the fruiting. If the plants are to be allowed to grow, the flowering tops are removed. Crop is protected from plant diseases, which otherwise lead to loss of potency of drug.

Collection of Digitalis leaves

The leaves are picked up in the afternoon during August and September in the first and the second year, when 2/3rd of the flowers are fully developed. The basal leaves and the leaves at the top are collected at the end. The discoloured leaves are rejected. While collecting the leaves, dry weather is specifically selected. After plucking, the leaves are immediately brought to the drying centre and dried in vacuum driers. It curtails exposure of fresh leaves to the atmospheric conditions. The dried leaves (containing not more than 5 per cent of moisture) are packed into the air-tight containers, with suitable dehydrating agent. The digitalis plants from which flowering stems are removed will continue to produce leaves for several years, which can be processes ferther.

The activity of the leaves is due to the glycoside. The presence of moisture and the enzymes (i.e. digipurpuridase and oxidase) cause deterioration of glycosides. If the leaves are dried above 60°C, potency is lost due to chemical degradation.

Macroscopic Characters

Colour : Dark greyish-green. Digitalis plant beares purple colored flowers.

Odour : Slight

Taste : Bitter

Size : 10 to 40 cm long and 4 to 20 cm wide

Shape : Ovate-lanceolate to broadly ovate; with irregularly crenate or serrate or occasionally dentate margin.

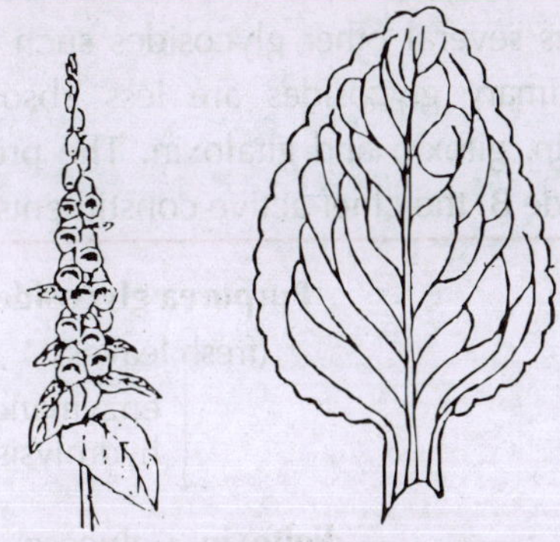

Fig. 9.14: The herb and Leaf of *Digitalis purpurea*

Extra Features : The leaves are slightly pubescent on both the surfaces with pinnate venation and prominent veinlets on the under surface. Generally, the leaves are sessile broken and crumpled.

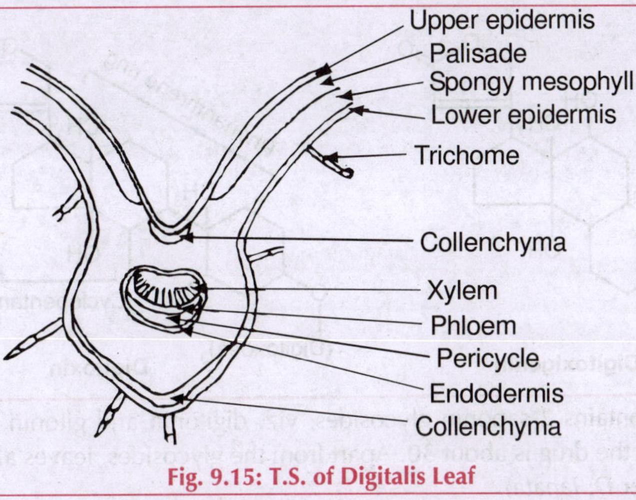

Fig. 9.15: T.S. of Digitalis Leaf

Microscopic Characters (Fig. 9.15)

Digitalis is a dorsiventral leaf. It has anomocytic stomata on both surfaces and water pores at the apex of most of the marginal teeth. The trichomes are uniseriate, multicellular (3 to 5 cells) and bluntly pointed. There are also glandular trichomes with unicellular stalk and unicellular or bicellular head. The glandular trichomes are generally located over the veins. Collapsed celled covering trichome is an important characteristics of digitalis. Digitalis is free of calcium oxalate and sclerenchyma. Starch grains are present in the endodermis. There is collenchyma at 3 different places i.e. at the upper epidermis, lower epidermis, and pericyclic part, which is also characteristic to digitalis.

Chemical Constituents

Important contributions to the studies pertaining to the chemical composition of the drug were made by Nativelle (1868), Kiliani (1891) and Stoll (1938).

Digitalis contains 0.2 to 0.45 per cent mixture of both primary and secondary cardiac glycosides (cardenolides). Purpurea glycosides A and B and glucogitaloxin are primary glycosides possessing at C-3 of the aglycone, a linear chain of 3 digitoxose moieties terminated by glucose. Digitalis also contains several other glycosides such as odoroside H, gitaloxin, verodoxin, and glucoverodoxin. The primary glycosides are less absorbed and less stable than secondary glycosides such as digitoxin, gitoxin and gitaloxin. The products of hydrolysis of purpurea glycoside A and purpurea glycoside B, the chief active constituents of the drug, are as under.

Purpurea glycoside A
(fresh leaves)
↓ enzymatic hydrolysis
digitoxin + glucose
↓ hydrolysis
digitoxigenin + 3 digitoxose

Purpurea glycoside B
(Fresh leaves)
↓ enzymatic hydrolysis
gitoxin + glucose
↓ hydrolysis
gitoxigenin + 3 digitoxose

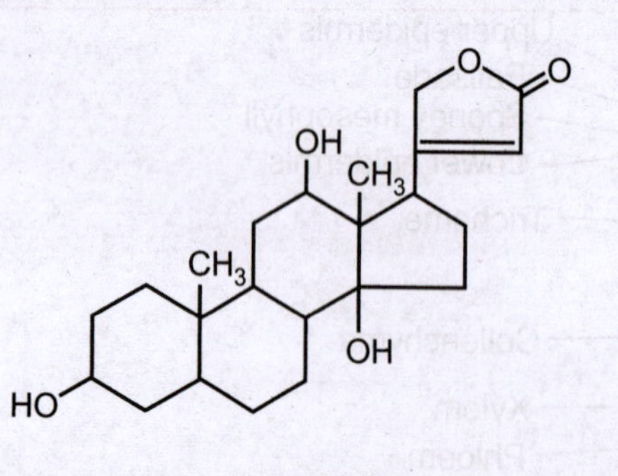

Digitoxigenin

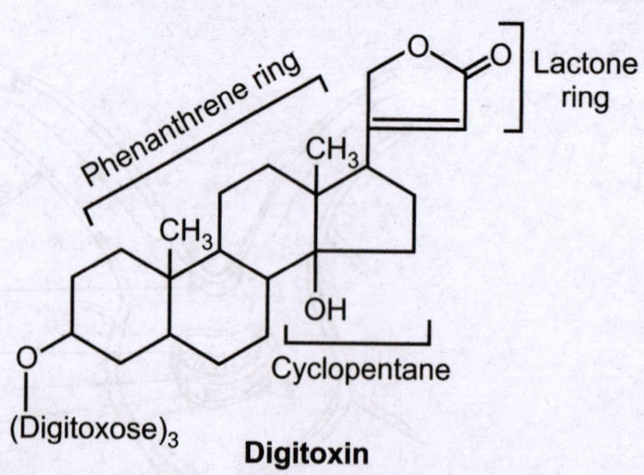

Digitoxin

Additionally, it contains 2 saponin glycosides, viz. digitonin and gitonin. The total number of glycosides reported in the drug is about 30. Apart from the glycosides, leaves also contain hydrolytic enzymes (Details under *D. lanata*).

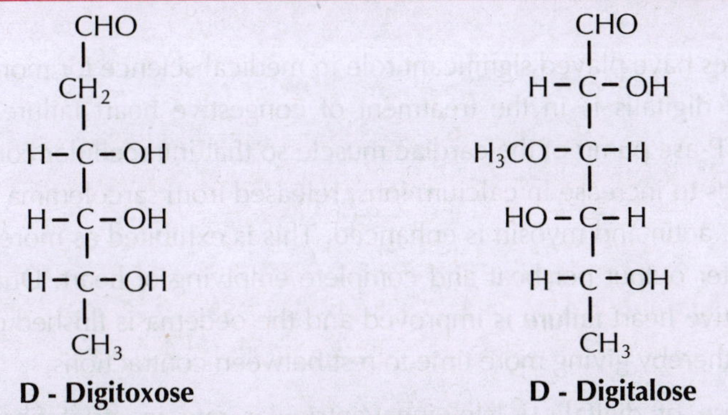

Structures of Some Cardenolide glycones from Digitalis

Chemical Tests

Keller-Kiliani test for digitoxose: The test consists of boiling about 1 g finely powdered digitalis with 10 ml 70 per cent alcohol for 2 to 3 minutes. The extract is filtered. To the filtrate is added, 5 ml water and 0.5 ml strong solution of lead acetate. Shake well and separate the filtrate. The clear filtrate is treated with equal volume of chloroform and evaporated to yield the extractive. The extractive is dissolved in glacial acetic acid and after cooling, 2 drops ferric chloride solution is added to it. These contents are transferred to a test tube containing 2 ml concentrated sulphuric acid. A reddish brown layer acquiring bluish-green colour after standing is observed due to the presence of digitoxose.

Legal test: The extract is dissolved in pyridine, sodium nitroprusside solution is added to it and made alkaline-pink or red colour is produced.

Baljet test: To a section of digitalis, sodium picrate solution is added. It shows yellow to orange colour.

Standards of Quality

Foreign organic matter	:	Not more than 2 per cent
Loss on drying	:	Not more than 5 per cent w/w, by drying to constant weight at 105°C.
Ash	:	Not more than 10.0 per cent
Acid-insoluble ash	:	Not more than 5 per cent

Assay

Digitalis is assayed both by chemical and biological methods. For bio-assay, Indian pharmacopoeia suggests both the use of pigeon or guinea pig. The assay is done for prepared digitalis or digitalis tincture. In both the animals, definite volumes of the diluted preparations are given at 5 minutes interval till cardiac arrest occurs. For guinea-pig, which is previously anaesthesized, the drug is slowly injected into jugular - vein. In anaesthesized pigeons, the jugular vein is exposed, canulated and the drug is injected.

Use

Digitalis glycosides have played significant role in medical science for more than 200 years. The most frequent use of digitalis is in the treatment of congestive heart failure. Digitalis blocks the sodium-potassium ATP-ase pump of the cardiac muscle so that intracellular concentration of sodium is increased. This leads to increase in calcium ions, released from sarcolemma and thereby brushing action of proteins viz. actin and myosin is enhanced. This is exhibited as more forceful contractions of myocardium, greater output per beat and complete emptying of heart. Due to this, the delayed circulation in congestive heart failure is improved and the oedema is flushed out. Digitalis shortens the length of systole, thereby giving more time to rest between contractions.

Another major use of digitalis is slowing ventricular rate in atrial fibrillation, atrial flutter, supraventricular tachycardia and premature extra systoles. This is due to negative chronotropic effect (slowing the heart rate) by directly suppressing the conduction of electric impulses at auriculo-ventricular node.

Digitalis has a cumulative effect in body and is slowly eliminated. Hence, dosage is important aspect. Generally, it is given by oral route except in emergency. As the digitalis dosage forms are prepared only from natural drug, variations in concentration and thereby exact content in different brands is possible. Hence, during therapy, physician relies on only one brand at a time.

Dose

Initial dose - 1 - 2 g in 24 - 48 hours
Maintenance - 100 mg daily

Prepared Digitalis

Powdered digitalis leaves are adjusted to the standard strength in prepared digitalis. It is an official preparation along with digitalis tablets, digitalis tincture and digitoxin injection.

Storage

Digitalis should be stored in a well-closed, well-filled container in cool place away from light. While storing care must be taken to ensure that, leaves do not contain more than 5 per cent of moisture, because it causes the destruction of the glycoside and ultimately loss of cardiac activity on storage. Therefore, digitalis powder is stored in containers with dehydrating agents like calcium chloride or silica gel to absorb excess of water present in the drug or in atmosphere.

Adulterants

Digitalis leaves are adulterated with the following:

1. The leaves of *Verbascum thapsus* (Scrophulariaceae) are mixed with genuine drug and can be distinguished microscopically by the presence of large woolly branched candelabra trichomes.

2. The primrose leaves: The leaves of *Primula vulgaris* (Primulaceae) are added to digitalis. They can be detected microscopically by the presence of uniseriate covering trichomes, which are 8 to 9 celled long. The lateral veins of the leaves of primrose are straight.

3. Comfrey leaves: These are the leaves of *Symphytum officinale* (Boraginaceae) and can be detected by the presence of multicellular trichomes forming hook at the top.

ALLIED DRUGS

1. Digitalis Lanata

Synonyms

Woolly fox glove leaf, Austrial Digitalis.

Biological Source

These are the dried leaves of *Digitalis lanata* Ehrhart, belonging to family Scrophulariaceae, and contain 1 to 1.4 per cent of a mixture of cardiac glycosides.

This drug is indigenous to Southern and Central Europe and currently it is cultivated in U.S.A., Holland and Eqneddor.

Macroscopic Characters

The leaves are oblong, lanceolate and sessile with entire margin. They are about 21 cm in length and 6 cm in width. The plant is a biennial herb about one metre in height (Fig. 9.16).

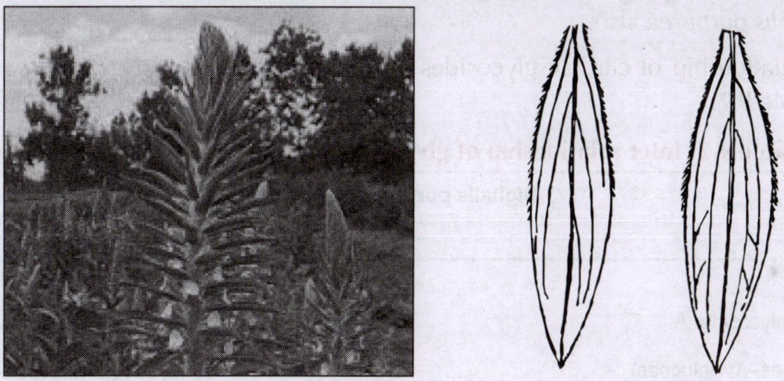

Fig. 9.16: *Digitalis lanata* herb and leaves

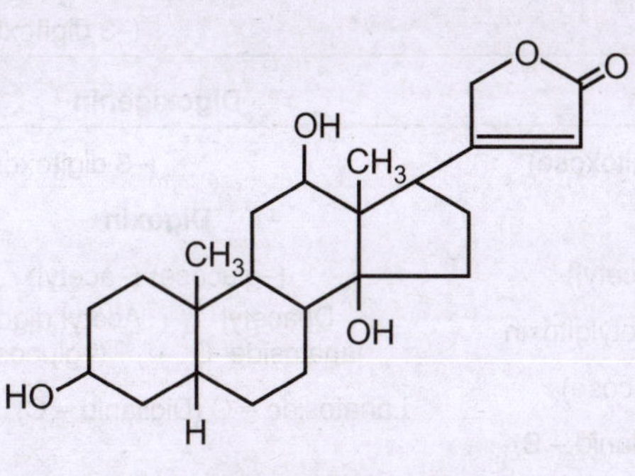

Digitoxin

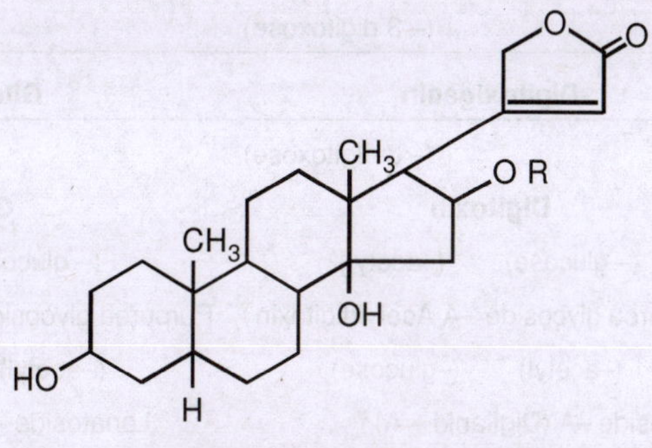

Gitoxigenin R = H

Gitaloxigenin R = CHO

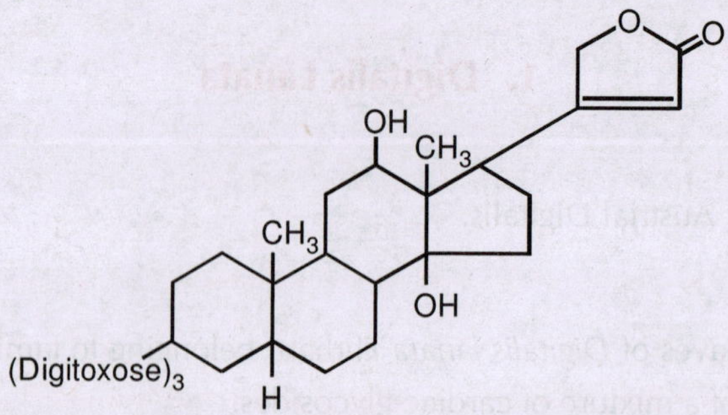

Digoxin

Chemical Constituents

It contains five primary glycosides and in all about 70 cardiac glycosides. The primary glycosides are identified as lanatosides A, B, C, D and E. It should be noted that glycone digitoxose and not aglycone, is acetylated. It is more potent than *Digitalis purpurea* and is used as a source for the manufacture of digoxin, lanatosides C and other cardiac glycosides.

The aglycones viz. digoxigenin and diginatigenin are specific to *Digitalis lanata*, while others are present in *Digitalis purpurea* also.

The inter-relationship of cardiac glycosides and their aglycones of *D. purpurea* and *D. lanata* are given in Table 9.2.

Table 9.2: Inter relationship of glycosides of *D. purpurea* and *D. lanata*

	Digitalis purpurea glycosides ↓	
↓		↓
Purpurea glycoside- A		Purpurea glycoside – B
(Deacetyl lanatosides–A) (-glucose)		(Deacetyl lanatosides–B) (-glucose)
↓ Enzymatic hydrolysis		↓
Digitoxin		**Gitoxin**
↓ (– 3 digitoxose)		↓ (–3 digitoxose)
Digitoxigenin	**Gitoxigenin**	**Digoxigenin**
↑ (– 3 digitoxose)	↑ (– 3digitoxose)	↑ (–3 digitoxose)
Digitoxin	**Gitoxin**	**Digoxin**
(– glucose) (–acetyl)	(– glucose) (– acetyl)	(–glucose) (–acetyl)
Purpurea glycoside –A Acetyldigitoxin	Purpurea glycoside–B Acetylgitoxin	Deacetyl lanatoside–C Acetyl digoxin
↑ (–acetyl) (–glucose)	(–acetyl)) ↑ (–glucose)	(–glucose)
Lanatoside –A (Digilanid – A)	Lanatoside – B (Digilanid – B)	Lanatoside – C (Digilanid – C)
↑	↑	↑
	↑ **Digitalis lanata glycosides**	

Uses

(1) *Digitalis lanata* is used as commercial source for digoxin, lanatosides C, lanatosides A and a mixture of lanatosides. Lanatosides C and digoxin have the same actions as that of digitalis. Lanatoside C is poorly absorbed from gastrointestinal tract and it is less cumulative. Digoxin, by oral route, shows effect in about 1 hour and maximum effect is reached within 6 hours. Digoxin is used for rapid digitalisation in the treatment of auricular fibrillation and congestive cardiac failure. It is preferred because of less cumulative effects.

(2) *Digitalis lutea* (Straw Foxglove): These are the dried leaves of *Digitalis lutea* (Scrophulariaceae) grown in U.S.A. and Europe. Leaves are sessile, oblanceolate with serrate or dentate margin. The size of the leaves is 28 × 6 cm, but most of the leaves reach only 50 per cent of their size. The drug is as potent as *Digitalis purpurea*. It is used for the same purpose but, is supposed to have less irritation. The chemical constituents of the drug are not thoroughly known, but it is free of calcium oxalate. It is used as a common **substitute** for official drug.

(3) *Digitalis thapsi* (Spanish Foxglove): It is found in Spain and Italy. The leaves are small (5 to 10 × 1.5 to 5 cm) in size. The leaves are yellowish-green in colour and lanceolate with crenate margin and decurrent lamina. Histologically, leaves can be identified by the absence of non-glandular trichomes and also due to the presence of very strongly striated cuticle. It contains calcium oxalate crystals. The drug is 1.25 to 3 times more potent than *Digitalis purpurea.*

The other species of Digitalis containing cardiac glycosides are *D. subalpina, D. dubia, D. grandiflora, D. ferruginea,* and *D. mertonensis.*

The various Digitalis preparations are:

(1) Acetyl digitoxin

(2) Acetyl digoxin

(3) Deslanoside injection (derivative of digoxigenin)

(4) Digitalin (standardised mixture of glycosides from the seeds of D. purpurea)

(5) Digitoxin

(6) Digoxin as elixir, injection, tablets

(7) Gitalin (mixture of glycosides from D. purpurea)

(8) Lanatoside A

(9) Lanatoside C

(10) Digitalis tincture, and

(11) Prepared digitalis.

EUROPEAN SQUILL

Synonyms

Scilla, Squill bulb, White squill, Urgenia squil.

Biological Source

Squill consists of the sliced white and dried fleshy, inner scalve leaves from the bulbs of *Urginea maritima* (Linn.) Baker, belonging to family Liliaceae.

Geographical Source

It is indigenous to the countries near the Mediterranean region like France, Italy, Malta, Spain, Greece, Algeria, and Morocco.

Collection and Preparation

White squill gives the bulbs which are about 12 - 15 cm in diameter and 18 - 20 cm in height. They grow in half submerged condition in sandy soil in the coastal region. They are collected in spring or in August when the plant sheds all aerial leaves and flowering stage is completed. After collection, the external thin scaly leaves and central portion are removed and the remaining fleshy bulb is cut into transverse slices and then dried by sun rays or artificially.

Macroscopic Characters

It contains concavo–convex slices which are thick at middle point, 3 - 6 cm in length and 4 - 8 mm in width. It has a slight odour. The drug appears as brittle, translucent and yellowish-white scales. It is hygroscopic in nature and becomes tough and flexible on storage.

Fig. : Urgenia maritima herb

Microscopic Characters

The transverse section shows epidermis, mesophyll and vascular bundles. Epidermis consists of polygonal, axially elongated rectangular cells and the occurrence of stomata is almost rare. Mesophyll is characterised by mucilage present in large polygonal parenchymatous cells. The mesophyll also shows raphides of calcium oxalate. Mesophyll is interrupted by small vascular bundles.

Chemical Constituents

Squill contains the cardiac glycosides of bufadienolide type. It contains a number of glycosides, but most important is scillaren A, about two third of total glycosides in quantity. It also contains scillaren B.

The hydrolysis of scillaren A shows that it is biglycoside (containing 2 sugars viz. glucose and rhamnose). The aglycone is called as scillaridin A (scillarenin A).

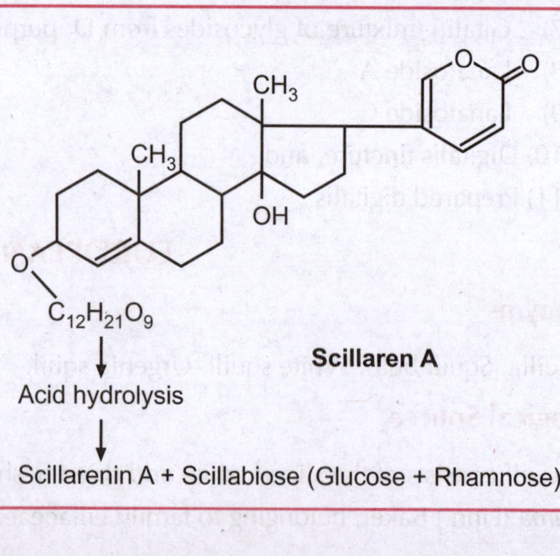

Scillaren A

The drug also contains other cardiac glycosides in small amounts like glucoscillaren A, which is a triglycoside and Proscillaridin A.

Glucoscillaren A : Scillarenin A + rhamnose + glucose + glucose and

Proscillaridin A: Scillarenin A + rhamnose

Along with cardiac glycosides, the drug also contains xanthoscillide, flavonoids, mucilage, calcium oxalate, sinistrin (a carbohydrate similar to inulin), and a volatile substance which causes irritation.

Standards of Quality

Acid insoluble ash : Not more than 1.5 per cent

Uses

It is a cardiotonic without cumulative effect. It is also used in chronic bronchitis as an expectorant in small doses, while in higher doses causes vomiting.

INDIAN SQUILL

Synonyms

Jangli pyaz, Sea onion, Scilla, Urginea.

Biological Source

Urginea consists of dried slices of the bulbs of *Urginea indica Kunth,* family *Liliaceae.*

Geographical Source

As the name indicates, it is found in India. It is grown on sea coasts including Konkan and Saurashtra and also in the dry hills of the lower Himalayas at an altitude of 1500 m. European squill is grown in Italy, Spain, Greece, France and Algeria.

Cultivation

It is not cultivated, scientifically. However, it is observed that the herb needs sandy soil and mean annual temperature of 15°C. Bulbs can be raised from seeds and the plant acquires full size within 5 years. Generally, the distance between two rows of plants is about 75 cm. Other agricultural practices followed are similar to that of the common bulbs like garlic and onion.

Collection of Squill Bulbs

The bulbs are collected after flowering in autumn. They are cleaned and cut into quarters which are further sliced. It looses about 80 per cent of its weight when the sun-dried. The bulbs are finally packed in bags or barrels.

Macroscopic Characters

Colour	:	The slices of squill bulbs are slightly yellowish to white.
Odour	:	Slight and characteristic.
Taste	:	Bitter, mucilaginous and acrid.
Size	:	30 to 60 mm in length and 3 to 8 mm broad.
Shape	:	The slices of the Indian squill are united in groups of 4 to 8, which are curved.

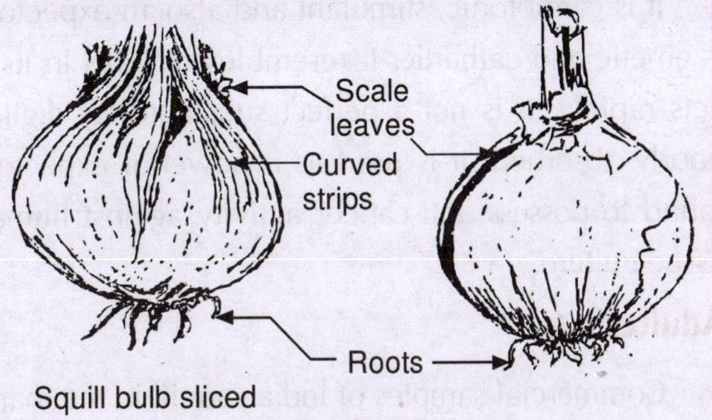

Fig. 9.17: Squill bulb

Extra Features

The entire bulb of the Indian squill is about 15 cm in diameter. The slices of the squill are translucent and become tough and flexible after absorbing the moisture.

Microscopic Characters

The transverse section shows presence of polygonal axially elongated epidermis, parenchymatous mucilaginous mesophyll with raphides of calcium oxalate crystals and is interrupted by small vascular bundles.

Chemical Constituents

Indian squill contains about 0.3 per cent of cardiac glycosides. It yields to alcohol about 20 to 40 per cent of extractive. The other contents of the drug are mucilage (about 40 per cent) and calcium oxalate.

Scillaren A and scillaren B are the major cardiac glycosides of the drug. Scillaren A on hydrolysis (by an enzyme) scillarenase yields proscillaridin A and on acid hydrolysis scillaridin A, whereas scillaren B yields proscillaridin B and scillaridin B respectively. The drug also contains glucoscillaren A and an enzyme scillarenase.

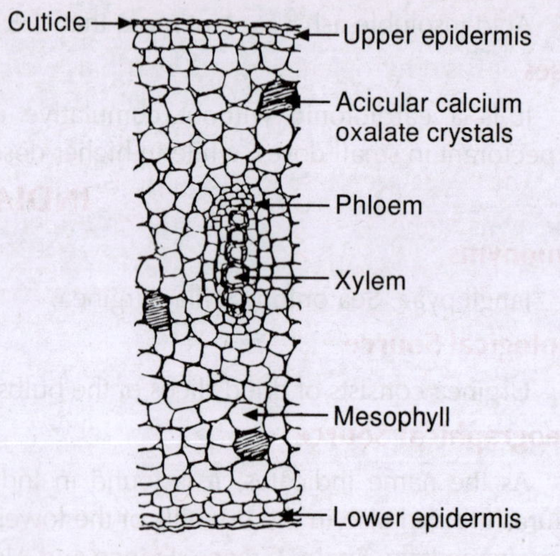

Fig. 9.18: Transverse Section of squill bulb

Chemical Test

Mesophyll stains red with alkaline colarin solution and reddish purple with 0.1 M Iodine solution.

Standards of Quality

Acid insoluble ash : Not more than 1.5 per cent

Uses

It is cardiotonic, stimulant and also an expectorant. It is diuretic in small doses. In large doses, it is emetic and cathartic. It resembles digitalis in its cardiotonic activities, but is less cumulative and acts rapidly. It is not a perfect substitute for digitalis, as it is associated with irritant effect and is poorly absorbed. It is used as a powerful expectorant in chronic bronchitis and asthma. It is also found to possess anti-cancer activity against human epidermoid carcinoma of the nasopharynx in tissue culture.

Adulterants

Commercial samples of Indian squill in the market, many a time, contain *Scilla hyacintoiana*.

STROPHANTHUS

Synonyms

Semino Strophanthi, Arrow poison.

Biological Source

These are the dried ripe seeds of *Strophanthus kombe* Oliver or of *S. hispidus* De Candolle, belonging to family Apocynaceae.

Geographical Source

Strophanthus is a climber indigenous to regions near the Shire river, Nyanza and Tanganyika lakes of Eastern tropical Africa. The tribals were using the extract of seeds as arrow poison.

Macroscopic Characters

The seeds are obtained from fruits with 2 follicles. The awns of the seeds are separated. The seeds are 10 - 20 mm in length, 3 - 5 mm in width and 2 mm in thickness. The seeds have bitter taste and unpleasant odour. The shape is lanceolate and awn is thread like extension of testa. The silky touch of the seeds is due to trichomes. The endosperm and embryo are quite oily in nature unpleasant odour.

Fig. 9.19: Strophanthus kombe climber

Chemical Constituents

The drug contains a mixture of glycosides called K-strophanthin. It contains K-strophanthoside (strophoside), K-strophanthoside β cymarin and cymarol. Except cymarol, all the 3 glycosides yield strophanthidin on hydrolysis. K-strophanthoside is considered as main primary glycoside which on hydrolysis gives aglycone K-strophanthidin along with 3 sugars cymarose, α and β glucose.

The drug also contains mucilage, resin, trigonelline, choline, and upto 30 per cent of fixed oil.

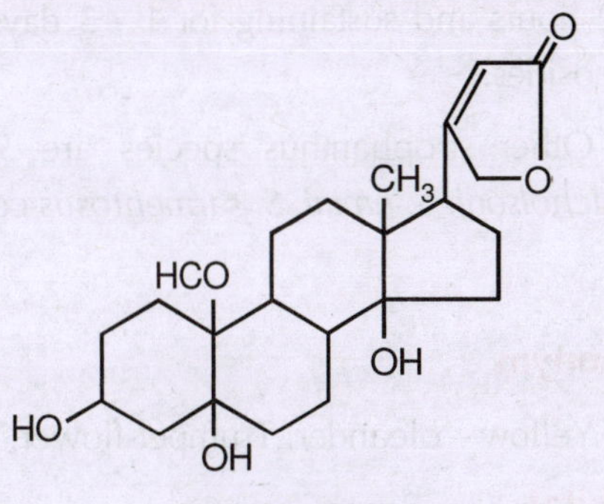

Strophanthidin

Chemical Test

The glycosides show emerald green colour on addition of 80 per cent sulphuric acid.

Uses

It is a cardiotonic. The drug is mainly considered as a source of K-strophanthin. These glycosides have less cumulative effect otherwise similar in actions as that of digitalis. It is given parenterally, because of poor absorption by oral route.

OUABAIN

Ouabain is a name of G-strophanthin and it is a cardiac glycoside obtained from the seeds of *Strophanthus gratus* or from the wood of *Acokanthera schimperi* or *A. ouabain*, belonging to family Apocynaceae.

It is poisonous substance and is present from 4 - 8 per cent in seeds. It is two times more potent than K-strophanthin.

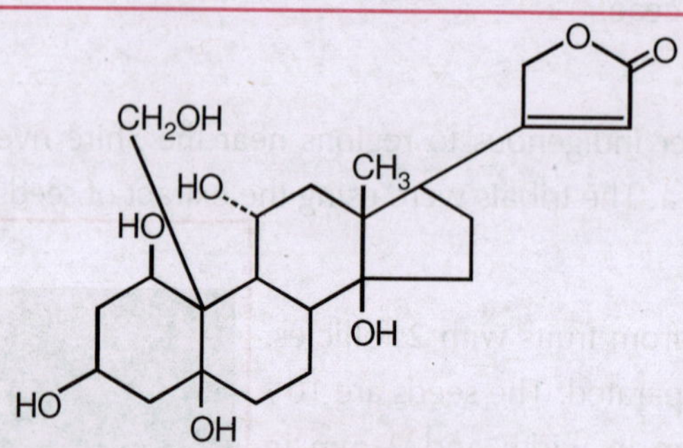

Ouabagenin

G-strophanthin on hydrolysis gives G-strophanthidin, which on further hydrolysis gives ouabagenin and rhamnose.

Uses

It is a cardiac glycoside, given intravenously in acute cardiac failure, because of its rapid onset of action. The onset of action occurs within 5 -10 minutes and reaches to peak effect in 30 minutes to 2 hours and sustaining for 1 - 3 days. It is used as a reference standard for bioassay of cardiac glycosides.

Other strophanthus species are *S. - courmontii* (one fourth active as *S. kombe* seeds); *S. nicholsoni, S. emini, S. sarmentosus* containing sarmentogenin.

THEVETIA

Synonym

Yellow – oleander, Trumpet-flower, Lucky-nut tree.

Habitat

A large, evergreen shrub 450 - 600 cm tall with scented bright yellow flowers in terminal cymes bears triangular fleshy drupes, containing 2 - 4 seeds. Leaves are about 10 - 15 cm in length linear and acute.

Biological Source

These are the dried seeds of *Thevetia - peruviana. Merrill syn. Thevetia nerifolia*, Juss. belonging to family Apocynaceae.

Geographical Source

West Indies, Florida, Hawaii, America and also found throughout India.

Macroscopic Characters

Colour : Seeds are green to greenish black

Odour : None

Taste : Thevetia seeds are very bitter, when chewed produce numbness

Shape : Oval, hard and pointed at one end.
Bilocular and endocarp is hard.

Size : 1.6 × 1.2 × 0.5 cm

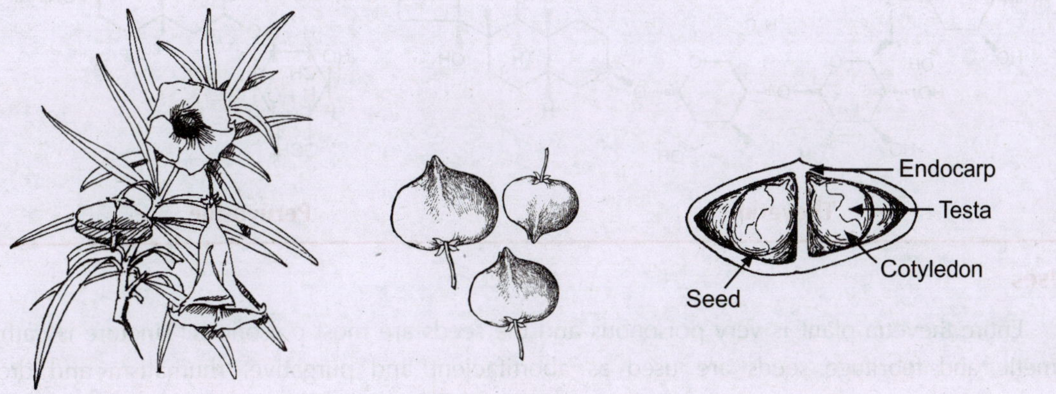

(a) Thevetia peruviana twig (b) Thevetia fruits (c) Fruit exposed to show seeds

Fig. 9.20: Thevetia

Chemical Constituents

Thevetia kernels are very rich in cardio-active glycosides, which are found to be triosides i.e. the aglycone of these glycosides is a combination of three sugar units.

Thevetin is the major and active constituent of Thevetia which is a mixture of two triosides i.e. Thevetin A and (Thevetin B) cereberoside.

The monosides separated from thevetia seeds are cerberin, neriifolin, peruvoside, theveneriin (ruvoside) and peruvosidic acid (Perusitin).

Table 9.3: The details of glycosides, aglycones and sugars of Thevetia are as under

Glycoside	Aglycone	Sugars
Cereberoside (Thevetin B)	Digitoxigenin	L-Thevetose + 2 mol. D-glucose
Thevetin A	Cannogenin	L-Thevetose + 2 mol. D-glucose
Peruvoside	Cannogenol	L-Thevetose
Neriifolin	Digitoxigenin	L-Thevetose
Theveneriin	Cannogenol	L-Thevetose
Peruvosidic acid	Cannogenic acid	L-Thevetose

Peruvoside shows a quick action with well absorption and tolerance. It has a very less accumulative tendency and can be given orally. Enzymatic biotransformation of peruvoside and nerrifolin from yellow oleander gives digoxin.

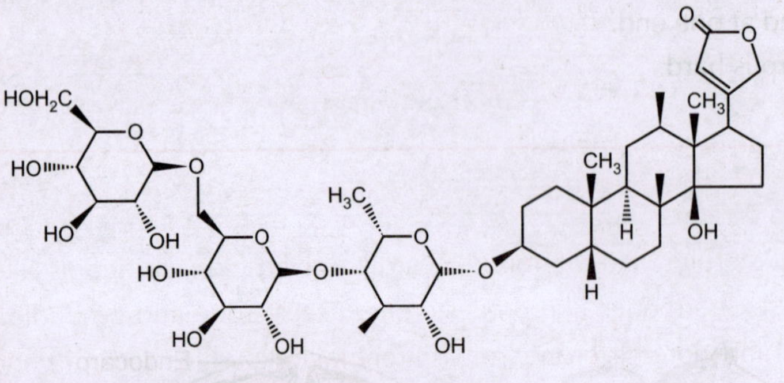

Thevetin

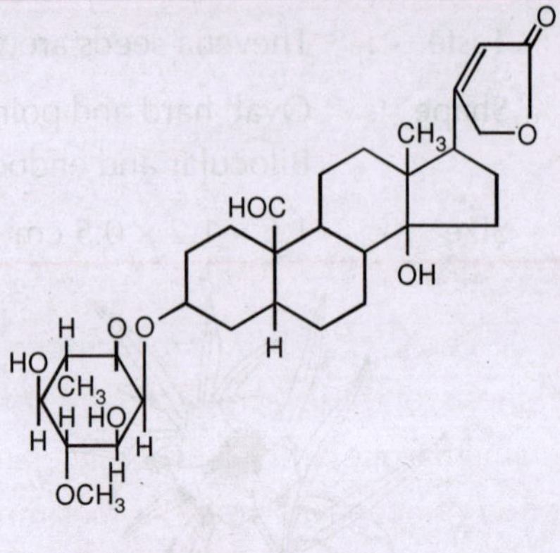

Peruvoside

Uses

Entire thevetia plant is very poisonous and the seeds are most poisonous. Tincture is cathartic emetic and febrifuge seeds are used as abortifacient and purgative, rhumatism and dropsy. Peruvoside is used in treatment of mild cardiac insufficiency and weak heart. This drug shows relatively high degree of therapeutic index compared to digoxin.

In West Germany, peruvoside is marketed under name *"Encordin"*.

III. SAPONIN GLYCOSIDES

Most of the plants containing saponin glycosides are medicinally and commercially important. As the name indicates, the aglycone part of these glycosides has soap like action. They exhibit some physical properties like foaming action by shaking with water and yielding colloidal solutions. They are generally considered as haemotoxic, because they cause haemolysis of erythrocytes. Due to this activity, some of them are used as fish poisons. Saponins have a bitter and acrid taste. They cause irritation of mucous membrane. They are mostly non-crystalline substances soluble in water and alcohol, but insoluble in non-polar organic solvents.

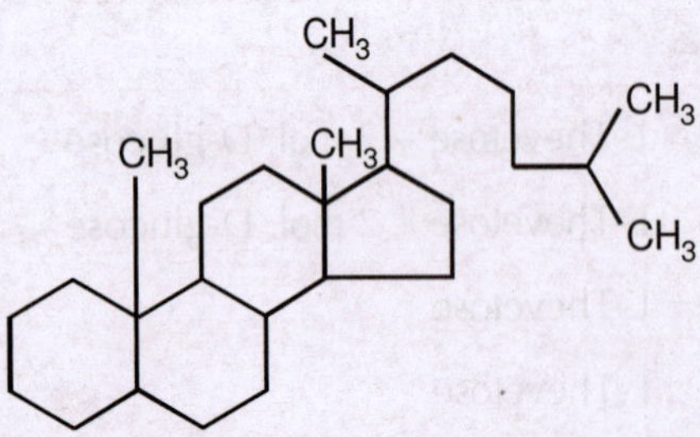

Steroid Pattern of Saponins
(Tetracyclic Triterpenoids)

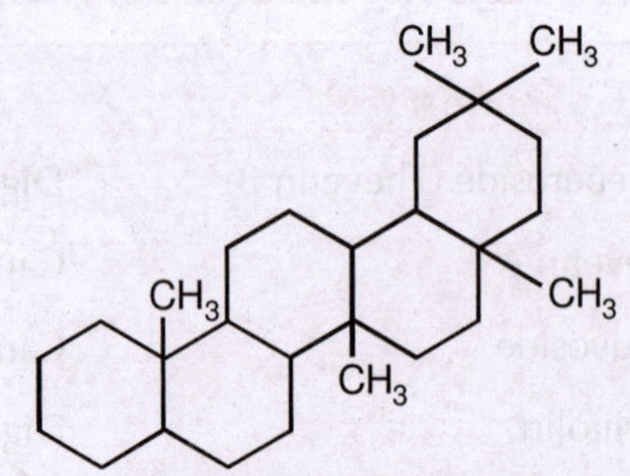

Pentacyclic Triterpenoids

Saponins are good emulsifiers, detergents and toxic compounds.

Chemically, they contain aglycone called as sapogenin. Sapogenins are high molecular weight substances which by acetylation give crystalline forms. The harmful sapogenins are called as sapotoxins. Depending on the nature of aglycone, saponins are categorized into 2 groups viz.

(i) Steroidal saponins (Tetracyclic triterpenoid saponins)

(ii) Pentacyclic triterpenoid saponins

Both types of aglycones are linked with different types of sugars and uronic acids.

1. Steroidal Saponins

Commercially, steroidal saponins are more important, as they are used as raw material for the synthesis of various medicinally useful steroids like vitamin D, cardiac glycosides, corticoids like betamethasone and cortisone acetate, sex hormones like progesterone, testosterone and oestradiol, oral contraceptives such as mestranol and norethisterone; and spironolactone which is a diuretic steroid. Steroidal sapogenins viz. diosgenin and hecogenin can be considered as a representative example of this group of saponins. Due to their pharmaceutical importance, many plants have been screened for detection of steroidal saponins. Their distribution is limited to plant kingdom. In dicot plants, important sources are from Leguminosae, Solanaceae, Apocynaceae, etc. They are mainly obtained from monocot plants like Liliaceae, Dioscoreaceae and Amaryllidaceae.

Digitonin, gitonine and digonin from digitalis seeds, while sarsaponim diosgenine from other sources are the examples.

2. Pentacyclic Triterpenoid Saponins

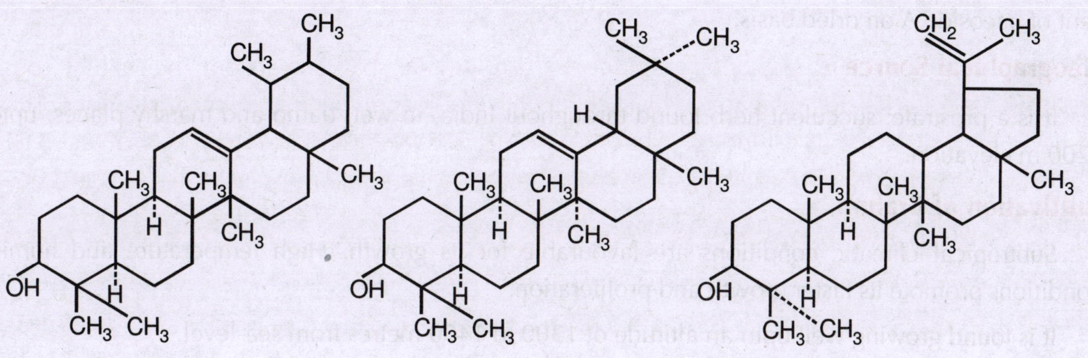

α–amyrin β–amyrin Lupeol

This group contains the sapogenin with pentacyclic triterpenoid nucleus, which is linked with sugars or uronic acids. The sapogenin is further differentiated into

(a) α - amyrin type; (b) β - amyrin type; and (c) Lupeol.

An important derivative of this group are triterpenoid acids. These acids are present in various drugs formed by substitution of carboxylic group at C_4, C_{17} and C_{20}.

	R'	R''
Gypsogenin	CHO	H
Hederagenin	$CH_{20}H$	H
Oleanolic acid	CH_3	H
Quillaic acid	CHO	OH

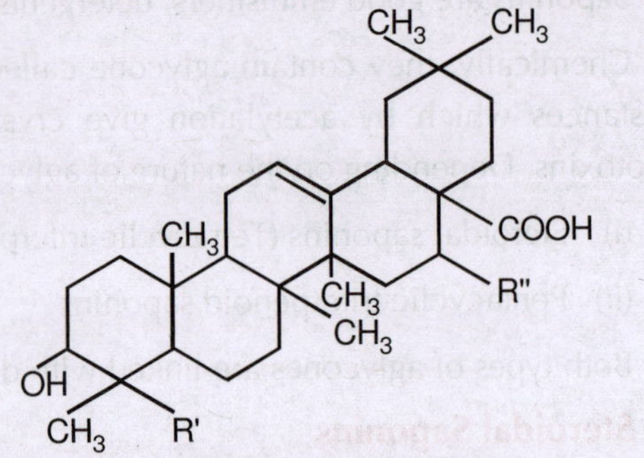

Triterpenoid acids

Besides the chemical structure, these types of saponins differ from steroidal saponins by way of their distribution. Pentacyclic triterpenoid saponins are available from various families of dicot plants like Polygalaceae, Caryophyllaceae, Berberidaceae, Umbelliferae, Rubiaceae, Compositae, Primulaceae, Rutaceae, Chenopodiaceae, etc. They are however scarcely found in monocot plants. Glycearrhiza, quillaia and senega are the drugs contianing these soponins.

BRAHMI

Synonym

Bacopa

Biological Source

It consists of the fresh leaves and the stems of the plant known as *Bacopa monniera* Linn. (*Herpestis moniera*), belonging to family Scrophulariaceae, and should contain not less than 2.5 per cent of bacoside A on dried basis.

Geographical Source

It is a prostrate, succulent herb found throughout India, in wet, damp and marshy places, upto 1200 m elevation.

Cultivation of Brahmi

Subtropical climatic conditions are favourable for its growth. High temperature and humid conditions promote its faster growth and proliferation.

It is found growing well upto an altitude of 1300 to 1400 metres from sea-level.

High temperature ranging from 30 to 40°C is favourable for growth.

50 - 100 cm rainfall which is well distributed.

It can be grown under varying soil conditions, but wet soil with poor water drainage, is found to be suitable for its growth.

Seeds can be used, but due to excessive loss of seeds, vegetative propagation (plant cuttings) is practised widely. Ideally transplanted during summer (March - May) or at the onset of monsoon (June - July). Its a perennial crop.

Fig. 9.21 (a): Enlarged Twig of Bacopa with leaves

Fig. 9.21 (b): Bacopa moniera herb

Land is ploughed twice or thrice thoroughly and made weed free. About 4 - 5 tonnes of well decomposed manure/hectare is added, again ploughed, levelled and irrigated a day before planting.

Transplanting material is obtained from the plant cutting raised in nursery bed. Mixture of soil, manure and sand is used for raising nursery bed. It is irrigated well to keep the soil moist. Then seeds alongwith portion of sand are sown in the nursery bed and irrigated regularly. Within a period of 2 to 3 months, the seedlings are ready for transplanting.

The mother plants are cut into small pieces alongwith roots, which are ideal transplanting material. At the onset of monsoon or during summer, they are transplanted at a spacing of 40 ∞ 40 cm in well prepared wet field; followed by flood irrigation.

About 100 kg plant cuttings are required for planting in an acre of land.

It is moist loving, marshy plant, hence requires adequate irrigation. First irrigation after transplanting is essential for establishment of plants, followed by the flood irrigations as per the requirement of soil. Usually during monsoon, no additional irrigation is required. On an average, 7 - 8 irrigations are sufficient.

Application of natural fertilizers like FYM or compost manure is sufficient for its growth. For its good quality and yield, application of nitrogenous fertilizers is advisable. A basal dose of 60 kg phosphorus and 60 kg of potash/hectare should be applied alongwith 100 kg nitrogen/hectare in three split doses.

During initial stages of plant growth, hand weeding is preferred at an interval of 10 - 15 days and once the plants establish themselves and proliferate with dense mass, weeding is carried out as and when required.

Collection of Brahmi

The crop can be maintained in perennial state and with two harvests a year, first in month of June and next after monsoon in month of October-November. Usually, after 120 - 150 days of planting, the crop is ready for harvesting. Cutting is done in such a way that the aerial part is removed and the basal portion is left in the soil for subsequent regeneration of crop.

Harvested material is washed and cleaned of the external matter. It is then spread in thin layer on ground under shade at room temperature. The product is tossed twice a day for uniform drying. The drying process takes about 8 - 10 days.

The total dry herb yield of plant in two harvesting seasons is approximately 100 quintal per hectare.

Macroscopic Characters

Colour : The entire herb is green in colour

Odour : Slight and characteristic

Taste : Bitter

Size : (Leaves) About 2 cm

Shape : The leaves are fleshy and obovate

Leaves are alternate, simple, entire, with broad apex, sessile and their lower surface is dotted. Flowers are solitary, stalked; corolla bluish white in colour and about 1 cm in size. One of the 5 sepals is larger than the others. Flowers arise in the axes of the leaves in April-June. Capsules are ovoid, seeds are minute and numerous. Roots are present at nodes. It is a creeping herb.

Microscopic Characters

Brahmi consists of whole plant and hence microscopy of leaf and stem are to be covered.

(a) Microscopy of leaf: Midrib in leaf is indistinct and mesophyll is isobilateral of spongy parenchymatous cells. Stomata are present on both the surfaces of leaf. Calcium oxalate in the form of prismatic crystals are reported in mesophyll only.

(b) Microscopic characters of stem: Cortex of the stem is with large air spaces xylem vessels are radially arranged, while the xylem rays are uniseriate. Pith consists of collapsed parenchymatous cells.

Chemical Constituents

Brahmi is found to contain the alkaloids brahmine, herpestine and the mixture of 3 other alkaloids. It contains saponins, namely bacosides A and B. Additionally it also contains betulic acid, stigmasterol, monnierin and hersaponin. Bacosides A and B on acid hydrolysis yield triterpenoid aglycone bacogenins A and B respectively. Asiatic acid and Brahmic acid are the final products of hydrolysis of saponins of Brahmi and centella.

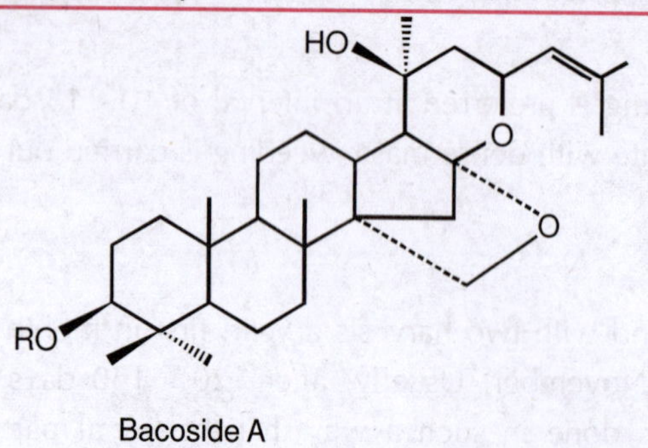

Bacoside A

R = Ara (3→1) Ara

Bacoside B

R = Glu (2→1) Ara (3→1) Glu

Structure of Bacoside A and B

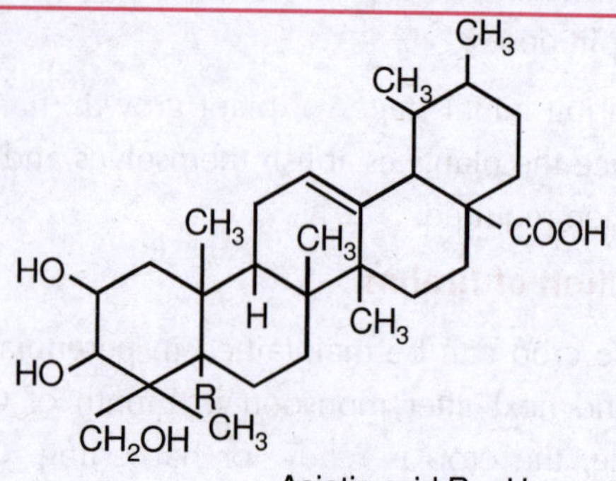

Asiatic acid R = H

Brahmic acid R = OH

Standards of Quality

Foreign organic matter	:	Not more than 2.0 per cent
Total ash	:	Not more than 18.0 per cent
Acid insoluble ash	:	Not more than 6.0 per cent
LOD	:	Not more than 12.0 per cent
Water soluble extractives	:	Not less than 22.0 per cent
Alcohol soluble extractives	:	Not less than 06.0 per cent

Uses

It is used as nervine tonic, in the treatment of asthma, epilepsy and insanity. It is aperient and also diuretic. The alcoholic extract of the entire plant is found to have anti-cancer activity.

Recently, it has been reported that bacosides A and B enhance protein kinase activity and new protein synthesis especially in those parts of cerebrum associated with long term memory, and consequently enhancement of recall levels, learning and treatment of dementia. It has significant effect on retention of new information and promotes clarity of thought memory and mental concentration.

DIOSCOREA

Synonyms

Yam, Rheumatism root, Chinese yam.

Biological Source

It consists of dried tubers of the plants, *Dioscorea deltoidea, D. composita*, and other species of *Dioscorea*, family Dioscoreaceae.

Geographical Source

D. deltoidea is found growing in North Western Himalayas from Kashmir and Punjab to Nepal and China upto an altitude of 1000 to 3000 m. It is cultivated in Jammu and Kashmir and in part of Himachal Pradesh. *D. deltoidea* is also found in U.S.A. and Mexico.

Cultivation and Collection

Climatic conditions: Following climatic conditions are more favourable for cultivation of Dioscorea rhizomes:

Altitude: 1000 – 3000 metres above sea level.

Soil : Sandy, loamy clay soil but preferably acid, neutral or slightly alkaline soil.

Temperature: Sunny situation or warm green house. Moist soil is also prefered.

It is a perennial. Climber growing upto 3 metres.

In view of the pharmaceutical significance of the drug, it is tried and successfully grown in various parts of India. Commercially, it is grown in Tamil Nadu, West Bengal, Maharashtra, Karnataka, and Jammu and Kashmir.

Cultivation of dioscorea

The crop can be raised from seeds, but variability in progeny and comparatively longer time for harvesting are the disadvantages with this method. Therefore, healthy tubers of about 70 - 80 g in weight with crown are selected for cultivation. For checking the tuber-rot, they are treated with fungicide and sown in nursery beds. It takes about 30 to 40 days for their sprouting.

After 2 to 3 months of growth, tubers are transplanted in the field, which are treated with the insecticide earlier. While planting, the tubers are placed at a distance of 30 x 60 cm. Initially, the veins are weak and tender and they need support for their optimum growth. Trellises of 2.5 m in height are provided for this purpose. Since the tubers are very exhaustive, a high dose of farmyard manure to the extent of 5 - 10 tones per hectare is applied in the beginning.

Fig. 9.22 (b): *Dioscorea deltoidea* Twig

Organic fertilizers should be applied subsequently in equal doses at an interval of one month. Irrigation should be done every 10 days. No major pests are reported in case of *Dioscorea*.

However, preventive measures should be taken against white-bugs and mites. Inter-cropping with legumes can be done. Fresh tuber-yield as high as, 18 tones per hectare, can be expected from two year crop.

Collection of dioscorea tubers

Harvesting is done by deep ploughing in the dormant season, as during this period, the diosgenin content is found to be high as compared to other seasons. Rhizomes loose about 50 per cent of their weight on drying.

Macroscopic Characters (Fig. 9.22)

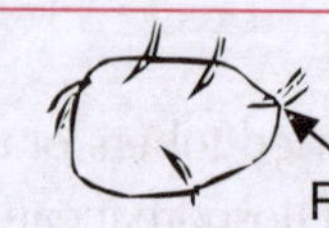

Fig. 9.22 (a): Dioscorea tuber

Colour	:	Tuberous roots are slightly brown
Odour	:	Odourless
Taste	:	Bitter
Size	:	Varies depending upon age of tubers 10 – 12 × 8 – 10 cm normally.

Extra Features:

It is a climber with alternate leaves (Fig. 9.22). Rhizomes are soft, horizontally arranged and are very close to the soil. Drug is covered with scattered roots. They weigh about 20-50 g. Every plant bears 10 – 15 rhizomes.

Microscopic Characters

Epidermis is normally absent in the transverse section of the drug. The cork consists of only few layers, followed by thin walled cortical parenchymatous tissue. Stele forms the major part of the drug and consists of several close collateral fibro-vascular bundles. Endodermis and pericycle are indistinguishable.

Chemical Constituents

Dioscorea rhizomes contain 75 per cent of starch. They are non-edible, since they are very bitter in taste. The chief active constituent of dioscorea is diosgenin, a steroidal sapogenin (4 to 6 per cent), and its glycosides, smilagenin and epismilagenin and β-isomer yammogenin. Rhizomes are also found to contain an enzyme sapogenase. Tubers are also rich in glycosides, and phenolic compounds. Diosgenin is the hydrolytic product of saponin-dioscin.

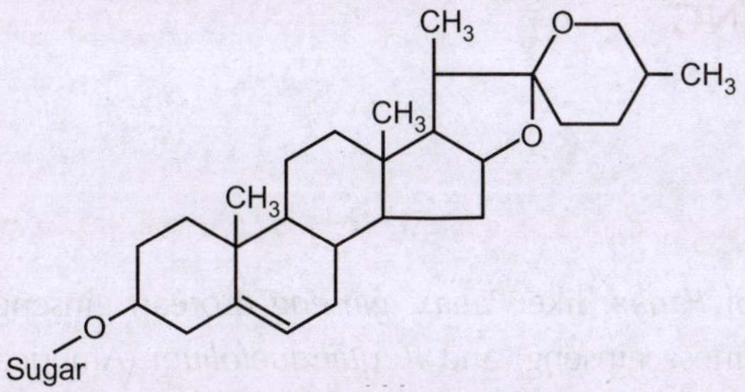
Dioscin

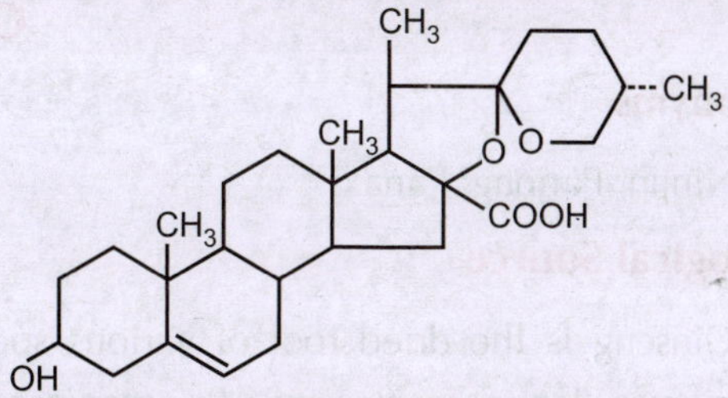
Diosgenin

Standards of Quality

Foreign organic matter	:	Not more than 2.0 per cent
Alcohol soluble extractives	:	Not less than 11.0 per cent
Water soluble extractives	:	Not less than 17.0 per cent
Total ash	:	Not more than 6.0 per cent
Acid insoluble ash	:	Not more than 1.0 per cent

Uses

Pharmaceutically, the rhizomes are used as rich source of diosgenin. Diosgenin being steroidal in nature is used as precursor for synthesis of several corticosteroids, sex-hormones, and oral-contraceptives. Dioscorea is used in the treatment of rheumatic arthritis.

Isolation

The tubers are dried, powdered and hydrolysis of dioscin is done by mineral acids. The liberated diosgenin is extracted with the help of non-polar solvents like benzene or solvent ether.

After isolation, diosgenin is degraded to 16-dehydro pregnenelone acetate. This latter molecule is used as precursor for synthesis of:

(1) Corticosteroids like cortisone, hydrocortisone, prednisolone,

(2) Pregnenes like progesterone, 17 - α - hydroxy progesterone,

(3) Androstanes like testosterone, methyl testosterone, etc. and

(4) 19 - NOR steroids like estrone, 17 - α ethynyl estradiol and norethisterone acetate.

The conversions are done with the help of microbial transformations.

Allied Species

Dioscorea flouribunda is cultivated in Central America and India (Karnataka State). It contains 3 to 5 per cent of diosgenin. *D. villosa* Linne mainly from Virginia and Carolina in U.S.A. is also very rich in diosgenin content. *D. villosa* is a twining perennial with beautiful yellow flowers and triangular capsules. *Sikkimensis* Prain and Burkill (syn. *D. deltoidea* wall var. *sikkimensis* Prain) occurs in Eastern Himalayas, Nepal, Sikkim, Bhutan, Assam, Bihar and Bengal upto an altitude of 1600 - 2000 m. It contains 2 to 2.8 per cent of diosgenin.

Costus speciosus is an alternative potential source for diosgenin (1.5 per cent) and can be used as a substitute for the genuine drug.

GINSENG

Synonyms

Ninjin, Pannag, Panax.

Biological Source

Ginseng is the dried root of various species of *Panax*, like *Panax. ginseng* (Korean ginseng), *P. japonica* (Japanese ginseng), *P. notoginseng* (Chinese ginseng) and *P. quinquefolium* (American ginseng), belonging to family Araliaceae. It contains not less than 0.4% of Ginesenosides calculated on dried baris.

Geographical Source

It grows widely in Korea, China and Russia. Presently, ginseng is commercially cultivated in Korea, China, Japan, Russia, Canada and United States of America. In India at present, it is cultivated in Kohima, Tunsang district of Nagaland. It is also available in Arunachal area.

Like some of the well known ayurvedic medicines, ginseng has generated large interest because of its novel pharmacological actions. An international symposium was arranged at Lugano in 1975 to discuss the analytical, pharmacological and clinical aspects of ginseng, which was sponsored by World Health Organisation.

Description

Ginseng is native of Manchuria, Chinese territory and other parts of eastern Asia, and is largely cultivated in Korea and Japan. It is a smooth perennial herb, with a large, fleshy, spindle-shaped, annulated very slow-growing root. It is pale yellow to a brownish colour, odourless with mucilaginous taste with slight bitterness. The stem is simple and erect, leaves with five finely-toothed leaflets, and a single, terminal umbel, with a few small, yellowish flowers. The fruit consists of cluster of bright red berries.

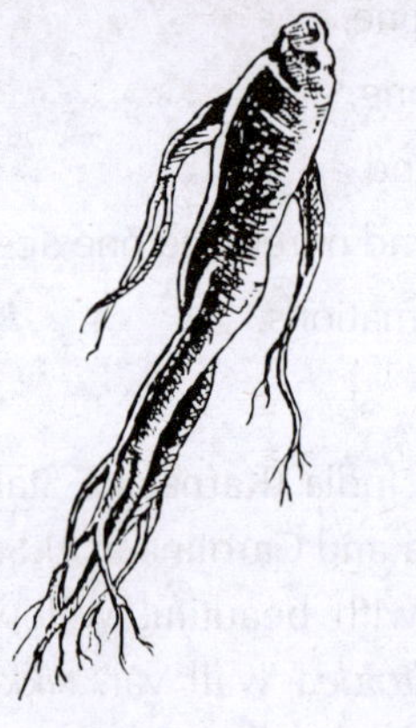

Fig. 9.23: (a) Ginseng Root

Fig. 9.23: (b) Ginseng Herb

The term *Panax* indicates curve all (*Pan and axos*), while *ginseng* is derived from Chinese words *shen sang*, which stands for manroot, because the shape of the root resembles the human body. The references about ginseng are found in ancient Chinese literature, stating its medicinal properties. Now-a-days, ginseng is considered as adaptogen. It increases non-specific resistance and defence mechanism of the body.

Cultivation of Ginseng

The cultivation technology adopted in Korea is briefly described.

Ginseng is propagated by means of seeds in nursery beds and then transplanted into open fields i.e. permanent beds. The ripe seeds are collected from four year old plants. They are sown in November in nursery beds. There are three types of nursery beds viz. Yang-Jik, To-Jik and Ban-Yang-Jik. The first type gives high quality seedlings. After attaining sufficient growth, the seedlings are dug up in the following May and transplanted to permanent beds for next 3 - 5 years. Ginseng requires clay loam or sandy loam soil. It grows at altitudes from 100 - 800 metres. The soil with high amount of potassium gives better results. About 10 - 15 seedlings are planted in one square metre. About 7 - 10 days after transplantation, shades are provided to plants to protect them from excessive sunlight. Generally, use of fertilizers is avoided, but before transplantation, the soil is mixed with large amount of green grass.

Collection of Ginseng roots

Periodically, weeding is done. The plants are harvested successively for 3 - 5 years after transplantation. Generally, they are harvested between July to October. White ginseng is obtained by removing the outer layers of the roots. Red ginseng is obtained by first steaming the roots and after that they are dried. But, removal of outer layers may lead to loss of active constituents.

Macroscopic Characters

Ginseng roots are tuberous corpulent roots. They are yellowish brown, white or red in colour depending on type. They are translucent and possess the stem scars.

Microscopic Characters

Pale yellow coloured roots exhibit following microscopic characters.

The cork cells of the root are polygonal followed by oval parenchymatous cortex cells which are full of starch grains. Yellow brown resinous canals are also present. Medullary rays are made up of rectangular cells. Spiral and reticulate vessels which are slightly lignified are also observed. Starch grains are 0.5 – 1 μm in size. Cluster crystals of calcium oxalate are also available.

Chemical Constituents

Ginseng contains a mixture of several saponin glycosides, belonging to triterpenoid group. They are grouped as follows:

(1) Ginsenosides; (2) Panaxosides; and (3) Chikusetsusaponin

Ginsenosides contain aglycone dammarol while panaxosides have oleanolic acid as aglycone. About 13 ginsenosides have been identified. Panaxosides give oleanolic acid, panaxadiol and panaxatriol on decomposition.

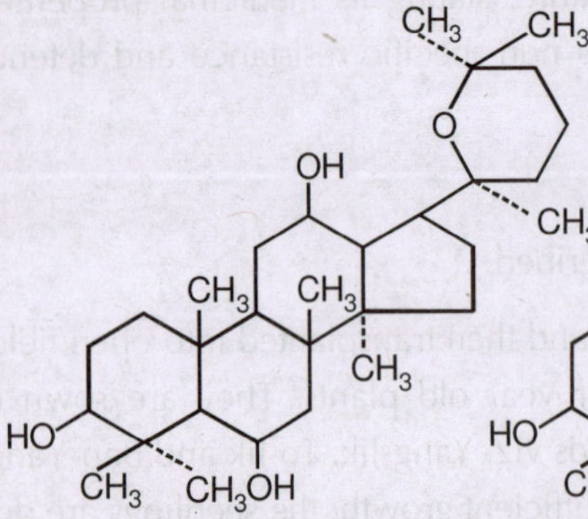

Oleanolic acid

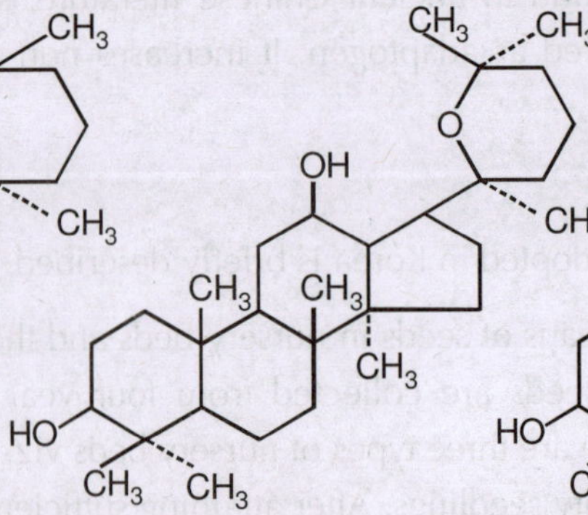

Panaxadiol

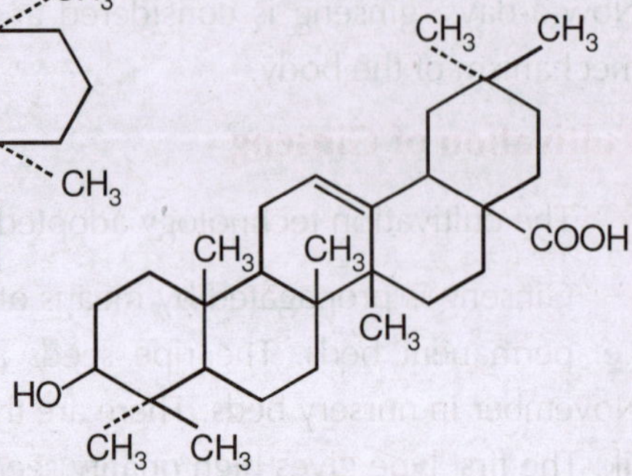

Panaxatriol

Glycosides of Ginseng

Standards of Quality

Foreign organic matter	:	Not more than 2.0 per cent
Total ash	:	Not more than 7.0 per cent
Acid insoluble ash	:	Not more than 1.0 per cent
Sulfated ash	:	Not more than 12.0 per cent
Alcohol soluble extractive	:	Not less than 22.0 per cent
Water soluble extractive	:	Not less than 27.0 per cent
Loss on drying	:	Not more than 10.0 per cent at 105ºC with 1.0 gm drug

Uses

Ginseng is an important immunomodulatory drug. It shows a wide range of activities. It increases the natural resistance (non-specific resistance) and enhances the power to overcome the illness or exhaustion. It has both stimulant and sedative properties. It is used as aphrodisiac. It is believed to be useful in adrenal and thyroid dysfunctioning. In old days, ginseng was used for a number of ailments, like curing the giddiness and prolonging life of elderly and diabetic persons. It is given as demulcent and in gastritis and anaemia.

Although, ginseng shows a low toxicity, long term use leads to poisoning, similar to that of corticosteroids.

Ginseng extracts are also used externally in cosmetics.

GOKHRU

Synonyms

Puncture vine, Tribulus.

Biological Source

It consists of dried fully ripen fruits of the plant *Tribulus terrestris* Linn, family Zygophyllaceae. It contains not less than 0.5 per cent diosgenin on dried basis.

Geographical Distribution

The drug is found widely distributed in warmer regions of the world. Gokhru occurs throughout India, Sri Lanka and in West Tibet.

Collection Gokhru

It flowers in rainy season and fruits are collected in October and November. It is a annual or biennial. It is collected from wild grown plants and not cultivated on commercial scale.

Macroscopic Characters

Colour	:	Fruits are greenish to gray
Odour	:	Odourless
Taste	:	Characteristic
Size	:	1.0 to 1.5 cm
Shape	:	Fruits are globose, consisting of 5 to 10 woody cocci, each with two pairs of hard and sharp divergent spines.

Fig. 9.24: Flowering twig of Gokhru

Chemical Constituents

The drug contains traces of alkaloids, harmine and harman. Additionally, they contain saponins Teresterosin A and Teresterosin E, Tribulosin, which on hydrolysis yield, steroidal sapogenins like diosgenin, gitogenin, chlorogenin and ruscogenin. The roots and leaves also contain saponin.

Flavonoids, kaemferol, tribuloside and its derivaties have also been isolated from the leaves and fruits. Fixed oil (3.5 per cent), resin, traces of essential oil and nitrates are the other constituents of the fruits. Fatty oil possesses excellent drying properties.

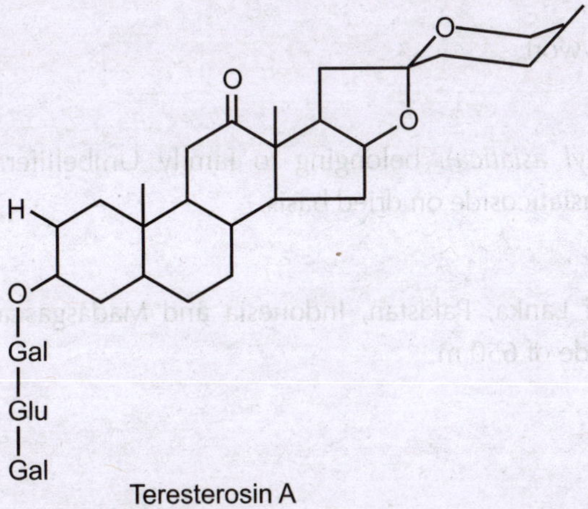

Teresterosin A

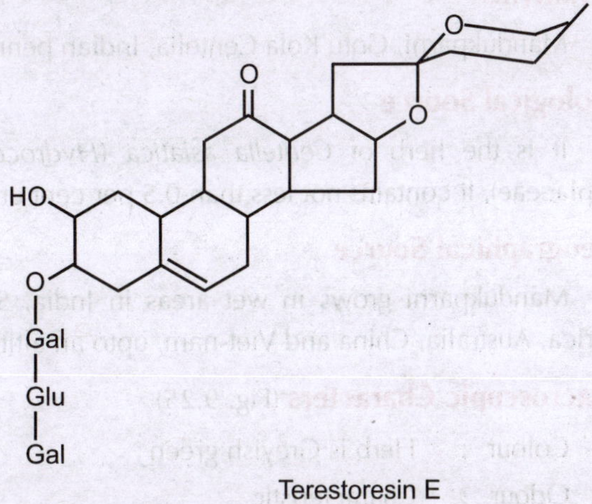

Terestoresin E

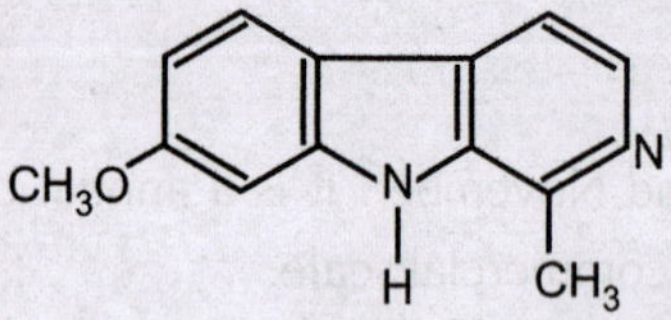

Harmine

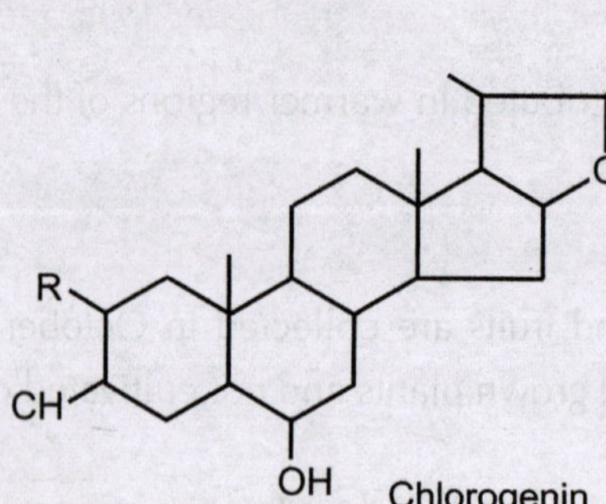

Chlorogenin

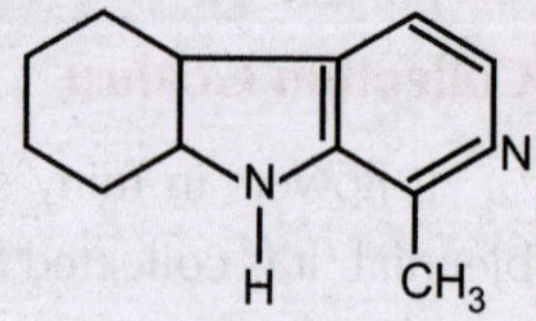

Harman

Standards of Quality

Foreign organic matter	:	Not more than 2.0 per cent
Total ash	:	Not more than 11.0 per cent
Acid insoluble ash	:	Not more than 1.0 per cent
Alcohol soluble extractives	:	Not less than 3.0 per cent
Water soluble extractives	:	Not less than 15.0 per cent
Loss on dryig	:	Not more than 5.0

Uses

Fruits are used as diuretic, tonic and also in the treatment of calculous affections and painful micturition. They are also used as aphrodisiac and in gout. Gokhru is a common ingredient of ayurvedic preparations, Dashamoolarishta and Chyavanprash.

The dried fruits of *Pedalium murex* (Pedaliaceae) are known as **BADA gokhru.** The plant is found near sea coast of India. The fruits contain fatty oil, mucilage, an alkaloid and resin. Bara gokhru is used as a diuretic and in dysuria and gonorrhoea.

JALBRAHMI

Synonyms

Mandukparni, Gotu Kola Centella, Indian pennywort.

Biological Source

It is the herb of *Centella asiatica* (*Hydrocotyl asiatica*), belonging to family Umbelliferae (Apiaceae). It contains not less than 0.5 per cent or asiaticoside on dried basis.

Geographical Source

Mandukparni grows in wet areas in India, Sri Lanka, Pakistan, Indonesia and Madasgascar, Africa, Australia, China and Viet-nam, upto an altitude of 650 m.

Macroscopic Characters (Fig. 9.25)

Colour	:	Herb is Greyish green
Odour	:	Characteristic
Taste	:	Bitter sweet

It is also a creeping plant, but the leaves are bigger long and petiolate. The leaves are entire, crenate, orbicular and reniform. Leaves are 1.5-6.5 cm in diameter, petioles 7.5-15 cm in length, narrow flattened stipules are short forming sheathing base. It bears an umbel inflorescence with 3 - 4 pink sessile flowers. The stems are 0.1 to 0.5 mm wide, twisted, flattened red and show long internodes. Like Brahmi, rooting occurs at nodes. Fruits are in umbels pale green to brown wrinkled, 2 mm long.

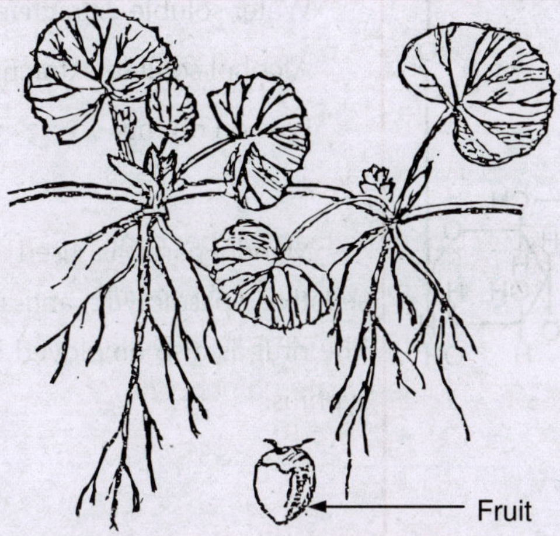

Fig. 9.25: Centella asiatica plant

Microscopic Characters

Leaf: All the typical characters of leaf are seen in transverse section i.e. cuticle, epidermis and cortex. Leaf has uneven cuticle covering the polygonal epidermal cells. Frequent paracytic stomata and trichomes on epidermis. Trichomes are narrow and twisted. Multicellular uniseriate trichomes are found occasionally.

Large, single prismatic crystals of calcium oxalate are present in epidermal polygonal cells of fruits. No vittae an present in fruits.

Chemical Constituents

Mandukparni mainly contains saponins in the form of α - amyrin derivatives i.e. triterpenoid called asiaticoside and madecassoside about 1.0 per cent. They yield asiatic acid and madecassic acid (triterpene acids) on hydrolysis respectively.

$$\text{Asiaticoside} \xrightarrow{\text{hydrolysis}} \text{Asiatic acid + 2 Glucose + Rhamnose}$$

$$\text{Madecassoside} \xrightarrow{\text{hydrolysis}} \text{Madecassic acid + 2 Glucose + Rhamnose}$$

The sugars are esterified at carboxyl group. Chennai and Lucknow varieties from India also contain brahmoside and bramhinoside, which on hydrolysis give brahmic acid isobrahmic acid arabinose, glucose and rhamnose.

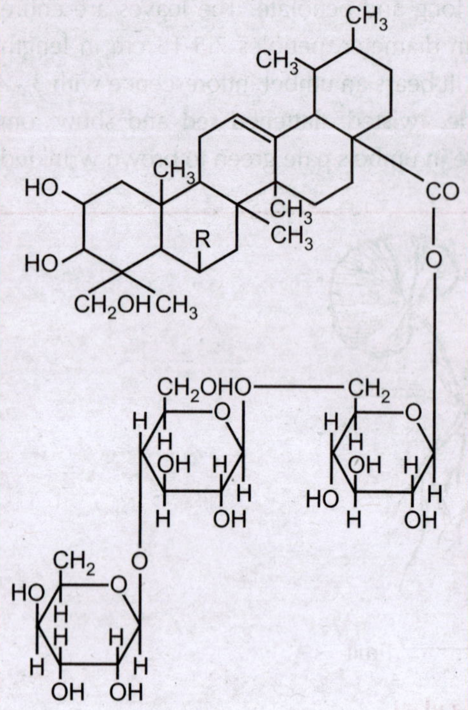

Asiaticoside (R = H) ;
Madecassoside (R = OH)

Standards of Quality

Foreign organic matter	Not more than 2.0%
Total ash	Not more than 19.0%
Acid insoluble ash	Not less than 6.0%
Water soluble extractives	Not more than 6.0%
Alcohol soluble extractives	Not more than 9.5%
Loss on drhing	Not more than 12.0 %

Uses

Mandukparni is used as nervine tonic. It shows sedative, spasmolytic, anti-anxiety and anti-stress action. The drug is also employed in skin diseases, leprosy and syphilis.

MOMORDICA

Synonyms

Bitter gourd, Karela, Bitter melon.

Biological Source

Karela consists of fresh green fruits of the plant known as *Momordica charantia* Linn.,: Cucurbitaceae.

Geographical source

It is cultivated through out India, China, Malaya, tropical Africa and America.

Cultivation

It is a genus of annual or perennial climbers found throughout India and regularly cultivated as an vegetable up to an altitude of 1500 m.

It is cultivated during warm season i.e. during April to July by using 2 - 3 seeds in a pit. The pits are prepared at a distance of half a meter and provided with manures. Only one plant is retained and seedlings are watered once or twice a week.

Collection of Karela

Plants begin to flower 30 - 35 days after sowing and the fruits are ready for harvesting 15 - 20 days after flowering.

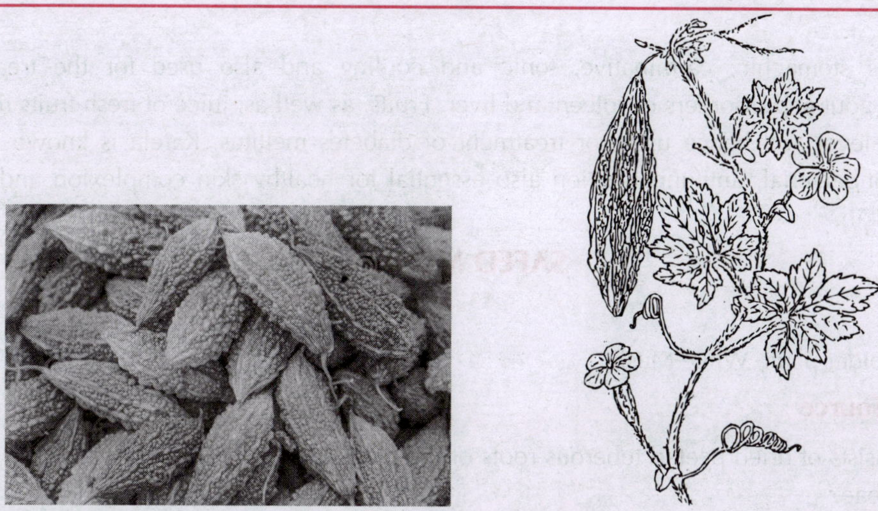

Fig. 9.26: Momordica - climber

Macroscopic Characters

- Colour : Fruits are dark green to whitish green
- Odour : Characteristic
- Taste : Intensely bitter
- Size : 0.5 - 25 cm long, 2 - 4 cm wide
- Shape : Pendulous, fusiform and beaked. It is ribbed with numerous tubercles; fruits contain numerous brownish seeds embedded in red pulp.

Stems are slender and pubescent. The leaves of karela are sub-orbicular, 5 - 7 lobed, pubescent or sub-glabrous; flowers are solitary.

Chemical Constituents

The fruits and leaves contain charantin, a steroidal saponin which shows blood sugar lowering activity. Karela fruits even contain a cathartic principle called momordicin. The drug also contains carbohydrates (10 per cent), mineral matter (1.5 per cent) and ascorbic acid (88 – 188 mg/100 gm). Additionally, alkaloids, glucoside, saponins, and mucilage are the other contents of Karela.

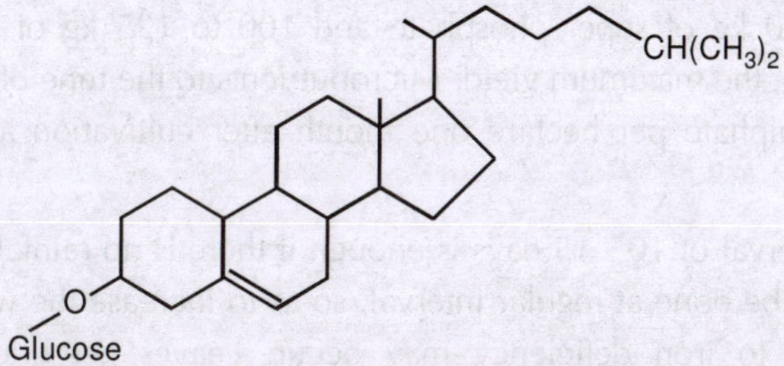

Charantin

Uses

Karela is stomachic, carminative, tonic and cooling and also used for the treatment of rheumatism, gout and disorders of spleen and liver. Fruits, as well as, juice of fresh fruits reduce the blood sugar level and hence used for treatment of diabetes mellitus. Karela is known for blood purifying, antibacterial, antifungal action also essential for healthy skin complexion and provides glow to the skin.

SAFED MUSALI

Synonyms

Indian spider plant, White Musli.

Biological Source

This consists of dried peeled tuberous roots of the plant known as *Chlorophytum borivillianum*, family Liliaceae.

Geographical Source

This is found in African countries and also in India. In India it is collected from wild grown plants from forests. Cuttently it is cultivated scientifically on commercial scale in Madhya Pradesh, Chikaldara in Maaravati district, Maharashtra, Southern Rajasthan, and North Gujarat.

Cultivation and Collection of Safed Musali

Cultivation: Safed musali is a crop of 210 - 240 days. Several commercially exploited varieties are available in the market, few of them are as under RC-2, RC-16, RC-36, RC-20, RC-23 and RC-37.

These varieties are being maintained at Rajasthan Agricultural University, Udaipur.

For cultivation purposes red loamy, gravelly soil or well drained soil is suitable. Desired pH of soil for cultivation is 6 - 7. Rainfall in the range of 50 - 75 cm is favourable for cultivation. Cultivation is done either by using seeds or tuberous roots, in the month of June or early in July. Approximately 100 - 125 kilogram of tuberous roots per hectare is required. The roots be treated with fungicides like Bavistin (in the proportion of 1.5 gm per litre of water) before sowing. Treatment of Humic acid (2 ml per litre of water to the roots) may be fruitful to initiate the root system.

Roots are placed at a distance of 15 × 15 cm or 20 × 20 cm for proper yield. 10 to 15 tones of farm yard manure per hectare before cultivation provides sufficient nutrient status for healthy crop. During sowing 250 - 300 kg of super phosphate and 100 to 125 kg of potassium sulphate per hectare be provided to get the maximum yield. Micronutrients to the tune of 20 to 25 kg and 100 to 120 kg of ammonium sulphate per hectare one month after cultivation are supplied for healthy growth.

Irrigation after an interval of 10 - 15 days is enough if there is no rainfall, drip irrigation is most suitable. Proper weeding be done at regular interval, so as to increase the weight of tuberous roots. Chlorosis of leaves due to iron deficiency may occur. Leaves get dried during November - December and are lost. Irrigation after leaf fall upto end of January is essential. Normally, no pesticides are required for pest-control.

Collection: After the second week of February, tuberous roots in the form of bunches are uprooted by digging method and no mechanical method is suggested as it may result in loss of roots. Roots are collected and spread in shade for 4 - 8 days. Then fleshy roots are separated from bunches. Small tuberous roots are separated, dried thoroughly and stored for cultivation. Fleshy roots are peeled to get inner milky white roots. Roots are cleaned and dried for 7 - 10 days. 5000 to 6200 kg of fleshy roots per hectare are obtained, during drying 80 per cent of water is lost.

Total cost of cultivation is approx. ₹ 22500 / hectare. Indian consumption of roots is estimated to be 300 - 500 tones, but it has high export potential. It is also cultivated as inter-crop by third community in Hilly region in Vidarbha region of Maharashtra.

Morphological Characters

Colour : Tuberous roots of safed musli are white to off coloured.

Odour : Slight characteristic.

Taste : Slightly bitter

Size : 2 to 4 cm in length and 0.5 cm in width.

Shape : Elongated, hard, tapering at both the ends, fracture is short.

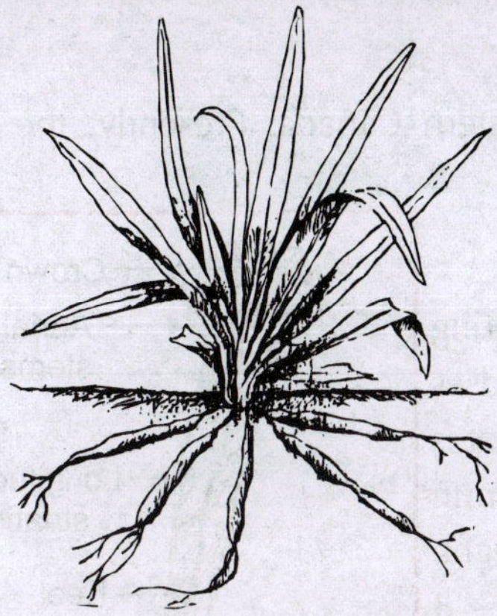

Fig. 9.28 (a): Safed musali herb

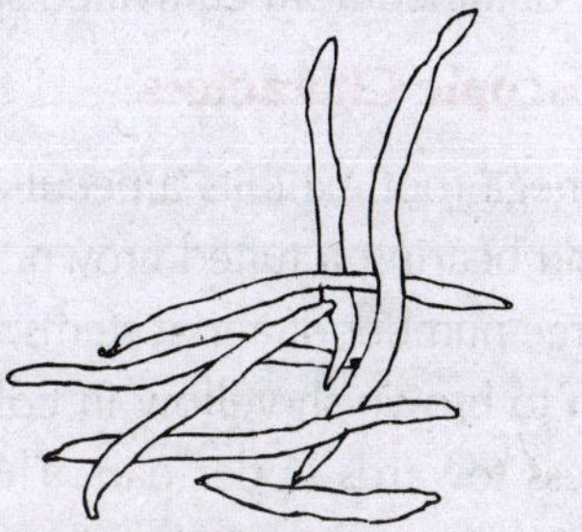

Fig. 9.28 (b): Dry tubers of Safed musali

Chemical Constituents

Safed musali contains 39 to 42.0 per cent of carbohydrates, proteins 8 - 12 per cent, saponins 2 - 4 per cent, sapogenin (hecogenin) 0.18 to 0.2 per cent, several rare elements like zinc, copper and phosphorus fibre content of the roots is about 3.0 to 5.00 per cent.

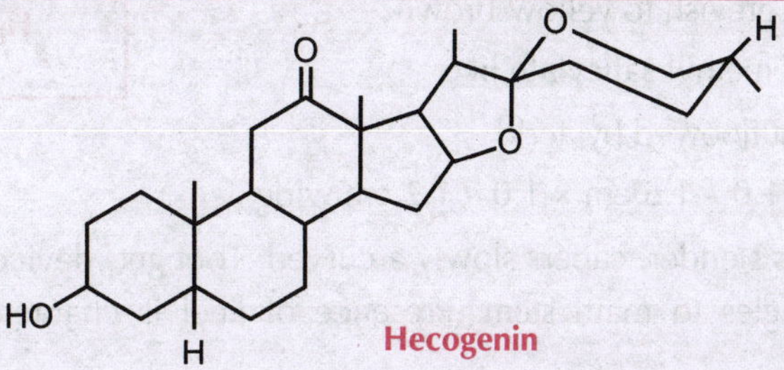

Hecogenin

Uses

In the Ayurvedic system of medicine, Safed musali is used as well known tonic for general debility and also as aphrodisiac.

Dose

2 - 3 gm of powdered Safed musali alongwith milk once in a day.

Storage

During storage tuberous roots are treated with thiram and captan in the proportion of 4.0 gm/kg of roots to reduce rotting of roots.

SENEGA

Synonyms

Senega root, Senega snake root, Radix Senegae, Rattle snake root.

Biological Source

Senega root is the dried root and rootstock of *Polygala senega* or *Polygata senega* var. *latifolia* Torret Gray, belonging to family Polygalaceae. It should contain not less than 29 per cent of alcohol soluble extractive.

Geographical Source

The plant is indigenous to Minnesota in U.S.A. and Eastern Canada. Presently, the drug is mainly obtained from cultivated species in Japan.

Macroscopic Characters

Senega root exhibits a peculiar shape as an elongated tapering root and bearing a tufted crown. The crown is linked to the bases of a large number of aerial stems, bud and small shoots. The root is greyish to brownish-yellow in colour, which is about 3 - 6 mm in thickness towards upper part. The crown is 1 - 1.5 cm in height, while 1 - 2 cm in width. The root is either simple or in 2 - 3 branches, divided just below the crown. Senega root shows a keel with a sharp spiral portion tapering downwards. The root has smooth and splintery fracture.

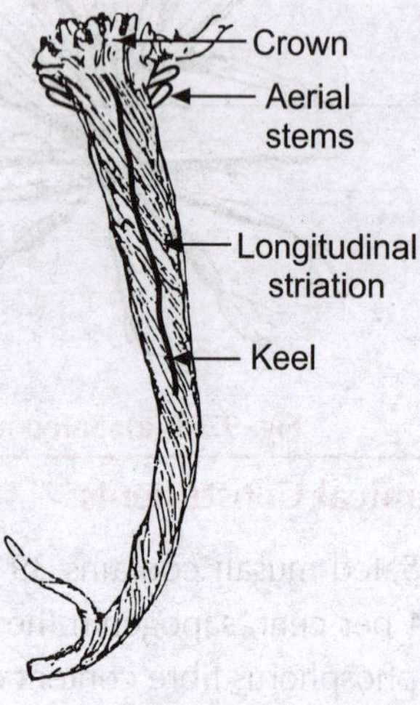

Fig. 9.29: Senega Root

Senega has first sweet and then acrid taste with characteristic odour of methyl salicylate.

Colour	:	Roots are greyish to yellow-brown
Odour	:	Flower of methyl salicylate like
Taste	:	Sweet first floowed by acrid
Size	:	Crown is 1.0 – 1.5 cm × 1.0 – 1.2 cm wide

Extra Features: Root is slender, capers slowly a curved. Toot gets devicded into two branches of which one is at right angles to main stem. Presence of keel is characteristic root is tranversly wrinckled.

Microscopic Characters

In general, the transverse section shows the presence of cork, phellogen, collenchymatous phelloderm, diarch primary xylem at the centre, surrounded by secondary xylem. In some portion of root, there is a band of enlarged phloem giving rise to keel. The medullary rays in wood portion are narrow, lignified and large in number. There are no phloem fibres. Rootstock shows central parenchymatous pith.

Chemical Constituents

Senega contains two saponin glycosides which are triterpenoid in nature. They are senegin (4 per cent) and polygalic acid (5.5 per cent). Senegin on hydrolysis gives senegenin, senegenic acid and presenegenin. It has been also reported that senega contains other derived forms of presenegenin called senegin II.

The sweet taste of the drug is due to polygalitol (1, 6 - anhydrosorbitol). The odour of drug is due to small amount of methyl salicylate which is formed as a result of breakdown of some unknown glycoside. Senega also contains fixed oil and sterol, but it does not contain starch.

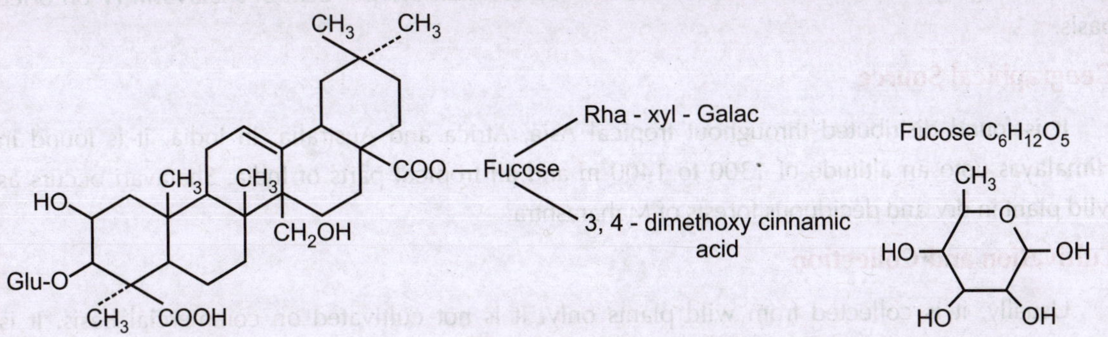

Senegin II

Standards of Quality

Ash value	:	About 4 per cent
Acid insoluble ash	:	Not more than 2.0 per cent
Alcohol soluble extractive value	:	Not less than 29 per cent

Uses

The saponin glycosides of senega are not absorbed, but cause irritation of gastric mucosa. This leads to secretion of mucous in bronchioles through reflex action and hence, used as stimulant expectorant. It is used in treatment of chronic bronchitis, but generally prescribed alongwith other common expectorants. Large doses of senega cause vomitting.

Allied Drugs

(1) Indian senega: It is the root and rootstock of *Polygala chinensis* Linn. collected from 3 - 4 years old plants. It is also called chinensis root and was official in Indian Pharmacopoeia 1955. It does not contain saponins and hence fails to show persistent frothing and haemolysis. It was further showed that, Indian senega is root of *Andrachne aspera* which is without keel.

(2) Spurious Indian senega: It is root of *Glinus oppositifolia* belonging to family Molluginaceae. It contains a saponin and starch. It shows several rings of vascular bundles.

(3) White senega: It is the root of *Polygala alba* and is obtained from southern states of U.S.A. It has less acrid taste and also less activity. It does not show keel.

SHATAVARI

Synonym

Shatmuli, Asparages racemosus root.

Biological Source

Shatavari consists of dried roots and the leaves of the plant known as *Asparagus racemosus* Wild, belonging to family Liliaceae. It contains not less than 0.1 per cent of shatavarin IV on dried basis.

Geographical Source

It is found distributed throughout tropical Asia, Africa and Australia. In India, it is found in Himalayas upto an altitude of 1300 to 1400 m and all tropical parts of India. Shatavari occurs as wild plant in dry and deciduous forests of Maharashtra.

Cultivation and Collection

Usually, it is collected from wild plants only. It is not cultivated on commercial basis. It is cultivated in gardens as a potted plant for its graceful feathery leaves and flowers. Shatavari bears white, fragrant flowers in the form of simple racemes. Fruits are red coloured berries.

It is a perennial climbing shrub. The stem is covered with recurved spines. The leaves are linear, green and needle-like It bears tufts of cladodes axillary to minute scales. Roots occur in a cluster or fascicle at the base of the stem. Cultivation yields ten tones of roots per hectare.

Macroscopic Characters

Colour	:	Roots are silver white or ash coloured externally and white internally.
Odour	:	None
Taste	:	Starchy slightly bitter followed by sweet taste
Size	:	5-15 cm × 2 cm in thickness
Shape	:	Spindle shaped
Extra features	:	Longitudinally wrinkled with yellow hand central core.

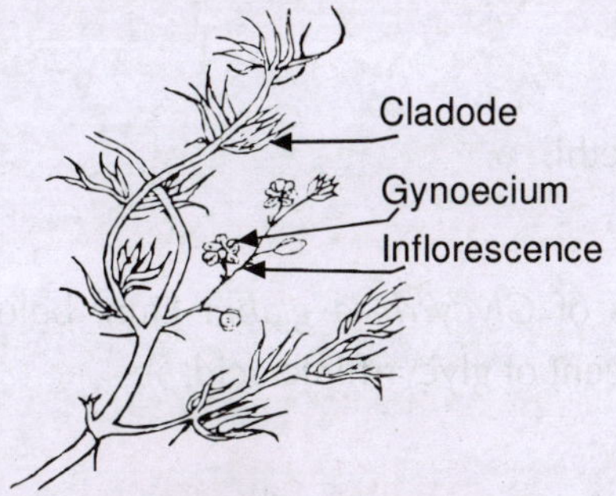

Fig. 9.30: Shatavari Plant

Fig. 9.31: Shatavari Roots

Chemical Constituents

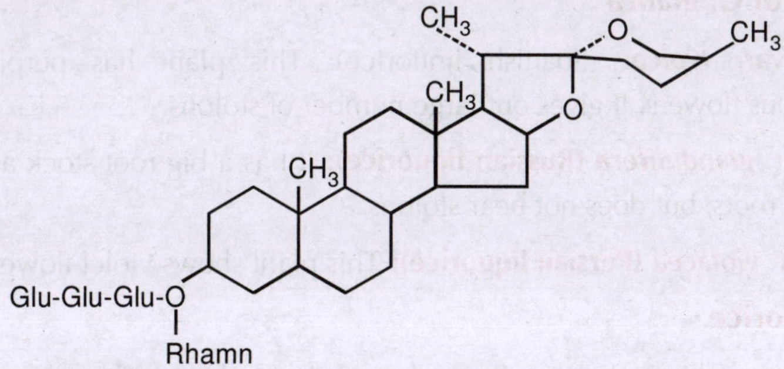

Shatavarin I

Shatavari roots contain 4 steroid saponins: shatavarin I - IV (0.2 per cent). Shatavarin I is the major glycoside with 3 glucose and rhamnose moieties attached to sarsapogenin, whereas in shatavarin IV two glucose and one rhamnose moieties are attached.

Flowers and fruits of shatavari contain quercetin, rutin and hyperoside, while leaves contain diosgenin and quercetin.

Standards of Quality

FOM	: Not more than 2.0 per cent
Ash	: Not more than 5.0 per cent
Acid insoluble ash	: Not more than 0.6 per cent
Alcohol soluble extractive	: Not more than 9.0 per cent
Water soluble extractive	: Not more than 26.0 per cent
L.O.D.	: Not more than 8.0 per cent

Uses

Shatavari roots are used as Galactogogue, tonic and diuretic. Shatavarin I is reported to possess antioxytocic property. Roots are largely used for medicinal oil recommended in the treatment of rheumatism and nervine disorders. Its use in Ayurveda in threatened abortion and safe delivery is justified by uterine blocking activity.

YASTI

Synonyms

Glycyrrhiza, Liquorice root, Glycyrrhizae radix, Mulethi.

Biological Source

Yasti consists of dried, unpeeled, roots and stolons of *Glycyrrhiza glabra* Linn, belonging to family Leguminosae. Yasti contains not less than 3.0 per cent of glycyrrhinic acid.

Geographical Source

The drug is commercially cultivated on a large scale in Spain, Sicily and England. *Glycyrrhiza glabra var. glandulifera* (Russian liquorice) grows in Russia and Glycyrrhiza *glabra* var. *violacea* comes from Iran.

Different varieties of *G. glabra*

1. ***G. glabra* var. *typica*:** (Spanish liquorice): This plant has purplish blue coloured papilionaceous flowers. It gives out large number of stolons.
2. ***G. glabra* var. *glandulifera* (Russian liquorice):** It has a big root stock along with a number of elongated roots, but does not bear stolons.
3. ***G. glabra* var. *violacea* (Persian liquorice):** This plant shows violet flowers.

Cultivation of Liquorice

Spanish liquorice is cultivated on a large scale in Spain and Italy. The propagation of the plant is done with young pieces of stolons. They are divided into young pieces, each piece should exhibit 2 - 3 buds of aerial shoot. The plant requires deep sandy soil which is well prepared and added with farmyard manures. The pieces of stolons are planted in March at 2' × 3' distance. The fertilizers are to be supplied when the green parts are developing. The crop is kept free of weeds. The roots are harvested 3 - 4 years after planting when they show sufficient growth.

Collection of Liquorice

Rhizomes and roots are dug up in October, preferably from the plants which have not borne the fruits. Buds and rootlets are removed, and the drug is washed. Some pieces are peeled and divided into small pieces. The drug is dried first under sun and then in shades, during which it loses about 50 per cent of its weight.

Macroscopic Characters

Colour	:	Unpeeled roots yellowish-brown or dark brown externally, and yellowish internally, while the peeled liquorice is pale yellow in colour.
Odour	:	Faint and characteristic.
Taste	:	Sweet.
Size	:	Length 20 to 50 cm and 2 cm in diameter.
Shape	:	Cylindrical pieces which are straight may be peeled or unpeeled. Peeled liquorice is angular.
Fracture	:	It is fibrous in the bark and splintery in wood.

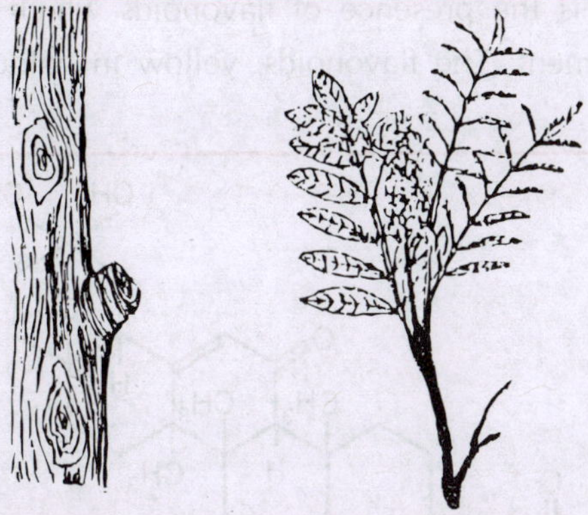

(a) Liquorice Root (b) *Glycyrrhiza glabra* Twig

Fig. 9.32

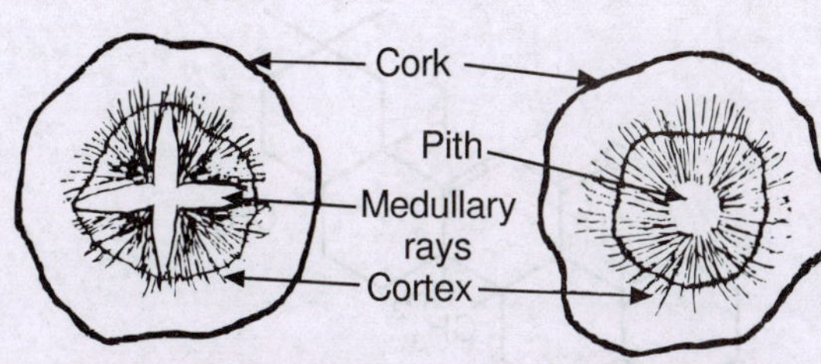

(a) T.S. of Root (b) T.S. of Stolon

Fig. 9.33: T.S. of Glycyrrhiza

Extra Features

Unpeeled pieces show the presence of small buds encircling scaly leaves and longitudinally the drug is wrinkled, while the peeled drug is fibrous without wrinkles.

Microscopic Characters

The important histological diagnostic characters of liquorice are given below:

(1) Unpeeled drug shows the presence of polyhedral tubular brownish cork cells.

(2) Fibres are thick, lignified or partially lignified, in the groups of 10 to 15 in phloem and xylem. Vessels are large and closely arranged with bordered pits. Starch and calcium oxalate crystals are present in parenchyma.

In case of stolons, the pith is present and is parenchymatous. The root is characterised by the presence of tetrarc xylem and absence of pith.

Chemical Constituents

The chief constituent of liquorice is a triterpenoid saponin known as glycyrrhizin (glycyrrhizic acid), which is a potassium and calcium salt of glycyrrhizinic acid. Glycyrrhizinic acid is a glycoside and on hydrolysis yields glycyrrhetinic acid (glycyrrhetic acid), which has a triterpenoid structure. The different varieties are found to contain varying amounts of glycyrrhizin (from 6 to 14). Spanish liquorice contains 5 to 10 per cent, Russian variety contains about 10 per cent, while Persian liquorice contains 7.5 to 13 per cent of glycyrrhizin. Other constituents of Liquorice are glucose (upto 4 per cent); sucrose (2.5 to 6.5 per cent); bitter principle glycyramarin resins, asparagin (2 to 4 per cent) and fat.

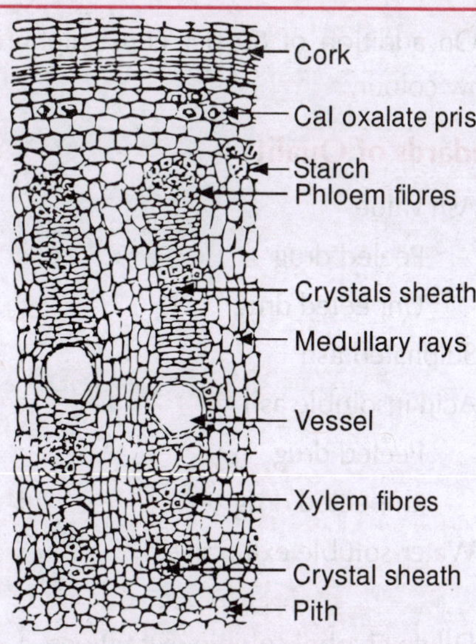

Fig. 9.34: T. S. of Liquorice stolon

Another important chemical aspect of liquorice is the presence of flavonoids which cause antigastric effect and are useful in peptic ulcer treatment. The flavonoids, yellow in colour, are liquiritin and isoliquiritin.

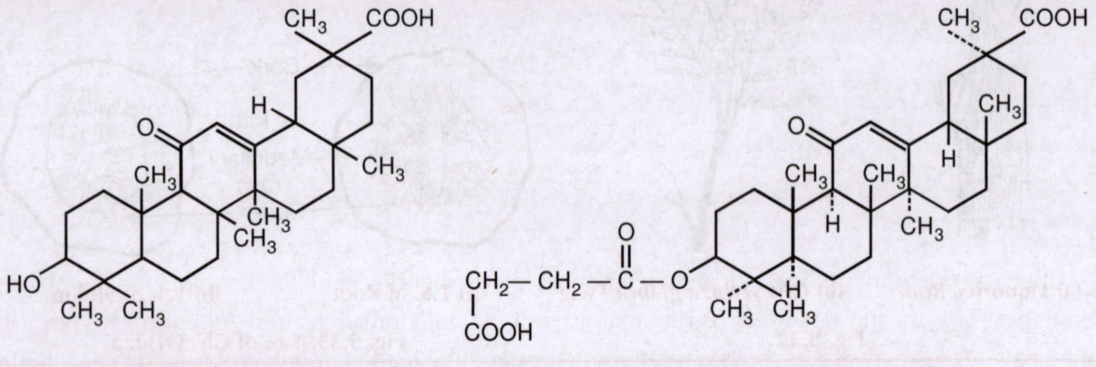

Glycyrrhetinic acid Carbenoxolone

The Indian liquorice roots have shown the presence of 2 - methylisoflavones and a coumarin viz. liquo-coumarin.

Carbenoxolone is an oleandane derivative prepared from glycyrrhiza and possesses significant minerocorticoid activity. It is used as an anti-ulcer drug. It changes the composition of mucous and increases mucosal barrier for the diffusion of acid. It is postulated that carbenoxolone inhibits the enzymes which inactivate prostaglandins and suppresses the activation of pepsinogen. This drug has marked anti-inflammatory effects. It is employed in treatment of gastric and duodenal ulcers. It is also used along with antacids for treatment of gastric reflux and reflux oesophagitis. It has also application in the form of gel for mouthwash in treatment of oral ulcers.

Chemical Test

On addition of 80 per cent sulphuric acid, the thick section of drug or powder shows deep yellow colour.

Standards of Quality

Ash value	
Peeled drug	: Not more than 6 per cent
Unpeeled drug	: Not more than 10 per cent
Sulphated ash	: Not more than 10.0 per cent
Acid insoluble ash	
Peeled drug	: Not more than 1 per cent
Unpeeled drug	: Not more than 2.0 per cent
Water-soluble extractives	: Not less than 20 per cent, including 7 per cent of glycyrrhizin
Dilute alcohol soluble extractives	: Not less than 25.0 per cent

Uses

Traditionally, liquorice has been used as an expectorant and demulcent. It is used in cough mixtures, and as a flavouring agent in formulations with nauseous drugs like, ammonium chloride, alkali iodides, quinine, cascara, etc.

Due to flavonoids content (isoliquiritin) with antigastric effects, it is used in peptic ulcer in the form of deglycyrrhized liquorice (DGL). This form has a reduced minerocorticoid activity and therefore used in treatment of peptic ulcer, for healing purposes.

The drug is also an antispasmodic. This is due to flavonoids glycoside viz. isoliquiritin, the aglycone part of this glycoside has antispasmodic effects.

Because of minerocorticoid activity (due to glycyrrhetinic acid), it is employed in place of corticosteroids for the treatment of rheumatoid arthritis, inflammations and Addision's disease. But, the heavy doses may cause sodium retention, consequently leading to hypertension, water retention and severe electrolyte imbalance. Glycyrrhizin is an established anti-inflammatory drug.

Liquorice is used most commonly as a flavouring agent for chewing tobacco and snuff tobacco. Ammoniated glycyrrhiza is used as a flavouring agent in beverages, confectionery and pharmaceuticals.

Residual matter remaining in the preparation of liquorice liquid extract is reported to have been used as a foam stabilizer in foam type of fire extinguisher. Liquorice is an ingredient of liquorice compound powder which is claimed to have a potentiating action of senna.

Adulterants and Substitutes

Manchurian liquorice is obtained from *Glycyrrhiza uralensis*. It is pale chocolate brown in colour with exfoliated cork and wavy medullary rays. It is free from sugar, but contains glycyrrhizin. Russian liquorice may be peeled and obtained from *Glycyrrhiza glabra* variety *glandulifera*. The drug is purplish in colour with numerous long roots, but no stolons.

SOAP POD

Synonym

Shikekai.

Biological Source

This consists of dried fruits of the plant *Acacia concinna*. Fam., Mimosac.

Geographical Source

Available in tropical moist mansoon forests very common in Konkan and north karnataka forest. Also available in China and Malaya islands.

Description

Shikekai is a climber plant where lenticells and hooked prickles are the characteristic features of this drug.

Leaves are 5 – 10 cm long, Bi-pinnate rachisalso with prickles and with glands. Pinne 4 – 6 pairs and ruchis ends in spine. Leaf-lets are in 10-20 pairs and overlapping.

Flowers are with globose heads, yellow in colour, **Fruits** are pod, fleshy when green, wrinkled when dry and are 7.15 – 12.5 × 2.0 - 2.2 cm. long **Seeds** are 6-10.

Chemical constituents

They are rch source of seponin.

Uses

Aqueous solutions are used as detergent as it produces froth/foam. Powdered drug removes dandraff and promotes growth of hairs hence hair tonic. Seeds are used to facilitate delivery in child birth.

Bark of the soap pod tree is used for dyeing and tanning nets in South India.

SOAP - NUT

Biological Source

Ritha

Biological Source

It is dried ripe fruit of *Sapiandus tripholiatus* (Sy: *Sapindus Laurifolias)* of fam: Sapindaceae.

Geographical Source

Evergreen Monsoon forests collected in West Bangal, Assam in western ghats from Konkan to south Karnataka. Panch Mahal and Kheda district of Gujarat. It is often planted, also available in Shri-Lanka wild in Himalayas above 1000 metres.

Habitate

It is a large handsome tree with straight trunk, girth around 9.1 to 9.2 metre with grey coloured bark with rough deciduous scales.

Morphological Characters

Leaves are ovate or oblong lanceolate and occuminate apex. Leaves are entire glabrous, shinning above and pubescent beneath. Petiole 2-3 cm. It bears numerous flowers in clusters, small dull whitish in colour. Calyx pubescent, petals covered with hairs, fruits are fleshy 3 lobed soft when ripe and brownish in colour. **Seeds** are round obovate. **Flowering** lime Octover to December fruiting time February – April evey year.

Chemical constituents

Fruit shell contains from 13 – 15% saponins. Seeds contains semi solid fixed oil but is very costly.

Uses

Ritha is used mostly externally as detergent. It is also a fish-poison. Roots and nuts are expectorant. Internally they are emetic and nauseant. It is also used in epilepsyl . Ritha is a substitute for soap.

SOAP BARK

Synonyms

Quillaia, Soap bark, Panama wood.

Biological Source

Quillaia is the dried inner part of the bark of *Quillaja saponaria* Molina, belonging to family Rosaceae.

Geographical Source

It is indigenous to Chile and Peru. It is grown in North India, Punjab and Uttar Pradesh.

The bark is stripped off from trunks of wildly grown trees, and dried.

Macroscopic Characters

Colour	:	(Bark) Outer surface brownish-white with occasional reddish or blackish patches. Inner surface yellowish-white.
Odour	:	Odourless.
Taste	:	Acrid and astringent.
Size	:	100 cm in length, 10 to 20 cm in breadth and 5 to 10 mm in thickness.
Shape	:	It is found in the form of hard, tough and flat pieces.

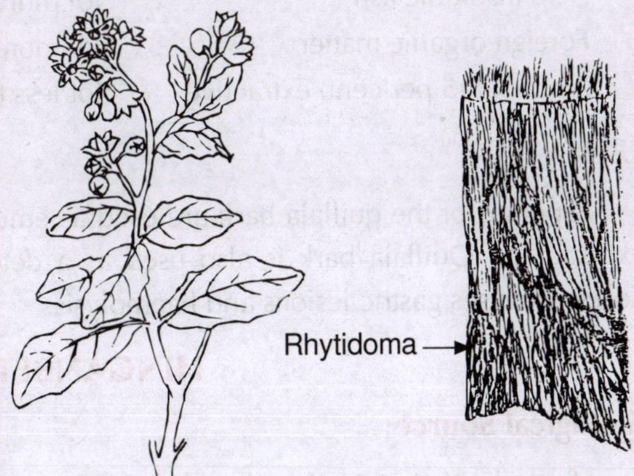

Fig. 9.27: *Quillaja saponaria* herb and bark

Extra Features

The outer surface shows partly removed rhytidoma. It is longitudinally striated. Internally, it is very smooth. Fracture is splintery.

Microscopic Characters

Rhytidoma of quillaia represents the cork cells. It is followed by parenchymatous tissues containing numerous groups of bast fibres and large prismatic crystals of calcium oxalate. Secondary phloem, phloem fibres and starch grains are found in the parenchyma. Medullary rays (multiseriate and pitted) and sclerides adjacent to phloem fibres are also present. The innermost part of the bark is unlignified and without fibres.

Chemical Constituents

Quillaia contains about 10 per cent of colourless amorphous triterpenoid saponin glycosides. Glycosides on hydrolysis yield quillaic acid and quillaia-sapotoxin. It also contains sucrose and tannin. The acrid taste of quillaia is due to quillaia-sapotoxin. It yields about 25 to 40 per cent of alcohol (45 per cent) soluble extractive. Quillaia responds to the tests of frothing and haemolysis.

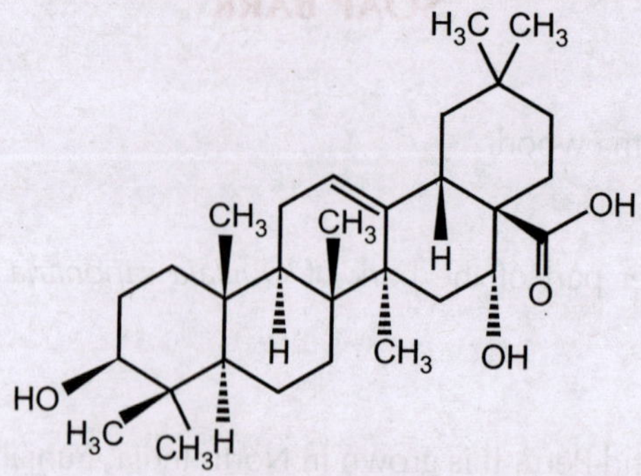

Quillaic acid

Standards of Quality

Acid insoluble ash	:	Not more than 1 per cent
Foreign organic matter	:	Not more than 2 per cent
Alcohol (45 per cent) extractive	:	Not less than 28 per cent

Uses

Saponins of the quillaia bark are used as emulsifying agents for tar and volatile oil. It is also an expectorant. Quillaia bark is also used as a detergent and in preparation of shampoo. In higher doses, it causes gastric lesions and haemolysis.

HINGANBET (BALANITE)

Biological Source

Hinganbet consists of bark, unripe fruits and leaves of the plant *Balanites aegyptiace* family Simarubaceae.

Geographical Source

It is available throughout the country specially in Bihar, Gujrat, Maharashtra, Punjab and in Deccan along with Sikkim.

Description

It is a plant of 5-6 metres in height. **Leaves** are rounded at the apex and bear very strong and sturdy spines neat the segments. It bears scented yellow coloured **flowers** in clusters. **Fruits** are egg shaped slighly enlarged hard seeds are pentagonal, seeds contain unpleasant penetirating fixed oil.

Chemical Constituents

Bark of the tree as well as fruits contins high amount of saponins. Kernel of the seeds yield yellow bland tasteless fixed oil about 42-45 %.

Uses

Balanite is bitter, antiseptic, expectornat, purgative and diuretic also.

Tribal community in forests is using the drug for washing clothes effectively due to Saponines.

SOAP-WART

This actually a weed known biologically Saponaria vaccaria fam Caryophyllaceae also knownn as **Sabuni**.

It is a tall robust annual plant, flowers in corymbose, sepals with five broad green vains. Rosy, flowers in Feb, also cultivated. Ornamental when grown in clumps contains high conent of saponins.

Use: Similar to soap-pod

IV. CYANOGENETIC GLYCOSIDES

They are also called as cyanophore glycosides, because of the presence of hydrocyanic acid in the aglycone moiety. The interest about such compounds was generated, while studying cyanides from different organisms. One of the prominent cyanogenetic glycosides from plants is amygdalin which was isolated long back in 1830.

Most of these glycosides contain derivatives of mandelonitrile i.e. nitrile of mandelic acid. This substance yields on hydrolysis, hydrocyanic acid and benzaldehyde. Because of this, they may be also considered as aldehyde glycosides. About 110 families of plant kingdom have been reported to contain cyanogenetic glycosides. Rosaceae being prominent amongst them. The cyanogenetic glycoside containing drugs do not exert a specific category of pharmacological activity, but are used for different pharmaceutical purposes. Many of them are used as flavouring agents. Some of these glycosides worth mentioning here are amygdalin from bitter almonds, prunasin from wild cherry bark, linamarin from linseed, manihotoxin from *Manihot utilissima*, sambunigrin from *Sambucus nigra*, etc.

Identification of cyanogenetic glycosides is based on liberation of hydrocyanic acid on hydrolysis which changes sodium picrate into sodium purpurate, as is evident from colour change from yellow to brick-red.

BITTER ALMOND

Synonym

Amygdala amara

Biological Source

Bitter almonds are dried ripe seeds of plants, *Prunus amygdalus* Batsch var *amara* (D.C.) (Focke), family Rosaceae.

Geographical Source

Bitter almond trees are native of Iran and Asia minor. At present, they are extensively cultivated in Sicily, Italy, Spain, Portugal, South France, and Morocco. The bitter almond trees cannot be easily distinguished botanically, from sweet almond trees (*Prunus communis*).

Macroscopic Characters

Colour : Bitter almond seeds are brown in colour.

Odour : Odourless

Taste : Bitter

Size : About 20 mm in length, 125 mm in width and 10 mm in thickness.

Shape : Flattened, oblong, ovoid in shape with markings on testa.

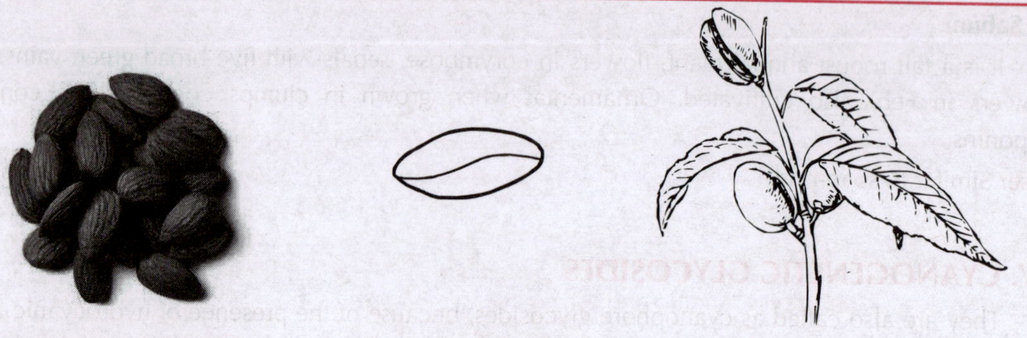

Entire bitten almond seeds   Kernel cut transversely   Bitter almond herb

Fig. 9.35: Bitter Almond

Chemical Constituents

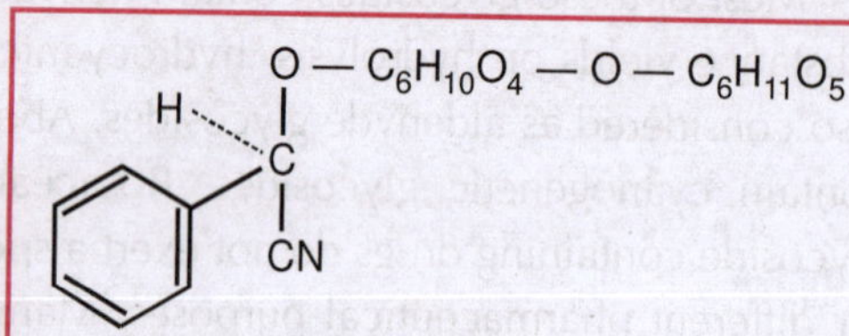

Amygdalin

Bitter almonds contain about 40 to 50 per cent of a bland fixed oil, 20 per cent of proteins, an enzyme emulsin and a colourless crystalline bitter glycoside known as amygdalin (1 to 3 per cent). Bitter almonds also contain 0.5 per cent of volatile oil. Amygdalin on hydrolysis due to emulsin in presence of water decomposes into benzaldehyde and hydrocyanic acid.

Hydrocyanic acid is very poisonous and makes the drug unsuitable for internal consumption. Sweet almonds do not contain amygdalin (bitter glycoside) and hence are not bitter in taste. Due to absence of amygdalin, they do not produce volatile oil on hydrolysis. Bitter almond oil contains 80 per cent of benzaldehyde and about 2 to 6 per cent of hydrocyanic acid.

Uses

Bitter almonds are used as sedative due to hydrocyanic acid content. The oil is used in demulcent skin lotion. It is also used in the preparation of amygdalin, bitter almond water, in perfumery and in the form of liquors. Bitter almond oil should not be used for flavouring of foods.

WILD CHERRY BARK

Synonyms

Virginian Prune Bark, Wild Black Cherry, Cortex Pruni.

Biological Source

It is the dried bark of *Prunus serotina* Ehrhart, belonging to family Rosaceae.

Geographical Source

It is indigenous to eastern states of U.S.A. and some parts of Canada. It is grown in Florida, Virginia, Dakota, Mississipi and North Carolina.

Collection

The bark shows maximum medicinal potency during autumn season. Hence, it is the proper period for collection of bark. The inner bark contains more amount of hydrocyanic acid. Therefore, after collection, generally the cork is removed from bark. After removal of cork and partly cortex, the exposed surface showing phloem appears uniform dark brown coloured product called rossed bark. It is dried and carefully preserved.

Macroscopic Characters

Colour	:	Bark is greenish brown to dark brown
Odour	:	Odourless, If treated with water gives odour resembling bitter almond.
Taste	:	Bitter aromatic and astringent
Size	:	10 - 12 × 4 - 5 × 1.5 - 2 cm.
Shape	:	Flat or Curved.

Fig. 9.36 (a): Wild cherry herb

Fig. 9.36(b): Wild cherry bark

Chemical Constituents

Wild cherry bark contains mandelonitrile glucoside, called prunasin, a cyanogenetic glycoside. Because of enzyme prunase present in the bark, prunasin in presence of water gives benzaldehyde, glucose and hydrocyanic acid. The bark collected in autumn contains about 0.12 to 0.16 per cent of hydrocyanic acid. The drug also contains p - coumaric acid, traces of benzoic acid, resin containing β - methyl aesculetin and trimethyl gallic acid.

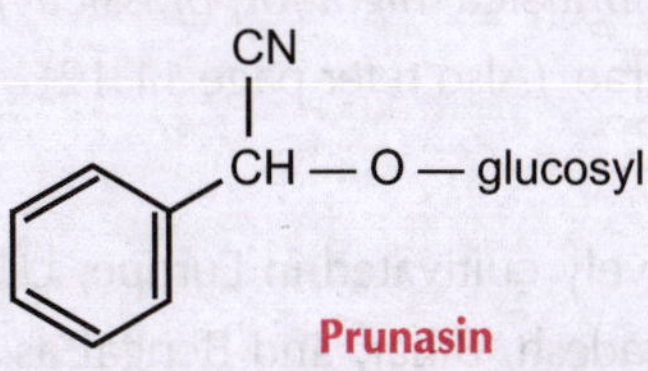

Prunasin

Standards of Quality

FOM	:	Not more than 2.0 per cent
Ash value	:	3 - 6 per cent
Acid insoluble ash	:	0.2 - 0.6 per cent
Water soluble extractives	:	Not less than 10 per cent
Hydrocyanic acid	:	Complies BPC requirements

Chemical Tests

(1) **Sodium picrate test:** The small pieces of bark are put in a flask containing water. A filter paper soaked in sodium picrate is suspended at the neck of the flask. After some time, yellow colour of the paper is turned to brick red due to liberated hydrocyanic acid, which turns sodium picrate into sodium isopurpurate.

(2) The hydrocyanic acid liberated on hydrolysis reduces mercurous nitrate to metallic mercury.

(3) A piece of paper is dipped in guaiacum resin and moistened with dilute copper sulphate. When it is exposed to freshly cut surface of drug, blue stain is produced.

Uses

It is used as expectorant and syrup is used as a flavouring agent.

V. ISOTHIOCYNATE GLYCOSIDES

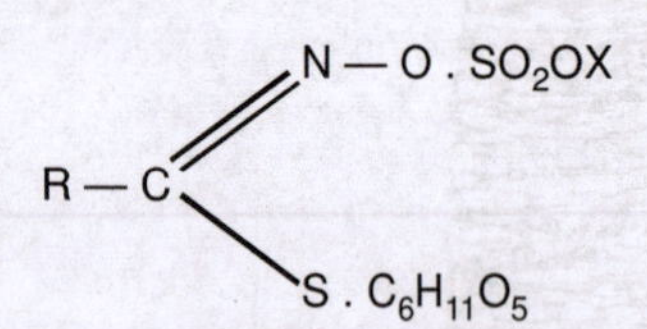

They are also called as glucosinolate compounds. The aglycone part of such glycoside contains isothiocynate, as it contains sulphur along with nitrogen. Taxonomical survey indicates that they are widely scattered in few higher families, prominent amongst them is Cruciferae. The other families are Resedaceae and Capparidaceae. Many of the plants belonging to such families contain myrosinase enzyme. The general structure of these aglycones may be represented as above.

Hence, in a glycosidic form, they are the glucosides of isothiocynates. The nature of R and X in the aglycone varies for respective drug.

MUSTARD SEED

Synonym

Black mustard, Synapsis

Biological Source

These are dried ripe seeds of *Brassica nigra* or *Brassica juncea* (Indian mustard) and their varieties. It belongs to family Cruciferae. (Also refer page 11.12)

Geographical Source

Black mustard seeds are extensively cultivated in Europe, U.S.A. and India. In India, the variety *Brassica juncea* is grown in Uttar Pradesh, Bihar, and Bengal as rabbi crop. Mustard is a herb with slender erect stems and yellow flowers. It has four sided siliquous fruits with very short stalk. After the maturity of the plant, the aerial part is cut out and partially dried. The partially dried herb is beaten to collect the seeds. The seeds are thoroughly dried in air.

Macroscopic Characters

Colour : Black or dark brown or reddish-brown.

Odour : None, crushed seeds have pungent odour.

Taste : Bitter.

Size : About 1 mm in diameter.

Shape : Seeds are nearly spherical in shape.

Extra Features

The weight of 10 seeds is about 150 to 170 mg. Seeds are covered with a brittle testa. The kernel is greenish-yellow and oily. It has two cotyledons and embryo.

Microscopic Characters

The seed coat i.e. testa, consists of two layers, outer layer being well developed, while the inner layer consists of collapsed cells. Epidermis is made up of polygonal tubular cells containing mucilage. Hypodermis consists of large polygonal cells. The endospermic cells contain aleurone grains and fixed oil. Embryo also contains fixed oil and aleurone grains in its polyhedral cells.

Fig. 9.37: Mustard plant and seeds

Chemical Constituents

Black mustard seeds contain about 30 per cent fixed oil, 20 per cent proteins and 0.7 to 1.3 per cent of volatile oil. The mustard seed contains about 4 per cent of isothiocynate glycoside called sinigrin (Potassium myronate). Because of enzyme myrosin, in presence of water, sinigrin is hydrolysed into allyl isothiocynate.

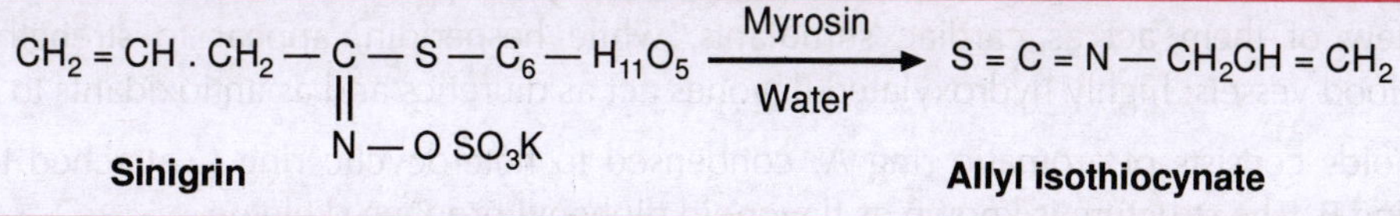

Allyl isothiocynate is also called as essence of mustard or **volatile mustard oil**. It is a pale yellow or colourless liquid having a very sharp, pungent and irritating odour with an acrid taste. Refer black-mustard oil in lipids. This oil is sparingly soluble in alcohol or water but miscible with organic solvents. It is a powerful vesicant and irritant substance. The seeds also contain mucilage and little quantity of acid sinapine sulphate.

Standards of Quality

Ash value	:	4.2 - 5.7 per cent
Moisture content	:	Not more than 5 per cent
Fixed oil	:	Not less than 25 per cent

Chemical Test

Mustard powder when treated with alkali shows bright yellow colour.

Uses

It is a condiment. It is emetic when used internally in higher doses. Externally, it is used as counter irritant and rubefacient in the form of poultice or plasters. All the activity of seed is due to sinigrin or consequent volatile mustard oil.

VI. FLAVONOL GLYCOSIDES / FLAVONOIDS

Flavonoids are the phytonutrients found in almost all plants and vegetables alongwith carotenoids are responsible for various colours in fruits, flowers and vegetables. Quercitin and kaempferol are the suitable examples of flavonoids. Flavonoid rich diets are powerful antioxidants, anti-allergic, antiviral and anti-inflammatory. They are also associated with prevention of cancer and cardiovascular disorders.

The examples of flavonoid rich foods which can be quoted are strawberries, grapes, brussels-sprouts parsley, onion and tea.

Polyphenols constitute largest group of phyto-chemicals distributed amongst plant kingdom. Polyphenols cover flavonoids, non-flavonoids and phenolic acids. Non-flavonoid polythenols are hydrolysable tannins while phenolic acids are hydroxyl cinnamates and hydroxyl benzoates.

Flavonoids

The flavonoid compounds are regarded as C_6-C_3-C_6 compounds, (fifteen carbon skeleton) in which each C_6 moiety is a benzene ring, the variation in the state of oxidation of the connecting C_3 moiety determines the properties and class of such compounds.

Flavonoid compounds usually occur in plants as **glycosides** in which one or more of the phenolic hydroxyl groups are combined with sugar residues. The hydroxyl groups are generally found in positions 5 and 7 in ring A, while ring B commonly carries hydroxyl or alkoxyl groups.

These occur in almost all parts of plants. Numerous physiological activities have been attributed to them; few of them act as cardiac stimulants; while hesperidin, appear to strengthen weak capillary blood vessels; highly hydroxylated flavones act as diuretics and as antioxidants to lipids.

Flavonoids consists of aromatic ring A, condensed to heterocyclic ring C attached to second aromatic ring B. The structure is known as flavonoid diphenyl propane skeleton.

Flavonoids are divided into two sub-classes i.e. authocyanins and anthoxanthins. Anthocyanins are glycosides of anthocyanidin and are the most important class of water soluble plant pigments responsible for various beautiful colours such as purple, blue and red to fruits, flowers and leaves. Anthoxanthins are white or yellowish or colourless compounds and includes flavonols, flavanols, isoflavones and flavanones.

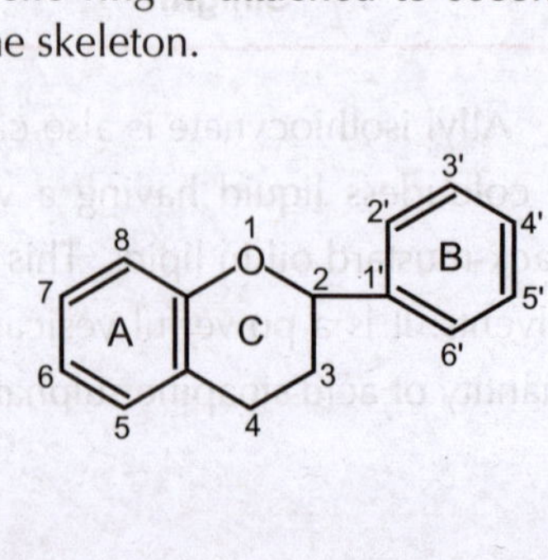

Table 9.4: Classification of Polyphenols

1. Flavonoides:

Type	Representative compounds	Example
1. Flavonols 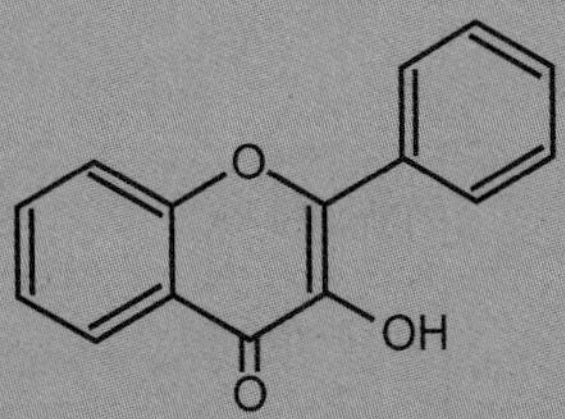	Myricitin Quercetin Kaempferol	Apple, olive Orange Black tea
2. Flavanols 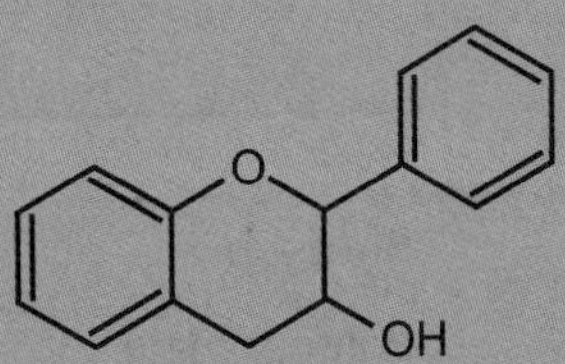	Catechin Epicatechin	Pear Green tea
3. Flavones 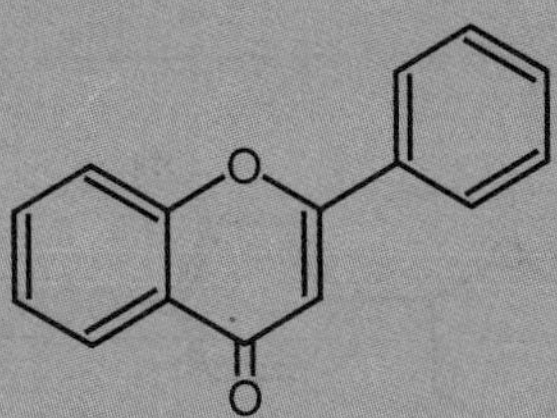	Apiginin Luteolin Fisetin	Celery Olives
4. Isoflavones 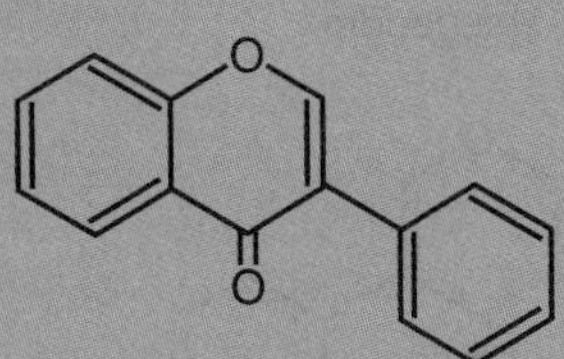	Genistein Daidzein	Soya-bean
5. Flavanones 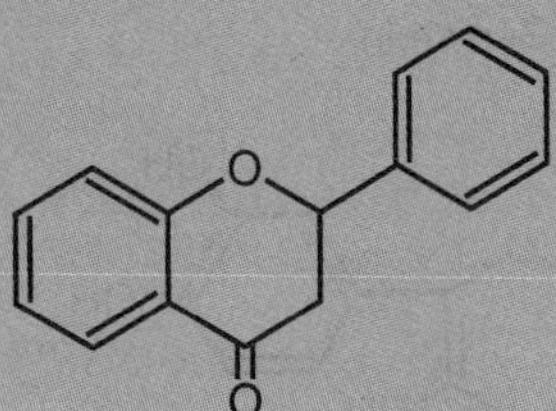	Hesperidin Naringin Hesperitin	Orange Grape fruit

Contd...

2. Phenolic acids

(a) Hydroxy cinnamic acid 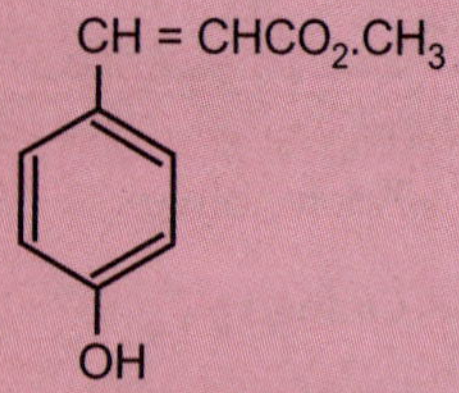	Caffeic acid Ferulic acid Gingerol	Coffee Ginger
(b) Hydroxy benzoic acid 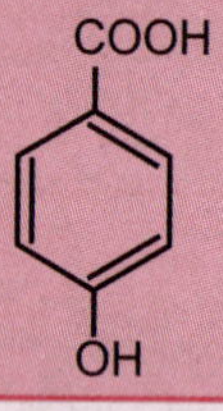	Ellagic acid Gallic acid	Raspberry Strawberry

3. Hydrolysable tannins

	Chebulic acid Gallo tannins	Rhubarb Pomogranate

Few important phyto-polyphenols are represented as:

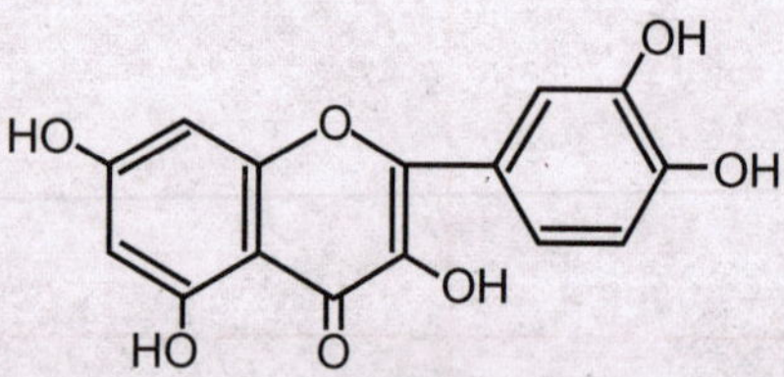

Quercetin

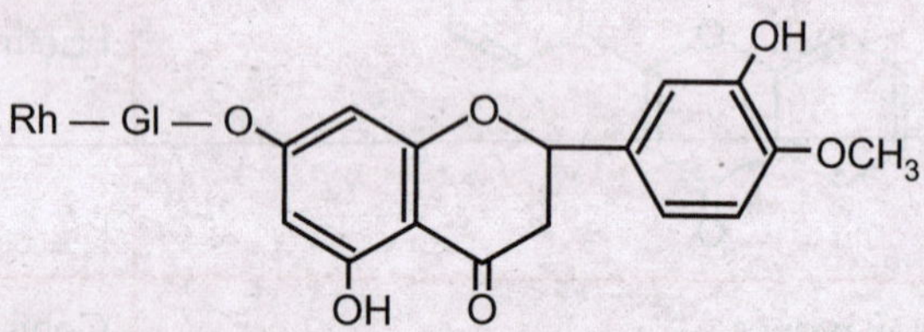

Hesperidin

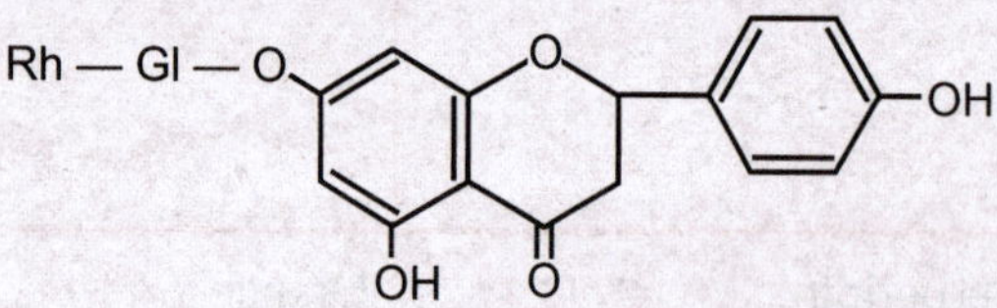

Naringin

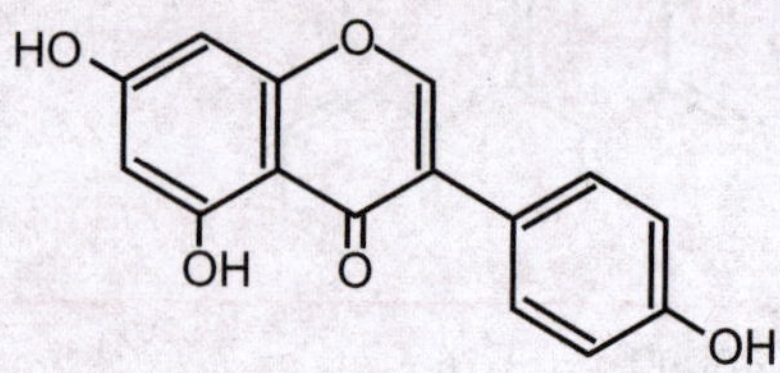

Genistein

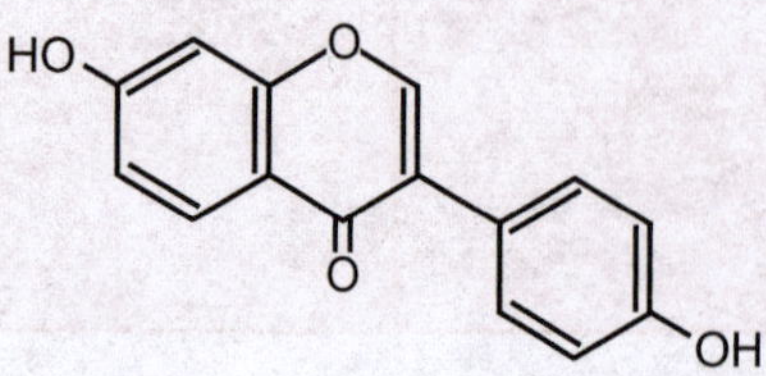

Diadzein

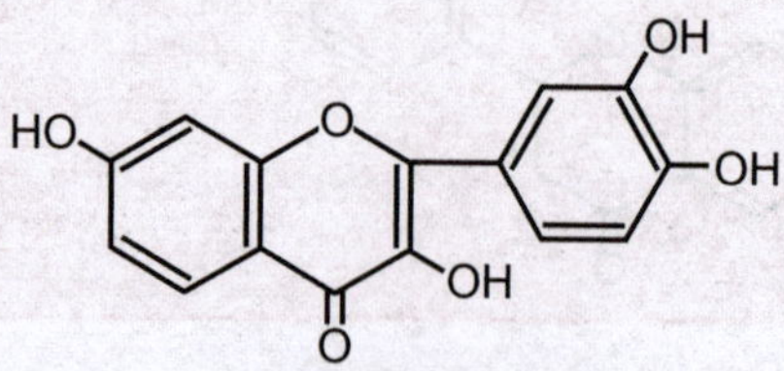

Fisetin

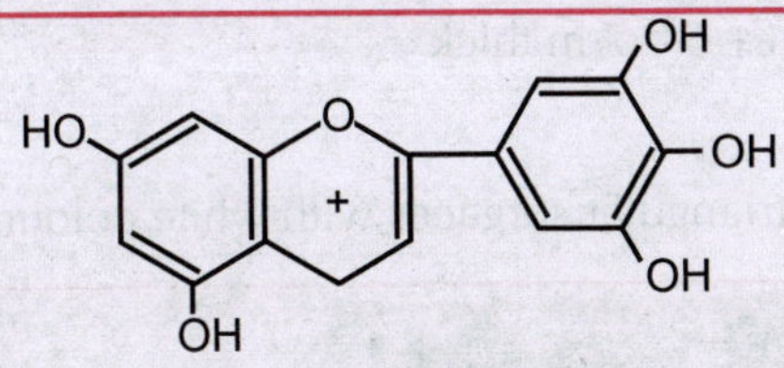

Anthocyanidin

The sugar-free pigments are called anthocyanidins.

The natural anthocyanidins are classified into three groups; pelargonidin, cyanidin, and delphinidin.

Several factors, such as the pH, complex-forming metals, and tannins, affect the colours of the anthocyanins.

Leucoanthocyanidins:

Leucoanthocyanidins are flavan-3, 4-diols. These are colourless substances but gives red solutions with acids. Leucoanthocyanidins are widely distributed in the plant kingdom: Leucopelargonidin and peltogynol are typical examples of this class.

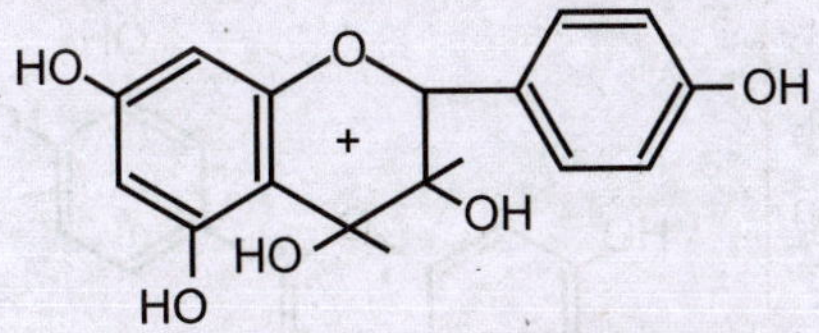

Leucopelargonidin

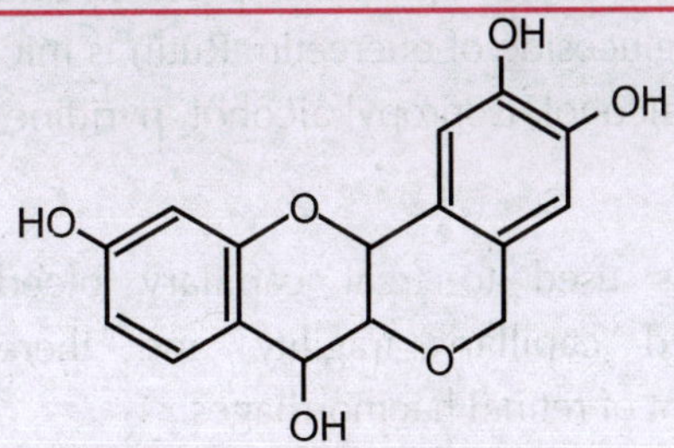

Peltogynol

BUCK WHEAT

Biological Source

It is the powder of dried food grains of *Fagopyrum esculentum* family Polygonaceae.

Geographical Source

It is a crop of cold climatic conditions and favourable temperature required is about 18ºC. It is cultivated commercially as food crop in USA, Russia and Japan. It is cultivated as main food crop in Sikkim. It is also under cultivation to Kerala.

Macroscopic Characters

These are brown coloured irregular shaped food grains with four triangular surface of the size of soyabeans 6 – 10 × 5.8 mm. The leaves of the plant are heart shaped, stems are hollow and slender. Plants are about one meter in hight and bear white coloured flowers.

Colour : Food grains of Brown colour.

Odour : Odourless

Taste : Taste-less

Size : 6 – 10 × 5.8 mm + 5 – 7 m thick

Shape : Irregular

Extra features: It is with four triangular surgaces with white coloured flowers.

Fig. 9.37 (a) : Buck Wheat food grains

Chemical Constituents

It contains 12 per cent proteins and 2.0 per cent fat. Buck wheat mainly contains rutin rhamnoglucoside of quercetin. Rutin is microcrystalline greenish yellow tasteless powder, soluble in methyl alcohol, isopropyl alcohol, pyridine and solutions of alkali hydroxides.

Uses

It is used to treat capillary bleeding alongwith increased capillary fragility and thereby useful in treatment of retinal haemorrhages.

It is also obtained from flower buds of Chinese pagoda tree i.e. *Sophora japonica*.

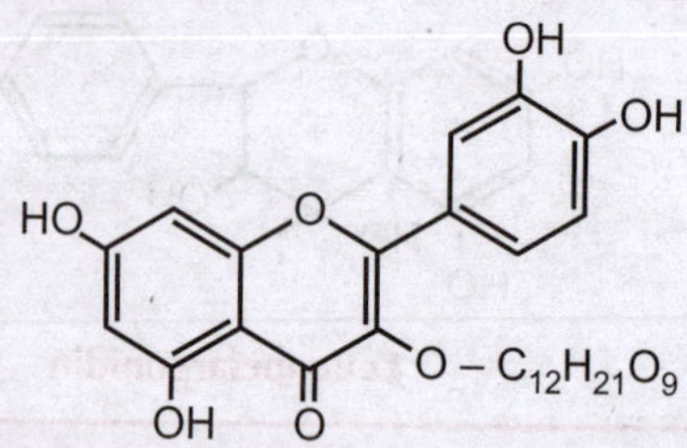

Rutin

GINKGO

Synonyms

Maiden hair Tree, Kew Tree.

History

Ginkgo is one of the oldest known trees, tracing its history upto Mezoic era, about 150 million years back. It was one of the most widely distributed trees. Chinese system of medicine also shows reference to Ginkgo in its first pharmacopoeia, called Shen Nong Pen T'Sao, 2,000 years ago.

Ginkgo occurs as a street and park tree. The leaves of ginkgo have an appearance of maiden hair fern, hence it is called as maiden hair tree. It grows upto 40 metres in height and is appreciably long lived upto 1,000 years. Ginkgo is known to be resistant to insects, bacteria and pollution.

Biological Source

The dried leaves of *Ginkgo biloba* Linn., the only living member of family Gingkoaceae.

Geographical Source

It is native to China cultivated in France and USA. As ornamental shade-tree in Japan, Europe and Austraila.

Macroscopic Characters

The plant bears bilobate leaves in young stage turning to deciduous form later on and in old age fan shaped bright yellow in appearance. Ginkgo is dioecious in nature in which female plant bears a single naked ovule on a stock while pairs of anthers in clustered form grow on male plants. Ovule develops into a drupe–shaped organ, yellow in colour, which falls on ground. This later material gives out a odour like rancid butter.

Fig. 9.38: (a) Ginkgo twig

Fig. 9.38: (b) Ginkgo biloba fruiting plant

Chemical Constituents

Medicinally, active ingredients of ginkgo leaf are various flavonol glycosides, which mainly include flavonol, mono–, di– and triglycosides of kaempferol, quercetin and isorhamnetin. Biflavones present are ginkgetin, isoginkgetin, gingkolic acid and bilobetin which occur in the form of coumaric esters of glucorhamnosides. Along with these ingredients, leaves also contain diterpene lactones namely ginkgolides A, B, C and J, anthocyanins, sitosterol and catechins. It also contains organic acids namely 4–hydroxybenzoic and shikimic acid. Total ash not more than 11.0%.

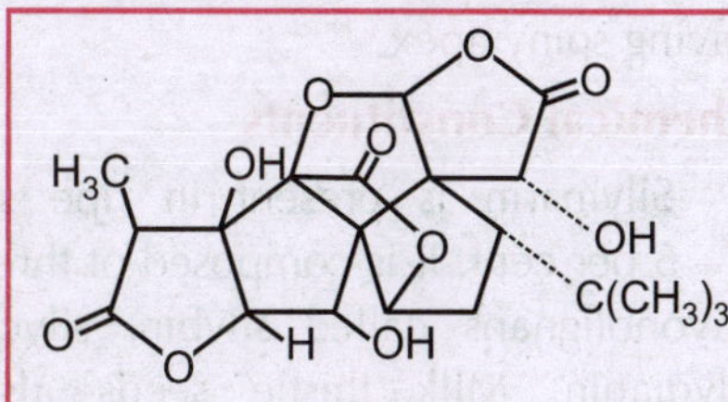

Ginkgolide B

Uses

Although a drug of ancient origin, ginkgo has been established in modern times to have a wide spectrum of activities. It is mainly used in treatment of metabolic and vascular disorders especially cerebral and peripheral vascular insufficiency and intermittent claudication.

It is used in treatment of diseases arising due to microcirculatory problems such as age–associated memory impairment, acrocyanosis and post–phlebitis syndrome, Raynaud's disease (a type of peripheral vascular disease), vertigo and senile macular degeneration. Ginikgolides are specific platelet activating factor antagonists.

Ginkgolide B is used in treatment of severe sepsis. Ginkgo extract is traditionally used in asthma. It has been reported that ginkgolide B is a potent inhibitor of PAF–induced thrombocytopenia and bronchoconstriction. It is also found to be effective against anaphylaxis, possibly because of which, ginkgo extract is used, as per modern Chinese pharmacopoeia, for treatment of chronic bronchitis. It is used as a source of drug especially in the treatment of various ailments owing to microcirculatory problems.

Commonly ginkgo is used in the form of water extract. Sugar coated tablets containing standardized extract (40 mg of active principle with 9 mg of flavone glycosides) of gingko are also available.

MILK-THISTLE

Synonyms

Marian Thistle, Our Lady's Thistle, The Wild Artichoke (Silymarin).

Biological Source

These are the ripe seeds of Milk thistle *Silybum marianum* (L.) Gaerth, belonging to family Asteraceae (Compositae).

Geographical Source

Although, it occurs wildly in Europe, Canada and S. America, it is known to be indigenous to Kashmir.

Macroscopic Characters (Fig. 9.39)

Milk thistle (Thorny) plants grow from five to ten feet in height and commonly found in hedge banks and waste places in Europe.

It bears dark–green, prickly large leaves showing milk–white veins. The plant shows reddish purple flowers having spiny apex.

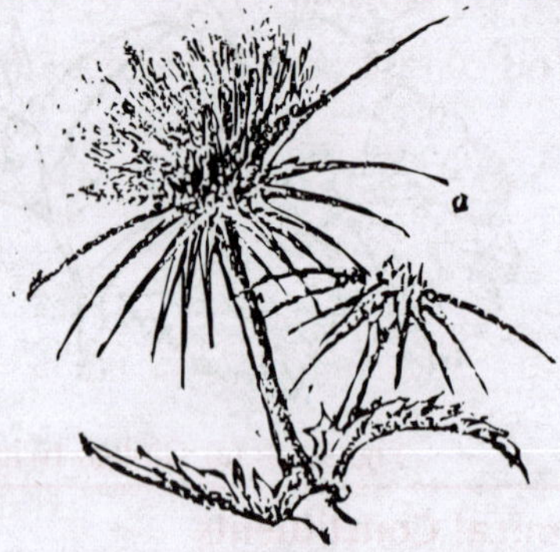

Fig. 9.39: Milk thistle herb

Chemical Constituents

Silymarin is present in ripe seeds from 4 – 6 per cent. It is composed of three isomeric flavonolignans called silybin, silycrystin and silydianin. Milk thistle seeds also contain other flavonalignans namely dehydrosilybin, desocysilycristin, desoxy silydianin, silyhermin, neosilyhermin, silybinome and silandrin. The other constituents present are betaine, apigenin and silybonol.

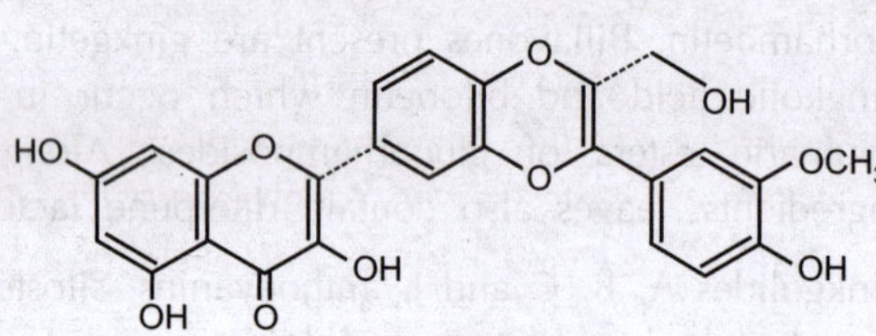

Silybin

Uses

Presently, silymarin is mainly used for its effects in liver diseases. It is used as hepatoprotective and in chronic inflammatory hepatic disorders including hepatitis, cirrhosis and fatty infiltration which occur due to industrial pollutants and alcohol. It has also been found to be effective against liver poisoning due to α – amanitin and phalloidin, which are the chief toxic constituents of lethal death cap fungus, *Amanita phalloides*. Besides these hepatotoxins, silymarin also shows hepato protective effects against D–galactosamine, carbon tetrachloride and thioacetamide. It has been reported that therapeutic utility of silymarin is due to stabilization of cell membrane, stimulation of protein synthesis and accelerating the process of regeneration of hepatic cells. Traditionally, it has been used in Europe as galactogogue in nursing mothers. In old times, it was also used as bitter tonic, anti–depressant and in liver problems. It has also a mention in homeopathy for treatments of jaundice, bronchitis etc.

The mechanism of hepatoprotective effective of silymarin has been suggested variously like anti–oxidant activity by trapping super oxide anions, stimulation of RNA synthesis and in case of *Amanita phalloides* poisoning, blocking the receptor sites of outer liver cell membranes.

Silymarin is preferably given by parenteral route, due to low water solubility of flavonolignans. If taken by oral route, only 20 – 50 per cent is absorbed.

RUE

Synonyms

Garden Rue, Herby Grass.

Biological Source

This consists of dried herb of *Ruta-graveolens* family Rutaceae, a perennial undershrub.

Geographical Source

The plant is indigenous to Southern Europe and often cultivated in Britain and Indian gardens.

Organolephic Characters

It is an erect, scented, glabrous herb about 30 to 40 cm in height

- Colour : Leaves are blue-green in colour.
- Odour : Characteristic and strongly aromatic.
- Taste : Pungent
- Leaves : Leaves are oblanecolate in shape often broken into pieces of 2 × 2 mm, fragrant, glabrous, spathulate.
- Flower : Flowers are greenish yellow in colour. Petals concave and incurred at the tip.
- Fruits : Fruits are sub cylindrical, brown greenish capsules 4 to 5 lobed, rough externally.
- Stem : Cylendrical branched 2-3 cm in length and 1 mm wide.

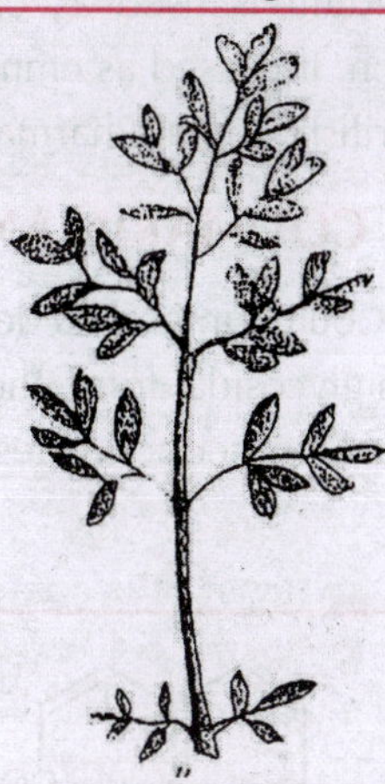

Fig. 9.40 (a): Branch of Ruta graveolens

Microscopic Characters

The leaves bear Anomocytic stomata on lower surface; mesophyll is large with clusters of calcium oxalate crystals and large oil glands. Polen grains are 35 mm in diameter.

Chemical Constituents

Rue contains various types of chemicals like glycosides, alkaloids, volatile oil and fixed oil (in the seeds). Rutin about 2.0%, psoralin, bergaptin, graveoline, naphthoharniarin, rutarin are the constituents of the drug. Rutin is a flavonoidal glycoside and is pale yellow crystalline compound.

The **alkaloids** rutalinium, 1-methyl - 2 - nonyl-4-quinolone, dictamine, rutamine, rubalinium are reported in the plant. **Fixed oil** contain linoleic, stearic and palmitic acid, while **volatile oil** contains. Limonene, cineole, ethyl valerate, methyl salicylate, caryophylline and myrecine.

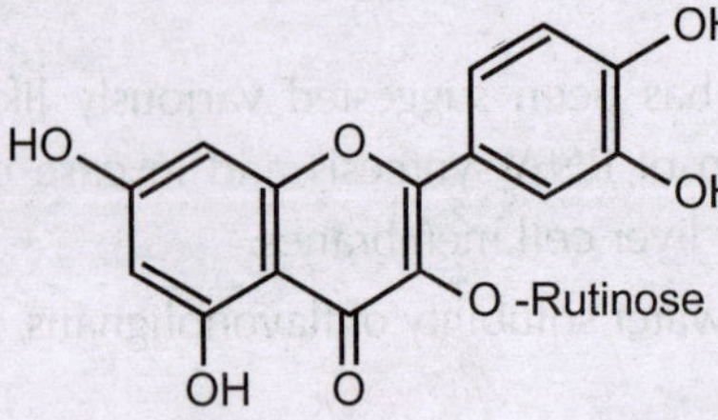

Rutin

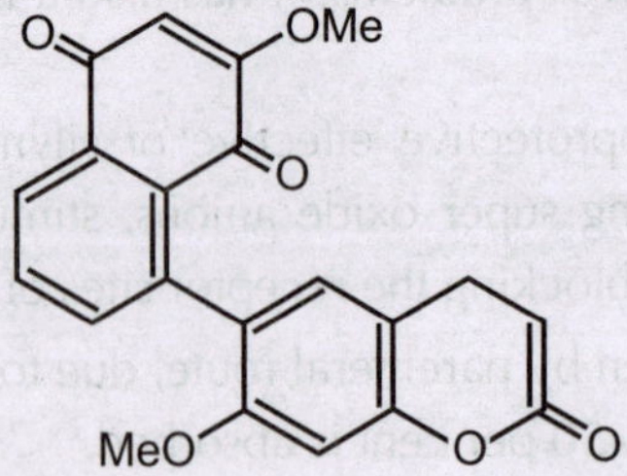

Naphthoherniarin

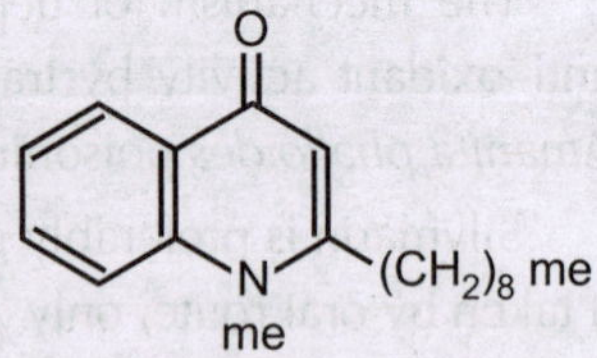

1 methyl-2-nonyl-4-quinolone

Standards of Quality

FOM	:	Not more than 2.0 per cent
Total ash	:	Not more than 10.0 per cent
Acid insoluble ash	:	Not more than 2.0 per cent

Uses

Rutin is used as capillary protectant therapeutically. Rutin has antitussive and spasmolytic action. It is used as emmenagogue and used in the form of liquid extract or infusion. Drug is official in British Herbal Pharmacopoeia 1983.

VII. COUMARIN AND FURANOCOUMARIN GLYCOSIDES

Coumarin and its derivatives are present in many plants. Coumarin is present either in free state or in glycosidal form, the former being most common. Coumarin and other forms like umbelliferone, scopoletin, scopolin, aesculetin aesculin, etc. are the derivatives of Benzo-α-pyrone.

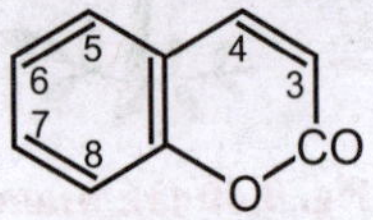

Coumarin

Furan

Coumarin derivatives

	Compound	Additional groupings
1.	Aesculetin	OH at 6, OH at 7
2.	Fraxin	OCH_3 at 6, OH at 7 O - glucose at 8
3.	Herniarin	OCH_3 at 7
4.	Scopoletin	OCH_3 at 6, OH at 7
5.	Umbelliferone	OH at 7

Coumarin glycosides are present in families like Leguminosae, Solanaceae, Rubiaceae, Caprifoliaceae, Umbelliferae, Oleaceae, etc. Furanocoumarins are formed by fusion of furan ring to coumarin at either 6th and 7th position or 7th and 8th position. The significant drugs containing furanocoumarins are visnaga, psoralea, *Ammi majus,* bergamot oil, bael fruit, etc.

PSORALEA FRUIT

Synonym

Bavchi, Babchi

Habitat

Bavchi is an erect annual plant about 60 to 125 cm in height. Stem, as well as, branches are grooved, studded with glands and white hairs. Leaves are 3.5 to 5 cm in size, broadly elliptic and rounded, white hairy and dotted with black glands on both surfaces. Lower glands are small, solitary or a bunch of 10 - 30 and arising in axils of leaves. They are bluish-purple in colour. The plant flowers from August to December.

Biological Source

These are the dried ripe fruits of the plants known as *Psoralea corylifolia* Linn., (Syn. *Cullen corylifolium*) belonging to family Leguminosae.

Geographical Source

It is found throughout India as a weed in waste places and also in Sri Lanka. It is grown in Maharashtra and Madhya Pradesh states of India.

Macroscopic Characters

Fig. 9.41: Psoralea herb

Colour	:	Dark chocolate to black.
Odour	:	None, pungent and characteristic after crushing.
Taste	:	Bitter, unpleasant and acrid.
Size	:	3 to 5 × 2 to 3 mm.
Shape	:	Pods are ovoid, oblong beam-shaped.

Compressed, glabrous closely pitted seeds are compilotropous, non-endospermic, smooth oily and free of starch.

Chemical Constituents

Psoralea fruits are found to contain essential oil (0.05 per cent), resin (8.9 per cent), brown fixed oil (10 per cent) a pigment, raffinose and coumarin compounds, psoralen, isopsoralidin, psoralidin and corylifolin.

Psoralens are one of the furocanmarins and are phyto-alexins chemically. Phyto alexins occur in several plant families and are the defensive chemicals of plants, produced in response to attacks by fungi and insects. Phytoalexins also exhibit photo sensitysing and phototoxic effects in animals and humans. They are also used in photo-chemo therapy in management of vitiligo psoriasis and microsis fungals.

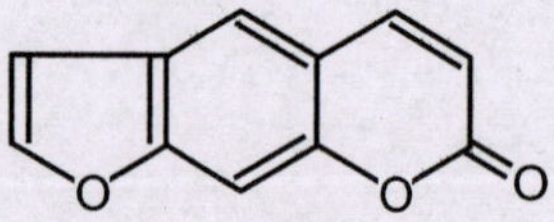

Psoralen

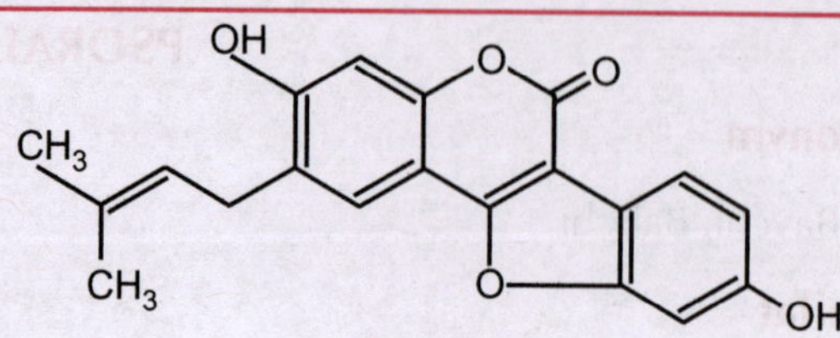

Psoralidin

Recently, two more compounds namely bavachromanol and psoralenol have been reported. The unsaponifiable matter of the fixed oil of seeds is found to contain stigmasterol.

Chemical Test (for psoralen)

(1) A small quantity is dissolved in alcohol to which 3 times propylene glycol, 5 times acetic acid and 40 times water is added. A blue fluorescence is observed under ultra violet light.

(2) The drug shows yellow fluorescence in ultra violet light when dissolved in alcohol and little sodium hydroxide solution is added to it.

Uses

Psoralea is recommended in treatment of leucoderma, leprosy, psoriasis and inflammatory diseases of the skin. It is taken by oral route of administration or applied externally in the form of ointment or paste. The seeds of psoralea are used as diuretic, anthelmintic, and laxative. Volatile oil is used as flanouring agent.

TONKA BEAN

Synonyms

Tonka bean camphor, Tonco seeds, Tonquin beans.

Biological Source

These are the dried seeds of i.e. *Dipteryx odorata* Wild and *Dipteryx oppositifolia,* belonging to family Leguminosae.

Geographical Source

Tonko beans are available in Brazil Mexico, Holland and British guina.

Description

Seeds are oblong, straight, curved flat with shining and brownish black in colour. Flavour similar to clove with aromatic acrid bitterish taste. Tonko beans are as good as pen in size.

Chemical Constituents

The seeds mainly contain coumarin. Coumarin is lactone of cis-o-coumarinic acid occurring as prismatic colourless crystals with persistent fragrant odour and bitter burning taste. It is sparingly soluble in water, but soluble in alkali hydroxides and different organic solvents.

Tonco seeds contain about 25 per cent fat and 1 - 3 per cent coumarin.

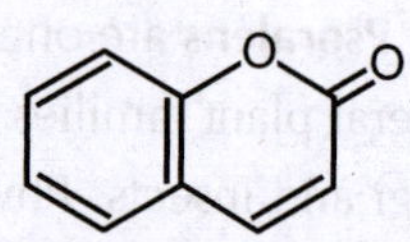

Coumarin

Uses

Coumarin is used as flavouring agent, fixative in perfumery and in tobacco manufacture. Previously, it was used as flavouring agent in food products, but it has been prohibited since 1965. Coumarin is also present in red clover *Trifolium pratense*), sweet clover ((*Melilotus albus*), and sweet scented bedstraw (*Galium trifolium*). Tonko seeds are used extensively in Homoeopathic practice.

AMMI

Biological Source

These are the fruits of *Ammi majus*. Linn, belonging to family Umbelliferae. *Ammi visnaga* is the other member knonw as visnaga, used for same purpose.

Geographical Source

It is a plant native to Egypt and is wildly grown in Europe, Abyssinia and West Africa. It is cultivated in Jammu and Kashmir state of India.

Cultivation and Collection

It is cultivated in North India mainly by some private pharmaceutical firms as well as Central Institute of Medicinal and Aromatic Plants.

The propagation is done by seeds which are mixed with soil and sown at a distance of 90 cm. The sowing is done in October. Germination takes place in a month. Before sowing, farmyard manure and super phosphate are added to soil.

The harvesting is done in April - May, when the plant bears immature green fruits. The fruits are dried and separated by thrashing.

Macroscopic and Microscopic Characters

The fruits resemble those of visnaga, except that they are small, and possess 4 prominent secondary ridges. No lacuna is present in primary ridges.

Chemical Constituents

The active ingredient of the fruits is xanthotoxin which is present from 0.4 to 1.9 per cent. The other ingredients are bergapten, isopimpilin and imperatorin. All these compounds are furanocoumarin derivatives.

Fig. 9.40 (b): Ammi visnaga herb

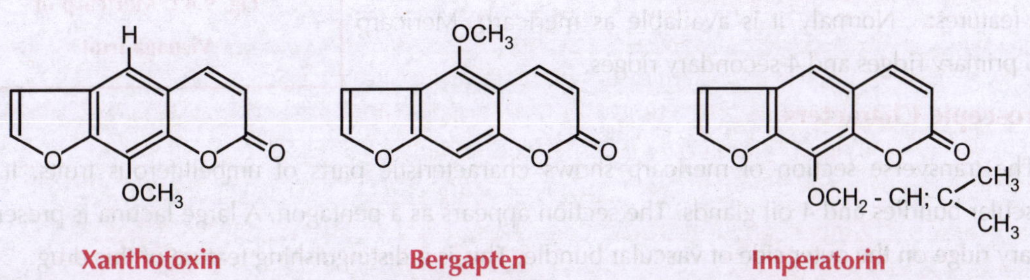

Uses

Xanthotoxin, the active ingredient of this drug, also known as methoxsalen in U.S.P., is used to increase the formation of melanin pigments in the skin, on exposure to ultra violet light. It is useful in the treatment of idiopathic vitiligo. It has not shown any useful effects in treatment of leucoderma due to trauma or leprosy.

It can be given orally. Skin lotion is also applied, after which exposure to ultraviolet light for one minute, is essential.

VISNAGA

Synonyms

Khella, Pick tooth fruit.

Biological Source

These are dried ripe fruits of *Ammi visnaga* Linn., belonging to family Umbelliferae.

Geographical Source

The drug is native of Nile delta in Egypt. It has been cultivated in Egypt and Chile in South America.

Cultivation and Collection

The plant which is an annual herb growing about 1 to 1.5 metres in height, is propagated by means of seeds in loamy soil of nursery beds in August. After attaining a height of about 6 - 7 cm, transplantation is done into open fields. After 7 - 8 months, the plant bears flowers and harvesting is done at the stage of ripening of first fertilized flowers. The plants are cut and preserved in heaps, during which all the fruits are ripened.

Macroscopic Characters

Colour : Friots are greenish brown in colour.

Odour : Slightly aromatic

Taste : Aromatic bitter and slightly pungent

Shape : Plano convex and ovoid Lanceolate

Size : 2.0 - 2.5 × 0.7 - 1.2 × 0.8 - 1.0 mm

Extra features: Normaly it is available as mericarp. Mericarp has 5 primary ridges and 4 secondary ridges.

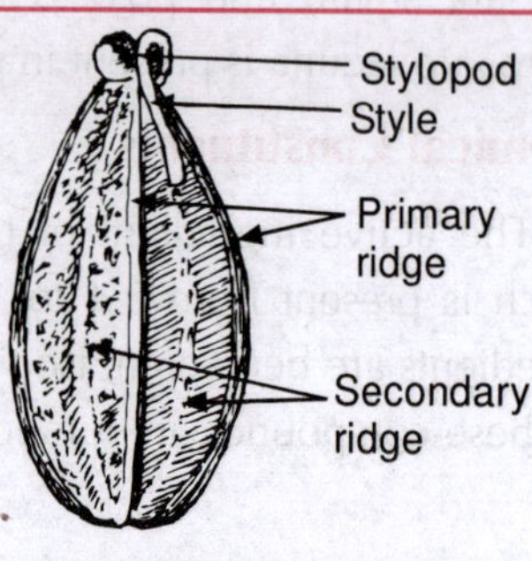

Fig. 9.42: Mericarp of Visnaga fruit

Microscopic Characters

The transverse section of mericarp shows characteristic parts of umbelliferous fruits. It has 5 vascular bundles and 4 oil glands. The section appears as a pentagon. A large lacuna is present in primary ridge on the outer side of vascular bundle. This is a distinguishing feature of this drug.

Chemical Constituents

The drug contains volatile oil about 1 per cent khellin, 0.1 per cent visnagin and 0.3 per cent. khelloside. They are furanocoumarin derivatives. Khellin occurs as odourless bitter tasting crystals and it is 2-methyl -5, 8 dimethoxy furanochrome. It is soluble in mineral acids and organic solvents except ether.

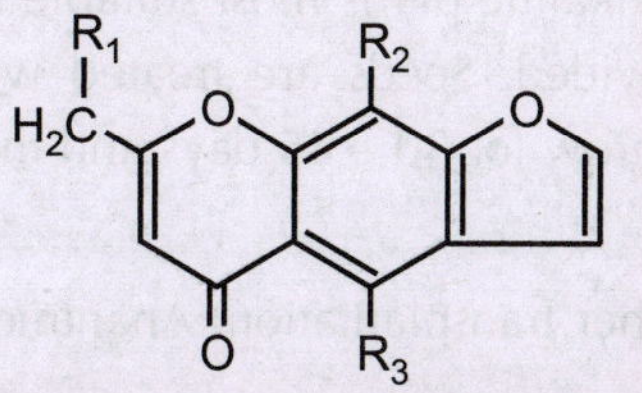

Glucoside	R_1	R_2	R_3
Visnagin	OCH_3	H	H
Khellin	OCH_3	OCH_3	H
Khelloside	OCH_3	H	O-glucoside

The oil present in drug contains samidine, dihydrosamidine, and visnadine.

Uses

Khellin is a smooth muscle relaxant and used as coronary vasodilator in angina pectoris, renal and uterine colic, bronchial asthma and whooping cough.

Samidine, dihydrosamidine and visuadine are strong vasodilators.

VIII. ALDEHYDE GLYCOSIDES

ANANTMUL

Synonyms

Anantmoola, Sariva, Indian Sarsaparilla.

Biological Source

It consists of dried roots of the climber known as *Hemidesmes indicus* Linn. belonging to family Asclepladaceae. The roots contain not less than 0.020 per cent of Iso-vanillin evaluated on dry basis.

Geographical Source

If is fond throught India and Sri Lanka.

Macroscopic Characters

Anantmul roots are woody, thickened at nodes, leaves are opposite short and petioled, variable in size and shape. Flowers are greenish yellow outside and purplish inside. Fruits are dehiscent follicles, about 10 cm long. Seeds are numerous, black and flattened. It is a lateciferous plant. Seeds are ciliated.

Colour : Roots are yellowish brown

Odour : Vanilline – like

Taste : Aromatic and sweetish

Size : 30 cm or more in length 3 – 6 mm in thickness.

Shape : Cylendrical, seldom branched, having corky bark, longitudinal furrows and transverse fissures. All plants parts have white milky juice.

Fracture : Short at periprery and fibrous in the centre.

Cultivation of Anantmul

Cultivation can be done by using seeds or by vegetative propogation using well developed root cuttings, well drained soil with enough carbon content and having alkaline pH is most suitable for cultivations. Soil is ploughed twice and organic manures are provided. Seeds are treated with Indole-buteric acid (20 ppm) before sowing. Cuttings are allowed to grow for 30 – 45 days and then transplanted at a distnace of 1 × 1 metre.

After 1.5 – 2 years, the roots of 1 cm diameter are used for further transplantation. Anantmool being climber and suitable arrangements are required for support.

Collection of Anantmul

Crop is ready for harvesting after 2 years. Roots are washed properly to remove debris and dried in shade for 6-8 days. About six to eight quintals of dried roots per acre is the normal yield.

Microscopic Characters

T.S. of root shows 2-3 layered cork cambium, having compressed cells full of brown contents. Xylem characterized by narrow medullary rays. Pith is absent several laticiferrous cells are present in secondary cortex, vessels show pitted thickening.

Fig. 9.43: Hemidesmes indicus plant

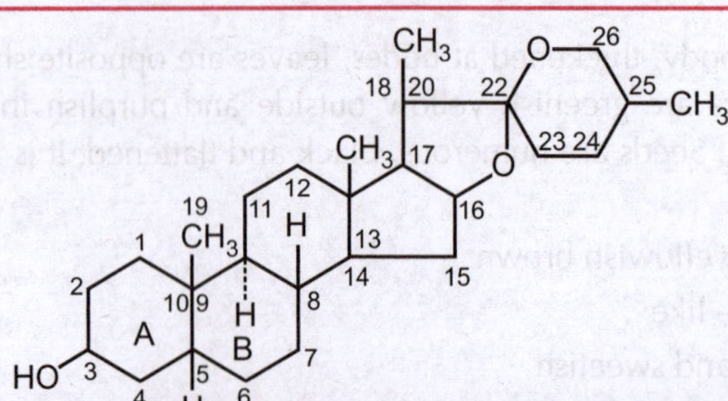

Sarsasapogenin

Chemical Constituents

Roots contain about 0.2 per cent vol. oil, β sitosterol, hemisterol, demidesmol, a & b amyrins, lupeol, tannins, and saponins. Sarsasapogenin contains 3 glucose and one rhamnose as sugar componants.

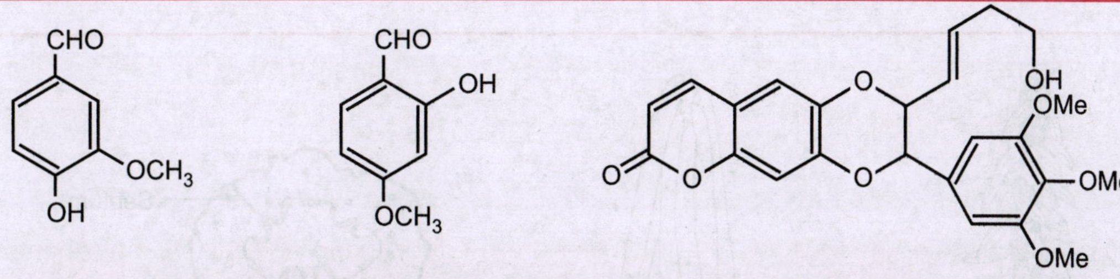

Vanillin p-methoxy salicylic aldehyde Hemidesminine

Volatile oil contains p-methoxy salicylic aldehyde as major constituent. Roots contain coumarino-lignoids, hemidesminine and hemidesmin I. Iso vanilline is the main constitutent responsible for pleasant flavour of the drug.

Standards of Quality

Foreign organic matter	:	Not more than 2.0 per cent
Alcohol soluble extractives	:	Not less than 8.0 per cent
Water soluble extractives	:	Not less than 12.0 per cent
Ash contents	:	Not more than 15.0 per cent
Acid insoluble ash	:	Not less than 02.0 per cent
Loss on drying	:	Not less than 12.0 per cent by using 5 gm in an oven at 105ºC.

Uses

Syrup of the root is used as flavouring agent and in preparation of herbal tea. It is also used as blood purifier, in rheumatism, anti-inflammatory tonic in urinary disorders. Powdered roots are used for pre and post natal care, and to increase lactation in cows. It is used in psoriasis and also as flavour in beverages.

Storage

Protect the drug against insect and rodent attack and keep away from moisture and heat.

VANILLA POD

Synonyms

Baunilha, Vanilla beans, Fructus Vanillae, Mexican vanilla.

Biological Source

Vanilla consists of cured, fully grown, unripe fruits of *Vanilla planifolia* Andrews (Madagascar, Mexican or Bourbon vanilla) or V. *tahitensis* J.W. Moore (Tahiti vanilla), family Orchidaceae.

Geographical Source

It is pereneal indigenous to east coast of Mexico. Now-a-days, the plant is cultivated in Mauritius, Tahiti, Java, Seychelles, Madagascar, Sri Lanka and in Lakshdweep island.

Being cultivated sucessfully in Kerala since three decades.

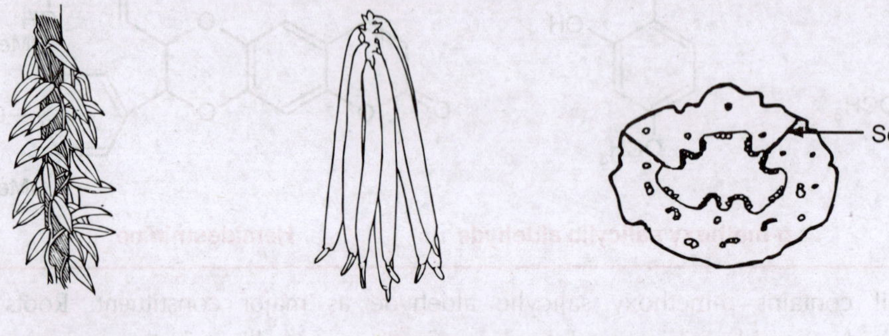

Fig. 9.44: (a) Vanilla plant (b) Vanilla pods (c) T.S. of Vanilla pod

Description

Vanilla plant is a perennial climbing epiphyte, attached to the trunk of the trees by means of aerial rootlets.

The fruits of vanilla are capsules, they are slender linear and 15 – 25 cm long and only 10.0 mm in diameter they are flat cylendirical and curved usually at stalk and fruits are black in colour, Odour and taste is aggreable, fragrant.

Fruits are longitudinally wrinkled and covered with minute crystals of vanilline. Capsule is unilocular with three carpels containing numerous black seeds, covered by dark balsamic fluid. The balasenic fluid is secreted by hair like cells on the inner epidermis of the pericarp. Fruit of vanilla breaks with two unequal parts very characteristic to the fruit.

Plant starts bearing fruits from fourth year onwards upto 50 years. Plant servives for about 70 years. Flowers of vanilla are hard pollinated also, fruits are collected when they are ripen and are yellow coloured. Fruits are curved by dipping in warm water and sweating is repeated between woolen blankets in sun, during day time while in nights-fruits are packed in wool-covred boxes, process takes about two months and the pods loose about 60 -80 per cent of thier weight. Pods take now desired yellow colour and which becomes black after few days. The pods are now put into bundler of 50 and packed in thin containers and sent to market.

Chemical Constituents

Vanilla pods contains mainly two glycosides about 2.0% namely glucovanillin (Avenein) and gluco vanellic alcohol. During the curing period.

These glycosides get hydrolysed as under:

$$\text{Glucovallin} \xrightarrow{\text{Hydrolysis}} \text{Glucose} + \text{Vanillin}$$

$$\text{Glucovanillic alcohol} \xrightarrow{\text{Hydrolysis}} \text{Glucose} + \text{Vanillic acid}$$

$$\text{Vanillic acid} \longrightarrow \text{Vanillic aldehyde (Vanillin)}$$

These glycosides are present in fluid secreted by inner epiderms of pericarp and it gets spreaded over the entire fruit. These glycosides are responsible for the aromatic pleasant flavour of vanilla-pods. Vanilla pods also contain 10.0% fixed oil, 10% sugars and calcium oxalate.

Vanilla plant is a peremial climbing epiphyte attached to the trunk of the tree by means of aerial rootlets.

Macroscopic Characters

Colour	:	Vanilla pods are black in colour
Odour	:	Pleasant due to vanilline. Pleasant and aromatic and flurour
Taste	:	Bitter and aromatic
Size	:	15.0 to 25.0 cm × 0.8 – 10 mm
Shape	:	Flattened cylindrical

Exatra features: These are capsuler fruits with longitudinally wrinkled, usually curved at stalk end.

The fruits of vanilla are capsular type which are flattened and cylindrical in shape. The fruits are 15 - 25 cm in length and about 8 -10 mm in diameter. The plant gives the fruits for about 30 - 40 years. They have a longitudinally wrinkled surface, and it is covered with crystals of vanillin.

Chemical Constituents

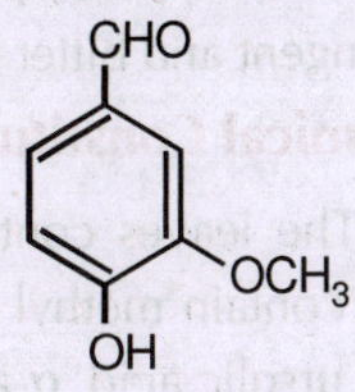

Vanillin

The main constituent of vanilla is vanillin, contained in the fluid, secreted by inner epidermis of pericarp.

The fruits contain about 2 per cent vanillin, (4-hydroxy-3 methoxy benzaldehyde) which is present in the form of glucovanillin. It also contains another glycoside called glucovanillic alcohol. Vanillin is obtained from glucovanillin by hydrolysis. On hydrolysis, glucovanillic alcohol gives vanillic alcohol and glucose. The oxidation of vanillic alcohol gives vanillin.

Vanillin occurs as white to cream coloured powder or crystals. It is soluble in fixed and volatile oils, alkali hydroxides, alcohol, chloroform and ether, but sparingly soluble in water.

The odour and flavour of vanilla are not totally due to vanillin, but collectively due to vanillin and other fragrant priniciples.

Uses

It is used as a flavouring agent, in perfumery and source of vanillin.

Other Sources of Vanillin

(1) It is present in small amounts in benzoin, balsam of peru and balsam of tolu.

(2) The waste sulphite liquors obtained in the manufacture of wood pulp is converted into vanillin.

(3) It is mainly obtained from eugenol present in clove oil (cinnamon oil contains very less quantity of eugenol, hence not used commercially).

IX. PHENOL GLYCOSIDES

BEARBERRY

Synonyms

Uva ursi, Bearberry leaves, Busserole, Arctostaphylous.

Biological Source

These are the dried leaves of *Arctostaphylous uva-ursi* (Linne) Sprengel, belonging to family Ericaceae. (Bears eat the fruits with relish and hence bear berry)

Fig. 9.45: Uva-ursi herb

Geographical Source

It is found in different parts of Central and North Europe, North America, Canada and Scotland.

Macroscopic Characters

Leaves are spatula like, 2 to 2.5 cm in size tapering towards base, with short stalk or petiole, brittle, margin is entire.

Fruits are scarlet red with calyx at base.

The leaves are collected in April to June and dried under sun. They show olive green upper surface and pale colour on lower surface. They are small coriaceous, shining leaves. They have astringent and bitter taste, without any specific odour.

Chemical Constituents

The leaves contain a glycoside called arbutin which contains phenolic aglycone. The leaves also contain methyl arbutin, quercetin, ursone, iriodoids, quinones, tannins (6 - 10 per cent), gallic acid ursolic acid, α-amyrin, β-amyrin and terpenoids.

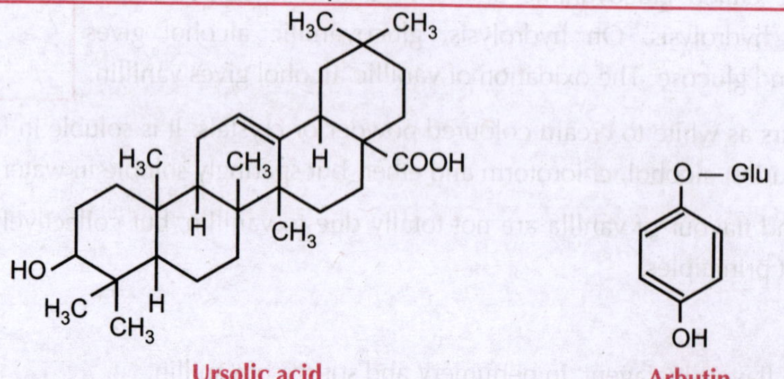

Ursolic acid Arbutin

Standards of Quality

Total ash	:	Not more than 4.0 per cent.
Acid insoluble ash	:	Not more than 1.0 per cent.
Foreign organic matter	:	Not more than 5.0 per cent.

Uses

The leaves have diuretic and astringent properties. As an infusion, it is used in urethritis and cystitis.

X. STEROIDAL GLYCO-ALKALOIDS

SOLANUM KHASIANUM

Habitat

A stout highly-branched shrub upto 1 - 1.5 metre tall with straight pricles, Leaves: ovate, lobed, lobes are triangular prickly on both surfaces. Flowers: white coloured, in lateral 1 - 4 flowered racemes.

Biological Source

It consists of dried berries of *Solanum khasianum* C.B. Clarke, belonging to family Solanaceae.

Geographical Source

The plant is found wildly growing at various altitudes in India right from coastal region upto 2000 metres. It is found in hilly regions of Assam, Manipur, NEFA, Sikkim, Nilgiris, Central India and also in Myanmar and China. Now-a-days it is cultivated on commercial scales in Maharashtra.

Cultivation

In view of its solasodine content, it has commercial significance. Solasodine, a steroidal glycoalkaloid, has similar applications as that of diosgenin. The cultivation of this plant is scientifically studied and the observations of those trials are given here in brief. The seeds are used for propagation, either through nursery beds or by direct broadcasting.

Fig. 9.46: *Solanum khasianum* plant with fruits

In February, the seeds are sown in nursery beds. The seed beds are covered with sand or farmyard manures and weeding is done periodically. When the seedlings show sufficient growth, they are transplanted into open fields.

The raising in nurseries is preferred to direct broadcasting. The plant grows in various climatic and agricultural conditions. The well drained soil and sunny atmosphere are preferred. The seedlings are transplanted in moist soil at 50 × 50 cm distance.

Urea, potash and superphosphate are given as fertilizers. In the initial period, irrigation is done once in a week and then in later stages as per requirement.

Collection of Berries

After six months, the plants are harvested for collection of berries. They are immediately dried in shade or artificially at low temperature to reduce the large content of moisture.

Macroscopic Characters

It bears yellowish to greenish berries which are globose and 2.5 cm in diameter with compressed smooth brown seeds.

Chemical Constituents

The berries contain about 3 per cent of steroidal glycoalkaloid called **solasodine**. Solasodine is hydrolysed product of solasonin and is used in partial synthesis of corticosteroids. A new glycoalkaloid solakhasianin having rhamnose and galactose as sugar components have been isolated. Mucilage surrounding part of the seeds contain highest amount of alkaloid. Immatured and over-ripe fruits contain neglegible content of alkaloid, while it is maximum when fruits change colour from green to yellow. Colour change of fruits takes place about two months after setting the fruits to the plants.

Isolation of solasodine on commercial scale is being done by one of the local manufacturer at Jalgaon in Maharashtra.

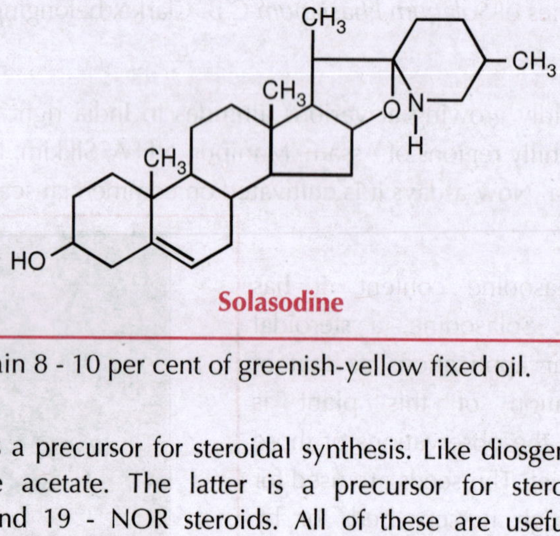

Solasodine

The berries also contain 8 - 10 per cent of greenish-yellow fixed oil.

Uses

Solasodine is used as a precursor for steroidal synthesis. Like diosgenin, it is first converted to 16-dehydro-pregnenelone acetate. The latter is a precursor for steroids, like corticosteroids, pregnane, androstanes and 19 - NOR steroids. All of these are useful as sex hormones, oral contraceptives, etc.

Isolation

For the isolation of solasodine, the berries are powdered and the oil is removed by defatting. The defatted material is extracted with ethanol. The extract is concentrated, treated with concentrated HCl and refluxed for 6 hours. After this, ammonia is added to basify the extract and again refluxed for 1 hour. It is filtered and the residue is washed, dried and dissolved in chloroform. This mixture is filtered and solasodine in the form of residue is obtained by evaporating the solvent.

Allied Species

Many other species of Solanum containing small quantities of solasodine are *S. eleatnifolium, S. inacanum, S. indicum, S. surratense, S. seaforthianum, S. mammosum, S. trilobatum, etc.*

STEVIA

Synonyms

Honey plant, Sweet herb of paraguay.

In future due to its high potency sweetening property, stevia is likely to be superior to chemical and synthetic sweetner, as well as sugar too. Its real wonder herb which is 300 times sweeter than sugar along with calorie **free** property.

Biological Source

These are dried roots and leaves of plant *Stevia rebaundiana* Bertoni belonging to family Asteraceae (Compositae).

Geographical Source

It is native of Paraguay and Brazil. Now cultivated in Canada, China, Japan, USA, Mexico, Indonesia, Korea and also to some extent in India.

Cultivation

Subtropical climatic conditions are favourable for its growth. High temperature during day time and semi humid condition promote its faster growth.

Grows well upto an altitude of 200 to 250 metres from sea-level. Minimum 10°C during night and between 30°C to 35°C during day time. Well distributed rainfall ranging between 100 to 150 cm is favourable for its growth.

Cultivation of Stevia

In India cultivation is done from April to June or August to September if there are heavy rains. Histance between two plants is maintained 25-30 cm and in between two rows 45-50 cm. Exposure to sunlight upto 40°C is essential.

It does not tolerate saline soil with water logging. Hence, should be grown on red laterite or calcareous soil with rich organic matter and well drained property with a pH between 4 - 5.

It can be propagated either by seeds or vegetatively by using rooted plantlets. But due to poor seed germination (only 10 per cent) vegetative propagation is prefered. It is a perennial plant. For cultivation 75000 to 100000 plantlets per hectare are used.

Collection of Stevia

After 2 - 3 months from transplanting it starts bearing flowers. The appropriate timing for harvesting this herb, is period just before it starts flowering, when the active constituents are at their maximum in the leaves. The plants are cut at their base leaving 10 to 15 cm stem for regeneration of the crop. After first harvesting, the crop is again supplied with top dressing of nitrogen and irrigated. The procedure is repeated after every harvest. The crop is then ready after 45 days of first harvest. In this way harvesting can be carried out after every 45 days for three consecutive years.

Whole plant is dried in shade for 3 - 4 days. Then the leaves are separated from the plant. Which are used for preparation of powder. Depending upon the number of plucking the leaves (3 to 6 times in a year) 2000 to 4000 kg dried leaves, can be collected per hectare.

Macroscopic Characters

Colour	:	Leavers are dark green
Odour	:	None
Taste	:	Sweet, liquorice-like
Size	:	5 cm in length, 3 cm in width
Shape	:	Ovate
Flowers	:	White, throats funnel form lobes 5 available in June to August every year.
Extra Features	:	Leaves, simple, sessile, petiolate, acuminate, crenate faces are glabrous, planted crosswise, facing each other pubescent.

Fig. 9.47: Stevia herb

Stavia is an erect, cauline, perennial, herbaceous plant about 50 – 100 cm in height. Dry weight of single plant about 70 gram.

Chemical Constituents

Total sweet glycoside concentration as Stevioside (5-10 per cent); Rebaudioside-A (2 - 4 per cent), Rebaudioside-C (1 - 2 per cent), Dulcoside - A (0.4 - 0.7 per cent), Amongst these stevioside is the major component, besides these it also contains good amount of sterols, triterpenes, flavonoids, tannins and also negligible amount of volatile oil.

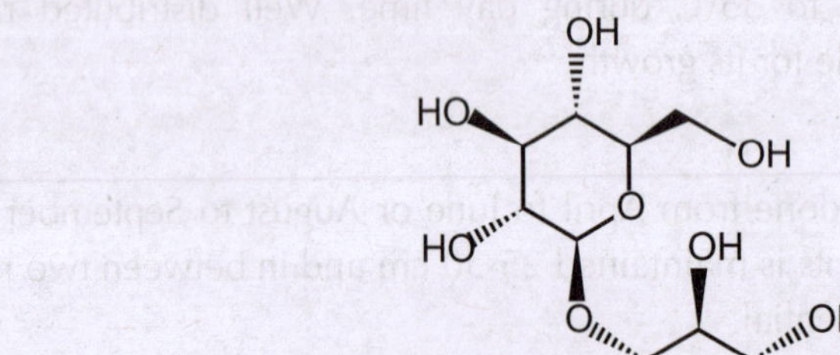

Stevioside

Uses

Several therapeutic uses of stevia are reported amongst them antidiabetic is the major. It also shows antibacterial, antifertility, antiinflammatory, antiseptic properties. It is also used as digestive tonic. It has shown good results in clearing up skin problems like, acne, seborrhea, dermatitis, eczema etc.

XI. GLYCOSIDAL BITTERS AND MISCELLANEOUS GLYCOSIDES

Bitters are the substances taken before meal to improve appetite. They include various drugs of natural origin. Bitter glycosides have the activities like bitter, stomachic, digestive and febrifuge. These type of glycosides do not belong chemically to same class, but prominent drugs among them contain the glycosides of monoterpene irridoids with pyran cyclopentane ring.

Physiologically, the bitters have the stimulant effects on gustatory nerves in the mouth and lead to the increase in gastric juice secretion.

CHIRATA

Synonyms

Chirayata, East Indian Balmony, Chretta.

Biological Source

Chirata is the dried entire herb of *Swertia chirata* Buch Ham, belonging to family Gentianaceae, collected in flowering stage and dried. It should contain not less than 1.3 per cent of bitter principle of chirata.

Geographical Source

Chirata is found in India, Nepal and Bhutan. In India, it grows in Kashmir, Meghalaya, Madhya Pradesh, and Khasi hills, at an altitude of 1200 - 1500 m.

Cultivation and Collection

At present, it is collected from wild grown plants. It can also be cultivated in temperate region. The plant flowers from July to October and is collected for medicinal purposes, when the capsules are fully formed. The dried plants are made into sheaves weighing 1 kg.

Fig. 9.48: Chirata Twig

Macroscopic Characters

Colour	:	Leaves, flowers and fruits: peculiar with yellowish tinge
Stems	:	Yellowish-brown to purple
Odour	:	Odourless
Taste	:	Extremely bitter
Size	:	Stems are about 6 mm broad and 1 m in length.
Shape	:	They are cylindrical, glabrous and quadrangular at the apex and with large pith. The leaves are placed opposite to each other, broad at the base, ovate or lanceolate in shape with entire margin and five nerved acuminate apex and glabrous. Flowers are 2 to 3 mm broad and ovoid. Fruits are superior, bicarpellary unilocular with numerous reticulated ovoid seeds. Roots are brown, twisted, tapering with root-lets and about 5 -10 cm in length.

Chemical Constituents

Chirata is found to contain bitter yellow acid known as ophelic acid and two bitter glycosides chiratin and amarogentin. Chiratin is not a pure substance. Amarogentin is a phenol carbonic acid ester of sweroside, a substance related to gentiopicrin. It also contains yellow crystalline substance used in dyeing. The two alkaloids reported in chirata are known as gentianine and gentiocrucine.

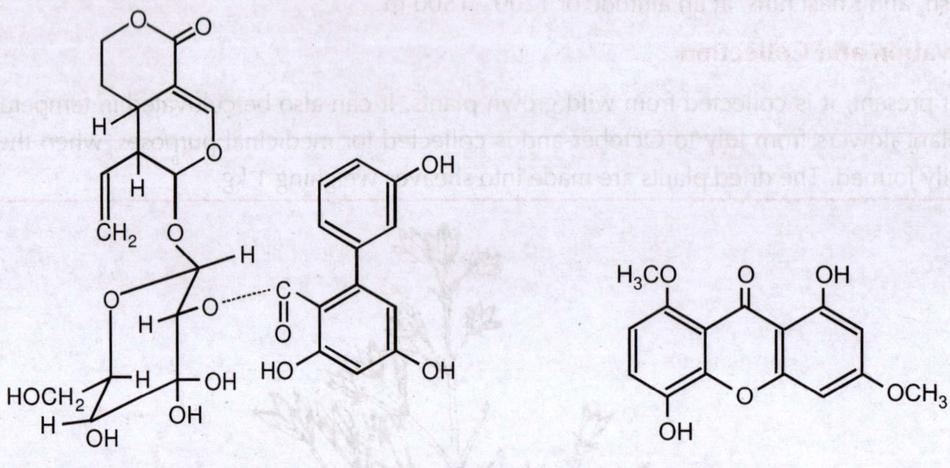

Amarogentin Chiratol

Standards of Quality

Foreign organic matter	:	Not more than 4.0 per cent.
Acid insoluble ash	:	Not more than 1.0 per cent.

Uses

Chirata is used as a bitter tonic. It is free of constipative effect. It is used since very early period as febrifuge. It is also prescribed in dyspepsia. It is a remedy for scanty urine and also used in epilepsy.

It is generally administered either as an infusion or a tincture. Industrially, it is used for dyeing cotton cloth. The present Indian requirement of chirata is said to be approximately 100 tones per year. However, small quantities are also exported from India regularly. It is an ingredient of ayurvedic preparation "Sudarshan Powder", used in chronic fever.

Concentrated compound chirata infusion contains chirata, orange peel, lemon peel and alcohol (25 per cent).

Substitutes

Following are the common substitutes for chirata.

(1) *S. densifolia* from Konkan. Plant is about 30 to 90 cm in height with decussate sessile leaves and white flowers.

(2) *S. ciliata* from Himalayas in Kashmir. It is a small herb with oblong or lanceolate leaves. Flowers are purple or dark red in colour.

(3) *S. paniculata.* It grows in Himalayas from Kashmir to Nepal and also in Mizoram. The herb is 50 to 90 cm in length. The leaves are oblong or lanceolate. Flowers are white in colour, but with two purple patches at the base and long pointed capsules.

GARCINIA

Synonym

Vilayati Imli, Amsul, Kokum rind.

Biological Source

It consists of dried de-seeded fruits of *Garcinia combogia* Desr. belonging to family Guttiferae. It should contain not less than 12.0 per cent of total hydroxy citric acid and hydroxy-citric acid lactone on dried basis.

Geographical Source

In India it is found in evergreen forests of Western Ghats in Nilgiris and also in Srilanka upto an altitude of 2000 metres.

Macroscopic Characters

Colour	:	Fruits are dark brown to blackish brown
Taste	:	Acidic
Odour	:	Characteristic
Shape	:	Ovoid and longitudinally grooved
Size	:	3 to 4 × 1 to 2 cm
Extra features	:	Mesocarp of fruit is very wide

Fig. 9.49: Garcinia cambogia – Flowering branch

Chemical Constituents

Fruits contain approximately 10.0 per cent tartaric acid, hydroxy citric acid, hydroxyl-citric acid-lactone reducing sugars 15 per cent and phosphoric acid in the form of calcium triphosphate. The seeds of fruit contain fat about 30 per cent kokum butter, seeds are also edible. Hydroxy citric acid is a 1, 2 dihydroxy – 1, 2, 3 propane tri carboxylic acid.

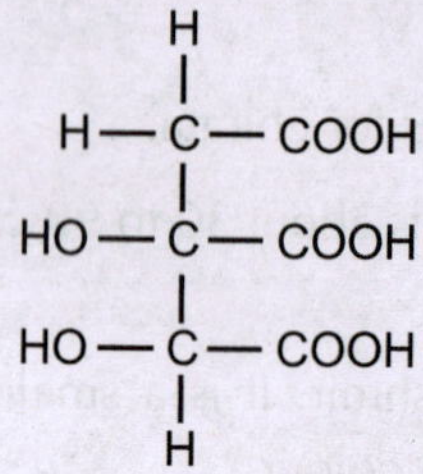

Hydroxy citric acid

Standards of Quality

Citric Acid	:	Not more than 2.0 per cent
Foreign organic matter	:	Not more than 5.0 per cent
Water soluble extractives	:	Not less than 40.0 per cent
Ash	:	Not more than 8.0 per cent
Acid insoluble ash	:	Not more than 1.5 per cent
Loss on drying	:	Not more than 12.0 per cent

It should also comply microbial limits.

Uses

Decoction of the fruits is given in the treatment of rhumatism and digestive disorders. It is also antiseptic. In food, it is used as condiment for flavour and in the preporation of **Solkadhi** (curry).

Dried rind due to acidic nature is used to polish golden and silver commodities. It is also used as a substitute for formic acid in coagulation of rubber-latex.

GENTIAN

Synonyms

Gentian root, Gentiana, Radix Gentianae.

Biological Source

Gentian is the dried partially fermented rhizome and root of yellow gentian i.e. *Gentiana lutea*, family Gentianaceae.

Geographical Source

Gentian is a perennial herbaceous tree native to hilly areas in Southern and Central Europe like Jura, Vosges mountains, and Yugoslavia.

Collection and Preparation

The large fleshy roots and erect rhizomes from about 2 - 5 year old plants are dug up. Collection is done in autumn, followed by cutting longitudinal into slices. At this stage, the pieces of roots and rhizomes are white in colour, without any odour. During slow drying or by keeping them in heaps and fermenting by slow heating, they become dark or yellow coloured and develop characteristic odour.

Macroscopic Characters

The rhizome is yellowish-brown and has transverse annulations and shows conical buds at the top. The root is narrower but continuous with rhizome. It is longitudinally wrinkled and has circular scars of rootlets. The drug has a peculiar odour. The drug first gives a sweet taste, followed by an intensely bitter taste. The fracture is short and smooth in dried drug, but tough and flexible in moist drug.

Colour	:	Roots and rhizonmes both brown
Odour	:	Characteristic and plesant
Taste	:	Sweet at first followed be intensely bitter.
Size	:	Rhizomes: 5 – 30 × 3 – 9 cm, Root : 1 – 2 cm in diameter
Shape	:	Roots and rhizomes both sub-cylendrical

Extra features: Rhizomes have tranrerae annulations and conical bud at the top. Rotts show presence of scars left by rootlets. Fracture is short and even for both.

Microscopic Characters

The transverse section of rhizome shows bark, cambium, wood and pith. The root shows these parts but no pith, in place of which a triarch primary xylem is present. The cork cells are thin walled. Cortex has parenchyma with oil globules and calcium oxalate. Phloem is present in small groups and phloem fibres are absent. The xylem contains spiral and annular vessels and also shows presence of interxylary phloem. The drug does not contain sclereids.

Fig. 9.50 (a): *Gentian luteа twig*

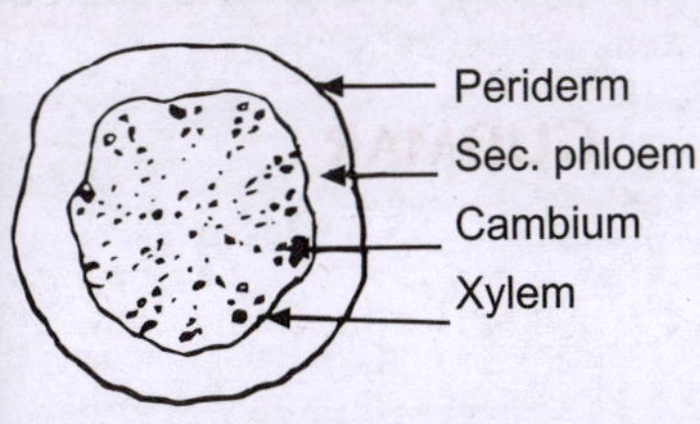

T.S. of Gentian root

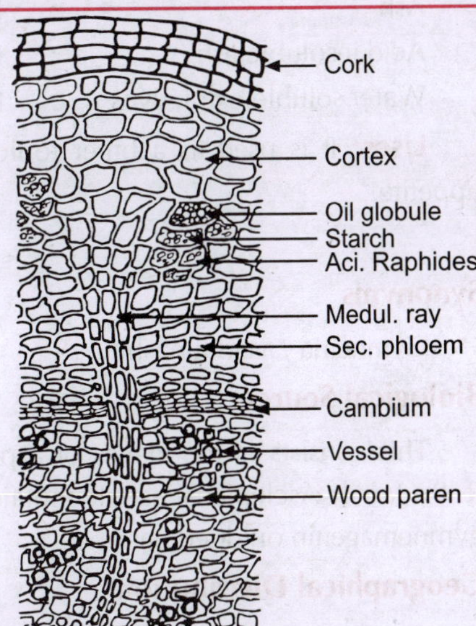

Fig. 9.50 (b): T.S. of Gentian root

Chemical Constituents

The drug contains bitter glycosides mainly gentiopicrin, which is also called gentiopicroside.

It is a water soluble, crystalline compound with a bitter value of 12,000. During fermentation and drying, it breaks down to gentiogenin and glucose.

The drug also contains amarogentin, amaroswerin, gentioside and a mixture of gentiopicrin and gentisin called gentinin.

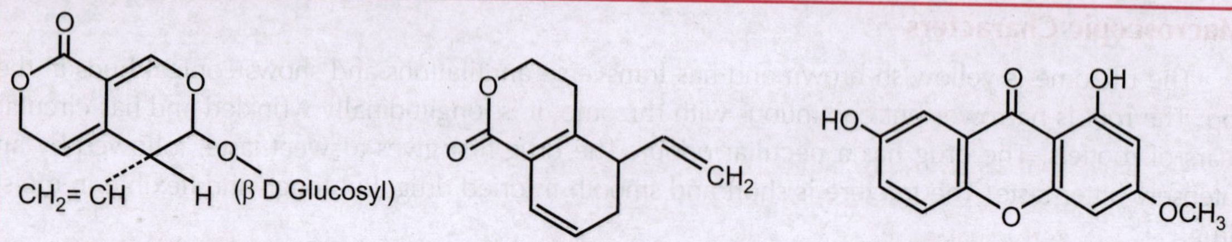

Gentiopicroside **Gentisin** **Gentianin (Gentisin)**

In the natural products, amarogentin is considered as bittermost substance and imparts a bitter taste in even 5.8 lakh times dilution. Although, most of the above mentioned constituents are bitter, the taste of the drug is mainly due to amarogentin. These glycosides contain monoterpene irridoids. The drug contains a flavonoid alkaloid called gentianine or gentisin. Gentisin gives yellow colour to drug.

The other constituents present in drug are gentisic acid, gentianose (a trisaccharide consisting two glucose and one fructose units), gentiobiose (disaccharide) and sucrose. The drug should yield not less than 33 per cent water soluble extractive value, due to sugars. It is decreased due to excessive fermentation.

Chemical Test: Under UV radiation gentian extract shows light-blue fluorescence.

Standards of Quality

Ash	:	Not more than 6.0 per cent
Acid insoluble ash	:	Not more than 3.0 per cent
Water soluble extractives	:	Not less than 33.0 per cent

Uses: It is used as a bitter tonic to stimulate the gastric secretion and hence improving the appetite.

GUDMAR

Synonyms

Gymnema , Madhunashini.

Biological Source

This consists of the leaves of a perennial woody climber plant known as *Gymnema sylvestre* R. Br. family Asclepiadaceae. It contains not less than 1.0 per cent of Gymnemic acids calculated on gymnemagenin on dried basis.

Geographical Distribution

It is a woody climber found in India, common in Deccan-peninsula and also in Northern and Western parts of India. It is occasionally cultivated for medicinal purposes.

Macroscopic Characters

Colour	:	Green
Odour	:	Pleasant and aromatic odour
Taste	:	Tasteless
Size	:	3 - 5 × 1 - 2 cm
Shape	:	Elliptic or ovate with acute or acuminate apex

Fig. 9.51: Gymnema plant

Extra Features

The leaves when chewed, have remarkable property of paralysing the taste glands for few hours against sweet and bitter taste.

Lower surface of leaf is more pubescent and base is rounded or cordate. Non-glandular-trichomes are present on both the surfaces of leaf. It bears small yellow coloured flowers in the form of cymes.

Cultivation

Climatic Conditions

It grows well in vivid agro-climatic conditions in tropical and subtropical regions, under shady situations.

It can grow upto 600 metres from sea level, favourable temperature is ranging from 25° to 30°C at rainfall of 50 - 100 cm per annum. It can grow on variety of soils, preferably loamy soil with rich organic matter and well drained property.

Method of Cultivation

Gymnema can be propagated from seeds and also vegetatively by using stem cuttings and apical portion of plants. For sowing seeds in nursery beds, December - January is appropriate season; and seedlings are transplanted in field during February - March. It is a perennial crop.

Stem cuttings are planted in field, at the onset of monsoon i.e. in months of June - July. The land is ploughed well, weeds are removed, levelled and during last ploughing, soil is mixed with farmyard manure.

The land is levelled well and pits of 45 × 45 × 45 cm are prepared at spacing of 2 × 2 metres. The pits are filled with soil, sand and FYM.

The seeds are collected from mature climbers during October - December. These are used for sowing. Nursery bed of 2 × 1 metre length are raised using mixture of soil, sand and farmyard manure. The seeds are sown followed by light irrigation. After a span of 2 to 3 months of sowing, the raised seedlings are transplanted in well prepared field, during monsoon and irrigated.

The stem cutting, young, as well as, woody cutting of about 15 – 20 cm are used for cultivation in well prepared land, at the onset of monsoon. These cuttings are planted at a spacing of

2 × 2 metres. In early stages, it is provided with shade and water logging in the soil is avoided. Once the plants are established in soil and attain height of about 15 to 20 cm, they are provided with support.

Periodic irrigation during early stages of growth is necessary for proper growth of plant. It is provided with weekly watering. During rainy season, it does not require special irrigation, but during dry periods, it is supplied with water.

During preparation of land for cultivation application of 20 tonnes of FYM/hect. is preferred. The soil is supplied with NPK as basal dose and nitrogen split into two doses for healthy yield. Usually, it is taken as a under crop, hence can grow well on the nutrients of main crop.

Since, it is taken as under crop, periodical weeding and hoeing is required so as to get sufficient nourishment to the crop. Usually, weeding after rainy season is preferred, alongwith the hoeing, as and when required.

Gymnema can also be planted between trees as undercrop. Planting of *Gymnema* under eucalyptus, coconut or mango can also be undertaken. 400 – 500 kg of Gymnema leaves can be obtained per year per hectare.

Collection of Gymnema

After a year of cultivation, leaves are ready for harvesting. Leaves are usually collected during winter season i.e. October to February. During the same period, roots are ready for harvesting, but they are collected or harvested during summer season i.e. during March - May.

The harvested leaves are cleaned of external matter by washing with water and are shade-dried. The roots are also cleaned, washed and cut into equal pieces and are shade-dried.

An average 5-6 kg leaves per plant can be harvested in 3-4 years i.e. about 10000 kg per hector per year.

Chemical Constituents

The leaves contain pentriacontane, hentriacontane, phytin, α and β chlorophylls, resin, tartaric acid, formic acid, butyric acid, mucilage inositol, d-quercitol, gymnemic acids (anti-sweet compounds), the mixture of triterpene saponines and anthraquinone derivatives.

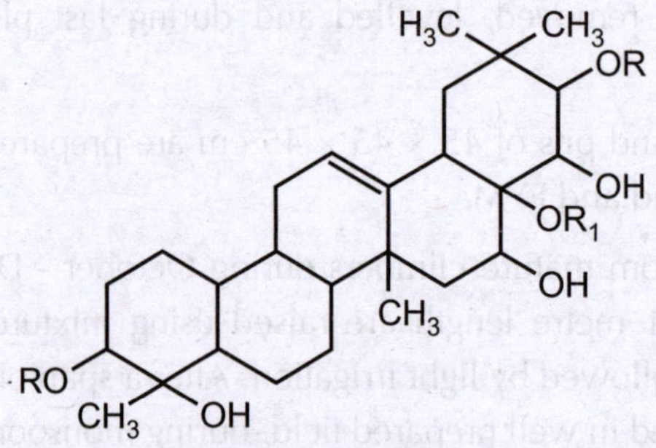

Gymnemic acid

	R	R_1	R_2
Gymnemic acid I	Tigloyl	Acetyl	Gluconic
Gymnemic acid II	Methylbutyryl	Acetyl	Gluconic
Gymnemic acid III	Methylbutyryl	H	Gluconic
Gymnemic acid IV	Tigloyl	H	Gluconic

The plant is reported to contain alkaloids, betain, choline and trimethylamine in the leaves. Lupeol, β-amyrin, stigmesterol are also present in the leaves.

Gymnemic acid is a mixture of at least nine closely related acidic glycosides fractionated by successive extraction with different solvents. Ethyl acetate fraction of such gymnemic acid mixture has a paralysing effect on taste glands. Gypenosides have also been isolated from leaves.

Chemical Tests

1. Dilute solution anaesthetises sweet taste buds.
2. Gives copious foam after shaking with water and on addition of dil. hydrochloric acid forms voluminous precipitate.

Standards of Quality

Foreign organic matter	:	Not mroe than 2.0 pe cent
Acid insoluble ash	:	Not mroe than 6.0 pe cent
Ash	:	Not more than 15.0 per cent (w/w)
Organic acids	:	Not less than 5.5 per cent (w/w)
Alcohol soluble extractives	:	Not less than 5.0 per cent (w/w)
Water soluble extractives	:	Not less than 20.0 per cent
Loss on drying	:	Not less than 14.0 per cent

Uses

It is used as antidiabetic, stomachic, stimulant, laxative and diuretic. The antidiabetic formulations of this drug are commercially available. Hypoglycemic effect is due to indirect stimulation of insulin secretion by pancreas. Dental plaque and caries are prevented by gymnemic acids. Recently the drug has received world attention due to its weight-lowering properties.

Indian exports of Gymnema extracts during 1996-1997 was reported to be Rs. 65 lakhs.

The leaves of *G. hirsutum* and *G. montanum* of Western ghats are the substitue for *Gymnema.*

G. tingens and *G. acuminatum* are other species of *Gymnema* and used also for other purposes.

GUDUCHI

Synonyms

Gulvel, Tinosopora, Giloe, Amrita.

Biological Source

These are the dried leaves and stem pieces of woody climber *Tinospora cordifolia*, Miers belonging to family Menispermaceae. It contains not less than 0.02 per cent of cordifolioside.

Geographical Source

It is found in deciduous and dry forests of India. It is spread throughout india from Kuman to Assam, Bihar, Kokan and also in Sri Lanka and Indonesia.

Cultivation of Guduchi

It usually grows in tropical dry areas and does not tolerate moist and humid climate. It can be grown under varying climatic conditions. It is habitated through out tropical India ascending to an altitude of 500 metres in the temperature range of 25 to 40°C. It thrives well in almost all types of soils. Sandy loam soil, rich in organic matter with good drainage is found to be good for higher yield of crop. Propagated by using stout stem cuttings. It is cultivated at onset of monsoon during May - June.

It can be easily intercropped with neem, mango and coconut.

As it is a perennial crop, periodical weeding is carried out so as to keep the crop weed free. One manual weeding should be done to keep weeds under check in early stage of growth i.e. after 20 – 30 days and during the remaining period of crop growth, 3 - 4 weedings may be sufficient to keep weeds under check.

Collection of Guduchi

The plants are harvested for stem and leaves. The leaves are collected when the plants are matured enough, and the stems (stout) are collected by cutting.

The collected stems are cut into small pieces. They are made free of adhered impurities, if any and then leaves and the stem-cuttings are shade-dried and are placed in dry places.

On an average, about 9 to 10 quintals of stem are obtained per hectare of land.

Macroscopic Characters

Colour	:	Grayish black
Odour	:	None
Taste	:	Bitter
Size	:	3 to 5 cm in length and 3 to 8 mm in diameter
Shape	:	Cylendrical
Fracture	:	Fibrous

(See coloured photograph at the end of the book.)

Chemical Constituents

It consists of tinosporine, tinosporic acid, tinosporol, giloin, gilonin, berberine, syringin, cordifolioside A ($C_{22}H_{32}O_{13}$) tinosporidine, tinosporoside, tinosporaside, etc. The bitter principles identified are chasmanthin, palmarin, and columbin. The stems are rich in proteins, starch, calcium and phosphorus.

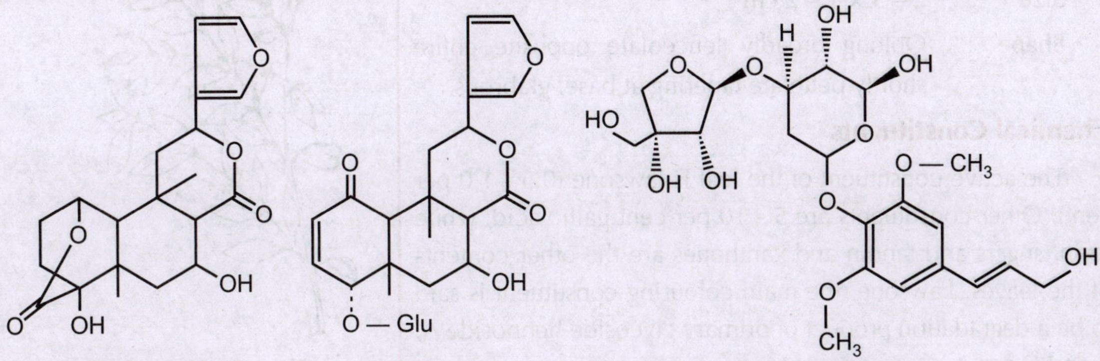

Tinosporoside Tinosporaside Cordifolioside A

Standards of Quality

Total ash	:	Not more than 7.0 per cent
Acid insoluble ash	:	Not more than 0.8 per cent
Water soluble extractives	:	Not less than 13.0 per cent
Alcohol soluble extractives	:	Not less than 6.0 per cent

Uses

It is used in general debility, pyrexia and skin diseases. Gulvel is effective in prevention of fibrosis. Also, used in rheumatoid arthritis, jaundice and in diabetes. It is effectively used in viral hepatitis.

Traditional Uses

The drug has been used in traditional medicine since ancient times for its ability to impart youthfulness, vitality and longevity. For these qualities, it is also referred to as *'Amrita'*. Now a days, the drug is gaining popularity due to its potent immunostimulant activity. It is also referred as *"Rasayan dravya"* in Ayurveda for these properties.

HENNA

Synonyms

Egyptian privet, *Lawsonia – alba*, Mehandi.

Biological Source

Henna consists of fresh or dried leaves of the plant *Lawsonia - inermis* Lam. family Lythraceae.

Geographical Source

Henna is indigenous to Africa and is largely cultivated in Egypt, Sudan, Caribbean islands, Florida, India and China.

Macroscopic Characters

Colour : Leaves are greenish brown

Odour : Characteristic

Taste : Bitter and astrignet

Size : 2 – 3 × 1 – 2 cm

Shape : Oblong broadly lenceolate opposite entire shortly petiolate tapering at base, glabrous.

Fig. 9.52: Henna herb

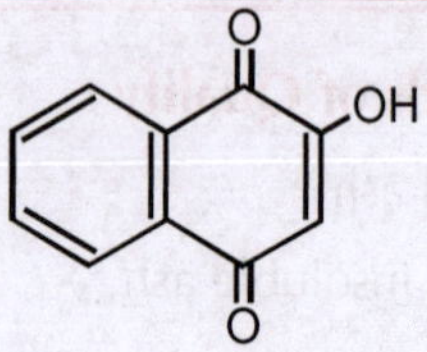

I - Lawsone

Chemical Constituents

The active constituent of the leaf is lawsone (0.5 - 1.0 per cent). Other constituents are 5 - 10 per cent gallic acid, white resin, sugars and tannin and xanthones are the other contents of the leaves. Lawsone, the main colouring constituent is said to be a degradation product of primary glycoside hennoside A, B and C.

Chemical Test

Extract henna leaves with water by boiling and filter, and cool. This decoction fades on addition of the a cid while deepens by addition of alkali.

Standards of Quality

Moisture content	: Not more than 9.0 per cent
Acid insoluble ash	: Not more than 4.0 per cent
Ash content	: Not more than 15.0 per cent
Tannin content	: Not less than 10 per cent
Water soluble extractives	: 25 – 33 per cent

Uses

Henna is used as a favourite hair dye either alone or in combination for the treatment of gray hair. It is used in several hair care products like rinses, conditioners, applications.

Henna imparts orange-red colour, which is more stable in the acidic media (pH 5.0).

Lawsone, the active constituent of the leaves is shown to have antibacterial and antifugal properties.

KALMEGH

Synonyms

Andrographis, Kirayat, Bhui-nimb.

Biological Source

This consists of dried leaves and tender shoots of the plant known as *Andrographis paniculata* Nees fam. Acanthaceae. It yields not less than 1.0 per cent of andrographolide calculated on dry basis.

Geographical Source

Kalmegh is an annual herb found in Shri Lanka and throughout India, specifically in Maharashtra, Karnataka, Uttar Pradesh, Tamil Nadu, Andhra Pradesh and Madhya Pradesh. It is cultivated to some extent in Assam and West Bengal. It flowers in the months of September to December and is collected from November to December. Yield per hectare is 2.5 tones.

Cultivation of Kalmegh

Climatic Conditions

It grows well in tropical and subtropical regions. However, it is seen as a shade loving plant; as it grows abundantly in moist climate with well distributed rainfall.

Altitude: It is grown upto 1500 metres.

Temperature: Ranging between 25° - 40°C is suitable.

Rainfall: Well distributed, in the range of 40-50 cm.

Soil: It can be grown in various types of soils, though sandy loamy soil with rich organic matter is suitable for its growth. It is also found growing on sandy clay soil. But soils with water logging condition are not suitable for its growth.

Cultivation

It can be propagated either by seeds or vegetatively by using cuttings. But, seed propagation is economical and easy for commercial purpose.

For raising nursery, period before onset of monsoon i.e. May - June is suitable and for transplantation in the main field, period after onset of monsoon i.e. the months of July - August is favourable. Total duration of crop is 100 - 120 days.

For nursery raising, either planting in polythene bags filled with soil, sand and organic manure in proportion of 1 : 1 : 1, or in beds of 3 × 1.5 × 0.15 metre size is carried out. The beds are mixed well with cattle manure at the rate of 25 kg/m^2. About 5 cm apart lines are marked on the bed and the seeds are sown. The beds are watered daily. After sowing, the seeds germinate within 8-10 days and are ready for transplanting after one month of sowing.

The land is prepared by ploughing twice or thrice and levelled. Then, soil is mixed well with farmyard manure at rate of 25 to 30 t/hectare at the time of last ploughing. Irrigation channels are prepared in the prepared soil.

At the onset of mansoon, the seedlings are transplanted at a spacing of 15 × 15 cm. If seeds are sown directly, then distance of 30 × 15 cm is kept for healthy yield. On an average, 450 gm/hect. of seeds are required for sowing.

It is mainly grown as rainfed crop. In areas with well distributed rainfall, no special technique for irrigation is required. But in case of dry weather, it is to be irrigated regularly in initial stages with an interval of 3 - 4 days, lateron after one week and once the crop stands in field, depending upon the requirement of crop.

To increase the yield and healthy crop production, kalmegh is fertilized even after application of 25 to 30 tonne/hectare of FYM during cultivation. Fertilizer dose of NPK is reported to be beneficial for good yield. 75 kg of nitrogen, 75 kg phosphorus and 50 kg of potash for one hectare are required. Out of this, 50% N and the entire dose of phosphorus and potash are applied as basal dose and the rest 50% nitrogen is used as a top dressing after 30 days of transplanting / sowing.

During its initial stage of growth, the land should be kept weed free to promote growth of the crop. For this, first weeding is done after one month of planting. After this, one or two more weedings are required.

Plant - Protection and Pesticides

No disease or pests of serious nature have been reported to attack kalmegh, as it is a hardy shrub.

Collection of Kalmegh

Total 2 - 3 harvestings can be done in a period of one year. The crop is ready for the first harvest after 4 - 5 months of sowing/planting. The appropriate period of collection is when it starts flowering. For harvesting, the plants are cut at its base leaving 15 to 20 cm stem for regeneration of the crop. After first harvesting, the crop is again supplied with top dressing of nitrogen and irrigated. The procedure is repeated after every harvest. The crop is then ready after 60 - 65 days of first harvest. In this way 2 - 3 harvesting can be made within 100-120 days.

The plant or the harvested material is dried in shade for 3 - 4 days and stored.

Macroscopic Characters (Fig. 9.53)

Colour	:	Leaves are dark green, while flowers are rose coloured.
Odour	:	Odourless
Taste	:	Intensely bitter
Size	:	Leaves 7 × 2.5 cm, flowers 1.8 cm in length.
Shape	:	Leaves are lanceolate and petiolate and with entire margin and acuminate apex.

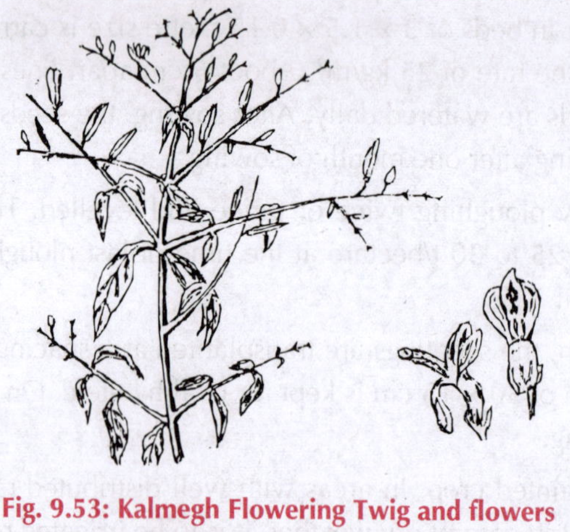

Fig. 9.53: Kalmegh Flowering Twig and flowers

Extra Features:

The venation of leaf is unicostate reticulate and mid-rib is ventrally grooved. The stem bears numerous branches and many of them bear flowers. The calyx is about 3 mm long, linear, lanceolate, glandular and pubescent; corrolla, 1 cm in length, rose coloured and hairy. Flowers are arranged in axillary and terminal racemes, which are 7.5 - 10 cm in length. The plant flowers in the months from September to December.

Chemical Constituents

Kalmegh contains bitter principle andrographolide 1.0 per cent, a bicyclic diterpenoid lactone 1.0 – 2.0 per cent and kalmeghin (0.85 - 2.5 per cent).

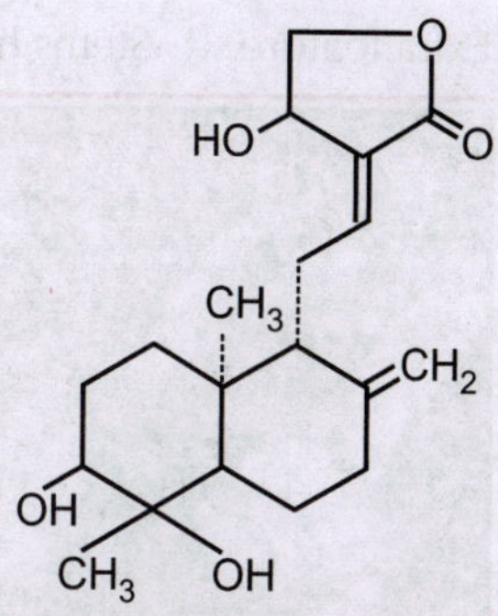

Andrographolide

Standards of Quality

Ash	:	Not more than 20.0 per cent
Acid-insoluble ash	:	Not more than 5.1 per cent
Foreign organic matter	:	Not more than 2.0 per cent
Alcohol-soluble extractives	:	Not less than 24 per cent
Water-soluble extractives	:	Not less than 20 per cent
Loss on drying	:	Not more than 12.0 per cent

Uses

Kalmegh is used as a bitter tonic, an alterative and anthelmentic. Green leaves are used as stomachic, anthelmintic and in the treatment of dysentery and dyspepsia. It possesses antityphoid and antibiotic activity. It has been proved to be hepatoprotective drug, mainly due to andrographolide.

Substitutes

Kalmegh is substituted with *Andrographis echioides,* found in tropical India and in dry districts of Maharashtra, Rajasthan and Tamil Nadu.

MANJISHTHA

Synonyms

Indian maddar, Rubia cordifolia, Rakta pushpi.

Biological Source

It consists of dried stems of a climber known as Rakta pushpi *Rubia cordifolia* Linn. sensu. Hook F. belonging to family Rubiaceae. It should contain not less than 0.02 per cent of Rubiadin on dried basis.

Geographical Source

It is found in Afganistan, Nepal and Iran upto an altitude of 2700 metres. It is also found abundant in Konkan.

Macroscopic Characters

Colour	:	Brown to purple coloured
Odour	:	None
Taste	:	Slightly bitter

Size : 4 to 5 cm × 0.5 cm

Shape : Stems are cylindrical, slender and wiry

Extra feature : Stems have longitudinal cracks.

Fig. 9.54: (a) Manjishtha herb (b) Manjishtha stems

Extra Features

Leaves are Heartshaped, 5-10 cm, pubescent, covered with trichomes, petiole double in length as compares to size of leaves, leaves in whorls, tow small and two big, flowers-bluish 0.5 to 2.5 cm in length, available in October-November. Roots reddish brown elongated about 4 – 8 cm in length.

Chemical Constituents

It contains glycosides manjisthin, purpurin, resin and red dye rubiadin (1: 3 dihydroxy – 2 methyl anthraquinone).

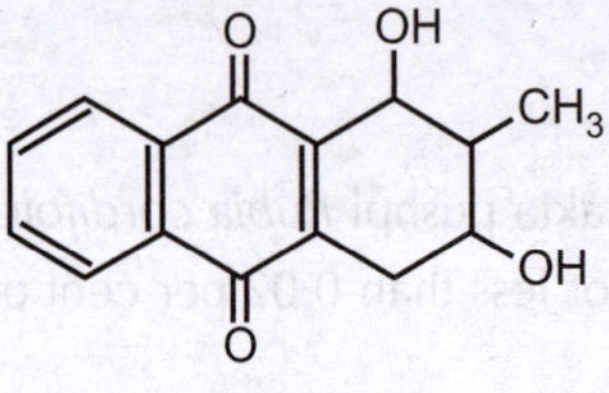

Rubiadin

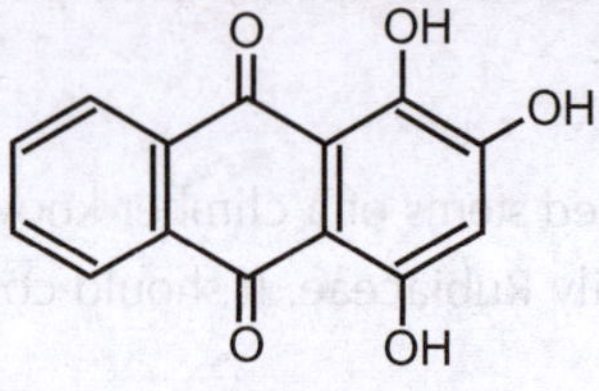

Purpurin

Standards of Quality

Water soluble extractives	:	Not less than 20.0 per cent
Alcohol soluble extractives	:	Not less than 04.0 per cent
Ash	:	Not more than 10.0 per cent
Acid insoluble ash	:	Not more than 0.5 per cent
Loss on drying	:	Not more than 5.0 per cent
Foreign organic mattter	:	Not more than 2.0 per cent

Uses

It is used in the treatment of leucoderma, gauty arthritis and skin pigmentation. It helps to gain lustre and glow of skin and aids to remove pimples, freckles and discolouration. In Ayurbedic medicine it is used as blood-purifier. It is used in textile industry for dyeing of fabrics.

PICRORRHIZA

Synonyms

Indian gentian, kutki.

Biological Source

Picrorrhiza consists of dried rhizomes of the plant *Picrorrhiza kurroa* (Scrophulariaceae), cut in small pieces and freed from attached root-lets. It contains not less than 5.0 per cent of Kutkin on dried basis.

Geographical Source

Picrorrhiza is a perennial herb well distributed in the upper Himalayas and also in China. It is found growing naturally in Kashmir and Sikkim. It is also reported in Punjab, Uttar Pradesh and Himachal Pradesh. The drug is collected from the naturally growing plants and may be cultivated at higher altitudes (2700 - 4500 m) in Himalayas. It is propagated by seeds and also from rhizomes. [Fig. 9.55 (a)].

Cultivation and Collection

These can grow well at higher altitudes in temperate or tropical climate. Partial shady areas are good for the growth of crop.

It can be cultivated at an altitude of 1800 to 2700 metres and also at higher altitude in Himalayas (2700 - 4500 metres).

Temperature ranging from 25° to 35°C is suitable.

These can be grown on variety of soil, though sandy loam soil with rich organic content and high moisture content are suitable for its cultivation. But, soil should not be water-logging.

Cultivation

It can be propagated from seeds and rhizomes.

Cultivation phase differs according to the range of altitude. At lower altitude, sowing is carried out during November - December, while at higher altitudes it is during April - May. Total duration of crop is two years.

For cultivation of Kutki, the land should be ploughed and harrowed, so that the soil is well pulverised. The field is provided with farmyard manure and it is levelled.

For raising nursery bed, mixture of soil, farmyard manure and sand is used. Healthy seeds collected are sown during November - December at lower altitude and during May at higher altitude. To increase seeds germination, the seed beds are covered with thin layer of moss so that it retains moisture content and water loss is avoided. After period of six months, when the seedlings are raised properly, they are transplanted on the well prepared field. At lower altitude, the seedlings are transplanted during winters and at higher altitude the plantation is carried out during spring.

The seedlings are planted at a distance of 30 × 30 cm followed by light irrigation immediately after transplanting. About 40,000 to 45,000 seedlings are cultivated in one acre of land.

Vegetative Propagation

For vegetative propagation, the cuttings of rhizomes are used. These cuttings are water dipped. At a suitable distance, high moisture trenches are prepared and the rhizome cuttings are planted. If kept covered with moss for two weeks, rhizome cutting also shows enhanced rooting.

It needs regular watering so as to decrease the mortality rate. At lower altitude, to keep land moist, irrigation is carried out every 24 hours during the initial stages of development. In rainy season, watering is done as per the requirement. During winter, watering should be done at every two days interval.

During land preparation or before transplanting, the field is supplied with sufficient amount of manure. In general, at lower altitude, maximum required manure varies from 55 to 70 q/acre to achieve healthy crop for three years. In areas rich with organic matter, the average requirement of manure for three years is 20 to 25 q/hect. No chemical fertilizers, other than manures, are required.

The land should be kept weed-free, so as to get maximum nourishment to the crop. Hence, during first year of cultivation, weeding is done at weekly intervals and at an interval of a month or two in the next years of crop till harvesting.

Inter-cropping with saunf *(Foeniculum vulgare)* or potato are most suitable. They also provide better microclimate for the growth of crop (*P. kurroa*) by retaining moisture in the soil for longer period. Inter-cropping with *Digitalis purpurea* has also been suggested, which also gives better yield.

No disease or pests of serious nature are found to attack the crop.

Harvesting:

To get maximum amount of bioactive ingredients, harvesting should be carried out at an appropriate period. After completion of reproductive phase, the crop matures and at this stage contains good amount of active ingredients, this is the period for harvesting. The crop is harvested by uprooting the plants manually, which is carried out during month of September at lower altitude and in the month of October at higher altitudes.

Collection of Picrorrhiza:

The collected roots and rhizomes are washed with plane water and are kept for shade drying. They are cut into small pieces and stored well.

Macroscopic Characters [Fig. 9.55 (b))]

Colour	:	The rhizomes are deep greyish-brown in colour, externally white, blackish internally with whitish-wood.
Odour	:	Slight and unpleasant.
Taste	:	Bitter.

Size	:	3 to 5 cm in length and 0.5 to 1 cm in diameter.
Shape	:	Cylindrical pieces with longitudinal wrinkles and annulations at the tip.
Extra Features	:	Conical, buds and stems along with the roots also constitute the drug. The roots are longitudinally wrinkled with transverse cracks. Roots are grey to brown in colour. Fracture is tough.

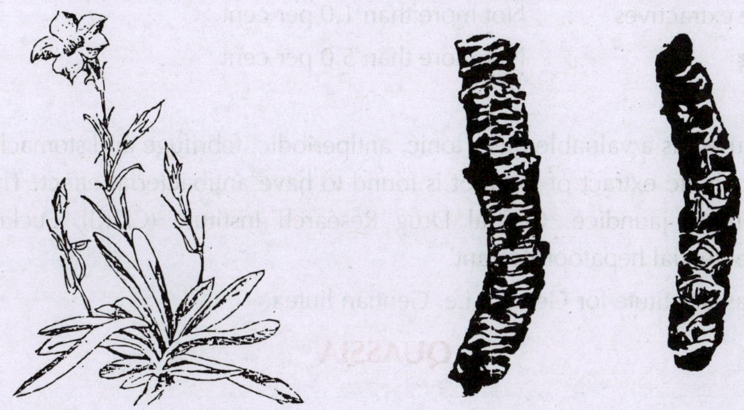

(a) (b)

Fig. 9.55: (a) Picrorrhiza Twig (b) Rhizomes of Picrorrhiza

Microscopic Characters

The periderm of the rhizomes constitutes 8 to 10 layers of thin walled cork. The phelloderm cells are full of brownish contents. The cortical cells are rounded with occasional inter-cellular spaces. The secondary phloem is many layered and xylem is made up of vessels, fibres and tracheids. Pith consists of thick parenchymatous cells.

Chemical Constituents

	R	R_1	R_2
Picroside I	H	H	trans-cinnamoyl
Picroside II	Vannilloyl	H	H
Kutkoside	H	Vannilloyl	H

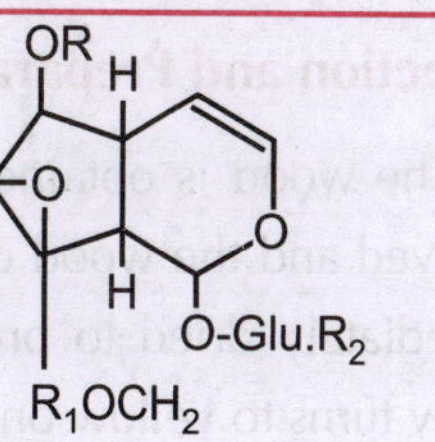

The drug is found to contain irridoid bitter substances, kutkin, picroside I, picroside II and kutkoside. Picrosides and kutkoside are C_9 monoterpene glycosides with an epoxy oxide in ring.

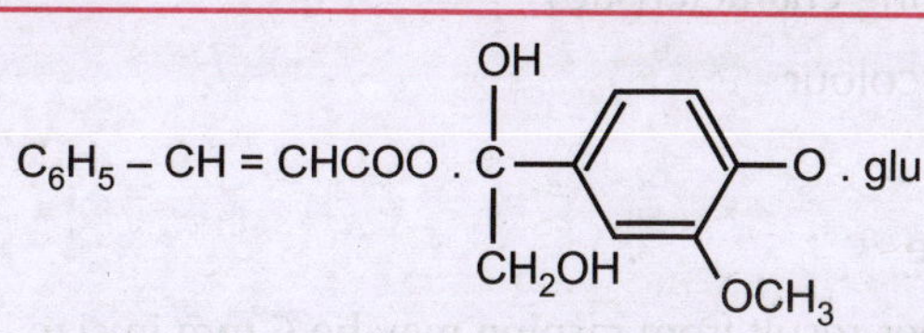

Kutkin

Standards of Quality

FOM	:	Not more than 2.0 per cent
Alcohol soluble extractives	:	Not less than 15.0 per cent
Wate soluble extractives	:	Not less than 25.0 per cent
Total Ash	:	Not more than 6.0 per cent
Acid insoluble extractives	:	Not more than 1.0 per cent
Loss on drying	:	Not more than 5.0 per cent

Uses

Picrorrhiza is used as a valuable bitter tonic, antiperiodic, febrifuge and stomachic. It is laxative in large doses. Alcoholic extract of the root is found to have antibacterial effect. The drug is found useful in treatment of jaundice. Central Drug Research Institute (CDRI), Lucknow has found kutkoside to be a potential hepatoprotectant.

It is regarded as substitute for Gentian i.e. Gentian liutea.

QUASSIA

Synonyms

Bitter wood, Quassiae Lignum, Jamaica Quassia, Picrasma.

Biological Source

It consists of the dried stem wood of *Picrasma excelsa* (S.W.) Planchon (*Picroena excelsa or Aeschrion excelsa*), belonging to family Simarubaceae. It contains about 0.2 per cent of the bitter lactone i.e. quassin and its hemiacetal neoquassin.

Geographical Source

The tree of quassia which grows to a height upto 30 metres, is native of Caribbean islands mainly Jamaica, Barbados and Martinique.

Collection and Preparation for Market

The wood is obtained from main trunk and other branches. After cutting the tree, the bark is removed and the wood of trunk and large branches are cut into small chips, raspings and logs. It is immediately dried to prevent the growth of mould. Naturally, the wood is white in colour, but slowly turns to yellow on drying.

Macroscopic Characters [Fig. 9.56 (a)]

Quassia wood has following characteristics

Colour	:	Brite yellow colour
Odour	:	Odourless
Taste	:	Intensity bitte
Size	:	Chips or cubes result from rasping may be 6 mm in size
Shape	:	Chips, raspings or logs

Extra-features: It is light in weight pieces of bark greyish bark in colour may be present. It is diffused porus wood and splits longitudianlly.

Microscopic Characters

The section shows secondary xylem consisting of xylem vessels, xylem parenchyma, fibres and medullary rays. In longitudinal section, the wood parenchyma appears as irregular concentric rings and in transverse section shows presence of thickened pitted walls. Quassia wood has large xylem vessels, present in groups with bordered pits. The wood fibres are present in abundant quantity. They are thick-walled with pointed ends and oblique pits. The medullary rays are bi-or multiseriate, the cells beings radially elongated with pitted cell walls [Fig. 9.56 (b)].

Fig. 9.56 (a): Quassia Herb

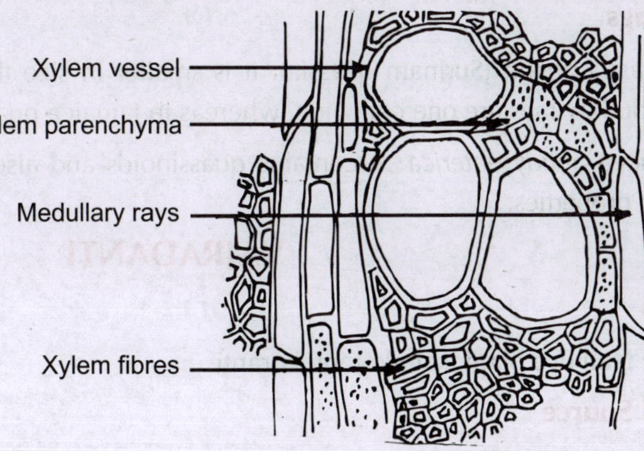

Fig. 9.56 (b): T.S. of Quassia Wood

Chemical Constituents

The active constituents of quassia wood are amaroid compounds like quassin, neoquassin, isoquassin (picrasmin) and 18 hydroxy quassin.

The drug does not contain tannins. Because of a yellow substance present, it shows blue fluorescence in acidified alcohol.

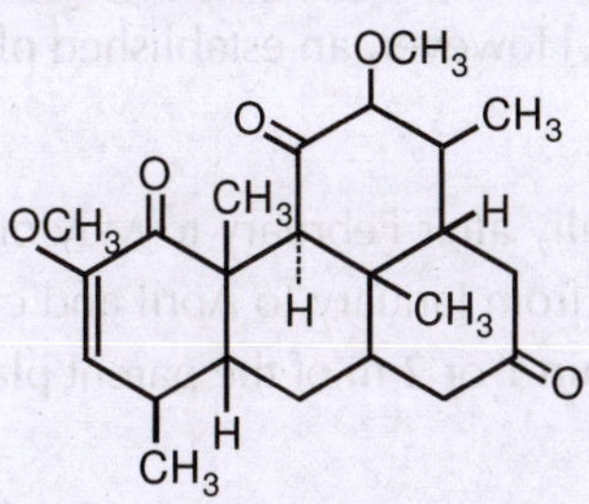

Quassin

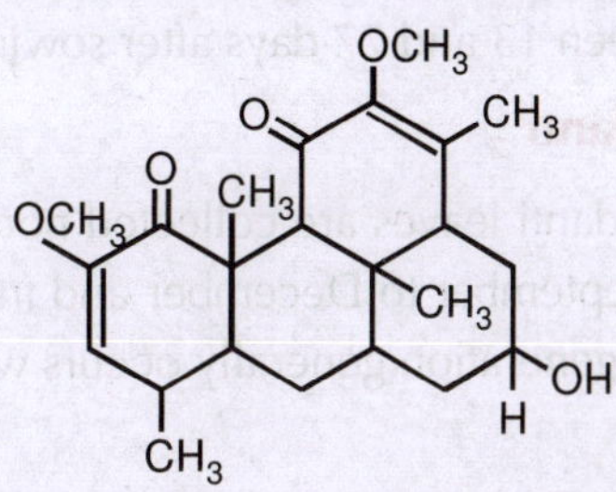

Neoquassin

Standards of Quality

Total Ash	:	Not more than 5.0 per cent
Acid insoluble ash	:	Not more than 1.0 per cent
Water soluble extractives	:	Not less than 4.0 per cent
FOM	:	Not more than 4.0 per cent

Uses

Quassia is used as bitter tonic. Because of absence of tannins, it has no astringent activity and also can be used with iron salts. It is given as an enema in the form of infusion to expel threadworms. The infusion is also used as a lotion for pediculosis. Quassia possesses insecticidal properties. The quassia preparations are used for denaturating alcohol.

Allied Drugs

(1) *Quassia amara* (Surinam quassia): It is smaller in size than Jamaica quassia. About 60 per cent of medullary rays are one cell wide, whereas in Jamaica quassia they are 2 - 3 cells wide.

(2) *Brucea antidysenterica:* It contains quassinoids and also bruceantin. The latter has shown anti-cancer properties.

VAJRADANTI

Synonyms

Yellow Nail dye plant, kate-korant, koranti.

Biological Source

It consists of dried whole plant of *Barleria prionitis* Linn family : Acanthaceae

Geographical Source

It is found throughout the hotter parts of India.

Cultivation of Vajrdanti

It is commonly grown as an ornamental plant in gardens.

The shrubs prefer a sunny situation on wide variety of well-drained moist soil. They tolerate temperature above at least 1°C. It best grows with average annual rainfall between 750 - 900 mm. The seeds are very light in weight about 33,000 seeds/kg. Seedlings grow slowly at first i.e. germinated between 13 and 77 days after sowing. However, an established plant grows fast.

Collection of Vajrdanti

Fully green vajradanti leaves are collected normally after February to May of every year. These shrubs flower from September to December and fruit from January to April and can be harvested for four years. Natural regeneration generally occurs within 1 or 2 m of the parent plant.

Description

It is evergreen, tender, erect, branching, evergreen shrub. Bushy shrub grows up to 1 - 2 m with spiny stem nodes, elliptic to oblong, mid-green leaves, 6 to 12 centimeters long, narrowed, and pointed at both ends. Flowers are yellow and axillary, with the upper ones in spikes. Bracts and calyx are green, with the outer bract usually foliaceous. Corolla is about 4 centimeters long, it flowers in summer.

The stems are terete, glabrous, much branched with cylindrical and tapering branchlet.

Leaves are smooth, opposite, ovate-elliptic to obovate, acuminate, tapering to base, entire, margin finely ciliate, bristle-tipped and about 6-15 cm long and 4-6 cm wide. The petioles are about 0.5-3 cm long, widely spreading spines present in axils.

The flowers are sessile, yellow in colour and often solitary in lower axils and spictate in the upper axils. Bracts are acute, linear-lanceolate, foliaceous, about 1-1.5 cm long and 0.2-0.8 cm wide with bristle tipped. The corolla is bright, golden yellow in colour with pubescent outside and glabrous inside and about 1.5 cm long. The filaments are hairy and about 2-2.5 cm long, yellowish in colour with 3mm long yellow anthers. The ovary is ovoid and sigma is long, linear, sticky and pinkish in colour.

The fruit is ovoid capsule containing 2 seeds. It is about 1.5-2 cm long and 0.6 - 0.8 cm wide. The seeds are oval-oblong, covered with silky copper-brown appressed hairs and measuring about 7.4-8.5 × 6-6.8 mm.

Roots are central tap type, with lateral roots branching in all directions.

Macroscopic Characters

Colour	:	Whole Vadjradanti plant is pale green to greenish
Odour	:	Characteristic slight
Taste	:	Bitter
Extra features	:	Crumbled pieces or coarse powder, brittle in nature.

Fig. 9.57 : Flowering Vajradanti herb

Chemical Constituents

Brassica prionitis shows presence of saponins, glycosides, flavonoids, steroids and tannins

- Glycosides are prioniside A, prioniside B, barlerinoside, verbascoside, shanzhiside methyl ester

- Saponins - lupulinoside, luteolin-7-O-β-D-glucoside
- Flavonoids - apigenin 7-O-glucoside, 6-hydroxyflavones
- Phenolic acid - *Melilotic acid*, syringic acid, vanillic acid, p-hydroxybenzoic acid

It also shows the presence of triterpenoids and steroids such as lupeol and β-sitosterol. Bark consists of 16% tannins and ellagic acid.

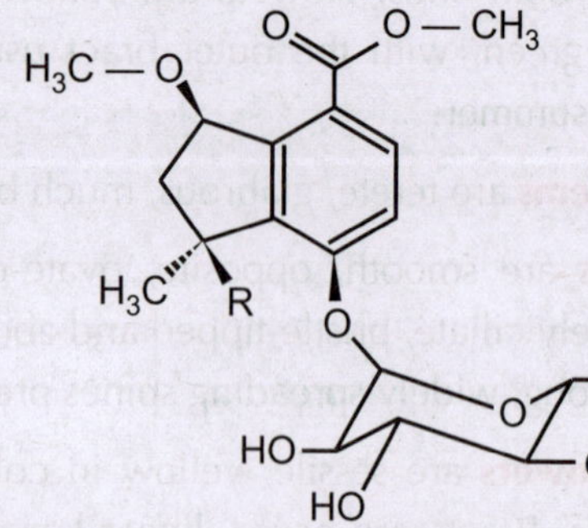

Prioniside

Standards of Quality

Foreign organic matter	:	Not more than 2 per cent
Total ash	:	Not more than 7 per cent
Acid-insoluble ash	:	Not more than 1 per cent
Alcohol-soluble extractive	:	Not less than 4 per cent
Water-soluble extractive	:	Not less than 10 per cent

Therapeutic Uses

Charaka recommended leaf paste as an ingredient of hot poultices and steam bath in the treatment of stiffness of limbs, enlargement of scrotum, sciatica and erysipelas. Leaf-powder is used in ayurvedic tooth-powder as abrassive (vicco-vajradanti).

Sushruta prescribed flowers internally in migraine, internal abscesses, oedema, haemoptysis, urethral discharges, seminal disorders and also to reduce obesity. The herb's extract in oil was prescribed for massage in arthritis and gout.

Traditionally, the plant parts have been recommended in different health aliments, such as astringent and recommended to strengthen the gums, also prescribed in diarrhoea.

Leaves are traditionally used as under

- Fresh leaf juice is used to control diarrhoea and urinary infection.
- The crushed leaves are applied on wounds and boils.
- The leaves are applied locally for joint pain and swelling.
- Fresh leaf juice applied on the scalp in fungal infection, furthermore it promotes hair growth.
- Leaves and flowering tops are recommended as diuretics and in urinary infection.
- Fruits are bitter, acrid, anthelmintic, haemolytic; also prescribed in gingivitis as expectorant.
- Seed powder is used as emetic and expectorant.
- Bark as astringent, useful in diarrhoea, whooping cough and expectorant.
- Roots are utilized in bleeding gums, loose teeth, dental cavities and sensitive teeth.
- Decoction of *Barleria prionitis* roots is prescribed for treatment of viral fevers.

- The Ayurvedic pharmacopoeia of India recommended it in goiter and psychological disorder.
- The whole plant is antiseptic, diuretic and applied over boils and glandular swelling.

Pharmacological Uses

It has antimicrobial activity over a wide pathological microorganism. Iridoid glycoside shown to have potent antiviral property. Whole plant extracts possess anthelmintic activity in-vitro studies. The Iridoid glycoside present in roots shown reduced spermatogenesis in male albino rats. The aerial parts of plant shown to possess strong antioxidant hepatoprotective, antidiabetic and cytoprotective ability in experimental animals. The plant also exhibited strong anti-inflammatory, anti-arthritic, analgesic and diuretic property.

KAVA-KAVA

Synonyms

Kava-kava, Fawa, Methysticum

Biological Source

These are the dried peeled rhizomes and roots of the plant *Piper methysticum* family: Piperaceae.

Geographical Source

Micronesia, Melanesia, Polynesia and Hawaii Islands. Polynesia is a group of more than 1000 islands scattered over central and Southern Pacific Ocean. Largest country in polynesia is New-Zealand.

Cultivations and Collections

Kawa is a perennial plant kawa shrubs thrive is loose well drained soils where the air reaches roots. It grows where plentiful rainfall is there about 2000 mm. Temperature of 70 – 95 F° and 70 – 100% relative huminity is available, too much sunlight is harmful to crop.

It can not reproduce sexually, female flowers are too less and do not produce fruits, cultivation is by stem cuttings only. Plants are harvested around four years of age. Roots reach to the depth of 60 cm.

Macroscopic Characters

Colour	:	Light yellow brown
Size	:	03-20 × 01-10 cn
Odour	:	Slight
Shape	:	Irregular piece, Transversely and longitudinally cut
Taste	:	Sweet then pungent followed by slight numbness.

Extra Features:

The piece of rhizomes are longitudinally wrinkled, with large circular white root scars. Fracture is short and granular. Inner surface is yellowish white, bark is thin, xylem is distinctly radiate and pith is large.

Fig. 9.58: Kava-kava herb

Microscopic Characters

Transverse section shows large oval cells of pith, numerous starch grains, which are spheroidal and upto 45 μ in size, with central hilum and cleft. Secretory canals with yellow to brown resin contents, medullary rays are with elongated cells, porons and lignified, vessels are lignified and reticulate, fibres are slightly lignified. Xylem parenchyma cells are elongated.

Calcium oxalate crystals and cork is absent.

Chemical Constituents

Kava contains pyrones, lactones, (Methysticin, dihydromythesticin) Yangonin, dihydrokawain, Kawain (Kavain) Flavonoids, alkaloids and glutathione. Kawalactones namely yangonin and 7, 8 epoxyangonin have also been isolated. Older plants have higher concentration of kawa lactone.

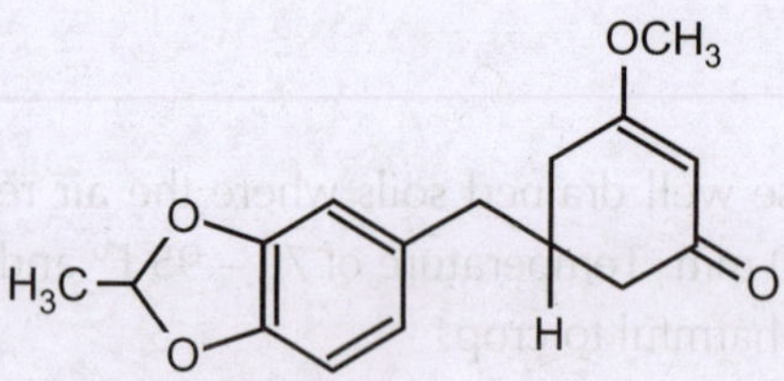

Dihydromythysticin

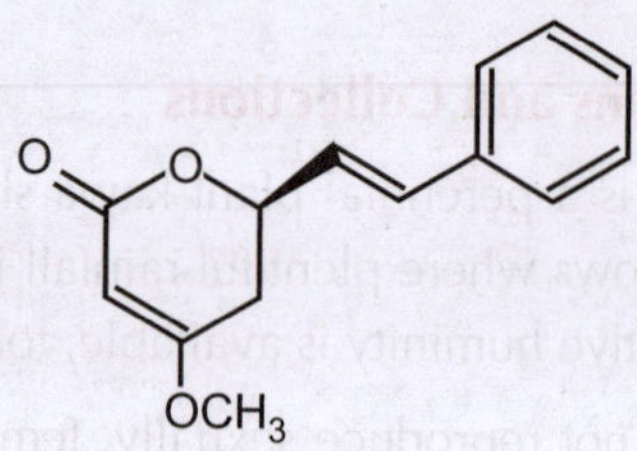

Kawain

Chemical Tests

(i) Powdered kava-kava when treated with sulphuric acid on tile produces red colour.

(ii) Mounted powdered kava-kava is absolute alcohol and partially evaporate the alcohol. Prismatic crystals of methysticin appear on the edge of covor slip. These crystal polarise red-violet to yellow colour and dissolve in sulphuric acid with the formation of violet solution.

Standards of Quality

Total Ash : Not more than 8.0 per cent

Uses

It has antiinflammatory, Anxylolytic (free of uneasyness) and neuroprotective properties.

BLACK-COHOSH

Synonyms

Black snake root, Bugwort, Raffle root, Black bugbane, Cimicifuga.

Biological Source

It consists of dried rhizomes and roots of the perennial plant known as *Cimicifuga racemosa* and also known as *Actaea racemosa* Family; *Ranunculaceae*.

It is available in BPC 1934 and British Herbal Pharma 1990.

Geological Source

Eastern North America and also cultivated in Europe.

Description

This perennial grows in moist heavy soil. It bears tall tapering racemes, flowers are white coloured on wiry black-purple stems. Flowres are produced on tall stem of 100 – 250 cm with racemes upto 50 cm. Flowers do not have sepals or petals but in clusters of 50 – 100, having long stemens and white stigma.

Flowers have sweet foetid flavour which attracts insects and beetles.

It has unpleasent medicinal smell. It bears deeply cut compound leaves which are burgundy dark red coloured. It resists summer heat and also drought. Basal leaves are upto one metre in length which are broad in the sets of three leaflets (i.e. Tripinnately compound leaves having serrated margin).

(a)

(b)

Fig. 9.59 (a) Black-cohosh herb (b) Enlarged leaf

Macroscopic Characters

Colour : Rhizomes are dark brown

Odour : Odourless

Taste : Bitter and acrid

Extra Features : Rhizomes are hard, subcylendrical, bearing several, shout ascending branches, marked with encircling leaf scars, terminating in cup-shaped scar. Under surface has long brittle roots, fracture is horny, exhibiting then dark bark.

Roots are quadrangular dark brown externally longitudinally wrinkled fracture in short.

Microscopic Characters

Distinct cambium and radiade xylem are characterisitc to rhizome. Xylem is 4 – 5 mm in diamter, dark medullary rays. Pith towards centre and 3 – 5 mm in diameter.

Chemical Constituents

Fukinolic acid is the main constituent of balck cohosh alongwith isoferulic acid and salicyclic acid in small amounts triterpine glycosides such as actein, 27 deoxyactein resin 15-20% cimicifugin, cimicifugosides, tannins cimigaside, volatile oil and actelna (resinous matter) are the other constituents of black cohosh roots.

Standards of Quality

Total Ash	:	Not more than 10.0 per cent
Acid insoluble ash	:	Not more than 4.0 per cent
Water Soluble Extractives	:	Not less than 10.0 per cent

Uses

For menopausal symptoms, premenstrual syndrome, painful menstruation, acne and for starting-labour in pregnant women.

Side Effects

Head ache, heaviness of limbs. Low blood pressure, perspiration, vomiting and weight gain.

OLEANDER

Synonyms

Nerium, Kanher.

Biological Sources

This consist of dried roots and leaves of the plant *Nerium indicum* or *Nerium oleander* fam Apocynaceae.

Description

It is large glabrous evergreen shrub with milky juice, leaves in three coriaceous 10 – 15 cm long, Linear – lanecolate, thick, tapering into short petiole, dark green in colour shining on upper surface. Flowers are red coloured or white, fragrant, funnel shaped. Fruits are follicle 10 – 120 cm in length, rigid flowers more or less throughout the year.

Geographical source

It is found in upper Gangetic plain, Himalayas, Baluchistan, Nepal, Central and South India and Cultivated through out India.

Chemical Constituents

Root, bark and seeds contain the toxic glucosides neriodorin, nenoderin and karabin. Leaves contain glucosides, oleandrin and neriin. Oleandrin is cytotoxic and acts similar to digitalis.

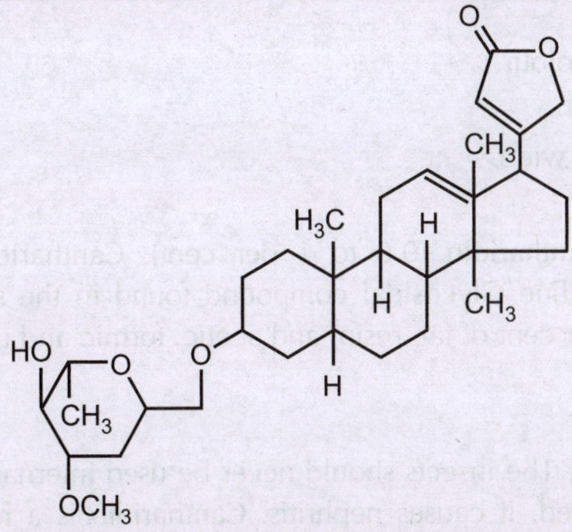

Oleandrin

Oleander flowering twig

Uses

Roots are very poisonous and used externally. Desacetyl olendrin is cardiotonic and diuretic.

ANIMAL DRUGS CONTINING GLYCOSIDES

CANTHARIDES

Synonyms

Spanish flies, Blistering beetle, Cantharis.

Biological Source

Cantharides consists of dried insects of *Cantharis vesicatoria* Linn. (*Lyatta vesicatoria*), belonging to family Meloidae. Cantharides contain not less than 0.6 per cent of cantharidin.

Geographical Source

These insects are found upon the plants of family Oleaceae and Caprifoliaceae in the Southern and Central Europe. The countries collecting cantharides are Russia, Romania, Spain, Italy and Sicily.

(a) Cantharide (Dorsal side)

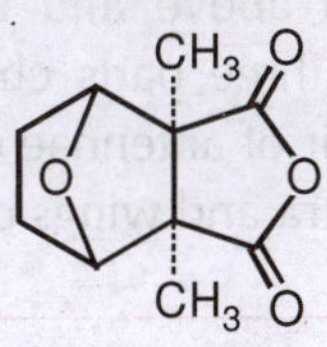

Cantharidin

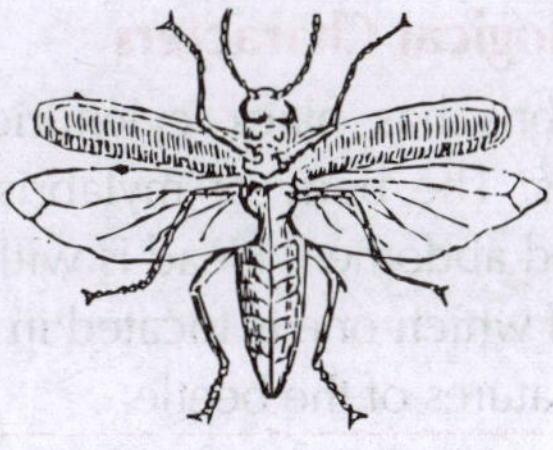

(b) Cantharide (Ventral side)

Fig. 9.60 : Cantharide insect

Collection

The mature insects, which are brilliant green in colour with a metallic lustre, are collected when they are inactive in the early morning by shaking the shrubs or beating with poles. The insects are killed either by plunging them into vinegar or exposing them to fumes of sulphur dioxide, ammonia or chloroform. The insects are dried artificially below 40°C.

Description

Colour : Insects are green or bronze green in colour.

Odour : Unpleasant.

Size : 10 to 20 mm in length and 3 to 6 mm wide.

Chemical Constituents

It contains a vesicating principle known as cantharidin (0.6 to 1 per cent). Cantharidin, anhydride of cantharidic acid, is a colourless crystalline Glycosidal compound found in the soft parts of insects. Insects also contain about 12 to 15 per cent of fat, resin, and acetic, formic and uric acids. Cantharidin is insoluble in water and alcohol.

Uses

Cantharide is an irritant, vesicant and rubefacient. The insects should never be used internally. They are used in veterinary practice. Internally if used, it causes nephritis. Cantharidin is a hair growth stimulant and is used in hair oils.

Storage

The insects are much susceptible to attack by mites and moths. They should be dried thoroughly and stored in well-closed containers, containing small amount of carbon-disulphide.

Substitutes

Cantharides insects [Fig. 9.60 (a), (b)] are substituted by Chinese cantharis, also known as Mylabris.

MYLABRIS

Synonym

Chinese cantharis

Biological Source

It consists of dried beetles of *Mylabris cichorii* or *Mylabris pustulata* family Meloidae.

Geographical Source

China and east India. Found in faily large quantities in the neighbourhood of Banglore and many parts of India.

Morphological Characters

Mylabris has black cylendrical body rounded above and flatish below. It is smaller than canthrides. The body of mylabris is divided in to three parts characteristics to insect, i.e. head, thorax and abdomen. Head is with characteristic pair of antennae of eleven joints. It has three pairs of legs, of which one is located in thorax region. Elytra and wings covering thorax and abdomen are special features of the beetle.

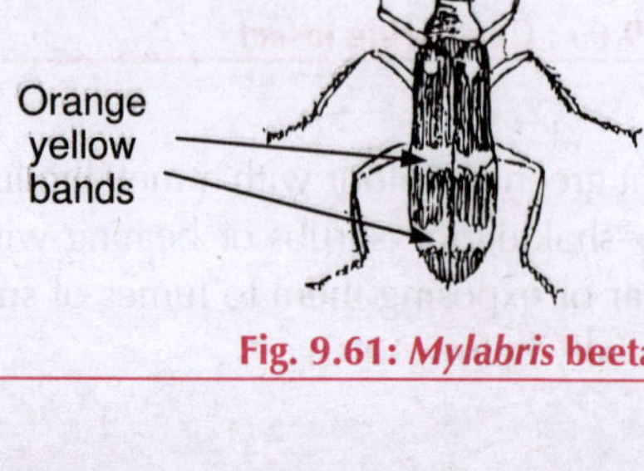

Fig. 9.61: *Mylabris* beetal

Description

Size : 12 - 30 mm × 5 - 10 mm.

Colour : Each Elytra is having orange yellow spot at its base, where it gets united to thorax. It is also marked with two wide transverse organge yellow bands.

Bands : Two bands and black back ground has stiff black hairs.

Chemical Constituents

The insects contains 1 to 1.2 per cent of cantharidin.

Uses

Similar to cantharides, these insects are also used as irritant, vesicant and rubefacient.

Storage

These beetles should be stored in well closed containers. Presence of moisture causes destruction of cantheridi by fermentation and hence needs to be preserved very carefully.

COCHINEAL

Synonyms

Coccus, Red scale insect.

Biological Source

Cochineal consists of dried full grown female insects, enclosing young larvae of *Coccus cacti* L. family Coccidae, order Hemiptera.

Geographical Source

These insects are indigenous to Central America and Mexico, but are now-a-days reared in Caribbean islands, Canary Islands, Algeria and Southern Spain.

Description

Colour : The insects are grey to black in colour depending upon the method of killing.

Size : 3.5 to 5.0 mm long and 2.5 to 3.5 mm wide.

Shape : Oval and plano-convex, strongly wrinkled transversely.

Extra Features:

Dorsal surface shows eleven segments, while ventral surface has head, thorax and abdomen, the typical parts of insects. Thorax shows three pairs of legs, while two antennae, proboscis, pair of eyes are present on the anterior end. Female insects do not have wings, while male have four wings. Each fully matured female insect encloses about 150 - 450 larvae. 70,000 insects weigh approximately 1 lb, one pound by weight.

Cultivation of Cochineal

The insects of cochineal are reared (raised) in Canary islands (on west coast of North Africa) on commercial basis on various species of the plant of cactaceae family.

The female insects suck their food through proboscis grow in size and develop larve within them, when fully grown.

Collection of Cochineal insects

The full grown insects are bruhsed from the plants on which they grow by using brooms and are killed by plunging in boiling water or by explosing to the fumes of sulphur or even by charcoal heat. Depending upon the method of application of heat to the insects various grades of insects are produced.

If heat is used to kill the insects, insects with purplish black colour are formed and they are known as **black-grain**. If fumes are used to kill the insects purple grey coloured insects are formed and are known as **silver grain**.

Immatured insects and larve are separated by sieves and are graded as shiftings.

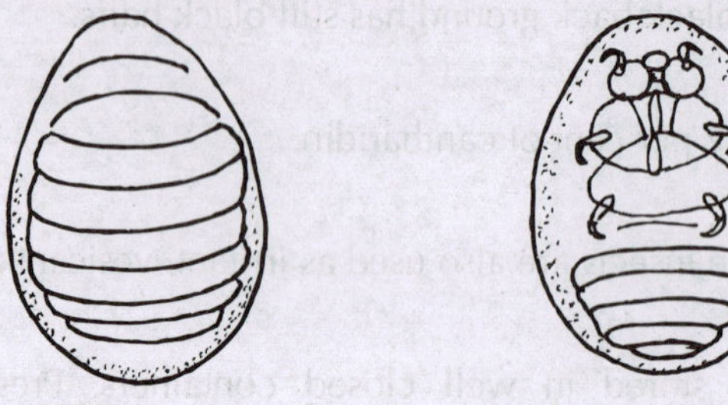

(a) (b)

Fig. 9.62: Cochineal insect. (a) Dorsal view, (b) Ventral view

Chemical Constituents

Insects contain about 10 per cent of carminic acid, a red colouring matter (glucosidal-anthraquinone), 10 per cent of fat, and 2.0 per cent of wax, coccerin. The colouring matter is present in the fatty parts of the insect and yolk of the eggs (larvae).

Carminic acid gives deep red colour in water, and yellow to violet colour in acidic solutions. It is soluble in alcohol. It occurs as small prismatic crystals.

Carminic acid is used to prepare **carmine**, which is a aluminium calcium lake produced by treating cochineal extracts with alum and lime salts.

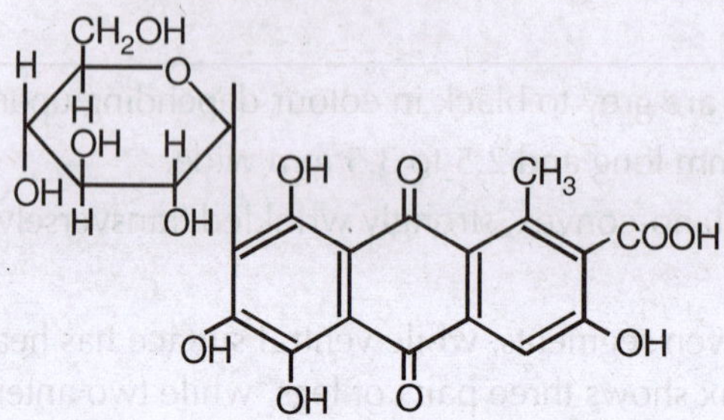

Carminic acid

Standards of Quality

Ash	: Not more than 2.0 per cent
Foreign organic matter	: Not more than 2.0 per cent
Alcohol 45 per cent soluble extractives	: Not less than 3.5 per cent

Uses

It is used as colouring agent for foods, drugs and for cosmetic products. It is FDA approved colour for foods and drugs.

It is utilized for preparation of industrial lakes, inks, acid-base indicator, microscopic stains, and also in photography.

Chapter 10...

DRUGS CONTAINING TANNINS

Tannins are one of the most widely occurring group of natural substances in different families of higher plants. These secondary metabolites are present in solution form in the cell sap and also in distinct vacuoles. Since long they are known as the astringent substances because they combine with tissue proteins and precipitate them, hence, they are used in medicines as mild antiseptics, in the treatment of diarrhoea, and to forestall minor haemorrhages. Commercially, they find extensive application in leather industry, where the hides are treated with tannins to prevent the putrefaction.

Chemically, they contain the mixture of complex organic substances wherein polyphenols are present with o-dihydroxy or o-trihydroxy groups on a phenyl ring. Normally, they have fairly high molecular weight and unlike alkaloids are devoid of nitrogen.

Tannins form colloidal solutions with water and are non-crystalline substances. In solution, they show acidic reaction due to phenols. They are also soluble in alcohol, glycerine, dilute alkalies, but practically insoluble in organic solvents except acetone. Tannins exhibit some specific chemical reactions.

(1) Solution of tannin precipitates gelatin, and alkaloids.

(2) Tannins are precipitated by salts of copper, tin, and lead.

(3) They are precipitated by strong potassium dichromate solution or chromic acid solution.

(4) They show colour reactions with iron salts. Ferric chloride gives bluish-black or brownish-green colour; potassium ferricyanide with ammonia gives deep red colour.

(5) **Goldbeater's skin test:** Goldbeater's skin is a prototype of untanned fresh skin of an animal and is obtained as a membrane from the intestine of ox. This membrane is treated with hydrochloric acid, rinsed with distilled water and then placed in tannin solution for 5 minutes. It is followed by washing with distilled water and keeping in ferrous sulphate solution. Tannin imparts brown or black colour to the skin.

(6) Tannins are precipitated by a 2 per cent solution of phenazone, the tannin solution being prepared with sodium acid phosphate.

CLASSIFICATION

The tannins are classified into three groups based on complexity of their chemical nature and according to their behaviour on dry distillation.

(1) Hydrolysable tannins: As the name indicates, these tannins are hydrolysed by acids or enzymes quickly and the products of hydrolysis are Gallic acid or ellagic acid. On dry distillation, Gallic acid and other components get converted to pyrogallol. They respond to ferric chloride solution, producing blue colour. The examples of hydrolysable tannins are gallotannin in nutgall, rhubarb, clove and chestnut; ellagitannin from oak, myrobalans and pomegranate bark.

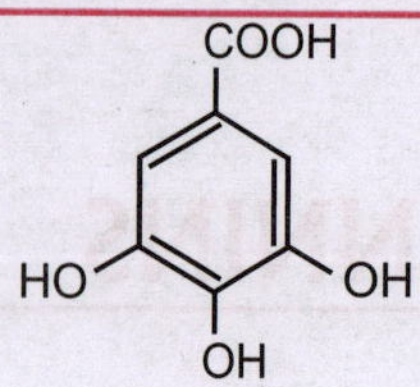

Gallic acid

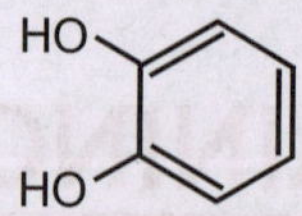

Catechol (Pyrocatechol)

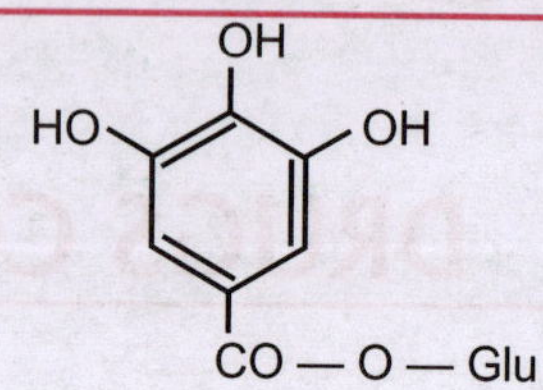

Glucogallin

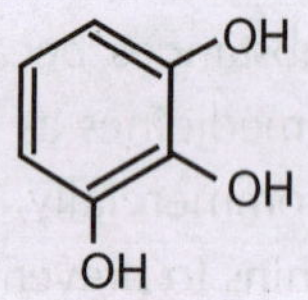

Pyrogallal

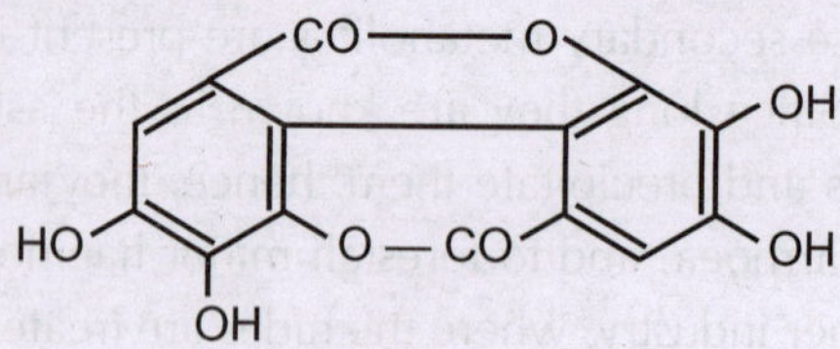

Ellagic acid

(2) Condensed tannins: They are also known as non-hydrolysable tannins, phlobatannins or proanthocyanidins. They are very resistant to hydrolysis. Being related to flavonoids pigments because they are formed via derivatives of flavones, like catechin or flavan-3-ol or flavan-3, 4-diols. Unlike the hydrolysable tannins, on treatment with enzymes or mineral acids, they are polymerised or decomposed into red coloured substances called phlobaphenes, which are insoluble in water and indicate the typical brownish-red colour of many plants and drugs. On dry distillation they yield catechol, with ferric chloride tannins produce brownish-green colour.

They are distributed in different parts of plants. The green tea and hamamelis leaves; cinchona, cinnamon and wild cherry bark; male fern rhizome; cocoa, cola and areca seeds; pale and black catechu are rich in condensed tannins.

Depending upon the source of the tannins, various modified methods are used for extraction. Methyl alcohol, hot water, acetone and ethyl acetate are the common solvents used for extraction; the extract is filtered and dried under vacuum.

(3) Phlorotannins:

Phlorotannins are present in dark brown algae and also in few red algae such as kerps and rock weeds. Chemically these are oligomers of phloroglucinol, and are based on arrangement of phloroglucinol monomers.

Phlorotannins are divided into sub-classes as fucols, phlorethols, fuco phlorethols, fuhalols and eckols.

As per linkage they are classified into four subclasses:

(1) Ether linkage i.e. fuhalos and Phlorethols

(2) Phenyl linkage i.e. Fucols.

(3) Ether and phenyl linkage i.e. Fucophlorethols.

(4) Dibenzodioxin linkage i.e. Eckols and carmalols.

Phlorotannins form precipitate with proteins and are reported in only alariaceae family.

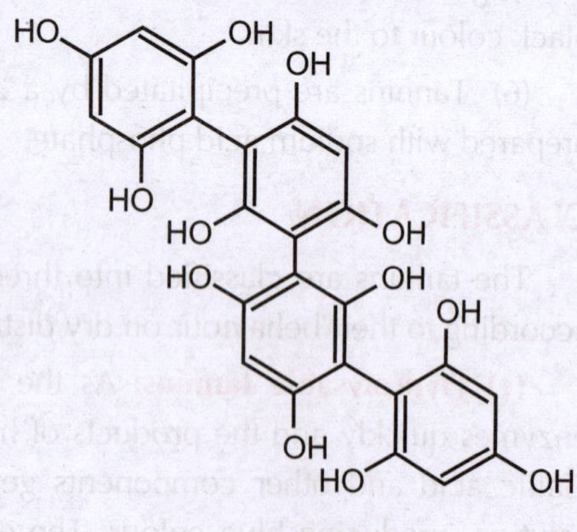

Phlorotannin

Pseudotannins:

These are not as such a separate group of tannins, but may be treated as subgroup because they do not obey to Goldbeater's skin test and are low molecular weight compounds. Chlorogenic acid in coffee and nux vomica, ipecacuanhic acid in ipecacuanha and catechins in cocoa are examples of pseudotannins. The detection test for chlorogenic acid is carried out by extracting the drug with water and treating this extract with ammonia solution, followed by exposure to air, which leads slowly to formation of green colour.

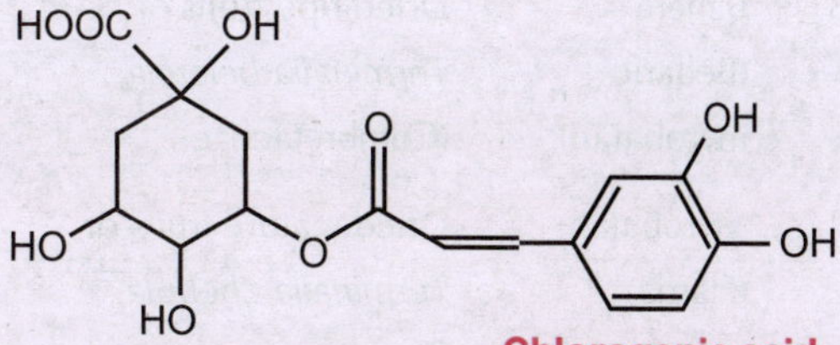

Chlorogenic acid

Isolation of Tannins

Myrobalan fruits are the rich source of tannic acid available in various parts of India. These fruits contain about 20 – 30% of hydrolysable tannins, i.e. Ellagitannins, which are soluble in ethyl alcohol and ethyl acetate.

For isolation of tannic acid fully matured and well dried fruits are selected. Seeds are removed from the fruits. The seed-free fruits are powdered and dried once again.

Then dried powdered myrobalan is subjected to the treatment of four times of its weight of ethyl alcohol on a shaker for 4 to 6 hours.

Extract is filtered and kept aside for one hour. The drug is re-extracted with alcohol as above for four hours and filtered.

Both the extracts are mixed together. The extract is vacuum concentrated to a semisolid mass. This semisolid mass is diluted with distilled water in the proportion of 1:40, preserve it over-night at 5.0°C.

Filter the extract and reject the flocculated precipitate. The clean filtrate is extracted three times with equal quantity of ethyl acetate. Concentrate the combined ethyl acetate extracts in vacuum.

The residue obtained is Tannic acid only. Dry it further in dessicator if necessary.

Table 10.1: Tannin Containing Drugs

Sr. No.	Name of Drug and Synonym	Biological Source	Active Constituents	Uses
I.	**Hydrolysable Tannins**			
1.	Amla (Indian goose berry)	Dried as well as fresh fruits of *Emblica officinalis* (Euphorbiaceae)	Vitamin C, phyllemblin, tannins	Diuretic, laxative, ingredient of *Triphala* and *Chyavanprash*
2.	Arjuna	Dried stem bark of *Terminalia arjuna* (Combretaceae)	Ellagic acid, β-sitosterol	Cardiotonic, hypotensive.

Contd...

3.	Bahera (Bellaric myrobalan)	Dried ripe fruits of *Terminalia belerica*, (Combretaceae)	Gallic acid, chebulagic acid	Astringent, ingredient of triphala churna
4.	Myrobalan (Harda)	Dried mature fruits of *Terminalia chebula*, (Combretaceae)	Chebulic acid, chebulagic acid, Gallic acid	Astringent, stomachic, purgative, ingredient of triphala churna
5.	Tannic acid	Fermented oak galls which are grown on young twigs of *Quircus infectoria* (Fagaceae)	Gallic acid and glucose	Astringent for mucous membrane
6.	Galls	Pathological out growth of *Quircus inpectorea* Fagaceae	Tannic acid	Astringent
7.	Witch-hazel	Leaves of *Hamamates verginiana N Hamamaliacea*	Hamamales tannins	Astringent, Hena stetic

II. Condensed Tannins

1.	Amra (mango)	Dried bark of *Mangifera indica* (Anacardiaceae)	Tannins, catechin mangiferin	Anti oxidant, antidiarrhoeal
2.	Ashoka bark	Dried stem bark of *Saraca indica*, (Leguminosae)	Catechol, ketosterol	Uterine tonic, oxytocic
3.	Black catechu (cutch)	Dried aqueous extract of heartwood of *Acacia catechu*, (Leguminosae)	Acacatechin, quercetin	Astringent for boils, skin eruptions, mainly used for commercial purposes
4.	Pale catechu, (Gambier)	Dried aqueous extract of leaves and young shoots of *Uncaria gambier*, (Rubiaceae)	Catechin, catechutannic acid	Astringent for treatment of diarrhoea
5.	Dhayati	Dried flowers of Woorfordia *floari bunda* and worfordia fruits Fam. Lithraceae	Tannins upto 2.5 %	Disorders of liver and dysentry.
6.	Pterocarpus (Indian kino)	Dried juice of the plant *Pterocarpus marsupium* (Leguminosae)	Kinotannic acid, kino-red, k-pyrocatechin	Hypoglycemic, powerful astringent

I. HYDROLYSABLE TANNINS

AMLA

Synonyms

Emblica, Indian goose berry, Amalki

Biological Source

This consists of dried, as well as fresh fruits of the plant *Emblica officinalis* Gaerth *Phyllanthus emblica* Linn. belonging to family *Euphorbiaceae.* It contains not less than 1.0 per cent w/w of gallic acid calculated on dry basis.

Geographical Source

It is a small or medium size tree found in all deciduous forests of India. It is also found in Sri Lanka and Myanmar. The leaves are feathery with small oblong pinnately arranged leaflets. The tree is characteristic greenish-grey with smooth bark.

Cultivation of Amla

Intolerant to frost or drought, it is grown by seed germination; besides, Amla can also be propagated by budding or cutting. It does not tolerate the frost or drought. It is normally found up to an altitude of 1500 m. Commercially, it is collected from wild grown plants.

Now a days the newly released varieties are selected for better yield. These are known as Banarasi, Kanchan, Anand-2, Balwant, NA6, NA7 and BS-1. Seeds or seedlings are placed at a distance of 4.5 × 4.5 metres in red loamy or coarse gravelly soil. Proper arrangement for irrigation is required, Drip irrigation is most suitable. Fertilizers in the dose-range of 750 - 900 gms of urea, 1 kg superphosphate and 1-1.5 kg of potash per annum depending upon the quality of soil are sufficient.

Fig. 10.1: Amla twig with fruits

The above dose is divided into two equal parts, one part is applied in September / October and the second in April - May every year. Pruning is done regularly and only 4 - 6 branches about 0.75 - 1.0 metre above the ground are retained. Plant bears male and female flowers separately. Male flowers are reported in the axil of the leaf, in bunches while the solitary female flowers are in the axil of the branches. The extent of fertilization is 25 - 30 per cent of flowers. Cultivated plants bear comparatively large fruits. The tree flowers in hot season and the fruits ripen during the winter.

Alternative crops to the extent of 7 - 8 years age of Amla trees can be undertaken. Black gram, tomato, gaur, sunflower, ground nut etc. are the common alternative crops of choice. Each plant can bear 175 - 300 kg of fruits and each of healthy fruit weights approximately 25 - 35 gm.

Plant hormones like Gibberlic acid or planofix in the range of 30 - 50 ppm are most useful to increase the yield per hectare.

Pesticides used are as under:

Diathase-78, DDT is useful to get rid of rust, blue mold or other fungal infections.

Macroscopic Characters

Colour : Fruits are the green colour changes to light yellow or brick red at maturity.

Odour : Odourless.

Taste : The taste of Amla is sore and astringent.

Size : The average size of an Amla is between 1.5 and 2.5 cm in diameter.

Shape : The fruits are depressed, globular.

Extra Features

Fruits are fleshy obscurely 4 lobed with 6-trygonus seeds. They are very hard and smooth in appearance.

Chemical Constituents

Amla fruit is a rich natural source of vitamin C (Ascorbic acid) and contains 600 - 750 mg per 100 g of the fresh pulp. Furthermore, fruits also contain about 0.5 per cent fat, phyllemblin and 5 per cent tannin. Amla fruits are also rich in mineral matters like phosphorus, iron and calcium. It contains appreciable amount of pectin. The fresh fruits contain about 75 per cent moisture. The fruits are dehydrated and stored. It is found that vitamin content of dried fruits is not lost considerably.

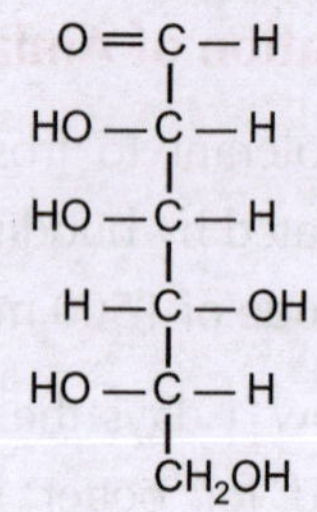

Vitamin C (Ascorbic acid)

It may be due to the presence of tannins, which retards oxidation of vitamin C.

Chemical Tests

1. Alcoholic or aqueous extract of the drug gives blue colour with ferric chloride solution.
2. Adding gelatin and sodium chloride in aqueous extract produces milky white colour.
3. In the aqueous extract of Amla add lead acetate to remove percipitate by filtration. To the filtrate add solution of 2 : 6 dichlorophenol – indophenols; the colour disappears.

Standards of Quality

FOM	: Not more than 3.0 per cent
Total ash	: Not more than 5.0 per cent
Acid insoluble ash	: Not more than 2.0 per cent
Water soluble extractives	: Not less than 40.0 per cent
Alcohol soluble extractives	: Not less than 30.0 per cent

Uses

Amla fruits are largely used in Indian medicines. It is used as an acrid, diuretic, refrigerant and laxative. Dried fruits are given in diarrhoea and dysentery. They are also administered in jaundice, dyspepsia and anaemia along with iron compound. Fruits are also used in preparation of inks, hair oils and shampoo. It is reported that fixed oil from fruits possesses the property of promoting hair growth. Seeds of the fruits are given in treatment of asthma and bronchitis.

The leaves are used as fodder. Alcoholic extract of the fruit is anti-viral. It is a popular ingredient of '*Triphala*' and '*Chyawanprash*''. Amla, being a rich source of vitamin C, is considered important to slow the ageing process. It improves skin health. Ageing is a cumulative result of damage to various cells and tissues, mainly by oxygen free radicals. Vitamin C is a scavenger of free radicals which breaks them down. It has an antioxidant synergism with vitamin E (which prevents peroxidation of lipids). Amla is a major ingredient of ancient Ayurvedic preparation Chyawanprash, believed to delay ageing process thereby adding to longevity.

ARJUNA

Synonyms

Arjun bark, Arjun

Biological Source

Arjuna consists of dried stem bark of the plant known as *Terminalia Arjuna* Rob, belonging to family Combretaceae. It contains not less than 0.02 per cent of arjungenin on dried basis.

Geographical Source

The tree is common in Indian peninsula. It is grown by the side of streams and very common in Chotta Nagpur region.

Cultivation and Collection

Arjuna plants are not cultivated on commercial scale. Arjuna is found as naturally growing plant in the dense forests. It is very common in Baitul in Madhya Pradesh and Dehradun, Uttrakhand. Arjuna can be successfully raised by sowing seeds or by means of stumps. The seeds take about 21 days for germination. It needs moist fertile alluvial loam and rainfall in the range of 75 - 190 cm. It grows satisfactorily up to 45° C. The bark is collected from wild growing plants and it is reported that yield per tree varies from 9 - 55 kg.

Macroscopic Characters

Colour	:	Colour of the outer side, is greyish brown while inner side of bark is reddish brown.
Odour	:	None
Taste	:	Astringent
Size	:	The pieces of various-sizes, about 15 × 10 ×1 cm.
Shape	:	Flats

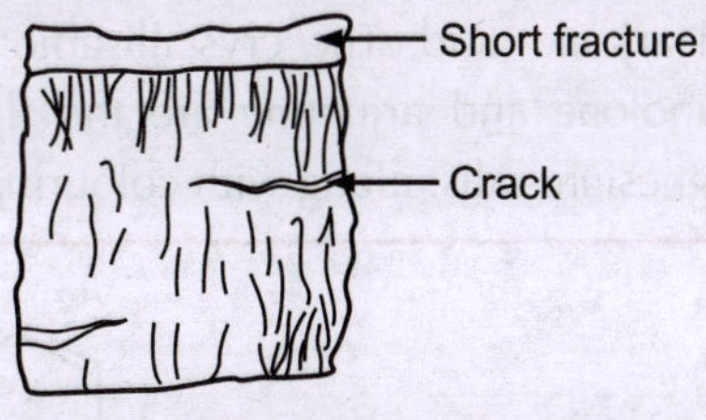

Fig. 10.2 (a): Arjuna-bark

Extra Features

The presence of the cork is not reported in the commercial drug. As Arjuna is collected from the old trees, the cork gets removed due to exfoliation. The appearance of the transversely cut surface is dark brown with characteristic greyish shining patches.

Microscopic Characters

Transverse section of the stembark shows outermost layer of cork of 9 - 10 layers of tangentially elongated cells. Outer layers show presence of brown matter. Cork cambium and secondary cortex not distinctive. The secondary phloem is wide and consists of phloem parenchyma, phloem fibres, sieve tubes and companion cells. The phloem is traversed by phloem rays, uniseriate (occasionally biseriate). Sieve tubes get collapsed and form **ceratenchyma** in the middle and outer phloem region. Phloem fibres in groups of 2-10 and distributed in rows. Phloem parenchyma contains rosette **calcium oxalate crystals** and also Idioblasts with crystals of calcium oxalate. Simple and sometimes compound **starch grains** distributed in the cortex.

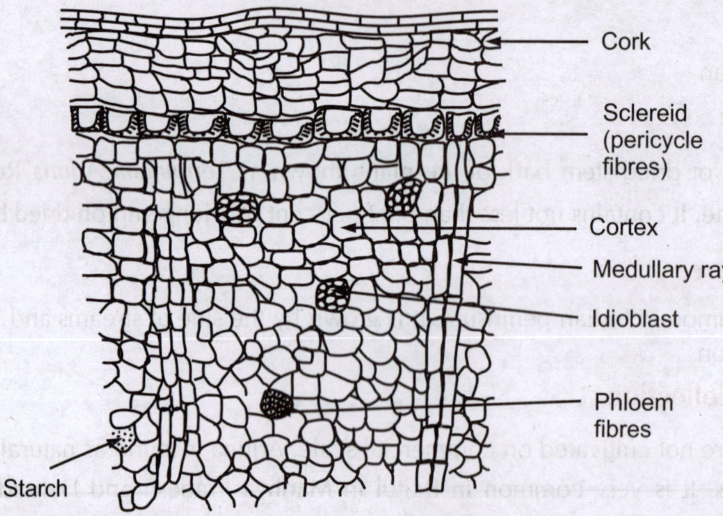

Fig. 10.2 (b) : T.S. of Arjuna Bark

Chemical Constituents

Arjuna contains about 15 per cent of tannins (hydrolysable). It also contains triterpenoid saponins, arjunolic acid, arjunic acid, arjungenin. In addition, it contains β-sitosterol, ellagic acid, and arjunic acid. The crystallisable compounds reported are arjunine and arjunetine. Arjunetin, arjunolone and arjunone are the flavonoids reported in Arjuna bark. Calcium, aluminium and magnesium salts, along with colouring matter and sugar are the other constituents of Arjuna.

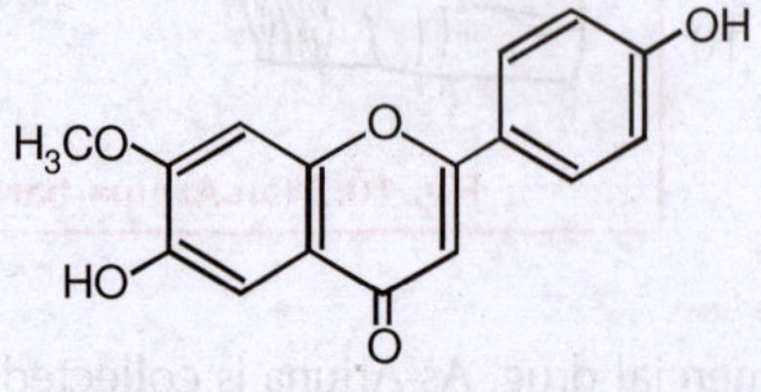

Arjunolone

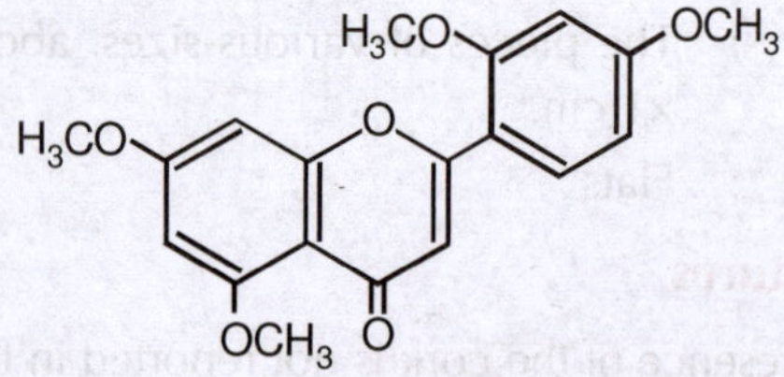

Arjunone

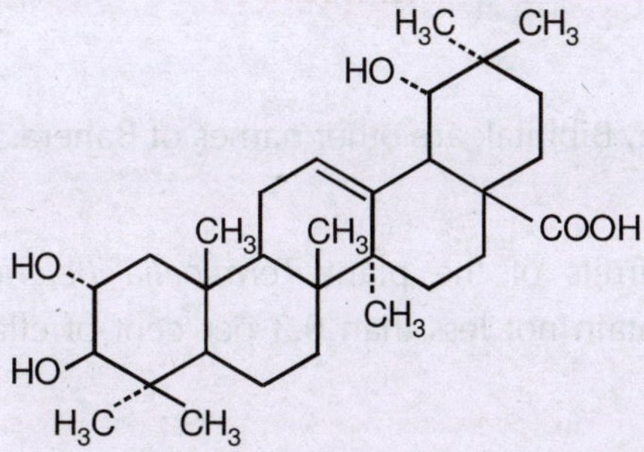

Arjunic acid

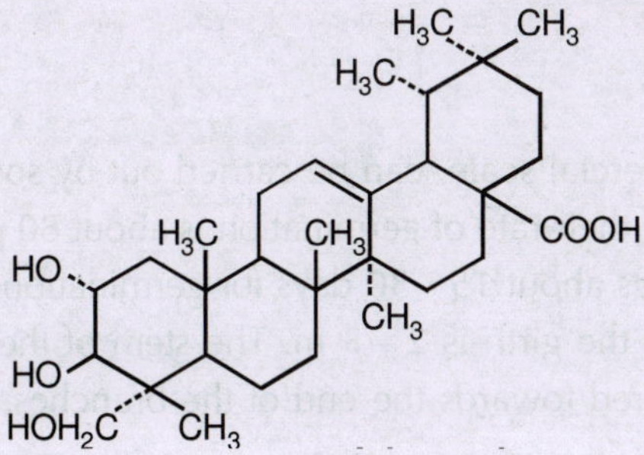

Arjungenin

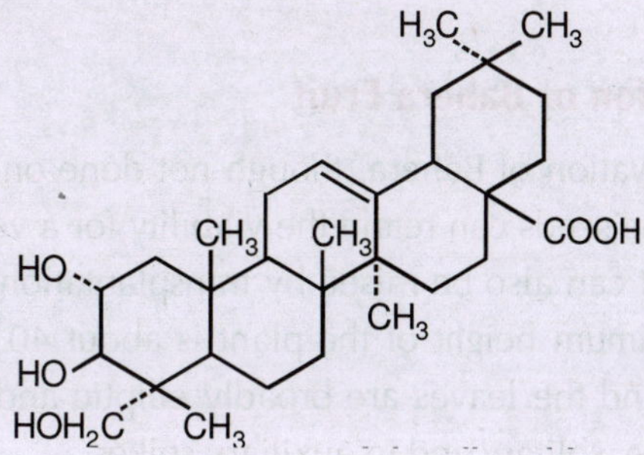

Arjunolic acid

Chemical Test

Ethereal extract of Arjuna shows pinkish fluorescence under ultra-violet light.

Standards of Quality

Total ash	:	Not more than 30.0 per cent
Acid insoluble ash	:	Not more than 2.0 per cent
Alcohol soluble extractives	:	Not less than 20.0 per cent
Water soluble extractives	:	Not less than 20.0 per cent
Loss on drying	:	Not more than 10.0 per cent

Uses

Arjuna bark is used as a cardiotonics. It reduces anginal frequency It is also styptic, febrifugel and antidysenteric. It possesses diuretic and tonic properties. The drug exhibits hypotensive action with vasodilation and decreased heart rate. The diuretic activity is due to arjunolic acid. Arjuna bark is used extensively in the tanneries and also as a timber.

Adulterants

The dried bark of the plant *Terminalia tomentosa* is used as an adulterant for the drug. However, it can be distinguished from Arjuna bark by fluorescence test. Ethereal extract of Arjuna gives pinkish fluorescence, while *Termenalia tomentosa* gives pale blue fluorescence.

BAHERA

Synonyms

Bellaric myrobalan, Baheda, Bibhitak are other names of Bahera.

Biological Source

It consists of dried ripe fruits of the plant *Terminalia belerica* Linn. belonging to family Combretaceae, and should contain not less than 0.3 per cent of ellagic acid and 0.75 per cent of gallic acid in dried form.

Geographical Source

The tree is found in all the decidous forests of India, up to an altitude of 1000 m. It is found in abundance in Madhya Pradesh, Uttar Pradesh, Punjab; Maharashtra and in Sri Lanka and Malaya also.

Cultivation of Bahera Fruit

Cultivation of Bahera, though not done on commercial scale, can be carried out by sowing the seeds. The seeds can retain the viability for a year and their rate of germination is about 80 per cent. The plant can also be raised by transplantation. It takes about 15 - 30 days for germination of seed. The maximum height of the plant is about 40 m and the girth is 2 - 3 m. The stem of the plant is straight and the leaves are broadly elliptic and clustered towards the end of the branches. Flowers are simple, solitary and in auxiliary spikes.

Collection of Bahera Fruit

Fully grown fruits are collected before the rainy season in March/April every year. Fully grown fruits fall on ground and are collected.

Macroscopic Characters (Fig. 10.3)

Colour : Fruits are dark brown to black

Odour : None

Taste : Astringent

Size : 1.3 - 2 cm in length

Shape : Fruits are globular and obscurely 5 angled

The fruits are pulpy with hard and stony seeds.

Fig. 10.3 (a): Bahera twig

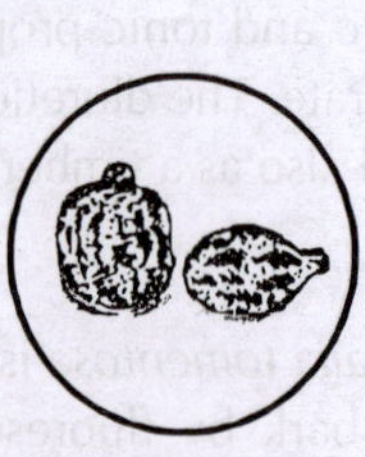

Fig. 10.3 (b): Bahera fruits

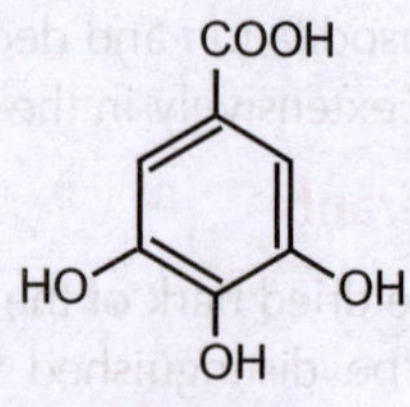

Gallic acid

Chemical Constituents

The fruits contain about 20 - 30 per cent of tannins and 40 - 45 per cent water-soluble extractives. It also contains colouring matter besides gallic acid, ellagic acid, phyllemblin, and ethyl gallate and galloyl glucose. The seeds contains non-edible oil. The plant produces a gum. It also contains most of the sugars as reported in myrobalan.

Standards of Quality

Total ash	:	Not more than 4.5%
LOD	:	Not more than 10.0%
Acid insoluble ash	:	Not more than 0.2 %
Alcohol soluble extractives	:	Not less than 17.0 %
Water soluble extractives	:	Not less than 26.0 %
Loss on drying	:	Not more than 10.0 %

Uses

Bahera is used as an astringent and in the treatment of dyspepsia and diarrhoea. It is a constituent of triphala. The purgative property of half ripe fruit is due to the presence of fixed oil. The oil on hydrolysis yields an irritant recipe. Gum is used as a demulcent and purgative. Oil is used for the manufacture of soap. Oil expressed from seeds is used for dressing of hairs.

MYROBALAN

Synonyms

Chebulic myrobalans, Harde, Haritaki.

Biological Source

It consists of dried, ripe, and fully matured fruits of *Terminalia chebula* Retzr belonging to family *Combretaceae*. It contains not less than 5.0 per cent of chebulagic acid and not less than 12.5 per cent of chebulinic acid. Ambegaon, Junnar, Khed and Akola (Nagar district(are the important centres of collection of myrobalaus in Maharashtra.

Geographical Source

Myrobalan tree is found in the sub-Himalayan tracks from Ravi to West Bengal, Assam and in all deciduous forests of India, specifically in Madhya Pradesh, Maharashtra, Bihar and Assam.

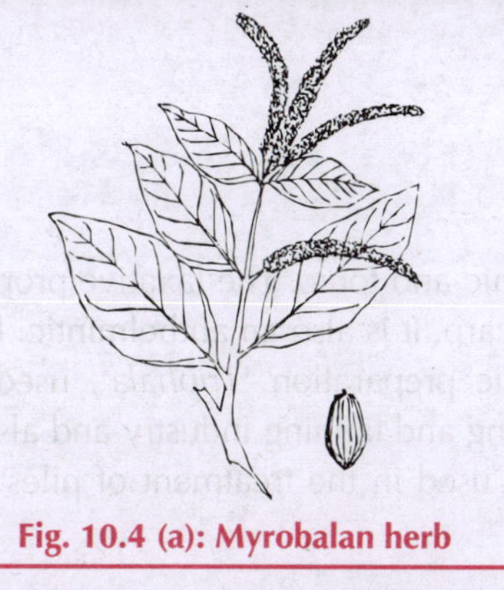

Fig. 10.4 (a): Myrobalan herb

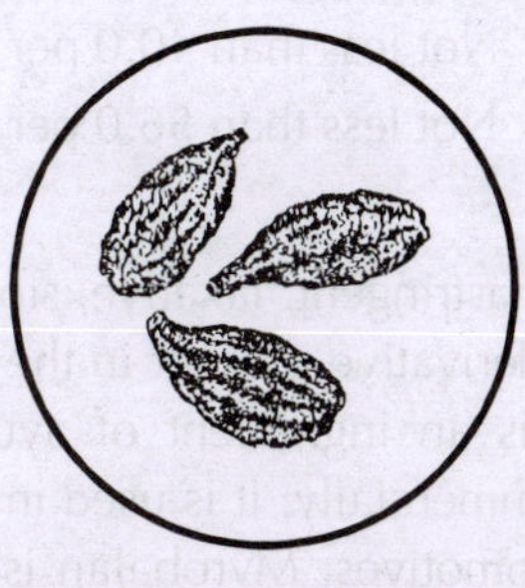

Fig. 10.4 (b): Myrobalan fruits

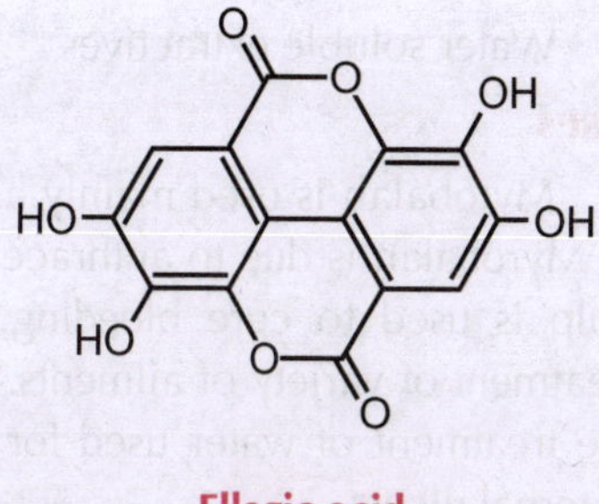

Ellagic acid

Cultivation and Collection

It grows at an altitude of 1800 m. It is not cultivated and fruits are collected from wild grown forest plants. It is a tree, 15 - 25 m in height, and 1.5 - 2.5 m in diameter. The tree is rounded, crowned with spreading branches and oxate leaves. It has yellowish-white flowers in the terminal spike. Fully grown fruits are collected before rainy season in April / May every year.

Macroscopic Characters

Colour	:	Fruits are yellowish-brown
Odour	:	Odourless
Taste	:	Astringent, slightly bitter and sweetish at the end
Size	:	20 to 25 mm long and 15 to 25 mm wide
Shape	:	Ovate and wrinkled longitudinally

Extra Features

The fruits are hard and stony with single seed which is light yellow in colour and 15 to 320 mm in length. The pulp of the fruit is non-adherent to the seed.

Chemical Constituents

Myrobalan fruits are an important source of tannin. Depending upon the geographical source, they vary in tannin content and the fruits collected from Chennai are very rich in tannin. The approximate analysis of the fruits is as follows:

Moisture - 10 per cent; Tannin - 25 - 32 per cent;

Water - insoluble matter - 40 - 50 per cent

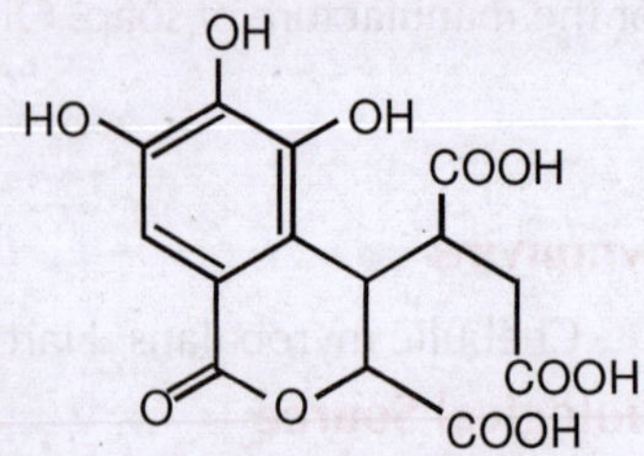

Chebulic acid

The tannins of myrobalans are of pyrogallol type (hydrolysable tannins), which yield chebulic acid and d-galloyl glucose on hydrolysis. Chebulagic, chebulinic, ellagic and gallic acids are the other contents of myrobalans. Myrobalan also contains glucose and sorbitol (about 3.5 per cent). During the maturation of the tree, the amount of tannin decreases, whereas the acidity of the fruits increases.

Standards of Quality

Total ash	:	Not more than 5.5 per cent
Acid insoluble ash	:	Not more than 0.5 per cent
LOD	:	Not more than 9.0 per cent
Alcohol soluble extractives	:	Not less than 40.0 per cent
Water soluble extractives	:	Not less than 56.0 per cent

Uses

Myrobalan is used mainly as an astringent, laxative, stomachic and tonic. The laxative property of Myrobalan is due to anthracene derivative present in the pericarp. It is also an anthelmintic. Fruit pulp is used to cure bleeding. It is an ingredient of ayurvedic preparation '*Triphala*', used for treatment of variety of ailments. Commercially, it is used in dyeing and tanning industry and also in the treatment of water used for locomotives. Myrobalan is also used in the treatment of piles and external ulcers.

GALLS

Synonyms

Tannic acid, Tannin, Gallotannic acid, Oak, Tamarix.

Biological Source

Galls are pathological out growth formed on young twigs of *Quircus infectoria* (Oak tree), belonging to family Fagaceae. The galls are formed due to egg deposition by gall wasp: *Adleria gallaetinctoriae*.

Tannic acid the main constituent of galls is also obtained from Chinese galls formed by an aphis, *Schlechtendalia chinensis* on the tree of *Rhus chinensis*, belonging to family Anacardiaceae.

Galls are collected in August and September every year.

Geological Source

Galls are obtained mainly from Aleppo in Asiatic Turkey (Turkesthan), Syria, Cyperans, Greece and Iran.

Description of Galls

Colour	:	Galls are bluish green or olive green externally and pale buff internally.
Odour	:	Odourless.
Taste	:	Astringent followed by sweet taste.
Size	:	Galls are 10 – 30 mm in size and
Shape	:	Sub spherical in shape and with short basal stalk.

Extra Features

They are hard to touch and sink in water. Galls absorb water, get enlarged and also are soft. Individual galls are 2 - 4 gm in weight, every gall has blunt and conical projections, which are about 10 - 15 in number. Solubility: Tannic acid is soluble in water, insoluble in organic solvent.

Isolation of Tannin acid

The galls are powdered and extracted with alcohol-ether mixture. The extract thus obtained is treated with water. Ethereal part separates out and evaporates. Alcohol-water part contains tannic acid. It is further concentrated into vacuum dryer to remove water at low temperature It is purified by several washings with organic solvents and vacuum dried.

Chemical Constituents

Galls contain about 50 – 70% of tannic acid, moisture and vegetable debris of the plant on which they are deposited. Tannic acid is not a single compound but is a mixture of esters of gallic acid with glucose and its exact composition varies with its source.

Tannic acid is hydrolysable tannin corresponding to complexity of pentadigalloyl glucose (C_{76} H_{52} O_{46}) composed of gallic acid and glucose. Tannic acid is incompatible with alkaloids, gelatin, albumin, alkalis and iron salts.

The main constituent of galls. Tannic acid is yellowish white to light brown shining powder. It has as astringent taste and no odour. It is amorphous and light in weight, it darkens on storage.

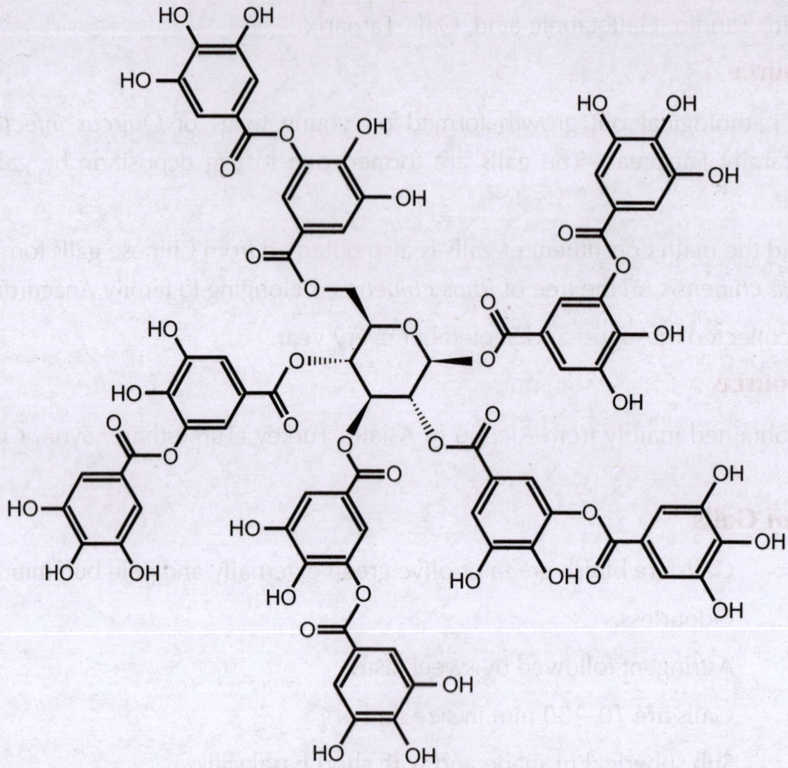

Tannic acid

Chemical Test

(i) With freshly prepared ferric chloride solution it produces bluish-black colour.

(ii) When boiled with sulphuric acid gives red brown colour.

Standards of Quality

Loss on Drying	:	Not more than 12.0 per cent
Ash	:	Not more than 0.2 per cent
Acid insoluble ash	:	Not more than 1.0 per cent
Alcohol soluble extractives	:	Not less than 10 per cent

Uses

It is used as an astringent for mucous membrane of mouth and throat hence employed for sore throat and receding gums. In the form of suppositories, it is effective in the treatment of piles. Tannic acid is also used as an antidote for poisoning due to alkaloids, heavy metals, and some glycosides.

Storage

It should be stored in well closed containers, away from light and in cool place.

WITCH-HAZEL

Synonym

Hamamelis leaves, Winier bloom.

Biological Source

These are dried leaves of Hamamelis. *Virginia;* Family Hamamelilaceae.

Geographical Source

Native of North America, available in Canada, Minnesotta, Louisiana, Virginia, North Curolina.

Morphological Characters

It is very common herb in USA and attains the height of about 3 meters.

Leaves are collected only Autumn and carefully dried in shade found in low damp woods. It flowers in fall season as the fruits of previous year ripen.

Leaves are dried shade in open air to pressure their green colour.

Organoleptic Characters

Colour	:	Leaves are dark green or brownish green
Odour	:	Slight, characteristic
Taste	:	Astringent and bitter
Size	:	4.0 – 15 × 3 -1 0 cm
Shape	:	Broadly ovate or rounded obvate

Extra features : Petiole is short, apex is acute.

Margin of leaf is coarsely crenate: Midrib on under surface is prominent. Branched trichomes (stellate) on midrib is characteristic.

Chemical Constituents

It mainly contains Hamamel-Tannin also gallic acid, phlobatannin and traces of volatile oil.

Commercial samples of witch hazel leaves contain approximately 12.- percent of Tannin Hamamalin available in the market is the mixture of substances obtained by extracting leaves.

Standards of Quality

Total ash	:	Not less than 5.0 per cent
Acid alcohol ash	:	Not less than 1.5 per cent

Uses

Hamalis leaves are astringent and also haemostatic

II. CONDENSED TANNINS

AMRA BARK

Synonym

Aam ka chhal

Biological Source

It consists of dried stem bark of *Mangifera indica* Linn. *family:* Anacardiaceae, a tree found wild or cultivated throughout the country. It contains not less than 1.5 per cent of mangiferin on dried basis.

Geographical Source

Abundant in India in Uttar Pradesh, Madhya Pradesh and Maharashtra.

Description

(a) Macroscopic:

Colour	:	Bark is greyish to dark brown externally and yellowish-white to reddish internally
Odour	:	Pleasant
Taste	:	Astringent
Size and Shape	:	Drug occurs in pieces of variable size and thickness
Extra features	:	Surface rough due to longitudinal cracks, fissures and scattered, raised lenticels.

Fig. 10.5: Fruiting mango tree

(b) Microscopic:

Mature bark has a wide cork consisting of tangentially elongated cells, few outer layers brown and inner lighter in colour, at few places lenticels; secondary cortex absent; secondary phloem wide, consists of sieve elements, parenchyma and phloem fibres, traversed by medullary rays, resin canals and yellow coloured elongated; tannin sacs abundantly scattered throughout phloem; stone cells thick walled, lignified, rectangular with wide lumen starch grains and prismatic crystals of calcium oxalate present in phloem cells; phloem fibres in groups of 2-15 or more cells, long and thick walled, phloem rays 1-3 seriate, containing crystals of calcium oxalate and starch grins.

Chemical Constituents

Mango bark contains 10-20 per cent tannins, namely protocatechuic acid and catechin. Additionally it also contains mangiferin, alanine, glycine, aminobutyric acid, kinic acid and Shikimic acid. Mangiferin is a polyphenolic compound and is an antimicrobial, analgesic and antioxidant.

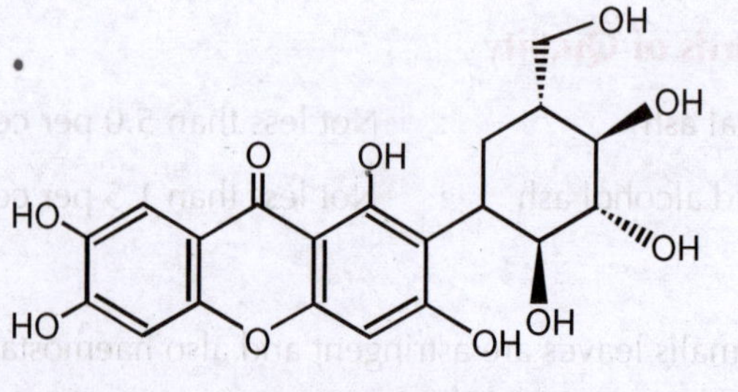

Mangiferin

Standards of Quality

Foreign organic matter	:	Not more than 2 per cent
Total ash	:	Not more than 9 per cent
Acid-insoluble ash	:	Not more than 2 per cent
Alcohol-soluble extractives	:	Not less than 20 per cent
Water-soluble extractives	:	Not less than 14 per cent

Identification by T.L.C.

TLC of the alcoholic extract on Silica gel 'G' plate using n-Butanol: Acetic acid: Water (4 : 1 : 5) shows under:

(1) UV (366 nm) three violet spots at R_f. 0.12, 0.73 and 0.87.

(2) On exposure to Iodine vapour four yellow coloured spots appear at R_f. 0.33, 0.51, 0.74 and 0.88.

(3) On spraying with 5 per cent Methanolic-sulphuric acid reagent and after heating the plate at 105ºC for ten minutes, three grey coloured spots appear at R_f. 0.49, 0.69 and 0.88.

Uses

Amra-bark is used as an astringent, antioxidant and also in the treatment of diarrhoea, dysentery and rheumatism. Fruits are edible, sweet, delicious and rich source of nutrients. Fixed oil obtained from mango-stone is valuable industrially.

ASHOKA-BARK

Synonyms

Ashoka, Sita ashok.

Biological Source

Ashoka consists of dried stem bark of the plant *Saraca indica* Linn. belonging to family Leguminosae.

Geographical Source

It is distributed in South Asia: Malaysia, Indonesia, Sri Lanka and India.

Cultivation of Ashoka

It is one of the most sacred trees of the Hindus. It is frequently grown as an ornamental and avenue tree in India. It is not found cultivated on commercial scale. It can be easily propagated from seeds. It is found growing suitably at an altitude of 750 m in the Himalayas, Khasi, Garo and Lushai hills. It is an evergreen tree, bearing dark **red** coloured **flowers'** reaching a maximum height of 9 m.

Collection of Ashoka

Bark is collected from the plant by making transverse and longitudinal incisions. Bark is dried in sun only which taken about 15 days as the bark is quite thick.

Macroscopic Characters

Colour	:	Outer side of bark is dark brown or almost black with warty surface. Internally, it is reddish-brown with fine longitudinal striations. It bears dark red coloured flowers.
Odour	:	Odourless
Taste	:	Astringent and bitter
Size and shape	:	It occurs in the form of channels of various sizes up to 50 cm long and 1 cm in thick. The bark is marked by bluish and ash white patches of lichens.

Fig. 10.6: Ashoka Twig and Fruit

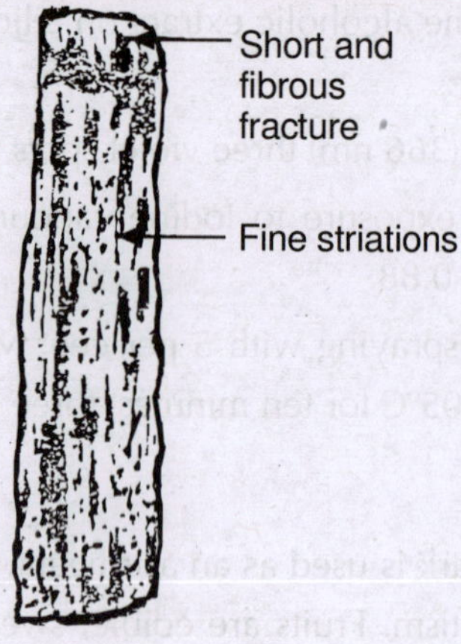

Fig. 10.7: Ashoka bark

Microscopic Characters

Phellem (cork) and phelloderm form the outer most layers. Cortex shows the presence of stone cells. Sclerenchymatous patches are reported in phloem. Medullary rays are multiseriate.

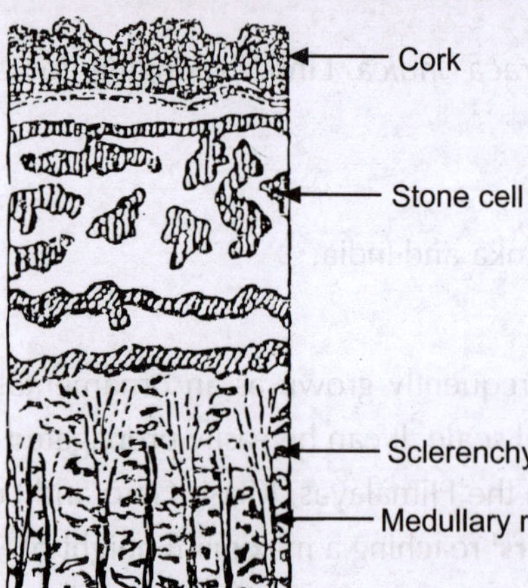

Fig. 10.8: T. S. of Ashoka bark

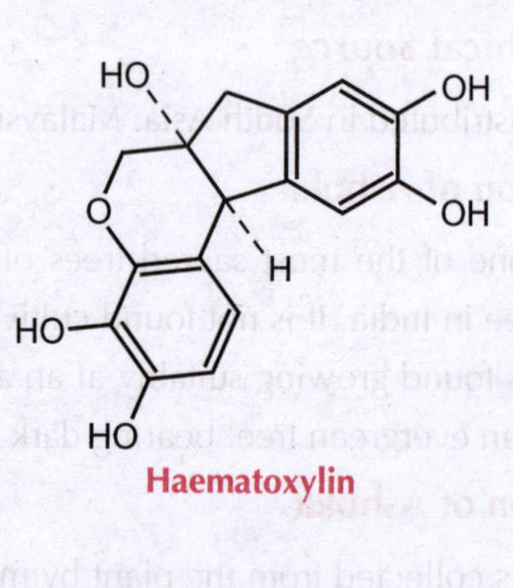

Haematoxylin

Chemical Constituents

Ashoka contains about 6 per cent tannin, haematoxylin, ketosterol, saponin and organic calcium and iron compounds. The tannins are found to be of condensed type. The ketosterol seems to be androgenic in nature. Leucopelargonidin and leucocyanidin have been isolated from the ashoka bark. The activity of drug is due to the presence of steroidal component and the calcium salt. Bark contain powerful oxytocic principle, a phenolic glycoside P_2.

Chemical Test

With alkaline solution it gives dark red colour.

Standards of Quality

Ash	:	10.8 per cent
Water soluble extractives	:	10.3 per cent
Acid soluble extractives	:	13.0 per cent

Uses

It is used as a uterine tonic and a sedative. It stimulates the uterus by the prolonged and frequent uterine contractions. It is also suggested in all cases of uterine bleeding, where ergot can also be used. It is reported to have a stimulant effect on the endometrial and ovarian tissue and useful in menorrhagia.

Note: This bark should not be confused with garden plant ashoka, which is botanically known as *Polyalthia longifolia* which bears white flowers.

BLACK CATECHU

Synonyms

Kattha, Cutch, Khadir-catechu, Catechu

History

Possibly, the use of black catechu could be traced back in history to the time of chewing betel leaf, wherein it has been used as adjuvant. In the bygone days, women used Kattha to colour thier feet. Since 15th century, this natural material has been exported to Europe.

The old information about Catechu is by a Portuguese writer Garcia de Orta in 1574. Dr. Wrath first used the scientific process to extract catechu, and showed that catechu consists of two parts: kattha and cutch.

Biological Source

It consists of dried aqueous extract prepared from the heartwood of *Acacia catechu* wild and *Acacia chundra* willd family Leguminosae.

Geographical Source

Plants used for preparation of catechu are grown in India and Myanmar.

Manufacture of Black Catechu

In the traditional method, the heartwood is boiled in earthen vessels, fired by sap wood, till all the soluble portion is extracted from it. It is cooled naturally till it is converted to semi-solid mass.

On cooling less soluble fraction separates out. The latter is removed as kattha and semi solid mass as cutch. It is transferred to rectangular pits, at the bottom of which sand and clay are placed. It is kept for several days, till the cutch part is absorbed by clay and solid mass, called kattha, is taken out and moulded into blocks.

In modern methods, the red heart wood obtained by felling the tree and separating the bark and sapwood, is cut into chips mechanically and put into extractors. The steam is passed through the drug for maximum extraction. The extract is concentrated under vacuum and is cooled by refrigeration. It is then centrifuged to isolate the cake of kattha. The cake is moulded in desired sizes and dried in proper condition. By this way, a good quality marketable kattha, ready for market is obtained. The mother liquor, left behind during centrifugation is concentrated, which on cooling gives cutch.

Fig. 10.9: Twig of catechu herb

Description

Colour	:	Black catechu is light brown to black
Odour	:	None
Taste	:	Very astringent
Size	:	About 2.5 - 5 cm
Shape	:	Cube or irregular fragments of broken cubes or brick shaped pieces.

Extra Feature

The cubes as well as brick shaped pieces of catechu show the presence of vegetable debris and break with a short fracture. The broken pieces are angular with pale cinnamon-brown colour. Catechu is friable and porous.

Chemical Constituents

Black catechu contains about 10 per cent of acacatechin. It is distereoisomer of 5, 7, 3', 4' tetrahydroxy flavan-3-ols. Acacatechin is also known as acacia catechin. Acacatechin undergoes oxidation to catechutannic acid in presence of water and the latter constitutes about 30 per cent of the drug. The other contents of black catechu are catechu red, quercetin, gum and quercitrin. Black catechu does not contain chlorophyll and also the fluorescent- substance present in pale catechu.

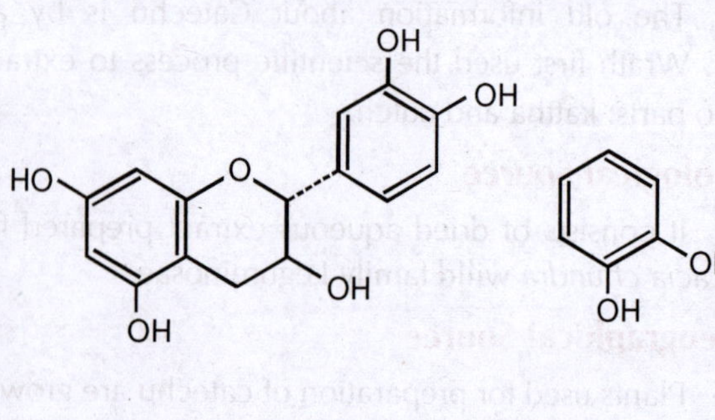

Catechin Catechol

Chemical Tests

(1) Because of the presence of catechin, black catechu gives pink or red colour with vanillin and hydrochloric acid.

(2) Catechin when treated with hydrochloric acid produces phloroglucinol, which burns along with lignin to give purple or magenta colour. For this purpose, tannin extract is taken on match stick dipped in hydrochloric acid and heated near the flame.

(3) Lime water when added to aqueous extract of black catechu gives brown colour, which turns to red precipitate on standing for some time.

(4) Green colour is produced when ferric ammonium sulphate is added to dilute solution of black catechu. By the addition of sodium hydroxide, the green colour turns to purple.

Standards of Quality

Ash value	:	Not more than 6 per cent w/w
Acid insoluble ash	:	Not more than 3 per cent w/w
Water insoluble residue	:	Not more than 25 per cent w/w
Alcohol insoluble residue	:	Not more than 90 per cent w/w

Uses

Kattha is used as an astringent externally for boils, skin eruptions and ulcers. It is also used in cough and diarrhoea. Kattha has cooling and digestive properties.

Cutch is not much used medicinally, but is used for other purposes like dyeing and colouring, water softening, reducing the viscosity of drill mud, removal of mercaptans from gasoline, protective agent for fishing nets and in the manufacture of ion-exchange resins.

PALE CATECHU

Synonyms

Gambier, Gambir Catechu.

Biological Source

It is a dried aqueous extract of the leaves and young shoots of *Uncaria gambier* Roxburgh, belonging to family Rubiaceae.

Geographical Source

The plant is indigenous to South East Asian regions like Arachipelago in Malaysia; presently, it is also cultivated in Singapore and Indonesia.

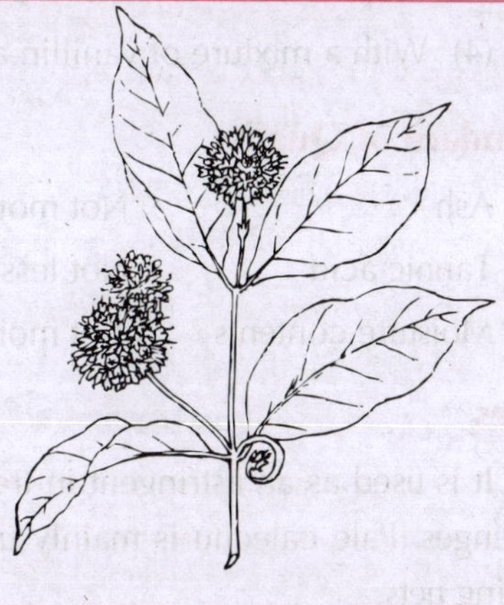

Fig. 10.10: Uncaria gambier (herb)

Cultivation and Collection

The cultivation is carried out in fields up to 170 m height and propagation is done by sowing seeds in damp soil. Nursery beds are raised and after 9 months, the seedlings are transplanted in open fields.

The first harvesting is done when the plant attains a height of 2 m. The plant yields the drug up to 20 years. The leaves and young shoots collected are boiled in pot called *Cauldron*, made up of wood and with iron bottom for 3 hours and decoction obtained is concentrated till it becomes a pasty mass with yellowish-green colour. This mass is moulded in cubes and dried.

Description

Pale catechu occurs as reddish-brown coloured cubical mass quite friable in nature. When broken, it shows cinnamon brown colour and porous nature. The drug has no odour, but highly astringent taste which first appears bitter and then sweet. When placed in water, it shows minute acicular crystals.

Chemical Constituents

The drug contains condensed tannins in the form of catechins 7-33 per cent, catechutannic acid 22-50 per cent and catechu red. The drug also contains quercetin and gambier fluorescin.

Chemical Tests

(1) It gives test for catechin by dipping a match stick in hydrochloric acid and warming it near flame similar to black catechu.

(2) Test for gambier-fluorescin: The drug is extracted with alcohol and sodium hydroxide is added to extract, followed by addition of a few drops of light petroleum. The mixture is shaken and kept for sometime. Green fluorescence is observed in light petroleum layer (distinction from black catechu).

(3) Small quantity of drug is warmed with chloroform and filtered in a porcelain dish and evaporated to dryness. Due to presence of chlorophyll, it shows greenish yellow colour.

(4) With a mixture of vanillin and hydrochloric acid, it shows pink or red colour.

Standard of Quality

Ash	: Not more than 5.0 per cent
Tannic acid	: Not less than 40.0 per cent
Moisture contents	: Not more than 25.0 per cent

Uses

It is used as an astringent in treatment of diarrhoea and also as a local astringent in the form of lozenges. Pale catechu is mainly used in dyeing and tanning industries, and also for protecting the fishing nets.

PTEROCARPUS

Synonyms

Bijasal; Indian kino tree, Red sandors wood, Malabar kino, Rakta chandan.

Biological Source

It consists of the dried juice of the plant *Pterocarpus marsupium* Linn. family Leguminosae obtained by making vertical incisions to the stem bark collected, and dried.

Geographical Distribution

It is found in hilly regions of Gujarat, Madhya Pradesh, Uttar Pradesh, Bihar and Orissa. It is also found in forests of Karnal, Kerala, West Bengal and Assam.

Macroscopic Characters

Colour : Dried juice is ruby-red

Odour : odourless

Taste : astringent

Size : 3 - 10 mm granules

Shape : angular grains

Fig. 10.11: *Pterocarpus* plant in flowering

Extra Features

The pieces of kino are angular, glistening, transparent, breaking with vitreous fracture. It is partly soluble in water (about 80 - 90 per cent), completely soluble in alcohol (90 per cent).

Chemical Constituents

Kino contains about 70 - 80 per cent of kinotannic acid, kino-red, k-pyrocatechin (catechol), resin and Gallic acid. Kinotannic acid is glucosidal tannin, while kino-red is anhydride of kinoin. Kinoin is an insoluble phlobaphene and is produced by the action of oxidase enzyme. It is darker in colour than kinotannic acid.

Crystalline principle santal, pterocarpin and homopterocarpin are also present in the wood.

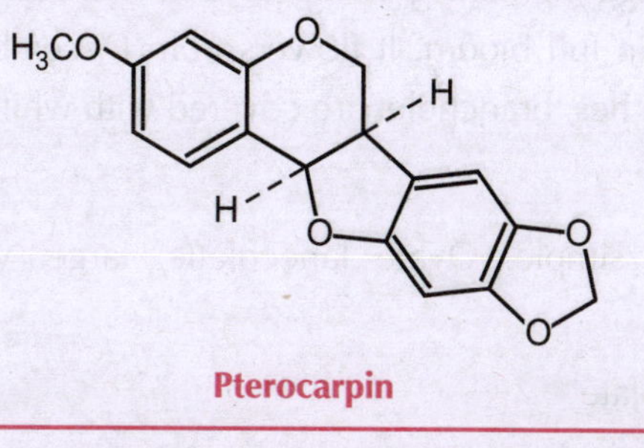

Pterocarpin

If the juice is boiled during drying, enzyme gets destroyed and thus, insolubilisation and darkening is prevented.

Identification

1. When the solution of drug is treated with ferrous sulphate, green colour is produced.
2. With alkali (like potassium hydroxide) violet colour is produced.
3. With mineral acid, a precipitate is obtained.

Standards of Quality

Water soluble extractives : Not less than 18.75 per cent

Alcohol insoluble extractives : Not more than 80.0 per cent

Uses

Kino is used as powerful astringent and also in the treatment of diarrhoea and dysentery, passive haemorrhage toothache and in diabetes. It is used in dyeing, tanning and printing.

The aqueous infusion of the wood is of considerable benefit in diabetes. Both the alcoholic and aqueous extracts of heartwood are known to possess hypoglycemic action. For the treatment of diabetes the cups made of wood are available with sales outlets of *Khadi and Gramodyog Commission*.

DHAYATI

Synonym

Dhataki.

Biological Source

It consists of the dried flowers of *woodfordia-flouribunda/Woodfordia fruticosa*. Fam. Lythraceae.

Geographical Distribution

It is found abundantly through out Indian, also in Sri Lanka, tropical, Africa, Madagaskar, China, Japan, Java and Sumatica.

It is very common on sea coast, Konkan and Mahabaleshwar in Maharashtra.

Description

It is large, deciduous, struggling herb growing wild in Monsoon forests, occasionally grown in Gardens, looks pretty, when in full bloom. It flowers from December to May every year. Dhayati plant has numerons long branches, branchelet are covered with white trichomes.

Organoleptic Characters

Leaves mostly opposite, simple. Ovate lanecolatie, large, velvety above. Nigropunctuate beneath.

Shape : Ovate lanceolate

Size : 5 - 9 × 1.3 - 2.5 cm

Colour : Dark green

Taste : Astrigent, characterstics.

Flowers : Scarlet red numerons

Fruits : Capsules 1.0 cm long irregular demiscent.

Seeds : Numerous wedge shaped brown smooth.

Extra features: India marginal nerve is preminant beneath, leaves have brown spots.

Chemical Constituents

Dhayati flowers are very rich in Tannin content and contain upto 2500% on dry basis. Leaves also contain a dye.

Use

Dhayati flowers are astringent used in Dysentry, disorders of livers and hacmorhoids.

Industrial Use

Used in manufacture of Ayurvedic Asava and Arishta to encourage / enhance fermentation processes. None of the asava or arishta can be prepared without dhyati flowers.

Shortage

Store in well closed and well filled containers away from moisture.

HAMAMELIES LEAF

Synonym

Witch Hazel leaf.

Biological Source

This consists of dried leaves of shrub of *Hamamelis Verginianae* fam. Hamamelidaceae.

Geographical Source

It is found in USA and Canada.

Description

Leaves have following characteristic

Colour : Dark greenish brown

Odour : Not marked

Taste : Bitter and astringent

Size : 5 to 12 cm long 3 - 10 cm wide.

Shape : Broadly oval to rhomboid, petiole is round and about 1.2 cm long. Margin of leaf is crenate to sinuate. Apex is acute, base chordate, unequal at base venation is pinnate.

Extra Features: Leaves are brittle and fibrous. Laterol veins straight, running to margin and also prominant on lower surface. Trichomes are slellate commercial samples are discolored and broken.

Chemical Constitutes

Harmomelis leaves contains 8 - 12 percent of Tannin (Hamamelin-Tannin), gallic acid calcium oxalate and traces of volatile oil and bitter principle.

Standards of Quality

Total ash : Not less than 7.0 per cent

Acid insoluble : Not more than 2.0 per cent

Content of stalks: Not more than 3.0 per cent

Uses

Astringent, haenostatic and sedative.

❖❖❖

Chapter 11

LIPIDS (FIXED OILS, FATS AND WAXES)

Lipids are the substances of animal or plant origin and comprise of fixed oils, fats and waxes. The basic function of oils and fats is storage of energy, apart from their several uses in medicine and industries. They are obtained by expression or extraction methods. Several tests and standards such as acid value, saponification value, iodine value, solubility in various solvents, and specific gravity are considered for determining their purity and identity.

FIXED OR FATTY OILS

These are the reserve food materials of plants and animals. Those, which are liquid at 15.5° to 16.5°, are called as fixed oils; while those which are solid or semi-solid at above temperature are termed as fats. Fixed oils, derived from plant sources, occur generally in seeds. They possess the following properties.

1. Fixed oils are thick, viscous, yellow-coloured liquids with characteristic odour.
2. They are non-volatile and cannot be distilled.
3. They do have food value and can be saponified.
4. They turn rancid on storage due to free acidity.
5. Fixed oils and fats are insoluble in water and ethyl alcohol, but soluble in organic solvents like chloroform; ether; petroleium-ether and benzene.
6. Fats and oils are esters of glycerol (three carbon trihydric alcohol) and various straight chained monocarboxylic acids, known as fatty acids. The fatty acids of natural fats have 4 - 24 carbon atoms (and any even number). These fatty acids may be saturated, monounsaturated, polyunsaturated or cyclic unsaturated. The cyclic unsaturated fatty acids are hydnocarpic, chaulmoogric and prostanoic.

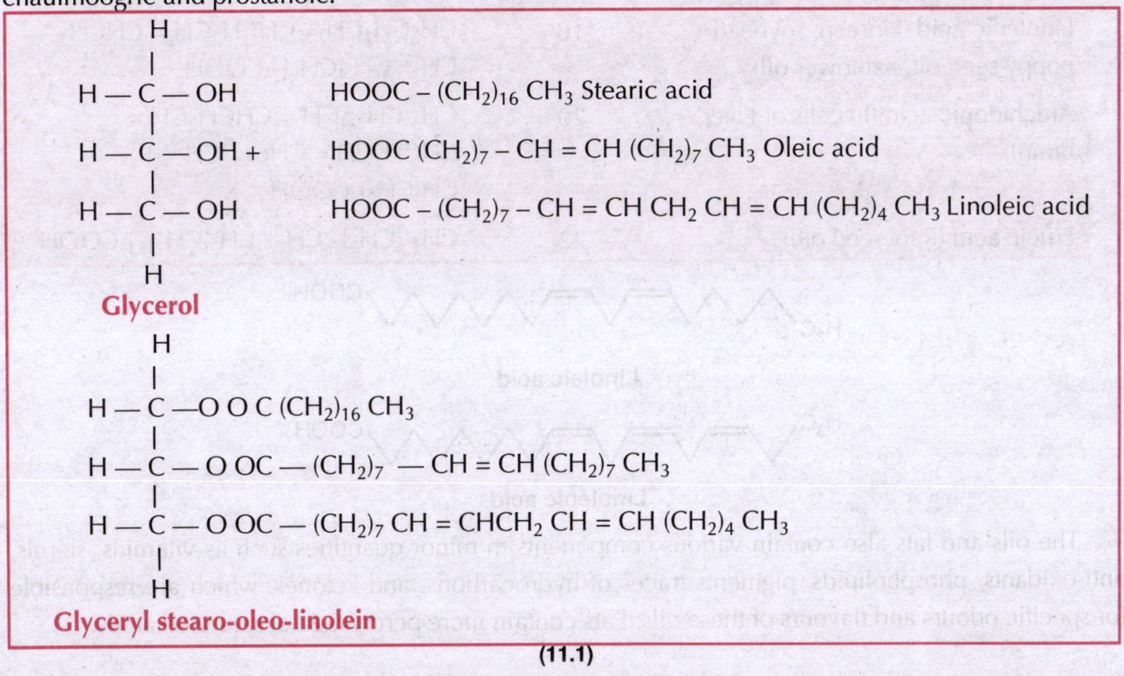

(11.1)

Table 11.1: Common Saturated Fatty Acids of Natural Origin

Fatty acids	No. of Carbon	Structural Formula	Source
Butyric acid	4	$CH_3 [CH_2]_2 COOH$	Butter fat
Caproic acid	6	$CH_3 [CH_2]_4 COOH$	Palm kernel oil
Caprylic acid	8	$CH_3 [CH_2]_6 COOH$	Coconut oil
Capric acid	10	$CH_3[CH_2]_8 COOH$	Palm oil
Lauric acid	12	$CH_3 [CH_2]_{10} COOH$	Coconut oil
Myristic acid	14	$CH_3 [CH_2]_{12} COOH$	Palm oil
Palmitic acid	16	$CH_3 [CH_2]_{14} COOH$	Arachis oil, sesame oil
Stearic acid	18	$CH_3 [CH_2]_{16} COOH$	Arachis oil, sesame oil
Arachidic acid	20	$CH_3 [CH_2]_{18} COOH$	Mustard, peanut oil
Behenic acid	22	$CH_3 [CH_2]_{20} COOH$	Mustard, peanut oil, rapeseed oil
Lignoceric acid	24	$C_{23} H_{47} COOH$	Peanut oil

Table 11.2: Some Unsaturated Fatty Acids of Natural Origin

Name and Source	No. of Carbon	Structure
Palmitoleic acid (Cotton seed oil)	16	$CH_3 [CH_2]_5 CH = CH [CH_2]_7 COOH$
Oleic acid (Safflower oil, corn oil)	18	$CH_3 [CH_2]_7 CH = CH [CH_2]_7 COOH$
Linoleic acid (Sesame, sunflower Linseed, corn oil, Safflower oil)	18	$CH_3 [CH_2]_4 CH = CHCH_2 CH = CH [CH_2]_7 COOH$
Linolenic acid (Linseed, soybean poppy seed oil, Safflower oil)	18	$CH_3CH_2CH = CHCH_2CH = CHCH_2 - CH = CH [CH_2]_7 COOH$
Arachidonic acid (In cells of Liver, brain)	20	$CH_3(CH_2)_4CH = CHCH_2CH = CHCH_2CH = CHCH_2CH = CH(CH_2)_3COOH$
Erucic acid (Rapeseed oil)	22	$CH_3 [CH_2]_7 CH = CH [CH_2]_{11} COOH$

Linoleic acid

Linolenic acid

The oils and fats also contain various components in minor quantities such as vitamins, sterols, anti-oxidants, phospholipids, pigments traces of hydrocarbons, and ketones, which are responsible for specific odours and flavours of these oils. Fats contain more percentage of saturated acids.

7. Physiologically, they are emmolients and demulcents.

The common examples of fixed oils obtained from vegetable sources are arachis oil, castor oil and sesame oil. The oils, obtained from animal sources are cod liver, and shark liver oil.

The unsaturated fatty acids namely lenoleic acid, linolenic acid are called ***Essential fatty acids,*** as they are not produced in the body and required to be provided in the diet.

Chemical Tests for Identification of Fixed Oils or Fats

Fixed oils or fats can be confirmed by chemical test for glycerine which is produced by their hydrolysis. The test is performed as under:

(a) Using Sodium Hydroxide: Mix 1 ml 1 per cent copper sulphate solution and 5 drops of the fixed oil or fat. Then add 5 drops of 10 per cent sodium hydroxide solution. A clear blue solution is obtained which shows glycerine is present in the sample. The cupric hydroxide formed in the reaction does not precipitate out as it is soluble in glycerine.

(b) Using Sodium Hydrogen Sulphate: Take 5 drops of the sample in a test tube and add a pinch of sodium hydrogen sulphate. Pungent odour emanates from the tube indicating glycerine is present in the sample. The pungent odour is due to the formation of acrolein.

Adulteration in Oils, Fats and Waxes

Adulteration of costly oils with cheaper one is quite common currently, especially if those are meant for edible purposes.

It is a normal practice to ascertain the purity of oils and fats by some routine physical or chemical test and by comparing with analytical constants such as iodine value, saponification value, refractive index, specific gravity, etc. But the range of values of these analytical constants is not very sharp and very frequently, these values for different oils overlap each other.

However, detection of adulterants based on the development of characteristic colour, turbidity or a precipitate is possible. These methods are qualitative and depending upon the skill of the analyst and can detect even as low as 0.1 per cent concentration of adulterant.

1. Detection of sesame oil as adulterant: To 2 ml oil, add 1 ml hydrochloric acid containing 1 per cent sucrose and keep aside for 5 minutes. The presence of pink colour in the acid layer indicates presence of sesame oil.

2. Detection of cotton seed oil as adulterant by Halphen's test: To 2 ml oil, add 2.5 ml alcohol and 2.5 ml solution of sulphur in carbon disulphide (1 per cent). Heat the mixture. Formation of pink or red colour indicates presence of cotton seed oil.

3. Detection of karanja oil as adulterant: Take 5 ml of sample oil in a test tube; add 10 ml of ethyl alcohol to it followed by addition of 5 drops of ferric chloride solution. Shake vigorously for two minutes and allow to stand for two hours. The development of light blue to dark colour in the alcoholic layer indicates the presence of karanja oil or any other oil containing phenolic group.

4. Detection of castor oil as adulterant: Take 1 ml of the sample to be tested in a clean test tube and to it add acidified petroleum ether (60 - 80°). Shake it thoroughly. Then to it add a drop of ammonium molybdate solution. Development of white turbidity indicates the presence of castor oil.

5. Detection of argemone oil as adulterant: It is an adulterant to mustard oil. Ferric chloride reacts with argemone oil in hydrochloric acid medium to produce small needle shaped reddish-brown crystals. However, if content of argemone oil is less than 1 per cent, needle-shaped crystals can be observed under microscope.

6. Detection of linseed oil as adulterant: About 1 ml sample is dissolved in 5 ml chloroform in a dry test-tube and one ml liquid bromine is added. The tube is stoppered and cooled in ice bath and to it added 1.5 ml rectified spirit and 10 ml ether. It is then shaken thoroughly again for 30 minutes. The appearance of precipitate (due to formation of hexabromide) indicates presence of linseed oil.

7. Detection of mineral oil as adulterant: To 1 ml oil or fat under test, add 25 ml alcoholic caustic potash. Boil in water bath, till it becomes quite clear. Transfer the contents of the flask to a wide mouth test tube and to it, add 25 ml distilled water along the side of the test tube, keeping the tube shaken during addition of water. The presence of turbidity indicates presence of mineral oil.

Classification of Fixed Oils

Depending upon sources and physical characteristics, fixed oils and fats can be sub-classified as follows.

Oils and Fats
(Vegetable)

Fats	Non-drying oils	Semi-drying oils	Drying oils
Cocoa butter	Olive oil	Castor oil	Linseed oil
Kokum butter	Peanut oil	Mustard oil	Poppy seed oil
Nutmeg butter	Almond oil	Sesame oil	Hemp oil
Coconut oil	Croton oil	Rapeseed oil	Walnut oil
Palm oil	Rice-bran oil	Cottonseed oil	
Mango kernel oil		Safflower oil	

Oils and Fats
(Animal)

Marine animals		Terrestrial animals	
Fats	**Oils**	**Fats**	**Oils**
Bone tallow	Cod liver oil	Lard	Lard oil
	Shark liver oil	Mutton-tallow	Neat-foot oil
	Whale oil	Butter Suet	
	Halibut liver oil		
	Tuna fish liver oil		

Classification of Lipids

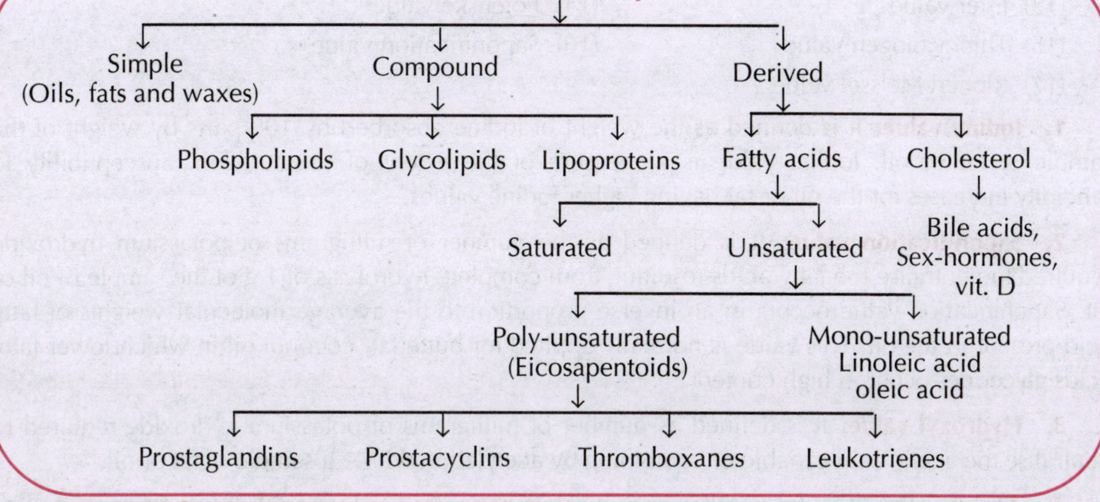

WAXES

Waxes are unctuous, fusible, variably viscous solid substances, with characteristic waxy lustre. These are esters of fatty acids with high weight **monohydric alcohol**, such as cholesterol, cetyl alcohol, melissyl alcohol, etc. They are insoluble in water, but soluble in most organic solvents. The esters in waxes are generally more resistant to saponification than the glycerides of oils and fats. They are obtained from vegetable and animal sources.

(a) Vegetable : Seasal wax, carnauba wax, Japan wax, and bayberry wax

(b) Animal : Spermaceti, bees wax, wool fat

The difference between fats and waxes is that fats may be saponified by either aqueous or alcoholic alkali, but waxes are only saponified by alcoholic alkali. Waxes are unsuitable for internal consumption since there is no enzyme in human body to hydrolyse them internally.

Analytical Parameters for Oils and Fats

The properties of oils and fats vary along with the degree of unsaturation, average molecular weight and also acidity from hydrolysis. A number of parameters are used for their analysis which are included under physical constants and chemical constants. Physical constants include viscosity, specific gravity, refractive index, solidification point etc. Under the group of chemical constants, conventionally the parameters like iodine value, acid value, peroxide value, saponification value, unsaponifiable matter are noted. In recent times, gas chromatographic determination of fatty acid profiles has been introduced for identification of fats and oils. Following is a brief idea about some of the analytical parameters grouped under chemical constants.

Normally, the following physical standards are covered for the complete study of Lipids, where so ever applicable:

(1) Specific gravity
(2) Refractive index
(3) Viscosity
(4) Optical rotation
(5) Free fatty acids
(6) Solidifying point
(7) Unsaponifiable matter
(8) Acid value
(9) Iodine value
(10) Acetyl value

(11) Hydroxyl value
(12) Peroxide value
(13) Ester value
(14) Polenske value
(15) Thiocyanogen value
(16) Saponification value
(17) Richert Meissel value

1. Iodine value: It is defined as the weight of iodine absorbed by 100 parts by weight of the sample of fat or oil. Iodine value is a measure of the extent of unsaturation. Susceptibility to rancidity increases for the oil or fat having higher iodine values.

2. Saponification value: It is defined as the number of milligrams of potassium hydroxide required to neutralise the fatty acids resulting from complete hydrolysis of 1 g of the sample of oil or fat. Saponification value occurs in an inverse proportion to the average molecular weights of fatty acid present in the oil. This value is normally applied for butterfat, coconut oil in which lower fatty acids glycerides occur in high content.

3. Hydroxyl value: It is defined as number of milligrams of potassium hydroxide required to neutralise the acetic acid capable of combining by acetylation with 1 g sample of fat or oil.

4. Acetyl value: It is the number of milligrams of potassium hydroxide required to neutralize the acetic acid obtained when 1 g of sample acetylated oil is saponified. Except castor oil (acetyl value 150), most of the oils and fats have low acetyl value (3 - 15).

5. Unsaponifiable matter: It is the matter present in fats and oil, which after saponification by caustic alkali and subsequent extraction with an organic solvent, remains non-volatile on drying at 80°C. It includes sterols (phytosterol and cholesterol), oil soluble vitamins, hydrocarbons and higher alcohols. Paraffin hydrocarbons can be detected by this method as adulterants.

6. Acid value: It is defined as the number of milligrams of potassium hydroxide required to neutralise the free acids present in 1 g sample of fat or oil. Generally, rancidity causes free fatty acids liberation, hence acid value is used as an indication of rancid state.

7. Peroxide value: It is a measure of peroxides present in oil. A peroxide value is generally less than 10 mEq/kg in fresh samples of oil. Due to temperature or storage, rancidity occurs causing increase in peroxide values.

8. Kreis test (rancidity index): Due to rancidity, epihydrin aldehyde or malonaldehyde are increased which are detected by Kreis test using phloroglucinol which produces red colour with the oxidized fat.

9. Ester value: It is defined as number of milligrams of potassium hydroxide required to combine with fatty acids which are present in glyceride form in 1 g sample of oil or fat. Difference between saponification value and acid value is ester value.

10. Reichert Meissle Value: This value is a measure of volatile water soluble acid contents of the fat. It is defined as number of millilitres N/10 potassium hydroxide solution required to neutralise the volatile water soluble fatty acids obtained by 5 g fat.

11. Polenski value: It is defined as the number of millilitres of N/10 potassium hydroxide solution required to neutralise water-insoluble, steam - distillable acids liberated by hydrolysis of 5 gm of fat.

The distribution of lipids in natural drugs, their chemical constituents and pharmaceutical uses are briefly summarized in Table 11.3.

Table 11.3: Summary of Lipid Drugs

Sr. No.	Name of drug and Synonym	Biological Source	Active Constituents	Uses
1.	Arachis oil (Peanut oil)	*Arachis hypogaea* (Leguminosae)	Glycerides of oleic, linoleic and palmitic acids	Base for oily injectables nutritive
2.	Black Mustard oil (Sarson oil)	*Brassica nigra, Brassica juncea* (Cruciferae)	Allyl isothiocynate and fatty acid glycerides, sinigrin	Rubefacient, counter irritant and condiment
3.	Carnauba wax (Brazil wax)	*Copernicia verifera* (Palmae)	Carnaubic and cerotic acid	Depilatories, cosmetic products
4.	Castor oil	*Ricinus communis* (Euphorbiaceae)	Ricinoleic and isoricinoleic acid	Laxative, lubricant, cosmetics, urethanes
5.	Hydrogenated castor oil (Castor wax)	*Ricinus communis* (Euphorbiaceae)	Triglyceride of hydroxy stearic acid	Ointment base
6.	Chaulmoogra oil (Hydnocarpus oil)	*Hydnocarpus wightiana* (Flacourtriaceae)	Hydnocarpic and chaulmoogric acid	Antileprotic
7.	Cocoa-butter (Theobroma oil)	*Theobroma cacao* (Sterculiaceae)	Glycerides of stearic, oleic and palmitic acids	Suppository base
8.	Corn oil (Maize oil)	*Zea-mays* (Gramineae)	Oleic and linoleic acids, phospholipids, vit. E	Dietary supplement, margarine
9.	Cotton seed oil	*Gossypium herbaceum* (Malvaceae)	Glycerides of saturated and unsaturated fatty acids	Pharmaceutical aid
10.	Evening primrose oil	*Oenothera biennis* (Onagraceae)	Gamma linoleic acid and linoleic acid	Nutritive and cosmetics
11.	Jojoba oil	*Simmondsia chinensis* (Buxaceae)	Mixture of long chain esters of fatty acids	Cosmetics and lubricant
12.	Karanja oil	*Pongamia glabra* (Papilionaceae)	Glycerides of fatty acids, a sitosterol	Non-edible oil, cosmetics, antiseptic
13.	Kokum butter (Goa-butter)	*Garcinia indica* (Guttiferae)	Glycerides of stearic, palmitic and linolenic acids	Demulcent, emollient, confectionary fat

Contd...

14.	Lecithin (Lecithol)	Soybean oil, egg yolk	Phosphates of phosphatidyl choline and phosphatidyl ethanolamine (cephaline)	Source of choline, also used in cosmetics, soaps
15.	Linseed oil	*Linum usitatissimum* (Linaceae)	Oleic, linoleic and linolenic acids	Lotions, liniments, ointments varnish and paint industry
16.	Neem oil (Margosa oil)	*Azadirachta indica* (Meliaceae)	Glycerides of saturated and unsaturated fatty acids	Non-edible oil, cosmetics, spermicidal
17.	Olive oil	*Olea europoea* (Oleaceae)	Olein, palmitin and linolein	Vehicle for oily injection and emolient
18.	Poppy seed oil (Poppy oil)	*Papaver somniferum* (Papaveraceae)	Glycerides of linoleic and oleic acids	Drying oil, iodised oil, paint and varnish industries
19.	Rice bran oil	*Oryza sativa* (Gramineae)	Glycerides of oleic and linoleic acids, tocopherol	Food oil, cosmetics
20.	Safflower oil	*Carthamus tinctorius* (Compositae)	Glycerides of linoleic and oleic acids	Edible oil, oleo margarine vegetable ghee
21.	Sesame oil (Gingelly oil)	*Sesamum indicum* (Pedaliaceae)	Glycerides of oleic, linoleic, palmitic acids and sesamolin	Demulcent, laxative, base for oily injectables
22.	Soya-bean oil	*Glycine soja* Papilionaceae	Omega-3-fatty acids (EPA + DHA)	Source of sterols.
23.	Wheat germ oil	*Triticum aestivum* (Graminae)	Glycerides of fatty acids	Nutritive and source of vit. E.
24.	Almond oil	*Prunus amygdalus* or *Amygdalus communis* Var. *amara* or mixture of both (sweet and bitter) family Rosaceae.	Saturated and unsaturated fatty acids.	For toiletry preparations such as soap, shampoo and hair oils.

Contd...

25.	Coconut oil	*Cocus nucifera* fam. Arecae.	Triglycerides of saturated fatty acids and caprylic acid, capric acid, lauric acid, myrestic acid.	Emollient and for edible purposes.
26.	Canola oil	*Brassia-napus* rapa or *Brassica compestris* or Brassica junsed fam. Erusiferie.	40.0% fixed oil and contain linoleic and linolenic fatty acids in proportion of 2:1, phytosterols like β-sitosterol campesterol.	Substitute for rape-seed oil for cooking and preparation of lipsticks.
27.	Mango kernel oil	*Mangifera indica Anacardiaceae*	olein, lenolenic acid tocopherols	substitute for cocoa-butter
28.	Sunflower oil	*Helianthus annus*	Arachidic acid, lenolenic acid	Emolliant, Humectant
29.	Bees wax (Yellow bees wax)	*Apis mellifica, A. indica* (Apidae)	Myricyl palmitate, cerotic acid	Hardening agent, ointment base
30.	Cod-liver oil	*Gadus morrhua* (Gadidae)	Vit. A and Vit. D, glycerides of fatty acids	Deficiency of vit. A and D, emollient, in treatment of rickets
31.	Halibut liver oil	*Psettodes specis*	Vit. A and Vit. D, glycerides of fatty acids	Deficiency of vit. A and D, emollient, in treatment of rickets
32.	Shark liver oil	*Hypoprion brevirostris*	Vit. A and glycerides of fatty acids	Deficiency of vit. A, in treatment of rickets
33.	Tuna fish liver oil	*Katsuwonus pelamis*	Omega-3 fatty acids	Dietory supplement
34.	Lard (Adeps)	*Sus scrofa* (Suidae)	Olein stearin	Ointment base
35.	Spermaceti	*Physeter macrocephalus* (Physeteridae)	Cetyl palmitate	Cosmetic creams, source of acetyl alcohol
36.	Suet	*Ovis aries* (Bovidae)	Palmatin, stearin	Ointment base
37.	Hydrous wool fat (Lanolin)	*Ovis aries* (Bovidae)	Esters of cholesterol and carnaubic and oleic acids	Absorbable ointment base

LIPIDS OF PLANT ORIGIN:

ARACHIS OIL

Synonyms

Groundnut oil, Peanut oil, Sweet oil.

Biological Source

It is fixed oil expressed from the seed kernels of cultivated varieties of *Arachis hypogaea* Linn., family Leguminosae.

Geographical Source

Arachis plant is indigenous to Brazil, but now extensively cultivated in Southern United States, China, Africa and India.

Method of Preparation

Peanut kernels contain about 40 - 50 per cent of fixed oil. The graded and cleaned kernels are crushed and the oil is separated by expression. Further quantity of oil is collected by hot expression method. Both fractions are mixed together. It is then allowed to stand for some time to separate stearin and filtered. The oil is then treated with fueller's earth or carbon for bleaching and filtered thoroughly by using filter presses.

Peanut cake or peanut oil meal is a rich source of proteins and contains about 39 - 45 per cent of crude proteins. In the crude form, it is used as cattle food or as raw material for preparation of protein isolate.

Fig. 11.1 (a): *Arachis hypogaea* plant with fruits

Fig. 11.1 (b): Peanuts Isolated

Description

Colour	:	Oil is pale yellow coloured liquid
Odour	:	Faint and characteristic
Taste	:	Bland and nut-like
Solubility	:	It is slightly soluble in alcohol and soluble in solvent ether, chloroform and light petroleum (40 – 60°).

Extra Features: On exposure to air, it thickens very gradually and may become rancid. It is sterilized by dry heat process. It is non-drying oil.

Chemical Constituents

Arachis kernels contain about 40 – 45% of fixed oil. Arachis oil consists of glycerides of fatty acids, chiefly oleic, linoleic, stearic and arachidic acid. Other acids are lignoceric and palmitic acids.

Oleic acid : $CH_3 (CH_2)_7 CH = CH (CH_2)_7 COOH$

Palmitic acid : $CH_3 (CH_2)_{14} COOH$

Arachidic acid : $CH_3 (CH_2)_{18} COOH$

Linoleic acid : $CH_3 (CH_2)_4 CH = CHCH_2CH = CH (CH_2)_7 COOH$

Raw oil may contain aflatoxin (a carcinogenic substance produced by the fungal growth) and colouring substances. Refining of the oil removes colour, odour, impurities and aflatoxin, too. The polyunsaturated fatty acid content of this oil is about 31.0 per cent.

Chemical Test

Under ultra-violet radiation Arachis oil shows blue fluorescence.

Standards of Quality

Unsaponifiable matter	: Not more than 1.5 per cent.
Acid value	: Not more than 0.5
Iodine value	: 85 - 105
Saponification value	: 185 - 195
Weight per ml	: 0.911 - 0.920
Refractive index	: 1.467 - 1.470
Hydroxyl value	: 2.5 – 9.5
Freezing point	: – 5ºC
Thiocyanogen value	: 67 – 73
Peroxide value	: Not more than 0.5

Arachis oil for parenteral dosage form should contain not more than 0.3 per cent of water.

Uses

Peanut oil is a solvent for intramuscular injections. Since it resembles to olive oil, it is used in the preparation of liniments, plasters and soap. Being a non-drying oil, it is valuable as a lubricant. It produces firm and excellent white soap. Principally, it is used as an edible oil.

Substitution and Adulteration

Peanut oil is adulterated with cotton seed oil or mixed with sesame oil. The presence of these oils is indicated by the following tests.

1. Take 2.5 ml of peanut oil and to it add 2.5 ml of alcohol and 2.5 ml of solution of sulphur in carbon disulphide (1 per cent). Heat the mixture. The presence of pink or red colour indicates cotton seed oil.

2. Shake 2 ml of arachis oil with 1 ml of hydrochloric acid containing sucrose (Baudouin's test). Keep it aside for 5 minutes. The presence of pink colour in the acid layer indicates the presence of sesame oil.
3. Foreign fixed oils can be confirmed by gas chromatographic methods.

BLACK MUSTARD OIL

Synonym

Sarson ka tel.

Biological Source

It is a fixed oil obtained from matured seeds of *Brassica nigra* (L) Koch or *Brassica juncea* L. Czern, family Cruciferae (Brassicaceae), also refer page 9.75.

Geographical Source

It is cultivated in India, China, Canada and England.

Description

The oil is prepared by expression. It is yellow coloured liquid of strong acrid odour until refined.

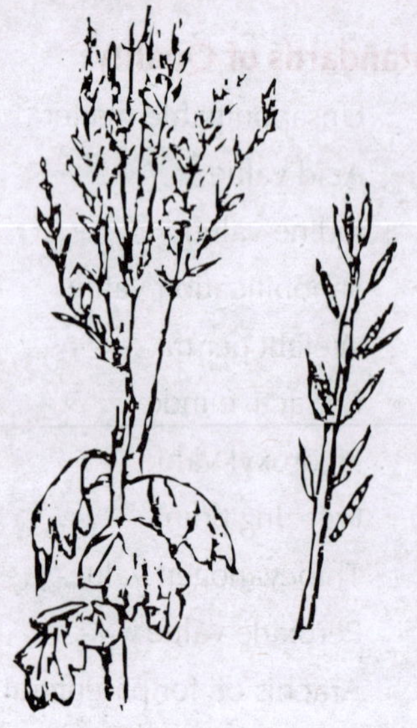

Fig. 11.2: Mustard plant

Chemical Constituents

Mustard oil contains glycerides of arachidic (0.5 per cent), behenic (2-3 per cent), eicosenoic (7-8 per cent), erucic (40-60 per cent), lignoceric (1-2 per cent), linoleic (14-18 per cent), linolenic (6.5-7.0 per cent), oleic (20-22 per cent) and myristic (0.5-10 per cent) acids.

(Black mustard seeds contain 35-40 per cent of fixed oil and a glycoside known as sinigrin (also known as potassium myronate) along with an enzyme myrosin. The myrosin acts on sinigrin in presence of water and produces potassium acid sulphate and allyl isothiocynate. Allyl isothiocynate is responsible for the strong acrid smell of volatile oil of mustard produced on hydrolysis of glycoside, which are not present in edible oil). Black mustard seeds also contain 0.75 to 1.35 per cent volatile oil.

Standards of Quality

Specific gravity	:	0.914 - 0.923
Unsaponifiable matter	:	0.9 - 1 per cent
Refractive index	:	1.472 - 1.4733
Solidifying point	:	–11 - 17°C
Iodine value	:	96 - 194
Saponification value	:	173 - 184

Uses

Fixed oil is used as edible oil after refining, but medicinal properties are due to allyl isothiocynate, which is a local irritant and emetic. If applied externally, it is rubefacient and vesicant. It is also used as condiment and in manufacture of soap. Refined mustard oil is used in preparation of vegetable ghee.

WHITE MUSTARD SEEDS

Biological Source

White Mustard seeds are obtained from *Brassice alba* herb; Fam; Crusiferae.

Geographical Source

India, China, Canada and England.

Description

White mustard seeds have:

Colour : Yellow

Odour : Odourless

Taste : Pungent taste

Size : 2 mm

Shape : Spherical

Extra features: When soaked in water become mucilaginous fruits are appressed and smooth. Each white mustard fruit contains only four to six seeds. Fruits are hairy.

Chemical Constituents

Seeds contans approximately 30.0% of fixed oil, 25% proteins and mucilage. Starch absent, crystalline glycosiae sinalbim and enzyme myrosin is also present in seeds.

CARNAUBA WAX

Synonym

Brazilian wax.

Biological Source

It is an exudate from pores of the leaves of the Brazilian wax-palm tree *Copernicia prunifera* (Mueller) HE Moore, *Copernicia cerifera* (Arrunda), da-camara (M artius), and family Palmae.

Geographical Source

Brazilian wax trees are found in North Brazil to Argentina in South America.

Preparation of Carnauba Wax

The leaves of Brazilian wax-palm are collected, dried, and then spread on a piece of cloth. By brushing and beating, the wax is separated. It is then melted, processed further to purify and poured into the moulds.

Fig. 11.3: *Carnauba* tree

Description

It is hard greenish solid wax with crystalline fracture.

It has sharp characteristic odour and bland taste. It is not soluble in water and is soluble in fat solvents.

Chemical Constituents

It contains esters of hydroxylated fatty acids, i.e. carnaubic, cerotic acid and melissyl cerotate.

Standards of Quality

Melting point	:	82 - 85.5ºC
Specific gravity	:	0.990 - 0.999
Acid value	:	04 - 07
Saponification value	:	78 - 89
Iodine value	:	13 - 15
Refractive index	:	1.45 - 1.459

Uses

Carnauba wax is used for preparation of cosmetic products, depilatories and deodorant sticks. It is also used for tablet coating. High quality shoe polishes and automobile waxes are other products made from carnauba wax. During 1995-96, India has exported carnauba-wax of worth ₹ 65.0 lakhs.

CASTOR OIL

Synonym

Ricinus oil.

Biological Source

Castor oil is the fixed oil obtained by the cold expression of the seeds of *Ricinus communis*, family Euphorbiaceae.

Geographical Source

Castor seeds are produced in almost all tropical and sub-tropical countries. In India, castor is one of the major oilseed crops, and India is the second largest producer of castor seeds in the world, producing about 2,80,000 tonnes per annum. Brazil, U.S.S.R., Thailand, U.S.A. and Romania are other countries producing this drug on large scale. In India, it is largely grown in Andhra Pradesh, Gujarat, and Karnataka. Andhra Pradesh produces about 60 per cent of the total crop in India. About 0.6 million hectares of land in India is under cultivation of castor.

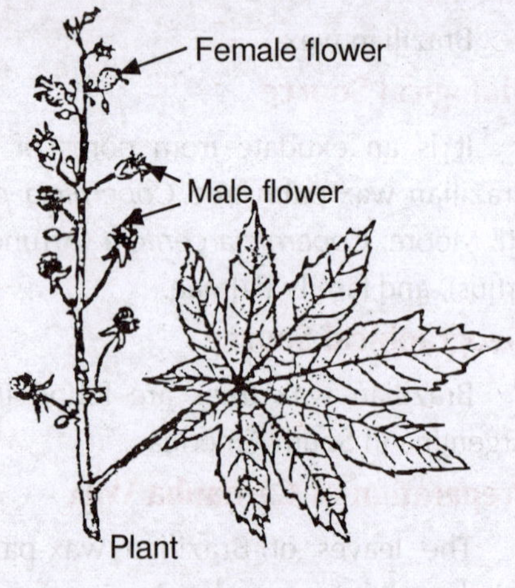

Fig. 11.4: *Ricinus communsis* tree in flowering

India exported, medicinal castor oil worth ₹ 376 crores during 1994-95. The export of medicinal caster oil during 1996-97 was to the tune of approximately ₹ 550 crores. The other two forms of castor oil i.e. hydrogenated and dehydrated are also getting exported regularly.

Composition of Seeds

Since castor oil is a fixed oil of therapeutic importance and purgative in action, it cannot be consumed freely. Apart from the oil, seeds also contain some of the toxic substances. Castor seeds consist of 75 per cent kernel and 25 per cent of hull. Seeds weigh from 0.1 to 1 g. Castor seeds are rich in phosphorus content and most of it is in the form of phytin. Hull is rich in mineral and also contains an alkaloid ricinine, resin, pigment, etc. The oil content of the kernel varies from 36 per cent to 60 per cent. Amongst different varieties, the Hyderabad muggelai variety is supposed to be the richest (about 48 per cent) in the oil content. Castor seeds contain several enzymes including lipase, maltase and invertase. The proteinous toxic principle ricin, constituting about 3 per cent of the whole seeds, is poisonous. Aruna and Gujarat castor hybrid-3 varieties are of shortest span (120 days).

Preparation of Medicinal Castor Oil

Castor oil can be prepared by two different methods, the first being the crushing of whole or decorticated seeds in power driven hydraulic presses and the second one known as Ghani, which consists of manually operated screw press driven by bullocks. For commercial extraction, the first method is adopted. The oil thus produced, is a non-medicinal castor oil.

The seeds are graded and freed of impurities like metallic pieces of iron and sand. The seeds are decorticated and hulls are removed. If the seeds are not decorticated, the manurial value of the cake increases. But, for medicinal purposes, it is desired that the seeds should be decorticated, as it improves the colour of the oil and also helps in controlling its acid value.

Decorticated seeds are pressed under hydraulic press with a pressure of 2 tonnes per square inch, which helps in extracting out 30 per cent of the oil present in the seeds at room temperature. The oil is known as *cold drown oil*. Rest of the oil from the seeds is removed by further increasing the pressure, and sometimes by hot pressing or even by solvent extraction process. The oil, thus processed, is not suitable for medicinal purposes. The cold drown oil is then steamed at 80°C, to destroy the enzyme lipase and ricin (toxic protein). It is then bleached and de-acidified with sodium carbonate to remove free fatty acid. If necessary, oil is washed with hot water before steaming to remove the mucilaginous matter present in the oil. Finally, it is treated with activated earth or animal charcoal to remove the impurities by adsorption and filled into the containers.

Description

Colour : Castor oil is pale yellow or almost colourless liquid.

Odour : Slight and characteristic.

Taste : First it is bland but afterwards slightly acrid, and usually nauseating.

It is a viscous and transparent liquid.

It is soluble in alcohol (an exception to the category of fixed oils); miscible in chloroform, solvent ether, glacial acetic acid and petroleum ether. It is insoluble in mineral oil.

Chemical Constituents

Castor seeds contains 35 – 50% of castor oil. Castor oil chiefly contains triglyceride of ricinoleic acid (about 80 per cent). Other glycerides are also present in the drug, where the fatty acids are represented by isoricinoleic, linoleic, stearic and isostearic acids. The viscosity of the castor oil is due to ricinoleic acid.

Ricinoleic acid: $CH_3\ (CH_2)_6\ CHOH\ CH_2\ CH = CH\ (CH_2)_7\ COOH$

Castor oil also contain heptaldehyde (heptanal) undecenoic acid and sebacic acid.

Identification

1. It mixes with half its volume of light petroleum ether (40 – 60°) but insoluble in double the volume of petroleum ether.
2. Add to the oil an equal volume of ethanol; clear liquid is obtained. On cooling at 0°C and on storage for three hours; the liquid remains clear (distinction from other fixed oils).

Standards of Quality

Weight per ml	:	0.945 to 0.965 g
Acid value	:	Not more than 2
Acetyl value	:	Not less than 143
Iodine value	:	Between 82 to 90
Saponification value	:	Between 176 and 187
Optical rotation	:	Between + 3.5° and + 6.0°
Solidifying point	:	– 10 to –18°C
Hydroxyl value	:	Not less than 150
Viscosity	:	6 – 8 poises
Refractive index	:	Between 1.4758 and 1.4798

Uses

Castor oil is cathartic. It is also used for lubrication commercially. Several other forms of the castor oil, such as dehydrated castor oil (DCO) or hydrogenated castor oil (HCO) are used industrially for several other purposes. The fatty acid like ricinoleic; heptaldehyde and undecenoic acid are the other substances commercially prepared out of the castor oil. Turkey red oil and soap are the other commercial products, extensively used in textile industry. It is used in preparation of paints, enamel, varnishes, grease, polishes, printing ink, hydraulic and brake spirit with little modifications.

Castor seed cake contains a very poisonous protein known as ricin. That's why castor cake is not used in fodder to cattles.

The cathartic property of the castor oil is due to irritant action of ricinoleic acid. Castor oil is often given orally or as aromatic castor oil or in the form of capsules. It is used in abortificient paste and ricinoleic acid is used in contraceptive creams and jellies. Atropine and cocaine for opthalmic purposes are suspended in castor oil.

Cosmetic Uses

Castor oil is used as emollient in the preparation of lip-sticks and in the form of **sulphorecinolate** being strong bactericide in tooth formulations. Other cosmetic formualtions in which it is used include perfumed hair oil and hair fixers. It imparts transparency to soaps hence used in cosmetic soaps.

Commercial Uses

For the commercial manufacture of sebacic acid ($HOOC(CH_2)_8COOH$), which is a raw material for synthesis of resins. Dehydrated castor oil is used in manufacture o linoleum and alkyl resins.

Castor oil imparts transparency to soaps. Dehydrated castor oil is used in manufacture of linoleum and alkyl resins.

Industrially, it is used in the production of rigid, semi-rigid foams and urethanes called elastomers, used for trolley wheels. It is a basic raw material in the manufacture of nylon - 11. Fungicide undecylenic acid is prepared from castor oil. Castor cake is used as a source of enzyme lipase. Castor oil does not freeze at lower temperature and hence, is valuable as lubricant.

HYDROGENATED CASTOR OIL

Synonyms

Castor wax, opal wax.

Biological Source

It is decolourised, deodourised, and hydrogenated castor oil. It mainly contains the triglycerides of hydroxy stearic acid.

Description

It is white to yellowish powder of uniform consistency and texture, it is unctuous to touch.

Solubility

It is soluble in ether, chloroform and petroleum ether but insoluble in water, alcohol and mineral oil.

Standards of Quality

Melting point	: 85 - 88°C
Acetyl value	: Not less than 193
Acid value	: Not more than 2.0
Iodine value	: Not more than 5.0
Peroxide value	: Not more than 5.0
Hydroxyl value	: 154 - 162
Saponification value	: 176 - 182

Additionally it should qualify test for heavy metals (Not more than 10 ppm) and free fatty acids.

Uses

It is used as substitute for spermaceti, bees wax and carnauba wax, in the preparation of ointments, creams and polishes.

Storage

It is stored in well closed, well-filled containers away from light and at temperature not exceeding 30ºC.

CHAULMOOGRA HERB

The unique well known herbal antiseptic plant available in Mahabaleshwar. It available in Western Ghats. It is 10 – 12 metres in height evergreen found at an attitude of 300 metres.

Biological Source

Chaulmoogra consists of plants known as *Hydnocanpus anthelmintica* or *Hydroearpus Wightina* or *Hydnocarpus heterophylla* and even other species of *Hydrocarpus* family *Flacourtriaceae.*

Geographical Source

It is native of Myanmar, Thailand and East India, Sri Lanka and Bangladesh are the other sources. Found in Ratnagiri and Sindhudurg forests. Plants are available near water streams or rivers and servive for 15 – 20 years only. Now they are very rear.

Description

The plant bears evergreen glabrons leaves. Plant is 10 – 12 metres in height. Leaves are pointed acuminate 10 to 15 cm long and 3.0 – 8.0 centimeters wide. Male and female plants are separate. Trunk of the tree is light brown and very much rough. It flowers in March or April and fruits are available in May onwards.

Fig. 11.5: Chaulmoogra herb with fruit

Fruits are hard and spherical and fruits-------- twelve month for maturity and bear 6 to 10 seeds. Fruits are edible and consumed by most of forest animals. Annual number of fruits per tree is about 200 – 800. Plants bears 7 to 8 kg seeds per annum.

Chemical Constituents

Seeds of chaulmoogra contains 40 – 45 percent fixed oil.

Uses

Oil is used as antileprotic and for other skin diseases, employed as fish poison.

CHAULMOOGRA OIL

Synonyms

Hydnocarpus oil, Gynocardia oil, Jangali badam ka tel.

Biological Source

Hydnocarpus oil is the fixed oil obtained by cold expression method from ripe seeds of the plant *Taraktogenos Kurzii* King, and *Hydnocarpus anthelmintic* Pierre, *Hydnocarpus heterophylla* Kurz *Hydnocarpus wightina Blume* and other species of the *Hydnocarpus,* family Flacourtriaceae.

Geographical Source

Chaulmoogra plant is native of Myanmar, Thailand, and East India. It is also found in Sri Lanka and Bangladesh. In India, it is grown in Assam and Tripura.

Fig. 11.5: *Hydnocarpus wightiana* twig

Fig. 11.6: Chaulmoogra seeds

Method of Preparation of Chaulmoogra Oil

Seeds are sub-ovoid, obtusely angular and 2 cm in length. Chaulmoogra seeds contain 40 - 45 per cent fixed oil. Seeds are decorticated by machine after grading the kernels are pressed with the hydraulic press and the oil obtained is filtered.

Description

Colour	:	It is yellow to brownish-yellow coloured liquid
Odour	:	Characteristic
Taste	:	Somewhat acrid
Solubility	:	Chaulmoogra slightly soluble in alcohol, soluble in chloroform, ether, benzene and carbon disulphide.

It is soft white solid below 25°C.

Chemical Constituents

It contains chemically esters of unsaturated fatty acids of chaulmoogric acid (27 per cent) and hydnocarpic acid (48 per cent), gorlic acid, about 20.0 per cent of proteins, cyanophoric glycosides and glycerides of palmitic acid and oleic acids.

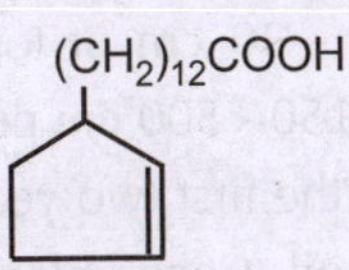

Chaulmoogric acid

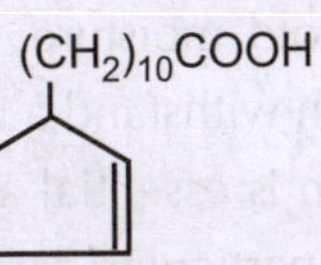

Hydnocarpic acid

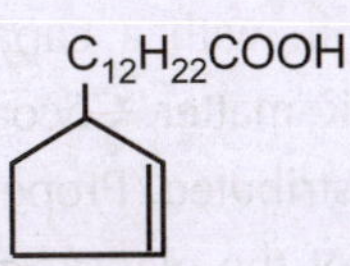

Gorlic acid

Standards of Quality

Weight per ml	:	0.935 - 0.960 g
Acid value	:	Not more than 10
Saponification value	:	195 - 213
Iodine value	:	93 - 104
Specific rotation	:	Not less than + 48° and not more than +60°
Refractive index	:	1.472 - 1.476

Uses

The unsaturated fatty acids of chaulmoogra oil possess strong bactericidal effect, against *Mycobacterium leprae*, and *M. tuberculosis*. It is found to be useful in the treatment of TB, leprosy, psoriasis and rheumatism. It is intended only for **external use**. Since, the seeds contain *cyanophoric glycosides,* oil cake is not edible and can not be used as food for animals.

Storage

It is stored in closed containers away from light and in cool place.

The plant is substituted in India by *Hydnocarpus wightiana* found abundantly in West Bengal, (Fig. 11.5) Kerala and Western Ghats and also by *Hydnocarpus* alpine occurring in Karnataka, Kerala and Tamil Nadu.

COCOA BUTTER

Synonyms

Theobroma oil; Cacao butter.

Biological Source

It is a fat obtained from roasted seeds of *Theobroma cacao* L, family Sterculiaceae.

Geographical Source

Cocoa is cultivated in most of the tropical and sub-tropical countries, It is native of South America and found in Sri Lanka, Philippines, Brazil; Curacao, Mexico, Ecuador; West Africa and parts of India.

Cocoa has been used by Mexicans since long time and even was known to Columbus and Cortez. Cocoa butter was prepared as early as 1695 A. D.

Cultivation of Cocoa

Cocoa is cultivated up to an elevation of 1000 metres above sea level. It needs well-drained good quality soil, with a capacity to hold moisture. About 15 - 30 cm of top soil should have sufficient organic matter. Cocoa plant can withstand a rainfall of 150 - 500 cm per annum, provided it is properly distributed. Proper irrigation is essential at least for the first two years of cultivation to enable taproot of the plant to penetrate particular depth in the soil. Cocoa plant is sun loving, but can not bear direct sunlight. Hence, permanent ever-green forest trees are grown to provide shade, well in advance before cocoa plants are planted. Inter-planting can be done successfully by raising coconut or arecanut trees between. Cocoa plants survive for 60 - 70 years and start bearing fruits after three years of planting. Hindustan Cocoa Products Limited in India has undertaken extensive cultivation of cocoa in Kerala.

Preparation of Cocoa butter

Cocoa seeds contain about 50 per cent of cocoa butter. The seeds are separated from pods and are allowed to ferment wherein the seeds change their colour from white to dark reddish-brown due to enzymatic reaction. The fermentation process takes place at 30 - 40°C. The process of fermentation is carried out in tubes, boxes or in the cavities made in the earth for 3 - 6 days. After fermentation, the seeds are roasted at 100 - 140°C, which looses water and acetic acid from the seeds and facilitates removal of seed coat. The seeds are then cooled immediately and are fed to nibbling machine to remove the shells followed by winnowing. The kernels are then fed to hot rollers which yield a pasty mass containing cocoa butter. This is further purified to give cocoa butter. The cocoa shells are processed further to yield an alkaloid.

Description

Cocoa butter is yellowish-white solid and brittle below 25°C. It has pleasant chocolate odour and taste.

It is insoluble in water, but soluble in ether, chloroform, benzene and petroleum ether.

Cocoa plant bears simple and alternate leaves.

Colour : Leaves are green

Odour : Aromatic agreeable

Taste : Bitter

Size : 15 cm in length.

Shape :

Fig. 11.7: *Theobroma cacao* Plant

Extra features: Leaves are linear, simple, alternate. Flowers yellowish-red, Flowers and fruits in clusters dividely attained to stems or branches.

Chemical Constituents

It consists of glycerides of stearic (34 per cent), palmitic (25 per cent), oleic (37 per cent) acids and small amount of arachidic and linoleic acids. The non-greasiness of product is due to its glyceride structure.

Standards of Quality

Specific gravity	:	0.858 - 0.864
Melting point	:	30 - 35°C
Refractive index	:	1.4637 - 1.4578
Saponification value	:	188 – 195
Iodine value	:	35 - 40

Chemical Test

Dissolve 1 gm of theobroma oil in 3 ml of ether in a test-tube at a temp of 17°C and immerse the tube in water having the temp of melting ice. The solution does not become turbid or deposit white flexes in less than three minutes after congealing raise the temp to 15°C, a clear liquid is gradually formed.

Uses

It is used as a base for suppositories and ointments, manufacture of creams and toilet soaps.

Substitute

Mango kernel oil, which is a solid fat at room temperature and has a melting point of 35°C, is used as substitute for cocoa butter.

CORN OIL

Synonyms

Maize oil, Mazola oil.

Biological Source

It is the fatty oil obtained from embryos of the grains of *Zea mays* Linn family Gramineae by Wet method.

Geographical Source

Maize is cultivated in almost all the countries in the world. It occupies third place among all cereal grains being cultivated at present. The USA, China, Brazil, India, U.S.S.R., Nigeria, and Kenya are the major producers of maize.

Method of Preparation

The maize grains or fruits (caryopsis) of maize are made up of 80 - 82 per cent endosperm, 10 per cent germ (embryo) and 5-7 per cent caryopsis. The germ (embryo) contains 40 per cent fixed oil.

During the manufacture of corn starch by wet milling, corn oil is obtained as a by-product. The oil-rich embryos (i.e. germs) are separated by flotation. The germs are washed thoroughly to make them free of starch, subjected to high pressure and heat to liberate the oil. The product, thus obtained, is filtered and refined by usual methods of settling, refrigeration and clarification.

Fig. 11.8: Maize fruit

Description

It is a light yellow coloured liquid with characteristic odour and taste. On prolonged exposure to air, it thickens and becomes rancid.

It is slightly soluble in alcohol, and fully soluble in chloroform, ether, benzene and petroleum ether.

Chemical Constituents

Maize oil contains the glycerides of linoleic (35 - 60 per cent), Linolenic (1 - 2 per cent), oleic (20 - 50 per cent), palmitic (12 - 18 per cent) and stearic acids (2.5 - 4 per cent). Unsaponifiable matter (1.3 per cent) contains γ-tocopherol and sitosterol. Crude oil contains about 2 per cent of phospholipids. It contains natural anti-oxidant in the form of vitamin E (tocopherol).

The stability of corn oil is supported by high level of natural antioxidant ferulic acid. Corn oil also contains ubiquinone (coenzyme Q), which is reported to have antioxidant activity.

Standards of Quality

Specific gravity	:	0.916 – 0.921
Refractive index	:	1.464 – 1.468
Acid value	:	2 – 6
Iodine value	:	109 – 133
Saponification value	:	187 – 196
Hydroxyl value	:	8 – 12
Thiocyanogen value	:	71 – 77
Viscosity	:	37 – 39

Uses

It is used as solvent for intra muscular injection and high-calorie dietary supplement. It is also used in cosmetics and preparations of margarine. The corn oil is a source of essential fatty acids. It is used in lowering blood cholesterol levels, besides being a source of vitamin E.

COTTON SEED OIL

Biological Source

It is fixed oil obtained from the dried, matured seeds (deprived of hairs) of cultivated varieties of *Gossypium herbaceum* or other species of *Gossypium* family Malvaceae.

For the manufacture, the seeds are steamed and placed in hydraulic presses, where about 145 kg of oil is expected from about 0.9 metric ton of the seeds. The defatted seeds and hulls are used as food for livestock.

Description

Cotton seed oil is a semi-drying pale yellow, practically odourless and oily in nature. It is slightly soluble in alcohol and miscible with ether, carbon disulphide, chloroform and petroleum ether. It solidifies between 0 to – 5°C.

Chemical Constituents

Cotton seeds contain 20 – 25% of fixed oil. It contains triglycerides of fatty acids, mainly palmitic, oleic Gossypol pigment and linoleic acid.

Chemical Test (Halphen reaction)

Mix equal quantities of cotton seed oil and ethyl alchol and 1.0 per cent solutions of sulphur in carbon disulphide in test-tube. Heat on boiling water bath for few minutes red colour is produced. This test is even used to detect cotton seed oil in other oils.

Standards of Quality

Iodine value	:	109 - 115
Saponification value	:	190 - 198.
Specific gravity	:	0.915 - 0.927
Refractive index	:	1.4645 - 1.4655
Viscosity	:	70.4 cp

Uses

Cotton seed oil is a pharmaceutical aid (solvent) in solid state as a substitute for Lard. It is used for preparation of cosmetics, hydrogenated fats, and emollients. It is also used as edible oil for cooking or preparation of margarine. It is employed as pediculicide, acaricide and laxative in veterinary medicine. In agricultural practices it is used as carrier oil for spraying insecticides. Oil cake is used as fodder for animals.

EVENING PRIMROSE OIL

Synonym

King's cureall.

Biological Source

This consists of fixed oil obtained from the ripe, dried seeds of the plant known *Oenothera biennis* L and other species of family Onagraceae.

Geographical Source

This plant occurs mainly in various parts of united kingdom.

Fig. 11.9: Evening primrose herb

Description

Colour	:	Golden yellow clear and transparent liquid
Odour	:	Characteristic peppery
Taste	:	Bland

Solubility

Insoluble in alcohol, but soluble in ether and chloroform.

Chemical Constituents

Evening primrose oil contains 72 per cent linoleic acid (LA) and 9 per cent γ-linolenic acid (GLA). It is known to be the richest natural source of GLA. Both LA and GLA are **essential fatty acids**, and essential in the formation of prostaglandins.

Standards of Quality

Density	:	0.9283
Refractive index	:	1.4782
Saponification No.	:	288
Iodine No.	:	155

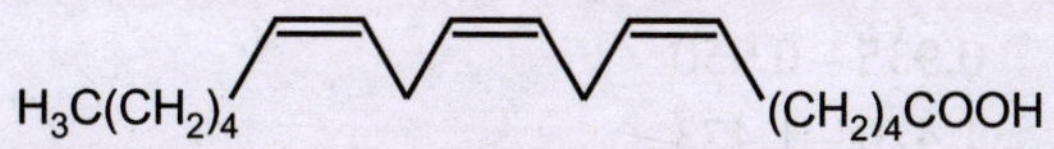

γ-Linolenic acid (GLA)

Uses

It is used in the form of dietary supplement and cosmetics.

Therapeutically, evening primrose oil is useful for those who lack the capacity to convert linoleic acid to γ-linolenic acid enzymatically. Hence, clinically it is useful as prostaglandin precursor. It has been also found to be useful in controlling severity of premenstrual syndrome and also in eczema. The other benefits reported are pain reduction in arthritis, controlling liver and kidney damage due to alcohol, controlling complications of diabetes, multiple sclerosis, etc.

Dose

6 - 8 g (adults) and

2 - 4 g (children)

JAMALGOTA

Synonyms

Purging nut, Croton seed.

Biological Source

These are the dried seeds of the plant *Croton tiglium,* fam: Euphorbiaceae.

Geographical Source

Naturalised in South-Kokan and cultivated in West Bengal, Shri Lanka, Malay Brahmdesh, China and Assam.

Description

It is small evergreen tree with young shoots, sparsley hairy **Leaves** which are thin, membranous ovate, serrate, acuminate 3-5 nerved, 20 cm long, bearing male and female flowers separately, small green in clusters. It bears capsular fruit 1.0 – 2.5 cm long, white, ovoid, obtusely trigonomos.

Morphological Characters

Seeds are 1.0 – 1.5 cm in size, compressed, ellipsoidal, with eight raised lines.

Croton oil is orange-yellow in colour and disagreeable odour.

Chemical Constituents

Seeds contains 30-45 per cent fixed oil, known as croton oil. While the seed kernels also contain 43-63 per cent fixed oil. The poisonous principle in the fatty acids is 2.0 – 3.0 only. Seeds also contain two toxic proteins i.e. croton globulin and croton albumin a glycoside crotonoside and croton resin. Fatty acids of the oil are normal fatty acids along with lauric acid, tiglic acid, valeric acid and butyric acid. Cotton oil contains **phorbol.**

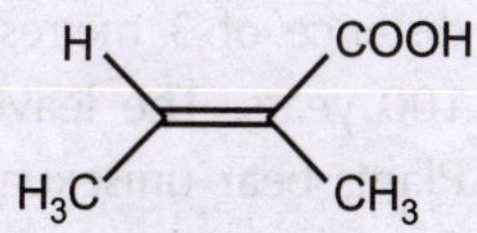

Tiglic acid

Standards of Quality

Density	:	0.935 – 0.950
Refractive index	:	1.470 – 1.473
Saponification no.	:	205 – 220
Iodine no.	:	102-115

Uses

Jamalgota seeds and oil both are drastic cathartic. Seeds in small doses are used as carminative. Oil applied externally produces vesicular eruptions. It is highly toxic and poisonous.

It also contains phorbol in fixed oil. Phorbol is a parent alcohol of tumour promoting compounds.

JOJOBA OIL

Synonym

Hohoba oil

Biological Source

Jojoba oil is a mixture of liquid wax ester from the seeds of *Simmondsia chinensis, S. californica* (family Buxaceae) obtained by extraction of ground seeds.

Geographical Source

Jojoba is an evergreen desert shrub indigenous to the southwestern United States (Arizona and California), and North-Western Mexico also called as Hohoba by the American Indians, it was described botanically and classified at the beginning of the 19th century.

In India its commercial cultivation has been tried in Rajasthan, Maharashtra, Orissa and in the coastal area.

Attempts to cultivate jojoba at Jodhpur, Pune, Ahmednagar, and Satara area have been successful.

Fig. 11.10: Male flowering Jojoba herb

Cultivation and Collection of Jojoba

Jojoba is a multi-stemmed, drought resistant, small, woody shrub, with tap-root upto 1 to 1.5 m, which is able to penetrate further deep into soil.

Jojoba plants are propagated directly by sowing the seeds or by transplanting the seedlings grown earlier. For cultivation arid land or sandy calcareous soil having a temperature range of 30 - 40 °C is most suitable. One acre accommodates 1300 to 1350 plants by maintaining the distance of 3 metres in between two plants and 2 metres in height and may live for more than 100 years. The leaves of the plant are simple, smooth edged elliptical, grayish green in colour. Plants bear unisexual flowers and it is needles to say that female plants bear only fruits. Female flowers are solitary and are greenish yellow in colour, while male flowers are in clusters and look like a bunch of small grapes. For proper pollination and fertilization proportion of 10 : 1 (female to male) is maintained.

Jojoba can be cultivated twice in a year that is in June and July or October and November. Alternative crops like black-gram, Safflower, can be taken for the first three years, as it starts bearing the fruits after four years. Jojoba bears splitting fruits and seeds need to be collected carefully. Cultivation becomes only economical after 5 years of cultivation. Normally, 2.0 - 2.5 kg of seeds are produced by each plant annually. No pesticides or insecticides are required for plant protection.

Description

Colour : Golden yellow coloured liquid

Odour : Aromatic, pleasant

Taste : Bland

Solubility : Soluble in organic solvents insoluble in water and alcohol.

Chemical Constituents

Jojoba seeds constitute about 45 - 55 per cent of liquid wax ester. Jojoba oil is not a triglyceride but a mixture of long chain esters and therefore referred to as wax. About 97-98 per cent of the oil consists of esters of C-18, C-20, C-22 and C-24 chain monounsaturated (monoethylene) acids and alcohols. These acids and alcohols are similar to those present in sperm whale oil. Two ester molecules containing 40 and 42 carbon atoms make up to 80 per cent of the oil. It is stable and not oxidised at room or high temperature. It contains small amounts of tocopherol which contribute to its stability in the air. The minor constituents include phytosterols.

Standards of Quality

Specific gravity	:	0.865
Refractive index	:	1.4648
Iodine value	:	80 – 81
Saponification value	:	90 – 92
Acid value	:	0.30 – 0.50

Quite stable, can be stored without rancidity for long time, resistant to bacterial degradation.

Uses

In the 1950s, jojoba oil was suggested as a substitute for sperm whale oil because of similarities between the oils. In 1971, the US Government banned whale hunting and also import and use of the whale. Due to this, jojoba oil has become valuable substitute for sperm whale oil in cosmetics, lubricants and various industrial chemicals.

Due to partial isomerisation or hydrogenation of double bonds of jojoba oil, it becomes a creamy product and is found to be a good carrier in dermatological preparations. Commercially jojoba oil is used as a lubricant. It is used with carnauba wax to minimise the stretch marks.

KARANJA OIL

Biological Source

It is a non-edible semi-drying fixed oil obtained from the seeds of *Pongamia glabra* Vent. family Papilionaceae.

Fruits of the plants are having a shape of Karanja, A fried sweet delicious items served in South India, hence the name.

Geographical Source

The plant is found throughout India along streams and rivers, mainly in Andhra Pradesh, Kerala, Madhya Pradesh; Maharashtra and Tamil Nadu. About 12,000 tones of oil per annum is produced in India.

Description

Fresh karanja oil is yellow-orange in colour, but darkens and becomes reddish-brown on storage. The oil has peculiar pungent and disagreeable odour and bitter taste.

Fig. 11.11 (a): Karanja Twig with fruits

Fig. 11.11 (b): Karanja Seeds

Chemical Constituents

Karanja seeds contain about 27 per cent of fixed oil. The fatty acid composition of the karanja oil is as follows:

Lignoceric acid	:	1.0 - 3.5 per cent
Oleic acid	:	44 - 75 per cent
Linoleic acid	:	10 - 18.5 per cent
Behenic acid	:	4.0 - 5.5 per cent
Palmitic acid	:	3.5 - 9 per cent
Stearic acid	:	2.5 - 9.0 per cent
Arachidic acid	:	2.2 - 5 per cent

Unsaponifiable matter of the oil contains β-sitosterol, and non-fatty components contain karanjin, karanja chromene, pongapin and pongamol.

Standards of Quality

Specific gravity	:	0.9273 - 0.930
Refractive index	:	1.4736 - 1.4739
Saponification value	:	181 - 182
Iodine value	:	89 - 90
Acid value	:	6 - 7
Acetyl value	:	21 - 22
Unsaponifiable matter	:	4.2 per cent

Uses

Karanja oil is used in the treatment of scabies, herpes, leucoderma and other cutaneous diseases. It is also used in rheumati sm. Now-a-days used in source of biodiesel.

KOKUM BUTTER

Synonyms

Goa butter, Kokum oil, Mangosteen oil.

Biological Source

It is the fat expressed from the seeds of *Garcinia indica* Chois, belonging to family Guttiferae. Commercial kokum butter from the market is melted; its free acidity is neutralised with the treatment of sodium carbonate solution and washed in hot water. It is decolourized, if necessary, by using suitable adsorbent.

Geographical Source

The tree (Fig. 11.11) is indigenous to Thailand, Cambodia, and China. It is also found in India. It is wildl plant that grows in Konkan, Western Ghats, Coorg dist and Nilgiri Hills. It is cultivated in Maharashtra and in Mauritius. Nearly, 1200 hectares of land is under cultivation in India producing 10,500 tones of drug per annum.

Cultivation of Kokum

Climatic Conditions

Kokum plant can grow under different agro-climatic conditions, though, warm humid climate is preferable for its healthy growth. It does not tolerate dry weather. It can be grown upto altitude of 400 to 500 metres. Temperature ranging from 10°C to 35°C is ideal for the growth and development of crop. Actually, it is a rainfed crop; well distributed rainfall ranging from 1000 to 2000 mm per annum is required.

It grows well in variety of soils ranging from rich loamy to laterite soil. It can be grown in sandy loam, red soil with adequate manuring. Water logging of soil should be avoided.

Propagated through soft-wood grafting. Usually planted at onset of monsoon during July - August. Its a perennial crop. The land is deep ploughed to remove weeds and the soil is well pulverised. At the time of last ploughing, compost farmyard manure is mixed, the land is levelled, and pits at a distance of 60 × 60 cm are prepared. It bears male and female plants separately. As female plants bear fruits, they are selected for grafting. Prepared grafts are planted carefully in the pits and watered.

The soft-wood grafts are planted in well prepared land at a distance of 60 × 60 cm at the onset of monsoon, followed by irrigation. About 150 to 200 plants are sufficient for one hectare plantation.

Government of Goa has undertaken the cultivation of Kokum in 15 hectares in association with NOVOD Board since 1995.

Normally, it is grown as rainfed crop. Hence, regular irrigation is not necessary though during dry seasons 2 to 3 irrigations are carried out.

For the better yield of crop, it is supplied with enough doses of fertilizers. It grows well with FYM in dose of 20 kg alongwith basal dose of 250 gm of phosphate. 500 gm of nitrogen, applied in split-doses.

Weeds are kept under check by regular weeding. Usually after 15 - 20 days, weeding is carried out. The procedure can be repeated as and when required.

It is a hardy crop, hence no serious pests or diseases have been reported to attack the crop. However leaf minor and pink disease, may sometimes appear on the plant which can be controlled by using appropriate pesticides.

Collection o Kokum Fruits

After 6 - 7 years of plantation, plants start bearing the fruits. It bears flowers, during November to February and fruit in the month of April - May. Fruits are collected manually. It is also harvested for the bark, which is removed during March or April. The harvested fruits and barks are dried in shade and are stored in dry place.

Total yield of fresh ripe fruits reported to be 8.5 to 9 quintals per hectare.

Method of Preparation of Kokum butter

The kernels from the seeds are churned and boiled in water. The melted fat is separated by skimming, washed twice with hot water and decolourized with animal charcoal or fueller's earth.

Description

Colour	:	Fat is light grey or yellowish.
Odour	:	Slight and characteristic.
Taste	:	Characteristic and slightly bitter.
Shape	:	It is marketed in the form of egg-shaped lumps.
Solubility	:	Soluble in organic solvents and insoluble in water and alcohol.

Fig. 11.12: Kokum Fruits

Chemical Constituents

Kokum butter contains glycerides of stearic (55 per cent), oleic (40 per cent), hydroxy capric acid (1 per cent), palmitic (2.5 per cent), and linoleic acids (1.5 per cent).

Standards of Quality

Weight per ml	:	0.895 - 0.899 g
Melting point	:	39 - 42°C
Acid value	:	Not more than 3
Sap. value	:	185 to 190

Iodine value	:	35 - 37
Refractive index	:	1.4565 - 1.4575
Unsaponifiable matter	:	2 - 3 per cent

Uses

It is used as a nutritive, demulcent, astringent and emollient. Locally, it is used in fissures of lips and hands. Cake, left after the extraction of the oil, is a manure. It is employed in the sizing of cotton yarn.

It is used in the preparation of ointments and suppositories. The dried rind of the fruit is called ***amsul*** and used as a substitute for tamarind.

LECITHIN

Synonyms

Vitellin, Lecithol, Phosphatidylcholine.

Source

Commercial lecithin consists of acetone-insoluble phosphates of phosphatidyl choline, phosphatidyl ethanolamine (Cephalin), phosphatidyl serine and triglycerides, fatty acids. It should contain not less than 50.0 per cent of acetone insoluble matter.

It is mainly obtained from soybean oil. It is also available in several vegetable seeds, egg-yolk, corn and from of animal brain and nervous tissue.

Description

Lecithins are waxy, white substances which become brown on exposure to air. They may also be thick pourable fluids depending upon acid value; characteristic in smell; insoluble in acetone, and soluble in other organic solvents. Lecithins swells in water forming colloidal solution, soluble in mineral oils and fatty acids. Lecithins decompose on heating.

Chemical Constituents

Chemically, lecithin contains glycerol, fatty acids, phosphoric acid and choline. Normally lecithins contain a saturated fatty acid at α-position and an unsaturated fatty acid at β-position.

When treated with sulphuric acid, choline separates forming phosphatidic acid. If boiled with alkalis or acids glycerophosphoric acid, choline and fatty acids are formed.

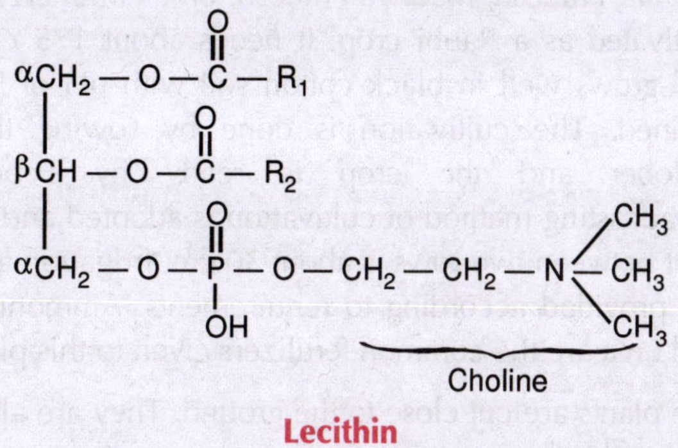

Lecithin

Standards of Quality

Density	:	1.305
Iodine value	:	95
Saponification value	:	1.96
Iso-electric point	:	3.5

Uses

With the combination of proteins it forms lipoproteins of plasma and cells; acetylcholine formed from choline part has very important role in transmission of nervous impulses across the synapses. It is very important lipotropic agent and can prevent formations of fatty liver, lowers the surface tension of lung alveoli.

It is an emulsifying agent of natural origin, dispersing agent, wetting agent, penetrating agent and also an antioxidant. It is used in the manufacture of cosmetics, soaps, candies, chocolates. It is industrially used as lubricant for textile fibres, petroleum industry and printing inks.

LINSEED

Synonyms

Flax seed; Linum are synonyms of linseed.

Biological Source

Linseed consists of dried ripe seeds of the plant known as *Linum usitatissimum* Linn. family Linaceae. It contains not less than 25 per cent of fixed oil and not more than 1 per cent of foreign organic matter.

Geographical Source

Linum is found in Russia, Canada, the U.S.A. and Argentina. In Egypt, Algeria, Italy and Greece, only fibre flax is cultivated, while in India it is cultivated for fibres as well as oil. Four species of linseed are recorded in India but only *Linum usitatissimum* is cultivated for its oil content.

Cultivation and Collection

In India, it is cultivated entirely for the seeds. It is cultivated in Uttar Pradesh, Madhya Pradesh, on a commercial scale. It is cultivated as a Rabbi crop. It needs about 175 cm of rainfall and grows well in black cotton soil with pH of 5.7 and well drained. The cultivation is done by sowing the seeds in October and the crop is ready by February/March. Broadcasting method of cultivation is adopted and the distance kept between two rows is about 30 cm. Irrigation and fertilizers are provided according to requirements. Ammonium sulphate and urea are the common fertilizers given to this plant.

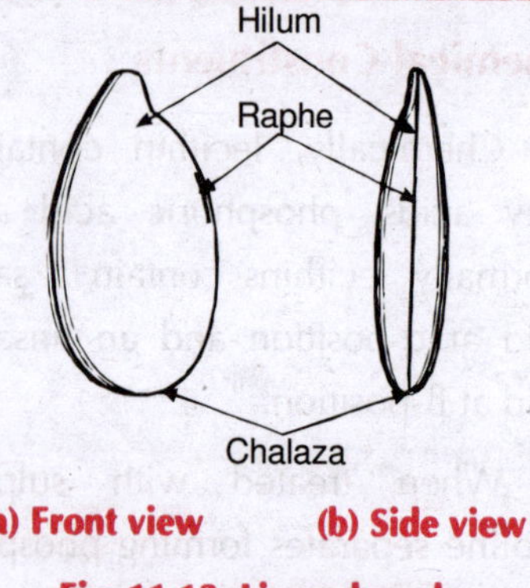

Fig. 11.13: Linseed seed

The plants are cut close to the ground. They are allowed to dry for 2 - 3 days and are then thrashed to break the capsules. Seeds coming out are separated by winnowing. The seeds are seived and packed in gunny bags.

Macroscopical Characters

Colour : Brown

Odour : None

Taste : Mucilaginous and bland

Size : Seeds are about 4 - 6 mm in length and 2 - 2.5 mm in width.

Shape : They are elongated, ovate, strongly flattened, rounded at one end and obliquely pointed at the other.

Extra Features

Seeds are glossy in appearance. One end has slight depression enclosed in hilum and micropyle. From the hilum yellowish raphe runs to the chalazas.

Chemical Constituents

Linseed contains about 20 - 40 per cent of fixed oil; 20 per cent of proteins. It also contains mucilage (2 - 7 per cent) and a cyanogenetic glycoside, linamarin. Additionally it contains pectin about 10 per cent and an enzyme linase.

Standards of Quality

Total ash : Not more than 5.0 per cent.

Acid insoluble ash : Not more than 1.0 per cent.

Uses

Linseed is used as a source of edible fixed oil and also as a demulcent. Linseed meal is used as poultice.

LINSEED OIL

Biological Source

This consists of fixed oil obtained from the dried fully ripe seeds of *Linum usitatissimum* Linn family Linaceae.

Geographical Source

Its origin is uncertain, is believed as indigenous to India. It is cultivated at present extensively as a source of fibres in Egypt, Algeria, Spain, Italy and Greece, while in Turkey, Afghanistan and India it is a source of oil. In Russia, it is cultivated for oil and fibre.

Method of Preparation

The variety yielding high percentage of oil is selected for extraction of oil. Seeds are sieved to make free of earthy matter and other materials.

Commercially, linseed oil is produced by use of expellers. Before the seeds are subjected to the expellers, they are rolled into meal, then moistened and heated by means of steam jacketed troughs filled over the expellers. An average yield of oil is 30 - 35 per cent. The expressed oil is tanked for a long period to settle the colouring matter and mucilage. The oil is then treated with alkali immediately after filtration. Alkali treatment helps to remove free fatty acids. Bleaching of the oil is done by using either charcoal or fueller's earth at elevated temperature. The refined oil so obtained, is chilled to separate wax.

Description

Colour : It is pale yellow coloured clear liquid.

Odour : Characteristic.

Taste : Pleasant.

Linseed oil gradually thickens on exposure to air forming a thin transparent film.

It is slightly soluble in alcohol, insoluble in water and miscible with ether, petroleum ether and chloroform.

Fig. 11.14 (a): A twig of Linseed plant

Fig. 11.14 (b): Linseed seeds

Chemical Constituents

It contains the glycerides of palmitic, stearic, oleic; linoleic and linolenic acids. The unsaponifiable matter of the oil contains considerable quantities of sterols, tocopherol and squalene. Linseed also contains a cyanogenetic glycoside linamarin and mucilage (5 per cent) in addition to fixed oil (20 - 40 per cent). Linseed oil also contain anti oxidants and 3-omega moega-6 fatty acids.

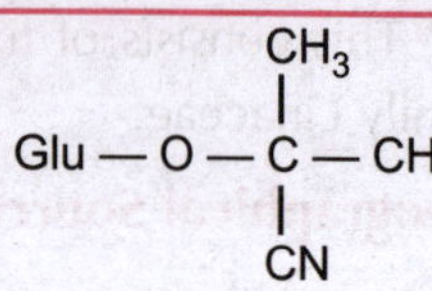

Linamarin

Standards of Quality

Specific gravity	:	0.927 - 0.931
Refractive index	:	1.4786 - 1.4815
Saponification value	:	188 - 195
Iodine value	:	160 - 200
Acid value	:	Not more than 4
Unsaponifiable matter	:	Not more than 1.5

Uses

It is the most important drying oil; hence, considerably large quantities are used for paints and varnishes. Medicinally, it is mainly recommended for external applications like lotions and liniments. It is used in treatment of scabies and other skin diseases along with sulphur. Since, it has very high iodine value; it is used in the preparation of non-staining iodine ointment and other products like cresol with soap. It is nutritive and an emollient, too.

Industrially, this important oil used for various purposes such as in the manufacture of soap, linoleum, greases, polishes and plasticisers, polymers, etc.

Caution

To meet the requirements of various industries, different grades and types of linseed oil are available in the market. Boiled linseed oil dries at a faster rate and forms a smooth and lustrous film. Linseed oil that has been boiled or treated in dryers such as linoleates or resinates of lead, manganese, cobalt or zinc should not be used in medicine.

Adulterants

Linseed oil is adulterated with boiled linseed oil, cotton seed oil; sunflower oil, rosin, mineral oils, fish oils, and mustard oil. Fish oil is detected by odour and rosin while the presence of mineral oils is identified by studying the composition of unsaponifiable matter.

NEEM OIL

Synonym

Margosa oil.

Biological Source

It is a non-edible fixed oil obtained from fully matured seeds of *Azadirachta indica* Juss, family Meliaceae, collected late in summer.

Geographical Source

The plant is found throughout India, Myanmar and other tropical countries. In India, it is very common in Maharashtra, Rajasthan, MP, UP and Tamil Nadu.

The present estimated production of the oil is about 6,00,000 tones per annum. The seeds contain about 20 per cent oil.

The export of neem oil was approximately ₹ 263 Lakhs during 1996-97.

Description

It is yellow coloured oil with specific odour and bitter taste. It is soluble in ether and chloroform.

Chemical Constituents

It contains glycerides of saturated and unsaturated fatty acids. The main fatty acids are oleic (50 per cent) and stearic (20 per cent) acids. The oil contains 2.0 per cent of bitters, which are sulphur containing compounds, nimbidin, nimbin, nimbinin and nimbidol. The unsaponifiable part contains nimbosterol (0.03 per cent).

Standards of Quality

Specific gravity	:	0.913 - 0.918
Refractive index	:	1.417 - 1.4627
Saponification value	:	195 - 205
Iodine value	:	68 - 75
Unsaponifiable matter	:	0.7 - 1.0 per cent
Acid value	:	Not more than 2.0

Uses

Nimbin, nimbidin and related compounds possess anti-viral activity. As non-edible oil, it is used for soap making and for manufacture of oleic and stearic acids. It is indicated in rheumatism and also as a pesticide and in medicated soaps for skin diseases. It is also spermicidal.

India has exported neem oil during 1994-95 and 95-96 to the tune of ₹ 356 lakhs and 217 lakhs respectively.

OLIVE OIL

Synonym

Oleum Olivae.

Biological Source

It is the fixed oil expressed from the ripe fruit of *Olea europoea* Linne, belonging to family Oleaceae.

Geographical Source

Olive tree, from which the oil is obtained, is native to Palestine and the countries near to Mediterranean sea. It is cultivated in United States and Southern Australia.

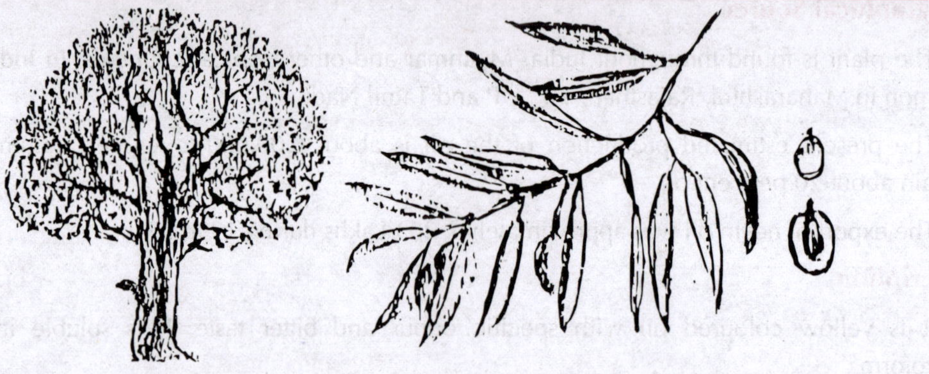

(a) (b)

Fig. 11.15: Olive tree and fruiting twig

Olive oil is very commonly used in various countries along the Mediterranean sea. Among the largest producers of olive oil, Italy, Spain and Greece are the leading countries. The others are United States, Turkey, Tunisia, Morocco and Syria.

Method of Preparation

It is prepared by crushing and pressing the ripe fruits called "Olives". The entire olive consists of 20 - 30 per cent oil and the fruit pulp has 60 - 80 per cent oil. The hydraulic presses are used to squeeze the oil out of fruit under low pressure. This type of technique is called cold pressing and it generates very little amount of heat, because of which it gives the best quality oil called 'virgin olive oil'. The further pressing gives low quality oil. The oil that comes from the last pressing is called olive residue and is generally used for cosmetics, medicines, etc.

Description

Colour : Olive oil is pale yellow or greenish-yellow coloured liquid.

Odour : Slight and characteristic

Taste : Bland, faintly acrid

Solubility : It is slightly soluble in alcohol and miscible with carbon disulphide, chloroform and ether.

Chemical Constituents

The olive oil contains the triglycerides mainly in the form of Olein, palmitin and linolein.

Chemical Test

Under ultra-violet radiation it gives deep golden-yellow colour, while refined oil at gives pale blue fluorescence Discolouring with charcoal removes fluorescence.

Standards of Quality

Specific gravity : 0.910 - 0.915 (25°C)

Iodine value : 79 - 88

Saponification value : 190 - 195

Free fatty acids : The free fatty acids in 10 g require not more than 5 ml 0.1 N sodium hydroxide for neutralisation.

Uses

Externally, it is an emollient and soothing agent for inflamed surfaces. It is used to soften the skin and crusts in eczema and psoriasis; it is also used as an ingredient of ear wax.

Internally, it is used as nutrient, demulcent and as mild laxative. It is used as a vehicle for oily suspensions for injection.

POPPY SEED OIL

Synonym

Poppy oil.

Biological Source

It is the fixed oil obtained from poppy seeds i.e. dried seeds of *Papaver somniferum*, family Papaveraceae.

The seeds are collected from the poppy capsules, the fruits of opium poppy. The countries wherein opium is collected, the seeds are produced commercially.

Geographical Source

It is collected in Yugoslavia, Bulgaria, Turkey, Iran, India and also in England.

Description

The seeds are very small, about 1 to 1.25 mm in diameter. Poppy oil is pale yellow coloured liquid with pleasant odour. It is a drying oil.

Chemical Constituents

Poppy seeds contain 50 – 55 % of fixed oil. The seeds are expressed for production of oil. The oil is free of all the narcotic alkaloids of opium. The oil is rich source of linoleic acid and contains about 73 per cent of it. In addition, it contains palmitic (10 per cent), arachidic (0.5 per cent) and oleic (12 per cent) acids.

Standards of Quality

Specific gravity	:	0.924 - 0.927
Refractive index	:	1.467 - 1.470
Acid value	:	3.0 - 13
Iodine value	:	132 - 158
Saponification value	:	188 - 196
Hydroxyl value	:	20 - 21
Thiocyanogen value	:	77 - 78
Unsaponifiable matter	:	0.4 - 1.2 per cent

Uses

Its taste is similar to olive and hence, used along with olive oil or alone in salads and also as edible oil. As iodine value is high, it is used in preparation of iodised oil. Being drying oil, it is very frequently used in the manufacture of varnish, soap and paints.

RICE BRAN OIL

Synonym

Rice oil.

Biological Source

Rice bran is the cuticle existing between the rice and the husk of the paddy and consists of embryo (germ) and endosperm of the seeds of *Oryza sativa*, family Gramineae.

It is obtained as a byproduct in rice mill during polishing of rice obtained after dehiscing of paddy.

Rice bran contains about 15 per cent of fixed oil and is obtained by solvent extraction method.

Geographical Source

It is produced in almost all states of India producing rice as main crop, Tamilnadu, Andhra Pradesh and Karnataka are leading.

Method of Preparation

The quality of rice bran oil depends upon the time which elapses between milling of the rice and removal of oil from the bran. Rice bran contains an active enzyme lipase, which raises the free fatty acid content on storage. The oil obtained from fresh bran is of good quality and has good flavour and low free fatty acid content. Therefore solvent extraction plant for rice bran oil should be set as nearer as possible to the rice milling so as to process out the rice bran oil quickly.

Rice bran is found in extremely small pieces. It is impermeable to solvents. Before solvent extraction, it is subjected to drying, cooking and flaking operations. The normal percolation method of solvent extraction does not serve the purpose with this type of material, but it is pressed and then extracted with solvent special continuous immersion extractors.

Description

It is golden yellow oil difficult to bleach, odourless, bland in tube not affected by temporary heating to 160°C. It is insoluble in water but soluble in common fat solvents.

Chemical Constituents

Rice bran oil contains 20 - 25 per cent of saturated and 80 - 85 per cent of unsaturated fatty acids as glycerides. Main fatty acids are oleic (40 - 50 per cent), linoleic (30 - 40 per cent) and palmitic acids (12 -18 per cent). The oil contains squalene and antioxidants like tocopherols.

Rice-bran oil contains a steroidal compound orizanol, which boosts good cholesterol and reduces bad cholesterol.

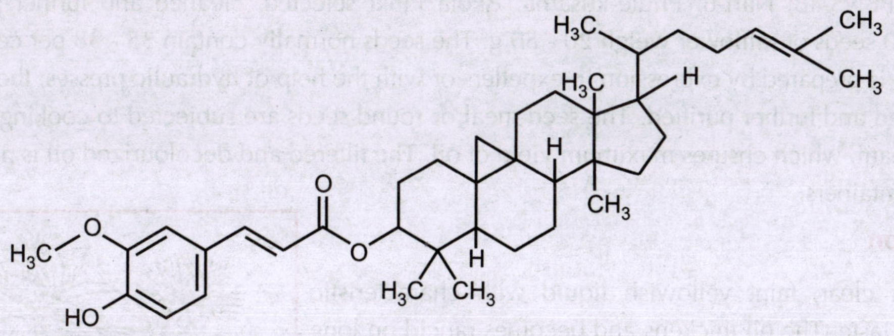

Orizanol-A

Standards of Quality

Acid value	:	04 - 05
Saponification value	:	181 - 189
Iodine value	:	99 - 108
Thiocyanogen value	:	69 - 76
Hydroxyl value	:	05 - 14
Refractive index	:	1.470 - 1.473
Specific gravity	:	0.916 - 0.921

Uses

Now-a-days it is used as edible oil. Since, it contains antioxidants, its keeping quality is very good. It is used in the manufacture of cosmetics as it is very effective skin protecting moisturizer and as an emollient. It is used in preparation of vegetable ghee.

It is preferably used in **baby cosmetic formulations, hair** and **skin products.**

SAFFLOWER OIL

Synonym

Kardi-ka tel.

Biological Source

It is a fixed oil obtained from the ripe and dry seeds of *Carthamus tinctorius* Linn family Compositae.

Geographical Source

This is one of the most ancient crops cultivated in Egypt as a dye-yielding herb. Now, it is cultivated as an oil seed plant and regarded as substitute for sunflower. It is cultivated in Mexico, India, the USA, Ethiopia and Australia.

Method of Preparation of oil

For expression of oil, the seeds from promising varieties in India namely A-300, S-7-13-3 and PBNS-12, PBNS-48, Nari-6, Phule-kusama, Akola Pinki selected, cleaned and further processed. About 1000 seeds of safflower weigh 20 - 50 g. The seeds normally contain 35 - 38 per cent of fixed oil. The oil is prepared by expression in expellers or with the help of hydraulic presses; thereafter the oil is filtered and further purified. The seed meal or round seeds are subjected to cooking by means of open steam, which ensures maximum yield of oil. The filtered and decolourized oil is packed into suitable containers.

Description

It is a clear, faint yellowish liquid with characteristic odour and taste. The oil thickens and becomes rancid on long exposure to air.

Safflower oil is slightly soluble in alcohol and freely soluble in ether, chloroform, benzene and petroleum ether.

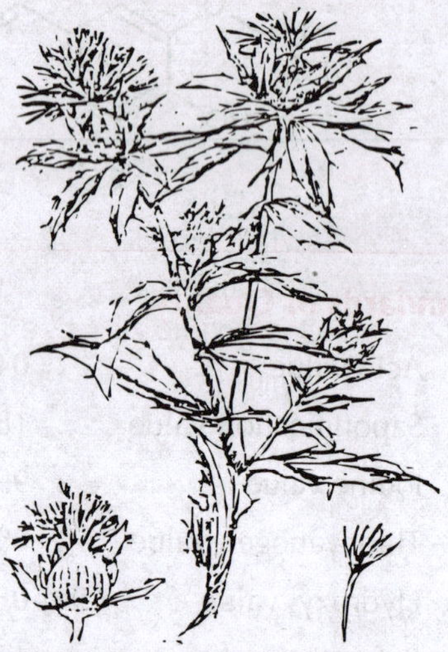

Fig. 11.16: Safflower-plant

Chemical Constituents

Safflower oil contains glycerides of palmitic (6.5 per cent), stearic (3.0 per cent), arachidic (0.296), oleic (13.0 per cent), linoleic (76 - 79 per cent) and linolenic acids (0.15 per cent). The polyunsaturated fatty acid content of the oil is the highest (75 per cent) and is said to be responsible to control cholesterol level in the blood, and thereby, reduces incidence of heart attacks.

Carthamin – Scarlet-red colouring matter and safflower yellow are obtained from flowers of carthamus. Carthamin is water insoluble while safflower yellow is water soluble.

Standards of Quality

Specific gravity	:	0.9211 - 0.9215
Acid value	:	01 – 9

Refractive index	:	1.472 - 1.475
Saponification value	:	188 - 194
Iodine value	:	140 - 150
Hydroxyl value	:	2.9 - 6.0
Unsaponifiable fraction	:	Not less than 1.5
Thiocyanogen value	:	82.5 - 86

Uses

The edible oil is used in the manufacture of oleomargarine, as a dietary supplement in hypercholesteremia and also in treatment of atherosclerosis. Due to its high linoleic acid content, it is consumed for preparation of vegetable ghee.

Industrially, it is used for preparation of soft soap varnishes, linoleum and water-proofing material. India has exported safflower oil to the tune of ₹ 699 lacs during the year 1995 - 96.

SESAME OIL

Synonyms

Teel oil, Gingelly oil and Benne oil.

Biological Source

It is fixed oil obtained by expression from the seeds of *Sesamum indicum* family Pedaliaceae.

Geographical Source

The plant (Fig. 11.17) is indigenous to India, and is cultivated in Caribbean islands, China, Japan, Africa and the United States.

Method of Preparation of oil

Sesame seeds contain about 50 per cent of fixed oil. Only white variety of seeds is used for pharmaceutical purposes. Seeds are very small in size. They are cleaned, if necessary, washed, sun-dried and expressed to yield oil at room temperature. Subsequently, the temperature and pressure are raised. The oil is purified by refining method and used.

Fig. 11.17: Flowering Sesame plant

Fig. 11.17: Sesame seeds

Description

Colour : Pale yellowish liquid

Odour : Slight, characteristic

Taste : Bland

Solubility : It is slightly soluble in alcohol, miscible with chloroform, solvent ether, light petroleum (40 - 60 per cent) and carbon disulphide.

Extra Features : It does not solidify at 0°C.

Chemical Constituents

Sesame seeds contain 45.50% of fixed oil. Gingelly oil contains glycerides of higher fatty acids mainly oleic, linoleic, palmitic, stearic, and arachidic acids. It contains about 5 per cent of olein, and a phenol known as sesamol which is responsible for stability of oil. It also contains lignan derivatives; sesamin and sesamolin, which are present in the unsaponifiable matter of oil. Sesamol is produced by hydrolysis of sesmolin.

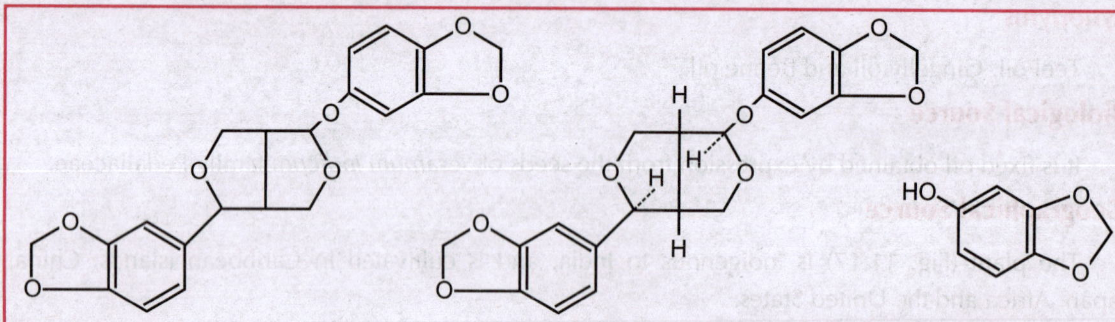

Sesamolin (+) Sesamin Sesamol

Identification: Badouin's test

Shake 2 ml sesame oil with 1 ml 1 per cent solution of sucrose in hydrochloric acid; a pink or red colour is produced due to sesamol.

Cool the sesame oil to 0ºC. It remains liquid only.

Standards of Quality

Weight per ml : 0.916 - 0.919 g

Refractive index : 1.472 - 1.476

Acid value : Not more than 2

Iodine value : 103 - 116

Saponification value : 188 - 195

Free fatty acids : Not more than 1 per cent

Unsaponifiable matter : Not more than 1.5 per cent

Uses

Sesame seeds are brain tonic and also improve complexion when applied to skin in the form of crushed meal. Oil has strong healing properties. Oil is nutritive, laxative and demulcent; it has got emollient properties. It is used in the preparation of liniments, plasters, ointments and soaps, similar to olive oil.

Pharmaceutically, it is used as a vehicle for intramuscular oily injections.

After burning, sesame oil yields high quality black ink.

SOYABEAN OIL

Biological Source

It is the fixed oil obtained from matured seeds of soya bean plants *Glycine soja* Sieb (*Glycine hispida* Maxim) family Papilionaceae (See also page 18.12) by mechanical pressing or by solvent extraction.

Fig. 11.18: Soyabean seeds

Geographical Source

It is an annual herb cultivated in Europe, China, Japan and India.

Description

Colour	:	It is golden yellow liquid
Odour	:	Characteristic
Taste	:	bland
Solubility	:	Insoluble in water and alcohol. Soluble in all organic solvents.

Chemical Constituents

Soya been seeds contain 20 – 25 per cent of fixed oil. It is a drying oil, and contains glycerides of saturated fatty acids namely linoleic acid 50%, oleic acid 30.0%, linolenic acid 7.0%. Stearic and palmitic acid 14.0%.

The unsaponifiable matter of the oil contains b-sitosterol, stigmasterol, campesterol and tocopherols. It also contains Lecithin and omega 3 fatty acids (EPA and DHA).

Fatty acid contents of the Soyabean oil are as under:

Saturated fatty acid	:	14 - 16 per cent
Mono unsaturated fatty acid	:	23 - 27 per cent
Poly unsaturated fatty acids	:	59 - 62 per cent
Omea-3 fatty acid	:	8.0 per cent
Omega-6 fatty acid	:	5.0 per cent

Standards of Quality

Specific gravity	:	0.922 - 0.926
Saponifcaiton value	:	190 - 195
Iodine value	:	130 - 142
Refractive index	:	1.4680
Hydroxyl value	:	4.0 - 8.0
Acid value	:	0.5 - 3.0

Uses

It is not only a cooking oil, but used extensively in manufacture of vanishes, paints, and insulators. Oil cake is used as valuable livestock food.

Stigmesterol obtained from oil is a basic raw material for synthesis of sterols.

BITTER ALMOND OIL

Biological Source

It is the volatile oil obtained from dried ripe kernels of sweet-almond *Prunus amygdalus* or *Prunus communis* var dulsis or Bitter almond *Prunus amygdalis* var amora, or mixture of both belonging to family Rosaceae and containing not less than 0.95 % benzaldehyde.

Other kernels containing amygdalin such as apricot, penches are also used.

Graphical Use

It is native of Iran (Persia) and asia minor. Commercially almond plants are culivated in Italy, France, Spain and North Afrcia. Now a days to some extent is also cultivated in Punjab and Kashmir.

Oil is obtained by macerating seeds with water and then steam distilling.

Description

The oil has

Colour : Colourless to yellowish

Odour : Characteristic

Taste : Of benzaldehyde

Solubility : Slightly soluble in water and solubie in alcohol, ether and pi------ oils.

Chemical Constituents

Contains not less than 95% benzaldehide, 2.4% benzaldehyde, Hydrocyanic acid and phenoxy aceto nitrile

Density : 1038 to 1060

Refractive index : 15425 to 15439

WHEAT GERM OIL

Biological Source

It is the fixed oil from wheat germ, *Triticum aestivum,* (Gramineae), obtained by solvent extraction or hydraulic expression. It constitutes about 2 per cent of wheat grain.

Description

It is bland yellow oil having a nut-like odour and taste. It is miscible in ether, benzene, petroleum ether and chloroform.

Chemical Constituents

Wheat germ oil contains saturated fatty acids (4.7 per cent), linoleic acid (44.1 per cent), linolenic acid (11.8 per cent), oleic acid (30 per cent), and unsaponifiable matter up to 4.7 per cent. Vitamin E tocophenol is the main constituent of unsaponifiable matter. The oil contains up to 0.5 per cent of vitamin E (2 International Units of vitamin E per gram of the oil.).

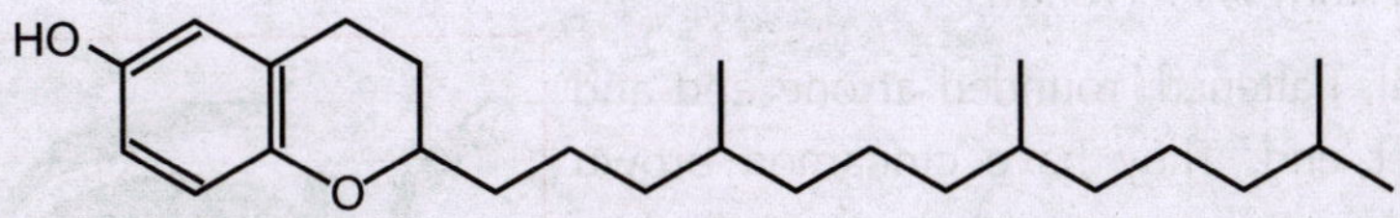

α tocopherol
(Vitamin E)

Standards of Quality

Specific gravity	:	0.925 - 0.935
Refractive index	:	1.469 - 1.478
Saponification value	:	179 - 190
Acid value	:	06 - 20
Iodine value	:	115 - 130
Hydroxyl value	:	10 - 48
Polenske value	:	0.4 - 2.1
Reichert - Meissel value	:	0.3 - 1.4
Thiocyanogen value	:	80 - 85

Uses

It is used as nutritional supplement and also used as source of natural vitamin E and unsaturated fatty acids (vitamin F). It also improves skin texture and firmness, it delays facial wrinkles and skin aging.

ALMOND OIL

Synonyms

Badam tail, Bitter Almond oil.

Biological Source

It is a fixed oil obtained by expression of seeds of *Prunus amygdalus* (*Amygdalus communis*) and/or *Prunus amygdalus* var. *amara* (bitter almond) or mixture of both belonging to family Rosaceae containing not less than 0.95 % benzaldehyde.

Commercially, bitter almonds are used as source of almond oil.

Geographical Source

It is native to Persia and Asia Minor.

Commercially, almond plants are cultivated and oil is produced on large scale in Mediterranean countries such as Italy, France, Spain and North Africa. In India, it is cultivated in some parts of Panjab and Kashmir.

Description

The almond is a deciduous tree, growing 5–10 m with large leaves, pinkish flowers and drupe fruits. There are two types of almonds that is sweet and bitter. The sweet almonds are used for edible purpose while bitter almonds are major source for almond oil. The fruits in unripe condition are green in colour, during ripening it changes colour with development of sclerenchymatous endocarp, which develops as stony shell (yellow).

Seeds are oval, flattened, rounded at one end and pointed at another end. They have cinnamon brown wrinkled papery testa. Below testa it contains two white cotyledons. Both verities of almonds are similar in morphology except the taste and size; seeds of the bitter almonds are bitter in taste while sweet variety is sweeter in taste. Seeds of the bitter variety are smaller in size (1.5 to 2 cm), while sweet varieties are 2 - 3 cm in size.

Fig. 11.19: Branch of almond herb with fruits

The oil is obtained by crushing and grinding the seeds and oil is filled in canvas bags. These bags are piled over the other and expressed between two slightly heated iron plates; oil is collected and furthermore, the oil is clarified by filtration.

Macroscopic Characters

Colour : It is colourless to Pale yellow coloured liquid.

Odour : Characteristic

Taste : Bland and characteristic, may be bitter or sweet, depending on variety.

Solubility : Insoluble in alcohol and soluble in chloroform, solvent ether and petroleum ether.

Chemical Constituents

Both the varieties of almond contain 40-50% of fixed oil, 20% of proteins, and mucilage. The fixed oil contains majorly unsaturated fatty acid such as 62-68% of linoleic acid and 20-30% linolenic acid; while saturated fatty acids as minor constituents contain 4% of palmitic acid and traces of other saturated acids.

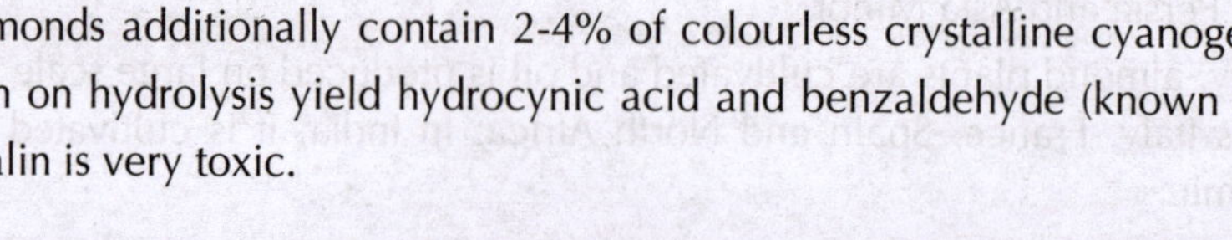

Benzaldehyde **Amygdalin**

The bitter almonds additionally contain 2-4% of colourless crystalline cyanogenetic glycoside, amygdalin, which on hydrolysis yield hydrocynic acid and benzaldehyde (known as volatile oil of almond). Amygdalin is very toxic.

Chemical Tests

1. Dissolve 100 ml of almond oil in cyclohyxane and dilute to 10.0 ml with same solvent. Absorbance is measured at maximum between 264 nm to 276 nm is not greater than 0.20.

2. Take 2 ml of oil with mixture of 1 ml of fuming Nitric acid and 1 ml of distilled water. Shake vigorously for five minutes. The mixture yields slight red colour (where as sesame oil or cotton oil yields a brown mixture when similarly treated).

Standards of Quality

Specific gravity	:	0.910 to 0.915
Acid value	:	1 to 3
Iodine value	:	97 to 105
Saponification value	:	188 to 197
Sterilization temperature	:	150ºC for one hour.
Refractive Index	:	1.5428 to 1.5489

Uses

It is used commercially for many toiletry preparations such as soap, shampoo, hair oil, etc.

Therapeutic Uses

Sweet oil of almond is nutritive and mild laxative, which is free of bitter taste as the name indicate. Sweet almonds are used in confectionary, almond milk and almond oil.

Pharmaceutically, it is used as vehicle for oily injection and volatile oil of almond is used as flavouring agent. Almond oil phenetinates deep into the skin and protects lips.

COCONUT OIL

Synonym

Copra oil.

Biological Source

It is a fixed oil obtained by expression from thoroughly dried kernels of the *Cocus nucifera* belonging to family *Palmae*.

Geographical Source

Western coastal area and specially Kerala and Goa. In Goa alone 18497 hectares of land is under cultivation and producing seven crores of coconuts per annum.

Indian Council of Agricultural Research (ICAR) at E.I.A., (Old Goa), has established research institute of coconut products.

Fig. 11.20: Copra-kernel fresh

Description

Colour	:	It is a white, semisolid fat.
Odour	:	Characteristic, and pleasant.
Taste	:	Bland and characteristic
Solubility	:	Insoluble in water and alcohol. Soluble in chloroform either and carbon tetra chloride. It has lowest level of unsaturated fatty acids and hence cost of hydrogenation is very less.
Extra Features	:	Two types of Copra are made in India, milling type and edible type. Milling copra is used to extract oil while edible copra is used as dry fruit.

Chemical Constituents

Copra contains 65 – 68% fat, 15 – 20 % carbohydrates, 9.0% proteins, 4 – 9% crude fibres. Carbohydrates glucose, fructose and raffinose. Coconut oil has lower level unsaturated fatty acids and hence the costs of Hydrogenatum is less as compared to other fixed oil.

Copra oil contains Trimyristine, trilaurin, tripalmetin, tristearin, and many other glycerides.

Standards of Quality

Density	:	0.903 – 0.910
Melting point	:	21 – 25ºC
Refractive index	:	1.4485 – 1.4495
Saphonification no.	:	255 – 258
Iodine no.	:	08.0 – 9.5
Acid no.	:	Not more than 6

Uses

Manufacture of soap, edible fats, chocolate candies, in baking instead of lard, in candles, in dyeing cotton, in ointment base in hair dressing in massage.

In the preparation of vegetable ghee.

- It does not elevate blood total cholesterol
- It increases blood HDL cholesterol.
- Lowers blood cholesterol if consumed alongwith coconut kernel.
- Decreases serum triglycerides.

It is used in food industry as it resists to oxidative rancidity and has pleasing flavour and good digestibility. It is preferred as dietary fat due to high content of lauric acid and low content of omega-6, fatty acid.

It is preferred as a fat in the preparation of filled milk, infant milk-powder, ice-cream, confectionery and bakery products.

Vergin coconut oil is prepared from coconut milk. Coconut milk is prepared by grating fresh coconut kernels which are then digested by healing and pressed to get the oil.

Vergin coconut oil helps in reducing body weight.

CANOLA OIL

Biological Source

It is edible oil obtained from the seeds of the plant *Brassica napus rapa, Brassica campestris* or *Brassica Juncea* family Erusiferie and should contain less than 2.0% Erucic acid and less than 30 micro moles of glucosinoletes. Canola oil is considered to be healthiest of plant derived oils.

Geographical Source

Canola is cultivated in Europian countries, Canada, China and India. It is third most widely consumed vegetable oil.

Plant Habitat

Canola plant reaches to a height of one to two metres with yellow flowers. It bears 60-100 pods per tree. Each pod contains 30-60 seeds. Seeds are black and round in shape and are about one mm in diameter. Canola plant is green in colour and may get changed to yellow due to climatic changes. It is a crop of 4 months only. It grows luxuriously in cool nights and hot day temperature.

Canola is also drought and disease resistant crop.

Fig. 11.21: Brassica – campestris herb with flowers

Description

Colour : It is light yellow coloured liquid.

Odour : Characteristic.

Taste : Bland.

Extra characters : Oil has very long shelf life.

Chemical Constituents

Canola oil has very good distribution of saturated, monosaturated and polysaturated fatty acids. Canola seeds contain approximately 40.0% fixed oil. It contains linoleic (Omega-6) and α-linolenic (Omega-3) fatty acid in the proportion of 2 : 1, which makes it one of the healthiest cooking oils.

Phytosterols namely β-sitosterol and compesterol are present in the oil, which competitively inhibit cholesterol absorption in the gut and there by reduce cholesterol levels by 10 per cent or more.

Erucic acid and glucosinolates are undesired contents and are removed to control the edibility of the oil. It contains no cholesterol and is good source of vitamin E.

Erucic acid

Fatty acid content of canola oil is Oleic acid 60.0% and Linoleic acid 20%, α-Linolenic acid (Omega-3) 10.0%; Palmilic acid 4.0%; Stearic acid 2.0% and saturated fatty acids 7.0%.

Standards of Quality

Specific gravity	:	0.916 – 9.921
Iodine value	:	110 - 120
Saponification value	:	188 - 198

Uses

It is a good substitute for rapeseed oil and used mainly for cooking purpose and in cosmetics for preparation of lipsticks.

Industrially it is used as lubricant, as bio-diesel, in the preparation candles, and also in news paper ink manufacture.

Storage

Keep it in well closed, well filled containers away from direct sun-light and in cool place.

MANGO KERNEL OIL

Synonym

Mango butter

Biological Source

It is a edible fat obtained from fully matured kernels of the seeds of *Mangifera Indica*, Family : *Anacardiaceae*.

Geographical Sources

Mango is indigenous to India and is the second largest producer of Mangoes in world and is producing about 15 million tons of mangoes per annum. China being first while Thialand is the third in the list.

Mango seed oil is of high quality due to low level of peroxide value and is of great impratnce due to unsaponifiable matter containing sterols (tocopherols) and squaline.

It is also rich in unsaturated fatty acids and also rich in anti oxidant activity, due to phenolics, betalains and carotenoids.

Preparation of Oil

Good quality mango seeds are collected and are cracked to obtain kernels, the seed coats are separated and the kernels are ground to reduce the particle size maximum to 500 nm. Yield of the oil is 8 – 8.5%.

For extraction hexane is used as it gives maximum yield of good quality oil. For industrial purposes oil expellers are to be used during extraction temp of 50ºC is maintained.

Description

Colour	:	Light yellow coloured wax and fat (solid)
Odour	:	Characteristics
Taste	:	Bland
M.P.	:	35 to 43ºC

Chemical Constituents

Approximate fatty acid composition of Mango kernel oil is as under

Palmitic acid	:	9 – 10 %
Stearic acid	:	37 – 39 %
Oleic acid	:	43 – 45 %
Lenoleic acid	:	06 – 07 %
Unsaponifiable matter	:	1.2 %
Lenolenic acid	:	02 – 03 %
Arachidic acid	:	02-2.3 %
Sterols	:	0.2 - 0.5 %

Kernels contain high percentage of starch 72.5% and fat 10.5% only. Kernels are also very rich in calcium, magnecium, potassium, phosphorus, vitamin C, E and A.

Iodine value of the oil represents unsaturated fatty acids, while saponication value indicates average molecular weight.

Mango fruits are also very rich in anti oxidant compounds like phenolics, betalans and carotenoids.

Standards of Quality

Specific gravity : 0.9991

Physical characteristics of Mango kernel oil.

Acid value	:	8 – 9
Saponification value	:	190 – 220
Iodine value	:	37 – 38
Peroxide value	:	8 – 9
Rechal michel value	:	0.12

Uses

Due to high content of unsaturated fatty acids oil is very stable and can be used as substitute for cocoa-butter, in chocolate manufacturing process. Kernels can also be used as feed for cattle and poultry.

Kernel powder is used as anthelmintic and as astringent in bleeding piles.

Mango seed kernel is good source of phenolic compounds including micro elements like zinc, copper, and seleum. The extract of mango seed kernel has heigh free radical scavenging and tyrosinase inhibition activities as compared to methyl gallate and phenolic compounds.

Sterols are represented as tocopherols in the mango kernal butter. Tocopherols are as anti oxidant 250 times more active than bytelaled hydroxytoluene or butylated hydroxy anisol.

One molecule of tocopherol can protect about 103 – 105 molecules of polyunsaturated fatty acids.

The use of mango kernel oil in body creams, suncare balms, hair products and other moisturising products is suggested. Mango kernel oil is free from any toxic principles.

Content of tocopherols in mg/kg of selected vegetable oils is as under:

	α tocopherol	γ tocopherol	δ tocopherol
1. Soyabean oil	144	16	870
2. Sunflower oil	608	17	041
3. Wheat gern oil	1330	71	160
4. Mango kernel oil	140	63	062
5. Rice-bran oil	347	-	089
6. Corn oil	191	-	942
7. Palm kernel oil	002	21	09

Storage

Store mango kernel oil in well closed containers in cool place, away from direct sunlight.

SUNFLOWER OIL

Synonym

Suryaphula oil.

Biological Source

It is the fixed oil obtained from the matured seeds of cultivated varieties of *Helianthus annus* Linn. belonging to family Compositae.

Geographical Sources

It is produced on large scale in USSR, Ukrain, India, Egypt, Canada, Argentina, Minisota and North Dakota in U.S.A.

Method of Preparation

Fully dried cleaned and graded seeds are selected for the production of oil. The seeds are decorticated by using machines and oil is obtained by expression. Oil is filtered and refined for pharmaceutical purposes.

Description

Colour : It is pale yellow in colour and liquid at ambient temperature.

Odour : Odourless

Taste : Bland agreeable taste

When pure sunflower oil is equally good to olive or almond oil for culinary and table purposes.

It is slightly soluble in alcohol and soluble in solvent ether. Chloroform and light petroleum (40 - 60), benzene and carbon tetra chleinde.

Fig. 11.22 : Flower of Sunflower

Chemical Constituents

Sunflower seeds contain 32 – 45% of fixed oil. Sunflower oil contains mixed triglycerides. The chief constituents are palmitic acid 6.5%, arachidic acid 4.0%, behenic acid about 1.0%, oleic acid 21.5%, linoleic acid 66.0% and Vitamin E (tocopherols) approximately 75 mg/100 gm.

Fatty acid contents in sunflower oil are as under:

Saturated fatty acid	:	12.0 per cent
Mono unsaturated fatty acid	:	25.0 per cent
Poly unsaturated fatty acids	:	62 per cent

$$CH_3(CH_2)_{18} \cdot COOH$$

Arachidic acid

Standards of Quality

Acid value	:	Not more than 0.5
Peroxide value	:	Not more than 10.0
Saponification value	:	188 – 194
Indine value	:	125 – 135
Thiocyanogen value	:	78
Hydroxyl value	:	14 – 16
Relchart – Meissle value	:	Below 0.5
Polenske value	:	Below 0.5
Unsaponifiable matter	:	Below 1.5%
Refractive index	:	1.472 – 1.474

On exposure to air it thickens gradually, It is a semi drying oil.

Uses

Industrially it is used for wool-dressing, lighting candle and soap making. It is said to be equal to olive oil for culinary and table purposes. It is an ingredient of dietary supplements marketed under the name "Travasorb", oil is used for culinary purposes mainly. Oil cake is a valuable food for cattle and poultry.

Sunflower oil is rich in vitamin E content and very low in saturated fatty acids and provides protection in Asthma and colon cancer. It is good source of vitamin A and hence it is aid in preventing cataract. It also contains lecithin and helps in lowering cholesterol level.

Being exceptionally light and non-greasy sunflower oil gets absorbed in the skin easily, without clogging. It is also excellent source of beta-carotene and makes the skin less sensitive to sunlight. The antioxidants present in this oil neutralise free radicals that penetrate the skin causing sunburn exposure to free radicals sunlight increases the rate of aging of skin causing wrinkles. Being naturally emollient it improves moisture retention capacity and is beneficial to dry skin.

RAPE SEED OIL

Synonyms

Colza oil, Sarson tel.

Biological Source

It is fixed oil obtained by expression of dried seeds of *Brassica compestris*, fam. Cruciferae.

Geographical Source

It is extensively cultivated in North Gujarat and also to some extent in Maharashtra.

Method of Preparations

Oil is prepared industrially on large scale by expression method only. The seeds are very small yellow or brown in colour. Subjected to high pressure by expression yield 35 per cent of oil.

Organoleptic Characteristics

Rape-seed oil is pale yellow in colour with characteristic smell and bland taste. Oil is viscous.

Chemical Constituents

It contains glycerides of fatty acid alongwith erucic acid.

Fig. 11.22 (a) : Brassica campestris (Rapeseed) herb

Standards of Quality

Specific gravity	:	0.913 – 0.917
Refractive index	:	1.4720 – 1.4752
Saponification no.	:	170 – 177
Iodine no.	:	097-105
Solidifying point	:	– 2.0 to – 10
Solubility	:	It is soluble in ether chloroform and carbon disulphide.

Uses

Used as food, in the manufacture of margarine, soft soap, vegetable ghee and as lubricant. Seeds are used in counter irritant poultice and preparation of comphor liniment.

CROTON OIL

Biological Source

It is a fixed oil obtained from dried and matured seeds of *Croton tiglium* fam Euphorbiaceae.

Geographical Source

Seeds producing croton oil are Sri-Lanka, Malaya, China and India. In India seeds are cultivated in West Bengal and Assam.

Description

Colour	:	Croton oil is a pale yellow to brownish yellow viscid liquid.
Odour	:	Slight and disagreeable
Solubility	:	Slightly soluble in alcohol and soluble in chloroform, ether, fixed oils, petrolium ether, carbon di-sulphate and glacial acetic acid.

Chemical Constituents

It contains croton resin, and glycerides of stearic, lauric, palmetic, myristic, tiglic acid, crotin, pharbolesters.

Standards of Quality

Density	:	0.935 – 0.950
Optical rotation	:	1.470 – 1.473
Saponification no.	:	205 – 220
Iodine no.	:	102 – 118

Uses

Formerly used as irritant. As carcinogen in cancer research. It is cathartic and counter irritant.

Caution

It is very poisonous. Over exposure casues hemorrhagic-gastroentritis Direct contact casues vesiculation, necrosis and sloughing at skin.

PALM OIL

Synonym

Red oil palm.

Biological Source

It is the fixed oil obtained from fleshy orange-red mesocarp of palm fruits *Elaesi-guincensis,* fam. Arecaceae (Palmae).

Geographical Source

West Africa, Malaysia and to some extent in India.

India has imported 8.85 million tonnes of palm oil in 2019 – 2020.

Habitat

Palm tree bears large bunches of plum-sized fruits. Palm tree is an erect, tall, handsome plant with stout stem and large pinnate leaves of dark green colour. Fruits are red in colour and some what obovate heads. At present, in India only 3 lakh hectares of land is under cultivation of palm trees. Cultivation produces four tonnes of oil per hectare.

Godrej agrowet has under taken cultivation on large commercial base to face the shortage of edible oil and also to encourage the Indian farmers about 5000 hectare area is planned for cultivation.

Description

Colour : Palm oil is a orange coloured semisolid.

Odour : Characteristic sweetish odour.

Taste : Bland and characteristic

Extra Features

Insoluble in water and ethenol.

Soluble in organic solvents.

Chemical Constituents

Plam oil contains several fatty acids, carotenoids and sterols. Fatty acids are palmitic acid – 45%, oleic acid 39.0%, lenoleic acid 10.0%. Carotenoids are α & β carotenes, tocopherols i.e. tocotrienol, β-sitostarol.

Standards of Quality

Melting Point : 27.5 ºC

Density : 0.89 to 0.91

Refractive index : 1.453 to 1.459

Saponification no. : 196 – 209

Uses

It is mainly used as edible oil used for cooking and frying after refining only. It is cocoa-butter substitute. It is used in preparation of margarines, soaps and oleo-chemicals, in cosmetics, in personal care products, in skin conditioners and also as solvent.

Palm kernal oil: Palm seeds are used to produce palm kernal oil by solvent extraction method or by mechanical solvent method. It is used similar to palm oil.

LIPIDS OF ANIMAL ORIGIN

YELLOW BEESWAX

Synonyms

Beeswax; Cera-flava.

Biological Source

Yellow bees wax is purified wax obtained by melting the walls of the honey comb of the bees *Apis mellifica* and other species of *Apis*, belonging to family Apidae.

Geographical Source

It is processed and commercially prepared in France, Italy, West Africa; Jamaica and India.

Processing and Preparation

The combs and capping of honeycomb are broken and boiled in soft water. These are then enclosed in a porous bag weighted to keep under water. The boiling causes oozing of the wax, which gets collected outside the bag and forms a cake after cooling. The debris on outer surface is removed by scraping. Beeswax is purified by heating in boiling water or dilute sulphuric acid and settling. The process is repeated several times and finally wax is skimmed off. Various techniques are adopted to bleach wax, such as treatment with hydrogen peroxide, chromic acid, ozone, etc.

Sometimes, treatment with charcoal, chlorine or potassium permanganate is also given to bleach the wax. Natural bleaching by exposing the wax to the sun-light in thin layers is also preferred. India exported 45 tones of beeswax worth ₹ 62 lakhs during 1996 - 97.

Description

Colour : Yellow to yellowish-brown, wax and solid.

Odour : Agreeable and honey-like

Extra Features

Yellow beeswax is non-crystalline solid. It is soft to touch and crumbles under the pressure of fingers to plastic mass. Under molten condition, it can be given any desired shape. It breaks with a granular fracture.

Fig. 11.23 (a): Honey-bee

Solubility

It is insoluble in water, but soluble in hot alcohol, ether, chloroform, carbon tetrachloride, fixed and volatile oils.

Chemical Constituents

It consists of esters of straight-chain monohydric alcohols with straight chain acids. The chief constituent of the bees wax is myricin i.e. myricyl palmitate (melissyl palinitale) (about 80 per cent). Free cerotic acid (about 15 per cent), small quantities of melissic acid and aromatic substance cerolein are the other constituents. Indian bees wax is characterised by its low acid value, while European bees wax has the acid value of 17 to 22.

Chemical Test

Saponification claud test: Boil 0.5 gm of bees wax with 20 ml of aqueous caustic soda solution for 10 minutes. cool it. No turbidity is produced.

Standards of Quality

Melting point	:	62° – 65°C
Specific gravity	:	0.958 – 0.967
Acid value	:	05 – 10
Saponification value	:	90 – 103
Ester value	:	80 – 95
Peroxide value	:	Not more than 8.0
Flash point	:	245 – 258°C

Uses

Bees wax is used in preparation of ointments, plasters and polishes. It is used in ointment for hardening purposes and the manufacture of candles, moulds and in dental and electronic industries.

It is also used in cosmetics for preparation of lip-sticks and face creams. Pharmaceutically, it is an ingredient of paraffin ointment IP.

Adulterants

Very frequently beeswax is adulterated with colophony, hard paraffin, stearic acid, Japan wax, spermaceti, carnauba wax and several other substances. Adulteration can be detected on the basis of solubility and melting point. The genuine wax should not give turbidity when 0.5 g of wax is boiled with 20 ml of aqueous caustic soda for 10 minutes and cooled.

White beeswax (White-wax): Obtained by bleaching yellow bees wax, and it should not be used for ophthalmic purposes.

COD LIVER OIL

Synonym

Oleum morrhi

Biological Source

It is processed from fresh liver of cod fish, *Gadus morrhua* and other species of *Gadus* (family - Gadidae).

Geographical Source

Large quantities of oil consignments are prepared in coastal regions of Norway, Scotland, Iceland, Germany, Denmark and Britain.

Method of Preparation

Single cod fish is 1.0 – 1.5 m in length and 15 – 25 cm in width. 15 – 25 kg in weight.

The fishes are caught by nets, opened, and livers are separated. The healthy livers free from gall bladders are washed, minced, steamed in steam jacketed vessels or 'kars' at a temperature not exceeding 85°C for half an hour, cooled and buried in snow for several days. Special barrels are used for this cooling process, which results in separation of stearin. The steaming of oil destroys enzyme lipase. The medicinal oil after filtration is kept in well-closed air tight containers in a cool place protected from light.

Description

Colour	:	Pale yellow coloured liquid
Odour	:	Slightly fishy
Taste	:	Fishy
Solubility	:	Soluble in chloroform, ether, petro-ether, slightly soluble in alcohol.

Five major steps involved in refining of medicinal cod-liver oil are (a) removal of impurities, (b) drying, (c) winterization, (d) deodorisation, (e) standardisation for vitamin content. The vitamin A content of the oil is determined spectrophotometrically.

Fig. 11.24: Cod Fish

Chemical Constituents

The medicinal value of oil is due to vitamin A and vitamin D. About 1 g of oil contains not less than 255 mcg of vitamin A and 2.125 mcg of vitamin D. The oil contains glyceryl esters of oleic, linoleic, gadoleic, myristic, palmitic and other acids. Cod liver oil also contains 7 per cent eicosapentaenoic acid and 7 per cent olecosohexanoic acid. (Both of them are omega-3 fatty acids).

Standards of Quality

Specific gravity	: 0.918 - 0.927
Refractive index	: 1.4705 - 1.4745
Acid value	: Less than 2
Saponification value	: 180 - 190
Iodine value	: 145 - 180

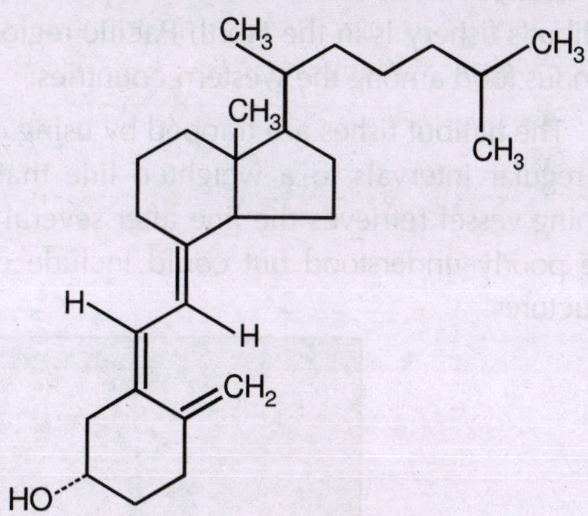

Vitamin D_3 (Cholecalciferol)

Uses

The oil is used as source of vitamins, as a nutritive and in treatment of rickets and TB.

As a result of competition from vitamin concentrates, the consumption of medicinal oil has substantially decreased in developed countries of Europe and the USA. The renewed interest in fish liver oils, particularly cod liver oil resulting from nutritional requirements for polyunsaturates in diet, coupled with blood cholesterol reducing property the vistas of trade seem to open ahead for commerce and industry of fish liver oil.

Non-destearinated cod liver oil is the entire oil that has not been chilled to separate stearin. The oil contains not more than 0.5 per cent by volume of water and liver tissues and it deposits stearin upon chilling.

Storage

In order to avoid loss of vitamins during storage, the oil should be kept in well-filled airtight containers, protected from light and in a cool place. The addition of small quantities of certain antioxidants (e.g. dodecyl gallate) is permitted. It may be bottled in containers from which air has been expelled by production of vacuum or by an inert gas like nitrogen.

HALIBUT LIVER OIL

The name is derived from haly (holy) and butt (flat fish), for its popularity on Catholic holy days.

Synonyms

Spiny turbot, Flat fish.

Biological Source

It is obtained from fresh as well as well preserved livers and viscera of flat fish *Psettodes species (Psettodes. erumei, Psettodes bennetti Psettodes belcheri)* and *Hippoglossus vulgaris* by solvent extraction or alkali digestion, belonging to family Psettodidae.

Geographical Source

It is found in the North Pacific and the North Atlantic oceans.

Collection and Extraction of Oil

Halibut fishes feed on almost any animal they can fit into their mouths. Commercially the largest halibut's fishery is in the North Pacific region from last century to till dates. The halibut meat is very famous food among the western countries.

The halibut fishes are trapped by using chunks of octopus or other bait on circle hooks attached at regular intervals to a weighted line that can extend for several miles across the bottom. The fishing vessel retrieves the line after several hours to a day. The effects of longline gear on habitats are poorly understood but could include disturbance of sediments, benthic structures, and other structures.

Fig. 11.25: One of the halibut fleet on voyage

The fishes are sent to processing unit where their livers and viscera are separated and stored separately. These are well preserved either by means of cool storage or addition of preservatives. Preservative treatment is necessary for preventing microbial decomposition of the oil and formation of free fatty acids rather than to protect vitamin A, which is quite stable within the liver.

Oil can be extracted by using any of the method suitable mentioned under.

Method 1

The process of extraction largely depends upon the oil content of the liver and the vitamin A potency. For the production of medicinal oil only sound, well preserved healthy livers are used. The livers which are relatively poor in oil content (but high Vitamin A content) are held tenaciously by proteins, will not be released completely during the normal method of extraction i.e. steaming. So in such cases the other effective methods are used. The digestion of liver protein with alkali has proved the most satisfactory method.

In this technique, the livers are minced and mixed with dilute caustic soda (1 - 2% by weight) or sodium carbonate (2 - 5%) and digested with live steam at temperature of 80 - 88°C with constant stirring for an hour. The digested material is processed through large centrifuge to separate the oil and tissues, further more the tissues are subjected for washing with other edible oil so as to remove the vitamin from tissue material.

Method 2

In another method i.e. enzyme alkali digestion process, minced livers are mixed with equal valume of water, and pH is adjusted to 1.2 - 1.5 with the help of mineral acid such as hydrochloric acid. This is followed by addition of pepsin and mixture is digested at temperature of 43 - 49°C. After the digestion, the pH is adjusted to 9.0 with the help of sodium carbonate, the mixture is heated to 80°C for an hour, which results in separation of oil which is filtered and separated.

Method 3

Solvent extraction method is also considered to be suitable for this oil. The yield of the oil by this process is always higher, but the oil obtained is darker with higher viscosity. The reason is, during solvent extraction process the free fatty acids also get extracted, thus it has to subject for refining process.

Refining of the oil:

The medicinal grade oil is refined by filtration, centrifugation or treatment with fullers earth. The oil contains the sufficient amount of stearin, which is separated by gradual cooling of oil between temperature 0 - 10°C where stearin congeal out and can be separated out by filtration at very low temperature.

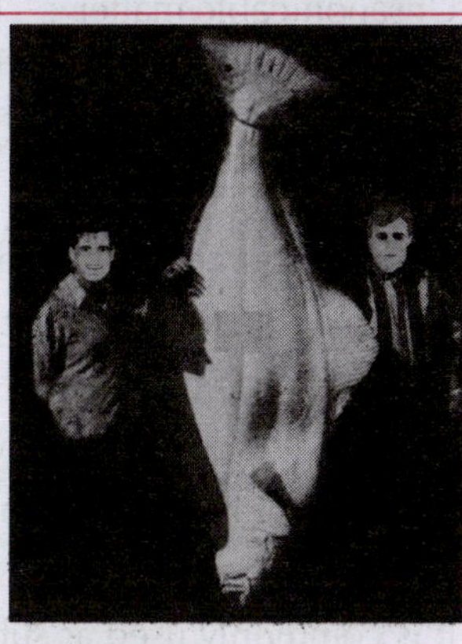

Fig. 11.26: Large Halibut fish

Description

Colour : Pale to golden yellow liquid.

Odour : Fishy

Taste : Fishy

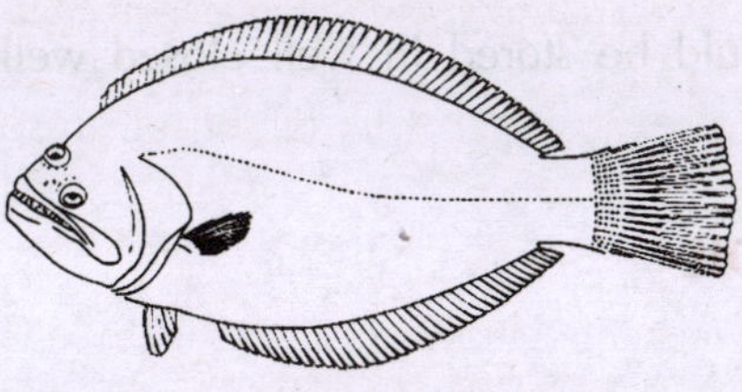

Psettodes belcheri fish

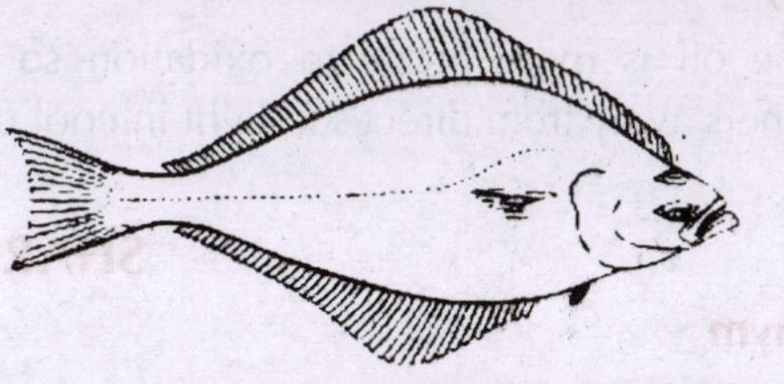

Hippoglassus vulgaris fish

Fig. 11.27 : Types of Halibut fishes

Chemical Constituents

The commercially important fish livers may be grouped into three classes such as (i) high oil content and low vitamin A potency, (ii) those with low oil content and high vitamin A potency and (iii) those with high oil content and high vitamin A potency.

The halibut fish comes into the second category, as the oil content is 4 - 28% with vitamin A potencies ranging from 25,000 - 6,00,000 I.U./g. The vitamin D potencies are upto the 5000 I.U./g.

The liver is the principal storage for vitamin A but in this flatfish the vitamin is concentrated in other organs as viscera.

Chemical Tests

1. 1 ml of halibut liver oil when treated with 1 ml of chloroform and 0.5 ml of sulphuric acid initially it turns to light purple followed by dark purple and finally turns to dark brown colour which indicates presence of Vitamin A.
2. 1 ml of halibut liver oil is dissolved in 10 ml of chloroform and this solution when treated with antimony trichloride in chloroform, after shaking, the mixture became blue coloured which indicates presence of vitamin A.

Standards of Quality

Specific gravity	:	0.915 - 0.925 at 20°C
Saponification value	:	Not more than 180
Iodine value	:	Not less than 112
Acid value	:	Not more than 2.0
Unsaponicable matter	:	Not less than 7.0
Vitamin A content	:	Not less than 30,000 IU/g
Vitamin D content	:	2500 – 3000 IU/g

Uses

1. As a source of Vitamins A and D.
2. It can be used in night blindness and rickets.
3. It can be employed as anti-arthritic agent and also finds its uses in osteo arthritis.
4. It is also claimed to lower the blood lipid content.
5. Halibut liver oil is a minor source of essential fatty acids.

Substitutes

It is commonly used as substituent for cod liver oil; the halibut liver oil is more concentrated for vitamin A content.

Storage

The oil is more prone to oxidation so the oil should be stored in well closed well filled containers away from direct sun light in cool place.

SHARK LIVER OIL

Synonym

Oleum Selachoids.

Biological Source

Shark liver oil is the fixed oil obtained from the fresh and carefully preserved livers of various species of the shark, mainly *Hypoprion brevirostris* and *Galeorhinus zyopterus*.

In India, *Scoliodon*, *Carcharias*, and *Sphyrna* are abundant among the species, and are generally utilised for the extraction purpose. According to *Indian Pharmacopoeia*, one gram of oil should contain not less than 6000 International Units of vitamin A activity.

Geographical Source

In India, the shark livers are processed and oil is obtained on commercial scale in Tamil Nadu, Maharashtra and Kerala. Most of the European countries are also producing shark liver oil on large scale.

Method of Preparation of oil

With a little variation, the principle involved in extraction of the oil from the livers is uniform in almost all cases. Government factories in Tamil Nadu and Maharashtra are processing livers for extracting oil. The livers are cleaned and minced. The minced mass is taken to a boiling pot, where the temperature of 80°C is maintained. The oil extracted is treated with dehydrating agent to remove traces of water.

The oil is then taken to the vacuum still for dehydration and chilled to separate stearin. Centrifuges are used to separate suspended materials in oil. The clear oil is manipulated to adjust the desired strength of vitamin A. The oil being sensitive to light and air, care is taken all the while to minimize its exposure to sunlight and air. Many a time, the livers are stored at very low temperature, until they are taken for processing.

Description

Colour : It is pale yellow to brownish-yellow liquid.

Odour : Characteristic fishy, but not rancid.

Taste : Bland or fishy.

Fig. 11.28: Shark fish (Carcharias sp.)

Solubility

Shark liver oil is soluble in solvent ether, chloroform, and light petroleum. However, it is insoluble in water and slightly soluble in ethyl alcohol.

Chemical Constituents

Shark liver oil contains vitamin A. The concentration of vitamin A in the oil varies from 15000 - 30000 International Units per gramme. Other constituents of the oil are the glycerides of the saturated and unsaturated fatty acids along with alkyl glycerol. Additionally it contains squalene and omega-3-fatty acids. (EPA & DHA).

Conversion of IU and Mcg for vitamin A activity:

700 mcg = 1500 IU

800 mcg = 2400 IU

900 mcg = 3000 IU

1100 mcg = 3300 IU

3000 mcg = 10000 IU

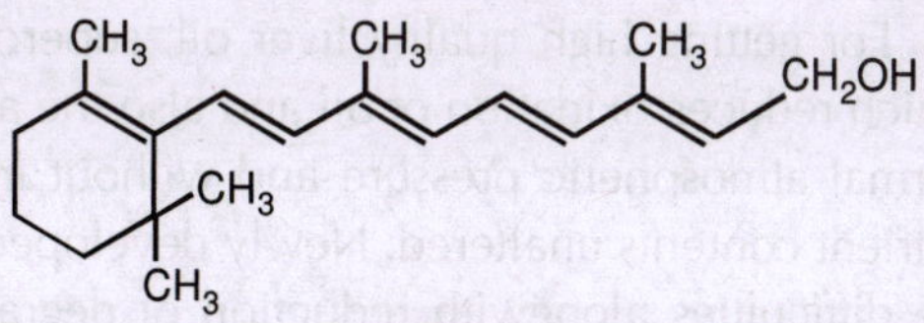

Vitamin A

Identification

1. Mix 1 g shark liver oil with 1 ml chloroform and treat with 0.5 ml sulphuric acid. It acquires light violet colour changing to purple and finally to brown (due to vitamin A).
2. Dissolve 1.0 ml shark liver oil in 10 ml chloroform and treat with saturated solution of antimony trichloride in chloroform. Shake it well. A blue colour is developed (due to vitamin A).

Standards of Quality

The shark liver oil should comply with following standards.

Specific gravity	:	0.912 - 0.916
Refractive index	:	1.459 - 1.477 at 40°
Acid value	:	Not more than 2
Iodine value	:	Not less than 90
Saponification value	:	Not more than 200.0
Unsaponifiable matter	:	Not more than 2.0 per cent.

Uses

It is used in the deficiency of vitamin A. It is also known as antixeropthalmic factor. (However, it should be noted that shark liver oil is free of vitamin D and is required to be fortified when necessary. Due to the absence of vitamin D, shark liver oil is not a genuine substitute for cod liver oil). It is also nutritive. Pharmaceutically, it is used in the preparation of dilute shark liver oil, shark liver oil emulsion (Indian N.F.) and shark liver oil with vitamin D. It is used in burn and sunburn ointments. Few species of shark are used for isolation of mercury from their tissue.

Storage

Shark liver oil is preserved in well-filled and well-closed containers protected from light.

TUNA FISH LEVER OIL

Biological Source

It is the fixed oil obtained from frozen marine fish levers of *Katsuwonus pelamis*, which is a sea food.

Geographical Source

Tuna fish is a sea food, and excellent precursor of marine oil and is found in China and Lakshadweep islands in India. In china, Tuna fish livers are discarded on large scale and hence used as source of marine oil and for livers. Output of tuna in China is about 4.6 million tones per annum.

Fig. 11.29: Tuna Fish

Method of Preparation

For getting high quality liver oil, supercritical carbon dioxide method is the ideal for tuna oil which reduces oxidation of oil and also the amount of impurities in the oil. This method is carried at normal atmospheric pressure and without freeze drying of the oil, though it is costly it keeps the nutrient contents unaltered. Newly developed subcritical dimethyl either method overcomes most of the difficulties alongwith reduction of degradation of bioactive components. The oil produced by this method possess a potential of low toxicity. The yield of tuna fish liver oil is approximately 17.5%.

Chemical Constituents

Tuna fish oil also contains omega 3 fatty acids EPA and DHA, but have a profile of around seven times as much omega-3 oils as omega-6 oil. Tuna fish contains lessor amount of omega fatty acids.

Uses

Mainly Tuna fish oil is used as dietary source, as emolient externally and in ointments.

SPERMACETI

Synonyms

Sperm wax; Cetaceum, Spermaceti. Acetyl-ester wax.

Biological Source

It is a waxy substance obtained from the head of sperm whale *Physeter macrocephalus* Linn. and other species of *Physeter*, family Physeteridae.

Geographical Source

Sperm whales are found mainly in Pacific, Atlantic and Indian oceans. The coastal regions of Mozambique, Zanzibar and Sri Lanka are the habitats for sperm whale.

Preparation

Sperm oil is secreted in a large special cavity above the right nostril in the jaw of the sperm whale. The cavity is emptied after the animal is captured. The oil on cooling deposits the crystalline mass known as spermaceti.

Description

Spermaceti is found in translucent pearly white masses, unctuous to touch and with crystalline fracture. It is odourless, tasteless but becomes yellow on exposure to air. It is soluble in oils, chloroform, carbon disulphide, boiling alcohol; but insoluble in petroleum ether and water.

Fig. 11.30: Sperm Whale Fish

Chemical Constituents

Spermaceti chiefly contains acetyl palmitate, free acetyl alcohol and esters of lauric, myristic and stearic acids.

Standards of Quality

Specific gravity	: 0.95 - 0.96
Melting point	: 42 - 50°C
Refractive index	: (at 80°F) - 1.433
Saponification value	: 125 - 136
Iodine value	: 03 - 05

Uses

It is mainly used as emollient and in preparation of ointments, especially cold-creams. Acetyl alcohol is a valuable emulsifying agent.

Synthetic Spermaceti or Acetyl ester wax

Spermaceti is scarce material in the market and hence mixture containing esters of saturated fatty alcohols and saturated fatty acids is used as substitute for spermaceti.

Guppy Fishes:

Fishes known as *Libistes reticularis* fam Cyprinodontidae are now-a-days used for biological control of Anopheles mosquito since they feed on anopheles mosquito which otherwise cause malaria.

LARD

Synonym

Adeps

Biological Source

It is the purified internal fat obtained from the abdomen of the hog (Pig) *Sus scrofa* Linn, family Suidae.

Preparation

The omentum and parts of peritoneum containing lard are washed thoroughly and minced to break the membranous vesicles. It is subjected to temperature of about 50 - 55°C to melt the lard. It is then separated by passing through muslin cloth. Cooling is done with proper stirring. The entrapping of air should be avoided which otherwise helps to develop rancidity on storage.

Description

It is a white, homogeneous fatty mass unctuous to touch. Lard has slight odour and bland taste. It is insoluble in alcohol and water, and soluble in benzene, carbon disulfide, ether and chloroform.

Fig. 11.31: Pig (Hog)

Chemical Constituents

Chemically, it contains about 60 per cent olein and 40 per cent of stearin and palmitin.

Chemical Test

Dissolve 1 gm lard in 5 ml ether. Allow to stand for 18 hrs. crystals will be deposited. It gives irregular groups of flat rhomboidal plates cut off obliquely at one end.

Standards of Quality

Melting point	:	34 - 41°C
Specific gravity	:	0.934 - 0.938
Acid value	:	01 - 02
Iodine value	:	52 - 56
Saponification value	:	192 - 198
Refractive index	:	1.4520 - 1.4550
Unsaponifiable matter	:	Not more than 0.5 per cent

Uses

It is used as an ointment base and in formulations where more effective absorption is desired. Benzoinated lard contains benzoin resin as preservative.

SUET

Synonyms

Sevum; Mutton suet.

Biological Source

Medicinal suet is the internal fat of abdomen of sheep *Ovis aries* Linn., family Bovidae.

The fat contained in the vesicles of the omentum and near the kidneys is used for suet. The omentum is crushed to break membranous vesicles. It is then melted and strained appropriately. Cooling is done by thorough stirring and then it is allowed to solidify.

Geographical Source

Australia, USA.

Description

It is a white solid fat, which is smooth, uniform and unctuous to touch. It has a slight odour and bland taste if fresh, but becomes rancid after long exposure to air. It is soluble in chloroform, ether and light petroleum and insoluble in alcohol and water.

Fig. 11.32: Sheep

Chemical Constituents

It mainly contains palmitin and stearin (about 75 - 80 per cent), along with olein (20 - 30 per cent).

Standards of Quality

Melting point	:	45 - 50°C
Specific gravity	:	0.948 - 0.953
Iodine value	:	33 - 45
Saponification value	:	192 - 195
Congealing point	:	37 - 40°
Refractive index	:	1.449 - 1.451

Uses

It is used as ointment base.

HYDROUS WOOL FAT

Synonyms

Lanolin, Adeps Lanae.

Biological Source

Hydrous wool fat is the purified fat-like substance obtained from the wool of the sheep *Ovis aries* Linn. family Bovidae.

It is the secretion of sebaceous glands of sheep deposited on the wool fibres. It contains between 25 - 30 per cent of water.

Geographical Source

Commercially, lanolin is manufactured in Australia, the USA. and to a very less extent in India.

Method of Preparation of Wool fat

Raw wool contains about 31 per cent wool fibres, suint or wool sweat (chemically potassium salts of fatty acids), about 32 per cent earthy matter and about 25 per cent wool grease or crude lanolin. Crude lanolin is separated by washing with sulphuric acid or suitable organic solvent or soap solution. It is further purified and bleached. The product is known as anhydrous lanolin or **wool fat**. The hydrous wool fat is produced by intimately mixing wool fat with 30.0 per cent of water.

Description

Colour : It is whitish-yellow, solid

Odour : Faint and characteristic

Taste : Bland

Extra Features

It is found in the form of ointment like mass and on heating in water bath, it separates into two layers.

Solubility

It is practically insoluble in water but soluble in chloroform and solvent ether with separation of water.

Chemical Constituents

It is a complex mixture of esters and polyesters of 33 high molecular weight alcohols and 36 fatty acids.

Hydrous wool fat contains mainly esters of cholesterol and isocholesterol with carnaubic, oleic, myristic, palmitic, lignoceric, and lanopalmitic acids. It also contains 50 per cent of water.

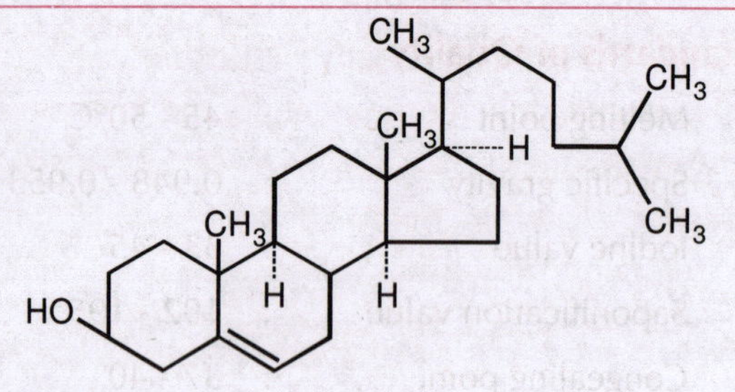

Cholesterol

Identification Test

Dissolve 0.5 g of hydrous wool fat in chloroform and to it add 1 ml of acetic anhydride and two drops of sulphuric acid. A deep green colour is produced, indicating the presence of cholesterol.

Standards of Quality

Anhydrous lanolin (wool fat) has following standards.

Melting point	34 - 44°C
Acid value	Not more than 1
Iodine value	18 - 36
Saponification value	90 - 105
Peroxide value	Not more than 20

Uses

The lanolin is mainly used as water absorbable ointment base. It is a common ingredient and base for several water soluble creams, cold creams and cosmetic preparations. It can be allergic also.

Anhydrons lonolin:

Lanolin which contains not more than 0.25 per cent of water is known as anhydrons lanolin. It is more readily absorbed through the skin than any other fat hence more valuable as a base for ointment.

White wax: is a bleached purified bees wax.

Chapter 12...

ENZYMES AND PROTEIN DRUGS

[A] ENZYMES

Enzymes are the proteins which act as biological catalysts. They play a vital role in the function of cells and activities of an organism. The enzymes show maximum activity between 35° - 40°C. They are practically inactive at 0°C and beyond 65°C get denaturated. Although, they are soluble in water and dilute alcohol, concentrated alcohol precipitates them. The factors like pH of medium has direct effect on their action. Enzymatic activity is reduced by formaldehyde, free iodine, heavy metals and tannins. The properties of enzymes which make them exceptional catalysts are as under.

(1) They catalyse only a specific range of reactions and in many cases only one reaction is catalysed by a given enzyme. Some of them have a low degree of specificity like pepsin, which hydrolyses almost all soluble native proteins. Hence, specificity is one of the most important characters of enzymes.

(2) As a group, they are exceptionally versatile catalysts. They effectively catalyse hydrolytic reactions, dehydrations, oxidation reduction reactions, acyl-transfer reactions, aldol condensations, polymerisations and free radical reactions.

(3) They are exceedingly efficient under optimal conditions. Most of the enzymatic reactions proceed 8 to 10 times more rapidly than the corresponding non-enzymatic reactions.

The enzymes are classified into the following categories:

(1) **Hydrolases** for catalysis of hydrolytic reactions.

(2) **Transferases** for the transfer of chemical group from one molecule to another.

(3) **Oxido-reductases** catalyse the oxidation-reduction reactions.

(4) **Lyses** catalyse the addition of groups to double bonds or *vice-versa*.

(5) **Isomerases** are responsible for intramolecular rearrangements.

(6) **Synthetases** catalyse the condensation of two molecules coupled with the cleavage of pyrophosphate bond of ATP or similar triphosphate.

Further, on the basis of site of action, enzymes can be grouped as under:

(a) Endoenzymes: Those which act only inside the cell are known as endoenzymes or intracellular enzymes. These involve in the synthesis of cell components, food reserves and bioenergetic, i.e. liberation of energy from foodstuffs. Since these processes are intracellular, enzymes involved are also intracellular. Examples are Synthetases, Isomerases, and Phosphorylases.

(b) Exoenzymes: The enzymes which are secreted outside the cell are known as exoenzymes or extracellular enzymes. Their function is to facilitate digestion. They hydrolyse very complex molecules into simpler compounds i.e. proteases, lipases, amylases acting on proteins, lipids or starch respectively.

Many of the enzymes also possess non-protein chemical groups. An enzyme moiety comprises a protein component 'apoenzyme', and a prosthetic group representing non-protein component. The latter is also called as cofactor or coenzyme. Certain metals and vitamins are coenzymes.

Table 12.1: Enzymes of Pharmaceutical Importance

Sr. No.	Name of drug and synonym	Biological source	Type of Enzyme	Uses
(A) Enzymes of Animal Origin				
1.	Chymotrypsin	Pancreas of Ox ,*Bos taurus*, (Bovidae)	Proteolytic enzyme	In ophthalmology, also as anti-inflammatory for soft tissues
2.	Deoxyribonuclease (Pancreatic dornase)	Beef pancreas	Nucleolytic enzyme	To reduce viscosity of bronchopulmonary secretions.
3.	Diastase (Amylase)	Human saliva or digestive tract of animals	Amylolytic enzyme	Enzymes of Animal origin Digestion of starch
4.	Fibrinolysin	Human plasminogen	Proteolytic enzyme	Treatment of thrombotic disorders
5.	Hyaluronidase	Mammalian (Bovine) testes	Amylolytic	To enhance absorption of intramuscular or subcutaneous, injections
6.	Muramidase (Lysozyme)	Tears, serum, lungs of animals	Mucolytic enzymes	Treatment of bacterial or viral infections
7.	Pancrealipase	Pancreas of hog *Sus scrofa* (Suidae)	Carbolytic, proteolytic and lipolytic enzymes	Treatment of chronic pancreatitis, cystic fibrosis,
8.	Pancreatin	Pancreas of hog *Sus scrofa* (Suidae)	Carbolytic, proteolytic and lipolytic enzymes	Digestive aid for starch, proteins and fats
9.	Pepsin	Glandular layer of fresh stomach of hog, *Sus scrofa* (Suidae)	Proteolytic enzyme	Conversion of protein into peptone and protease
10.	Rennin	Glandular layer of true digesting stomach of calf *Bos taurus* (Bovidae)	Proteolytic enzyme	Preparation of cheese and junkets.

Sr. No.	Name of drug and synonym	Biological source	Type of Enzyme	Uses
11.	Trypsin	Pancreas of Ox, *Bos taurus*, (Bovidae)	Proteolytic enzyme	Proteolysis of blood clot, necrotic tissue
12.	Urokinase	Human urine or kidney tissue cultures	Fibronolysis activating enzyme	Lyses of blood clots or fibrin in pulmonary embolism and anterior chambers of eye
(B) Enzymes of Herbal Origin				
1.	Bromelain (Plant protease concentrate)	Stem of pineapple plant *Ananas comosus* (Bromeliaceae)	Mixture of proteolytic enzymes	Anti-inflammatory for soft tissues
2.	Collagenase	Fermentation of *Clostridium histolyticum*	Proteolytic enzyme	Debridement of dermal ulcers and burns
3.	Fungal lipase	Asperigillus sp.	Lipolytic enzyme	Digestive
4.	Hemicellulase	Controlled fermentations of Trichoderma	Cellulose digestion	Digestive
5.	L-asparginase	*From Escherichia coli* bacteria	L-asparginase	Treatment of acute lymphocytic leukemia
6.	Papain	Latex of unripe fruit of tropical melon tree *Carica papaya* (Caricaceae)	Proleolytic enzyme	Meat tenderiser, clarification of beverages
7.	Ficin	Succulant fruits of *Ficus carica* Morazeae	Proteolytic enzyme	Anti-inflammatory anti-oxidant.
8.	Serratio-peptidase	Bacteria belonging to genus *Serratia*	Proteolytic enzyme	Anti-inflammatory especially to enhance antibiotic effects
9.	Streptokinase	Culture filtrates of β-hemolytic Streptococci group c.	Plasminogen activator enzyme	Treatment of thromboembolic disorders.
10.	Sutilains	*Bacillus subtilis*	Proteolytic enzyme	Treatment of burns

Following is the broad classification of enzymes depending on substrates on which they act.

* Amylolytic enzymes or Carbohydrases	* Proteolytic Enzymes or Proteases	* Lypolytic Enzymes
(1) Amylopsin or Diastase	(1) Trypsin	(1) Lypase
(2) Ptyalin	(2) Chymotrypsin	(2) Steapsin
(3) Malt-diastase	(3) Pepsin	(3) Pancreatin
(4) Invertase	(4) Renin	(4) Fungal lipase
(5) Maltase	(5) Muramidase	
(6) Zymase	(6) Fibrinolysin	
(7) Emulsin	(7) Eurokinase	
(8) Myrosin	(8) Streptokinase	
(9) Hemicellulase	(9) Papain	
(10) Hyaluronidase	(10) Bromelain	

Penicillinase

Bacterial enzymes produced by many bacteria which distroy penicillin and cephalosporins by catalysing the hydrolyris of amide bond in the β-lactam ring *Bacillus substilis – mesentericus and bacillus anthrasis* are good producers of penicellinase.

Use

Penicillinase is used in culture medic to autogonise antibacterial activity of penicillion, also used is treatment of ellergic reactionto penicillin.

(A) Enzymes of Animal Origin

CHYMOTRYPSIN

It is a proteolytic enzyme obtained from the extract of pancreas of ox *Bos Taurus,* family Bovidae. Chymotrypsin is a crystalline or amorphous odourless powder with white to yellowish colour.

It is soluble in water and shows maximum activity at pH 8. Each mg of Chymotrypsin contains not less than 1000 USP units.

This enzyme is used in ophthalmology for the dissection of zonule of lens for intracapsular cataract extraction. It is also used topically to reduce soft tissue inflammation from abscesses, ulcers, fistulas and necrotic wounds.

DEOXY-RIBONUCLEASE

Synonym

Pancreatic dornase

Biological Source

It is a Nucleolytic enzyme prepared by fractional precipitation of aqueous acid extracts of beef pancreas, which is followed by dialysis, sterilisation by filtration and freeze-drying.

It is available as white powder, soluble in water. It is stable in dry form, but losses its activity in solution. The maximum activity is exhibited at pH 6-7. Magnesium ions are necessary for its activation.

It is used in treatment of local abscesses and haematomas. It is also used in the form of aerosols to reduce the viscosity of pulmonary secretions and to enhance the expectoration of sputum in bronchopulmonary infections.

DIASTASE

Synonym

Amylase

Biological Source

It is one of the amylolytic enzymes present in saliva: i.e. salivary diastase or *Ptyalin* and pancreatic diastase or *amylopsin,* found in the digestive tract of the animals (Also known as *animal-diastase).*

It is also formed during the germination of barley grains and known as *Malt diastase.*

Several other amylolytic enzymes or carbohydrates of commercial importance are known, many of them are also used in therapeutics.

Zymase, maltase, sucrase, cellulase, invertase, hyaluronidase, lysozyme are amongst the important amylolytic enzymes.

Description

It is a yellowish white, amorphous powder obtained from the infusion of malt. Diastase has faint characteristic odour.

Solubility

It is soluble in water forming colloidal solution, it is precipitated by alcohol.

Properties

Like other enzymes, it is thermolabile and sensitive to a temperature more than 45°C. Diastase is active between 35° - 40°C. It is active in solution having pH range of 6 - 7. Acidic range of pH up to 4 causes its deactivation.

Diastase catalyses the hydrolysis of 1, 4-glycosidic linkages of polysaccharides: such as starch and glycogen. Diastase can convert 50 times its weight of potato starch into sugars.

Uses

Diastase from various sources is used as digestant. It is used in the production of predigested starchy foods and also for the conversion of starch to fermentable sugars in fermentation and brewing industries.

FIBRINOLYSIN

Synonym

Plasmin

Biological Source

It is a proteolytic enzyme obtained from activation of human plasminogen by highly purified streptokinase. It is also present in human serum.

It is available as white powder, soluble in water. It is stable in dried form and loses its activity in solution within 4-8 hours.

It is used in treatment of thrombotic disorders due to its fibrinolytic nature.

HYALURONIDASE

It is mucolytic enzyme and also prosent in heads of Leeches and snake venum.

It is a sterile, dry, soluble enzyme preparation, which is prepared from mammalian (bovine) testes and semen by fractional precipitation of aqueous extract and further refined by dialysis, sterilisation by filtration and lyophilisation. It is capable of hydrolysing mucopolysaccharides of hyaluronic acid type. Hyaluronidase injection is spreading agent.

This enzyme is odourless, white to yellow in colour, and is highly soluble in water, but insoluble in most of the organic solvents. Hyaluronidase acts on hyaluronic acid which is a mucopolysaccharide component of mucoprotein ground substance or the tissue cement of tissue spaces. It reduces its viscosity so that tissues become easily permeable for injected fluids. Because of this property, this enzyme is used to enhance the absorption rate and to reduce the discomfort caused by intramuscular or subcutaneous injections. It is also used in hypodermoclysis as an aid to subcutaneous administration of large volume of parenteral dose.

MURAMIDASE

Synonym

Lysozyme

Biological Source

It is a crystalline mucolytic enzyme present in tears, lungs, serum, leucocytes, different tissues and secretions of animals. It contains mainly arginine, aspartic acid, tryptophan, and traces of tyrosine.

It hydrolyses mucopolysaccharides and is active in transforming insoluble polysaccharides of cell wall to soluble mucopeptides, especially in gram positive bacteria. Due to this activity, it is given intravenously for the treatment of bacterial or viral infections.

PANCREALIPASE

It is a substance which contains enzymes mainly lipase along with protease and amylase. It is obtained from pancreas of the hog, *Sus scrofa* var. *domesticus* belonging to family Suidae.

Pancrealipase is more concentrated form of Pancreatin wherein the lipase activity is increased. It occurs as cream coloured amorphous powder with a characteristic odour. Each mg contains not less than 24 units of lipase activity, 100 units of protease activity and 100 units of amylase activity.

Pancrealipase is used in the treatment of chronic pancreatitis, cystic fibrosis and partial or complete surgical pancreatectomy.

PANCREATIN

It is a preparation extracted from pancreas of certain animals like hog (Pig) (*Sus scrofa var. domesticus*, family Suidae) or ox (*Bos Taurus,* family Bovidae). Pancreatin contains enzymes namely amylase, lipase and protease. Pancreatin is amorphous powder with white or buff colour. It is without any characteristic taste and odour. It is soluble in water, but insoluble in alcohol and ether. It shows maximum activity in faintly alkaline medium. It is used to treat pancreatic deficiency like pancreatitis and also in fibrocystic diseases of pancreas. It is also used for preparing peptonized or predigested food. Pancreatin is employed as digestive aid for converting starch into dextrin and sugar, proteins into proteases and derived substances, and fats into glycerol and fatty acids.

Each gramme of Pancreatin contains not less than 12,000 units of amylase activity, 1000 units of protease activity and 15,000 units of lipase activity.

PEPSIN

It is a substance containing proteolytic enzyme and is present in the gastric juice of animals. It is obtained from the glandular layer (mucous membranes) of fresh stomach of hog, *Sus scrofa* var–*domesticus,* belonging to family Suidae.

Pepsin is light buff or white coloured amorphous powder. It also occurs as translucent scales. It has a little acidic or saline taste with slightly meaty odour. It is soluble in water, but insoluble in alcohol, ether and chloroform. If pepsin is heated with alkali or pancreatic enzymes, its biological activity is lost. It shows maximum activity at pH 1.8. Pepsin has the capacity to digest 2500 times its weight of coagulated egg albumin. It is also available in other forms which may digest even up to 10,000 times their weight of coagulated egg albumin.

For preparation, the minced stomach linings are digested with hydrochloric acid followed by clarification, controlled evaporation, dialysis and concentration of the digested solution. When processed, solution is subjected carefully to vacuum evaporation, spongy pepsin is obtained.

Pepsin degrades proteins into peptones and proteases.

RENNIN

It is a partially purified, milk-curdling, proteolytic enzyme from glandular layer of the fourth or true digesting stomach of the calf, *Bos Taurus* (Bovidae).

Rennin has peculiar odour and characteristic saline taste. It is available as scales or powder. It has hygroscopic nature, yellowish-white or greyish-white colour.

For production of rennin, the glandular layers of digesting stomach of calf are minced and macerated in 0.5 per cent sodium chloride solution, followed by filtration. The filtrate is acidified with hydrochloric acid and rennin is precipitated by saturating the filtrate with sodium chloride. Further, rennin is separated, dried and powdered. Rennin can also be prepared microbiologically by controlled fermentation of *Endothia parasitica, Mucor pusillus and Bacillus cereus.*

Rennin is used to prepare junkets and cheese. It is also used to coagulate milk and hence making the milk easily digestible for physically weak patients.

TRYPSIN (CRYSTALLINE TRYPSIN)

It is a proteolytic enzyme obtained from mammalian pancreas like ox (*Bos Taurus,* family Bovidae). It is obtained by alcoholic or aqueous acid extraction of its precursor trypsinogen and further conversion to crystalline trypsin.

It is a crystalline or amorphous, yellowish-white powder without any odour. It is soluble in water, but insoluble in alcohol, glycerine, chloroform and ether. It shows maximum enzymatic activity at pH 7.6 - 8. It is stable in dry air, but the solution should be freshly prepared because of degradation. In the presence of calcium ions, its stability is increased.

Trypsin is used to enhance the proteolysis of different protein substrates, like blood clot, necrotic tissue, purulent exudates, etc. It does not act on living tissue due to presence of inhibitors. Each mg of trypsin should contain not less than 2500 USP trypsin units.

UROKINASE

It is an enzyme produced by kidney and obtained from human urine or kidney tissue cultures.

It is a lyophilised white powder, soluble in water. It is an activator of endogenous fibrinolytic system, which converts plasminogen to plasmin and degrades fibrinogen, fibrin clots and other plasma proteins.

It is used to dissolve (lyse) fibrin or blood clots in anterior chamber of eye and in acute massive pulmonary emboli. As it is derived from human source, it is less antigenic than enzymes with similar actions like streptokinase.

(B) Herbal Enzymes

BROMELAIN

Synonyms

Bromelin, Ananase.

Biological Source

Bromelain is a mixture of proteolytic enzymes from the stem and ripe and unripe fruits of pineapple plant *Ananas comosus*, belonging to family Bromeliaceae. Unlike papain it does not disappear as the fruit ripens.

Geographic Source

Pine-apple fruits used for isolation of Bromelain are produced commercially in south and central America, Philippines Guatemala, Mexico, Panama, Hawaii along with India. Costa-Rica is the world centre for marketing of pineapple. In India it is cultivated in Goa, Maharashtra in Konkan.

Cultivation of Pineapple:

Several varieties of pineapple fruits are available for commercial cultivation. Variety Q, Qeen morisons and Jayant are preferred. Because fruits of Q varieties are given preference for canning, while fruits of Queen variety are selected for table use. Q variety pine-apples are about 1.5 – 2.5 kg in weight while Queen variety fruits are 1.0 – 1.5 kg in weight. Pineapple is normally cultivated in coconut plantation in Konkan and Kerala by using cuttings or apical tops of fruits.

Fig. 12.1: Pineapple Fruit

Cuttings of 450 gm each are used for cultivation and are given prior treatment with solution of fungicides.

Deep ploughing of soil (Approx. 30 – 40 cm deep) is done. Distance of 60 cm in two lines and distance of 30 to 45 cm in between two plants is maintained. To get the bigger size of fruits distance of 45 cm in plants is essential 18 to 22 months are required to get the fruits. Any type of soil with proper irrigation is necessary for cultivation.

About 40 tonnes of manures per hectare are provided to the plants or chemical fetilizers of 12 gm Nitrogen, 06 gm of phosphorus and 12 gm of potash per plant per annum are used, in two or three installments.

First installment of fertilizers is given after three months of sowing and the last (or third) installment before completion of one year of plantation.

In winter, plants are irrigated after every six days or after 8 to 10 days in Summer. Plants are supported with soil frequently and weeding is done at regular intervals.

Preparation of Bromelain

Juice of ripe or unripe fruits is obtained by crushing and filtered to clear solution at low temperature. Transparent juice is precipitated with acetone by constant stirring at low temperature. The precipitate containing bromelain is collected and washed several times with acetone. Washings are rejected and the precipitate is dried at low temperature (below 40ºC) in vacuum dryer and filled in air-tight containers.

Description

It is available as odourless to slightly putrid buff coloured powder with irritating taste.

Solubility

It has slight solubility in water. It is insoluble in organic solvents like ether, chloroform, alcohol etc. Fruit Bromelain is acidic while stem bromelain is alkaline.

Uses

It is used in treatment of soft tissue inflammation and oedema due to surgery and injury. Used as meal tenderising, in chil-proof beer and in production of protein hydrolysate. It also reduces toxicity in diet.

COLLAGENASE

It is an enzyme derived from fermentation of *Clostridium histolyticum*. It has the capacity to digest native and denatured collagen.

The potency of collagenase is expressed in terms of its ability to digest indentured bovine collagen *in vitro*. It shows maximum activity between pH 7 and 8.

It is used for debridement of dermal ulcers and burns and other necrotic lesions. It is used in the form of ointment.

FUNGAL LIPASE

It is manufactured from *Asperigillus* sp. by fermentation process. The fermented fluid is filtered by sophisticated process to produce the filtrate suitable for spray-drying.

It is light brown to brown coloured amorphous powder with characteristic odour. It soluble in water but insoluble in organic solvents.

Incomplete digestion of fat in the diet may result in diarrhoea and deficiency of essential fatty acids.

The role of essential fatty acid can not be neglected in body, they are essential to maintain membrane structure in cells and also for healthy skin.

With wide range of temperature and pH fungal lipase digests fats and fat bodies. It is also effective in regulating the levels of cholesterol, triglycerides and for weight management.

Use: Digestive (for hydrolysis of lipids).

Storage: Store in well-filled well-closed containers away from light in coot place. It is stable for three years.

HEMICELLULASE

It is an enzyme complex necessary for hydrolyzing cellulose and various hemicellulosic materials including gums and plant fibres.

Hemicellulose is manufactured by controlled fermentation of *Trichoderma*.

The fermented material is passed through several fine filtration processes and concentrated. Finally, it is spray-dried to get fine free flowing material.

It is white to off white coloured powder, soluble in water and insoluble in organic solvents.

Uses: Used as digestive especially in the consumption of cereal food grains containing high percentage of non-soluble polysaccharides.

Highly viscous polysaccharides form complex with digestive enzymes and reduce their penetration to the substrates which result in poor digestion. This disorder is solved by prescribing hemicelluloses with diet. It is also used in the treatment of excessive gas and bloating, constipation and heart burn.

It should be stored in well filled well closed containers in cool and dry place. It is stable for three years.

L-ASPARGINASE

This enzyme is obtained from *Escherichia coli.* It also contains asparginase without antileukemic activity and it is removed during the purification of enzyme. L - Asparginase is also obtained from plant and animal tissues, fungi and yeast in pure form. This enzyme catalyzes hydrolysis of L-asparagines to L-aspartate and ammonia.

It is available as white and crystalline powder, soluble in water. Each mg of L-asparginase contains 250 units.

This enzyme interferes with the growth of malignant cells, which are not capable of synthesizing L-aspargine for their metabolism and hence, it is used in chemotherapy of acute lymphocytic leukemia in sequential combination with other drugs. It is also useful for induction of remission in children with relapse of acute lymphocytic lymphoma. It also shows immuno-suppressive activity.

PAPAIN

Biological Source

It is a mixture of proteolytic enzymes derived from the latex of unripe fruit of tropical melon tree, *Carica papaya,* belonging to family Caricaceae (Fig. 12.2).

Geographical Source

Aperennial plant indigenous tropical America and extensively cultivated in Shri-Lanka, Tanzania, Hawaii, Flurida and India.

Method of Preparation

For processing of papain, the latex of these fruits is collected in aluminium trays. To the collected latex, potassium metabisulphite (5 g/kg of latex) is added. The extraneous matter is cleared out by passing through sieves and latex is dried in vacuum shelf drier at 55 - 60°C. It is also processed by spray-drying method. This dried latex is called papain.

Description

Papain is available as light brown or white coloured amorphous powder with typical odour and taste. It shows maximum proteolytic activity between pH 5 - 6. It is soluble in water and glycerine.

Fig. 12.2: Papaya tree with fruits

Chemical Nature

The different proteolytic enzymes present in papain are the mixture of papain and chymopapain, proteolytic enzymes act on polypeptides and amides.

Identification

1. It decolourises aqueous potassium permanganate solutions.
2. It causes curdling of milk (Proteolytic activity).

Standards of Quality

Loss on Drying	:	Not less than 7.0%
Drying on	:	Not more than 60.0 %

Uses

It is used in clarification of beverages and as a meat tenderiser. It is employed in cheese manufacture as a substitute of rennin. It is also used for degumming of silk fabrics in textile industry and in leather industry for removing hairs of skins and hides.

Medicinally, it is used as an anti-inflammatory agent. It has shown relieving symptoms of episiotomy.

One NF unit of papain represents the activity which releases equivalent of 1 μg of tyrosine from a standard casein substrate.

ANJIR

Synonym

Anjir fruit, Common Fig.

Biological Source

This consists of nearly ripe and sun dried succulent fruits of the plant *Ficus-carica* family Moraccae.

Geographical Source

Figs are believed to be indigenous to Western Asia and have been distributed to mediterranean area. Turkey is the leading producer of figs in the world and produces 27% of world production, While Egypt, Algeria, Morocco, Iran, Syria and USA are the other producers on commercial scale. Maharashtra is the only state in India, producing figs on commercial base. About 500 hectares of land is under cultivation in Maharashtra. It is cultivated in Pandhanpur and Baramati.

Plant Habitate

Fig tree is about 3 to 9 metres in height with numerous spreading branches, trunk of tree is 15 cm. It contains copious milky latex, roots may descend 5 metres, leaves are deciduous and palmate with 3-7 main lobes and are irregularly toothed on the margin. Leaf blades are 20-25 cm hairy on lower side, flowers are all female and need no pollination for fertilization.

(a) Branch of fruiting fig herb

(b) L.S. of fig fruit

Fig. 12.3: Fig fruit

Cultivation and Collection of Figs

Trees thrive well at an attitude of 800 - 1800 metres and can tolerate the temperature of 10ºC - 20ºC. Even if the trees are irrigated, very hot and dry climatic conditions cause fruit-drop.

Fig cultivation is done in dry and sunny locations. Special care is required to limit the excessive growth of the plants to promote the fruiting. It thrives equally good in both sandy and rocky soil. Shade is to be avoided essentially for the crop. Two crops of figs are potentially produced every year. The first crop is in spring, propogation by using seeds is not preferred because other methods are quicker and more reliable. Second crop propogation starts in January.

pH of the soil for cultivation of figs should be in between 6.0 - 6.5. Highly acidic soils are unsuitable and cause damage to the crop.

In vegetative propogation, shoots with healthy buds at their ends of 8 to 12 months age and 15-25 cm and less than 1.5 cm in size are selected. The shoots are set into moist mix of soil and peat-moss located in the shade are burried 3/4th of their length. Distance of 30 cm in every fig plant is to be provided and also with 50 kg farm yard mannure 2.5 kg urea, 3.25 gm single supuer phosphate and 425 gm potash per annum. Irrigation should be done at the interval of 4 – 6 days under normal conditions, until winter and then they are planted in the desired location.

California produces about 80.0% of US production. Each fig tree bears 180-300 fruits per year. About 7000 hectares of land is under cultivation of fig in California. Fig trees remain productive up to 15 years of their age. Soybean, mung (phaseolus mungo), Udid (Phaseolus radiatus can be taken as inter crop for first 2-3 years. Plantation is quite resistant to pests hence cultivation is economical.

Macroscopic Characters

Colour	:	Fruits are Amber or more or less pink, rose-red or purple.
Odour	:	Pleasantly fruity.
Taste	:	Sweet.
Size	:	5 cm in length and breadth.

Figs are compound fruits, soft and fleshy. At the summit fruits have a small opening surrounded by scales and at the base a short stalk-like prolongation. The hollow receptacle bears numerous drupelets. Each drupelet containing stone seeds and have endosperm and curved embryo.

Chemical Constituents

Various polyphenols like gallic acid, chlorogenic acid, syringic acid, catechin, epicatechin and rutin are present in fig fruit. Anthocyanin, cyanidin and – 3-0 rutinoside in high concentration are also present in drug. Carbohydrates to the tune of about 20.0%, dietary fibres 2.9% various vitamins like B_1, B_2, B_3, B_5, B_6, vitamin C and vitamin K are also present in the figs. Proteolytic enzyme Ficin in small quantity is present in the fruits.

Standards of Quality

Water soluble extractives	:	Not less than 60.0%
Ash	:	Not more than 2.5%
Loss on Drying	:	Not more than 25.0 %

Microscopical Characters

Being a fruit epicarp consists of radially elongated cells with mucilaginous outer walls. Mesocarp is delicate, endocarp contains radially elongated sclereids with pitted walls. Endosperm and embryo is with small cells containing aleuronic grains and fixed oil. Starch is absent.

Receptacle : Epidermal cells are polyhedral with stomata and thick walled unicellular trichomes upto 300 µm. Hypodermis consists of rounded polyhedral cells, and contain rosette crystals of calcium oxalate.

Aerenchyma is made up of large irregular cells, containing large rosette crystals of cal oxalate, numerous latex tubes upto 50 mm wide and slender vascular bundles.

Uses

Figs are used as dietetic and as mild laxative. Traditionally, it is used for anti-carcinogenic properties also for anti-inflammatory and anti-oxidant activities. Figs are used traditionally for treatment of warts. Milky juice from fresh fruits destroys warts.

Storage

Fresh figs are perishable. If frozen whole, keeping quality can be maintained for several months. They should be stored in a dry place.

SERRATIOPEPTIDASE

It is a proteolytic enzyme derived from the bacteria belonging to genus *Serratia*, present in the gut of silk worm. Originally, it was discovered in Serratia E_{15} species. Now-a-days, it is produced by fermentation bio-technology.

Serratiopeptidase is considered as very effective bacterial enzyme and it is found to have better effects than trypsin and chymotrypsin, with negligible toxicity and side effects.

Given orally, it enters systemic circulation in unchanged form and can penetrate into all tissues, especially inflamed areas. It exerts histamine and bradykinen hydrolysing and proteolytic effects. Hence, it reduces capillary permeability and also breaks down proteins and exudates and hence, supports wound healing. Unlike chymotrypsin, serratiopeptidase, being a bacterial enzyme, shows rarely the allergic reactions.

Serratiopeptidase has a number of therapeutic applications, like (a) resolution of inflammation, (b) sputum liquefaction due to lyses of various proteins in sputum and hence lowering viscosity,

(c) enhancement of antibiotic effects due to removal of inflammatory barrier and hence, increasing antibiotic transfer to infected areas.

STREPTOKINASE

It is proteolytic enzyme obtained from culture filtrates of β-hemolytic *Streptococci* group C. This enzyme has the property of activating human plasminogen to plasmin.

It is available as a sterile, friable solid or white powder. It is soluble in water with maximum activity at pH 7. The solution at higher concentrations is stable for 6 hours at 4°C, otherwise dilute solutions are unstable. It is used in the treatment of thromboembolic disorders for the lysis of pulmonary emboli, arterial thrombus, deep vein thrombus and acute coronary artery thrombosis. The activity of this enzyme is due to activation of plasminogen to a proteolytic enzyme namely plasmin which degrades fibrin clots, fibrinogen and other plasma proteins.

SUTILAINS

Sutilains contain the proteolytic enzymes derived from *Bacillus subtilis*. It is available as cream coloured odourless hygroscopic powder. Each gramme of sutilains contain 2,50,000 NF casein units of proteolytic activity.

It is used for biochemical debridement of second and third degree burns, incision traumatic and pyrogenic wounds removed of injured tissues and ulcers occurring due to peripheral vascular diseases. It is used in the form of ointment can not be given by intravenous injection in it contains prototypic enzyme.

[B] PROTEINS AND PROTEIN DRUGS

Proteins are complex nitrogenous organic substances of plant and animal origin. They are of great importance in the functioning of living cells. They are produced by and associated with living matter. Apart from the fact that they are essential food stuffs like carbohydrates and fats, they also provide very important group of therapeutically active compounds such as hormones, enzymes, sera, antitoxins, etc. They are easily extractable from plant sources and are generally stored in the form of aleurone grains in plants.

In animals, they are present as structural material in the form of collagen (connective tissue), keratin (hair, wool, nail, feathers and horns), elastin (epithelial connective tissue), casein (milk) and plasma proteins. Casein, gelatin, heparin and hemoglobin are pharmaceutically important proteins of animal origin.

Proteins contain carbon, hydrogen, oxygen, nitrogen and rarely sulphur. The ultimate products of complete hydrolysis of proteins, either by chemical reagents or enzymes, are amino acids. Proteins are the compounds of high molecular weight, forming colloidal solution in water. Proteins are amphoteric in nature and get easily denatured due to heat, changes in pH, treatment of organic solvents or by ultraviolet radiation. Depending upon the products of hydrolysis, they are classified as simple (containing only amino acids), conjugated (amino acids and non-amino acids like nucleic acids, lipids, metals, and phosphoric acid) and derived proteins.

Proteins are broadly classified as under:

I. Simple proteins: They yield only amino acids on hydrolysis.

1. **Albumins** are soluble in water and are coagulated by heat. Examples are *egg albumin* and *lactalbumin*.
2. **Globulins** are insoluble in water but are soluble in dilute salt solution. They are precipitated by half-saturating their solution with ammonium sulphate, and are also coagulated by heat. Examples are *ovoglobulin, myosin, arachin, amandin*, and *serum globulin*.
3. **Glutamines** are soluble in dilute acids and alkalis, and insoluble in neutral solvents. Examples are *glutenin of wheat* and *oryzenin of rice*.
4. **Prolamines** are soluble in 70-80 per cent alcohol and insoluble in water, dilute salt solution, or absolute alcohol. Examples are *zein of corn, gliadin of wheat*, and *hordein of barley*.
5. **Scleroproteins** are insoluble in water or salt solution, but are soluble in strong acids or alkalis. Examples are *keratins of hair, horns, and hoofs, elastin of connective tissues, collagen of bones, fibroin and sericin of silk*.
6. **Histones** are soluble in water and insoluble in dilute ammonia. They are readily soluble in dilute acids and alkalis. Examples are *globins'* and *gadus histone of codfish sperm*.
7. **Protamines** are the simplest of the proteins. Soluble in water and not coagulated by heat. They are basic in nature and yield mainly basic amino acids upon hydrolysis. The protamines are soluble in water, dilute ammonia, acids, and alkalis. Examples are *Salmine of salmon sperm* and *clupeine of herring sperm* used as antidote in heparin over dosage administered intravenously. Neutralises anti-coagulant action of heparin. Used in the preparation of several long acting insulin preparations.

II. Conjugated proteins: They are composed of a simple protein combined with a non-protein group, known as the prosthetic group.

1. **Chromoproteins** are proteins united with coloured prosthetic groups such as haemoglobin or chlorophyll.
2. **Lipoproteins** are the combination of proteins with lipids such as lecithin of fatty acids. They are found in blood, milk, egg yolk, and the chloroplasts.
3. **Metalloproteins** are proteins which contain heavy metals such as Fe, Co, Mn, Zn, Cu, Mg, etc. Many enzymes belong to this group.
4. **Mucoproteins** are proteins and mucopolysaccharides. They are found in serum, human urine, and albumin.
5. **Nucleoproteins** as the name indicate are proteins and nucleic acids. The plant and animal viruses are ribonucleoproteins, since they consist chiefly of protein and RNA. The *tobacco mosaic virus* is best known nucleoprotein.
6. **Phosphoproteins** contain phosphoric acid. They are available in casein and egg yolk.

Some of the animal and plant proteins are very poisonous. Toxalbumin, ricin from castor seeds, and abrin from *Abrus precatorius* are very toxic to human beings.

Table 12.2: Protein drugs

Sr. No.	Name of drug and synonym	Biological source	Main chemical contents	Uses
1.	Casein	Milk	Phosphoprotein	Dietary supplement
2.	Collagen	White fibres of connective tissues	Glycine and prolene	Preparation of sutures.
3.	Collagen microfibrillar	Skin, connective tissue and bones of animals	Polypeptides	To check bleeding in surgery
4.	Gelatin	Skin, ligaments, tendons and bones of animals	Protein (Gluten)	Manufacture of hard and soft capsule shells
5.	Leech	*Hirudo medicinalis* Fam. Hirudinidae	Hirudin	Anti coagulant
6.	Cobra Venum	Secretion of poison gland of *Naja-naja* Calubridae	Neurotoxins	Local haemostatic
7.	Heparin sodium (Soluble heparin)	Mammalian tissues like lungs and intestinal mucosa	Glycosamino-glycans	Anticoagulant in blood transfusions and vascular surgery
8.	Protamine sulphate	Sperms or mature testes of fish of family Salmonidae or Clupeidae	Various simple proteins	Heparin antagonist
9.	Kavach	Seeds of *Mucuna-prurines* (Leguminosae)	l – dopa	Treatment of parkinsonism
10.	Lectins	Obtained from plants, bacteria, and fungal growth.	Proteins or glycoproteins chemically	Micro-chemical studies
11.	Malt extract	Barley grains, *Hordeum vulgare* (Graminae)	Proteins, maltose and amylotytic enzymes	Nutritive
12.	Protein hydrolysate	Sterile acid or enzyme hydrolised proteins.	Short chain peptides and amino acids	Severe protein deficiency
13.	Thaumatin	Fruits of *Thaumatococcus denielli* (Marantaceae)	Protein named as thaumatin	Sweetening agent.
14.	Yeast	Fungus *Saccharomyces cerevisiae* (Saccharomycetaceae)	Proteins, glycogen, vitamins of B group	Nutritive, also in brewing industry

KAVACH

Synonyms

Cow Itch plant, cow harge, Kuheli.

Biological Source

This consists of dried seeds of *Mucuna pruriens* Linn. (M. prurita Hook) family Fagaceae (Leguminosae).

Geographical Source

Found throughout India, commercially cultivated in tropical parts especially in Maharashtra. It produces 12 tonnes per hectare of seeds on commercial scale.

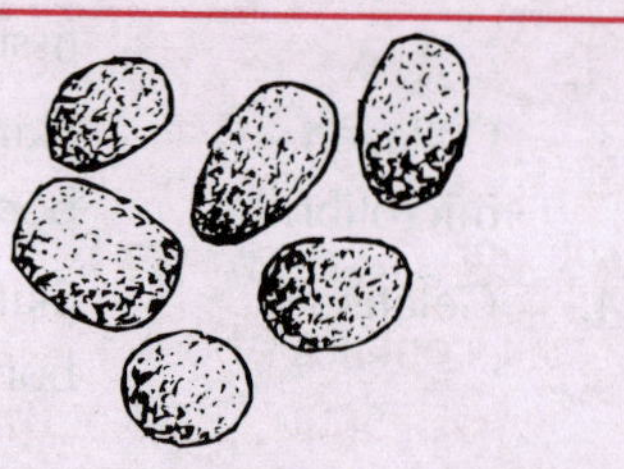

Fig. 12.7: Kavach seeds

Macroscopic Characters

Seeds are black in colour, odourless and bitter in taste; seeds are reniform in shape and 10 - 20 × 7 - 15 mm in size.

Extra Features

Seeds are hard, thick and glossy with prominent funicular hilum and cellular pit growth near the hilum.

Chemical Constituents

Seeds contain good number of amino acids including L-dopa (L 3.4, dihydroxy phenyl alanine) about 1.5 per cent, Licithin 11.5 per cent, alkaloids mucunine, mucunadine, prurienine, prurieninine and fat.

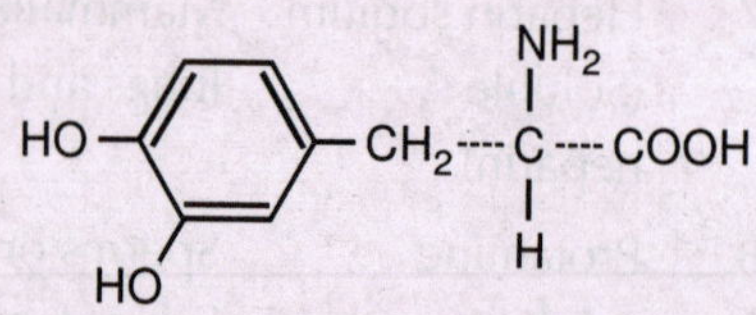

***l* - Dopa**

Uses

Levodopa is the amino acid of prime importance in the treatment of incapacitating paralysis agitans (parkinsonism). Levodopa shows maximum effects in rigidity and hypokinesia. It eliminates or suppresses the conditions like dysphagia, sialorrhea, seborrhea, postural instability, speech difficulties, etc. The side effects of Levodopa are nausea, vomiting, anorexia, epigastric pain, flatulence and dry mouth.

It elevates sperm count and hence used in stress related infertility. It is also used as an aphrodisiac.

CASEIN

Biological Source

Casein is a principal phospho-protein in milk and constitutes 3.0 per cent milk.

It comprises about 80 per cent total protein content of milk. There are two types of casein in the market.

Acid Casein: Warm skimmed milk is acidified with dilute acid, the whey is separated, curd is washed several times, dried and pulverised.

Rennet Casein: Skimmed milk is treated with an enzyme, rennet extract; product is separated, and purified.

Description

Casein is white, slightly yellow, tasteless, odourless, amorphous solid, hygroscopic, stable when dry but deteriorates rapidly when damp.

Solubility

It is insoluble in water, soluble in dilute alkalies, concentrated acids, but precipitates from dilute acid solutions.

Chemistry of Casein

Casein is a phosphoprotein containing about 0.85 per cent phosphorus and 0.75 per cent sulphur.

It contains about 15 amino acids and also rich in essential amino acids. Molecular weight of casein is 75000 - 3,70,000, Isoelectric point 4.7, Nitrogen content 15 - 16 per cent.

Standards of Quality

Loss on drying	:	Not more than 6.0 per cent
Sulphated ash	:	Not more than 1.5 per cent
Specific gravity	:	1.25 - 1.31.

Uses

It is useful dietary supplement and source of protein in pre and post operative care; as a base in the standardisation of proteolytic enzymes and as emulsifying agent.

Industrially, it is used in sizing of textile and paper, as an adhesive, in preparation of casein plastic and casein paints.

COLLAGEN

Synonym

Ossein.

Source

It is the protein which consists of major portion of white fibres of connective tissues of the animal body specifically, from the tendons, skin, bones and teeth.

Description

The molecule of collagen is similar to three strand rope, each strand consisting of polypeptide chain with molecular weight of 10,000. These three strands are left-handed helices and are wrapped together in a right-handed superhelix. The strands are held together by hydrogen bonds, which give the molecule its strength.

Collagen fibres range from 10 - 100 μm in diameter and visible by microscope as banded structure in the extracellular matrix of connective tissues.

Chemical Constituents

Glycine and proline are the important amino acids in the central core of the triple helical molecule of collagen. It can be differentiated from other accompanying fibrous proteins like elastin and reticulin. Elastin is highly crossed linked hydrophobic protein.

Collagen is characterised by the presence of glycine, proline, hydroxyproline, and hydroxylysine and low tyrosine and sulphur contents, while elastin contains non-polar amino acids like valine, isoleucine and leucine. Various types of collagen exist depending upon the amino acid sequence.

Collagen is converted to gelatin by boiling with water.

Uses

It is used in the preparation of sutures, as a gel in food casings.

Collagen is also used in photographic emulsions.

COLLAGEN MICROFIBRILLAR

It is an animal originated preparation containing polypeptide substance which occurs as the main constituent of skin, connective tissue and organic substances of bones.

It is used to stop bleeding in surgery. It is the character of collagen that platelets adhere to it and in turn, they secrete a substance which further enhances aggregation. It is absorbed within 84 days. Normally, it is used on those wounds where surgical ligatures cannot be used.

Absorbable Surgical Suture

It is also called as **surgical catgut** or catgut suture. It is a sterile strand prepared from collagen, which is derived from healthy mammals or from a synthetic polymer. It is available as flexible strand which varies in diameter, size, colour and capable of getting absorbed into living mammalian tissue. The extent of absorption is varied by certain modifications. It is coated with a suitable antimicrobial substance and is coloured with suitable colouring agent approved under Drug Act.

GELATIN

Synonyms

Gelatina, Gel foam, puragel.

Biological Source

Gelatin is a protein extracted by partial hydrolysis of animal collagenous tissue like skins, tendons, ligaments and bones with in boiling water. (Fig. 12.3)

Description

This protein product is available in the form of flakes, sheets, shreds or a coarse or fine powder. It has a characteristic odour and faintly yellow to amber colour.

Solubility

It is insoluble in cold water, but soluble in hot water. In cold water, it swells, softens and absorbs about 5 - 10 times its weight of water. With hot water, it forms a jelly on cooling. It is soluble in a mixture of glycerine and water, but insoluble in fixed and volatile oils, alcohol, chloroform and ether.

In dried condition, gelatin remains stable in air, but when moist may be degraded due to microbial attack.

The quality of gelatin is expressed as 'bloom strength'. It is the weight in gramme, which when applied to a plunger, 11.7 mm in diameter, under controlled conditions shall produce a depression exactly 4 mm deep in a jelly matured at 10°C and containing 6.66 per cent w/w gelatin in water.

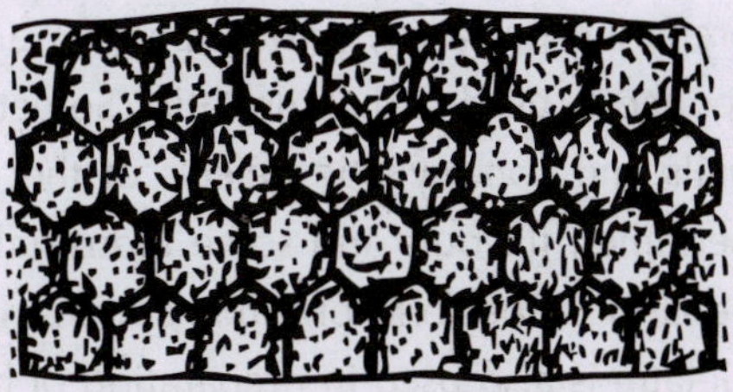

Fig. 12.4: Market Sample of Gelatin Sheet

The jelly strength of not less than 150 blooms is recommended by British Pharmacopoeia for pharmaceutical uses such as coating of pills, as vehicle for suppositories and as an emulsifying agent. But gelatin for capsule manufacture and microbial culture media, higher jelly strengths are used. The jelly strength is designated by bloom gelometer number.

Preparation of Gelatin

For the manufacture of gelatin, the bones are to be defatted and decalcified with organic solvent and mineral acid respectively. The material obtained by this treatment is treated with water at 85°C in successive quantities, due to which collagen dissolves into gelatin. It is further bleached and concentrated under reduced pressure to specific gelatin content and allowed to set in shallow trays. Such moulded gelatin is dried in drying room to eliminate moisture.

Chemical Constituents

As a protein, chemically, it contains different amino acids out of which major is lysine, an essential amino acid, but does not contain tryptophan. Gelatin is composed of gluten protein.

Standards of Quality

Ash	:	Not more than 3.2 per cent
Gel strength	:	150 - 250
L.O.D.	:	Not more than 15 per cent
pH 1.0 per cent soln.	:	3.6 - 7.6
Microbial limits	:	It should comply for absence of *E. coli* and for *Salmonella*. Total bacterial count should be less than 1000 per g.

Solubility

For confirming the identity of gelatin, the following chemical tests are applied.

Identification

1. It evolves ammonia when heated with soda lime.
2. It is precipitated by trinitrophenol and solution of tannic acid, but not with alum, lead acetate or acids which indicates that it does not contain chondrin.
3. It gives a white precipitate with mercuric nitrate and on warming turns to brick red colour.

Uses

Gelatin is mainly used in manufacture of hard and flexible capsule shells. It is also used for preparing pessaries, pastes, pastiles and suppositories. Gelatin in the form of absorbable gelatin sponge is used as haemostatic. Sometimes, it is also recommended for treatment of brittle finger nails and non-mycotic defects of the nails. Gelatin is employed for microencapsulation of drugs, perfumes, flavours and some industrial materials. It is used as a vehicle for certain injections, like heparin in the form of Pitkin's menstrum which contains gelatin, dextrose, acetic acid and water. Gelatin is also used in preparation of bacteriological culture media, absorbable gelatin sponge and gelatin film.

Absorbable Gelatin Sponge

Absorbable gelatin sponge is a sterile, white, tough and finely porous spongy material, which is absorbable and water insoluble.

For the preparation of this material, the warm solution of gelatin is whisked to foam of uniform porosity and then it is dried, cut into pieces of specific size and finally, sterilised at 150°C for one hour.

Though, it is insoluble in water, it is absorbed in body fluids. It takes up not less than 30 times its weight of water. About 9 gm of absorbable gelatin sponge takes upto 45 times its weight of well agitated oxalate whole blood.

It is used as a haemostatic. It is moistened with sterile sodium chloride solution and put within a surgical incision where it gets absorbed in 4-6 weeks. It is also used as a local coagulant and haemostatic.

Absorbable Gelatin Film

It is sterile, light amber, pliable, non antigenic gelatin film obtained from a specially prepared gelatin formaldehyde solution by drying, followed by sterilisation.

Description

It is a transparent film which becomes rubbery on moistening, but it is insoluble in water. It is used in the form of saline soaked sheets in surgical repair of defects in membranes like, pleura and dura mater, where it serves as mechanical protective, replacement matrix and temporary supportive structure.

Non-Absorbable Surgical Suture

It is sterile or non-sterile strand of material that is suitably resistant to action of living mammalian tissue. It may be modified to reduce capillarity, suitably bleached and also coloured with approved colouring agents. It is also coated with an antimicrobial agent. The various materials used in preparation of non-absorbable surgical suture are /silk, synthetic fibres of monofilament twisted or braided construction, cotton or linen fibres, coated natural or synthetic fibres, and monofilament or multifilament metal wire.

LEECH

Synonym

Hirudo

Introduction

The British Physiologist **John Berry** reported in 1884, that Leech secretes a powerful anticoagulant and named it as **Hirudin**. It was isolated in pure form in 1950 and its structure was revealed only in 1976.

Biological Source

Leeches are the living annelid worms known as *Hirudo medicinalis* Linn., scientifically, belonging to family Hirudinidae.

Geographical Source

There are three varieties of medicinal leeches namely.

(1) German or speckled

(2) Hungarian or Green variety and

(3) Australian variety.

As their variety indicate they are grown in Germany, Hungery and Australia only.

Cultivation and Collection of Leech

Cultivation of leeches is undertaken in special leech farms near **Hanover** in Germany. The ponds are ablong and about one metre in depth. The bottom and banks of the leech farms are covered with clay.

The leeches which are five to seven years of an age start laying eggs from June onwards upto the cold winter. The life span of leeches is about 12 to 15 years. Leeches are fed on fresh blood provided in linen bags twice yearly. The bags are suspended in the water for easy availability of food for their survival. Leeches are exported in barrels covered with stout canvas to expose to air, every barrel accomodates approximately 2000 leeches.

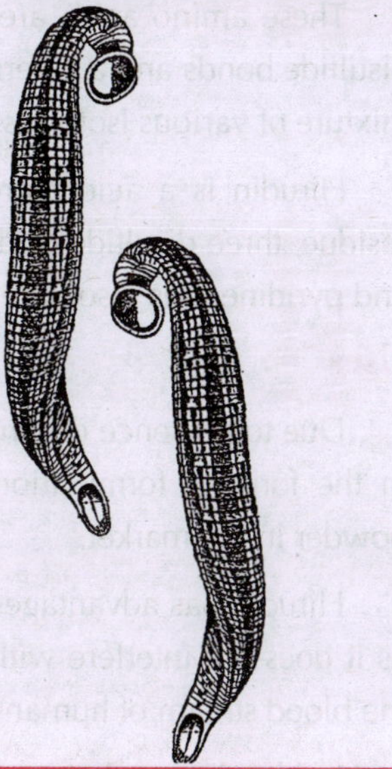

Fig. 12.5: Annelid worms Leeches

Physical Characteristics of Worms

Colour:

(1) Hungerian or green leech : Olive green

(2) German or speekled leech : Greenish yellow with black spots.

(3) Australian leech : Yellowish brown

Size:

(a) Fully extended : 10 - 12 cm × 0.8 - 1.0 cm

(b) Contracted : 3 - 3.5 cm × 1.0 - 1.8 cm

Shape

Made up of 90 - 100 ring shaped segments, tapering at both the ends. Each end of variety has disc shaped sucker with six longitudinal stripes on dorsal surface, while Astralian leech has five longitudinal stripes.

Extra Features

Anterior disc contains three jaws with minute teeth resembling circular saw.

Leeches get attached to skin by means of anterior sucker and suck the blood.

Each leech sucks about 4 to 8 ml of blood and it is sufficient for it for several months, as diet.

Chemical Constituents

Salivary glands of leeches contain most potent blood coagulant known as Hirudin which is a peptide chemically. Its isoelectric point is at pH 3.5 to 4.0.

Hirudin is a peptide and is made up of 65 amino acids.

These amino acids are organized into compact form, i.e. N-terminal domain containing three disulfide bonds and a C-terminal domain which is completely disordered. Natural hirudin contains a mixture of various isoforms of the proteins.

Hirudin is a acidic protein containing 65 amino acid residues. It contains a sulfated tyrosyl residue, three disulfide bridges and high proportion of dicarboxylic acids. Hirudin is soluble in water and pyridine. It is insoluble in ether, ethyl alcohol, acetone and benzene.

Uses

Due to presence of hirudin Leeches are used to reduce inflammation by drawing blood. Hirudin in the form of formulation is used in the treatment of thrombosis. It is available as lyophilized powder in the market.

Hirudin has advantages over commonly used anticoagulants and thrombolytics such as heparin, as it does not interfere with biological activity of the serum proteins. Hirudin is not metabolized in the blood stream of humans and is eliminated unchanged via kidney filtration.

In blood coagulation, conversion of fibrinogen into fibrin takes place by the action of enzyme thrombin. Prothrombinase enzyme produces thrombin from prothrombin. Fibrin is cross linked to form blood clot. Hirudin has inhibitory action on thrombin and thus produces anti-coagulation effect on blood.

Market Products

Lepirudin (Refludan R and Disirudin Iprivask) are hirudin based anti-coagulant pharmaceutical products in the market.

Storage

Leeches are preserved in the earthenware pans full of rain water with coating of fine sand and wood charcoal at the bottom. The pans are kept in shade at 10 - 20ºC temperature covered with muslin cloth. Leeches are in habit of changing their skin, and the fine sand at the bottom of earthenwares helps them in shading the skin. Hirudin should be stored at 2 - 8ºC.

COBRA VENUM

Biological Source

It is the dried secretion obtained from the 'poison gland' of *Naja-naja* and other species of Naja belonging to family Colubridae. It contains not less than 50 mouse-units per mg.

Immediately after extraction, the poison is freez-dried. It is then pooled, mixed, dissolved in ice-cold water for injection and filtered through bacteria-proof filter to give stock solutions.

Physical Characteristics

Colour	:	Venum is white or very light yellow dry powder.
Odour	:	Odourless
Solubility	:	Dissolves in water to give a clear solution with some insoluble matter.

Fig. 12.6: Cobra, Source of Corbra-venum

Identification: (1) It causes heamolysis of RBC, suspended in saline solution, in presence of lecithin.
(2) Mix 0.5 mg of soluble function with 1 ml of polyvalent anti-snake venum serum and incubate at 37ºC for one hour. Inject 0.5 ml of this mixture intravenously to group of mice weighing 17 to 20 g. No animal dies during 24 hours of observations.

Chemical Nature

Cobras belong to subgroup known as elapids. Elapids venom contain post synaptic neurotoxins that spread rapidly in the victim's blood stream, causing resperatory failure and eventually death.

Lucien-Bonaparte was the first to establish the proteinous nature of snake-venom in 1843.

Cobratoxin is the main toxic protein in cobra-venom which is a crystalline compound and is composed of single peptide chain, with 62 amino acid residues, intramolecularly cross-linked by four disulfide bonds.

Proteins constitute about 90 - 95 % of dry weight and are responsible for biological effects. Several proteins found in venom consists of toxins, neurotoxins and innumerable enzymes, polypeptide toxins in cytotoxins, cardiotoxins and post syanaptic neurotoxins.

There are four types of venom that act on the body:

Proteolytic venom

Hemotoxic venom

Neurotoxic venom

Cytotoxic venom

Viper venom acts on vascular system and brings coagulation of the blood and clotting of pulmonary arteries. The pain of wound is severe it is of burning character and speedily followed by swelling of limb and discolouration.

Use: It is used as local haemostatic.

Dose: 1 to 3 mouse - units, initial dose and subsequently increased to 5 to 25 mouse - units.

Storage: It should be stored in light resistant **single dose** containers only.

HEPARIN SODIUM

Heparin, also known as Soluble Heparin is a polymer of D-glucoronic acid and D-glucosamine. The amino groups and some of the hydroxyls are sulfated-molecular weight 6000 – 30000 DG.

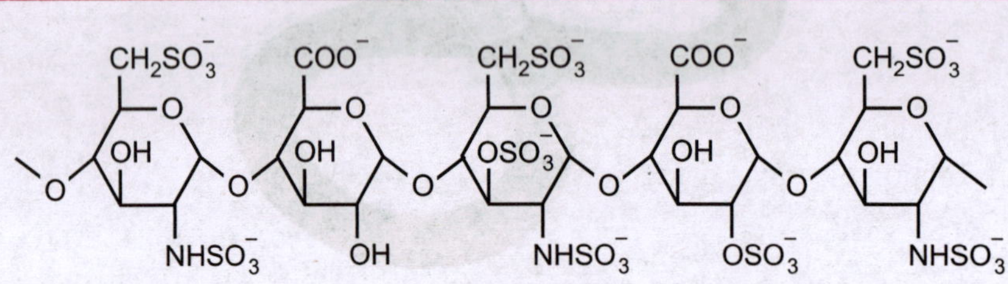

Heparin

It is a sterile preparation of active glycosaminoglycans, which are present in the mammalian tissues of oxen, pigs or sheep like lungs and intestinal mucosa.

Heparin has a significant role in mammalian body as a natural anticoagulant, thereby maintaining the fluidity of blood. Each mg of heparin (lung) contains not less than 110 units and (mucosa or other tissues) not less than 130 units of activity of delaying the clotting of shed blood. It is obtained as a white, hygroscopic powder without any colour or odour. It is soluble in water, saline, alcohol, glacial acetic acid and acetone.

Heparin is mainly used as an anticoagulant in vascular surgery and sometimes in blood transfusions. It is used in treatment of arterial and venous thrombosis. It is also recommended for treating primary and post-operative thrombophlebitis, non-fatal pulmonary embolism, diffuse intravascular coagulation and immune thrombocytopenia.

LECTINS

Lectins are proteins or glycoproteins without an immune origin. They are isolated from various natural sources like seeds, roots, bark; bacteria, fungi, fish eggs; body fluids of invertebrates lower vertebrates, mammalian cell membranes; seaweed and sponges. They can agglutinate the cells and precipitate complex carbohydrates. The agglutination activity of these highly specific carbohydrate binding molecules is usually inhibited by a simple monosaccharide, but for some, di, tri and polysaccharides are required. Lectins are not directly used in medicines but have following utilities.

1. For blood grouping and erythrocyte polyagglutination studies.
2. For isolation, purification and structural studies of carbohydrate containing molecules.
3. For histochemical studies of normal and pathological conditions.
4. For mitogenic stimulation of lymphocytes.

Some of the lectins containing natural sources abrin (*Abrus precatorius*), concanavalin A (*Canavalic ensiformis*), green marine algae (*Codium fragile*), horse gram (*Dolichos biflorus*) and red kidney bean (*Phaseolus vulgaris*).

MALT EXTRACT

It is a product obtained by extracting malt or malted barley, which is the partially and artificially germinated grain of one or more varieties of barley grain, i.e *Hordeum distichon* or *Hordeum vulgare*, family Graminae, or it is a mixture of malted barley with not more than 33 per cent of malted wheat, viz. *Triticum aestivum*, belonging to same family - Graminae.

Malt extract occurs as yellowish-brown or amber coloured viscous liquid with a characteristic odour and sweet taste. It contains nitrogen equivalent to not less than 4 per cent w/w of protein. Along with proteins, it also contains maltose (50 per cent), dextrin, glucose and amylolytic enzymes.

It is used as nutritive, and a flavouring agent for masking bitter taste. It is mainly used as a vehicle for preparations containing cod liver oil and halibut liver oil.

PROTAMINE SULPHATE

It is a purified mixture of different simple proteins obtained from the sperms or mature testes of fish, belonging to the family Clupeidae or Salmonidae.

It occurs as a hygroscopic crystalline powder with white or grey-yellow colour and an astringent taste. It is sparingly soluble in water and alcohol, while insoluble in ether and chloroform.

For the preparation of protamine sulphate, the testes of salmon are isolated, frozen and then minced, washed with water, centrifuged and then dehydrated using solvents and dried under vacuum. It is further subjected to extraction by sulphuric acid, filtered and protamine sulphate rich fraction is precipitated by adding cold alcohol. By dissolving in water, the fraction separates out as an oily layer on cooling. This layer is separated and dissolved in hot water and refractionated using cold alcohol. The fraction is further solvent dehydrated and finally dried under vacuum.

Protamine sulphate is a heparin antagonist and used to counteract the effects due to heparin overdosage like haemorrhage.

PROTEIN HYDROLYSATE

It is a sterile solution of short chain peptides and amino acids, which represent approximate nutritive equivalent to plasma, fibrin, lactalbumin, casein or other suitable protein, from which it is derived by acid, enzymatic or other method of hydrolysis.

Protein hydrolysate occurs as a yellowish to reddish amber transparent liquid and contains not less than 50 per cent total nitrogen in the form of α-amino nitrogen.

It is used to maintain the positive nitrogen balance in the conditions of severe protein deficiency, which arise due to starvation, surgical operations of GI (Gastro-Intestinal) tract or severe illness. It is given along with glucose or other carbohydrates which act as source of energy. Protein hydrolysate injection is given by infusion into superficial vein or by catheter into the large central veins.

Plant Protein Sweeteners:

Wide variety of sweetening agents or sweeteners are used in Pharmacy are basically carbohydrates like sucrose, glucose, honey etc. or glycosides such as glycyrrhizin, sterioside or even synthetic i.e. saccharine.

However the sweetening agents other than above chemical nature and from natural source are really very rare.

But plants containing very potent protein sweeteners are reported in the last two-three decades. Thaumatin, menelline and miraculin are well-known plant protein sweeteners.

THAUMATIN

Biological Source

Thaumatin is a highly potent protein sweetener obtained from the fruits of the plant known as *Thaumatococcus danielli* family: Marantaceae.

Geographical Source

The plant is found in Africa, especially in Ghana, Ivory Coast, Sierra Leone and Togo.

Macroscopic Characters

The plant bears reddish trigonal fruits with 2-3 seeds. Seeds of the plant are covered with fleshy aril. Arils contain active constituents and are sweet in taste.

Chemical Constituents

The aqueous extract of the arils is a mixture of two water soluble proteins thaumatin I and thaumatin II. These proteins are quite stable and their sweet taste is not lost by heating, it is stable at pH 2.7 to 3.0.

Sweet taste of the thaumatin can be detected even in concentration of 10^{-8}. Thaumatin is non-Toxic and non-carcinogenic. It gives flavour even at low concentration.

Uses

Thaumatin is used as a sweetening agent for consumer products such as chewing gums and branch fresheners.

YEAST

Yeast is a collective term for fungi which under normal conditions of growth contain a vegetative body of simple individual cell. It is a unicellular fungus rarely forming mycelia containing small but, definite nuclei and reproducing by budding and ascospore foramen or otherwise. About 350 species are known until now. Yeasts are classified according to their uses whereto they are put or based on their morphological characters. Wine yeast, baker's yeast, brewer's yeast; distiller's yeast are the common examples of various types of yeast. In pharmaceutical industry, either the brewers' yeast or the baker's yeast is used. Yeasts can also be classified as per their morphological character based on formation of ascospore; formation of true or pseudomycelium and reproduction by budding or fission.

Yeast forming ascospore is known as Ascomycetes and that which does not form ascospore is called as *fungi imperfecti*.

Synonyms

Brewer's yeast, Baker's yeast.

Biological Source

It consists of unicellular fungal microorganisms *Saccharomyces cerevisiae*, belonging to family Saccharomycetaceae.

The natural habitat of this microorganism is food juice, bread and several fermented, or fermenting media. On large scale, yeast is produced by using citrus, molasses, molasses wort, grain wort, etc.

Description (Fig. 12.8)

Colour	:	Yeast is available as whitish powder
Odour	:	Characteristic
Size	:	Unicellular microorganism measuring less than 1.5 μ.

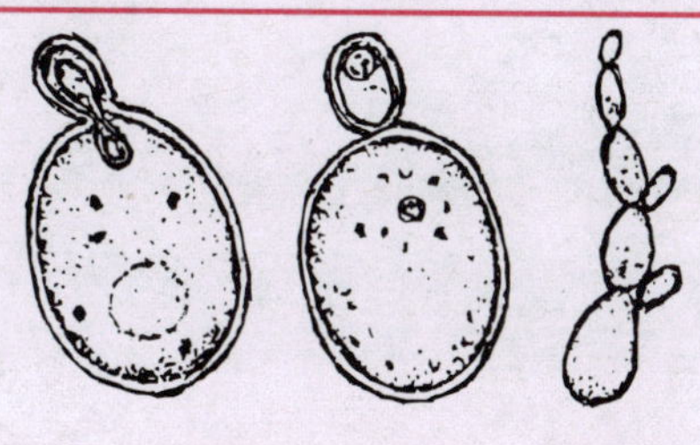

Fig. 12.8: Budding yeast microorganism

Microbial Limits

One gram should comply for absence of *E.coli* and *Salmonella*.

Manufacture of Yeast

Various substrates are used for the commercial production of yeast as mentioned above. But common and industrially used are beet and cane-molasses. A suitable strain like *S. cerevisiae* is grown and it is transferred to sterilized substrate. The equipment is sterilized to get rid of bacteria and other microorganisms. The nutrient medium on which the yeast is grown is continuously added to the culture medium.

It is confirmed that the sugar content of nutrient medium is not more than 1 per cent. The pH is kept in between 4 to 4.5 and temperature at 25 to 30°C. Nitrogenous nutrients like ammonium sulphate and super phosphate are added to the nutrient medium and fermentation is allowed to take place for about 11 hours. Yeast is separated in centrifuge, washed with sterilised cold water, press filtered and finally dried. The moisture content of the final product is maintained at 7.5 per cent and it is filled in containers and stored at low temperature.

Chemical Constituents

The yeast contains about 65 to 85 per cent of moisture, nitrogenous compounds (proteins), glycogen, fat, vitamins and ash. Vitamins reported are from vit B group i.e. thiamine, riboflavine, nicotinic acid, pantothenic acid, folic acid, biotin, etc. The enzymes invertase, diastase, zymase and maltase are present in yeast.

Uses

The yeast is used in manufacture of alcohol, beer and various wines and in the bread industry to raise dough. Irradiated yeast has been used as a source of Vitamin D. Glutathione and invertase are also manufactured from yeast. It is a good source of protein.

❖❖❖

Chapter 13...

DRUGS OF MINERAL ORIGIN

Eventhough, majority of natural drugs are derived from plant and animal origins, a few of them obtained from mineral sources are of paramount significance as pharmaceutical aids. The drugs of herbomineral origin like shilajit and momia are very effective tonics and stimulants. The summary of mineral drugs is given in Table 13.1.

Table 13.1: Mineral Drugs

Name	Main Constituents	Uses
1. Asbestos (Amianthus)	Double silicate of calcium and magnesium with little iron montmorillonite.	In bacterial filters and for filtration of caustic alkalis, heat resistant insulator fireproof gloves and clothing.
2. Bentonite (Whilkinite)	Zinc oxide and 0.5 per cent of ferric oxide.	Suspending agent, emulsifier, and base for plasters, lipsticks, rouges and depilatories.
3. Calamine	Zinc oxide and 0.5 per cent of ferric oxide.	Skin protectant and astringent locally.
4. Fueller's earth (Multani mitti)	Aluminium magnesium, silicate (Non-plastic).	Dusting powder for decolouring of oils, filter aid, cleansing woolen fabrics and as filler for rubber.
5. Kaolin (China-clay, porcelain clay)	Hydrated aluminium silicate.	Adsorbent, as filter aid, pill excipients in poultices.
6. Kieselguhr (Diatomaceous-earth)	Aluminium silicate.	Filter aid for clarifying liquids.
7. Paraffins	Saturated hydrocarbons	Ointment bases
8. Prepared Chalk (Whiting, chalk)	98 per cent of calcium carbonate and 0.5 per cent of magnesium carbonate.	Antacid, in toothpowder, dusting powder, in diarrhoea and a dietary supplement.
9. Shilajit	Herbomineral.	General tonic, aphrodisiac.
10. Talc (Talcum, French chalk)	Hydrated magnesium silicate.	Lubricant, dusting powder and as filter aid.
11. Gopichandan	Magnecium, Iron and Aluminium silicate	Absorbent and dessicant.

Contd...

12.	Rock Salt (Saindhava)	Naturally occurring sodium chloride in solid state.	Used in dyspepsia and as emetic.
13.	Borax (Sodium borate) (Takan khar)	As natural deposit on shores of dried lakes.	Diuretic, astringent emmanoguage and antacid.
14.	Sulphur (Gandhak)	In the beds of Gypsum and in volcanoes.	Laxative and diuretic

Clays are products of composition of rocks rich in aluminium silicate. They possess characteristic plasticity and can be easily moulded into any shape. Clays are made up of mineral grains, some of which are very small in size; and are classified into following three groups.

1. **Kaolin group:** It includes kaolinite; nacrite; dickite; annauxite and halloysite. All of them are hydrated aluminium silicate ($Al_2O_3 \cdot 2SiO_3 \cdot 3H_2O$)

2. **Montmorillonite group:** It includes montmorillonite. (Mg · Ca) Al_2O_3 ; $5SiO_3$ $2H_2O$; beidellite and saponite.

3. **Alkali-bearing clay minerals:** Like clay, mica-clays can also be classified according to their uses.

a) These minerals include, white-ware clays (non-plastic and plastic), kaolin (porcelain or China clay) ball-clay and paper clay.

b) Refractory or fire clay: Glass pot clay, flint clay, plastic fire clay, and graphitic fire clay.

c) Besides Pottery or stoneware clays are alkali bearing clay minerals.

d) Furthermore, medicinal clays, i.e. Bentonite also belong to this category.

ASBESTOS

Synonym

Amianthus

Source

It is a fibrous variety of mineral silicates found in two forms; serpentine and amphibole. The most common fibrous form of serpentine is chrysotile [Mg_6 (S_{14} O_{10}) OH_8] while amphibole is subdivided into four types:

1. Anthophyllite [($Mg \cdot Fe_7$) Si_8 O_{22} $(OH)_2$]
2. Amosite [Fe_5 Mg_2 (Si_8 O_{22}) (OH_2)]
3. Actinolite [Ca_2 $(Mg \cdot Fe)_3$ (Si_8 O_{22}) $(OH)_2$]
4. Tremolite [$Ca_2 \cdot Mg_5$ (Si_8 O_{22}) $(OH)_2$]

Geographical Source

It is found in Canada, Italy, Australia and Rhodesia. Asbestos is mined in open quarries.

Description

Asbestos is white to slightly greenish mineral. It is odourless and is easily broken into long, fine and flexible fibres. It can be spun and woven to form a coarse cloth. It is unaffected by heat. Chrysotile is attacked by acid, while amphiboles are acid resistant. It is also unaffected by alkalis.

Chemical Constituents

It is a double silicate of calcium-magnesium with little amount of iron which gives colour to asbestos.

Uses

It is used as filtering medium for caustic alkalis, for preparation of bacterial filters, heat resistant insulators, fireproof gloves, brake lining, and fireproof clothing.

Caution

Occupational exposure to asbestos is carcinogenic.

BENTONITE

Synonym

Whilkinite

Biological Source

It is a natural colloidal hydrated aluminium silicate (clay).

Geographical Source

It is found in many parts of world, especially Italy, Canada, South Africa and Mid-West USA.

The finest variety of bentonite comes from Fort Benton in USA and hence named as bentonite.

Preparation

Bentonite is found in the form of soapy lumps in the mines. Lumps are steam-dried, followed by pulverisation and sifting.

Description

It is cream to pale buff (yellowish or pinkish tint) coloured very fine powder, free of grittiness. It is odourless and earthy in taste. Bentonite is insoluble in water and organic solvents. When added to water, it swells about 12 times of its volume.

Chemical Constituents

It contains about 80.0 per cent montmorillonite ($Al_2O_3 . 4SiO_2 \cdot H_2O$). Further analysis reveals about 60 per cent SiO_2; 20 per cent Al_2O_3, 2 - 3 per cent magnesium oxide, besides 3.0 per cent ferric oxide.

Chemical Tests

1. Take 2 g of bentonite with 4.0 g of anhydrous sodium carbonate; warm with water, and filter. Acidify the filtrate with hydrochloric acid and evaporate to complete dryness. Warm the residue with dilute hydrochloric acid. The acid solution when treated with solution of ammonium chloride gives a white gelatinuous precipitate, soluble in acetic acid and also in hydrochloric acid.

2. Mount little bentonite in safranin solution (1.0 per cent w/v safranin in 70 per cent alcohol). It acquires permanent deep red colour.

3. When mounted in methylene blue solution (0.1 per cent solution of methylene blue in 95.0 per cent alcohol), it produces deep-blue colour.

Standards of Quality

Loss on drying	:	5 - 12 per cent
pH of 2.0 per cent solution	:	9 - 10.5
Viscosity (dynamic)	:	75 – 225 centipoise for 5.5 per cent w/v aqueous suspension
Swelling power	:	Not less than 12 times

Uses

Bentonite is used as suspending and emulsifying agent and base for plasters. Bentonite should comply tests for sedimentation volume and swelling power.

In cosmetics, it is used as base for creams, lipsticks, depilatories etc.

India is a regular exporter of bentonite.

CALAMINE

Synonym

Prepared Calamine

Source

Calamine is an ore and chemically it contains zinc oxide with a small amount of ferric oxide and contains after ignition not less than 98.0 per cent of zinc oxide.

Description

It is a odourless colourless powder, pink in colour and very fine. Calamine is insoluble in water and soluble in mineral acids.

Chemical Constituents

It contains 99 per cent zinc oxide and 0.5 per cent of ferric oxide. The colour of calamine is due to ferric oxide only. It should not contain calcium for pharmaceutical purposes.

Chemical Tests

1. Mix 1 g of calamine in 10 ml of dilute hydrochloric acid and filter. By adding ammonium sulphide to this solution, a white precipitate is formed, percipitate is soluble in hydrochloric acid, but insoluble in acetic acid.

2. Dissolve 1 g of calamine in 10 ml of dilute hydrochloric acid, boil and filter. To the filtrate, add solution of ammonium thiocyanate; red colour is produced.

Standards of Quality

Acid insoluble substances	:	Not more than 2.0 per cent
Loss on ignition	:	Not more than 1.0 per cent
Arsenic	:	Not more than 10 ppm
pH	:	Slightly alkaline

Use

It is used topically as astringent and skin protectant. It is an ingredient of lotions and cosmetics.

FUELLER'S EARTH

Synonyms

Floridin, **Multani mitti**, Armenian Bole.

The word Fueller's earth is derived from the ancient process of cleaning or pulling wool to remove oil and dust particles with water slurry of Fuller's earth.

Source

Fueller's earth is mined in open quarry.

It is a non-plastic type of kaolin, containing aluminium magnesium silicate.

Geographical Source

It is found in Hampshire, Surrey, Somerset, Dorset and Glaucestershine.

Description

It is white to yellowish grey in colour, odourless and tasteless powder. If put into water, it swells and acquires non-plastic texture.

Chemical Constituents

The approximate composition of Fueller's earth is as follows:

SiO_2	:	55 per cent
Al_2O_3	:	16 per cent
CaO	:	3.5 per cent
MgO	:	2.0 per cent
Fe_2O_3	:	6 per cent
Water	:	10 per cent representing montmorillonite 56 per cent and silica 18 per cent, calcite 90 per cent, magnesite 4.0 per cent, iron oxide 6.0 per cent.

Chemical test

Mount in cresol and examine in polarized light in dark field; it shines brightly.

Uses

It is used as decolorizer for oils and other liquids, as clarifying and filtering agent and for cleansing of woolen fabrics. Due to absorbent property, it is used in the preparation of dusting powders.

It is also used in cosmetic products since it soften and smoothen the skin and improves the blood circulation. It adds glow to the skin as it is natural cleansing agent.

KAOLIN

Synonyms

China clay, Kaolinum proderesum, Porcelain clay.

Source

Kaolin is a purified native hydrated aluminium silicate free from gritty particles. It is obtained by powdering the native kaolin, elutriating and collecting the fraction, which complies with the requirements of particle size. The native clay is derived from decomposition of the feldspar (potassium aluminosilicate) or granite rock and contains silica (47 per cent), alumina (40 per cent) and water (13 per cent).

Geographical Source

It is mainly found in South Eastern United States, England, France, and India.

Description

The different types of kaolin are as follows:

1. **Heavy kaolin:** It is purified hydrated aluminium silicate powdered so as to make free from gritty particles by elutriation.
2. **Light kaolin:** It is identical to heavy kaolin, but is limited to certain particle size and contains suitable dispersing agent.
3. **Natural light kaolin:** It is also a native hydrated aluminium silicate free from gritty particles, but devoid of dispersing agent.

Colour	:	Normally, it is white, but often gets tinged to grey, yellow or brown due to impurities.
Odour	:	It is almost odourless.
Taste	:	Tasteless
Size	:	10 μ - 60 μ in size.

Solubility

It is insoluble in water and in mineral acids. It is not affected by dilute hydrochloric acid, but decomposed by concentrated sulphuric acid after prolonged boiling. It absorbs small amount of water.

Identification

Heat kaolin on charcoal block with cobalt nitrate; it forms blue mass due to alumina.

Standards of Quality

Fusion point	:	1700°C - 1800°C.
Density	:	2.3
Loss on Ignition	:	Not more than 15.0 per cent
pH	:	4.0 – 7.5 (20.0 per cent w/v aqueous slurry).

Uses

It is used as an adsorbent by oral administration, in the treatment of enteritis, dysentery and in alkaloidal and food poisoning. It is also applied externally as a dusting powder and also as clarifying agent during the filtration. Mostly, light kaolin with a particle size less than 10 μ is used in pharmaceutical preparations.

Heavy kaolin with particle size up to 60 μ is only used in the preparation of kaolin poultice.

It is used as filler in paper, rubber, ceramics, cement and fertilizer industries. It is used in anticaking preparations, cosmetics, insecticides, paints and as source of alumina.

Currently, India is exporting kaolin.

KIESELGUHR

Synonyms

Diatomaceous earth, Celite super-cel, Industrial Earth.

Source

It is a natural diatomaceous earth consisting of siliceous skeletons of fossils, family Bacillariceae (subdivision of the algae), purified by treating with dilute hydrochloric acid, washings with water and drying.

Geographical Source

Huge quantity of this earth is available in West Germany, Denmark, Algeria, Kenya, the USA (California and Virginia), Scotland and Ireland.

Preparation

Kieselguhr is normally mined in an open quarry wherein large blocks containing moisture to the extent of 30 - 40 per cent are arranged and dried in air. The blocks containing 5 - 10 per cent of moisture are then pulverised to produce fine powder and subsequently graded. The powder is then subjected to acid treatment, washed thoroughly with water and finally dried.

Description

Kieselguhr is a brownish-grey to white coloured light powder. It is odourless and tasteless. Kieselguhr is very smooth, adheres to the skin after rubbing. It is not slippery. It absorbs moisture, but does not swell when mounted in cresol. It is invisible in polarised light with crossed nicols. Diatoms vary in size from 5 to100 to 500 μ and exhibit two shapes, elongated and circular or triangular known as discoid.

Chemical Constituents

Diatomite contains 75 to 90 per cent of silica, 1 to 5 per cent of aluminium oxide; calcium oxide (1.5 per cent), magnesium oxide (1.5 per cent) and iron oxide (5 per cent).

Standards of Quality

Loss on drying	: Not more than 10.0 per cent
Acid insoluble substances	: Not more than 8 mg
Particle size	: 5 – 100 – 500 μ

Uses

It is used as a filter aid, and for clarification and decolourization of liquids. It is used for the manufacture of tooth powder, face powder and nail polishes. Now-a-days, filter candles are made from diatomite.

PARAFFINS

(A) HARD PARAFFIN

Synonyms

Solid paraffin, paraffin wax.

Definition

Hard paraffin is a mixture of solid hydrocarbons obtained from either petroleum or from shale oil.

Description

Hard paraffin is colourless, tasteless, odourless, translucent wax like white solid, showing crystalline structure. It is slightly unctuous to the touch and burns with a luminous flame.

Chemical Nature

Hard Paraffin is a purified mixture of solid saturated hydrocarbons and can be represented as C_nH_{2n+2}.

Standards of Quality

Density	:	0.84 - 0.89 g/cm^2 (20ºC).
Melting point	:	50 - 57ºC.
Congealing point	:	It congeals at 50 - 57ºC.
Acidity or alkalinity	:	Hot alcohol shaken with an equal volume of melted paraffin, remains neutral to litmus paper.
Solubility	:	Practically insoluble in water, alcohol and acetone while soluble in chloroform, ether, warmed fixed oils and vegetable oils.

Chemical Tests

1. Paraffin burns with luminous flame and deposits carbon.
2. When paraffin is heated with an equal weight of sulphur, paraffin evolves hydrogen sulfide and becomes black from liberation of carbon.

Uses

In topical pharmaceutical formulations, paraffin is used as component of ointments and creams. In ointment it may be used to stiffen and increase melting point of formulation. Additionally it is used as coating agent for tablets and capsules.

Storage

Paraffin should be stored at a temperature not exceeding 40ºC in a well closed container.

(B) LIQUID PARAFFIN

Synonyms

Liquid petrolatum, white mineral oil.

Source

It is a mixture of purified liquid saturated hydrocarbons obtained from petrolatum mines located deep under the sea.

Description

It is colourless, odourless, tasteless transparent oily liquid free from florescence in daylight. Liquid paraffin is odourless when cold but develops a faint odour when heated.

Solubility

Liquid paraffin is more or less insoluble in water, however, it is miscible with chloroform, ether fixed and valuable oils and hydrocarbons besides being sparingly soluble in ethanol (95 per cent). Miscible with fixed oils.

Light Absorption

Not more than 0.1 in the range of 240-280 nm of 2.0 per cent solution in 2, 2, 4 trimethyl pentane. It should pass the test for sulphur compounds readily corbonisable substances and solid paraffin.

Chemical Nature

It is a mixture of liquid saturated hydrocarbons.

Chemical Test

In a test tube cautiously boil 1 ml liquid paraffin with 1 ml of 0.1 M sodium hydroxide with continuous shaking for about 30 sec. On cooling to room temperature two phases separate. To aqueous phase add 0.1 ml of phenolphthalein solution. The colour turns to red.

Standards of Quality

Weight per ml	:	0.860 - 0.904
Viscosity	:	Between 110 and 230 mPas at 20ºC.

Uses

Internally it is used as laxative and faecal softner as formulations aid in foods. It is used as excipients in topical preparations. It is used in ophthalmic ointments (less than 15 per cent), in topical ointments (0.22–0.23 per cent), topical emulsions (1–20 per cent), topical lotions (7–16 per cent), optic preparations (less than 50 per cent). It is also used in cosmetic and food products. In cosmetics it is used in the preparation of vanishing creams, all purpose cream, cleansing cream, men's hairdressing formulations (30 - 40 per cent), nail bleaches and nail white edging cream (46 per cent w/w), nail extenders, wooden eye pencils, moulded wax type (15 per cent w/w) and faecal softener.

Storage

It should be stored in airtight containers in a cool dry place. It should be protected from light. It can undergo autoxidation on storage forming aldehyde and organic acids. Antioxidants butylated hydroxyanisol, butylated hydroxytoluene and α–tocopherol can be used (10 ppm) to preserve the same.

(C) LIGHT LIQUID PARAFFIN

It is colourless, odourless, tasteless transparent liquid hydrocarbon and free from fluoresence by daylight. It is soluble in chloroform, ether but insoluble in alcohol and water. It is miscible with fixed oils. It is mixture of saturated hydrocarbons.

Chemical Nature

Identical with liquid paraffin.

Standards of Quality

Weight per ml	:	0.82 – 0.86
Viscosity	:	25 – 80 mp

Uses

Used as vehicle, other uses are identical to liquid paraffin only.

(D) YELLOW SOFT PARAFFIN

Synonyms

Yellow petroleum jelly. Paraffin Jelly.

Source

It is purified semi-solid mixture of hydrocarbons obtained per petroleum.

Description

It is pale yellow translucent soft unctuous mass. It is slightly fluorescent to day light. It is odourless soluble in either light petroleum (b.p. 40 - 60) but insoluble in water and alcohol (m.p. 38 - 56°C).

Use

Ointment base.

(E) WHITE SOFT PARAFFIN

Synonyms

White petroleum jelly, white petroleum.

It is obtained by bleaching yellow soft paraffin.

Source

It is semi-solid white translucent unctuous mass similar to yellow soft paraffin.

White soft paraffin is soluble in ether, chloroform but insoluble in water and alcohol.

Use

White soft paraffin is used as ointment base. It is unsuitable for ophthalmic use due to its bleaching and may cause irritation to eyes.

PREPARED CHALK

Synonyms

Chalk; Creta; Paris-white; **Whiting**; English white.

Source

Chalk is a native form of calcium carbonate, freed from most of the impurities by elutriations. It contains not less than 97.0 per cent w/w of calcium carbonate ($CaCO_3$), when dried at 100°C.

Collection and Preparation

Chalk is mined in open quarries, pulverized and then purified by elutriation. The water is removed and the insoluble chalk settle forming flat cakes, known as 'whiting'. It is purified further for pharmaceutical use.

Description

Chalk is colourless, odourless, white, earthy and soft to the touch. It is amorphous and insoluble in water. When treacted with acids, it effervesces. Prepared chalk has various shapes and sizes of foraminiferous shells, (the shells of bicellular animals of *Foraminifera*) as observed under microscope using cresol as mountant.

Chemical Constituents

It contains calcium carbonate (96 per cent), magnesium carbonate (0.5 per cent), 0.5 to 1.0 per cent of silica, traces of iron, manganese and aluminium oxides, and also sulphates, phosphates and organic matter.

Standards of Quality

Not more than 0.2 percent silica, oxides of iron and aluminium.

Uses

Prepared chalk is used as an antacid, a dietary supplement, a dusting powder, and an anti-diarrhoeal. It is used in face powders, and as an abrasive in tooth powders and toothpastes. It is also used in the manufacture of antibiotics and other pharmaceuticals.

Precipitated Chalk

This variety can easily be detected from the prepared chalk by microscopic studies. The foraminiferous shells are absent in precipitated chalk, as it is obtained by chemical reactions.

TALC

Synonyms

French Talc, Talcum, Soap stone, Purified Talc.

Source

It is natural hydrated magnesium silicate. It is purified by boiling with dilute hydrochloric acid, decanting the liquid, washing the residue several times with water and drying at 110°C.

Geographical Source

It is plentifully available in France, New York State, Italy and also in India.

Description

It is a very fine white or greyish-white powder, without odour and taste. It is free of gritty particles, greasy to touch and adheres to skin. Talc is insoluble in water, dilute acids and alkali hydroxides. It is a fine variety of steatite occurring as massive mineral with pearly white lustre and non-plastic thus, distinguished from mica.

Chemical Constituents

Chemically, talc is hydrated magnesium silicate represented by formula $Mg_6\ (Si_2O_5)_4\ (OH)_4$. Normally, it contains 1 - 2 per cent of iron oxide and only traces of aluminium oxide. Colour of the talcum powder is due to iron oxide content.

Standard of Quality

Density : 2.2 - 2.8

Chemical Tests

1. Fuse about 0.5 g talc with 0.2 g each of anhydrous sodium carbonate and potassium carbonate in a platinum crucible. Dissolve the fused mixture into 50 ml of water and to it add hydrochloric acid to it until it ceases to effervesce. Add little more acid and evaporate the contents to dryness on water bath. Cool it; dissolve in 20 ml of water, boil and filter. To the filtrate, add about 2 g of ammonium chloride and 5 ml of dil. ammonia solutions. If any precipitate is formed, remove it by filtration. To the filtrate, add sodium phosphate, white crystalline precipitate of magnesium ammonium carbonate will be formed.

2. Yields the reactions characteristic of silicate.

Uses

Talc is used as a filter aid for filtration and clarification of cloudy liquids; as lubricant in preparations of pills and tablets and as a dusting powder.

It is used in paint and varnish industry, as a filler for paper and as heat and electric insulator.

GOPICHANDAN

Biological Source

Gopichandan is a type of clay, like multani mitti.

Geographical Source

It is available near Dwarka in Gujrat state, near the lake GOPI (Hence, named as Gopichandan).

Description

It is light yellow coloured powder and its small particles resembles multani mitti.

Being clay contains mangnesium, iron and aluminium silicates chemically. It absorbs water put on it immediately.

Uses

It is used as absorbent, powder is dusted on unhealthy ulcers and wounds. It is cooling and desiccant in action. When applied with rose water in the form of paste to the forehead it relieves headache it is also used to treat inflamed boils.

ROCK-SALT

Synonyms

Saindhava, Sea salt, Bay salt, **Sendhalon.**

Source

Naturally available transparent sodium chloride in solid state is known as Rock salt.

Large beds of rock salt associated with clay are available and to get them holes are drilled in rocks which get filled with salt water. The resulting salt water is evaporated producing saindhava.

Geographical Source

This is available near sea-shores only.

Description

Rock salt is found in the form of whitish crystalline granules or even in transparent cubes. It is brownish white externally and white internally. It is with slight odour and salty taste.

Identification

It burns with yellow flame.

Uses

It is used in indigestion (dyspepsia) and abdominal disorders. Warm water made with rock salt is emetic and is used when needed.

BORAX

Synonyms

Suhagoor, **Takan khar**, Sodium borate.

Source

Borax is a sodium salt of Boric acid and occurs naturally. Crude borox occurs as natural deposit by evaporation of water on the shores of dried up lakes.

Geographical Source

It is available in India, Tibet and also from mud of lakes, surrounded by hills in Nepal.

Purification

It is purified by dissolving in water, straining through cloth and evaporating to dryness. It can also be purified by steeping (soaking) in whey and dried in Sun. It is recrystallised from water.

Description

Crude borax is dirty white in colour and is found in translucent and irregular masses on exposure to atmosphere it becomes opaque. It has balsamic odour and saline (salty) taste. Its aqueous solution is alkalene to litmus paper.

Identification

(1) The acidified solution of borax gives pink or brownish red colour with turmeric paper, dry it, and add dilute solution of ammonia, paper turns blue or greenish black.

(2) The mixture of borax, sulphuric and alcohol burns with a flame tinged with green.

Uses

It is diuretic, astringent, emmenagogue, antacid and antipyretic.

SULPHUR

Synonyms

Sublimed sulphur, Gandhak.

Source

Sulphur is a non-metalic element found free in beds of Gypsum and also in the form of sublimed sulphur in regions of volcanoes. It is also available in the form of ores known as pyrutes as sulphates and sulphides of Iron, Copper, Lead, Zinc etc.

Geographical Source

It occurs in Nepal, Kashmir, Afghanisthan in Bangladesh also.

Description

In market four types of crude Gandhaka are available. They are as under:

(1) **Vitreous** or **precipited sulphur** which is yellow in colour. Semi-transparent crystals.

(2) **White sulphur**, found in the form of pieces which are 5 × 7.5 cm in diameter, having nauseous smell and bitter and astringent in taste. Pieces are very brittle and gummy to touch. This variety is preferred for external use.

(3) **Lal-Gandhak**, and as the name indicates it is red in colour. Occurs as crystalline pieces and are shining, may also be purple or brick red in colour. Their taste is acrid and bitter. It burns with blue flame.

(4) **Black coloured**, it is regarded as pure sulphur and prepared by washing with milk. It is light yellow powder with a typical smell and bitter astringent in taste.

Identification

(1) It burns with blue producing sulphur di-oxide.

(2) It melts at 115°C to yellow liquid.

(3) It becomes dark and viscid at about 160°C.

Uses

Sulphur is having peculiar smell due to which it has several actions. It is laxative in action, diuretic and also insecticide. It can be used internally as well as externally.

In Ayurvedic practice normally it is used alongwith metallic mercury, as it combines with mercury very easily.

SHILAJIT

Synonyms

Mumie, shilaras.

Source

It is a herbomineral drug ejected out of fissures in iron rich rocks, during hot weather.

Geological Source

It is found naturally in mountainous areas, especially Himalayas, Vindhya and other mountains in India and also in Nepal. It may also be found as a tar in the earth crust, formed due to decomposition of vegetable substances.

It is observed that 'Momia', which occurs in Arabia and Persian mountains, resembles Shilajit.

Description

Colour : Brownish black to dark black

Odour : Characteristic

Taste : Distinctive

Extra features : Hard, greasy, resinous

The following forms of this drug have been reported in ancient Hindu literature:

1. Iron shilajit (blackish-brown variety)
2. Copper shilajit (blue variety)
3. Silver shilajit (white variety)
4. Gold shilajit (red variety)

Chemical Nature

Purified shilajit contains albuminoids, fatty acids, trace elements, waxes in addition to minerals. Fulvic acid-7-methoxy carboxyl biphenyl isolated from Shilajit has potent antistress action.

Uses

It is used as a general tonic stimulant and aphrodisiac Shilajit of authentic quality is an important ingredient of ayurvedic preparations meant for strengthening immunising system of human body. It is a unique blend of herbal and mineral ingredients formed by a unique natural process.

Chapter 14...

TERPENOIDS

Introduction:

Terpenes are hydrocarbons and components of resins and also turpentine produced from resin. Word terpene is derived from turpentine. When terpenes are modified chemically such as by oxidation or rearrangement of carbon skelton, the resulting compounds are known as **terpenoids**.

Carterpillars also emit terpenes in their osmeteria. (Osmeterium is a fleshy organ found in the prothoracic segement of larve of butterflies.)

Terpenoids are regarded as derivatives of polymers of **isoprene**, $H_2C = \underset{CH_3}{\underset{|}{C}} - CH = CH_2$ (i.e. C_5H_8) usually joined head to tail and widely distributed in the plant kingdom.

It can also be written as:

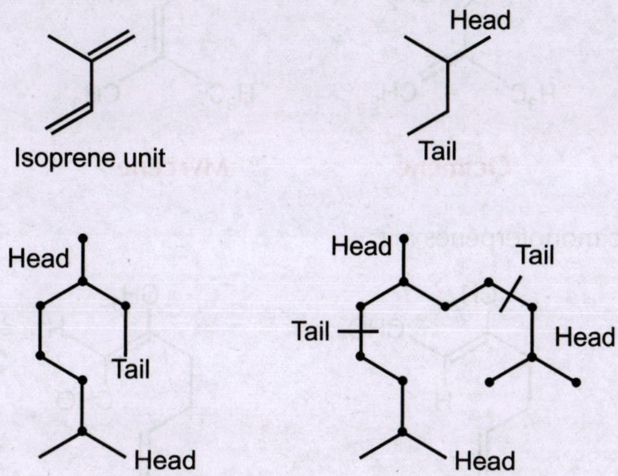

Terpene hydrocarbons are classified as under:

Isoprene* /Hemiterpene	C_5H_8
Monoterpenes	$C_{10}H_{16}$
Sesquiterpene**	$C_{15}H_{24}$
Diterpene	$C_{20}H_{32}$
Triterpenes	$C_{30}H_{48}$
Tetraterpenes	$C_{40}H_{64}$
Polyterpenes	$(C_5H_8)_n$

Terpenoids are abundantly available in volatile oils. They consist of a complex mixture of terpenes or sesquiterpene, alcohols, aldehydes, ketones, acids, and esters. The separation of individual components is accomplished by vacuum fractionation and by chromatographic methods.

The unsaturated hydrocarbons are conveniently separated as their crystalline addition products with hydrochloric acid or hydrobromic acid, or nitrosyl chloride.

* This does not occur free in nature. Isoprene itself is considered as only hemiterpene, while prenol (3 methyl-2 buten-1-ol) and isovalaric acid (3 methyl butanoic acid) are the hemiterpenoids.

** Prefix sequin means one and half.

MONOTERPENES

These are subdivided into three groups: acyclic, monocyclic and bicyclic.

Acyclic monoterpenes

Important hydrocarbons are ocimene and myrcene.

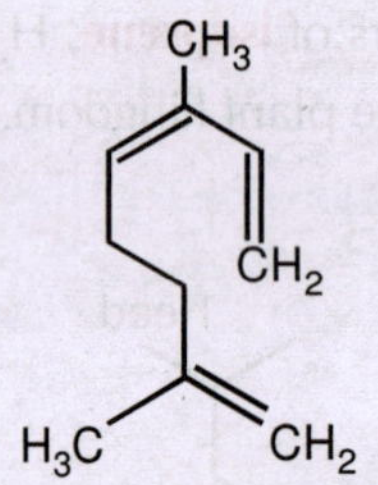

Ocimene

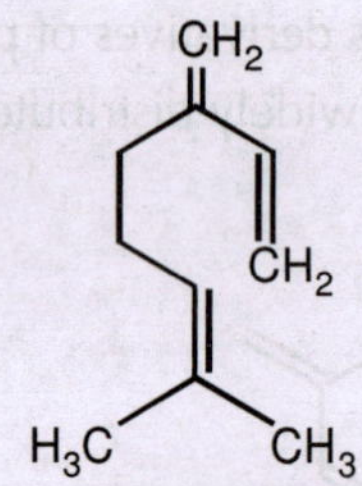

Myrcene

Examples of aldehydic monoterpenes are:

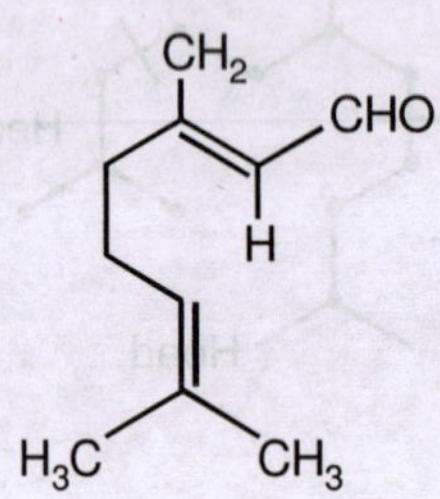

Geranial

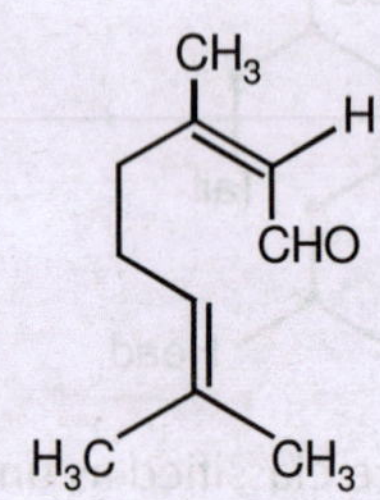

Neral

Examples of alcoholic monoterpenes are:

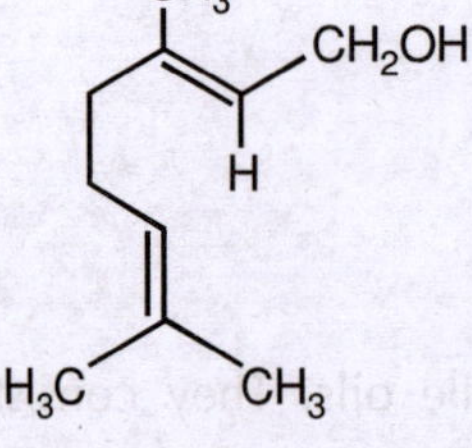

Geraniol

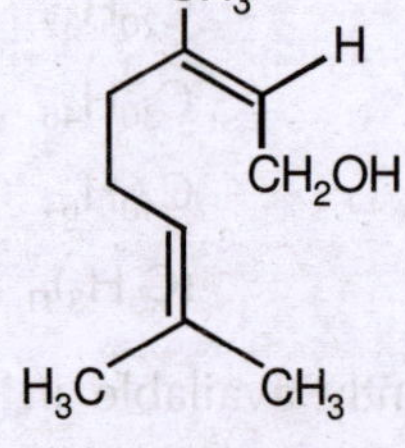

Nerol

Monocyclic monoterpenes

Containing only hydrocarbons are:

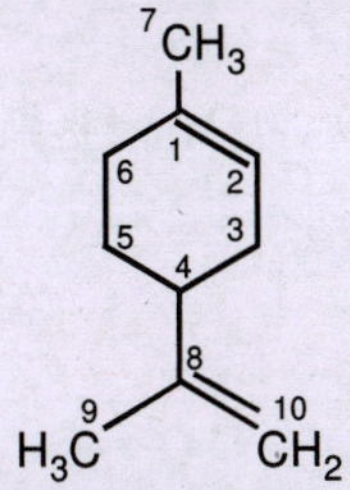

d-Limonene

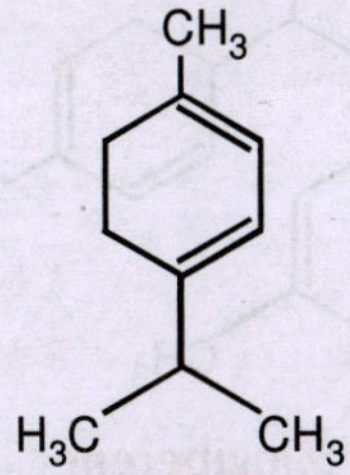

α-Terpinene

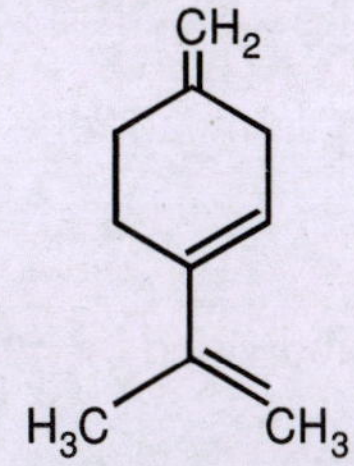

β-Terpinene

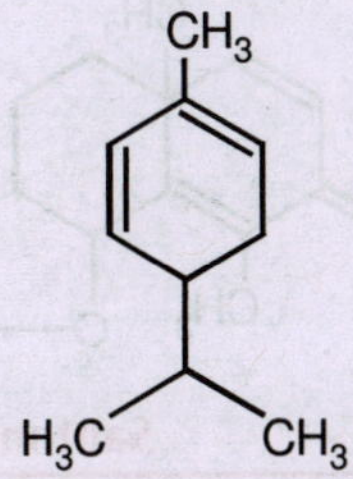

α-Phellandrene

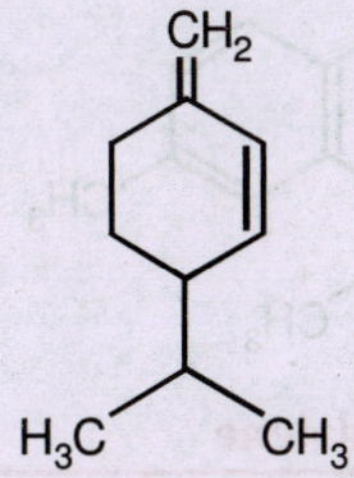

β-Phellandrene

Monocyclic monoterpenes containing alcohol are

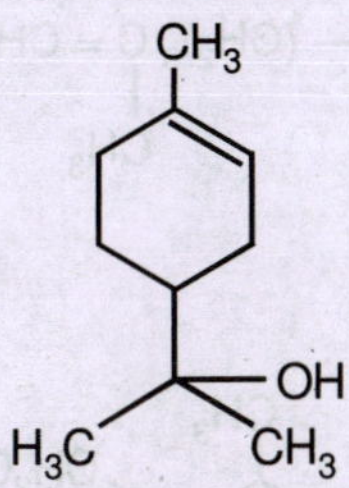

a-terpineol

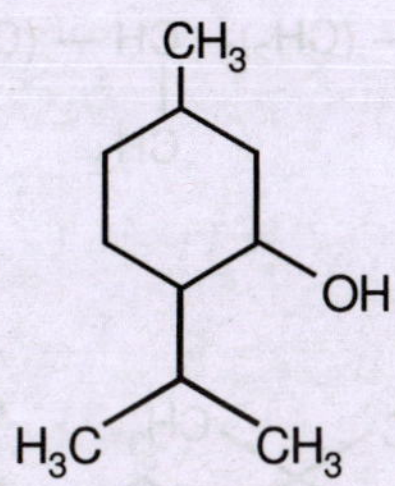

Menthol

Monocyclic monoterpenes containing aldehydes are

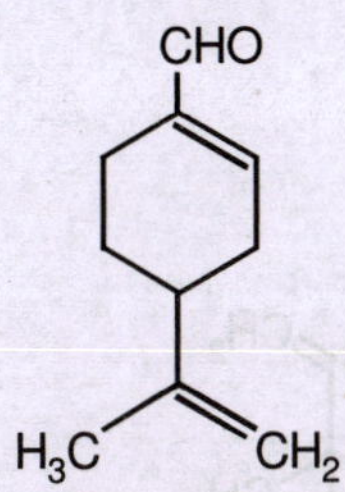

Perillaldehyde

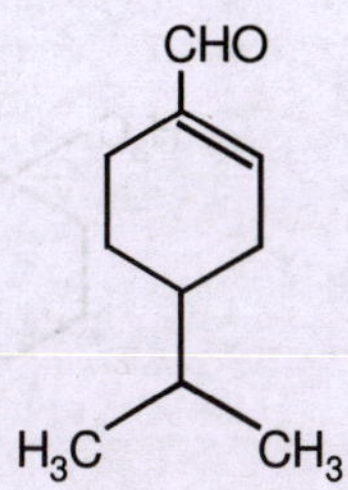

Phellandral

SESQUITERPENES

Monocyclic sesquiterpenes

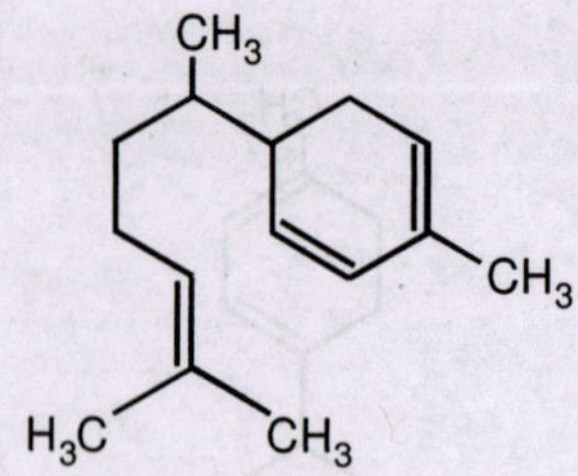

Zingiberene

Bicyclic sesquiterpenes

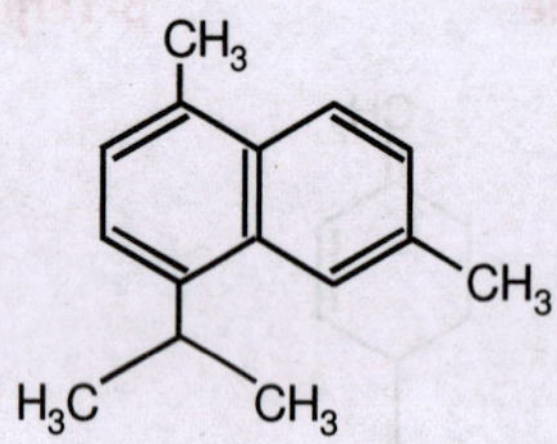

Cadalene

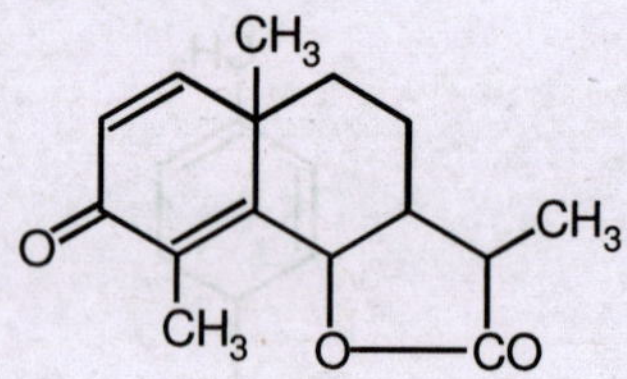

Santonin

DITERPENES

These are formed by the union of four isoprene units and are found mainly in the plant resins.

Acyclic diterpene

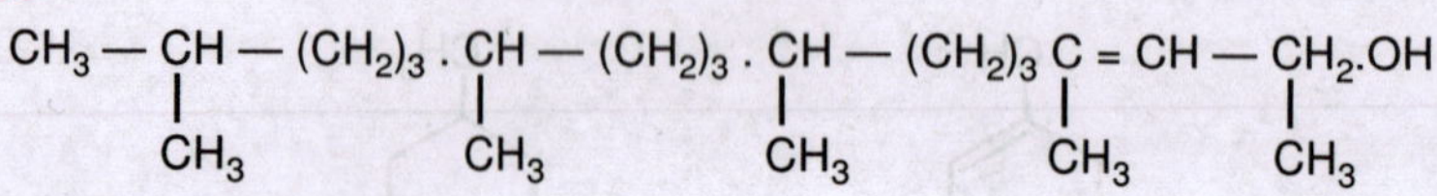

Phytol

Monocyclic diterpene

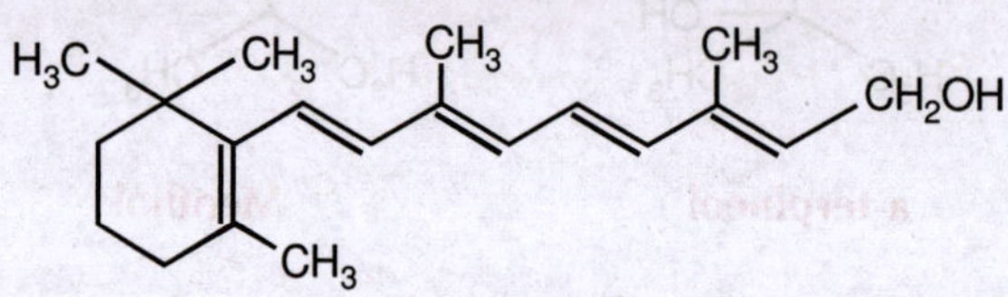

Vitamin A₁

Bicyclic diterpene

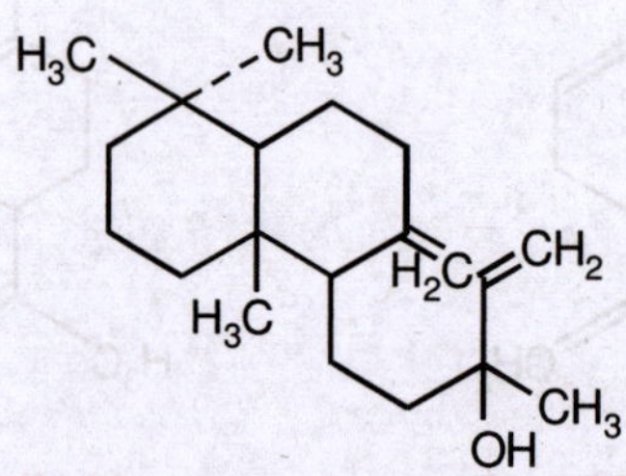

Manool

Tricyclic diterpenes

The resin acids form the major nonvolatile part of many natural resins, such as

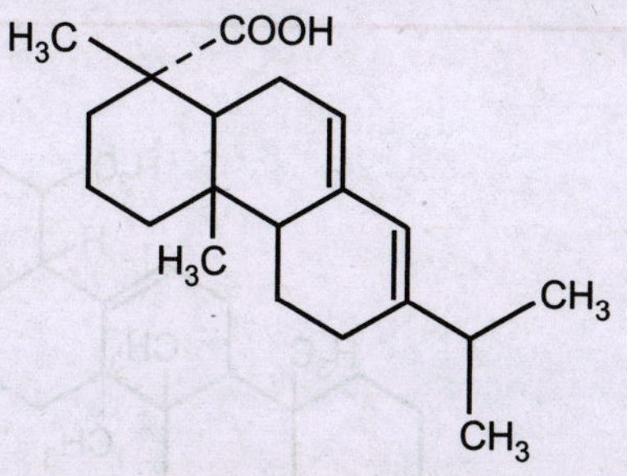

Abietic acid

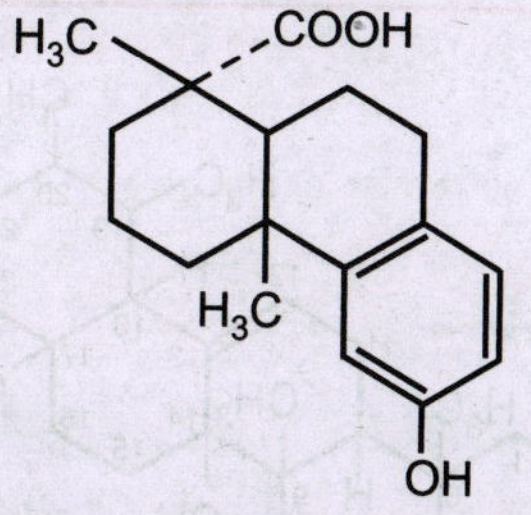

Podocarpic acid

TRITERPENES

The triterpenes are also widely distributed in the plants and animals kingdom. They are classified into three groups:

1. Acyclic triterpene; 2. Tetracyclic triterpene; 3. Pentacyclic triterpene.

1. Acyclic triterpene

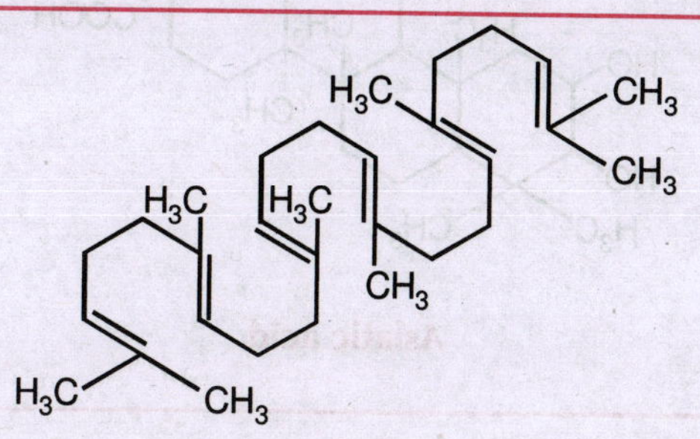

Squalene

2. Tetracyclic triterpenes

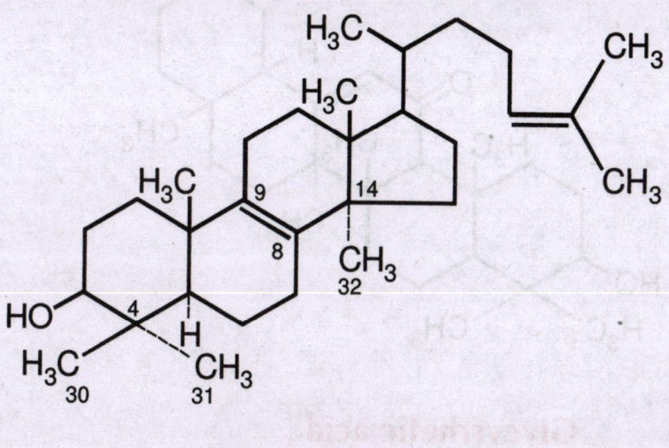

Lanosterol

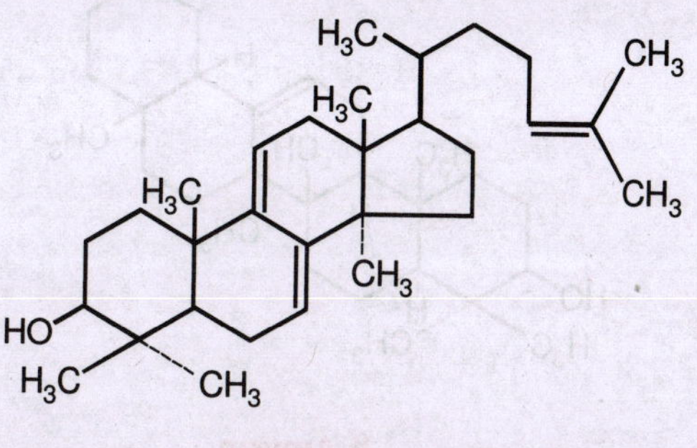

Agnosterol

3. Pentacyclic triterpenes

These compounds are subdivided into three subclasses.

α-Amyrin class

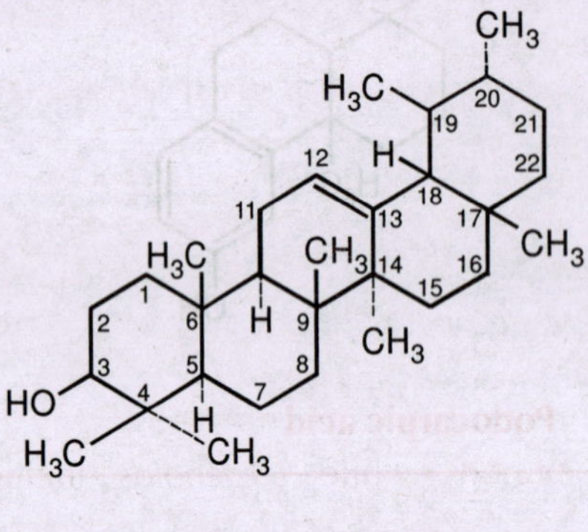

a-Amyrin

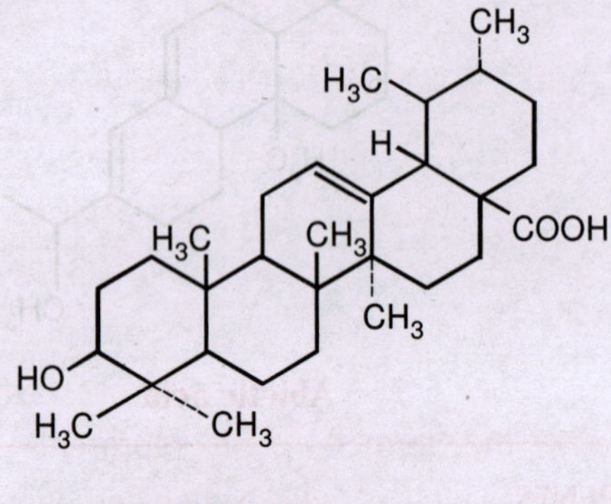

Ursolic acid

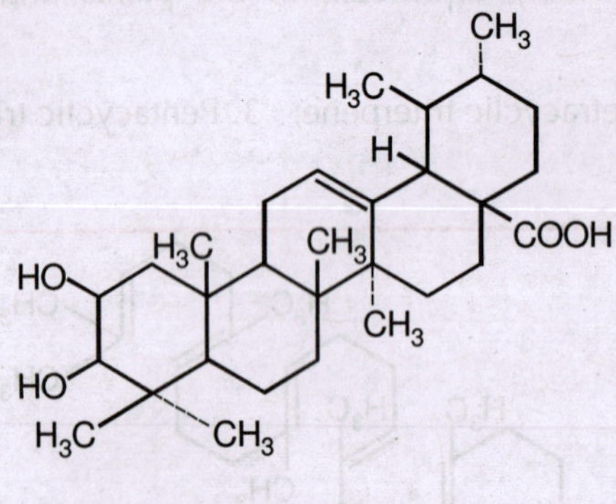

Asiatic acid

β-Amyrin class

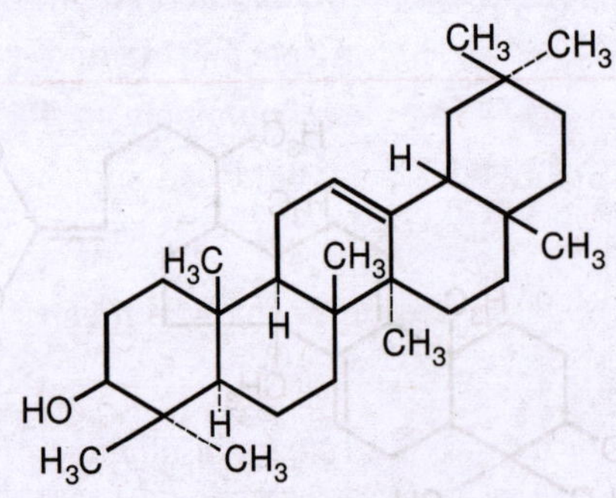

β-Amyrin

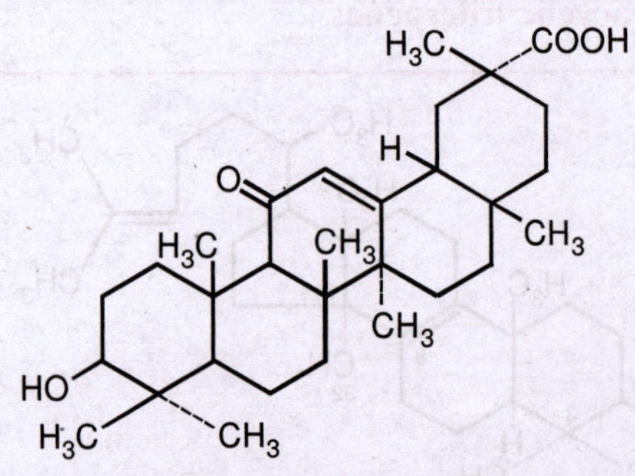

Glycyrrhetic acid

Lupeol group

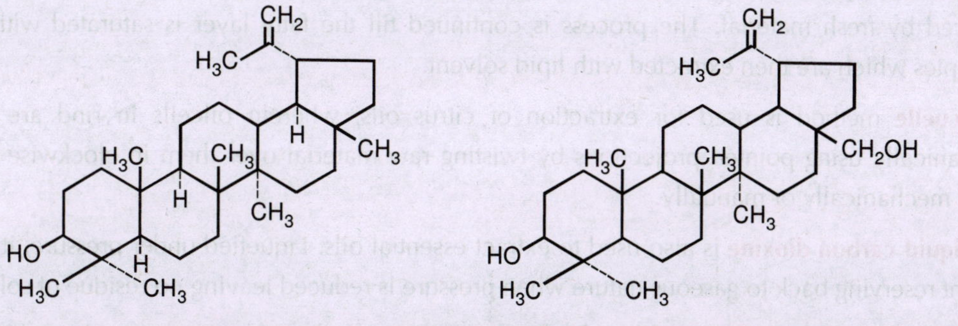

Lupeol **Betulin**

Actually, the term *terpene* represents the hydrocarbons while *terpenoids* include the hydrocarbons, as well as their oxygenated derivatives.

Terpenes and terpenoids are found in all volatile oils, resins and resin combinations of plant or animal origin. Drug containing terpenoids are known as Terpenoidal drugs.

(I) VOLATILE OILS

The odorous, volatile principles of plant and animal sources are known as ***volatile oils***. As they evaporate when exposed to air at ordinary temperatures, they are also called as "ethereal oils." They represent essence or active constituent of plant, or animal hence they are also known as "essential oils".

Chemically, they are derived from terpenes and their oxygenated compounds. They are made up of isoprene units (C_6H_8) and are usually mono-, sesqui and diterpenes with empirical formulae as C_5H_8, $C_{10}H_{16}$, $C_{15}H_{24}$ and $C_{20}H_{32}$ respectively.

Volatile oils are soluble in alcohol, ether and other lipid solvents and practically insoluble in water. They are usually lighter than water. They possess characteristic odours and they have high refractive index. Most of them are optically active. They are secreted in special structures such as duct, cell, schizogenous or lysigenous glands, trichomes, etc. They are commonly found in the species of Labiatae, Rutaceae, Piperaceae, Zingiberaceae, Umbelliferae, Myrtaceae and Lauraceae. They are present in entire plant or in any part of the plant.

Volatile oils are extracted by steam-distillation, solvent extraction or mechanical means such as ecuelle and effleurage techniques.

Hydro-distillation method comprising of water distillation, water and steam distillation, or steam distillation is used for extraction of volatile oil from herbal drugs. The fresh material is subjected to hydro-distillation in case of the leaf drugs. Air-dried subterranean parts are extracted by steam distillation.

Effleurage method is used for extraction of delicate perfumes. The fresh flower petals are mechanically spread on layer of fatty material, allowed to imbibing and the exhausted petals are replaced by fresh material. The process is continued till the fatty layer is saturated with volatile principles which are then extracted with lipid solvent.

Ecuelle method is used for extraction of citrus oils, wherein oilcells in rind are ruptured mechanically using pointed projections by twisting raw material over them in clockwise direction either mechanically or manually.

Liquid carbon dioxide is also used to extract essential oils. Liquefied under pressure, it acts as a solvent reserving back to gaseous nature when pressure is reduced leaving no residue of solvent.

Drugs containing volatile oils are classified as follows.

Table 14.1: Types of volatile oils

Type	Examples
1. Alcohol volatile oils	Peppermint, cardamom, coriander, orange flower oil, rose oil, sandalwood
2. Aldehyde volatile oils	Cinnamon, lemon peel, orange peel, citronella oil lemon-grass, bitter almond
3. Ester volatile oils	Gaultheria, lavender, mustard
4. Hydrocarbon volatile oils	Turpentine, black pepper
5. Ketone volatile oils	Caraway, spearmint, buchu, camphor, musk, civet oil
6. Oxide volatile oils	Chenopodium, eucalyptus
7. Phenolic ether volatile oils	Anise, fennel, nutmeg
8. Phenol volatile oils	Clove, thyme, creosote

Volatile oils are used as flavouring and perfuming agents in pharmaceutical formulations, foods, beverages and cosmetics. Besides, they are also used as medicinal agents such as carminatives (umbelliferous fruits, cinnamon, etc.), anthelmintics (chinopodium), diuretic (Juniper), antiseptic (eucalyptus), counter-irritant (oil of wintergreen), local anaesthetic (clove), sedative (jatamansi), local irritant (turpentine), etc. Citronella oil is an insect repellant, whereas lemon-grass oil is used for citral content, and subsequently for synthesis of vitamin A. Terpeneless oils are valuable perfuming and flavouring agents.

The sophisticated perfume industry utilizes solvent extraction technique for production of volatile oils using lipid solvents like ether or benzene, as it is quick, quality assured and economical method. The deterioration of volatile oils is less than hydro-distillation. The solvent distilled off may be re-used for extraction.

Terpeneless volatile oils fetch more prices in perfumery industry because of their specificity and stability. They are processed by removing hydrocarbons and undesired components by fractional distillation and other techniques which are usually kept as guarded secrets by manufacturers. They are used in manufacture of high priced cosmetics and perfumes. During 1993-94 and 1994-95, India's total exports of volatile oil were ₹ 3407.2 and ₹ 4426.3 lakhs, respectively.

Chemical tests for Volatile Oils

Presence of volatile oil in the natural drugs can be detected by the following tests:

1. To the thin section of the drug, add alcoholic solution of Sudan III. Red colour obtained by globules indicates the presence of volatile oil.

2. To the thin section of the drug, add a drop of tincture alkane. Red colour indicates the presence of volatile oil.

Storage of Volatile Oils

Certain volatile oils are liable to deteriorate on keeping. The deterioration is accompanied by change in colour, or increase in viscosity of the oil, or change in odour of the oil. Therefore, volatile-oils should be preserved properly in well-closed, well filled containers away from light and in cool place. Following are few examples of volatile oils alongwith their terpenoids.

Table 14.2: Occurance of terpenoids

Volatile oils	Terpenoids present
Almond oil	benzaldehyde
Caraway oil	carvone, and limonene
Citronella oil	geraniol, citronellal, farnesol
Eucalyptus oil	cineole
Geranium oil	geraniol
Peppermint oil	menthol
Ginger oil	zingiberene
Fennel oil	anethole
Sandalwood oil	santalol
Clove oil	eugenol

TERPENOIDS

Depending upon type and number of rings reported in terpenoids (except hemiterpene), they are named as

1. Acyclic terpenoids : have open-chain structure
2. Monocyclic terpenoids : have one-ring in structure
3. Bicyclic terpenoids : have two-rings in structure
4. Tricyclic terpenoids : have three-rings in structure

General properties of Terpenoids

Since terpenoids occur only in the volatile oils, they are normally colourless liquids or solids with pleasant smell and are insoluble in water, but they are soluble in alcohol, organic solvents and fixed oils. They readily volatilize in steam. Most of them are optically active. Terpenoids normally contain one or more doule bonds and form additive compounds with halogens, nitrosyl chloride and nitrosyl bromide. They get oxidised by oxidising agents.

Toxicity of Volatile Oils

Toxicity of volatile oils by oral route in very low though they are used frequently.

Isolation of Terpenoids

Volatile oils obtained in first step of distillation contain number of mono and sesqui-terpenoids. During fractional distillation of volatile oils, monoterpenoid hydrocarbons get distilled first followed by their oxygenated derivatives. Physical methods, chemical methods (treating with sodium bisulphite, phenylhydrazine, Tilden's reagent, Grignard's reagent), distillation under reduced pressure and several types of chromatographic methods are used for their isolation. Proper adsorbents are used and then the chromatogram is eluted suitably.

Uses

Terpenoids by virtue of their pleasant flavour are compounds of tremendous importance in perfumery, cosmetics, soaps, incense - sticks, foods, pharmaceuticals and beverage industries.

Therapeutically, they exert wide spectrum of activities like antiseptic, stimulant, carminative, diuretic, anthelmintic, analgesic, antirheumatic, aromatic, counter-irritant, etc.

After the nutritives like carbohydrates proteins and fats, condiments and spices are the major compounds consumed by the society on large scale, every where in India and abroad to make the food delicious and tasty. About 75 per cent of terpenoidal compounds are covered under condiments and spices.

To differentiate both of them, it is stated that condiments are mild in action and are used exclusively for sweetpreparation i.e. Nutmeg, Lavender, Mogra, Oragne, cardamom can be quoted as suitable examples.

Where as spices are strong, powerful in action and are pungent, penetrating and spicy in taste such as clove, ginger, cumin, cinnamon, black papper etc.

Apart from flavour in food and pharmaceutical industries, they are also used as insect repellant, insecticides, pesticides and deodorants.

Classification of Terpenoids

These are classified on the basis of isoprene units i.e. (C_5H_8) they contain.

Table 14.3: Classification of Terpenoids

Sr. No.	Name of the class of terpenoids	No. of Isoprene units	Molecular formulae
1.	Monoterpenes or terpenes	2	$C_{10} H_{16}$
2.	Sesquiterpenes	3	$C_{15} H_{24}$
3.	Diterpenes	4	$C_{20} H_{32}$
4.	Triterpenes	6	$C_{30} H_{48}$
5.	Tetraterpenes or Carotenoids	8	$C_{40} H_{64}$
6.	Pentaterpenes	10	$C_{50} H_{80}$
7.	Polyterpenes or rubber	n	$(C_5 H_8)_n$

All the above referred terpenoids are originated only from isoprene units. In nature, there are very few chemicals, which are found to be composed of both isoprenoid and non-isoprenoid units. They are called as **meroterpenoids**. Such meroterpenoid structures of natural origin are reported in quinine, phylloquinone (vitamin K), tocopherols (vitamin E), cannabinoids and various ergot alkaloids.

The distribution of terpenoids in natural drugs, their sources, chemical constituents and uses are briefly summarised in Table 14.4.

Table 14.4: Terpenoidal Drugs

1. Drugs Containing Monoterpenoids of Herbal Origin

Sr. No.	Name of Crude Drug	Synonyms	Scientific Name and Part used	Active Constituents	Uses
1	2	3	4	5	6
1.	Ajowan	Carum copticum Bishop's weed	*Fruits of Trychyspermum ammi* Umbelliferae	Volatile oil, proteins, carbohydrates, thymol, p-cymene, terpinine	Antispasmodic, stimulant, carminative, antiseptic, antifungal, insecticide, anthelmintic
2.	Anise	Anise-fruit Aniseed	*Fruits of Pimpinella anisum* Umbelliferae	Anethal, methyl chavicol	Stimulant, flavouring agent, carminative, expectorant, condiment
3.	Benafsha	Sweet violet	*Aerial parts of Viola odorata violaence Epicarp*	Violene, odoratine	Flavouring agent
4.	Bitter Orange peel	Orange cortex	*Citrus aurantium* Rutaceae	Hesperidin, isohesperidin, neohesperidin, Vit. C, limonene, citral	Stomachic, aromatic carminative, flavour
5.	Black pepper	Pepper	*Fruits of Piper nigrum* Piperaceae	Piperine, starch piperidine, *l*-phellandrene, Caryophylling	Aromatic, stimulant, stomachic, carminative, condiment, stimulate the taste buds with gastric juice
6.	Camphor oil	Rectified oil of camphor	Wood of *Cinnamomum comphora* Lauraceae	Camphor, eugenol, eucalyptol	Rubefacient, counter irritant

Contd...

1	2	3	4	5	6
7.	Caraway	Carum, caraway seed	*Fruits of Carum carvi* Umbelliferae	Carvone, carvacrol	Aromatic, stimulant, carminative, flavour
8.	Cardamom	Cardamom fruits	*Fruits of Elettearia cardamomum* Zingiberaceae	α-terpineol, borneol, cineole	Aromatic, carminative, stimulant, flavour
9.	Cassia cinnamon	Cassia bark, Chinese cinnamon	*Bark of Cinnamomum cassia* Lauraceae	Mucilage, coumarin, cinnamic aldehyde	Carminative, stimulant, flavour, aromatic, spices
10.	Celery	Apium, celery seeds	*Seeds of Apium graveolens* Umbelliferae	d-limonene, d-selinine, sedanoic acid, sedanolide	Stimulant, carminative, nervine sedative and tonic, rheumatism
11.	Chenopodium	American wormseed oil, cheno-podiol, chenoposan	*Flowering tops of Chinopodium ambrosioides* Chenopodiaceae	Asconidole	Anthelminitic, to expel hook worms and intestinal amoebae
12.	Cinnamon	Kalmi-Dalchini, Cinnamon bark	Bark of *Cinnamomum zeylanicum* Lauraceae	Eugenol, benzaldehyde Cuminaldehyde	Carminative, stomachic, flavour, stimulant, aromatic, antiseptic, astringent
13.	Citronella oil	Ceylon citronella	*Leaves of Cymbopogon nardus* Graminae	Geraniol, citronellal	Flavour, perfume for soap, in mosquito repellant cream
14.	Coriander	Coriander fruits	*Fruits of Coriandrum sativum* Umbelliferae	Coriandrol, coriandryl acetate, L-borneol, geraniol, pinene	Aromatic, flavour, carminative, stimulant
15.	Cummin	Jira	*Fruits of Cuminum cyminum* Umbelliferae	Cuminaldehyde, α-pinene, β-pinene, Phellandrene	Stimulant, carminative and used in dyspepsia
16.	Dill	European dill, Dill fruits, Anethum	*Fruits of Anethum graveolens* Umbelliferae	Dill - apiole, carvone	Aromatic, stimulant, carminative flavour, used in gripe water.

contd...

1	2	3	4	5	6
17.	Eucalyptus oil	Dinkum oil eucalyptus	*Leaves of Eucalyptus globulus* Myrtaceae	Cineole camphene phellandrene	Counter-irritant antiseptic, expectorant, used in cough, chronic bronchitis
18.	Fennel	Fennel fruits, fructus foeniculum	*Fruits of Foeniculum vulgare* Umbelliferae	Fenchene, anethal, Ketone phellandrene limonene	Carminative, flavour, aromatic, stimulant, expectorant
19.	Garlic	Allium	*Bulbs of Allium sativum* Liliaceae	Allicin, alliin, iron, phosphorus, copper, propyl disulphide	Carminative, expectorant, stimulant, disinfectant, condiment, rubefacient, anthelmintic
20.	Gaultheria oil	Bectula oil, oil of wintergreen	*Leaves of Gaultheria procumbens* Ericaceae	Gaultherin, methyl-salicylate	Rheumatism, flavour, Vermicide [Hook worms]
21.	Geranium oil	Algerian geranium	*Leaves of Pelargonium graveolens Geraniaceae*	Geraneol, citronellol	Flavouring agent
22.	Jatamansi	Indian spike nard, Nard	*Rhizomes of Nardostachys jatamansi* valerianaceae	Jatamanson Nardostachone	Sedative, diuretic, anti-spasmodic, emmunagogue, stomachic, stimulant, epilepsy, palpitation of heart
23.	Lavender oil	Common Lavender	*Flowering tops of Lavendula officinalis* Labiateae	Esters linalyl acetate linalool, pinene, geraniol, cineol	Aromatic, carminative, flavour, in perfumery
24.	Lemon peel	Cortex limonis	*Epicarp of Citrus limonis* Rutaceae	Limonene, hesperidin, citral geranyl acetate terpineol	Carminative stimulant
25.	Lemon grass oil	East-Indian lemon grass oil, Indian melissa oil	*Leaves of Cymbopogon flexuausus* Graminae	Citral, nerol, citronella, methyl heptanal	Flavour

contd...

1	2	3	4	5	6
26.	Long pepper	Catkins	*Climbing vine Piper longum Pipevaceae*	Piperine and volatile oil	Bronchitis
27.	Nutmeg	Myristicci, Nux-Moschata	*Seeds of Myristica fragrans* Myristicaceae	Myristicine, safrole, elemicin	Aromatic, flavour, stimulant, carminative used in soap industry treat rheumatism
28.	Palmarosa oil	Rosha oil, Geranium oil	*Leaves of Cymbopogon martini* Graminae	Geraniol, linalool, citronella dipentene	Flavour, treatment of rheumatism and skin diseases
29.	Peppermint oil	Mentha oil, Oleum mentha piperites	*Leaves of Mentha piperita* Labiatae	Menthol, pulegon menthone, methofuram	Carminative, stimulant, flavour, antiseptic and perfumes
30.	Pippala mool	Pippala root	*Roots of Piper longum Pipevaceae*	Piperine and volatile oil	Bronchitis
31.	Rasna	Galanga, East India root, Chinese ginger	*Rhizomes of Alpinia officinarum* Zingiberaceae	Cinnamates, cineole, pinene, camphor, galangin	Aromatic, stimulant, carminative, antibacterial, treatment of rheumatism
32.	Rosemary oil	–	*Flowering tops of Rosemarimus officinalis* Lab. Labiatae	Volatile oil, resin, Borneol, ursolic acid	Carminative, flavour, stimulant, rubefacient
33.	Shati (Kapur Kachri)	Spiked ginger lily	*Rhizomes of Hedychium spicatum* Zingiberaceae	Paramethoxy cinnamic acid ester, cineole, limonene	Stomachic, carminative, stimulant tonic, flavour, preparation of ABIT.
34.	Spearmint	Mentha viridis	*Leaves and flowering tops of Mentha spicata Labiatae*	l carvone	Flavouring agent
35.	Thyme	Garden thyme, mother of thyme	*Leaves and flowering tops of Thymus vulgaris* Labiateae	Thymol, linalool, carvacrol	Carminative, antispasmodic flavour, expectorant

Contd...

1	2	3	4	5	6
36.	Thyme oil	-	Flowering tops of *Thymus vulgaris* *Labiatae*	Thymol, Carvacrol	Anthelmenth antiseptic
37.	Tulsi	Sacred basil, holy basil	*Leaves and flowering tops of Ocimum sanctum* Labiatae	Eugenol, methyl eugenol, carvacrol, Caryophylline	Antibacterial, insecticidal, stimulant, aromatic, anticatarrhal, spasmodic, diaphoretic, antiperiodic, stomachic, good-immunomodulatory agent
38.	Turpentine oil	Oleum terbenthae	*Oleo-resin of Pinus roxburghi* Pinaceae	Carene, limonene, turpentine	Counter-irritant rubefacient, expectorant, antiseptic
39.	Vetiver oil	Khus oil	*Roots of Vetiveria zizanioides* Gramineae	Vetivone (α and β) vetiverol	Stimulant, aromatic, refrigerant stomachic, used as flavour, preparation of soap, sharbat etc.
40.	Vidang	Embelia	*Dried fruits of Embelia ribes*	Embelin, vilangin	Anthelmintic

2. Drugs Containing Monoterpenoids of Animal Origin

41.	Castoreum	Castor Canadian beaver	*Caster faber* Castoridae	Volatile oil, resin, castorin, benzyl alcohol, methyl phenol, *l*-borneol.	Flavour in beverages and in cosmetics.
42.	Civet	Zibeth	*Scent-glands of Viverra zibetha* Viverridae	Civetone, Indole civetole ethylamine, propylamine	Fixative for flavours
43.	Musk	Kasturi, Moschus	*Moschus moschiferus* Cervical	Musckone, Cholesterine, albuminoids and resin	Perfumery used in treatment of hysteria

3. Drugs Containing Sesquiterpenoids

1.	Acorus	Calamus, Bach, vaj, sweet flag	*Rhizomes of Acorus calamus* Araceae	Asarone, eugenol, aconine	Carminative stimulant, vermifuge perfume

Contd...

1	2	3	4	5	6
2.	Arnica	Arnica flowers, Mountain tobacco, Wolf's bane	*Flowering tops of Arnica montana* Compositae	Volatile oil, flavonoids, terpenoids	Counter-irritant, cosmetic, rheumatism, spinal paralysis and blindness
3.	Artemisia	Santonica, worm seed	*Flower-buds of Artemisia china, A. brevifolia A. aritime* Compositae	Santonin, artemisin, cineole, pinene, resin	Anthelmintic, cumulative in action
4.	Artemisia annua	Quinghao	*Artemisia annua* Compositae	Artemisinin, deoxy artemicinine	Antimalerial
5.	Clove	Clove-flower, Clove-buds	*Flower-buds of Eugenia caryophyllus* Myrtaceae	Eugenol, ester eugenin	Dental analgesic, carminative, flavour, stimulant, aromatic and antiseptic
6.	Cubeb	Kababchini	*Unripe fruits Piper cubela* Piperaceae	Cubebin, cubebic acid, wax	Stimulant, aromatic, urinary antiseptic used
7.	Davana oil	–	*Flowering herb Artemisia pollens* Compositae	Devanone, argemone, non Devanone	Perfume, flavour
8.	Feverfew	Feather few, Feather foil Midsummer daisy Altamisia	*Dried leaves of Tanacetum parthenium* Compositae	Sesquiterpene lactones, chrysanthemin A and B, parthenolide, Partholide, chrysanthemonin	Used in treatment of migraine, vertigo, fever, arthritis and menstrual disorders
9.	Hops	Humulus	*Female flowers of Humulus lupulus* Cannabinaceae	α, β-acid-Humulone, cohumulone, adhumulone, lupulone, colupulone, adlupalone	Sedative, spasmolytic
10.	Sandal wood	-	*Heart wood of Santalam album* Santalaceae	Sandalwood oil	Isolatoin of oil, in perfume

Contd...

11.	Sandal wood oil	East Indian Sandal wood oil	Volatile oil of *Santalum album* *Santalaceae*	α, β-santalol, santene, santenone, teresantol, standalone, santalene	Treatment of dysuria, used in perfume
12.	Saussurea	Costus, kuth	*Roots of Saussurea lappa* Compositae	Resinous, inulin costuslactone dihydrolactone camphene	Used in asthma, insecticide, antiseptic disinfectant, carminative, stimulant.
13.	Valerian	Indian valerian	*Rhizomes of Valeriana wallichii* Valerianaceae	Chatanene, valerine, borneol formate, acetate, valrate, valranic acid	In cough, antispasmodic depress CNS, stimulant, carminative, tonic
14.	Valerian European		*Rhizomes of Valeriana officienatis* Valerianaceae	Chatanene, valerine, borneol formate, acetate, valrate, valranic acid	In cough, antispasmodic depress CNS, stimulant, carminative, tonic
4.	**Drugs Containing Diterpenoids**				
1.	Coleus	Forskohlii	*Roots of Coleus forskohlii* Labiatae	Forskolin (coleonol)	Vasodilator, Cardio-stimulant, lowering blood pressure and intra-occular pressure used in glaucoma
2.	Taxus	Yew, Himalayan yew, Talis-patra	*Leaves and bark of Taxus baccata, T. cuspidata, T. brevifolia, T. canadensis.* Taxaceae	Taxol, cephalo mannine, 10-deacetyl baccatin	Anti-cancer
5.	**Drugs Containing Triterpenoids**				
1.	Ambergris	Grey-amber	*Physeter catoden* Physeteridae	Ambrein, epicoprostemol, coprostemone	Perfume
6.	**Drugs Containing Tetraterpenoids and Polyterpenoids**				
1.	Annatto	Annota, Arnota	*Dried seeds of Bixa orellana* Bixaceae	Bixin	Colouring agent for foods, cosmetics.

contd...

2.	Crocus	Saffron, Kesar	*Dried stigma of Crocus sativa* Iridaceae	Crocin, crocetin, picrocrocin	Flavour, antispasmodic, emmenagogue, stimulant.
3.	Chlorophyll	-	Pigment from all green leaves	-	Colouring agent
4.	Gossypol	-	*Pigment from* seed oil of *Gossypium berbaceum*	Malvaceae	Male contraceptive

General description of various types of terpenoids has been given here. This general description has been followed, under respective categories, with the pharmacognosy of the drugs containing such terpenoids.

(A) MONOTERPENOIDS

They are derived from geranyl pyrophosphate and are considered to be main ingredients of volatile oils. Monoterpenoid occur in plants some animals, microorganisms as well as in insects also. They occur both in terrestrial and marine plants and animals. Among plant kingdom, they are especially present in higher plants and their parts like seeds, roots, leaves, inflorescence and heartwood, etc. Prominent families containing monoterpenoid are Umbelliferae, Lauraceae, Labiatae, Gramineae and Myrtaceae. The families like Burseraceae and Pinaceae contain high percentage of monoterpenoid in the form of resin or oleo-resin. Various marine algae also contain monoterpenoid but they have not been found in fungi. Different terpenes also occur in different parts of the same plant. Most commonly occurring monoterpenoid are limonene, cineole and α-pinene.

Many of these structures have been found to have biological roles in some plants and animals. In insects, they show sex attractant and defence activities. Iridomyrmecin and citronellal, both monoterpenes, are defence chemicals of some arthropods. Sex attractant monoterpenes are present in male boll weevil. Many monoterpenoids are antimicrobial due to which they protect the plants.

HERBAL DRUGS CONTAINING MONO TERPENOIDS

AJOWAN

Synonyms

Trychyspermum copticum Link, *Carum copticum* Hieren, Bishop's weed, own.

Biological Source

Ajowan consists of dried ripe fruits of the plant *Trychyspermum ammi* Sprague (Umbelliferae). It contains not less than 1.0 per cent (w/w) of thymol on dried basis.

Geographical Distribution

It is said to be native of Egypt and is cultivated in Iraq, Iran, Afghanistan, Pakistan and India. In India, it is extensively cultivated in Madhya Pradesh, Uttar Pradesh, Gujarat, Rajasthan, Maharashtra, Bihar and West Bengal.

Cultivation and Collection

It is cultivated as a rabbi crop by broadcasting method in October-November. Thinning, weeding, and manuring, are done as in case of Umbelliferous crop. Irrigation is essential for cultivation and harvesting is done in February-March. About 300 to 350 kg fruits per hectare are obtained on commercial scale. Cultivation expenses 7000/- to 8000/- per acre. Selling price ₹ 5,000/quintal, yield 4-5 quintal/acre.

Organoleptic Characters

Colour : yellowish-brown

Odour : agreeable

Taste : aromatic, warm like thymol

Size : 1.7 - 3 mm long, 1.5 - 2.4 mm broad and 0.5 - 1.5 mm thick.

Shape : It occurs in the form of cremocarps or separated mericarps with bifid stylopod, and five light coloured ridges. Apical half to 2/3 parts of cremocarps have purplish tinge.

Fig. 14.1: Ajowan Fruits

Chemical Constituents

Ajowan fruits contain 2 - 4 per cent of volatile oil, about 21 per cent fat, 17 per cent proteins and 25 per cent carbohydrates. Traces of tannin, glycoside and steroidal substances have been reported. Volatile oil mainly contains thymol (35 - 50 per cent), p-cymene (50 - 55 per cent), terpinene (30 - 35 per cent). Pinene, dipentenes, etc. are other constituents of the oil. The flavour and taste of drug is due to thymol and volatile oil.

Standards of Quality

Foreign organic matter : Not more than 2.0 per cent

Ash : Not more than 20 per cent

Uses

Ajowan is used as antispasmodic, stimulant and carminative. It is also recommended in sore throat and bronchitis. Ajowan is used in preparation of lotions and ointments for checking chronic discharge. Ajowan is used as substitute for parsley fruits.

Ajowan oil and dethymolised ajowan oil is used as antiseptic, antifungal, insecticide and anthelmintic. The oil is also preferred as deodarant in mouth washes, tooth pastes and gargles and as a flavouring agent for disinfectant soaps.

ANISE

Synonyms

Anise fruit; Aniseed, Pimpinella.

Biological Source

It consists of dried ripe fruits of *Pimpinella anisum*, Linn. family Umbelliferae.

Geographical Source

Aniseed is indigenous to Egypt, Greece and Western Asia. It is cultivated in all warm climates, especially in Russia, France, Spain, Italy, Bulgaria, Mexico and India. In India, it is cultivated in Uttar Pradesh, Punjab and Orissa. Anise was known to Pleny and Dioscorides.

Macroscopic Characters

Colour	:	Greenish-grey or brown
Odour	:	Characteristic, aromatic, more, if crushed
Taste	:	Sweet, aromatic
Size	:	Cremocarps are 3 - 5 mm long and 1 to 3 mm wide.
Shape	:	Ovoid, conical, compressed laterally, rough to touch; primary ridges are slender, pale and uniform in width. Short bifurcate stylopod at the apex.

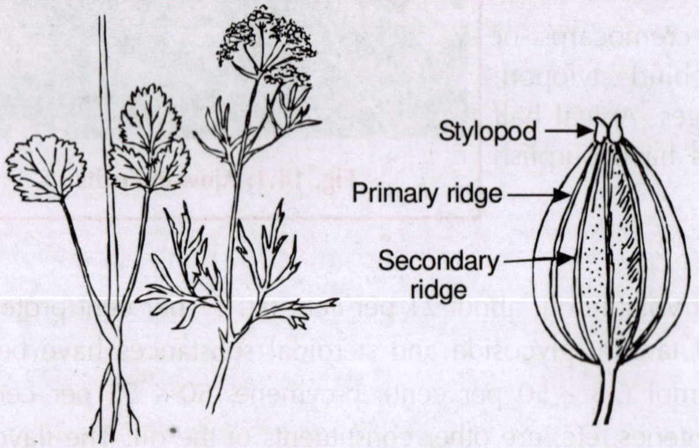

Fig. 14.2(a): Pimpinella anisum herb and Fruit

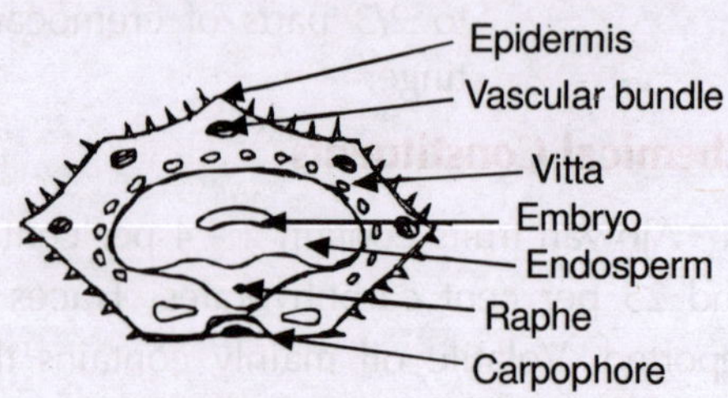

Fig. 14.2 (b): T.S. of Anise mericarp

Microscopic Characters

The epidermis of fruits is covered by numerous, unicellular conical thick walled warty trichomes. About 30 - 50 vittae are present in each mesocarp on dorsal surface, while two large vittae are seen on commissural surface. Small aleurone grains, rosette crystals of calcium oxalate and fixed oil are present in endospermic cells.

Chemical Constituents

Anise contains 1.5 - 3.5 per cent volatile oil, 10 per cent fixed oil, proteins, mucilage, and starch. The volatile oil mainly contains anethole (about 90 per cent), methylchavicol (13 - 15 per cent) and anisaldehyde.

Volatile oil of anise is colourless to yellow liquid with following constants.

Specific gravity	-	0.978 - 0.988
Optical rotation	-	+1 to - 2°
Refractive index	-	1.553-1.560

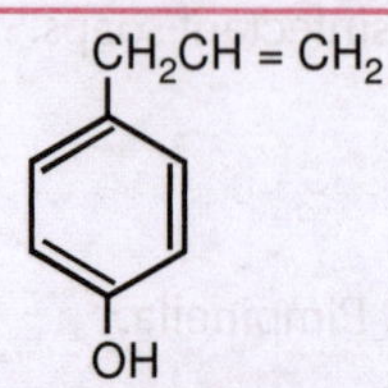

Chavicol

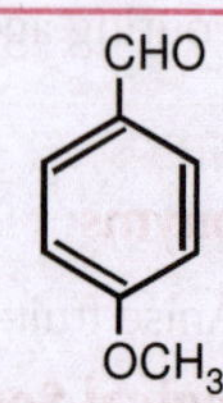

Anisaldehyde

Standards of Quality

Foreign organic matter	:	not more than 1.0 per cent
Total ash	:	not more than 10.0 per cent
Acid insoluble ash	:	not more than 1.5 per cent
Volatile oil	:	not less than 2.0 per cent

Uses

It is used as stimulant, carminative, expectorant, condiment and flavour in foods or beverages and also in manufacture of dentrifices.

STAR ANISE

Biological Source

This consists of fully ripe dried fruits of *Illicium verum* fam Magnoliaceae.

Geographical Source

It is found in Southern and south-western China,Spain, Turkey, Egypt, Russia abd Poland.

Morphological Character

One seeded follicles is the characteristic of fruit, about Eight follicles are present follicles are adherant at bases.

Colour	:	Carpels are reddish brwn
Odour	:	Agreeable aromatic odour
Taste	:	Sweet spicy taste
Size	:	1.5 – 1.7 × 4-5 mm
Shape	:	Boat shaped and bluntly beaked at apex.

Extra features : Carpels are woody and seeds are hard with shining, wrinkled externally and smooth and glossy internally.

Fig. 14.3: Star anise

Chemical Constitutents

Star anise fruits contain volatile oil about 10.0 per cent. It closely resembles to Anise (*Pimpenella anisum*) and contain 80 - 90 per cent of anethol, anisaldehyde and estragole.

Uses

Used as carminative, stimulatnt and flavouring agent for cough mixtures, specially.

BENAFSHA

Synonym

Sweet violet, Viola odorata.

Biological Source

It consists of dried aerial parts obtained from *Viola odorata* Linn, belonging to family Violaceae.

Geographical Source

Sweet violet is indigenous to India and found in Kashmir (Kangra), Himachal Pradesh (Chamba), and Kumaon hills.

Benafsha is cultivated only in gardens. Its cultivation is also undertaken in hilly regions of North India. It grows quite satisfactorily in cool and moist climatic conditions. It does not survive on exposure to heavy rains. Its propagations can be done either by cuttings or with seeds. Drug is naturally found at an altitude of 1500 - 1800 m.

Macroscopic Characters (Fig. 14.4)

Benafsha is a glabrous or pubescent herb, about 15 cm in height. Its root stocks are very stout and stolons are cylindrical. Leaves are dark green, tough, broadly ovate or cordate in shape with crenate margin. They are 1.5 - 5 cm in size. Flowers are solitary, auxillary forming central flowering rosettes. Flowers are very beautiful in colour. They are deep violet in shade with bluish-white base. Flowers are sweet, scented and which is why the plant is cultivated in gardens as an ornamental crop. Fruits are in the form of capsules, round, three angled and often purplish in colour. The plant blooms in second year.

Fig. 14.4: Viola odorata Herb

The drug benafsha in the market is available in different forms, which constitute various aerial parts of plant.

1. Kashmiri benafsha : Aerial parts like stems, leaves and flowers
2. Gul-i-benafsha : Only dried flowers
3. Berg benafsha : Aerial parts without flowers

For medicinal purposes, plant is collected from April to July.

Chemical Constituents

The leaves are found to contain an essential oil, alkaloid and a colouring matter. It has a very agreeable flavour and is used for high grade perfumes. The root stocks contain saponins, a glycoside of methyl salicylate responsible for expectorant property, an essential oil, and alkaloid odoratine. Flowers contain a substance known as violine, volatile oil, rutin and cyanin. It contains 0.1 per cent concentrate responsible for highly pleasant odour of the drug.

Standards of Quality

Total ash	:	not more than 14.0 per cent
Acid insoluble ash	:	not more than 3.0 per cent
Foreign organic matter	:	not more than 2.0 per cent
Water soluble extractives	:	not more than 20 per cent

Uses

The herb is used as expectorant, diaphoretic and antipyretic. The herb shows antibacterial and antifungal activities, and hence used in the treatment of eczema. It is used in the form of *sarbat*. The flowers are emollient, demulcent and are said to relieve pain due to cancerous growth. The leaves in large doses and seeds are cathartic. Violine is an emetic.

BITTER ORANGE PEEL

Synonym

Orange cortex

Biological Source

Orange peel is fresh outer part of the pericarp of the ripe or nearly ripe fruits of *Citrus aurantium* Linn, family Rutaceae. It contains not less than 2.5 per cent of volatile oil.

Geographical Source

It is indigenous to India, but is cultivated commercially in South Spain (Seville), Caribbean islands, USA, Morocco and Sicily. In India, it is grown in Maharashtra, Assam, Kurg, Karnataka, Andhra Pradesh, Punjab, West Bengal and Kashmir. In India, about 1,35,000 hectare of land is under cultivation of organge. Amaravati and Nagpur districts are leading in cultivation.

Cultivation and Collection

Orange trees are evergreen plants of 7 to 8 m in height. In India, it is specifically cultivated in the warmer and moist regions like Khasia Hills in Assam and Central India around Nagpur. Temparative of 15 – 37ºC, 350 per cent 600 mm of rain and relative humidity of 50 – 55 per cent are fourable for highest growth of plants. It grows in loamy soil, with adequate arrangements of irrigation. The plants are cultivated by sowing the seeds or by grafting or cutting. Generally, grafting method is adopted for cultivation. The plant starts bearing after 5 to 6 years and when the fruits are just ripe with orange spots, they are picked up with hands, taking care that oil glands are not ruptured. The juice is taken out and used in the food industries and the peels are treated to extract the oil or pectin and are dried and preserved for further use. At present about 1.5 lack hectates area is under cultivation of orange and 7.0 lakh tonnes of oranges are produced annualy in Maharashtra only.

Fig. 14.5: Orange Fruits   Entire and decorticated

Macroscopic Characters (Fig.14.5)

Colour : Peels are dark orange red

Odour : Aromatic

Taste : Aromatic and bitter

Size : The strips or the ribbons are of varying size and shape. They are about 3 - 4 mm in thickness.

The strips are brittle, with little inner white portion of the rind. The outer surface is rough due to the presence of numerous small pits of oil glands. The fracture is short.

Chemical Constituents

Bitter orange peel contains about 3.5 per cent of volatile oil (Bitter orange oil), which is an active constituent of the drug. Apart from the volatile oil, it also contains several other compounds like hesperidin, iso-hesperidin, neohesperidin, vitamin C, and pectin. The bitter substances present in the drug are glycosidal compounds known as aurantiamarin and aurantimaric acid. Volatile oil of the drug contains about 90 per cent of limonene and small percentage of aldehydes, chiefly citral and citronellal.

Characterisitcs of BITTER ORANGE OIL:

Bitter orange oil is obtained from fresh bitter peels only.

Colour : It is pale yellow liquid.

Odour : Orange flavour

Taste : Bitter in taste.

Density : 0.842 – 0.848.

Opitical rotation : +88 to _98

Solubility : Very slightly soluble in water, Soluble in 4 volumes of alcohol.

Bitter orange oil and orange oil are two different oils and are not identifical.

Chemical Test

Solution of recently expressed orange oil in dehydrated alcohol is neutral to moistened litmus paper.

Uses

Bitter orange peels are used as stomachic, aromatic and carminative. Pharmaceutically, used as a flavouring agent for several oral liquid preparations and also for the extraction of pectin.

Now-a-days, terpeneless orange oil is preferred commercially as a flavouring agent, as it has stronger flavour and is more readily soluble than natural oil. It has about 25 times stronger flavour as compared to the natural orange oil.

Substitute for Bitter Orange Peels

Two types of substitutes are reported.

(1) Peels of sweet-orange are usually mixed with bitter orange peels, which are thin, more yellow in colour and less bitter in taste. These are giving more volatile oil contents.

(2) Indian orange peels obtained from *Citrus auroutium* locally available are used frequently. These pieces are irregular in size and shape and dark in colour.

ORANGE OIL

Synonym

Oil of orange (Sweet orange oil)

Biological Source

This is the volatile oil expressed from fresh peel of ripe fruit of the orange (Citrus aurantium var: sinensis fam Rutaceae.

Morphological Characters

It is a yellow to deep orange coloured oil with its characterstic orange odour and taste. Slightly soluble in water two volumes of 90.0% alcohol.

Density	:	0.842 to 0.846
Optical rotation	:	+ 94° to +99°
Refractive index	:	1.4723 to 1.4737

Chemical Constituents

It contains 90.0% of d-lemonine, cetral, decylaldehyide, methyl anthranilate along with terpenoid and linalool.

Uses

Therapeutically it is expectorant and as flavour and perfume.

BLACK PEPPER

Synonyms

Pepper, Piper nigrum, Maricha.

Biological Source

Pepper is the dried unripe fruit of perennial climbing vine *Piper nigrum* Linn., family Piperaceae. It contains not less than 2.5 per cent of piperine on dried basis.

Geographical Source

It is indigenous and cultivated in South India. It is also cultivated in Indonesia, Brazil, Malaysia and Sri Lanka. India ranks first in cultivation of this drug. About 50,000 tones of pepper is being produced, which is 30 per cent of total world production. Nagarjun Aushadhi vanaspati Sahakari Utpadak Company at Anjangoan – Surji, Dist. Amaraoti in Maharashtra is also cultivating Black-pepper in 50 acres of land since several years. The export of pepper oleo-resin during 1996-97 was 519 tones, costing approximately ₹ 3240 lakhs.

Cultivation and Collection

Large scale cultivation of crop is mostly confined to sub-mountanious tracks of western ghats. i.e. Kerala, Karnataka and part of Maharashtra. Approximately, 1,60,000 hectares of land is under cultivation of pepper. Well distributed annual rainfall of 250 cm is ideal for successful cultivation of drug. It can be cultivated right from sea level upto an altitude of 1200 metres. Temperature variation of 10°C to 40°C is tolerated by the plant. Pepper thrives best in virgin soil rich in humous content.

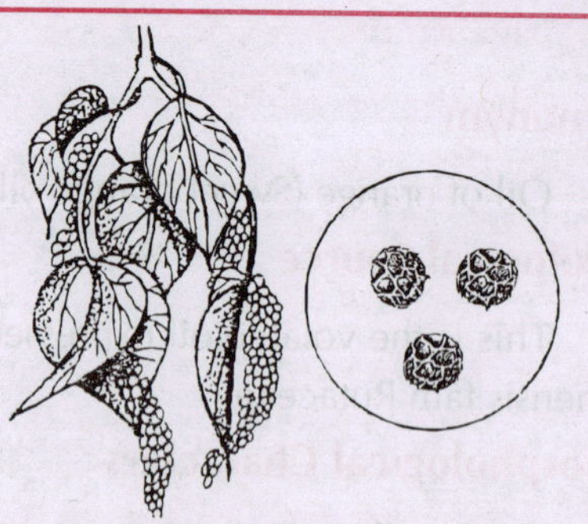

Fig. 14.6: *Piper nigrum* Vine Fruiting

Clay is best suited for the plant, but still it can be grown on red loams, sandy loams and even alluvial soil. When grown on slopy land, slopes facing south are to be avoided, as it is to be protected from severe sunlight.

Pepper is cultivated by sowing the seeds or by propagating cuttings. The plants raised from cutting start bearing from second year and survive upto 15 years. The seed-raised plant starts fruiting after 7 - 8 year and can even survive upto 60 years. The cuttings are planted in March-April by keeping the distance of 3 - 4 metres in either direction. The supports or stands for every vine are also raised at the time of planting the cuttings. The plant climbs well on silver oak, coconut arecanut, palms and *Erythrina indica*. Since it is a perennial vine and may cause depletion of soil fertility, the doses of the followings are most suitable for economical cultivation. An application of dose of about 100 g of potash, 500 g of ammonium sulphate, 1 kg of superphosphate and 10 kg of rotten cattle compost per year for each tree is desirable. Flowering occurs in May/June and harvesting of berries which are orange or red coloured in the spikes is done in February/March by hand plucking. The berries are dried in air. The drug develops wrinkles and also becomes black in colour during drying. A plantation of one hectare containing 1000 vines yields about 100 kg of dry pepper per annum.

Preparation for the market of black pepper

Black pepper: The green spikes turn to yellow or orange colour and are suitable for further processing. They are dried on pavement in sunlight.

To get the quality product spikes are subjected to special treatment for uniform colour spikes are put into boiling water for one mininute which reduces drying time and do not develop fungal growth. This type of product fetches high price in the market. Fruits should not certain more than 12.0 per cent moisture. This process causes a loss of about 70 per cent of weight.

White pepper: White variety is also prepared from black pepper. The process involves boiling of fully matured dried black-pepper fruits in water for half an hour or steaming the black pepper fruits. The outer black pericarp is removed by using pulping machine. After removing the pericarp. they are treated with bleaching powder solution for some time and drying thoroughly!

The black coloured pericarp powder is used for isolation of volatile oil, which is used in food, spices and condiments.

Macroscopic Characters

The colour of drug is blackish-brown or greyish-black. It is aromatic and pungent. The berries are 3.5 - 6 mm in diameter; globular and coarsely reticulately wrinkled with remains of stigma at apex. The pericarp is thin with a single white kernel. The kernel is hollow at the centre, entirely consisting of perisperm and a small endosperm and embryo.

Microscopic Characters

The transverse section of drug shows tubular epidermal cells followed by thin walled parenchymatous hypodermis with rectangular stone cells. The inner pericarpic layer is brown coloured and is made up of sclerenchyma. Seed coat layer is attached to it and is reddish-brown. Pericarp and perisperm contain oil glands and abundant starch grains are also present.

Chemical Constituents

Pepper contains an alkaloid piperine (5 - 9 per cent), volatile oil (1 - 2.5 per cent), pungent resin (6.0 per cent), piperidine and starch (about 30 per cent). The volatile oil which is yellowish in colour contains mainly *l*-phellandrene and caryophyllene.

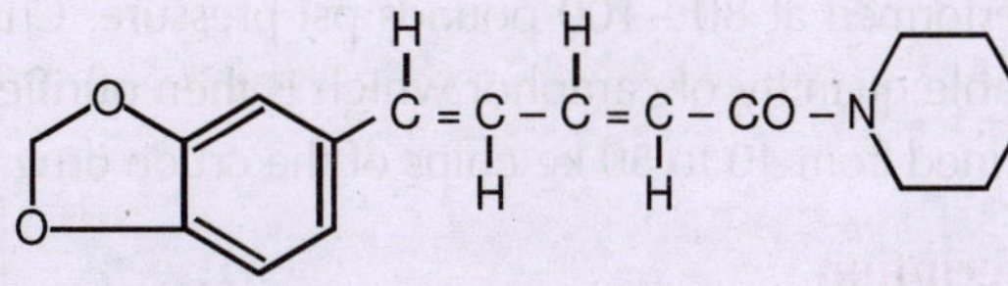

Piperine

It has specific gravity of 0.898 - 0.900, optical rotation - 3° to - 5°, and refractive index of 1.4539 - 1.4977.

Standards of Quality

Foreign organic matter	:	Not more than 2.0 per cent
Ash	:	Not more than 6.0 per cent
Acid insoluble ash	:	Not more than 1.0 per cent
Crude fibre	:	Not more than 15 per cent
Alcohol soluble extractive	:	Not less than 60 per cent
Water soluble extractive	:	Not less than 6.0 per cent

Uses

The fruits are used as aromatic, stimulant, stomachic and carminative. It causes feeling of warmth and used as condiment. It also stimulates taste-buds, with increase in gastric juice. The oil is mainly used as spice due to pungent taste. It is reported to enhance the bio-availability of certain drugs.

Substitutes

(1) *Piper attenuatum* is found in Vishakhapatnam district of Andhra Pradesh and Madurai and Tirunnelveli district of Tamil Nadu.

(2) *Piper brachystachyum* is found in Tirunelveli and Nilgiri districts of Tamil Nadu.

(3) *Piper longum* is available in Tamil Nadu, Andhra Pradesh and Kerala States. It is also known as PIPALMAL in Hindi. It is an important substitute for the official drug.

CAMPHOR OIL

Synonyms

Formosa oil of camphor, Rectified oil of camphor.

Biological Source

Camphor oil is obtained from the wood by steam distillation of *Cinnamomum camphora* (Chinese-camphor) (Nees and Eberm), belonging to family Lauraceae.

Geographical Source

Camphor trees are large, evergreen and indigenous to Eastern Asia and are abundant in Taiwan, Formosa, Japan and China.

Method of Preparation of Camphor Oil

The wood is reduced into small pieces of chips and subjected to steam destillation. Distillation is performed at 80 - 100 pounds psi pressure. Crude natural camphor oil, thus obtained contains a variable quantity of camphor which is then purified by sublimation. About 1 kg of crude camphor is obtained from 40 to 50 kg chips of the crude drug.

Description

It is colourless or yellowish liquid with camphoraceous odour. It is soluble in 3 volumes of alcohol, chloroform and ether.

Fig. 14.7: Camphor herb

Chemical Constituents

It mainly contains safrole, acetaldehyde, dipentene, camphor eugenol, d-pinene, eucalyptol, phellandrene and cineole.

Standards of Quality

Specific gravity	:	0.875 - 0.90
Refractive index	:	1.465 to 1.470
Optical rotation	:	+ 9° to + 27°

Camphor

Camphor is isolated from camphor oil by chilling or synthesised from pinene chemically.

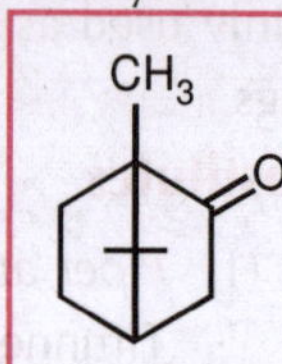

Camphor

Camphor is colourless, crystalline solid granular mass known as flower of camphor. It has penetrating characteristic odour and aromatic pungent taste followed by sensation of cold. It is translucent and readily pulverisable in presence of alcohol or chloroform. Camphor gets readily liquefied if triturated with phenol, menthol, thymol, chloral hydrate and resorcinol. It slowly vaporises at room temperature. It burns with a bright smoky flame. Natural camphor is dextrorotatory whereas synthetic camphor is racemic and has sp. gr. 0.990; and m.p. 178-179°C.

Chemical Test for Camphor

A drop of solution of vanillin (1: 100) and sulphuric acid when added to powdered natural camphor, produces immediately a yellow colour, changing to red, violet, and finally blue.

Synthetic camphor does not respond to this chemical test.

Uses

Camphor oil and camphor are used as a rubefacient, counter irritant, antifungal and as flavour in soaps and tooth powder and cosmetic products. It is used for isolation of camphor. The oil should be stored in well closed container away from light and in cool conditions.

CARAWAY

Synonyms

Carum, Caraway seed

Biological Source

Caraway consists of the dried ripe fruits of *Carm carvi* Linn., family Umbelliferae. It contains not less than 3.5 per cent (v/w) of volatile oil.

Geographical Source

It is indigenous to Holland and Central Europe. It is cultivated in Central Asia, Europe and many other countries. In India, it is found wild in north Himalayan region, and is cultivated in Kashmir, Kumaon, Garhwal and Chamba, at an altitude of 3000 to 4000 metres.

Cultivation and Collection

It is grown in dry and temperate climate. It needs humus soil. The total height of the plant is not more than a metre. For cultivation, the dried seeds are sown by broadcasting method in rows. The fruits are collected before ripening. The entire plants are dried and fruits are thrashed out. The yield is about 0.5 tonne per hectare.

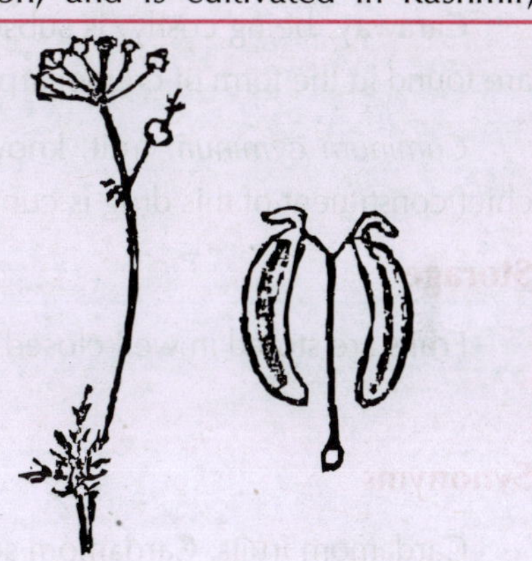

Fig. 14.8: Caraway Flower-head and Fruit

Macroscopic Characters

It is found in the form of a mericarp (Fig. 14.8).

Colour : Yellowish-brown

Odour : Characteristic and aromatic

Taste : Characteristic and aromatic

Size : 7 × 2 mm

Shape : The mericarps are elongated, 5-sided, narrow, tapering at the ends and are glabrous. It shows the presence of 5 slightly yellow primary ridges. Mericarps are usually separate and free from pedicel and carpophores. It is an orthospermous fruit.

Chemical Constituents

Caraway contains 2.5 to 8 per cent volatile oil, about 10 per cent of fixed oil, 15 per cent of proteins and resin. Volatile oil contains 45 to 65 per cent of carvone, limonene, dihydro-carvone and traces of carvacrol.

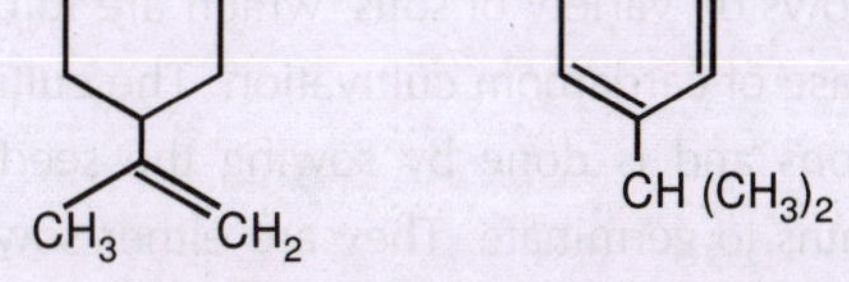

Carvone Carvacrol

Standards of Quality

Foreign organic matters	:	Not more than 2.0 per cent
Acid insoluble ash	:	Not more than 1.5 per cent
Volatile oil	:	Not less than 3.5 per cent

Uses

It is used as an aromatic, a stimulant, and carminative. It is used as a spice for culinary purposes and also for flavouring the bread, biscuits, cheese, and cakes.

Substitutes

Caraway, being costly, is substituted by Indian dill fruits. The fruits are yellowish in colour and are found in the form of cremocarps. The fruits are 4 to 6 × 2 mm in size.

Cuminum cyminum fruit, known as jeera in the market, is another substitute of the drug. The chief constituent of this drug is cuminic aldehyde and not carvone.

Storage

Fruits are stored in well closed containers in a cool place.

CARDAMOM

Synonyms

Cardamom fruits, Cardamom seeds, Cardamon, Small cardamom.

Biological Source

Cardamom consists of the dried ripe fruits of *Elettaria cardamomum* Maton var, *minuscula* Burkill, family Zingiberaceae, the seeds of which are removed as and when required for use. The seeds should contain not less than 4 per cent of volatile oil actually seeds constitute the drug.

Geographical Source

It occurs wild in Sri Lanka (partly in Ratnapura and Lunugala districts) and also in Myanmar and Malaysia. In India it is cultivated scientifically in Karnataka, Tamil Nadu and Kerala. In Karnataka, it is chiefly cultivated in Mysore and Kurg districts. It also grows wild on Malabar hills. Guatemala is the largest producer of cardamom in the world.

Cultivation of Cardamom

For cultivation, cardamom needs about 150 to 600 cm of rainfall. The altitude, which suits the drug favourably is about 600 - 1600 m and the temperature range required is between 50° - 100° F. It grows on variety of soils, which are full of moisture and also well-drained. Mulching is necessary in case of cardamom cultivation. The cultivation of the drug varies to a very great extent in different regions and is done by sowing the seeds. The seeds of cardamom are very hard and take four months to germinate. They are either sown by broadcasting method or selected seeds are sown in prepared seed-beds. Cardamom-71 and cardamom-81 are the varieties giving highest yield per hectare. The seedlings are transplanted into nursery beds, and finally in the fields. The germination of the seeds is poor (30 - 40 per cent). Fresh seeds should always be used for cultivation. The seed-beds of 1 × 10 m are prepared on the raised grounds. The seeds are sown and covered with soft earth. Sowing is done from August to October and it takes 7 - 8 weeks for germination. In Tamil Nadu, the cultivation is done in the months of February and March. About 3 - 4 months old seedlings are transplanted into nursery bed by keeping the distance of 15 - 40 cm in between. The seedlings, being delicate, are provided with sheds in the beginning. When they are about 130 cm in height or bear about 8 - 10 leaves and are 1 - 1.5 year old, they are transplanted in the forest fields just before the setting in of monsoon.

Collection of Cardamom Fruits

The fruits loose about 70 - 80 per cent of their weight during drying. The fruits are hand-rubbed and sieved to remove foreign matter. The average life of cardamom plant is about 20 - 25 years and yield per hectare in the beginning is about 25 - 50 kg and may reach 50 - 70 kg in successive years.

The cardamom plants start bearing from third year of their planting. The flowering commences in May and continues upto August. The fruits are picked up just before their ripening, so as to prevent the capsules from splitting on the floors. The collected capsules are dried after harvesting either in sun or in artificial dryers. During drying, the fruits shrink and the shrinked appearance persists in the final product. The artificial drying takes about 48 hours, while the natural drying in sun requires more time. In sun-drying, the bleaching of the fruits takes place to some extent, while in the artificial drying the green colour of the capsule is maintained. The sun-dried fruits are preferred, because the seeds remain intact and also the characteristic sweetish aroma can be maintained.

Preparation for Market

Many a time, the fruits are subjected to the treatment of bleaching with sulphur dioxide to improve the colour. Therefore, they are bagged, stored and transported as needed. The Cardamom Act of 1965 was passed by the Government to encourage the cultivation of drug in Karnataka.

Macroscopic Characters [Fig. 14.9 (a)]

Colour	:	The natural colour of cardamom without processing is green or pale buff. If treated with sulphur dioxide, it changes to white.
Odour	:	Aromatic, agreeable and pleasant.
Taste	:	Strongly aromatic
Size	:	The cardamom fruits, i.e. capsules are about 2 cm in length.
Shape	:	They are ovoid or oblong, plump, 3-sided, sharply beaked at the top, with smooth or longitudinally striated surface.

Each capsule of the fruit contains three chambers. Each chamber consists of two rows of seeds, about 5 - 10 in number. The seeds are 3 - 4 mm in size, very hard to touch, reddish-brown in colour, and are covered by membranous arid. Seeds are derived from anatropous ovules. They are irregularly triangular in shape, covered with transverse wrinkles, known as rugae. The seeds are strongly aromatic in taste and colour. The proportion of pericarp to seeds is 1: 3 w/w.

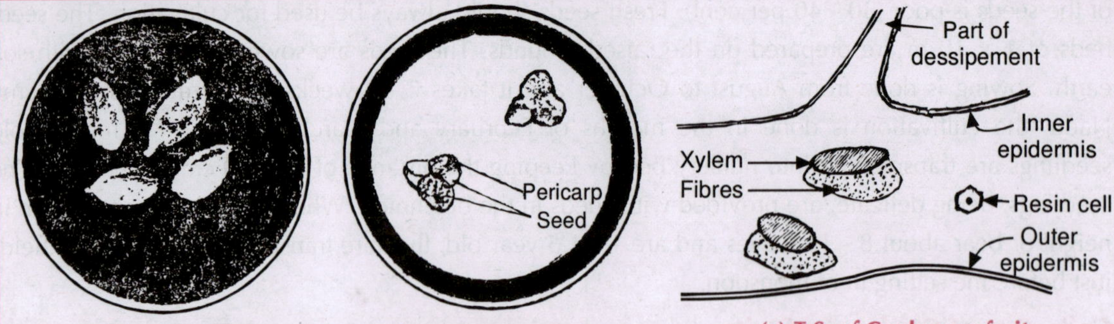

(a) Cardamom Fruits

(b) Cardamom Fruit (Cut transversely)

(c) T.S. of Cardamom fruit

Fig. 14.9: Cardamom fruit

Microscopic Characters

The outer as well as inner epidermis of the pericarp are made up of polygonal tubular cells. Mesocarp is parenchymatous in nature. Mesocarp includes few brown to yellow coloured resinous cells (scattered throughout the mesocarp), alongwith partially lignified fibrovascular bundles.

Transverse section of the seed shows the presence of outermost layer of arillus, followed by two layers of testa which contain highly aromatic volatile oil.

Perisperm, endosperm and embryo characteristic to the seed are also observed in the section. Calcium oxalate crystals and starch grains are present in the perisperm.

Chemical Constituents

Cardamom seeds contain volatile oil to the extent of 2 per cent to 8 per cent. The active constituent of the volatile oil is cineole. Other aromatic compounds present are terpinyl acetate, terpineol, borneol, terpinene, etc. The other constituents of the cardamom seeds are fixed oil, starch and proteins.

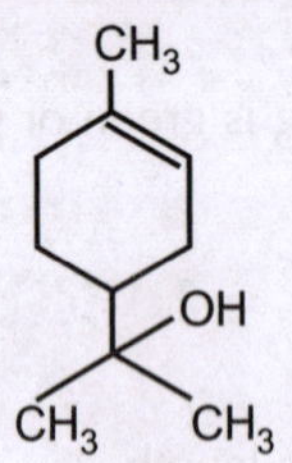

α-terpineol

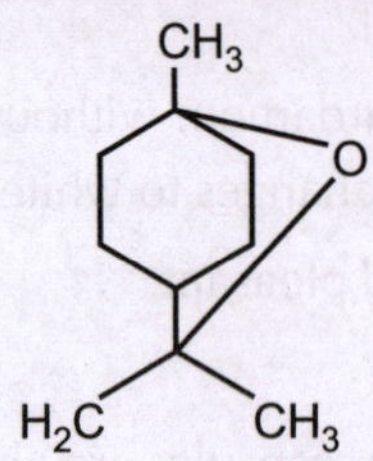

Cineole

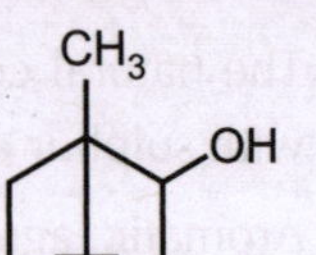

Bomeol

CARDAMOM OIL

Volatile oil of cardamom is obtained by steam distillation from caradamom seed and yield is 6.5 – 10.0 per cent. Oil is a complex mixture of terpenes mainly containg α termpenyl acetate 25 – 42 per cent 1,8 cineole 20-40%, linalyl acetate, limonene, linalool.

Chemical Constituents

Colour : It is colourless, transparent, mobile liquid with.

Odour : Warm and spicy

Taste : Spicy, comphoraccous, lemony.

Soluble in diethyl phthalate, Benzyl alcohol, fixed oils, mineral oil but insoluble in glycerin and propylene glyceol.

It has folloiwng characteristics

Specific Gravity : 0.917 – 0.947

Optical rotaiton : +22° to + 44°

Refractive index : 1.462 to 1.466

Standards of Quality

Total ash	:	Not more than 9.0 %
Acid insoluble ash	:	Not more than 3.0 %
ASE	:	Not less than 05.0
WSE	:	Not less than 10.0%
LOD	:	Not more than 15.0%

Uses

Cardamom is used as an aromatic, carminative and stimulant. It is also a good flavouring agent. It is used in the form of compound tincture. It is also a flavour for confectionary, tea, coffee, curry powder, Garam masala. As perfume for soaps, lotions

Storage

Cardamom fruits should be stored in well-closed, well filled container in cool places. As and when needed, the seeds should be removed from the fruits and used. It has been observed that if the seeds are stored after dehusking, about 30 per cent volatile oil is lost within 18 months of storage.

Substitutes and Adulterants

Cardamom fruits are often adulterated with orange seeds and unroasted coffee grains. The adulterants of the seeds are even small pebbles and other aromatic seeds.

Very common adulterants of cardamom are mentioned below:

Fig. 14.10: Long Wild Cardamoms

1. **Long wild native cardamom:** The long native cardamom **(big cardamom)** is substitute to the genuine drug. It is obtained from *Elettaria cardamom Var. major* widely grown in Shri Lanka. The fruits of this drug are about 4 cm in length, 1 cm in diameter, dark brown in colour and coarsely striated. The seeds of this drug are less aromatic, bitter in taste and with inferior aroma. The oil of the seeds has different composition.

2. **Korarima Cardamom:** These are 4 × 2 cm in size, red-brown in colour. Seeds do not show the presence of rugae, which is a characteristic of the genuine drug.

3. The loose seeds of cardamom or fully ripe seeds are also mixed. Due to the long exposure or due to ripening, they contain less volatile oil as compared to authentic drug.

4. Cardamom husk is also used to adulterate the genuine drug and can be distinguished by its ash content. Genuine drug should not have more than 6 per cent of total ash and 3.5 per cent acid-insoluble ash.

Varieties

Depending upon the geographical sources, four species are available and these species can be distinguished from one another by the morphological characters. Under the histological characters, special reference is available regarding number of sclerenchymatous cells present per sq. mm. which is supposed to be constant for a species, details of which are given below.

Table 14.5: Characteristics of Cardamom Varieties

Variety	Size, mm	Description	Sclerenchymatous cells/sq. m
Alleppy	8 - 20 4 - 10	striated longitudinally shrunken, green or pale buff	3790
Mysore	4 - 10	smooth, stout, bleached or cream coloured	3310
Malabar	4 - 10	shorter, ovoid less smooth, striated longitudinally	4610
Mangalore	larger than Alleppy type	globular, transverse wrinkles rugae 6 to 8 for all species	4960

CASSIA CINNAMON

Synonyms

Cassia bark, Chinese cinnamon

Biological Source

This is dried stem bark of the plant *Cinnamomum cassia* Blume, belonging to family Lauraceae. It contains not less than 1.0 per cent of volatile oil.

Geographical Source

The plant producing cassia bark is native of China. It is also found in Myanmar and Sri Lanka. At present, it is cultivated in southeastern province of China.

Macroscopic Characters

Colour	:	Dirty brown (both surfaces)
Odour	:	Characteristic, sweet, aromatic
Taste	:	Slightly sweet, aromatic
Size	:	5 - 40 × 12 - 18 × 1 - 3 mm
Shape	:	Single quills
Fracture	:	Short granular on outer part, occasional greyish patches of cork

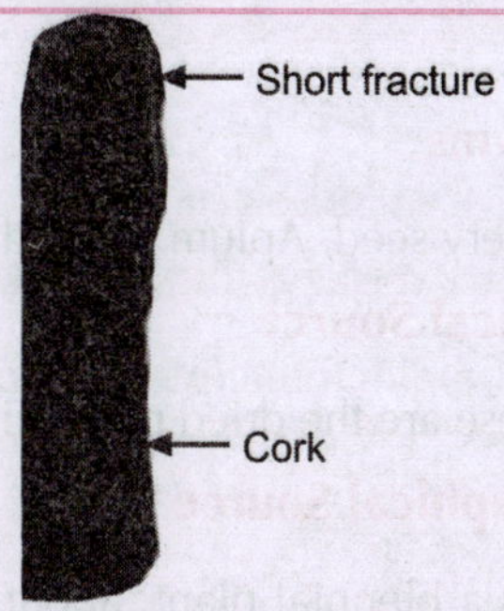

Fig. 14.11: Cassia Bark (Quill)

Microscopic Characters

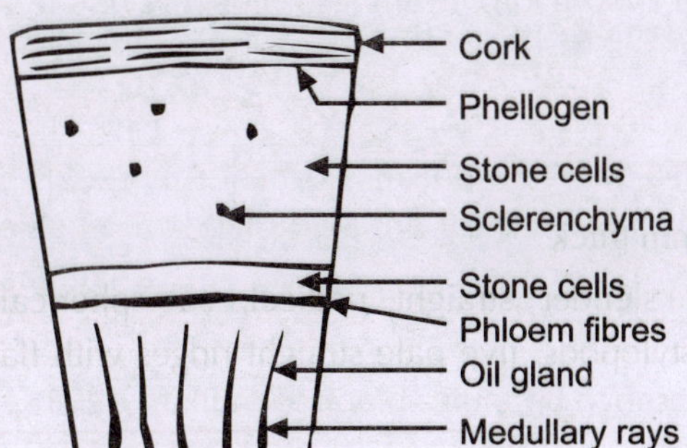

Fig. 14.12: T.S. of Cassia Bark

The cork consists of thick walled and thin walled alternate layers of polygonal tubular- cells. Few cells of cork are lignified. Phellogen is followed by scattered sclerides in cortical tissues. Pericycle is made up of continuous band of stone cells. Lignified pericyclic fibres are embedded in the sheath of stone cells. Oil glands, mucilage cells and bast-fibres are located in the phloem parenchyma. Medullary rays are multiseriate in pericyclic region, while very narrow towards centre. Abundant starch grains and acicular crystals of calcium oxalate are also present.

Chemical Constituents

Cassia bark contains 1 - 2 per cent of volatile oil and is known as cassia oil. It also contains mucilage, starch, calcium oxalate and tannins. Cassia oil contains not less than 85 per cent of cinnamic aldehyde and very small amount of eugenol.

It also contains cinnamyl acetate. The volatile oil is brown in colour, with sp. gr. 1.045 - 1.063; optical rotation – 1° to + 1° and refractive index 1.6020 - 1.6060. Cassia bark also contain coumarin not found in Cinnamon bark.

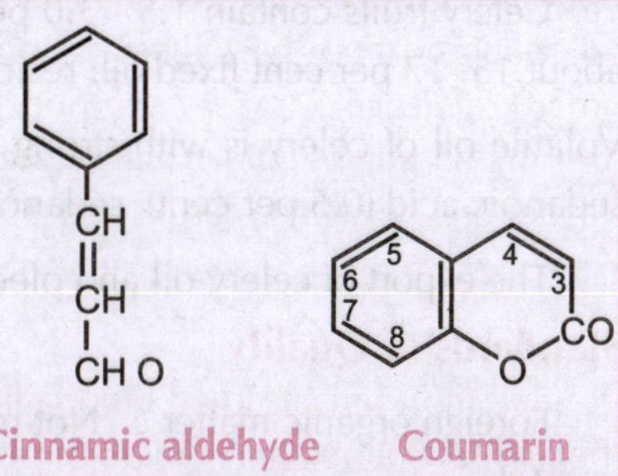

Cinnamic aldehyde Coumarin

Standards of Quality

1.	F.O.M.	:	Not more than 1.0 per cent
2.	Total ash	:	Not more than 5.0 per cent
3.	Acid insoluble ash	:	Not more than 2.0 per cent
4.	Sulfated ash	:	Not more than 6.0 per cent

Uses

The drug is used as a carminative, stimulant, flavouring agent and an aromatic. In India, it is also used in spices.

CELERY

Synonyms

Celery seed, Apium, Ajmoda.

Biological Source

These are the dried ripe fruits of the cultivated plants of *Apium graveolens* Linn, Umbelliferae.

Geographical Source

It is a biennial plant, about 25 - 250 cm in height, found wild or cultivated. It is wild in North Western Himalayas, hills of Punjab and in North Africa. It is also found as garden plant. It is cultivated on large scale in Europe, America and India. In India, it grows abundantly in Punjab. The yield per hectare is about 750 - 850 kg.

Macroscopic Characters (Fig. 14.13)

Colour	:	Fruits are dark brown in colour
Odour	:	Characteristic and aromatic
Taste	:	Aromatic and camphoraceous
Size	:	1 - 1.5 mm long, 1.5 mm wide and 0.5 mm thick
Shape	:	Found as mericarps, crescent-shaped, slender straight pedicel, sub-spherical, laterally compressed having two small stylopods, five pale straight ridges with flat commissural surface.

Microscopic Characters

Polyhedral epidermal cells with slightly wavy anticlinal walls. Outer walls are radially striated and frequently papillose. Endocarpous tissue is narrow, thin walled, lignified and tangentially arranged. Mesocarp consists of 6 to 9 oval vittae, of which 2 vittae are on commissural surface and 1 to 3 in each groove of the dorsal surface. Endosperm is dense, composed of thick walled polyhedral cells, containing aleurone grains, fixed oil and rosette crystals of calcium oxalate.

Fig. 14.13: Celery herb

Chemical Constituents

Celery fruits contain 1.5 - 3.0 per cent pale yellow volatile oil, about 15 -17 per cent fixed oil, resins and bitter extractives.

Volatile oil of celery is with strong aromatic flavour. It contains d-limonene (60 per cent), sedanoic acid (0.5 per cent), sedanolide (2.5 to 3.0 per cent) and d-selinine (10 per cent).

The export of celery oil and oleo-resin together was approximately ₹ 510 lakhs during 1996-97.

Standards of Quality

Foreign organic matter	:	Not more than 1.0 per cent
Ash	:	Not more than 10.0 per cent
Acid-insoluble ash	:	Not more than 2.0 per cent

Volatile oil	:	Not less than 1.5 per cent
Specific gravity	:	0.870 to 0.895
Optical rotation	:	+ 67° to + 79°

Uses

Celery fruits are used as stimulant, carminative, nervine sedative and tonic. These are also used in the treatment of rheumatism and as a flavour for soft drinks and unpleasant medicaments.

Storage

The fruits should be stored in cool dry place.

CHENOPODIUM OIL

Synonyms

American Worm seed oil, Chenopodiol, Chenoposan.

Biological Source

Chinopodium oil is a volatile oil obtained by steam distillation from the fresh flowering and fruiting plants (except roots) of *Chenopodium ambrosioides* var *anthelminticum* Linn., family Chenopodiaceae. It contains not less than 65 per cent of Ascaridole.

Geographical Source

The plant chinopodium is very common in Eastern and Central U.S.A. It is indigenous to Caribbean islands. It is extensively cultivated in Europe, Maryland, Mexico and Canada. In India, the other species are found in West Bengal, Tamil Nadu, Assam and in Karnataka. The plants are cultivated in Eastern America.

Description

Colour	:	Colourless to light yellow coloured liquid.
Odour	:	Unpleasant and characteristic.
Taste	:	Bitter and burning.
Solubility	:	It is soluble in 8 volumes of alcohol (70 per cent) and insoluble in water.

Chemical Constituents

The herb contains 0.5 to 1.0 per cent volatile oil, while fruits contain 1 to 4 per cent of oil. The oil contains chiefly ascaridole which is an unsaturated terpene peroxide (70 to 80 per cent), about 20 per cent p-cymene, myrcene, 1-limonene and camphor.

Fig. 14.14: Chinopodium herb

Chemical Test

Heat 1 ml chinopodium oil in a test tube with a piece of porcelein. A deep golden yellow liquid is produced (Caution: Perform the test very carefully as the oil explodes when heated).

Standards of Quality

Weight per ml : 0.955 to 0.975 g

Refractive index : 1.474 to 1.480

Optical rotation : – 3° to – 8°

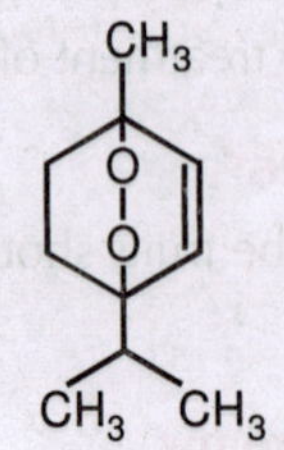

Ascaridole

Uses

Chinopodium oil is a very valuable anthelmintic, specifically for round-worms (*Ascaris*). It is also used to expel hook-worms and dwarf tape-worms. It has no action on large tape-worms. Chenopodium oil is used against intestinal amoebae and in veterinary practice. It is contraindicated in pregnancy and in patients with impaired kidney or liver.

Substitutes

In India, the plant chinopodium is substituted by *Chinopodium ambrosoides* Linn. and *Chinopodium album* Linn. found in Uttar Pradesh, Tamil Nadu, Karnataka and West Bengal.

Storage

Chinopodium oil is stored in closed containers, in cool places away from light.

CINNAMON

Synonyms

Cinnamon bark; Kalmi-Dalchini, Ceylon cinnamon

Biological Source

Cinnamon consists of the dried inner bark of the shoots of coppiced trees of *Cinnamomum zeylanicum* Nees., (Syn. *Cinnamomum verum* J. S. Presl) belonging to family Lauraceae. It should not contain less than 1.0 per cent of volatile oil.

Geographical Source

Cinnamon, the evergreen tree of tropical area, is considered to be native of Sri Lanka and Malabar coast of India. It is also found in Jamaica and Brazil. However, most of the world requirements are met by Sri Lanka and hence true cinnamon is known as Sri Lanka cinnamon.

Cultivation, Collection and Preparation for the Market

Eventhough, cinnamon can be propagated by planting the cuttings and layers commercially, it is generally propagated by seed, as it is the easiest method to adopt. Cinnamon is a crop of tropical countries. It needs sandy or siliconous soils with an admixture of humus. The altitude at which it grows favourably is 800 to 1000 metres. Sheltered situation with an annual rainfall of 200 - 250 cm is ideal for cultivation of cinnamon. The seeds are sown in well prepared nursery beds located at suitable places in June and July. The seeds are sown at a distance of 10 cm and covered with small layer of soil and watered properly. It takes approximately 20 days for germination of seeds.

Seedlings are provided with the shades and are allowed to grow for about 10 - 12 months. Transplantation is done in Oct./Nov. or in rainy season by keeping the distance of 2 meters in between two plants. Shades are provided with the pendals of coconut leaves. The field is weeded 2 - 4 times in a year. For healthy growth, each plant should be manured with 100 g each of ammonium sulphate and superphosphate in the first year. Subsequently, it is increased, depending upon the age of the plant. Fertilizers are applied in two equal doses, first in monsoon and second in October-November to encourage the growth of side shoots.

Fig. 14.15: Young cinnamon plant

Trees are coppiced to induce the formation of shoots. The trees are allowed to grow further unless they turn to uniform brown by formation of cork. Harvesting is done in rainy season when it is easy to peel off the bark from shoots. Longitudinal incisions are made on the shoots and transverse markings are given so as to form the rings at the nodes, which also connect the longitudinal incisions producing the strips, which are then peeled off. Strips, thus formed, are made into bundles, wrapped in coir matting and allowed to ferment for 24 hours. Fermentation results in losening of the outer cork and cortex, which is then removed by scrapping with curved brass knives. During drying, the bark contracts and gets converted into quill. The smaller quills are inserted into larger quills to form compound quills. Quills are arranged end to end, upto the length of 90 cm when soft and fresh quills are rolled by hand and slightly pressed so as to avoid swelling and splitting into pieces. The drug is dried in shade over the mats. The quills are collected, packed into bundles of different grades and sold. Small pieces and debris produced during the handlings of the quills are known as quilling and feathering. These are also used for the manufacture of cinnamon oil.

One hectare plantation of cinnamon, on an average, produces 200 - 300 kg bark and 2 - 3 kg of leaf oil per year.

Macroscopic Characters (Fig.14.16)

Colour	:	The outer surface is dull yellowish-brown, while the inner surface is dark yellowish-brown.
Odour	:	Fragrant
Shape	:	Found in the form of compound quills.
Size	:	About 1 m in length and 1 cm in diameter. The thickness of the bark is approximately 0.5 mm.
Taste	:	Aromatic and sweet followed by warm sensation.
Fracture	:	Splintery

The outer surface of the bark is marked by wavy longitudinal striations with small holes of scars left by the branches. The inner surfaces also shows the longitudinal striations. Bark is free of cork.

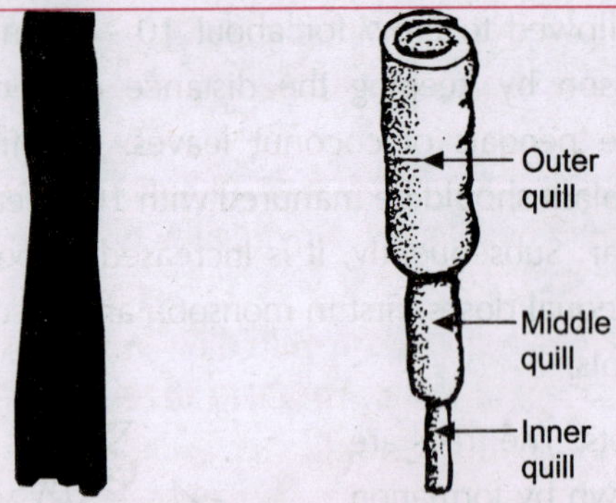

(a) Single quill (b) Compound Quill of Cinnamon

Fig. 14.16: Cinnamon bark

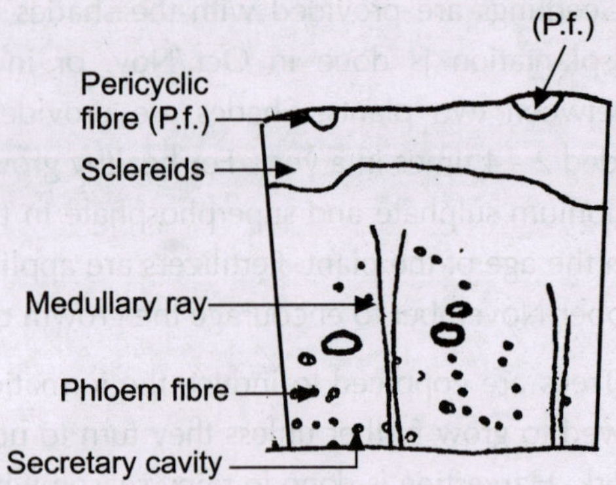

Fig. 14.17: T.S. of Cinnamon bark

Microscopic Characters

Being an inner bark, the cork and primary cortex are absent. Rarely, patches of primary cortex may be present. Sclerenchymatous pericycle is prominent. The stelar part shows phloem, phloem fibres, biseriate medullary rays and secretory cavities containing volatile oil and mucilage. Starch grains in cortical parenchyma and medullary rays and calcium oxalate crystals in parenchymatous cells are also present.

Chemical Constituents

Cinnamon bark contains about 0.5 - 1.0 per cent of volatile oil, 1.2 per cent of tannins (phlobatannins), mucilage, calcium oxalate, starch and a sweet substance known as mannitol.

The volatile oil is the active constituent of the drug. It is light yellow (when freshly distilled) in colour and changes to red on storage. Bark yields 14 - 16 per cent of 90.0 per cent alcohol-soluble extractive.

Cinnamon oil contains 60 - 70 per cent of cinnamaldehyde, 5 - 10 per cent eugenol, benzaldehyde, cuminaldehyde and other terpenes like phellandrene, pinene, cymene, caryophyllene, etc.

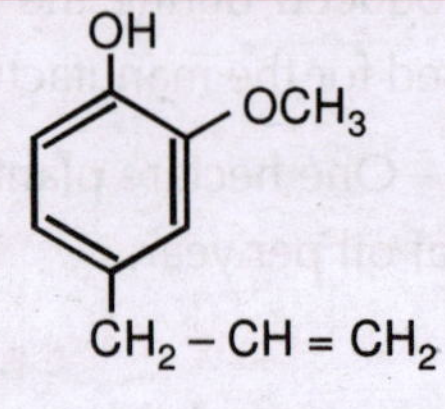

Eugenol

Colour	:	Cinnamon oil is yellow to red in colour
Sp. gr.	:	1.00 - 1.030;
Optical rotation	:	0 to –2;
Refractive index	:	1.562 - 1.582˚.

Standards of Quality

1.	Foreign organic matter	:	Not more than 2.0 per cent
2.	Total ash	:	Not more than 6.0 per cent
3.	Acid insoluble ash	:	Not more than 4.0 per cent
4.	Sulfated ash	:	Not more than 6.0 per cent

Chemical Test

On addition of a drop of ferric chloride solution to a drop of volatile oil, produces a pale green colour. With ferric chloride, cinnamic aldehyde gives brown colour and eugenol gives blue colour, resulting in the formation of pale green colour.

In cassia oil, brown colour is obtained, as it contains only cinnamic aldehyde.

Uses

Bark is used as a carminative, stomachic and mild astringent. It is also used as a flavouring agent, stimulant, an aromatic, and antiseptic. Commercially, it is used as a spice and condiment, and also in the preparation of candy, dentrifices and perfumes.

Substitutes and Adulterants

Jungle cinnamon: It is the bark obtained from wild growing trees, which is dark in colour, less aromatic than the cultivated trees, and slightly bitter.

Cinnamon chips: These are pieces of untrimmed bark. They can be distinguished from genuine drug by the presence of abundant cork cells and by poor yield to 90 per cent alcohol.

Cultivated Saigon cinnamon: It consists of the bark of the trees of *Cinnamomum loureirii* (Lauraceae). It is exported from the port of Saigon. It is in China and Japan. The bark is greyish-brown in colour with light patches and sweet taste. Quills are 30 × 4 × 0.7 cm, unpeeled and contain 2.5 per cent of volatile oil.

Java Cinnamon: It is derived from *Cinnamomum burmanii* (Lauraceae). Bark is less aromatic, peeled and found in the form of double quills. Histologically, medullary rays contain small tubular crystals of calcium oxalate, not found in *C. zeylanicum*. It contains about 75 per cent of cinnamaldehyde in the oil. It also gives poor yield to 90 per cent alcohol as compared to Sri Lanka cinnamon.

Storage: Keep it in well closed containers in cool place, away from light.

CITRONELLA OIL

Biological Source

It is obtained by steam distillation from the fresh leaves of *Cymbopogon nardus* (Linn.) Rendle, family Graminae.

Geographical Source

It is probably indigenous to Sri Lanka and cultivated in Myanmar, Malaysia, Indonesia, Fiji and Sri Lanka. In India, it is cultivated in Kerala.

Cultivation and Collection of Citronella grass

It is tall perennial, throwing dense fascicles of leaves from a short rhizome. The plant is quite stout, erect, about 1.75 metres in height. The leaves are linear and tapering, upto 60 - 70 cm long, glabrous green and lower ribs are red in colour. Flowers are atheate panicles, 30 - 50 cm long and nodding. The grass has pleasant aromatic smell. It is cultivated by vegetative propagation from slips or even by sowing seeds. It needs to be irrigated regularly during winter and summer. The crop is ready for harvesting after eight months of growth. It can be harvested several times with regular intervals. About 20 tonnes of grass can be obtained in a hectare. The grass contains 0.4 to 0.5 per cent of volatile oil, with an average yield of 90 kg of oil per hectare.

Fig. 14.18: Flowering Twig of *Cymbopogen nardus* Linn.

Description

The oil is pale-greenish yellow in colour with pungent taste. It is insoluble in water, soluble in 80 per cent alcohol (1:10) and also in fixed oils. Two commercial varieties of oil exhibit following constants.

Chemical Constituents

Ceylon citronella grass contains 0.3 - 0.9 per cent of strongly aromatic and pleasantly flavoured volatile oil. It mainly contains geraniol (40 - 60 per cent) and citronellal (5 - 20 per cent). Other constituents of the oil are d-camphene, limonene, dipentene, borneol, linalool, and methyl eugenol. Recently, it is found to contain cadinene and elmicin.

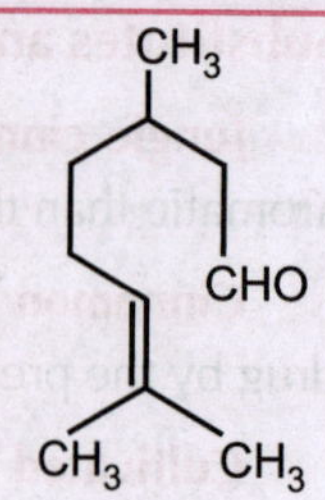

Citronallal

Java variety contains citronellal (25 - 50 per cent), geraniol (25 - 40 per cent), camphene, limonene and dipentene.

Standards of Quality

	Sri Lanka variety	Java variety
Specific gravity	0.898 - 0.920	0.895 - 0.900
Optical rotation	– 7° to – 14°	– 2° to –5°
Refractive index	1.4784 - 1.485	1.468 - 1.473
Ester value	175	250

Uses

It is used as perfume for soaps, brilliantine, in mosquito repellent creams and sprays and as a flavouring agent for liniments and lotions.

It should be stored in well closed container, protected from light and in a cool place.

Exports of citronella and its concentrates to the extent of 8.87 × 9.11 million US dollars are on record during 2017 – 2018 and 2018 - 2019 respectively.

CORIANDER

Synonyms

Coriander fruits, Cilanthro leaves.

Biological Source

These are the fully dried ripe fruits of the plant known as *Coriandrum sativum* Linn. (Family - Umbelliferae). The fruits should contain not less than 0.3 per cent of the volatile oil.

Geographical Source

Plant is cultivated throughout European countries, principally in Russia, Hungary and Holland. It is also cultivated in India, Egypt and Morocco. In India, it is widely cultivated in Andhra Pradesh (Guntur, Anantpur), Maharashtra (Jalgaon and Satara), West Bengal (Howrah and 24-Paragana districts), Uttar Pradesh, Rajasthan and Jammu and Kashmir.

Cultivation and Collection

Coriander is cultivated as kharif, as well as, rabi crop. It needs light to heavy black soil. About 15 to 20 kg of fruits per hectare are required for cultivation. It is sown by dibbling method. It is rotated with wheat, grain, jowar and onion and also grown with cotton, sugarcane and brinjal as mixed crop. Crop is ready for harvesting after 100 days of growth. J-16, J-214, K-45 and New Pusa are few of the improved varieties of coriander. About 2 lakhs hectares of land is under cultivation in India. India produces about 70 per cent of global requirement of coriander. Rajasthan alone accounts for 70.0 per cent of production in India.

Macroscopic Characters

Colour : Yellowish - brown to brown

Odour : Aromatic

Taste : Spicy and characteristic

Size : Fruits are 2 - 4 mm in diameter and 4 - 30 mm in length

Shape : Coriander is a sub-globular cremocarpous fruit.

About 10 primary ridges and 8 secondary ridges are present.

Primary ridges are wavy and inconspicuous, while secondary ridges are straight. It is further described as an endospermic and a coelospermic fruit. The weight of 100 fruits is approximately 1 g.

Microscopic Characters

The epidermis of the pericarp is made up of polygonal tubular cells with stomata. Several epidermal cells contain prisms of calcium oxalate. Mesocarp consist of inner and outer layer of parenchyma with a layer of sclerenchyma in between them. Inner epidermis of pericarp consists of parquetry cells.

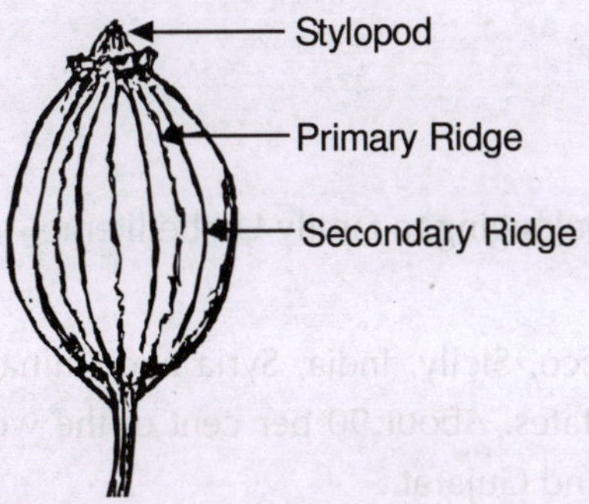

Fig. 14.19 (a): Coriander Fruit

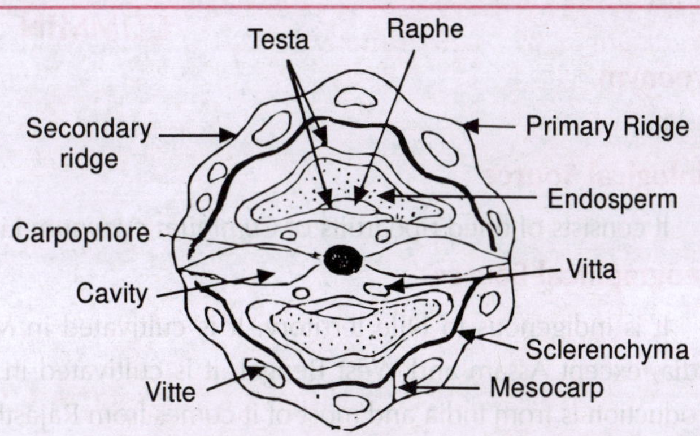

14.19 (b): T.S. of Coriander fruit (Cremocarp)

The seed is characteristic to umbelliferous fruits. Starch grains, trichomes and lignified reticulate parenchyma absent. Fixed oil globules are present in the endosperm, while volatile oil in the vittae. Aleurone grains are present in polygonal thick walled cellulose parenchyma of endosperm.

Chemical Constituents

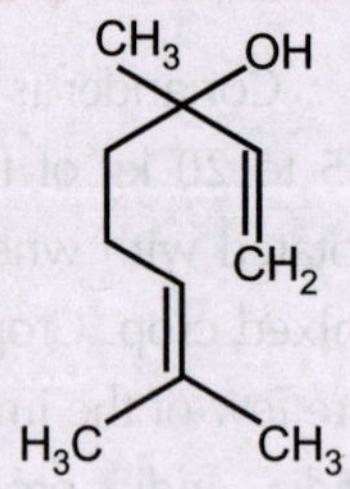
Coriandrol

Coriander yields from 0.3 - 1 per cent of volatile oil. The fixed oil (13 per cent) and proteins (20 per cent) are the other contents of the drug. Volatile oil of the drug contains 90 per cent of D-linalool (coriandrol) and coriandryl acetate, and small quantities of L-borneol, geraniol and pinene. Coriander leaves are rich in vitamin A content. The fruit yields 5 - 7 per cent ash. During ordinary storage of crude drug, the volatile oil composition alters considerably.

Coriander oil is pale yellow liquid having specific gravity of 0.863 - 0.875, refractive index 1.462 - 1.472, and optical rotation of + 8° to + 15°.

Standards of Quality

Ash	:	Not more than 7.0 per cent
Acid insoluble ash	:	Not more than 1.5 per cent
Volatile oil	:	Not less than 0.3 per cent
Foreign orgnaic matter	:	Not more than 2.0 per cent

Uses

The fruits, as well as, volatile oil, are used as an aromatic, carminative, stimulant and flavouring agent. Coriander oil is used alongwith purgatives to prevent gripping. It is an ingredient of compound spirit of orange and cascara elixir.

Storage

Coriander fruits are thoroughly dried and stored in well-closed containers as they are highly prone to insects.

Substitutes

It is substituted by Bombay coriander fruits, which contain less volatile oil and are ellipsoidal in shape.

CUMMIN

Synonym

Jira.

Biological Source

It consists of dried ripe fruits of *Cuminum cyminum* Linn, belonging to family Umbelliferae.

Geographical Source

It is indigenous to Nile territory. It is cultivated in Morocco, Sicily, India, Syria and China. In India, except Assam and West Bengal, it is cultivated in all states. About 90 per cent of the world production is from India and most of it comes from Rajasthan and Gujarat.

Macroscopic Characters

Colour	:	Brown coloured, ridges are light in colour.
Odour	:	Characteristic and aromatic.
Taste	:	Characteristic and aromatic
Size	:	4 - 6 mm in length and about 2 mm thick

Shape : Elongated and tapering at both ends. Cremocarps generally separate. Each mericarp has fine longitudinal ridges. Alternating with these are secondary ridges which are flat and bear conspicuous emergences.

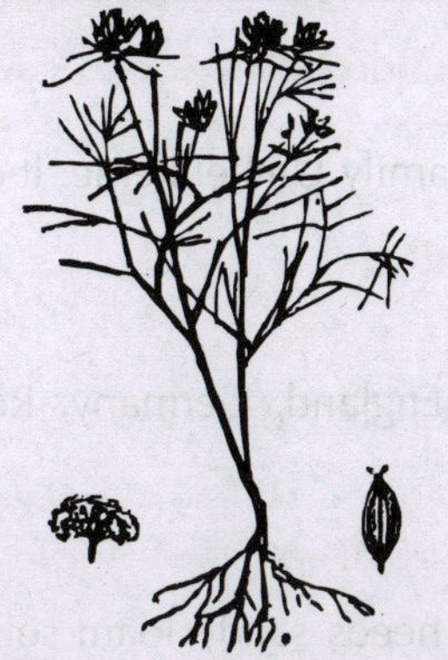

Fig. 14.20 (a): Cummin plant (*Cuminum cyminum* Linn.)

Fig. 14.20 (b): Cummin fruits

The transverse section of mericarp exhibits an oily endosperm and six vittae, of which four are on dorsal surface and two on ventral surface. The large pluriserial hairs, characteristic to it, are present.

Chemical Constituents

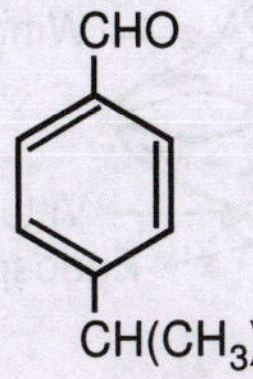

Cuminaldehyde

Cummin fruits contain 2.5 - 4 per cent volatile oil, 10 per cent fixed oil and proteins. Volatile oil mainly consists of 30 - 50 per cent cuminaldehyde, small quantities of α -pinene, β-pinene, phellandrene, cuminic alcohol, hydrated cuminaldehyde and hydro-cuminine.

Volatile oil is colourless to pale yellow, with following constants:

Specific gravity - 0.900-0.935

Optical rotation - + 4 to + 8°

Refractive index - 1.4950 - 1.5090

The export of Cummin during 2008 – 09 and 2009 – 10 was to the tune of ₹ 5440 & 5482 lakh respectively.

Standards of Quality

Foreign organic matter	:	Not more than 2.0 per cent
Ash	:	Not more than 8.0 per cent
Acid insoluble ash	:	Not more than 01.0 per cent
Alcohol soluble extractives	:	Not less than 8.0 per cent
Water soluble extractives	:	Not less than 16.0 per cent

Uses

It is used as stimulant, carminative, and also in diarrhoea and dyspepsia. India is exporting 1.5 tonn of cummin per annum which is about 2.5% of its total production in India. At present it costs Rs. 35,000/- per tonn.

DILL

Synonyms

Dill fruits, Anethum, European Dill.

Biological Source

Dill consists of dried ripe fruits of *Anethum graveolens* Linn. family Umbelliferae. It contains not less than 2.5 per cent of volatile oil.

Geographical Source

Anethum is indigenous to South Europe and cultivated in England, Germany, Romania and Mediterranean countries.

Cultivation and Collection

The cultivation of the drug is done by sowing the fruits. It needs sandy loam soil, devoid of gravels, with slightly acidic nature. It is a cold weather crop and is grown during Rabi season. Its sowing is done in October and the crop is harvested in March/April. The umbels are collected after full maturity, dried thoroughly and shaken to separate the fruits. Fruits are sieved, excess of foreign organic matter is removed and then packed into the bags.

Macroscopical Characters (Fig. 14.21)

Colour	:	The fruits are yellowish or pale brown with narrow wings.
Odour and Taste	:	Aromatic and characteristic
Size	:	The fruits are about 4 mm in length, 2 to 3 mm in width and 1 mm in thickness.
Shape	:	Broadly ovate, strongly compressed dorsally

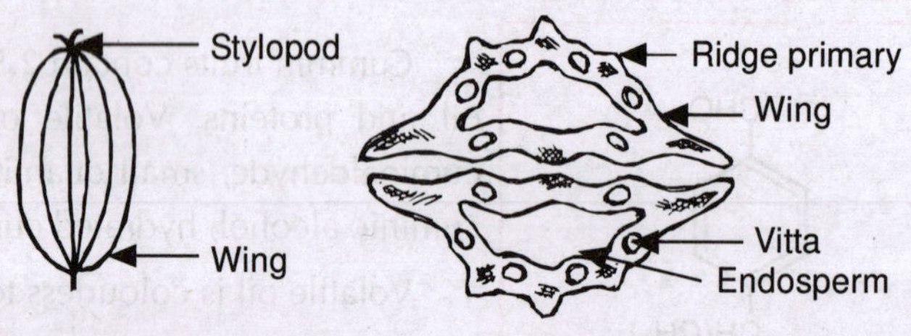

(A) Entire Fruit (B) T.S. of Cremocarp

Fig. 14.21: Dill Fruit

The fruits are found in the form of mericarps usually separated and free from pedicel. Each mericarp is yellow in colour, lateral ridges of mericarp are extended as wings. Carpophore is usually adherent to the commissural surface. Endosperm is orthospermous.

Chemical Constituents

Dill contains about 2.4 to 4 per cent of volatile oil; about 20 per cent each of fixed oil and proteins. The chief constituent of the volatile oil is aromatic liquid, known as carvone (43 to 63 per cent). It also contains dihydrocarvone, D-limonene, phellandrene.

Indian dill from *Anelhum sova* contains dill-apiole which is poisonus.

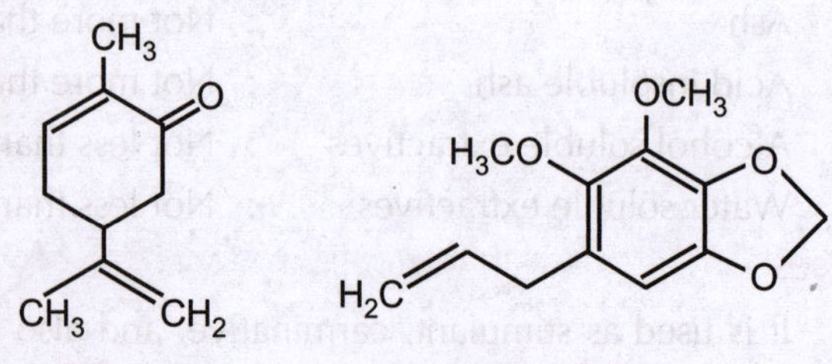

Carvone Dill-apiole

Volatile oil of dill is colourless liquid with following characteristics:

Specific gravity	:	0.900 - 0.915
Refractive Index	:	1.481 - 1.492
Optical rotation	:	+ 70° to + 80°

Standards of Quality of Dill

Total ash	:	Not more than 10.0 per cent
Acid insoluble ash	:	Not more than 3.0 per cent
F.O.M.	:	Not more than 2.0 per cent

Uses

It is used as an aromatic, a stimulant, carminative and also in gastric disturbances of infants and children. The oil of dill is used in the preparation of dill water, gripe water and also as flavouring agent. Indian dill is used in veterinary medicines.

Substitutes

It is substituted by *Indian dill* obtained from *Anethum sowa*. It contains whole cremocarps. They are narrower and less compressed as compared to European dill. India exported Dill oil worth of ₹ 27 lakhs during 1996-97. The volatile oil of Indian dill contains dill-apiole. Indian dill is generally used in the veterinary practices, since the dill-apiole, is regarded as a poisonous substance. It is cultivated in Andhra Pradesh, Punjab, Uttar Pradesh, Maharashtra and Gujarat.

Storage

Dill fruits are stored in well closed containers and in a cool place.

EUCALYPTUS OIL

Synonyms

Eucalyptus, Dinkum oil, Lemon gum tree, Nilgiri oil.

Biological Source

Eucalyptus oil is the volatile oil obtained by the distillation of the fresh leaves of *Eucalyptus globulus* and other species of *Eucalyptus*, belonging to family Myrtaceae. It is rectified, if necessary.

It should contain not less than 65 per cent of cineole.

Fig. 14.22: Eucalyptus leaves

Geographical Source

It is indigenous to Australia and Tasmania. It is cultivated in the United States (California), Spain, Portugal, and in India. E. citriodora, known as citron scented or lemon scented gum, is grown on large scale in Kerala, Tamil Nadu and other states.

Method of Preparation

In India, the oil is commercially produced on very large scale in Nilgiri. The trees, about 16 years of age are preferred for the collection of leaves. About 750 to 1200 kg of leaves are collected per hectare. The leaves are dried in shade for about 3 days and subjected to steam distillation. The distillation unit (Fig.14.22) consists of false perforated bottom and is made up of copper. Every charge for distillation consists of about 350 kg of leaves. Sufficient quantity of water is also charged in the still and the steam under pressure is passed through it. It takes about 6 hours for complete distillation. Eucalyptus oil (0.75 - 1 per cent) produced is collected in the receiver.

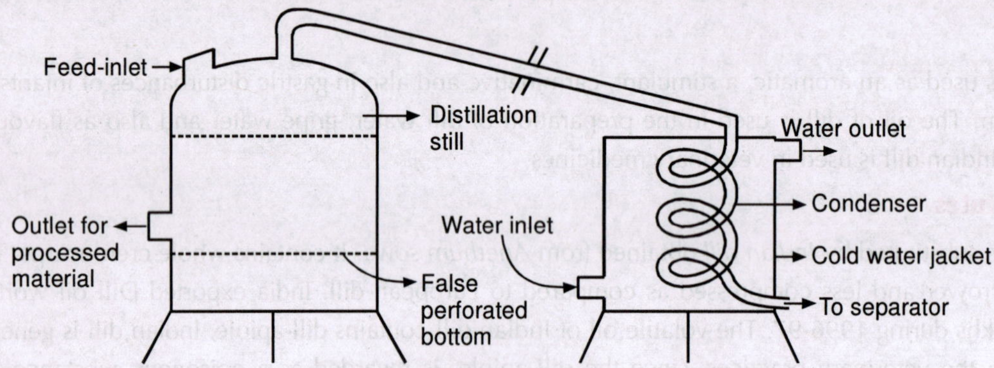

Fig. 14.23: Distillation Unit for Volatile Oil

The crude oil, thus produced, is rectified again after treatment with sodium hydroxide. The oil is filtered and filled in suitable containers.

Description

Colour : It is a colourless or pale yellow liquid.

Odour : Aromatic and camphoraceous.

Taste : Pungent and camphorous followed by the sensation of cold.

It is soluble in 90 per cent alcohol, fixed oils, and fats and in paraffin, but it is insoluble in water.

Chemical Constituents

Eucalyptus oil chiefly contains cineole, also known as eucalyptol (about 80 per cent). It also contains pinene, camphene and traces of phellandrene, citronellal, geranyl acetate.

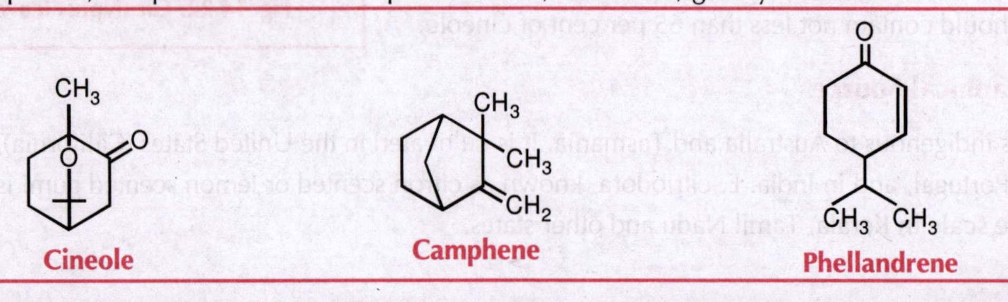

Cineole **Camphene** **Phellandrene**

Chemical Test

Mix 2.5 ml. of eucalyptus oil is with 5 ml of purified petroleum benzins add 5 ml of solution of sodium nitrite (5 gm of sodium-nitrite and 8 ml of purified water). Then add 5 ml of glacial acetic acid crystals of phellandrne nitrite do not form in the mixture within ten minutes.

Standards of Quality

Weight per ml	:	0.897 - 0.916 g
Refractive index	:	1.457 - 1.469
Optical rotation	:	0° - +10°

It should be free from heavy metals and should contain only traces of phellandrene.

Uses

Eucalyptus oil is used as a counter-irritant, an antiseptic and expectorant. It is used to relieve cough and in chronic bronchitis in the form of inhalations. It is an ingredient of several liniments and ointments. Solution of eucalyptus oil is used in nasal drops.

Storage

Keep in well closed container away from light and in cool place.

FENNEL

Synonyms

Fennel fruits, Fructus foeniculum.

Biological Source

Fennel consists of dried ripe fruits of the plant known as *Foeniculum vulgare* Miller, family Umbelliferae, obtained by cultivation. It should contain not less than 0.6 per cent of anethole calculated on dried basis.

Geographical Source

It is indigenous to mediterranean countries and largely cultivated in Romania, Russia, Germany, France, India and Japan. In India, it is cultivated in Gujarat, Punjab, Maharashtra, Rajasthan, Uttar Pradesh and West Bengal.

Cultivation and Collection

Fennel is cultivated by dibbling method. Quality fruits of good germination rate are sown just before the spring. Free branching of the herb and specific arrangement of leaves (in hair-like segments) on the stems require plenty of space in between two plants and rows. Four to five seeds are put, at a time, in holes at a distance of 25 cm in between them. Well drained and calcareous soil in sunny situation is found to be favourable for cultivation of fennel. In India, nearly 90 per cent of the fennel production comes from Gujarat alone.

The crop is kept free of weeds and provided with suitable fertilizers. When the fruits are ripe, crop is harvested and dried in the sun. Fruits are separated by thrashing.

Plants respond well to the use of mannures and fertilizers. Mostly organic mannures are prefers. NPK are used the increase the healthy crop.

Altitude : 1000 – 1800 m

Temperature : 25 – 45°C

Rain fall : 700 – 900 mm per annum

Type of soil : Well drained celceious soil and slury situation is desired.

pH of soil : 6 – 7.5

Duration of crop : Annual

Frequent mulching is unnecessary.

Macroscopic Characters (Fig. 14.24)

Colour : Green to yellowish-brown

Odour : Sweet aromatic

Taste : Strongly aromatic

Size : 5 - 10 × 2 - 4mm

Shape : Straight or slightly curved

It is five-sided fruit in the form of cremocarps with pedicels and rarely found in the form of mericarps. Fruits are glabrous with straight, prominent, yellow coloured five primary ridges and a bifid stylopod at the top. It is an orthospermous fruit. Transversely cut surface shows 2 commissural vittae and 4 dorsal vittae. The embryo is small embedded in upper end with abundant oily endosperm. Commissural surfaces of the endosperm are not grooved.

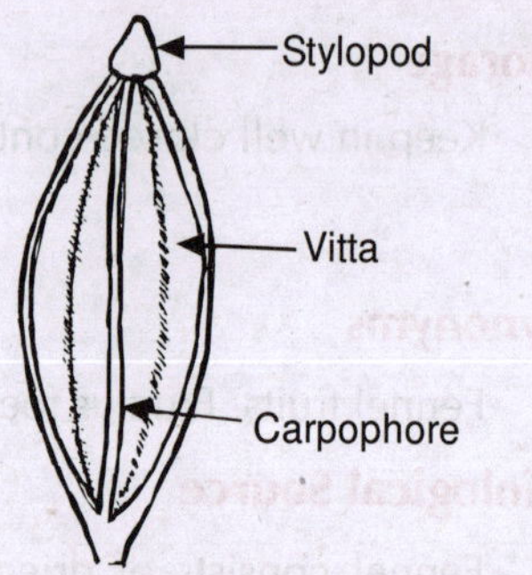

Fig. 14.24: Commissural surface of mericarp of fennel

Microscopic Characters (Fig. 14.25)

The special features of the histological characters of fennel are the presence of anomyocytic stomata on the epidermis of pericarp and mesocarp containing lignified and reticulate parenchyma. Parquetry arrangement of cells on the inner epidermis of pericarp is another feature of the drug and common to all umbelliferous fruits. Vittae, the secretory canals, contain volatile oil and are brown in colour. Endosperm is made up of polyhedral thick-walled cells containing fixed oil and aleurone grains and minute rosette crystals of calcium oxalate. Trichomes and starch grains are absent.

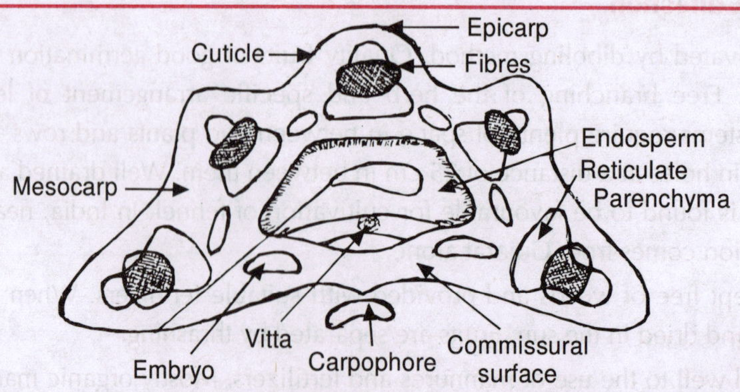

Fig. 14.25: T.S. of Mericarp of Fennel

Chemical Constituents

Fennel consists of 3 to 7 per cent of volatile oil, about 20 per cent each of proteins and fixed oil. The chief active constituent of the volatile oil is a ketone, fenchone (about 20 per cent) and a phenolic ether anethole (about 50 per cent). The other constituents are phellandrene, limonene, methyl chavicol, anisic aldehyde, etc. Fenchone is a colourless pungent liquid with aromatic odour. The anethole is sweet in odour and taste. Oil of fennel is pale yellow liquid with sp. gr. 0.953 to 0.973, refractive index of 1.526 - 1.538 and optical rotation of + 12° to + 24°.

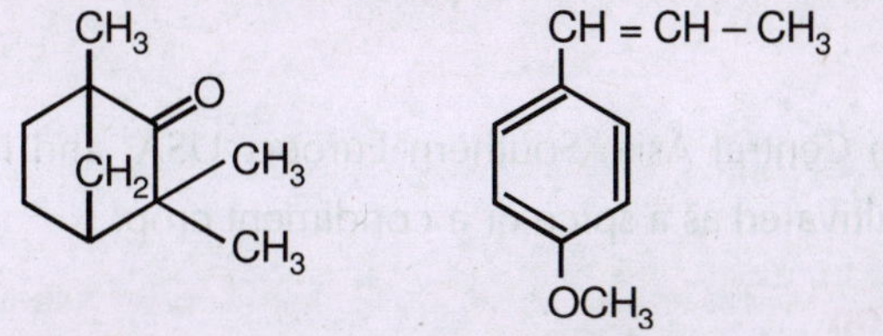

Fenchone Anethole

Standards of Quality

Total Ash	:	Not more than 11.0 per cent
Acid insoluble ash	:	Not more than 1.5 per cent
Water soluble extractives	:	Not more than 10 per cent
Foreign organic matter	:	Not more than 1.5 per cent

Uses

It is used as a carminative, aromatic and stimulant. It is also an expectorant. Pharmaceutically, it is used as flavouring agent.

Indian exports of fennel oil during 1995-96 and 96-97 were ₹ 17.5 lakhs and 14.00 lakhs respectively.

Adulterants

Fennel is commonly adulterated with exhausted fennel fruits which can be distinguished by the following tests.

1. The fruits from which volatile oil is removed by treating with alcohol, contain less percentage of volatile oil and have a typical odour of fusel oil. Such fruits do not contain fenchone.
2. If the fruits are exhausted by the application of steam, they look dark greenish-brown in colour and contain only traces of volatile oil and sink in water.

Substitutes

Table 14.6: Characteristics of Different varieties of Fennel Used as Substitutes

Variety	Size in mm	Taste	Volatile oil content (per cent)	Fenchone content of volatile oil (per cent)
1. Saxony	10 × 4	aromatic	4.76	22.00
2. Russia or Rumanian	4 to 6 ×1 to 2	camphoraceous	4.50	18.00
3. French sweet or Roman	7 to 8 × 2 to 3	sweet aromatic	2.1	nil
4. Indian	4 to 7	camphoraceous	0.720	6.70
5. Japanese	3 to 4 × 2 to 3	very sweet	2.70	10.20

GARLIC

Synonyms

Allium.

Biological Source

This consists of dried bulbs of the plant known as *Allium sativum* Linn. family Liliaceae. It contains not less than 0.2 per cent of alliin on dried basis.

Geographical Source

Lahsun is cultivated in Central Asia, Southern Europe, USA, and India. In India, it is found in almost all the states and cultivated as a spice or a condiment crop.

Cultivation and Collection

Garlic is cultivated in well-drained moderately clay loamy soil. It needs cool moist climatic conditions during the growth and dry period during maturity. Garlic is a hardy plant with narrow flat leaves and bears white small flowers and bulbils. The cultivation of drug is done by planting bulbs generally in the month of September to late in October. It takes about four months for harvesting. It is also taken as an alternate crop with many other vegetables. For cultivation, about 300 kg of bulbs per hectare are required, and yield per hectare is about 8,000 kg.

Macroscopic Characters (Fig.14.26)

Colour	:	Bulbs are white to pink in colour.
Odour	:	Characteristic and aromatic.
Taste	:	Aromatic and pungent.
Size	:	1.5 to 2.5 cm

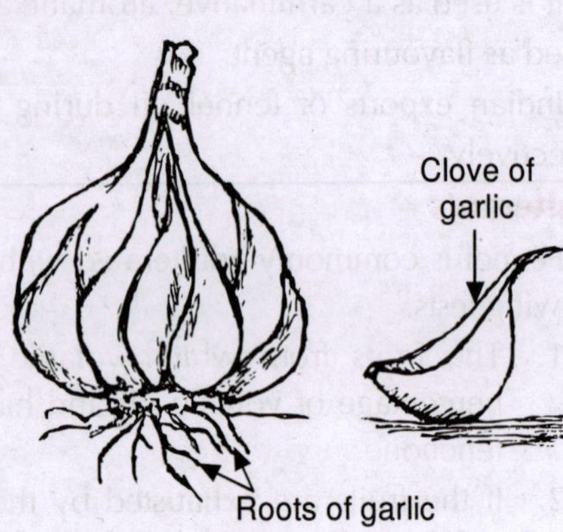

Fig. 14.26: Garlic bulb

Chemical Constituents

Garlic bulbs contain 29 per cent of carbohydrates, about 56 per cent of proteins (albumin), 0.1 per cent of fat, mucilage, and 0.06 to 0.1 per cent of volatile oil. It also contains phosphorus, iron and copper. Volatile oil of the drug is the chief active constituent, and contains allyl propyl disulphide, diallyl disulphide, alliin and allicin. Alliin by action of enzyme allinlyase is converted into allicin. Garlic oil is yellow in colour and has specific gravity of 1.046. It is optically inactive.

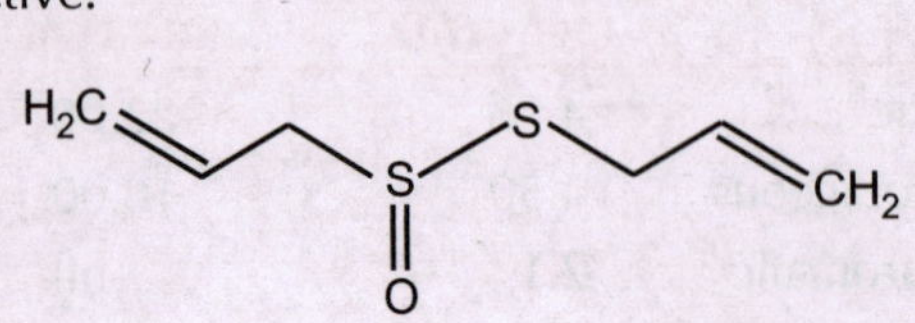

Allicin

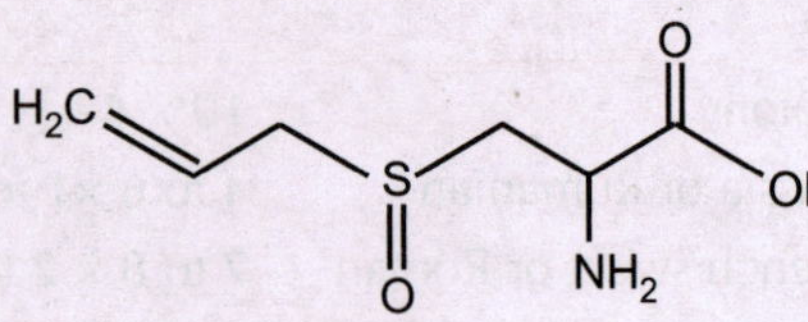

Allin

Standards of Quality

Foreign organic matter	:	Nore more than 2.0 per cent
Ash	:	Nore more than 5.0 per cent
Acid insoluble ash	:	Nore more than 1.0 per cent
Loss on drying	:	Nore more than 65.0 per cent

Uses

Garlic is used as carminative, aphrodisiac, expectorant, stimulant, and disinfectant in the treatment of pulmonary conditions. It is largely used as condiment. Oil of garlic is used as anthelmintic and rubefacient. Allicin is antibacterial. Garlic oil is useful in high blood pressure and atherosclerosis. Fresh garlic is prophylactic against amoebic dysentery. It has strong anti-oxidant effect. Mainly responsible to regularise blood flow in blood vessels and also to heart. It helps to reduce formation of blood clot (Thrembosis) alongwith anti-platelet aggregation effect. Lasun also protects blood vessels from high blood lipid levels. It is also described on intestinal disunfectant.

Storage

It undergoes degradation and hence, should be carefully stored at zero degree centigrade and 65 per cent of relative humidity.

GAULTHERIA OIL

Synonyms

Betula oil, Sweet birch oil, Oil of wintergreen, Tea berry oil.

Biological Source

Gaultheria oil is obtained by steam distillation from the leaves of *Gaultheria procumbens* Linn, family Ericaceae.

Geographical Source

The plant is found throughout the Eastern United States of America and Canada.

Cultivation and Processing of Wintergreen Oil

Gaultheria is a shrub like perennial with slender creeping subterranean stems and branches. It is an evergreen plant of 5 -15 cm height, with white axillary flowers, and bright red globular aromatic berries.

The oil is also obtained from the birch bark i.e. the bark of the plant *Betula lenta* Linn, family Betulaceae.

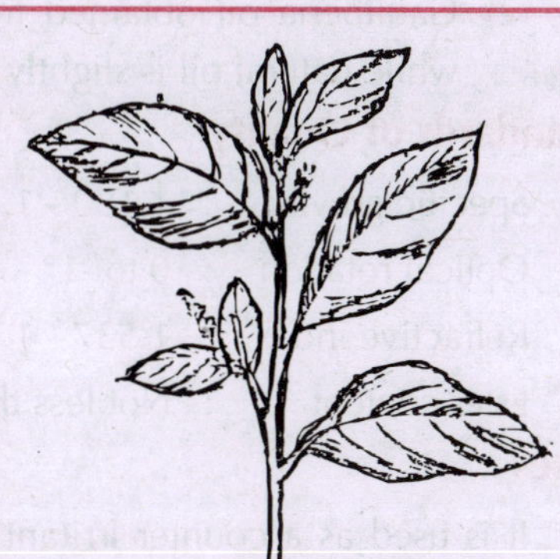

Fig. 14.27: Gaultheria Plant

In India, it is obtained from the leaves of *Gaultheria fragrantissima* Wall, family Ericaceae found growing in Khasia hills, Western ghats, Nilgiris and near Travancore at an altitude of 1600 metres.

Description

Colour : colourless or pale yellow liquid

Odour : characteristic, aromatic and strong

Taste : sweet, aromatic, pungent

Solubility : slightly soluble in water, soluble in six parts of 70 per cent alcohol.

Chemical Constituents

The drug yields about 1.5 per cent of oil.

The oil contains more than 98 per cent of methyl salicylate, formed by the hydrolysis of a glycoside gaultherin in presence of water by the naturally occurring enzyme gaultherase. Additionally, the oil also contains an ester that splits into enanthic alcohol and an acid.

Enanthic alcohol (n-heptyl alcohol, $CH_3\ (CH_2)_5\ CH_2OH$) and its ester show characteristic odour due to which natural gaultheria oil can be distinguished from synthetic methyl salicylate.

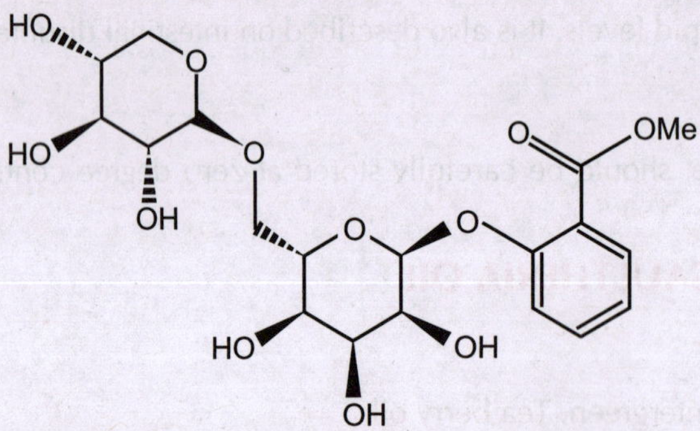

Gaultherin

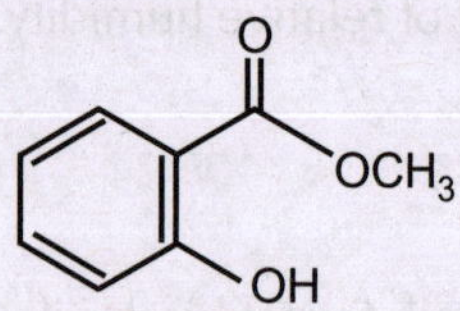

Methyl salicylate

Chemical Tests

(1) To 5 ml oil, add 0.5 ml 5.0 per cent solution of vanillin in 90 per cent alcohol, 2 ml alcohol and shake. A blood-red colour is produced.

(2) Gaultheria oil obtained from betula bark or from synthetic source are optically inactive, while natural oil is slightly laevo rotatory.

Standards of Quality

Specific gravity : 1.180 - 1.187

Optical rotation : 0 to -1°

Refractive index : 1.537 - 1.539

Ester content : Not less than 98.0 per cent w/v, calculated as methyl salicylate.

Uses

It is used as a counter irritant and in the treatment of rheumatism. It is used as very effective vermicide against hook worms. Gaultheria oil can be used internally also. It is employed as a flavouring agent for candies and also in perfumery.

It should be stored in well closed containers away from light and in cool place.

Storage: Keep in well closed, well filled containers.

GERANIUM OIL

Synonyms

Moroccan geranium oil, Algerian geranium oil.

Biological Source

Geranium oil is obtained from fresh leaves and stems of *Pelargonium graveolens* L. family Geraniaceae by steam distillation.

Geographical Source

Geranium is indigenous to South Africa and cultivated in Algeria, Morocco, Spain, France and Italy. Indian geranium oil obtained from other species and is known *Palma Rosa-oil.* It is raised in Jammu to some extent.

Cultivation and Collection of Geranium

Geranium is a perennial plant and is cultivated in Jammu only. It can be cultivated in Maharashtra also. It is cultivated by using slips. Manures and single super phosphate is used before sowing. Cuttings of 20 cm size are prefered with enough number of buds. Antifungal agent and plant growth hormones like IBA, NAA or NAD are used. The cuttings are planted inthe nursery beds at a distance of 10 to 15 cm. After the full growth they are transplanted at a distance of 75 × 60 cm for further growth. Fertilizers like urea, potash and phosphate at the rate of 40 kg each per hectare are provided in the first year. Second dose of fertilizers is provided after 60 days again. Frequent weeding is a must. If necessary the crop may be irrigated.

After nine months when the leaves become pale green in colour and start producing pleasant smell and bear flowers are harvested. Leaves are dried in the shade. Harvesting (cutting of leaves) is done thrice in a year. 25 – 30 tonnes of green leaves are produced, which are processed to give volatile oil. The yield of oil is about 22 – 30 litres per hectare. Geranium herb at present is sold at the rate of Rs. 7000/- per tonne. India has exported Germanium oil worth of 36.25 and 40.34 US million dollars 7000/- in 2017 – 18 and 2018 – 19 respectively.

Geranium Flower

Geranium Cultivation

Description

It is colourless to slightly yellowish coloured liquid, with characteristic pleasant flavour. Insoluble in water, soluble in alcohol.

Chemical Constituents

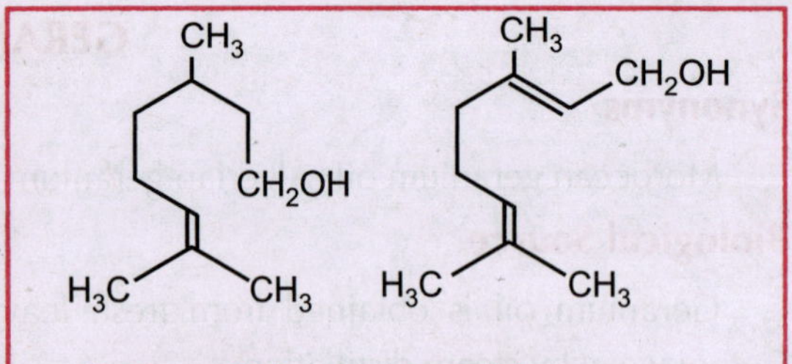

β-Citronellol **Geranlol**

All varieties of geranium generally contain 0.08 to 0.4 per cent of fragrant volatile oil. Geranium oil contains two types of constituents i.e. alcohols and esters. The alcohols are β-citranellol and geraniol about 60 - 70 per cent. While esters namely geranyl acetate, geranyl-tiglate, citranellyl formate and citranellyl acetate contribute about 20–30 per cent of the oil. Several sesquiterpene hydrocarbons, sesquiterpene alcohol are also reported in the oil and are responsible for the pleasant fragrance. It is regarded on costiliest flavour for food and pharma products.

Standards of Quality

Specific gravity	:	0.8830 – 0.9000
Refractive index	:	1.4600 – 1.4700
Total alcohol (geraniol)	:	66 – 78 per cent

Uses

Geranium oil is used as flavouring agent for creams, lotions, soaps and other cosmetic products (to the tune of 1.0 per cent). Where as 0.001 per cent in alcoholic, non-alcoholic beverages, candy, puddings and other dairy products.

JATAMANSI

Synonyms

Nard, Indian spike nard.

Biological Source

Jatamansi consists of dried rhizomes of *Nardostachys jatamansi,* D.C. family Valerianaceae.

Geographical Source

These plants are found in the Alpine Himalayas at an altitude of 3000 - 5000 m. It is cultivated in Punjab, Sikkim and in Bhutan.

Cultivation and Collection

Jatamansi is a perennial herb propagated by cuttings of the underground parts. The favourable altitude for the luxurious growth of the plant is 3000 - 5000 m. The rhizomes are collected from the wild grown plants only. The plant is about 10 - 60 cm in height and with stout and long woody root stocks. The leaves of the plant are sessile, a few, and oblong-ovate in shape. Flowers are rosy, slightly pink or blue in dense cymose. 1300 kg/hactare of drug is produced after cultivation.

Macroscopic Characters (Fig. 14.28)

Colour	:	Dark grey rhizomes are crowned with reddish-brown tufted fibres.
Odour	:	Highly aggreeable, aromatic.
Taste	:	Acrid, slightly bitter and aromatic.
Size	:	Rhizomes are 2.5 - 7.5 cm in length.
Shape	:	Elongated and cylindrical.

The fibres present on the rhizomes are the remnants of leaf bases. Rhizomes break easily and internally they are reddish-brown in colour.

Fig. 14.28: Jatamansi herb with rhizome

Microscopic Characters

Transverse section of the rhizome is characterised by the presence of outermost 2 - 5 layers of cork cells containing oil globules. The cortex is marked by the presence of schizogenous canals. Phloem is in the form of patches and xylem is characterized by vessels. Vessels have typical scalariform thickening. Cambium is continuous and well marked.

Chemical Constituents

Jatamansi contains 1 - 2 per cent of pale yellow volatile oil, resin, sugar, starch bitter principle, an alcohol and its isovaleric ester. It also contains jatamansic acid and ketones jatamansone and nardostachone.

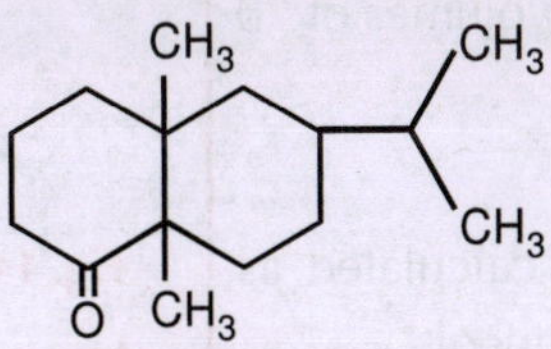

Jatamansone

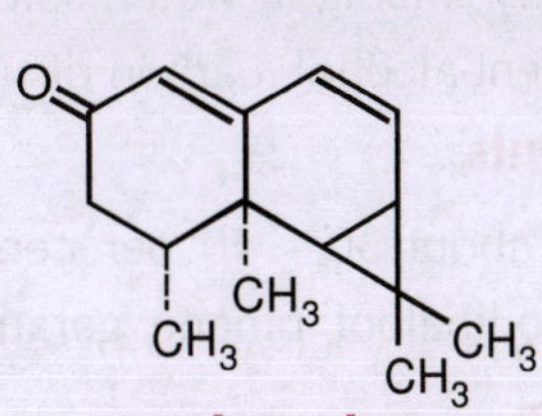

Nardostachone

Chemical Test

The 80 per cent alcoholic extract of drug shows bluish-white fluorescence under ultraviolet light.

Standards of Quality

FOM	:	Not more than 5.0 per cent
Ash	:	Not more than 9.0 per cent
Acid insoluble ash	:	Not more than 5.0 per cent
Alcohol soluble extractives	:	Not less than 2.0 per cent
Water soluble extractives	:	Not less than 5.0 per cent

Uses

Jatamansi is used as a sedative, anti-spasmodic, diuretic, emmenagogue, and stomachic. It is stimulant in small doses and also useful in epilepsy, hysteria, and palpitation of heart. The oil possess antiarrhythmic activity and also used as a flavouring agent in the preparation of medicinal oil. The oil is used as hair tonic, since it is reported to promote the growth of hair and it also imparts blackness to the hairs.

LAVENDER OIL

Synonym

Comman Lavendor. Gavati-chaha tel.

Biological Source

It is the volatile oil obtained by steam distillation of fresh flowering tops of *Lavandula officinalis* Chaix (L. Vera D.C.), family Labiatae.

This plant is also known as true lavender or common lavender. It is indigenous to mountainous or hilly regions of Europe and is supposed to be the most superior in all varieties of lavender. The other variety known as *Lavandula stoechas* is also used for the isolation of oil. In commerce, it is known as French lavender.

Geographical Source

It is found in Portugal and eastwards throughout the Mediterranean region. Lavender is cultivated to some extent in India.

Fig. 14.29: Lavender Plant (*Lavandula officinalis* Chaix)

Description

Colour : Colourless or yellow liquid.

Odour : Characteristic pleasant aroma.

Solubility : Slightly soluble in water, soluble in 4 volumes of 70 per cent alcohol, carbon disulphide.

Chemical Constituents

The oil contains about 30 - 40 per cent esters, calculated as linalyl acetate and also linalool, pinene, geraniol and cineol.

Standards of Quality

Density	: 0.878 – 0.892
Refracted index	: 1.445 – 1.466
Optical rotation	: 12.5° to – 7.0°
Acid value	: Not more than 1.0 per cent

Uses

It is used as aromatic, carminative and as flavouring agent in perfumery and cosmetics.

LEMON GRASS OIL

Synonyms

East Indian Lemon - Grass Oil, Indian Melissa oil.

Biological Source

The lemon grass oil is a volatile oil obtained by steam distillation from the leaves and aerial parts of the plants *Cymbopogon flexuous* or *Cymbopogon citratis*, etc. belonging to family Graminae. It contains not less than 75 per cent of aldehyde calculated as citral.

Geographical Source

It is indigenous to India and found wild, as well as, cultivated throughout India. It occurs in West Africa, Guatemala and East Africa. In India, it is grown in Kerala, Tamil Nadu, Karnataka, and

Maharashtra. The West Indian lemon grass oil is distilled from *Cymbopogon citratis*. This is mainly collected from plants grown in Gujarat, Maharashtra and Punjab.

India produces about 900 tonnes oil per year. There is stiff competition from Guatemala in the trade of oil.

Indian export of lemon-grass oil during 1995-96 & 1996-97 was to the tune of ₹ 180 & 234 lakhs respectively.

Fig. 14.30: Lemon-grass plant with flowers

Fig. 14.31: Lemon grass leaves

Description

Colour : Yellowish-brown liquid

Odour : Resembling to lemon oil

Taste : Similar to lemon oil

Solubility

Entirely soluble in 3 parts of 70 per cent alcohol, chloroform and in fixed oil, and insoluble in water. It is also soluble in mineral oil and propylene glycol.

Chemical Constituents

Lemon grass oil chiefly contains citral, in addition to methylheptenol, nerol, citronellal, dipentene and geraniol.

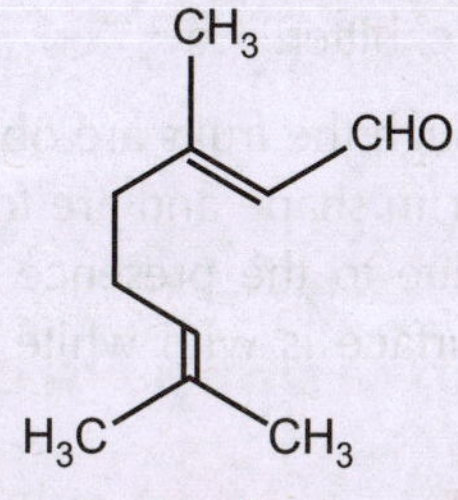

Citral

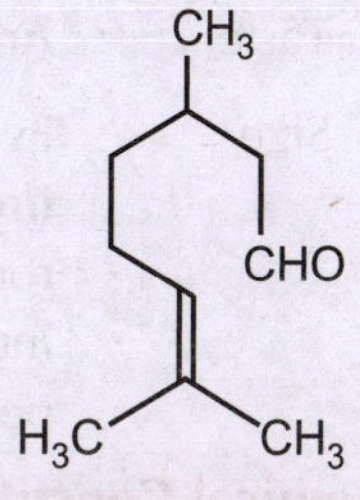

Citronellal

Standards of Quality

Weight per ml : 0.892 to 0.909

Refractive index : 1.4774 – 1.4784

Optical rotation : –3° to + 1°

Relative density : 0.886 – 0.906

Uses

It is used as a flavouring agent and in perfumery. It is employed as a source of citral from which β-ionine is prepared. β-ionine is starting material for the synthesis of vitamin A. Kerala State Drugs and Pharmaceutical Co. has commissioned vitamin A plant at Alleppey.

Indian exports of 7.68 and 10.92 US million dollars in the year 2017 – 18 and 2018 – 2019 of Lemongrass oil were made.

Storage

It should be kept in well closed containers in cool place and away from light.

LEMON PEEL

Synonym

Cortex limonis

Biological Source

Lemon peel is the outer part of the pericarp of the ripe or nearly ripe fruits of *Citrus limonis* (L) Burm, family Rutaceae. It contains not less than 2.5 per cent of volatile oil.

Geographical Source

It is indigenous to North India, but is cultivated on a very large scale in countries like Spain, Italy, and Sicily. It is also cultivated in California, Florida, Jamaica, Australia and India. In India, the cultivation is carried out in Uttar Pradesh, Maharashtra, Karnataka, Madhya Pradesh and Punjab.

Macroscopic Characters (Fig.14.32)

Colour : (i) Fresh lemon peel: Outer surface is bright green yellow, internally white.

(ii) Dried peels: The outer surface is yellow and inner surface pithy white.

Odour : Strong, fragnant, aromatic and characteristic.

Taste : Aromatic, bitter.

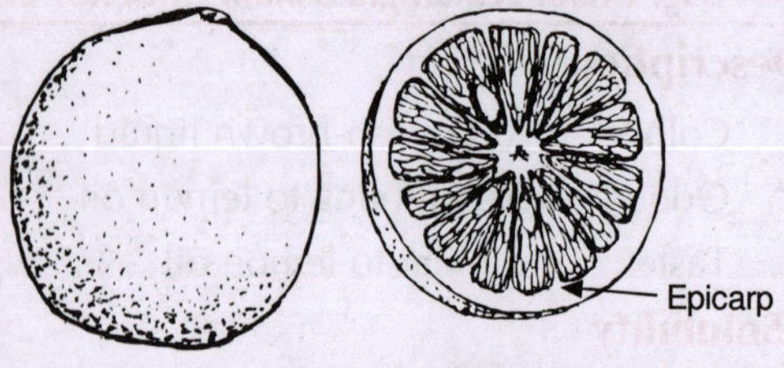

Fig. 14.32: Lemon Fruit

Size : Eventhough the fruits are oblong ovoid of about 5 to 12 cm in length, the peels are irregular in shape and are found in the form of trips or pieces. The outer surface is rough due to the presence of numerous minute pits representing oil glands. The inner surface is with white spongy pericarp. The fracture is short in dried lemon peel.

Chemical Constituents

Flavour of lemon oil is due to neral and geraniol only.

Lemon peel contains volatile oil from 2 - 4 per cent. The other constituents of the peels are hesperidin, pectin, calcium oxalate and bitter substances.

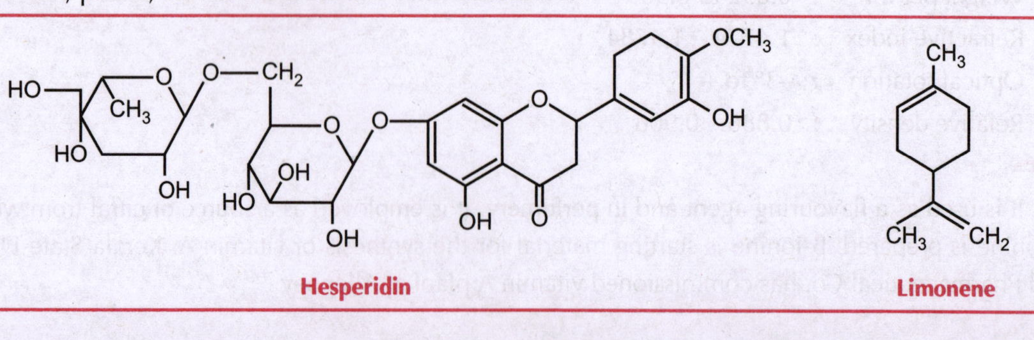

Hesperidin **Limonene**

Standards of Quality

The volatile oil of the drug contains mainly limonene (about 90 per cent), citral (about 4 per cent) and other aromatic compounds like geranyl acetate and terpineol. The lemon oil has pale yellow colour and also following standards:

Special gravity	:	0.849 - 0.855;
Optical rotation	:	+ 57 to + 65°; and
Refractive index	:	1.4742 - 1.4755.

Chemical Test

Solution of recently expressed oil is added to three volumes of alcohol which is neutral or only slightly acidic to litmus paper.

Uses

It is a carminative and stimulant and oil is used as perfuming and flavouring agent. It is also used for extraction of pectin and volatile oil.

Adulterants

The peels which have been rejected by machines after drying are mixed with the drug. They can be identified by the presence of scarification lines on the outer surfaces of the peels.

LONG PEPPER

Synonyms

Pippali Large, Catkins (Big), Pan-pimpali.

Biological Source

This consists of dried fruiting spikes of climbing vine, called as *Piper longum* belonging to the family Piperaceae. It contains not less than 1.0 per cent of piperine on dried basis.

Varieties like Gajapippali, Cheemthipalli, Sihali pippali and Viswam are also known.

Geographical Source

In India, it is found in Bengal, Bihar, Annamaliai district of Tamilnadu and Cherapunji in Assam. Also in parts of East Nepal. It is also found in Singapore, Sri Lanka, Indonesia and Malaysia. It is native of Indo-Malaya region.

Cultivation of Long pepper

Area with high humidity and high rainfall is suitable for its growth. It is shade loving plant.

It can be cultivated from sea level upto an altitude of 1000 metres. It is perennial crop. Temp variation from 10° to 35°C is tolerated by plant. High rainfall from 200 to 400 cm is favourable for healthy growth.

Well drained forest soil, rich in organic matter or laterite soil, with high organic content and good moisture holding capacity is also suitable for cultivation. It is propagated by using seeds, suckers or rooted vine cuttings.

Fig. 14.33 (a): Climber of Long pepper

Fig. 14.33 (b): Piper longum Fruits

Collection of Long pepper

The spikes are harvested when they are full grown but unripe (i.e. 7 - 8 months after planting) during months of February / March. If they are left without picking, they ripe and their pungency is lost to a great extent. The spikes are collected and dried in sun for 4 - 5 days and stored.

The yield of dried spikes is 400 kg per hectare during first year and increases upto 1000 kg per hectare till third year and decreases thereafter.

Macroscopic Characters

Colour	:	Pale brown to dark brown
Odour	:	Aromatic spicy
Taste	:	Hot and sweet taste
Size	:	2 to 5 cm in length 0.4 to 0.5 cm in diameter
Shape	:	Circular and elongated

Spikes are cylindrical, erect and blunt.

In case of Small variety spikes are 1.0 – 1.9 cm long and 0.2 to 0.30 m in diameter. Colour, odour and taste are identical to large variety.

Chemical Constituents

It consists of alkaloids piperine, piplartine and piplasterol (about 6 per cent), 1 per cent essential oil and pungent resin. Large variety contains not less than 1.0 per cent while small variety contains not less than 0.4 per cent of piperine.

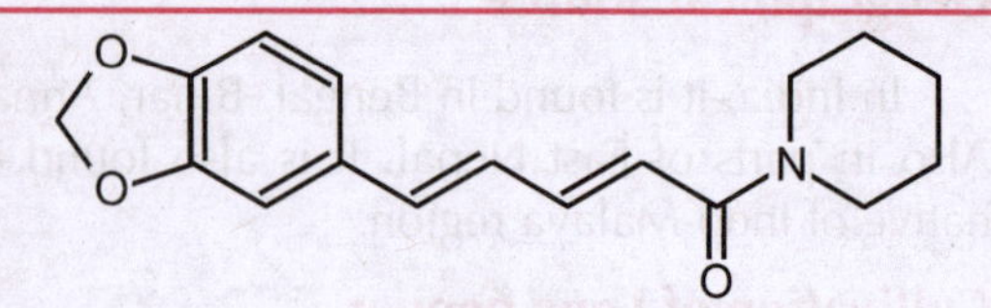

Piperine

Standards of Quality

Total ash	:	Not more than 5.0 per cent
Acid insoluble ash	:	Not more than 0.6 per cent
WSE	:	Not less than 11.0 per cent
ASE	:	Not less than 12.0 per cent
LOD	:	Not more than 13.0 per cent

Uses

It is widely used in Ayurvedic and Unani medicines, especially in diseases of respiratory tract. The roots are used for bronchitis, stomach-ache, diseases of spleen and tumours.

It improves appetite. It is also used in pickles.

NUTMEG

Synonyms

Myristica, Nux Moschata.

Biological Source

Nutmeg consists of dried kernels of the seeds of *Myristica fragrans* Houtt. family Myristicaceae. The kernels are freed from their arillus and seed-coat. It contains not less than 5 per cent (v/w) volatile oil.

Geographical Source

Nutmeg is indigenous to Molucca islands and cultivated in Indonesia, Caribbean islands and other tropical countries. In India, it is cultivated in Kerala and Tamil Nadu.

Cultivation and Collection

The plant is a dioecious tree bearing female and male flowers separately. The drug is obtained from female plants. Nutmeg thrives well in warm humid climate and grown in locations from sea level to an altitude of 900 m. The requirement of the rainfall is about 200 - 250 cm. Sandy loam, clay loam and red lateritic soils are satisfactory for cultivation. The plant is cultivated by sowing the seeds. Only freshly collected seeds are suitable for cultivation. Vegetative propagation of nutmeg is also possible. The seedlings are raised in the nursery beds. The seeds take about 2 - 3 months for germination. When the seedlings are about 16 - 18 months old, they are transplanted by keeping the distance of 8 - 9 m in between. The nutmeg essentially needs shade for its protection. If sufficient rain is not available, irrigation is necessary. Fertilizers and manures are provided to the plants from time to time. The sex of the plant is determined and male trees are reduced to 10 per cent of the total population. Fruits are produced by the plants throughout the year, but the number is maximum from December to May. Plants start bearing fruitsafter 10 years of age of normal 400 – 500 fruits are born by a tree per year depending upon the age normally. The fruits are suitable for picking when the husk of the fully ripe nut splits, or else they are allowed to drop on the ground after dehusking. The red feathery arilus (mace) covering the seed is detached from the seed shell by hands and flattened out in the sun. It takes about 4 - 8 weeks for drying of the drug in sun. Alternatively, it is dried by artificial heat. Shells are broken up and nutmegs are removed. Normally, the full grown female tree, under favourable conditions, produces 2 - 3 thousand fruits per year, weighing 10 kg and 1.5 - 3 kg of mace. Thus, 100 female plants located in a hectare can produce 1000 kg of nutmegs and 150 - 300 kg of mace. Commercially, the plantation is economical only when the plants are more than 10 years old.

Volatile oil of nutmeg is highly fragrant yellow coloured liquid having sp. gr. 0.855 - 0.9241, optical rotation + 10° to + 30° and refractive index 1.470 - 1.4880.

Leaves of plant are also used medicinally as they also cantains volatile oil and are very costly.

Macroscopical Characters (Fig. 14.34)

Colour	:	Externally, the kernels are greenish-brown or brown
Odour	:	Strongly aromatic
Taste	:	Pungent and aromatic
Size	:	Kernels are about 20 - 30 mm in length and 20 mm broad
Shape	:	Ellipsoidal

The kernels are externally covered with small irregular patches or lines and cup shaped mace, 1 mm thick.

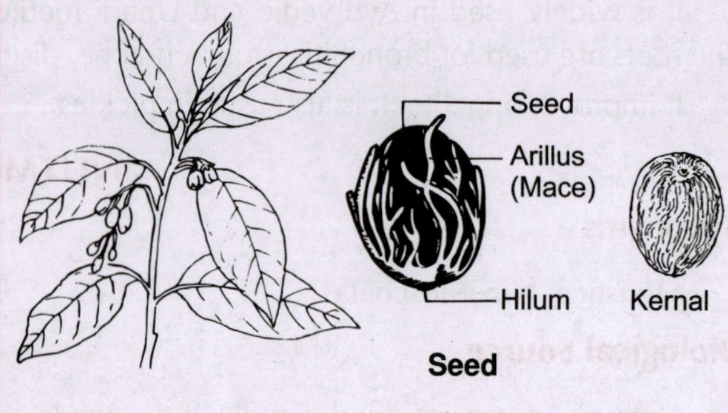

Fig. 14.34: Myristica fragrans herb & kernel

Chemical Constituents

Nutmeg contains 5 - 16 per cent of volatile oil, and about 30 per cent of fat. The volatile oil contains about 4 - 8 per cent myristicin, elimicin and saffrole. The fatty acid constituents of the fixed oil are myristic (about 60 per cent), palmitic, oleic, lauric and other acids. The fat of nutmeg is also known as nutmeg butter. The other constituents of the drug are protein and starch.

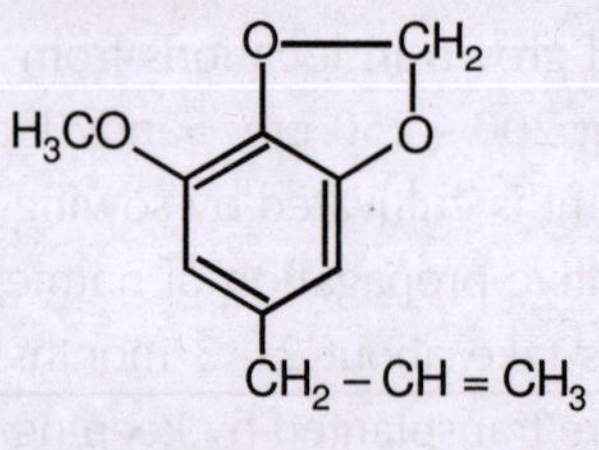

Myristicin

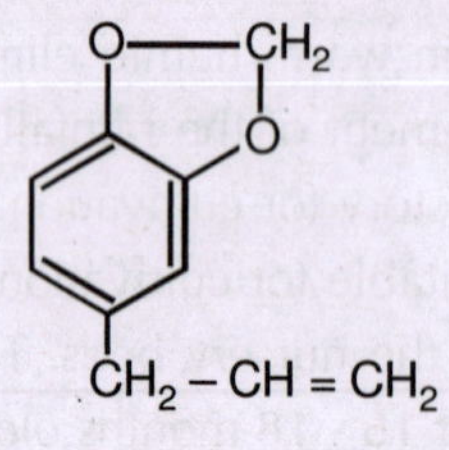

Saffrole

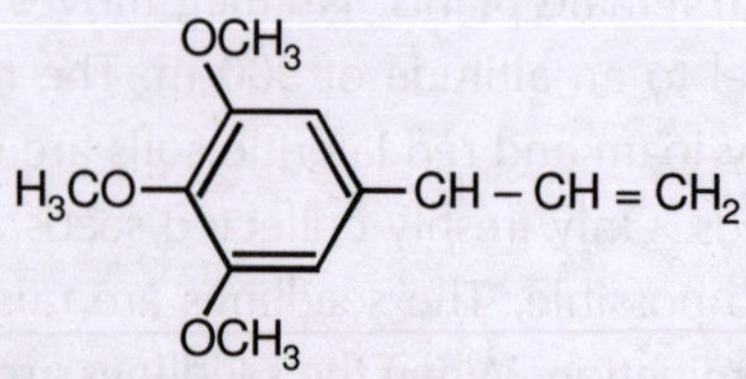

Elimicin

Myristicin (methoxy saffrole) is very poisonous and the narcotic effect of the drug is said to be due to myristicin. Geraniol, terpineol, d-camphene, etc. are the other constituents of volatile oil of nutmeg. Nutmeg butter contains trimyristin.

Standards of Quality

Total ash	:	Not more than 2.0%
Acid insoluble ash	:	Not more than 0.1 %
Alcohol soluble extractives	:	Not less than 20.5 %
Water soluble extractives	:	Not less than 04.5 %
LOD	:	Not more than 13.0%

Uses

Nutmeg is used as an aromatic, stimulant, and carminative. It is used as a flavouring agent, too. The fat (nutmeg butter) also known as **banda soap**, is used in soap industries. The fat and volatile oil of nutmeg are used in the treatment of rheumatism.

MACE

The arillus of the seeds of myristica is known as mace which arises in the region of hilum, before the flowers open and fertilization takes place. The mace is an orange coloured modification of nutmeg seed. Mace consists of Flattene lobed pieces of 20 mm in length and 1 mm in thickness. Mace is transluecent and brittle. Its odour and taste is similar to kernels only.

Chemical Constituents

Mace contains about 20 per cent fixed oil and 10 - 15 per cent volatile oil and amylodextrin. Macilocic acid and macelenic acid chemically

Uses

Mace is used as a condiment, aphrodisiae aromatic and a carminative. It is used as a flavouring agent, specifically for biscuits, pickles, meat and fish preparations.

Chemical Test

Mace gives red colour with Iodine due to amylodextrine.

Substitutes and Adulterants

Being costly, nutmegs are adulterated with inferior varieties or with the exhausted drug. *Myristica malabarica* found in evergreen forests of Konkan in Maharashtra and in Kerala and *Myristica argentea* are the two substitutes to the authentic drug. The first is devoid of odour, while the latter is acrid in taste. The imitation of nutmeg, made by moulding the exhausted powder and flavouring, can be distinguished by putting it into water wherein the imitations break down quickly.

Myristica beddomei found upto an altitude of 1200 metres in Kerala, Karnataka, and Tamil Nadu and *Myristica attenuata* cultivated in Gujarat and Karnataka are the other substitutes for nutmeg.

PALMAROSA OIL

Synonyms

Rosha oil, Geranium oil, Motia-variety.

Biological Source

It is obtained from leaves and flowering tops of *Cymbopogon martini* Stapf, family Graminae.

Geographical Source

The grass is found wild in Punjab, Kashmir, Almora and Garhwal regions of Uttar Pradesh, Madhya Pradesh, Gujarat, Maharashtra, Andhra Pradesh and Tamil Nadu. In Maharashtra, it is cultivated in Jalgaon, Dhule and Nasik districts.

Two varieties of grass are commercially available. The **Motia** variety contains highest percentage of geraniol in oil (about 90 per cent) and is known as palmarosa or rosha oil, or East - Indian geranium oil. Another variety called as **Sofia oil** is poor in geraniol content and is known as Ginger - grass oil.

Cultivation, Collection and Processing of Palmarosa grass

This consists of tall, perennial and sweet scented grass, about 1.5 - 2.5 metres in height. The stem is straw coloured and leafy. The leaves are usually broad, subcordate or rounded at the base, about 20 × 1 to 2.5 cm in size. Flowers are in the form of spikes, 12 - 18 mm long, oblique or divaricate. The grass gives flowers in the months from October to December.

It can be cultivated by sowing the seeds or vegetatively propagated. The seedlings are raised in nursery beds and then planted by keeping the distance of 1 metre in between two rows. It needs light, well drained calcarious soil with sunny situation and rainfall of about 70 - 80 cm per year. The grass is irrigated when necessary. First crop is harvested when the grass flowers. One hectare needs about 10 kg of seeds for cultivation.

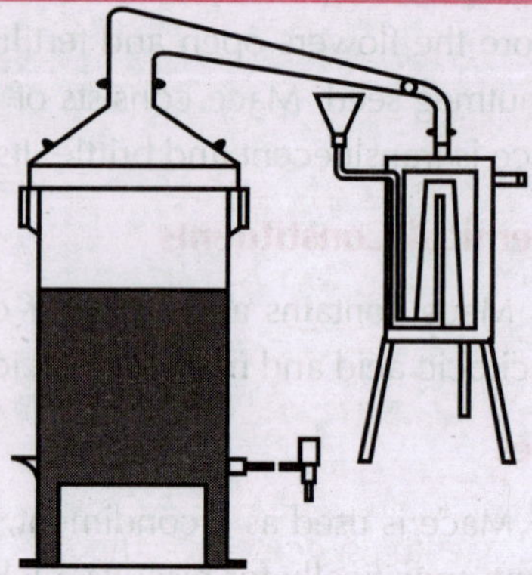

Fig. 14.35 : Distillation assembly for palmarosa oil

Rosha is perenneal plant. The leaves of which are 80 – 90 cm in length and flowering stem is about 30 cm. Leaves are yellowish green and have pleasant aromatic flavour reseulting to rose flowers.

The grass yields about 7.5 - 22 kg of palmarosha oil per hectare.

Description

Colour : Colourless to light brown coloured liquid

Odour : Aromatic resembling to rose

Taste : Slight

Solubility : Insoluble in water, soluble in 90 per cent alcohol

Chemical Constituents

The palma rosha grass contains about 1.25 - 1.5 per cent of oil. The flowers contain maximum oil, while the stems contain minimum. Rosha oil contains 85 - 95 per cent of geraniol, citronellal dipentene, linalool and terpenes.

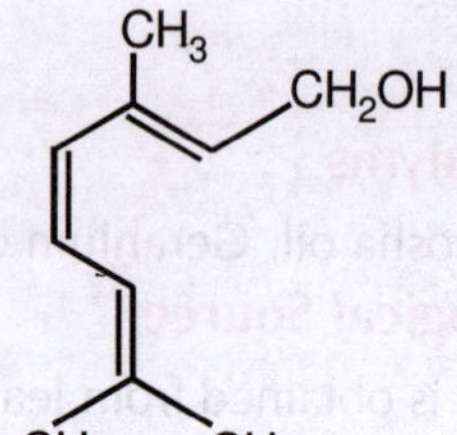

Geraniol

Standards of Quality

Specific gravity : 0.885 to 0.896

Optical rotation : + 1.4° to – 2°

Refractive index : 1.4760 - 1.4805

Uses

Palmarosa oil is used in perfumery, cosmetics, soaps and in the treatment of rheumatism and skin diseases. It possesses insect repellent properties.

Storage

Store in well closed container in cool place.

PEPPERMINT OIL

Synonyms

Oleum mentha piperita, Colpermin.

Biological Source

The oil is obtained by steam distillation of the fresh flowering tops of the plants known as *Mentha piperita* Linn. belonging to family Labiatae. If necessary, it is rectified. Mentha oil contains not less than 4.5.0 per cent and not more than 10.0 per cent w/w of esters calculated as menthyl acetate and not less than 44.0 per cent of free alcohols calculated as menthol and not less than 15 per cent and not more than 32.0 per cent of ketone calculated as menthone.

Geographical Source

Mentha species are cultivated in various parts of the world. Mentha grows wild in Europe, while it is cultivated in Japan, England, France, Italy, USA, Bulgaria and Russia. In India, it is cultivated near Jammu and in Tarai region of Uttar Pradesh.

Cultivation and Collection of Mentha

For the cultivation of mint, well-drained fertile, sandy-loam soil with neutral pH is required. The rainfall in the range of 95 to 105 cm and temperature in between 15 to 25°C are the other requirements for cultivation. The altitude of 250 to 400 m is supposed to be ideal location. The cultivation is done by vegetative propagation method by using suckers. The treatment of mercury compounds as fungicides encourages better sprouting. Suckers are placed at a distance of 10 cm in January/February. The distance maintained between two rows is approximately 50 to 60 cm. The plants should be kept free of weeds by hoeing frequently. Folliar spray of urea (2 to 3 per cent) alongwith other fertilizers like superphosphate, potash etc. are found to be advantageous. Harvesting is done when crop reaches the flowering stage. The mentha plant is very susceptible to pests. The treatment with thiodan (1 : 700), endrine 20 EC (1 : 700), BHC (10 per cent), sulphur (0.5 per cent), and malathion (5 per cent) are found to be very effective for pest control.

Fig. 14.36: Mentha Twig (Mentha piperita)

Preparation of Oil

To reduce the bulk, mentha plants (Fig. 14.36) are dried to 1/4th of their weight. It helps in saving the time required for distillation and reduces its cost of production. Drying directly in sunlight results in loss of volatile oil. Fermentation should not take place during drying of the herb, as it adversely affects the quality and quantity of the oil. The air-dried material is charged into galvanised iron or mild-steel still. The still designed for this purpose has a false perforated bottom. The steam under pressure, generated with the help of boiler, is then passed through the drug. It takes about 3 to 4 hours for distillation. More than 80 per cent of the oil is distilled off during the first half of distillation. Distillation should be completed carefully, as menthol of medicinal and commercial importance comes over in the latter part of distillation. The condenser should be made up of either aluminium or stainless-steel and should be coiled, so as to increase the area of condensation. The distillate i.e. mentha oil is collected in separating can. Mentha oil is insoluble in water and also lighter than water, which keeps it floating in a separating can. The oil is then decanted and filtered.

The average oil content of the herb is found to be 0.5 to 1 per cent (v/w) approximately. About 100 kg of the oil per hectare of the crop per year is supposed to be satisfactory for commercial purposes.

Description

Colour : Peppermint oil is colourless to yellow.

Odour : Characteristic and pleasant.

Taste : Pungent followed by cooling sensation.

Macroscopic Characters

Peppermint Leaves *(Mentha-peperita)*	Mentha Leaves *(Mentha arvensis)*
(1) Leaves more or less crumpled and detached from stems. Stems are quandrangular.	(1) Leaves mixed with stem and aerial parts. Stems circular.
(2) Leaves are petioled, oblong lanceolate or ovate, oblong, opposite. Leaf size : 1.0 – 5 × 2.0 – 2.5 cm.	(2) Leaves smooth sessile lanceolate to oblong, acute coarsly dentate, petiole 0.8 to 1.2 cm. Smooth above glandular below. Leaf size : 2.0 – 3.0 × 1.5 – 2 cm
(3) Light green to purplish brown surface glabrous, lower surface with amber coloured trichomes on midrib and vein. Odour : Aromatic Taste : Aromatic, followed by cooling sensation on drawing breath.	(3) Coarsly dentate, smooth above and glandular below. Odour : Aromatic Taste : Aromatic, followed by cooling sensation on drawing breath.

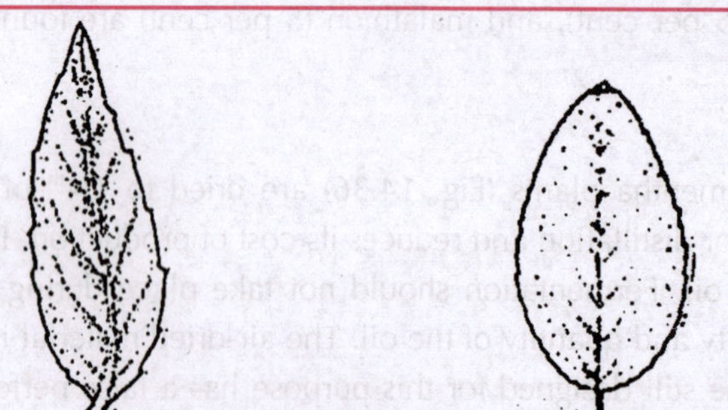

Fig. 14.36 (a): (I) Leaf of Mentha piperita (II) Mentha arvensis

Solubility

It is soluble in 70 per cent alcohol, ether and chloroform and insoluble in water.

Chemical Constituents

Peppermint herb contains 0.8 to 9.5 per cent of volatile oil and 5.0 to 10.0 percent tannins.

Peppermint oil contains chiefly *l* – menthol to the extent of 70 per cent in free, as well as, in the form of esters, depending upon variety (like American, Japanese, Indian). American peppermint oil contains 80 per cent menthol while Japanese oil contains 70 – 90 per cent. Other important constituents of the peppermint oil are menthone, menthofuran, jasmone, menthyl isovalerate, menthyl acetate and several other terpene derivatives. The other terpenes include *l*-limonene, isopulegone, cineole, pinene, camphene, etc. Jasmone and esters are responsible for pleasant flavour, while menthofuran causes resinification and develops dirty smell.

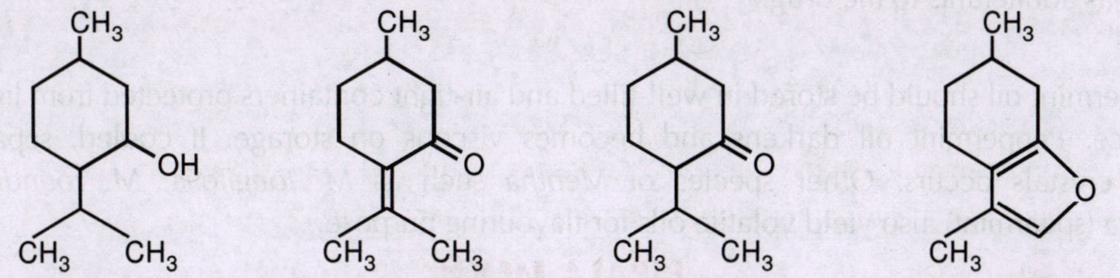

Menthol Pulegone Menthone Menthofuran

MENTHA OIL

Mentha oil is obtained by steam distillation from *Mentha arvensis* var-*piperascens* (Japanese Mint), commercially cultivated in U.P., Himachal Pradesh, Punjab, Haryana, Jammu and Kashmir and Central India. It contains about 80 per cent *l*- menthol. It is also cultivated in Japan, Brazil and California. The yield of oil is upto 1.2 per cent v/w, as compared to 0.8 per cent v/w in *M. piperita*.

Chemical Test

A few drops of peppermint oil are mixed with 5 ml of nitric acid solution (prepared by adding 1 ml of nitric acid to 300 ml of glacial acetic acid, heat on water bath. Within five minutes liquid develops – blue colour, which on further heating deepens and shows copper colours fluorescence after some time it becomes golden yellow.

Standards of Quality

(1) It is neutral to litmus paper.

(2) Weight per ml is 0.900 to 0.912 g.

(3) Optical rotation at 25°C is – 16° to – 30°.

(4) Refractive index – 1.4590 to 1.4650.

Uses

Peppermint or Mentha oil is used as carminative, stimulant, and flavouring agent. It has mild antiseptic properties too. It is used in toothpaste, tooth powders, shaving creams, and different pharmaceutical dosage forms. It is also consumed in the preparation of chewing gums, candies, jellies, perfumes and essences. Menthol is manufactured in India by S. H. Kelkar and Co., Bhavna Chemicals, Procter and Gamble Ltd. And several others. India produces about 500 tonnes of menthol annually.

Both mentha oil and menthol have calcium channel blocking activity causing spasmolytic and smooth muscle relaxant effects, and hence useful in irritable bowel syndrome. They show better pharmacokinetic profile when given by enteric coated capsules for release in large intestine. Muscle relaxant activity is employed to reduce spasm during endoscopy of colon. For this purpose, emulsified oil is injected through biopsy channel of the endoscope. Mentha oil shows digestant activity by stimulating bile flow. This is also supported by flavonoid pigments of leaf. Azulene from the leaf is believed to be anti-inflammatory and anti-ulcer in activity. For the nasal decogestant effect, menthol is considered by U.S.F.D.A. as generally regarded as safe (GRAS). Mentha oil is used for inhalation in steam, as well as topical products and lozenges for its antitussive effects.

Peppermint oil and menthol are being exported from India since a decade. The export values during 1994-95 were ₹ 2890 lakhs and ₹ 623 lakhs for menthol and peppermint oil respectively.

Substitutes and Adulterants

Several species of *Mentha* are known to contain oil. Many a time, these oils are de-mentholized and used as adulterants to the drug.

Storage

Peppermint oil should be stored in well-filled and air-tight containers protected from light and in cool place. Peppermint oil darkens and becomes viscous on storage. If cooled, separation of menthol crystals occurs. Other species of *Mentha* such as *M. longifolia, M. roundifolia* and *M. spicata* (spearmint) also yield volatile oils for flavouring purpose.

PIPALA MOOL

Synonym

Pipala-roots, Pipala mool.

Biological Source

The roots of *Piper-longum* fam.: Piperaceae are known as pipala mool or pipala roots in the market.

After collecting the fruits for three successive years, the whole plant is harvested by cutting the stems close to the ground and the roots are dug up. The roots and stems are cleaned and cut into small pieces. Roots are dried in the shade.

Geographical Source

Indonesia, Phillipines, Sri-Lanka, Brazil, Malaysia, and India. In India it is cultivated in Bengal, Kerala and Karnataka.

Macroscopic Characters

Roots are cylindrical, slightly curved with swollen internodes, covered with rootless scars.

Roots are 2.5 – 5.0 cm in length and 2 – 5 mm in thickness. Dark brown in colour with pungent bitter taste, producing numbeness to tongue.

The average yield of dried roots per hectare is about 500 kg.

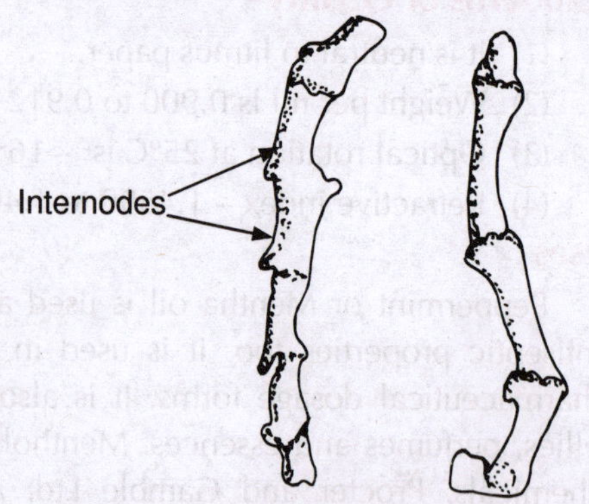

Fig. 14.37: Pipala-mool (Pipala-roots)

Chemical Constituents

Pipala roots contain 0.15 to 0.18 per cent piperine, along with piplartine, triacontane and dihydro stigmesterol.

Standards of Quality

Foreign organic matter	:	Not more than 2.0 per cent
Ash	:	Not more than 8.0 per cent
Acid insoluble ash	:	Not more than 1.0 per cent
Alcohol soluble extractives	:	Not less than 8.0 per cent
Water soluble extractives	:	Not less than 11.0 per cent

Uses

Pipala roots are used similar to long pepper. Additioally in Ayurvedic medical practice they are also used as emmenagogue abortifacient, anthelmintic and especially for cough, bronchitis and asthma.

RASNA

Synonyms

Galanga, East India root, Chinese ginger, Lesser galangal, **Koshta kolingan**.

Biological Source

These are the dried rhizomes of *Alpinia officinarum*, belonging to family Zingiberaceae. The rhizomes are trimmed, washed, cut into segments and dried. They contain not less than 0.5 per cent of volatile oil. Commercially, these are known as lesser galangal rhizomes, while the greater rhizomes are obtained from *Alpinia galanga*. (Syn. *Amomum galange*)

Geographical Source

The drug is found throughout the Eastern Himalayas and in South West India. It is also cultivated. It is a perennial herb, about 1 – 2 m in height bearing rhizomes. The drug is collected in autumn.

Macroscopic Characters (Fig.14.38)

Colour	:	The rhizomes are reddish-brown externally and light orange brown.
Odour	:	Characteristically aromatic.
Taste	:	Aromatic and pungent.
Size	:	The rhizomes are about 2 – 8 cm in length and 2 cm in thickness.
Shape	:	They are irregularly branched and marked with fine annulations.

Rhizomes are tough and have a fibrous fracture.

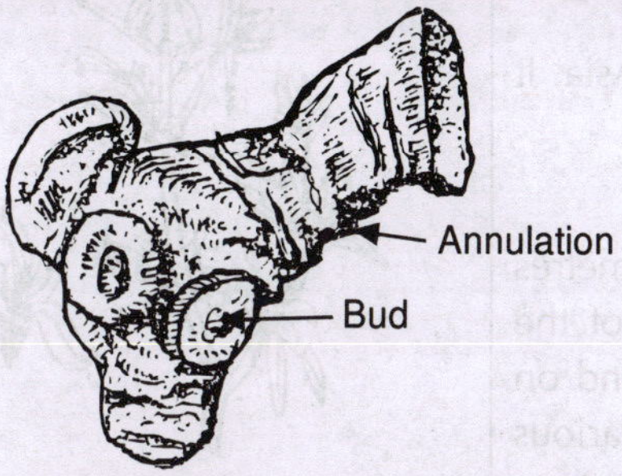

Fig. 14.38 (a): Rhizome of Rasna

Fig. 14.38 (b): Galanga herb

Chemical Constituents

The green rhizomes contain 0.6 per cent to 1.5 per cent of volatile oil. The oil contains methyl cinnamate (about 48 per cent), cineole (about 25 per cent), camphor and pinene. It also contains resin, oily pungent galangol, alpinol, yellow crystalline substance known as galangin, and di-oxyflavanol. The antiinflammatory triterpene compounds reported are sorghumol and bochmerd.

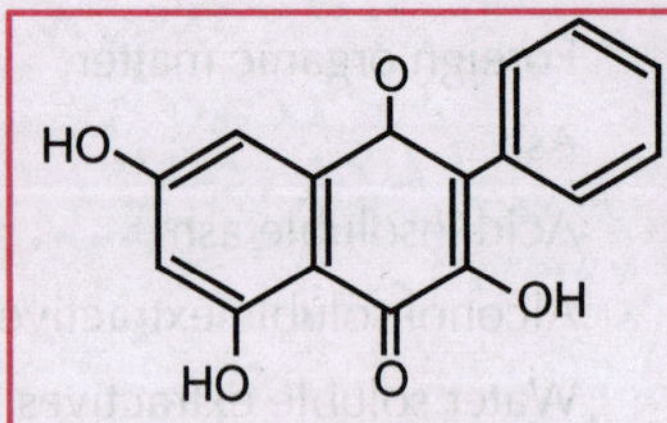

Galangin

Standards of Quality

Foreign organic matter	:	Not more than 2 per cent
Alcohol soluble extractives	:	Not less than 6.0 per cent
Water soluble extractives	:	Not less than 13 per cent
Total ash	:	Not more than 3.5 per cent
Acid-insoluble ash	:	Not more than 1 per cent

Chemical Test

Shake about 1 g finely powdered rasna with 60 per cent alcohol for 10 minutes and filter. Put a drop on filter paper and dry it. Examine the same under ultraviolet light, a bright bluish-white fluorescence is observed.

Uses

Galanga is used as an aromatic, a stimulant and carminative similar to ginger. It has antibacterial properties. In India, it is used in treatment of rheumatism and catarrhal affections.

Substitute

Alpinia galanga (Java galanga or Greater galangal) is similar to the official varieties, but, less pungent. It does not contain flavonoids. The alcoholic extract does not show any fluorescence under ultraviolet light.

ROSEMARY OIL

Biological Source

It is the volatile oil obtained from fresh flowering tops of the plant *Rosmarinus officinalis* Linn, family *Labiatae*.

Geographical Source

Rosemary is indigenous to South Europe and South Asia. It is cultivated in Mediterranean basin and India.

Cultivation and Collection of Rosemary

Rosemary is an evergreen plant (*Fig. 14.39*) about 2 metres in height. It bears blue coloured racemes. The leaves of the plant are about 2.5 cm tall. It is cultivated in gardens and on slopes. Its cultivation is found successful on light calcarious soil. It is cultivated by sowing the seeds or with slips. After flowering, the plants are cut about 10 cm above the ground and steam distilled.

Fig. 14.39: *Rosmarinus officinalis* Plant

Description

Colour : Colourless or pale yellow liquid
Odour : Characteristic rosemary flavour
Taste : Camphoraceous

Chemical Constituents

The flowers contain about 1.0 per cent of volatile oil, resin, ursolic acid and bitter principle. While leaves, in addition to oil, also contain tannins. Volatile oil contains mainly 10 -15 per cent of borneol, 2.5 – 3 per cent of bornyl acetate, camphor, eucalyptol, pinene, d-camphene, cineol and 45 per cent of terpenes.

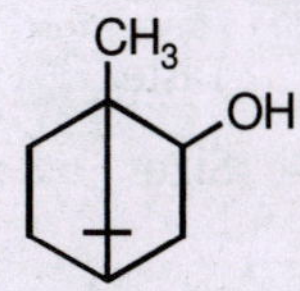
Borneol

Standards of Quality

Specific gravity : 0.894 – 0.912
Refractive index : 1.464 – 1.476
Optical rotation : – 5 to 10°
Solubility : Oil is insoluble in water, soluble in 10 volumes of 80 per cent alcohol.

Uses

It is used as carminative, rubefacient, stimulant and as flavouring agent for liniments, hair lotions, inhaler, soaps and cosmetics.

Storage

Keep in well closed containers, protected from light and in cool place.

KAPUR KACHARI

Synonyms

Spiked ginger lily, *Sontakka*, Shati.

Biological Source

This consists of dried sliced rhizomes of the plant known as *Hedychium spicatum,* belonging to family Zingiberaceae. It contains not less than 0.80 per cent of p-methoxy cinnammic acid ethyl ester on dried basis.

Geographical Source

Kapur kachari is found in tropical and sub-tropical Asia. In India, it is found in Western and Central Himalayas at an altitude of 1000 to 2500 m.

Cultivation and Collection

Ginger lily is cultivated for the ornamental foliage and agreeable sweet flowers. It is propagated by divisions of rhizomes. It needs rich loamy soil with plenty of water. Kapur kachari is grown in gardens as a potted plant. It thrives well in semi-shady situation. It is perennial herb, about 1 m in height with glabrous leaves and white ascending flowers, found in the form of dense terminal spikes. It yeilds 12 tones of rhizomes per hectare.

Macroscopic Characters

Colour	:	Rhizomes are white and starchy internally, while rough and reddish-brown externally.
Odour	:	Very strong aromatic.
Taste	:	Camphoraceous, bitter.
Size	:	1.5 to 2 cm in diameter and about 0.5 cm in thickness. The slices are covered with occasional rootlets.
Shape	:	They are flat and spherical.

Fig. 14.40: Kapur kachari plant

Chemical Constituents

Kapur kachari rhizomes contain 4 per cent of volatile oil, 52 per cent of starch, glycosidal organic acid and ash (4.5 per cent). The volatile oil is responsible for flavour of the drug and chiefly contains ethyl paramethoxycinnamic acid ester (about 68 per cent). It also contains cineole limonene, 8-caryophyllene, ethyl cinnamate, and traces of cinnamic aldehyde.

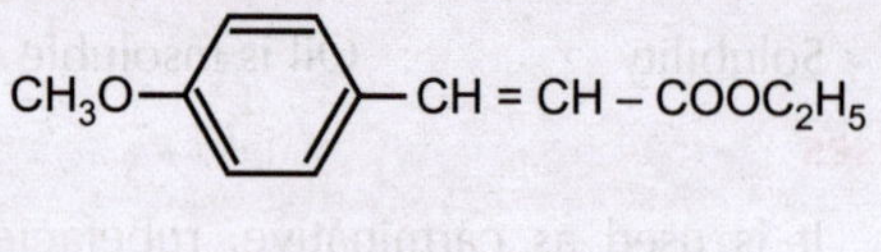

Ethyl p-methoxycinnamate

Standards of Quality

Foreign orgnaic matter	:	Not more than 02 per cent
Ash	:	Not more than 08.0 per cent
Acid insoluble ash	:	Not more than 03.0 per cent
LOD	:	Not more than 14.0 per cent
Alcohol soluble extractives	:	Not less than 0.7 per cent
Water soluble extractives	:	Not less than 12.0 per cent

Uses

Rhizomes are used as stomachic, stimulant, carminative and tonic. They are either used in the form of powder or decoction. Being aromatic, it is used as a flavouring agent for several cosmetic preparations and also to flavour tobacco. The rhizomes are used in preparation of 'abit', a fragrant highly perfumed powder used during the religious functions. It is also known to possess insect repellant properties.

Storage

Rhizomes are thoroughly dried and stored in well closed containers.

SPEARMINT

Synonyms

Spearmint leaves, Mint, Mentha viridis Stotch – spearmint, common spearmint.

Biological Source

This consists of dried leaves and flowering tops of the plant known as *Mentha spicate* Linn. (*Mentha viridis* Linn.) or *Mentha cardiaca* Geranol ex. Baker belonging to family Labiatae.

Geographical Source

Spearmint is indigenous to Europe, but extensively cultivated in North America and Asia. More than 30,000 acres of land is under cultivation of spearmint in Washington State alone, Michigan, Indiana, Idaho, Yakima are the other places where it is cultivated commercially.

Fig. 14.41: Spearmint – herb

Macroscopic Characters

Colour	:	Green to dark green, stems are purple
Odour	:	Aromatic, and characteristic
Taste	:	Aromatic (but not followed by cooling sensation as in peppermint) leaves are sessile, flowers are arranged in spikes.

Chemical Constituents

It contains mainly volatile oil about 0.5 to 1.0 per cent resin and tannins. Spearmint oil contains not less than 50 per cent of earvone, small amount of linalool, pinene, cineole and phellandrine.

Spearmint oil is obtained by steam distillation of the herb and is yellowish in colour and having characteristic aromatic taste of mint.

Standards of Qualtiy

Specific gravity	:	0.930 – 0.940
Optical rotation	:	48° to – 59°
Refractive index	:	1.4820 – 1.4900

Spearmint oil contains not less than 50 per cent of *l*-carvone, and small amount of linalool, pinene, cineole and phellandrene.

Uses

Spearmint is used as carminative, flavour and source of spearmint oil is used as flavour mainly for tooth-pastes, mouth-washes, chewing gums, and also for sauces and cosmetic products.

THYME

Synonyms

Garden thyme, French-thyme, Mother of thyme, Red thyme.

Biological Source

This consists of dried or partially dried leaves and flowering tops of the plant *Thymus vulgaris* Linn, family Labiateae (Lamiaceae).

Geographical Source

It is indigenous to Mediterranean region i.e. Asia, Australia, Greece, Italy, Spain and extensively cultivated in France, Spain, Portugal and USA.

Macroscopic Characters

It is an aromatic perennial shrub about 20 – 45 cm in height, with quadrangular, greyish brown to purple brown lignified twisted stems.

Leaves are 3 – 10 × 3 mm in size, sessile, or with very short petiole, entire, lanceolate to ovate; lamina is tough, greyish green in colour, pubescent on the lower surface.

Calyx is green with violet spots, calyx tube is covered with long stiff hairs. Corolla is twice in length as compared to calyx, bilobiate, pinkish or whitish in colour and verticillaster.

Fig. 14.42: Thyme plant

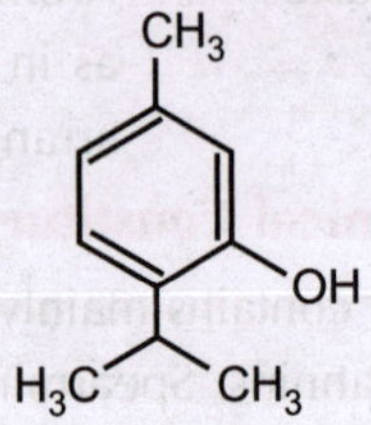

Thymol

Chemical Constituents

Thyme contains 1.0 - 2.5 per cent of volatile oil, flavonoids, caffeic acid, labiatic acid, ursolic acid, resins and tannins. Volatile oil contains mainly thymol (20 - 80 per cent), carvacrol, terpineol and linalool.

Standards of Quality

Total ash	:	Not more than 12.0 per cent
Acid insoluble ash	:	Not more than 4.0 per cent
Foreign organic matter	:	Not more than 2.0 per cent
Content of stalks	:	Not more than 15.0 per cent
Specific gravity	:	0.910 – 0.935
Optical rotation	:	Slightly leavorotatary
Refractive index	:	1.4950 – 1.5050
Angle of rotaiton	:	Not more than –3 at 25ºC

Uses

It is used as expectorant, antispasmodic and carminative. Thymol possesses antifungal, antibacterial and anthelmintic properties. It is used as culinary herb and also for flavouring. The labiatic acid has anti-oxidative properties.

Thymol is a counter-irritant and very strong anthelmintic (specially against hook worm). Both oil and thymol are used in tooth paste, detergents and soaps as flavour and also in perfumes. Thyme oil is used in cough syrup, liniments, lotions and mouth washes.

Substitutes

Wild thyme: *Thymus - serpyllum* (Labiateae) is a substitute for thyme. It contains 0.5 to 2.5 per cent volatile oil and constitutes about 45 - 75 per cent of thymol. It also contains carvacrol.

THYME OIL

Biological Source

It is the volatile oil distilled from flowering plant. *Thymus Vulgaris,* family - Labiateae.

Description:

It is colourless to readdish brown liquid with pleasant thymal odour, Sharp taste. Specific gravity : 0.894 – 0.930.

Chemical Constituent

It contains 30 – 40 per cent thymol and carvecrol also contains bomyl acetate, cymene, pinene and Linalooe.

Uses

Thyme oil is antiseptic, anthelmentic and flavouring agent. It is also rubifacient counter irritant and carminative.

TULSI

Synonyms

Sacred basil, Holy basil.

Biological Source

Tulsi consists of fresh and dried leaves of *Ocimum sanctum* Linn. (Syn. *Ocimum tenuiflorum)* family Lamiaceae, and contains not less than 0.40 per cent eugenol on dried basis.

Geographical Source

It is a herbaceous, multi branched annual plant found throughout India. It is considered as sacred by Hindus. The plant is commonly cultivated in garden and also grown near temples. It is propagated by seeds. Currently Tulsi is cultivated commercially for its volatile oil.

Macroscopic Characters

It is multi branched small herb and 30 - 75 cm in height. All parts of tulsi are used in medicine, especially fresh and dried leaves. Leaves are oblong, acute with entire or serrate margin, pubescent on both sides and minutely gland-dotted. The leaves are green in colour with aromatic flavour and slightly pungent taste. Flowers are purplish in colour in the form of racemes. Nutlets are subglobose, slightly compressed, pale brown or red in colour. Seeds are reddish-black and subglobose.

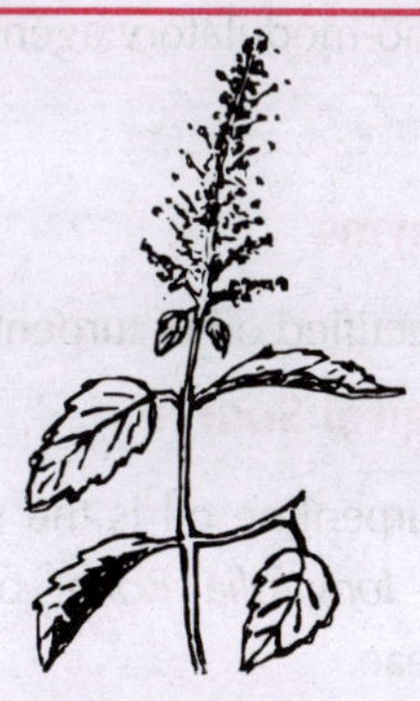

Fig. 14.43: Tulsi herb

Varieties of Tulsi

Three varieties of Tulsi are cultivated in India everywhere.

(1) Ram Tulsi: has broad, bright green leaves with slightly sweet taste.

(2) Krishna Tulsi: has purplish green leaves.

(3) Ran Tulsi: is wild rare variety.

Chemical Constituents

Tulsi leaves contain bright, yellow coloured and pleasant volatile oil (0.1 - 0.9 per cent). The oil content of the drug varies depending upon the type, the place of cultivation and season of its collection. The oil is collected by steam distillation method from the leaves and flowering tops. It contains approximately 70 per cent eugenol, carvacrol (3 per cent) and eugenol-methyl-ether (20 per cent). It also contains caryophyllin and anthocyanins. Seeds contain fixed oil with good drying properties.

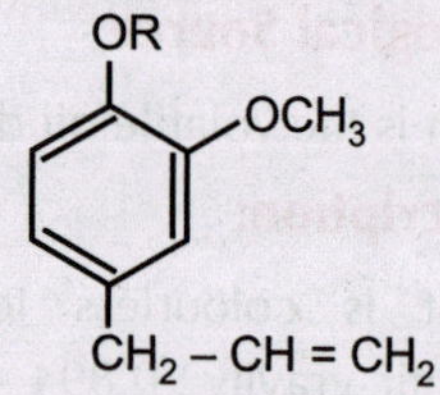

R = H, Eugenol
R = CH_3, Methyl eugenol

The plant is also reported to contain alkaloids, glycosides, saponin, tannins, an appreciable amount of vitamin C, and traces of maleic, citric and tartaric acid.

Standards of Quality

Foreign organic matter	:	Not more than 2.0 per cent
Ash	:	Not more than 15.0 per cent
Acid insoluble ash	:	Not more than 5.0 per cent
LOD	:	Not more than 12.0 per cent
ASE	:	Not less than 3.0 per cent
WSE	:	Not less than 10.0 per cent

Uses

The fresh leaves, its juice and volatile oil are used for various purposes. The oil is antibacterial and insecticidal. The leaves are used as stimulant, aromatic, anticatarrhal, spasmolytic, and diaphoretic. The juice is used as an antiperiodic and as a constituent of several preparations for skin diseases and also to cure ear-ache. Infusion of the leaves is used as a stomachic. The drug is a good immuno-modulatory agent. Tulsi has expectorant and anti-inflammatory properties.

TURPENTINE OIL

Synonyms

Rectified oil of turpentine; Oleum terbenthae.

Biological Source

Turpentine oil is the volatile oil obtained by distillation from oleoresin of *Pinus roxburghi Sarg* (*Pinus longifolia Roxb*) or *Pinus palustris* or from other species of *Pinus*, belonging to family Pinaceae.

Geographical Source

It is commercially manufactured in India, U.S.A., France, and Russia.

Cultivation and Collection of Turpentine Oil

Different species are cultivated for the commercial production of turpentine oil. *Pinus echinata*, *Pinus maritima* and *Pinus sylvestris* are grown in America, France and Russia respectively.

In India, several species like *Pinus insularis* and *Pinus roxburghi* (also known as *Pinus longifolia*) are used commercially. *Pinus longifolia* is found growing naturally in Himalayas from Kashmir to Bhutan. It is also cultivated and found naturally in Himachal Pradesh, Jammu, Punjab and Uttar Pradesh. Climatic conditions suitable for cultivation of the drug are rainfall between 100 to 175 cm, altitude 600 to 2400 m, and temperature 30°C to 38°C. It grows on variety of soils like limestone, band of quartz and clay, sand stone and even on bare rocks. It grows naturally by seed germination. In the nursery, plants are grown by sowing the seeds in March and April. The seedlings are transplanted at a distance of 1.5 m in July **and once again, after two years just** before the beginning of monsoon. The plants of 20 to 25 years of age are utilised for the collection of the resin (colophony) and oil.

Method of Preparation of Oleo-resin

Two methods are employed in India for tapping the plant. In the first method, suitable number of blazes are made in the stems of tree, while in the second heavy tapping i.e. "tapping to death" is done and resin is collected. In several cases, **it is continued for 8 to 10 years**. The tapping produces schizogenous ducts. The oleo-resin which comes out of the plant is collected in earthenware or in suitable containers. The crude oleo-resin is purified by heating in a steam jacketed vessel with spiral mixer. Vegetable debris and materials like sand and other particles are removed by settling or floating. The clarified resin is then taken to the distillation plant. The oil is redistilled after treatment with sodium hydroxide or lime water to remove resins, acids, phenols, etc. It is then dehydrated with common salt or sodium sulphate and stored in suitable containers. The oleo-resin produces approximately 25 per cent of turpentine oil and about 70 per cent of colophony. The turpentine oil produced after chemical treatment by alkalis is known as rectified turpentine oil. About 70 per cent of total world production of turpentine is produced in the USA.

Description

Colour : It is colourless to slightly yellow.

Odour : Strong, characteristic and becomes much stronger on storage with less pleasantness.

Taste : Pungent and bitter. It is a clear, transparent liquid.

Fig. 14.44 : Pinus – roxburghi tree with fruit

Solubility

The oil is insoluble in water but soluble in alcohol. It is miscible with chloroform, carbon-disulphide, glacial acetic acid, benzene, petroleum ether, and fixed oils.

Chemical Constituents

About 40 terpenes are reported in turpentine oil. Mainly turpentine oil contains α - pinene (20 - 30 per cent), β - pinene (5 - 10 per cent) camphene, β - phellandrene, delta-3-carene, limonene, etc. Indian turpentine oil contains less proportion of pinene and high content of carene. American oil contains 65 per cent of α - pinene, 30 per cent of β - pinene and 10 per cent of other terpenes.

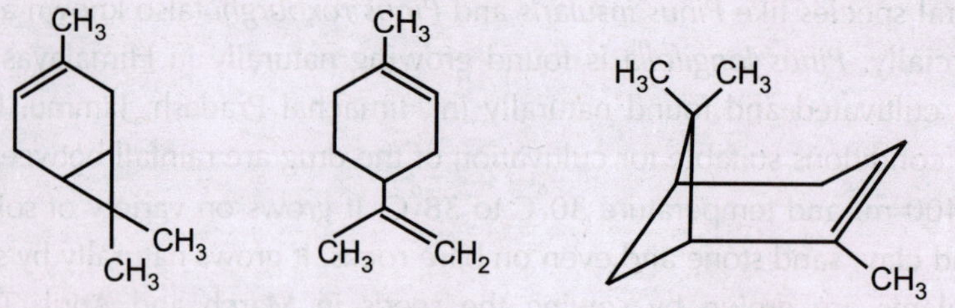

Carene Limonene α-Pinene

Standards of Quality

Weight per ml	:	0.854 to 0.868 g
Refractive index	:	1.467 to 1.477

Uses

Turpentine oil is chiefly used as a counter-irritant and rubefacient in the form of liniments to be used externally. It is also a weak antiseptic and expectorant. It is used in the treatment of chronic bronchitis. Industrially, it is used in the manufacture of synthetic pine oil, disinfectants, denaturants, insecticides, paints and varnish.

Adulterants

Resin oil, wood turpentine and petroleum jelly are the common adulterants of turpentine. Petroleum jelly can be detected by low weight per ml of the oil and resin oil leaves the stain of fatty substances, if stained on paper.

Storage

The oil is stored in well-closed bottles away from light and in cool place. If the oil is exposed to atmospheric oxygen, it undergoes oxidation, becomes viscous, yellowish and acidic in action and its weight per ml, as well as, boiling point increases.

VETIVER OIL

Synonym

Khus oil

Biological Source

It is obtained by steam distillation from roots of the plant *Vetiveria zizanioides* Stapf, family Gramineae.

The medicinal and commercial importance of oil is known since ancient times. Vetiver oil was known to Charaka as described in Ayurveda. The commercial importance of roots is confirmed from the incidence of leving duty on vetiver roots by the king of Kanauj in twelfth century.

Geographical Source

The plant is found growing in India, Myanmar, Sri Lanka and East and West Africa. It is cultivated in Indonesia, Caribbean islands, Malasiya, Philippines and USA.

In India, it is found abundant in Punjab, UP, Rajasthan, Kerala, Karnataka and Tamil Nadu.

Cultivation and Collection of Vetiver

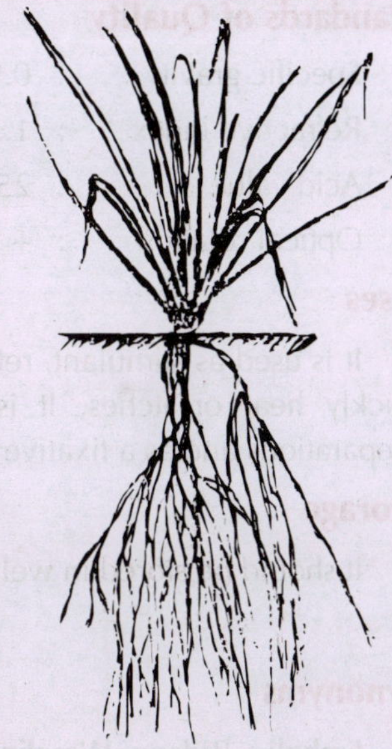

Fig. 14.45: Vetiver plant with roots

Cultivation of vetiver grass (*Fig. 14.45*) is done by sowing the seeds or from slips. A well-drained sandy loam is most suitable for cultivation. The temperature ranging from 25° - 38°C and rainfall of 100 - 200 cm are desired. It thrives best in marshy places and humid climate. About 6000 seedlings per hectare are required for commercial cultivation of vetiver grass. 3-4 tonnes of dried roots can be obtained by cultivation per hectare per year of vetiver roots.

Planting of slips is done just before the outbreak of monsoon. The distance between two plants and between two rows is approximately 22.5 cm. Proper arrangements for irrigation must be made after rainy season is over. Fertilizers and manures are provided to produce sturdy grass and roots. The grass attains the height of 1 - 1.5 metres above the ground. When the plant is about 15 -18 months of age, the roots are collected by uprooting in dry months of the year. If necessary, digging is done for collection of roots. The drug is slashed, cut into small pieces and used for extraction of oil.

Preparation of Oil

The small pieces of roots are chipped, put into the still and are subjected to steam distillation. Since, the specific gravity of the oil is 0.990 - 1.032, it is very difficult to separate the oil from condensate. Hence, the condenser is always kept warm, so that the warm oil after cooling separates very quickly without forming any emulsion. The salt (sodium chloride) may be used to increase density of water, so as to separate it from the condensate. The traces of water in oil are removed by treating it with anhydrous sodium sulphate. The yield of oil is 1.2 - 2 per cent v/w.

Description

The oil is light brown to deep brown or green in colour with characteristic odour. It is soluble in fixed oil and alcohol.

Chemical Constituents

The vetiver oil mainly contains alcohols (45 - 60 per cent), i.e. vetivenol, vetiverol, and 8 - 35 per cent ketones namely α and β-vetivones. Indian vetiver oil contains khusal, khusitol and khusinol.

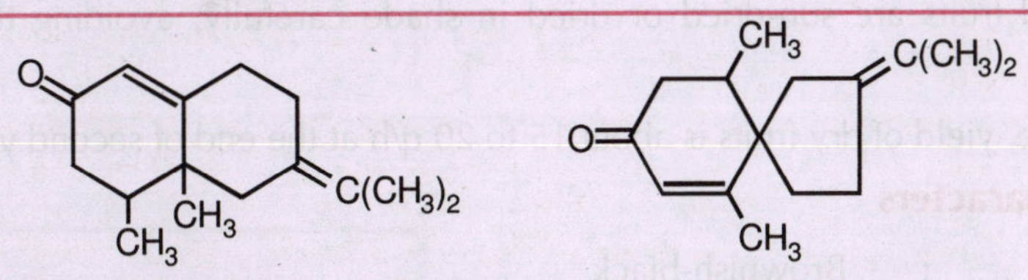

α-vetivone β-vetivone

Standards of Quality

Specific gravity	:	0.990 - 1.032
Refractive index	:	1.512 - 1.523
Acid value	:	25 - 80
Optical rotation	:	+ 15° to + 45°

Uses

It is used as stimulant, refrigerant, flavouring agent, aromatic, stomachic and in the treatment of prickly heat or itches. It is also used in preparation of *sharbat*, soap, perfumery and toilet preparations and as a fixative of volatile oils.

Storage

It should be stored in well filled, well closed containers and in a cool place.

VIDANG

Synonyms

Embelia, Bidang, **Wavding**, Briberang.

Biological Source

This consists of dried fruits, of the plant *Embelia ribes,* belonging to family Myrsinaceae.

Geographical Source

It is distributed in hilly parts of India, found in Assam, Bihar, Bengal, Orissa, Chhatisgarh, Madhya Pradesh and throughout northern parts of India. It is also grown in Sri Lanka.

Cultivation and Collection

It grows in temperate to tropical regions. It needs moist soil and partial shady region for favourable growth. It can be grown under different agro-climatic conditions. It grows well upto altitude of 900 to 1700 metres. Temperature ranging from 25°C to 30°C is suitable for its growth. Rainfall in the range of 50 - 150 cm is favourable for cultivation.

It thrives well on variety of soil ranging from black cotton soil to sandy or rocky soils. Moist soil with rich organic matter is preferable for its healthy growth.

It is propagated by using seeds (nursery rising). It is a perennial crop.

After two years of plantation, the crop is ready for harvesting. Generally, after two years, it bears fruits during August - September. When the fruits attain maturity (ripes), they are harvested manually in the months of December - January.

The harvested fruits are sun-dried or dried in shade carefully, avoiding the attack of fungal infection.

On an average, yield of dry fruits is about 15 to 20 q/h at the end of second year of plantation.

Macroscopic Characters

Colour	:	Brownish-black
Odour and taste	:	Aromatic

Shape : Globular with beak-like projection at the apex

Size : 2.5 mm in diameter

Fruits are with pedicel and 3 – 5 sepals. The fruit contains horny red coloured seed. The seed is covered by thin membrane and has yellowish spots containing embelin. The weight of 100 seeds is about 0.5 g.

Vidang-plant

Vidang berries

Fig. 14.46: Vidang

Microscopic Characters

Epicarp of the fruit is the outermost layer consisting of one celled epidermis. In surface view the epidermal cells appear round with inter cellular spaces. The outer 4-6 layers of mesocarp are parenchymatous with intercellular space and contain tannins mesocarp, vascular bundles, encircled by a pericyclic ring of pseudofibres, are seen; xylem vessels show scalariform to spiral thickenings; phloem indistinct. Endocarp is sclerenchymatous consisting of palisade like stone cells, which are much thickened and having prominent pits. Testa consists of 1 or 2 layers of rectangular cells with brown pigments. Endosperm has thick hemicellulosic walls and contain masses of embelin. Embryo consists of thin walled rectangular cells with fixed oil and protein.

Chemical Constituents

Vidang fruits contain embelin (2 per cent) and its dimer vilangin, quircitol (1.0 per cent), fatty material (5.0 per cent), 1 per cent tannin, neglegible amount of volatile oil, colouring matter and an alkaloid christembin.

Embelin - 2.5 - 3.1 per cent, is an inseparable mixture of three components i.e. homoebelin, embelin and rapanone.

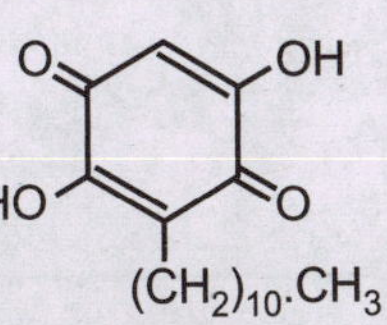

Embelin

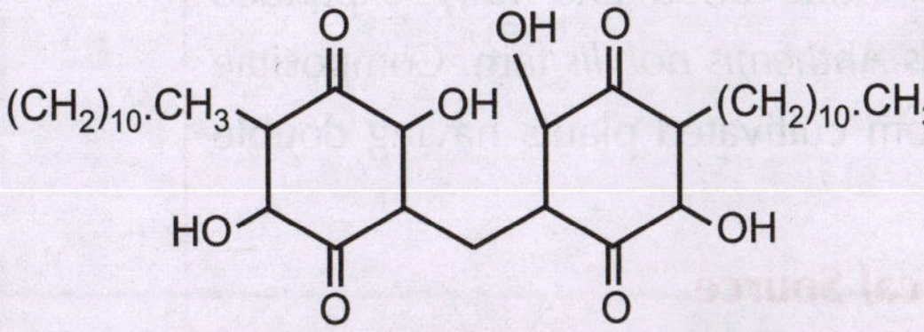

Vilangin

Chemical Tests

1. A 5.0 per cent ethereal extract of the drug gives deep-violet colour with sodium hydroxide solution.
2. A 5.0 per cent ethereal extract gives bluish-violet precipitate with dilute ammonia solution. (Test for embelin).
3. Extract 5 g of powdered vidang with 25 ml of alcohol and filter. Divide the filtrate in two equal parts.
   (a) To one portion, add solution of lead acetate. A dirty green precipitate is produced.
   (b) To the other portion, add solution of ferric chloride. A reddish brown precipitate is produced.

Uses

Vidang is considered to be anthelmintic, astringent, carminative, stimulant and alternative. The fruits are very effective against tapeworms but have no effect on hook worms and round worms. Aqueous extract of vidang fruits is found to have anti-fertility action.

Vidang fruits are used as an ingredient in applications for ring worm.

This climbing shrub is known for its medicinal values since the period of Charaka and Sushruta. It is widely used for its antihelmintic property.

Storage

Dried fruits are stored in clean porous jute bags in dry place.

Substitutes and Adulterants

Commercial samples are substituted by *Embelia robusta* fruits which are bigger in size, reddish in colour and longitudinally wrinkled. They contain volatile oil glands in mesocarp and stone cells in inner mesocarp.

CHAMOMILE ROMAN

Synonyms

Roman chamomile, Camomilla, Chamaemelum.

Biological Source

These are the dried and fully expanded flower-heads *Anthemis nobilis* fam. Compositae collected from cultivated plants having double capilula.

Fig. Chamomile Roman

Geographical Source

It is cultivated mainly in France, Belgium, Saxony, Hunary and to some extent in England.

Chemical Constituents

Volatile oil about 0.8 – 1.0 is per cent in Roman chamomile flowers. Yellow colouring matter i.e. flavonoids like Apigenin, luteolin and quercetin are also reported.

Angelic acid, tiglic acid esters, nobilin chamazulene, farnesol are the other constituents of volatile oil.

Anthemic acid is the biter principle reported in chamomile.

Uses

It is used as flavouring of food products and essential ingredient of herbal tea.

Carminative, antiemetic and antispasmodic are the other properties of the oil for which it is used suitably nobelion of the drug in allergic.

Roman chamomile is reputed to be abortifacient and also affect menstrual cycle.

It should be used with all precautions.

CHAMOMILE GERMAN

Synonyms

Wild chamomile, Hungarian chamomile.

Biological Source

This consists of dried *flower heads* of *chamomilla-reticulata* or *Mantricaria retuculata* fam: (Compositae) Asteraecae.

Fig. Chamomile German

Description

Flowers are conical, 3-10 mm in breadth narrow, hollow receptacle, with short, brown in colour with white margins ray flovets 13 mm in length, corolla white disc florets are 2 – 5 mm, corolla tubes with 1 narrow lobes, odour is aromatic, taste is bitter.

Chemical Constituents

Contains *volatile oil* 0.2 to 2.0 per cent containing and bisabold and chamazulene. *Flavonoids* like apiginine, luteolin quercitin are also reported. Anthemic acid, tannin are other constituent of the drug.

Uses

It is used as carminative, antispasmodic, anti inflammatory, anti-septic.

Chamomile flowers are recommended for haemorrhoids and leg ulcers.

It is used as food flavouring agent and also in herbal tea formulation.

MONOTERPENOIDS OF ANIMAL ORIGIN

CASTOREUM

Synonyms

Castor, Canadian beaver.

Biological Source

A reddish brown resinous substance obtained from dried perpetual follicles of the beaver *Castor faber* L. family Castoridae.

The aromatic compound is obtained from two pear-shaped membranous sacs found in male and female and are situated between the anus and the sexual organs of the animal.

Two varieties North American *castor canadensis* and Russian are known in the market.

Fig. 14.47 : North American Beaver (Castor canadensis)

Description

When fresh it is creamy secretion but turns to brown after drying. It has a strong odour but becomes more pleasant on dilutions. It is soluble in alcohol.

Chemical Constituents

It contains volatile oil (1 - 2 per cent), resin (80 per cent), a crystalline compound known as castorin. Benzyl alcohol, methyl phenol and l-borneol are other contents.

Uses

It is used as flavouring agent in alcoholic and non-alcoholic beverages and as a fragrance in cosmetics.

CIVET

Synonym

Zibeth, Civet cat.

Biological Source

Civet is the odorous secretion obtained from highly specialised scent glands in the region of the external generative organs of male and female civet cat, *Viverra zibetha* Linn family Viverridae.

Geographical Source

Civet cats are found all over South East Asia, Africa, Madagascar and Southern Europe. Three genera of these cats are also found in India.

These cats can be distinguished from ordinary cats by their five toed hind feet, longer muscles and shorter limbs and by the presence of scent glands. Civets are omnivorous in diet.

Description

Colour : Civet is pale yellow semi-liquid

Odour : Obnoxious odour, when diluted, becomes very pleasant.

Taste : Bitter

Solubility : Partly soluble in hot alcohol, ether and insoluble in water

Fig. 14.48: Indian civet cat (*Viverra zibetha*)

Chemical Constituents

Civet mainly contains civetone, civetol, indole, ethylamine, propylamine and few unidentified free acids.

$$CO(CH_2)_7\,CH = CH(CH_2)_7$$

Civetone

Standards of Quality

It is fusible and burns without leaving any residue.

Loss on drying at 100°C	:	Not exceeding 25 per cent
Mineral Matter	:	Not more than 2 per cent
Acetone extract	:	65 - 80 per cent
Alcohol soluble extract	:	45 - 65 per cent
Ether soluble extract	:	11 - 24 per cent

Uses

It is used as a flavouring agent in cosmetics and foods and also as fixative in perfumery.

MUSK

Synonyms

Moschus, Kasturi.

Biological Source

Musk is a dried secretion obtained from the prenaptial follicles of male musk-deer *Moschus moschiferus* Linn family Cervidae.

Geographical Source

The animal musk deer (*Fig 14.49*) is found in the mountainous regions of the Himalayas and in China. It is also reported in Russia.

Fig. 14.49: Musk-deer

Processing of Musk

Musk deer is a small animal of 50 cm height and differs from other deers in respect that there are no antlers in either sex. It is a wary animal of iron-grey colour.

The fragrant musk is contained in an oval hairy projecting sac found only in the male animals, situated between the umbilicus and the prepuce. The size of sac is about 3 - 7 cm long and 3 - 5 cm broad which weighs about 25 to 35 g and contains half its weight of musk secretion. The sac has a small hairy orifice in its anterior part and is marked posterior by a ***furrow***, which corresponds with the opening of the prepuce.

Description

Colour : Secretion is dark brown or brownish-red

Odour : Very strong characteristic and penetrating

Taste : It has bitter taste

Musk occurs as viscid mass or coarse granular powder.

Chemical Constituents

When distilled, musk yields about 1.5 per cent w/w of dark brown volatile oil. It also contains fat, wax, cholesterin, albuminoids and resins.

The volatile oil chiefly contains ketonic substances, of which muskone is the main constituent.

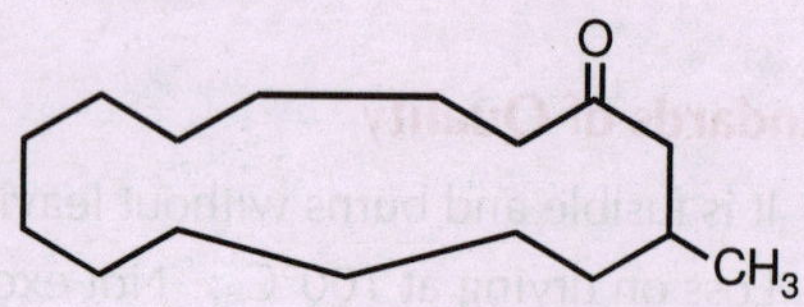

Muskone

Standards of Quality

Loss on drying	:	20 - 30 per cent
Alcohol soluble extractive	:	20 - 30 per cent
Water soluble extractive	:	55 - 70 per cent
Ash	:	Not more than 8 per cent

Uses

It retains its odour for very long time and the smell is perceptible even if diluted to 3000 times. Due to this property, it is used in perfumery. Therapeutically, it is administered as powerful stimulant in the treatment of hysteria.

Substitutes

There are several animals that secrete substances with strong odour, more or less similar to musk. Beaver (*Castor fiber*), Civet (*Viuerra zibetha*) and American musk (*Fiber zibeythicus*) are few important examples.

The herbal source, known as Musk mallow (*Abelmoschus moschatus*, family Malvaceae) is found abundant in hotter plains of India. This plant is cultivated in Maharashtra, Gujarat, and M.P. The seeds of this plant contain volatile oil, which has resembling flavour to that of musk. Synthetic musk is a yellowish white crystalline compound and has very strong persistent odour, somewhat similar to, but distinct from natural musk.

(B) SESQUITERPENOIDS

Sesquiterpenoids are derived from three isoprene units. Hence, they are C_{15} compounds and biosynthesized from farnesyl pyrophosphate. They have wide occurrence in nature and mainly found in plants and fungi. The carbon skeleton of sesquiterpene is found in highly diverse forms as compared to other terpenes. Among such various forms, sesquiterpenoid lactones are considered more significant, due to their marked biological activity. They contain α-methylene - γ-lactone system. Some members of sesquiterpenoid lactones also contain α, β-unsaturated carbonyls and epoxides. Sesquiterpenoid lactones are further categorised into three groups viz.

(i) Germacranolides (Ten membered ring),

(ii) Eudesmanolides (Two fused six-membered rings) and

(iii) Guainolides (Five membered and seven membered ring fused).

ACORUS

Synonyms

Calamus, Bach, Ghoda Vaj, Vaj, Sweet flag, Vekhand.

Biological Source

These are the dried rhizomes of the plant known as *Acorus calamus* Linn, belonging to family Araceae. It contains not less than 1.5 per cent of volatile oil.

Geographical Source

Calamus is native of Eastern Europe and is found wild in England. It is chiefly collected in Holland, North America, most of the European countries and in Central Asia. In India, it is cultivated, as well as, found wild in Himalayas and also in Sikkim. In Himalayas, it is found upto an altitude of 2000 m.

The rhizomes are collected during autumn when volatile oil content is highest.

Macroscopic Characters

Colour : Rhizomes are brown in colour.

Odour : Characteristic and aromatic.

Taste : Bitter, pungent and disagreeable.

Size : 5 - 15 cm in length and 1 to 2 cm in thickness.

Shape : The rhizomes are cylindrical and branched. They are somewhat shrunken and with deep longitudinal wrinkles. The leaf scars are more prominent on upper surface and encircle the rhizomes. The under surface of rhizomes bear very small, but raised circular root scars. The lateral branching is occasional. The fracture is short.

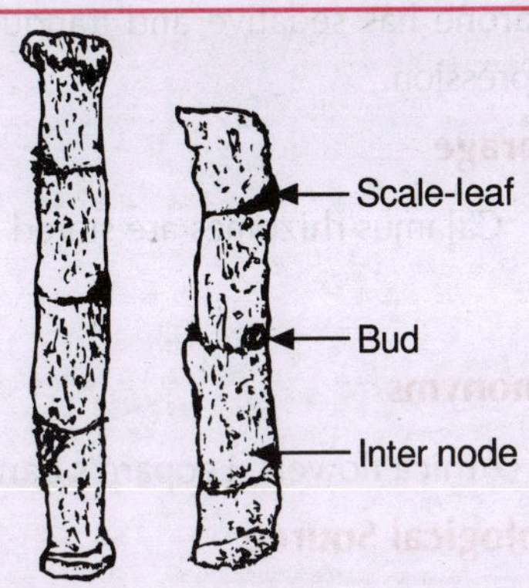

Fig. 14.50 (a): Acorus Rhizomes

Microscopic Characters

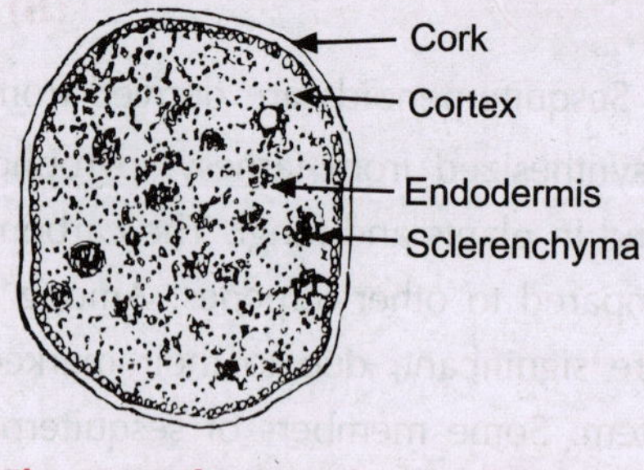

Fig. 14.50 (b): T.S. of Acorus Rhizome

The transverse section of the unpeeled acorus rhizome shows rarely cork followed by the cortex. The large stele separated by a prominent endodermis is exhibited. Numerous, small, ovate vascular bundles are scattered throughout the section. The oleoresinous cells and abundant starch grains are also found. Calcium oxalate prisms are seen accompanying the fibres. Tannin cells are found in parenchyma of the drug.

Outer part of cortex contains smaller cells and sometimes, may be collenchymatous. Stele is composed of rounded parenchyma enclosing large air spaces and vascular bundles. Vascular bundles are endodermis and are of concentric type. Secretive cells are present in parenchyma.

Chemical Constituents

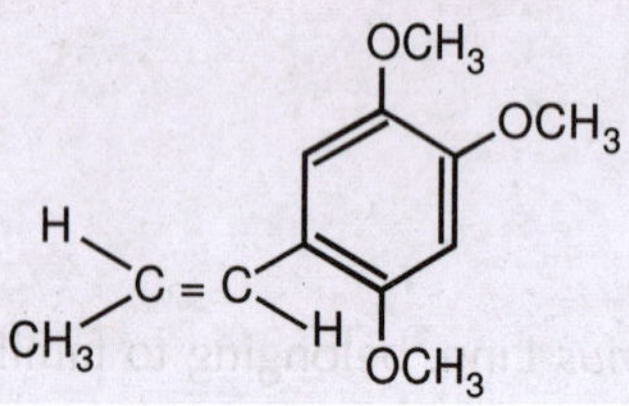

Asarone

The drug contains 1.5 - 3.5 per cent of volatile oil, starch, resin (2.5 per cent) and tannin (1.5 per cent). Volatile oil contains asaraldehyde. The other contents of the oil are asarone and eugenol. Calamus also contains a bitter amorphous principle known as acorine. The oil from North American race is devoid of asarone, whereas, European race contains about 10 per cent asarone.

Standards of Quality

Total ash	:	Not more than 6.0 per cent
Acid insoluble ash	:	Not more than 1.0 per cent

Uses

Calamus is used as carminative, bitter stimulant and vermifuge. Volatile oil is used in perfumery, as insect-repellant and for flavouring gin, beer, etc. It is also used for preservation of food grains. Asarone has sedative and tranquillising properties. In India, the drug is also used in epilepsy and depression.

Storage

Calamus rhizomes are stored in well closed containers.

ARNICA

Synonyms

Arnica flowers, Leopard's bane, Mountain tobacco.

Biological Source

The dried flower heads, of *Arnica montana* Linn, belonging to family Compositae, are a source of medicine which is used in external form only.

Geographical Source

This species occurs growing as a perennial herb, mainly in calcareous soil of mountainous regions in Central Europe, England, Southern Scotland and in USA, Canada as a garden cultivated plant. The other species known to occur are *A. fulgens, A. sororia* and *A. cordifolia*. This plant has been named Arnika from the Greek word *ptarmikos,* meaning sternutatory, causing sneezing.

Macroscopic Characters

Arnica occurs as rhizomatose perennial herb which grows upto 20 – 30 cm in height. The rhizome grows in a horizontal direction. The stem is stiff, branched or unbranched. The leaves, less in number, are tiny, opposite, elliptical, sessile and have a leathery texture. Arnica bears daisy like flower heads, 8 cm in diameter and central flowers are deep yellow in colour. They occur solitary or in a cluster of 3 – 4 heads. The peripheral flowers are orange yellow and have 3–toothed tongue. The root is dark–brown with wiry rootlets on lower side. July – August (summer) is a better period of harvesting leaves, while roots are harvested after autumn.

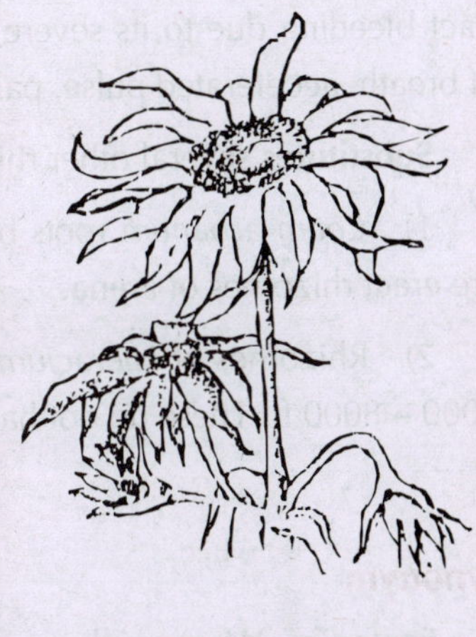

Fig. 14.51: Arnica flower

Odour : Pleasant

Taste : Slightly irritant

Chemical Constituents

Flowers contain 1.0 per cent volatile oil, flavonoids, terpenoids and bitter principles.

The flowers contain sesquiterpene lactones collectively called helenanolides which include helenalin; 11, 13 – dehydrohelenalin, epoxyhelanalin. Arnifolin is also present as a sesquiterpene lactone in flowers. Arnica flowers and rhizomes contain a bitter principle called arnicin which includes various chemically unrelated constituents. Flowers and rhizomes also contain arnidiol and faradiol and they are counter irritants. The flavonoid principles present in flowers are apigenin, quercetin, tricin and kaempferol. In total, seven flavones and nine flavonols have been identified. Volatile oil is present in flowers, rhizomes and roots. 4–hydroxythymoldimethyl ether is the major constituent of volatile oil, along with thymol.

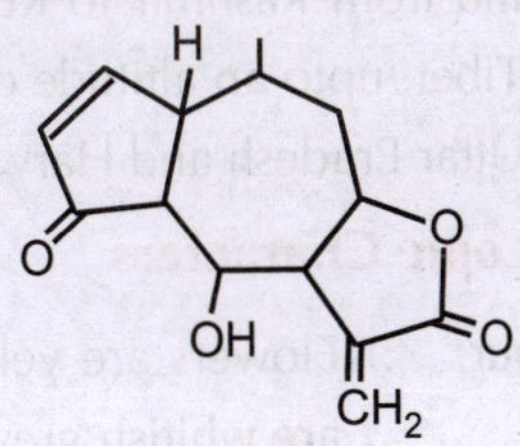

Helenalin

Standards of Quality

Ash : Not more than 8.0 per cent

Foreign orgnaic matter : Not more than 2.0 per cent

Uses

Arnica tincture is used as topical counter irritant for bruises, sprains and aches. Arnica is used in cosmetic and hair shampoos. It is also used in treatment of chronic rheumatism, spinal paralysis and amourosis (blindness). Arnica is mainly used in homeopathy for various purposes. Sesquiterpene lactones from helenanolide group like helenalin and other ester derivatives have been reported to possess antihyperlipidemic, antitumor, anthelmentic, antifungal immune stimulating properties. It is not suitable for internal use. Helanine is potent anti-inflamatory and antitumour compund.

Arnica is recommended only for external applications. Toxic symptoms, both in internal and external applications, are known. It is believed that activity and toxicity are due to sesquiterpene lactones, due to their strong sensitization capacities. Although used externally, in some cases, they cause allergic contact dermatitis.

When taken internally, helenalin causes fatal gastroenteritis with vomitting, diarrhoea and G. I. tract bleeding due to its severe irritant activity. It also causes respiratory stimulations and shortness of breath, accelerated pulse, palpitation and paralysis of heart.

Substitutes several other rhizomes are present in drug but only following two are prominant

1) *Ceum arbanum* roots belonging to fam Rosaceae which have flavour similar to cloves and are erect rhizomes of arena.

2) Rhizomes of *Hieracium virosum* fam. Compositae found in Spain and kashmir at altitude of 7000 – 8000 ft. These do not have oleo-resin ducts and said to be poisonous.

ARTEMISIA

Synonym

Santonica, Wormseed.

Biological Source

These are the unexpanded flower-heads of *Artemisia cina Berg, Artemisia brevifolia Wall, Artemisia maritima* Linn. and other species of Artemisia, belonging to family Asteraceae. It is collected late in spring or early in summer when the flower-heads are young. It should contain not more than 2 per cent of stems, and not less than 0.75 per cent of santonin.

Geographical Source

It is found growing wild in the Kurran valley in Pakistan, Turkey and from Kashmir to Kumaon in Himalayas, as well as, in West Tibet, upto an altitude of 4000 metres. It is also found in Punjab, Uttar Pradesh and Haryana.

Fig. 14.52: Artemisia cina herb

Macroscopic Characters

Colour	:	Flowers are yellow in colour, while other parts are whitish-grey
Odour	:	Aromatic and sweet
Taste	:	Bitter and camphoraceous

The drug consists of yellowish or brownish flower-heads, which are oval in shape. Flowers are fertile with tubular corolla and short cylindrical tube and narrow limb. Calyx is absent.

Chemical Constituents

Santonica contains essential oil and two crystalline substances i.e. santonin and artemisin. The volatile oil content varies from 1.0 - 2 per cent, while the percentage of santonin is about 2.0 per cent. The volatile oil contains cineole, pinene and resin. The chief active constituent of the drug is santonin. Santonin is a sesquiterpene lactone which is anhydride of santonic acid. The amount of santonin present varies considerably depending upon the type of species and time and nature of collection. If the flower-heads are unexpanded and quickly dried, they yield over 3.0 per cent of santonin

Artemisin   Santonin

Identification

Boil 1 g finely powdered drug with 10 ml alcohol and filter. To the filtrate, add sodium hydroxide and heat again. The liquid develops red colour.

Uses

Santonica is used as a strong anthelmintic, especially for round worms. It has less or no effect on hook worms and tape worms. The crude drug, at present, is rarely used in therapeutic and has been totally replaced by santonin. Santonin is found to create remarkable disturbances of vision. Santonin is cumulative in action. The B.P. variety of the drug is obtained from *Artemisia cina*.

Substitutes

Artemisia is substituted by a tall aromatic shrub, known as *Artemisia vulgaris* Linn. (Compositae) found in all mountaneous districts of India. It is found at Mount Abu in Rajasthan, in Konkan, Uttar Pradesh, Tamil Nadu and Jammu and Kashmir in Himalayas, up to an altitude of 4000 metres. It occurs in Sikkim and Sri Lanka too.

ARTEMISIA ANNUA

Synonyms

Quinghao, Sweet-annie, Worm weed.

Biological Source

This consists of Chinese traditional herb *Artemisia annua* Linn. Family Asteraceae and should contain not less than 0.8 per cent of artemisinin on dried basis.

Geographical Source

It is endemic to China and grows wild in Europe and America. It is cultivated in Vietnam, China, Iran, Turkey and Australia. It is also cultivated on experimental basis in Gujarat, Jammu & Kashmir, Uttar Pradesh, Himachal Pradesh and Karnataka.

Cultivation and Collection

CIMAP at Lucknow has developed uniform and high yielding genetically improved variety of this plant and is being cultivated for the purpose of Aretemisinin. It gives about 29 quintals per hectare of dry leaves and flowering tops containing 0.11 per cent of Artemiesinin. The yield per hectare achieved is 3.19 kg of Artemisinin.

Fig. 14.53: Twig of *Artemisia annua* herb

Macroscopic Characters

It is annual herb 0.7 - 1.75 m in height with deeply grooved branches. The leaves are bipinnate compound and dark green in colour, strongly aromatic in flavour and bitter in taste. Leaves are 0.25 - 0.5, 5-7 cm in size, 1-2 pinnately divided into lanceolate oblong deeply cut segments and subglobose. It bears tiny green yellow flowers in clusters.

Stems are longitudinal, cylindrical, about 2-5 mm in diameter and 30 to 80 cm in length.

Microscopic Characters

Histologically leaves are characterised by isobilateral arrangement. Anomocytic stomata alongwith glandular trichomes abundant in number on both the surfaces with very short stalk and multicellular head. T shaped covering trichomes are the special character of leaves.

Chemical Constituents

Out of seven species of *Artemisia,* only *A. annua* contains Artemisinin, deoxyartemisinin 1.0 per cent. It is a sesquiterpene lactone with an internal peroxide linkage. Artemisinic acid, arteannuin A and B and artemether are the other constituents of drug.

Amyrin, luteolin, β-sitosterol, stigmasterol are also reported in the drug.

Unlike other known antimalarials, artemisinin does not contain a heterocyclic ring with nitrogen. Hence, it is a new type of structure in the group of antimalarials. It is sparingly soluble in water while it shows decomposition in other solvents. Volatile oil 0.3 to 0.4 per cent with artemisia alcohol, artemisia ketone, camphor, Caryophyllene and myrcene are also present in the drug.

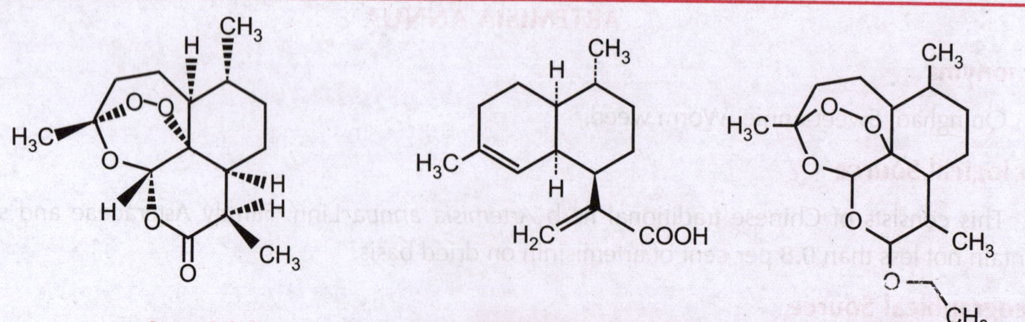

Artemisinin **Artemisinic acid** **Artemether**

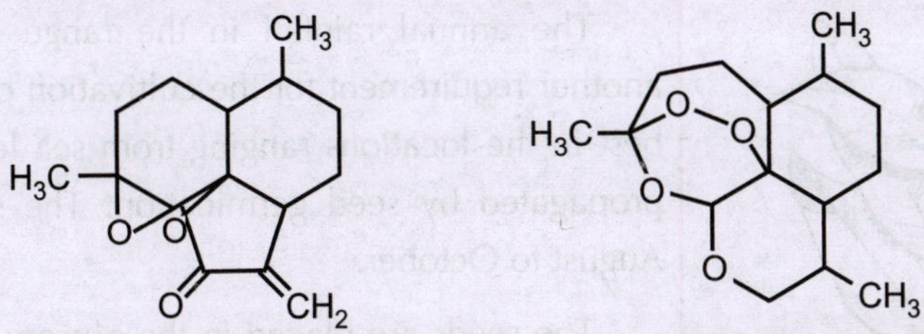

Arteannuin B **(+) – Deoxyartemisinin**

Standards of Quality

Ash content	:	Not more than 11.0 per cent
Acid insoluble ash	:	Not more than 3.0 per cent
Foreign organic matter	:	Not more than 2.0 per cent

Uses

Artemisinin shows antimalarial effects by its rapid blood schizonticidal activity. The trophozite development is arrested and pathological consequences are not observed. It acts both against Chloroquine sensitive and resistant *P. falciparum* and *P. vivax* malarial parasites. Presently, artemisinin is the most potent drug for treating cerebral malaria. Artesunate and artemether, which are the derivatives of artemisinin, have also been found to be highly effective. Artemether has been developed by C.D.R.I., Luknow. Artemisic acid is reported to have not only antibacterial but also cytotoxic and anti-inflammatory actions. Artemisinin has shown anti-HIV activity with an IC_{50} of 100 μm and EC_{50} of 100 μm.

CLOVE

Synonyms

Caryophyllum, Clove flower, Clove buds.

Biological Source

Clove consists of dried flower buds of *Eugenia caryophyllus*, family Myrtaceae. It should contain not less than 7.0 per cent (w/w) of eugenol calculated on dried basis.

Geographical Source

It is indigenous to Amboyna and Molucca islands. It is now cultivated chiefly in Zanzibar, Pemba, Penang, Madagascar, Caribbean islands, Sri Lanka and India. In India, cloves are grown in Nilgiri, Tenkasi-hills and in Kanyakumari district of Tamil Nadu state. It is also cultivated in Kottayam and Quilon districts of Kerala.

Cultivation and Collection

Deep rich loamy soil with high humus content is suitable for clove cultivation. It is also found growing favourably in open sandy loam and laterite soils of South Kerala region. Water logging of soil must be avoided. It needs warm humid climate and grows well in the vicinity of sea.

Fig. 14.54: Flowering Twig of *Eugenia caryophyllus* Linn

The annual rainfall in the range of 150 - 250 cm is another requirement for the cultivation of clove. Clove thrives best in the locations ranging from sea level upto 900 m. It is propagated by seed germination. The seeds are sown from August to October.

The seeds are placed in the nursery beds at a distance of 10 cm. It takes from four to five weeks for seed germination. The seedlings of clove are slender and delicate, with poor growing rate. After six months, they are transplanted to the pots where they are allowed to grow for a year.

Thereafter, they are again transferred to the field and are provided with the shade in initial stages of growth. Clove can also be grown with arecanut, coconut or nutmeg plants.

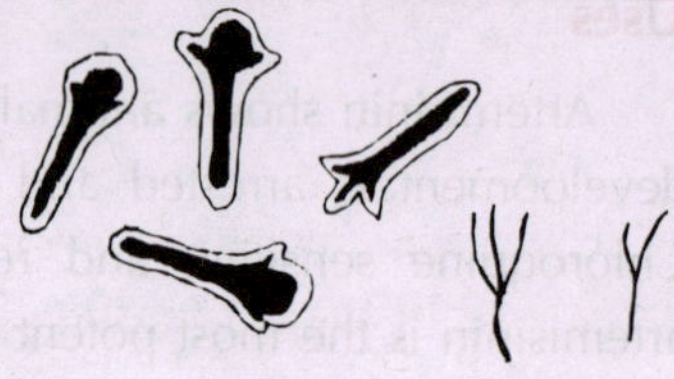

Fig. 14.55: Cloves and Clove Stalks

The plants are provided with suitable fertilizers like ammonium sulphate, super phosphate and potash. Generally, the fertilizers are given in two doses; first in May/June and second in October. The plant starts bearing after 7 - 8 years and a satisfactory yield per hectare of the drug is achieved only after 15 - 20 years of growth. Under normal condition of soil, clove tree produces on an average, 3 kg of the drug. Cloves are handpicked or collected by beating with bamboos. This operation commences when the cloves start changing their colour from green to slightly pink. When the tree is tall and cloves are beyond reach, platform ladders are used for collection. The cloves are dried in the sun and freed from foreign material and graded. The cloves on drying become perfectly crimson or brownish-black in colour.

Apart from cloves and clove powder as spices, India exported 4.8 tones of clove oil costing approximately ₹ 20 lakhs during 1994-95, and clove oleoresins of ₹ 25 lakhs.

Macroscopic Characters

- Colour : Crimson to dark brown.
- Odour : Slightly aromatic
- Taste : Pungent and aromatic followed by numbness
- Size : About 10 to 17.5 mm in length, 4 mm in width, and 2 mm thick.
- Shape : Hypanthium is surmounted with 4 thick acute divergent sepals surrounded by dome shaped corolla. The corolla consists of unexpanded membranous petals with several stamens and single stiff prominent style. Cloves are heavier than water.

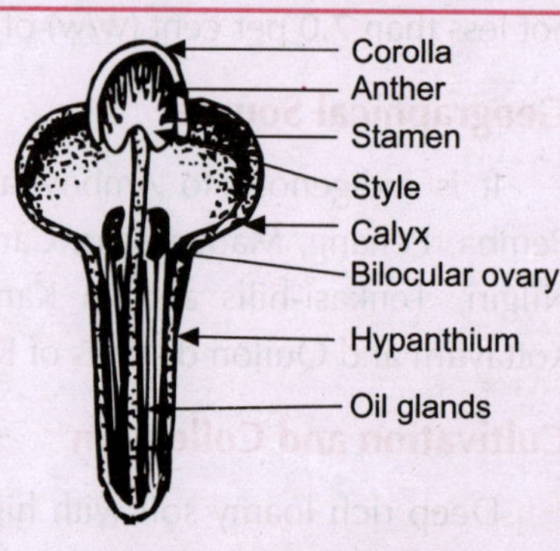

Fig. 14.56: L.S. of Clove

Microscopic Characters

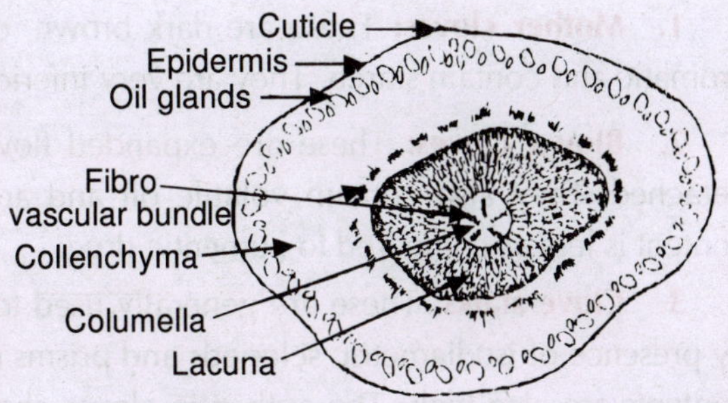

Fig. 14.57: Transverse Section of Clove

The epidermis of the clove is covered with thick cuticle. The epidermis itself consists of straight walled cells and large anomocytic stomata. The oil glands, which are ovoid and schizolysigenous are found in all parts of the drug.

Phloem fibres, which are isolated, are occasionally found in the spongy tissue. Cluster crystals of calcium oxalate and small number of stone cells are found in the drug. Clove does not contain starch.

Chemical Constituents

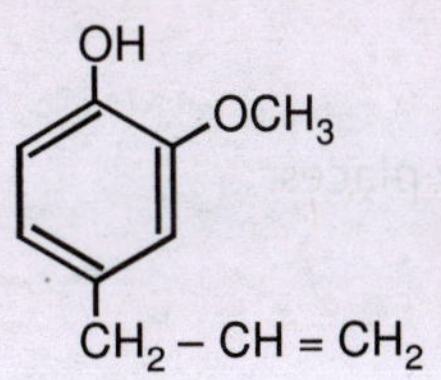

Eugenol

Clove contains about 15 to 20 per cent of volatile oil; 10 per cent to 13 per cent of tannin (gallotannic acid), resin, chromone and eugenin. The volatile oil of the drug contains eugenol (about 70 to 90 per cent), eugenol acetate, Caryophyllenes and small quantities of esters, ketones and alcohols.

Oil of clove is colourless to pale yellow in colour. It becomes thick and darker in colour on storage. It has specific gravity of 1.038 - 1.06, refractive index is 1.527 to 1.535 and it boils at 250°C.

Chemical Test

If the transverse section of clove is treated with strong potassium hydroxide solution, the needle-shaped crystals of potassium eugenate are observed.

Standards of Quality

Loss on drying	:	Not more than 12.0 per cent
Ash	:	Not more than 07.0 per cent
Acid insoluble ash	:	Not more than 10 per cent
Acid soluble extractives	:	Not less than 7.0 per cent
Water soluble extractives	:	Not less than 20.0 per cent

Uses

Due to presence of Tannins, it tightens the gums hence used in dental products. Clove is used as a dental analgesic, carminative, stimulant, flavouring agent, an aromatic and antiseptic. It is also used in the preparation of cigarettes. The oil is used in perfumery and also in the manufacture of vanillin.

In searcity of cloves, clove stalks and blown cloves are used as substitute to the cloves.

Adulterants

1. **Mother cloves:** These are dark brown, ovate ripened fruits of clove tree. They are slightly aromatic and contain starch. They are very inferior in volatile oil content.

2. **Blown cloves:** These are expanded flowers of the clove tree. The stamens generally get detached. They also contain volatile oil and are similar in colour to the cloves. The volatile oil content is less as compared to authentic drug.

3. **Clove stalks:** These are generally used to adulterate the powdered cloves and are detected by presence of isodiametric sclerieds and prisms of calcium oxalate. Their ash value and crude fibre contents are also high. The authentic cloves should not contain more than 5 per cent of stalks to pass the pharmacopoeial limit. Due to similarity in colour, odour and taste, clove stalks are mixed with the cloves. Clove stalks contain only 5 per cent of oil.

4. **Exhausted cloves:** These are the cloves from which oil has been extracted by distillation. They are dark in colour, more shrunken and when pressed with finger nails, do not show the presence of oil. Exhausted cloves float on water.

Storage

Clove and its powder should be stored in air-tight containers in cool and dry places.

CUBEB

Synonyms

Tailed pepper; Kababchini.

Biological Source

Cubeb consists of dried fully grown unripe fruits of *Piper cubeba* Linn, family Piperaceae.

Geographical Source

It is indigenous to Indonesia. It is cultivated in Sri Lanka, India, Indonesia and Caribbean islands.

Macroscopic Characters

Colour : Fruits are dark brown

Odour : Strong spicy

Taste : Bitter and spicy

Size : About 3 - 4 mm in diameter

Shape : Globular, depressed at the base, pericarp is reticular wrinkled, bears triradiate stigma, with slender stalk about 4 mm long. It contains only one seed. Internally, it is white and hollow.

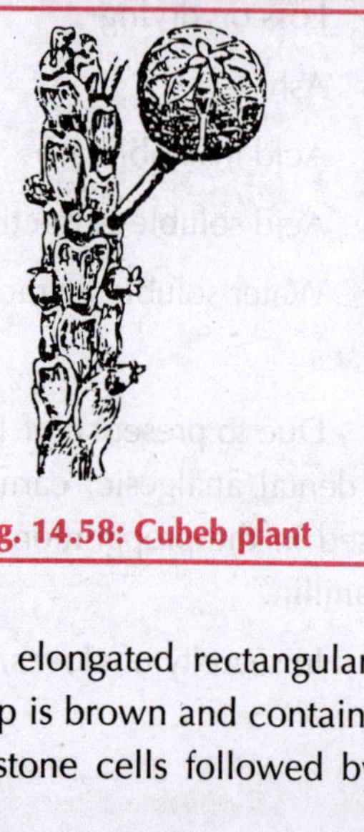

Fig. 14.58: Cubeb plant

Microscopic Characters

The transverse section shows presence of a single layer of radially elongated rectangular, sclerenchymatous cells in the inner layer of pericarp. Outer layer of pericarp is brown and contains crystals of calcium oxalate. Hypodermis is made up of small group of stone cells followed by parenchymatous mesocarp. Mesocarp contains large oil cells, round starch grains and calcium oxalate crystals. Perisperm consists of large polyhedral thin walled cells. It does not contain beaker cells as reported in black pepper fruits.

Chemical Constituents

Cubeb contains 5 - 20 per cent of volatile oil, resins, fatty oil, gum, and cubebic acid, cubebin and wax.

Cubeb oil is pale green or yellow liquid, contains dipentene, cadinene, and cubeb camphor. Volatile oil of cubeb has the following standards.

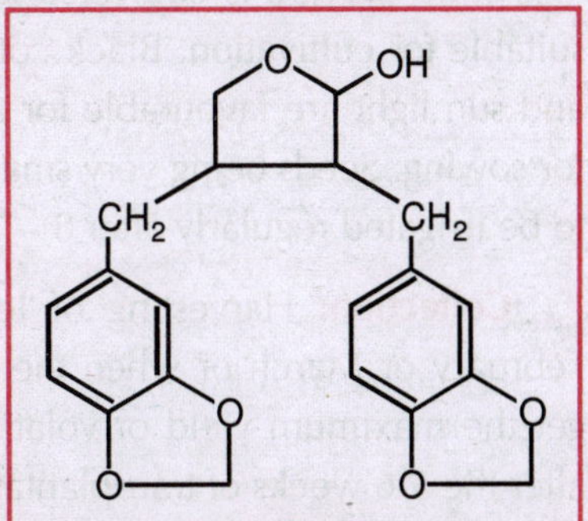

Cubebin

Specific gravity	:	0.905 - 0.925
Optical rotation	:	20 to - 40°
Refractive index	:	1.4800 - 1.5020

Standards of Quality

Foreign organic matter	:	Not more than 2.0 per cent
Immature fruits	:	Not more than 10.0 per cent
Ash	:	Not more than 8.0 per cent
Acid insoluble ash	:	Not more than 2.0 per cent

Chemical Test

To powdered cubeb fruits, add drop of sulphuric acid, purple colour is produced.

Uses

It is used as stimulant and aromatic. Cubebin which is ether soluble constituent of the fruit, is used as urinary antiseptic. It is also recommended in the treatment of cough.

DAVANA OIL

Biological Source

It is the volatile oil obtained by steam distillation of flowering herbs *Artemisia pallens* wall ex. DC belonging to family Asteraceae.

Geographical Source

The plant *Artemisia pallens* called *Davana* is cultivated in Maharashtra, Andhra Pradesh, Karnataka and Tamil Nadu. Davana oil is produced on commercial basis in the South Indian states especially in Kolar and Chittoor districts.

Cultivation and Collection

Davana is cultivated in Pune district of Maharashtra and more specifically in Saswad, Jejuri, Sangali and Kolhapur also. Approximately 200 hectares of land is under cultivation of Davana.

Description

It is a Rabi crop and reaches the height of about 45 – 60 cm. It bears number of branches and the plant has yellowish green colour leaves are arranged atternately on rachis, plant bears very minute oval shaped brown coloured seeds. Seeds are also sweet smelling and pleasant about 6000 seeds weigh one gram.

Cultivation: It is cultivated late in October or early in November. Well drained gravelly soil is suitable for cultivation. Black cotton or fertile soil is not suitable for cultivation. Humid atmosphere and sun light are favourable for the growth 1.5 kg of seeds mixed with 10 kg of fine sand are used for sowing. Seeds being very small they are also covered with fine sand before irrigation. Crop needs to be irrigated regularly after 8 – 10 days. Enough fertilisers are provided as required.

Collection: Harvesting of leaves and flowers is done in February or March or when the crop is in flowering stage. To get the maximum yield of volatile oil. The plants are dug out after the 5-6 weeks of transplantations in the farm.

The entire plant is cut about 1 cm above the soil either in the morning or evening. Plants are dried in the shade for 2 to 3 days. Yield is about 10 tonnes and can produce 12 kg of oil from leaves.

Fig. 14.59: Davana - herb

Macroscopic Characters

The plant bears characteristic leaves and tiny yellow flowers. The leaves show dissected and petiolate nature. It gives erect, branched and woolly stem. The flowers have compound raceme inflorescent character. The seeds are also sweet smelling.

Discription

Davana oil: It is pale yellow to yellowish brown in colour. It is transparent, light and has a strong agreeable flavour.

Solubility

It is insoluble in water and soluble in alcohol.

Chemical Constituents

Davana oil contains sesquiterpene called davanone, artemone and non-davanone. It contains davana furans which are responsible for its aroma. The oil also contains minor constituents in the form of cineol, borneol, geraniol, linalool, eugenol etc.

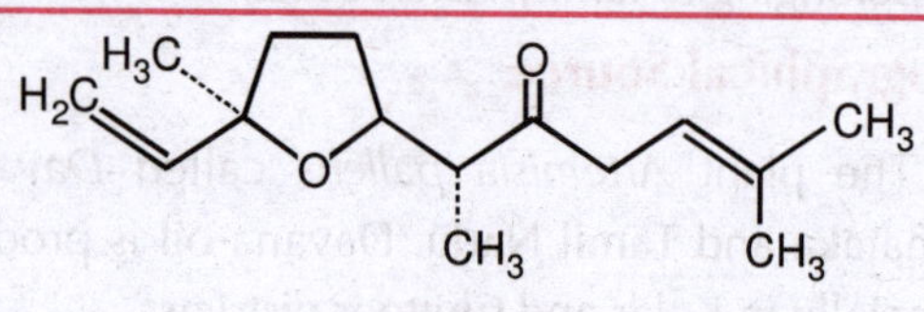

Davanone

Standards of Quality

Specific gravity	:	0.942 to 0.970
Refractive index	:	1.479 to 1.491
Optical rotation	:	+ 15 to + 60º
Flash point	:	93.3ºC

Uses

Used as pleasant flavouring agent and in perfumes.

Davana has about 0.2 per cent volatile oil. The yield of this oil is about 7.5 kg/hectare. The flowers have higher content of oil. During the year 2017 – 2018 and 2018 – 2019 export of 5.26 and 3.4 US million dollars of Davana oil was made.

FEVERFEW

Synonyms

Featherfew, Featherfoil, Midsummer daisy, Altamisia, *Chrysanthemum parthenium, Pyrethrum parthnium.*

Biological Source

Feverfew consists of fresh or dried leaves and aerial parts of the plant *Tanacetum parthenium* (L) Bernh belonging to family Asteraceae (Compositae), when the plant is in flowering stage.

Geographical Source

Found in Britain, Balkan peninsula and also cultivated in Britain.

Macroscopic Characters

Colour : Yellowish green

Odour : Strongly aromatic

Taste : Bitter and acrid

(It may cause blisters on the lip and soft palate.

Stems: Quadrangular, branched and about 1 - 5 mm wide, bearing pinnate or bipinnate, alternate leaves.

Leaves: Leaves are petiolate, ovate 5 - 7 segmented, 2.5 - 10 cm in size, margin is crenate, apex is obtuse, mid-rib is very much prominant on under surface.

Flowers: In the form of corymb, from the axils of the leaves about 2.5 cm in size, capitulum with downy bracts, numerous yellow disc-florets, ray florets in single row which are white and short, receptacle is hemispherical.

Fig. 14.60: Feverfew herb

Microscopic Characters

Leaves: Striated cuticle, epidermal cells with wavy anticlinal walls, anomocytic stomata on lower epidermis, trichomes of both-types i.e. multicellular, uniseriate covering trichomes and also biseriate glandular trichomes with biseriate multicellular head.

Stems: Thin walled epidermal cells, anomocytic stomata, uniseriate covering trichomes. The cortex is parenchymatous and narrow. Collenchyma in ridges, xylem with spirally and thickened vessels are present. Pith is parenchymatous.

Chemical Constituents

Feverfew leaves are reported to contain sesquiterpens lactones, chrysanthemin A and B, parthenolide, chrysanthemonin, chrysanthemolide and partholide. The volatile oil found in traces (0.02 to 0.07 per cent) contains borneol, camphor, farnesene and their esters, while pyrethrin flavonoids are the other constituents of the herb.

Standards of Quality

Foreign organic matter	:	Not more than 2.0 per cent
Water soluble extractives	:	Not less than 15.0 per cent
Tolal ash	:	Not more than 15.0 per cent
Acid insoluble ash	:	Not more than 03.0 per cent

Identification

TLC method of identification as described below is to be applied. Use solvent system A:

Prepare Test solution as under:

1. Extract 1 g of powdered feverfew with 10 ml methanol on water bath for 15 min. Cool and filter. Apply 20 µl of this solution to plate and
2. 0.025 per cent rutin in methanol and apply 20 µl of this solution to plate.
3. Spray the plate with spray reagent A and examine in ultra-violet light at 366 nm.

Major bands relative to rutin are whitish-blue 2.25; greenish blue 2.0; yellowish green 1.65, yellowish orange 1.5. There should be neither green-band at 2.3, nor blue-band at 1.1, (distinguishing from Matricaria and from Roman chamomile respectively).

Uses

Traditionally, feverfew is being used in the treatment of migraine, vertigo, arthritis, fever and menstrual disorders.

HOPS

Synonyms

Humulus, Lupulus.

Biological Source

These are the female dried fruits (strawbiles) of *Humulus lupulus* of *Cannabinaceae.*

Geographical Source

It is cultivated on large scale in California, England, Germany and Russia and found all over Europe.

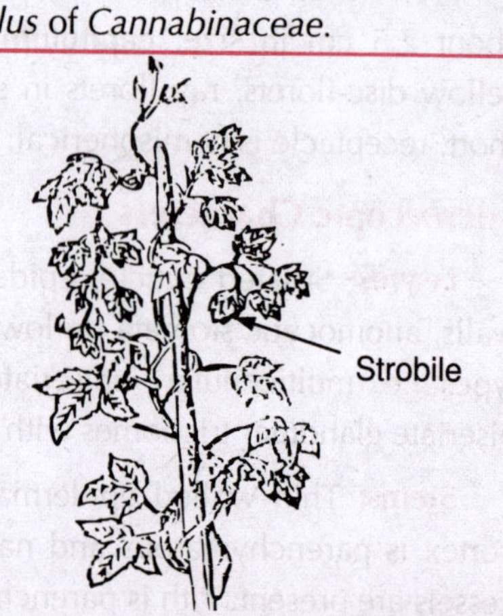

Fig. 14.61: Climber of Humulus

Description

It is climbing plant and perennial in habit. It is grown by cuttings. Plants are placed at a distance of 2 metres. Poles of 2.5 - 3.5 m are provided to support the plant. It starts bearing the fruits in second year only. Dried fruits are exposed to sulphur-di-oxide fumes to bleach the colour. Hop is a compound fruit.

Chemical Constituents

Hops contain volatile oil about 0.3 – 1.0 per cent tannin, resins, fatty acids. The chief constituent is humulene. Bitterness is due to humulol.

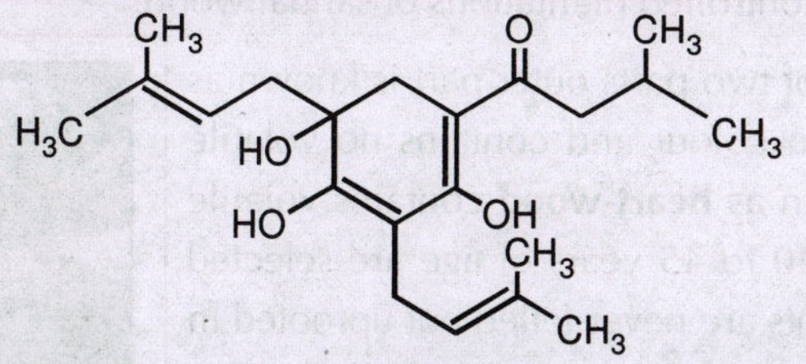

Humulone

The unsaturated alcohol viz. 2 - methyl - 3 - butene - 2 - ol exerts sedative and hypnotic activity. Humulone and lupulone, being keto-enol structures are sensitive to oxidation and hence, believed to get broken down to this unsaturated alcohol on prolonged storage. The drug shows smooth muscle relaxant effects on G.I. tract. It also shows in-vitro fungistatic and antibacterial activity. A novel flavone structure, viz. 3 - isopentenyl phlorisovalerophenone has a strong fungistatic activity against *staphylococcus, mucor, candida, trichophyton* and *fusarium*.

Standards of Quality

Ash	:	Not more than 10.0 per cent
Acid insoluble ash	:	Not more than 5.0 per cent
Water soluble extractives	:	Not less than 10.0 per cent
Foreign organic matter	:	Not more than 2.0 per cent

Uses

Hops are used as sedative and spasmolytic. Flowers are given to treat G.I. tract problems like irritable bowel syndrome. Lupulin is used for checking ejaculation in hyperexcitable males, due to its estrogenic and antiandrogenic effects. Hops creams and lotions are used for softening of skin.

Since old days, this plant has been known for its use in brewing industry and also in medicine. Sedative effects of its strobiles have been known since long. It was used as herbal tea, liquid extract and especially as hop pillow. It is believed that volatile oils of hops cause the sedative effects by inhalation when used in pillow form. The cones have also been known to cause toothache relieving and diuretic effects.

Volatile oil of hops is sedative, Bitter substances are stomatic and tonic, it also promotes sleep.

SANDAL WOOD

Synonym

Yellow sandal wood.

Biological Source

Sandal wood consists of heart wood of the plant known as ***Santalam album*** family Santalaceae.

Geographical Source

Exclusively this plant grows in forests of South India and specifically in Karnataka. It is also grown in Tamilnadu, Kerala, Andhra Pradesh and in Timor of Indonesia. Now a days Vietnam, New caledonia are also having controlled plantations of sandal wood.

Sandal wood consists of two parts outer part is known as **Sapwood** which is white in colour and contains no volatile oil. The inner part is known as **heart-wood** contains volatile oil. The plants which are 40 to 45 years of age are selected for isolation of oil. The plants are never felled but uprooted in rainy season. Plants are 12 – 20 meters in height and about 1.5 to 2.0 meteres in girth.

Fig. 14.62: Sandal wood twig

The flowers of the plant are red, green or violet in colour, while fruits are fleshy, globose indehiscent and purple to black-coloured. Sandal wood plant flowers in March to April.

Cultivation and Collection of Sandal Wood

It is the oldest drug known and has been in use since 4000 years. It is ever green plant and flurishes well from sea level upto 1800 metres in sandy and lateritic loamy soil. It is cultivated by sowing seeds in nursery beds. Seedlings produced and transplanted in the field. Sandal wood plants are parasitic and their successful regeneration depend upon providing suitable host. The roots of sandal seedling attach themselves to the root of the host plant through haustorea.

Germination of seeds is up to 80% and seeds take about 40 – 90 days to germinate after sowing. Total duration of the crop is about 30 – 40 years.

The systematic cultivation of these plants is undertaken mainly in South Indian states of Karnataka and Tamil Nadu. The leaves and other parts of the tree are free of volatile oil. The plant bears very beautiful flowers and fruits. Trees more than 25 years of age are normally selected for the collection of the oil. All parts of the wood contain volatile oil.

Forest department of Karnataka, TN, AP have undertaken commercial cultivation of sandal on Government lands, since it has high export potential.

Types of Propagation:

Natural regeneration: Most of the sandal wood is obtained from natural forests as it is reporduces from seeds easily, which are dispersed by birds and wind freely. The seeds germinate profusely after monsoon. But mortality rate of such seedlings is equally high for the withstand drought and excessive heat.

For successful regeneration of sandal trees naturally, following measures are to be taken -

- Providing suitable host.
- Avoiding damage to seedlings from torrid condition and during drought.
- Making available facilities for the free and natural growth of the plant.
- Looking after carefully the suppressed trees should be looked after carefully.

Artificial Regeneration: The systematic cultivation for artificial regeneration of these plants is undertaken in South Indian states of Karnataka, Tamilnadu and also in Andhra Pradesh. The method tried for regeneration is dibbling of seeds, raising seedlings on nursery-beds or in polythene bags and then transplanting in field. For successful artificial regeneration, the fact of root parasitism should be understood carefully and appropriate and suitable selection of host plant should be carried out.

Parasitism:

One of the most characteristic features of sandal is that, it is an obligatory root parasite. The roots of sandal seedlings attach themselves to the root of host plant through haustoria.

It has been observed that sandal obtains nutrition mainly of calcium and potash from soil through its roots, while part of nutrition, mainly nitrogen and phosphorus, is drawn from host plant. The hosts are classified as good, medium or poor hosts depending upon the number of host parasite connections.

Sandal can parasite almost all species including sandal itself, amongst the favoured one are *Cassia, Acacia, Bambusa, Gossypium, Pongamia,* Neem, *Semecarpus, Strychnos, Thespesia, Vitex, Eucalyptus*, etc.

Seed Treatment:

The treatment of seeds with gibberellic acid has been found to be effective in inducing quick and uniform germination.

After sowing seeds, the period, just after first monsoon showers, is suitable for growth. For transplanting seedlings from nursery bed to field period between May to October preferable. Total duration of crop is 30 to 40 years.

Cultivation:

The most favoured method is by dibbling the seeds. For this purpose, the seeds are collected during January-March. Then, the forest floor beneath selected host plant is cleaned and just after first showers of monsoon the sandal seeds are dibbled in a batch of 4 - 5 under selected host and are scattered over the entire area. Germination of these seeds is upto 80% and the seeds take about 40 - 90 days to germinate after sowing.

In some areas, sandal seeds are dibbled under thorny bushes. These thorny bushes provide protection against grazing animals and shade to protect from excessive heat. The method of cultivation is cheap and successful.

The land is prepared by ploughing and made clear of all weeds. The pits of 45 × 45 × 45 cm are dug at a distance of 3.5 × 3.5 m to 5 × 5 m and the sandal seedlings with the host seedlings are planted in the pits. On an average, 600 to 800 sandal seedlings are planted per hectare.

Nursery Raising and Transplantation:

In some areas especially Tamilnadu where sandal trees are raised in small area, the seedlings are raised on nursery beds in polythene bags and then transplanted into the field. The procedure for this is to collect fresh seeds from fruiting trees in October, the fruits are depulped and dried well. Then, they are sown in well prepared nursery beds, alongwith host plants at an equal distance.

The seedlings can also be raised in polybags. After a period of 30 - 35 days i.e. when the seedlings are 25 - 35 cm in height and basal portion turns darker, they are ready for transplantation. The appropriate timing for transplanting these seedlings is from May to October.

20 tonnes farmyard manure/hectare is required for good growth of plants.

It should be kept weed free. To keep the weeds under control, weeding is carried out in the first year at regular intervals. It is also done in the second year, if necessary.

In case sandal seedlings are overtopped by the host plant, the host is lopped so as to give enough sunlight to the seedlings of sandal. If there are any other climbers, they are cut and removed.

Few diseases are reported to affect the plant, amongst them spike disease is the most lethal in nature. Once it affects the crop, it is easily transmitted by insect vectors.

Symptoms of the disease are as follows:

- reduction in the size of leaves.
- internodes of the affected branches are shortened.
- reproductive phase is shortened in affected plants.
- leaves turn yellow and finally red, leading to death.
- also causes death of root-ends and haustorial connections.

However, no definite cure for sandal spike has been discovered so far. Disease control through vector host plant control is ideal, practical and lasting. The only control measure suggested is treating with antibiotics of tetracycline group. They are effective in suppressing the symptoms, though they cannot completely eradicate the disease. Application of terramycine or ledermycine alongwith fungicide like bentate to the diseased plant brings about temporary remission of symptoms within 20 – 30 days. The recovery lasts for 100 – 120 days, after which second treatment is necessary.

The other diseases of sandal reported are diseases of seedlings, leaf-spot disease alongwith certain diseases affecting the stored logs of sandal.

It is a rainfed crop and no special arrangements for irrigation are required to be made; though, during dry weather it should be irrigated once or twice till the crop is well established.

Sandal is susceptible to virulent disease caused by fungal and nematode infection. Hence, seed beds are spread with fungicide dithane Z-78 (0.25%) once in 15 days to avoid fungal attack and with Ekalux solution (0.02%) once in a month, to avoid nematode attack.

The sandal tree starts flowering at an early age of 2-4 years. But selection for the collection of seeds can be done only when they are 20 - 25 years old and the quality of seeds is excellent. The tree blooms twice a year once, prior to onset of monsoon and the next after monsoon. The fruits mature between July - September when they are collected.

For the harvesting of sapwood, heart wood which is the actual part containing sandal oil, should be in the matured state or its formation should be maximum. The roots of sandal are also scented, hence harvesting is carried out by uprooting and not by cutting. The heartwood formation in the tree starts when it is of 20 years old and it is at its maximum when the trees are 35 to 60 years old and have attained the girth of 40 - 60 cm. The heartwood has clear demarcation.

The fully grown plants are uprooted and the bark from roots and stems is removed alongwith little sap wood. The wood is yellowish or pale red in colour. It is very much dense, hard, heavy and splits easily. The wood surface shows darker and higher zones. It is cut in small pieces and subjected to steam distillation. It contains 2.5 per cent oil. India's exports of sandal wood oil were ₹ 16 crores during 1994-95 and ₹ 23.5 crores during 1995-96.

Morphological Characters

Colours : Sandal wood is yellowish to pale brown

Odour : Strong and fragrant

Taste : Slightly bitter

Size : Logs are about one metre in length and about 15 – 20 cm in diameter.

Extra features: Wood is hard, heavy and dense and splints easily.

Chemical Constituents

Sandal wood contains 2.0 – 5.0 per cent of sandal wood oil, which mainly contains santalol, sesque terprenes, sentane and santenone.

Uses

The volatile oil of chandan calms the mind, nervous system, mood it awakens courage, strength and intelligence. Oil is stimulant also anti-pyretic, diuretic and expectorant. It is used technically for manufacture of boxes, carvings and it is largely used in perfumary as flavouring agent.

SANDAL WOOD OIL

Synonym

East Indian sandal wood oil.

Biological Source

It is obtained by distillation from the heart-wood of *Santalum album* Linn, family Santalaceae.

Geographical Source

The plants are found in India and Malaysia. Sandalwood tree is an evergreen plant of 10 - 12 m height found widely distributed in India.

Description of Sandal Wood Oil

Colour : It is pale yellow to colourless viscid liquid

Odour : Characteristic, persistant

Taste : Unpleasant.

Solubility : It is very slightly soluble in water and soluble in alcohol and chloroform.

Chemical Constituents

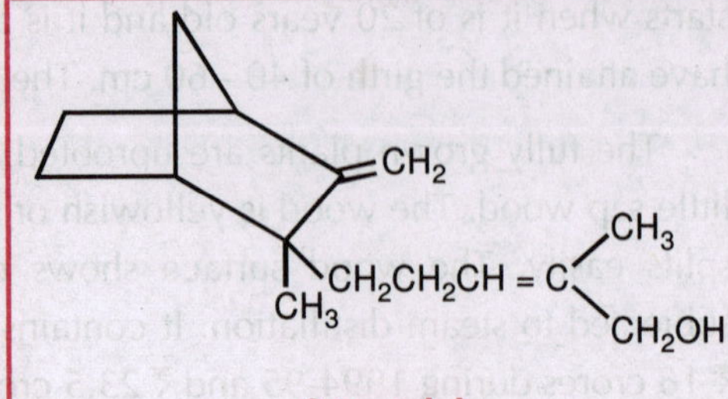

β-Santalol

Oil of sandal contains about 95 per cent of two isomeric sesquiterpene alcohols, α - santalol (b.p. 300 - 301°C) and β - santalol (b.p. 170 - 171°C). Additionally, the oil contains an aldehyde santalol $C_{16}H_{22}O$, santene, santenone, teresantol, santalone and santalene. The oil is present in elements of wood. It is not present in or secreted by any special types of cells or glands.

Standards of Quality

Specific gravity	:	0.973 - 0.985
Optical rotations	:	15° - –20°
Refractive index	:	1.500 - 1.510
Esters	:	Not less than 2 per cent w/w as santolyl acetate
Alcohols	:	Not less than 90 per cent w/w as santalol

Uses

Sandal wood oil is used for symptomatic treatment of dysurea and in diminishing the frequency of micturition marked in the tuberculosis of the bladder. It is mainly used as a perfume in preparation of cosmetics and incense sticks. Export of sandal wood oil worth of 0.55 ml 0.1.61 US million dollars was made during 2017 – 18 and 2018 – 19 respectively.

Substitutes

The volatile oil obtained from wood of plant *Eucarya spicata*, found in Western Australia known as Australian sandal wood oil is a substitute for the oil. It contains sesquiterpene alcohols known as fusanols.

Being very costly, the volatile oil is very often adulterated with fixed oils.

West Indian Sandalwood oil is obtained from *Amyris balsamifera* (Rutaceae) and used as substitute. Its specific gravity is 0.900 to 0.967 and optical rotation + 24 to + 29°. It contains 50 per cent amyrol.

Storage

It should be stored in well-filled, well-closed containers, away from light and in cool place.

SAUSSUREA

Synonyms

Costus, Kuth.

Biological Source

Saussurea consists of dried roots of *Saussurea lappa* Clarke, belonging to family Compositae.

Geographical Source

Saussurea is found at an altitude of 2500 to 3000 m in Kashmir and neighbouring Himalayan regions. In Jammu, Chenab and Kishenganga Valley it grows in the natural environment. It has been successfully cultivated in Kashmir and Uttar Pradesh.

Fig. 14.63 (a): Saussurea lappa herb with Roots

Cultivation and Collection

For medicinal purpose, it is cultivated in Kashmir and other Himalayan regions. It prefers deep rich porous soil at an altitude of 2600 - 3000 m. Saussurea is propagated either by root cutting or by seeds, but propagation by seed is found to be economical. Seeds for propagation are collected in September.

They retain their viability for a period of one year. The sowing of seeds is done in spring and the seedlings are transplanted at a spacing of 90 × 90 cm. The roots are collected after three years of vegetative growth. They are cut into pieces of 10 cm and dried in the sun. The yield of 2 - 2.5 tones per hectare can be obtained.

Fig. 14.63 (b) : *Saussurea lappa* Root

Macroscopic Characters (Fig. 14.63 (b))

Colour	:	Grey to dull brown
Odour	:	Aromatic and strong
Taste	:	Bitter
Shape	:	Cylindrical, thick and stout
Size	:	7 - 15 cm in length and 1 - 1.5 cm in breadth.

The roots are occasionally ridged and wrinkles are longitudinal and anastomosed rootlets are rarely present. The cut surfaces of the root shows two distinct regions. Fracture of the root is short and horny.

Microscopic Characters

The transverse section of the root shows the presence of periderm followed by broad zone of phloem and xylem. Cork is 3 - 5 layered, and phloem fibres are also exhibited. Medullary rays are multiseriate and wider in phloem region than in the xylem. Oleoresin cavities and inulin in storage parenchyma are also present.

Chemical Constituents

Kuth roots contain about 6 per cent resinoids, 1.5 per cent of essential oil, 18 per cent of inulin, 0.5 per cent of an alkaloid, fixed oil, traces of tannins and sugar. The constituents of the volatile oil of the kuth are costuslactone, dihydrolactone, camphene and several sesquiterpene and sesquiterpene alcohols. The alkaloid is known as saussurine. Kusthin, a liquid having formula $C_{10}H_{26}O_8$, has been also isolated.

Standards of Quality

Foreign organic matter	:	Not more than 02.0 per cent
Ash	:	Not more than 09.0 per cent
Acid insoluble ash	:	Not more than 03.0 per cent
ASE	:	Not less than 12.0 per cent
WSE	:	Not less than 28.0 per cent

Uses

Saussurea is useful in the treatment of bronchial asthma and is categorised as an expectorant. The oil of the kuth is used as an essence. Saussurea is also a powerful insecticide. The volatile oil of the saussurea has antiseptic and disinfectant properties, especially against *Streptococcus* and *Staphylococcus*. It is used as a carminative and stimulant. Drug is also used in chronic skin diseases and in the treatment of rheumatism. India is exporting saussurea on a very large scale to several countries including Japan, France, Hongkong and others. At present, about 500 tonnes of the drug is being exported.

Adulterants

The oil of saussurea, being very costly is frequently adulterated with vetiver oil.

VALERIAN

Synonyms

Indian valerian or Valerian jatamansi.

Biological Source

It consists of dried rhizomes, stolons and roots of *Valeriana wallichii* DC, family Valerianaceae.

Geographical Source

It is found abundant in the Western Himalayas and is reported in Afghanistan and Pakistan. Indian valerian is collected in Kashmir, Garhwal and other places in North-West Himalaya in India.

Cultivation and Collection

It is found in Himalayas in wild condition. It is propagated from root stocks. It needs deep rich soil and flourishes in moist and shady locations. The roots and rhizomes are dug out in autumn.

Fig. 14.64: Indian Valerian Plant

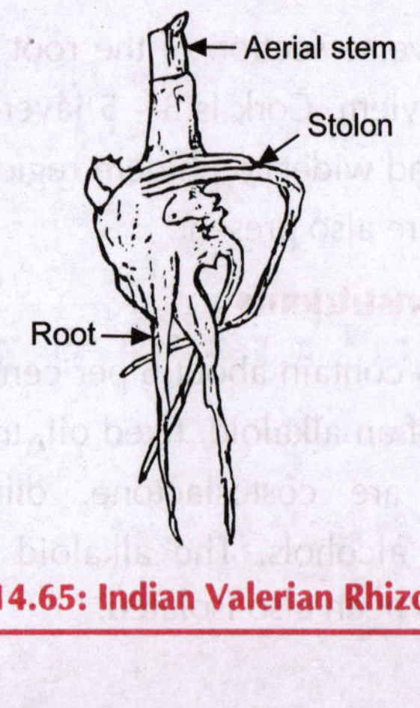

Fig. 14.65: Indian Valerian Rhizome

Macroscopic Characters

Colour : Rhizomes - Dull yellowish-brown

Stolons - Yellowish-grey

Roots - Yellowish-brown

Plant bears white or rosy flowers

Odour : Characteristic, penetrating and resembling isovaleric acid.

Taste : Bitter and camphoraceous

Size : Rhizomes - About 4 to 8 cm long and 4 to 20 mm in thickness

Stolons - About 5 mm in length and 2 to 4 mm in thickness

Roots - About 3 to 5 cm in length and 1 mm in diameter

Shape : Rhizomes are sub-cylindrical and flattened, dorsiventrally curved and unbranched.

Rhizomes: The presence of leaf scars on the upper surface, and small circular root scars on the lower surface. Rootlets, as well as few scars being remains of aerial parts, are also observed on the rhizomes.

Stolons: They are connecting the rhizomes and are characterised by the presence of nodes, internodes and root. Fracture is short and horny.

Chemical Constituents

Valerian contains pale brown to amber coloured oil about 1 per cent, and alkaloids chatanene and valerine. Volatile oil contains borneol formate, borneol acetate, camphene, borneol isovalerianate, sesquiterpenes, etc. Valepotriates or valtrate, responsible for the therapeutic activity of the drug are present to the extent of 2 per cent as compared to 0.5 per cent in European variety. The oil has sweet odour, which is more pleasant in the fresh rhizomes of cultivated plants.

6-methyl epigenin is the active compound of valerian and acts as a benzodiazepine binding ligand.

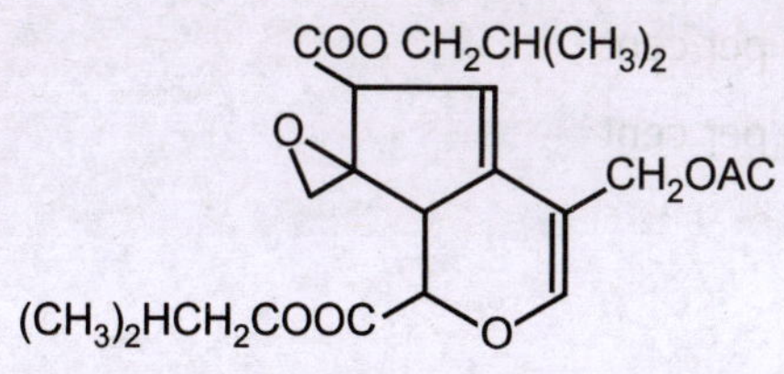

Valtrate

$CH_3\,(CH_2)_3\,COOH$

Valeric acid

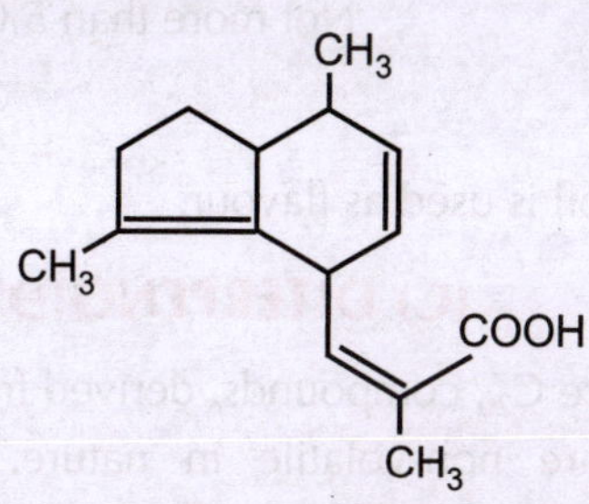

Valerenic acid

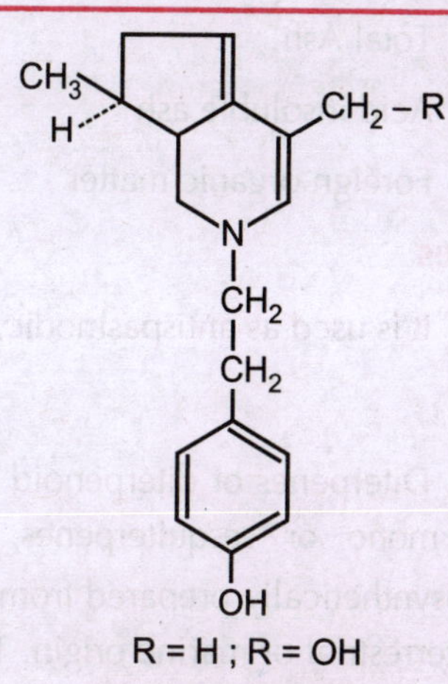

R = H ; R = OH

Valerian alkaloids (Chatinine and valerine)

Standards of Quality

Foreign organic matter	:	Not more than 02.0 per cent
Ash	:	Not more than 12.0 per cent
Alcohol (60.0%) soluble extractives	:	Not more than 30.0 per cent

Uses

Valerian roots have antispasmodic and depressant action on the central nervous system. It can be used as stimulant and carminative. The oil is used as a tonic and stimulant in pharmaceutical preparations and also in perfumery. Valerian roots are used in preparation of tinctures, extracts, infusions, medicinal tea and extraction of volatile oil. Steam distilled volatile oil, as well as, solvent extracted absolute of valerian are used for flavouring of tobacco and beer. Freeze dried preparations of valepotriates are manufactured. Valepotriates possess tranquillising activity in animals.

VALERIAN EUROPEAN

Biological Source

It consists of thick dried rhizomes and roots of *Valeriana officinalis*, Family Valerianaceae collected in autumn.

Geographical Source

It is obtained from wild and cultivated plants in England, Holland, Germany, Belgium, France and Japan.

Chemical Constiuents

It yields 0.5 - 1 per cent of volatile oil. This contains esters (bornyl acetate, bornyl formate, bornyl isovalerianate), terpenes, alcohols and a sesquiterpene alcohol (valerianol).

Standards of Quality

Total Ash	:	Not more than 15.0 per cent
Acid insoluble ash	:	Not more than 7.0 per cent
Foreign organic matter	:	Not more than 5.0 per cent

Uses

It is used as antispasmodic, oil is used as flavour.

(C) DITERPENOIDS

Diterpenes or diterpenoid are C_{20} compounds, derived from four isoprene units, and in contrast to mono or sesquiterpenes, are non-volatile in nature. These C_{20} natural compounds are biosynthetically prepared from geranyl pyrophosphate. They are present both in plants and animals of terrestrial or marine origin. The plant families rich in diterpenoids are Pinaceae, Podosocarpaceae and Taxodiaceae (all conifer resins); Leguminosae, Cistaceae and Burseraceae (all angiosperm resins). The other families containing appreciable quantities of diterpenes are Labiatae,

Ranunculaceae and Euphorbiaceae. They have also been found in marine animals (Coelentrates) like soft corals and sea fans. Different parts of the same plant may contain different diterpenes; and percentage of diterpenes varies alongwith seasons.

Most of the diterpenes appear to be stress compounds, produced as a response to environmental stress. They are also considered as defence chemicals prepared against parasites and mechanical injury. Apart from these activities, certain diterpenes have specific roles. Gibberellins are plant growth regulators, phytol is an important part of chlorophyll with crucial role in photosynthetic process while retinol having essential role in mammalian visual function. Besides their active biological role in plants and animals, many diterpenes have important pharmacological activities.

COLEUS

Synonym

Forskohlii, Maeenmool.

Biological Source

It consists of roots of *Coleus forskohlii,* Syn. *Plectranthus barbalus* belonging to family *Labiatae* (Lamiaceae), and should contain not less than 0.4 per cent of forskolin on dried basis.

Geographical Source

The plant grows perennially over the tropical and subtropical regions including India, Pakistan, Sri Lanka, Brazil and Tropical East Asia. In India, it occurs at high altitudes upto the height of 2500 metres in Himalayas, in the regions from Garhwal, Kumaon to Nepal. It also occurs in Deccan Peninsula and Parasnath hills in Bihar. It is common on dry and barren hills.

Cultivation and Collection

About 2500 tonnes of leaves and roots are harvested every year from commercial cultivation in Gujarat and Kanchipuram district of Tamil Nadu.

It is propagated by using stem cuttings in spring. All types of soils are suitable for cultivation. Coleus is harvested in autumn. Irrigation once in ten days is enough. It is highly disease resistant crop. It is a crop of 6-7 months duration. It fetches the value of Rs. 1.5 lakh per tonne. Yield per hectare is about 1 to 1.5 tonnes.

Traditionally, it has been used in India for pickles and for this purpose it is cultivated to some extent.

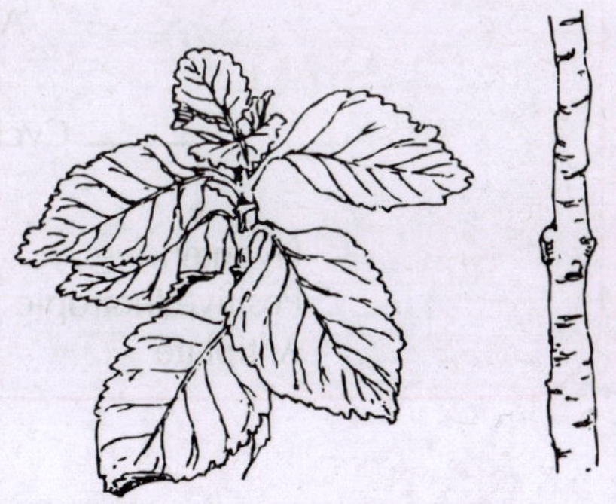

Fig. 14.66: Coleus forskohlii twig and root

Macroscopic Characters

It is a branched, aromatic perennial herb 1 - 2 feet in height leaves have camphor-like flavour with several, fasciculated, succulent, radially spread roots.

Chemical Constituents

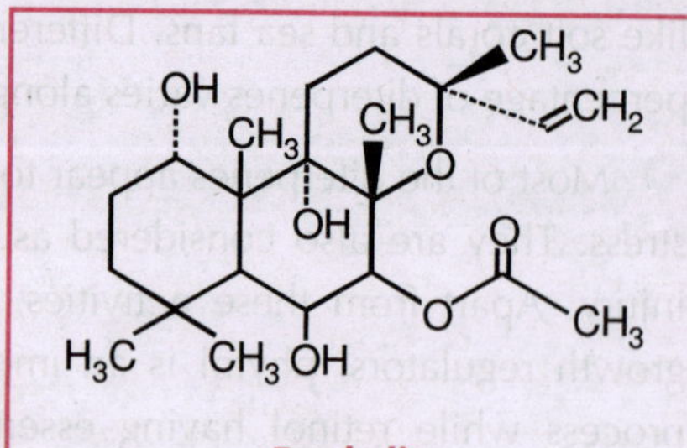

Forskolin

It contains the diterpene Forskolin about 0.6 per cent (coleonol) was discovered at the Hoechst Research Centre, Mumbai, India and Central Drug Research Institute (CDRI), Lucknow, India. It also contains various diterpenoid derivatives. The leaves give coleon E (methylenequinone), barbatusin and coleon F. The roots contain coleonol B, coleonol C, deoxycoleonol and labdane diterpenoids (I, II and III).

Standards of Quality

Loss on drying	:	Not more than 12.0 per cent
Foreign organic matter	:	Not more than 2.0 per cent
Ash	:	Not more than 15.0 per cent
Acid insoluble ash	:	Not more than 05.0 per cent
Alcohol soluble extractives	:	Not less than 15.0 per cent
Water soluble extractives	:	Not less than 18.0 per cent

Uses

Forskolin has vasodilator and cardio stimulant effects. Major biological effects of forskolin include activation of adenylate cyclase, positive inotropic action, blood pressure lowering, and decreasing intra-occular pressure. Forskolin inhibits human platelet aggregation induced by epinephrine or collagen. It reduces intra occular pressure and hence useful in glaucoma.

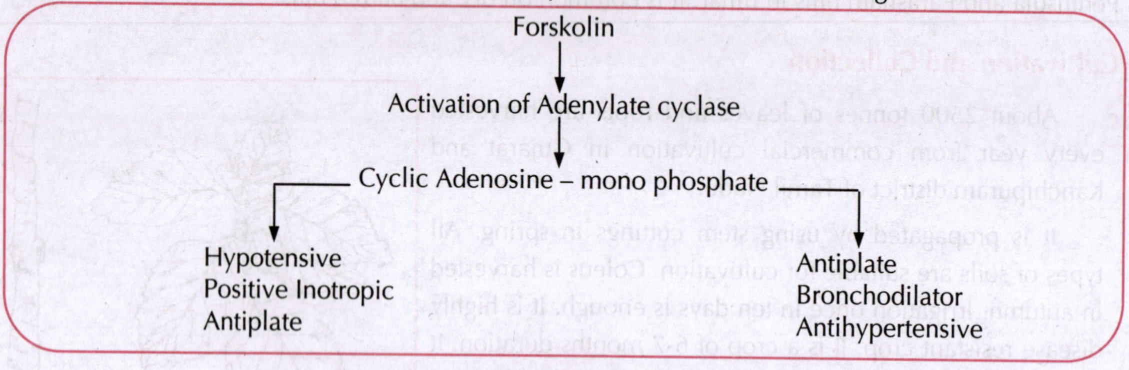

TAXUS

Synonyms

Yew, Talispatra, Himalayan yew, Birmi.

Biological Source

This consists of dried leaves, bark and roots of various species of *Taxus*, belonging to family Taxaceae. The four important species with parts used are as under.

1. *Taxus baccata* (English or European yew) mainly leaves.
2. *Taxus brevifolia* (Pacific yew) mainly stem bark.

3. *Taxus Canadensis* (Canadian or American yew) Leaves and roots.
4. *Taxus cuspidate* (Japanese yew) leaves.
5. *Taxux wallichiana* :

It contains not less than 0.04 per cent of Paclitaxel (Taxol) culculated on dried basis.

Fig. 14.67: Yew tree and bark

Fig. 14.68: Twig of Yew tree

Geographical Source

It is very slow growing evergreen gymnospermous tree. Found in India, Canada and America. It is reported in temperate Himalayan region of India upto an altitude of 2000 – 3500 metres.

Morphological Characters

Leaves

Colour	:	Dark green
Taste	:	Bitter
Size	:	1 - 3 cm × 1 - 2 cm
Shape	:	Lanceolate, flat Leaves are arranged spirally on the stem, leaf bases are twisted to allign the leaves in two flat raws on either side of stem except on erect leading shoots.
Bark	:	Thin and scaly brown
Seed cones	:	Each contain one seed which is 4 - 7 mm surrounded by aril, Arils get matured after 6 - 9 months.

Chemical Constituents

The main constituent paclitaxel or taxol is present in all parts of the plant especially in leaves, roots and bark.

Taxanes are the most important group of chemical constituents and uptil now 40 different Taxane compounds have been found, all of which are diterpenoid structures. Among them three most important members are Taxol, cephalomannine and 10 - deacetyl baccatin. In all species, with little variations, taxol occurs from 0.007 per cent to 0.01 per cent. Presently, it is mainly obtained from stem bark of *T. brevifolia*. But the method of isolation is tedious and like vinca, yields are also less. It needs at least 60 years old 3 - 4 trees to get 1 gm of taxol. Yields 50-150 mg of taxol are obtained from 1 kg dried yew bark. About 10 kg bark is available from an average tree. Considering the slow growing nature of yew tree, it has been already predicted that this will lead to severe ecological problem. Obviously alternative sources, like total synthesis or biotechnological route through tissue culture are also presently under development. It has been reported that leaves contain 10-deacetyl baccatin III, which can be comparatively easily converted to Taxol. The leaves can be quickly regenerated after harvesting. Among the various yew species, most of the compounds are anticancer. The most potent compounds include Taxol (containing a rare oxetane ring and amide side chain), cephalo mannine (0.031 per cent), baccatin – III (0.084 per cent), and 10 - deacetyl baccatin III.

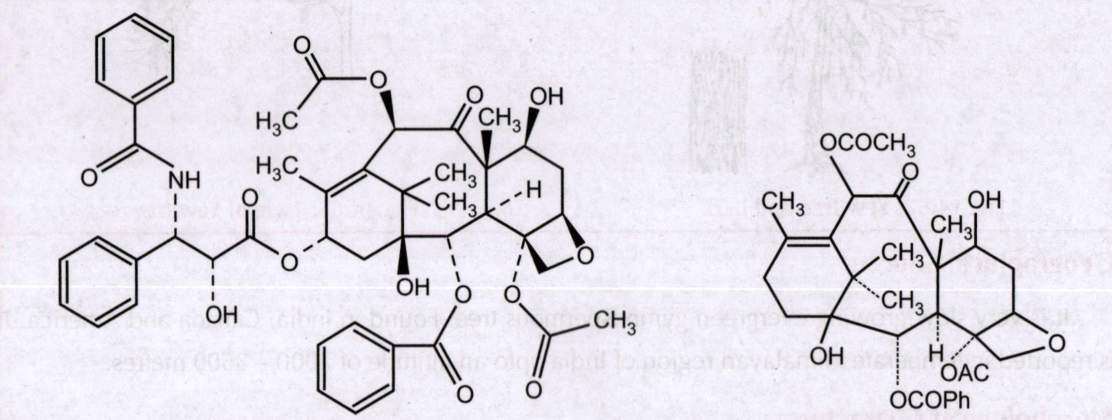

Paclitaxel (Taxol) 10-Deacetyl baccatin III

A derivative of Taxol, called **taxotere** has been reported to have better bio-availability and pharmacological properties and has been claimed as a promising anticancer agent.

Standards of Quality

Alcohol soluble extractives	:	Not less than 11.0%
Water soluble extractives	:	Not less than 17.0%
Total ash	:	Not more than 4.0%
Acid insoluble ash	:	Not more than 0.7%
Loss on drying	:	Not more than 7.0 % at 105ºC

Uses

Used as anticancer, analgesic, anti-inflammatory, antipyretic and anti-convulgant. The biological target of Taxol is microtubules produced from α and β-tubulin. The microtubules are responsible for the formation of mitotic spindle necessary for cell division. α and β-tubulin polymerise to give microtubules and for this process microtubule associated proteins (MAP) and guanosine triphosphate (GTP) are necessary. Taxol brings out the polymerisation to microtubules in absence of MAP and GTP. Due to this, microtubule formation is much enhanced which causes detrimental effects on dividing cells which leads to blockade of cell cycle. Eventually, multiple abnormal esters are formed from microtubules and get distributed in cytoplasm. These structures are non-functional. Taxol also inhibits cell migration thus, preventing spread of metastatic cancer cells. Taxol has been approved by USFDA for treatment of refractory ovarian cancer. It has also a promising role against non-small cell lung carcinoma, gastric and cervical cancers and also carcinomas of head, neck, prostate and colon.

(D) TRITERPENOIDS

Among various terpenoid groups, this is the largest one. Triterpenoids are widely distributed in plant kingdom but few in animals. They are present in either free state or as esters or glycosides. Biosynthetically, triterpenoids are C_{30} compounds prepared from six isoprene units. They are further classified as: (i) Tetracyclic triterpenoids and (ii) Pentacyclic triterpenoids.

Steroids are degraded triterpenes.

Triterpenes are very commonly present in most of the dicotyledonous plants. The plant families producing appreciable quantity of triterpenoids are Cucurbitaceae, Caryophyllaceae, Leguminosae, Apocynaceae etc. Azadirachtin from *Azadirachta indica* is a prominent triterpenoid from plant kingdom. They are also present in marine sources like squalene, ambrein and lanosterol from shark, ambergris and wool fat respectively.

They are mostly stress compounds (phytoalexins) produced by the plants as a response to environmental stress and ecological interaction.

AMBERGRIS

Synonyms

Grey amber, Amber grease.

Biological Source

It is the pathological product formed in the intestinal tract of the sperm whale *Physeter catodon* Linn, family Physeteridae.

It is found floating in tropical seas or near sea-shores. The sperm whale feeds on squid or cuttle fish. It is said that indigestible beaks of these animals cause irritation of whales, stomach and in turn stimulates the formation of ambergris. Sometimes, the masses weighing 1 kg to 10 kg or more are found floating at a time.

Description

Colour : Grey to black waxy mass.

Odour : Agreeable, characteristic, persistent odour, recalling musk. It is brittle, flammable and completely volatile by heat. It volatalises at 100ºC into white vapour.

Solubility : Insoluble in water and even in alkali hydroxides. Ambergris is soluble in hot alcohol, chloroform, ether, fats and volatile oils.

Chemical Constituents

It contains a triterpene alcohol, ambrein (about 25 per cent) which is white and crystalline epicoprostanol and coprostanone.

Standards of Quality

Special gravity : 0.780 - 0.926

Melting Point : 62 - 63°C

Uses

It is used in perfumery and also as fixative for delicate flavours.

(E) TETRATERPENOIDS AND CAROTENOIDS

Carotenoids are produced from 8 isoprene molecules and contains 40 carbon atoms. Thus, carotenoids are **tetraterpenoids** and consist of organic pigments. These are biosynthetically prepared by tail to tail condensation of geranyl geraneol. They contain long sequence of conjugated double bonds.

Carotenoids are prominant group of natural colouring matters (pigments) exhibiting purple, red, yellow or orange colours. They are present in plants and animals. Highest amount of these compounds are present in carrots, fruits and hence the name carotene.

They are found in chloroplast and chromoplasts of plants and other photosynthetic organisms including bacteria and fungi.

Carotenoids are produced from fats and other basic metabolic building blocks by these organisms.

In animal, they are produced by zephids and spider mites.

Carotenoids from diets are stored in fatty-tissues of animals exclusively carnivorus animals obtain these compounds from animal fats.

Carotenoids are deep yellow, orange or red. Carotenoids absorb light energy for photosynthesis and protect chlorophyll from photo damage.

Carotenoids contain unsubstituted beta-ionone rings and have vitamin A activity, that is they can be converted to retinol and these act as antioxidants.

Pink colour of flamengos, seales of wall lizards are due to carotenoids only. CRT is the gene cluster responsible for biosynthesis of carotenoids.

Carotenoids are responsible for the yellow colour of egg yoke, bananas, daffodils, carrots and corn.

Lutein, Capsanthen, Bixin, Crocetin and lycopersene are the naturally occuring carotenoids.

Annato, saffron, chlorophyll and gossypol are the examples of drug containing tetraterpenoids.

ANNATTO

Synonyms

Arnotta, Annotta, Lipstic pods, Kumkum, Shendri.

Biological Source

This consists of dried seeds of the plant *Bixa orellana* L. family Bixaceae.

Geographical Source

The plant is cultivated in Peru, Jamaica, Mexico, Kenya, India, Brazil. About 40 per cent of the world requirement is met by Peru while India contributes only 10 per cent.

Cultivation of Annato

Cultivation of Bixa is done either by sowing the seeds or from cuttings. It is a tall, perennial herb about 2-3 metres in height. Plant is quite stout, erect, decorative and ever green.

It is cultivated in well drained fertile loamy soil full of organic matter. It also needs tropical setting at an alternate of 3000 metres. Bixa can be cultivated at any time in the year provided, if there is sufficient arrangement for irrigations. Sowing of seeds is done just before the raining season. Sowing is done by drilling method and distance of one metre is maintained in between two plants and two roots. Plants should be watered immediately after planting.

Fig. 14.69: Annatto herb

Organic mannures and fertilisers are provided to the plant at the time of planting only.

August to October is the flowering season of the crop and it starts bearing fruits. Fruits get matured during December to February and ready for harvesting.

Fruits are prickly heart shaped pods and each fruit contains about 50 brick red coloured seeds. Before breaking fruits are collected and are dried.

Preparation for the Market

Bixa is commercially grown for dye and also for the seeds which are used as spice.

Fully matured prickly pods are collected and dried. The fruits are then macerated in water for a while. The dye settles, is collected and dried, made into cakes.

The seeds from pulp are separated and washed thoroughly for further processing and use.

Macroscopic Characters

Colour	:	Seeds are brick-red
Odour	:	Slightly sweet and peppery
Taste	:	Hotness scale 1 to 2
Size	:	3 to 5 mm
Shape	:	Triangular .

Chemical Constituents

Annato seeds are covered with aril and contains bixin dye. Annato seeds are found to contain about 12 per cent of annato oleo resin of which 50 per cent is water soluble. Volalile oil covers only 0.3 - 0.8 per cent, while pigment covers 4 - 5 per cent.

The main constituent of the pigment is known as Bixin which constitutes about 70 - 80 per cent of pigment. Bixin is a carotenoid carboxylic acid and is responsible for yellow colour. Isobixin, trans-bixin, nor-bixin are the other isomers of it. Bixin is naturally present in *cis form and gets* converted to *trans-bixin* during extraction. Bixin, in pure form, is insoluble in water but soluble in organic solvents and alkaline aqueous solutions. Oil soluble annatto is available in the market for commercial use. Seeds also contain crocetin and abscisic acid.

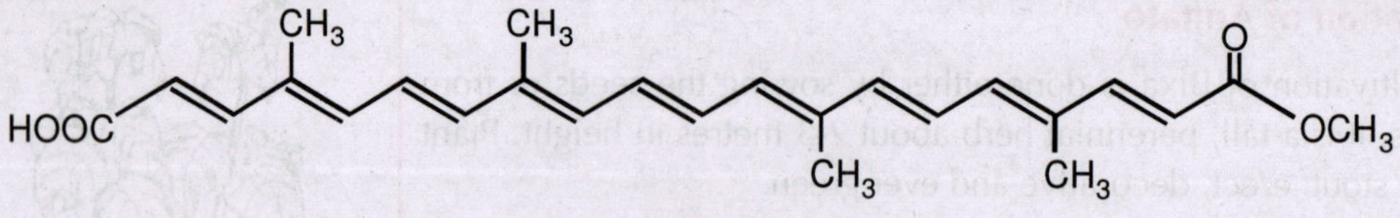

Bixin

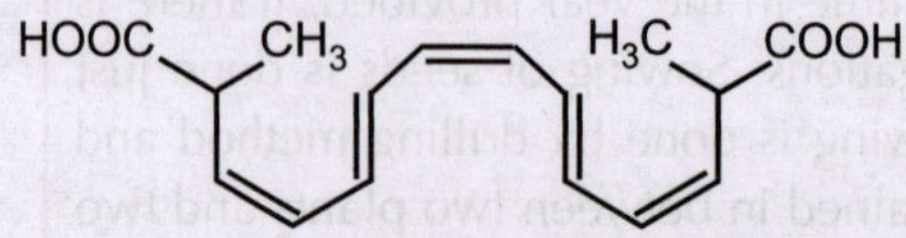

Crocetin

Bixin is a monomethyl ester of nor-bixin (a dicarboxylic acid) and gets readily hydrolysed to dicarboxylic acid by alkali during extraction. Annatto extracts are the potassium salts of norbixin.

When compared to β-carotene in the tinctorial strength, bixin is quite stable. However, it looses its tinctorial strength on storage. Light and heat has adverse effect on its tinctorial power. Annatto is most stable at pH 8.

Uses

Bixin is antioxidant and protects against ultra violet light. It also has liver protective properties.

Annatto is used as colouring agent for foods, cosmetics, alcoholic and non-alcoholic beverages, dairy desserts, fats and oils and in margarine.

CROCUS

Synonyms

Saffron, Hay saffron, Kesar, Keshar.

Biological Source

Crocus consists of dried stigmas and upper parts of styles of flowers of plant known as *Crocus sativus* Linn, family Iridaceae.

Geographical Source

It is cultivated in India (Kashmir), Spain, Iran, France and Greece. Now under controlled condition it is cultivated successfully in PUNE, in Maharashtra and total production is about 3.5 tonnes against 100 tonnes requirement of India.

Cultivation and Collection

The plants are raised from corms planted during July/August. Cultivation is undertaken at PAMPORE and KIRATWAD in area of Kashmir. Which is about 2000 – 2500 metres above sea-level. Since, the temperature requiement of 10ºC is fulfilled at this place. Corms are placed at the distance of one foot apart in cultivation. The flowering takes about one year. Flowers are collected in early morning in November and December, stigmas and upper style portions are detached manually, and the drug is dried by artificial heat. It is stored in dry place.

India's share in global supply is 50 tones and is about 10 per cent of world requirement. Approximately, 5000 hectares of land is under cultivation in Kashmir and Himachal Pradesh.

After the sucessful cultivation of strawberry and apples, experimental and trial cultivation of saffron is now taken at Mahabaleshwar in Maharashtra. Saffron costs ₹ 5.0 lakh per kg.

Macroscopic Characters (Fig. 14.70)

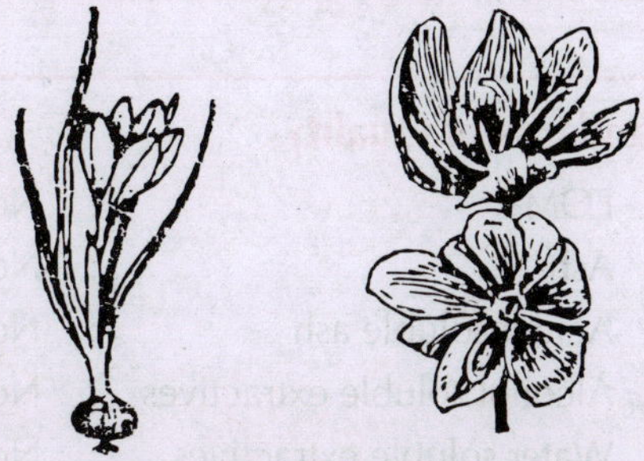

Fig. 14.70: Saffron Plant and Flower

Colour	:	Stigma dark red to reddish-brown. Style, yellowish-brown to yellowish-orange.
Odour	:	Strong, characteristic and aromatic.
Taste	:	Characteristic and bitter.
Size	:	Stigmas are 25 mm long, and styles about 10 mm long.
Shape	:	Stigma trifid and styles cylindrical.

Stigmas may be attached to the apex of styles or may be separated. About 12000 flowers weigh 100 g.

Chemical Constituents

Saffron contains red colouring matter known as Crocin and crocetin, bitter principle picrocrocin and traces of volatile oil. Protocrocin, a carotenoid glycoside splits during drying into two molecules of picrocrocin and one molecule of crocin. Crocin further hydrolyses to crocetin and gentiobiose while picrocrocin yields glucose and safranal. It yields about 50 to 60 per cent of extractives to cold water. It also contains non-reducing substances and crystalline hydrocarbon. The crocins are responsible for the blue colour when treated with concentrated sulphuric acid.

Chemical Tests

1. Add a drop of sulphuric acid to dry stigma. It turns blue, gradually changing to purple and finally purplish-red.
2. Saffron imparts yellowish orange brown colour to water.

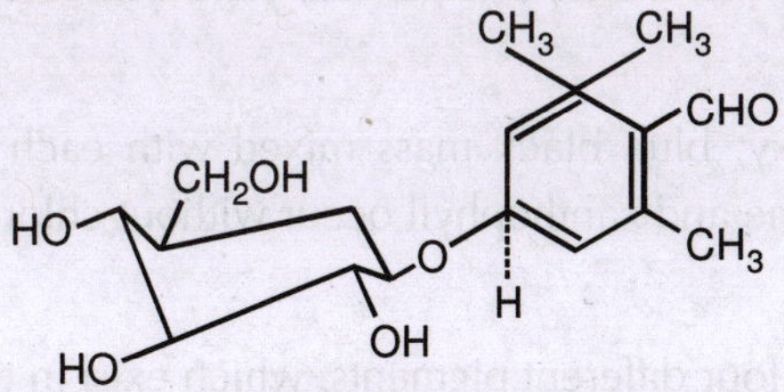

Picrocrocin

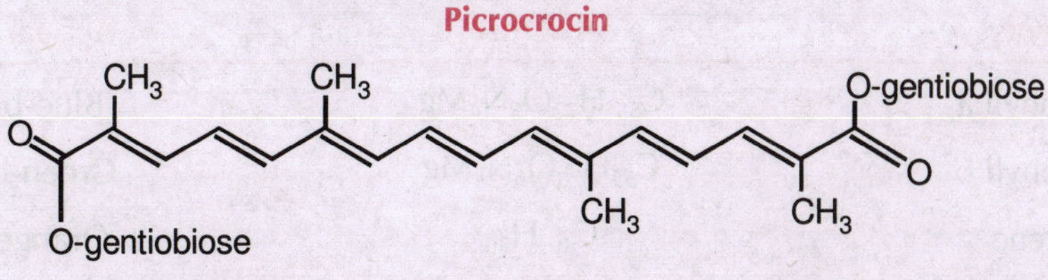

Crocin

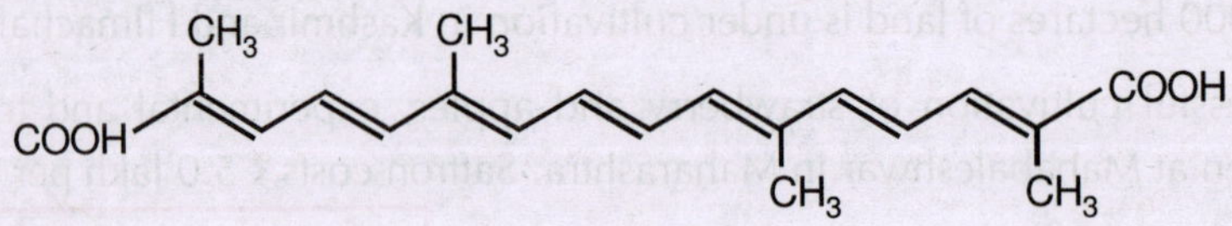

Crocetin

Standards of Quality

FOM	:	Not more than 2.0 per cent
Ash	:	Not more than 8.0 per cent
Acid insoluble ash	:	Not more than 2.0 per cent
Alcohol soluble extractives	:	Not less than 8.0 per cent
Water soluble extractives	:	Not less than 45.0 per cent

Uses

Saffron is used as a colouring (food dye) and flavouring agent. It is also as antispasmodic, emmenagogue and a stimulant.

Adulterants

Due to high price of drug in the market, saffron is adulterated. In India, it is commonly adulterated with florets of safflower, botanically known as *Carthamus tinctorius* Linn, family Compositae, which are orange in colour. Safflower florets impart orange colour to alcohol, whereas no colour is observed with authentic drug. The weight of the drug is sometimes, increased by the addition of glycerine and ammonium nitrate. Such type of adulteration can be detected by determining water-soluble extractive.

CHLOROPHYLL

Biological Source

Chlorophyll the natural pigment is obtained from the green leaves of higher plants and green algae. It is a mixture of four different pigments i.e. Chlorophyll a, Chlorophyll b, Carotene and Xanthophylls, i.e. blue–black, green–black, orange–red, yellow in colour.

Description

Chlorophyll occurs as waxy, blue black mass mixed with each other, in the green plasmids called chloroplast while carotene and xanthophyll occur without chloroplast in any part of plants.

Chemical Constituents

Chlorophyll is a mixture of four different pigments, which exist in the chloroplast.

Pigments	Structure	Appearance
Chlorophyll a	$C_{55}H_{72}O_5N_4Mg$	Blue-black
Chlorophyll b	$C_{55}H_{70}O_6N_4Mg$	Green-black
Carotene	$C_{40}H_{56}$	Orange-red
Xanthophyll	$C_{40}H_{56}O_2$	Yellow

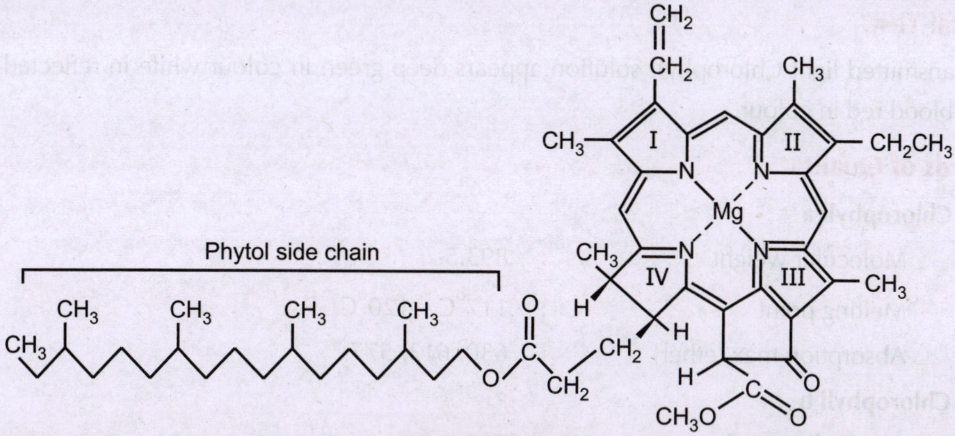

Chlorophyll a

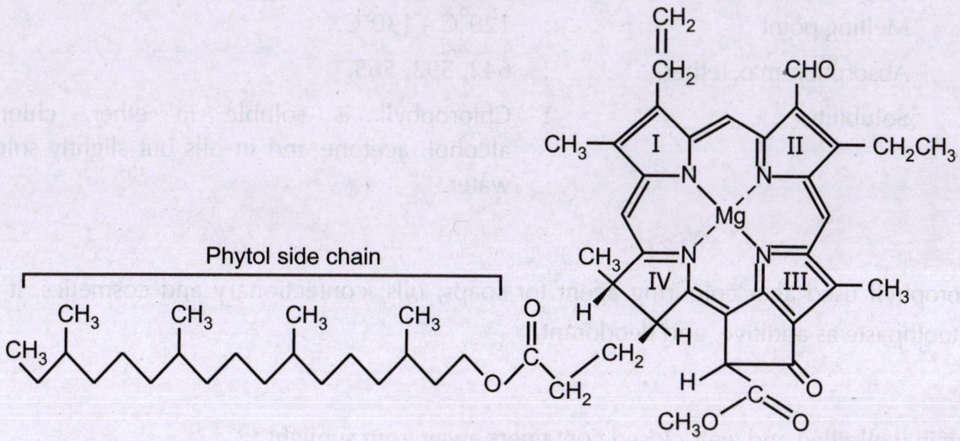

Chlorophyll b

Isolation of Chlorophyll

Pieces of green leaves are treated with acetone. Then shaken for a few minutes and allowed to stand, the acetone appears blue green in colour which is then filtered. The solutions show fluorescence. The solution is poured into a separating funnel with petroleum ether and methyl alcohol, shaken and allowed to settle within 2 or 3 minutes. The solution separates into two layers.

(A) The upper layer appears deep green in colour containing chlorophyll a and carotene which are separated by adding potassium hydroxide solution and methyl alcohol, again shaken for a few minutes and allowed to stand, immediately liquid separates out into two layers, upper layer is deep green containing chlorophyll a, and the lower layer is yellowish brown containing carotene.

(B) Lower layer appears yellowish brown light greenish yellow containing chlorophyll b and xanthophyll which are separated by adding potassium hydroxide solution and methyl alcohol, shake it for a few minutes. Allow it to stand for a few minutes, immediately each liquid separates out into two layers, upper layer is greenish containing chlorophyll b and lower is light yellow containing xanthophyll.

Chemical Test

In transmitted light Chlorophyll solution appears deep green in colour while in reflected light it appears blood red in colour.

Standards of Quality

Chlorophyll a

Molecular weight	:	893.5.
Melting point	:	$117^{o}C - 120^{o}C$.
Absorption max (ether)	:	630, 613, 577.

Chlorophyll b

Molecular weight	:	907.4.
Melting point	:	$120^{o}C - 130^{o}C$.
Absorption max (ether)	:	642, 593, 565.
Solubility	:	Chlorophyll is soluble in ether, chloroform, alcohol, acetone and in oils but slightly soluble in water.

Uses

Chlorophyll used as a colouring agent for soaps, oils, confectionary and cosmetics. It is also used in toothpaste as additive, and deodorant.

Storage

Store in well-filled and well-closed containers away from sunlight.

GOSSYPOL

Biological Source

It is a pigment found in the oil and cottong seed cake of seeds of cotton *Gossypium herbaceum,* family Malvaceae, to the extent of 0.5 per cent. Also found in ceiba (*Ceiba-pentandre*) family: *Bombacaceae*.

Description

It is insoluble in water but soluble in alcohol, ether and chloroform. It is a brown solid with Melting Point 177 – 180ºC.

Chemical Nature

It is polyphenolic aldehyde with molecular formula $C_{30}H_{30}O_8$.

Uses

Gossypol is a male contraceptive agent due to its antispermatogenic activity. Degossypolised cotton seed cake is now available in the market. It is used on poultry feed also as food for fishes.

It has been well studied in China for this purpose. But it is poisonous in nature as it reduces oxygen carrying capacity of blood in non-ruminant animals. It also shows insecticidal properties.

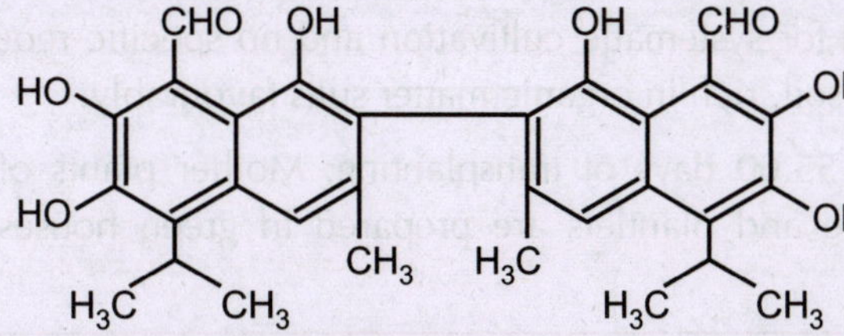

Gossypol

Gossypol has been investigated for spermicidal, antiviral, antiprotozoal and antitumour activities. Side effects include persistent oligospermia and much loss of libido.

STRAWBERRY

Synonym

Alpine-strawberry.

Biological Source

Strawberry consists of Prostrate runner of *Fragaria vises* family Rosaceae.

Geographical Source

It is commercially cultivated in USA, Italy, India and Sikkim in about 5000 hectares. In Himalayan territory and Sikkim it is cultivated at an ultitutde of 2000 – 4000 metres. In India now it is cultivated to some extent in Maharashtra also at Mahabaleshwar. Other places where it is cultivated are Satara, Jawali, Rai, Koregoan and Khatav area. Recently its cultivation Nashik has been found successful and is cultivated in surgoan Tq. In green houses.

Fig. 14.71 (a): Strawberry Plant

Cultivation of Strawberry

Strawberry needs cold climatic conditions, and the varieties which are cultivated are sweet-charli, selva and winter dawn.

White poly mulching paper is used in the cultivation for survival of these delicate plants and also to save the water less. Strawberry is mainly cultivated by tribal-community and have become experts.

Nersery beds are used for systematic cultivation and no specific requirement of soil. It can be cultivated on sandy loamy soil, rich in organic matter suits favourably.

It shorts fruiting after 55.60 days of transplanting. Mother plants of strawberry are are from America and Italy. In June and plantlets are prepared in green houses which take about three months.

Fig. 14.71 (b): Strawberry Cultivation and Strawberry Plant

Microscopic Characters

Strawberry is slender proslrate with internodes creping the ground rooting at infernodes. It arises as axillary bud and creeps. Some distance away from mother plant and grow into new plant. It also spreads fruits are dark red coloured with sweet taste.

Strawberry fruits are indehiscent, many seeded fleshy fruits, epicarp and meso carp can be distinguished.

Chemical Nature

Red colur of strawberry fruts is due to pelargonidin-3-galactoside. Leaves of the plant contain ellagitannin.

Uses

Strawberry fruits are astringent and diuretic root stock is used as coffee-substitute in Kashmir. Also used as colouring for several products and also in medicines.

Other berry fruits of commercial importance and used in food industry on large scale are:

* Goose-berry (*Physalis minima* fam Solanaceae)
* Black berry (*Syzysium cumini* fam. Myrtaceae)
* Mulberry (*Morus alba* fam Moraceae)
* Bear-berry (*Arctastaphylus uva-ursil* fam Ericaceae)

[II] RESINS AND RESIN COMBINATIONS

Resins are amorphous products of complex chemical nature. These are mixtures of essential oils, oxygenated products of terpene and carboxylic acids found as exudations from the trunk of various trees. They are transparent or translucent solids, semi-solids or liquid substances containing large number of carbon atoms. Most of the resins are heavier than water. They are insoluble in water, but soluble in alcohol, volatile oils, fixed oils, chloral hydrate and non-polar organic solvents like benzene or ether. They are hard, electrically non-conductive and combustible masses. When heated, they soften and ultimately melt. They are usually formed in schizogenous or schizolysigenous cavities or ducts as end products of metabolism. Chemically, they contain organic acids, alcohols, esters, and neutral resins. Depending upon the type of the constituents of the resin, they are further classified as: (a) Acid resin, (b) Ester resin, and (c) Resin alcohols.

1. Acid resins

Following are few examples of this type of resins alongwith their acids; colophony (abietic acid), safetid (sandracolic acid), copaiba (copaivic and oxycopaivic acids), myrrh (commiphoric acid) and shellac (alleuritic acid).

2. Ester resins

This group contains esters as the chief constituents of the resins, e.g. benzoin and storax. Benzoin contains coniferyl benzoate and storax contains cinnamyl cinnamate.

10. Resin alcohols

The contents are the complex alcohols of high molecular weight. They are either found in free state or as esters. The examples are balsam of peru with peruresinotannol, gurjan balsam with gurjuresinol and guaiacum resin with guaic-resinol.

Resins and oils in homogenous mixtures are called as **oleoresins**, e.g. copaiba, Canada balsam, capsicum, etc. **Oleo-gum resins** are the homogenous mixtures of volatile oil, gum and resin, e.g. myrrh, guggul and safetida. **Glycoresins** are made up of resins and sugars and are present in jalap and ipomoea. If the resin contains benzoic acid and / or cinnamic acid, it is called as a **balsam**, e.g. balsam of tolu, storax, balsam of peru, etc.

Resenes: These are the complex natural substances without any specific chemical properties. They are inert chemically. They neither form any salt nor they get hydrolysed. Examples of the drug containing resenes are the gum copal, gutta purcha, asofoetida, colophony and dammar.

Isolation of Resins

Pharmaceutical resins are obtained from the plants and animals by one of the following methods.

1. By extraction with alcohol and precipitation with water, e.g. jalap, podophyllum, ipomoea, etc.
2. By distillation for separation of oil, e.g. copaiba, colophony, etc.
3. By heating the plant part, e.g. guaiacum.
4. As plant exudates by incisions, e.g. myrrh, asafoetida, balsams, etc.
5. By collecting fossil resins, e.g. copal, kauri, etc.
6. By processing the encrustations i.e. shellac.

Table 14.7: Natural drugs containing resins and resin combinations

Sr. No.	Name of crude drug	Synonym	Biological Source	Active Constituents	Uses
(A)	Vegitable Drugs				
1.	Asafoetida	Devils drug	*Ferula foetida* (Umbelliferae)	Resin, gum, volatile oil, asaresinotannol	Carminative, nervine stimulant, intestinal flatulance
2.	Balsam of peru	Indian black balsam	*Myroxylon balsamum* (Leguminosae)	Resin, volatile oil, esters of cinnamic and benzoic acid	Scabies, wounds, flavouring agent
3.	Benzoin	Loban	*Styrax benzoin,* (Sumatra benzoin) *Styrax tonkinesis* (Siam benzoin) Styraceae	Benzoic and cinnamic acids and their esters, sumaresinolic and siaresinolic acid, coniferyl acetate	Expectorant, carminative, antiseptic
4.	Boswellia	Sallaki	*Boswellia serrata* (Burseraceae)	Resin, boswellic acid, volatile oil, sesquiterpenes	Rheumatiod arthritis
5.	Cannabis	Ganja	*Cannabis sativa* (Cannabinaceae)	Resin, tetrahydro-cannabinol	Narcotic analgesic psychotropic
6.	Capsicum	Chillies	*Capsicum annum* (Solanaceae)	Capsaicin, capsanthin, carotene, pigments	Stomachic, counter irritant in rheumatism, lumbago
7.	Capsicum oleo resin	Capsicum extract	*Capsicum annum* (Solanaceae)	Capsanthin, capsaicin	Powerful irritant, ingredient of capsicum plaster
8.	Colocynth	Bitter apple	*Citrullus colosynthis* (Cucurbitaceae)	Resin, alkaloids, cucurbitacin E	Purgative
9.	Colophony	Rosina	*Pinus species* (Pinaceae)	Resin acids, abietic acid resene	Stimulant diuretic
10.	Ginger	Zinziber	*Zingiber officinale* (Zingiberaceae)	Volatile oil, zingiberene, curcumene, resin, gingerol, shogaols, gingediols	Aomatic carminative, flavouring agent, motion-in treating sickness

Contd...

Sr. No.	Name of crude drug	Synonym	Biological Source	Active Constituents	Uses
11.	Ginger oleo resin	Gingerin	*Zingiber officinale* (Zingiberaceae)	Gingerol, shogaol, zingerone	Flavour for carbonated beverages
12.	Guggul	Commiphora	*Commiphora weightii* (Burseraceae)	Gum resin, steroids, guggulosterones	Anti-inflammatory, hypolipidemic
13.	Haridra	Haldi (Turmeric)	*Curcuma longa* (Zingiberaceae)	Volatile oil, curcuminoids	Anti-inflammatory, condiment, spice
14.	Ipomoea	Orizaba jalap	*Ipomoea orizabensis* (Convolvulaceae)	Resin, jalapin	Cathartic
15.	Jalap	Jalap radix	*Ipomoea purga* (Convolvulaceae)	Resin, volatile oil, jalapin convolvulin	Powerful cathartic
16.	Kaladana	Pharbitis	*Ipomoea hederaceae* (Convolvulaceae)	Resin (pharbiticin), saponin	Cathartic
17.	Male fern	Aspidium	*Dryoptiris filix-mas* (Polypodiaceae)	Oleo-resin, phluroglucinol derivatives, filicic acid, filicinic acid, flavispidic acid	Anthelmintic for tape worms
18.	Myrrh	Gum-Myrrh	*Commiphora molmol* (Burserraceae)	Volatile oil, gum, resin, commiphoric acids	Antiseptic, stimulant
19.	Nagli	Ragi	*Eleusine corocana* (Gramineae)	Volatile oil Carbohydrates	Nutritive Tonic
20.	Podophyllum	Indian podophyllum	*Podophyllum hexandrum* *P. emodi* (Berberidaceae)	Resin, podophyllin peltatins	Purgative treatment of veneral worts, anticancer
21.	Podophyllum resin	Podophyllin	*Podophyllum hexandrum,* *P. emodi* (Berberidaceae)	Resin, podophyllotoxin	Drastic purgative, treatment of veneral worts
22.	Storax	Prepared storax	*Liquidamber orientalis* (Hemamelidaceae)	Resin (storesin) free and cinnamic ester	Stimulant antiseptic and in perfumes
23.	Tar	Pine tar	*Pinus sylvestris* (Pinaceae)	Hydrocarbons, terpenoids resinous substances	Expectorant, antipruritic

Contd...

Sr. No.	Name of crude drug	Synonym	Biological Source	Active Constituents	Uses
24.	Tolu balsam	Balsam of Tolu	*Myroxylon balsamum* (Leguminasae)	Cinnamic acid, benzoic acid, benzoyl benzoate, toluresinotannol, volatile oil	Expectorant, flavouring agent, antiseptic
25.	Nirgudi	Monks pepper	*Vitex Negundo*	Volatile oil	Antiinflamatory
26.	Serpentary	-	*Aristolochia Serpentaria* *Aristolochicaeae*	Volatile oil	Stimulent
(B)	Animal Drug				
27.	Shellac	Sealing wax	*Lacifer lacca* (Lacciferidae)	Resin, shelloic acid, aleuritic acid	Sustained release medicaments

ASAFETIDA

Synonyms

Asafetida, Gum Asafoetida, Devil's dung.

Biological Source

Asafetida is the oleo-gum resin obtained by incision from the rhizomes and roots of *Ferula foetida* Regel, *Ferula rubricaulis* Regel and other species of *Ferula*, belonging to family Umbelliferae.

It contains not less than 0.10 per cent of Trans Fernlic acids calculated on dry basis.

Geographical Source

It is a genus of perennial herbs distributed from Mediterranean region to Central Asia, specifically Iran and Afghanistan. Three species of *Ferula* are found in India. *Ferula narthex* occurs in Kashmir.

Collection and Preparation for the Market

In Afghanistan, the resin is obtained from carrot shaped massive roots and rhizomes of the plants which are about 4 - 5 years of age and 12 - 15 cm in diameter. In the month of March-April, just before the flowering season of the plant, the upper part of the roots, very close to the crown, is cut off. The milky juice oozes out of the cut surface and starts coagulating. This cut surface is covered by the dome-shaped device made up of leaves and the branches to avoid the contamination with sand and foreign organic matter. After few days, the coagulated matter is scrapped off and the fresh cuts are given to collect more exudate. This is continued for about three months or until the plants cease to produce latex. On average, plant yields about 1 kg of the oleo-gum resin. After collection, it is dried thoroughly and packed in suitable containers.

Description

Colour : It is yellowish-white changing to reddish-brown

Odour : Intense, persistent, penetrating and alliaceous

Taste : Bitter, alliaceous and acid

Size : The tears are 0.5 - 3 cm in diameter

Shape : It occurs in 2 different forms i.e. tears and masses. Tears are rounded or flattened.

Extra Features

Fresh tears are tough but when dried they become hard and brittle. Tears are internally milky whitish-yellow, translucent or opaque. Mass of asafoetida is agglutinated and mixed with root fragment, foreign material and other impurities.

Chemical Constituents

Asafoetida contains resin (40-65 per cent), gum (20-25 per cent), and volatile oil (4-20 per cent). The resin of the drug consists chiefly asaresinotannol in the free or combined form with ferulic acid. Free umbelliferone is absent in the drug (distinction from galbanum). Ferulic acid on treatment with hydrochloric acid is converted into umbellic acid, which looses water to form umbelliferone.

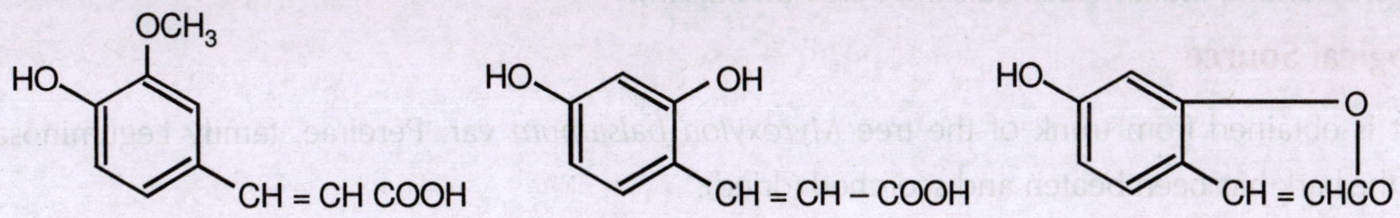

Ferulic acid **Umbellic acid** **Umbelliferone**

Oil of asafoetida is obtained by steam distillation of the oleo-gum resin. The chief constituents of the oil are secondary butyl propanyl disulphide. Other constituents of oil are di and trisulphides, pinene and other terpenes. The specific odour of the drug is due to sulphur compounds of the formulae $C_7 H_{14} S_2$, $C_{16}H_{20}S_2$ and $C_{10}H_{18}S_2$.

Standards of Quality

Foreign organic matter	:	Not more than 02.0%
Alcohol soluble extractives	:	Not less than 12.0%
Water soluble extractives	:	Not less than 15.0%
Total ash	:	Not more than 10.0%
Acid insoluble ash	:	Not more than 02.0%
Moisture	:	Not more than 10.0%

Identification

1. Fractured surface of the drug, if treated with sulphuric acid forms red or reddish-brown colour.
2. When treated with 50 per cent nitric acid, the drug gives green colour.

3. When triturated with water, it forms the yellowish-orange emulsion.
4. Triturate about 0.5 g of the drug with sand and 5 ml of hydrochloric acid. To it, add little quantity of water, filter and to the filtrate add equal volume of ammonia. A blue fluorescence is produced due to presence of umbelliferone.

Galbanum, an oleo-gum resin obtained from *Ferula galbaniflua*, is an allied drug to asafoetida. Galbanum contains free umbelliferone.

Uses

It is used as carminative, nervine stimulant, or in intestinal flatulence, as flavouring agent for curries, sauces and pickles and in veterinary medicine. Ferulic acid is used in food preservative.

Adulterants

Asafoetida is adulterated with gum arabic, rosin, gypsum, red clay, chalk and barley or wheat flour.

BALSAM OF PERU

Synonyms

Peru balsam, Indian black balsam, Peruvian balsam.

Biological Source

It is obtained from trunk of the tree *Myroxylon balsamum* var. Pereirae, family Leguminosae, after the bark has been beaten and scorched (dried).

Geographical Source

It is produced in Central America, San Salvadoer, Honduras and Guatemala. The plants have been naturalized in Sri Lanka and Florida.

Preparation of Balsam of Peru

The balsam trees are about 25 metres in height with ducts in bark. The balsam is a pathological product and is formed by injury to the plant. The process of collection starts in the months of November and December. The strips of the bark about 15 × 30 cm are beaten with blunt end of an axe and scorched with torch to injure the plant. The balsam is produced about one week after the injury and 2 weeks later, the bark is pulled and removed. The bark is then boiled in water to get the product. The trunk also exudes the balsam due to injury and is collected by wrapping with rags. The rags are saturated with the balsam, which is then collected by boiling in water. The rags are squeezed and balsam sinks to the bottom, water is decanted and balsam is collected.

Fig. 14.71 : *Myroxylon balsamu* var. Pereirae herb

Description

Colour : Pern balsam is dark-brown to reddish brown viscous liquid and transparent thin layers.

Odour : Aromatic vanilla like.

Taste : Bitter acrid taste.

Solubility : Insoluble in water, soluble in alcohol, glacial acetic acid, partly soluble in ether and petroleum ether.

Chemical Constituents

50 - 65 per cent high boiling volatile oil with cinnamein (Benzye cinnamate) and 25 - 28 per cent of resin, traces of styrene, vanillin and coumarin. Volatile oil contains esters of cinnamic acid and benzoic acid like, cinnamyl cinnamate, cinnamyl benzoate, benzyl benzoate, and benzyl cinnamate, with small amounts of free benzyl alcohol, free benzoic and cinnamic acids.

Standards of Quality

Density	:	1.150 – 1.170
LOD	:	Not more than 40%
Sap value	:	170 - 230

Uses

It is extensively used in topical preparations for scabies, treatment of wounds, ulcers and bedsores; in feminine hygiene sprays soaps, and cosmetics.

It is used as flavouring agent and also for masking. As fixative of flavours, maximum permissible limit is 0.8 per cent.

It is also used as flavour for food items like candy, dairy desserts, alcoholic and non-alcoholic beverages to the extent of 15 ppm.

BENZOIN

Synonyms

Sumatra Benzoin, Loban.

Biological Source

Benzoin is a balsamic resin obtained from *Styrax benzoin* Dryand or *Styrax paralleloneurus* Perkins and other species of *Styrax* known in the market as **Sumatra Benzoin** or it may also contain the balsamic resin from *Styrax tonkinesis* and other species commercially known as **Siam Benzoin**, belonging to family Styraceae. It should contain not less than 25 per cent of total balsamic acids, with reference to dry alcohol soluble matter.

Geographical Source

Sumatra benzoin is produced from trees grown in South Eastern Asia, chiefly Sumatra. While Siam benzoin is obtained from trees grown in Thailand and Vietnam.

Collection and Preparation for the Market

Fig. 14.72: Styrax benzoin herb

Benzoin trees are not grown in India. The drug is totally imported from Indonesia. Benzoin is a pathological resin and is collected from about 6 years old wild or cultivated trees. It is collected by making incisions near the base of the tree. The plant is reported to contain neither secretory nor the resinous constituents. These are only produced after incising the plant. The initial produce, which is amorphous and yellow, is not used in the medicine. The next flow which oozes out of the plant is collected and utilised for medicinal purposes. This resinous material is allowed to solidify and sundried. Approximately, a single tree produces about 10 kg of balsamic resin.

Description

Sumatra Benzoin

Colour : It is greyish-brown or grey.

Odour : Aromatic and characteristic.

Taste : Sweetish and slightly acrid.

It occurs in the form of lumps of varying sizes or tears. Tears are externally yellowish, milky white.

The surface is uneven. When heated, fumes of benzoic and cinnamic acids are produced.

Siam Benzoin

Colour : Yellowish-brown to rusty-brown.

Odour : Agreeable and vanilla-like.

Taste : It is sweetish and slightly acrid.

It occurs as hard and brittle masses. When heated, it is softened and becomes plastic.

Chemical Constituents

Sumatra benzoin contains free balsamic acids (benzoic and cinnamic acids) and esters derived from them. Triterpenoid acids such as summaresinolic and siaresinolic acids are also present. The major constituent of Siam benzoin is an ester coniferyl benzoate (about 76 per cent). The drug also contains styrol, vanillin and phenyl propyl cinnamate.

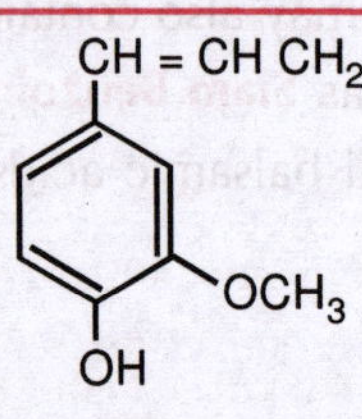

Coniferyl alcohol

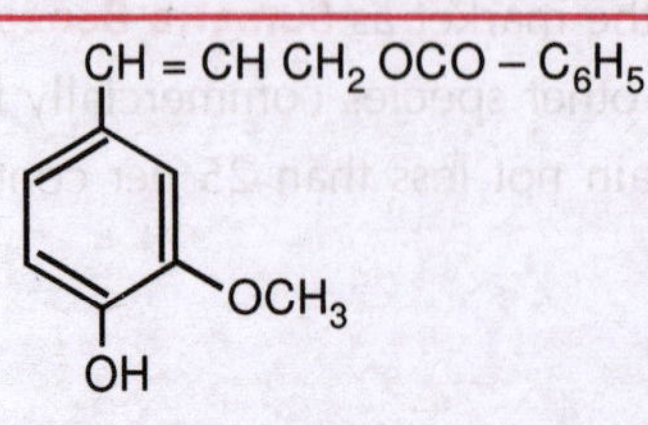

Coniferyl benzoate

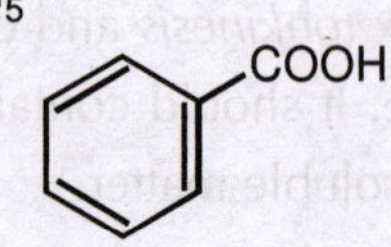

Benzoic acid

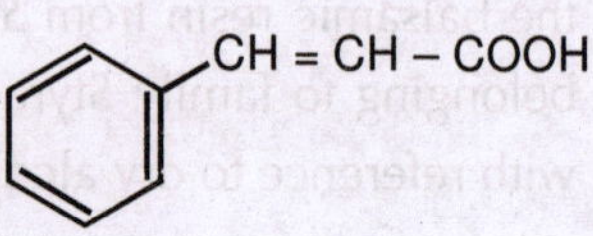

Cinnamic acid

Siam benzoin differs from Sumatra variety in that it contains insufficient cinnamic acid to give an odour of benzaldehyde, when warmed with potassium permanganate solution.

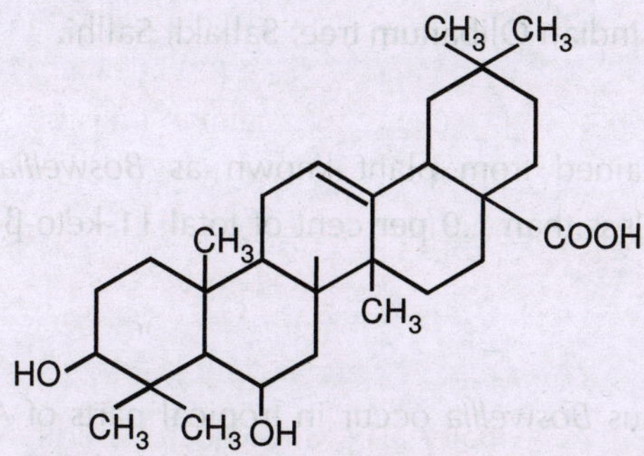

Summaresinolic acid

Identification

1. Alcoholic solution of benzoin with water gives milky white solution.
2. Heat small quantity of benzoin in a test tube covering opening of test tube with a glass plate. Cool the contents of the test tube. Examine the glass plate under microscope. The crystals of cinnamic acid shall be observed.
3. To 2.5 g benzoin, add 10 ml ether, shake it well and pour 2 to 3 ml of this extract in a porcelain dish, add 2 to 3 drops sulphuric acid. A deep brown colour is produced in case of Sumatra benzoin and a deep purplish red colour is obtained in case of Siam benzoin.
4. Add 4 ml solution of potassium permanganate to 1g benzoin and warm. Odour of benzaldehyde is produced in case of Sumatra benzoin.

Standards of Quality

1.	Benzoic acid content	:	Not less than 60 per cent (Sumatra)
			Not less than 12 per cent (Siam)
2.	Alcohol soluble matter	:	Not less than 76 per cent (Sumatra)
			Not less than 90 per cent (Siam)
3.	Loss on drying	:	Not more than 10 per cent
4.	Acid insoluble ash	:	Not more than 10 per cent

Uses

Benzoin is an irritating expectorant, a carminative and diuretic. It is also used externally as an antiseptic and a protective. It is used in the form of compound tincture of benzoin, and as an inhalation, specifically in the treatment of upper respiratory tract infection. It is preferred to retard the rancidity of fats and oil in the preparation of benzoate lard. Industrially, it is used to fix the odour of incense sticks, soaps, perfumes and several other cosmetics and to mask the taste of pharmaceutical preparations.

BOSWELLIA

Synonyms

Kunduru, Sallaki guggul, Indian Olibanum tree, Sallaki, Sallhi.

Biological Source

The oleo-gum-resin obtained from plant known as *Boswellia serrata* belonging to family Burseraceae. It contains not less than 1.0 per cent of total 11-keto-β-boswellic acid and acetyl-11-keto-β-boswellic acid.

Geographical Source

About 10 species of genus *Boswellia* occur in tropical parts of Asia and Africa. *B. serrata,* the species found in dry hilly areas of Bihar, Madhya Pradesh, Chhatisgad and Gujarat. It is a medium sized but highly branching tree. It grows upto 12-15 feet in height. The type of leaves distinguishes Indian olibanum into two varieties called *var serrata,* having pubescent and serrate leaves and *var. glabra* having glabrous and entire leaves.

Method of Preparation

Oleo-gum-resin is obtained by tapping process, between November and June. The average yield per tree per annum is maximum up to 1 kg. The trees with more girth and stunted growth have low yields. For exudation, shaving is done, at 2.5 feet height, after 4-5 days.

Description

The oleo-gum-resin of Indian olibanum has following characters and compositions.

Colour : Golden to dark brown

Odour : Turpentine like, agreeable on burning

Taste : Aromatic-bitter

Size : Agglomerates of various size and shapes tears are 2 × 5 cm. Translucent, brittle, club or pear shaped.

Fig. 14.73: *Boswellia serrata* twig

Chemical Constituents

Oleo-gum resin of bosewellia approximately consists of:

Volatile oil : 8 - 9 per cent

Gum : 20 - 23 per cent

Resin : 55 per cent

Volatile oil is composed of sesquiterpene alcohols, anisaldehyde, d-α-thujone, α-pinene, d-α-phellandrene and phenolic compounds. Similar to turpentine oil, this oil is soluble in colophony and dammer, but more volatile in nature. Acidic components of volatile oil are known as compholenic acid and compholytic acid.

Gum is mainly composed of arabinose with small amounts of xylose and galactose. Gum also contains oxidizing and diastatic enzymes. The highly brittle resin is soluble in various organic solvents.

Boswellia resin: Boswellia mainly contains boswellic acids. Boswellic acids belong to pentacyclic triterpenoids which are the active therapeutic contents. Important of them are β-boswellic acid, acetyl-β-boswellic acid, acetyl-11-keto-β-boswellic acid and 11-keto-β-boswellic acid of which acetyl-11-keto-β-boswellic acid is the main anti-inflammatory compound on which drug is standardized. Additionally it also contains diterpenic alcohol serratol and α and β amyrin.

Tetracyclic triterpenic acids present in resin are known as elemolic acid and elemonic acid.

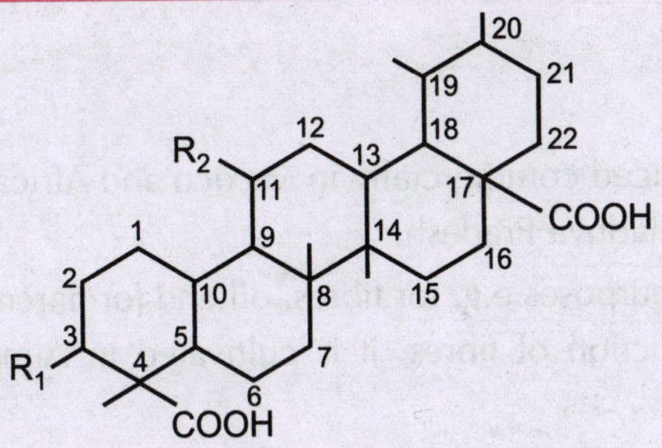

β-Bosewellic acid

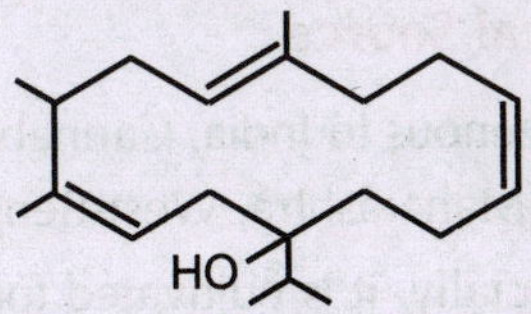

Serratol

Standards of Quality

Moisture	:	Not more than 12 per cent
FOM	:	Not more than 2.0 per cent
Alcohol soluble extractives	:	Not more than 35 per cent
Ash	:	Not more than 10 per cent
Acid insoluble ash	:	Not more than 2.0 per cent
Melting point	:	73 - 78°C
Heavy metals	:	Not more than 20 ppm
Water	:	Not more than 12.0 ppm

Table 14.8: Chemical nature of Boswellic acids

Sr. No.	Type	R_1	R_2
1.	β-boswellic acid	— OH	— H
2.	Acetyl-β-boswellic acid	$O—CO—CH_3$	— H
3.	Acetyl-11-keto-β-boswellic acid	$O—CO—CH_3$	= O
4.	11-keto-β-boswellic acid	— OH	= O

Uses

It is anti-inflammatory and antiarthritic. Indian olibanum is mainly used in treatment of rheumatoid arthritis. It is known to regain integrity of vessels in joints from damage or spasm. It is also used in preparation of incence sticks and as a fixative in perfumes.

CANNABIS

Synonyms

Cannabis Indica, Indian hemp; Ganja; Marihuana

Biological Source

Cannabis consists of dried flowering tops of the cultivated female plants of *Cannabis sativa* Linn, family Cannabinaceae (Moraceae).

Geographical Source

It is indigenous to India. Cannabis is produced commercially in Mexico and Africa. In India, it is cultivated in Maharashtra, West-Bengal, and Madhya Pradesh.

Commercially, it is cultivated for various purposes e.g. for fibres, oil and for narcotic substances like charas, bhang, ganja, etc. For the production of fibres, it is cultivated in Almora, Garhwal, Nainital, also in Kashmir, Nepal and Kerala.

Cultivation and Collection

The cultivation of cannabis is undertaken only under license from the Government. Since it is a *narcotic drug*, its cultivation is regulated by Government. It needs light loamy or sandy soil and humid tropical climate. It is cultivated as a kharif crop and sowing of the seeds is done in the month of June or July. The distance between two rows is about 1 metre and it needs about 6 - 9 kg of seeds per hectare. The plant is kept free of weeds and, if necessary, thinning is done. It flowers in the month of November or December harvesting is done in the month of December or January. The average yield of ganja per hectare is 275 kg, but may vary depending upon suitability of climatic conditions. The average crop per hectare is 1100 - 1700 kg.

Macroscopic Characters

Colour : Drug is dull green in colour.

Odour : Strong, characteristic and narcotic

Taste : Somewhat acrid and pungent

Shape : Flattened or cylindrical masses consisting of branches, upper part of the stems with bracts, bractioles, pistillate flowers, fruits and seeds. Stems which are not more than 3 mm in diameter also constitute the drug.

Stems are thin, straight, cylindrical and longitudinally furrowed. Bracts are 1.5 - 2 cm long, simple or lobed with subulate stipules. Bractioles are in pairs, present in the axil of bract.

Fig. 14.74: Cannabis Sativa herb

Flowers are formed in axil of each bractiole with 2 long brownish-red hairy stigmas. Achene type of fruit is about 5 - 6 mm in length and 4 mm in width, ovate glossy green or yellowish-green in colour with single seed.

Narcotic Products from Cannabis

The narcotic substances obtained from the hemp plant are the bhang or hashis, ganja and charas. The method of preparation and types of plants used for extracting the drug vary and only the female plants are used for preparation. Male plants are not selected, because the resinous material is formed only in female unfertilized plants.

Bhang or Siddhi

Bhang and siddhi are two different names of the same product. Bhang is prepared by cutting the leaves and flowering tops of the plant, exposing them to sun and dew, drying and pressing the drug. The product is stored in earthenware vessels. The resinous matter present in glandular trichomes contains hallucinogenic matter.

Ganja

Ganja consists of the dried flowering or fruiting tops of the female plants from which no resin has been removed. It is collected only from the cultivated plants. For the preparation of ganja, plants are collected when the lower leaves fall on the ground and flower stalks begin to turn yellow. The floral shoots are cut off and are spread out in ridges and furrows. The ridges are levelled down and crushed to press the floral shoots into compact sheaves. The turning of the material is done at regular intervals. After that, the material is collected and arranged in flat heaps. It is, then, subjected to pressure under the press. The heaps are turned over and broken up, so as to form thick layers. After 3 to 4 days, ganja is ready for storage. Two types of ganjas are available in the market. The first variety is flat or Bombay ganja, while the other is known as round or Bengal ganja. In case of flat ganja, the individual pieces of the plants are pressed by threading, wherein the resinous material of the flowering tops sticks together to form the flat mass. The flat ganja coming from Ahmednagar district of Maharashtra is well known in commerce. Round ganja is regarded as better quality where the resin free parts of the plants are removed and every piece (flowering top) is rolled individually, so as to form cylindrical mass. It is prepared in Bengal.

Charas

Charas is the resinous exudation collected from the leaves of the hemp plants. The resinous secretion, appearing just before flowering of the plant, is collected by rubbing the fresh tops between hands or by beating them on cloth or carpet. The adhering material is scrapped off to yield, charas of the market. It is collected even by walking through the cultivated plants after wearing the leather aprons. The resinous secretion, which sticks to the leather aprons is scrapped off and collected.

Chemical Constituents

Indian hemp comprises 15 - 20 per cent of resin (present in glandular trichomes) which contains the major active euphoric principle 1-3-4 trans tetra-hydrocannabinol (commonly known as Δ^1 THC). It contains volatile oil, trigonelline, and choline. The resin also contains cannabinol, cannabidiol, cannabidiolic acid, cannabichromene and cannabigerol. Indian hemp seeds contain about 20 per cent of fixed oil.

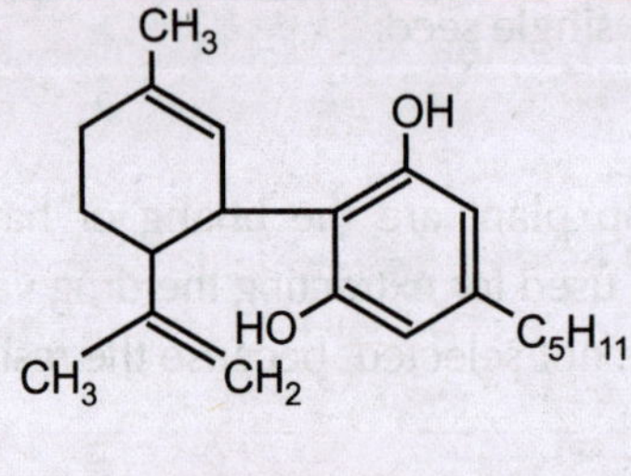

Cannabidiol

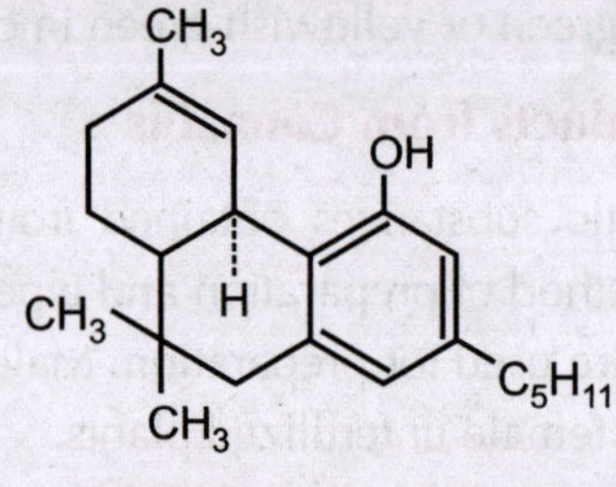

Tetrahydrocannabinol

Standards of Quality

Alcohol soluble extractives	:	10 – 18 per cent
Ash	:	Not more than 15.0 per cent

Identification

Shake about 0.1 g drug with 5 ml light petroleum (60 - 80°) and filter. To 1 ml filtrate, add 2 ml 15 per cent solution of hydrogen chloride in ethyl alcohol. At the junction of the two liquids, a red colouration appears. After shaking, the upper layer becomes colourless and lower layer acquires pink colour, which disappears on addition of water.

Uses

It is a narcotic, sedative and analgesic. It has psychotropic properties due to tetrahydrocannabinol. At present, it is very little used as a drug. It causes intoxication, euphoria and later mental disturbances.

Cannabis is notified as dangerous drug by government, its cultivation, sale, storage is controlled by government. It is very poisonous and should be used carefully under medical (consultation) supervision only.

Storage

It should be stored in a well closed container after thorough drying.

CAPSICUM

Synonyms

Chillies, Cayenne pepper.

Biological Source

Capsicum consists of dried ripe fruits (Lal-Mirchi) of *Capsicum annum* Linn. var: *minimum,* belonging to family Solanaceae. It contains not less than 0.25% of capsaicin on dried basis. Another variety capsicum frutescens known as **Lavangi mirchi** is also available.

Geographical Source

Capsicum is cultivated and collected in almost all the tropical countries. East Africa, West Africa and India are the regions producing the drug on commercial scale. In India, it is grown in Andhra Pradesh, Uttar Pradesh, Gujarat, Maharashtra, Assam and Tamil Nadu.

Cultivation and Collection

Chillies are very common throughout India. About 9.3 lach hectares of land is under capsicum cultivation in India. Cultivation of chillies is done by sowing seeds on the raised beds. The seedlings which are about one month old are transplanted into well prepared fields. As an irrigated crop, it is normally rotated with sugarcane and turmeric. The crop is ready for harvesting after six months. The fully ripened fruits are picked up by hands and are dried in the sun on the floors. After drying and grading, they are packed into the gunny bags and sent to the market. About 51.900 tanns of dry chillies are being exported to USA, Korea, Bangladesh, while large quantities of oleo resins of chillies are exported to USA, Japan, Germany, UK and France.

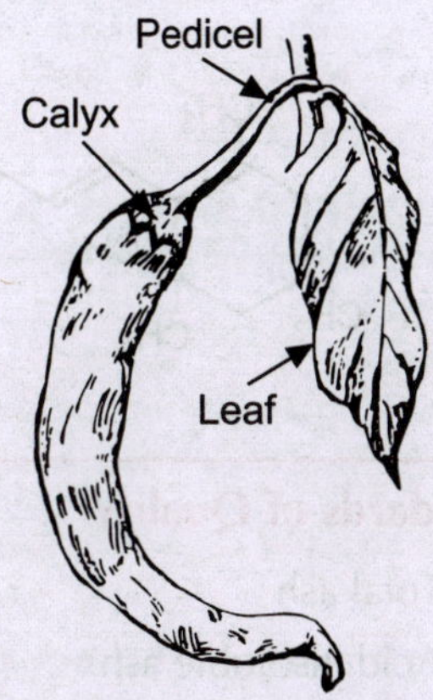

Fig. 14.75: Capsicum Fruit

Chillies are best grown in warm and humid climate with dry weather during maturity period.

They thrive well on light to heavy deep soil or well drained loamy soil. Chillies are grown as *Kharif*, as well as, *Rabi* crop. NP-34, NP-41, NP-46, and NP-51 are high yielding pungent varieties of capsicum. About 8.1 lakhs tones of chillies are produced every year in India. About 50.0 percent of world requirement of capsicum is produced in India.

Macroscopical Characters

Colour	:	Dull orange red to brownish-red.
Odour	:	Characteristic.
Taste	:	Pungent.
Size	:	About 12 to 25 cm in length and up to 7 mm in width.
Shape	:	It is conical or oblong.

Extra Features

Pedicel is attached to 5 toothed calyx, about 2 - 3 cm in length. Surface is shrivelled, glabrous, and leathery. The fruit contains 10 - 20 brownish-yellow flat seeds, which are about 3 - 4 mm in length. The seeds are either loose or attached to thin reddish dissepiment. The fruits are divided into 2 or 3 loculi. The fruits should not contain more than 3 per cent of calyx and pedicel.

Chemical Constituents

Capsicum contains about 0.5 - 0.9 per cent colourless, crystalline and pungent principle, known as capsaicin which is volatile above 65°C. Capsicum also contains volatile oil 1.5 per cent, fixed oil (4 - 16 per cent), proteins and pigments, viz. capsanthin and carotene. Pigments are responsible for the red colour. Thiamine and ascorbic acid are the other contents of the drug.

Unlike ginger, the pungency of capsicum is not destroyed by boiling it with 2 per cent solution of sodium hydroxide. But, it is destroyed by oxidizing agents like potassium permanganate.

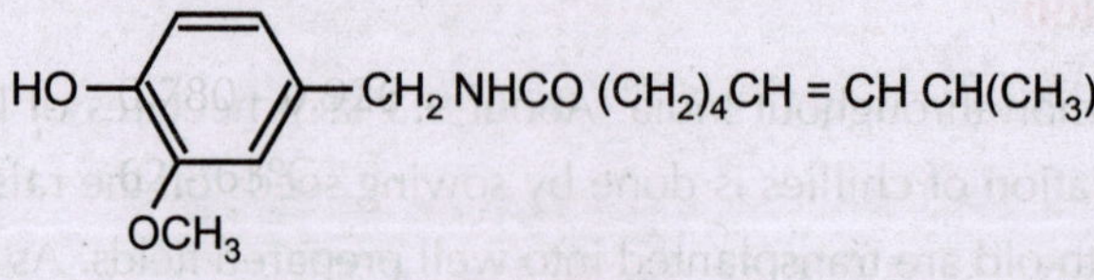

Capsaicin

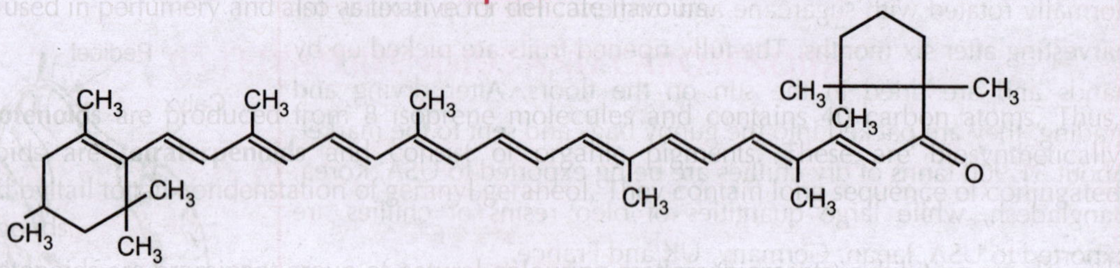

Capsanthin

Standards of Quality

Total ash	:	Not more than 10.0%
Acid insoluble ash	:	Not more than 1.2%
Loss on drying	:	Not more than 11.0% at 105ºC
Ether soluble extractives	:	Not less than 12.0%.

Uses

It is used as a carminative, an appetizer and a stomachic. Externally, it is used as a counter irritant in the treatment of rheumatism, lumbago and neuralgia. It is also used in spices.

CAPSICUM OLEO-RESIN

Synonym

Capsicum extract

Biological Source

It is the oleo-resin of Red-chillies obtained by percolation of dried ripe fruits of *capsium annum* van. *minimum* family solanaceae.

Method of Preparation

Capsicum oleo-resin is prepared by extracting the crushed capsicum with either hot alcohol or hot acetone by percolation method. The solvent (hot alcohol or acetone) is evaporated and the residue is extracted again with successive quantities of cold alcohol or acetone until the marc is free from pungency. It contains not less than 8 per cent of capsaicin.

Description

It is a thick dark reddish brown with intensely pungent taste. It is soluble in alcohol, acetone, ether, chloroform, volatile oils and fats. It gets solubilized with opalescence in fixed oils and benzene.

Chemical Constituents

Apart from a red colouring matter capsanthin, it contains not less than 8 per cent capsaicin. Capsaicin is present to the maximum extent in the longitudinal dissepiments. Its pungency is not affected by alkali, but destroyed by oxidizing agents.

Uses

Capsicum oleo resin is a powerful irritant and a carminative. It is used as a counter-irritant in lumbago and neuralgia. It is an ingredient of capsicum plaster, capsicum cotton wool and capsicum ointment.

Capsicum oleo-resin, being a very powerful irritant, even in minute quantity, produces burning sensation in contact with eyes and tender parts of skin.

Caution

To get rid of the skin irritation due to capsicum, a dilute solution of potassium permanganate may be applied. Cocaine eye drops may be used to relieve the pains in the eyes.

COLOCYNTH

Synonyms

Bitter apple, Bitter cucumber.

Biological Source

It consists of dried pithy pulp of ripe fruits of *Citrullus colocynthis* Schrador, family Cucurbitaceae, containing not more than 2 per cent of epicarp and 5 per cent of seeds.

Geographical Source

It is native of Asia, and Africa found in Syria, and Egypt. It is cultivated in Spain, Sicily and Morocco. It occurs throughout India growing wild, particularly in Tamil Nadu, Gujarat, Punjab and Andhra Pradesh.

Cultivation and Collection

It is not cultivated commercially. It is an annual or perennial herb with climbing stem. It bears spherical fruits which are green in colour and turn yellow when matured. Fruits are about 8 to 10 cm in diameter. The fruits are collected in autumn, peeled with sharp instruments and dried in the sun. The dried pulp of the full grown fruits, free of rind, constitutes the drug. It is marketed either as broken pieces which are light and spongy or in the form of flakey powder, which is yellowish-orange or yellowish-green in colour.

Macroscopical Characters

Colour	:	The fruit is yellow
Odour	:	Characteristic
Taste	:	Intensely bitter
Size	:	Entire fruits are upto 10 cm in diameter
Shape	:	Sub-spherical

The berries contain number of seeds which are about 6 × 4 × 2 mm in size, ovoid, flattened and yellowish-brown in colour.

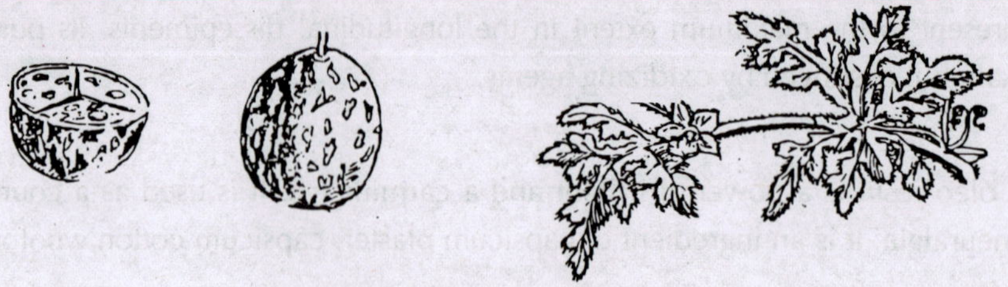

Fig. 14.76: Colocynth Twig and Fruit

Chemical Constituents

The principles of colocynth include bitter, amorphous, alkaloids and resin. Both of them are responsible for its purgative action. The pulp also contains α - elaterin (also known as Cucurbitacin E) glycosidal compounds colocynthin, colocynthidin, phytosterol and fatty acids.

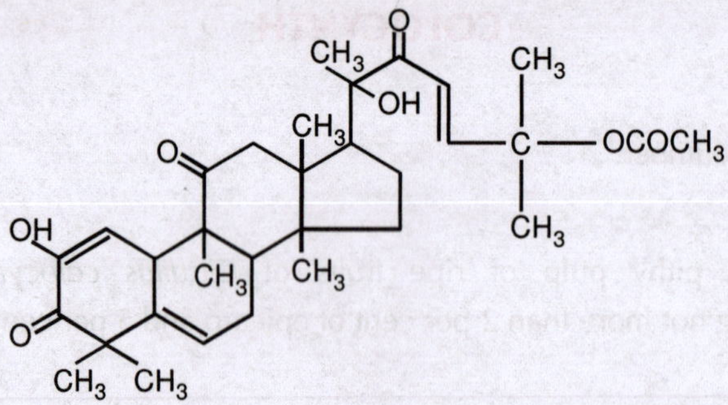

α-elaterin

Standards of Quality

Acid insoluble ash	:	Not more than 6.0 per cent
Light petroleum ether	:	50 – 60
Soluble matter	:	Not more than 3.0 per cent
Seeds	:	Not more than 5.0 per cent

Uses

It is used as powerful purgative along with carminatives. The constituents also exhibit necrotic action.

COLOPHONY

Synonyms

Rosin; Rosina; Colophonium; Resin; Amber-resin, Gum rosin, pine-resin.

Biological Source

Colophony is the residue left after the distillation of the oil of turpentine from the crude-oleo-resin obtained from various species of *Pinus* belong to family Pinaceae.

Geographical Source

It is prepared in North America, Northern Europe, Pakistan and India.

Description

Colour : Pale yellow to yellowish brown or amber colour.

Odour : Faint and terebinthinate.

Taste : It occurs in the form of angular, translucent masses of various sizes.

Extra Features

Colophony is brittle and readily fusible with glossy appearance.

Solubility

It is insoluble in water but soluble in alcohol, carbon disulphide, chloroform, ether, many fixed and volatile oils, glacial acetic acid and in light petroleum.

Chemical Constituents

Colophony contains mainly resin acids which are unsaturated and hence reactive. About 90 per cent of this is abietic acid. The composition of the drug varies according to the source and storage conditions. The drug also contains esters of oleic acid and resin acids. It also contains 0.5 per cent volatile oil; 5 to 6 per cent resenes. Other acids in the colophony are sapinic acid and pimaric acid. Unsaponifiable matter of colophony, which constitutes about 3 - 10 per cent contains hydrocarbons and high molecular weight alcohols.

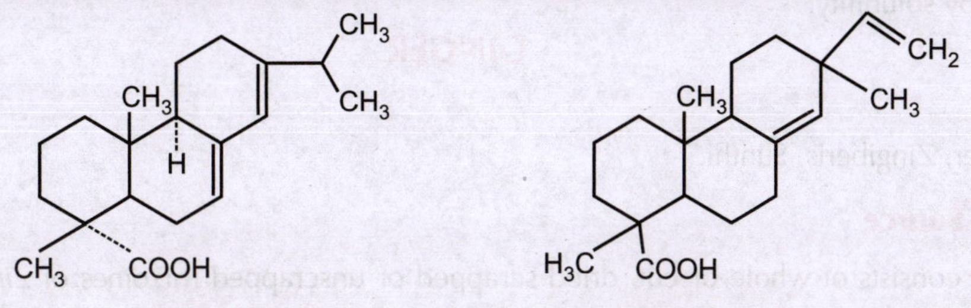

Abietic acid **Pimeric acid**

Identification

(1) Take 1 gm of Colophony and dissolve in 10 ml of acetic anhydride by gentle heat and add a drop of sulphuric acid, bright red colour changing to violet is produced.

(2) Alcoholic solution of colophony is acidic to litmus.

(3) Dissolve colophony in light petroleum ether and double the volume with dilute solution of copper acetate, petroleum layer takes emerald green colour (due to the copper salt of abietic acid).

Standards of Quality

Colophony complies with following standards:

Melting point	: 75°C to 85°C
Acid value	: 130 to 180

Saponification value	:	188 to 192
Ash value	:	Not more than 0.125 per cent
Specific gravity	:	1.08

Uses

Colophony is a stimulant and diuretic. It is ingredient of certain plaster masses and collodions. It is also used in the preparation of ointments. Rapin-esters are used as protective coatings like polymers.

Industrially, it is used in the preparation of varnishes, insulating compounds, soaps, printing inks, paper sizing in the preparation of floorings, soldering compounds, mastics and pressure sensitive's adhesives.

Storage

Colophony should be stored in the ungrounded condition in well filled, well closed containers away from light and in cool places. In the powder form, it gets oxidized quickly and looses its solubility in light petroleum and gets increased in the weight and hence, it should be stored properly.

Adulterants

Colophony is adulterated with black resin or apic resin. However, both of them can be confirmed by solubility.

GINGER

Synonyms

Zingiber, Zingiberis, Sunthi.

Biological Source

Ginger consists of whole or cut, dried scrapped or unscrapped rhizomes of *Zingiber officinale* Roscoe, family Zingiberaceae. It contains not less than 0.8 per cent of total gingerols on dried basis.

Geographical Source

It is said to be native of South East Asia, but is cultivated in Caribbean islands, Africa, Australia, Mauritius, China, Jamaica, Taiwan and India. More than 35 per cent of the world's production is from India. Assam and Maharashtra are the largest giner producing states of India.

Cultivation and Collection

Approximately, 25,000 hectares of land is under cultivation in India for the production of about 25,000 tones of dry ginger annually. In almost all states of India, ginger is cultivated, especially in Kerala, Assam, Himachal Pradesh, Orissa, West Bengal and Karnataka. Ginger needs warm humid climate and is cultivated in areas with heavy rainfall. It is cultivated even at sea level, but still it thrives best at an altitude of 1000 - 1500 m. If no sufficient rainfall is available, proper arrangements for irrigation are necessary. Sandy or clay or red loamy soils are suitable for ginger. Ginger is cultivated by sowing rhizomes in the month of June. Carefully preserved seed-rhizomes are cut into small pieces and, at least one living bud is allowed in each piece. About 1200 - 1400 kg ginger seed-rhizomes are necessary per hectare. Ginger is a soil exhausting crop and being a rhizome, needs to be supplemented with good quantity of manures and fertilizers. Superphosphate, ammonium sulphate and potash are the common fertilizers used for ginger. Ginger is ready for harvesting in about six months, when its leaves become yellow. Harvesting of ginger is done by digging the rhizomes. They are washed properly and then dried to improve the colour and to prevent its further growth. The rhizomes are scrapped, dried and coated with inert material like calcium sulphate. The yield of 1500 kg per hectare of green ginger is possible by cultivation.

Ginger is produced in almost all the states of India and ranks second among ginger producing countries of the world. There are one dozen large scale oleo resin producing industries in India at present with total installed capacity of 900 tones. About 38.5 lakh tonnes of giner is produced in India, and Maharashtra is the second largest ginger producing state, after Assam.

Preparation of Dry Ginger

Full grown ginger rhizomes after digging from soil are wahsed throughly to remove earthy matter with ample. Then they are kept in water for 6-8 hours and are subjected to remove cork by scrabing. After removal of cork they are are washed again and are ketp in 2.0% lime water for about six hours.

Then they are shifted to small closed rooms and spread uniformly and are exposed to fumes of sulpher for 12 hours. The process is repeated three times consecutively to get best results.

Now the rhizomes are taken to the perforated galvenised cages of 50 × 75 cm size and further three vessels containing each of 20%, 25% and 50% solution of sodium hydroxide w/v are prepared. Now the vessels containing sodium hydroxides solution are heated, to boil the solution.

The cages containing ginger are dipped in each solution in asending order of strength of sodium hydroxide solution for not more five minutes in each solutuion.

Thus, the alkali treated rhizomes are then passed in another vessels having 4.0% (w/v) citric acid solution, where they are kept for two hours. Then the rhizomes are washed throughly and kept on pavements to dry. About 14.6 lakh tonnes of ginger is produced in Maharashtra every year.

Either the sun dried method or drying in oven is followed and then the dried ginger is sent to market in suitable packing.

Macroscopic Characters

Colour : Externally, it is buff coloured.

Odour : Agreeable and aromatic.

Taste : Agreeable and pungent.

Size : Rhizomes of ginger are about 5 - 15 × 1.5 - 6.5 cm.

Shape : The rhizomes are laterally compressed, bearing short flat, ovate and oblique branches on the upper side, with bud at the apex.

Fracture : Short and fibrous.

Fig. 14.77: Ginger Rhizome with Buds

Extra Features

Longitudinal striations and the occasional projecting fibres are present on the surface of ginger. Transversely cut surface shows well marked endodermis and stele.

Microscopic Characters

Cork consists of irregularly arranged cells, followed by cortex. Cortex is made up of thin walled parenchymatous tissue. Well marked endodermis distinguishes the stele and the cortex. Cortical tissue encloses several closed collateral fibro-vascular bundles. Vascular bundles just inside the endodermis are free of fibres. Oleo-resinous cells and starch grains are found throughout the ground tissue. Endodermis is free of starch.

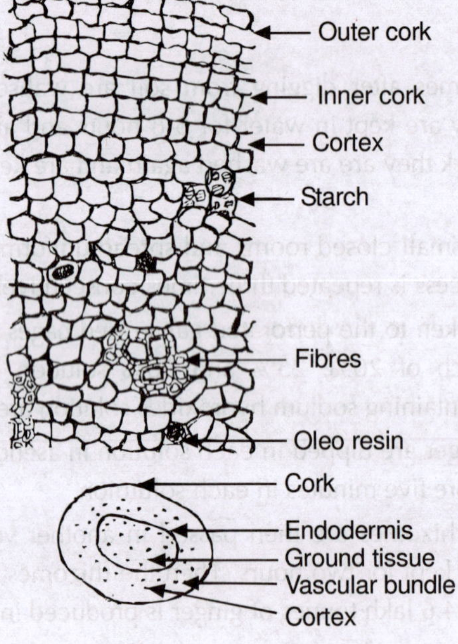

Fig. 14.78: T.S. of Ginger Rhizome

Chemical Constituents

Ginger consists of volatile oil (1 - 4 per cent), starch (40 - 60 per cent), fat (10 per cent), fibre (5 per cent), inorganic material (6 per cent), residual moisture (10 per cent) and acrid resinous matter (5 - 8 per cent). Ginger oil is constituted of monoterpene hydrocarbons, sesquiterpene hydrocarbons, oxygenated mono and sesquiterpenes, and phenyl propanoids.

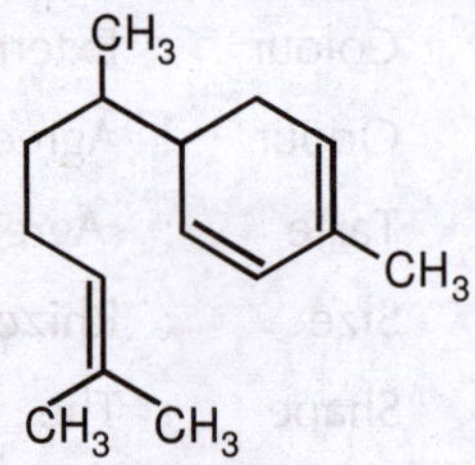

Zingiberene

Sesquiterpene hydrocarbon content of all types of ginger oil from different countries is found to be same and includes α-zingiberene, β-bisabolene, α-farnesene, β-sesquiphellandrene and α-curcumene.

Aroma and flavour are the main characters of ginger. Aroma is due to fragrant principles of volatile oil while the flavour, pungency and pharmacological action is exerted by phenolic ketones of oleo-resin. Various components of volatile oil like isometric terpenic aldehydes like geranial and citral, which cause the delicate and lemony aroma. Few sesquiterpene hydrocarbons are believed to exert spicy note.

Phenolic ketones of oleo resin include gingerols like shogaols, zingerone, paradols, gingediols, hexahydrocurcumin and also o-methyl ethers of these compounds.

Standards of Quality

Foreign organic matter	:	Not more than 2.0 per cent
Water soluble extractives	:	Not less than 10 per cent
Alcohol (90 per cent) soluble extractive	:	Not less than 4.5 per cent
Total ash	:	Not more than 6.0 per cent
Water soluble ash	:	Not less than 1.7 per cent
Acid insoluble ash	:	Not more than 2.0 per cent
Water	:	Not more than 12.0 per cent

Uses

Ginger is used as a stomachic, an aromatic, a carminative, stimulant and flavouring agent. Ginger oil is used in mouth washes, ginger beverages and liquors.

Ginger powder has been reported to be effective in motion sickness. It has been suggested that adsorbent, aromatic and carminative properties of ginger on G. I. tract cause adsorption of toxins and acid enhanced gastric motility. These may have probably blocking effects of G. I. reactions and nausea.

Z. officinale (Methanolic extract) has molluscicidal effects, possessing efficacy to control the parasitic infection viz. schistosomiasis. U.S. Food and Drug administration has included ginger as product that is generally regarded as safe (GRAS).

Storage

Ginger is coated with lime to improve its colour and quality and hence this particular variety is known as limed ginger.

Adulterants

Ginger is adulterated with exhausted ginger, but it can be detected by determination of water-soluble ash, volatile oil content and alcohol and water soluble extractives.

GINGER OLEO/RESIN

Synonym

Gingerin

Biological Source

It is the oleo-resin obtained by percolating the powdered rhizomes of *Zingiber officinale* (Zingiberaceae).

Method of Preparation

Coarsely powdered dried ginger is extracted with acetone or ethyl ether or ethylene dichloride by percolation method. The drug is extracted until exhausted and the extracts are mixed together. The solvent is removed by distillation under reduced pressure. Alcohol gives the maximum yield of oleo-resin. The yield of gingerin may vary from 3.5 to 9.0 per cent, depending upon the source of the material and the method of preparation. An average yield of oleo resin is 6.5 per cent.

Description

It is dark brown, aromatic and pungent viscous liquid.

Chemical Constituents

Gingerin contains approximately 18 - 35 per cent of volatile oil of ginger. It contains pungent, as well as, non-pungent principles of ginger. Fresh samples of gingerin may contain 30.0 per cent of gingerol (the main pungent substance of ginger), shogaol and zingerone. **Shzgoal** is formed from gingernol by loss of one molecule of water.

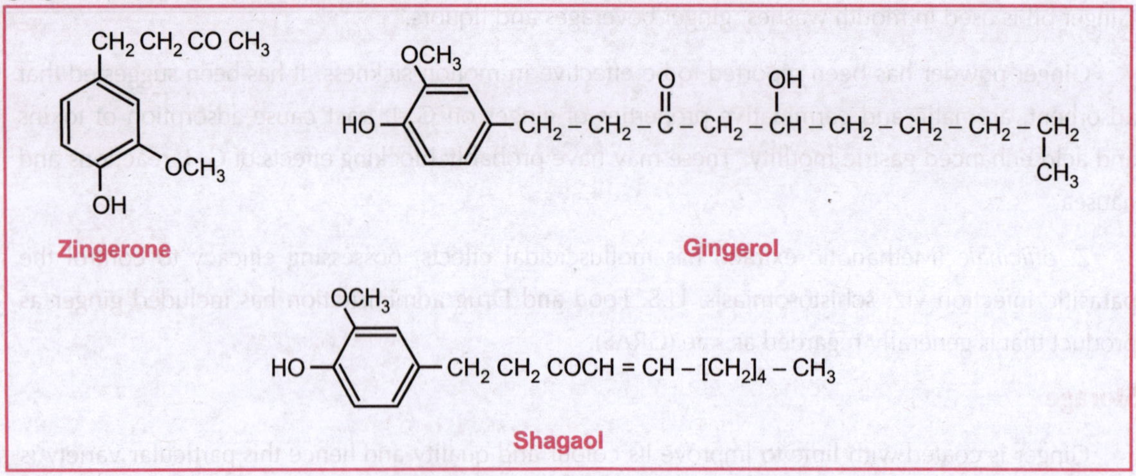

Uses

Since, the oleo-resin can be standardized, it is prepared in industry for its desired use. Oleo-resin of ginger is mainly used as flavour for carbonated beverages in the spices and condiments.

Indian export of ginger oleo-resin during 1996-97 was found to be ₹ 451 lakhs.

GUGGUL RESIN

Synonyms

Scented Bdellium, Gum guggul, Commiphora.

Biological Source

Guggul is the oleo-gum-resin obtained by making deep incisions at the basal part of stem bark of *Commiphora weightii* belonging to family Burseraceae, and should contain not less than 1.0 per cent and not more than 1.5 per cent Guggulsterones (Z & E).

Guggul is one of the noted drugs from Ayurveda and Unani system. In recent times, its demand in therapeutics has been substantially increased.

Geographical Source

Guggul plant is native to Africa especially in its arid zones like Ethiopia, Somalia, Kenya, Zaire and Zimbabwe. It grows widely in various subtropical regions. In India, guggul plant is mainly found in Rajasthan and Gujarat States. Ajmer and Jaisalmer districts of Rajasthan are the prominent habitats.

Cultivation and Collection of Guggul

It grows upto 2 - 3 metres as a woody tree and shows spinescent branches on pale yellow to brownish stem. It has characteristic silvery and paper like bark peelings. It bears compound leaves with ovale subsessile leaflets and they are serrated with smooth upper surface. The guggul tree grows well in sandy loam soil with more gypsum content, with a pH 7.5 - 9. In general, it requires low land use soils.

Fig. 14.79: Guggul plant

Guggul plant can be propagated both by seeds and stem cuttings. Seeds are a natural mode of propagation. In the arid or semi-arid zones, slopy well drained highly degraded lands are preferred for this purpose.

The seeds are collected from matured red berries in July-September when the viability is more. The plants are raised through nursery beds and transplanted after six months. For vegetative propagation, 25 - 30 cm long stem cuttings are planted in June or October-November.

CIMAP has identified a clone of guggul plant called *'Maru - Sudha'* which contains higher percentage of E and Z Guggulosterones.

Oleo-gum-resin is collected from at least 5 years old plants. It is tapped from main stem with 7.5 cm diameter on which deep circular incisions are made. It should be noted that the resin ducts occur only in bark portion near cambial layer. Guggul oozes out as yellowish white aromatic latex like matter. It has been reported that dose of 400 ml ethaphon (2 - chloro-ethyl phosphoric acid) three times a year enhances upto 22 times the secretion of guggul. Thick branches of tree gives best grade of guggul. Each plant gives from 0.5 - 1 kg of guggul per year.

Description

Colour : Brown to pale yellow or dull green, solid in nature.

Odour : Agreeable, aromatic and balsamic.

Taste : Characteristic bitter.

Size : 0.5 to 2.5 cm in diameter.

Shape : Rounded or irregular masses or agglomerated tears. Tears are somewhat transparent, with waxy surface and brittle in nature. Guggul is gummy to touch and tears are normally with fractured surface.

Solubility : When triturated with water, it forms white emulsion. It is partly soluble in alcohol.

Chemical Constituents

The gum-resin portion of guggul contains 30.0% resin and 60% gum alongwith steroids, diterpenoids, carbohydrates and aliphatic esters. They are present in a complex mixture form, but the resin does not contain free cinnamic acid, benzoic acid or their esters which are normally found in many oleo-gum-resin of pharmaceutical significance. The purified gum gives pentosan, pentose and furfural. Steam distillation of guggul gives pale yellow volatile oil, containing the terpenes like myrcene and caryophylline. Guggul contains Z - guggulosterone, E - guggulosterone and three new sterols viz. guggulosterol I, II and III.

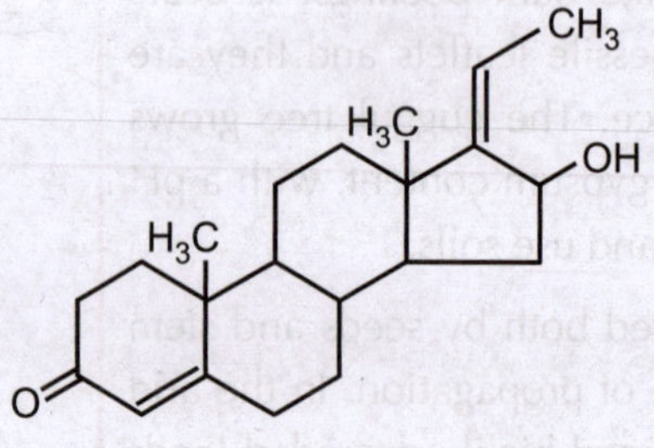

Guggulosterone-Z

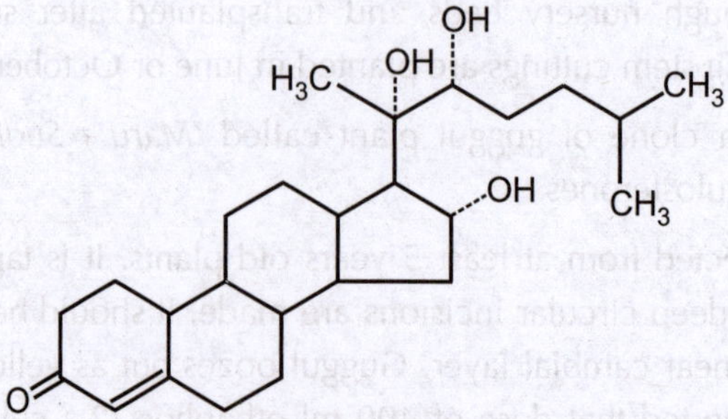

Guggulosterol I

Guggul contains therepeutically active steroids shown as under:

Oleo-gum-resin (Guggul)

↓ 120-130° C (Hot-ethyl acetate)

- Gum - 60 per cent (Insoluble)
- Resin - 30 per cent (Soluble)
  - Neutral fraction 90-95 per cent
    - Ketonic-part (12 per cent)
      Hypolipidemic and Hypocholesteremic contains
      Z - Guggulosterone
      E - Guggulosterone
    - Non-ketonic part (88 per cent)
  - Acidic fraction 4 - 5 per cent (Anti-inflammatory)

Chemical Test

To the ethyl acetate extract of guggal add acetic anhydride boil, cool and add two ml of sulphuric acid, green colour develops at the junction due to sterols.

Standards of Quality

Water soluble extractives	:	Not more than 48 per cent
Alcohol soluble extractives	:	Not less than 35.0 per cent
Ethyl acetate soluble extractives	:	Not less than 23.0 per cent
Ash value	:	Not more than 5.5 per cent
Sulphated ash	:	Not more than 10.0 per cent
Loss on drying	:	Not more than 11.8 per cent

Uses

It is used as anti-inflammatory, anti-rheumatic hypolipidemic and hypo-cholesteremic drug.

It lowers low density lipoproteins and supports weight contact. 'Guggulip' developed from *Commiphora mukul* is an antihyperlipidaemic product. Guggul provides protection from artherosclerosis.

Adulterants

Guggul is found as adulterated with resins of various *Commiphora* species like *C. abyssinica, C. roxburghii, C. molmol* and also with *Boswellia serrata*.

HARIDRA

Synonyms

Indian saffron, Curcuma, Turmeric, Haldi.

Biological Source

Turmeric consists of dried, as well as, fresh rhizomes of the plant known as *Curcuma longa* Linn. (*C. domestica*), belonging to family Zingiberaceae. It contains not less than 1.5 per cent of curcumin.

Geographical Source

India accounts for as much as, 90 per cent of the total output of the world. Tamil Nadu and Andhra Pradesh together contribute about 70 per cent of the Indian production. Kerala also produces large quantity of turmeric. It is very superior in quality and is exported on large scale. At present, about 1,07,800 hectares of land is under cultivation of turmeric with 2,94,900 tones of production. In the year 1996-97, India exported 155 tones of turmeric-oleo resin of the value of ₹ 868 lakhs.

Curcuma is a genus of about 70 species of rhizomatous herbs distributed in South East Asia and especially India, China, Thailand, Italy, Malaysia, Archipelago and N. Australia. Commercially, *C. amada, C. angustifolia, C. aromatica, C. caesia, C. zedoaria* and *C. longa* are important. Out of these, *C. longa* (turmeric) is more important due to its uses like spice, condiment, antiseptic in bruises, anti-inflammatory and in sprains. It has long been known traditionally as a natural dye-stuff for dyeing wool and silk. Most of the species are perennial herbs which grow 2 - 3 ft. high with a short stem and tufted leaves. Rhizome is the product of commerce.

Cultivation and Collection

C. longa (turmeric) is the main species of commerce and is cultivated for its rhizomes in India, China and also in Sri Lanka, Indonesia, Jamaica, Peru. India is the major grower with almost 175,000 hectares under this crop producing 12,40,000 tones per (2020 - 2021) annum and is the major, exporter to entire world. The plants are grown for 7 - 9 months after which the rhizomes (both mother and finger) are harvested, cooked, dried and then processed for powder, oleo resin and curcumin. The extraction of powder is carried out by using solvents, water or both. Diseases and insects are known for which proper methods of control are available. Genetic improvements have been attempted and five high yielding varieties have been developed. High yielding curcumin varieties have been evolved through tissue culture techniques; clonal propagation has been successfully developed in case of *C. longa*.

Macroscopic Characters

Fig. 14.80: Turmeric rhizomes

Colur	:	Yellowish-brown externally and orange internally.
Odour	:	Characteristic
Taste	:	Slightly bitter, aromatic
Size	:	5 - 10 × 2 - 4 cm
Shape	:	Oblong, cylindrical

Extra features: Presence of root scars annulatins. Fractures – fibrous.

Microscopic Characters

The transverse section of turmeric rhizome shows the outermost 4 to 6 layers of brick shaped parenchymatous cork, followed by cork cambium. The cortex consists of thin walled rounded parenchymatous cells containing scattered vascular bundles. Oleo resin cells with brownish contents are also observed throughout the ground tissue. Oil cells have suberised cell-walls. Vascular bundles are present in cortex and are collateral. Vascular bundles in pith region are scattered forming discontinuous ring under endodermis. Endodermis is well marked and starch grains (5 to 15 in diameter) are abundant.

Chemical Constituents

Turmeric contains about 5 per cent of volatile oil, resin, abundant zingiberaceous starch grains and yellow colouring substances known as curcuminoids. The chief component of curcuminoids is known as curcumin (50 - 60 per cent). Chemically, *Curcuma* species contain volatile oil, starch and curcumin. Curcumin and other related curcuminoids such as Demethoxy curcumin and Bis Demethoxy curcumin are reported to be responsible for the yellow colour in some species. Volatile oil content ranges from 1.0 - 6.5 per cent and is composed of mono and sesquiterpene such as α and β pinene, α-phellandrene, camphor, camphene, DL-ar-turmerone zingiberene and α, β curcumenes. Species like *C. angustifolia* and *C. caulina* have high starch content and are used as a substitute for arrow root.

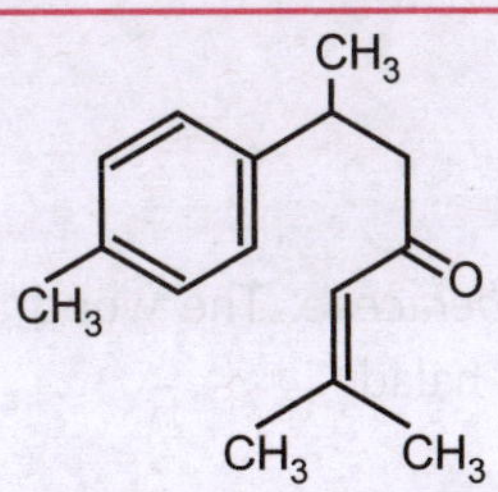

DL-ar Turmerone

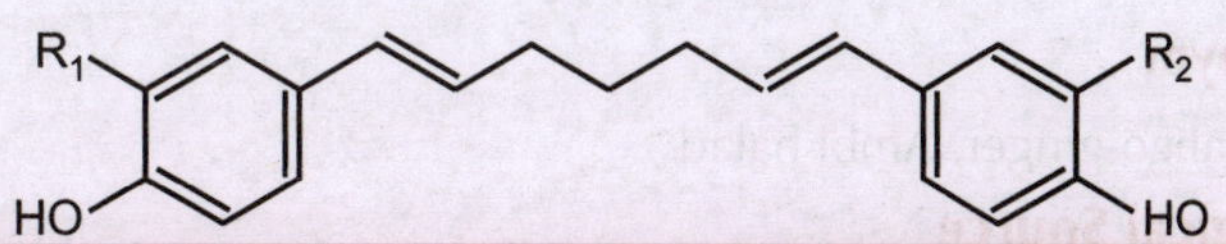

Constituents	R_1	R_2
Curcumin	OCH_3	OCH_3
De-methoxy curcumin	OCH_3	H
Bis-de-methoxy curcumin	H	H

Chemical Tests

1. Powdered drug with sulphuric acid gives crimson colour.
2. The aqueous solution of turmeric with boric acid gives reddish-brown colour which on addition of alkali changes to greenish-blue.

3. With acetic anhydride and concentrated sulphuric acid, it gives violet colour. When this test is observed under ultraviolet light, red fluorescence is seen.

Standards of Quality

Foreign organic matter	:	Not more than 2.0 per cent
Ash	:	Not more than 10.0 per cent
Acid insoluble ash	:	Not more than 2.0 per cent
Water soluble extractives	:	Not less than 12.0 per cent
Alcohol-soluble extractives	:	Not less than 6.0 per cent
Moisture	:	Not more than 12.0 per cent

Uses

Turmeric is used as a condiment or spice, and colouring agent, especially for ointments and creams impvoved complexion and Chemically, it is used for the detection of boric acid. It is antiseptic and anti-inflammatory too. Curcumin is also powerful antioxident.

Turmeric curcumin are official in various pharmacopoeias. Apart from traditional uses, curcumin has been proved as antiinflammatory drug. Antiarthritic agent has been isolated from *C. aromatica*. In China, *C. wenyjuin* (*C. aromatica*) has been used in cervical cancer. Curcumin has been defined by International Standards Organisation (ISO 5562-1983) and British Standards (BS 6147: 1983). It is estimated both by colorimetry and HPLC. G.L.C. and T.L.C. methods are also reported for various constituents. 30th report of WHO/FAO expert committee on food additives has included curcumin.

Curcuminoids isolated from ethyl acetate extract of turmeric have shown modest HIV-1 and HIV-2 protease inhibitory activity.

Substitute

Curcuma amada Roxbby is a common substitute for turmeric. It is found wild in Andhra Pradesh, Orissa and West Bengal.

AMBA-HALAD

Synonym

Mango-ginger, Ambi halad.

Biological Source

This consists of dried rhizomes of *Curcuma amada,* family - *Zinziberaceae*. The wood amba means mango. Since it has flvour of raw-mangoes so it is known as amba-halad.

Geographical Source

The plant grows wild in Bengal, Andhra Pradesh, Konkan and Tamil Nadu and also cultivated throughout India.

Description

A rhizomatous aromatic herb, rhizomes are buff coloured, laterally flattened, oval, cuticle thick, long unicellular hairs, submerged cork cells, cylindrical, pale yellow inside.

Flavour similar to raw-mangoes. Thsta aromatic.

Rhizomes are 2.0 - 6.0 cm long and 0.5 to 2 cm in diameter, remnant of scaly leaves which are arranged circularly giving appearance of growth rings, circular scars on the surface of rhizomes, branching sympodial, horizontal, roots thread like and short, odour raw mango like, taste pungent.

Leaves are long petioled, in tufts, oblong, lanecolate and 30 - 45 × 7.5 - 12.5 cm in size.

Flowers in antumnal spikes, size of spikes 7.5 - 15.0 × 3.0 - 8.5 cm, bracts are green, colour red to pink. Calyx white or very pale yellow, flowering time - September every year.

Microscopic Characters

Transverse section of rhizome is circular with thick cuticle and rectangular-oval epidermal cells. Root hairs are unicellular and cork cells are suberised. Lysigenous, oil glands containing volatile oil. Cortex is quite wide and scattered vascular bundles (V.B.). Each V.B. has fibrous sheeth. Marked endodermis followed by pericycle V.B. without bundle sheath are also present in ring just below pericycle. Thin walled and lignified fibres with narrow central lumen are also present. Cortex contains numerous oil cells. Being rhizome most of the parenchymatous cells are filled with starch grains which are oval or polygonal in shape, their size is 10 to 80 μ and are simple.

(a) (b)

Fig. 14.81: (a) Amba-Halad plantation, (b) Rhizomes of Amba-Halad

Chemical Constituents

Ambahalad is found to contain 1.0 per cent volatile oil and contains 1.8 cineole and myrecene mainly. It also contains pinene, camphor α and β curcumine, turmerone and a phytosterol.

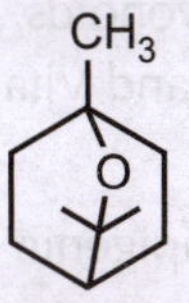

1,8 cineole

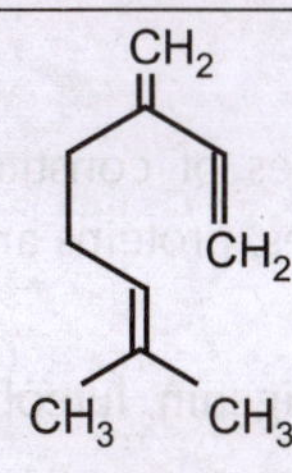

Myrcene

Standards of Quality

Total Ash	:	Not more than 12.0 per cent
Acid insoluble ash	:	Not more than 2.0 per cent
Water soluble extractives	:	Not less than 14.0 per cent

Alcohol-soluble extractives	:	Not less than 9.0 per cent
Foreign organic matter	:	Not more than 1.0 per cent
Loss on drying	:	Not more than 5.0 per cent

Uses

It is used as Hypolipidaemic mainly and also as carminative, stomachic, appetiser and antipyretic.

Storage

It should be stored in well closed and well filled containers in cool place.

PARSLEY

Synonyms

Apium petroselium; Carcum petroselium.

Biological Source

It is obtained from the plant known as *Petroselium sativa* fam. Umbellifarae (Apiaceae).

Geographical Source

It is native of Sardinia, but frequently cultivated throughout India in gardens. Now in Maharashtra, it is cultivated in Shirur, Dist. Pune.

Description

It is glabrous biennial or short lived perennial 45-75 centimeters in height. The stem of the herb is much branched. **Leaves** are decompound ultimate, segments are punctuate-ovate. Leaves are also stalked, deeply cut, curled and crisped. Parsley bears small greenish yellow compound umbels of **flowers**.

Fruits are very small and ribbed. The drug is official in British herbal Pharamacopoeia 1990.

Fig. 14.81 (c): Parsley Cultivation

Chemical Constituents

Parsley contains various types of constituents like flavonoids, fumarocoumarins, volatile oil, fixed oil, oleo-resin, carbohydrates, proteins and vitamins A and Vitamin D. The details of which are as under.

Flavonoids: Glycosides of Apigenin, luteolin (i.e. apiin, apigenin 7 - 9 incoside only in leaf.

Furanocoumarins are bergapten and oxy-penceilanin 0.02 and 0.01% respectively, 8-methoxy-psoralen, psoralen and xanthotoxin up to 0.003 per cent.

Volatile oil in seed 2.0 - 7.0 per cent and 0.05 per cent in leaves. Leaf contains myresticin upto 85% and apiol.

Apiole is also present in Dill which is different and is known as Apiole (dill), while the apiole present in parselay is known as Apiole-Parsley.

Apiole is very poisonous and it should be used carefully.

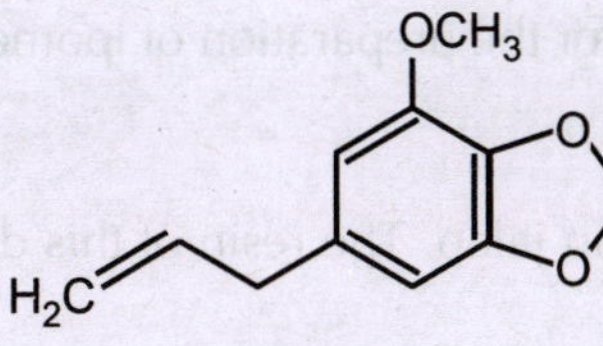

Apiole (Dill) m.p. 29.5°

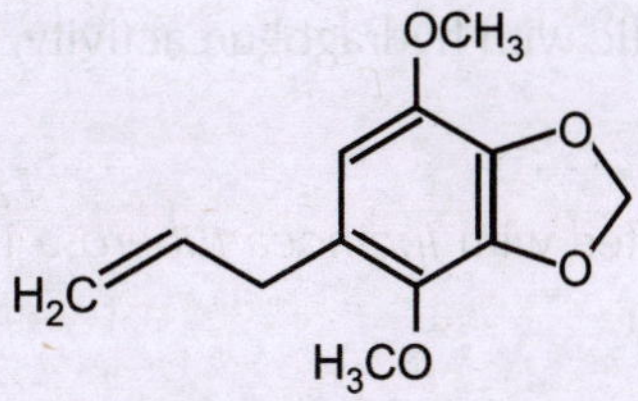

Apiole (Parsley) m.p. 29.5°, b.p. 254

Uses

Parsley fruits are used as flavouring agent in food and also in pharmaceuticals.

IPOMOEA

Synonyms

Orizaba jalap roots, Ipomoea root, Mexican scamony.

Biological Source

Ipomoea consists of dried tuberous roots of the plant known as *Ipomoea orizabensis*. family Convolvulaceae. Ipomoea yields not less than 15 per cent of resins.

Geographical Source

It is collected in Mexico (Orizaba) and Andes mountain.

Macroscopical Characters

Colour	:	Tubercles outer surface is greyish brown due to cork.
Odour	:	Characteristic and slight.
Taste	:	Faintly acrid
Size	:	3 to 10 cm wide and 2 to 4 cm thick.
Shape	:	Transverse or oblique slices, large roots some times cut longitudinally may result in rectangular pieces. It is wrinkled, and transversely cut surface is greyish in colour with 3 to 6 concentric rings of fibrous vascular bundles.
Fracture	:	Short, irregular fibrous.

Chemical Constituents

It yields about 10 to 20 per cent of complex resinous matter with 90 per cent alcohol. 95 per cent of this resinous matter is soluble in ether. Chief constituent of the resin is jalapin i.e. mixture of acidic glycocides, ipuranol and ipurganol. Other consitutents of the drug are volatile oil, scopoletin, cetyl alcohol, and several other fatty acids.

Chemical Test

Fluorescence in ultraviolet light shows deep bluish violet colour.

Uses

It has cathartic with hydragogue activity. It is used for the preparation of Ipomoea resin.

Adulterants

It is adulterated with *Ipomoea tuberosa* i.e. Brazilian jalap. The resin of this drug is very slightly insoluble in ether.

JALAP

Synonyms

Mexican Jalap, Jalap Radix.

Biological Source

Jalap consists of dried tubercles of *Ipomoea* purga Hayne, family Convolvulaceae.

Geographical Source

It is found in Jamaica, South America and India.

Cultivation and Collection

It is a perennial herb with horizontal runner, nodes and inter nodes. The nodes are characterised by the formation of adventitious roots. Jalap has tuberous roots, some of them swell due to the storage of reserve food material and get converted into tubercles. These tubercles are collected by digging and are dried. Sometimes, the wood-fire may also be necessary for drying and hence the drug gets smoky odour.

Macroscopical Characters (Fig. 14.82)

Colour	:	Tuberous roots are dark brown.
Odour	:	Distinctive and smokey.
Taste	:	It is sweetish, starchy and acrid.
Size	:	Slices are about 14 to 15 cm in length and 4 to 10 cm in diameter.
Shape	:	Fusiform, irregularly ovoid or nepiform.

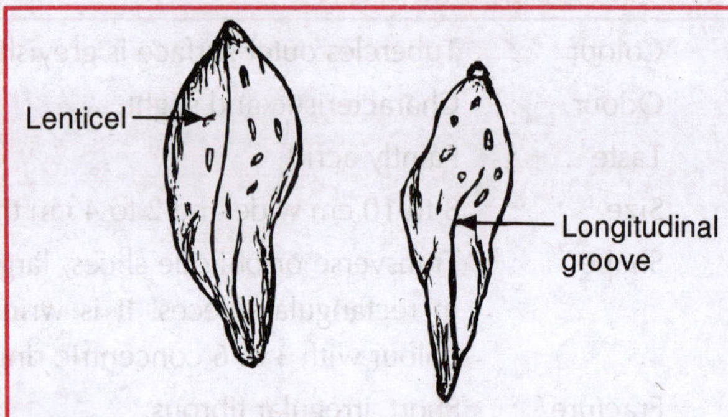

Fig. 14.82: Jalap Tubercles

Extra Features

It is irregularly furrowed longitudinally, sometimes wrinkled with circular rootlets and scars. The presence of lenticels is common. Transversely cut surface shows weak to pale brown concentric cambium zones.

Chemical Constituents

Jalap contains 8 - 18 per cent of resin, volatile oil, starch, gum and sugar. Resin is soluble in ether upto 10 per cent and the ether soluble portion is known as jalapin. Ether insoluble part is known as convolvulin. Convolvulin on hydrolysis yields rhamnoconvolvulic acid about 74 per cent, tiglic acid about 9 per cent and exogonic acid 7 per cent. The jalapin (scammonin), ether soluble part of the drug, is the main active constituents of jalap. In addition to all above constituents, resin also contains ipurganol, methyl aesculetin, palmitic and stearic acids.

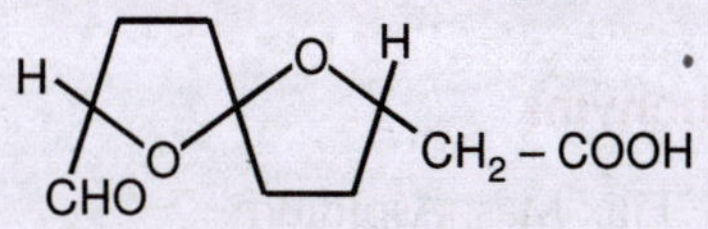

Exogonic acid

Uses

Jalap is used as a powerful cathartic. It acts in presence of bile. Jalap resin is prepared from the powdered jalap by precipitation with alcohol and used for the same purpose as that of jalap.

KALADANA

Synonyms

Pharbitis seeds, Kaladana seeds.

Biological Source

It consists of dried seeds of *Ipomoea hederaceae* Jacq, family Convolvulaceae.

Geographical Source

It is believed to be of American origin and is found in tropical and subtropical regions. It is also found in India.

Collection of Kaladana Seeds

It is found wild in the Himalayas upto 2000 m. This white flowered plant is found as weed in sugar-cane. It is also grown in Bihar.

Macroscopical Characters (Fig. 14.83)

Colour	:	Seeds are brown or black.
Odour	:	None.
Taste	:	Sweetish first and then acrid.
Size	:	About 5 mm in length and 3 - 5 mm in width.
Shape	:	Triangular and resembling quarter of the orange.

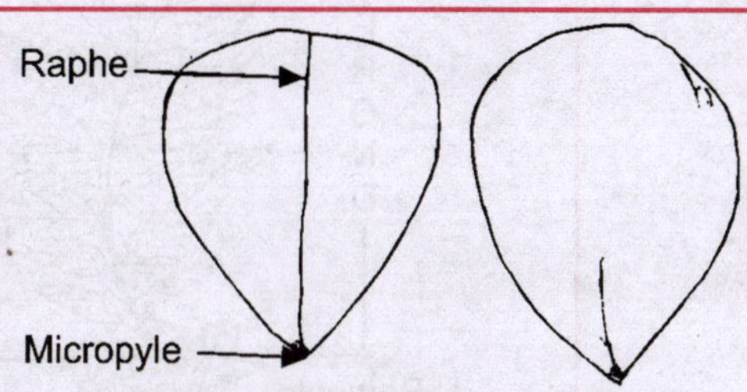

Fig. 14.83: Kaladana seeds

Extra Features

Every seed has two flat faces, united form a triangle exhibiting cordate hilum.

Chemical Constituents

Kaladana contains a cathartic resin known as pharbiticin about 11 per cent; fixed oil 15 per cent; mucilaginous glycoside (saponin) and resinous substances.

Uses

It is used as cathartic and in excessive doses, it causes irritation of small and large intestine.

MALE FERN

Synonyms

Filix Mas, Aspidium.

Biological Source

Male fern consists of rhizomes and frond bases of *Dryopteris filix-mas*, *D. odontoloma*, *D. marginata* and other species of *Dryopteris*, family Polypodiaceae, collected late in autumn, divested of roots and dead portion, and carefully dried retaining their internal green colour. It contains not less than 1.5 per cent of filicin.

Geographical Source

Male fern is indigenous to Great Britain and widely distributed in Europe, South America and India. It is found throughout India, in shaded forests upto 3000 m elevation. It occurs on Nepal border, and in Sikkim, and Himalayan area.

Macroscopic Characters (Fig. 14.84)

Colour	:	Rhizomes are dark brown or reddish-brown externally.
Odour	:	Slight and characteristic.
Taste	:	Somewhat sweet, mucilaginous and astringent.
Size	:	Several varieties available in the market, average 3 to 25 cm in length and about 5 cm in width.
Shape	:	Rhizomes are cylindrical to conical in shape.

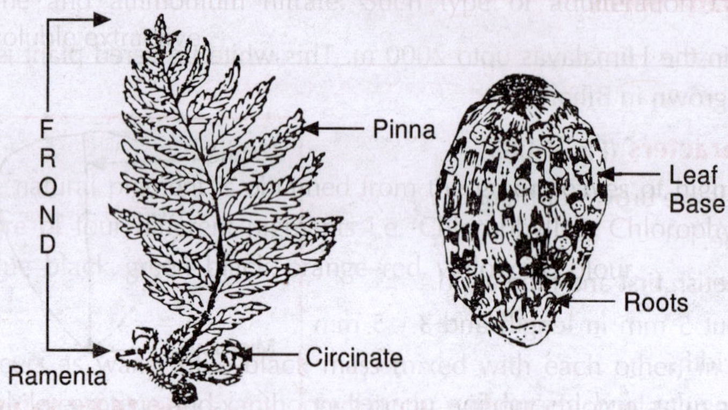

Fig. 14.84: Male Fern Plant and Rhizome

Extra Features

Rhizomes are covered with the frond bases which are about 3 to 5 cm long and one cm thick. Rhizomes, as well as, the frond bases are covered with ramenta. Rhizomes and frond bases show a circle of 8 to 9 yellowish vascular strands, if cut transversely.

Chemical Constituents

Male fern contains number of ether-soluble phloroglucirnol derivatives contributing to anthelmintic property. Oleo resin constituting an active constituent of the drug is about 6 to

15 per cent. Indian male fern yields 8 - 10 per cent ethereal extractive of which 30 per cent is filicin. Filicic acid, filicinic acid, aspidinol, albaspidin and flavaspidic acid are the components of resin. Tannin (filicotannic acid), volatile and fixed oils are the other constituents of drug. (Filicin is the name given to the mixture of all ether-soluble substances obtained from the drug).

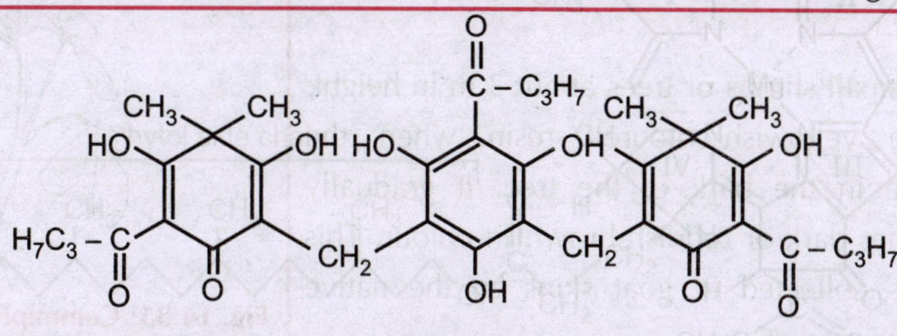

Filicic acid

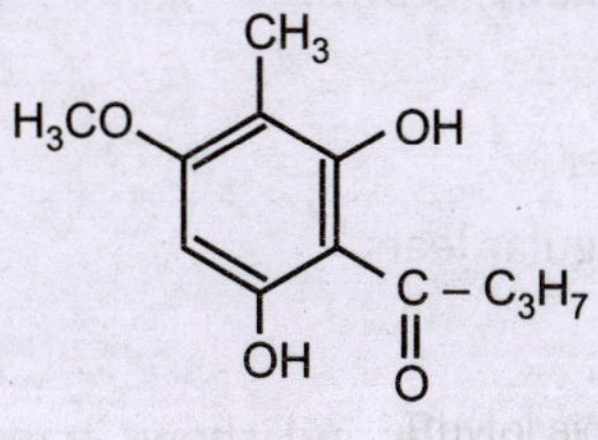

Aspidinol

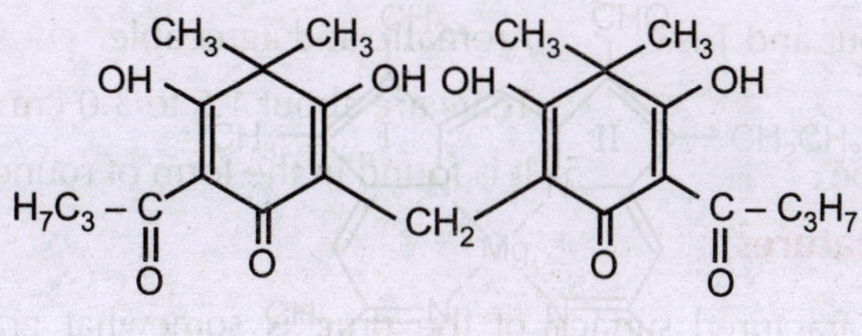

Albaspidin

Standard of Quality

Acid insoluble ash : Not more than 2.0 per cent

Uses

Male fern and its constituents are used as anthelmintic, especially as taeniafuge. Male fern is used in the form of male fern extract for the expulsion of tapeworms.

Substitute

Rhizomes and frond bases of **lady fern** *Athyrium filix*-foemina are mixed with genuine drug. Only dumb bell shaped vascular bundles are present in each frond base and internal glandular trichomes are absent in lady fern.

MYRRH

Synonyms

Gum Myrrh, Bol, Myrrha.

Biological Source

Myrrh is an oleo-gum-resin obtained from *Commiphora molmol* Engler and from other *Commiphora* species. It belongs to family Burseraceae.

Geographical Source

It is found in North East Africa and Southern Arabia.

Collection and Preparation for the Market

Fig. 14.85: Commiphora molmol herb

One of the characteristics of the family Burseraceae is that the plants possess oleo-resinous canals in their conducting tissues.

The plants are small shrubs or trees about 3 m in height. The plants exude yellowish-coloured resin when the incisions are made in the bark of the tree. It gradually hardens and becomes dark or reddish-brown in colour. This coagulated mass is collected in goat skins by the native tribals and sent to the market.

Description

Colour	:	Kesin is externally, reddish-brown, internally, brown.
Odour and Taste	:	Aromatic and agreeable.
Size	:	Tears are about 1.5 to 3.0 cm in diameter.
Shape	:	It is found in the form of rounded or irregular tears.

Extra Features

The fractured surface of the drug is somewhat granular. It is brittle and shows translucent surface. Whitish spots on broken pieces are also seen.

Chemical Constituents

Myrrh contains about 10 per cent of yellowish thick volatile oil; 60 per cent of gum; 25 to 40 per cent of resin, and bitter principle (about 3 to 4 per cent). It contains several impurities constituting about 5 per cent of the drug. Resin contains ether-soluble resin acids, α, β and γ commiphoric acids. The α and β heerabomyrrholic acids are ether-insoluble acids present in resin. The volatile oil contains terpenes, cuminic aldehyde, eugenol, etc. The gum is associated with oxidase enzyme. It yields not more than 70 per cent of alcohol-insolube matter.

Identification

1. When about 0.1 g of the substance and 0.5 g of sand are triturated with solvent ether, filtered and allowed to evaporate, the thin film formed gives violet colour on contact with bromine vapours.
2. When triturated with water, it forms yellowish-brown emulsion.

Standards of Quality

FOM	:	Not more than 4.0 per cent
Ash	:	Not more than 9.0 per cent

Uses

It is used as a stimulant and an antiseptic. It is also protective. Myrrh is astringent to the mucous membrane and hence its tincture is used in the mouth washes and gargles.

Adulteration

The drug is substituted by several species like Arabian myrrh, Yemen myrrh, etc. Both of them are less fragrant and less aromatic. In India, myrrh is substituted by *Balsamodendron mukul*, known as Indian bdellium.

PATCHOULI OIL

Synonym

Pucha-pat

Biological Source

Patchouli oil is obtained from cultivated species of leaves of the plant *Pogostemon patcholi*, *Pogostemon cablin Benth*, *Pogostemon patchouli* pellet var. Suavis Hook. Family : Labiatae.

Geographical Source

In India it is available at Western ghats, Karnataka and Nilgiris.

Fig. 14.85 (a) : Aerial top of Patchouli herb

Description

A yellow to dark brown viscous liquid. resembling that of sandalwood oil and the flavour is like rose with mild sweet herbaceous odour along with aromatic spicy fragrance which improves after ageing. It is strongly leavo rotatory.

Chemical Constituents

Patchouli oil contains benzaldehyde, eugenol, cinnamic aldehyde, alcohol, ketone, Patchouli azulene and sesquiterpene alcohol. Sesquiterpenes constitute to 40-45% of the oil, patchouli camphor or patcholi alcohol (patchoulol) constitute 35-40% of the oil. Recently, a crystalline fraction of norsesqiuterpinic alcohol called norpatchoulenol is isolated, which is true odour carrier of oil.

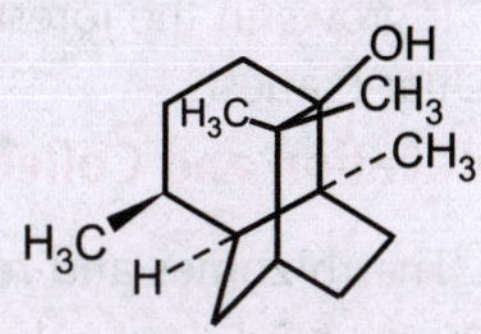

Patchouli alcohol

Standards of Quality

Specific gravity	:	0.9500 – 0.9750 at 25°C.
Optical rotation	:	– 48° to – 70°.
Saponification No.	:	3.3 to 9.9
Refractive index	:	1.5050 to 1.5130 at 25°C.
Acid value	:	Max. 5.
Ester value	:	Max. 10.
Solubility	:	Insoluble in water, soluble in alcohol and ether.

Uses

Patchouli essential oil is used in aromatherapy to treat skin complaints. It is thought to have a regenerative effect on skin tone and to help clear conditions such as eczema and acne. The oil may be used for varicose veins and hemorrhoids. The oil has strong fixative properties and helps to prevent rapid evaporation of perfume. It is also employed in cosmetics, soaps, toilet soaps and scents, body lotions, pre-shave and after-shave lotions, detergents. Very low concentration of oil serves as flavoring agent in alcoholic and non-alcoholic beverages, baked goods, candy, gelatin and frozen dairy desserts. Especially oil is used in preparation of lavender soap, lavender spray perfume and face powder perfume.

Storage

Patchouli oil is stored in clean, dry aluminium containers which should be free from any chemicals and odours. It should be filled up to the brim of the containers in order to expel all air, otherwise it will deteriorate the quality of the oil. The containers are made airtight.

Rubber cork spoils the quality of the oil, hence only metal cones be used and the oil containers should be kept in cool place.

PODOPHYLLUM

Synonyms

Indian Podophyllum, Podophyllum Radix, Himalayan May-apple.

Biological Source

Podophyllum consists of dried rhizomes and roots of *Podophyllum hexandrum* Royal or *Podophyllum emodi* Well, belonging to family Berberidaceae. American podophyllum consists of dried rhizomes and roots of *Podophyllum-peltatum.*

Geographical Source

It grows in the forests of the Himalayas from Kashmir to Sikkim in Himachal Pradesh and parts of Uttar Pradesh.

Cultivation and Collection

The rhizomes and roots are obtained from wild grown plants growing at an altitude of 3000 to 4000 m. Underground rhizomes remain dormant in winter and produce aerial shoots in April to May. Shoots flower during summer and die in November. Rhizomes and roots are dug up in spring or autumn, washed, cleaned and dried in the sun. The drug collected in the month of May has higher resin content than that collected in November. Actually, the roots contain more resin than the rhizomes and hence, the roots are preferred.

Macroscopical Characters

Colour	:	The rhizomes are yellowish-brown to earthy brown.
Odour	:	Slight and characteristic
Taste	:	Bitter and acrid.
Size	:	Rhizomes are 2.5 cm in length and 1 - 2 cm in thickness. Roots are about 7 cm in length and 2.5 mm in thickness.
Fracture	:	Both roots and rhizomes are brittle and even.

Extra Features

Rhizomes are irregular and knotty, somewhat flat and dorsiventral, upper surface of the root has presence of 3 to 4 cup shaped scars.

(a) Podophyllum – hexandrum plant

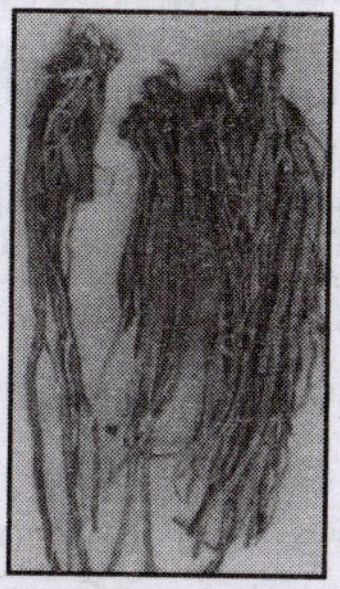
(b) Podophyllum – hexandrum roots

Fig. 14.86: Indian Podophyllum Rhizome (Natural Size)

Microscopic Characters

Rhizome: Outermost layer of rhizome consists of six layers of thin walled polygonal, tabular cork cells, while the cortex is made up of cellulosic parenchyma containing compound starch grains, calcium oxalate crystals (cluster type) are also present in the parenchyma. On the inner side of the vascular bundles short sclerides are present, xylem vessels are short and irregular.

Roots: T. S. of root exhibits the presence of epiblema which are strongly thickened, cortex is similar to rhizome but without calcium oxalate crystals. Exodermis and endodermis contain suberin. Stele consists of four to nine arch arrangements. Pith shows the presence of pitted-sclerides.

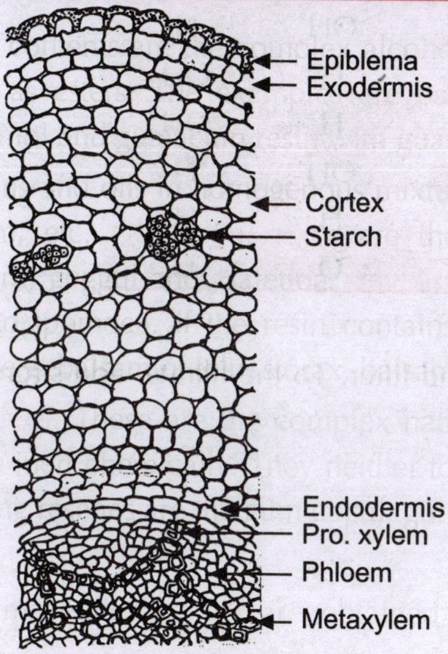

Fig. 14.87: T. S. of Podophyllum root

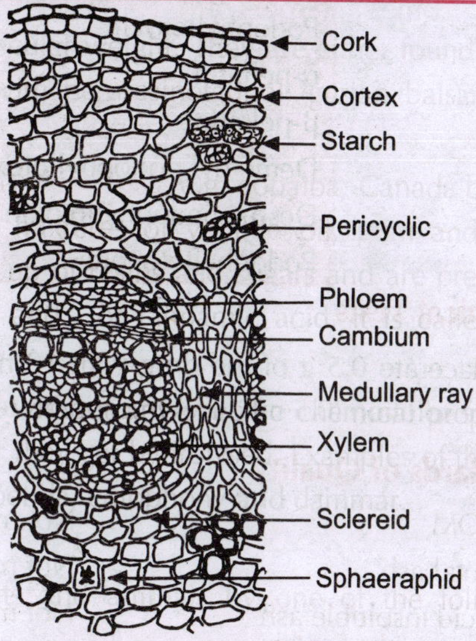

Fig. 14.88: T. S. of Podophyllum rhizome

Chemical Constituents

Indian Podophyllum contains from 7 - 15 per cent of resin known as podophyllin. Resin content varies in roots and rhizomes with the area and the season of collection. Roots contain more resin than the rhizomes. The active principle in podophyllin resin is known as podophyllotoxin which constitutes at least 40 per cent in Indian variety. α - and β-peltatins are present only in the American *Podophyllum*, but absent in *Podophyllum hexandrum*. The flavonoid astragalin and several other glucoside of podophyllotoxin are the other constituents of Indian podophyllum. Quercetin, about 8 per cent kaempferol, asiragalin and essential oil (responsible for odour of the drug) are the other contents of it. Chemically, Podophyllin is a **lignan** compound i.e. substance whose, chemical structure is based on 2 - 3 di-benzyl butane. Etoposide (4-demethylepipodophyllotoxin ethylidene-glucoside) is semi-synthetically processed and used in testicular and lung cancer.

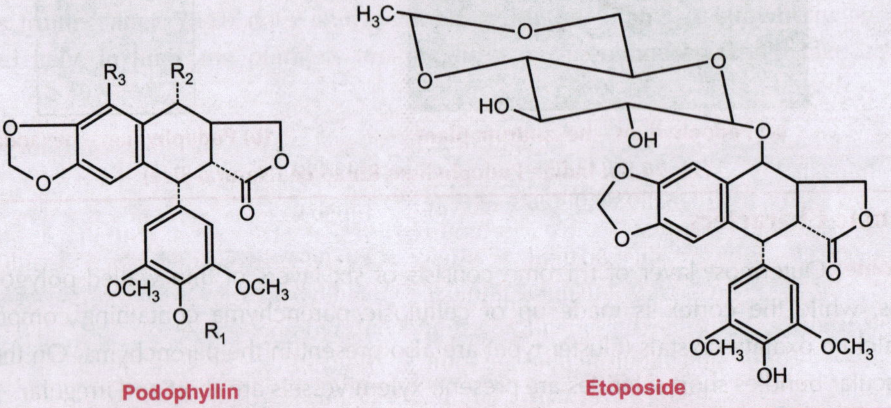

Podophyllin Etoposide

Table 14.9: Chemical nature of podophyllin

	R_1	R_2	R_3
Podophyllotoxin	CH_3	OH	H
α-peltatin	H	H	OH
β-peltatin	CH_3	H	OH
Demethyl podophyllotoxin	H	OH	H
Desoxypodophyllotoxin	CH_3	H	H
Podophyllotoxone	CH_3	O	H

Chemical Tests

Macerate 0.5 g of the drug with 10 ml of alcohol and filter. To the filtrate, add strong copper acetate solution (0.5 ml), brown precipitate is produced.

Standards of Quality

FOM	:	Not more than 2.0%
Total ash	:	Not more than 6.0%
Acid insoluble ash	:	Not more than 2.0%
Alcohol soluble extractives	:	Not less than 17.0%
Water soluble extractives	:	Not less than 1.0%

Uses

Because of cytotoxic action, podophyllum is used in the treatment of veneral and other warts. It is also purgative, cholagague and bitter tonic. The active constituents of the drug, viz. podophyllin is a resin and has recently been successfully used in controlling some forms of cancer. Its cytotoxic effect i.e. mitotic arrest is similar to colchicine. It is used in semi-synthetic production of etoposide.

PODOPHYLLUM RESIN

Synonyms

Indian podophyllum resin, Podophyllin.

Biological Source

It is a mixture of resin obtained from dried roots and rhizomes of plants *Podophyllum hexandrum* or *P. emodi*, belonging to family Berberidaceae. Podophyllum resin contains not less than 40 per cent and not more than 50 per cent of podophyllotoxin.

The dried rhizomes and roots of Indian podophyllum are obtained from slopes of Himalaya in Kashmir and Tibet. It is also grown in Afghanistan.

Method of Preparation

It is prepared by percolating powdered podophyllum with alcohol and followed by precipitation with acidulated water. The precipitated resin is washed several times with water and dried thoroughly.

Description

Podophyllum resin is found as amorphous powder of light brown to greenish-yellow or brownish green colour. If it is subjected to temperature of 25°C or more, it becomes dark in colour. It possesses peculiar bitter and acrid taste. It is also irritating to eyes and mucous membrane.

Podophyllum resin is insoluble in cold water, partly soluble in hot water and is fully soluble in alcohol 90 per cent. It is partly soluble in chloroform and ether.

Chemical Constituents

The composition of resin is made up of lignans which contain at least 40 per cent of podophyllotoxin. American podophyllum resin contains only 10 per cent of podophyllotoxin. Podophyllotoxin is ter–anhydronapthalene derivative with OH and lactone group in trans position which are essential for anti-mitotic activity. In cis position, it possesses only purgative activity. Apart from podophyllotoxin, picropodophyllin, quercetin, astragalin and resinous substances are also found in podophyllin.

Identification

Take 0.5 g powdered resin. To it add 5 ml of 60 per cent alcohol and 5 ml of 1 N potassium hydroxide solution. Shake it well. A stiff jelly is produced, in case of Indian podophyllum resin, while the American podophyllum resin does not produce any precipitate.

Uses

Due to cytotoxic action, podophyllum resin is used as a paint in the veneral and other warts. It is also caustic. It is used as a drastic purgative and hydro-gauge cathartic. It should not be given to children and in pregnancy.

STORAX

Synonyms

Prepared Storax; Purified Storax.

Biological Source

Storax is a balsam obtained from the wounded trunk of *Liquidambar orientalis* belonging to family Hamamelidaceae Miller and subsequently purified by solution in alcohol, filtration and evaporation of solvent. It contains not less than 30 per cent of total balsamic acids.

Geographical Source

It is indigenous to South West Asia and Turkey.

Collection

It is collected from dried bark, which is about 3 to 4 years old. The bark of the plant is injured by beating. The balsam comes out into the inner bark. The total bark is then peeled, inner bark is stripped off and boiled in water. Balsam floats over water and is separated, to yield crude storax. It is then processed to meet the pharmacopoeial requirements.

Description

Colour	:	Balsam is greyish or greenish-brown. It is semi-fluid, visicid substance.
Odour	:	Agreeable and balsamic.
Taste	:	Agreeable, balsamic and bitter.
Solubility	:	It is soluble in 90 per cent alcohol, chloroform, ether and carbon disulphide. Insoluble in water.

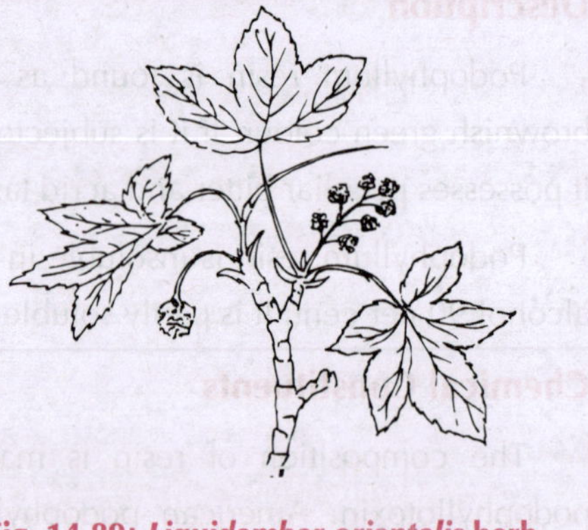

Fig. 14.89: *Liquidambar orientalis* herb

Extra Features

It is semi-solid and obtained in the form of thin transparent layers after processing.

Chemical Constituents

Purified storax is composed of alcoholic resin known as storesin, constituting 33 - 50 per cent of the drug and occurring as free and as cinnamic ester. It also contains free cinnamic acid (15 - 10 per cent), cinnamoyl cinnamates (5 - 10 per cent), and small amount of ethyl benzyl cinnamates and traces of vanillin. Steam distilled storax yields a dark brown volatile oil known as oil of storax.

Identification

1. Mix about 1 g storax with 5 g sand and 5 ml potassium permanganate solution. Warm it gently. The odour of benzaldehyde is produced.
2. About 1 g storax, if shaken with 10 per cent solution of potassium chromate and 1 ml sulphuric acid, odour of the benzaldehyde is produced.

Standards of Quality

Acid value	:	80.0
Ester value	:	100 - 133
Loss on drying	:	Not more than 5 per cent
Saponification value	:	170 - 200
Alcohol soluble extractives	:	Not less than 70 per cent

Uses

Storax is used as a stimulant and an antiseptic. It is also used for scenting the soap, cosmetics and as fixatives for heavy perfumes. It is a flavouring agent for tobacco and several pharmaceutical preparations.

Substitutes and Adulterants

It is adulterated with rosin, olive oil and other resins. **Red Gum** or **Sweet Gum** or **American Storax** is a common substitute available in the market. It is obtained from *Liquidamabar styraciflua* which is transparent, viscous and yellowish liquid.

TAR

Synonyms

Chir Tar, Stockholm Tar, Pine Tar.

Biological Source

It is a bituminous liquid obtained from the wood of Scots pine *Pinus sylvestris* P. *palustris*, P. *strobes* belonging to family Pinaceae and other species of *Pinus* by destructive distillation.

Geographical Source

Canada, USA, Northern Europe and India.

Description

Colour	:	Dark brown or almost black, transparent liquid, found in very thin layers.
Odour	:	Strong characteristic and naphthalene like.
Taste	:	Bitter and pungent.
Solubility	:	It is insoluble in water, partly soluble in alcohol and soluble in chloroform, ether, fixed oils and volatile oils.

Extra Features

If stored, it separates into a layer which is granular in character due to minute crystallization of catechol, resin acids, etc. It is acidic in reaction.

Chemical Constituents

Tar contains many substances including hydrocarbons, which are terpenoids in nature and also phenols and their methyl ethers. It also contains resinous substances, traces of terpentine and other substances.

Identification

1. Take 1 g of the tar, add 20 ml of water. Shake it well. To the filtrate add 3 drops of 0.1 per cent solution of ferric chloride; red colour is produced.
2. Its aqueous solution is acidic to litmus.

Standards of Quality

Weight per ml - 1.07 per g

Ash content - 0.47 per cent

Uses

In the form of syrup, it is used as an expectorant. Pine tar is used as antipruritic and antibacterial. It is also local irritant. It is applied as an ointment in the treatment of chronic skin diseases and in eczema.

Tar differs from the industrially, as well as, pharmaceutically important, coal tar. Coal tar is obtained by destructive distillation of bituminous coal below 1000°C. Coal tar also differs in the chemical contents, as it contains benzene, naphthalene, pyridine, phenols, etc. It is alkaline to litmus and becomes more viscous on exposure to air. Coal tar is more irritating to skin and more disinfectant as compared to pine tar.

TOLU BALSAM

Synonym

Balsam of Tolu

Biological Source

Balsam of tolu is solid or semi-solid balsam obtained from the trunk of trees *Myroxylon balsamum* Linn., belonging to family Leguminosae. It contains not less than 35 per cent and not more than 50.0 per cent of total balsamic acids, calculated as cinnamic acid on dry ethanol soluble matter.

Geographical Source

It is indigenous to Columbia and collected in Caribbean islands and Cuba. Tolu is the name of place on the North Coast of Columbia near Cartagena, hence the name of the drug.

Method of Preparation

It is a pathological product secreted in cortex of young twigs. It is collected by making V-shaped incisions in bark. The drug is collected in calabash cups and transferred to tin containers.

Description

Colour : Yellowish or brownish-yellow to brown

Odour : Aromatic and vanilla-like

Taste : Aromatic

Extra Features

It is a tenacious mass becoming harder and finally brittle. It darkens on keeping.

Fig. 14.90: Myroxylon balsamum herb

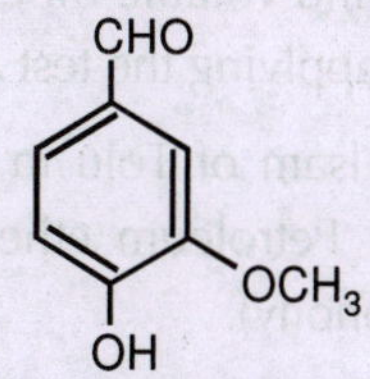

Vanillin

Solubility

It is soluble in alcohol 90 per cent, ether, chloroform and in glacial acetic acid.

Chemical Constituents

Balsam of tolu contains 12 per cent to 16 per cent of free cinnamic acid, about 8 per cent of free benzoic acid, 7.6 per cent of oily liquid (cinnamicin) containing benzyl benzoate and benzyl cinnamate. The resinous matter, which constitutes about 80 per cent of the drug, contains chiefly, ester of toluresinotannol. Small quantities of vanillin and styrol are also reported in the drug. The volatile oil obtained on distillation from balsam of tolu does not exceed 3 per cent. The tolu oil is yellow liquid with highly fragrant flavour.

Identification

1. When heated and pressed in between two glass slides and examined under microscope, it exhibits crystals of cinnamic acid.
2. To alcoholic solution of balsam of tolu, add solution of ferric chloride; green colour is produced.
3. Warm gently about 1 g drug with 5 ml potassium permanganate solution; the odour of benzaldehyde is produced.

Standards of Quality

For pharmaceutical purposes, it should comply with the following standards.

Balsamic acid contents	:	35 – 50 per cent
Loss on drying	:	Not more than 4 per cent
Alcohol insoluble matter	:	Not more than 5 per cent
Acidity	:	Solution in 90 per cent alcohol is acidic to litmus.

Uses

Balsam of tolu is used as an expectorant and as flavouring agent. It is also an antiseptic. It is used as a common ingredient of cough mixtures. It is used in the preparation of confectionery, chewing gums and perfumery.

Adulterants

Many a time, exhausted balsam of tolu is added to the drug. Fictitious tolu balsam prepared by adding benzoin and cinnamic acids to the exhausted balsam can be detected by the lack of its natural flavour and volatile oil content. Colophony is also added to balsam of tolu. Its presence can be detected by applying the test as described here.

Dissolve balsam of Tolu in petroleum ether and add double the volume of dilute solution of copper acetate. Petroleum ether takes emerald green colour (due to copper salt of abietic acid present in colophony).

ALLSPICE

Synonyms

Piminto, Jamaica pepper, Fourt in one.

Flavour of mixture of cinnomon, clove, black-pepper and nutmeg together resembles. The flavour of this fruit hence it is known as allspice.

Biological Source

This consists of full grown but unripe fruits of *Pimento officianalis* or *Pimento dio* fam. Myrtaceae.

Geographical Source

Mainly cultivated in Jamaica. It is native of West-Indies and also cultivated in South America, Central American, Mexico.

Now-a-days in Kerala also, it is cultivated. Kallar and Burliyar are the centres of its cultivation in Tamil Nadu.

Description

Fig. 14.91: Allspice

Fruits have following characteristics

Colour : Leaves are green in colour and get change to reddish brown during drying of leaves

Odour : Agreeably aromatic

Taste : Warm and aromatic

Shape : Berreis are subspherical

Extra features : The berries (fruits) are collected while green in colour and dried. During drying fruits are likely to loose the flavour and hence are dried when they are green only.

Berries are subspherical and 5 to 8 mm is diameter. Berries are surmounted by four sepalous calyx. Their surface is rough. The leaves of pimento are thick and having shining and tapering at apex. The lower side of leaves have oil glands. The plant is unisexual. Female plants bear berries.

Chemical Constituents

Pimento fruits contains 3.5 to 4.5% of volatile oil and tannin volalite oil contans chiefly engenol to extent of 65%. Also contain quereetin, glycosides, catechin, fixed oil, sugar and vitamins.

Volatile oil has

Density	:	1.035 – 1.050
Refractive index	:	1.535 to 1.536
Optical rotaion	:	–2° to +0.5°

Uses

Flavouring agent for cosmetics, soaps, perfumes. It is carminative also.

NIRGUDI

Synonyms

Monks pepper, Chosteberry.

Biological Source

This consists of dried leaves and flowers of *Vitex negundo* family : Verbenaceae.

Geographical Source

It is indigenous to the Mediterranean counties and central Asia also found in India and Myanmar. It is found on waste land upto 2000 metres in Himalayas.

Cultivation

It is very sturdy plant and its cultivation is done by using seeds. It prefers light and medium soils and can be grown in nutritionally poor soil. All types of soils are suitable for cultivation. It cannot graw in shade. It needs dry or moist soil.

Fig. 14.91: Nirgudi herb

Macroscopic Characters

Colour	:	Herb is grayish green
Odour	:	Pepper like aromatic
Taste	:	Bitter

It is a small shrub about 2 - 3 metres is height. Bark is reddish brown.

Leaves with five lanccolate leaflets each leaflet 4 to 10 cm central leaflet being largest and possessing stalk leaf edges are toothed or serrated and bottom surface is covered with hairs. It bears numerous flowers in panicles about 10-20 cm in size and white to blue in colour. It flowers in September and October, flowers are hermaphrodite.

Corola and calyx is covered with dense hairs. Fruits are succlant drupe about 4 mm in diameter, egg shaped and become black or purple when ripe.

Chemical Constituents

Nirgudi contains 0.05 per cent of very fragrant volatile oil. Vitexin and isovitexin are isolated from methanolic extract of nirgudi. Leaf juice contains casticin, isoorientin, chrysophenol, luteolin, and hydroxyl benzoic acid.

Volatile oil of nirgudi contains sabinene, linalool, terpineol, caryophyllin, and globulol about 60%.

Uses

Leaves are useful in dispersing swelling of joints from acute rheumatism. Roots are useful in colds and rheumatic ailments. Drug is known to be used for various female disorders. Even Hippocratis, Dioscorides and Theophrastus have mentioned the use of the plant for haemorrhage after the child birth.

Leaves are anti parasitical, aromatic, analgesic, vermifuge, antifungal and anti-inflammatory. Leaves have shown anti-tumor activity. Nirgudi is also used in the preparation of herbal tea. It is very good insect repellant.

NAGARMOTHA

Synonyms

Mustake, Motha, Nut-grass.

Biological Source

These are dried tubers of *Cypercus-rotundus* Fam. *Cyperaceae.*

Geographical Source

Distributed through out the plains of India upto 1800 meters of elevation. It grows in Shri-Lanka, in moist areas, rice fields alongwith water courses.

Description

It is a glabrous herb stolons are alongated bearing hard, ovoid, tunicate fragrant tubers, stems sub solidory 10.0 – 75 cm. Leaves longer than stems, and are linear.

Colour : Outer surface of tubers is dark brown and white inside.

Odour : Pleasant aromatic

Taste : Aromatic fragrant

Size : 0.8 – 2.5 cm

Shape : Tubers are ovoid

Extra features: Tubers are elongated, glabrous.

Fig. 14.92: Nagarmotha

Chemical Constituents

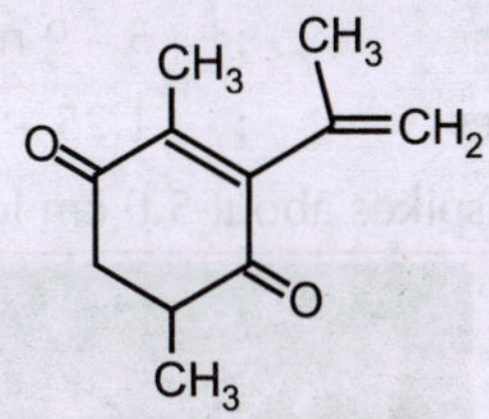

α cyperone

Rhizomes mainly contains volatile oil as active constituent and is present to the extent of only 0.1 to 0.3 per cent on dy basis. Volatile oil contains mainly and cyperone, rotundone and b-rotunol. Cyperenone content of oil is about 25-32 per cent.

Standards of Quality

Foreign organic matter	:	Not more than 2.0 per cent
Ash	:	Not more than 6.5 per cent
Acid insoluble ash	:	Not more than 2.5 per cent
Alcohol soluble extractive	:	Not more than 14.0 per cent
Water soluble extractives	:	Not more than 20.0 per cent
Loss on drying	:	Not more than 11.5 per cent

Use

Anti inflammatory and antimalerial. It is also muscle relaxant a antipyretic. Roots diaphoretic, astringent used for ulcers sores and in flavouring agent.

BETEL LEAF

Synonyms

Betel, Nagwel, Nagvalli.

Biological Source

Betal leaves consist of dried and fresh leaves of *Piper betle* fam. Piperaceae.

Geographical Source

It is indigenous to India and cultivated in all hotter parts of India. Malaya and Sri-Lanka. It is common in Maharashtra, West-bengal, and Karnataka. In Maharashtra it is cultivated in Thane, Solapur, Satara, Pune and to some extent in Vidarbha and Marathwada.

Description

This is a climber and its stems are semi-woody with many adventiatous roots which are stout, thickened at nodes young parts are glabrous. It is unisexual plant.

Morpholgy

The leaves are:

Colour	:	Green, slightly yellow
Odour	:	Aromatic
Taste	:	Slightly warm and aromatic
Size	:	10 – 18 × 5 – 10 cm
Shape	:	Broadly ovate, acuminate, entire slightly cordate.

Leaves : 5 – 9 nerved base usually round alternately arranted.

Petioles : 1.25 – 2.5 cm, thick, shining on both the surfaces.

Bears spikes about 5.0 cm long.

Fig. 14.93: Betel Leaf (Nagwel)

Chemical Constituents

Betal leaves chemically contains about 0.2 to 1.0 percent volatile oil, tannin, sugars, starch and enzyme diastase about 0.8 to 1.9 per cent.

Volatiel oil is slightly, laevo-toratory.Oil contains chavical, cadenene, chavlbetol and segqueterpines along with isomerde of eugenol.

Burning taste of leaves is only due to volatile oil. Leaves are rich of vitamin B and Vitamin C.

Uses

Leaves are stimulant antiseptic and carminative. Due to diastase conents leaves are used for masticatory purpose for digestion of starch. Leaves are also silogouge.

AVOCADO

Synonyms

Green gold, Loni-phal.

Biological Source

This consists of dried fruits and roots stocks of *Persea – americana* Fam. Lauraceae. Malona and Haas are the two important species being cultivated, while chokvet, green lvta, purple green, flurt-pullor are the other species.

Geographical Source

Avocado is cultivated on large scale in Mexico, Chilley, Indonesia, Columbia, Keniya, Brazil, Gwatemala, China, Israil, Haitee, Australiya, USA and Venuzuala.

In India it is under cultivation in Tamilnadu, Kerala, Karnataka and Sikkim

Project of cultivation of this plant durg is sponsered by Balasaheb Sawant Agricultural University at its farm in Wakwali, Tal. Dapodi, Dist. Ratnagiri.

Description

The creamy coloured avocado fruits are preferred for medicinal use. Fruits are harvested earlier and allowed to ripen, where upon fruits become soft and move slightly on pressing.

Avocado Plant

Avocado Fruit

Fig. 14.94

Cultivated plants starts fruiting after three years. The average weight of avocado fruit is about 500 gm and are very soft like butter. Hence, named as Loni-phal and develop flavour like butter. They are acidic sweet in taste.

Chemical Constituents

Fruits of avocado contains vitamin B, C, E and K. Lipids and salls of potassium calcium, magnesium, sodium, sulphur and chlorine.

Uses

Since the fruits are rich source of several vitamins, nutritive constituents, have high demand in pharmaceuticals, cosmetics products and also in food products.

Avocado have high export potentials vale.

Avocado fruits are rich source of anti-oxidants, lutein, and zeo xanthin covotinoids are responsible for filtering out blue-liquid. Which can damage eye and help in presserrina healthy eye cells. Therefore it reduces the chance of cataract and molecules degeneration in adults.

SERENDIPITY RHIZOME

Ornamental Onion

It is propagated from purple coloured rhizomes. Rhizomes are dug-out and spit by dividing into 7 - 8 pieces. Every piece sprouts into new plant. It gets propagated in fall season only.

It grows into soil of any fertility even in clay soil. It is a drought resistant crop grows in full sun and also in partial shade. Height of plant is about 10 - 20 cm and spreads into area of 25 - 35 cm. It is a perrineal and develops with attractive nice blue-green foliage. Plant is deer and rabbit resistant due to bitter taste. Serendipity crop is pest and disease resistant. It bears globe like rosy purple flowers in mid summer or late in summer.

Serendipity plant has average water-requirement. When the serendipity plant is crushed it emits smell of onion.

SERPENTARY RHIZOME

Synonyms

Virginian snake root, Rhizoma serpentaria, Texan snake root.

Biological Source

This consists of dried rhizomes and roots of *Aristolochia serpentaria* OR *Texan: Aristotochia reticulata,* fam. Aristolochiaceae, collected in autumn and dried.

Geographical Source

USA in the basin of Mississippi river.

Description

Texan snake-root is a short slender and horizontal or some times oblique rhizome.

Macroscopic Characters of Root

Colour : Rhizomes are yellowish brown in colour.

Odour : Characterstic camphoraceous odour

Taste : Strong disagreeably bitter taste

Size : About 3 cm long and 3 to 4 mm thick.

Extra featrues : Sides and lower surface of rhizome is covered with numerous slightly curved roots (10 cm × 0.2 mm).

Lower surface bears number of short bases of aerial roots.

Upper surface bears short bases of slender aerial stems.

Rhizomes and roots are brittle and break with short fracture. Bark is yellowish brown and thin.

Chemical Constituents

Rhizomes contains volatile oil, about 1.0 per cent, tannin a bitter principle and alkaloid aristolochine.

Uses

It is used as local and general stimulent and also as tonic.

RASPBERRY LEAVES

Synonyms

Red rasberry, Rubus, Raspberry leaves.

Biological Source

These are the dried chopped leaves of *Rubus idaeus* frm. Rosaceae.

Geographical Source

These leaves are cultivated commercially in United Kingdom, Canada, Europe and to some extent in India.

Microscopic Characters

Colour	:	Green on upper surface
Odour	:	Faint and characteristic
Taste	:	Bland
Size	:	5 to 12 cm in length
Shape	:	Ovale to ovale lanceolate, acuminate and rounded at base, dentate

Extra features : Rasp berry is a deciduons woody bush, leaves are having 2.5 cm long petiole, cylendrieal terminal leaflet are longer than others.

Chemical Constituents

It contains Anthocyanins, ellagitannins, terpenoids, flavonoids, sugar, malic acid and citric acids. Flavonoids contains rutin.

Standards of Quality

Total ash	:	Not more than 8.0 per cent
Acid insolube ahs	:	Not more than 1.5 per cent
Water soluble extractives	:	Not less than 15.0 per cent

Uses

Used in Jams, desserts, sauces and beverages. It is used as flavouring agents in Pharmaceutical formulations. It is astringent also. It is used to faciliate the particritions.

ZANTHOXYLUM

Synonyms

Tumru.

Biological Source

This consists of dried entire herb of *Zanthaxylum alatum* fam Rutaceae.

Geographical Source

Available in sultropical area of Himalya, Trans-indus, Panjab, Kumaon, Bhutan, Khasia Hills at about 3,000 – 5,000 ft.

Chemical Constituents

Part of entire plant are used medicinally fruits contain volatile oil about 1.5 per cent. Bark contains bitter crystalline compound, identical with berberine. Aqueous extract of fruits contain saponins tannins. Alcoholic extract contain zygophyllin. Leaves, stems also contain zygophyllin of quonovic acid.

Uses

Zanthoxylum is used as aromatic. Carmunative, stomachic and in dyspepsla. In the treatment of fever and also as fish-poison. It is remedy for tooth ache and used in the preparation Dant-Kanti tooth paste, a pantented product of Patanjali Ayurved Ltd., Haridwar.

TEA-TREE OIL

Synonym

Melaleuca oil.

Biological Source

It is a volatile oil obtained from leaves and terminal branchlets of Tea-tree. *Melaleuca alternifolia* family Myrtaceae.

Geographical Source

The plant producing tea tree oil is native to south east Queensland, North east, New South Wales, and Australia.

Method of Preparations

Tea-tree oil is a essential oil and is prepared by steam distillation. Put the tea-tree leaves in distillation pot and cover with water. Boil the water to steam the leaves. Place the lid on the pot upside down. Provide cooling arrangement on top of the upside down pot lid to hasten the steam condensation. Oil collected in the container is to be collected carefully.

Fig. 11.23: Flowering branch of Tea-tree herb

Description

Colour	:	Colorless to pale yellow liquid
Odour	:	Terpene like odour associated with nutmeg.
Solubility	:	Soluble in 80.0% alcohol. Insoluble in water.

Chemical Constituents

There are more than 100 components in tea-tree oil and mostly terpene hydrocarbons, i.e. Mono-terpenes, sesquiterpenes and their alcohols.

The important components and their concentration in the oil are as under.

Table 11.4: Chemical constituents of Tea-tree oil

Sr. No.	Components	Concentration
1.	Terpeniol	30 – 48%
2.	γ-terpinene	10 – 28%
3.	α-terpinene	05 – 13%
4.	1-8 cineole	00 – 15%
5.	Terpinolene	1.5 – 5%
6.	α-terpineol	1.5 – 8%
7.	a-pinene	01 – 6%
8.	p-cymene	0.5 – 8%

Standards of Quality

Refractive index : 1.4760 – 1.4810

Optical rotation : + 6.48 to +9.48

Density : 0.8950 to 0.9050

Uses

Tea tree oil is poisonous if taken internally. It is to be used only externally. Adverse effects such as skin irritation, allergic contact dermatitis, systemic hypersensitivity reactions are observed. Tea tree oil heals skin infections and soothe sunburns, relieves ichiness from rashes. α-terpineol is used for denaturation of fats for soap industry. It is very useful in acne, athlete's foot, bad breath, warts, insect bites, psoriasis. It is antimicrobial and also used in deodorants and household cleaners.

Additionally it is good flavouring agent for tooth-paste and for cosmeceuticals.

Storage

Oil is sensitive to light, heat exposure, air and moisture, store it in airtight containers in dark, cool and dry place.

Varieties

Several other species of tree tea herb are cultivated and their oil is extracted. *Melaleuca styphelioides* in Tunisia and Egypt. *Melaleuca leucadendra* in Malasia, Vietnam and Egypt, *Melaleuca acumenata* in Tunisia, *Melaleuaca ericifolia* in Egypt *Melaleuca quinquenervia* in USA.

ANIMAL DRUGS CONTAINING TERPENOIDS

SHELLAC

Synonym

Lac, Sealing wax.

Biological Source

Lac is found in India, the resinous protective secretion of the tiny lac insect, *Lacifer lacca Kerr*, family Laccidae and order *Hemiptera*, which produces the bulk of Commercial lac in India.

Geographical Source

It is also produced in Sri Lanka, Thailand, China, Vietnam, Indonesia, Philippines, Malaya and Pakistan.

L. lacca is a pest on number of plants either wild or cultivated. The minute red coloured larvae of the insect settle on the young fleshy shoots of the host plant and with their long proboscis suck nutrient from the sap. The insects secrete a thick resinous fluid which envelopes their bodies, and the secretions from individual insect coalesce and form a hard continuous envelop over the twigs. After the completion of life cycle and just around the time larvae of the next generation begin to emerge, the twigs are harvested and the encrustations scrapped off, dried and further processed to produce the lac of the commerce. This crude lac is known as stick lac and is not used in the industry after refining.

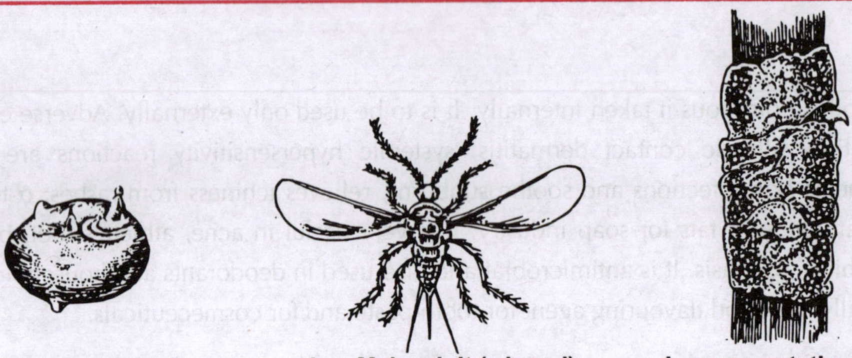

Fig. 14.92: Lac insects

Cultivation, Collection and Preparation

Although Shellac is not exactly a vegetable product but it is largely dependent on vegetation for its production. It is a resinous substance, secreted by a species of scale insects by sucking the juice of the tree, and forming on these a continuous waxy incrustation. Naturally these insects live under braches of trees or shrubs and producing heavy secretions on them. Uncultivated trees were at one time the only source of lac supply. But now, however, selected species are cultivated for the purpose and systematically infested with the lac insects.

Method of Purification

The resinous secretion of the lac insect and red colouring matter are known to Indians since early times. India held virtual monopoly of lac till 1950 and accounted for nearly 85 per cent of the world's production.

The stick lac is powdered and washed in water several times and then dried on floorings away from sunlight. It is in the form of grains and known as seedlac, which is further processed to remove the adhering 3-10 per cent impurities. Various trade names such as shellac, button-lac and dewaxed lac are given to them depending upon the type of processing.

Description

Colour : Shellac is available as yellowish transparent sheet or powder.

Odour : Characteristic.

Shellac is very brittle in nature and heavier than water.

Solubility : Very slowly soluble in 80 - 85 per cent alcohol; only 13 –15 per cent in ether, 10 - 20 per cent in benzene and 2 - 5 per cent in petroleum ether. It is soluble in alkali solutions, and insoluble in water.

Chemical Constituents

Stick lac contains 70 - 80 per cent of resin, sugars, proteins, colouring matter (1 - 2 per cent), wax (4 - 6 per cent), extraneous matter (8 - 12 per cent), and volatile oil in traces.

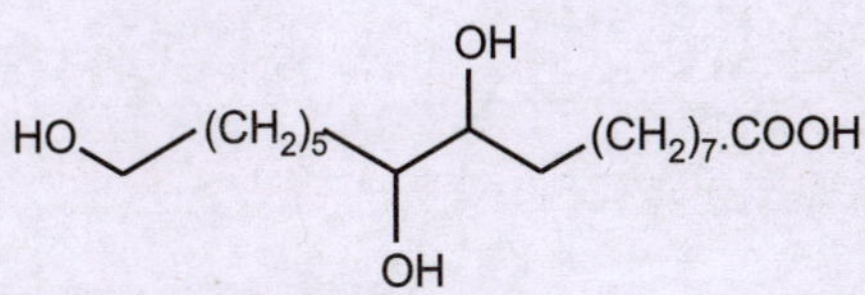

Aleuritic acid

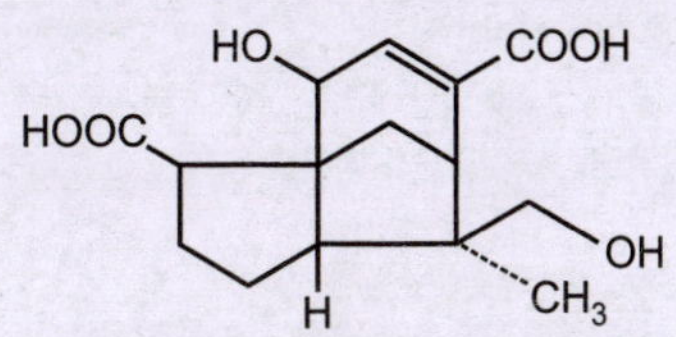

Shellolic acid

Lac resin consists of hydroxy fatty acid derivatives. Aleuritic acid is the main constituent and about 35 per cent of resin, while shellolic acid and its isomers along with kerrolic acid and butolic acid are present to a small extent. It resembles carnauba wax to a great extent. The colouring matter contains laccaic acid, which is water soluble.

Chemical Test

Add few drops of mixture of ammonium molybolate and 3 ml of sulphuric acid to 50 mg of shellac, a green colour is produced and it becomes lilac (pinkish-purple) on standing for five minutes.

Standards of Quality

Melting point	:	72 - 82°C
Density	:	1.035 - 1.140
Sulphated ash	:	Not more than 1.0 per cent
Saponification value	:	185 - 210
Iodine value	:	10 – 18
Acid value	:	65 – 95
Ethenol insoluble matter	:	Not more than 2.0 per cent
Arsenic	:	Less than 3.0 ppm.

Uses

Lac is used for the manufacture of sustained release medicaments.

Industrially, it is used in preparation of phonographic records, in lacquers and varnishes, in electrical machines as sealing wax, and in inks. India has exported Shellac worth of ₹ 36 crore and ₹ 33 crore during 1995 - 96 and 1996 - 97 respectively.

Chapter 15...

ALKALOIDAL DRUGS

Alkaloids are a chemically heterogeneous group of natural substances and comprise more than 6000 basic nitrogen containing organic compounds which occur in about 15 per cent of all vascular terrestrial plants and in more than 150 different plant families. The alkaloids exhibit diversity of structures and also show an extraordinary spectrum of pharmacological activities. Because of these characters, they are important for chemical, physiological, taxonomic and biogenetic studies.

HISTORY

Fig. 15.1: Friedrich Wilhelm Adam Serturner

The term 'alkaloid' or 'Pflanzenlkalien' was coined by **Meissner**, a German pharmacist, in 1819. The mankind has been using alkaloids for various purposes like poisons, medicines, poultices, teas, etc. The French chemist, **Derosne** in 1803, isolated narcotine. In the same year, morphine from opium was isolated by **Serturner**, a pharmacist of Paderborn near Hannover in 1803. **Pelletier** and **Calverton** from the Faculty of Pharmacy in Paris isolated emetine in 1817 and colchicine in 1819. This was followed by isolation of series of alkaloids from vegetable drugs, like strychnine (1817); brucine, piperine and caffeine (1819); quinine, colchicine and cinchonine (1820); coniine (1826), papaverine (1821) and thebaine (1835). By 1884, about 25 alkaloids were reported to be isolated from cinchona bark alone, but in 1870, a landmark in the domain of alkaloids was achieved by determining the structure of coniine which also became the first synthesized alkaloid in 1889. From the beginning of 19th century till date, it has proved to be a perpetual work to discover new alkaloids from plants and animals. In the present century, the proper structures were assigned to various alkaloids with the help of chromatographic and other sophisticated physical methods of analysis. As per a Russian review in 1973, the number of known alkaloids had reached up to 4959, amongst which, the structures of 3293 alkaloids were elucidated. At present, the number of alkaloids discovered has exceeded 6000.

Definition

In view of their chemical and physiological diversity, there is no comprehensive definition of alkaloids. The term is derived from the word 'alkali-like' and hence, they resemble some of the characters of naturally occuring complex amines. The term alkaloid also covers proto alkaloids and pseudoalkaloids. In view of all such variations, the only definition that brings all such compounds under one title is as follows: **These are the organic products of natural or synthetic origin which are basic in nature and contain one or more nitrogen atoms, normally of heterocyclic nature, and possess specific physiological actions on human or animal body, when used in small quantities.** The true alkaloids are toxic in nature, contain heterocyclic nitrogen which is derived from amino acids and always basic in nature. True alkaloids are normally present in plants as salts of organic acids. The 'proto alkaloids' or 'amino alkaloids' are simple amines in which the nitrogen is not in a heterocyclic ring. Some times, they are considered as biological amines. They are basic in nature and prepared in plants from amino acids. Some of the examples of these alkaloids are mescaline, N, N-dimethyl tryptamine, colchicine and ephedrine. The term 'pseudoalkaloids' includes mainly steroidal and terpenoid alkaloids and purines. They are not derived from amino acids. They do not show many of the typical characters of alkaloids, but give the standard qualitative tests for alkaloids. The examples of pseudoalkaloids are conessine and caffeine.

Sources of Alkaloids

They are obtained from plant, animal synthetic and tissue culture sources.

PROPERTIES

1. Physical Properties

With few exceptions, all the alkaloids are colourless, crystalline solids with a sharp melting point or decomposition range. Some alkaloids are amorphous gums, while others like coniine, sparteine, nicotine etc. are liquid and volatile in nature. Some alkaloids are coloured in nature, e.g. betanidin is red, berberine is yellow and salts of sanguinarine are copper red in colour.

In general, the free bases of alkaloids are soluble in organic non-polar, immiscible solvents. The salts of most alkaloids are soluble in water. In contrast, free bases are insoluble in water and their salts are also very sparingly soluble in organic solvents. The alkaloids containing quaternary bases are only water soluble. Some of the pseudoalkaloids and protoalkaloids show higher solubility in water. For example, colchicine is soluble in alkaline water, acid or water and caffeine (free base) is freely soluble in water. Quinine hydrochloride is highly soluble in water i.e. 1 part of quinine hydrochloride is soluble in less than 1 part of water, while only 1 part of quinine sulphate is soluble in 1000 parts of water.

The solubility of alkaloids and their salts is useful in pharmaceutical industry for the extraction and formulation of final pharmaceutical preparations.

2. Chemical Properties

Most of the alkaloids are basic in reaction, due to the availability of lone pair of electrons on nitrogen. The basic character of the alkaloidal compound is enhanced if the adjacent functional groups are electron releasing. The alkaloid turns to be neutral or acidic when the adjacent functional groups are electron withdrawing like amide group which reduces the availability of the lone pair of electrons. But, alkaloids exhibiting basic character are very much sensitive to decomposition and cause a problem during their storage. Their salt formation with an inorganic acid prevents many a time their decomposition.

The alkaloids may contain one or more number of nitrogen and it may exist in the form as primary ($R - NH_2$), e.g. mescaline; secondary amine ($R_2 - NH$), e.g. ephedrine; tertiary amine (R_3N) e.g. atropine; and quaternary ammonium compounds [R_4N^+X] e.g. tubocurarine chloride. In the last type, their properties vary from other alkaloids, owing to quaternary nature of nitrogen.

In the natural form, the alkaloids exist either in free state, as amine or as salt with acid or alkaloid N-oxides.

CHEMICAL TESTS FOR ALKALOIDS

The qualitative chemical tests used for detection of alkaloids are dependent on the characters of alkaloids to give precipitates as salts of organic acids or with compounds of heavy metals, like mercury, gold, platinum, etc.

The different reagents used are **Mayer's reagent** (potassium mercuric iodide solution) giving cream coloured precipitate; **Dragendorff's reagent** (potassium bismuth iodide solution) giving reddish brown precipitate; and **Wagner's reagent** (iodine-potassium iodide solution) yielding reddish brown precipitate. Some alkaloids also give yellow coloured precipitates with picric acid called as **Hager's reagent** and picrolonic acid. Individual alkaloid gives colour or precipitate with certain specific reagent.

The chemical tests with heavy metals are not solely limited to alkaloids. Proteins, coumarins and α- pyrone also give precipitates with these reagents. It may be also noted that some alkaloids do not give such tests, like caffeine which is highly water soluble. Hence, the tests with heavy metals are in some cases false positive reactions or false negative reactions. For this purpose, the specific tests for individual alkaloids are more important for qualitative evaluation of crude drugs. These tests are covered under individual drugs.

ISOLATION AND EXTRACTION OF ALKALOIDS

The extraction of alkaloids is based on their basic character and solubility pattern. The normal procedures followed are to treat the moistened drug with alkali so as to set free the base as it exists in salt form and then to separate free base with organic solvent. This is known as Stas Otto process. Though the methods of extraction vary, generally following procedure is applied for small scale extraction of alkaloids. First, the plant is defatted with petroleum ether, especially in case of seed and leaf forms of drugs. Before applying this treatment, the alkaloid should be tested for its solubility in petroleum ether. Otherwise, the drug should be pretreated with acid to convert alkaloids into their salts. This happens in case of extraction of ergotamine from ergot. In the second stage, the drug may be extracted with polar solvents like water, ethanol, methanol, aqueous alcohol mixtures or with acidified aqueous solutions. By this treatment, the alkaloidal salts are transferred to polar solvent. It also helps in removing pigments, sugars and other organic secondary constituents. In the following stage, the alcohol solution is evaporated to thick syrup and is subjected to partition between aqueous acid solution and an organic solvent. After continuous extraction with organic solvent for some time, the aqueous phase is made alkaline with either sodium carbonate or ammonia. The basic aqueous solution is then extracted with convenient organic solvent followed by drying of alkaloid containing solution, normally with sodium sulphate, filtered and evaporated to yield alkaloid residue.

The other method meant for extraction of alkaloids employs the treatment of drug with ammonia so as to convert the alkaloidal salts into their free bases. Such liberated alkaloids in free base form are conveniently extracted with organic solvents like ether, benzene, chloroform, etc. The method is not useful for isolating alkaloids with quaternary nitrogen.

The further purification of crude extract of alkaloids is done by following ways, which may, however, vary for individual alkaloid.

1. Direct Crystallisation from Solvent

It is a very simple method of isolation and may not be useful in case of complex mixtures.

2. Steam Distillation

This method is specially employed for volatile liquid alkaloids like coniine, sparteine and nicotine, but otherwise this process is not suitable for alkaloids with high molecular weights.

3. Chromatographic Techniques

This method has proved to be ideal for separation of a vast number of plant alkaloids. The different techniques of chromatography (thin layer, column, gas, liquid, ion exchange chromatography, HPTLC etc.) are used for separation of individual alkaloids from complex mixtures.

4. Gradient pH Technique

Though alkaloids are basic in nature, there are variations in the extent of basicity of various alkaloids of the same plant. Depending on this character, the crude alkaloidal mixture is dissolved in 2 per cent tartaric acid solution and extracted with benzene so that the first fraction contains neutral and/or very weakly basic alkaloids. pH of the aqueous solution is increased gradually by 0.5 increments up to pH 9 and extraction is carried out at each pH level with organic solvent. By this way, the alkaloids with different basicity are extracted. Strongly basic alkaloids are extracted at the end. The general scheme for extraction of alkaloids can be summarized as follows.

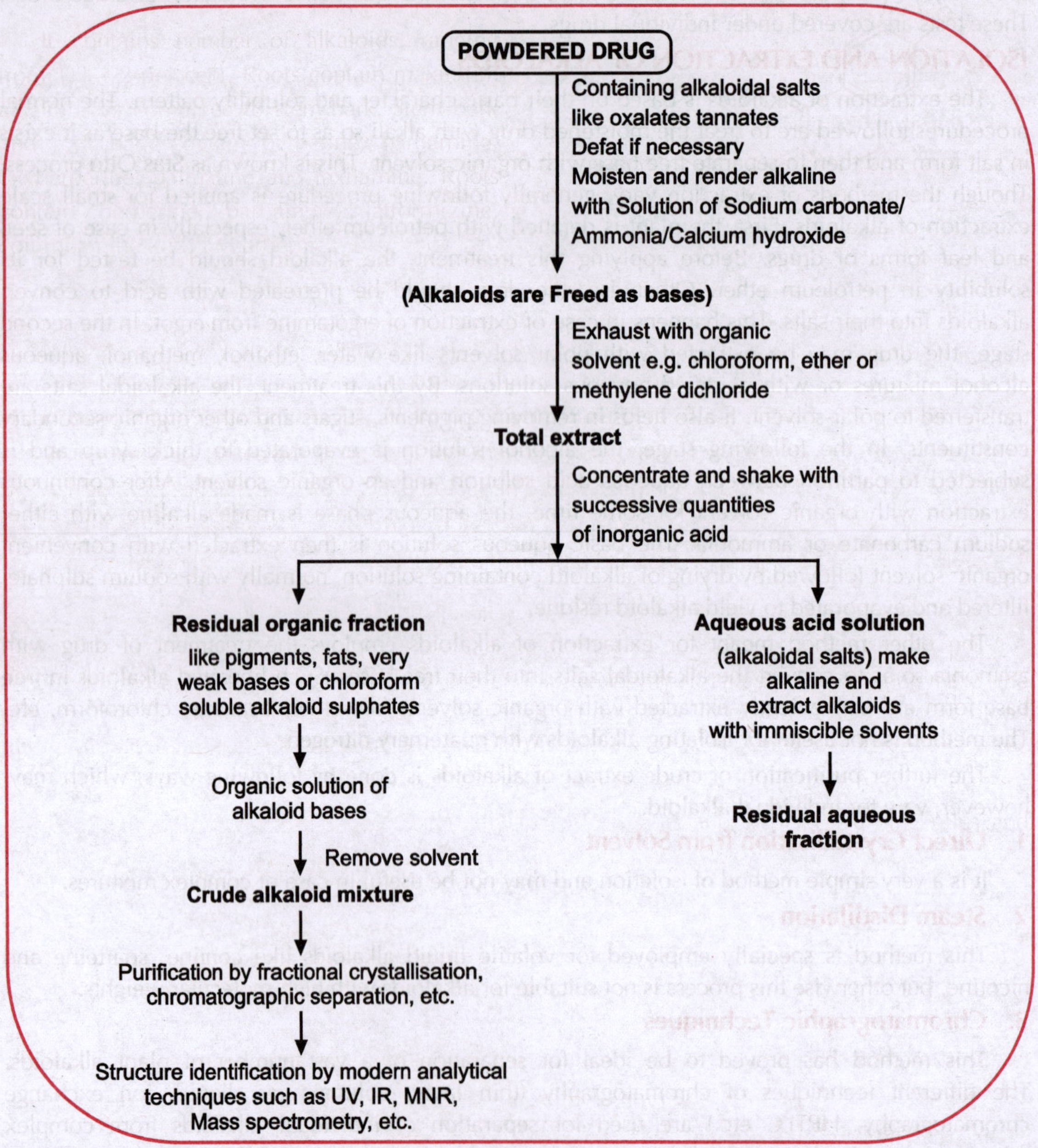

CLASSIFICATION OF ALKALOIDS

The various methods proposed for classification of alkaloids are as follows:

1. Pharmacological Classification

Depending on the physiological response, the alkaloids are classified under various pharmacological categories, like central nervous system stimulants or depressants, sympathomimetics, analgesics, purgatives, etc. This method does not take into account chemical nature of crude drugs. Within the same drug, the individual alkaloid may exhibit different action e.g. morphine is narcotic analgesic, while codeine is mainly antitussive. In cinchona, quinine is antimalarial, while quinidine is cardiac depressant.

2. Taxonomic Classification

This method classifies the vast number of alkaloids based on their distribution in various plant families, like solanaceous or papillionaceous alkaloids. Preferably, they are grouped as per the name of the genus in which they occur, e.g. ephedra, cinchona, etc. From this classification, the chemotaxonomic classification has been further derived.

3. Biosynthetic Classification

This method gives significance to the precursor from which the alkaloids are biosynthesized in the plant. Hence, the variety of alkaloids with different taxonomic distribution and physiological activities can be brought under same group, if they are derived from same precursor. E.g. all indole alkaloids from tryptophan are grouped together. The alkaloidal drugs are categorised on the fact whether they are derived from amino acid precursor as ornithine, lysine, tyrosine, phenylalanine, tryptophan, etc.

4. Chemical Classification

This is the most accepted way of classification of alkaloids. The main criterion for chemical classification is the type of fundamental (normally heterocyclic) ring structure present in alkaloid. The alkaloidal drugs are broadly categorised into two divisions.

(a) Heterocyclic alkaloids (True alkaloids) are divided into twelve groups according to nature of their heterocyclic ring.

(b) Non-hetero-cyclic alkaloids or proto-alkaloids or biological amines or pseudoalkaloids.

The following chart indicates type of alkaloids and their occurrence in various plants along with basic chemical ring.

Table 15.1: A. Heterocyclic Alkaloids (True Alkaloids)

Type	Basic ring structure	Examples
1. Pyrrole and Pyrrolidine	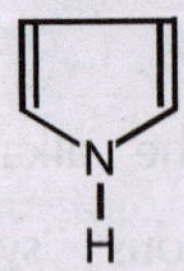 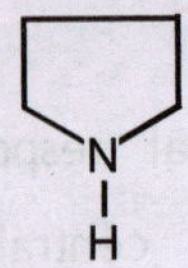	Hygrine, coca species
2. Pyridine and Piperidine	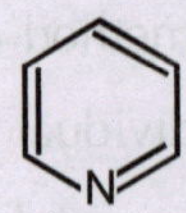 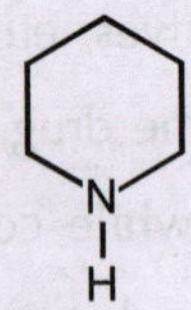	Arecoline, anabasine, coniine, lobeline, pelletierine, trigonelline
3. Pyrrolizidine	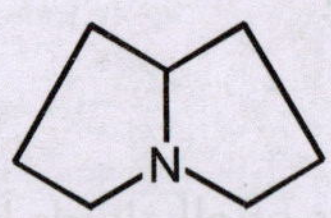	Echimidine, senecionine, seneciphylline, symphitine
4. Tropane (piperidine. N-methyl pyrrolidine)	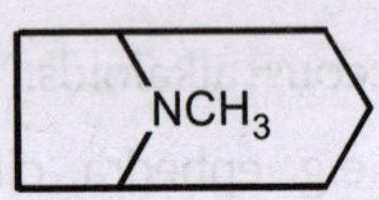	Atropine, hyoscine, hyoscyamine, cocaine, pseudo- pelletierine, meteloidine
5. Quinoline	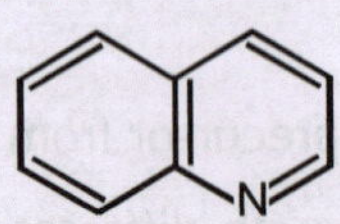	Quinine, quinidine, cinchonine, cinchonidine cupreine, camptothecin.
6. Isoquinoline	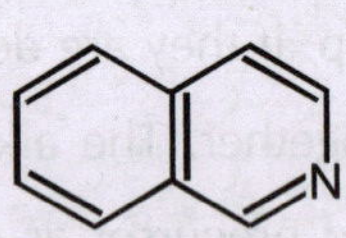	d-tubocurarine, berberine, emetine, cephaeline, papaverine, narcotine, narceine
7. Phenanthrene	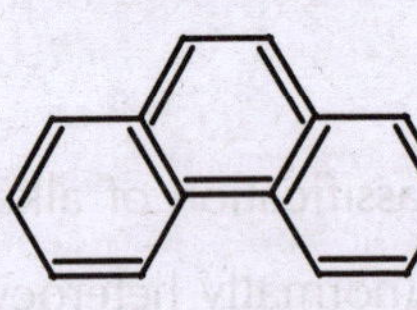	Morphine, codeine, hydrastine
8. Aporphine (reduced isoquinoline-naphthalene)	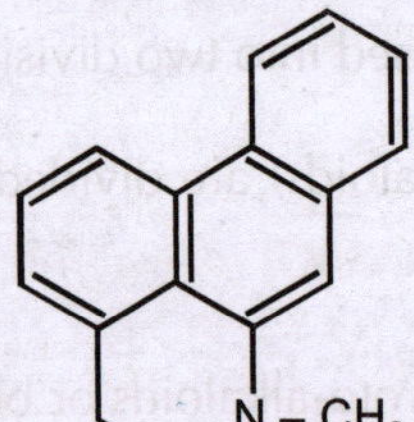	Boldine

Condt...

Type	Basic ring structure	Examples
9. Indole (Benz pyrrole)	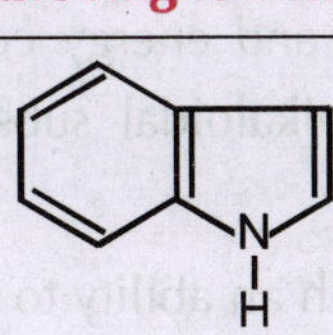	Ergometrine, ergotamine, reserpine, vincristine, vinblastine, strychnine, brucine, physostigmine
10. Imidazole	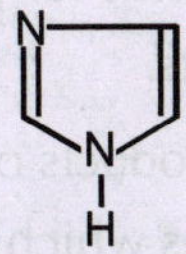	Pilocarpine, Isopilocarpine, Pilosine
11. Norlupinane	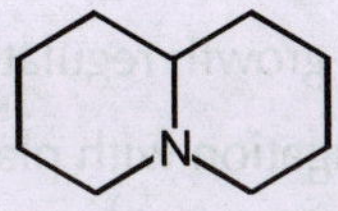	Cytisine, laburnine, lupanine, sparteine
12. Purine (pyrimidine/imidazole)	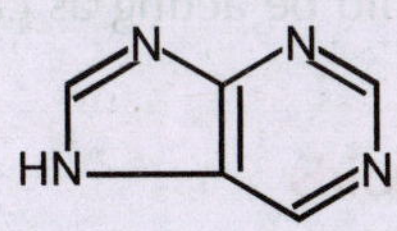	Caffeine, theobromine, theophylline
13. Steroidal (cyclopentano-perhydrophenanthrene ring)	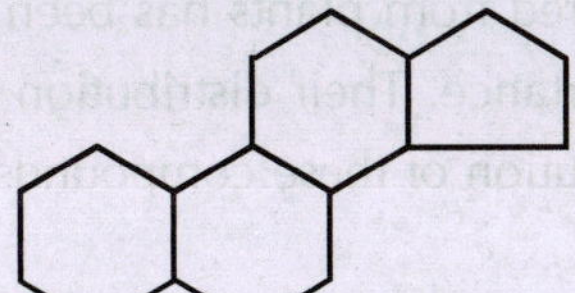	Protoveratrine, solanidine, conessine, funtumine

Table 15.1: B. Non-heterocyclic alkaloids Pseudo alkaloids

Type	Basic ring structure	Examples
14. Diterpene	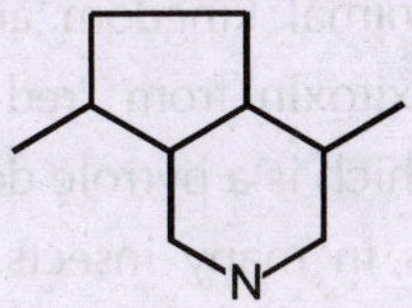	Aconitine, Aconine, hypoaconitine

Proto alkaloids (Amino – alkaloids)

Type	Basic ring structure	Examples
15. Alkylamine (Amino alkaloid)	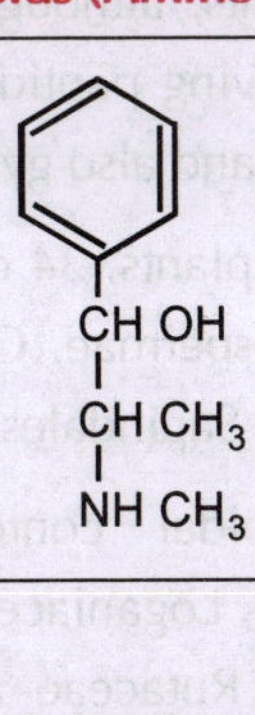	Ephedrine, pseudoephedrine. Mescaline, colchicine

ROLE OF ALKALOIDS IN PLANTS

The alkaloids are poisonous in nature, but when used in small quantities, exert useful physiological effects on animals and human beings and hence they have secured significant place in medicine. Their exact role in nature and functions in the plants, if any, are still a topic of ambiguity. Only one aspect is clearly understood that they are synthesized by a particular, stereospecific, many a time complicated, and energy consuming pathways and further they are metabolized to other alkaloidal or non alkaloidal substances. Some of the predicted roles of alkaloids in the plants are discussed below.

1. They are the reserve substances with an ability to supply nitrogen.
2. They might be the defensive mechanisms for plants growing in dry regions to protect from grazing animals, herbivores and insects.
3. It is also possible that they are end products of detoxification mechanism in plants, and by this way check formation of substances which may prove to be harmful to the plants.
4. They might have a possible role as growth regulatory factors in the plants.
5. They are present normally in conjugation with plant acids, like meconic acid, cinchotannic acid, etc. Therefore, alkaloids could be acting as carriers within plants for transportation of such acids.

Occurrence and Distribution of Alkaloids

The number of alkaloids discovered from plants has been continuously increasing. It is noted that alkaloids are of taxonomic importance. Their distribution in nature appears to be restricted or specific. Hence, the pattern of distribution of these compounds and their biological variability is of chemotaxonomic interest.

In general, the major distribution of alkaloids occurs in the angiosperms. But their presence is also detected in microorganisms, marine organisms, insects, animals and some of the lower plants. Some of the alkaloids reported from animal kingdom are castoramine from Canadian beaver, muscopyridine from musk deer and saxitoxin from "red tide". *Gonyaulax catenella* which has neurotoxic activity and a sex hormone which is a pyrrole derivative called 2, 3-dihydro – 7 methyl-1 H – pyrrolizin -1- one which occurs in many insects. It has been reported that among the bacteria, about 47 per cent species indicate the presence of alkaloids, e.g. pyocyanine from *Pseudomonas aeruginosa*. In the lower plants, although the alkaloids are found in less number, some important sources are ergot fungus giving peptide alkaloids, ergometrine, ergotamine, etc., lycopodine from lycopodium – a club moss, and also gymnosperms like ephedra alkaloids.

Out of the 60 different orders in higher plants, 34 orders contain alkaloids. Amongst them the prominent orders are Campanulales, Centrospermae, Gentianales, Geraniales, Liliflorae, Ranales, Rhoedales, Rosales, Rubiales, Tubiflorae and Sapindales.

The promising families with alkaloidal content are Amaryllidaceae, Apocynaceae, Berberidaceae, Euphorbiaceae, Leguminosae, Loganiaceae, Liliaceae, Lauraceae, Menispermaceae, Papaveraceae, Ranunculaceae, Rubiaceae, Rutaceae and Solanaceae. On the other hand, the orders like Curcurbitales, Fagales, Oleales, Salicales and the families like Labiatae and Rosaceae are practically devoid of alkaloids.

Sources of Alkaloids

Sources of alkaloids can be summarised as under:

Solanacerus alkaloids are:

1. Hyoscyamine
2. Hyoscine (sopalanio)
3. Atropine
4. Apo atropine
5. Cuskhygrine
6. Cusk hygrine

(a) From Animals

(i) Canadean beaver — Castoramine

(ii) Musk dear — Muskopyrine

(iii) Red tide — Soxitokin

Adrenalin from Adrenal cortex and bufotenine from toads (toad - frog)

(Adronaline is a harmone but it chemcally complies all the requirement of alkaloidal compounds.)

(b) From Lower plants

(i) Bacteria — Pyocyamine

(ii) Fungi (Ergot) — Ergotonine, Ergometrine

Erythromycine and chloremphenicol are artibiotic alkaloids.

(iii) Gymnosperm — Ephedra

(c) From Higher plants

(i) Leaves — --- (---) vnca (vincrystin); Lea (caffine)

(ii) Barks — Cinolona (Quinein), Kureti (conessine)

(iii) Seeds — Nuxomica (strychain), Betal nut (Arceolin)

(iv) Fruits — Hemolock

(v) Underground part — Colchicum, Glorisosa

(vi) Roots — Rauwolfia, Daruhalad, Ashwagandha (Withanine)

(vii) Entire organisms — Belladonna (Atropine), Bhringraj (Ectiptine)

Bhringraj (Ecliptines)

(viii) Latex — Pappy – Morphul, codem

(d) From Synthetic sources

Heroin and homotropine are the synthetic alkaloids and are not reported naturally.

Within a plant, in most of the cases, the alkaloids are highly localized and concentrated in certain morphological parts only e.g. seeds (nux vomica, areca, physostigma), roots (rauwolfia, belladonna, ashwagandha), underground stems (sanguinaria), barks (kurchi, cinchona), leaves (coca leaf, lobelia, duboisia) and fruits (conium). In some cases, like vinca and ephedra, practically every part of plant contains alkaloids. The list of crude drugs containing alkaloids as active constituents is presented here.

Table 15.2

1. Indole Alkaloids

Sr. No.	Name of drug and synonym	Biological source	Active constituents	Uses
1.	Ergot (Ergot of Rye)	A fungal sclerotium of *Claviceps purpurea* Family Hypocraceae in ovary of rye plant *Secale cereale.* Family Graminae	ergometrine, ergotamine	oxytocic, prevents postpartum haemorrhage ; used in treatment of migraine
2.	Nux vomica (Crow Fig)	Seeds of *Strychnos nux vomica,* Loganiaceae	strychnine, brucine	CNS stimulant, bitter, stomachic, tonic
3.	Phyostigma (Calabar beans)	Seeds of *Physostigma venonosum,* Loganiaceae	Physostigmine, physovenine	cholinergic, (ophthalmic) in glaucoma
4.	Sarpagandha (Rauwolfia)	Roots and rhizomes of *Rauwolfia serpentina,* Apocynaceae.	Reserpine, rescinnamine	hypotensive tranquilliser
5.	Vinca (Catharanthus)	Entire plant of *Catharanthus roseus,* Apocynaceae	vincristine, vinblastine	anticancer (treatment of Hodgkin's disease and leukemia)

2. Isoquinoline Alkaloids

Sr. No.	Name of drug and synonym	Biological source	Active constituents	Uses
6.	Curare (South American arrow poison)	Dried extract of stems & leaves of various plants from family Loganiaceae and Menispermaceae	d-tubocurarine chloride	skeletal muscle relaxant
7.	Daruhaldi (Berberis)	Roots and rhizomes of *Berberis aristata,* and other *Berberis* species, Berberidaceae	berberine	astrigent in inflammation of mucous membranes
8.	Ipecacunha (Ipecac)	Roots and rhizomes of *Cephaelis ipecacuanha* and *C. acuminata Rubiaceae*	emetine, cephaeline	antiamoebic, emetic, expectorant
9.	Opium (Raw opium)	Dried latex from the capsules of *Papaver somniferum,* Papaveraceae	narcotine, papaverine	narcotic analgesic, in diarrhoea

Contd...

3. Tropane Alkaloids

Sr. No.	Name of drug and synonym	Biological source	Active constituents	Uses
10.	Belladonna herb (Deadly night shade Leaf)	Dried leaves and flowering tops of *Atropa belladonna,* Solanaceae	*l*-hyoscyamine, atropine	anticholinergic, antispasmodic
11.	Coca leaves	Dried leaves of *Erythroxylon coca,* Erythroxylaceae	cocaine, cinnamyl-cocaine, α-truxilline	local anaesthetic
12.	Datura (Angel's trumpet)	Dried leaves and flowering tops of *Datura metel* var. *fastuosa*, Solanaceae	scopolamine, hyoscyamine, atropine	anticholinergic, in deodenal ulcers
13.	Duboisia (Cork-tree)	Dried leaves of *Duboisia myoporoides,* Solanaceae	scopolamine	anticholinergic
14.	Hyoscyamus (Henbane)	Dried leaf and flowering top of *Hyoscyamus niger*, Solanaceae	*l*-hyoscyamine, hyoscine	anticholinergic, antispasmodic
15.	Stramonium (Thornapple leaves)	Dried leaves and flowering tops of *Datura stramonium*, Solanaceae	*l*-hyoscyamine, hyoscine, and atropine	anticholinergic, mydriatic, control motion, sickness

4. Quinoline Alkaloids

Sr. No.	Name of drug and synonym	Biological source	Active constituents	Uses
16.	Camptotheca (Cancer tree)	Dried stem wood *Camptotheca acuminata* Nyssaceae	Camptothecin	Antitumour
17.	Cinchona (Peruvian bark)	Dried root or stem bark of *Cinchona calisaya, C. officinalis, C. ledgeriana, C. succirubra,* Rubiaceae	quinine, quinidine, cinchonine, cinchonidine.	Antimalarial, bitter tonic.

5. Amino Alkaloids (Proto-Alkaloids)

Sr. No.	Name of drug and synonym	Biological source	Active constituents	Uses
18.	Colchicum (Meadow saffron seed)	Seed and corm of *Colchicum autumnale,* Liliaceae	Colchicine, demecolcine	treatment of gout, induction of polyploidy

Contd...

Sr. No.	Name of drug and synonym	Biological source	Active constituents	Uses
19.	Ephedra (Ma-Huang)	Dried stems of *Ephedra gerardiana, E. equisetina, E. sinica etc.,* Ephedraceae	ephedrine, pseudoephedrine (alkaloidal amines)	sympathomimetic antiasthmatic, treatment of hay fever
20.	Gloriosa (Glory lily)	Dried rhizomes & roots of *Gloriosa-superba,* Liliaceae	Colchicine	In the trearment of gout & cancer
		6. Pyridine alkaloids		
21.	Areca nut (Betel nut)	Dried ripe seed of *Areca catechu,* Palmae	arecoline, arecaidine	respiratory stimulant
22.	Lobelia (Indian tobacco)	Dried leaves and tops of *Lobelia nicotianefolia,* Campanulaceae	lobeline, lobelanidine	respiratory stimulant
		7. Purine Alkaloids		
23.	Cocoa seed (cocoa beans)	Seeds of *Theobroma cocoa,* Sterculiaceae	theobromine, caffeine	diuretic
24.	Coffee seed	Dried ripe seeds of *Coffea arabica,* Rubiaceae	caffeine trigonelline	Stimulant, to counter effect over dosage of CNS depressant
25.	Kola (Cola)	Seeds of *Cola nitida* Sterculiaceae	caffeine, theobromine	CNS stimulant
26.	Tea Leaves (Camellia Thea)	Leaves and leaf buds of *Thea sinensis,* Theaceae	caffeine, theobromine	CNS stimulant and diuretic
		8. Imidazole alkaloids		
27.	Pilocarus (Jaborandi)	Dried leaves of *Pilocarpus jaborandi,* Rutaceae	pilocarpine, pilosine	cholinergic (ophthalmic), used in treatment of glaucoma
		9. Steroidal Alkaloids		
28.	Ashwagandha (Asgandh)	Dried roots of *Withania somnifera,* Solanaceae,	withanine, somniferine, withanolide (steroid)	sedative and also as antirheumatic

Contd...

Sr. No.	Name of drug and synonym	Biological source	Active constituents	Uses
29.	Kurchi (Hollarrhena)	Dried bark of *Holarrhena antidysenterica*, Apocynaceae	conessine, isoconessine	Antiamoebic
30.	Veratrum (white and green hellebore)	Dried rhizomes of *Veratrum album* and *V. viride*, Liliaceae	germidine, protoveratrine A and B	Hypotensive, cardiac depressant
		10. Diterpene alkaloids		
31.	Aconite (Monkshood)	Dried roots of *Aconitum napellus*, Ranunculaceae	aconitine, neopelline	In treatment of rheumatism and sciatica
		11. Quinazoline alkaloids		
32.	Vasaka (Adulsa)	Leaves of *Adhatoda vasica*, Acanthaceae	vasicine and vasicinone	Antitussives, expectorant
		12. Miscellaneous alkaloidal drugs		
33.	Bhringraj (Eclipta) Maka	Entire herb of *Ecilipta-alba*, Asteraceae	Wedilolactone Ecliptine	Brain tonic, improves complexion, hepato protective
34.	Bhuiamla (Phyllanthus)	Aerial parts of *Phyllauthus amarus*, Euphorbiaceae	Phyllanthin	Anti-HIV, hepato protective
35.	Punarnava (Rakta-punarnava)	Dried herb of *Boerhaavia diffuse* Fam: Nictaginaceae	Punernavoside, boervinone	Jaundice, diuretic
36.	Shankhpushpi (Shankhini)	Aerial parts of *Canscora decussata*, Gentianaceae	Bitter substances, triterpenes	Nervous debility
37.	Dwarf morning glory (Aloe-weed)	Aerial parts of *Evolvulus alsinoides*, Convolvulaceae	Shankhpushpi β-sitosterol	Brain tonic, hair-growth stimulent
38.	Hemlock	Unripe fruits of *Conium maculatum*, Umbelliferae	Coniine conhydrine	Sedative antispasmodics

INDOLE ALKALOIDS

Most of the indole alkaloids are biosynthesized in the plants from amino acid tryptophan (see chapter 4). It is considered as one of the most important groups of alkaloids, as they yield the drugs with very useful therapeutic effects. The indole alkaloids normally contain two nitrogens, out of which one is present as indolic nitrogen and the other one is present in the position created by removal of two carbons from the p-position of the indole ring.

- The crude drugs containing indole nucleus are discussed here.

ERGOT

Synonyms

Ergot of Rye, Ergota.

Biological Source

Ergot is the dried sclerotium of a fungus, *Claviceps purpurea* Tulasne (Clavicipitaceae or Hypocraceae) developed in ovary of rye plant, *Secale cereale* Linne (Graminae). It contains not less than 0.19 per cent of the total alkaloids of ergot, calculated as ergotoxine, of which not less than 15 per cent consists of water soluble alkaloids of ergot, calculated as ergometrine.

Geographical Source

Switzerland, Yugoslavia, Hungary and Czechoslovakia.

History

The name of this drug is originated from a French word 'Argot,' which means fur and indicates the shape and attachment of the sclerotia to the infected rye spikes, like the fur which is attached to the body of the birds. Even in old days, ergot fungus was known to be a pathogen, infecting the rye fields in European countries and Russia. It is known that the toxic effects were observed owing to contamination of ergot with rye grains. The toxic symptoms were gangrene in the extremities and convulsions. In middle ages, such symptoms were reported and called as St. Antony's fire. After knowing the cause, it was called as ergotism which had severely occured in sixteenth century in Germany. Further, it was discovered that ergot has specific uses in obstetrics and came into wide use from nineteenth century onwards. In 1836, it was introduced in London Pharmacopoeia. The life history of fungus was studied and the name *Claviceps purpurea* was first coined by Tulasne in 1853.

Collection and Preparation of Ergot

Presently, ergot is produced by natural way i.e. cultivation of rye plants and subsequently infecting with this fungus, as well as, by artificial way i.e. saprophytic production.

For the natural way of production, rye plant is host and ergot is a parasite. It is known that more than 600 plants from different families of wild and cultivated grasses act as hosts for ergot fungus as a parasite or pathogen. The various other known species of this fungus are *C. microcephala, C. nigricans* and *C. paspali,* which can produce ergot. Among all the hosts, rye is the better host for the large scale production of ergot by way of quality and quantity.

Among the various stages of development of this fungus, sclerotial stage or a dormant stage contains the maximum amount of drug. For a systematic study, it is necessary to know the other developmental stages of fungus or more precisely the life history of ergot (Fig. 15.1).

Life Cycle of Ergot

The ovary of the rye plant at its base, gets infected by ascospores of the fungus in spring or summer season. The spread of ascospores to ovaries is influenced by wind and insects. After infecting, the ascospore germinate in the favourable conditions, like moisture and damp climate.

The germination of ascospores leads to formation of hyphal strands which go on invading the wall of ovary with the help of an enzyme. By this way, the hyphae form a soft, white mass of tissue over the surface of the ovary which is called as mycelium. The mycelium secrets a viscous and sugary fluid called honey-dew. At the same time, the hyphal strands produce asexual spores called conidiospores, which remain in a suspended form in honey-dew. Due to the sugary fluid i.e. honey-dew, the insects and ants are attracted which further help in the spread of the fungus to other host plants. This developmental stage is the sexual stage and called as **Sphacelial stage**.

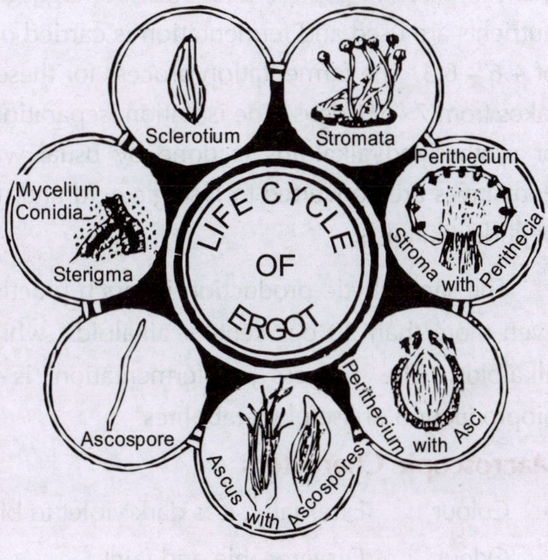

Fig. 15.1(a): Life Cycle of Ergot

The hyphae further invade into the deeper parts of ovary and slowly replace the entire tissue of ovary by a compact tissue called pseudoparenchyma which is hard and dark purple. It is called as **Sclerotium stage** and is considered as resting or dormant stage of the fungus and contains maximum amount of ergot alkaloids. If this sclerotium is left uncollected, it eventually falls on the ground and in the favourable season, i.e. spring gives out 'stromata' which are in the elongated form. Each stromatum has a globular head and a stalk. The head portion contains a large number of perithecia and every perithecium is like a flask shaped structure which contains a number of sacs, each sac containing the ascospores which are thread like in appearance. Ascospores are the sexual spores capable of inducing fresh life cycle of fungus by infecting the ovary of rye plant.

Selection of a correct strain of fungus (Claviceps purpurea), appropriate containers for preparing large scale ergot inoculum and an ideal nutrient medium are important requirements for commercial production of ergot. The various chemical races of fungus can produce only specific ergot alkaloids like ergotamine, ergometrine and ergotoxin in appreciable quantities from their sclerotia. The ascospores of this species with the specific chemical race are germinated on nutritive medium and by this way large bulk of conidiospores are formed. The suspension of this strain of fungus is sprayed on rye plants in large cultivated areas.

Apart from field cultivation, other method which is much practised is saprophytic production of ergot. This process was initiated in Japan by Prof. Abe. Saprophytic production is convenient in many ways as it eliminates the variation in yield due to weather conditions and production can be achieved throughout the year. For this method, various strains of ergot are used depending on the type of ergot alkaloid to be obtained. Claviceps paspali gives clavines and simple lysergic acid derivatives.

It is much easier to manufacture clavines and simple lysergic acid derivatives and then convert them to different peptide alkaloids, i.e. ergot alkaloids. For nutrition of cultures of fungus, specific nutrients are used and fermentation is carried out in temperature range of 20 C – 30 C and in a pH of 4.6 – 6.3. The fermentation process for these submerged cultures in shaking flasks or fermenters takes from 7 – 21 days. The isolation, separation and purification of simple lysergic acid derivatives or synthesized alkaloids is done by usual ways applied for other alkaloids. The lysergic acid derivatives are converted to lysergic acid and further partly synthesized into ergometrine and other peptide alkaloids.

The saprophytic production is much practised now-a-days, because mycelial dry weight gives even more than 20 per cent of alkaloids, while natural sclerotia contain less than 1 per cent of alkaloids. The process of fermentation is properly regulated or controlled for optimum bioproduction of useful metabolites.

Macroscopic Characters

Colour	:	Externally, it is dark violet to black. Internally, it is whitish or pinkish white.
Odour	:	Disagreeable and faint.
Taste	:	Unpleasant.
Size	:	The sclerotia are 1 – 3 cm in length and 1 – 5 mm in width.
Shape	:	Sclerotia are fusiform, triangular and usually tapering on both the ends.
Fracture	:	It is brittle with short fracture.

Extra Features

Longitudinal furrows and transverse cracks are present on each surface. Sclerotia are highly susceptible for fungal growth.

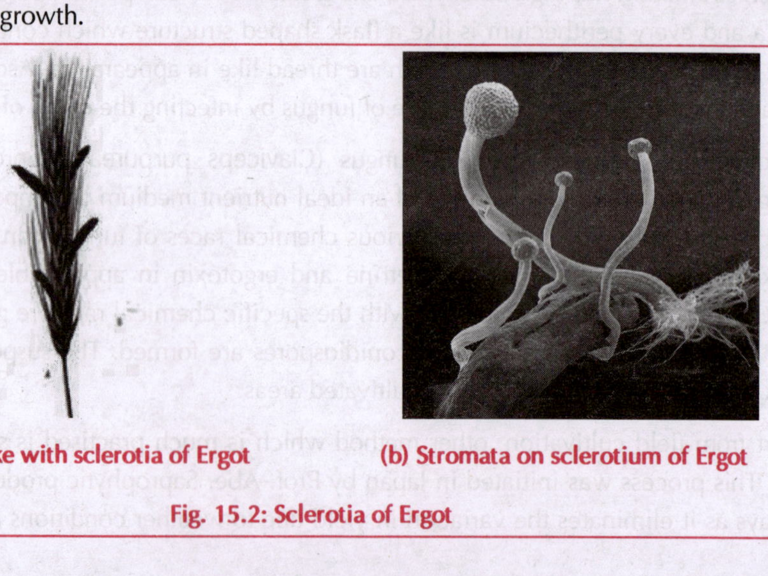

(a) Rye spike with sclerotia of Ergot (b) Stromata on sclerotium of Ergot

Fig. 15.2: Sclerotia of Ergot

Microscopic Characters (Fig. 15.3)

The outermost layer of the sclerotium is made up of few thin, flattened, polygonal cells of purple to dark brown colour, while inner part is made up of dense pseudo parenchymatous cells composed of chitin. The mycelial cells (central region) are round or oval, thick and with high refractive walls. They also contain cells with fixed oil. Sclerotium does not contain starch, calcium oxalate or any of the lignified tissue.

(a) T.S. (b) Longitudinal section

Fig. 15.3: Microscopic characters of Sclerotium of Ergot

Chemical Constituents

Ergot contains large number of potent indole alkaloids (0.1 – 0.25 per cent), which are derivatives of lysergic acid. Lysergic acid is present in its peptide derivative form and hence the alkaloids are also called as peptide alkaloids. The six pairs of alkaloids are broadly grouped into water soluble and water insoluble categories. Each pair contains laevo-form which is medicinally active, while dextro form is inert in action. The water soluble pair contains (–) ergometrine and its dextro part as ergometrinine. The water insoluble group is further divided into Ergotamine and ergotoxine group.

Table 15.3: Ergot alkaloids

(–) Laevorotatory alkaloids		(+) Dextrorotatory alkaloids
Ergometrine	Water-Soluble	Ergometrinine
Ergotamine	Water insoluble	Ergotaminine
Ergosine		Ergosinine
Ergocristine		Ergocristinine
Ergocryptine		Ergocryptinine
Ergocornine		Ergocorninine

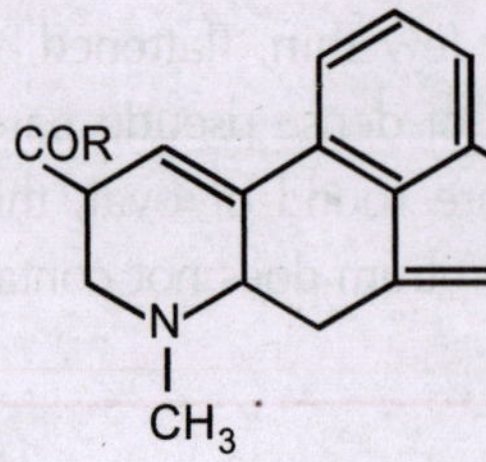

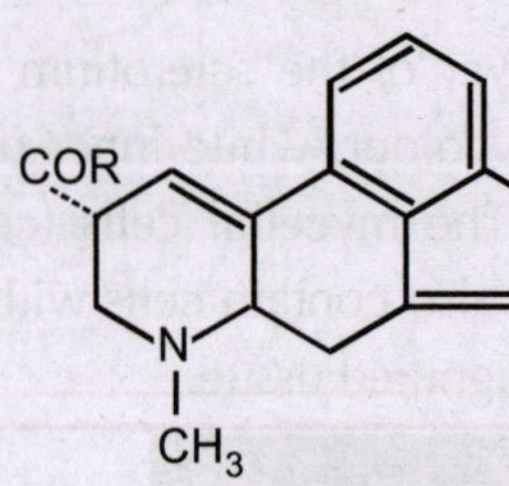

R		
R = – OH	Lysergic acid	Isolysergic acid
R = $-NH_2$	Lysergic acid amide **(Ergine)**	Isolysergic acid amide **(Erginine)**
R = – NH — C(H)(CH_2OH) — CH_3	Ergometrine	Ergometrinine
R = – N $(C_2H_5)_2$	Lysergic acid diethylamide (LSD)	

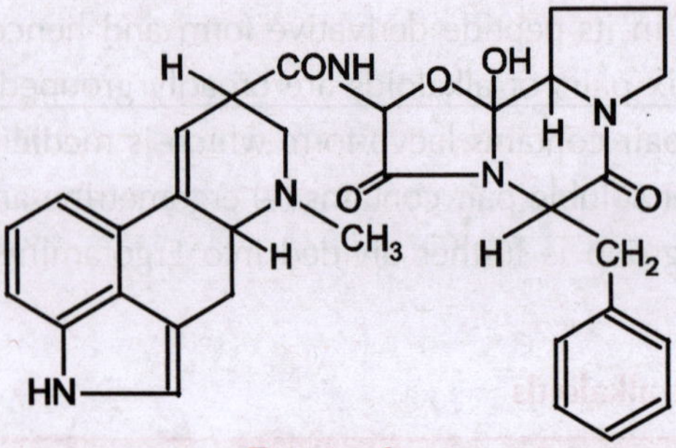

Ergotamine

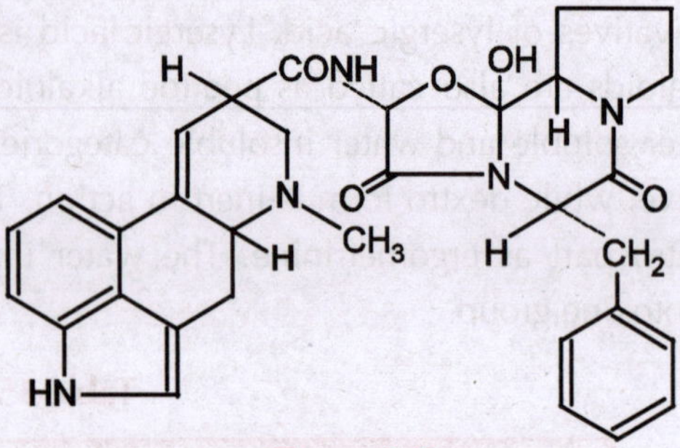

Ergocristine

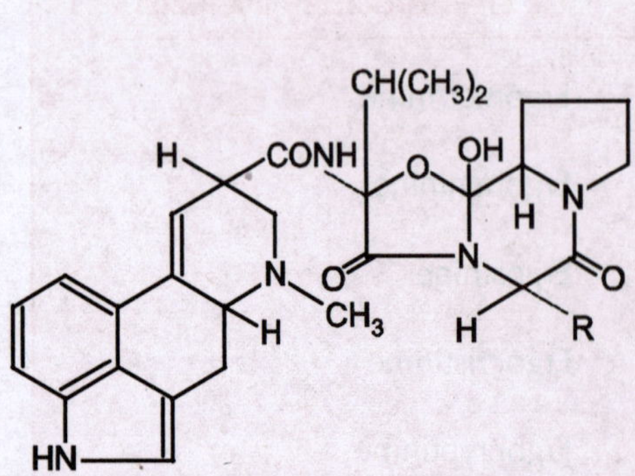

Ergocryptine

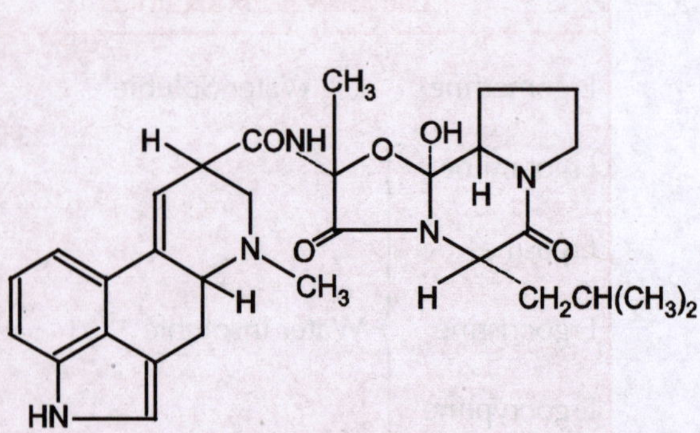

Ergosine

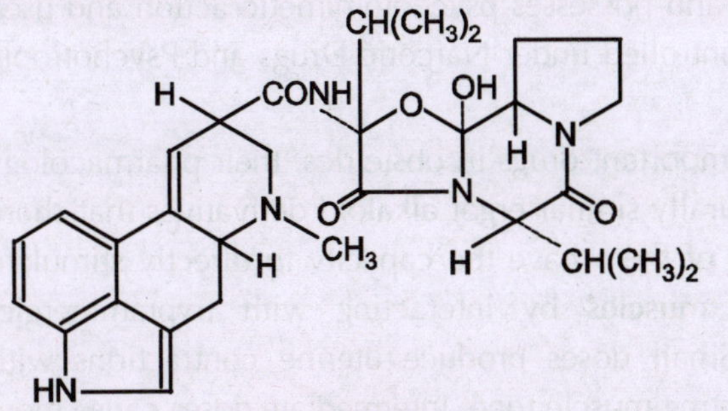

Ergocornine

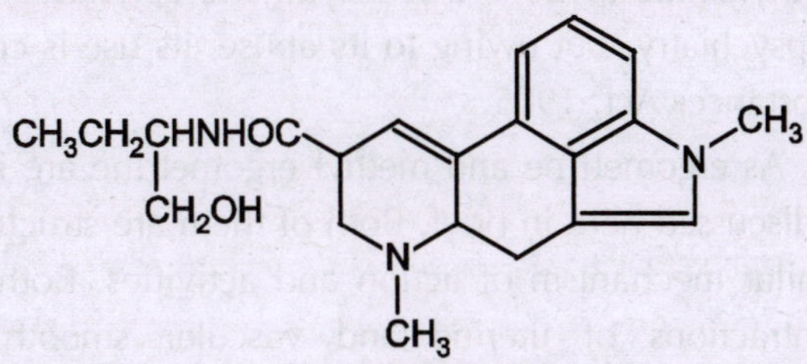

Methysergide

α-Ergocryptine R = $CH_2CH(CH_3)_2$

β-Ergocryptine R = $CH(CH_3)CH_2CH_3$

(Derivative of Lysergic acid used as migraine prophylactic)

Besides the alkaloids, ergot also contains pigments, ergosterol and fungisterol, histamine, tyramine, amino acids, acetyl choline, chitin, up to 30 per cent fixed oil and 8 per cent moisture.

Chemical Tests

1. Ergot powder gives a blue colour with p-dimethylaminobenzaldehyde (Van-Urk's reagent).
2. Ergot is treated with solvent ether and sulphuric acid and the filtrate obtained shows red violet colour in its aqueous layer, when treated with saturated solution of sodium bicarbonate.
3. Ergometrine gives a blue fluorescence in water.
4. Ergotamine responds to a specific test. Little quantity of ergotamine is dissolved in glacial acetic acid and ethyl acetate. A small portion of this is treated with sulphuric acid and shaken well by which blue colour with red tinge appears. By addition of ferric chloride, blue colour deepens, while red tinge becomes faint.

Standards of Quality

Moisture : Not more than 8.0 per cent

Ash : Not more than 7.0 per cent

Total alkaloids : Not less than 0.1 per cent as ergotoxin, of which

Water soluble alkaloids Ergometrine : Not less than 7.0 per cent

Standardisation

Ergot and its alkaloids are analysed by spectrocolorimetric method and bio assay is carried out for oxytocic activity.

Uses

Ergot and its alkaloids have many different uses. Now-a-days, ergot is not used as a whole, but the isolated alkaloids are used in therapeutics. Ergot and ergometrine maleate (in the United States, ergometrine is called as ergonovine) are used as oxytocic and sometimes used to enhance the labour pains in delivery cases and also to prevent the post partum haemorrhage. Ergotamine tartarate is used as a specific analgesic in treatment of migraine. It is given along with caffeine. Ergotoxine methanesulphonates (mesylates) are used in geriatric patients. Lysergic acid diethylamide (LSD) is a semisynthetic derivative, and possesses psychotomimetic action and used in psychiatry, but owing to its abuse, its use is controlled under Narcotic Drugs and Psychotropic Substances Act, 1985.

As ergometrine and methyl ergometrine are important drugs in obstetrics, their pharmacology is discussed here in brief. Both of them are structurally similar ergot alkaloid derivatives that share similar mechanism of action and activities. Both of them have the capacity to directly stimulate contractions of uterine and vascular smooth muscles by interacting with tryptaminergic, dopaminergic and alpha-adrenergic receptors. Small doses produce uterine contractions with increased force and frequency and with normal resting muscle tone. Intermediate doses cause more forceful and prolonged contractions with an elevated resting muscle tone, while large doses cause sustained contractions and tetany. Small doses of these drugs are used after delivery to control bleeding and maintain uterine firmness. Methyl ergometrine is generally preferred, because it causes less hypertension.

Dose

1. Ergometrine maleate and methyl ergometrine maleate, oral 200 – 400 g, 2 – 4 times a day, i.m./i.v. → 200 g every 6-8 hours.
2. Ergotamine tartarate → 1 – 2 mg sublingual; 250 – 500 g intramuscular/subcutaneous.
3. Ergotoxine methanesulphonates → 0.5 mg sublingual, 4 - 6 times a day

Prepared Ergot

The powdered and immediately defatted ergot is called as prepared ergot. It is required to contain 0.19 per cent of total ergot alkaloids, calculated as ergotoxine of which 15 per cent should be ergometrine. Its dose is 0.15 - 0.5 g by oral route.

Storage

Ergot should be dried thoroughly and kept in entire form in cool place. It should be stored in well closed containers. Ergot alkaloids are very sensitive to moisture and hence thorough drying of ergot is necessary. The broken sclerotia are very susceptible to the fungal growth and hence broken pieces should not be stored. Ergot alkaloids are sensitive to light and temperature. Hence, the drug is stored at low temperature in cool place away from light. If powdered ergot is required to be stored, it should be defatted first and then stored or, otherwise, decomposition of active constituents take place.

NUX VOMICA

Synonyms

Crow-fig, Semen strychni, Nux vomica seed.

Biological Source

Nux Vomica consists of dried ripe seeds of Strychnos nux vomica Linn, family Loganiaceae. It should contain not less than 1.2 per cent of total alkaloids calculated as strychnine.

Geographical Source

It is indigenous to East India and is largely collected from forests in Sri Lanka, Northern Australia and India. It is found abundantly in South India i.e. in Tamil Nadu, Kerala and on Malabar Coast. It is also available in the forests of Bihar, Orissa, Konkan, Mysore and Gorakhpur.

History

Because of its poisonous nature, this drug was used in sixteenth century to kill the animals. It is derived from a Greek word Strychnos, meaning poisonous and nux vomica indicates a nut with vomiting effects. Strychnine and brucine, the alkaloids present in this drug, are some of the few alkaloids which were first isolated. It was introduced in medical practice from 1640 onwards.

Collection

In India, the entire drug is collected from wild grown plants by the local tribal community. The nux vomica tree is found throughout the tropical area, 1300 m above the sea level. The plants are about 10 - 12 metres in height with a crooked trunk and several branches. The leaves are orange, oppositely arranged, with oval shape, entire margin and acute apex. The flowers are greenish-white and the bark is greyish to yellow. Fruits of the plants are orange yellow, berries of normal size. Each fruit contains about 4 - 5 seeds and heavy bitter pulp. The ripened fruits are collected and seeds are freed of the pulp. They are washed with water thoroughly. Unripened seeds are separated by the floating test in water. The seeds are dried on mat and packed in gunny bags for marketing. The collection of the fruit and seeds is carried out from November to February. In India, about 15,000 tones of seeds are collected annually. Seeds, pure and crude alkaloids of Nux-vomica are regularly exported from India.

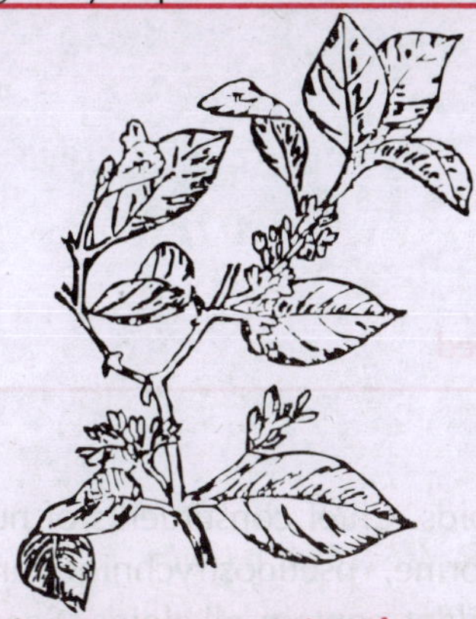

Fig. 15.4: Nux Vomica Plant

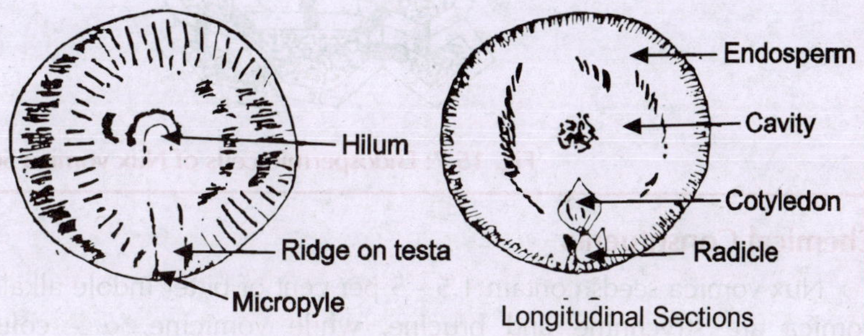

Fig. 15.5: Nux Vomica Seed

Macroscopic Characters

Colour	:	Seeds are greenish-brown
Odour	:	None
Taste	:	Intensely bitter
Size	:	Seeds are 10 to 30 mm in diameter and 4 to 6 mm in thickness.
Shape	:	The seeds are disc shaped, somewhat flat or irregularly bent and concavo-convex. Margin of the seeds is rounded.

Extra Features

Surface of the seeds is silky due to the radially arranged, densely covered, closely appressed unicellular lignified covering trichomes. The presence of endosperm, embryo and cotyledons can be confirmed in the L. S. of the seed.

Microscopic Characters

The epidermis consists of strongly thickened, pitted and lignified trichomes. Epidermis is followed by a layer of collapsed cells. Endosperm is characterised by thick walled polyhedral unlignified cells with plasmodesma, aleurone grains and oil globules. Calcium oxalate crystals and starch grains are absent in drug (Figs. 15.6 and 15.7).

Nux vomica Seeds

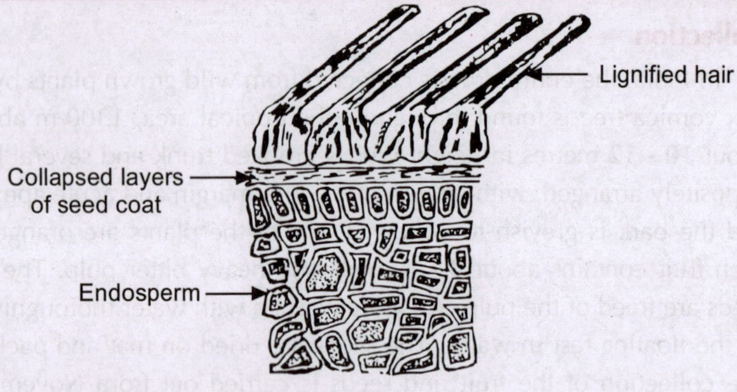

Fig. 15.6: T.S. of Nux vomica

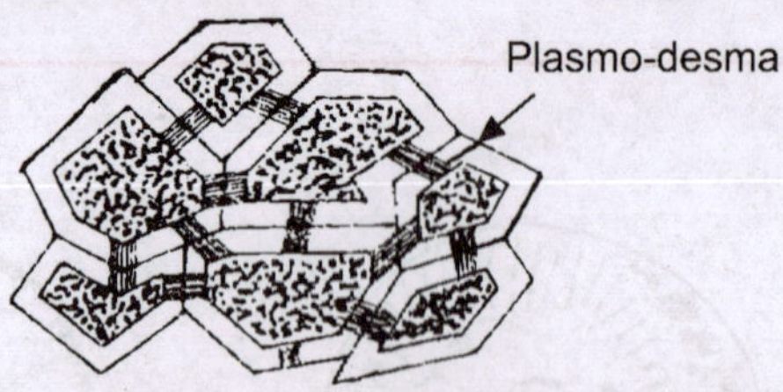

Fig. 15.7: Endospermic cells of Nux vomica Seed

Chemical Constituents

Nux vomica seeds contain 1.5 - 5 per cent of bitter indole alkaloids. Chief constituents of nux vomica are strychnine and brucine, while vomicine, α - colubrine, pseudostrychnine and strychnicine are also present. Apart from seeds, other parts of the plant contain alkaloids. Seeds also contain 3.0 per cent of fat. Bark contains brucine and traces of strychnine. Wood and root of the plant also contain strychnine.

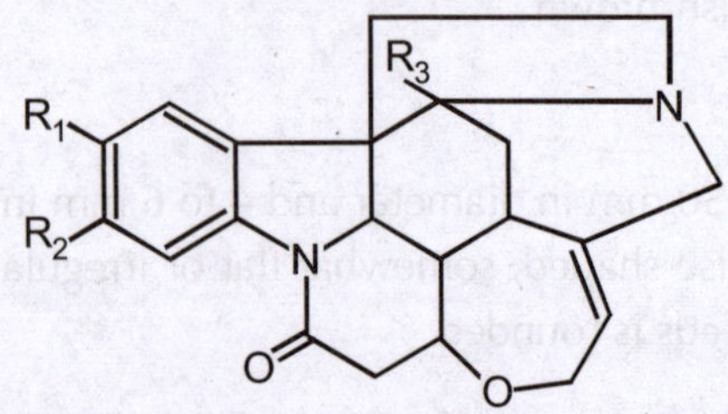

Strychnine	$R_1 = R_2 = R_3 = H$
Brucine	$R_1 = R_2 = ROCH_3$, $R_3 = H$
α - colubrine	$R_1 = H$, $R_2 = OCH_3$, $R_3 = H$
β - colubrine	$R_1 = OCH_3$, $R_2 = R_3 = H$

The other minor, but, chemically related alkaloids are isostrychnine, N-oxystrychnine, protostrychnine, β-colubrine, and novacine.

Nux vomica also contains a glycoside viz. loganin, chlorogenic acid and fixed oil.

The alkaloids can be isolated with the use of dilute sulphuric acid and lime. Strychnine sulphate is meagerly soluble in water and alcohol.

Chemical Tests

The thin sections of nux vomica seed are defatted and the following tests are performed.

1. Stain the transverse section of nux vomica with ammonium vanadate and sulphuric acid Manddin's reagent. The endospermic cells become purple due to the presence of strychnine.
2. Stain the transverse section of nux vomica with concentrated nitric acid. Endospermic cells take yellow colour due to the presence of brucine.
3. Strychnine with sulphuric acid and potassium dichromate gives violet colour which turns to red and finally yellow.

Standards of Quality

Foreign organic matter	: Not more than 1.0 per cent
Ash	: Not more than 3.0 per cent
Strychnine content	: Not less than 1.2 per cent

Uses

Due to its bitter taste, nux vomica is used as bitter stomachic and tonic. It is a stimulant to central nervous system. It increases the blood pressure and is recommended in certain forms of cardiac failure. It stimulates respiratory and cardiovascular systems. Brucine possesses very less physiological actions and is about one-sixth in potency as compared to strychnine. But as far as the bitterness is concerned, it is four times bitter than strychnine. Brucine is used for denaturing alcohol and inedible fats, as a standard for bitterness and as a dog poison.

Adulterants

1. Dried seeds of Strychnos nuxblanda Hill, are used as adulterant to nux vomica seeds. These are similar in size, pale in colour with a distinct ridge on the edge of the seeds. Nuxblanda seeds are regular in shape and contain traces of alkaloids. Seeds are tree of bitterness.

2. Dried seeds of Strychnos potatorum are another adulterant to authentic drug. The seeds are also known as clearing nuts. They are smaller and thicker with yellowish buff colour. Seeds contain diaboline and traces of strychnine and brucine. Seeds are tree of bitterness.

Allied Drugs

1. The seeds of Strychnos wallichiana are used as substitute to nux vomica, as their alkaloidal content and composition are comparable to the genuine drug.

2. The dried seed of Strychnos ignatii is another allied drug. Seeds are about 2.5 cm in diameter, ovoid in shape, dark green in colour with unlignified detached trichomes. It contains 2.5 per cent to 3 per cent alkaloids of which 60 per cent is strychnine. The seeds are used for the manufacture of strychnine.

PHYSOSTIGMA

Synonyms

Calabar bean, Ordeal bean

Biological Source

It is the dried ripe seed of Physostigma venenosum Balfour, belonging to family Leguminosae. It contains not less than 0.15 per cent of alkaloids of physostigma.

Geographical Source

West Africa, Gulf of Guinea

Macroscopic Characters (Fig. 15.8)

The drug is obtained from the legumes of perennial woody climber. The beans are flattened and reniform in shape with extremely hard, brown coloured. The testa of the seeds has a grooved hilum.

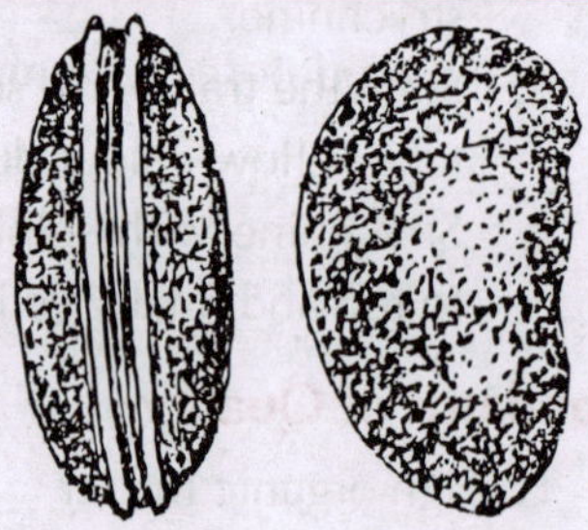

Fig. 15.8: Physostigma seeds

Chemical Constituents

The drug contains indole alkaloids, amongst which physostigmine, also called as eserine, is most important and it is present in the cotyledons.

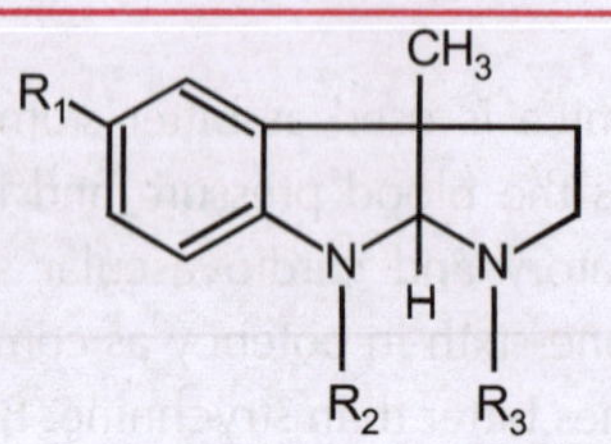

Physostigmine	$R_1 = CH_3NHCOO$,	$R_2 = CH_3$,	$R_3 = CH_3$
Eseramine	$R_1 = CH_3NHCOO$,	$R_2 = CH_3$,	$R_3 = CONHCH_3$
8-norphysostigmine	$R_1 = CH_3NHCOO$,	$R_2 = H$,	$R_3 = CH_3$
Eseroline	$R_1 = OH$,	$R_2 = CH_3$,	$R_3 = CH_3$

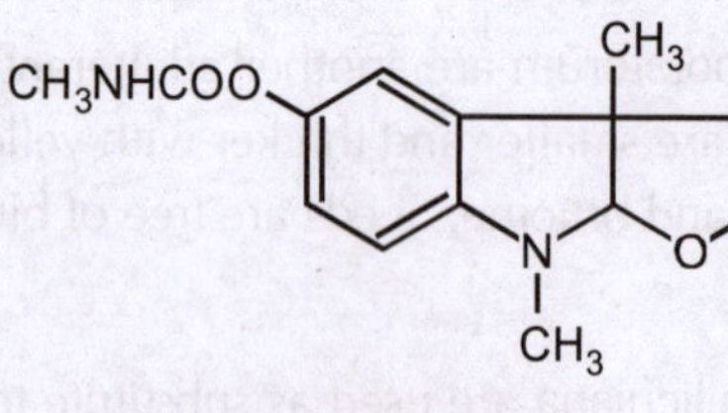

Physovenine

It is to be preserved properly from air and light as it is oxidized into rubeserine, a red compound, when exposed to air. Along with physostigmine, the drug also contains the alkaloids eseramine, geneserine, physovenine, calabatine, calabasine, isophysostigmine and N - 8 - norphysostigmine.

Uses

Physostigmine shows parasympathomimetic (ophthalmic) activity. It is categorised as an anticholinesterase agent. It enhances the cholinergic activity by inhibiting cholinesterase, and hence parasympathetic activity is sustained by physostigmine administration. It is used for contracting the pupils of the eye to counteract the poisoning caused by anticholinergic agents. It is also used as an antidote for reversing most of the cardiovascular (tachycardia and arrhythmia) and CNS effects due to overdosage with tricyclic antidepressants.

Dose

For ophthalmic purpose, topically 0.1 ml of 0.25 - 0.5 per cent solution.

For other purposes, intramuscularly or intravenously 1 - 4 mg slowly.

RAUWOLFIA

Synonyms

Rauwolfia root, Serpentina root, Chhotachand, **Sarpgandha**

Biological Source

Rauwolfia consists of dried roots of the plant known as Rauwolfia serpentina Benth, belonging to family Apocynaceae. Serpgandha contains not less than 0.15 per cent of reserpine and ajmalcine, calculated on dried basis.

Geographical Source

Several species of Rauwolfia are found distributed in the tropical regions of Asia, America and Africa. Commercially, it is produced in India, Sri Lanka, Myanmar, Thailand and America. In India, it is cultivated in Uttar Pradesh, Bihar, Orissa, Tamil Nadu, West Bengal, Karnataka, Maharashtra, and Gujarat.

History

This drug is known to Indian System of Medicine since last many centuries. Because of snake like shape of the drug, it has been known as 'Sarpagandha'. It has found its place as an important drug in treatment of insanity and snake bite since traditional times. But the drug came into limelight only after the isolation of reserpine, its most significant alkaloid, in 1952 by Mueller. Since then a large interest has been generated regarding the activity of this drug.

Cultivation and Collection of Rauwolfia

Under wide range of climatic conditions, rauwolfia grows luxuriantly. However, it flourishes in hot humid condition and grows satisfactorily in shade. In wild state, it grows in variety of soils. But for cultivation, clay loamy soil with large amount of humus and good drainage are supposed to be ideal. The pH of the soil should be acidic and around 4. The temperature range for cultivation is 10 to 38 C. Rainfall should be in the range of 250 - 500 cm. Soils containing large amount of sand make the plants more susceptible to diseases.

In can be propagated by various methods, such as by seeds, roots, cutting, root stumps, etc. The propagation from seeds is usually the method of choice. The healthy seeds are sown into the nursery beds. The rate of germination of seeds is very low, hence sufficient quantity of the seeds are sown. Sowing is done in the month of May or at the break of monsoon. The seedlings are then transplanted in the month of August at a distance of 16 to 30 cm. The plants are provided with various chemical fertilizers and manures. The chemical fertilizers include ammonium sulphate, urea; while the manures include, generally, the bone-meal. The plants are kept free from weeds. When the plants are about 3 to 4 years old, they are uprooted. The roots are cut properly, washed so as to remove the earthy matter and dried in air.

It needs about 5 kg of seeds to produce the seedlings sufficient to cover the area of one hectare after transplantation. The average yield of roots per hectare is 1200 kg. It may vary, depending upon the soil, climatic conditions and age of the plant.

Macroscopic Characters

Colour	:	Roots and root bark are greyish yellow to brown and wood, pale yellow in colour.
Odour	:	Odourless
Taste	:	Bitter
Size	:	About 10 to 18 cm long and from 1 to 3 cm in diameter.
Shape	:	Roots are sub-cylindrical, slightly tapering and tortuous.

Fracture is short and irregular. The transversely cut surface is white, dense with finely radiating xylem.

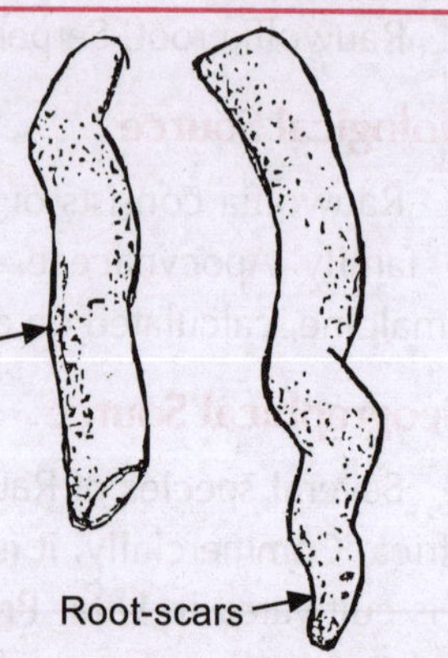

Fig. 15.9: Pieces of Rauwolfia Roots

Extra Features

Roots are rough with longitudinal marking and slightly wrinkled surface. Rootlets are usually absent, but few small circular root scars with tetrastichous arrangements are seen.

Microscopic Characters

Fig. 15.10: Rauwolfia Twig

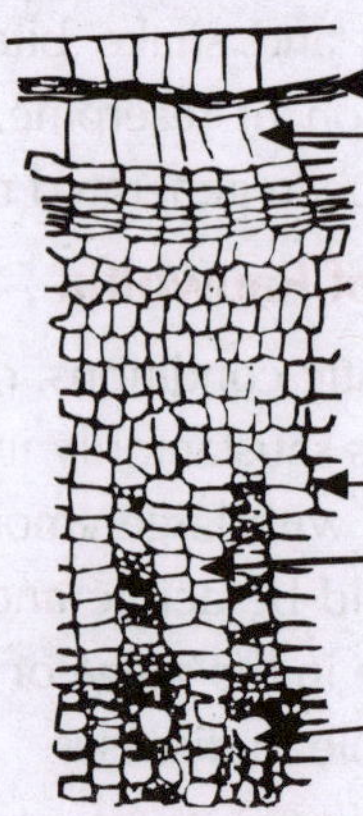

Fig. 15.11: T.S. of Rauwolfia Root

The cork is made up of stratified cells followed by phelloderm of few rows of parenchyma. Phloem is narrow, parenchymatous with small scattered sieve tissue. Parenchyma contains starch grains and few latex cells, with brown resinous matter. Secondary phloem contains calcium oxalate crystals. Xylem is about 4/5th of the diameter of the root and consists of vessels, tracheids, wood parenchyma and wood fibres. Xylem vessels are elongated up to 350 in length and 50 in width and contain simple or bordered pits. Stone cells and phloem fibres are absent.

Chemical Constituents

About 30 indole alkaloids have been reported in drug and total alkaloidal content of rauwolfia roots ranges from 0.7 - 3 per cent, depending upon the source. Alkaloids are concentrated mostly in the bark of the roots. The alkaloids of rauwolfia are broadly classified into the following types, (a) indole alkaloids, (b) indoline alkaloids, (c) indolenine alkaloids (d) oxyindole alkaloids and (e) pseudo indoxyl alkaloids. The important alkaloid of rauwolfia is reserpine. Apart from the alkaloids, it also contains oleo-resin, phytosterol, fatty acids, alcohol and sugars. The other alkaloids present in the drug are ajmaline, ajmalicine, rauwolfinine, rescinnamine, reserpinine, yohimbine, serpentine and serpentinine. The major alkaloids reserpine and rescinnamine are esters derived from methyl reserpate and trimethoxybenzoic acid in reserpine and trimethoxycinnamic acid in case of rescinnamine. Syrosingopine is methyl carbethoxy syringoyl reserpate.

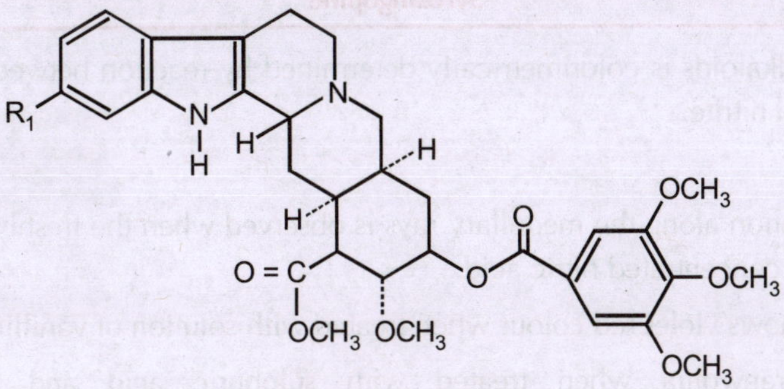

Deserpidine R_1 = H

Reserpine R_1 = OCH_3

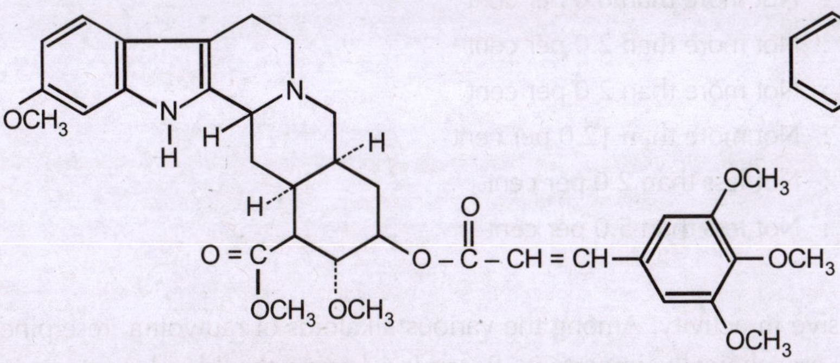

Rescinnamine

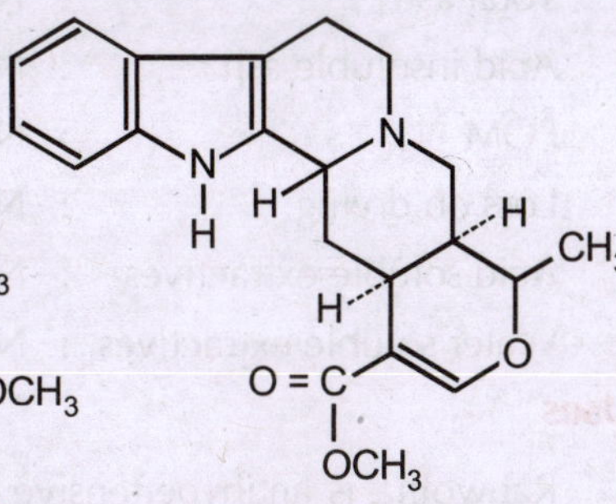

Ajmalicine

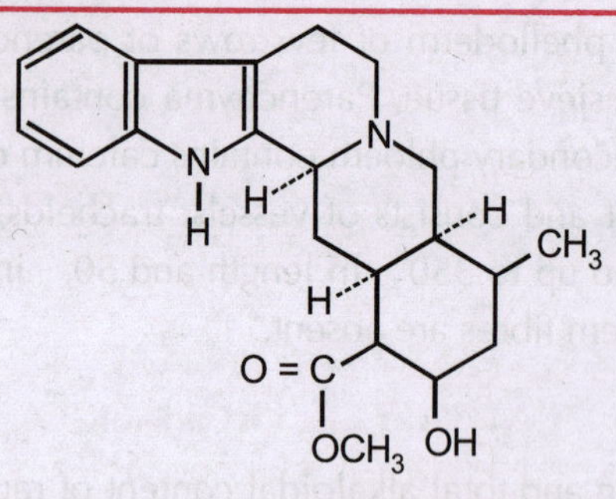

Yohimbine

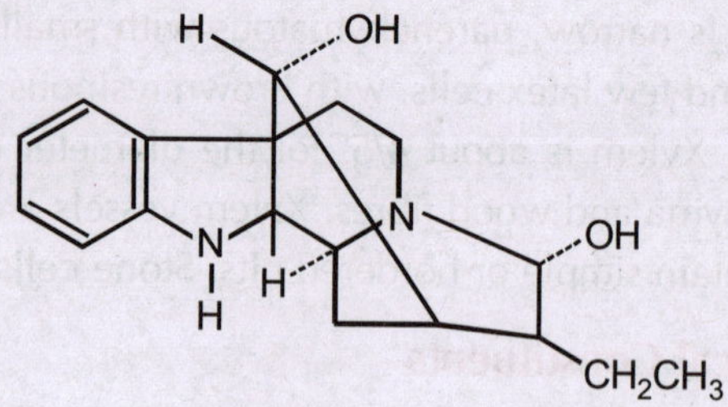

Ajmaline

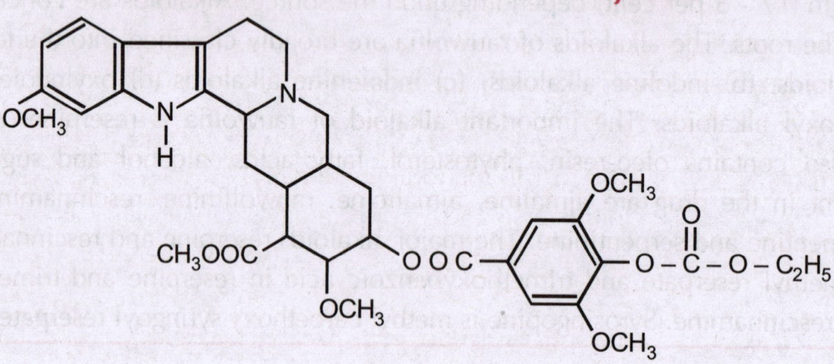

Syrosingopine

Reserpine like alkaloids is colorimetrically determined by reaction between acidic solution of alkaloids and sodium nitrite.

Chemical Tests

1. A red coloration along the medullary rays is observed when the freshly fractured surface is treated with concentrated nitric acid.
2. Reserpine shows violet red colour when treated with solution of vanillin in acetic acid.
3. Powdered ranwolfia when treated with sulphuric acid and p-di-methyl amino benzaldehyde, develops violet to red colour. (Test is for modle alkaloids).

Standards of Quality

Total ash	:	Not more than 8.0 per cent
Acid insoluble ash	:	Not more than 2.0 per cent
FOM	:	Not more than 2.0 per cent
Loss on drying	:	Not more than 12.0 per cent
Acid soluble extractives	:	Not less than 2.0 per cent
Water soluble extractives	:	Not less than 5.0 per cent

Uses

Rauwolfia is antihypertensive in activity. Among the various alkaloids of rauwolfia, reserpine, rescinnamine and ajmalicine are clinically important. Reserpine lowers the blood pressure by depleting stores of catecholamines at nerve endings. It prevents re-uptake of nor epinephrine at storage sites, allowing enzymatic destruction of neuronal transmitter. It is used to treat mild essential hypertension and may be an effective adjunct to the treatment of more severe hypertension.

Because of the tranquillising effects, the drug is used in mild anxiety conditions and reserpine in some of the neuropsychiatric disorders.

Rescinnamine is also used as antihypertensive, but it causes mental depression in higher doses.

Deserpidine is used as antihypertensive and tranquilliser. It shows very less side effects.

Ajmalicine, though less in quantity, has the uses in treatment of circulatory diseases, in relief of obstruction of normal cerebral blood flow.

Syrosingopine shows peripheral effects similar to reserpine. It has less sedative actions and it is used for the treatment of mild or moderate hypertension.

Dose

Rauwolfia roots	:	100 to 150 mg (oral twice daily)
Reserpine	:	Initial dose 250 g once a day (oral)
		Maintenance dose 100 - 250 g once a day
Rescinnamine	:	500 g oral twice a day (initial dose); 250 g oral daily maintenance dose.

Allied Drugs and Substitutes

The rauwolfia species are not limited only to South East Asian region, but also found in Africa, Central and South America, New Guinea, Hawaii, New Caledonia, Australia and far east regions. It is reported that rauwolfia has about 86 different species. From the medicinal point, the most pertinent to mention here is R. vomitoria, which is known as African rauwolfia. It is used as a commercial source for the preparation of reserpine. The other known species of rauwolfia from Africa are R. caffra, R. cumminsfi, R. mombasiana, R. oreogiton, R. obscura, R. rosea and R. volkensii. All of them contain reserpine.

The other rauwolfia species with reserpine content are R. tetraphylla and R. nitida. Catharanthus roses contain ajmalicine.

Pausinystalia yohimba, known as yohimbe bark, contains yohimbine, which is structurally related to reserpine.

The root bark of Alstonia venenata and A. constricta also contain reserpine. The various species of Aspidosperma genus contain indole alkaloids which resemble to those from rauwolfia.

The various other species with which rauwolfia is found to be substituted are Rauwolfia tetraphylla, R. densiflora and R. vomitoria (African rauwolfia). R. densiflora contains sclerenchyma, while R. tetraphylla has uniform cork, abundant sclereids and fibres, but devoid of rescinnamine. The root of R. vomitoria has 5 discontinued bands of sclerenchyma and very large vessels.

VINCA

Synonyms

Catharanthus, Periwinkle, Vinca minor.

Biological Source

It is the dried whole plant of Catharanthus roseus, belonging to family Apocynaceae. It is also known as Vinca rosea.

Geographical Source

It is probably indigenous to Madagascar. It is cultivated in South Africa, India, U.S.A., Europe, Australia and Caribbean islands as an ornamental plant, as well as, for its medicinal properties.

History

Probably the use of vinca has been known since B.C. 50 in European countries as antidysentric, antihaemorrhagic, diuretic and wound healing. This plant was used in the form of 'tea' for treatment of diabetes in Jamaica and in Brazil for toothache. This plant was first scientifically investigated by Canadian workers Noble, Beer and Cutts. During these studies, it was found that it does not have any oral hypoglycemic principle, but contains alkaloid possessing antileukemic principle and the alkaloid was named as vincaleucoblastine. Because of such activity, the plant was thoroughly investigated at M/s Eli Lilly by Svoboda and his colleagues and they reported four dimeric indole alkaloids, viz. vinca-leucoblastine, leurosine, leurosidine and leurocristine. All these compounds exhibited anticancer activities. Some changes were later reported for the names of these compounds such as vincaleucoblastine to vinblastine and leurosine, leurosidine and leurocristine to vinleurosine, vinrosidine and vincristine respectively.

Cultivation, Collection and Preparation for the Market

Vinca grows all over India up to 500 metres. It is grown well in tropical and subtropical area in South Indian and North Eastern States of India. Except the highly alkaline or water-logged soil, vinca does not require any special conditions of soil. It favourably grows in light sandy soils, rich in humus. The rainfall of about 100 cm is most suitable for it. For the propagation, the fresh seeds are used and sown in nurseries or some times direct sowing is also done. For direct sowing, about 2.5 kg of seeds per hectare are required. They are mixed with 10 times quantity of sand and sown in monsoon in rows of 45 cm apart. When the plants are sufficiently grown up, they are thinned out and a distance of about 30 cm is left between two plants. Nursery sowing is found to be economical. In February or March, they are sown in nursery and transplanted in open fields after 2 months when they achieve 6 - 7 cm height. They are transplanted in open fields at 45 cm 30 cm distance and about 74,000 plants per hectare are necessary. The plants do not need much water supply and are drought resistant. Though, the plant does not require any special supply of fertilizers, a mixture of nitrogen, phosphorus and potassium gives favourable results. Farmyard manure is also sometimes used. Weeding is done periodically and the leaf strippings are done 6 and 9 months after sowing. The stems are cut about 7 - 8 cm above the ground level after one year of growth and the leaves, stems and seeds are separated and air dried. For collection of roots, the field is profusely irrigated and roots are dug out by ploughing, which are further washed, dried in shades and packed in bales. The seeds are collected from matured fruits for next propagation. The yield of dried roots, stems and leaves per hectare in irrigated land is 1-5, 1 and 3 tones respectively.

As the plant contains very less percentage of alkaloids, it is not used as a galenical, but for extraction of the alkaloids. The extraction and separation procedure for alkaloids is based on their separation into soluble and insoluble tartarates in other solvents. Due to this vinblastine, vincristine and other weak bases are separated and then fractionated with the help of column chromatography, using alumina as adsorbent.

By using tissue culture technique, some of the vinca dimeric indole alkaloids like catharanthine and ajmalicine have been isolated. But yet, it has not, become the procedure for commercial utilisation. India is exporting vinca roots regularly.

Macroscopic Characters

Colour : Vinca leaves are green in colour.

Odour : Characteristic

Taste : Bitter.

Size : 2 – 3 1.0 – 1.5 cm

Shape : Ovate or Oblong

Extra Features : Leaves are shiny with rounded mucronate apex, simple glabrous. Acute apex, petiolate unicostate, reticulate, entire leaves are also brittle with glossy appearance.

Fruits are follicular with many black seeds.

Flowers are bractate, pediculate hermophrodite, normally 2-3 cm in cymose, axillary clusters violet-pink in colour.

Vinca-alba bears white flowers.

Vinca is an errect, pubescent herb with branched tap roots. Roots are pale grey in colour. Herb is upto one metre in length and 20 – 25 cm wide.

Microscopic Characters

Fig. 15.12: Vinca plant

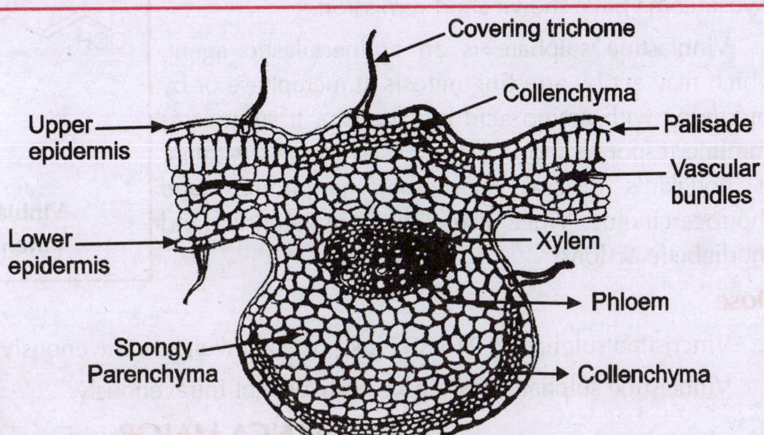

Fig. 15.13: T.S. of Vinca Leaf

Upper surface shows presence of single layer of rectangular celled epidermis with unicellular covering trichomes. Palisade is made up of single layer beneath upper epidermis and contains compact elongated cells. Spongy parenchyma is 5 - 8 layered with intercellular spaces. Midrib shows presence of collenchyma below the upper epidermis and above the lower epidermis. Xylem and phloem are present in the centre. Cruciferous stomata are present more frequently on lower epidermis. Calcium oxalate crystals are absent.

Chemical Constituents

A large number of indole alkaloids are present in vinca. Out of them, about 20 dimeric indole-dihydroindole alkaloids possess oncolytic activity, and among them, vincristine and vinblastine are most significant. Vinblastine contains indole alkaloid part called catharanthine and dihydroindole alkaloid part called vindoline. The other alkaloids present in vinca are ajmalicine, lochnerine, serpentine and tetrahydroalstonine. It requires about 500 kg crude drugs to extract out 1 g of vincristine, because of its extreme low content, viz. 0.0002 per cent. This makes these alkaloids very costlier and hence, the efforts for their synthesis are under attempts. From their structure, the five-ring dihydroindole system is present in few other natural drugs. Therefore, the attempts towards the synthesis of four-ring indole system are going on presently. Such two systems can be further coupled.

A new variety under the name vinca major has reached the market in all vinca producing countries, which is with big leaf size.

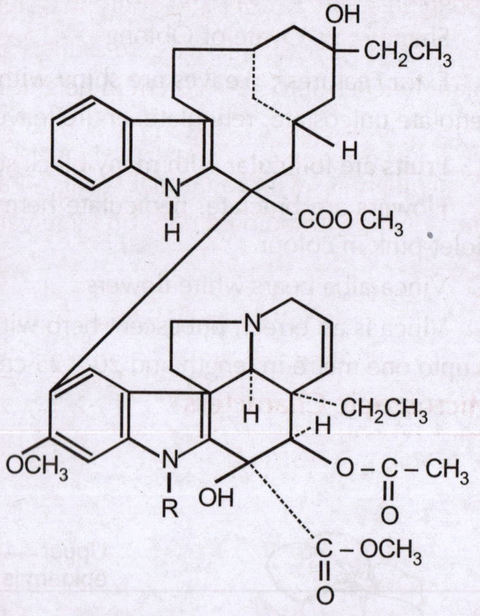

Vinblastine **R = CH_3**

Vincristine **R = CHO**

Uses

Vinca is used to extract vincristine, vinblastine and ajmalicine. Vincristine sulphate is an antineoplastic agent which may act by arresting mitosis at the metaphase. It is given intravenously in the treatment of acute leukemia of children; some childhood leukemias are also responded. In adults, hodgkin's diseases, reticulum cell sarcoma, lymphosarcoma and myosarcoma have shown short remission.

Vinblastine sulphate is an antineoplastic agent, which may act by arresting mitosis at metaphase or by interfering with amino acid metabolism. It suppresses immune response and is mainly used in the treatment of hodgkin's disease and other lymphomas and choriocarcinoma. Vinca also exhibits hypotensive and antidiabetic actions.

Dose

1. Vincristine sulphate: 10 to 30 g/kg of body weight intravenously, but maximum upto 2 mg.
2. Vinblastine sulphate: 100 g/kg body weight intravenously.

VINCA MAJOR

Synonyms

Blue periwinkle, large periwinkle, big leaf periwinkle.

Biological Source

It is the flowering plant vinca-rosea family apocyanaceae. The species name major refers to the large size. Compared to vinca rosea, which is a normal garden flowering plant.

Geographical Source

It is found in Southern Europe, Northern Africa, Spain and France also in Turkey Western Caucacus and Himalaya ranges. Plant grows well in fall sun and deep shade. It is an invasive species.

Macroscopical Characters

It is an ever green, perennial, frequently used in cultivation as ground cover.

Colour : Leaves are green.

Odour : Odourless

Taste : Bitter

Size : 3 - 9 2 - 4 cm

Extra features : **Leaves** are opposite, orbicular at the base of stems and laneolate at apex, glossy, leathery texture and entire in shape. Distinctly ciliate margin and hairy petiole, hairs short.

Flowers are harmaphrodite axillary, solitary violet purple 3 – 5 cm long. Corolla five-lobed plant reaches upto 50 – 70 cm in height. Flowering from spring to autumn.

Microscopical Characters

Epidermis with aniclinal walls. Parxoylic stomata on lower surface oil, trichomes are unicellular with pointed ends on large vains. Calcium oxalate crystals absent.

Chemical Constituents

It contains alkaloids and tannins.

Standards of Quality

Total ash	:	Not more than 10.0 per cent
Acid insoluble ash	:	Not more than 3.0 per cent

Uses

Similar to vinca rosea.

ISOQUINOLINE ALKALOIDS

CURARE

Synonym

South American arrowroot poison.

History

It is reported by Sir Walter Raleigh (1596) and Pedro de Ceiza de Leon (1553) that the plant preparation, called by various names 'curari', 'wourari', 'urari', etc. were used in battle for arrows and spears and also for killing animals. Sir Benjamin Brodie in 1812 showed that this extract causes asphyxation and Claude Bernard in 1844 suggested that curare had a capacity to interfere with the passage of nerve impulse from brain to muscles. Tubocurarine was isolated by Boehm in 1898.

Previously, the extract was described by different names depending on the method used for their storage like calabash curare (stored in the shell of a gourd), pot curare (earthenware containers) and tube curare (bamboo tuber.)

Biological Source

Some of the plants found in Amazon region like Brazil and Peru, contain highly poisonous substances. The extract of all such plants is called curare.

These plants belong to family Loganiaceae and Menispermaceae. Some of the prominent members yielding curare are Chondrodendron tomentosum, Strychnos castelnaea, S. crevauxii, S. toxifera, etc.

All these plants contain alkaloid tubocurarine. Similar alkaloids, but slightly differing from tubocurarine, are obtained from genera like Stephania, Menispermum, Cyclea, Pleogyne, Cocculus, Anomospermus from family Menispermaceae and Daphandra (Monimiaceae), Magnolia (Magnoliaceae) and Berberis (Berberidaceae).

Geographical Source

All such plants are available from geographical regions like Brazil, Peru, Columbia, Guiana, Venenzuela, etc.

Description

Curare appears as black or brown, resinous material having a bitter taste. Curare is soluble in cold water and dilute alcohol.

Chemical Constituents

The extract contains a wide range of alkaloids like (+) tubocurarine, curine, curarine, isochondrodendrin, cycleanine, chondrocurine, tomentocurine, etc. Amongst all of them (+) tubocurarine is most potent in activity and has bisbenzylisoquinoline moiety with quaternary structure.

(+) Tubocurarine chloride (C_{38} H_{44} Cl_2 N_2 O_6 $5H_2O$) is a quaternary base. It is insoluble in organic solvents, but soluble in water. For the extraction of this alkaloid, aqueous acid solution containing the extract is treated with reinecke acid H $[(CrCNH_3)_2]$ by which reineckate, a precipitated complex is obtained, which is filtered and dissolved in dry acetone. This is followed by addition of silver sulphate to precipitate silver reineckate and tubocurarine sulphate. The later is converted to tubocurarine chloride by addition of barium chloride. The filtrate of mixture contains this alkaloid which is crystallized by mild evaporation.

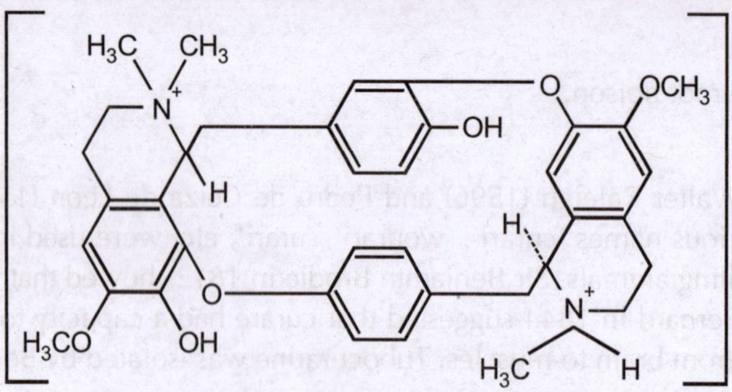

(+) Tubocurarine chloride

Chemical Tests

(1) The saturated solution of (+) tubocurarine gives a green colour with ferric chloride.

(2) By treatment with mercuric nitrate solution, (+) tubocurarine gives cherry red colour.

Uses

Because of the neuromuscular blocking actions, the main alkaloid of curare i.e (+) tubocurarine is used as skeletal muscle relaxant. It is used in surgical operations, to relax abdominal muscles due to which lower and safer dose of general anaesthesia is applied. It is also occasionally used to aid in diagnosis of myesthenia gravis.

Dose

15 - 30 mg i.m. as initial dose.

DARUHALDI

Synonyms

Darvi, Rasavanti, Berberis, Chitra, Indian berbery.

Biological Source

This consists of the dried roots and root bark of Berberis aristata DC belonging to family Berberidaceae and contain not less than 0.70 per cent of berberine on dry basis.

Geographical Source

It occurs Assam, Bihar and Himalayas and also in Nilgiri Hills, at an altitude of 1000 – 3000 m.

Cultivation and Collection

It grows well in temperate climate. It flourishes better under moist and humid climatic conditions. The plant does not tolerate hot climate. It grows well at higher altitude between 2000 - 3000 metres above the sea-level. Well distributed rainfall in the range of 60-75 cm; and range of temperature from 15 C - 30 C appears to be well suited to the plant. It mainly grows as rainfed crop. It grows on variety of soils, ranging from sandy alluvial loam to red lateritic loam or even dark loam soil is suitable for its growth. Propagation is carried out during spring season.

Self sown seeds are the major propagation source in nature. It can be successfully propagated vegetatively by stem cuttings. The method of raising seedlings on nursery bed and then transplanting can also be practiced. It is a perennial plant.

It is an erect, deciduons shrub about 1.75 to 3.5 metre in height. Leaves are 4 8 4 cm in size and obovate, deciduous. With basal spines and serrate.

Flowers are racemic and golden yellow in colour and fruits are blue.

Fig. 15.14: Berberis aristata – Flowering and fruiting branch

The plants are ready for harvesting after two years of plantation. The root bark is removed after its maturity. It is cut into small pieces, sun dried and stored in well closed container in dry place.

Macroscopic Characters

Colour : Roots are pale yellowish brown

Taste : Bitter

Odour : None

Roots are cylindrical, knotty, much branched, easily separable, wood is yellow.

Pieces of roots are variable in length and thickness, bark is 0.5 to 0.8 cm in thickness fracture is short in bark and splintery in xylem.

Chemical Contituents

It contains number of alkaloids, ranging from 1.2 – 2 per cent. Roots contain maximum alkaloids as compared to stem bark. Stem bark and root bark contain berberine, barbamine, oxyberberine, palmatine and taxilamine. Roots contain berberine, barbamine, Jatrorrhizine, columbamine and oxyberberine.

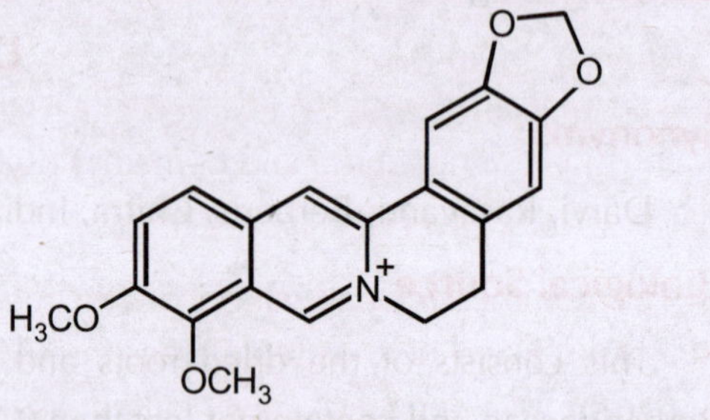

Berberine

Standards of Quality

For Roots

Foreign organic matter	:	Not more than 2.0 per cent
Ethanol soluble extractives	:	Not more than 2.0 per cent
Water soluble extractives	:	Not less than 6.0 per cent
Ash soluble extractives	:	Not less than 5.0 per cent
Acid insoluble ash	:	Not more than 1.0 per cent
Loss on Drying	:	Not more than 10.0 per cent
Microbial limits	:	Should comply IP limits.

Uses

It mainly possesses anti-inflammatory and antibiotic activity. It is also used as antipyretic, the plant extract is valuable in malarial fever. It is used as purgative in children; also used as blood purifier, tonic, febrifuge, anti bacterial and antidiarrhoeal.

Berberis is contraindicated in diarrhoea and pregnancy.

The plant extract **Rasaut** is used externally in eye diseases and also for washing piles and swellings.

IPECACUANHA

Synonym

Ipecac

Biological Source

It consists of the dried roots, or the rhizomes and roots of Cephaelis ipecacuanha (Brot.) A. Rich. or of Cephaelis acuminata Karsten, both belonging to family Rubiaceae. It should contain not less than 2 per cent of the total alkaloids, in which at least 50 per cent should be emetine.

Geographical Source

Cephaelis ipecacuanha is called Rio or Brazilian ipecac, which is obtained from Brazil, India, Myanmar, and Malaysia. C. acuminata is called Panama or Cartagena ipecac, which is procured from Columbia, Panama, Nicaragua and India.

History

Ipecacuanha means the small wayside plant with vomiting effects. Even before it was introduced in Europe as a medicine, ipecac was known in Brazil and used as an antidysenteric drug. Helvetius, a Dutch physician, launched use of this drug in Europe in 1688 under the name Brazilian root. In 1817, Pelletier and Magnedi first isolated emetine in crude form and later in 1894, Paul and Cownley separated emetine in pure form. Pyman succeeded in isolating two more alkaloids, viz. emetamine and o–methyl psychotrine in 1917.

Cultivation, Collection and Preparation for Market of Ipecacuanha

The commercial cultivation of Rio ipecac (also called Brazilian or Matto Grosso ipecac) in India is discussed here as large areas from West Bengal are under the cultivation of Rio ipecac. The drug from W. Bengal is also called as Johore ipecac. Better results have been obtained by cultivating ipecac at the lower foot hills of Eastern Himalayas. But the cultivation needs special attention and precautions.

The propagation is done by sowing the seeds in mid January to mid February. The germination is improved by treatment of seeds with lime water or hydrogen peroxide. The two months old seedlings are transplanted at a spacing of 10   10 cm. It favourably grows in the temperature range of 23 - 38 C, with a rainfall of 300 cm. The humid atmosphere helps in the growth of plant. The nitrogenous fertilizers have significant effect in increasing the quantity of emetine. The percentage of all alkaloids is maximum in third year and hence, harvesting of roots is done after three years of vegetative growth.

Macroscopic Characters

1. Brazilian ipecacuanha

(a) Roots

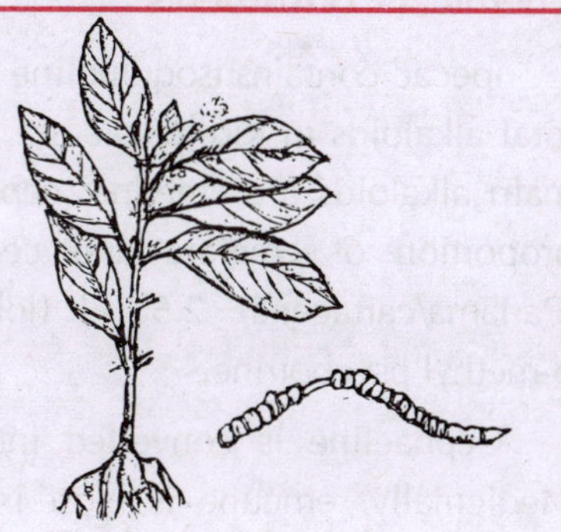

Fig. 15.15: Ipecacuanha herb & root

Colour : Dark brick red to dark brown

Odour : Faint

Taste : Bitter

Size : Up to 150 mm in length and 6 mm in thickness

Shape : Roots are found in tortuous pieces.

Extra Features : They are closely annulated externally, ridges rounded and completely encircling roots. Fracture is short in the bark and splintery in the wood. Pith is absent.

(b) Rhizomes

Colour : Brick red to dark brown.

Size : About 2 mm in diameter and short, attached to the roots

Shape : Cylindrical.

Extra Features : They are wrinkled longitudinally and show the presence of prominent pith.

2. Panama ipecacuanha

Colour : Greyish-brown to reddish-brown

Odour and taste : Same as in Brazilian variety

Size : Characterized by large size upto 9 mm in thickness.

Shape : Cylindrical

Extra Features : They are characterised by the absence of annulations and the presence of transverse ridges at an interval of 1 to 3 mm. They partially encircle the roots.

Microscopic Characters

Roots: Transverse section of root shows the presence of cork layer with brown contents, which is followed by phelloderm composed of thin walled parenchyma. It is full of starch grains and acicular crystals of calcium oxalate. Xylem consists of tracheids and small vessels. Secondary medullary rays with starch grains are also present. Size of the starch grains is up to 15 microns.

Rhizomes: Pericycle is with thick walled sclereids, protoxylem and spiral vessels. Columbian variety differs only in one aspect i.e. the starch grains are larger in size and up to 22 microns.

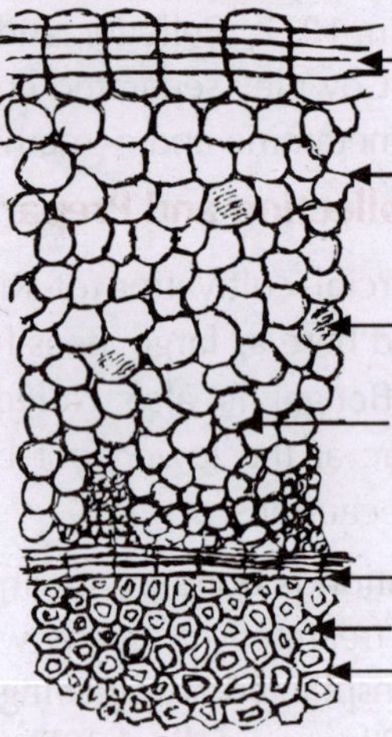

Fig. 15.16: T. S. of Ipecacuanha Root

Chemical Constituents

Ipecac contains isoquinoline alkaloids which belong to phenolic and non phenolic groups. The total alkaloids in Rio-ipecac are upto 2 per cent, and in Panama ipecac, about 2.2 per cent. The main alkaloids are emetine, cephaeline, psychotrine, o-methyl psychotrine and emetamine. The proportion of emetine and cephaeline in different varieties is 4 : 1 (Rio-ipecac), 1 : 1 (Panama/cartagena), 2.5 : 1 (Johore), alongwith minor bases like psychotrine, emetamine and o-methyl psychotrine.

Cephaeline is converted into emetine by methylation of phenolic C (6) hydroxyl group. Medicinally, emetine is most important. Non-phenolic alkaloidal group includes emetine and o-methyl psychotrine, while phenolic alkaloidal group includes cephaeline and psychotrine. Ipecac also contains ipecacuanhic acid, glycoside ipecacuanhin, starch and calcium oxalate.

Chemical Tests

(1) To about 2.5 g powdered drug, add 20 ml hydrochloric acid and 5 ml water. Shake it well and filter. To the filtrate, add 0.5 g potassium chlorate. The presence of yellow colour gradually changing to red, after standing, is due to emetine.

(2) The addition of sulphuric acid and sodium molybdate (Frohde's reagent) to small quantity of emetine gives bright green colour.

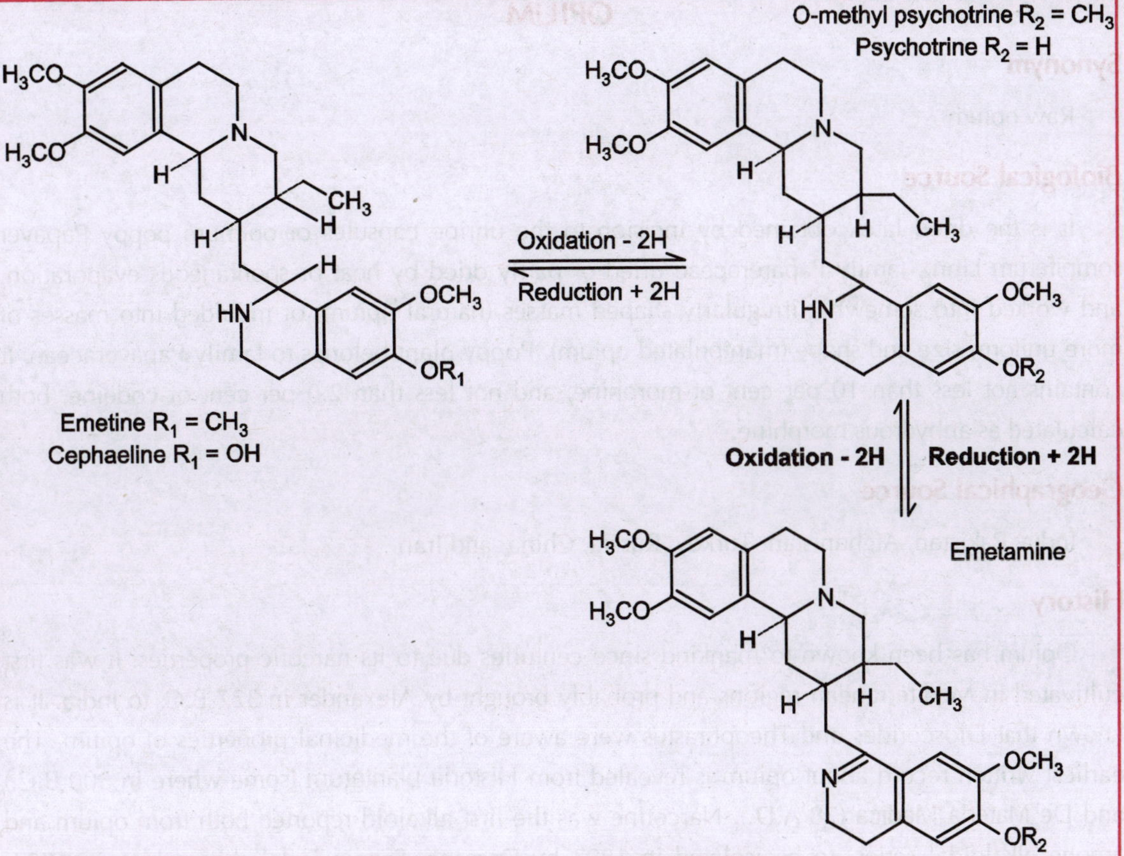

Ipecac Alkaloids

Standards of Quality

Total ash	:	Not more than 5 per cent
Acid-insoluble ash	:	Not more than 2 per cent
Foreign organic matter	:	Not more than 1 per cent

Uses

Ipecacuanha is expectorant in small doses and emetic in higher doses. Cephaeline has more emetic and less expectorant action as compared to emetine.

Emetine hydrochloride is used as antiprotozoal, as it is highly toxic to amoeba i.e. Entamoeba histolytica even in very low concentrations like 1 in 6 millions. Hence, it is used by administering parenterally, in treatment of amoebic dysentery.

Ipecacuanha is used for isolation of emetine and cephaeline. It is also reported that emetine has anti-tumour properties.

Dose

Emetine hydrochloride: 1 mg per kg body weight, subcutaneously or intramuscularly, but should not exceed 60 mg and maximum for 5 days.

Allied Plants

Various genera belonging to family Rubiaceae contain emetine. Some of them are Alangium, Hillia, Manettia Psychotria, Borreria, Remijia and Ferdinandusa.

OPIUM

Synonym

Raw opium

Biological Source

It is the dried latex obtained by incision to the unripe capsules of opimum poppy Papaver somniferum Linn., family Papareraceae dried or partly dried by heat or spontaneous evaporation, and worked into somewhat irregularly shaped masses (natural opium) or moulded into masses of more uniform size and shape (manipulated opium). Poppy plant belongs to family Papaveraceae. It contains not less than 10 per cent of morphine, and not less than 2.0 per cent of codeine, both calculated as anhydrous morphine.

Geographical Source

India, Pakistan, Afghanistan, Turkey, Russia, China, and Iran.

History

Opium has been known to mankind since centuries due to its narcotic properties. It was first cultivated in Mediterranean regions and probably brought by Alexander in 327 B.C. to India. It is known that Dioscorides and Theophrastus were aware of the medicinal properties of opium. The earliest written record about opium is revealed from Historia plantarum (some where in 300 B.C.) and De Materia Medica (78 A.D.). Narcotine was the first alkaloid reported both from opium and among alkaloidal series, to be isolated in 1803 by Derosne. Segnin isolated morphine in 1804. Magendi and Bally first introduced it in medical practice in 1818. Gulland and Robinson elucidated the structure of morphine in 1923. In 1833, Robiquet isolated codeine from opium, and in 1881, Grimaux reported that codeine is o-methyl derivative of morphine. Merck company isolated papaverine in 1848.

Cultivation, Collection and Preparation

Being a potent narcotic drug, the cultivation and other aspects of opium are governed by respective governments in different countries, including India. In India, all the activities about opium and its derivatives are controlled under Narcotic Drugs and Psychotropic Substances Act, 1985.

The genus Papaver has 50 different species, of which six species are found in India, viz. P. somniferum (Opium poppy), P. nudicaule (Iceland poppy), P. rhoeas (corn poppy), P. orientale, P. argemone, and P. dubium.

Poppy is an erect plant attaining 60 - 120 cm height. It is rarely branched. The leaves are linear, oblong or ovate oblong and have a dentate or serrate margin. It bears bluish white, purple or violet coloured large flowers. Accordingly, the varieties P. somniferum var. glabrum, P. somniferum var. album, P. somniferum var. nigrum are described. The second variety is cultivated in India.

Indian opium is considered as the only legal source of opium to many countries including United States of America and Britain.

In India, about 54 thousand hectares of land is under opium poppy cultivation. It is under government control, and cultivation of poppy is restricted to Madhya Pradesh, Rajasthan and Uttar Pradesh.

The weather conditions affect, up to a large extent, the yield of opium. Although, temperate climate is the natural requirement of opium poppy, it can be grown with success under subtropical climate in winter season, as there is a favourable effect on yield by cold weather. But, extreme cold conditions, including frost, adversely affect the plant and ultimately yield of opium. In short, the best climatic conditions for opium poppy are cool weather without freezing temperature and cloudiness, and sufficient sunshine.

Opium poppy is grown from November to March. Propagation is done by sowing the seeds, for which 3 - 4 kg of seeds per hectare are necessary. The seeds admixed with about 3 - 4 parts of sand are sown. Opium poppy requires highly fertile, well drained loamy soil with fine sand. The soil should contain organic matter, nitrogen and should have a pH around 7. The distance between two plants maintained is usually 25 cm and the plant reaches maximum height of one metre.

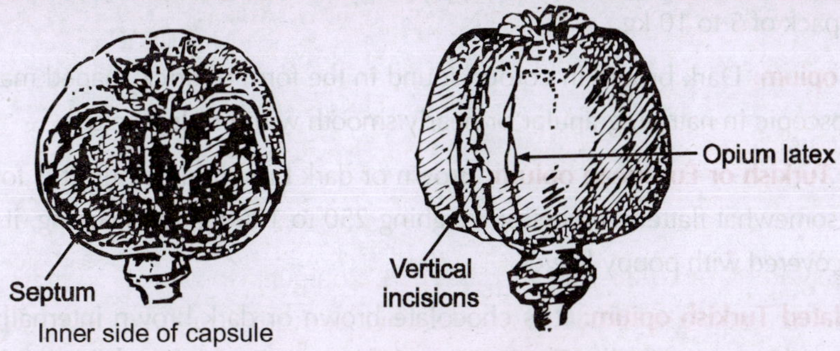

Fig. 15.17: Poppy Capsules

Periodically, the thinning of plants is done to get uniform growth and better development. The plants are kept totally free from weeds with the use of suitable weedicides. The plant should be protected from various insect pests like cut worms, leaf minor and poppy borer. The use of manures and fertilizers markedly improve the quality and yield of opium poppy. Especially, nitrogen and phosphorus have remarkable effects on growth of plant.

After sowing, within 3 - 4 months, the plant bears flowers, which are converted to capsules within few days and attain maturity after 15 - 20 days. During the maturity period, the capsule exudes maximum latex which shows a colour change from dark green to light green. Such capsules (Fig. 15.17) are incised vertically in the afternoon with the help of specific needle like apparatus called 'nushtur'. It penetrates maximum up to 2 mm into the capsule. Because of incisions, latex exudes out and thickens due to cold weather in night which is eventually scrapped and collected next morning by an iron scoop called Charpala. The incising process is repeated for about 4 times on the same capsule with 2 days interval. The incisions must remain superficial, so as to maintain the external exudation of latex. The latex is collected in plastic containers. Then, capsules are collected and dried in open areas and further the seeds are separated by beating. The average yield of opium is about 25-26 kg per hectare and for seeds, it is from 4-5 quintals per hectare. Opium is exported traditionally from India. The exports for 95-96 and 96-97 were ₹ 2365.5 lakhs and 4102 lakhs respectively.

The opium collected by this way is either exported or some of the part is further processed at Government opium factory at Ghazipur. A generalised process is outlined to cover the industrial method for extraction of alkaloids of opium.

Macroscopic Characters

Colour	:	Dried latex is brown or dark brown.
Odour	:	Strong characteristic
Taste	:	Bitter
Shape and size	:	Found in the form of powder or masses weighing 250 – 2000 gm.

1. **Indian opium:** Dark brown in colour. It is found in the form of cubical pieces weighing about 900 g for marketing purposes. It is enclosed in tissue paper and is brittle and plastic in nature. Internally, it is homogenous. Depending upon the requirement, the powdered form is available in the pack of 5 to 10 kg.

2. **Persian opium:** Dark brown in colour, found in the form of brick shaped masses, weighing 450 g. It is hygroscopic in nature, granular or nearly smooth with brittle fracture.

3. **Natural Turkish or European opium:** Brown or dark brown in colour. It is found in conical or rounded and somewhat flattened masses, weighing 250 to 1000 g. On keeping, it becomes hard and brittle. It is covered with poppy leaves.

4. **Manipulated Turkish opium:** It is chocolate brown or dark brown internally and covered with broken poppy leaves externally. The masses of this type are oval and flattened on upper and lower surface weighing about 2000 g. It is somewhat plastic or even brittle.

5. **Manipulated European opium:** It is dark brown in colour internally and covered with broken leaves. It is found in the form of elongated masses with rounded ends weighing 150 to 500 g. It is firm, plastic and with brittle fracture.

Chemical Constituents

The latex contains mainly the alkaloids derived from amino acids phenylalanine and tyrosine. Chemically, they are placed under benzylisoquinoline and phenanthrene types.

Narcotine (also called noscapine), narceine and papaverine belong to the former, while morphine, codeine and thebaine represent latter category.

Fruits of poppy contain numerous off white coloured and minute seeds. These seeds contain 30-35 per cent drying fixed oil. Which is used commercially in oil paint industry, which is colourless, tasteless and transparent.

ISOLATION OF ALKALOIDS

Powdered Opium
(shaken with calcium chloride solution and filtered)
↓
Filtrate (concentrated)
↓
Add 10% sodium hydroxide solution
↓

- **Filtrate** (morphine, codeine, narceine)
  Extract with chloroform
  - **Chloroform**
  - **Aqueous solution** (morphine, narceine). Acidify, then make slightly alkaline with ammonia
    - aqueous solution (narceine)
    - Precipitate (morphine)
- **Precipitate** (narcotine, papaverine, thebaine)
  1. Dissolve ppt. in alcohol
  2. Acidify with acetic acid
  3. Add 3 volumes of boiling water
  - aqueous solution (thebaine)
  - Precipitate (narcotine)
    1. Dissolve in boiling 0.3 % oxalic acid solution
    2. Cool
    - **Crystals** (Papaverine oxalate)
    - **Aqueous solution** (narcotine oxalate)

OPIUM ALKALOIDS

Morphine is monoacidic, laevorotatory phenolic alkaloid and also contains an alcoholic hydroxyl group at C (6) position. Due to presence of phenolic hydroxyl group, it is soluble in alkali hydroxides, except ammonium hydroxide. Morphine is very less soluble in different solvents like ether (1 in 600), chloroform (1 in 1200), alcohol (1 in 200) and water (1 - 3000). Diacetyl derivative of morphine is heroin.

Codeine (methyl morphine) is a strong monoacidic base and laevorotatory. It is soluble in water and organic solvents.

Papaverine is a weak monoacidic base and inactive optically. It is slightly soluble in organic solvents, but insoluble in water.

The other important benzylisoquinoline alkaloid narcotine is also a weak monoacidic base and is laevorotatory, while its salts are dextrorotatory. Narcotine is soluble in acetone, benzene, chloroform, but insoluble in water, alcohol and ether.

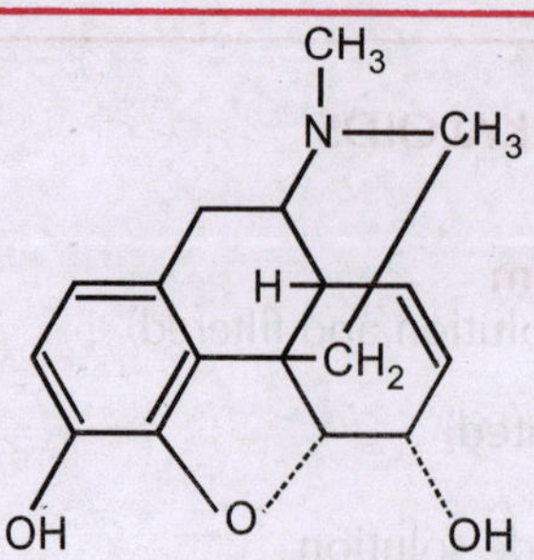

Morphine

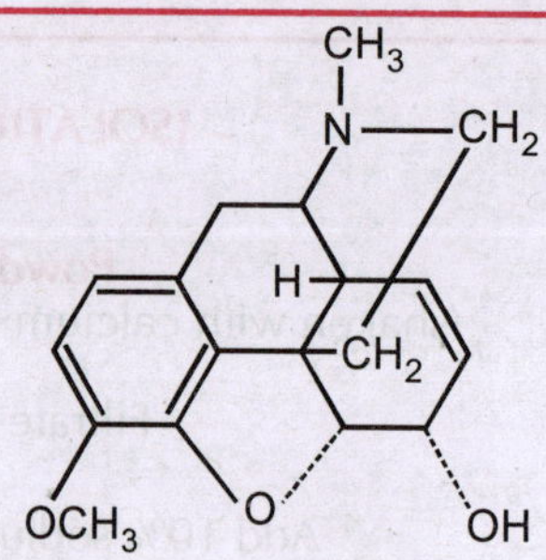

Codeine

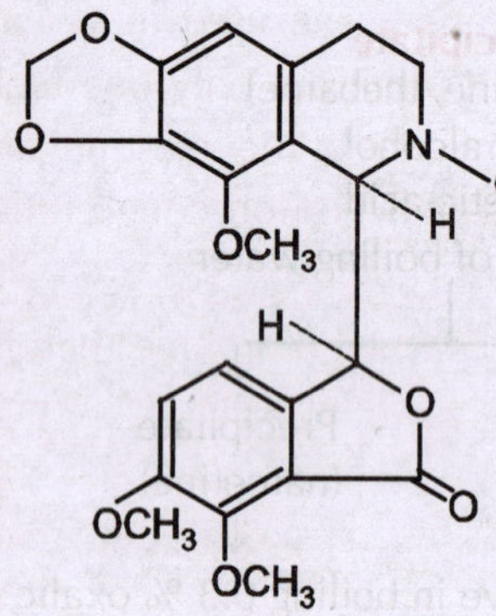

Narcotine

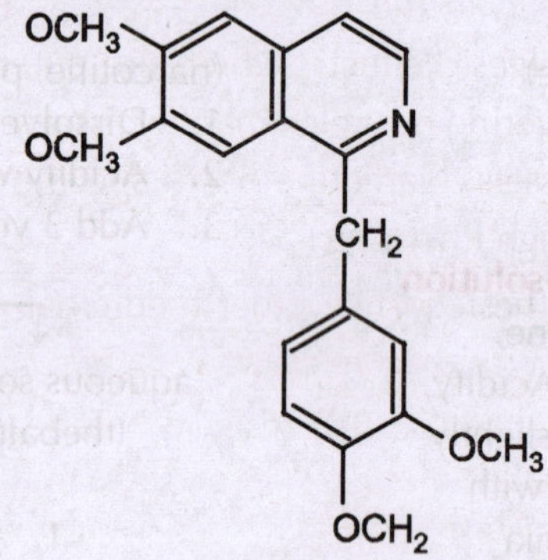

Papaverine

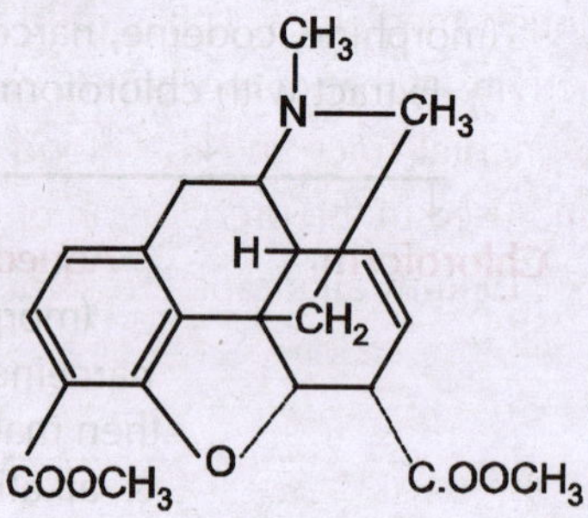

Heroin

The opium alkaloids are present as salts of meconic acid.

Protopine and hydrocotarnine are the minor alkaloids of opium. Opium also contains sugar, wax, mucilage and salts of calcium, potassium and magnesium. Opium does not contain tannins, starch and calcium oxalate.

Chemical Tests

(1) The general test to detect opium is by testing presence of meconic acid. The alkaloids are present as the salts of meconic acid. Opium is dissolved in water and to the filtrate, ferric chloride solution is added by which deep reddish purple colour is obtained, which persists even on addition of hydrochloric acid.

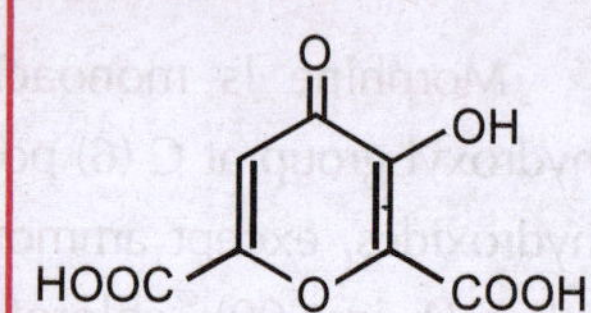

Meconic acid

(2) Morphine when sprinkled on nitric acid gives orange red colour. Codeine does not respond to this test.

(3) The treatment of morphine solution with potassium ferricyanide and ferric chloride solutions gives bluish green colour. Codeine does not respond to this test.

(4) Papaverine solution in hydrochloric acid gives a lemon yellow colour with potassium ferricyanide solution.

Standards of Qualtiy

Total ash	:	Not more than 6.0 per cent
LOD	:	Not more than 15.0 per cent
Morphine contents	:	Not less than 10.0 per cent

Uses

Opium belongs to the category of hypnotic sedative and analgesic in which the action is mainly due to morphine. Morphine is a potent analgesic. Due to its central narcotic effects, it causes addiction. Hence, it is given only in severe pains and in those cases, when patient does not show response to other analgesics. Morphine has a biphasic action on central nervous system. It sedates the cerebrum and has a mixture of stimulation and sedation on the medulla. In the medulla, it sedates the respiratory centre, emetic centre and the cough reflex. It also stimulates chemoreceptor trigger zone in the medulla, which leads to nausea and vomitting and is considered as a side effect. Morphine also produces respiratory depression and constipation.

Codeine relieves local irritation in the bronchial tract and as an antitussive used in various cough medicines. It has mild analgesic effects, which are potent than aspirin, but only one tenth activity of that of morphine. Papaverine has relaxant effects on smooth muscles of the intestinal and bronchial tract and the blood vessels. Narcotine has a specific depressant action on cough reflex and used in the preparation of cough linctuses.

Opium alkaloids are semisynthesized like other medicinal agents. Diacetyl morphine (heroin) has more narcotic, analgesic property than morphine. By losing one molecule of water, morphine gives apomorphine which is emetic and used subcutaneously to treat poisoning cases. Hydromorphone is formed by replacing one of the hydroxyl groups and also removal of adjacent double bond. It is also a potent narcotic analgesic, but habit forming tendencies are less.

The synthetic morphine like compounds are called 'opioids', which are non habit forming, but possess the medicinal activity of morphine.

Dose

(1) Morphine Sulphate	:	10 mg, 6 times a day parenterally.
(2) Codeine sulphate/phosphate	:	10 - 20 mg every 4 - 6 hours, orally.
(3) Narcotine (noscapine)	:	15 mg, 4 times a day, orally.
(4) Papaverine hydrochloride	:	150 mg orally and 30 mg parenterally.

Commercial Varieties of Opium

(1) Indian opium: It is dark brown in colour and found in the form of cubical pieces weighing 900 g. It is brittle and plastic in nature. The powdered form is available as 5 - 10 kg packs. It contains 10 per cent anhydrous morphine.

(2) Persian opium: It is dark brown in colour and available as brick shaped masses of 450 g. It is hygroscopic, granular or smooth.

(3) Turkish opium: It is commonly called as druggists opium or soft opium. It is brown or dark brown in colour and available as conical rounded or flattened masses.

(4) Chinese opium: It comes in market in the form of flat globular cakes and contains 4 - 11 per cent morphine.

Different Forms of Opium

(1) Powdered opium: It contains 10 per cent anhydrous morphine with lactose, caramel and powdered cocoa husks.

(2) Opium concentratum: It contains different alkaloid hydrochlorides of opium in following proportions:

Anhydrous morphine	:	47.5 - 52.5 per cent
Codeine	:	2.5 - 5 per cent
Narcotine	:	16 - 22 per cent
Papaverine	:	2.5 - 7 per cent

(3) Camphorated opium tincture: It contains alcoholic solution of opium, benzoic acid, camphor, anise oil and the formulation is prepared in alcohol. It is used in treatment of diarrhoea as antiperistaltic.

Storage

Opium is preserved in a well closed container to prevent loss of morphine.

Adulteration

Since the production of opium is under government control, it is not found to be adulterated. The adulterated forms show presence of opium capsules in powdered form, gum and sugary fruits.

Allied Plants

The various other species of poppy, which do not contain morphine are Papaver argemone, P. dubium, P. orientate, P. bracteatum, P. strigosum, P. intermedia, P. paeoniflorum, hybrid of P. somniferum and P. orientate, P. psendo orientale, and plants from genera Argemone and Eschscholzia (both belonging to family Papaveraceae).

Among all these species, P. bracteatum has scored more importance, as it does not contain morphine, which causes addiction. The amount of total alkaloids and consequent percentage of thebaine is also very high. Because of such morphine free contents, this species is more significant as a potential new source of opiates.

TROPANE ALKALOIDS

Tropane molecule represents the fusion of pyrrolidine and piperidine ring with common methylated nitrogen. The alkaloids containing with methylated tropane nucleus are chemotaxonomic characters of family Solanaceae. Out of 10-12 such alkaloids of this family, therapeutic value is present only in l-hyoscyamine. Hyoscine and racemic form of hyoscyamine viz. atropine. They have anticholinergic effects. They are employed for different purposes. Hyoscyamine is used in parkinsonism. Hyoscine is useful as preanaesthetic in surgery and also in motion sickness. Atropine is employed to achieve paralysis of parasympathetic nerves like in treatment of eye diseases.

Family Solanaceae includes 72 genera, out of which only 8 genera viz. Datura, Atropa, Duboisia, Hyoscyamus, Scopolia, Physoclaina, Mandrogora and Solandra contain l-hyoscyamine, hyoscine and atropine.

The species, with such alkaloids, under these genera are grouped in the following way.

1. Hyoscyamine as main alkaloid: *Datura stramonium, Atropa belladonna, A. acuminata, Duboisia myoporoides* (South Australian Strain).
2. Hyoscine as main alkaloid: *D. metel, Duboisia myoporoides* (North Australian Strain)
3. Hyoscyamine and Hyoscine (both in low quantities): *Hyoscyamus niger, Mandrogora officinarum.*

BELLADONNA HERB

Synonyms

Belladonna Leaf; Belladonnae Folium; Deadly night shade leaf.

Biological Source

Belladonna herb consists of dried leaves or the leaves and other aerial parts of *Atropa belladonna* Linn. (European belladonna) or *Atropa acuminata* Royle ex-Lindley (Indian belladonna) or mixture of both the species collected when the plants are in flowering condition. It belongs to family Solanaceae. It contains not less than 0.3 per cent of the alkaloids of belladonna herb, calculated as *l*-hyoscyamine.

Geographical Source

It is indigenous to and cultivated in England and other European countries. In India, it is found in the Western Himalayas from Simla to Kashmir and adjoining areas of Himachal Pradesh. Its chief habitat is Jammu and in forests of Sindh, and Chinab valley.

History

Because of the hallucinogenic effect of this plant, it was used as witch craft in the middle ages. In ancient times, the juice of this plant was used as a cosmetic, because of its dilatory effect on the pupil of the eye. This drug was first introduced in the London Pharmacopoeia in 1809.

Cultivation and Collection

Cultivation of belladonna at an altitude of 1400 m above the sea level is found to be satisfactory, if proper irrigation facilities are provided. It is observed that the yield per hectare can be increased substantially by proper cultivation technology. The experimental trials of applications of several fungicides and insecticides right from the treatment of the seeds up to the foliar sprays were very encouraging. Its cultivation in Jammu and Kashmir is found to be successful.

Belladonna berries are crushed to get the seeds for cultivation. Proper processing like washing and sieving is performed. Only healthy seeds are used for cultivation. Seeds are sown by broadcasting method in well prepared beds with the application of fungicide like diathon.

Sowing is done in May and July. The seedlings are ready for transplantation by the end of September. Transplanting is done by keeping certain distance between two plants and the seedlings are irrigated carefully. Fertilizers like urea, potash and superphosphate are given as per the needs. Insecticidal sprays like sevin are also tried when the plant reaches maturity. The leaves, as well as, the flowering tops are cut and sundried or dried in shade. During drying, care is taken to retain the green colour. While grading and packing for market, woolly stems and foreign organic matter are rejected. The yield per hectare is found to be 200 to 600 kg.

Macroscopic Characters (Fig. 15.18)

Colour : Leaves - Green to brownish-green
Flowers - Purple to yellowish-brown
Fruits - Green to brown

Odour : Slight and characteristic

Taste : Bitter and acrid

Size : Leaves - 5 - 25 cm long and 2.5 - 12 cm wide
Flowers - Corolla 2.5 cm long and 1.5 cm wide
Fruits - About 10 cm in diameter.

Shape : Leaves - Ovate, lanceolate to broadly ovate, with acuminate apex, decurrent lamina, entire margin, petiolate, brittle and transversely broken.
Flowers - Campanulate, 5, small reflexed lobes of corolla.
Fruits - Berries, sub-globular in shape with numerous flat seeds.

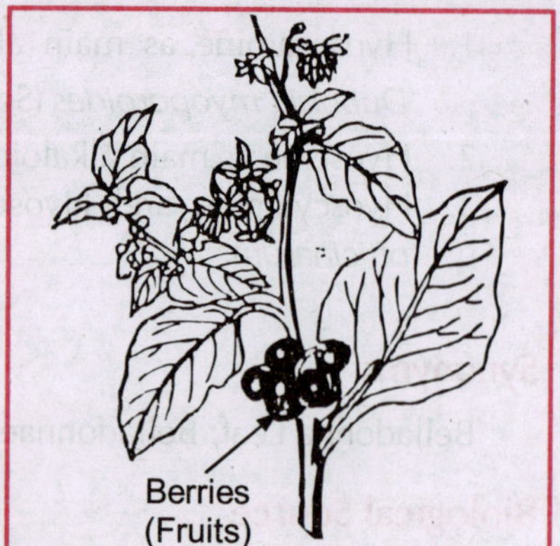

Fig. 15.18: Belladonna Herb

Extra Features

In general, the entire drug is seen as crumpled and twisted. The dropping flowers are associated with as many pairs of leaves. The flowers are with 5 stamens, superior bilocular ovary with numerous seeds.

Microscopic Characters (Fig. 15.19)

Epidermal cells with slightly sinuous anticlinal wall and striated cuticle, anisocytic stomata and occasionally uniseriate multicellular covering trichomes are present. There are glandular trichomes which are uniseriate and with unicellular heads. The palisade ratio is 5 to 7.

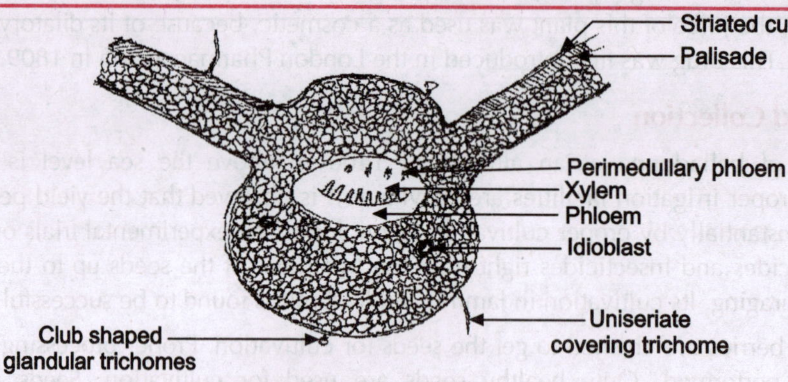

Fig. 15.19: T. S. of Belladonna Leaf

Chemical Constituents

The total alkaloidal content of drug is 0.4 - 1 per cent and varies in different parts of plant, roots (0.6 per cent), stems (0.05 per cent), leaves (0.4 per cent), unripe and ripe berries (0.19 - 0.21 per cent) and seeds (0.33 per cent).

The main alkaloids are l-hyoscyamine and its racemic form is atropine. The drug also contains belladonine, scopoletin (I - methyl aesculetin), hyoscine, pyridine and N - methyl pyrroline. The later two are the volatile bases. Homotropine is a synthetic compound and is preferred in the medical profesion as the synthetic process of atropine and hyoscyamine is very costly.

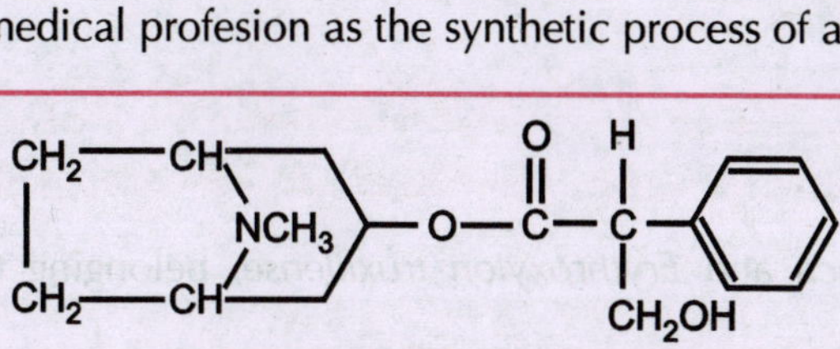

Atropine (Tropiine (±)-Tropate)

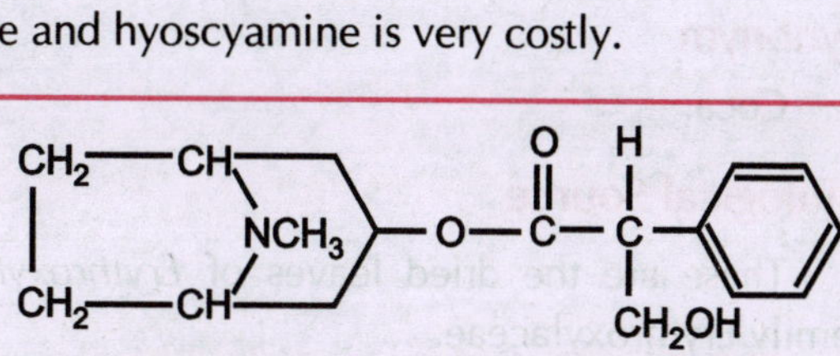

Hyoscyamine (Tropine (–)-Tropate)

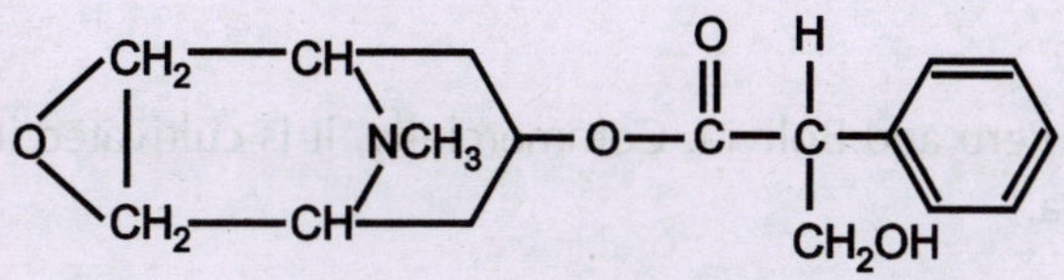

Hyoscine (Scopalamine)

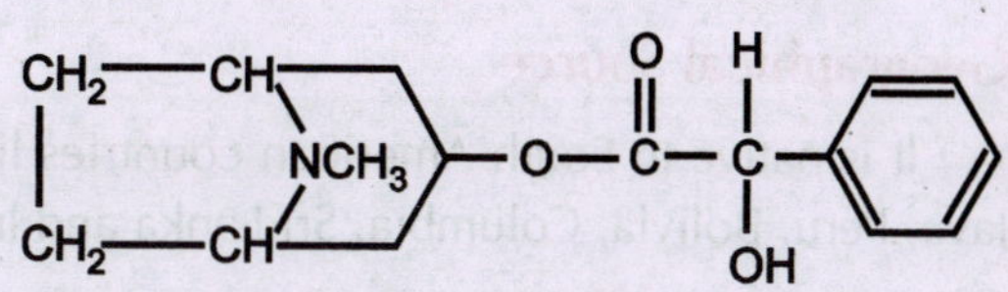

Homotropine

Standards of Quality

Total ash	: Not more than 14 per cent
Acid-insoluble ash	: Not more than 3 per cent
Foreign organic matter	: Not more than 3 per cent
Stems	: Not more than 20.0 per cent
Alkaloids	: Not less than 0.3 per cent
Calcualted as Hyoscyamine	
It gives Vitali - Morin reaction	: Positive

Uses

It is the parasympatholytic drug with anticholinergic properties. It is used to reduce the secretions such as sweat, saliva and gastric juice and also to reduce spasm in cases of intestinal gripping due to strong purgatives. It is also used as an antidote in opium and chloral hydrate poisoning.

Dose

0.6 to 1 ml in the form of belladonna tincture - 4 times a day.

Adulterants and Substitutes

The drug is adulterated with the leaves of Phytolacca americana, Solanum nigrum, and Ailanthus glandulosa. Each of them is distinguished by their histological character. Idioblast is present in Phytolacca leaves, lamina is denser, needle shaped crystals are present and anomocytic stomata are distinct. Palisade ratio is from 2 - 4 in S. nigrum leaves. Clustered crystals of calcium oxalate near the veins are present in Ailanthus leaves. The leaves also show the presence of unicellular lignified trichomes.

COCA LEAVES

Synonym

Coca.

Biological Source

These are the dried leaves of *Erythroxylon coca* and *Erythroxylon truxillense*, belonging to family Erythroxylaceae.

Commercially, *E. coca* is known as Bolivian or Huanuco coca and *E. truxillense* is known as Peruvian or Truxillo coca.

Geographical Source

It is native to South American countries like Peru and Bolivia. Commercially, it is cultivated in Java, Peru, Bolivia, Columbia, Sri Lanka and India.

History

Since ancient times, the coca leaves have been used by South Americans as a masticatory and were reserved for only native chiefs and Incas. It was considered as 'divine plant'. The first report about coca leaves was prepared by **Monardes** in 1569. The plant was brought to Europe in 1688. Cocaine was isolated in 1860 and its local anaesthetic effects were discovered in 1882 by **Koeller** from Vienna. It was introduced into medicine some time in the later half of nineteenth century.

Cultivation, Collection and Preparation

Coca leaves contain the tropane alkaloids, mainly cocaine. Though it has local anaesthetic effects, it has hallucinogenic actions, leading to addiction. Because of this, all activities about this plant are strictly governed by Narcotic Drugs and Psychotropic Substances Act 1985 in India and by relevant acts in other countries.

Coca plant is mainly cultivated in Java, Peru and Bolivia. The open fields at an altitude of 500 - 2000 metres are suitable for cultivation of this drug. Propagation is done by sowing the seeds in nursery beds. The seedlings on attaining a height of about 15-20 cm are transplanted to open fields by keeping a distance of 2 metres. A thatched roof is provided in initial days of cultivation. The height of the plant is maintained upto 2 metres by regular prunning. The drug is collected over a period of 3 years, at an interval of one year. The leaves are collected in dry weather and dried in the shed or by artificial means.

From the crude cocaine obtained from leaves, the pure cocaine is isolated. The total alkaloids are hydrolysed to get ecgonine, from which cocaine is synthesised. Large scale manufacture of cocaine is mostly done in U.K, where the processes are protected by patents.

Macroscopic Characters [Figs. 15.20 (a) and 15.20 (b)]

Peru variety gives the elliptical leaves which are 1.5 - 5 cm long with very thin texture and pale green colour. Bolivian variety contains greenish-brown leaves, which are oval in shape and measure 2.5 - 7.5 cm in length and 1.5 - 4 cm in width. It bears a prominent midrib. In Bolivian variety, the lower surface shows two curved lines, one each on both sides of midrib.

Both the varieties have a characteristic odour with a bitter and aromatic taste.

Fig. 15.20 (a): *Erythroxylm coca* herb and leaf

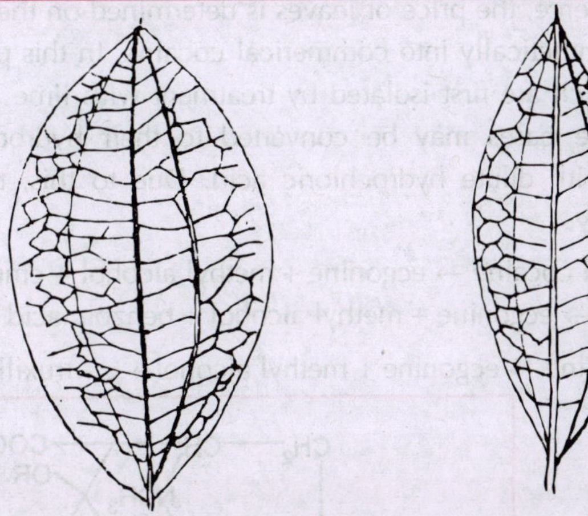

Fig. 15.20 (b): A leaf of *E. truxillense*

Microscopic Characters

Midrib is typical and is partly encircled by considerable amount of collenchyma. The lower epidermis shows presence of papillae and numerous paracytic stomata. The drug does not show trichomes, but some of the cells of epidermis contain mucilage. Numerous lignified idioblasts are present near the veins. Starch grains, stone cells and rarely calcium oxalate crystals are present in epidermal cells.

Microscopically, both the varieties have similar characters. They are isobilateral leaves and show presence of parenchyma under both epidermal layers.

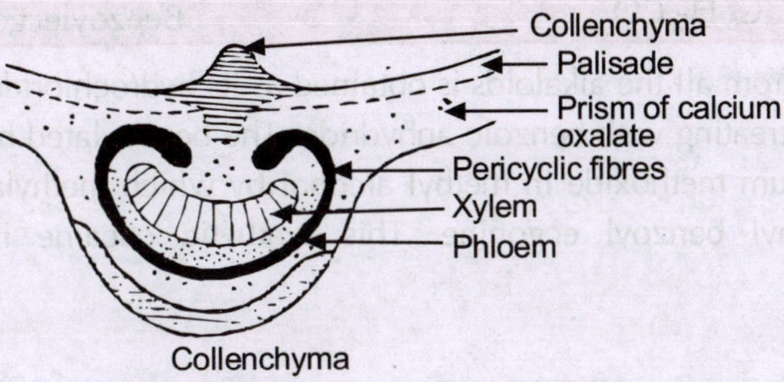

Fig. 15.21: Coca Leaf in T.S.

Chemical Constituents

The drug contains 0.7 - 1.5 per cent of total alkaloids. The majority of alkaloids are tropane esters. Both Peruvian and Bolivian coca contain less alkaloids, but higher proportion of cocaine among them. Java coca has higher percentage (upto 2 per cent) of alkaloids, but low amount of cocaine.

The coca plant yields the tropane derived alkaloids such as cocaine (methyl benzoyl ecgonine), cinnamyl cocaine (methyl cinnamoylecgonine), α-truxilline (methyl-α-truxilloylecgonine), tropocaine, benzoyltropine, dihydroxytropane and benzoylecgonine.

In commerce, the price of leaves is determined on the basis of ecgonine content, because it is converted synthetically into commerical cocaine. In this process, ecgonine is obtained from crude alkaloids which are first isolated by treatment with lime and organic solvents or dilute sulphuric acid. All free bases may be converted to their hydrochlorides. Further, they are hydrolysed by boiling with dilute hydrochloric acid. Due to this, the different bases liberate ecgonine in following way.

Cinnamyl cocaine → ecgonine + methyl alcohol + cinnamic acid
Cocaine → ecgonine + methyl alcohol + benzoic acid

α - truxilline → ecgonine + methyl alcohol + α - truxillic acid

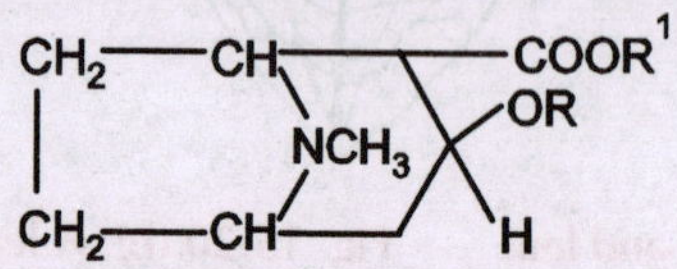

Ecgonine Derivatives

R	R′	Name of Alkaloid
H	H	Ecgonine
CH_3	C_6H_5CO (benzoyl)	Cocaine
H	CH_3	Methyl ecgonine
H	$C_6H_5CH = CHCO$ (cinnamoyl)	Cinnamoyl ecgonine
H	C_6H_5CO	Benzoylecgonine

Ecgonine liberated from all the alkaloids is obtained as its hydrochloride. It is further converted to benzoylecgonine by treating with benzoic anhydride. The benzoylated base is next treated with methyl iodide and sodium methoxide in methyl alcohol by which methylation occurs and brings out cocaine i.e. methyl benzoyl ecgonine. This synthetic cocaine is isolated as cocaine hydrochloride.

Chemical Test

Cocaine powder is treated with sulphuric acid, heated, followed by addition and mixing of water. It gives the characteristic smell of methyl benzoate.

Uses

Cocaine is a local anaesthetic. It is the first known local anaesthetic from which various other synthetic substitutes with similar activity have been prepared.

In general, coca leaves are used as stimulant, restorative and also in convulsions. Cocaine reduces the sedative and respiratory depressant effects of morphine and allied drugs, due to CNS stimulant properties.

Owing to hallucinogenic and addictive effects of cocaine, it has become the drug of abuse and hence, its uses are limited to ophthalmic surgery and surgery of ear, nose and throat.

DATURA

Synonyms

Datura herb, Angel's trumpet.

Biological Source

Datura consists of the dried leaves and flowering tops of Datura metel and D. metel var. fastuosa Safford. It belongs to family Solanaceae.

It should contain not less than 0.20 per cent of total alkaloids of Datura, calculated as l-hyoscyamine.

Geographical Source

It is found in India, England and other tropical and subtropical regions.

Cultivation and Collection

The drug is cultivated of Datura rarely by sowing the seeds. Mainly collected from wild grown plants. The germination is normally very slow. If the seeds are soaked in water and kept overnight, the rate of germination increases. About 7 - 8 kg of seeds per hectare are required for sowing purpose. The seeds require about 15 - 20 days for germination. Weeding and thinning are necessary and performed when are 10 - 15 cm tall. The distance kept in between 2 plants is about 75 - 100 cm. The plants should be supplied with organic fertilizers and proper irrigation. The drug is collected after 4 months of its cultivation. The leaves and branches are removed, drug is dried in the sun and marketed by packing in gunny bags.

Macroscopic Characters

The drug has a characteristic but unpleasant odour and a bitter taste. The drug contains entire, broken wrinkled, crushed leaves along with stem fragments and floral parts.

The entire leaf has length of 9 - 18 cm and width of 8 - 13 cm. Normally, the margin is of entire, but in some cases sinuated with rounded or acute 2 - 4 broad lobes. The leaf is covered with minute hairs, lower surface is slightly pale in colour and the leaf has a thin texture. The leaf is unequal at the base with acute apex and glabrous lamina. Each leaf has 3 - 4 coarse veins on each side and 4 - 6 secondary veins on either side of the midrib. Flowers are reddish-purple on outer side and whitish on inner side. Corolla is thin, acuminate, triangular to circular in shape. Flowers are funnel shaped with pedicel which is never erect. The stems, as well as, branches of drug are purple coloured. Brown coloured seeds are triangular and are found in the thorny capsule.

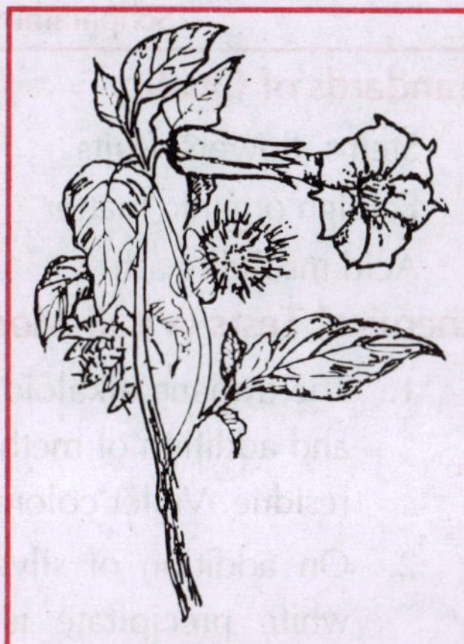

Fig. 15.22: *Datura metel* in Fruiting Stage

Microscopic Characters

Through the transverse section, it shows its dorsiventral character. The epidermal cells of both sides show anisocytic or cruciferous stomata. The cells are covered with thin cuticle and glandular and non-glandular simple trichomes. About 40 per cent of the lamina is occupied by single layer of palisade cells. Approximately 6 - 8 layers of spongy parenchymatous cells are present. The midrib shows vascular tissue with protoxylem and metaxylem. Trichomes are more on the midrib region. Stomatal index is 12.7 - 19.5 for upper surface and 21.2 - 24 for lower surface. Palisade ratio is 3.5 - 6.5. The spongy parenchyma contains calcium oxalate crystals.

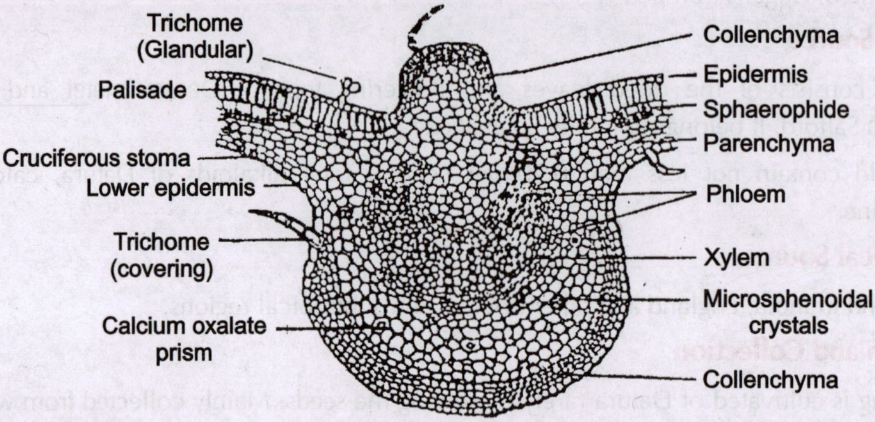

Fig. 15.23: T.S. of Datura leaf

Chemical Constituents

Datura herb contains upto 0.5 per cent of total alkaloids, among which hyoscine (scopolamine) is the main alkaloid, while *l*-hyoscyamine (scopoline) and atropine are present in very less quantities (see belladonna herb).

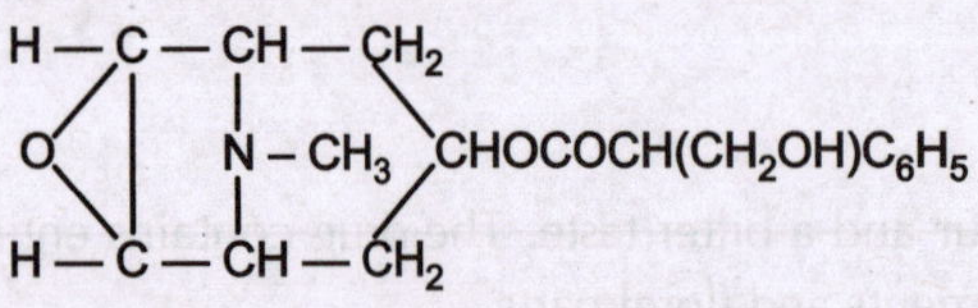

Scopalamine

Hyoscine ($C_{17}H_{21}O_4N$) is an ester of tropic acid and scopine. It is the principal alkaloid of *Datura, Scopolia* and *Duboisia* species. The occurrence of hyoscine is restricted only to Solanaceae family.

Standards of Quality

Stems, flowers, fruits	:	Not more than 20 per cent
Foreign organic matter	:	Not more than 2 per cent
Acid-insoluble ash	:	Not more than 4 per cent

Chemical Tests (Vitali-Morin reaction)

1. The tropane alkaloid is treated with fuming nitric acid, followed by evaporation to dryness and addition of methanolic potassium hydroxide solution to an acetone solution of nitrated residue. Violet coloration takes place due to tropane derivative.
2. On addition of silver nitrate solution to solution of hyoscine hydrobromide, yellowish white precipitate is formed, which is insoluble in nitric acid, but soluble in dilute ammonia.

Uses

Datura herb and its main alkaloid hyoscine are parasympatholytic with anticholinergic and central nervous system depressant effects. The drug is used in cerebral excitement. Along with morphine, it is used as preoperative medication. It is also used in treatment of asthma and cough.

Hyoscine hydrobromide is used in motion sickness, gastric or duodenal ulcers.

DUBOISIA

Synonyms

Cork wood, cork tree

Biological Source

Duboisia consists of the dried leaves of Duboisia myoporoides R. Brown, and D. leichhardtii. It belongs to family Solanaceae.

Geographical Source

It a Australian herb found around Sydney along the coastal area tablelands of North Quinceland, New South Wales and New Caledonia island.

Macroscopic Characters

Duboisia is perennial shrub up to 3.0 metres in height with brown to purple colour bark on young stage while a corky older bark at later stage

Leaves are 3-10 cm into 1-2 cm. Leaves are tapering at both the ends and alternatively arranged.

Colour is pale green.

Taste is intensely bitter.

Fig. 15.24: Duboisia herb

Chemical Constituents

Duboisia is considered as the chief commercial source of scopolamine and atropine. It contains l–hyoscyamine which is converted to atropine during extraction. Along with it, the drug also contains nor-hyoscyamine, tigloidine, valtropine, tiglyoxytropine.

The synthetic process for scopolamine and atropine is very costly and hence, much reliance is placed on its natural source.

Atropine (C_{17} H_{23} O_3 N) occurs as colourless crystals, with a bitter taste and no odour. It is soluble in chloroform and alcohol. It is a racemic form of hyoscyamine.

Chemical Tests

1. The addition of gold chloride solution to atropine in water and hydrochloric acid gives lemon yellow precipitate. (2) It gives positive Vitali Morin reaction.

Uses

Duboisia leaves are the main source of atropine and scopolamine.

Atropine is the parasympatholytic drug. It also causes stimulant action on central, medullary and higher nerve centres. Atropine has many different therapeutic uses. It is used as an antidote for pilocarpine, physostigmine and other choline esters. It relieves bronchial spasms in asthma. As it suppresses the gastric secretions, it is used in peptic ulcer. It has applications in ophthalmic practice, because of its dilatory effects on pupil of the eye. It is also used to reduce tremor and rigidity in parkinsonism.

HYOSCYAMUS

Synonyms

Henbane, Hyoscyamus herb, Hyoscyamus leaves.

Biological Source

It consists of the dried leaves, or leaves and flowering tops of Hyoscyamus niger, belonging to family Solanaceae.

It should contain not less than 0.05 per cent of alkaloids of hyoscyamus, calculated as l-hyoscyamine. It is grown as a biennial herb.

Geographical Source

Hyoscyamus is a native of Western Asia, North Africa, Europe and India. It is cultivated in Russia, Belgium, Hungary and India.

History

Hyoscyamus is known since ancient times and was used as household drug. Dioscorides has also mentioned about this drug in his literature. The references about this drug are found in 'Arabian Nights' and also in Anglo-Saxon works on medicine. The drug was reintroduced in 1809 in the London Pharmacopoeia.

Cultivation and Collection

The cultivation is done in temperate region at an altitude of 2400 to 3300 m. The drug is cultivated on commercial scale in England, Egypt, U.S.S.R., and Hungary. In India, it is cultivated in Kashmir to a limited extent. The method of propagation is by seeds. The small seed beds are raised and seeds are sown. The seeds require about two weeks for germination. The seedlings are transplanted in field in the month of May by keeping a distance of half metre between them and about 75 cm between two rows. The plants are kept free of weeds and occasional hoeing is also done. The crop is harvested when it reaches maturity. Under all favourable conditions, the yield of the drug per hectare is 1000 kg - 1500 kg.

Macroscopic Chracters (Fig. 15.25)

The fresh drug has characteristic and strong odour with a bitter and acrid taste.

The leaves have about 25 cm long lamina and they are pale greyish green in colour. The shape is ovate-oblong to triangular ovate. In few cases, they may contain short petiole, otherwise they are sessile. The margin has acute triangular lobes, which are irregularly dentate. The apex is acute, the lamina is covered with glandular hair. There is the prominent midrib with pinnate venation. The flowers are funnel shaped and yellow in colour, showing purplish veins.

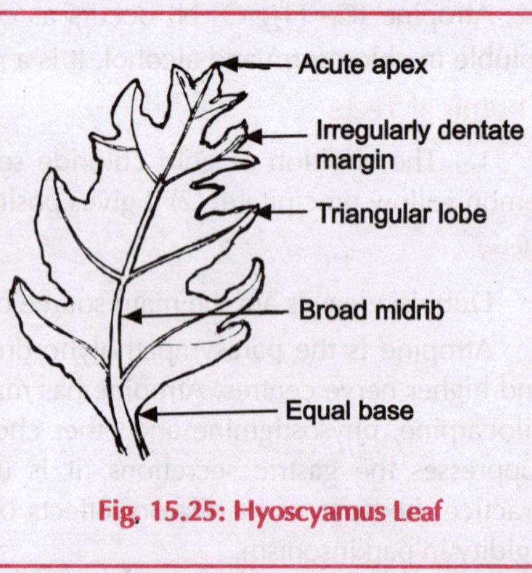

Fig. 15.25: Hyoscyamus Leaf

Microscopic Characters (Fig. 15.26)

The leaf is dorsiventral. Epidermis is covered with smooth cuticle and numerous glandular trichomes. Anisocytic stomata are present in epidermal layer. Palisade cells are present in a single layer but all cells contain prismatic or cluster crystals of calcium oxalate. Near the veins, some times, the idioblasts are present containing microspheroidal crystals. Midrib shows many bicollateral vascular bundles arranged in an arc.

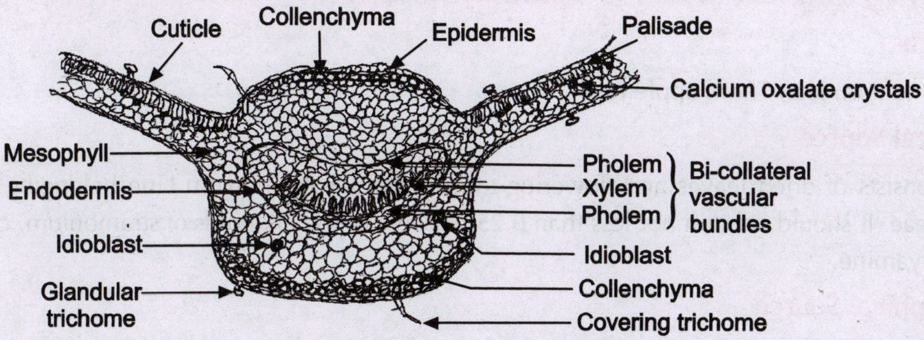

Fig. 15.26: T.S. of Hyoscyamus leaf

Chemical Constituents

The total alkaloids present in the drug range from 0.05 - 0.15 per cent in which about 75 per cent is hyoscyamine. Atropine and hyoscine are present in fewer amount. The alkaloidal percentage is more in petiole than in the stem or lamina.

The principal alkaloid of this drug (-) hyoscyamine (C_{17} H_{23} NO_3) is an ester of tropic acid and tropine, and is more active than the racemic form i.e. atropine. During the extraction, it is racemized to atropine (see belladonna herb).

Chemical Test

It gives Vitali-Morin test positive in case of datura herb.

Standards of Quality

Acid insoluble ash	:	Not more than 12.0 per cent
Foreign organic matter	:	Not more than 2.0 per cent
Stems	:	Not more than 3.0 per cent having a width greater than 5.0 cm

Uses

It is used to counteract gripping due to purgatives and also to relieve spasms of urinary tract. It is also sedative and used to check salivary secretion. It is an expectorant too. It is an antispasmodic and antiasthmatic.

Dose

Hyoscyamine sulphate - 125 – 250 m, 3 - 4 times a day.

Allied Drugs

The other species of this drug viz. Egyptian Henbane (H. muticus) is used in the form of dried leaves and flowering tops and grows wildly in Egypt, Arabia, Iran. The leaves contain high proportion of total alkaloids upto 1.7 per cent, in which about 75 per cent is l-hyoscyamine. The other alkaloids are apoatropine, l-hyoscine and small quantities of noratropine and norphyoscine. It is mainly used as a commercial source of atropine.

STRAMONIUM

Synonyms

Stramonium leaf, Thorn apple leaf.

Biological Source

It consists of dried leaves and flowering tops of Datura stramonium Linn belonging to family Solanaceae. It should contain not less than 0.25 per cent of the alkaloids of stramonium, calculated as hyoscyamine.

Geographical Source

It is indigenous to Caspian region and is cultivated in United States, South America, France, Germany and Hungary.

History

Both for datura and stramonium, the name of the genus owes its origin to Sanskrit and Arabian words. **Gerarde** brought the seeds of drug from Constantinople and introduced this drug in Europe in sixteenth century.

Cultivation and Collection

Stramonium is an annual herb. It is propagated by seeds, which are sown by keeping a distance of 1 metre between two plants and 60 m between two rows. It needs calcarious soil. It flowers in September. The collection of the leaves is done twice, once before flowering and another during the flowering stage.

Macroscopic Characters (Fig. 15.27)

Stramonium leaves are greyish green in colour, unpleasant in odour and bitter in taste. Because of drying, they are shrunken and brittle. They are 8 - 25 cm in length, 7 - 16 cm in breadth and ovate or triangularly ovate in shape. The leaves have a short petiole with unequal base. Stramonium bears funnel shaped, white or purple coloured flowers, with 5 stamens and superior ovary. The ovary is bilocular at the upper part and quadrangular on lower portion.

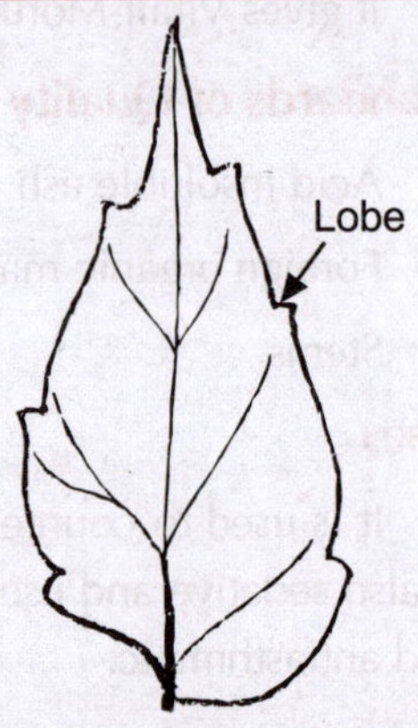

Fig. 15.27: Stramonium Leaf

Extra Features

Leaves are very thin, brittle, shortly petiolate and unequal at the base. The apex is acuminate and margin is dentately lobed or irregularly serrate. Flowers are solitary, about 7.5 cm in length. They arise in the axil, where from the stem forks, and are erect with a short pedicel. Calyx is about 3.5 cm and sharply folded. Corolla is funnel shaped. Five stamens and superior ovary, which is bilocular in the upper part and quadrangular in lower part are present.

Microscopic Characters (Fig. 15.28)

Stramonium leaves have smooth cuticle followed by a layer of wavy epidermal cells. Palisade in single layer is present on upper surface only. At the base of the palisade tissue, crystal layer (each cell containing cluster or prism of calcium oxalate) is present. The midrib shows a bicolateral structure. Upper and lower hypodermal collenchyma is characteristic to midrib. Epidermis shows

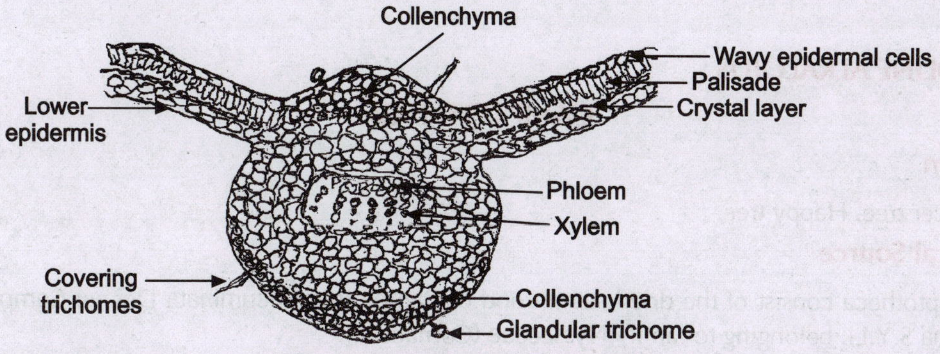

Fig. 15.28: Stramonium leaf in T.S.

numerous multicellular conical uniseriate trichomes. Short glandular trichomes are occasionally present on both the surfaces. Anisocytic stomata are more numerous on the lower epidermis than the upper one. Stem in the transverse section shows the presence of uniseriate multicellular covering trichomes, pericyclic fibres, wood fibres (i.e. xylem) and large vessels.

Chemical Constituents

The amount of total alkaloids found in drug is between 0.2 - 0.5 per cent, in which l-hyoscyamine and hyoscine are principal alkaloids along with small quantity of atropine. Hyoscine occurs in high quantity in young leaves. However, l-hyoscyamine content of fully grown plant is more i.e. up to 2/3rd of total alkaloids. The percentage of alkaloids is very less in stems and hence pharmacopoeia limits its content to 3 per cent. The seeds contain 0.2 per cent of alkaloids and about 25 per cent of fixed oil.

It gives Vitali Morin reaction due to tropane alkaloids.

Standards of Quality

Stems	:	Not more than 3 per cent and maximum diameter 5 mm
Foreign organic matter	:	Not more than 2 per cent
Acid-insoluble ash	:	Not more than 5 per cent
Sulphated ash	:	Not more than 20 per cent

Uses

Stramonium leaves are used in the treatment of asthma. They are similar in action to belladonna and hyoscyamus. It is used to control salivation, muscular rigidity and the tremors in parkinsonism. Atropine depresses the nerve endings to the secretory glands and plant muscles.

Hyoscine lacks the central stimulant action of atropine. Its sedative properties enable it to be used in the control of motion sickness.

Substitution and Adulterants

Different substitutes used to adulterate stramonium are as follows.

1. Xanthium strumarium (Compositae): Leaves are pale green, rhomboid-ovate and with ciliated margin.
2. Solanum nigrum (Solanaceae): The leaves are ovate, petiolate and coarsely toothed. They resemble small stramonium leaves, but contain no calcium oxalate crystals.

QUINOLINE ALKALOIDS

CAMPTOTHECA

Synonym

Cancer tree, Happy tree.

Biological Source

Camptotheca consist of the dried stem wood of Camptotheca acuminata Decne; Camptotheca lowreyona S.Y.Li. belonging to family Nyssaceae (Cornaceae).

Geographical Source

It is distributed in China and Tibet. It is indigenous to southern China. Camptothecin was discovered by National Institute of Health, U.S.A. during the antitumor screening programme for this plant.

Cultivation and Collection

It is a medium sized deciduous tree growing up to 10 meters and its plants can be raised by sowing the seeds or by means of vegetative propagation i.e. using the cuttings. Cultivation is done in autumn. Shading of whole seedlings increases the concentration of alkaloid in the leaves and lowers in the roots. Drought conditions increase the concentration of alkaloids in leaves.

Fruits are collected late in autumn, while leaves, twigs, and bark is collected around the year. All plant parts are dried in the sunlight and then used.

Fig. 15.29: Camptotheca herb

Macroscopic Characters

Colour	:	Leaves are dark green in colour with reddish petiole.
Odour	:	Odourless
Shape	:	Leaves are entire, accuminate, ovate and lanceolate.
Size	:	8-10 cm in length and 3-5 cm in width. Flowers are red coloured.
Taste	:	Bitter.

Chemical Constituents

Camptotheca plant contains 0.004 - 0.03 per cent of quinoline alkaloid camptothecin. Other minor constituents of the drug are Irinotecan, topotecan, 9-aminocamptothecin, 10-hydroxy camptothecin and 10-methoxy camptothecin. All plant parts leaves, bark, fruits, and twigs also contain camptothecin, young leaves have highest concentration.

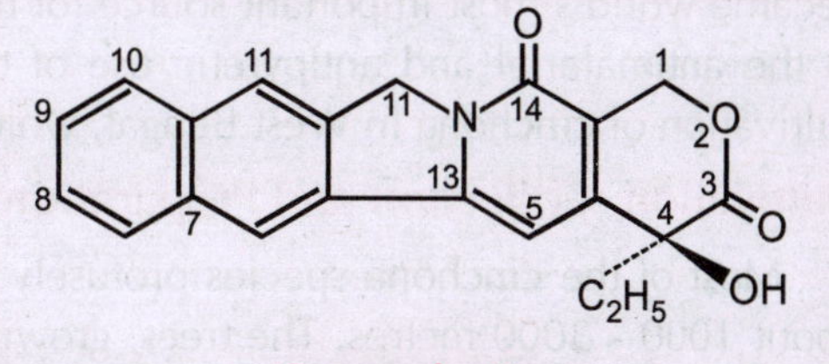

Camptothecine

Camptothecin has yellow needle shaped crystals. It is a very weak basic alkaloid and does not form stable salts with mineral acids. It does not respond Dragendorff's and Mayer's reagent test. The other minor constituents are 10-hydroxycamptothecin and 10-methoxycamptothecin.

Uses

Camptothecin is prototype DNA topoisomerase I inhibitor. It has shown a broad spectrum antitumour activity in animal tests. Its sodium salt shows pronounced antileukemic activities. Preliminary and Phase II clinical studies of sodium salt for gastrointestinal cancer have given promising results, but show extremely high toxicity in both animals and human beings.

Camptothecin inhibits the replication of DNA viruses such as herpes virus, adenovirus and vaccinia virus, but shows no effect on RNA virus like poliovirus.

Camptothecin and 10-hydroxycamptothecin are used in China, clinically, for treatment of leukemia and liver cancer. Their toxicity is less than sodium salt of camptothecin.

CINCHONA

Synonyms

Jesuit's bark, Peruvian bark

Biological Source

It is the dried bark of the cultivated trees of Cinchona calisaya Wedd., C. ledgeriana Moens, C. officinalis Linn, C. succirubra Pav. ex-klotzsch, or of hybrids of either of the last two species with either of the first two. Cinchona belongs to family Rubiaceae. It contains not less than 6 per cent of total alkaloids of cinchona.

Geographical Source

India, Bolivia, Columbia, Ecuador, Peru, Tanzania, Guatemala, Indonesia and Sri Lanka are the countries where cinchona is found. In India, it is cultivated in Annamalai hills (Coimbatore district) and Nilgiri hills (Nilgiri district) in Tamil Nadu and in Darjeeling area of West Bengal.

History

Cinchona is native to Eastern slopes of the Andes at high altitudes (1500 - 2500 metres). It is known that the bark was first used as an antipyretic in 1630 by Jesuits, although it was discovered in 1513 in Peru. Owing to the efforts of Viceroy of Peru, Count Chinchon it was introduced as a drug in Europe around 1655. It was officially reported as an infusion in London Pharmacopoeia in 1677. In the honour of viceroy, the genus was described by Linnaeus as Cinchona in 1742. After the isolation of quinine and cinchonine in 1820 by Pelletier and Canventon, the alkaloids or their mixtures came into use as a medicine. In 1860, C. calisaya, C. micrantha and C. succirubra were introduced in India by Markham. Dutch, introduced C. ledgeriana in Java and sooner the country became world's most important source for this drug with a high alkaloidal content. In India, owing to the antimalarial and antipyretic use of this drug; right from 1880, a large area was taken for cultivation of cinchona in West Bengal, which eventually shifted to South India.

Cultivation, Collection and Preparation for Market

Most of the cinchona species profusely grow in sub-tropical or tropical climates at a height of about 1000 - 3000 metres. The trees, growing below this height are found to have less percentage of quinine. The rainfall conditions required are uniform (from 250 - 380 cm in a year). The favourable growth is achieved between an atmospheric temperature of 60 - 75 F. Cinchona requires light, well drained forest soil which is rich in organic matter. The acidic soil having a pH of 4.2 - 5.6 and a small amount of nitrogen are found to be most favourable for growth. Cinchona needs slopping situation, high humidity and protection from wind.

The propagation is done with either seeds or budding or layering. In West Bengal, only budding is practised and in Tamil Nadu, the budding and layering methods are applied. The seeds of cinchona are very small and light in weight. About one gramme of cinchona seeds contain 3500 seeds. They are admixed with soil during sowing. The maintenance of genetic purity causes a problem as high cross fertilization occurs in cinchona plants. This affects the yield, like in high alkaloid content giving species, such as C. ledgeriana, the average alkaloid content is reduced. The seeds should be immediately used for propagation as on storage they lose their viability. The germination takes place in 3 - 6 weeks. The seedlings with 2 pairs of leaves are transplanted and space of 6 - 10 cm is maintained inbetween two seedlings and 2 rows. The young seedlings are protected from direct sunlight. In forest soil, they are transplanted after 15 months of growth and preferably before heavy rainfall. A distance of 2 2 metres is maintained between two plants. As cinchona consists of stem, as well as root bark, the plants from 4 - 20 years of age are selected for harvesting, but the maximum alkaloidal content is found to 6 - 10 years old plants.

Collection of Cultivated Cinchona

The bark is collected by coppicing method. For this purpose, vertical incisions are made on branches, trunk of tree and these incisions are connected by horizontal circles. The bark is then stripped off and dried in sun light and further by artificial heat. The drying is done below 175 F. During drying, the bark loses up to 70 per cent of its weight. The care should be taken to avoid molding or fermentation during drying. The quills of drug are packed in gunny bags and marketed. The root bark is collected by uprooting the trees and bark is separated manually.

During the two world wars, Java and Indonesia lost their positions as potential producers of cinchona. After that, India has gained the prime position as producer and supplier of cinchona and quinine. By 1985 - 86, the production had reached upto 10 lakh kg of bark and about 26,000 kg of quinine salts.

Extraction of Quinine

For extraction of quinine, the bark is powdered and extracted with benzene or toluene in presence of alkali. Further, the alkaloids are extracted with dilute sulphuric acid. By bringing the acid extract to neutrality, quinine sulphate separates, as it is sparingly soluble.

Macroscopic Characters (Fig. 15.28)

Cinchona bark has a slight and characteristic odour, but somewhat astringent and intensely bitter taste. In general, the bark is available in the form of quills and curved pieces.

Flowering bunch of Cinchona

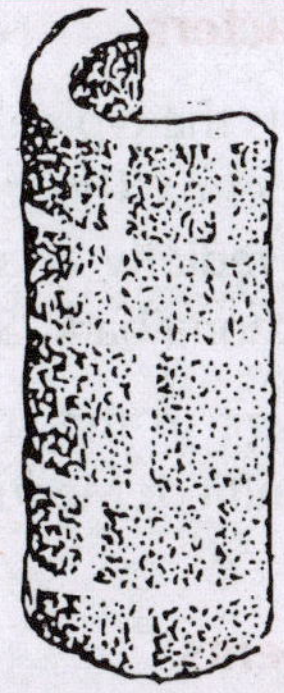
Piece of Cinchona Bark

Fig. 15.30

Stem bark: It is up to 30 cm in length and about 2 to 6 mm in thickness. The outer surface shows dull brown grey or grey colour and many a time, shows presence of mosses and lichens owing to its growth in heavy rainfall areas. The bark is rough and has transverse fissures. These fissures are different in different species. It is furrowed or wrinkled longitudinally. The outer bark in some varieties shows exfoliation. The inner surface is pale yellowish-brown to deep reddish-brown and the colour depends on the species. The fracture is short in external layers and fibrous in the inner portion.

Root bark: It occurs in length of 2 - 7 cm. The bark is curved, twisted or irregularly channelled. The outer and inner surfaces are similar in colour. The outer surface is scaly and shows depressions. The inner surface is striated.

The different commercial varieties have some special characters. C. succirubra is also called as red bark, while C. ledgeriana is referred to as yellow bark. C. robusta is the hybrid between C. succirubra and C. officinalis.

Table 15.4: The typical characters of 4 main species of cinchona

Characters	*C. calisaya*	*C. ledgeriana*	*C. officinalis*	*C. succirubra*
Size	Diameter is from 12 - 25 mm and thickness from 2 - 5 mm	Diameter is 12-25 mm and thickness varies from 2 - 5 mm	Diameter is upto 12 mm and thickness is upto 1.5 mm	Diameter is from 20 - 40 mm and thickness from 2 - 5 mm
Other features	Broad longitudinal fissure with transverse cracks.	Broad longitudinal fissures and cracks more in number, but less deep. Some pieces show longitudinal wrinkles and reddish warts	It shows a number of transverse cracks	Well marked longitudinal wrinkles, but less number of transverse cracks. Only some pieces show reddishwarts
Powder	Cinnamon brown	Cinnamon brown	Yellow	Reddish brown

Microscopic Characters

Cinchona exhibits the typical histological characters of the bark. The cork cells are thin-walled, followed by phelloderm. The cortex consists of several secretory channels and phloem fibres. Medullary rays with radially arranged cells are present. Idioblast of calcium oxalate is the specific characteristic of cinchona bark. Starch grains are present in the parenchymatous tissues. Stone cells are rarely present in the structure. A few of the cork cells are lignified. Medullary rays are 2 to 3 cells wide.

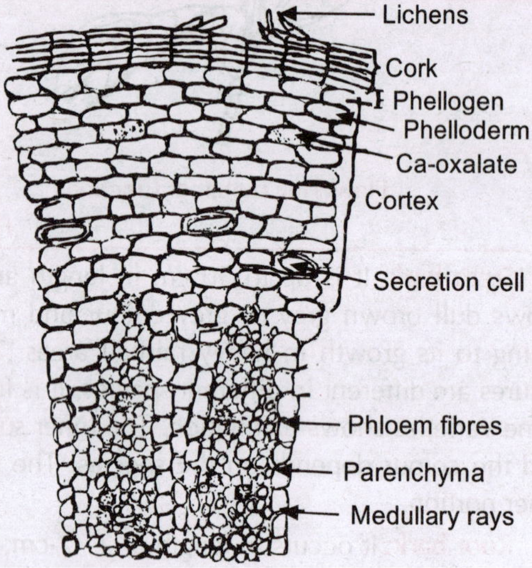

Fig. 15.31: T.S. of Cinchona Bark

Chemical Constituents

Cinchona bark contains about 25 alkaloids, which belong to quinoline group. The important alkaloids are quinine, quinidine, cinchonine and cinchonidine. The alkaloids of lesser importance are quinicine, cinchonicine hydroquinine, hydrocinchonidine and homocinchonidine. C. succirubra contains 5 - 7 per cent of total alkaloids, of which 30 per cent is quinine. C. ledgeriana yields from 6 - 10 per cent and, in some cases, upto 14 per cent of total alkaloids, with upto 75 per cent is quinine. C. salisaya has 6 - 8 per cent total alkaloids (about 50 per cent quinine).

Quinine and Quinidine are stereoisomers of each other. Quinidine is also obtained commercially from cuprea bark i.e. Ramijia pendunculata Fluckiger belonging to family Rubiaceae, or by isomerization of quinine.

Quinine and Quinidine form many salts, but medicinally their sulphates are more significant. Cinchonine and cinchonidine are also isomers of each other.

Apart from alkaloids, cinchona also contains quinic acid and cinchotannic acid. In the plant, the alkaloids are present as salts of these acids. Cinchotannic acid decomposes into insoluble cinchona red, due to its phlobatannin nature. Cinchona bark also contains a glycoside called quinovin, tannins and bitter essential oil.

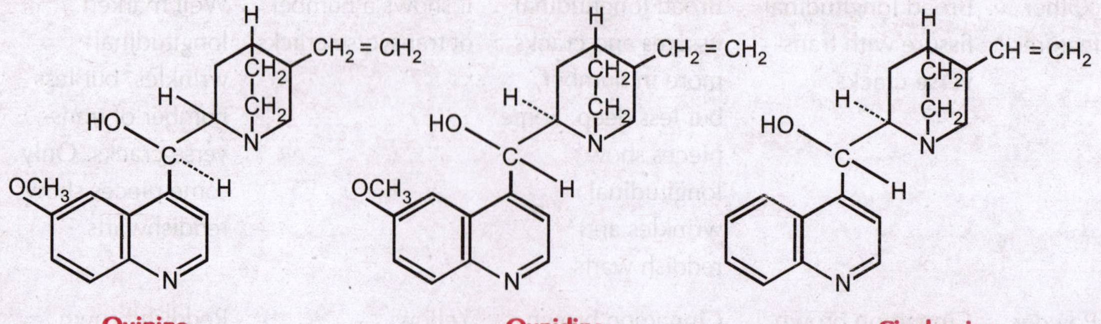

Quinine **Qunidine** **Cinchonine**

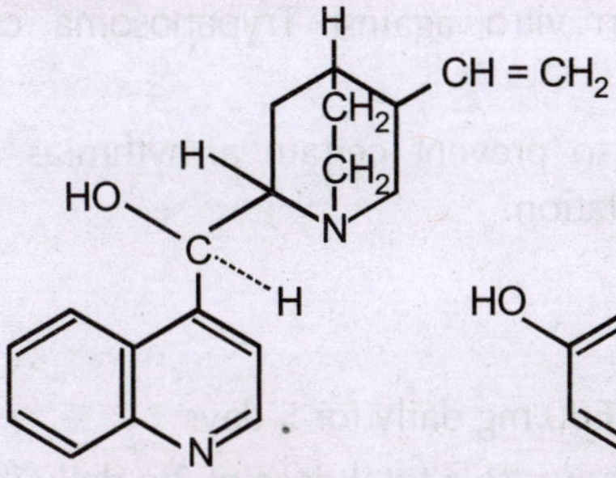

Cinchonidine

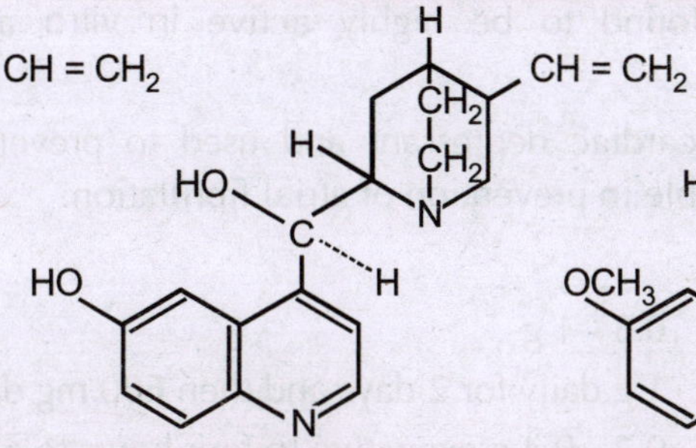

Cupreine

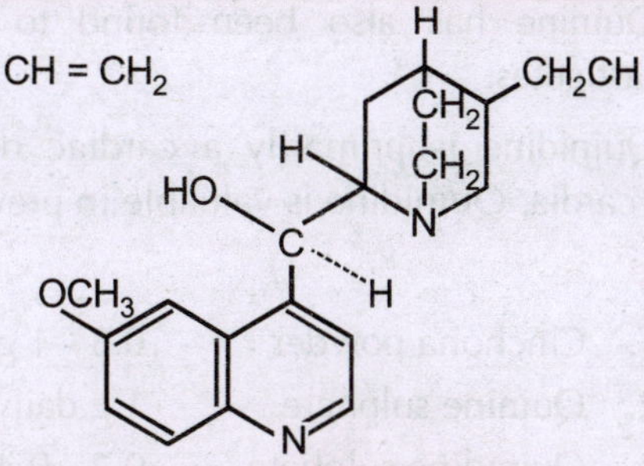

Hydroquinine

The alkaloid quinine occurs as bitter white crystals and it darkens when exposed to light and has fluorescent properties. It shows a strong blue fluorescence in ultra-violet light. This fluorescence is enhanced in presence of dilute sulphuric acid. Quinine forms salts with different acids. Quinine sulphate $(C_{20}H_{24}N_2O_2)_2 . H_2SO_4 . 2H_2O$ is important from pharmaceutical point of view. It has very less solubility in water (1 in 810 parts of water), due to which, it is suitable for oral use.

Quinidine $(C_{20}H_{24}N_2O_2)$ is similar to quinine in its physical and chemical properties and has higher water solubility. The free base is soluble in water, ethyl alcohol, methyl alcohol and chloroform.

Chemical Tests

1. Heat the powdered drug in a dried test tube with little glacial acetic acid, purple vapours are produced at the upper part of test tube.
2. Thalleoquin test: The powdered drug gives emerald green colour with bromine water and dilute ammonia solution.
3. Quinidine solution gives a white precipitate with silver nitrate solution, which is soluble in nitric acid.

Standards of Quality

Total ash	:	Not more than 4 per cent
Acid insoluble ash	:	Not more than 2 per cent
Foreign organic matter	:	Not more than 2 per cent
Water soluble extractives	:	Not less than 32.0 per cent
Acid soluble extractives	:	Not less than 10.0 per cent
Loss of Drying	:	Not more than 8.0 per cent

Uses

Cinchona bark is antimalarial in nature. The cinchona preparations like cinchona extract, compound cinchona tincture etc. are also employed as bitter stomachics and antipyretics. Quinine and its salts are used in the treatment of malaria. Quinine is a protoplasmic poison, especially for protozoa like Plasmodium vivax, P. falciparum, P. malarie and P. fatal, and hence, used as powerful antimalarial drug.

Recently the pharmacokinetic studies on quinine have shown that it can be better used in other forms. Infusion of quinine rather than intravenous injection eliminates the risk of sudden death. Secondly, quinine in microencapsulated form has been reported to give better bioavailability.

Quinine has also been found to be highly active in vitro against Trypanosoma cruzi epimastigotes.

Quinidine is primarily a cardiac depressant and used to prevent certain arrhythmias and tachycardia. Quinidine is valuable in prevention of atrial fibrillation.

Dose

1. Cinchona powder - 0.3 - 1 g
2. Quinine sulphate - 1 g daily for 2 days and then 600 mg daily for 5 days
3. Quinidine sulphate - 0.2 - 0.4 g every two to four hours to a total dose of 3 g daily in atrial fibrillation.

Substitutes

Cuprea bark (Remijia pedunculata), a coppery red coloured drug, contains quinine, quinidine and other alkaloids which resemble to those from cinchona bark. The bark contains numerous stone cells. Along with cinchona alkaloids, it also contains cupreine. False cuprea bark (R. purdiena) contains an alkaloid called cusconidine, traces of cinchonine, cinchonamine, but no quinine.

AMINO ALKALOIDS

COLCHICUM

Synonyms

Meadow saffron seeds, autumn crocus

History

Though, it was known from the time of Dioscorides, it was not much used, owing to its toxic nature. In the medieaval times, Arabian people were using it for treatment of gout. It was reintroduced in European countries towards beginning of seventeenth century and first appeared in the London Pharmacopoeia in 1616. Pelletier and Caventou isolated colchicine in 1820.

Fig. 15.32 (a): Colchicum autumnale herb

Biological Source

It consists of the dried ripe seeds of Colchicum luteum Baker and Colchicum autumnale Linn, belonging to family Liliaceae. Colchicum corm is also used medicinally.

Geographical Source

It is found and cultivated in various parts of Europe, like England, Czechoslovakia, Holland, Poland and Yugoslavia. It is also cultivated in India (in Western Himalaya and Kashmir regions).

Cultivation, Collection and Preparation

In nature, colchicum propagates by repeating the life cycle with the corm which is present as a swollen underground stem with sheathing leaves. Towards the end of summer, the fully grown corm develops daughter corms in the axil of scaly leaf near the base. These daughter corms develop parasitically on parent corm and subsequently, the parent corm withers away. After this, the daughter corms develop into new plants.

In Jammu and Kashmir and different parts of Europe and Africa, the drug is obtained by propagation with seeds. The propagation is done by sowing the seeds in boxes at an altitude of 1000 - 3000 m. The seedlings are transplanted in open fields at a distance of 1 m. The plants bear the capsular fruits after one year of vegetative growth. The fruits are collected before dehiscence and dark seeds are separated, processed and graded. The corms are isolated and the adhering scales and coats are removed. The corms are sliced transversely and dried below 65 C.

Macroscopic Characters

(a) Colchicum Seeds

Colchicum seeds are very hard in nature and show a reddish-brown testa. The seeds have a projection at the hilum and from there develops strophiole, which is an outgrowth of testa. The seeds are 2 - 3 mm in diameter, having bitter and acrid taste and no odour. The corms are 2 - 3 cm in diameter and used in the sliced form which are reniform and ovate in shape, with 2 - 5 mm, thickness. They have a short fracture, bitter taste and no odour.

(b) Colchicum Corms

Colour : Yellowish-brown

Odour : None

Taste : Bitter and acrid

Size : Slices are about 2 - 5 mm in thickness

Shape : Sub-reniform or ovate in outline or

Extra features : Fracture is short and cut surfaces are white and starchy, showing greyish points.

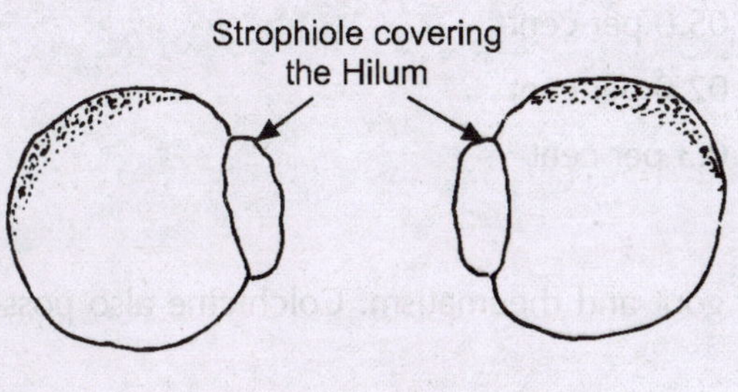

Fig. 15.33 (a): Colchicum Seeds

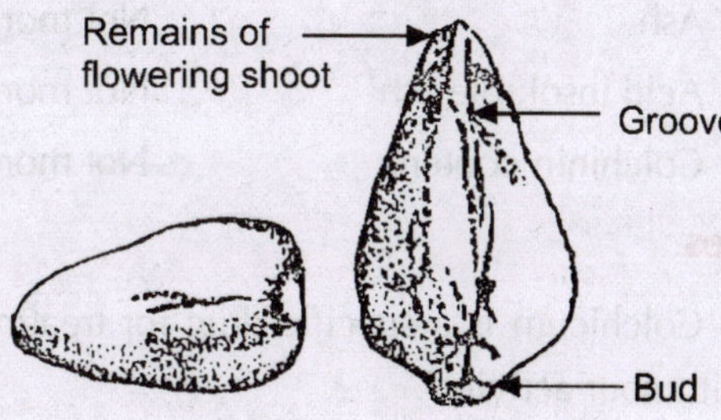

Fig. 15.33 (b): Colchicum Corms

Microscopic Characters

The section of seed shows parenchyma and endosperm. The parenchymatous cells are reddish-brown with thick walls. The endospermic cells show pitted walls and contain aleurone grains and fixed oil. The strophiole portion of seed contains starch.

The corm has epidermis, parenchyma and vascular tissue. The parenchymatous cells have abundant starch grains. The epidermis has circular stomata. In vascular part, the xylem vessels are spiral or annular.

Chemical Constituents

Colchicum seed contains 0.2 - 1 per cent of amino alkaloids of which colchicine is the main constituent. The seeds contain upto 0.8 per cent of colchicine and in corms, it is upto 0.6 per cent. Colchicum also contains demecolcine. Both the alkaloids contain tropolone or cycloheptatrien-ol-one ring structure.

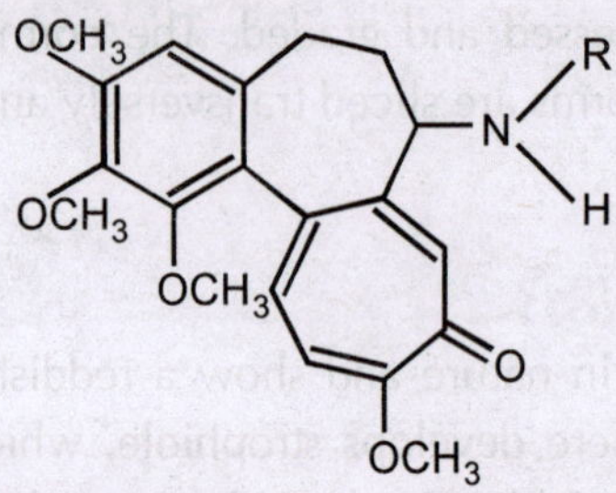

Colchicine, R = $COCH_3$, Demecolcine, R = CH_3

Colchicine ($C_{22}H_{25}O_6N$) is obtained as pale yellow crystals, amorphous or in powder form. It has a bitter taste and is odourless. It darkens on exposure to air. Colchicine is freely soluble in alcohol and chloroform, soluble in 25 parts of water and in 220 parts of solvent ether.

Chemical Tests

(1) Colchicine gives yellow colour with 70 per cent sulphuric acid.

(2) Alcoholic solution of colchicine, when treated with ferric chloride gives red colour.

Standards of Quality

Foreign organic matter	:	Not more than 2.0 per cent
Ash	:	Not more than 05.0 per cent
Acid insoluble ash	:	Not more than 02.0 per cent
Colchinin content	:	Not more than 0.3 per cent

Uses

Colchicum is a specific drug for treatment of gout and rheumatism. Colchicine also possesses antitumour activity.

Non Medicinal Uses

Apart from medicinal use, colchicine is widely accepted and practised as a chemical agent for bringing the polyploidy (increase in number of chromosomes) and hence used in horticulture and cultivation of medicinal plants.

Dose

Colchicine: 500 – 650 g, orally 1 - 3 times a day, orally

500 g to 1 mg,1 - 2 times a day, intravenously

EPHEDRA

Synonym

Ma-Huang

Biological Source

It consists of the dried young stems of Ephedra gerardiana (wall.) Stapf, and E. nebrodensis (Tineo.) Stapf, belonging to family Gnetaceae (Ephedraceae). Ephedra should contain not less than 1 per cent of total alkaloids, calculated as ephedrine.

Geographical Source

The main source of ephedra is from China, Pakistan, North-West parts of India, Australia, Kenya, Spain and Yugoslavia.

History

The drug originally belongs to Chinese System of Medicine. It has been used in China since last 5000 years. In Chinese language, it is called as Ma-Huang where 'Ma' denotes astrigent taste and 'Huang' is for yellow colour of drug. The references about this drug are found in the herbal of the emperor Shen Nung (2700 B.C.) and in 'Chinese medicinal plants' (1596 A.D.) by Pen T'sao Kang Mu. In those times, it was used for treatment of respiratory problems, fever and also for improving circulation. The drug was for the first time explored chemically by Yamanishi in Japan and he isolated ephedrine in crude form in 1885, which was further obtained in pure form by Nagai and Hari. Merck of Darmstadt, a German firm, carried out detailed search on Ephedra helvectica and isolated ephedrine in 1888. K. K. Chen and C. F. Schmidt are mainly credited for introduction of ephedra and ephedrine in modern therapeutics.

Cultivation, Collection and Preparation

Ephedra can be cultivated at an altitude of 2500 to 3000 m. Annual rainfall should not exceed 50 cm. It can be propagated by seeds or by layers or divisions of the root stock. Seeds are sown early in the spring at a distance of 5 cm, keeping the distance of one meter between 2 rows. The plants are collected after attaining the age of 4 years for the extraction of alkaloid. During this period, proper irrigation and weeding are necessary. The alkaloidal content of the drug varies from season to season. It is found to be maximum in autumn; when plants and twigs are dark in colour. Twigs are generally dried in sun or even by artificial ways. After drying, they are stored in dry and well closed containers, away from light.

Macroscopic Characters

Ephedra is a gymnospermous plant bearing thin stems which are woody, cylindrical and grey to greenish in colour (about 5 mm in diameter).

It shows the internodes at a distance of about 3 to 3.5 cm. Ephedra bears the scaly leaves from the nodes in a whorl of 2. The bases of the leaves are dark brown and they are joined on all sides of the node-forming a sheath. It bears a terminal bud, which is short and usually constricted at base. The male spikes are solitary, ovate, sessile and crowded.

Microscopic Characters

The T.S. of ephedra shows the following characteristics.

(i) unicellular epidermis made up of quadrangular cells along with thick-walled cuticle,

(ii) vertical rows of sunken stomata and papillae on the ridges,

(iii) chlorenchymatous cortex,

(iv) non lignified, hypodermal fibres,

(v) lignified pericyclic fibres,

(vi) crystals of calcium oxalate in the cortex, and

(vii) parenchymatous dark brown coloured pith.

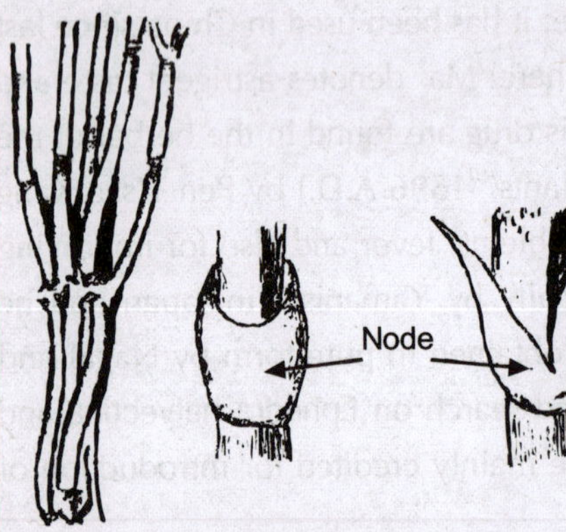

Fig. 15.34: Ephedra Twig

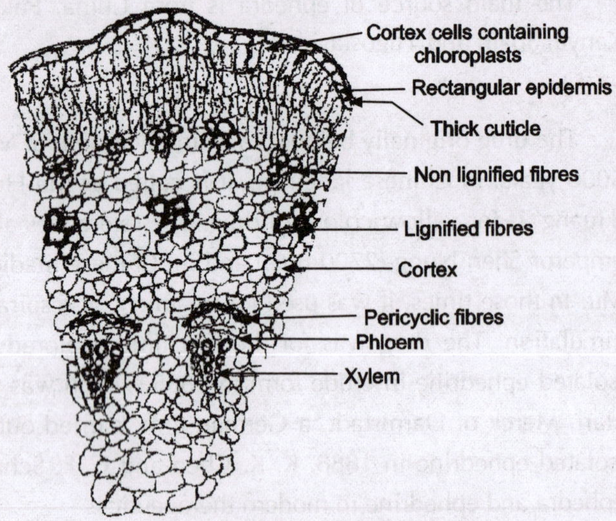

Fig. 15.35: T.S. of Ephedra Stem

Chemical Constituents

Ephedra contains the amino alkaloids. They are ephedrine, nor-ephedrine, n-methyl ephedrine, pseudo-ephedrine, etc.

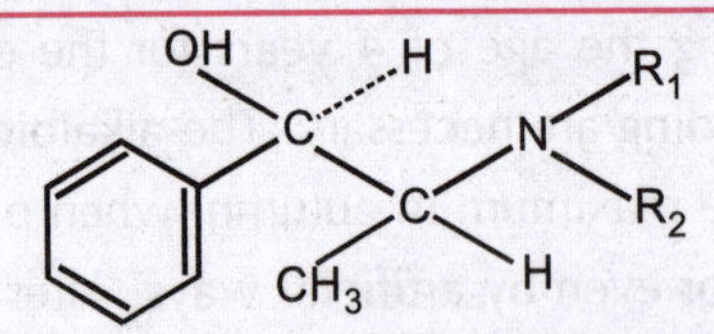

(–) Ephedrine	R_1 = H,	R_2 = CH_3
(–) Nor-ephedrine	R_1 = H,	R_2 = H
(–) n-methyl ephedrine	R_1 = CH_3,	R_2 = CH_3

Chemically, ephedrine ($C_{10}H_{15}$ NO) is 1-phenyl - 1 - hydroxy - 2 - methylaminopropane and is soluble in water, alcohol, organic solvents and oils. It is odourless and colourless, deliquescent and decomposes when exposed to air.

Along with the amino alkaloids, macrocyclic alkaloids called ephedradines are present in roots. The drug also contains oxazolidone.

Standard of Quality

Total ash : Not more than 10.0 per cent.

Alkaloidal contents : Not less than 2.25 per cent.

Chemical Test

Ephedrine is dissolved in water and dilute hydrochloric acid and then treated separately with copper sulphate and sodium hydroxide. The solution gives violet colour. If shaken with solvent ether, the organic layer shows purple and aqueous layer shows blue colour.

Uses

Ephedra and its alkaloids show sympathomimetic effects. Hence, it is used as a bronchodilator in asthma and also in the treatment of allergic conditions like hay fever. As compared to adrenaline, the onset of action for ephedrine is slow, but the effect is much prolonged, as it is not quickly hydrolysed by mono amino oxidase in the body. Ephedrine is also used to correct the low blood pressure conditions, because of its peripheral contraction of arterioles. Ephedradines have hypotensive effects.

In 1995-96 and 1996-97, India has exported total ephedrine salts of worth ₹ 820.8 lakhs and ₹ 1555 lakhs respectively.

Dose

Ephedrine hydrochloride or sulphate 25 - 50 mg, 6 - 8 times a day, orally or parenterally and 0.1 ml 1 - 3 per cent solution, 2 - 3 times a day intravenously.

Other Species and Allied Drugs

The genus Ephedra has about 45 species, amongst which nearly 25 species contain ephedrine. The prominent species containing ephedrine are *E. equisetina and E. sinica* (both Chinese). *E. intermedia, E. major, E. helryetica and E. alata.* They contain from 35 - 87 per cent of ephedrine in total alkaloids. The other plants containing ephedrine are *Aconitum napelles* (Ranunculaceae); *Sida cordifolia*; and *S. rhombifolia* (Malvaceae); *Roemeria refracta* (Papaveraceae); and *Taxas baccata* (Taxaceae).

ALLIED DRUGS

The plants having chemical contents similar to colchicine type of alkaloids belong to different genera, such as *Dipidax, Gloriosa, Androcybium, Camptorrhiza,* etc.

The other genera in which colchicine is present are from the same family Liliaceae, like *Androcymbium, Bulbocodium, Dipidax, Littonia, Fritillaria, Ornithoglossum,* etc. and *Narcissus* and *Crocus* from families Amaryllidaceae and Iridaceae respectively.

GLORIOSA

Synonym

Glory Lily, Climbing lily.

Biological Source

These are the dried tubers of *Gloriosa superba* Linn, belonging to family Liliaceae.

Geographical Source

This drug is an ornamental plant indigenous to tropical Asia and Africa. The drug is obtained on a large scale from Tamil Nadu (India).

Gloriosa has gained the importance in medicine in recent years only. Due to heavy demand for colchicine, many other related plants have been screened. Some of the plants which showed potential are Iphigenia, Morendera, Androcymbium and Gloriosa. Out of these, Gloriosa from India is indicated to be the promising drug for the production of colchicine on commercial scale.

Macroscopic Characters

Glory lily occurs as a branched herbaceous **climber** common in low jungles.

Its characteristics leaves and the attractive coloured flowers are the points for identification. The tubers are perennial and fleshy. They are cylindrical bifurcated, usually 'V' shaped and with the 2 limbs. They are 30 cm long and about 5 cm in diameter.

Fig. 15.36: ***Gloriosa superba*** **flowering twig**

Isolation

The conventional method for extraction of colchicine by alcohol and chloroform extract yields less percentage of colchicine type alkaloids from G. superba. But by present aqueous extraction method, more than 80 per cent colchicine is recovered and yield is also comparatively better.

Chemical Nature

It contains colchicine along with its different derivatives in minor quantities.

Uses

As a source of colchicine, and in the treatment of gout and inflammation.

PYRIDINE ALKALOIDS

ARECA NUT

Synonym

Betel nut, Supari.

Biological Source

These are the dried ripe seeds of Areca catechu, belonging to family Palmae. It should contain not less than 0.25 per cent of alkaloids, calculated as arecoline.

Fig. 15.37: Areca plant with fruits

Geographical Source

Areca is cultivated in different parts of world such as India, Sri Lanka, South Eastern Asia, Philippines, East Africa, etc.

Macroscopic Characters (Fig. 15.37)

The plant of Areca is a tall palm which bears the fruits of nut type, each containing a single seed, thin seed coat and a large ruminant endosperm. The testa is deep-brown coloured and colourless exhibits a network of depressed, fawn coloured lines.

The astringent seed is very hard and towards the flattened end, a small embryo is present.

Chemical Constituents

Areca contains a number of alkaloids, belonging to pyridine piperidine group and derived from amino-acid lysine. These alkaloids are reduced pyridine derivatives. The various alkaloids are arecoline, arecaidine, guvacine (tetrahydronicotinic acid) and guvacoline. Arecoline is methyl ester of arecaidine, while the latter is N-methylated derivative of guvacine. The drug also contains lipids 150 per cent, volatile oil, tannins and gum. Only arecoline possesses physiological activity. Arecoline is a liquid alkaloid.

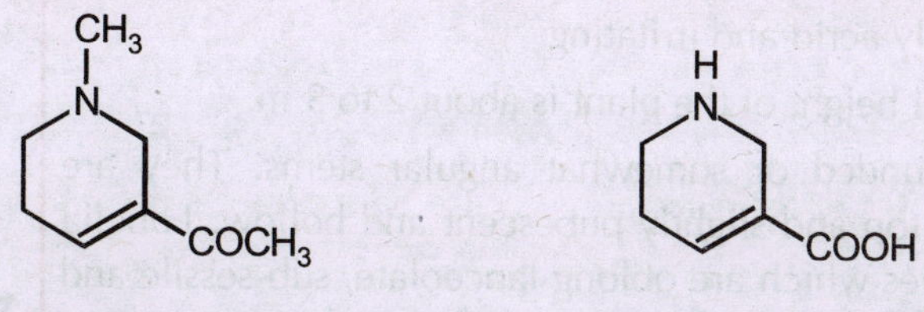

Arecoline Guvacine

Standards of Quality

Foreign organic matter	:	Not more than 2 per cent
Content of stalks	:	Not more than 60 per cent
Ash	:	Not more than 5 per cent
Water soluble extractives	:	Not less than 20 per cent
Acid-insoluble ash	:	Not more than 2 per cent

Uses

Arecoline is parasympathomimetic. Areca has sialogogue properties and is consumed as masticatory in India and other eastern countries. But, the habit of chewing areca may cause oral leukoplakia. Areca is an anthelmentic drug and used as vermicide and taenifuge in veterinary practice. **It is not used in human medicine.**

LOBELIA

Synonyms

Indian tobacco, Asthma weed

Biological Source

It consists of the dried aerial parts of Lobelia nicotianaefolia Heyne, collected in October-November and dried in the shade. It belongs to family Campanulaceae.

Lobelia should contain not less than 0.55 per cent of total alkaloids, calculated as lobeline.

The other species viz. L. inflata is official in British Pharmacopoeia and United States Pharmacopoeia and consists of herbs of this species.

Geographical Source

It is a biennial or perennial herb growing on the Western Ghats of Maharashtra to Tryare in Kerala at an altitude of 700 to 2300 m. It is also found growing in South Maharashtra Nilgiri and in Karnataka. Lobelia inflata is indigenous to eastern and central U.S.A. and Holland.

Cultivation, Collection and Preparation

Lobelia IP is collected from wild sources, while Lobelia BP/USP is collected both from wild and cultivated varieties. Lobeline, the major alkaloid of this drug is obtained by synthesis also.

Macroscopic Characters

Colour	:	Stems are green with purplish tinge
Odour	:	None
Taste	:	Extremely acrid and irritating
Size	:	The total height of the plant is about 2 to 3 m.

The plants have rounded or somewhat angular stems. They are usually branched at the top and slightly pubescent and hollow. Lobelia stems bear alternate leaves which are oblong lanceolate, sub-sessile and finely serrulate. The upper leaves are shorter and the lower leaves are bigger in size. The leaves are 5 to 25 cm in length, and 1 to 5 cm in breadth. Flowers are white with 3 lobes, terminal recemes are from 30 to 100 cm in length. Fruits are bilocular inflatted capsules which are sub-globose, about 5 to 10 cm in diameter and with longitudinal ribs. Seeds are small, ellipsoidal and compressed.

Fig. 15.38: Lobelia Herb

Chemical Constituents

Lobelia contains 0.5 to 1.2 of total alkaloids, important of them is lobeline. Lobelidine, Lobelanine, isolobelanine and lobelanidine are the other alkaloids present in the drug. It also contains pungent volatile oil, resin, gum and fixed oil.

Lobeline R_1 = H, OH, R_2 = O

Lobelanine R_1 = O, R_2 = O

Lobelanidine R_1 = H, OH, R_2 = H, OH

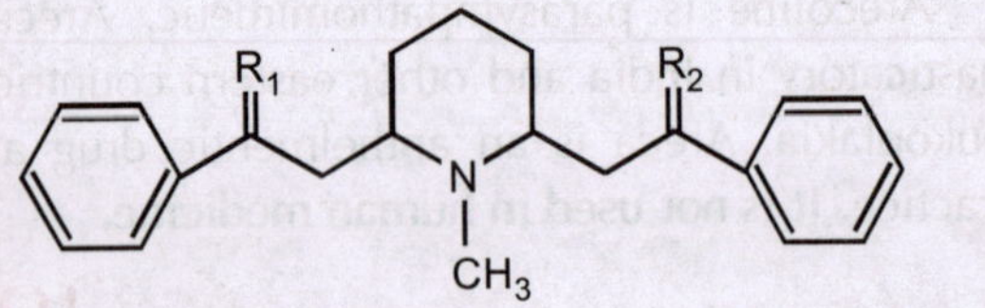

Lobeline $C_{22}H_{27}O_2N$ occurs as colourless crystals and soluble in hot alcohol. It forms hydrobromide, hydrochloride and methiodide salts.

Nobel prize winner H.O. Weiland and co-workers have been credited with the structural determination of lobelia alkaloids.

Chemical Tests

1. Solution of lobeline in sulphuric acid gives red colour with formaldehyde.
2. Lobeline solution on boiling produces acetophenone which is recognised by smell.

Standards of Qualtiy

FOM	:	Not more than 2.0 per cent
Ash	:	Not more than 5.0 per cent
Acid insoluble ash	:	Not more than 2.0 per cent
Content of stalks	:	Not more than 60.0 per cent
Water soluble extractives	:	Not less than 20.0 per cent

Uses

Lobelia is mainly used in the treatment of asthma and as respiratory stimulant. It is used for bronchial asthma and chronic bronchitis. Previously, the fumes of powdered lobelia were used for inhalation. It is also claimed that lobelia is useful as a smoking deterrent. In parenteral form, lobeline hydrochloride is given for resuscitation of new born infants, through umbilical vein.

Lobeline has similar physiological effects as that of nicotine.

It is a cardiovascular stimulant having the effects through carotid chemoreceptors. Cough and pain receptors are stimulated by lobeline.

PURINE ALKALOIDS

Various plants like tea, coffee, cola, cocoa, guavana and mate contain the derivatives of xanthine, generally called purine alkaloids. They occur in various parts of world and are used in stimulant beverages because of caffeine - a C.N.S. stimulant principle.

This group of alkaloids mainly includes caffeine, theobromine and theophylline.

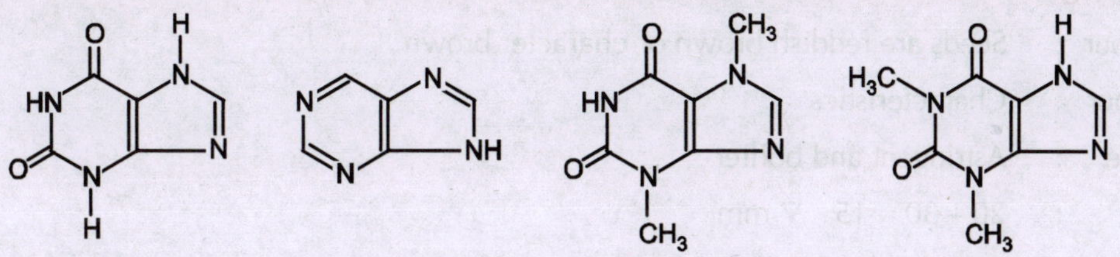

Xanthine Purine Theobromine Theophylline

These alkaloids contain purine base which is a fusion of an imidazole and pyrimidine rings. Purine base is derived into xanthine and latter is further differentiated as caffeine (1, 3, 7 trimethyl xanthine), theobromine (3, 7 dimethyl xanthine) and theophylline (1, 3 dimethyl xanthine). Xanthine does not occur free in nature.

It is pertinent to note that even though they are called alkaloids, none of them respond to common alkaloidal reagents. They give the murexide colour reaction positive. They are also sublimatary in nature.

All the main three alkaloids have stimulant and diuretic properties. Caffeine is mainly stimulant and theobromine has diuretic effects. Theophylline has both the actions, but it has also smooth muscle relaxant properties.

COCOA SEED

Synonym

Cocoa bean.

Biological Source

It is obtained from seeds of *Theobroma cocoa*, belonging to family Sterculiaceae.

Geographical Source

The seeds are obtained from the cocoa plants cultivated in Sri Lanka, Java, South American countries like Brazil, Ecuador, Guiana and Caribbean islands.

Macroscopical Characters

Cocoa plant is indigenous to Mexico and it is a tree of about 10 – 12 metres in height.

It bears the fruits directly on trunk (Similar to jack fruit) and fruit are yellow or reddish in colour. They are quite large in size, fleshy, avoid in shape cocoa fruits are 10-furrowed longitudinally. Fruits contain five-rows of seeds having 10 – 12 seeds in each row. Seeds are separated from the pod and allowed to fermetn. During fermentation they charge their colour from white to dark-red colour. Then the seeds are roasted, but the temperature are not allowed to cross 140ºC. During this process seeds loose their water and develop their characterisitc flavour and taste. Seeds roasted are subjected to ribbling machine to cracks the seed coat (cacao-shells). Then they are winnowed. Broken pieces are known as nibs in market. Nibs are then ground between the hot rollers whereby they yield pasty mass containing 50.0 percent cocao butter. It is allowed cool coco buter congeals and is separated.

Morphological Characters

Colour	:	Seeds are reddish brown or character brown
Odour	:	Characteristics
Taste	:	Astringent and boilter
Size	:	20 – 30 15 7 mm
Shape	:	Ovoid-flattened

Extra features: Sees coat is thin and brittle.

Cocoa fruits

Fig. 15.39 (a) Cocoa Seeds

Chemical Constituents

Cocoa beans contain mainly theobromine upto 1.0 per cent and cocoa butter (theobrom oil) 30-50 per cent. It contains less percentage of caffeine, various volatile compounds giving flavour to the seeds, polyphenols, etc.

Cocoa fruits are 15-20 7-10 cm in size bearing numerous cocoa seeds contains protein 15.0 per cent, starch – 15 per cent.

Uses

The stimulant and diuretic properties of the drug are due to theobromine. Cocoa is also used as nutritive. Cocoa butter is largely used as a suppository base in pharmaceuticals.

COFFEE SEED

Synonym

Coffee bean.

Biological Source

For highest yield of coffee seeds plant needs warm and humid climate, well distributed rains of 250 cm and 800 – 1650 meters of altitude are favourable for coffee plantation.

It is the dried ripe seed of Coffea arabica Linne or C. liberica Hiern, and deprived of most of the seed coat. It belongs to family Rubiaceae.

Geographical Source

It is found in Ethiopia, Brazil, India, Vietnam, Mexico, Guatemala, Indonesia and Sri Lanka.

Coffee beverage is popular in different parts of world. The amount of caffeine present is also more than in tea. In middle east countries, it is known by name 'qahuah'. It is a strong decoction of coffee seed powder. The word has its origin in Turkish and Arabic languages. The major suppliers of coffee, now-a-days, are Brazil and India. Karnataka, Kerala and Tamil Nadu grow large plantation of coffee. At present it is cultivated in Maharashtra also. Totally 125000 hactares of land is under cultivation of coffee in India at present.

Habitat

Coffee plant is an evergreen shrub which bears drupe type of fruit with an ellipsoidal or spheroidal shape. Each fruit has 2 locules, with one seed in each chamber. Each plant gives about 2 - 3 kg of coffee seeds. The fruits are dried to separate the seeds. The seeds are separated by wetting them and mechanically separating, followed by drying in heaps, which causes fermentation.

Morphological Characters

The separated seeds or beans are green in colour. They are roasted by which the colour and odour is effected. The seeds acquire dark brown colour and possess an agreeable odour and bitter taste. India stands third in coffee production in world.

Fig. 15.39: Coffee Plant with fruits

More than 70 per cent coffee plantation in India grow coffee arabica only producing caffeine.

India produced 1.65 and 2.35 lakh tones of coffee during 1998 - 99 and 99 - 2000 respectively.

Chemical Constituents

The main constituents of coffee bean are caffeine, tannin, fixed oil and proteins. It contains 2 - 3 per cent caffeine, 3 - 5 per cent tannins, 13 per cent proteins, 10 - 15 per cent fixed oils, chlorogenic or caffeotannic acid and sugars in the form of dextrin, glucose, etc. In the seeds, caffeine is present as a salt of chlorogenic acid. During roasting process, the agreeable smell of coffee is developed which is due to an oil called caffeol composed of mainly furfural alongwith minor quantities of phenol, pyridine and valerianic acid.

Extraction of Caffeine

Caffeine is prepared either by synthesis (from urea or uric acid) or extracted from natural sources. Coffee bean is one of the major sources for it. For the extraction, coffee roasters are used in which caffeine sublimed during roasting is recovered. It is the commercial method for extraction of caffeine.

Uses

It is used as a source of caffeine. The main effects of coffee i.e. stimulant and diuretic actions are due to caffeine. It is, sometimes, used to combat the toxic effects due to CNS depressant drugs.

Other Preparations

Decaffeinated coffee: This product has been introduced in developed countries. It has been specially designed, because of addiction effects of coffee. It has all the agreeable odour of coffee beans, but contains a meagre amount (about 0.08 per cent) of caffeine.

KOLA

Synonyms

Bissy or Gooroo seeds, cola nut, COLA.

Biological Source

It is also called as Kola nut and is obtained as the cotyledons of *Cola nitida, cola acumenata* belonging to family Sterculiaceae.

Fig. 15.40: Cola acuminata herb with seeds and pod

Macroscopic Characters

Seeds are chocolate brown in colour with aromatic and rose like flavour, bitterish-sweet in taste. Seeds are round or square shaped.

The drug is indigenous to West Africa and is cultivated in Brazil, Jamaica, Sri Lanka, Indonesia and East Africa.

Chemical Constituents

Kola nuts contain the purine alkaloids viz. caffeine and theobromine in combination with tannin called kolacatechin. They are separated by drying the kola nut. The drug contains about 1-3 per cent caffeine, and upto 1 per cent theobromine.

Uses

Kola nuts are used in preparation of aerated beverages, and as stimulant.

TEA LEAVES

Synonym

Camellia thea.

Biological Source

It contains the prepared leaves and leaf buds of Thea sinensis (Linne) O. kuntze, belonging to family Theaceae (Ternstroemiaceae).

Geographical Source

Large areas of land are put under cultivation of tea in India, Sri Lanka, China, Indonesia and Japan. It is available as **black tea** from India and Sri Lanka and **green tea** from China and Japan.

Black tea is obtained by fermenting the heap of fresh tea leaves and further drying with artificial heat. Green tea is obtained by putting tea leaves in copper pans and then drying by artificial heat.

Assam, West Bengal, Tamilnadu and Kerala are leading in cultivation of tea herb.

India is the biggest tea-growing country in the World and also exporter too! More than half the world's output of tea is from India.

Cultivation of Tea

In India tea is mainly cultivated in Assam, Darjeeling, Nilgiris.

Tea herb prefers shady and humid climate. Hot dry weather and hot winds are unfavourable to the plant. Rid soil is very favourble and given stimulus to successful plantation. It is cultivated on slopes only. Plant are raised from seeds and spaced out in plantations at about 5 feet above each way.The yield of leaves in obtained at the end of third year and full yield continuous to about 6th year. Harvesting of leaves is done after every three months.

For cultivation of tea tea-plants need high humidiy and high rain fall. Altitude of 2000 meters above sea level is a must. pH of soil in the range of 4.5 to 6.5 is quite tavourable. About 4,36000 hectare of land is under cultivation of Tea in India and 8,70,000 tonns of leaves are also produced annually.

Harvesting of leaves is done only after the age of three years of tea plants.

Macroscopic Characters

It is a small evergreen shrub when cultivated reaches to the height of 1.0 - 1.5 metres, while wild growing plants reach upto 6.0 metres. Plant is much branched and bears grey bark.

Leaves: Dark green, lanceolate or elliptical, blunt at apex, base is tapering margin shortly serrate. Young leaves are hairy while matured leaves are glabrous.

Flowers are solitary or in groups of 2 or 3 in the leaf axils, and drooping.

Colour	:	Fresh leaves are green, after processing become dark brown colour
Odour	:	Characteristics
Taste	:	Bitter and astrigent
Size	:	5 – 10 2 – 3
Shape	:	Lanceolate blunt and acumunate at apex and tapering at base, stalk in short margin is serrate.

Extra features: Young leaves are covered with trichomes but trichomes get lost when the leaf matures.

Fig. 15.41 (a) : Tea-leaf

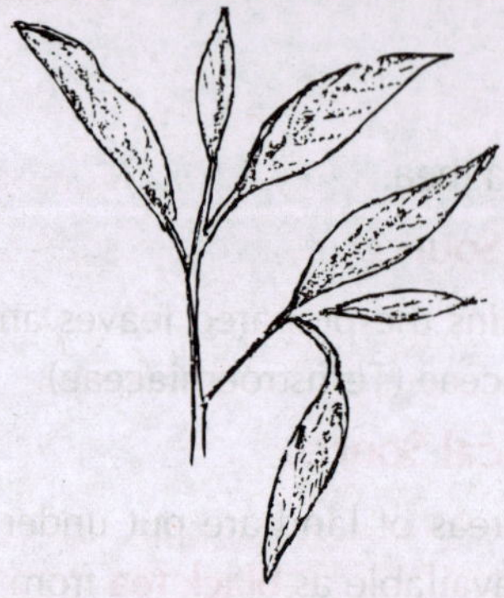

Fig. 15.41 (b) : Twig of tea-herb

Preparation of Green Tea

It is prepared by exposing the freshly collected leaves to the air until most of the moisture is removed. Then they are roasted and stirred continuously until leaves become moist and flaccid. Then they are passed to rolling table and rolled into balls and subjected to a pressure which removes the moisture. Then the leaves are shaken out on the copper pans and roasted again till the leaves assume dull green colour. Then the leaves are winnowed, screened and graded into various varieties.

Chemical Constituents

Tea leaves are considered as a rich source of caffeine (1 - 3 per cent). It is extracted from tea dust and tea leaf waste or sweepings. It also contains tannin 10 – 25 per cent theobromine and theophylline in minor quantities. The colour of tea leaves is due to gallotannic acid (15 per cent). The agreeable odour is due to presence of a yellow volatile oil. Tea leaves also contains an enzymatic mixture called thease responsible for colour change during preparation for market.

Standards of Quality

Alkaloidal contents : 1.0 – 3.0 per cent

Use

Tea is useful as a CNS stimulant in the form of beverage besides, it is a diuretic as well. Tea and coffee both are CNS stimulans, if their stimulant actions are compared to each other, tea sets the actions quickly and also diminishes quickly. Coffee offers stimulation title late but lasts longer.

Caffeine

Caffeine [$C_8H_{10}O_2N_4$] occurs as a white powder without any odour but possesses bitter taste.

It is a weak base and feebly soluble in water, alcohol, chloroform and solvent ether. But, its solubility in water is sharply increased in the presence of benzoates, bromides, citric acid and salicylates.

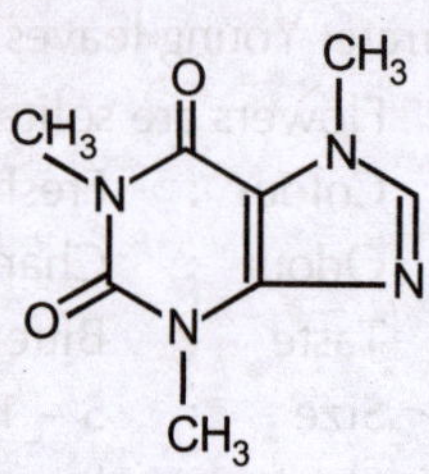

Caffeine

Chemical Tests

1. Caffeine (also the other purine alkaloids) gives **murexide colour reaction.** Caffeine is taken in a petridish to which hydrochloric acid and potassium chlorate are added and heated to dryness. A purple colour is obtained by exposing the residue to vapours of dilute ammonia. The purple colour is lost on addition of fixed alkali.
2. Caffeine also produces white precipitate with tannic acid solution.

Uses

Caffeine is widely accepted and used as a central nervous system stimulant, due to its cerebral vasoconstrictor effect. It is also given along with ergotamine tartarate to potentiate the action of latter as a specific analgesic in migraine.

IMIDAZOLE ALKALOIDS

PILOCARPUS

Synonym

Jaborandi

Biological Source

The drug consists of the leaves of closely related plants of the genus Pilocarpus, belonging to family Rutaceae. The genus includes various species known by different names like Pilocarpus jaborandi (Pernambuco jaborandi), P. pennatifolius (Paraguay jaborandi), P. microphyllus (Maranham jaborandi), P. trachylophus (Ceara jaborandi), P. selloanus (Rio jaborandi), P. spicatus (Aracati jaborandi), P. heterophyllus (Barqui simento jaborandi) and P. racemosus (Guadeloupe).

Now-a-days, P. microphyllus i.e. Maranham jaborandi is the main source of this drug.

Geographical Source

It is indigenous to South America and especially grown in Brazil. It is found in Venenzuela, Caribbean islands and Central America.

Macroscopic Characters

The characters of main commercial variety, Maranham jaborandi are discussed here.

Jaborandi leaves are greyish green to greenish, brown in colour and have a slight aromatic odour along with bitter taste of alkaloids. When tested, the leaf causes increase in salivary secretion. It is an, imparipinnate compound leaf with 7 leaflets. The leaflets are asymmetrical, obovate and sessile in shape and 2 - 6 cm long and 1 - 3 cm wide in size. The terminal leaflet is symmetrical and oval in shape. The leaflets are present on the winged and glabrous rachis. The leaflet shows pinnate venation. They contain numerous oil cells.

Fig. 15.42: Pilocarpus (jaborandi leaf)

Chemical Constituents

The leaflets contain imidazole alkaloids among which pilocarpine is most important. Other alkaloids are isopilocarpine, pilocarpidine, pilosine, pseudopilocarpine, and isopilosine.

The range of total alkaloids in different species is between 0.5 to 1 per cent.

The oil cells observed in the leaf give volatile oil containing different monoterpenes like limonene, α - pinene, sabinene and sesquiterpenes.

The main alkaloid pilocarpine occurs as viscous oil or hygroscopic crystalline solid without any colour and odour. It is soluble in water and organic solvents, but insoluble in petroleum ether. It is the lactonic derivative of pilocarpic acid. The hydrolysis of pilocarpine results into pilocarpic acid, with loss of optical activity. Treatment with alkali produces isopilocarpic acid which is the stable stereoisomer. The alkaloids are totally deteriorated on long storage.

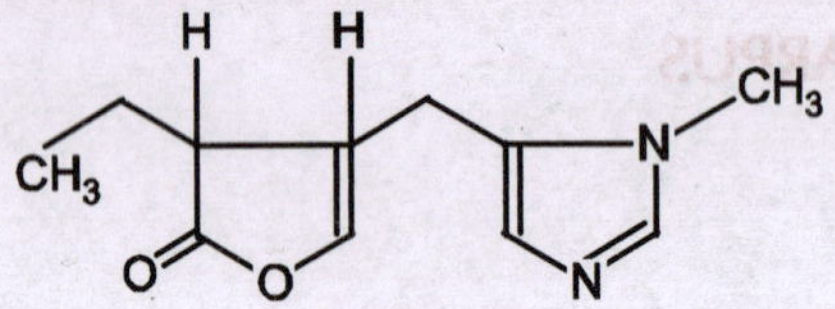

Pilocarpine

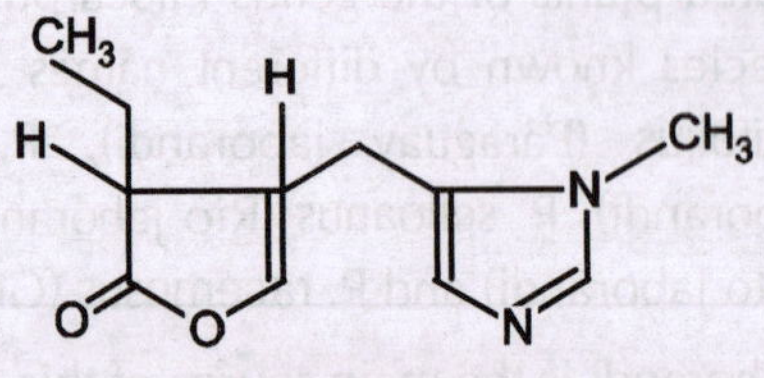

Isopilocarpine

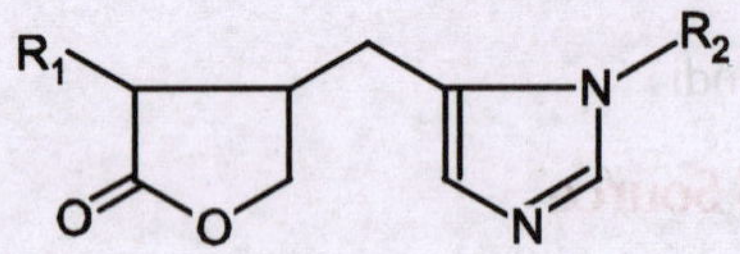

Pilocarpidine, $R_1 = C_6H_5$, $R_2 = H$
Pilosine, $R_1 = C_6H_5CHOH$,
$R_2 = CH_3$

Jaborandi Alkaloids

Chemical Test

To the pilocarpine solution, small quantity each of dilute sulphuric acid, hydrogen peroxide solution, benzene and potassium chromate solution is added. On shaking, organic layer gives bluish-violet colour and yellow colour appears in aqueous layer.

Uses

The drug is mainly used in the form of pilocarpine hydrochloride. It is a physiological antagonist of atropine. It acts directly on the autonomic effector cells of those structures innervated by post-ganglionic cholinergic nerves. Hence, it causes contraction of the pupil of the eye, and increase in sweating and salivation.

It is used in ophthalmology for the treatment of glaucoma.

Dose

Its recommended dosage is 0.05 - 0.1 ml of 10 per cent solution, topically.

STEROIDAL ALKALOIDS

Some of the important drugs covered under this class are Solanum species; kurchi bark, and veratrum. The active chemical principles of such drugs contain mainly steroidal (cyclopenteno phenanthrene) entity, along with basic nitrogen. Either they are used as medicines, as in case of kurchi and veratrum, or as a precursor for synthesis of various other steroids as in Solanum species. They belong either to C_{27} or C_{21} group of steroids.

ASHWAGANDHA

Synonyms

Withania root, Asgandh, Winter cherry.

Biological Source

It consists of dried roots and stem bases of Withania somnifera (Linn.) Dunal, belonging to family Solanaceae and should contain not less than 0.02 per cent of total withanolide A and withaferin A on dried basis.

Geographical Source

This plant grows wildly in all dry parts and subtropical India. It occurs in Madhya Pradesh, Uttar Pradesh, Punjab plains and North Western parts of India like Gujarat and Rajasthan. It is also found in Congo, South Africa, Egypt, Morocco, Jordan, Pakistan and Afghanistan.

Cultivation of Ashwagandha

It is reported that the plants from different sources vary in their morphological and therapeutic properties. Now-a-days, the cultivation is mainly done in Madhya Pradesh (Manasa plantations), where, about 2000 hectares are under cultivation. The propagation is done by seeds, for which about 4 - 5 kg of seeds are required per hectare. The seeds are sown in the soil which is unsuitable for other crops. The sowing is done towards June-July and during growth, no special arrangements are made for irrigation. Even the nitrogenous fertilizers lead to formation of small roots, but large foliage.

Collection of Ashwagandha Roots

Towards December or January, the plants bear flowers and fruits and during January, harvesting is initiated which lasts upto March. The roots are collected by uprooting the plant and either entire roots or the pieces thereof are dried immediately.

Macroscopic Characters

The roots show buff to grey yellow outer colour with longitudinal wrinkles. They are unbranched, straight, conical and some of them bear a crown. The root crown possesses a number of bud scars. Roots are bitter in taste and fresh roots smell similar to urine of horse (hence ashwagandha). The fracture is smooth and powdery.

Microscopic Characters

The transverse section of root shows exfoliated cork which is non-lignified with 2 - 4 layers of phellogen and about 15 - 20 rows of phelloderm. It prominently shows parts of vascular tissue like cambium, consisting of 3 - 5 layers of tangentially elongated cells, phloem region with parenchyma, sieve tubes and companion cells. Secondary xylem is hard which forms a continuous vascular ring interrupted by medullary rays. The transverse section of stem base shows pith, pericyclic fibres, xylem with tracheids, fibres, and starch grains.

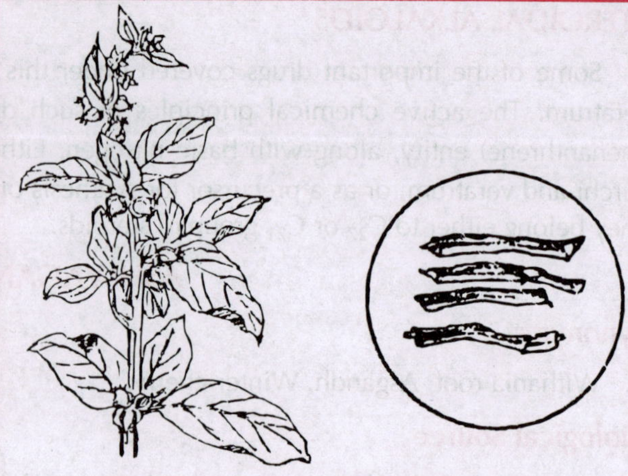

Fig. 15.43: Ashwagandha plant Roots

Chemical Constituents

The main constituents of ashwagandha are alkaloids and steroidal lactones. Among the various alkaloids, withanine is the main constituent. The other alkaloids are somniferine, somnine, somniferinine, withananine, pseudo-withanine, tropine, pseudo tropine, 3-α-gloyloxytropane, choline, cuscohygrine, isopelletierine, anaferine and anahydrine. Two acyl steryl glucosides viz. Sitoindoside VII and sitoindoside VIII have been isolated from roots.

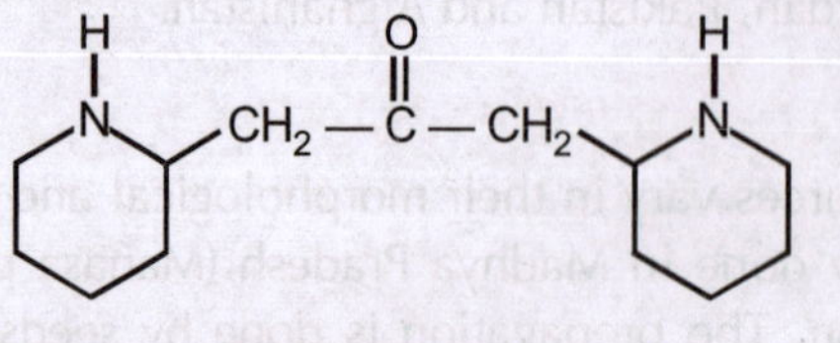

Anaferine

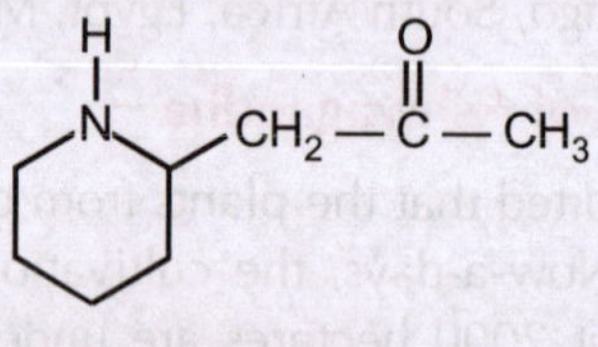

Dl-isopelletierine

The leaves contain steroidal lactones, which are commonly called as "withanolides". The withanolides have C_{28} steroidal nucleus with C_9 side chain, having six membered lactone rings. Lavie et.al, have isolated such compounds from plants grown in Israel, India and S. Africa.

The various withanolides reported are as follows:

Steroids (withanolides) of Ashwagandha

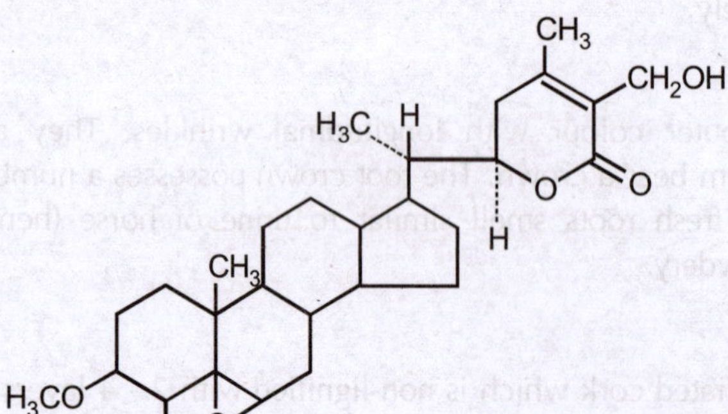

Withaferin

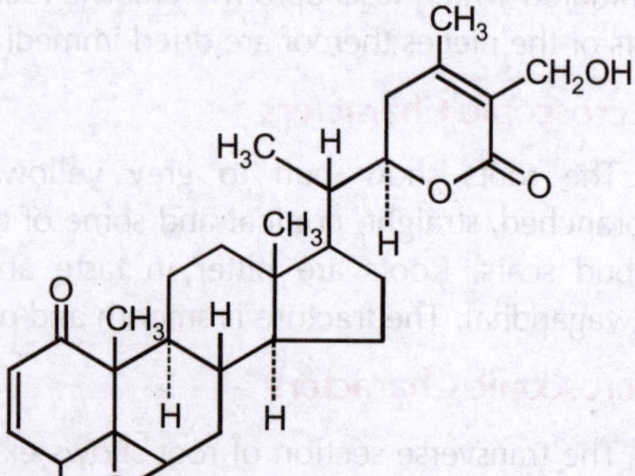

Withaferin A

These compounds have been obtained from Wilhania somnifera chemotype I.

Another series of steroidal lactones viz. withanolide E to M have been obtained from chemotype III.

The drug also contains two monohydric alcohols called somnitol and somnirol; withanic acid; a phytosterol and ipuranol; and a mixture of fatty acids containing cerotic acid, oleic acid, palmitic acid and stearic acid.

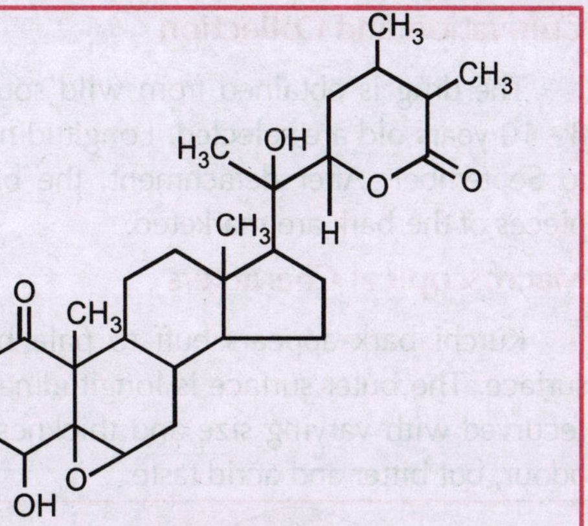

Withanolide A

Standards of Quality

Foreign organic matter	:	Not more than 2 per cent
Total ash	:	Not more than 7 per cent
Acid-insoluble ash	:	Not more than 1.2 per cent
Alcohol soluble matter	:	Not less than 16 per cent
Water soluble extractives	:	Not more than 15 per cent
Loss of drying	:	Not more than 12 per cent

Uses

It is well known valuable adoptogen, nourishes nerves and tissues. Essential to increase endurence and energy. Additionally it reduces auxiety and stress.

Ashwagandha has sedative and hypnotic effects. It has hypotensive, respiratory, stimulant actions alongwith bradycardia. It is an immuno-modulatory agent. Sitoindoside (VII and VIII) have been shown to possess anti-stress activity. It acts as mood stabilizer revives mind and body.

Traditionally, it has been used in the treatment of rheumatism, gout, hypertension, nervine and skin-diseases. This drug prevents bony degenerative changes in arthritic conditions. It has been widely used as sex stimulant and rejuvenator and is considered as strength and vigour promoting drug especially in geriatric cases.

The leaf extracts shows activity against Staphylococcus aureus and Ranikhet virus.

KURCHI

Synonym

Holarrhena

Biological Source

It is the dried stem bark of Holarrhena antidysenterica Wall belonging to family Apocynaceae. It is collected from 8 - 10 years old plant and freed from attached wood, and peeled into small pieces. It should contain not less than 2 per cent of total alkaloids of kurchi.

Geographical Source

Kurchi is indigenous to India and found throughout India in parts ascending up to 1000 metres in Himalayan region. It is also found in Orissa, Assam, Uttar Pradesh and Maharashtra.

Cultivation and Collection

The drug is obtained from wild source only. For the collection of bark, the plants which are 8 - 10 years old are selected. Longitudinal and transverse incisions are made on the trunk from July to September. After detachment, the bark is separated from the wood and dried. The recurved pieces of the bark are marketed.

Macroscopical Characters

Kurchi bark appears buff to pale brown on outer surface, while slightly brownish on inner surface. The outer surface is longitudinally wrinkled and bears horizontal lenticels. The pieces are recurved with varying size and thickness. The drug shows a short and granular fracture. It has no odour, but bitter and acrid taste.

(a) Kurchi twig fruiting (b) Pieces of Kurchi Bark (c) T. S. of Kurchi

Fig. 15.44: Kurchi herb and bark

Microscopic Characters

In the periderm portion, cork has both tangentially and radially elongated cells. Cork cambium has several layers. Stone cells are present in cortex, either singly or in groups or in horizontal layers. It does not show phloem fibres. Phloem contains sieve tubes, companion cells, phloem parenchyma and stone cells. Medullary rays are multiseriate. In many stone cells, prismatic calcium oxalate crystals are present.

Chemical Constituents

Kurchi contains about 25 total alkaloids (1.5 to 3 per cent). They are C_{21} group steroidal alkaloids. The active alkaloids are conessine (kurchicine), norconessine, isoconessine, dioxyconessine, conessimine, holarrhimine and holarrhidine.

Conessine is also present in root bark alongwith some other steroidal alkaloids.

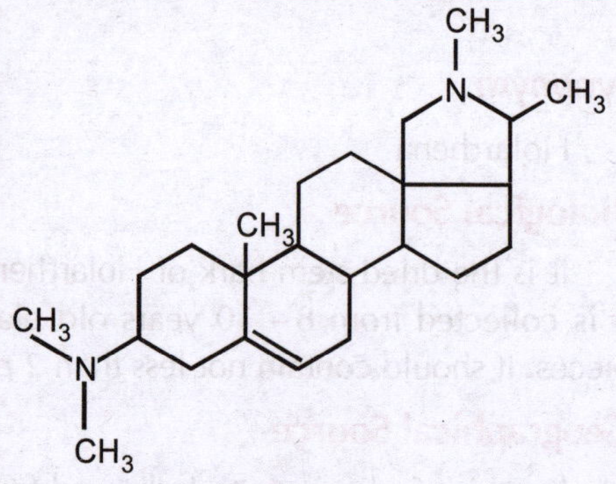

Conessine

Standards of Quality

Total ash	:	Not more than 7.5 per cent
Acid insoluble ash	:	Not more than 1 per cent
Alcohol (60 per cent) soluble extractives	:	4 - 6 per cent
Foreign organic matter	:	Not more than 5 per cent
Loss on drying	:	Not more than 8 per cent

Uses

Kurchi is antiprotozoal in activity and used to treat amoebic dysentry. Conessine is highly active against Entamoeba histolytica. A traditional preparation of kurchi bark, viz. "kutajarishta" is commonly used, especially for chronic amoebiasis.

VERATRUM

Two important species of genus Veratrum are Veratrum viride and Veratrum album. Both contain protoveratrine as active principle.

Veratrum contains two groups of alkaloids called jeveratrum and ceveratrum alkaloids. The alkamine part of both these groups is polyhydroxylated $C_{27}N$ fused polycyclic. The alkamines of jeveratrum group contain only 2 or 3 oxygen atoms while in ceveratrum, they have 7 - 9 oxygen atoms. The members containing ceveratrum alkaloids are only therapeutically active, and their examples are cevadine, germerine, veratridine, protoveratrine A and protoveratrine B.

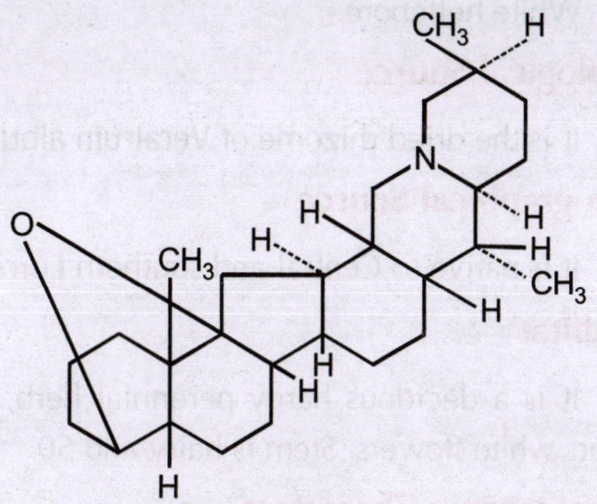

Ceveratrum nucleus

(a) American Veratrum

Synonym

Green hellebore.

Biological Source

It consists of dried rhizome and roots of Veratrum viride Aiton, belonging to family Liliaceae.

Geographical Source

The drug is obtained from wild growing plants in many parts of United States like states of New York, North Carolina, Georgia, Tennessee, etc. It is also found in some parts of Canada.

Macroscopic Characters

Colour	:	Brown
Odour	:	Unpleasant

Taste : Acrid

Size : Rhizomes are 5 - 8 cm in length and 2 - 3.5 cm in diameter

Shape : Subcylindrical with numerous stout yellowish brown roots.

It is a perennial about 1 - 2 m in height, large leaves, which are oral and strongly ribbed, star shaped many flowers in panicle.

Chemical Constituents

It contains two alkaloids namely, germidine and germitrine. Calcium oxalate and starch are the other constituents of durgs.

Uses

The medicinal action is due to its many alkaloidal constituents. It lowers blood pressure and decreases heart rate. It is used as a liquid extract or tincture in pregnancy associated hypertension. It is used as insecticide also.

(b) European hellebore

Synonym

White hellebore.

Biological Source

It is the dried rhizome of Veratrum album Linne, belonging to family Liliaceae.

Geographical Source

It is native to Central and southern Europe, China and Japan.

Habitat

It is a decidous hardy perennial herb, flowers in June - July bears white flowers. Stem is hairy and 50 - 125 cm.

Fig. 15.45: *Veratrum album* plant

Macroscopic Characters

Colour : Brown

Odour : Unpleasant characteristic when fresh, dried with no odour

Taste : Burning, acrid and bitterish

Size : 5 - 15 cm in length and 2 - 3 cm in diameter

Shape : Tuberous, fleshy with number of long white fibres at the end of the roots.

Chemical Constituents

It contains mainly veratrine and also germidine, germitrine, protoveratrine, cevadine, veratrine pseudojervijne, veratrosine, etc.

Both Protoveratrine A and B are soluble in chloroform, but insoluble in water and light petroleum.

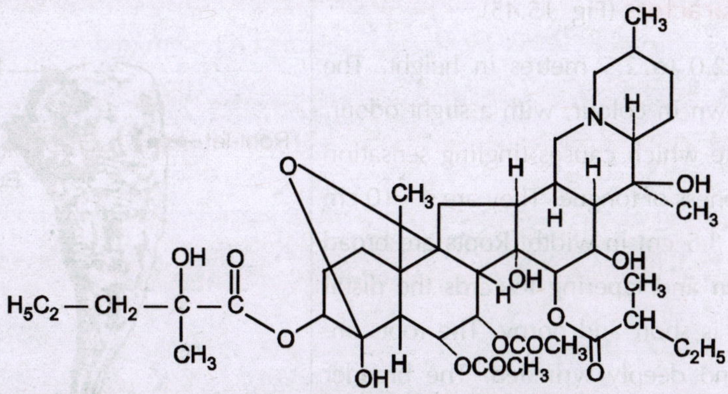

Protoveratrine A

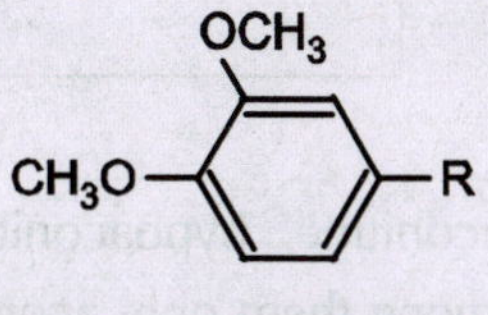

Veratridine

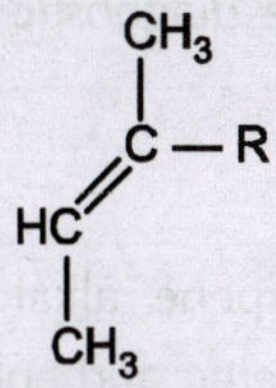

Cevadine

Chemical consistuents of veratrum herb

Uses

Veratrum album is mainly used as a source of protoveratrine A and B. Among these, protoveratrine A is medicinally more potent. They are used for the management of hypertension in pregnancy, especially in preclampsia and exclampsia conditions. It is not used therapeutically, however it is used as insecticide only.

DITERPINE ALKALOIDS

ACONITE

Synonyms

Aconite root, Bachnag, Monkshood, Wolfsbane.

Biological Source

It is the dried root of Aconitum napellus Linn belonging to family Ranunculaceae. It should contain not less than 0.6 per cent of alkaloids of aconite, of which not less than 30 per cent should be aconitine.

Geographical Source

Aconite is found is Hungary, Germany, Spain and Switzerland. It is cultivated in England. In India, it is collected from wild or cultivated plants and the drug collected is A. chasmanthum or A. ferox.

Macroscopic Characters (Fig. 15.45)

The plant is 2.0 to 2.5 metres in height. The roots are dark brown in colour, with a slight odour. It has a slight taste which causes tingling sensation followed by numbness of tongue. They are 4 - 10 cm in length and 1 - 3.5 cm in width. Roots are broad towards the crown and tapering towards the distal end. The fracture is short and horny. The roots are slightly twisted and deeply wrinkled. The broader end of the root bears numerous rootlets and scars. The daughter roots are the diagnostic characters of parent roots.

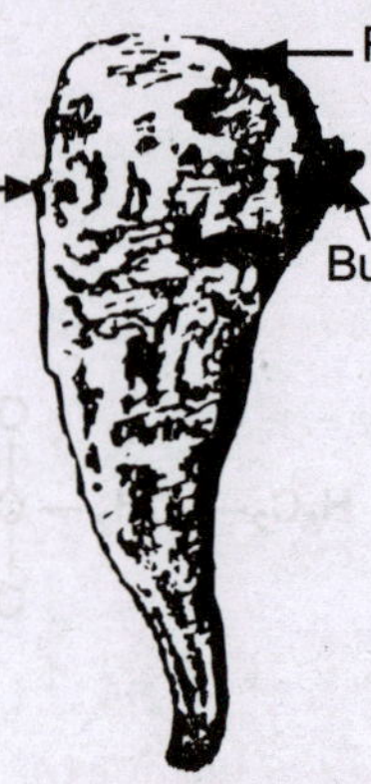

Fig. 15.46: Aconite root

Chemical Constituents

Aconite contains diterpene alkaloids like aconitine, hypoaconitine, neopelline, napelline, neoline and traces of sparteine and ephedrine. Among them only aconitine is the most important, having poisonous action. The chief constituent of drug is aconitine and is ether-soluble. The ether-soluble alkaloids vary in different species of aconite. The drug also contains aconitic acid, succinic acid and starch.

$$\text{Aconitine} \xrightarrow{\text{Hydrolysis}} \text{Benzoyl aconine + Acetic acid}$$

$$\text{Benzoyl aconine} \xrightarrow{\text{Hydrolysis}} \text{Aconine + Benzoic acid}$$

The products of hydrolysis are less active.

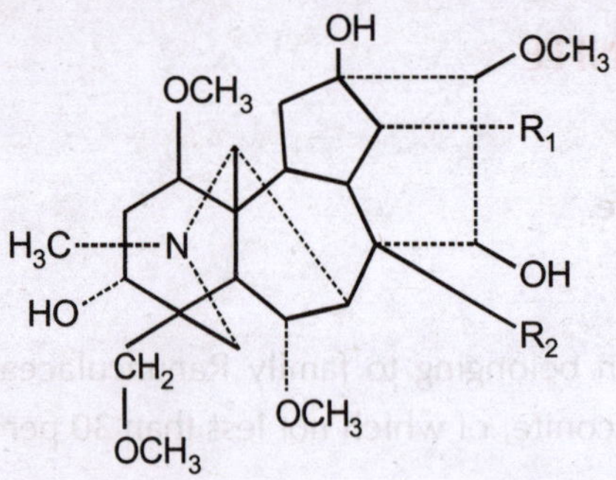

$R_1 = -OCOC_6H_5$, $R_2 = -OCOCH_3$ } Aconitine

$R_1 = R_2 = -OH$ } Aconine

Standards of Quality

Ash	:	Not more than 5 per cent
Acid-insoluble ash	:	Not more than 1 per cent
Aerial stem	:	Not more than 5 per cent
Foreign organic matter	:	Not more than 2 per cent

Uses

It is a highly poisonous drug. It is used externally (in the form of liniment) in the treatment of neuralgia, sciatica, rheumatism and inflammation. It is also analgesic and cardiac depressant. Now-a-days, its use is restricted only to homeopathic medicines.

Substitutes

1. Japanese aconite: These are the rosts of Aconitum uncinatum var. Japonicum. Roots taper gradually are dark grey to brwon in colour, smooth marked with pale longitudinal ridges on parent root. Daughter roots are smaller and more steady, less wrinkled. Trunsverese section of root shows circular cambium. It contains japaconitine, aconitine.

2. Indian aconite: These are the roots of Aconitum chasmanthum. Roots are smaller about 2.0   0.75 cm in size, black in colour externally less wrinkled, fracture in lighter less tough, rootlets are clustered at busal extremity. Starch is gelatinised.

QUINAZOLINE ALKALOIDS

VASAKA

Synonyms

Adhatoda, Adulsa, Malabar nut.

Biological Source

It consists of dried, as well as, fresh leaves of the plant Adhatoda vasica Nees, belonging to family Acanthaceae, and contains not less than 0.6 per cent of vasicine on dried basis.

Geographical Source

Vasaka is indigenous to India, where it is found in sub-Himalayan track up to an altitude of 1000 m, and in Maharashtra especially, in Konkan region. Besides India, it is found in Myanmar, Sri Lanka and Malaya.

Cultivation and Collection

The uses of vasaka have been known since old times and it is included in different formulations of ayurveda.

The plant is not cultivated on commercial scale. It is obtained from garden plants or wild sources. It can be easily propagated by stem cuttings and by seed germination. The plant is obtained in all seasons of the year. It reaches to a height of 2 - 3 metres. It is also observed that the plant favourably grows in loamy soil.

Macroscopic Characters

The drug contains stem leaf, fruit and seeds. The leaves have 10 - 30 cm length and width of 4 - 10 cm. They are petiolate and exstipulate. The shape is lanceolate. The margin is crenate with acuminate apex. There are 8 - 10 pairs of lateral veins. Taste is bitter and odour is characteristic.

Fig. 15.47 (a): Vasaka herb

Microscopic Characters

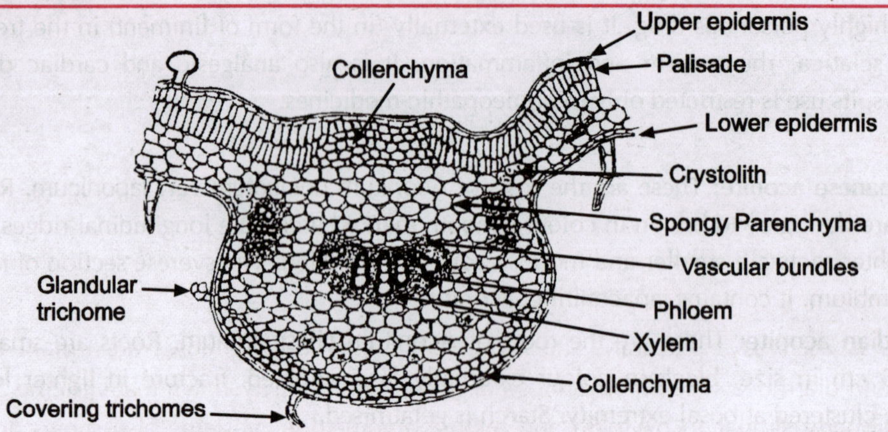

Fig. 15.47 (b): T.S. of Vasaka leaf

The epidermis shows caryophyllaceous stomata with sinuous epidermal cells, and covering and glandular trichomes. It is a dorsiventral leaf with palisade having 2 layers of cells. 2 - 3 bicollateral vascular bundles are seen in midrib. Mesophyll contains prismatic and acicular crystals of calcium oxalate. Stomatal index is from 10.8 - 18.2 and palisade ratio from 5 - 8.5.

Chemical Constituents

Vasaka leaves contain quinazoline derivatives such as vasicine 2.0 to 2.5 per cent, vasicinone and 6-hydroxy vasicine. Biochemically, vasicine is oxidised to its ketonic derivative vasicinone and the latter exerts main activity as bronchodilator. The drug also contains volatile oil, betain and vasakin. It is also reported that vasaka contains adhatodic acid.

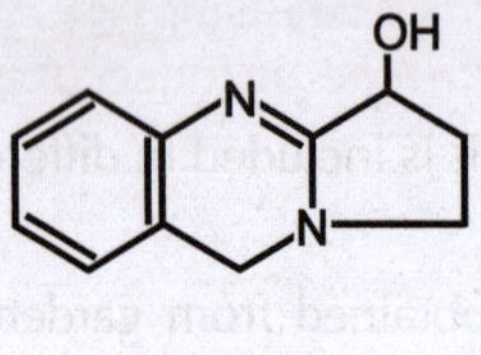

Vasicine

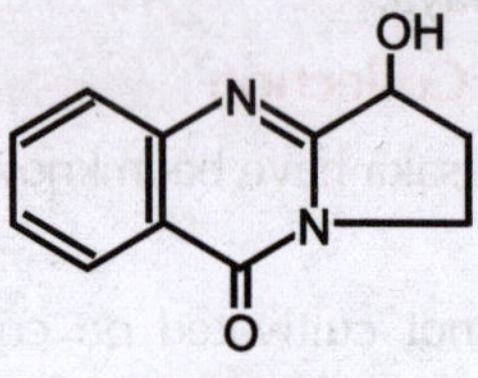

Vasicinone

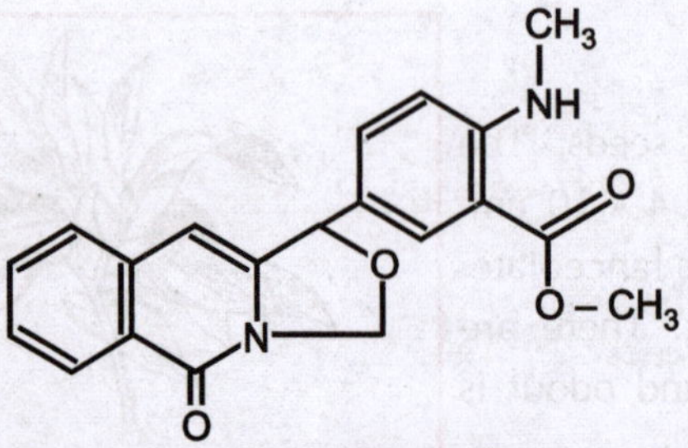

Adhatodine

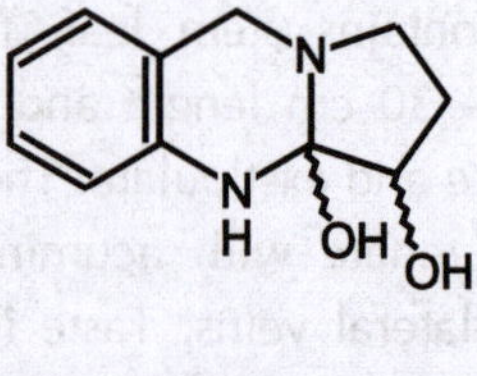

Vasicol

Standards of Quality

Foreign organic matter	:	Not more than 2.0 per cent.
Total ash	:	Not more than 21.0 per cent.
Acid insoluble ash	:	Not more than 2.0 per cent.
Loss on drying	:	Not more than 12.0 per cent.
Water soluble extractives	:	Not less than 22.0 per cent.
Alcohol soluble extractives	:	Not less than 3.0 per cent.

Uses

Vasaka is used as expectorant and bronchodilator. The large doses are irritant and cause vomitting and diarrhoea. The pharmacological investigations have shown that vasicine also shows oxytocic property similar to oxytocin and methyl ergometrine. Vasicine also shows abortificient action and both the actions are due to release of prostaglandins. Bromhexine HCl is a synthetic derivative of vasicine which changes the structure of bronchial secretions and reduces viscosity of sputum. It does not cause drowsiness or dependance.

MISCELLENEOUS ALKALOIDAL DRUGS

BHRINGRAJ

Synonyms

Eclipta, Bhangra, Maka, Wadelia.

Biological Source

This consists of whole plant botanically known as Eclipta Alba (L) Hassk of family Asteraceae (compositae) it contains not less than 0.1 per cent wedelolactone on dried basis. (Variety with yellow flowers is botanically knonwn as Wedella-chinensis).

Geographical Source

In tropical parts of world and all over India, abundant in marshy places and available in all seasons.

Description

Colour	:	Green to greenish brown
Odour	:	None
Taste	:	Bitter

It is a perrinial herb about 10-15 cm in height or a prostate plant.

Root : Cylindrical up to 7 mm in diameter, greyish with number of secondary branches.

Fig. 15.48: Bhringraj herb

Stem : Cylindrical or flat, node dishnet; rooting at nodes, appressed white trichomes, brownish in colour.

Leaves : 3.5 cm in length. Opposite, sessile, entire, oblong, lanceolate and trichomes on both the surfaces.

Flowers : Solitary on unequal axillary peduncles, ray flowers ligulate, white disc flowers tubular, corolla 4 toothed, stamen 5, anthers united, ovary inferior. It flowers in October November every year!

Fruit : Achenial, one seeded with narrow wing and brown in colour.

Seed : Dark brown hairy and non-endospermic.

Chemical Constituents

Bioactive steroidal alkaloids, glycosides, resin, alkaloid ecliptine, nicotine, and flavonoid wedelo-lactone; wedelic acid, epigenin and luteolin.

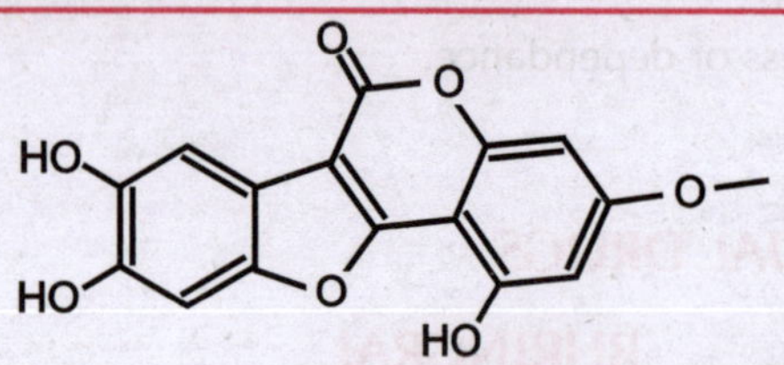

Wedelolactone

Standards of Quality

Foreign organic matter	:	Not more than 5.0 per cent
Ash	:	Not more than 22.0 per cent
Acid insoluble ash	:	Not more than 11.0 per cent
Loss on drying	:	Not more than 15 per cent
Water soluble extractives	:	Not less than 15 per cent
Alcohol soluble extractives	:	Not less than 05 per cent

Uses

Bhringraj is brain tonic, used for liver-disorders, viral hepititis, skin and hair care disorder, improves complexion. It is valuable to calm the mind and memory disorders. Bhringraj strengthens spleen and is general tonic.

BHUI-AMLA

Introduction

It is a small annual herb, known since the ancient period of **Charaka** for its medicinal use. The fresh juice of whole plant is found to be useful in various liver disorders.

Synonym

Phyllanthus

Biological Source

It consists of all aerial parts of the plant, Phyllanthus niruri belonging to family Euphorbiaceae. Indian variety available is Phyllanthus fraternus and Phyllanthus armarus. It contains not less than 0.25 per cent of total phyllanthin and hypophyllanthin on dried basis.

Geographical Source

The plant is found growing abundantly throughout India, mainly in the states of Maharashtra, Uttar Pradesh, Punjab, Bihar, Orissa, Andhra Pradesh and some parts of Madhya Pradesh, Karnataka and Bengal. Also found in Sri Lanka.

Cultivation and Collection

It grows well under tropical climatic conditions. However, very dry temperature or very low temperature conditions are not preferred. It can grow up to 700 - 800 metres above the sea level. Suitable temperature ranging from 25 - 40 C with average humidity is desired. Well distributed rainfall in the range of 40 - 50 cm is suitable.

Bhuiamla grows in variety of soils, preferably well drained rich organic and light textured soils. It also grows well on sandy loam to calcareous soil. Soil pH should be alkaline to acidic. Even water logging soil does not show adverse effects on its growth.

Propagated well through seeds or by raising seedlings and transplanting.

The plants mature within 3 to 4 months. They are usually harvested after rainy season is over, when the amount of green leaves is in abundance. Since, the active constituent is present in leaves, higher leaf mass is expected during harvesting. For harvesting, the whole plants are uprooted.

The collected whole plants are cleared of external matter and are allowed to dry in shade. The fresh juice of plant is also used in medicine.

Total yield of fresh herb per hectare is about 2 - 3 tones.

Macroscopic Characters

- Colour : Leaves are green to yellowish green in colour.
- Taste : Bitter
- Odour : None
- Stems : 1 - 1.5 cm in length and 1.4 mm in diameter stems are stout.
- Leaves : Short-stalked, oblong in shape and about 5 3 mm in size.

Fig. 15.49: Bhui-amla herb

Chemical Constituents

It contains, leucodelphimidin alkaloids, flavonoids as quercetin, astralgin, nerurin quercitrin, isoquercitrin and rutin. It also contains $\nless$ 0.25 percent of total phyllanthine and hypophyllanthine.

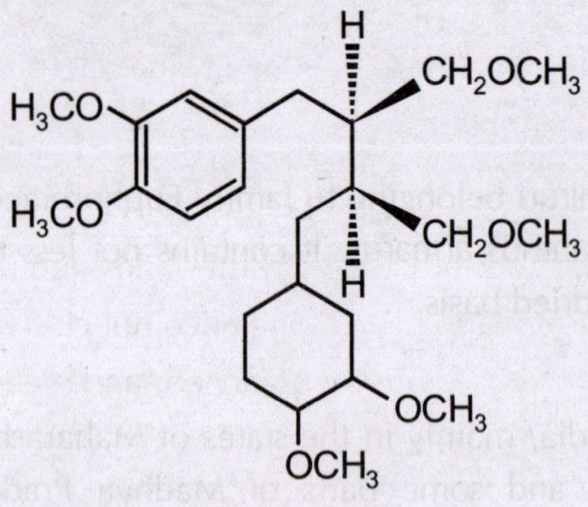

Phyllanthin

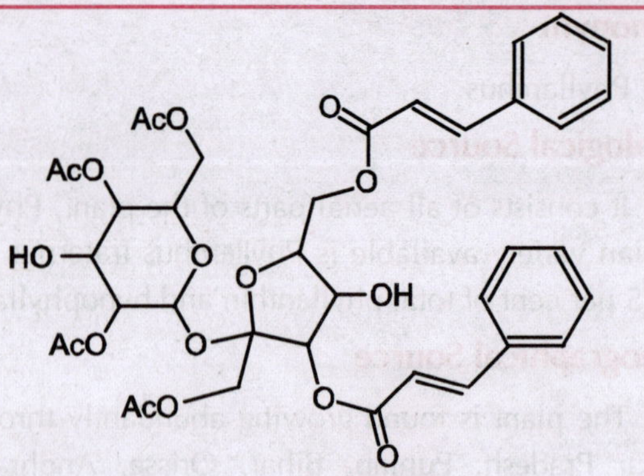

Niruriside

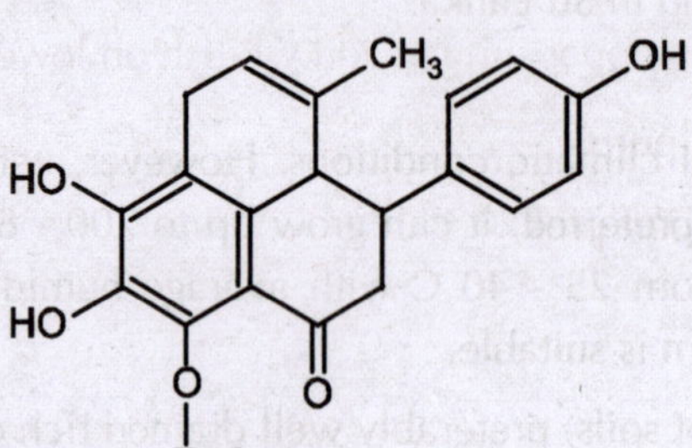

Rutinase

Nerurin

Niruriside, a carbohydrate from methanolic extract of Phyllunthus niruri has shown anti-HIV activity. It is specific inhibitor of REV protein, with an IC_{50} value of 3.3 µm.

Standards of Quality

Foreign organic matter	:	Not more than 2.0 per cent
Total ash	:	Not more than 8.0 per cent
Acid insoluble ash	:	Not more than 5.0 per cent
Water soluble extractives	:	Not less than 15.0 per cent
Alcohol soluble extractives	:	Not less than 6.0 per cent
Loss on drying	:	Not more than 12.0 per cent

Uses

The drug is used as a hepatoprotective mainly, used in the treatment of viral hepatitis and various other liver disorders. It is also used as diuretic in oedema. Externally, it is used to relieve inflammation, also used as good an appetiser.

PUNARNAVA

Synonyms

Rakta Punarnava, Hog weed

Biological Source

It consists of fresh, as well as, dried herb Boerhaavia diffussa (B. repens; B. procumbens) Linn. Nyctaginaceae. It should contain not less than 0.005 per cent of boervinone on dried basis.

Geographical Source

It is found wild throughout India and Sri Lanka. Punarnava is found in Himalayan valleys up to 2000-2500 m. The weed also grows in Malaysia, China and Africa.

Collection of Punarnava

Punarnava is very common and most troublesome weed of sandy tracts, waste-lands and road-sides. It is collected from wild grown plants. It is abundant during rainy season hence not cultivated scientifically and is collected in the flowering stage. It is variable, diffusely branched, and creeping herbaceous perennial.

Macroscopic Characters

Colour : Leaves: green on upper surface and whitish on lower surface
Stems: greenish-purple
Flowers: upper part pink

Odour : Odourless

Taste : Bitter

Size : Leaves are 25 - 30 mm long. Small leaves are 12 - 20 mm in length, ovate, oblong sub-orbicular and apex slightly pointed with rounded or sub-cordate base. Leaves are glabrous, rather thick in texture and petiolate. Petioles are double the length of the blade. Margin is entire.

Stems are cylindrical, stiff, and thick at the nodes. They are branched and about 1 m in length. Stems are light reddish-brown above and pale greenish below.

Flowers are internally sessile in small umbels, about 10 - 25 mm in length.

Fruits are one sided, glandular and about 0.5 cm in size.

Roots are elongated, fusiform, tapering and somewhat tuberous. Roots grow vertically downwards striking deep into the soil. They are cream or light brownish-yellow, with very soft skin. Old roots are often marked with knotty scars of fallen rootlets.

Microscopic Characters

Leaf: Epidermis shows the presence of multicellular uniseriate glandular trichomes. Cortex consists of one or two layers of collenchyma followed by few layers of parenchyma. Endodermis is distinct. Stellar part is represented by small vascular bundles, which are scattered throughout. Leaves have stomata on both the surfaces. Palisade is one layered and with small air-sac in spongy parenchyma. Calcium oxalate clusters and idioblasts are present in the mesophyll.

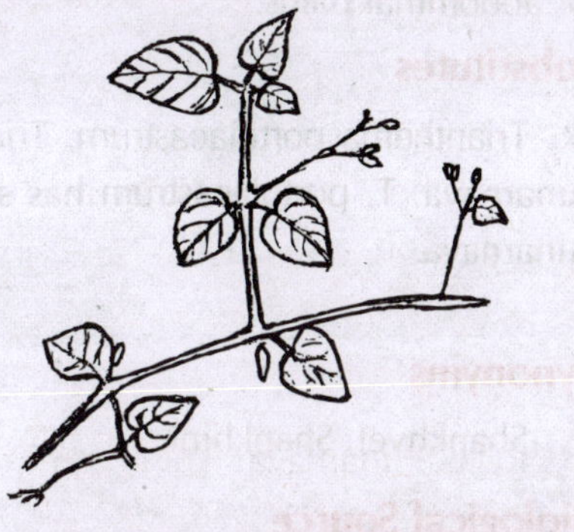

Fig. 15.50: Punarnava twig

Chemical Constituents

Punarnava contains about 0.04 - 0.1 per cent of alkaloid known as punarnavine (M.P. 235ºC), phenolic glycoside, punernavoside, an antifibrinolytic agent and rotenoids, boeravinone A, B, C and D. It also contains about 6 per cent of potassium nitrate, an oily substance, and ursolic acid.

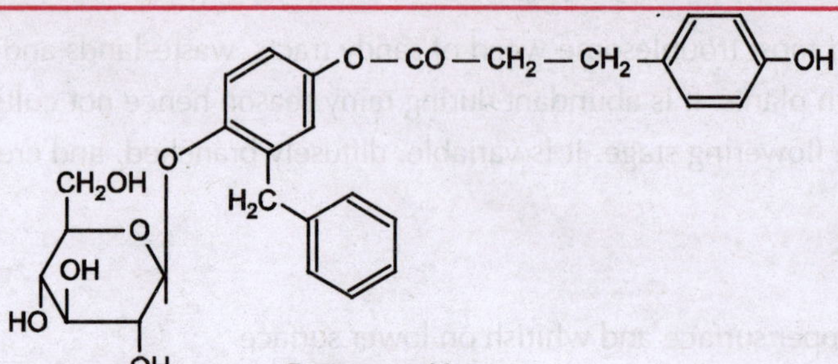

Punarnavoside

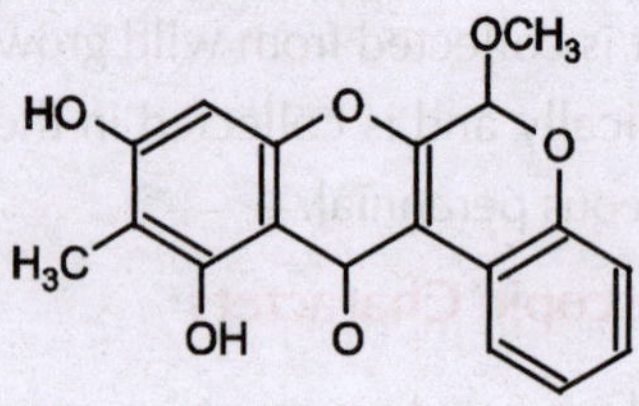

Boeravinone A

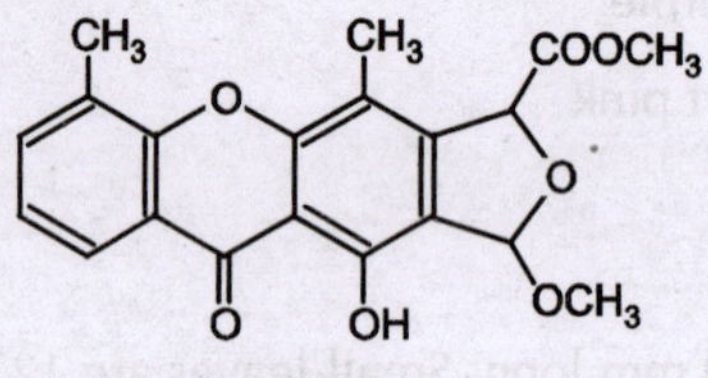

Borhavine

Standards of Quality

Foreign organic matter	:	Not more than 2.0 per cent
Water soluble extractives	:	Not less than 9.0 per cent
Alcohol soluble extractives	:	Not less than 0.5 per cent
Acid insoluble	:	Not more than 03.0 per cent
Ash content	:	Not more than 10.0 per cent
Loss on drying	:	Not more than 10.0 per cent

Uses

The herb is used as diuretic and as an expectorant. Punarnava is stomachic and is prescribed in the treatment of jaundice. It is also given in the loss of digestive power, enlargement of spleen and for abdominal pains.

Substitutes

Trianthema portulacastrum, Trianthema obcordata and T. decandra are usually substituted for punarnava. T. portulacastrum has similar foliage to that of punarnava and is also known as shwet punarnava.

SHANKHPUSHPI

Synonyms

Shankhvel, Shankhini.

Biological Source

This consists of the aerial parts of the plant known as Canscora decussata, family Gentianaceae.

Geographical Source

Shankhpushpi is found throughout India up to an altitude of 1300 m. It is also grown in Sri Lanka and Myanmar.

Macroscopic Characters

It is much branched, annual plant about 10-30 cm in height propagated from seeds. The flowering season of this plant is from October to December. The plant is cultivated in the gardens as ornamental plant for its flowers. This is an erect annual with four winged stems and half a metre in length with decussate branches. It grows well in moist situations.

Fig. 15.51: Shankhpushpi: *Canscora decussata herb*

Leaves are sessile, 2.5 - 4.0 cm in length, lanceolate, decussate with three prominent verticle lines; flowers are funnel shaped axilliary, solitary, white or yellowish or light pink in colour.

Fruits are glabrous, upto 2.5 mm in size, stem are woody at the base.

Chemical Constituents

Drug is found to contain bitter substance and an oleo-resin. Two crystalline compounds have been isolated from the aqueous and alcoholic extracts of the plant. Shankhpushpi is found to contain triterpenes, alkaloids, xanthones, volatile oil, β sitosterol and n-triacontane. Alkaloids are known as shankhpushpine and convoluine.

Uses

Entire plant, as well as, fresh juice is used in medicine. It is regarded as bitter alternative and **nervine tonic**. The fresh juice of the plant is prescribed in insanity, epilepsy, and nervous debility. Alcoholic extract possesses anti-viral activity against Ranikhet disease virus (chicks).

Substitutes

Canscora diffusa fam: Gentianaceae distributed throughout India is used as substitute for Shankhpushpi.

There is controversy in correct identity of shankhpushpi and some scientists have identified Dwarf morning glory i.e. Evolvulus alsinoides fam: Convolvulaceae and also Convolulus pruricalis family Convolvulaceae as Shankhpushpi.

DWARF MORNING GLORY

Synonym

Aloe-weed.

Biological Source

This consists of dried aerial parts of the plant known as Evolvulus alsinoides fam. Convolvulaceae. This drug is also considered as **shankhpushpi**.

Macroscopic Characters

It is a procumbent or sub-erect shrub. Stems are woody at base. Leaves are about one inch long, simple, linear, oblong or mucronate, alteranate, lanceolate, obtuse, subsessile with tapering base branches arise from woody base (root-stock). Flowers are auxillary, solitary or 2 - 3 together, pedunculate bluish or white coloured, while fruits are four angled capsules flowering season is from July to November. Roots are 5 - 15 cm in length, white with penetrating odour.

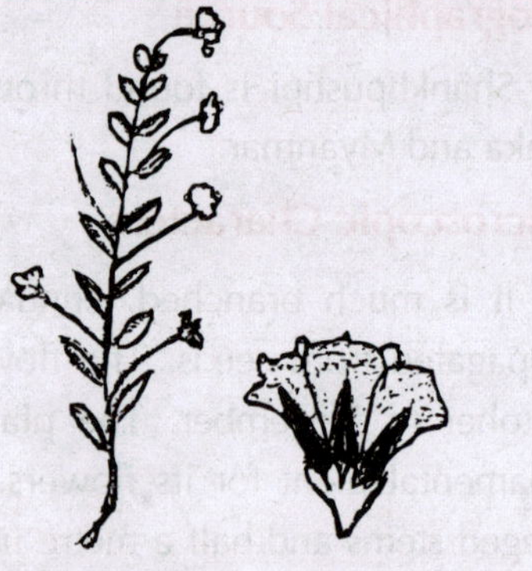

Fig. 15.52: *Evolvulus alsinoides* herb

Chemical Constituents

The drug contains an alkaloid shankhpushpine, betain, evolvine, volatile oil, yellow neutral fat and organic acids. In addition to potassium chloride kaempferol, B-sitosterol, triacontinol.

Uses

It is used as brain tonic and sedative. It is also used to treat chronic bronchitis, asthma and to promote growth of hairs.

HEMLOCK

Synonyms

Poison Hemlock, Spotted hemlock, Poison pursley.

Biological Source

This consists of dried full grown but unripe fruits of the plant Conium maculatum family Apiaccae (Umbelliferae).

Geographical Source

It is native of temperate regions of Europe, West Asia and North Africa. It has been naturalized in Australia. North America and New Zealand. It is found on drained soils, near streams, ditches and other waste places.

Cultivation and Collection

Hemlock fruits are collected from wild grown plants only, when they are fully grown, but before they ripen i.e. before they change their colour from green to yellow fruits are dried thoroughly and stored.

Macroscopic Characters

Hemlock is a biennial plant reaching the hight of 1.5 – 2.5 metres, stems are hollow green in colour and are usually spotted or streaked with red or purple on lower half of the stem.

Fig. 15.53 (a): Hemlock herb

Fig. 15.53 (b): Fruity top of Hemlock

Leaves are roughly triangular and about 50 40 cm in size. Flowers are small white and in umbels from 10 to 15 cm across.

Microscopic Characters

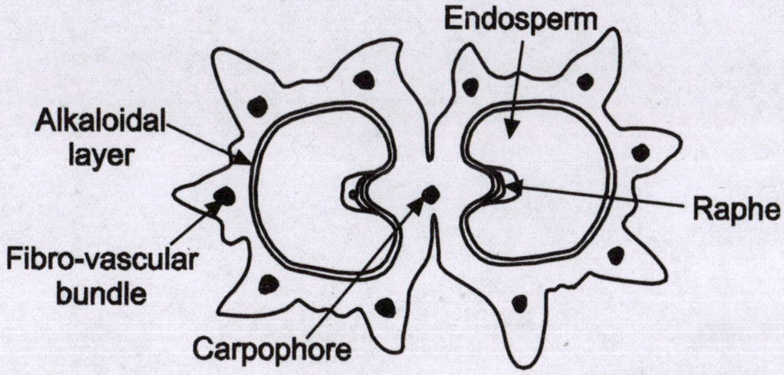

Fig. 15.54: T.S. of Hemlock Fruit (cremocarp)

Chemical Constituents

Pyridine-piperidinic, alkaloids are the active constituents of hemlock. About eight alkaloids have been reported and are named as coniine, N-methyl coniine, conhydrine, pseudo conhydrine and gamma-conicrine. Coniine is a liquid and volatile alkaloid. Their actions are similar to Nicotine. All parts of plant contain alkaloids but maximum in fruits from 0.05 – 2.5 per cent. Hemlock also contain volatile oil and fixed oil.

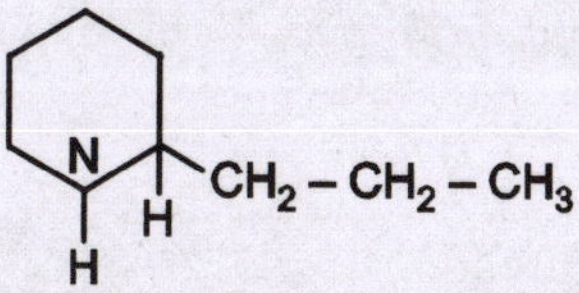

Coniine

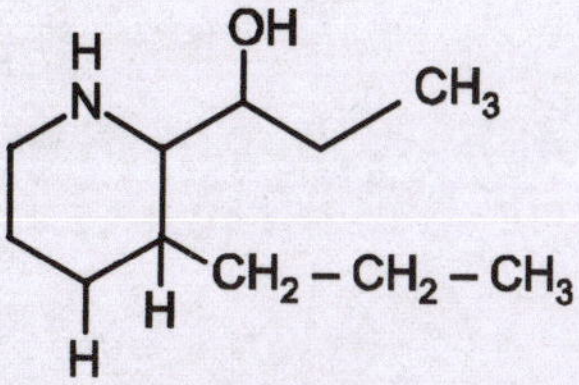

Conhydrine

Chemical Test

Crush the dried fruits and add dilute solution of potassium hydroxide. Strong mouse-like odour is developed (due to liberation of coniine alkaloid).

Action of Mechanism of Coniine

The alkaloid coniine disrupts the working of central nervous system through action on nicotinic acetylcholine receptors. Alkaloids are highly potent and even in small doses cause respiratory collapse and even death. The death is due to blocking the neuromuscular junction similar to curare, due to lack of oxygen to heart and brain. Overdoses produce paralysis and loss of speech, depression and respiratory system and then death.

Uses

Hemlock is highly potent and very poisonous, needs to be used with lot of care.

Poison hemlock has been used as sedative and anti-spasmodic.

❖❖❖

Chapter 16

FIBRES, SUTURES AND SURGICAL DRESSINGS

Fibres, as referred to in Pharmacognosy, are elongated thick walled cells with pointed ends; cell walls of which consist of cellulose and may or may not contain lignin. In medical practice, they are used as surgical dressings made up of natural or artificial materials. Apart from three natural sources, such as, plant, animal and mineral, fibres are now synthesised chemically from various materials. Fibres obtained from various sources can be categorised as follows:

(a) Plant fibres - Jute, flax, banana, cotton, hemp

(b) Animal fibres - Silk, wool

(c) Regenerated and Synthetic fibres - Nylon, terylene, orlon

(1) Fibres regenerated from carbohydrate materials - Alginate yarn, artificial silk or rayon or regenerated cellulose

(2) Fibres regenerated from protein materials - Aridil from groundnut protein and fibrolin from milk casein

(3) Synthetic - Nylon, terylene, orlon

(d) Mineral fibres - Glass, asbestos

Various chemical tests can be applied for the identification of fibres. The microscopical examination is the main criterion to confirm the identity of fibres.

Commercially, fibres are used in the textile industry for weaving the cloth, as filtering medium, and also for insulation purpose.

Natural fibres of surgical importance are listed in Table 16.1.

Table 16.1: Surgical fibres

Sr. No.	Name and synonym of fibre	Biological source	Active constituents	Uses
1.	Cotton (Absorbent cotton, Surgical cotton, Medicinal cotton)	Epidermal trichomes of seeds of *Gossypium* species (Malvaceae)	93 to 94 per cent cellulose and moisture – 5 - 7 per cent	Surgical dressing Filtering media and insulation.

Contd...

2.	Jute (Gunny bag fibres)	Phloem fibres of *Corchorus sp* Family: Tiliaceae	Cellulose hemicellulose and lignin	Manufacture of tows and Gunny-bags Straining Filtration media.
3.	Flax	Pericyclic fibres of stem of *Linum – usitatissimum* Family: Linaceae	Pecto-cellulose	Straining and filtering media, preparations of course bags.
4.	Silk (Silk fibres)	Fibres obtained from silk worm cocoons of *Bombyx mori* Family: Bombycidae	Protein known as Fibroin	Sutures, ligatures
5.	Wool (Wool fibres)	Fibres from flees of sheep *Ovis aries* Family: Bovidae	Protein known as keratin	In the manufacture of surgical dressings like domette, crepe bandange, flannel and blankets for warmth.
6.	Cellulose	Found in all plant cells.	Polysaccharide linked with glucose	Preparation of carboxy methyl cellulose (CMC) and other cellulose products.

A. PLANT FIBRES

ABSORBENT COTTON

Synonyms

Raw cotton, Cotton wool, Purified cotton, Surgical cotton.

Biological Source

Cotton consists of the epidermal trichomes or hairs of the seeds of cultivated species of the *Gossypium* (*Gossypium herbaceurre, Gossypium barbadense*), belonging to family Malvaceae. Purified cotton or absorbent cotton consists of the trichomes as mentioned above, but freed from fatty matter and adhering impurities. It is also bleached and sterilized.

Geographical Source

Cotton is produced commercially in USA, Egypt and India. It is also cultivated in various parts of Africa and South America. In India, seven million hectares of land is under cultivation of cotton, of which 30 per cent is irrigated and 70 per cent rainfed.

Preparation of Absorbent Cotton

Cotton is commonly grown for the purpose of fibres in the tropical countries. The plant after flowering, bears fruits known as capsules. The fruits are 3 to 5 celled. Each capsule contains numerous seeds. The seeds covered with the hairs are known as bolls. The bolls (Fig. 16.1) are collected, dried and taken to the ginning press, wherein the trichomes are separated from the seeds. Various devices are used to separate the hairs. The short and long hairs are separated from each other. The hairs with short length are known as linters and are used for the manufacture of absorbent cotton, while long hairs are used for the preparation of cloth. The raw cotton obtained by this way is full of impurities, like wax, fat, colouring matter, vegetable debris, etc. It is processed to get rid of most of the impurities. It is taken to the machine known as cotton opener and followed by treatment with dilute soda solution or soda ash solution under pressure for about 10 - 15 hours. The wax, fatty material and colouring matter are removed by this treatment. It is then washed with water and treated with suitable bleaching agent. It is again washed with water, dried and carded into flat sheets. It is finally packed in wrappers and sterilised by means of Gamma radiations.

Fig. 16.1: Cotton Plant with Fruit

Description

Colour : White (due to bleaching)

Odour : Odourless

Taste : Tasteless

Size : Cotton fibres are 2.5 - 4.5 cm. in length and 25 - 35 micron in diameter.

It is free from pieces of leaves, seed coat, foreign matter and dust.

It may be slightly off-white in colour, if sterilized.

Microscopic Characters (Fig. 16.2)

The trichomes are unicellular, flattened and ribbon-like with slightly thickened and rounded apex. They are tubular and hollow.

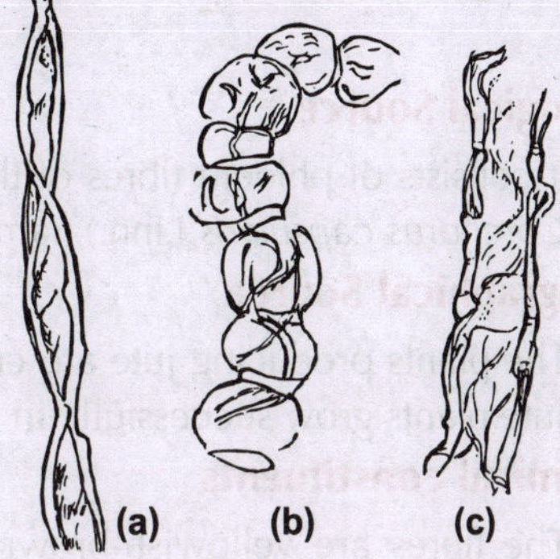

Fig. 16.2: (a) Cotton Fibre (b) Raw cotton mounted in cuoxam reagent (c) Absorbent cotton mounted in cuoxam reagent

Chemical Constituents

Raw cotton contains about 90 per cent of cellulose, 7 to 8 per cent of moisture, wax, fat and remains of protoplasm.

Purified cotton or absorbent cotton is entirely cellulose, with 6 - 7 per cent of moisture.

Standards of Quality

Absorbent cotton wool I.P. has the following standards:

Length of staples	:	Not less than 15 mm.
Water soluble extractive	:	Not more than 0.5 per cent
Sulphated Ash	:	Not more than 0.5 per cent

It should comply with the test for neps, fluorescence, acidity, absorbency and oxidising substances.

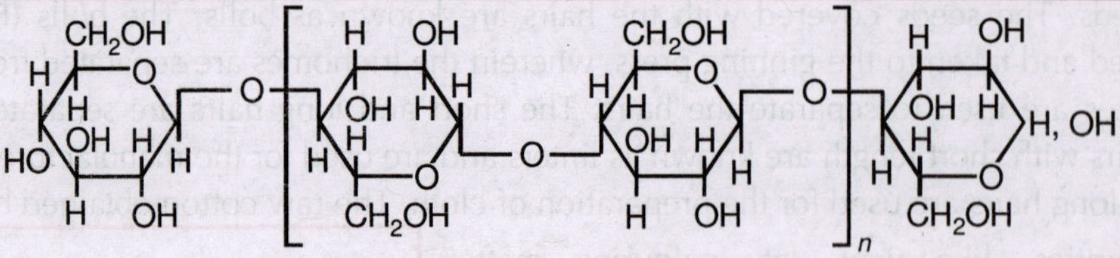

Cellulose chain

Chemical Tests

(1) Soak cotton fibres in iodine water and dry. Add few ml. of 80 per cent sulphuric acid-trichomes assume purplish-blue or bluish-green colour (distinction from jute, hemp, wool, silk, nylon, alginate yarn and acetate rayon).

(2) Ammonical copper oxide solution (cuoxam-regent) dissolves raw cotton fibres with the formation of balloons, while absorbent cotton dissolves completely with uniform swelling.

(3) Cotton is insoluble in dilute sodium hydroxide solution and hydrochloric acid (distinction from silk). It is soluble in 66 per cent of sulphuric acid.

Uses

Cotton is used as a filtering medium and in surgical dressings. It is also used as an insulating material. Absorbent cotton absorbs blood, mucus, pus and prevents the wounds from infections, hence indispensable part of dressing.

Storage

Absorbent cotton should be stored in cool place. Bacterial contamination makes the cotton friable and brittle. The absorbent cotton should be wrapped in wrappers so as to prevent the dust and microbial contamination. Heat and long storage make absorbent cotton non-absorbent.

JUTE

Biological Source

It consists of phloem fibres of the stem of various species of the *Corchorus* (*Corchorus olitorius* and *Corchorus capsularis* Linn.), family Tiliaceae.

Geographical Source

The plants producing jute are cultivated in West Bengal, in the basins of Ganges and in Assam. The Jute plants grow successfully in areas having loamy alluvial soil with pH values of 6 to 8.

Chemical constituents

The fibres are yellowish-brown in colour and contain cellulose (53 per cent), hemicellulose (20 per cent) and lignin (10 per cent).

Chemical test

The middle lamella is highly lignified and gives red colour with phloroglucinol and hydrochloric acid.

Uses

It is used in the manufacture of tows (stupa)*, padding splints, filtering and straining medium. Jute fibres are used for the preparation of coarse bags (Gunny bags).

---

* Stupa: Wet cloth or sponge charged with medication for external application.

FLAX

Biological Source

These are pericylic fibres obtained from the stem of the plant *Linum usitatissimum*, family Linaceae.

Geographical Source

The plant is an annual herb and is cultivated in many countries, such as Argentina, Russia, Canada, India, Holland and USA.

Description

Fibres are pale yellow in colour. Fibres are lustrous with sufficient tensile strength. Their length is about 20 - 50 mm. and are with fine pointed ends.

The seeds of the plant are good source of fixed oil mucilage and also used medicinally.

The chemical nature of the fibres is known to be pecto-cellulose.

Uses

The fibres were earlier used in the manufacture of lint, but at present cotton has replaced these fibres. It is used in preparation of rugs, lace, lawn and filtering medium.

B. ANIMAL FIBRES

SILK

Synonym

Silk fibres.

Biological Source

These are the fibres obtained from the cocoons of *Bombyx mori* (Mulberry silkworm) and other species of *Bombyx* and also from *Antheraea* species family, Bombycidae (Order Lepidoptera).

Geographical Source

Fine quality of silk is manufactured in Japan, Italy, China and France. Now-a-days, a large quantity of silk is also produced in India.

Preparation of Silk

The larvae of the silkworm produce silk fibroin fibres from the glands in their mouth. This fibroin gets united with a gum-like secretion known as sericin and forms cocoon. These cocoons are not allowed to grow further into an insect, but are heated to 60 - 80°C by exposing them to steam. The exposed cocoons are put into hot water to dissolve the gum and to separate the fibres. *(Fig. 16.3)*

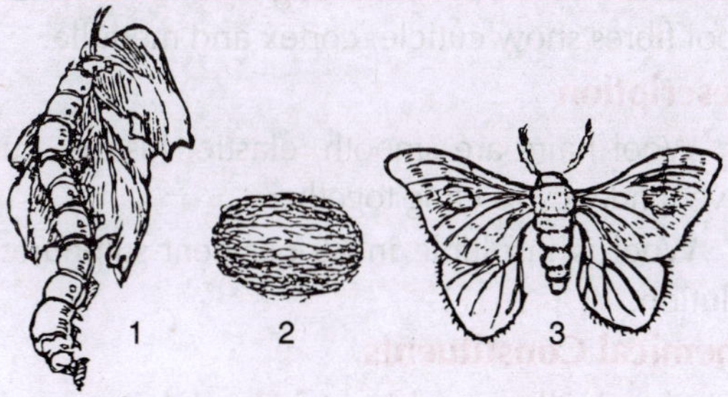

Fig. 16.3: Silk worm moth - *Bombyx mori* Linn.
1. Larva, 2. Cocoon, 3. Moth

Description

Silk fibres are 5 to 25 microns in diameter and about 1200 metres in length. Silk threads are very fine, solid, smooth.

Colour : Colour is usually yellowish

Odour : Odourless

Solubility

Insoluble in water and alcohol, silk is soluble in cuoxam solution, sulphuric acid (66 per cent) and concentrated hydrochloric acid (distinction from wool).

Chemical Constituents

Silk contains a protein known as fibroin. Fibroin on hydrolysis yields amino-acids glycine and alanine.

Chemical test

It does not contain sulphur and hence, the test with lead acetate solution is negative.

Uses

Special types of sutures, sieves and ligatures are prepared from the silk.

WOOL

Synonym

Wool fibres.

Biological Source

Wool or wool fibres are obtained from the fleece of sheep *Ovis aries*, family Bovidae, (Order Ungulata.)

Geographical Source

Large quantities of wool are manufactured in Australia, Russia, Argentina, U.S.A. and also in India.

Preparation of Wool

The hairs, forming the fleece of the sheep are removed at shearing time. They are then processed to remove the wool fat and dirt. The clean and defatted wool is subjected to bleaching, washed again and dried.

When observed under high power of microscope wool fibres show cuticle, cortex and medulla.

Fig. 16.4: A - The Sheep, B - Wool fibre

Description

Wool hairs are smooth, elastic, lustrus, curly, hygroscopic and slippery to touch. Wool hairs have tendency to cling together.

Wool is insoluble in 66 per cent sulphuric acid, concentrated hydrochloric acid and cuoxam solution.

Chemical Constituents

Chemically, wool contains sulphur containing protein known as keratin. Keratin is rich in sulphur containing amino-acid cystine.

Chemical Test

When lead acetate is added to a solution of wool in caustic soda, a black precipitate is formed owing to high sulphur content (distinction from silk).

Uses

Wool is used as: (i) filtering and straining medium and (ii) in the manufacture of dressings like domette and crepe bandages and also flannel.

C. REGENERATED AND SYNTHETIC FIBRES

RAYON

Synonyms

Regenerated cellulose, viscose rayon

Source

It is an artificial fibre, composed of regenerated cellulose in which substituents have replaced not more than 15 per cent of hydrogens of hydroxyl groups. Now, it is produced exclusively by the viscose process.

The cellulose of coniferous wood in the form of pulp is processed to give viscose rayon. The pulp contains about 80 - 90 per cent of cellulose and hemicellulose. The cellulose is treated with sodium hydroxide to get sodium cellulosate. It is further treated with carbon disulphide and sodium hydroxide to produce viscous solution of sodium cellulose xanthate.

After ripening, this solution is forced through the fine nozzles into the bath of sodium sulphate and sulphuric acid to give continuous filaments. The filaments are drawn together to form the yarn. The yarn is desulphurized, bleached, washed twisted and then dried.

Rayon contains about 10 per cent moisture and does not loose the absorbency on storage like cotton.

Uses

It is mainly used for making fabrics, preparation of surgical dressings and viscose rayon absorbent wool.

SYNTHETIC FIBRES

1. Nylon (Polyamide)

It is the polymer of adipic acid and hexamethylene diamine. The fibres are highly lustrous to dull, white or coloured. When applied to flame, the fibre melts with formation of bead. It is soluble in 5 M hydrochloric acid and insoluble in acetone. It is soluble in 90 per cent formic acid and phenol (90 per cent), which distinguishes it from fibres of biological origin.

Uses

It is used for preparation of filter cloths, sieves and non-absorbable sutures.

2. Terylene

It is a polymer of ethylene glycol and terephthalic acid, (whereas Orlon is a polymerised acrylonitrile.) Terylene may be distinguished from nylon that it retains its structure on boiling with phosphoric acid.

Uses

It is being used for artificial grafts and also as non-absorbable suture.

D. MINERAL FIBRES

Glass Fibres

The fibres consists of sand (silica), mixed with oxides of aluminium, calcium, boron and magnesium. They are unaffected by all usual reagents used in identification of fibres. They melt at high temperature and form a transparent bead. They are used in filter fabrics for insulation and splinting material.

Asbestos

It contains mainly hydrated magnesium silicates. Rock asbestos is white, yellow or green in colour. It is highly refractive and do not fuse when heated.

Uses

It is used for preparation of a filtering medium and for bacterial filters.

SURGICAL DRESSINGS

The word surgical dressing is used to include all the materials either used alone or in combination to cover the wound. The purpose of application of dressing is to protect the wound and favour its proper healing. A material which holds the dressing in desired position is known as bandage. Fibres are used for the preparation of surgical dressings.

Dressings are meant for the following functions:

(1) To reduce or prevent infections.

(2) To offer protection to healing wound.

(3) To offer mechanical support to the tissues.

Surgical dressings should comply the following requirements as mentioned in Pharmacopoeia.

(1) They should be sterilised before use.

(2) They should be stored in dry well-ventilated place at a temperature not exceeding 25°C.

(3) They should be used with permitted antiseptics in prescribed concentrations only.

(4) They should not be dyed unless mentioned in the monograph.

(5) Adhesive products should not be allowed to freeze.

(6) There should not be any loose threads, fibre-ends in dressings.

Surgical dressings are classified as:

(A) Fibres

(a) Non-medicated fibres

Absorbent cotton, wool, rayon, silk etc.

(b) Medicated fibres

Boric acid wool, Capsicum wool.

(B) Fabric

These are woven materials. These may be medicated or non-mediated. Examples are: Absorbent lint, Absorbent ribbon gauze, Boric acid lint, Absorbent gauze, X-ray detectable.

(C) Bandages

These are the products used to retain dressings in place and provide support for applications of medication to the skin. They are water proof, Bandages are of two types i.e. they may be elastic, or non-elastic, adhesive or non-adhesive.

(a) Non-medicated bandages:

Crepe bandage, Domette bandage, Calico bandage, Cotton and rubber elastic bandage.

(b) Medicated bandages:

Plaster of Paris bandage, Zinc paste bandage, Zinc paste Ichthamol bandage.

(D) Rubber and Oil Impregnated Materials

Belladonna self-adhesive plaster, Zinc oxide self-adhesive plaster, etc.

British Pharmacopoeia recognizes the following types of surgical dressings.

(1) Fibres and related materials	i.e. Viscose, animal, wood cellulose wadding
(2) Carded products	i.e. Absorbent cotton, viscose wadding, etc.
(3) Non-extensible, non-adhesive woven products	i.e. Absorbent muslin; Domette bandage, etc.
(4) Non-extensible adhesive woven products	i.e. Belladonna adhesive plater, Zinc oxide surgical adhesive tape,

(5) **Extensible non-adhesive woven products** i.e. Cotton bandage, elastic web bandage, etc.

(6) **Extensible adhesive woven products** i.e. Elastic adhesive bandage, extension plaster

(7) **Non-woven products** i.e. Impermeable plastic surgical adhesive tape, permeable plastic surgical adhesive tape, etc.

Surgical dressings are broadly classified as woven dressings and non-woven dressings.

WOVEN DRESSINGS

Cotton, silk and wool are used to prepare a number of woven dressings. They should also comply the pharmacopoeial requirements. Following are the few woven-dressings available in the market.

1. **Absorbent Lint:** It is cotton cloth of plain weave from the warp yarn of which a nap has been raised.
2. **Flannel Bandage:** It consists of a raised fabric of plain weave and is made up of entirely wool.
3. **Domette Bandage:** It consists of a combination of fabric of plain weave, where the warp yarn are of cotton and weft yarns are entirely of wool.
4. **Crepe - Bandage:** It consists of fabric of plain weave, wherein warp threads are of cotton and wool, whereas the weft threads are entirely of cotton only.
5. **Gauge and cellulose wadding tissue:** Cellulose tissue consists of thick layer of cellulose wadding enclosed in tubular absorbent gauze.

Few technical terms used to describe surgical dressings:

(i) **Staple:** It is an alternative term for fibre.

(ii) **Neps:** Neps are knots or tangled masses of fibres visible to naked eye. They may be produced in cotton during growth or when it becomes damp.

(iii) **Yarn:** It is term applied to thread spun from the fibres.

(iv) **Yarn Count:** It is the weight in grams of 1 kilometer of yarn. It is used as a measure of thickness.

(v) **Union fabric:** It is the fabric composed of two kinds of yarn e.g. cotton and wool.

(vi) **Warp and weft:** The threads in fabric running lengthwise are referred as 'warp' or 'ends', while those crossing them at right angles are 'weft'.

(vii) **Plain weave:** A weave in which the threads pass alternatively over and under the threads running at right angle.

(viii) **Twist:** It refers to the direction of twisting the fibres while spinning the yarn. Twist to the right is called " z " twist whereas, twist to the left is called " s " twist or reverse twist.

(ix) **Naps:** This refers to loose fluffy fibres of cotton cloth of plain weave by cutting and scraping the threads. These weakened threads enable the material to be easily torn. A fabric on which naps are produced is called 'Raised fabric'.

SUTURES

These are the sterile threads, strings or strands specially prepared for use in surgery meant for sewing tissues together. Sutures must possess the following properties.

(1) They must be sterile and should cause no irritation.

(2) They should have finest possible guage and adequate strength.

(3) If absorbable, their time of absorption must be known.

(4) They are intended to be used for one occasion only.

Sutures may be prepared from intestinal tissues and tendons of animals and birds; vegetable fibres; camel hair; human hair; synthetic threads or metallic wires. Depending upon their characteristic of absorption, i.e. digestion in the tissues of the body, sutures are known as absorbable or non-absorbable sutures.

Various methods are used to sterilize sutures. A few of the important methods are chemical sterilisation or sterilisation by irradiation, and sterilisation by heat.

Heat sterilisation involves two processes and sutures are classified into two categories depending upon heat resistance, (a) boilable (in anhydrous liquid) and (b) non-boilable.

The sterilisation should not affect the properties of sutures or their utility. In irradiation method, sterilisation is carried out by electron particles or by gamma rays from cobalt 60. The recognized dose of gamma rays treatment is 2.5 megarad.

Sutures may be of the following types:

(1) Absorbable sutures

(a) Sterile catgut (small intestine of sheep, ox and appendix of deer)

(b) Sterile reconstituted collagen suture

(2) Non-absorbable sutures

(a) Sterile non-absorbable sutures (silk and cotton)

(b) Sterile linen suture

(c) Sterile polyamide sutures (nylon)

(d) Sterile polyester sutures (terylene)

(e) Sterile braided sutures

(f) Sterile stainless and silver sutures

(3) Haemostatics

(i) Oxidised cellulose

(ii) Absorbable gelatin sponge

Uses

Depending upon the physical and chemical properties and also the desired effect, various sutures are effectively used under different conditions.

Silk sutures are strong, smooth inactive and also available in various diameters. The braided form is compact one and has a special advantage of not getting twisted when drawn through the body tissues. Nylon sutures are strong enough and enjoy the merit of their utility in skin and plastic surgery. Cotton sutures due to their low order of tissue reactivity are preferred, but suffer from poor tensile strength. Linen, though very much economical and quite effective, lacks in uniform diameter for practical purposes. Kangaroo tendons are specially used in harnia, while metallic sutures are used in surgery in general.

Chapter 17...

MARINE DRUGS

In the domain of drugs and medicines, it is somewhat contradictory to note that most of them are obtained either by synthesis or from terrestrial organisms. Even though, about seventy per cent of the earth's surface is covered by oceans and seas, rich in fauna and flora, it has not been properly explored for the search of biomedicinals. It is necessary to tap the marine sources for commercial utility of novel drugs in the treatment of human ailments.

The drugs obtained from marine organisms which are being conventionally used like shark and cod-liver oils, sodium alginate, agar-agar, pearls, prawal and chitin have been already discussed elsewhere in this book in accordance with chemical nature of constituents. Even today, they constitute an important group of marine drugs.

The present chapter is a compilation of some of the marine biologicals with proven chemical nature and pharmacological actions. About 5 lacs species of marine organisms have been reported from the oceans and seas from various parts of the world. Some of these organisms are antimicrobial, antiviral, antibiotic, anticancer/cytotoxic, enzyme inhibitors, anti-inflammatory, prostaglandins, neurophysiological and cardiovascular agents. Many of the species contain toxic compounds.

Classification

Broad classification of marine drugs is as under:

Table 17.1: Classification of Marine Drugs

Sr. No.	Class	Examples
1.	Antibiotic compounds	Cycloeudesmol, variabilin, Ircinin-1
2.	Anti-inflammatory and Antispasmodic agents	Manoalide, Tetrado toxin
3.	Antimicrobial compounds	Holotoxin ABC, Tholepin, eunicin acanthelin
4.	Cardio vascular active substances	Anthopleurins, Laminine, Eptatretin, saxitoxin, spongosine, eledosin, ATX-II, Autonomium
5.	Cytotoxic (Anti cancer) compounds	Ara-c, crassin acetate simularin.
6.	Marine toxins	Ciguatoxin, Palytoxin, Saxitoxin, Brevetoxin
7.	Miscellaneous compounds	Kainic acid, Domoic acid, Aplysinopsin.

1. ANTIBIOTIC COMPOUNDS

Some of the marine antibiotics and their respective sources have been listed below (Table 17.2).

Table 17.2: Antibiotic Principles from Marine Sources

Organisms	Antibiotic Substance
1. Marine bacterium, *Pseudomonas bromutilis*	2, 4 - Dibromo - 6 - (3, 4, 5 tribromopyrrole - 2 - yl) phenol
2. Red algae, *Chondria oppositiclada*	Cycloeudesmol
3. Sponge, *Ircinia strobilina* and *Ircinia oros*	Variabilin, Ircinin - 1
4. Sponge, *Verongia archery*	3, 5 - dibromo - 4 - hydroxy benzene-1 acetamide

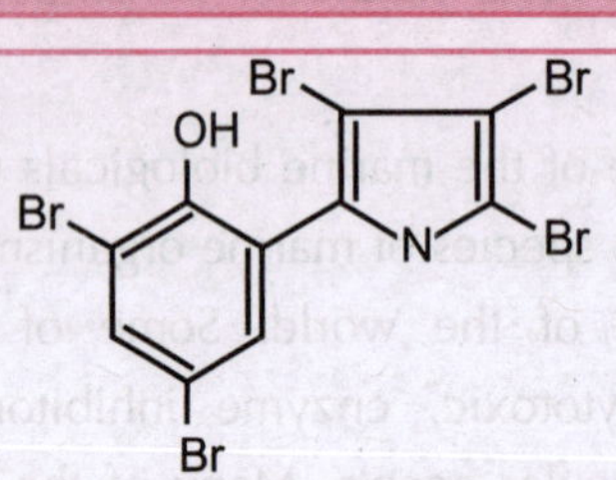

2, 4 - Dibromo - 6 - (3, 4, 5 - tribromopyrrole - 2 - yl) phenol

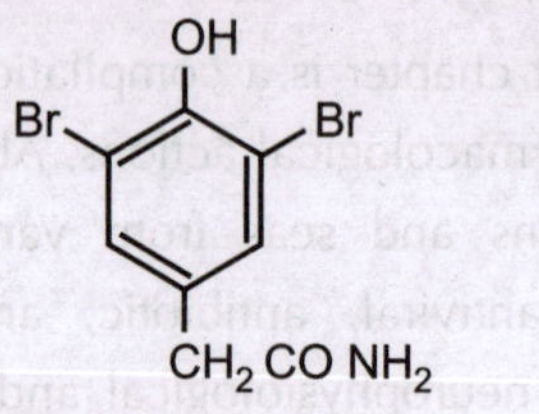

3, 5 dibromo - 4 - hydroxy - benzene - 1 - acetamide

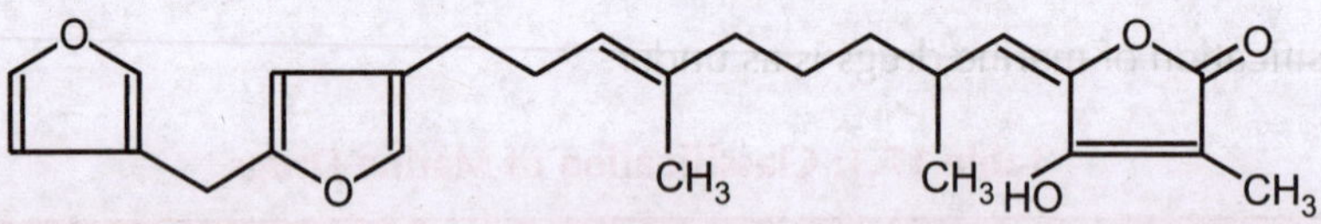

Ircinin - 1

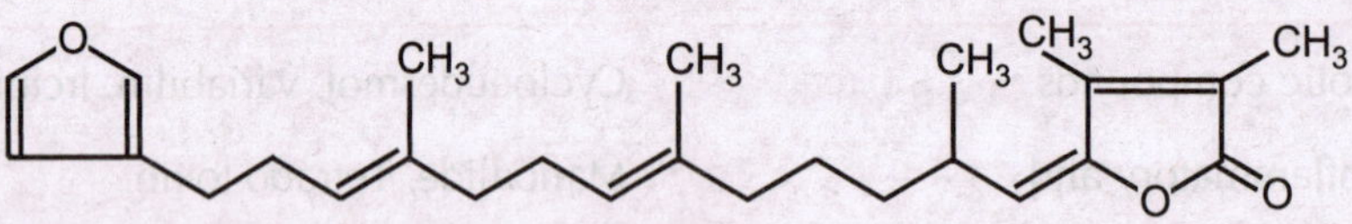

Variabilin

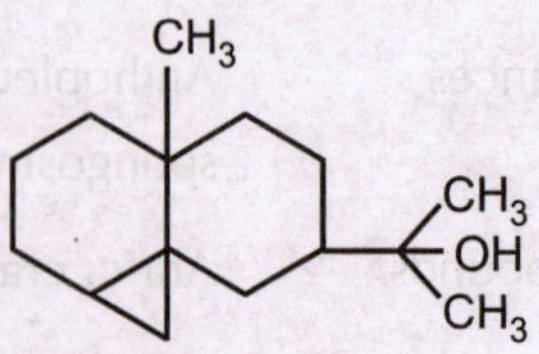

Cycloeudesmol

2. ANTI-INFLAMMATORY AND ANTISPASMODIC AGENTS

Manoalide, a non-steroidal anti-inflammatory compound, has been obtained from the sponge *Luffariella variabilis*. Probably, it acts by direct inactivation of phospholipase A_2, which is an enzyme present in some neurotoxins and also has a role in synthesis of prostaglandins in humans. Manoalide has analgesic and selective anti-inflammatory properties.

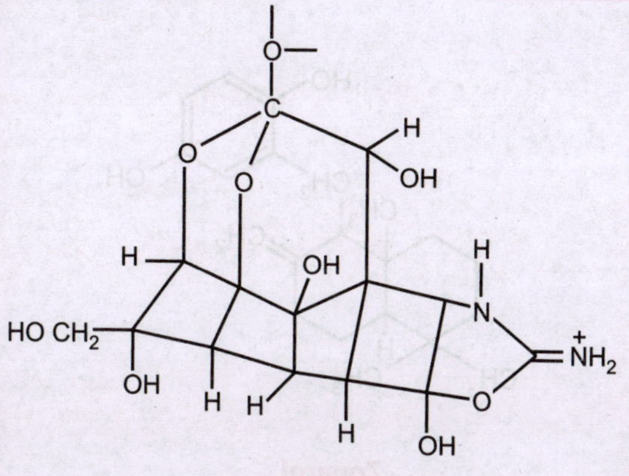

Tetradotoxin

The other anti-inflammatory agents reported are dendalone 3hydroxy butyrate from a sponge, *Phyllospongia dendyi*; flexibilide from a soft coral, *Sinularia flexibilis* and is chemically diterpenoid cembrane; 6-n-tridecyl salicylic acid from the brown algae, *Caulocystis cephalornithos*.

The muscle relaxant action is associated with flustramine A and B which are present in *Flustra foliaceae*, a Swedish marine moss. Tetradotoxin obtained from liver and ovaries of puffer fishes, is a strong antispasmodic.

3. ANTIMICROBIAL COMPOUNDS

A large number of antimicrobial compounds have been detected from marine organisms like sea cucumber, marine algae, sponges, sea hare, gorgonian corals and red algae.

The organisms, and their antimicrobial principles are listed below: (Table 17.3)

Table 17.3: Antimicrobial Principles From Marine Sources

Marine Organisms	Antimicrobial Compounds
1. Sea cucumber, *Stichopus japonicus*	Holotoxin A, B and C (steroida glycoside) antifungal in nature Zonarol and iso-zonarol
2. Brown algae, *Dictyopteris* zonaroides	Tetrabromoheptanone
3. Red algae, *Bonnemaisonia hemifera*	Four isomers of polyhalo 3 butene - 2 - one and seven isomers of polythaloacetones.
4. Red algae, *Asparogopsis taxiformis.*	Aeroplysinin-1 (+)
5. Sponge, *Verongia aerophoba*	Aeroplysinin-1 (-)
6. Sponge, *Agelas oroides*	2 - cyano - 4, 5 – dibromo pyrrole
7. Red algae, *Laurencia pacifica* and *L. filformis.*	Prepacifenol
8. Annelida, *Thelepsus setosus*	Tholpin
9. Sea hare, *Aplysia californica*	Debromolaurenterol
10. Gorgonian corals, *Eunicia mammosa*	Eunicin
11. *Acanthella acuta*	Acanthelin -1 (active against *Mycobacterium*)

The structural formulae of some of the above referred compounds are:

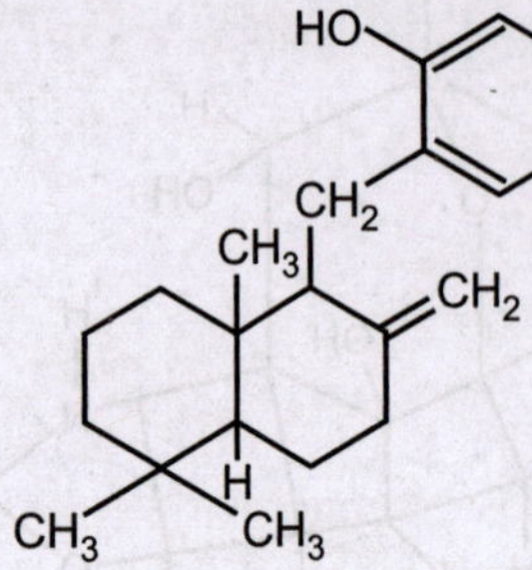

Zonarol

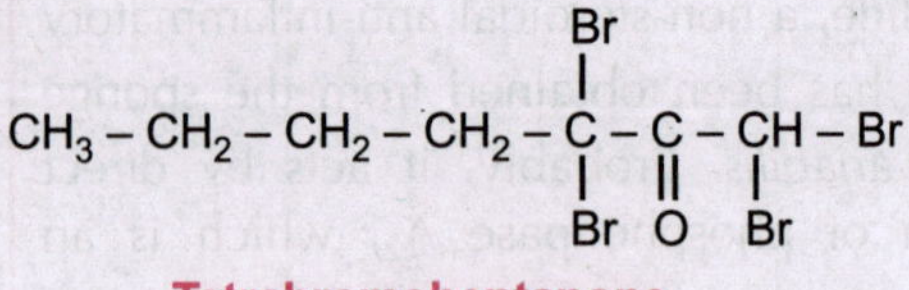

Tetrabromoheptanone

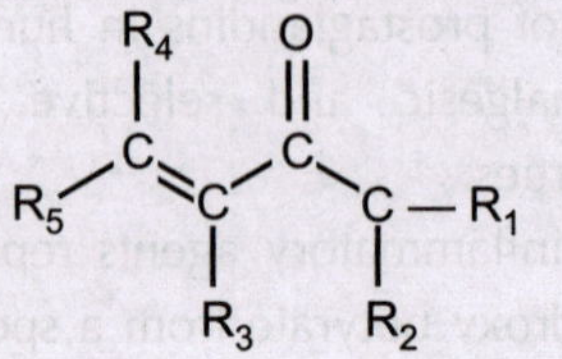

Polyhalo - 3 - butene - 2 one

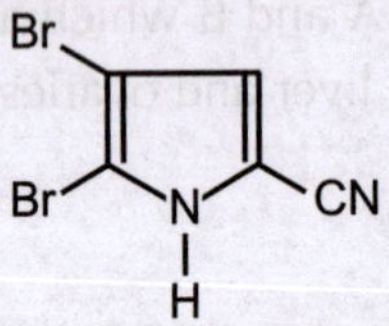

2, - Cyano - 4, 5 - dibromopyrrole

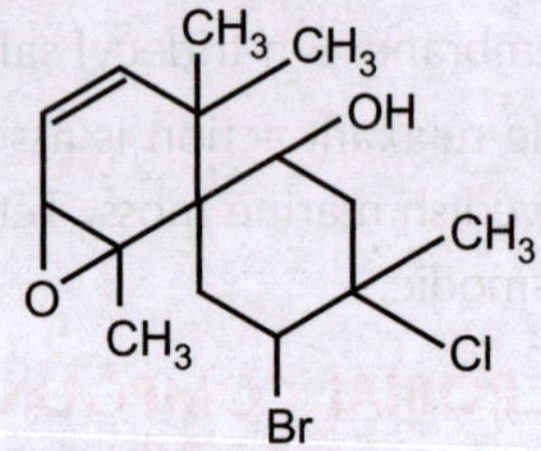

Prepacifenol

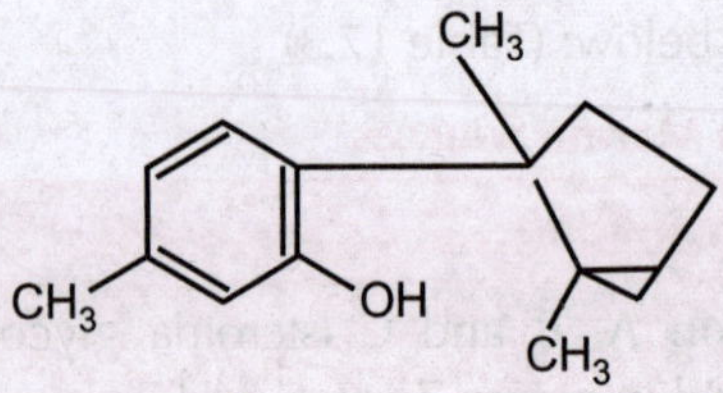

Debromolaurenterol

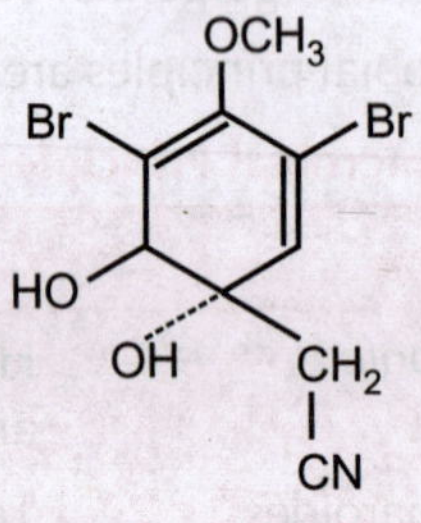

Aeroplysinin - 1(+)

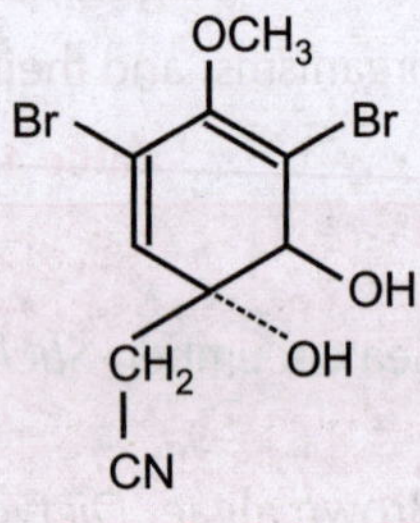

Aeroplysinin - 1(–)

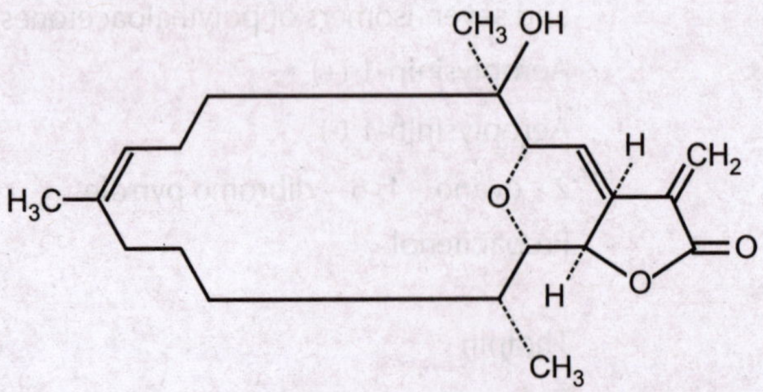

Eunicin

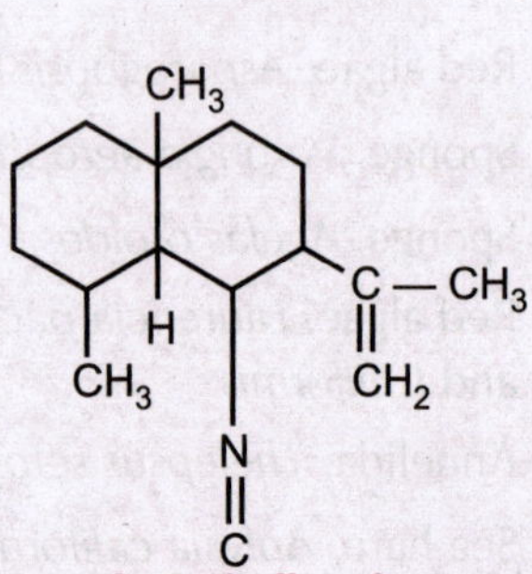

Acanthelin - 1

4. CARDIOVASCULAR ACTIVE SUBSTANCES

(i) Anthopleurins: These are the group of peptides obtained from coelenterates viz. *Anthropleura xanthogrammica* giving type A and B (AP - A and AP - B); and *A. elegantissima* giving type C (AP - C) anthopleurins AP - A shows strong positive inotropic action without an effect on Na^+, K^+ ATPase, pump heart rate and blood pressure in various animals. It produces cardiotonic effect in conscious dog, which is 35 times potent as compared to digoxin with very less toxicity.

(ii) Eptatretin: It is found in the aneural bronchial hearts of pacific hogfish viz. *Eptatretus stoutii*. It is a potent cardiac stimulant with direct stimulant action on mammalian myocardium, which is comparable to adrenaline.

(iii) Laminine: A marine algae, *Laminaria angustata* gives this basic amino acid compound with hypotensive effects.

(iv) D (–) Octopamine: It produces adrenergic and cardiovascular response and is found in *Octopus macropus, O. vulgaris and Eledone moschata.*

(v) Saxitoxin: It shows a hypotensive effect and is found in butter clam i.e. *Saxidomus giganteus,* and *California mussel, Mytilus californianus.*

(vi) Autonomium: This compound has isosteric structure of adrenaline and acetylcholine and is found in *Verongia fistularis*. It produces both α and β-adrenergic effects and also exhibits cholinergic activity. This compound shows CNS stimulant effects in mice exhibited by significant increase in spontaneous motor activity. Due to both adrenergic and cholinergic effects, this compound may prove to be useful in regulating the behaviour of heart.

(vii) ATX - II: This is one of the polypeptides found in sea anemones. Like anthopleurins, it also shows potent and dose dependent cardiotonic response in mammalian heart preparations. It is suggested that the cardiotonic effect is due to indirect action on the pump activity induced by increased Na^+ transport as a result of a delayed inactivation of fast sodium current.

(viii) Besides ATX - II and anthopleurins, various polypeptides have been obtained from different sea anemones like *Condylactis gigantea* with 195 amino-acid residue peptide and haemolytic action in rabbits; *Parasicyonis actionstoloides* with a neurotoxic peptides and very less number of amino acids; and *Actinia equina* with 147 amino-acids exhibiting bradycardia, rapid hypotension and respiratory arrest in the intact rat.

(ix) Holothurins and asterosaponins: These are marine saponins. Holothurins contain triterpenoid moiety as aglycone and asterosaponins have steroidal aglycone. Holothurins are present in sea cucumbers, belonging to family Holothuroidae of phyllum Echinodermata. Holothurins have cardiotonic and ichthyotoxic actions. They also show haemolytic activity. Holoxins like holotoxin A, B and C are desulphated holothurins and possess antifungal activity.

Asterosaponins which are obtained from star fishes of family Asteroidae are toxic to various marine organisms. They exert different pharmacological actions like haemolytic, anti-inflammatory, hypotensive and analgesic effects and also cytolytic action due to interference in protein metabolism and neuromuscular blocking effects. They are inhibitors of ATPase, but holothurins are more potent in this comparison.

(x) Spongosine: Chemically, it is a nucleoside (methoxy derivative of adenosine). It is found in the extract of Caribbean sponge *Crypotethia crypta*. It reduces both the rate and the force of contraction of heart.

Like spongosine, many nucleosides have been reported from marine sources like doridosine (hypotensive and hypothermic) from *Anisodoris nobilis*, 5'-deoxy-5-iodotubercidin (muscle relaxant, hypothermic and potent inhibitor of adenosine kinase) from red algae *Hypnea valentiae*.

(xi) Eledosin: It is a powerful hypotensive compound obtained from posterior salivary glands of cephalopod *Eledone moschata*. It also shows strong vasodilator effects.

The structural formulae of some of the compounds mentioned above are as follows:

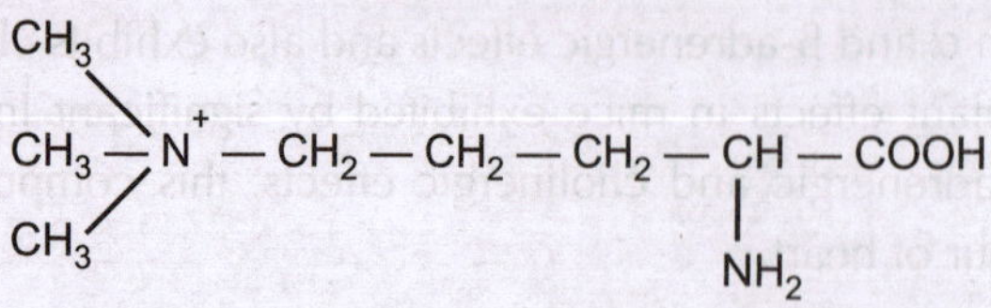

Laminine

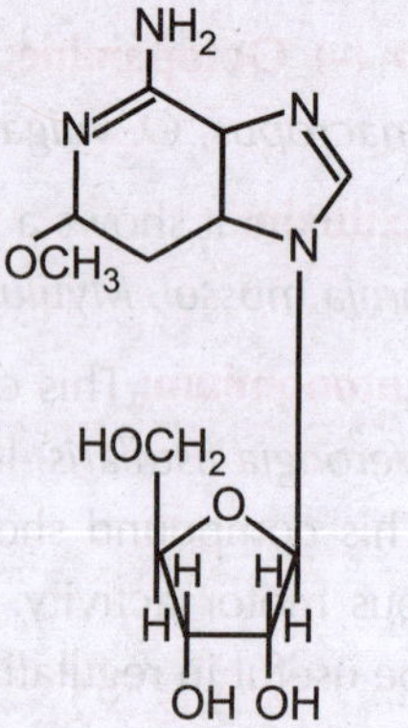

Spongosine

5. CYTOTOXIC (ANTICANCER) COMPOUNDS

1. Ara-C

It is 1 - α - D - arabinofuranosyl cytosine or cytosine arabinoside. It is obtained as a synthetic compound based on the knowledge of naturally occurring moieties present in Caribbean sponges, viz. spongosine and spongouridine. Ara-C is used in treatment of acute myelogenous leukemia and human acute leukemia, for therapeutic purposes. It is active against Erlich carcinoma, sarcoma -180 and L-1210 leukemia in mice.

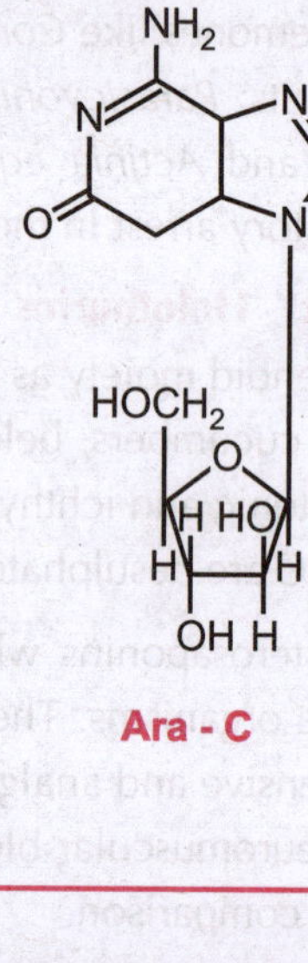

Ara - C

2. Crassin acetate

It is a member of cembranoids which are cyclic diterpene obtained from soft corals. Crassin acetate is obtained from Caribbean gorgonian, Pseudoplexaura porosa, which is cytotoxic to human leukemic and HeLa cells in-vitro.

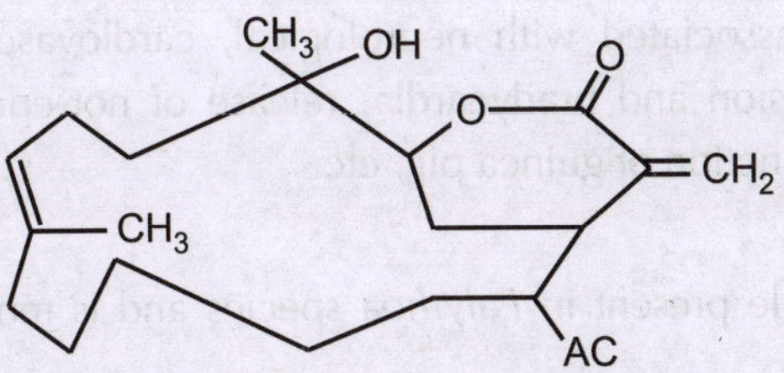

Crassin acetate

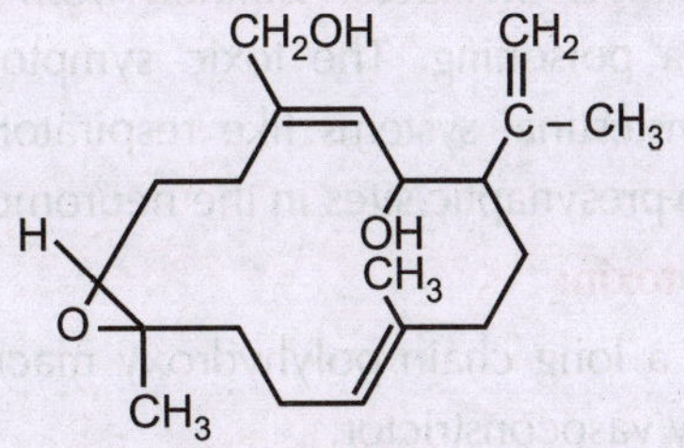

Asperdiol

Among the various cembranoids, a soft coral *Sinularia flexibilis* gives simularin, dihydrosimularin and simulariolide.

Asperdiol, a non-lactonic cembranoid from gorgonian coral, is cytotoxic.

Many compounds with antitumour and immuno-suppressant effects have been reported e.g. geranyl hydroquinone (cytotoxic to leukemia and mammary carcinoma) from chloroform extract of *Aplidium* species; aplysistatin (antileukemic) from sea hare *Aplysia angasi*; halitoxin (antitumour) from a sponge Haliclona *viridis*; didemnins A, B and C (potent inhibitors of L - 1210 leukemic cells *in-vitro* and P - 388 leukemia and B - 16 melanoma *In-vivo*) from Caribbean tunicate *Trididemnum* species; majusculamine - C (inhibitor of X - 5563 myeloma) from algae, *Lyngbya majuscula*; and extracts and purified fractions of sea sqirt, *Ecteinascidia turbinata*.

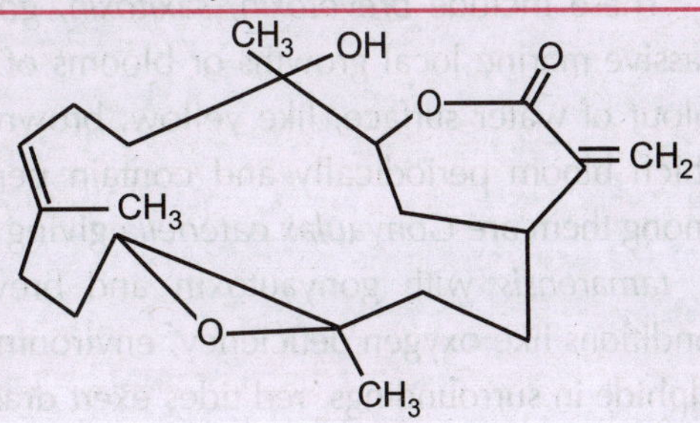

Simularin

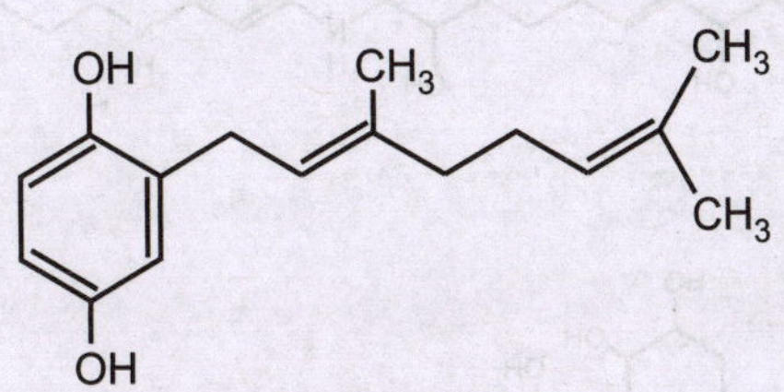

Geranyl hydroquinone

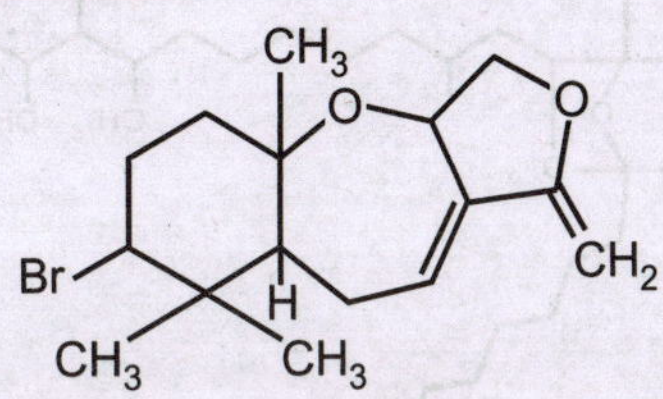

Aplysistatin

6. MARINE TOXINS

Some of the microorganisms from the sea which are autotrophic, also act as a source of toxins. These toxins are both ectocrine or external metabolites and endotoxins. Such endotoxins are some of the most potent materials. Some of these toxins have been discussed here in brief.

The best known marine biotoxins are tetradotoxin, paralytic shell fish toxin and ciguatera toxin.

(i) Ciguatoxin:

It is present in red tide dinoflagellate *Gambier discus-toxicus*. The term "ciguatoxin" was first used to describe an ailment associated with the ingestion of marine snails. Ciguatera poisoning is also thought to be due to various intermediates in the food, mainly species of blue green algae.

Ciguatoxin is the factor isolated from pacific red snapper and considered to exert toxicity i.e. ciguatara poisoning. The toxic symptoms are associated with neurological, cardiovascular and gastro instestinal systems like respiratory depression and bradycardia, release of nor-epinephrine from the presynaptic sites in the neuromuscular junction of guinea pig, etc.

(ii) Palytoxin:

It is a long chain polyhydroxy macromolecule present in *Palythoa* species and is most potent coronary vasoconstrictor.

Palytoxin can exert full effect on coronary arteries in an isolated guinea-pig heart even at very low concentration. Besides such effect, it also has actions on nerve tissues and muscles. It elevates T wave and S-T segment in ECG of rat, dog and guinea pig.

(iii) Red tide toxins (Paralytic shell fish toxins):

These include *brevetoxin, saxitoxin, gonyautoxins*, etc. The "red water" or "red tides" are the massive marine local growths or blooms of various algae or protozoa, and they cause a change in colour of water surface, like yellow, brown or green and not always red. They are dinoflagellates which bloom periodically and contain peridinin, a red coloured pigment. The most deleterious among them are *Gonyaulax catenella* giving saxitoxin,
G. tamarensis with gonyautoxin and brevetoxin from *Ptychodiscus breuis*. Mainly in adverse conditions like oxygen deficiency, environmental imbalance of nutrients, accumulation of hydrogen sulphide in surroundings, red tides exert drastic effects due to their toxins and may turn a rich fauna into biological desert. The different brevetoxins exert either neurotoxic or haemolytic effects. Brevetoxin - B shows positive inotropic and arrhythmogenic activitities on rat and guinea - pig hearts accompained by AV block.

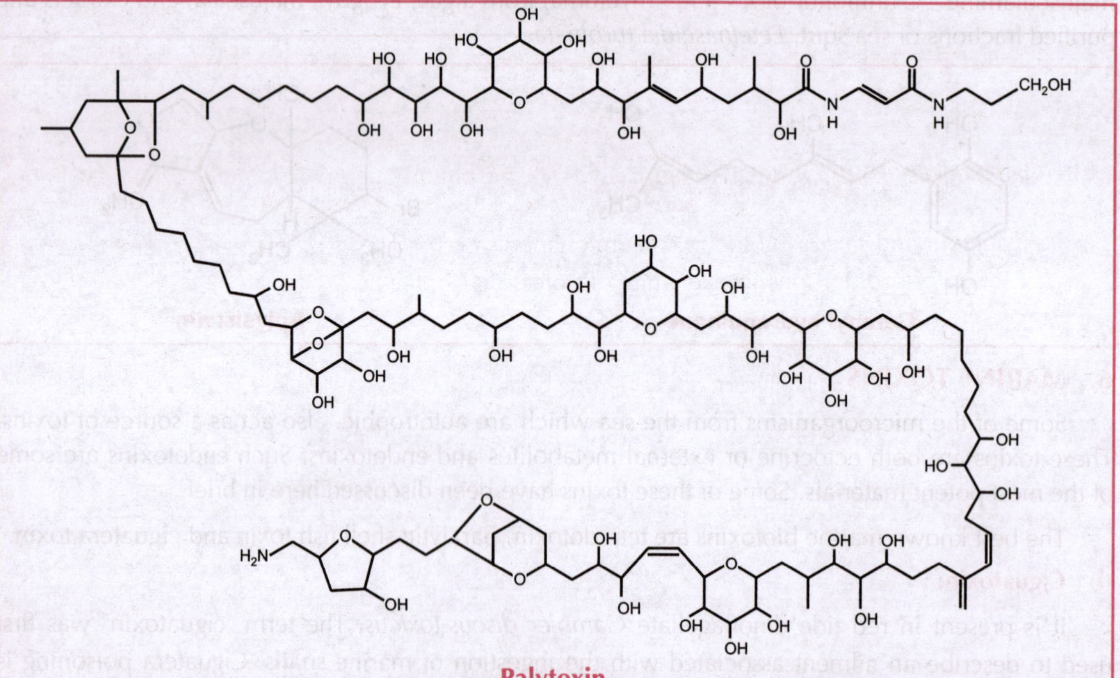

Palytoxin

Table 17.4: Marine Toxins and their Physiological Effects

Organisms	Physiological Effects
1. *Gymnodinium breve*	depolarises membranes, RBC haemolysis and neurotoxic
2. *Gymnopodinium ueneficum*	depolarises membranes, permeability alteration to Na^+
3. *Gonyaulax catenella (saxitoxin)* and *G. tamarensis*	block membrane Na^+ conductance
4. *Haliclona rubens* (holotoxin)	potent neuromuscular blocker
5. *Amphidinium carters*	acetyl-choline like activity
6. *Lophogorgia rigida* (lophotoxin)	neuromuscular blocker
7. *Anacystis cyanea*	neurotoxin, fast death factor
8. *Gambier discus toxicus* (maitotoxin)	stimulates calcium channels in insect skeletal muscles
9. *Prymnesium parka*	neurotoxin, blocks end plate transmission, haemolytic agent
10. *Palythoa species* (palytoxin)	most potent coronary vasoconstrictor
11. *Ptychodiscus brews* (brevetoxin)	positive inotropic and arrhythmogenic
12. *Babylonia japonica* (neosurugatoxin)	potent blocker of sympathetic ganglia and a specific blocker of nicotinic Ach receptors
13. *Gambier discus toxicus* (ciguatoxin)	respiratory depression and bradycardia

Gonyautoxin and *saxitoxin* bind to sodium channels on the outside of excitable membranes and allow an influx of sodium ions in exchange of potassium ions reflux with the attendant membrane depolarisation.

Although, the marine toxins have not yet been introduced in clinical medicine, they might prove to be useful as research tools in investigating the processes like biolectrogenesis, chemical transmission and to study excitable properties in nerve and muscle cells.

The research workers in biochemistry from Ottawa in Canada have found a protein in sea-tortoise. This specific protein is present in blood cells of the animal. Because of this protein, which is yet to be named, sea-tortoise can lower down the temperature of its body fluids below freezing point. When the fluids in the vicinity of blood cells go into freezing state, they absorb water from the tissues and also lead to expansion of fluids in tissues. Due to this, tissue is protected from any injury. During summer, the fluid melts and the water absorbed by them is released which re-enters the tissues. This phenomenon was unfold owing to the observation that the sea tortoise is not harmed for prolonged period in winter, although it does not enter the hibernation stage. Such protein may be employed to facilitate certain surgical operations like cardiac surgery, where it is essential to lower the body temperature to an appreciable level.

7. MISCELLANEOUS COMPOUNDS

The marine products like insecticides, anthelmentic, anticoagulants and compounds with special activities have been included in this group.

Kainic acid, a potent convulsant, has been isolated from red algae, *Digenea simplex* which shows antiascariatic activity. Also, domoic acid has been found in a red algae *Chondria armata* with activity against ascaris and pinworm.

A marine annelid, *Lumbriconeris heteropoda* gives nereistoxin found to be useful as an insecticide due to its ganglion blocking effects. A semisynthetic derivative viz. cartap from nereistoxin is commercially used as insecticide.

Didemnins A, B and C have a protective action against *Herpes simplex* type-2 virus in mice. Ara - A, a derivative of adenine arabinoside (from a sponge) is active against *Herpes encephalitis*.

A yellow sponge, *Verongia spengeli* gives aplysinopsin which has antidepressant action in animals and also cytotoxic activity.

Antilipemic activity is shown by laminarin. The different species of *Eucheuma, Chondrus* and *Gigartina* are reported to contain carrageenan, an anticoagulant agent.

A number of mammalian prostaglandins, like PGA_2, PGE_2, etc. have been obtained from the gorgonian, *Plexaura homomalla*, which may be used directly or in a simple derivative form as drugs for treating asthma, peptic ulcer and also in birth control.

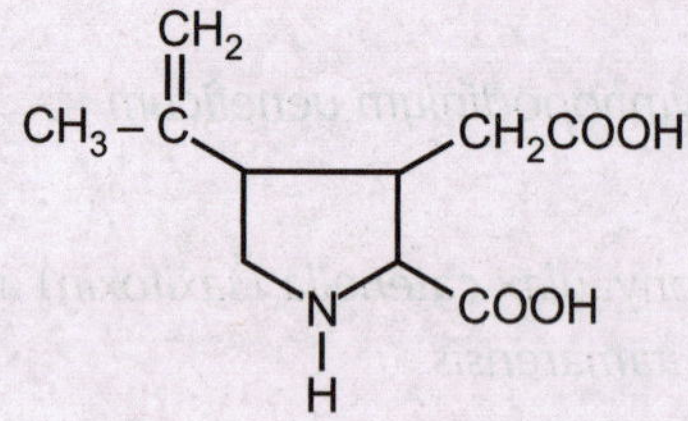

Kainic acid

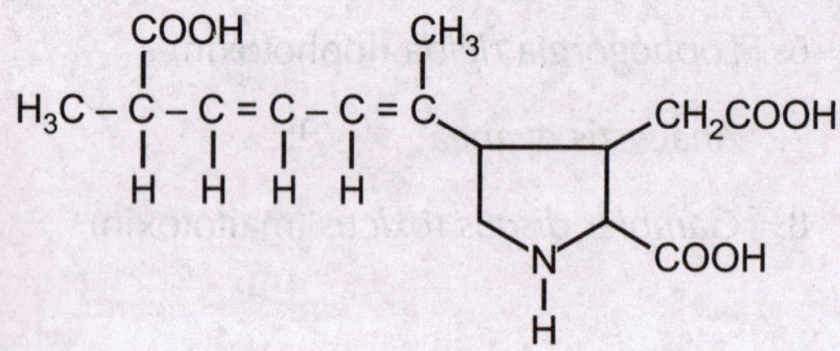

Domoic acid

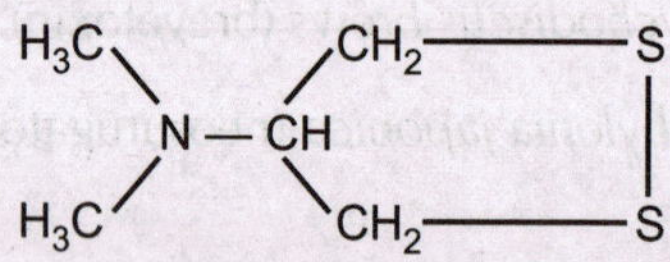

Nereistoxin

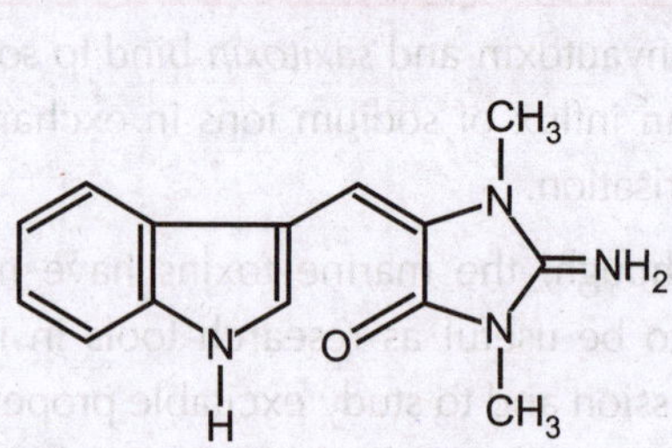

Aplysinopsin

Acanthifolicin is polyether antibiotic, found in a sponge *Pandaros acanthifolium*, which also possesses cytotoxic activity.

❖❖❖

Chapter 18...

NUTRACEUTICALS AND COSMECEUTICALS

The concept of generating utility of food as health promoting factor beyond its nutritional value is gaining acceptance within public arena and among scientific community. The nutraceuticals have evolved from the recognition of the link between food and health.

NUTRACEUTICALS:

Nutraceutical is defined as a substance which can be considered as food which, in addition to its normal nutritional value provides health benefits including prevention of diseases or promotion of health.

The knowledge about the nutritive value of various food items and their basic chemical roles has much increased. Due to this, the values of menu that various societies have adopted around the world have been appreciated. The western nutritional experts believe that the food which offers best value to bodily health is "Mediterranean Cuisine". It includes fish, whole wheat bread, tomato, olive oil, salad, garlic, onion, fruits and red wine. East Asian food scores high in its nutritional value and includes glutinous rice, soybean curd, fibrous fruits and tea.

Neutraceuticals are available in the following market products for day today use –

(1) Fruit powders, (2) Vegetable powders, (3) Herbal powders, (4) Spices powders, (5) Health supplements, (6) Food supplement, (7) Dietary supplements, (8) Baby food.

The major diseases for the prevention and or treatment whereof, nutraceuticals have been associated are heart disease, cancer, hypertension and diabetes. The other diseases, related with role of nutraceuticals are osteoporosis, arthritis and neural tube defects. Certain dietary habits can enhance cancer risk.

Designing a proper food to maintain proper health has gained recognition and acceptance world-wide. Due to this, the food industry in many countries is modifying their products as a response to consumer demands. In recent years, many of the natural products from India are gaining popularity in USA and Europe as Nutraceuticals. The present US market of nutraceuticals is three billion US dollars.

Classification of Nutraceuticals

Depending upon various characteristics, nutraceuticals can be classified by several methods. Their major classes are:

1. Nutraceuticals from food source.
2. Nutraceuticals indicating their action of mechanism.
3. Nutraceuticals according to their chemical nature.
4. Nutraceuticals according to their higher contents in specific foods items.

1. **Nutraceuticals from Food Source:** Nutraceuticals are obtained from plants, animals and microbial resources and are restricted.

However, the demerit of the system is that few nutraceuticals are found in plants and animals while others are reported in all the three resources. Linolenic acid is found in animal flesh and also synthesized in plants. Choline and phosphotidyl choline is available in plants, animals and also in microbes.

Table 18.1: Nutraceuticals from Food Source

Vegetable	Animal	Microbial
Allicin, α-Limonene	Choline	Lactobaillus acidophilus
α-tocopherol, Anthocyanates	Conjugated lenoleic acid (CLA)	Streptococcus salvaricus
Ascorbic acid, β-carotene	DHA	Yeast
Capsaicinoids, Catechins	EPA	
Cellulose, Curcumin	Lecithin	
Ellegic acid, Geraniol	Spingolipids	
Lycopene, Pectin	Ubiquinone	
Quercetin		

2. **Nutraceuticals indicating their pharmacological activities:**

Table 18.2: Nutraceuticals according to their Pharmacological action

Anti-oxidant	Anti inflammatory	Anticancer	Bone Protectives	Antibacterial
Ascorbic acid	Curcumin	Genestein	(Osteogenetic)	Garlic
β-carotene	Quercetin	Limonene	Soy-protein	Curcumin
Catechins	Capsaicin	Glycyrrhizin	Genestein	
Ellagic acid	Linolenic acid	Daidzein	Daidzein	
Gingeroot	EPA	Diallyl sulfide	Calcium compounds	
Glutathione	DHA	CLA		
Luteolin		Sphingolipids		
Lycopene		α–tocopherol		
Polyphenolics				
Tocopherols				

3. **Nutraceuticals according to Chemical Nature:**

Table 18.3: Chemical nature of nutraceuticals

Phenolic compounds	Protein based	Isoprenoids	Carbo-hydrate derivatives	Fatty acids and structural lipids	Microbial	Minerals
Tannins Anthocyanins Isoflavones Flavonones Flavonols Coumarins Lignin	Choline Isothio-cyanates Capsaici-noids Amino acids Indoles Folate Allyl thio-compounds	Carotenoids Saponines Tocopherols Simple-terpenes Tocotrienols	Ascorbic acid, Oligo-saccharides Non-starch products	CLA Sphingolipids Lecithin n 3 PUFA MUFA	Prebiotics Probiotics	Ca Se K Cn Zn

4. **Nutraceuticals according to their higher contents in specific food items**

Table 18.4: Nutraceuticals from specific foods

Nutraceuticals	Name of specific food
EPA & DHA	Fish oils
Lycopene	Tomatoes and tomato products
Isothiocyanates	Vegetables of crusiferae family
CLA	Beef and dairy products
Isoflavones	Soyabean and legumes
β-carotenes	Carrots, pumpkin
Curcumin	Turmeric
Catechin	Tea, berries
Quercitin	Citrus fruits, red grapes
Allyl sulfer compounds	Garlic, onion

The food substances used as nutraceuticals contain antioxidants, prebiotics, probotics, omega-3 fatty acids, certain phytochemicals and dietary fibres. Except probiotics, all these components are present in fruits, vegetables and different types of foods.

(1) ANTIOXIDANTS

Antioxidant nutraceuticals are those which contain Vitamin E, Vitamin C, Vitamin A and beta-carotene. They are present in some fixed oils, fruits, vegetables and fishes. Antioxidants present in such foods are those compounds which either prevent the formation of oxygen free radicals or trap them (scavenging effect).

Anti-oxidants or inhibitors of oxidation are compounds which retard or prevent the oxidation and in general prolong the life of the oxidizable matter. The reactive oxygen species (ROS) in the body include superoxide anion, singlet oxygen, hydroxyl radical and hydrogen peroxide. The oxidative damage initiated by these is propagated by lipid peroxidation which may cause further damage to DNA. The body defence system against the oxidative damage consists of enzymes such as superoxide dismutases, glutathione peroxidase, catalase and the reducing agents such as glutathione, ascorbate and iron.

Various theories have put forward to explain the role of antioxidants in treating disease. In general the reactive oxygen species circulating in the body tends to react with the electrons of other molecules in the body and these also affect various enzyme systems thereby causing damage which may further contribute to conditions such as cancer, ischemia, aging, adult respiratory distress syndrome, rheumatoid arthritis etc.

Through various studies it has been reported that the free radical is the cause for cytotoxicity, alternation of enzymes and nucleic acids and peroxidation of lipids which altogether result in loss of cell membrane integrity and in turn cause aging.

Some of the naturally occurring anti-oxidants which could be of therapeutic use are super oxide dismutase (isolated or recombinant), tocopherol, ascorbic acid, adenosine, transferrin, lactoferrin, glutathione and its precursors, carotenoids and other plant pigments, deferoxamine.

Antioxidants play a vital role in the life of a living being. S-allyl cysteine sylphoxide (SACS) from garlic and disulphides may act by scavenging free radicals. They have also shown hypolipodemic action. Various organic compounds with phenolic hydroxy groups have shown antioxidant activity by scavenging free radical or by metal chelating.

Vitamin E (Tocopherol) is a major radical trapper in lipid membranes and is found clinically useful in cardiac damage and carcinogenicity. Selenium is an important dietary anticarcinogen. It is present in the active site of glutathione peroxidase enzyme, which is essential for destroying lipid hydro peroxides and thus prevents oxygen radical induced lipid peroxidation. Dietary ascorbic acid is an effective scavenger of superoxide radicals and is found in high concentrations in the eye and ECF (Extra cellular fluid) of lungs.

Generally, peroxidation of lipids leads to formation of free radicals which cause detrimental effects on various tissues and organs of body. They may cause mutation and consequently cancer. They are also known to be involved in aetiological sequences of diseases like atherosclerosis, arthritis and aging process. Besides peroxidation of lipids, there are some other factors leading to formation of free radicals, including excessive alcohol consumption, environmental pollutants, excessive sunlight, etc.

The oxidation of LDL-cholesterol increases the chances of atherosclerosis. Antioxidants like tocopherol in vitamin E lower the susceptibility of LDL for oxidation. Tocopherol also reduce platelets' role in thrombus formation. Corn oil, wheat germ oil are rich sources of vitamin E. Various plant materials like amla, myrobalan, lemon contain an antioxidant in the form of ascorbic acid (vitamin C). It prevents both the formation and scavenging of oxygen free radicals. Beta-carotene and vitamin A have shown to cause antioxidant effects and immune enhancement.

(2) POLY-UNSATURATED FATTY ACIDS (PUFA)

They are present in various vegetable oils and marine animals. These sources include safflower oil, corn oil, mustard oil, soybean oil, etc. They help to reduce cholesterol formation/deposition. These vegetable oils mainly contain PUFA belonging to linoleic group (omega-6 type). Some marine fishes contain PUFA belonging to linolenic group (omega-3 type). The later are found to appreciably reduce thromboxane formation and hence useful as preventive measures for atherosclerosis. (Omega-3 fatty acids have been described in later part of this chapter).

In spite of high consumption of fats, Eskimos have rarely shown any evidence of atherosclerosis. The reason identified has been their consumption of such cold water fish oils which contain appreciable quantities of omega-3 fatty acids.

(3) PROBIOTICS

These are the living microorganisms, which when taken with or without food, improve the intestinal microbial balance, and in turn, functioning of the large intestine. Probiotics include *Bifidobacterium* and *Lactobacilli* species such as *L. acidophillus*. These microorganisms exert their effects by producing substances and conditions which inhibit the growth of harmful bacteria in the large intestine. The dairy products like sour milk and A/B-culture yoghurts contain these probiotics.

(4) PREBIOTICS

These are the nutraceuticals which promote the flourishing of probiotics. Before reaching to colonic region, the probiotic microorganisms have to survive the digestive enzymes and acids in the upper gut. To overcome this problem, nutraceuticals in the form of prebiotics are available. Prebiotics are the food substances which reach to colon in intact form, without getting depleted by gastric pH and digestive acids. Prebiotics also selectively promote the growth of colonic probiotic bacteria. At present, the best known prebiotic is inulin. It is a polyfructose obtained mainly from raw chicory (roots of *Cichorium intibus*) or Jeruslem Artichoke. Chicory is reach in fibrous polysaccharide inulin. It is a soluble dietary fibre and resistant to digestive enzyme. It reaches to large intestine or colon essentially intact, where it is fermented by the resident bacteria. *Lactobacilli* and *Bifido bacteria* digest inulin and feed themselves on it. Hence, prebiotics act as fertilizers for these symbiotic bacteria. Inulin also serves the role of dietary fibre. Safety of inulin has been evaluated and accepted by FDA of United States.

(5) DIETARY FIBERS

Fibers are non-digestible polysaccharides found in plant cell walls. They are present in food including fruits, vegetables, grains and legumes. Thus fibers which we eat are called **Dietary fibres.**

There are two types of fibers: the **soluble** and **insoluble** dietary fibres. Soluble fibers are partially soluble in water and form gel while insoluble fibers are insoluble in water and pass through the digestive tract largely intact. Both types of fibers are very important in the diet and provide several benefits to the digestive tract by helping to maintain regularity. Soluble fibers are more beneficial since they reduce blood cholesterol levels and also reduce the risk of heart attach.

Dietary fibers can be distinguished by their viscosity, water binding properties (fecal bulking) and fermentability. Viscosity is essential to regularise the blood glucose and also to reduce the quantity of bile acids re-absorbed. Insoluble fibers due to water-binding property reduce their transit time in large-intestine, which promotes regularly and reduces the chance or colon cancer by decreasing the time of exposure of colonocytes to potential carcinogenic metabolites. Both types of fibers are subject to fermentability, but soluble fibers are more susceptible as compared to insoluble. Fermentation is advantageous since it favours microbial proliferation. Fast developing proliferation will consume nitrogenous products which otherwise are carcinogenic if accumulated in the colon.

Insufficient ingestion of dietary fibres cause constipation, bowel irregularities, colorectal cancer and hemorrhoids. Food with insufficient fruits, nuts, whole grains or vegetables can not provide desired quantity of fibers and may lead to deficiency.

Following table indicates the sources of dietary fibers.

Table 18.5: Sources of dietary fibers

Soluble dietary fibers	Insoluble dietary fibers
Oat-meal	Whole-wheat
Oat-bran	Wheat-bran
Nuts	Carrots
Seeds	Cucumbers
Legumes	Celery
Lentils (peas, beans)	Tomatoes
Apples	Barley
Pears	Brown rice
Straw-berries	Bulgur
Blue berries	Whole-grains
Black berries	

Recommended dose of fibers for adult per day is about 30 gm. For children. Upto age of 5 + 5 gm. i.e. for a child of 9 years of age 9 + 5 = 14 gm per day.

Table 18.6: Dietary fiber content of selected foods

Type	Total (g)	Soluble (g)	Insoluble (g)
Legumes per 125 (g) cooked	5.8	2.9	2.9
Kidney beans	5.1	2.3	2.8
Chick beans	6.2	1.3	4.9
Navy beans	5.8	2.2	3.6
Northern beans	5.6	1.4	4.2
Pinto beans	7.4	1.9	5.5
Cereal grains per 125 g cooked			
Barley	4.2	0.9	3.3
Bulgur	2.9	0.5	2.4
Millets	3.3	0.6	2.7
Rice-brown	1.7	0.1	1.6
Rice white	0.2	-	0.2
Oat meal	2.0	1.0	1.0
Cornflakes	0.5	0.0	0.5
Popcorn	0.5	0.0	0.5
Fruits (Medium size)			
Banana	2.0	0.5	1.5
Apple with skin	3.0	0.5	2.5
Pear with skin	4.5	0.5	4.0
Orange with skin	2.0	0.5	1.5
Mango with skin	3.7	1.5	2.2
Peach with skin	3.2	1.3	1.9
Pineapple with skin	1.0	0.1	0.9
Plum with skin	1.7	0.9	0.8
Vegetables per 125 g.			
Carrots	2.8	1.4	1.4
Cauliflower	1.3	0.5	0.8
Beets	1.5	0.7	0.8
Peas	2.5	0.9	1.6
Spinach	2.7	0.5	2.2
Tomatoes	0.9	0.0	0.9
Sweet potato	3.8	1.4	2.4
Pumpkin	3.6	0.5	3.1

HEALTH BENEFITS OF OMEGA-3-FATTY ACIDS

A fatty acid with one double bond at third carbon atom from methyl end is known as **Omega-3 fatty acid**.

Biological Source

These are found in cold-water fishes like cod, salmon, tuna, sardines, shark fish, blue fish, marc kevel and herring. Additionally these are also reported in cold weather bean-oil plants like flax-seed, canola, walnuts, soyabean, and freshly ground wheat germ.

Chemical Nature

Omega refers to the location of the first double bond along the carbon chain counting from the methyl end. Parent fatty acid of omega-3 family is **lenolenic acid**. Omega-3-fatty acids are known as Eicosapentaenoic acid (EPA) and Docosahexaenoic acid (DHA). 15 – 30 per cent of fats in brain is DHA, while 30-50 per cent of fat in the eye is DHA.

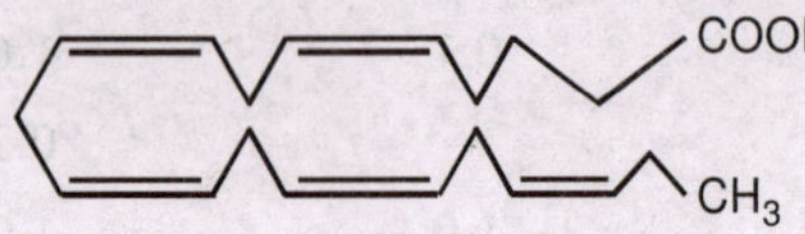

Eicosapentaenoic acid (EPA)

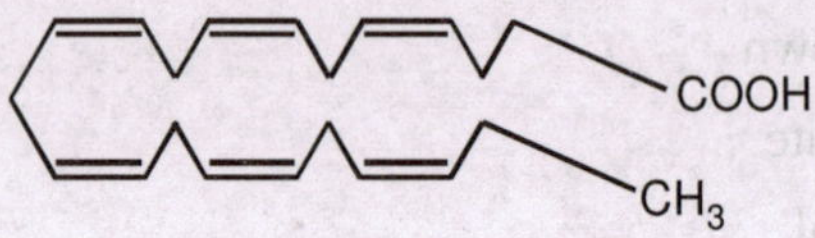

Decosahexaenoic acid (DHA)

EPA has an extra double bond between 3rd and 4th carbon atom from methyl end of molecule. These are reported in wild animals but deficient in domestic live stock.

Omega-3 fatty acids are important components of all cell membranes. Their presence in the cell membrane increases the physicochemical stability and functional integrity of the cell. They are essential for normal growth and development at all stages of life. Omega-3 fatty acids make the cells less susceptible to oxidative damage as well as they decrease the formation of lipid peroxides.

Omega-6 fatty acids are commonly found in seeds and oils of most plant. They are abundantly available in diet and hence are not supplemented. Parent fatty acids of omega-6 family are **linoleic acid**.

Biological Role of Omega-3-Fatty Acids

It has been found that omega-3 fatty acids have significant role in atherosclerosis. The major risk factor is increased levels of low-density lipoprotein (LDL), which carries 70 per cent of total serum cholesterol. Contrary to LDL, high-density lipoprotein (HDL) helps to clear cholesterol from arterial wall lesions, and hence reduces the risk. Atherosclerosis is characterized by injury to endothelium (*tunica intima*), normally by excessive free radicals. Increased permeability of this layer causes adhesion and entry of monocytes. Platelets start aggregating on the injured layer. PDGF (platelet-derived growth factor) and macrophages cause proliferation of smooth muscle cells which migrate to endothelial layer. Most significantly, LDL starts accumulating within the migrated smooth muscle cells and extracellular also. Adhesion of platelets, proliferation of smooth muscle cells and deposited cholesterol makes the artery rigid and thick structured, and causes decreased lumen.

Uses

Omega-3 fatty acids have been found to be useful due to their following activities:

(1) Suppression of smooth muscle cell proliferation and migration. (2) Reduction of LDL and VLDL levels. Decrease in hypercholesteremia and triglyceridaemia. (3) Increase in HDL levels. (4) Reduction in thromboxane formation and increased fibrinolysis.

This causes increase in blood flow property. Due to thromboxane lowering, they have anti-inflammatory action and used in arthritis.

Although they are essential fatty acids, their consumption in last 50 years has been decreased owing to changes in dietary habits. This has led to certain non-communicable diseases, particularly mental disorders like Alzheimer's disease, since omega-3 fatty acids are selectively enriched in brain.

SPIRULINA

Habitat

It is a microscopic plant which grows in fresh water (planktonic form). Its thallus form is filamentous, unbranched and non-differentiated, and looks like tiny green spiral coils. It possesses ability to thrive in conditions, which are much harsh for other algae.

Spirulina is a concentrated source of food containing nutraceuticals and contains antioxidants, probiotics and phytonutrients.

Biological Source

Spirulina is a blue green algae *Spirulina platensis* or *Spirulina maxima* family Oscillatoriaceae

This group of algae is considered to be one of the remarkable groups of photosynthetic simple plant forms. It represents a link between green plants and bacteria. It has a soft cell wall made up of complex sugars and proteins and is different from most algae, that it is easily digested.

Fig. 18.1 (a): Spirulina Plant

Geographical Source

Current world production of spirulina for human consumption is more than 1 thousand metric tones. It is being cultivated in United States of America and leads world production which is followed by Thailand, Mexico, India and China.

Structure of Spirulina

Cell wall is multilayered and chemically it is made up of mucopolymer and pectic compounds. Generally there is no sheath, when sheath is present, it is made up of polysaccharides. There are no chloroplast. The thylakoid lamellae are distributed randomly in the cell. Chlorophyll-a is present and chlorophyll-b is absent. Other pigments are carotenoid, phycocyanin and phycoerthrin. Gas vacuoles help to regulate buoyancy. Nucleus is prokaryotic.

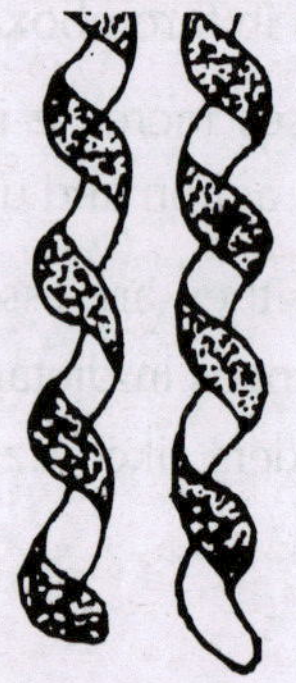

Fig. 18.1 (b) : Spiral coils of Spirulina

Spriulina is a blue green **nitrogen fixing alga**. Some of current researches indicate that *spirulina* holds a great promise as a source of single cell protein (SCP). SCP is the protein obtained from unicellular organisms. *Spirulina* contains 60 – 65 per cent **proteins**. It is cultured in clean as well as in salty water.

Chemical Composition

Spirulina contains proteins (50-70 per cent), proteinous nitrogen (11.36 per cent), total organic nitrogen (13.35 per cent), nitrogen from nucleic acids (1.9 per cent). It has net protein utilisation (NPU) upto 62 per cent.

It contains lipids (5-6 per cent) having mostly essential fatty acids (Vitamin F), composed of oleic, linoleic, gamma linoleic, palmitic, palmitoleic, heptadecanoic acids. About 40 per cent of the fats include glycolipids including sulpholipids (2 - 5 per cent) which have significant anti-HIV activity. Spirulina provides 8-14 per cent of recommended daily allowance (RDA) of fats.

Spirulina contains the carbohydrates in the form of glycogen and rhamnose which are easily digestible and require less insulin.

Among the vitamin content, it mainly possesses natural beta carotene with 9-cis-carotenoid isomer, which has more anti-oxidant capacity. The other vitamins present are B_1, B_2, B_3, B_6, B_{12} and E_3.

The mineral content (3-6 per cent) mainly includes iron which is reported to be better absorbed than other natural iron, because of its soluble complexes with **phycocyanin**, (phycobiliprotein) which is an algal protein having the linear tetrapyrrole i.e. phycocyanobilin and resembles haemogobin. Phycocyanin, which is a blue-green pigment is believed to enhance general immunity and useful in lymphocytic activity against cancer. Spirulina has an enzyme content in the form of super oxide dismutaes (SOD). This enzyme is known for its free radical scavanging effects and plays vital role in pathophysiological conditions like atherosclerosis, arthritis, cataract, diabetes and also in emotional stress and aging process. Spirulina also contains crude fibres (0.8 per cent) and ash (6 per cent).

Biological Role

Spirulina has been subjected to thorough screening for its biological role. Some of the findings are promising.

It has immuno-stimaulant activities. It stimulates the production and activity of bone marrow stem cells, macrophages and T-cells. Spleen and thymus gland shows enhanced function

In-vitro studies on spirulina indicate that it enhances cell nucleus enzyme activity and DNA repair and hence it has possible role in cancer treatment.

Water extract of spirulina inhibits HIV-1 replication in human derived T cell lines and in human peripheral blood mononuclear cells. Calcium spirulan inhibits in-vitro replication of HIV-1, Herpes simplex, Human cytomegalovirus, Influenza virus, Mumps and Measles virus.

Gamma linolenic acid of spirulina helps to reduce cholesterol levels. It has appetite suppressing activity.

Uses

(1) Spirulina is simple and has fast growth rate hence cultivation of spirulina can be undertaken even in **waste water**, this helps to solve the problems of waste water pollution.

(2) Spirulina grows well in **sewage water** which is best material for bio-degradations.

(3) Spirulina can fix atmospheric nitrogen during its growth, and can be used as a source of nitrogenous fertilizer.

Substitutes: Spirulina resembles *Chlorella* and *Aphanizomenon flos-aquae*, which are also blue-green algae. Chlorella is nutritious but lacks the antiviral and immune stimulant activities. Cell wall of Chlorella is made up of indigestible cellulose. *Aphanizomenon flos-aquae* contain potent nerve toxins.

ROYAL JELLY

Synonym

Queen bee jelly.

Biological Source

It is yellowish transluscent highly viscous liquid produced by fermentation of the secretion of the salivary gland of worker bees namely *Apis mellifiera* family Apidae alongwith honey and pollen grains.

Chemical Nature

Constitution of royal jelly reveals that it contains **moisture** 65 - 70 per cent, **proteins** 15 - 20 per cent, **carbohydrates** 10 - 15 per cent, and **lipids** 1.5 - 6.0 per cent.

Royal jelly is rich source of Vitamins i.e. vitamin A, B_1, B_2, B_6, B_{12}, pentothenic acid nicotinic acid, Vitamin C, and Vitamin E. It also contains various essential amino acids like histidine, methionine, leucine, lysine, phenylalanine, threonine, and tryptophan. Essential fatty acid namely 10, hydroxy - 2 - decinoic acid is also present in royal jelly.

Biological Role:

It is believed that the longevity of more than five years of the Queen-bee owes to Royal jelly. Now a days it is mainly used in China and also in USA as an important nutraceutical.

Uses

Royal jelly is used as general tonic and food supplement. Worker bees prepare it for the nutrition of Queen-bee and the larvae.

Preparation: Rejoov caps Rhone poulene – Mumbai.

PROPOLIS

Synonyms

Bee glue, Propolis wax.

Biological Source

It is the resinous brownish material collected by honey bee from the buds of tall coniferous trees and used by bees to fill cracks or gaps of the bee hive. It is available in the market in the form of capsules.

It is known as natural antibiotic that can build body healthy and protect against harmful bacteria.

Description

It is yellow green to dark brown in colour and having aromatic odour. It becomes hard when cold, but soften on warming and becomes sticky. Bees themselves use it to seal the openings in honey comb.

Extraction or propolis with alcohol, produces insoluble propolis wax and alcohol soluble propolis balsam.

Fig. 18.2: Propolis resin

Chemical Composition

Propolis contains 50 per cent resin, 30 per cent wax, 10 per cent essential oil and pollen grains.

Uses

It possesses antibacterial and antifungal properties. Product containing propolis are very effective in healing of wounds as propolis contains numerous antibiotic compounds.

SOYA BEANS

Synonym

Wonder bean.

Biological Source

These are the fully matured dried seeds obtained from pods of the plant *Glycene-soja* and *Glycine max* family Leguminosae. (See also Page 1.40)

Chemical Composition

It is rich source of carbohydrates, isoflavones, fats, vitamins and minerals. Soya has high contents of high quality proteins. Along with calcium it also contains iron, magnesium and potassium.

Biological Role of Soyabean

Soya contains low proportion of saturated fat, but is a rare source amongst plants containing omega-3-fatty acids (EPA & DHA). It contains no cholesterol but helps to lower blood cholesteral levels. Soya additionally contains isoflavones i.e. Genestein and Daidzein. They are believed to have preventive role towards breast and prostate as well as other carcinomas. These are also known to prevent osteoporosis.

Fig. 18.3: Soya bean herb with pods

It has been accepted by FDA of United States of America that soyabean has a role in reducing the risk of coronary heart disease. The nutritional value of soyabean curd (Tofu) has been long realized by Chinese people. Low incidence of cancer and cardio vascular diseases in Chinese and Japanese people has been linked to their traditional soya-rich diets.

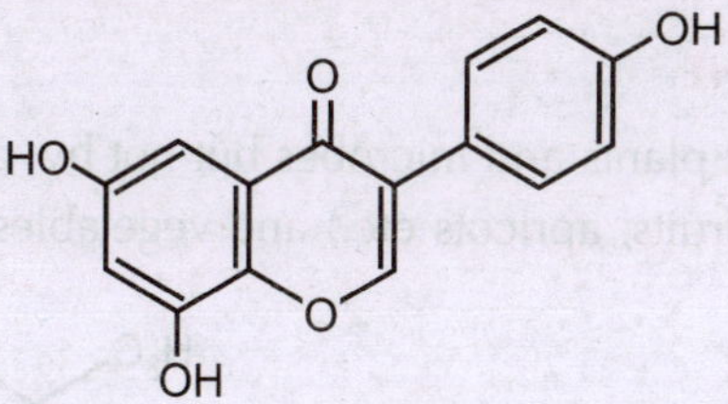

Genistein

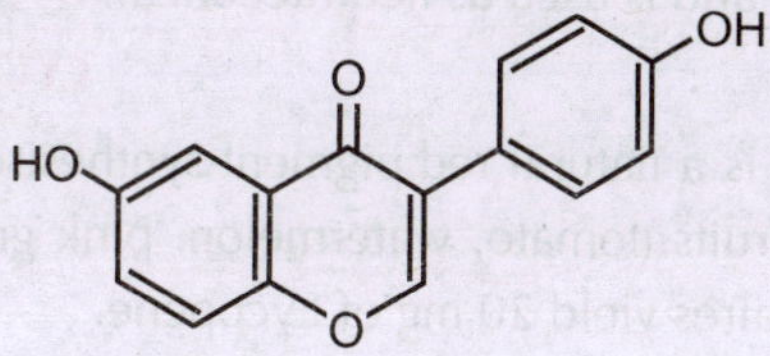

Daidzein

GARLIC

Garlic, the bulbs of *Allium sativum* (Family Liliaceae), has been associated with humans and their food since ancient times. It is grown and used as food and medicine in all temperate climatic regions of the world. Garlic contains carbohydrates (31 per cent), proteins (5 - 6 per cent), fat (0.2 per cent) and high amounts of phosphorus, potassium and calcium. Garlic contains a sulphur based compound called alliin (present in cell vacuoles). When the cells are broken, it is converted to allicin and finally diallyl sulphide, both of them are strong smelling and fiery tasting compounds.

Garlic reduces serum lipid levels because it causes, (i) reduction or inhibition of lipogenesis and (ii) enhancing break down and excretion of lipids. It increases HDL (High density lipoproteins) and reduces LDL (low density lipoproteins). Overall, garlic is used to reduce serum cholesterol and also in treatment of atherosclerosis. Garlic has also been found to reduce platelet aggregation.

Allicin from garlic shows antibiotic activity against *Mycobacterium tuberculosis, Staphylococcus aureus and S. faecalis.* Garlic is useful in the treatment of amoebic dysentery and parasites like tapeworm and hook worm.

Garlic exerts strong antioxidant effect, prevents lipid peroxidation and hence protects liver cells from various toxins including mutagenic chemicals.

NEUTRACEUTICALS

GLUCOSAMINE

It is found in chitin mucoproteins and mucopolyschharides. It is an amino monosaccharide made up of glucose and glutamic acid.

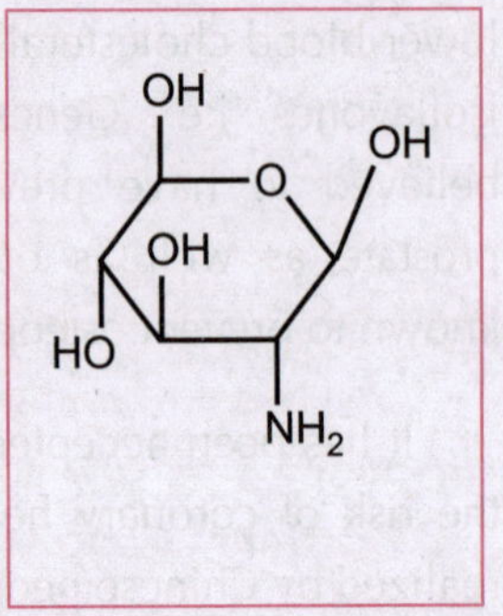

It is found naturally in human body especially in cartilage, tendons and ligament tissue. Glucosamine is not found in significant amounts in the usual diet and must be synthesized by the body.

But this ability declines with age and makes the body vulnerable to degenerative joint disease or arthritis. It has been found that fish oil and others as Evening primrose oil prove to be most useful for such joint diseases (being natural source of glucosamine). It is antiarthritic and pharmaceutical acid.

The marine herbal *Eclonia Maxima* of South Africa is extracted with water to get the biostimulatant and is used as neutraceutical.

LYCOPENE

Lycopene is a natural red pigment synthesised by plants and microbes but not by animals. It is found in red fruits (tomato, watermelon, pink grape fruits, apricots etc.) and vegetables. One kg of fresh ripe tomalres yield 20 mg of Lycopene.

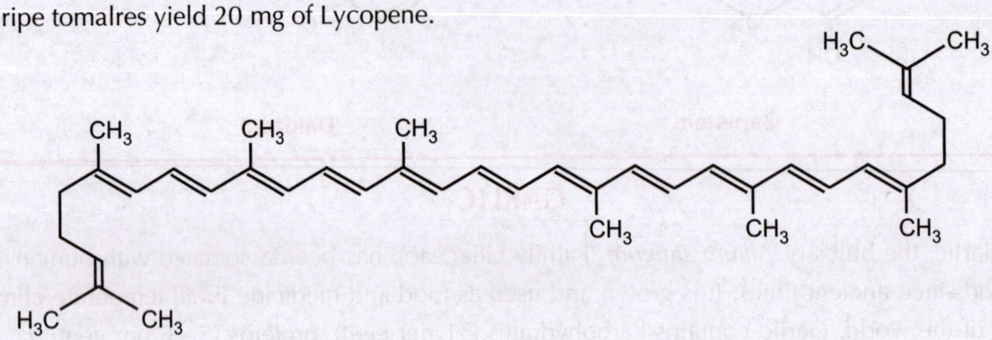

Lycopene

Lycopene is a linear carotenoid and acyclic isomer of betacarotene. It is insoluble in water. A lipid rich diet increases bioavailability of lycopene. Processed tomato products like tomato ketchup, paste and juice are all better sources of lycopene than raw tomatoes. Because heat processing is known to increase lycopene bioavailability.

Uses of Lycopene:

Lycopene is a strong antioxidant. Overall, antioxidant capacity of hycopene is double of betacarotene and upto 100 times that of α-tocopherol (vitamin E).

Beneficial anticancer effects of lycopene have been shown in cancer of lung, bladder, pancreas, prostate and breast.

Inverse associations between tomato/lycopene intake or serum lycopene and risk of cancer have been shown.

Lycopene containing fruits (carotenoids) have shown to exert beneficial effects in coronary vascular disease and age related macular degeneration (irreversible blindness due to damange of macula – the central part of retina).

Proanthocyanidins and Grape products

Proanthocyanidins (PACs) are a group of phytochemicals which are considered as one type of Bioflavonoids. PACs occur naturally in many fruits, vegetables, nuts, seeds as well as in flowers, roots and bark of many plants. Diet high in fruits and vegetables is known to given high protection against a variety of diseases. It is the increase in antioxidant status of body due to the fruits and vegetables.

Grape seed proanthocyanidin extract (GSPE) has been shown to prevent the development of certain diseases by acting as antioxidant.

PACs are also known as vegetable tannins, polyphenols or condensed tannins. They are the main precursors of blue-violet or red colouring matter in plants. The main sources are apples, grapes, black berries, raspberries, cranberries, rosehips and handthorn. They have been used in foods, beverages (wine and tea) and herbal preparations.

Most commonly they are present as polyphenol structures (derived from flavan-3-ol unit) which is either trans (catechin) or cis (epicatechin).

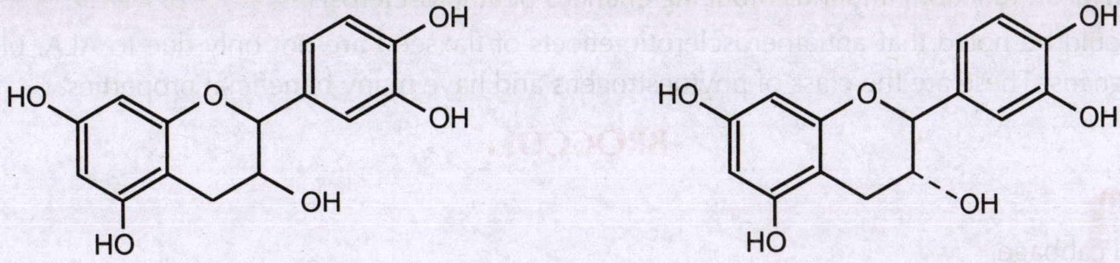

2, 3 trans (+)-catechin 2, 3 cis (–)-epicatechin

As these monomers (catechin and epicatechin) join together to form dimers, trimers and other oligomers, many different structures are formed with changes in stereochemistry. Usually there are five to seven aromatic rings with 12-16 phenolic groups. Dimeric proanthocyanidin B_1 is also called epicatechin – (4b – 8) and is formed in grapes and cranberries.

Uses:

Most of the pharmacological uses of PACs are related to their antioxidant properties. Especially grape seed extract has been shown to be a potent antioxidant. Studies performed on isolated rat heart perfused with red wine extract before ischaemic heart have provided evidence that PACs from red wine are effertive cardio protective agents. Red wine PACs have also been shown to be potential antioxidant, preventing the oxidation of LDL-C. PACs are also useful in vascular disorders like varicose veins, venous insufficiency and microvascular problems such as retinopathies.

Alpha-Linoleic Acid

Alpha-Linoleic Acid (ALA) is an essential fatty acid. ALA contains 18 carbon atoms; and three unsaturated bonds, starting at third carbon, ALA is also known as Omega-6 fatty acid.

Various sources of ALA and their contents are :

Source of oil	% of ALA
Flax seed (Linseed)	50%
Hemp seed	30%
Pumpkin seed	20%
Walnut	3-11%
Soyabean	5-7%

ALA is also polyunsaturated fatty acid (PUFA). But most of the oils with PUFA are omega-6 (double bond at 6th carbon). Ideally the diet should contain equal proportion of Omega-6 and Omega-3 PUFA. Because both of them are precursors of longer chain eicosanoids, like prostaglandins, thromboxanes and leucotriens. The eicosanoids derived from Omega-6 have opposing properties to those of Omega-3 and hence a balance is required. A diet rich in Omega-6 tends to cause thrombi, blood platelet aggregation, cardiovascular diseases, inflammation and diabetes.

Uses

ALA (especially Flaxseed or Linseed) is known to have beneficial (lowering cholesterol and improving heart function) and thus reducing chances of atherosclerosis.

It should be noted that antiatherosclerotic effects of flaxseed are not only due to ALA, but also due to lignans. These are the class of phytoestrogens and have many beneficial properties.

BROCCOLI

Synonym

Wild cabbage.

Biological Source

This consists of edible green plant of *Brassica-oleracea* var: Italica family: Cruciferae.

Geographical Source

Though found in many countries it is mainly cultivated in China and India constituting about 73% of world requirement about 26 million tonnes. It is a type of turnip only.

Plant is known for its vegetable and medicinal values since sixth century and it is cultivated on commercial scale in mediterraneans countries. Calfornia produces more than 92.0% of broccoli production of USA. Now-a-days it is cultivated in Mizoram near Aizwal on commercial scale.

Cultivation

Broccoli is cultivated by using seeds only, seeds are very small and brown in colour. It is cultivated in nursery beds as cool-weather crop and needs 20-25°C temperature, it is harvested before it flowers, flowers are yellow in colour. It can be cultivated in all seasons, it is a crop of 3 to 4 months only. It is to be watered after every 8 – 10 days if no rains are available. Inexperimental trial producing only 165 quintals per hectare. Brocoli c1 × 3512 variety is preferred. Cultivated at distance of 45 – 50 cm, keeping the distance of 8 – 10 cm in between two lines and using only 800 – 850 seeds. hoeing is done it an interval of 20 – 30 days poultry manure of 2.5 tonns per hectare are used.

Morphological Characters

Colour : Flower heads are dark green in colour while the stalk is light green.

Odour : Characteristic

Taste : Characteristic and slightly bitter due to sulphoraphane content.

Size : Pieces of broccoli are about 10 – 15 cm in length.

Shape : It has tree-like shape branching from thick stalk.

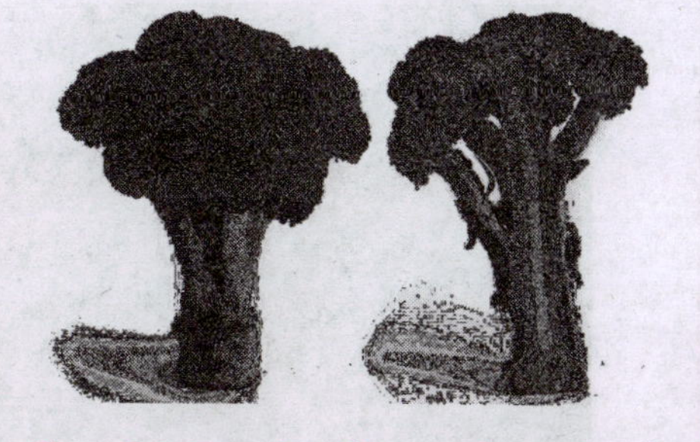

Fig. 18.4: *Brassica oleracea* Flower heads

Mass of flower heads are surrounded by leaves and resembles cauliflower.

Chemical Constituents

Broccoli contains most of the vitamins and is very rich in vitamin C and vitamin K which is more than contents reported in Amla-fruit.

Additionally it also contains minerals like Calcium, Iron, Magnesium, Manganese and Phosphorous, Sulphur containing glucosinolate compounds, Isothiocyanides and Sulforaphane. These are diminished by boiling, but are preserved by steaming, microwaving or stirr frying.

100 gm of broccoli gives 90 mg of vitamin C and 100 μg of Vitamin A.

Uses

Broccoli decreases the risk of obesity, diabetes, and heart disorders. It also promotes healthy complexion and hairs. It has also been reported that sulforaphane content of broccoli inhibits the histone deacetylase enzyme, responsible for the progression of cancer cells. Vitamin C plays vital role in formation of collagen and hence reduces wrinkles. Broccoli is a nutritive diet and also good detoxification agent.

NAGLI

Synonyms

Ragi, Nachni, Fox-tail millet, Finger millet.

Biological Source

It is a food grain obtained from the plant *Eleusine corocana* family: Gramineae. Fruiting top of Nagli looks like folded hand finger. Hence, it is known as finger millet.

Geographical Source

Africa, Malasia, China Japan are the major countries cultivating Nagli. Heavy crops of Nagli are taken in Ahmedabad and Khaira districts of Gujrat. It is also cultivated in Ratnagiri district of Maharashtra. About 5 lakh acres of land is under cultivation. Nashik, Nandurbar and Dhulia districts are also centers of cultivation of Nagli. 1.2 lakhs hectares of land is under cultivation of Nagli.

Cultivation and Collection of Nagli

Cultivation process of Nagli commences from first or second week of May. First of all soil is deep ploughed to remove the weeds, which is done by ploughing twice or thrice.

Fig. 18.5 (a): Nagali

Fig. 18.5 (b): Nagli grains

Nagli can be grown in various-types of soil, though sandy loamy soil rich in organic matter is most suitable. Water logging soils are not suitable.

It is cultivated by sowing the seeds only. Soils is mixed with farm yard manure before ploughing, at the rate of 25 – 30 tonns per hectare.

It is cultivated during first or second week of June. Seeds are sown in nursery beds at the rate of 2 to 4 kg seeds per acre. Depending upon the total area of cultivation quantity of seeds may vary.

After three weeks of sowing seedlings are ready for transplantation. Two varieties of Nagli are available in the market i.e. Black and White SHITULI from white variety is normally preferred by farmers. Black variety is also having two types of i.e. Garve and Nimgarve. For maximum yield of crop good quality namely BENDRI is the choice of farmers.

Transplantation of seedlings in the field is done in first or second week of July. Manures or organic fertilizers (i.e. Castor seed cake, Neem seed cake, Karanj seed cakes) are normally used by avoiding chemical fertilizers. Total crop duration is 120 days while harvesting is done during October.

Cultivation of Nagli

Nagli requires temp zone of 12 – 30°C, and rainfall in the range of 50 – 100 cm, where as pH of soil 5.5. – 8.5.

Nagli plants grow to the height of 150 – 170 cm. Nagli is cultivated twice in a year i.e. Kharip and Rubbi crop as well. Its out come is approximately 1070 kg per hectare.

Expenses for cultivation

₹ 6000/- per acre. Yield per acre 10 quintals. ₹ 15000/- income per acre

Morphological Characters

Nagli are very small, hard grains to touch and have shining. They are some what spherical 1.5 – 2.5 mm in size pale green, odourless, slightly mucilaginous in taste.

Chemical Constituents

Nagli grains contain vitamin B, and are very rich in carbohydrates and minerals. Nagli is free of gluten.

Nutritive Value of Nagli (Per 100 gm)

Energy - 336 calorie

Carbohydtates	-	72.0 per cent
Proteins	-	07.7 per cent
Fibres	-	03.6 per cent
Fat	-	01.3 per cent
Calcium	-	344 mg
Iron	-	06.4 mg
Phosphorus	-	283.0 mg
Naicin	-	02.1 mg
Thymin	-	00.42 mg
Riboflarin	-	00.19 mg

CEREALS AND MILLETS

Cereals and millets are part and parcel of food. Since both of them are equally good and important for health. Cereals like rice, wheat, maize, jawar provide us carbohydrates (Main source of energy) alongwith proteins and fat.

Millet are rich source of fibres, vitamins, antioxidants. They get digested slowly and very light for digestion. They are ideal food for seniors. World Health Organisation has declared year 2023 as the year of millets.

Locally millets are dry-land crops and can serve in shortage of rains or even water. Millets have higher nutritive value. Millets are commercially cultivated on large scale in various states.

Details of their dietic and nutritive value are as under:

	Percentage of Millets					
Millet	Proteins	Fats	Fibre content	Minerals	Iron per 100 mg	Calcium per 100 mg
1. Bajra	10.5	4.75	1.3	2.3	16.5	38
2. Jawar	10.5	3.0	2.0	1.6	5.5	330
3. Nachani (Nagli)	07.3	1.25	3.5	2.7	3.9	345
4. Congani (Raale)	12.25	4.0	7.5	3.25	3.0	30
5. Kodra	0.25	3.5	9.0	2.5	0.5	25
6. Sawa	7.65	5.25	7.5	1.5	9.0	18
7. China (Bhagar)	11.0	4.0	10.0	4.5	15.0	11
8. Rajgira	12.5	2.1	2.3	2.0	0.75	14

Biologically it is known as *Pennisetum glaucum;* Fam. Guaninae.

1. Bajra (Pearl millet): Its a crop of Africa but has been naturalised in India since long as Kharip crop and cultivated on very large scale in the area of short of rains.

2. Jawar: A crop of dry-land known as *Sorghum bicolour* fam. Gramineae. **Geo Source:** Available in Maharashtra, Karnataka, Tamilnadu and Rajasthan. Rich source of carbohydrate and important food grain cultivated on large scale.

3. Nachani (Finger Millet): Scientifically known as *Elusine. Corveana* fam. Gramineae. **Geo Source:** Available in Rajasthan, Karnataka, Andhra Pradesh and Tamilnadu.

It is rich in vitamins and also free of gluten, easily get digested and crop of dry land. It is highly nutritive.

4. Kangani (Fox-tail millet): It is type of Rice cultivated in Maharashtra known as **Raale** available in Andhra Pardesh, Karnakata, Telangana and Rajasthan. Easy to digest. Rich in iron and calcium. Botanically known as *Setaria Italica* fam. *Graminae.*

5. Kodra: Kodo millet: Botanically it is *Paspalum scroticuluro;* Fam. Gromineae. **Geo Source:** Found in MP, chhatisgar, Maharastra and Tamil Nadu.

6. Sawa: Barnyard millet known as *Echinochloa frumentacea or Panicum milliaceum.* Fam. Gramineae. **Geo Source:** Available in Tamil Nadu, Andhra Pardesh, Karnakata and Rajasthan. Cultivated over greater part of India and Himalayin region upto 2100 metres. Sawa is very useful in biliovsness (Feeling sick)

7. Chinna: Proso-millet: Known scientifically *Panicum-maliaecum* Fam.: Gramineae, cultivated in Tamil Nadu, Karnataka, Andhra Pradesh and Uttarakhand. Good source of fibres and easy to digest. Ideal for seniors.

8. Rajgira: This is botanically *Amaranthus candatus* Fam. Amaranthaceae also known as *Chawlai* in Hindi. Cultivated in Kerala, Tamil Nadu and Maharashtra. Flowering tops are red coloured and very compact and thick. The fruits of this plant are used in religeons fasts (Upaas) in Maharashtra. Gets digested very fast and rich in carohydrate and proteins.

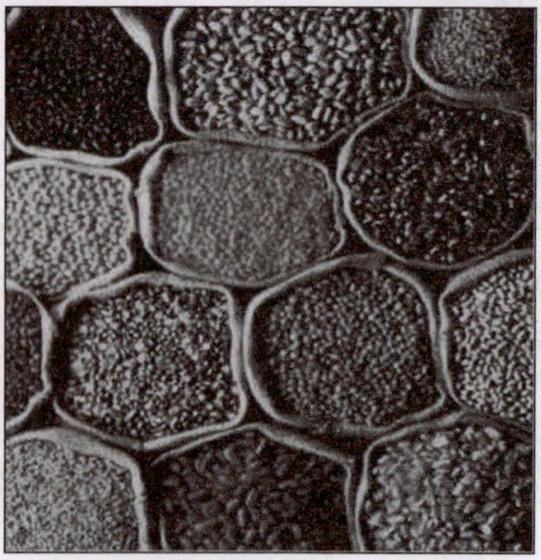

Fig. 18.6: Selected Food grains, Rich sources of nutraceuticals

9. Kurri: Brown top millet: One of the important millet botanically known as *Brachaiaria ramosa* Fam. Gramineae. **Geo Source:** Available in Karnataka, Andhra Pradesh.

10. Kutki: Little millet: Known botanically as *Panicum sunantness,* Fam. Gramineae. Available abundantly in Madhya Pradesh, Orisa, Czarkhand, Uttar Pardesh, Rajasthan, Assam and Khasia hills.

Uses

Grains are tonic, cooling useful in biliousness (Feeling sick) and astringent said to be very nutritious and also used in the preparation of bread and papad. According to Ayuveda, nagli grains are used in "Tridosha" and in blood diseases.

FAN-PALM

Freshly drawn juice of fruits of fan-pam is known in Toddy. Toddy is full of nutrients, vitamins, minerals and equally good to coconut nature.

Synonyms

Fan palm, Desert palm.

Biological Source

Fan-palm is botanically known as *Borassus-flabelti formis* Fam. Palmae.

Fig. 18.7: Fruits of Fan-palm

Geographical Source

Available in all dry parts of India, specially on sea-shore of peninsula, very common in Ratnagiri, Thane and Sindhudurg, Bihar, West bangal and Uttar Pradesh. It is cultivated and also self-swan. Sri-Lanka, Iran and Myanmar are the countries where it is available abundantly. It is native of Africa.

Description

Fruits are covered with green, very thick coating, interally grey are very smooth and juicy and white in colour. Sweet in Taste. Odourless fruits are subspherical, flat in shape and about 8 – 10 cm in size.

Chemical Constituents

Funils contains minerals, iron, potassium, phosphorus and good amount of vitamins A, B, and C.

Uses

It is expectorant, allays thivst, tonic. It is described to be aphrodisiac, intoxicalting and blood purifier.

COSMECEUTICALS

"Plants are indispensible to human being for survival" has been proved since antiquity. They are the source of food, clothing and shelter.

Plants used as food, were differentiated as therapeutic agents later on by experience and as on today they are equally important even as source of drugs.

In drug industry, herbs are used for the followings:

(1)	Source of phyto-pharmaceuticals	: To isolate pure compounds
(2)	Herbal raw materials	: To formulate herbal, ayurvedic, homoeopathic and unani medicines.
(3)	Medicinal herbs	: To prepare plant extracts and single ingredient powders.
(4)	Dietary supplements	: Sources of vitamins and enzymes.
(5)	As nutraceuticals	: Foods containing supplements of herbal origin for prevention or to treat ailments.
(6)	As cosmeceuticals	: Cosmetic products having medicinal value.

The major part of herbs is being consumed in the cosmeceuticals industry and is realised by going through their estimated demand in the US market.

At present United States annual market for cosmeceuticals has been estimated to be 2.5 US billion dollars, where as 5.0 million dollars for European countries.

Cosmeceuticals are the cosmetic products, which contain biologically active principles or ingredients of *plant origin* having effect on users, or they are combination products of cosmetics and pharmaceuticals, intended to enhance the health and beauty of the skin.

They differ from cosmetics, since cosmetics are the inert substances which cleanse or enhance the appearance of the skin without any *therapeutic benefit.* Where as pharmaceuticals are used to treat or prevent the diseases or are intended to affect the physiological structure or function of the body.

Pharmaceuticals are scrutinized and also reviewed by FDA, however one should note that cosmetics and cosmetic ingredients are tested for safety but evidence is rarely available for their beneficial claims.

Aging of the skin is cumulative phenomenon and is influenced by several factors, such as

(i) Temperature fluctuations

(ii) Exposure to the chemicals and gases

(iii) Exposure to sunlight

(iv) Environmental pollution.

UV radiations from sunlight penetrate the skin and accelerate damage due to free radicals, which includes inflammation, wrinkling and hyper pigmentation.

Due to prolonged exposure to UV radiation the collagen and elastin fibres of the skin are broken down by the enzymes collagenase elastase and texture of the skin deteriorates. Collagen and elastin are responsible for maintaining the elasticity and integrity of the skin. Several plant extracts and antioxidants obtained from natural sources are able to prevent the aging and also improving the appearance of the skin.

HIBISCUS

Synonym

Shoe flower, Jaswand.

Biological Source

This consists of roots, leaves and flowers of *Hibiscus, rosa sinensis,* fam. *Malvaceae.*

Geographical Source

Cultivated in Gardens through out India. Native country probably China.

Description of Plant

It is a evergreen woody perennial shrub, with short petioled leaves. Leaves are ovate lanecolite and irregularly serrate margin at the end. Leaves are entire at base and globrous. **Flowers** are axillary solitory, involucral bracts 5 – 7. Calyx is 7.0 – 7.5 cm in diameter tubular below, petals are red, 3 to 4 times long as calyx, staminal tube extented for beyond the petals. No fruits. Flowering throughout the year. There are single or double forms with different colour shades; may be orange crimson, bright red or magenta. Red flowers of this plant are specially offered to lord Shri Ganesha (Ganapati).

Cultivation

It is cultivated in garden everywhere by vegetative method (cutting) in shady places, during rainy season. Any type of soil is suitable for its cultivation.

Chemical Constituents

Flowers contain thimin, riboflavin, niacin, ascorbic acid, anthocyanin pigment, cyanidin diglucoside.

Uses

It is used to treat jaundice and irregular menstruation. Also used to arrest hair fall and promote hair growth.

Several herbal drugs (Ayurvedic drugs) are used in medical practice, in the formulation of cosmeceuticals since long. They are used in dental products, various disorders of skin and hair. Few of them are as under.

COCONUT WATER

Biological Source

The water of tender coconuts. (Biologically known as *Cocus-nucifera* Fam. Palmae) is sterile, nutritions and thirst quenching health drink from 6.7 months old fresh fruits.

Geographical Source

Coconut plants are cultivated in hot and damp region near sea shores in India, Kerla, Goa and also Maharashtra are the leading states.

Description

Coconut palm are about 25 – 30 metres in height and have tuff of leaves at the top and bears about 100 cocunuts per annum. Dried kernels are richest source of edible oil and edible copra which is popular dry-fruit in India. Coconut water is transparent colourless and with pleasing flavour.

Table 18.7: Herbal drugs used in formulation of cosmeceuticals

Sr. No.	Desired action	Names of herbal drugs used
(A)	Skin-Disorders	
1.	Anti-acne	Arjuna, Cucumber gel, Manjishta, Vetiver.
2.	Anti-inflammatory	Aloe-vera, Boswellia, Chandan, Raktachandan, Rose.
3.	Anti-oxident	Amla, Ginkgo-biloba, Rosemary, Turmeric, Green tea
4.	Anti-wrinkle	Apricot, Aloe-vera, Ashwagandha, Ginsing, Liquorice, Papaya
5.	Anti-aging	Amla, Aloe-vera, Curcuma.
6.	Astringent	Chammomile, Galls, Lemon, Purslay, Sage, Thyme
7.	Depigmentation	Papaya, Cucumber, Turmeric, Manjistha
8.	Moisturising	Almond oil, Aloe-vera, Fenugreek, Neem, Rose, Vetiver
9.	Bleaching	Marigold, Raktachandan, Aloe, Turmeric
10.	Cleansing	Cabbage, Citrus peels, Amla, Rose, Cucumber, Geranium Sandalwood, Papaya, Aloes
11.	Antiseptic	Neem, Tulsi, Citrus peels, Turmeric, Lavender oil, Witch-hazel
12.	Cooling effect	Vetiver, Lemon, Neem, Rose, Fenugreek, Mentha, Cucumber
13.	Sun Screen	Calundulla, Carrat, Chammomile, Cucumber, Aloe-vera, Red-sanders
14.	Smoothening of skin	Ashwagandha, Fenugreek, Almond oil, Sesame oil, Vetiver, Aloe-vera gel, Orange oil
15.	Nourishing to skin	Wheat-germ oil, Honey, Peach, Carrot, Apricot.

(B) Hair-Disorders		
1.	Anti-dandruff	Soap nut, Shikekai, Arnica, Thyme, Rosemary, Aloe-vera, Parsley, Kapur Kachri
2.	Anti-lice	Kachur-Sugandhi, Sitaphal
3.	Hair tonic	Jatamansi, Cocount oil, Sesame oil, Brahmi, Amla, Hibiscus, Rosemary.
4.	Hair wash	Amla, Soap-nut, Shikekai, Marigold
5.	Hair dye	Eclipta, Amla, Henna, Hibiscus, Sage, Blood root, Chammomile
6.	Hair conditioner	Eclipta, Rosemary, Fenugreek, Henna, Amla, Hibiscus
7.	Shining of hairs	Soap-nut, Shikekai, Sage, Henna coconut oil
8.	Reduce hair loss	Amla, Arnica

Following are the few naturally occurring agents frequently used in cosmeceutical formulations.

1. Retinoic acid (Tritinoin)

Naturally occurring form of vitamin A is known as retinoic acid which is very commonly used topically for the treatment of photo-aged skin (sun-damaged skin) and also for acne i.e. skin eruption with red pimples. Retinoic acid is said to increase the rate of cell-division and turnover. It is also useful in reducing the wrinkles, hyperpigmentation, and roughness due to over exposure to sunlight.

2. Alpha-hydroxy acids

Glycolic acid and lactic acid are the commonly used alpha hydroxy acids in cosmeceutical formulations. These weaken the linkage of cells in dry skin and favour its exfoliation (shedding). The flexibility of skin is increased due to topical applications. Aleuritic acid from shellac is now-a-days preferred as a substitute for α-hydroxy acids.

3. Boswellic acids

These are present in the resin of *Boswellia serrata* and are said to inhibit the enzymes responsible for inflammation (5-lipoxygenase) and damage (due to elastin) of the skin hence incorporated in anti-inflammatory creams and lotions usually.

4. Vitamins C and E

A well known antioxidant vitamin C from amla frequently used in the form of Ayurvedic formulation internally as well as externally. It is described to be the scavenger of oxygen free radicals which are responsible for aging process of skin. Exposure of the skin to UV light induces the formation of free radicals causing the skin to age rapidly. Vitamin C has synergistic effect with Vitamin E and is used as protective against UV radiation.

5. Co-enzyme Q 10 (Ubiquinone)

(Q stands for quinone group while 10 refers to isoprenyl units)

UV radiations not only damage the skin but also have reported to reduce its natural ubiquinone content. Experimentally also in the cultured human cells, the results with ubiquinone were found to coincide; additionally ubiquinone reduces the activity of enzymes which degrade the collagen.

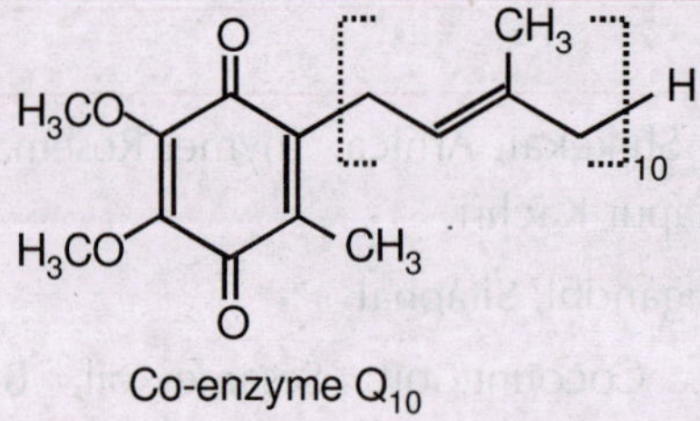

Co-enzyme Q_{10}

Mol. formula $C_{59}H_{90}O_4$

Mol. wt. 863.36

6. Miscellaneous

Tetrahydrocurcuminoids: White (colour free) curcuminoids of turmeric (*curcuma longa*) *carnonic acid, coumarinic acid, ursolic acid* from rosemary extract *Rosemarimus officinalis* as antioxidants are the other compounds which are used to facilitate the tissue damage and restoring the healthy status of skin.

Table 18.8: Natural Drugs used in Formulation of Cosmeceuticals and Neutraceuticals

Herbal Cosmeceuticals	Herbal Neutraceuticals
1. Aloe	1. Apples
2. Amla	2. Beets
3. Carrot	3. Cucumber
4. Coconut	4. Oat meal
5. Eclipta	5. Orange skin
6. Fenugreek	6. Peach-skin
7. Heena	7. Pears
8. Hibiscus	8. Pumpkin
9. Raktachandan	9. Spinach
10. Sesame	10. Tomato
11. Soat nut	
12. Turmeric	

❖❖❖

Chapter 19

NATURAL PESTICIDES, ANTIBIOTICS AND ALLERGENIC EXTRACTS

A. NATURAL PESTICIDES

Along with time, many advances have been reported in scientific field, but still the attitude towards medicinal crop protection is somewhat in a primitive condition. Medicinal crops are treated with highly lethal synthetic insecticides, so that harmful and harmless or beneficial insects are also killed. These insecticides also leave deleterious effects on the plants. Development of modern pesticides along with their indiscriminate use have resulted in number of harmful side effects in man and also development of secondary pests or resistance among insects. In this view, judicious use of selective pesticides is being mentioned as a part of integrated control of pests. Due to this, more emphasis is being given to natural products which were ignored for a long time. Many plants have been reported to exhibit insecticidal or insect-repellant effects and some of them are *Azadirachta indica, Chrysanthemum cinerariaefolium, Acorus calamus, Derris elliptica, Nicotiana tobacum Pistia stratiotes, Nigella sativa, Cyperus rotandus* etc. Some of them have been described here.

Table 19.1: Natural pesticides

Sr. No.	Name and synonym	Biological source	Active constituents	Uses
1.	Derris (Tuba-root)	Rolts and rhizomes *Derris-elliptica* Leguminosee	Rotenone tephrosin	Insecticide for vegetables (contact poison)
2.	Neem (Margosa)	Aerial parts of *Azadiracta* indica Meliaceae	Azadirechtin Nimbin Nimbidin	Insecticide, antifeedant nematicide
3.	Pyrethrum (Insect-flower)	Dried flowers of *Chrysanhemum cinerarifolium Compositae*	Pyrethrins	Domestic insect repellant (contact – poison)
4.	Ryania	Roots and stems *Ryania speriost* Flacourtriaceae	Ryanodine	Insecticide
5.	Sabadilla (Caustic barley)	Ripe seeds of *Schoenoeauion offioinale* Liliacae	Sabadine, cevadine	Domestic insecticides
6.	Tobacco	Dried leaves of *Nicotiana tobacum* Solanaccae	Nicotin nor-nicotine anabasine	Very potent insecticide
7.	Citronella oil	Cymbopogon nardus Graminae	Geraniol, Citronellal	Insect repellant

DERRIS

Synonym

Tuba root

Fig. 19.1: Derris herb

Biological Source

This consists of dried roots and rhizomes of *Derris elliptica* Roxb. and *D. malaccensis* Prain, belonging to family *Leguminosae* and Cube root of *Lonchocarpus utilis* and *L. urucu*, belonging to family Leguminosae. Contains not less than 2.0 % rotenone.

Geographical Source

Derris (also called Tuba root) is indigenous to East Indies and Myanmar. Cube root is available from Peru, Brazil and British Guiana.

Macroscopic Characters

Derris roots occur as slender pieces having dark red brown or grey brown colour. They have longitudinal furrows. They are flexible, tough with a fibrous fracture. Taste is bitter.

Chemical Constituents

Rotenone (previously called tubatoxin) is present in derris and cube roots from 2 - 10%, and is an isoflavonoid derivative.

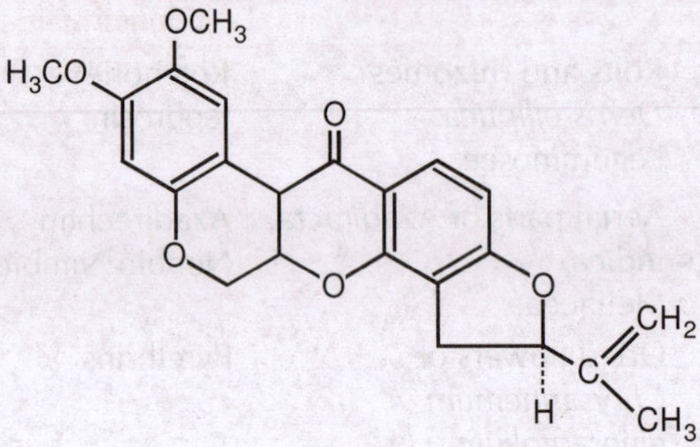

Rotenone

Alongwith rotenone, derris also contains tephrosin, toxicarol and degnelin. Rotenone occurs as colourless to brownish crystals without any odour and taste. It is insoluble in water, but sparingly soluble in alcohol, acetone and chloroform.

Standards of Quality

Ash	:	Not more than 6.0 per cent
Acid insoluble ash	:	Not more than 2.0 per cent

Uses

Rotenone is a contact poison and used in the form of sprays for killing vegetable insects during harvesting times, such as leaf hoppers, Mexican bean beetle, cater pillars and aphides. For veterinary purpose, it is used to control cattle grubs, fleas and chicken lice.

NEEM

Synonym

Margosa.

Biological Source

It consists of all aerial parts of plant known as *Azadirachta indica*, family Meliaceae.

Geographical Source

It is found in India, Pakistan, Banglades, Shri Lanka, Thialand, Malaysia, Mauritius, Fiji, South Africa and East Africa.

Macroscopic Characters

Leaves: Alternate, exstipulate, imparipinnate leaflets 5.0 - 10 cm in length lanceolate closely clustered towards the ends of branches. The leaves have serrate margin, green colour and bitter to taste.

Bark: Moderately thick, rough, brown in colour longitudinally and obliquely furrowed. Internally starchy white, laminated with characteristic smell of neem and bitter in taste.

Chemical Constituents

Good number of chemicals isolated from the plant belong to the classes diterpenes (sugiol), nimbiol (bark), triterpenes : β-sitosterol, stigmasterol (leaf), **Limonoids:** Maliantriol (seed oil) nimbidinine (seed oil), Nimbendiol (seed oil) nimbin and azadiractin (seed), chemically azadiractin is tetra nor tritarpinoid. **sulphurous compounds:** Number of cyclic tri and tetrasulphides (leaves), flavonol glycosides : Nimaton, quercetin, myrecetin, kaempferol. Neem leaves contain not less than 1.0% w/w of Rutin.

Fig. 19.2: Neem Twig

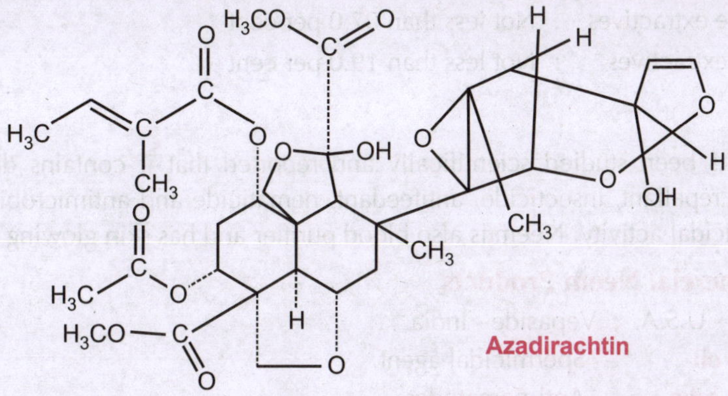

Azadirachtin

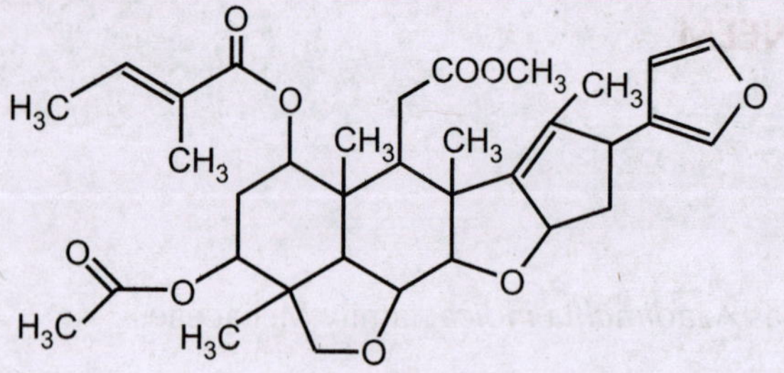

Salanin

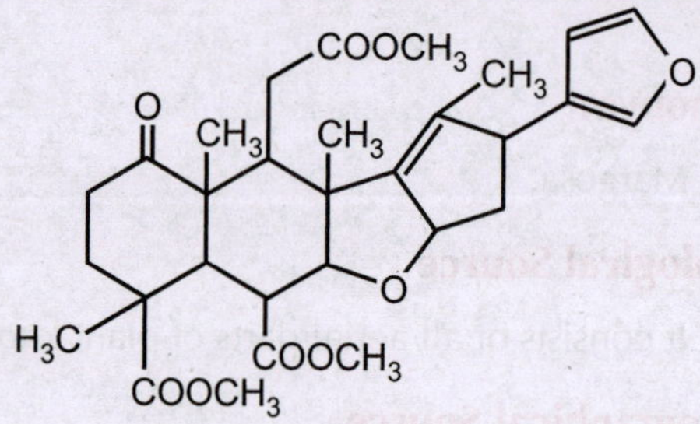

Nimbin

Leaves	:	Azadirachtin	-	insect repellant
		Meliantriol	-	anti-feedant
		Salanin	-	anti-feedant
Seeds	:	Nimbin	-	Antiviral action
		Nimbidin	-	Antiviral action
		Azadirachtin	-	Insect repellant
Flowers	:	Nimbosterol, Myricitin, Kaempferol	-	Insecticidal
Fruits	:	Deacetyl azadirachtinol	-	Paralyzes insects swallowing mechanisms
Bark	:	Nimbin, Nimbinin, Nimbidin	-	Antiviral
		Margolone, Margolonone	-	Antibacterial
Roots	:	Excellent for reforestation		
		Compounds with antibacterial and antifungal properties.		

The neem oil contains 2% of bitters, which are sulphur containing compounds nimbin, nimbidin, nimbinin and nimbidol. Azadirachtin-k, new tetraterpenoid has been isolated from seed kernels of neem alongwith other compounds such as nimbolide, olichinolide B, nimbin, 6-deacetyl-nimbin, salanin and azadiradione.

Standards of Quality

Foreign organic matter	: Not more than 02.0 per cent
Ash	: Not more than 12.0 per cent
Acid insoluble ash	: Not more than 02.0 per cent
Loss on drying	: Not more than 12.0 per cent
Alcohol soluble extractives	: Not less than 07.0 per cent
Water soluble extractives	: Not less than 19.0 per cent

Uses

Recently, it has been studied scientifically and reported that it contains different chemicals which have insect repellant, insecticide, antifeedant, nematicide and antimicrobial properties. The seed oil has spermicidal activity. Neem is also blood purifier and has skin glowing property.

Important Commercial Neem Products

Margosan - O - U.S.A. ; Vepaside - India

Neem oil - Spermicidal agent.

Neem cake - Anti-nematodes

PYRETHRUM

Synonyms

Insect flowers, Dalmation Insect Flowers.

Biological Source

Pyrethrum flowers are expanded flower heads of *Chrysanthemum cinerariaefolium*, family Compositae. It should contain not less than 0.7 per cent of total pyrethrins.

Geographical Source

Pyrethrum is cultivated in Yugoslavia (Dalmatia), Japan, Brazil, Kenya, Zaire, Tanzania and India. Jammu and Kashmir are the major areas of cultivation in India for pyrethrum.

Cultivation and Collection

Pyrethrum is cultivated in tropical zones at an altitude of 1500 to 3500 m, depending upon the distance from equator. The soil requirements depend upon rain fall and other climatic conditions in a particular area. A rainfall between 800 to 1300 mm is considered suitable for pyrethrum cultivation. Since, the plant is very sensitive to frost, sunny periods interrupting rainfalls, are desired conditions for cultivation. It needs a temperature between 15°C to 25°C. For higher yields per hectare, not only climatic conditions and soil, but also method of cultivation matters. Cultivation of the drug is done by seed germination and vegetative propagation. The plants produced by sowing the seeds do not grow satisfactorily. The vegetative method of cultivation is generally adopted. The splits are planted in well prepared fields. The plants reach maturity within four months. The crop should be kept free of weed for maximum yield. The pyrethrum flowers are collected by hands and dried immediately therefore, to avoid loss in the pyrethrin content. Sun-drying of the drug is generally preferred. After drying, flowers are packed into sacks and sent to the market. Average yield of the drug per hectare is 300 to 400 kg.

Macroscopic Characters

Colour : Cream to straw coloured

Odour : Characteristic, aromatic

Taste : Bitter followed by numbness

Size : 10 to 15 mm in diameter

Shape : Flower-head is flat with convex receptacle.

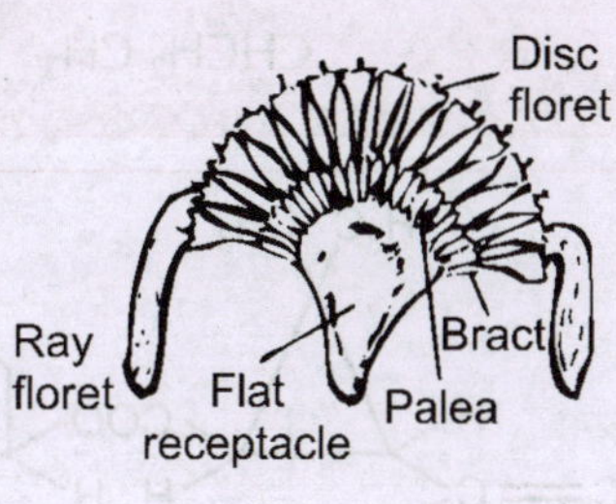

Fig. 19.3: Pyrethrum flower

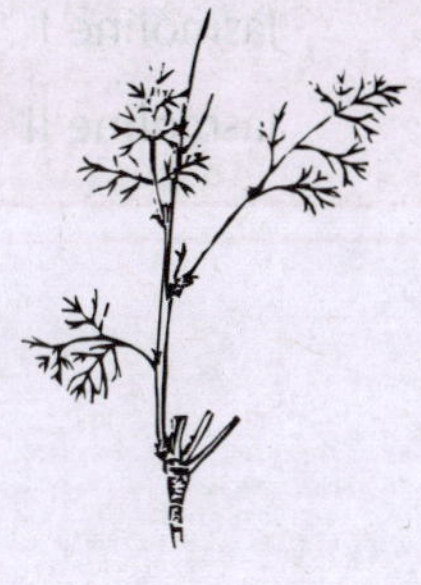
Fig. 19.4: Pyrethrum herb

Extra Features

Peduncle is short and longitudinal striated. Involucre of three yellowish green bracts and keel are pronounced. Ray-florets (female) are 15 - 23 in number. Disc-florets are 200 – 300 (hermaphrodite), each with 5 lobed, yellow coloured tubular corolla.

Chemical Constituents

The insecticidal principles of pyrethrum are located in the oleoresin secretion of floral parts (achenes) of partially open or closed flowers. Although, pyrethrin I and pyrethrin II are the main active constituents, it also contains other active compounds called cinerin I, cinerin II, Jasmoline I and Jasmoline II. All these constituents are esters.

Table 19.2: Pyrethrum Esters

Ester	Alcohol part	Acidic part
1. Pyrethrin I	Keto alcohol pyrethrolone	Chrysanthemum mono-carboxylic acid
2. Pyrethrin II	Keto alcohol pyrethrolone	Monomethyl ester of chrysanthemum dicarboxylic acid
3. Cinerin I	Keto alcohol cinerolone	Chrysanthemum mono-carboxylic acid
4. Cinerin II	Keto alcohol cinerolone	Monomethyl ester of chrysanthemum dicarboxylic acid
5. Jasmoline I	Jasmolone	Chrysanthemum mono-carboxylic acid
6. Jasmoline II	Jasmolone	Monomethyl ester of chrysanthemum dicarboxylic acid

Table 19.3: Nature of R and R_1 attachments to esters

Ester	R	R_1
Pyrethrin I	$CHCH = CH_2$	CH_3
Pyrethrin II	$CHCH = CH_2$	$COOCH_3$
Cinerin I	$CHCH_3$	CH_3
Cinerin II	$CHCH_3$	$COOCH_3$
Jasmoline I	$CHCH_2\ CH_3$	CH_3
Jasmoline II	$CHCH_2\ CH_3$	$COOCH_3$

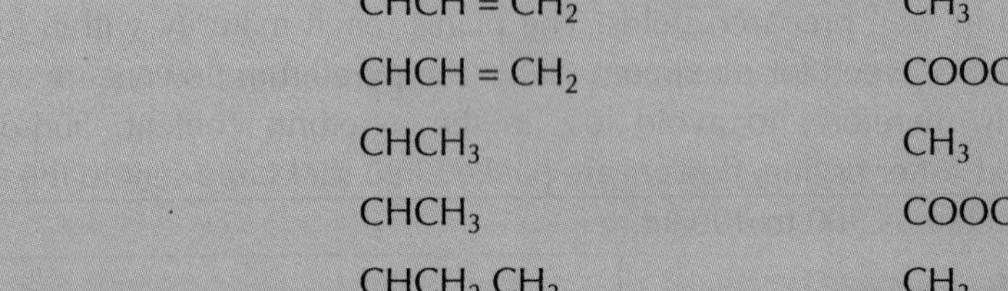

Allethrin

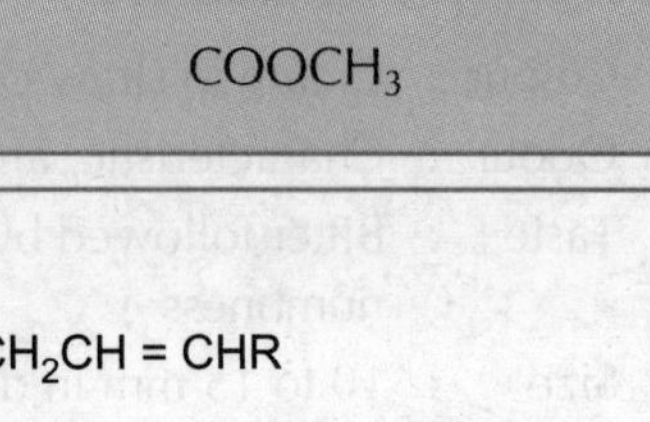

Pyrethrosin

Pyrethrum also contains pyrethrosin, pyrethrol and sesquiterpene lactones.

Allethrins are the synthetic analogues of naturally occuring insecticides i.e. pyrethrin, cinerin and jasmoline.

Where R = CH_3 ⟶ Allethrin I

and R = $COOCH_3$ ⟶ Allethrin II

Standards of Quality

Pyrethrum should contain not more than 5 per cent of naturally adhering stems.

Ash: Not more than 8 per cent

Acid-insoluble ash: Not more than 1 per cent

Uses

Pyrethrum has been used since long as an insecticide and it is a contact poison. It paralyses, different types of insects in short time. It shows rapid toxicity towards different pests, but has very low toxicity in mammals. Previously pyrethrum was used in the form of powder or extracted with odourless kerosene. Presently, the extracts containing about 20 per cent of pyrethrins, are used in the form of dust, wettable powder, emulsion concentrate kerosene solution and aerosol. Many a time, pyrethrum extract is mixed with other insecticides which enhance the action of pyrethrum by synergistic effect. Some of the compounds used for this purpose are sulfoxides, piperonyl butoxide, sesamin and DDT.

Pyrethrum exerts a knock down effect on insects. Either singly or in synergistic form, it is used to kill different plant insects, and also flies and mosquitoes. The pyrethroids are also prepared synthetically and used. They are allethrin derivatives, S-bioallethrin and resmethrin.

Other Varieties

Chrysanthemum coccineum, C. marschalli.

RYANIA

Synonym

Ryanicide

Biological Source

It consists of roots and stems of *Ryania speciosa,* belonging to family Salicaceae (Flacourtriaceae). It is indigenous to South America (Colombia).

Morphology

Ryania is a shrub (Fig. 19.5). It has ovate leaves and white flowers.

Chemical Constituents

Ryania contains alkaloids of which ryanodine is important which constituents only 0.16 – 0.2%.

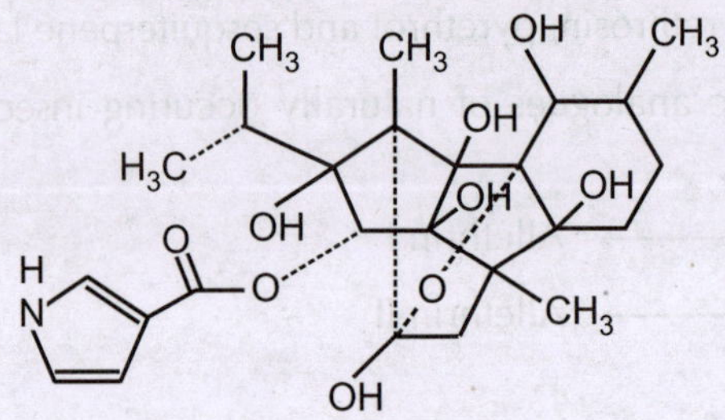

Ryanodine

Uses

Ryania extracts are used as insecticide to control codling moths corn borer and many others, but ineffective for cabbage and cauliflower worms. It is very low toxic to mammals.

SABADILLA

Synonyms

Cevadilla, Caustic barley.

Biological Source

These are the dried ripe seeds of *Schoenocaulon officinale*, belonging to family Liliaceae.

Geographical Source

It is a herbaceous, tall plant indigenous to Venenzuela, Guatemala and Mexico.

Macroscopic Character

The seeds are long, narrow, tapering to acute angle and dark brown to nearly black in colour. They have bitter and acrid taste and the seed powder causes intense sneezing.

Fig. 19.5: *Ryania* speciosa herb

Fig. 19.6: Sabadilla herb

Chemical Constituents

Sabadilla contains a number of alkaloids like cevadine (which is also called crystalline veratrine) veratridine, sabadine, sabadilline and among them the first two alkaloids are more potent. The mixture of these alkaloids is called *'veratrine'*. It also contain veratonic acid fatty oil and resin.

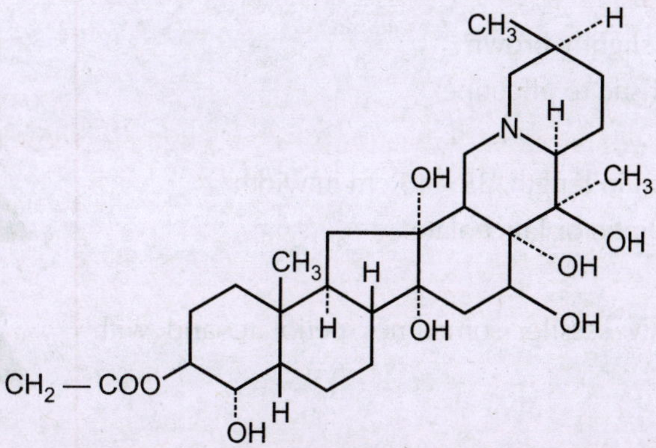

Sabadine

Uses

Powdered sabadilla is an insecticide used to kill house flies, thrips and bugs in the form of spray or dust.

TOBACCO

Biological Source

This consists of dried leaves of *Nicotiana tobacum*, belonging to family Solanaceae.

Geographical Source

Tobacco is cultivated on a commercial scale to a very large extent in China, United States and India. China produces annually 22 laks metric tonnes, while India produces about 5 laks metric tonnes of tobacco in a year. The other tobacco producing countries are Brazil, Russia, Turkey and Italy. In India, it is produced mainly in Andhra Pradesh, Gujrat, Karnataka, Orissa and Bihar.

Description

Tobacco, as a whole, is stout, ever green and viscid annual, 1 to 3 metres in height. It has thick erect stem and few branches. It bears about 20 leaves which are approximately 80 cm in length. Flowers are light-red, white or pink in colour; fruits are capsules, elliptic, ovoid, 1.5 to 2.0 cm in size. Seeds are spherical and 0.5 mm in diameter and brown in colour. Various varieties are known to be cultivated in India. Few of them are bidi tobacco, cigar tobacco, chewing and hookah tobacco.

Cultivation and Collection

Depending upon the type of tobacco, the requirements of soil and climate also vary. Warm climate, and well drained fertile land are favourable for its cultivation. Seeds are used for cultivation. The seeds are sown on the seed beds in winter or early spring. When the seedlings are about 12 weeks old, they are transplanted. In the flowering season, the flowering tops are cut so as to encourage the growth of foliage. Harvesting is done after 70 - 90 days of transplantation. The leaves are then subjected to processing by air curing, fire curing or fuel curing. During this process, the chemical changes occur and lead to development of flavour and aroma.

Macroscopic Characters

Colour	:	Green or slightly brown
Odour	:	Characteristic to Nicotine
Taste	:	Bitter
Size	:	60 - 80 cm in length, 35 - 45 cm in width
Shape	:	Ovate, elliptic or lanceolate

Extra Features

The leaves are usually sessile, sometimes petiolate and with frilled wing.

Fig. 19.7: Tobacco plant

Chemical Constituents

The tobacco contains pyridine-piperidine type of alkaloids, among which the most prominent is nicotine. The other alkaloids are nornicotine and anabasine upto 08 – 0 per cent nicotine.

Nicotine Nor-Nicotine

Uses

Nicotine exerts stimulant effects on heart and nervous system. It is not used medicinally. It is powerful quick acting poison. Even 40 mg dose orally is fatal to humans. Rectified tobacco seed oil is used as edible oil in European countries. Nicotine is used in the manufacture of nicotinic acid and nicotinamide.

Tobacco and nicotine are known insecticides for last three centuries. Nicotine controls a wide range of insects. It is mainly used against soft bodied insects like aphides. It acts as a contact poison. It is also effective against white flies, moths, termites, butterfly-larvae, red-spider mites, etc. Nicotine is sprayed on crops in the form of nicotine sulphate. It has certain advantages over synthetic insecticides that it is safer, easier to handle and much less toxic to warm blooded animals. Because of its volatility, it evaporates earlier and leaves no harmful residue on the marketable products.

B. ANTIBIOTICS

Although, antibiotics occupy the prominent position in today's world of drugs, it is considered that they were known to human beings since pretty old days. The people were aware about the antiinfective character of moulds and fungi grown on foods like bread, etc. But the real impetus to the field of antibiotics was given by Pasteur and also by extraction of pyocyanase from *Pseudomanas aeruginosa* which was introduced in medical practice as an antibiotic. Penicillin was extracted by Alexander Fleming in 1929, but the phenomenal antimicrobial activity of *Penicillium notatum* was pointed out by Chain Florey some time ago.

At present, antibiotics comprise of a large number of metabolites from various microbes, which exhibit inhibitory or killing response on the growth of other microbes, when used in very little concentrations. Production of antibiotic by a micro-organism is considered as microbial antagonism.

The antibiotics are produced either by natural way through fermentation or by semi-synthetic or synthetic method. The fermentation process is either surface or submerged one. The latter process is highly efficient one and employed for industrial purposes. A large number of microbial strains have been developed and tried for their antibiotic producing capacity and also for efficiency in their extraction.

The chemical classification of antibiotics covers the following groups.

(a) Aminoglycosides (b) Penicillins

(c) Cephalosporins (d) Macrolides

(e) Tetracyclines (f) Polypeptides (g) Miscellaneous

(a) Aminoglycosides

All the aminoglycoside antibiotics chemically, contain a basic 6 membered carbon ring (amino or guanidino) to which one or more amino sugars are attached through a glycosidic linkage. The examples of these antibiotics include gentamicin, kanamycin, neomycin, streptomycin, etc. These antibiotics combine with bacterial ribosomes and block the protein synthesis. They have broad spectrum activity against gram-negative bacteria.

(b) Penicillins

The penicillins are obtained from various species and strains of *Penicillium* and chemically, they are a - carboxamido derivatives of penicillanic acid. The natural penicillins are Penicillin G (benzyl), Pencillin F (2 - pentenyl), Pencillin K (n - heptyl), Penicillin V (phenoxy methyl), etc. The semisynthetic penicillins are derived from 6 - amino-penicillanic acid, e.g. ampicillin, amoxicillin, carbenicillin, cloxacillin, cyclacillin, etc.

Natural and semisynthetic penicillins prevent the formation of bacterial cell wall by interfering with the synthesis of N-acetylmuramic acid, mucopeptides and teichoic acids. Penicillns are effective mainly against gram - positive bacteria.

(c) Cephalosporins

They are derivatives of cephalosporanic acid, which is a close analogue of penicillanic acid. Cephalosporanic acid contains dihydrometathiazine ring Cephalosporins include cephalexin, cephaloglycin, cephapirin, cephradine, cephalothin, cefazolin, etc. Like penicillins, cephalosporins also interfere with synthesis of bacterial cell wall by blocking the cross-linking of the peptidoglycan units. But their effects are more pronounced in-vitro against gram-positive bacteria.

(d) Macrolides

Chemically, they are macrocyclic lactones with 12 to 20 carbon atoms in the primary ring. Among all the macrolides, except erythromycin, no other member has any significant therapeutic value. Erythromycin is used in the forms like estolate, stearate, ethyl succinate, lactobionate, etc. Some of the other members of macrolides are troleandomycin, mitasamycin and spiramycin.

Macrolides block the translocation step in protein synthesis by binding to 50s sub-unit of the bacterial ribosomes. The only clinically used member erythromycin is used for treatment of gram-positive bacterial infections like *Enterococci*, *Pneumococci* and *Staphylococcus aureus*.

(e) Tetracyclines

Tetracyclines comprise a group of antibiotics which have octahydronaphthacene skeleton (hydrocarbon made up of four fused rings). The different members of this group used in clinically are chlortetracycline, doxycycline, oxytetracycline, demeclocycline, methacycline and minocycline.

The tetracyclines inhibit the protein synthesis by binding to 30s ribosmoes and preventing t-RNA from combining with m-RNA. They are active against a wide range of gram-positive and gram-negative bacteria, rickettsia and viruses and hence, used in treatment of variety of infections.

(f) Polypeptides

Unlike other group of antibiotics mentioned earlier, members of this group differ in mechanism of action and their anti-bacterial spectrum. Only correlating part for all of them is that they are composed of a number of amino acids joined through peptide linkage and have a cyclic structure. They contain unnatural D-configuration amino acids. Most of them are obtained from genus *Bacillus*. Polypeptide antibiotics are in many cases, used for topical applications. The examples of polypeptide antibiotics include Polymyxin B sulphate, Bacitracin zinc, colistin sulphate, gramicidin and capreomycin sulphate.

(g) Miscellaneous antibiotics

The chemical nature and mechanism of action of the members of this divergent group are varying, but some of them are the products of bacteria, which have unique biosynthetic patterns. The miscellaneous antiobiotic group includes rifampin, cycloserine, vancomycin, chloramphenicol, clindamycin, novobiocin, fusidate sodium, etc. Among them, rifampin is a drug of choice in treatment of *Mycobacterium tuberculosis and M. laprae.*

Besides chemical mode of classification, antibiotics are also classified depending upon their mechanism of action. The generalized modes have been mentioned below:

(a) **Inhibition of protein synthesis**: The antibiotics with such effects are gentamicin, tetracyclines, erythromycin, chloramphenicol, neomycin, streptomycin, etc.

(b) **Inhibition of cell wall formation:** This type of activity is shown by penicillins, celphalosporins, bacitracin, vancomycin, etc.

(c) **Disruption of DNA metabolism:** The antibiotics included under this group are rifampin, actinomycin D, doxorubicin, mitomycin etc.

(d) **Alteration incellular membrane function:** The antibiotics included are polymyxin, nystatin, candicidin, etc.

C. ALLERGENIC EXTRACTS

Allergy is the unusual response to fubstances or conditions by a human-being, which is harmless to others.

Three different words, allergy, allergic and allergenic are very important and are explained as under. Allergy it is a condition causing unfavourable bodily reacts to contain sustenous. Allergic or Allergenic : substances causing allergic reactions.

ALLERGENIC EXTRACTS

Allergy is defined as unusual response to substances or conditions by a human being and is harmless to others.

While the substances causing allergy and known as avergens or allergenic substances.

Under mentioned are the various types of allergies known so far.

(1) Contactants (2) Infectants (3) Intestants

(4) Ingestants (5) Injectants

(1) Contactants: Substances of this type when come in contact with human body (specially skin) cause allergy, such as asporagns, lobelia, deffodils, marigold. English Ivy, trichomes of various types of leaves a flowers, pollen grains are the common examples.

Urushiol from poison-ivy plant is very harmful poisonous and dangerous, which is constituent of poisonous and dengerous, which is constituent of irritant oil of the plant and causes watery blisters, alongwith pruritus and (ittiching) poisonaly plant (Poisun vine, mark weed) is botanically known as *Toxicodendron radicans* Fam. Anacardiacere Urshiol is non-volatile phenolic compound and is allergic.

(2) Infectants: Certain living organisms cause allergy by the products they produce during metabolism in the human body.

(3) Infestants: Parasitic organisms like tape worm, hook warm, pinworm, thread worm when enter the body of human being cause serious allergic sysptoms due to their growth products, or metabolic wastes of these parasites.

(4) Ingestants: Cold liver oil, orange joule or even cow's milk is also allergic, even the chocolate and shell fish may also be the cause of allergy in it is reported milk allergy is another example of allergy reported to infants.

(5) Injectants: Injection of some antibodies also cause allergy. Penicillin injection is very serious and known to many, glandular products antitoxins, liver extracts are also other suitable examples of injectant allergens.

(6) Miscellaneous: Apart from the several ---- mentioned types some peopel get allergy of listening specific voice or by observation of some animals or the bites of animals. Bees ane wasps stings also cause allergy.

Allergenic extracts are the substances used for diagnosis and treatment of allergic diseases and are concentrated solutions or suspensions of allergens.

Allergy or hypersensitivity is considered as an untoward immunological reaction towards an antigen, which is called allergen. In some cases allergic reaction leads to a disease, otherwise in most of the cases remains cause-effect relationship. It is untoward, in the sense that, it causes tissue damage during the immunological response exhibited by body.

The allergans and allergenic extracts are treated as crude drugs. A number of natural substances or their extracts have allergic reactions and they are employed in diagnosis and treatment of allergy.

Dust and pollens are the most common allergens and the other known allergens are from various fungi, insects, food items, etc. The chemical nature of allergens appears to be diversified and also not yet correctly identified. In general, the chemical nature of those known allergens seems to be proteinous or glycoproteinous, and normally, their molecular weight ranges between 5,000 to 40,000. Many allergenic substances constitute more than one allergen and depending on such content, vary in their allergic potency and called as major or minor allergens.

A number of allergenic extracts are used for the diagnosis of allergy, by employing various tests. They are normally administered in the form of aqueous or glycerinated products, using normal saline or 50% glycerine as diluent. These extracts are added with preservatives like phenol (0.4%) and adjusted to pH 8.

The types of allergenic extracts have been summarised below:

1. Pollen Extracts

Pollen grains are produced in higher plants and in many cases are allergy causing. They get mixed in atmosphere due to wind flow and are inhaled by humans, which leads in many cases to pollinosis, also called hay fever. As this allergy is caused due to pollens, it is naturally limited to certain seasons only and also to certain geographical areas viz. temperate climatic zones. Some of the plants producing such allergic pollens from which diagnostic extracts are prepared have been listed here.

Table 19.4: Allergic pollen extracts

Alfalfa	Almond	Apple	Acacia	Barley	Blue grass
Canary grass	Cherry	Corn	Dandelion	Eucalyptus	Gram grass
Gladiolus	Hazelnut	Hemp	Juniper	Mulberry	Mustard
Lily	Orchard grass	Oats	Pigweed	Palm	Rye grass
Sunflower	Tulip	Worm seed			

2. Fungal Extracts

Alongwith pathogenicity, certain fungi are also responsible for allergic manifestations. Fungi and moulds grow in various places, more commonly in saprophytic form and cause rhinitis, asthma and cutaneous reactions. Few of the fungal and mould extracts used are listed below.

Table 19.5: Allergic fungal extracts

Alternaria	*Aspergillus species*	*Candida*	*Cephalosporium*
Cryptococcus	*Epidermophyton*	*Fusarium*	*Helminthotsporium*
Mucor	*Neurospora*	*Penicillium species*	*Rhizopus*
Saccharomyces	*Torula*	*Trichoderma*	*Verticillum*

3. Dust Extracts

Dust ranks second to pollens, as a causative factor for allergy. Dust occurs practically every where and comprises various ingredients of innumerable origin like food, fungi, hair, feathers, cleansing agents, etc., to name only a few. Hence, it poses problems to locate exact allergenic part. It is believed that dust mite (Dermatophagoides species) is mostly, the allergy causing ingredient from dust, as it is universally spread and normally present on vegetable fibres like cotton. For sensitivity testing, home dust extracts are prepared and tested.

4. Insect Allergen Extracts

The types and species of insects are innumerable and some of them cause allergy in humans. The insect allergy occurs due to either bite/sting or inhalation of insect parts like scales, hair, etc. Insects belong to order Hymenoptera, like honey bee, fire-ant, wasp, hornet etc. The symptoms are asthma, systemic anaphylaxis, urticaria and finally, leading to respiratory failure and/or cardiac arrest. The following part is the list of few allergy causing insects.

Table 19.6: Allergic insect extracts

Fire ant	Red ant	Black ant	Honey bee
Butterfly	Caddis fly	Citrus mealy bugs	Wasps
Cockroach	Fruit fly	House fly	Lady bug beet
Mites	Mosquitoes	Moth	Spiders

5. Food Extracts

Like dust allergy, it is also difficult to detect the food allergy. Precisely, it is necessary for a physician to diagnose out food allergy by eradicating possibility of gastro-intestinal tract disturbances. The various symptoms of food allergy are asthma, systemic anaphylaxis, rhinitis, GI disturbances, urticaria, headache, etc. Although infants and children are more prone to food allergy, owing to underdeveloped GI towards certain foods, it is believed that, normally, digestion or processing of food items causes allergy. This type of allergy is difficult to treat and hence, the way out is to eliminate such food items from diet, which is also not much feasible. The various types of foods causing allergy are listed below.

Table 19.7: Food materials causing allergy

Apricot	Banana	Berries	Lactalbumin	Cow milk
Buckwheat	Arrowroot	Chocolate	Gelatin	Cucumbers
Safflower seed	Yeast	Fig	Chicken	Honey
Lentil	Egg white,	Shellfish	Papaya	Beef liver
Radish	Turnip	Spinach		

6. Miscellaneous Inhalant Extracts

Besides various natural allergens listed, a number of miscellaneous substances are found to be responsible for allergy due to their inhalation.

Some of the commonly used substances in food, cosmetics, drugs, insecticides and other products and causing allergy are listed below.

- Various hairs like those from human, horse, camel, goat, dog, cat, rabbit fur and wool.
- Various bird feathers like those from chicken, pigeon, turkey and duck.
- Algae, gum acacia, cotton linters, fish glue, jute, lycopodium, pyrethrum, silk, tobacco in various forms.

Preparation of Allergenic Extracts

After properly identifying the allergen, various stages involved in the preparation of allergenic extracts are:

(a) grinding (b) defatting

(c) extraction and (d) standardization.

Strict aseptic procedures are followed during extraction. The process is carried out in cold-room, owing to thermolabile nature of many allergens sterilisation is done by filtration. After this, sterility testing is done both for aerobic and anaerobic micro-organisms.

The standards of allergenic extracts are expressed in the form like weight/volume, pollen units, protein-nitrogen units, total nitrogen units, etc.

Chapter 20

HARBALS USED IN AYURVEDIC PHARMACY

INTRODUCTION

Ayurveda is regarded as ancient science of life and is based on principle of "maintaining the health of a healthy person and relieving the patient from the diseased condition". Ayurveda states it in following way:

स्वस्थस्य स्वास्थ्यरक्षणं व्याधितस्य व्याधिपरिमोक्षः ।

Ayurveda is the science of health and healing practised by ancient Aryans which is based on Atharvaveda, one of the oldest scriptures of Hindus, about 3000 years old. The object of Ayurveda is to counteract the imbalance of three essential elements, *vaata, pitta* and *kapha* (air, bile and phlegm respectively), which constitute the *Tridosh* from which the body originates. It is the *Tridosh* which regularises the normal working of the human body.

An Ayurvedic medicine, as defined in the Drugs and Cosmetic Act 1940, includes all medicines intended for internal or external use, for or in the diagnosis, treatment, mitigation or prevention of disease or disorder in human beings or animals and manufactured exclusively in accordance with the formulae described in the authoritative books of Ayurvedic systems of medicine specified in the First Schedule of the Act.

Authoritative books of Ayurvedic system as mentioned in First Schedule of Drugs and Cosmetic Act 1940 are as below.

Arogya Kalpadruma, Arka Prakasha, Arya Bishak, Ashtanga Hirdaya, Ashtanga Samgraha, Ayurveda Kalpadruma, Ayurveda Prakasha, Ayurveda Samgraha, Bhaisajya Ratnavali, Bharati, Bhava Prakasha, Brihat Nighantu Ratnakara, Charaka Samhita, Chakra Dutta, Gada Nigraha, Kupi Pakva Rasayana, Nighantu Ratnakara, Rasa Chandanshu, Ras Raja Sundara, Rasaratna Samuchya, Rasatantra Sara Siddha Proyaga Samgraha, Rasa Trangini, Rasa Yoga Sagara, Rasa Yoga Ratnakara, Rasa Yoga Samgraha, Rasendra Sar Samgraha, Rasa Pradipka, Sahasrayoga, Sarvayoga Chikitsa Ratnam, Sharangadhara Samhita, Siddha Bhaishajya Manimala, Siddha Yoga Samgraha, Sushruta Samhita, Vaidhya Chintamani, Vaidyaka Shabda Sindhu, Vaidyaka Chikitsa Sar, Vasaya Jiwash, Vasaua Rajeeyam, Yoga Ratnakara, Yoga Tarangini, Yoga Chintamani, Kashyapa samhita, Bhela Samhita, Vishwantha Chikitsa, Virada Chikitsa, Ayurveda Chintamani, Abhinava Chintamani, Ayurveda Ratnakar, Yoga ratna samgraha, Ramamrsita, Dravyaguna nighantu, Rasamanjari and Bangasena.

According to hypothesis of ayurveda, the entire universe is composed of five basic elements called *panchmahabhuta* and include space (*akasha*), air (*vayu*) energy (*teja*), liquid (*jal*) and solid (*prithvi*). The variable combinations of these principles make *tridosha* in human body. These combinations are *vaata, pitta* and *kapha*. They have been interpreted in the following way.

Vaata is composed of space and air *(akasha* and *vayu*). It is concerned with the physical and mental activities which are activating or dynamic in nature. Vaata represents nervous system, or *nadichakra* as per ayurvedic terminology.

Pitta is composed of liquid and energy, and indicates biochemical / energy systems. It dominates nutritional and thermogenetic activities and also includes hormonal, digestive and metabolic systems.

Kapha is composed of solid and liquid and indicates various tissues and organs (colloidoplastic system). It is concerned with integration of structural elements of the body.

Ayurvedic therapeutics is based on five pharmacological principles (*panchsheel*) of the drug. These principles are *rasa, guna, virya, vipaka* and *prabhava*.

1. Rasa: It gives not only taste of drug (*dravya*), but may indicate properties and actions of drug. There are six types of *rasa* which have been given below.

Table 20.1: Types of Rasa

Type	Composition	Character
(i) *Madhur* (sweet)	Solid + liquid (*prithvi + jal*)	*pitta* increasing, *kapha* decreasing
(ii) *Amla* (sour)	Solid + energy (*prithvi + tej*)	*pitta* increasing
(iii) *Lavan* (saline)	liquid + energy (*jal + tej*)	*Kapha* and *pitta* increasing
(iv) *Katu* (bitter)	air + energy (*vayu + tej*)	*vaata* increasing
(v) *Tikta* (pungent)	air + space (*vayu + akasha*)	*vaata, pitta* increasing
(vi) *Kasaya* (astringent)	air + solid (*vayu + prithvi*)	*vaata* increasing

2. Guna: They are certain physical attributes of drug, which affect the *tridosha*. Effect of *guna* of drug supersedes its *rasa*. There are, in total, ten pairs of *guna*, each pair includes activity reciprocal to each other.

(i) *Guru* (heavy) ⟶ *Laghu* (light)
(ii) *Sukshma* (subtle) ⟶ *Sthula* (gross)
(iii) *Sthira* (stable) ⟶ *Sara* (unstable)
(iv) *Snigdha* (unctuous) ⟶ *Rooksha* (ununctuous)
(v) *Sandra* (dense) ⟶ *dense* (liquid)
(vi) *Sheeta* (cold) ⟶ *ushna* (hot)
(vii) *Manda* (dull) ⟶ *tikshna* (sharp)
(viii) *Mridu* (soft) ⟶ *Kathina* (hard)
(ix) *Visada* (nonslime) ⟶ *piccila* (slime)
(x) *Slaksan* (smooth) ⟶ *Khara* (rough)

3. Vipaka: The end product of all digestive transformations of drug is *vipaka*. It is also called *nisthapaka* and differs from *awasthapaka* (stages of digestion) or prapka (initial transformation). Types of *vipaka* are madhura, *amla* and *katu*. These types affect *kapha*, pitta and *vaata* respectively.

4. Virya: It indicates potency of drug and shows two intrinsic properties, including *sita virya* and *ushna virya*. *Virya* influences the balance of *tridosha*. *Sita virya* indicates *Kapha varag* and *ushna virya* indicates *pitta varag,* and *vaata* remains buffer (*Yoga-vaha*). Among all the *guna*, only eight of them reach to the state of *virya* and include *guru, laghu, sita, ushna snigdha, ruksha, mridu* and *tiksna*.

5. Prabhava: This pharmacological principle indicates specific power of a drug. *Rasa, guna, virya* and *vipaka* of various drugs may be similar, but differ in their *prabhava*. This is due to specific chemical composition *(bhautika ansha)* of a drug. *Prabhava* is characterised by chemical composition and site of action of a drug. Thus, *prabhava* can be compared with pharmacotherapeutic actions of a drug.

Ayurvedic Dosage Forms

Different solvents (menstrua) used in preparations of ayurvedic formulations are water, oils, milk, ghee, cow's urine, etc. The use of sweetening agents, binding agents, colourants, flavouring principles and other adjuvents is also very common in ayurvedic preparations. With an objective of obtaining maximum therapeutic benefit and making the formulation palatable, different dosage forms are prescribed in ayurveda. The important ones are as follows dosage forms:

1. Arka

It is the liquid preparation obtained by distillation of certain crude drugs soaked in water using the distillation unit (Arkayantra).

The coarsely powdered crude drug is soaked in adequate quantity of water and kept overnight, which softens the drug and releases the volatile principles during distillation.

Arka is a suspension of the distillate in water with slight turbidity and colour, depending upon the nature of crude drug undergoing the distillation. It has characteristic aromatic odour.

Examples: Ajmodarka, Karpurady arka, Jatamansy arka.

2. Asava and Arishta

Asavas and Aristas are the medicinal preparations processed by soaking the drugs in the powdered forms or in the form of their decoction (known as *'Kasaya* in ayurveda), in a solution of sugar or jaggery (Gur), for a specified period of time. During soaking, it undergoes fermentation generating alcohol and in process facilitating extraction of active constituents contained in the drugs. Alcohol so generated also serves as a preservative in the product.

Examples: Kumariasava, Madhukasava, Punarnavasava, Arvindasava, Chandanasava, Kanakasava, Lohasava, Kutarista, Draksharista, Dashmularista, Vidangarista, Ashokarista, Khadirarista.

Absolute cleanliness is maintained during the preparation of Arishta and Asava. The wooden pots (vessels) are fumigated with pippali churna and also smeared with ghee before the fermentation liquids are poured into them.

In the processing of arishta, the crude drugs mentioned in the formula (*Patha*), are coarsely powdered and decoction (*Kashaya*) is prepared, filtered and transferred to wooden vats. Sugar, honey or jaggery is added to it, dissolved and boiled. Dravyas, other finely powdered ingredients and Dhataki pushpa (*Woodfordia fruticosa*) are added to it. The vessel is closed with lid and edges are sealed with clay smeared cloth wrapped in seven consecutive layers. Normally the vessels used for processing of *arishta* and *asava* are placed in cellar (basement) for a specific period in order to facilitate *Sandhana* process (fermentation). The contents are examined for completion of *Sandhana* process and the *arishta* is filtered and bottled.

In the preparation of *asava*, required quantity of water and jaggery or sugar is taken, boiled, cooled and transferred to fermentation vessel or barrel. Finely powdered crude drugs and other ingredients as mentioned in the formula (*Patha*) are then added to it. The container is covered with the lid and edges are sealed with clay smeared cloth, wrapped in seven consecutive layers. Further processing is identical to that of *arishta*.

Filtered asava or aristha should be clear and without any froth (foam) at the top. It should not become sour on standing. It has characteristic, aromatic and alcoholic odour.

3. Avaleha or Leha and Paka

Aveleha or Leha is a semisolid preparation of drugs prepared by addition of sugar, jaggery (gur) or sugar candy and boiled with prescribed drug-juice or decoction.

Jaggery/gur or sugar candy is dissolved in liquid, boiled and strained. The powdered drugs in small quantities are added and stirred continuously to form the homogeneous mass. Ghee or oil is added while preparation is hot.

Examples: Kutajavaleha, Draksavaleha, Vasavaleha, Bilvadileha, and Surnava leha.

4. Bhasma

The powdered form of the substance, obtained by calcination of metals, minerals or animal products by a special process in closed crucibles or in pits covered with cow dung cakes (puta), is known as Bhasma.

The method of preparation of Bhasma consists of two steps, the first being *sodhana* and the second *marana*. Both the processess are time consuming and need special attention all the while.

Bhasmas are grey, whitish, yellowish or black coloured powders. They are stored in glass containers. Bhasmas are quite stable products and maintain their potency for a long time.

Examples: Tamra-bhasma, Godanti bhasma, Pravala bhasma, Mukta bhasma, Lauha bhasma, suvarna bhasma and shankha bhasma.

5. Churna (Curna)

Fine powder of drug or drugs is known as churna. Drugs mentioned in *patha*, are cleaned properly, dried thoroughly, pulverised and then sieved. The churna is free flowing and retains potency for one year, if preserved in air-tight cotainers.

Examples: Triphala curna, Trikatu curna, Drakshadi curna and Sudarshan curna.

6. Dravaka

The liquid preparations obtained from *lavanas* or *ksharas* are known as *Dravakas.*

These are prepared by distillation process with or without additions of any fluids. These should be stored in glass bottles and are **quite stable products.**

Example: Sankha-dravaka

7. Fant (fanta)

These are prepared by pouring hot water over the dry crude-drugs. Normally crude drugs are leaves but roots and fruits may also be used at times. The pulverised crude-drugs are placed in the clean and dry metal vessels and the hot water (just to boil) is poured and the lid is placed. The drug is strained through fine muslin cloth and the filtrate is used as fanta.

8. Ghana-saar

The decoction obtained in preparation of Kadha or Nikadha is further evaporated to semi-solid mass is known as Ghana saar. (Ghan means solid).

9. Ghrita (Snehkalpa)

Ghritas are the preparations in which ghee is boiled with the prescribed quantity of the decoction (Kasaya) and fine paste (Kalka) of the drug as specified in the formula.

The process of preparation of ghrita ensures the absorption of the therapeutically active constituents of the drugs used for the preparation.

Ghrita solidifies when cooled. It has colour, odour and taste of the ingredients used in the preparation. Ghritas are the preparations for internal consumption and are stable for about sixteen months. Normally they are taken alongwith warm vehicle (water or milk).

Examples: Asoka ghrita, Nirgundi ghrita, Brahmi ghrita, Sukumara ghrita and Pippalyadi ghrita.

10. Kalka

When fresh or dry crude-drug is pounded with water hot or cold to give a very fine paste then it is termed Kalka.

11. Kajjali

An amalgum used in Ayurvedic system of medicine as a vehicle for other medicines is prepared by mixing equal quantity of mercury and sulphur. Before the sulphur and mercury are mixed together every care is taken to see that both the ingredients are of highest purity. Stone mortar and pestal are used for uniform rubbing and converting them into very fine blackish grey powder.

12. Ksharas

Alkaline (basic) substances obtained from the ash of drugs are known as ksaras. The crude drug is cut into pieces and burnt to ash. The ash is dissolved in water, strained and again evaporated to remove water while salty solid obtained is known as *kshar*.

Kshar becomes moist, if exposed to atmospheric conditions. It is quite stable for years together. The taste of kshar is salty.

Examples: Apamarga ksara, Palasa ksara, Vasa ksara, and Yava ksara.

13. Kvatha (Decoction)

This is prepared by boiling the powdered crude drugs in water. According to the hardness of crude drug, either four, eight or sixteen times quantity of water is added to the drug and boiled to one fourth volume. The decoction is filtered and used. Sometimes butter, honey, sugar, jaggary or oils are added before intake.

Examples: Mahamanjishtadi Kvatha; Sariwadi Kvatha.

14. Kupipakva Rasayana

The minerals and drugs of metallic origin in the powdered form are mixed together and placed in a glass flask (Kachkupi i.e. glass vessel), occupying about 1/3rd of volume. The glass flask is closed with clay smeared pieces of cloth around the bottle in seven consecutive layers. The flask is dried and burried in sand (Valuka yantra) upto the neck. The flask in valukas yantra is then heated gradually in three stages i.e. Mridu-agni, Mandagni and Tiksnagni for a specified period of time as mentioned in process. Then a red hot iron rod about 5 cm in diameter is inserted into glass flask (*kach kupi*) through the opening and stirred properly, so that opening of the flask is not choked due to the coating formed by sublimed sulphur.

When the process is over, the glass-flask is cooled and broken to collect the contents carefully without the contamination of any glass pieces therein. Kupipakwa-rasayana has different physical characteristics depending upon the contents of formula.

Generally these are yellow, red or dark coloured. Kupipakva-rasayana are quite stable-products and can be preserved idefinitely.

Examples: Rasa karpura, Rasa sindura, Makaradhvaja, Swarna sindura, Samir pannag rasa, and Vatagni kumar rasa.

15. Lepa

The preparation in the form of paste meant for external application on the body is known as lepa. The crude-drugs in the powdered form are mixed with cows urine, water, oil or ghee and made into paste.

Lepas prepared from vegetable powders are stable **only for one month**, while lepas containing minerals or metals last long.

Examples: Sinduradi lepa, Pathyadi lepa, Dasanga lepa and Dosaghna lepa.

16. Matras

Highly potent, **stick like** preparations are known as Matras. They are rubbed over the sand-stone in a vehicle like honey or milk. The rubbing on the sand stone is termed as turn. Turns decide the dose of the medicament. The outcome of turns is collected on the finger tips and is applied to the tongue wherefrom the drug gets absorbed and offers its therapeutic activity.

17. Nassyas

These are the nasal preparations and may be very fine powders, aqueous or oily preparations. They are either directly blown into the nostrils or applied by dropper if in the liquid-form.

18. Netrabindu and Anjana

Netrabindu is processed by dissolving the specified crude drugs in water or kasaya or honey and used as eye drops.

Anjanas are very fine powders of medicaments to be applied with netrasalaka. Their potency depends upon the drugs used in their preparation. Normally, they are used **within one year.**

Examples: Muktadi mahanjana, Chandroday vartti.

19. Nikadha

The residue or marc obtained in Kadha or Kashaya is treated with boiling water and strained. The filtrate thus obtained is known as Nikatha.

20. Parpati Kalpas

The mineral or herbal drugs are processed by special techniques of roasting and are converted into flattened scales or thin layers.

Parpati-Kalpas are claimed to have a **long shelf life**.

21. Pisti

Pisti is obtained by triturating the drug with the specified liquids and exposing the same to sun or moon light.

Pisties are as fine as bhasma, and retain their **potency indefinitely**. They are stored in bottles.

Examples: Praval pisti, Mukta pisti, Manikya pisti

22. Praash

They are semi-solid preparations like jams. These are highly palatable preparations used for refreshing or rejuvination of the body. i.e. **chyavanpraash**.

23. Rasa-Yoga

The medicinal preparations containing mineral drugs as their main ingredients, in the form of powder or pills are known as Rasa-yoga.

The physical properties of Rasa-yoga vary to a great extent, depending upon their contents. They are stored in glass bottles and can be **preserved indefinitely**.

Examples: Karpura-rasa, Icchabhedi-rasa, Tribhuvana kirti-rasa, Prabhakara-rasa and Laghu malini vasant-rasa.

24. Sattva

Water extractable solid substance obtained from a drug is known as sattva. Crude drug is cut into pieces, soaked in water, kept overnight and strained. It is allowed to settle. The supernatant liquid is decanted and *sattva* (extract) is dried and powdered.

Example: Guduchi (Gulvel) Sattva.

25. Siddha-milks

The drops obtained by burning medicinal substances like roots or fruits (marking nut) or wicks impregnated with medicinal extracts (sulphur preparation) are allowed to fall in the boiling milk and are thoroughly mixed by stirring. The drops getting mixed with milk is enriched with medicinal activity are further used for treatment. Siddha milk are especially useful for invalids as correctives or tonics.

26. Swarasa

When the Kalka is pressed in between the palms of the hand or is expressed through the layers of cloth to produce a thick juice then it is called as swarasa. (swa-rasa). Swarasa is generally prepared from crushed leaves or the roots of the herbs.

27. Taila (Medicated oils)

They are called sneha kalpa/paka and prepared by cooking oil with the juice or the decoction and paste of drugs. Unless otherwise specified, paste of drug should be one fourth part of the oil and the liquid (drava) should be four times of oil. If no liquid is specified in recipi, water should be used.

For preparing medicated oil, the fine paste of the drug, liquid and oil together, cooked, stirred constantly to the paste at the bottom and prevented from getting charred. When medicated taila gets property cooked, large amount of foam appears at the surface of the oil. Therefore the formulation should be strained prior to packing. If salt or any alkali preparation is added to the recipi, it should be after the oil is strained and mixed thoroughly.

Tailas can be used internally and topically. They retain potency for about **sixteen months**. They are taken internally with warm water or warm milk.

Examples: Bhringaraja taila, Maha Narayan taila, Laghu Visagarbha taila, Anu-taila, and Jyotismati taila.

28. Vatika and Gutika

Medicaments in the form of tablets or pills are known vatika and gutika respectively. These are made by using single or combinations of vegetable, mineral or animal drugs.

These preparations can be used upto **two years**. Pills with mineral ingredients can be used indefinitely.

These formulations should not loose their original colour, odour, taste, and form on standing. They should be kept away from moisture, if they contain salt, ksara or sugar.

Examples: Gandhaka vati, Sankha vati, Sanjivani vati and Pranda gutika, Lasunadi gutika, Marma gutika.

Ayurvedic profiles of selected herbal drugs are given in the following table.

Table 20.2: Ayurvedic Profiles of Herbal Drugs

Ayurvedic drug	Part used	Ayurvedic properties				Important use
		Rasa	Guna	Veerya	Vipak	
1. ADULSA *Adhatoda vasica* (Acanthaceae)	leaves	kasaya tikta	laghu rooksha	sheet	katu	expectorant
2. ALOE *Aloe-vera* *(Liliaceae)*	dried juice	katu	guru snigdha	sheet	madhur	laxatiae
3. AMLA *Emblica officinalis* (Euphorbiaceae)	fresh and dried fruits	All rasas	laghu rooksha	sheet	Madhur	vit. C deficiency
4. APAMARGA *Achyranthus aspera* (Amaranthaceae)	aerial parts of herb	tikta	laghu rooksha	ushna	katu	astringent, diuretic
5. ARJUNA *Terminalia arjuna* (Combretaceae)	bark	kasaya	laghu rooksha	sheet	katu	cardiotonic
6. ASHOKA *Saraca indica* (Leguminosae)	bark-seeds	kasaya tikta	laghu rooksha	sheet	katu	uterine disorders
7. ASHWAGANDHA *Withania somnifera* (Solanaceae)	roots, leaves	Madhur kasaya	laghu snigdha	ushna	madhur	general debility
8. BACH *Acorus calamus* (Araceae)	rhizomes	tikta	laghu rooksha	ushna	katu	carminative, vermifuge
9. BEHRA *Terminalia bellerica* (Combretaceae)	fruits	kasaya	laghu rooksha	ushna	madhur	astringent, styptic
10. BHALATAK *Semicarpus anacardium* (Anacardiaceae)	fruits, oil	Madhur kasaya	Laghu tikshna	ushna	madhur	anti-inflammatory
11. BHRINGA RAJ *Eclipta alba* (Compositae)	leaves and roots	katu tikta	laghu rooksha	ushna	katu	hair-growth tonic
12. BRAMHI *Bacopa moniera* (Scrophulariaceae)	aerial parts	Tikta	laghu snigdha	ushna	katu	brain tonic
13. CURCUMA *Curcuma-longa* *(Zinziberaceae)*	rhizomes	tikta katu	rooksha laghu	ushna	katu	expectorant

Contd...

Ayurvedic drug	Part used	Ayurvedic properties Fruit				Important use
		Rasa	Guna	Veerya	Vipak	
14. DATURA *Datura metel* *(solanaceae)*	Aerial parts	tikta katu	laghu rooksha	ushna	madhur	Antispasmodic
15. GINGER *Zingiber officinalis* (Zingiberaceae)	rhizomes	katu	laghu snigdha	ushna	madhur	carminative
16. GOKSHURA *Tribulus terrestris* (Zygophyllaceae)	fruits	Madhur	guru snigdha	sheet	madhur	diuretic urinary-disorders
17. GUGGUL *Commiphora mukul* (Burseraceae)	oleo-resin	katu kasaya	laghu snigdha	ushna	katu	anti-rheumatic
18. GULVEL *Tinospora cordifolia* (Menispermaceae)	stem, roots	tikta kasaya	guru snigdha	ushna	madhur	diuretic bitter tonic aphrodisiac
19. GYNNEMA *Gymnema sylvestre* (Asclepiadaceae)	roots, leaves	kasaya, katu	laghu rooksha	ushna	katu	oral antidiabetic
20. HARITAKI *Terminalia chebula* (Combretaceae)	fruits	kasaya	laghu rooksha	ushna	madhur	astringent
21. KANTKARI *Solanum xanthocarpum* (Solanaceae)	entire plant	tikta katu	laghu tikshna	ushna	katu	expectorant, febrifuge carminative
22. KALIJIRI *Centratherum anthelminticum* (Composite)	all parts	katu tikta	laghu tikshna	ushna	Madhura	anthelmintic, tonic, stomachic
23. LAHSUN *Allium sativum* (Liliaceae)	bulbs	madhur lavan kasaya	snigdha tikshna guru	ushna	katu	carminative
24. NAGARMOTHA *Cyperus rotundus* (Cyperaceae)	rhizomes	katu kasaya	laghu rooksha	sheet	katu	perfume, flavour
25. NEEM *Azadiracta indica* (Meliaceae)	bark, leaves and fruit	Tikta kasaya	laghu	sheet	katu	bitter tonic, insecticide
26. OPIUM-POPPY *Papaver somniferum* (Papaveraceae)	latex, capsules	sukshma rooksha	kasaya	ushna	katu	narcotic

contd...

Ayurvedic drug	Part used	Ayurvedic properties				Important use
		Rasa	Guna	Veerya	Vipak	
27. PALASH *Butea monosperma* (Papilionaceae)	dried seeds, fruits and gum	Katu tikta kasaya	laghu snigdha	ushna	Madhura	astringent for treatment of diaorrhoea and dysentery
28. PUNARNAVA *Boerhaavia diffusa* (Nyctaginaceae)	entire-herb	Madhur tikta kasaya	laghu rooksha	ushna	Madhura	diuretic
29. RASNA *Alpinia galanga* (Zingibaraceae)	entire herb	tikta	guru	ushna	madhur	antibacterial rhumatism carminative
30. SHANKHPUSHPI *Evolvulus alsinoides* (Euphorbiaceae)	entire herb	kasaya katu tikta	guru snigdha	sheet	madhur	nervine tonic
31. SHATAVARI *Asparagus racemosus* (Liliaceae)	leaves and tubers	Madhur tikta	guru snigdha	sheet	Madhur	for treatment of habitual abortion
32. TULSI *Ocimum sanctum* (Labiateae)	entire herb	katu tikta	laghu rooksha	ushna	katu	expectorant, antiseptic

ADDITIONAL DRUGS:

Ayurvedic drug	Part used	Ayurvedic properties				Important use
		Rasa	Guna	Veerya	Vipak	
1. BABCHI *Psoralia-corylifolia luguminosae*	Seeds Leaves	katu tikta	Laghu rooksha	ushna	Katu	In the treatment lencoderma
2. BLACK-PEPPER *Piper nigra* (Piperacaeum)	Half rite fruits	Katu	Laghu tikshna	ushna	Katu	Carminative condiment
3. PIPAL Ficus-religiosa (Moraceae)	Bark fruits leaves	Kasaya	Guru rooksha	Sheet	Katu	Stimulant
4. Gloriosa *Gloriosa superba Liliaceane*	Tubers	Katu tikta	Laghu Tikshna	Ushna	Katu	In Goat and Rhumatigm
5. Gudmar *Gymnema-sylvestra* (Apclzpilleace)	Root and Leaves	Kasaya Katu	Rooksha	Ushna	Katu	Anti-diabetic

contd...

Ayurvedic drug	Part used	Ayurvedic properties				Important use
		Rasa	Guna	Veerya	Vipak	
6. Kastur bhendi Hibiscus Abelmoschas	Flowers	Madhur	Laghu	Sheet	Katu	Colouring agent
7. ANANT MUL *Hemidesinus indica* Asclpiacacoae	Roots	Madhur / tikta	Snigha guru	Sheet	Madhur	Anti-inflammatory
8. MAHUA *Madhuea-longi-folia* Sapotaceae	All, except root	Snigdha	Madhur	Sheet	Madhu	In the treatment of Rhumatism
9. CHITRAK *Plumbago---- indica* Plumbaginaceae	Roots	Katu	Laghu, Rooksh	Ushna	Katu	In Rhumatism
10. PTEROCARPUS *Pterocarpus-marsupium* Leguminosae	Gum, Bark, Leaves	Kasaya	Laghu, Rooksh	Sheet	Katu	Anti-diabetic
11. SARPGANDHA *Rauwolfia serpentina* apocyanaceal	Roots	Tikta	Rooksh	Ushna	Katu	Anti-hypertensive
12. MANJISHTA *Rubia-cordifolia* *Rubiaceae*	Root	Kasaya Katu	Guru Rooksh	Ushna	Katu	Leuesdema treatment
13. SAUSSUREA *Saussura lappa* Compositae	Root and Rhizome	Tikta Katu	Rooksha	Ushna	Katu	Anti-asthematic
14. RINGANI *Solanum-indicum* SOLANAECE	Fruits	Katu	Laghu, Rooksh	Ushna	Katu	Mainly as Expectorant
15. CHIRATA *Swertia-chirata* Gentianaceae	Entire drug	Tikta	Laghu	Sheet	Katu	Stomachic antipyretic
16. LODHRA *Symplocos receunosa* Symplocaceae	Bark	Kasaya	Laghu	Sheet	Katu	Cooling and astringent
17. TAXUS *Taxus-baccata* Taxaceae	Leaves	Tikta	Laghu Tikshna	Ushna	Madhur	In anti-cancer treatment
18. VALERIAN *Velerian wallichi* Valeriarecae	Roots	Tikta	Laghu	Ushna	Katu	Anti-spasmodic and cough

❖❖❖

Chapter 21...

IMMUNOMODULATORS, RASAYANA AND ADAPTOGENS

Immunomodulatory medicinal plants is comparatively a recent concept in pharmacognosy. In many of the diseased conditions, immune response is impaired. Hence to maintain a disease free state, modulation of immune response by either its stimulation or suppression, can be a helpful therapy. Immunomodulators can provide supportive therapy to the chemotherapy. Although human immune system is quite able to maintain immunity or fight against various infections, still certain bacteria and viruses can affect it adversely. Adding to these infectious agents, other factors like unhealthy living conditions, chronic illness, environmental pollution and also the stress are harmful to immune response.

Immunosuppressants have already been employed in transplantation surgery. The non-specific immune stimulation or immunopotentiation can be caused with the use of some medicinal plants. Ayurvedic concepts of preventive health care have shown certain links with non-specific immunostimulation by such medicinal plants.

Ayurvedic system of medicine describes the concept called **'Rasayana'** under which **Rejuvenating drugs** have been described. These rejuvenating drugs have been indicated for various uses like delaying the ageing process, maintenance of youth, improvement of mental health and also removal of diseased conditions caused due to infectious organisms and also mental or physical stress. 'Rasayana' promote the quality of all body tissues. The precise meaning of rasayana is the therapeutics which promotes dhatus - tissues of body and increases strength and immunity against diseases. It is one of the eight clinical specialities from Ayurveda. The medicinal plants mentioned under rasayana promote the quality of body tissues through three ways, which include the followings :

(i) *Poshaka rasa* (ii) *Agni* (iii) Srotasas

Shatavari, milk, ghee cause the enrichment of nutritional quality of nutrient plasma i.e. *Poshaka rasa*. Plants like bhallataka and pippali promote the nutrition by improving the digestive processes and metabolic activities i.e. *agni*. Increasing the efficiency of micro-circulatory channels is caused by *Srotasas*. They cause better availablity of nutrients to the tissues. *Guggulu* is a typical example of srotasa.

Recently much work has been done on these related issues in western countries, due to which a new branch of medical science called **'psychoneuroendocrino-immunology'** has been developed. In this branch of medical science, the function of brain and peripheral system have been linked so as to explain the etiopathogenesis of various diseases. Based on a number of experiments, it has been made evident that immune system can be influenced by brain which then sends impulses to the brain by means of secreting hormones and neuropeptides. By this way, brain endocrine system and immune system have been linked together to fight against diseases.

Adaptogens are those natural substances which increase the non-specific resistance of organisms against different type of stressors. The drug induced state of resistance against aversive stimuli is termed as adaptogenic activity and such drugs are called adaptogens. They are the metabolic regulators which enhance the ability of the organism to adapt the environmental factors. The concept and practice of adaptogens have been accepted by medical scientists and health authorities in many developed countries.

The natural products described as immunomodulator, rasayana and adaptogenic agents have been mentioned below in Table 21.1.

Table 21.1: Immunomodulating agents, Rasayana and Adoptogens

	Natural products		
1. Immunomodulators	*Arnica montana* (Arnica)		
	Calendula officinalis (Calendula)		
	Matricaria recutita (Chamomile)		
	Echinacea purpurea (Echinacea)		
	Panax ginseng (Ginseng)		
	Serrenoa serrulata (Saw palmetto)		
	Tinospora cordifolia (Gulwel)		
	Withania somnifera (Ashwagandha)		
	Azadirachta indica (Neem)		
	Asparagus racemosus (Shatavari)		
	Allium sativum (Garlic)		
2. Rasayana	**(a) Acting on Saptadhatus (tissues)**		
		rasa	– Shatavari, bala
		rakta	– Lohadi rasayana
		mamsa	– Ashwagandha, Shatavari
		meda	– Shilajit
		asthi	– Praval
		majja	– Shankhapushpi, Brahmi
		shukra	– Ashwagandha
		all dhatus	– Amalaki
	(b) Srotasa (Microcirculatory channels)		
		rasavaha	– Gulwel
		raktavaha	– Suvarna makshika
		shukravaha	– Makaradhwaja
		pranavaha	– Pippali, abhraka
		mamsavaha	– Shatavari
	(c) Agni (digestive and metabolic activities)		
			– pippali, bhallataka

3. Adaptogens	*Withania somnifera* *Ocimum sanctum* *Sida cordifolia* *Panax ginseng* *Eleutherococcus senticosus* *Schizandra chinensis*

Many of these plants have been studied for such effects. Through these studies, some of their ingredients have been corelated to immunomodulator (more to be immunostimulant), rasayana or adaptogenic activities. Immunostimulant activities have been attributed to high molecular weight polysaccharide components and in some cases glycolipids and peptides. Low molecular weight compounds with high percentage of immunostimulant activities include macrocyclic lactones, quinones, alkaloids, alkylamides, sesquiterpenes, di-and triterpenoids. Some of the plants, extensively studied for immunostimulant activity are *Tinospora cordifolia, Withania somnifera, Asparagus racemosus,* etc.

The activation of mononuclear cells to release cytokines and interleukin-1 has been documented for *Tinospora cordifolia*. Syringin and cordiol, the active constituents of this plant show anticomplimentary and immunomodulatory activity. They significantly increase Ig G antibodies in serum. *T. cordifolia* shows dose- dependant stimulation of humoral and cell-mediated immunity. The role of *T. cordifolia* as a supportive (adjuant) therapy to chemotherapy for pulmonary tuberculosis has been justified.

Azadirachta indica leaf extract has been shown to cause higher Ig G and Ig M levels in ovalbumin immunized mice. Enhancement of macrophage migration inhibition and foot pad thickness have also been observed which indicate cell mediated response.

Roots of *Withania somnifera* have been found to contain sitoindoside VII and sitoindoside VIII for which antistress activity has been established. Sitoindoside IX and X have shown significant immunomodulatory and CNS effects (antistress, learning abilities and memory).

Echinacea purpurea and other species of *Echinacea* have been extensively studied for their immunomodulatory properties. Immunostimulant activity of Echinacea has been associated with its polysaccharide and polyacetylene fractions, in both *in-vivo* and *in-vitro* studies. In animal studies, *in-vivo* immunostimulant activity has been observed by the way of phagocytosis enhancement and by an increase in the serum elimination of carbon particles (Carbon clearanace test). Phagocytosis enhancement of TNF (Tumour necrosis factor) - secretion, stimulation in human macrophages and lymphocytes is indicative of non-specific T-cell activation. The reported immunostimulant property is considered to be partly mediated via the binding of polysaccharide fraction of Echinacea to carbohydrate receptors on the cell surface of T-cell lymphocytes, resulting in non-specific T-cell lymphocytes resulting in non-specific T-cell activation.

Saw palmetto causes the immunostimulant activity due to its polysaccharide viz. glucuronic acid. For immunostimulant activity of calendula (using granulocyte and carbon clearance test), high molecular weight polysaccharide fractions have been found to be responsible.

Chapter 22

NATURAL EXCIPIENTS (Pharmaceutical Aids)

The substances which are of little or no therapeutic value but are essentially used in formulations or compounding of various pharmaceutical dosage forms are known as **Excipients** or **Pharmaceutical aids**.

Excipients (adjuvants or Pharmaceutical necessities) are obtained from synthetic and various natural sources, they are used in the preparations of dosage forms.

For their adequate study it is essential to arrange them in scientific and systematic manner. Their large number, varied occurance, and several uses make them difficult to put them in single uniform pattern.

CLASSIFICATION OF NATURAL EXCIPIENTS

To know them thoroughly they can be arranged in one of the following classes according:

1. To the source of their natural origin i.e. Animal, vegetable and mineral

   Table 22.1 given below illustrates suitably.

Table 22.1: Natural Excipients : Sources of Origin

Animal	Vegetable	Mineral
Beeswax	Agar	Asbestos
Cantheride	Cardamom	Bentonite
Cochineal	Guargum	Calamine
Gelatin	Kokum butter	Chalk
Honey	Orange oil	Fueller's earth
Lactose	Pectin	Kaolin
Lanolin	Peppermint	Kieselguhr
Lard	Saffron	Paraffins
Musk	Starch	Shilajit
Shellac	Turmeric	Talc
Spermaceti	Vanilla	
Suet		

Many other excipients used very frequently in pharmaceutical industries such as saccharin, sorbitol, glycols, boric acid, parabenes CMC are not covered in this chapter, since they are obtained totally from the synthetic sources.

2. According to the function they perform in formulation of dosage forms, they can be classified as given in Table 22.2.

Table 22.2: Natural Excipients According to Role they Perform

Sr. No.	Type of Natural excipient		Examples
1.	Acidulents	:	Tamarind, lemon juice
2.	Adsorbents	:	Bentonite, kaolin, pectin
3.	Anti-oxidants	:	Citric acid, tocopherol, lecithin, vit. C
4.	Anti-adherants	:	Talc, corn-starch, magnesium-stearate
5.	Anti-bacterials	:	Benzoin, neem, curcuma, myrrh.
6.	Astringents	:	Catechu, tannic acid, bahera
7.	Binding agents	:	Acacia, gelatin, tragacanth, starch
8.	Colourants	:	Chlorophyll, caramel, amaranth, cochineal
9.	Desiccants	:	Silica, fused calcium chloride
10.	Diluents	:	Lactose, starch, mannitol, sucrose
11.	Disintegrating agents	:	Starch, psyllum husk, CMC
12.	Emollients	:	Glycerin, glycol, olive oil
13.	Emulsifying agents	:	Acacia, agar, guar gum, methyl cellulose
14.	Filter aids	:	Talc, bentonite, kieselguhr
15.	Flavours	:	Cardamom, vanilla, lemon oil, orange oil
16.	Glidants	:	Corn starch, talc, aerosil
17.	Granulating agents	:	Acacia, guar-gum
18.	Hardening agents	:	Bees wax, hard paraffin, castor oil hydrogenated, suet
19.	Humectants	:	Glycerin, polyethylene glycol
20.	Lubricants	:	Cocoa butter, stearic acid
21.	Ointment bases	:	Lanolin, bees wax, petroleum jelly
22.	Solvents	:	Glycerin, mineral oil, peanut oil
23.	Surgical fibres	:	Absorbent cotton, silk, wool
24.	Suspending agents	:	Alginic acid, gelatin, guar-gum
25.	Suppository bases	:	Cocoa butter, glycero gelatin
26.	Sweetening agents	:	Glycyrrhiza, honey stevia
27.	Thickening agents	:	Pectin, tragacanth, methyl cellulose
28.	Tonicity agents	:	Dextrose, sodium chloride

3. Excipients can also be classified according to their chemical nature such as Carbohydrates, fixed oil, volatile oils, resins, proteins etc. Table 22.3 given below elaborates the same further.

Table 22.3: Natural excipients and their Chemical nature

Sr. No.	Chemical nature	Examples
1.	Carbohydrates	Acacia, Agar, Cellulose, Dextrin, Guar gum, Carrageenan, Pectin, Honey, Tragacanth.
2.	Fixed oils	Almond oil, Arachis oil, Castor oil, Sesame oil, Corn oil, Olive oil, Soyabean oil, Cod liver oil.
3.	Fats	Cocoa butter, Kokam butter, Palm oil, Coconut oil, Lard, Suet.
4.	Waxes	Bees wax, Wool fat, Jojoba oil, Carnauba wax, Spermaceti.
5.	Volatile oils	Cardamom oil, Clove oil, Caraway oil, Peppermint oil, Eucalyptus oil, Sandalwood oil, Lemon oil, Orange oil, Lavender oil, Vetiver oil, Cinnamon oil.
6.	Resins	Balsam tolu, Balsam peru, Shellac, Asafoetida, Benzoin, Myrrh, Guggul, Storax, Sallaki, Podophyllum, Ginger oleo resin, Capsicum oleo resin.
7.	Proteins	Gelatin, Casein, Thaumatin
8.	Enzymes	Papain, Diastase, Pepsin, Pancreatin, Renin.
9.	Tannins	Amla, Catechu, Bahera, Myrobalan, Tannic acid.

4. **Alphabetical classification:**

English names of excipients are put into alphabetical order for this purpose of classification such as :

Acacia, Aloe
Bael, Bees wax
Caraway, Cardamom
Dextrin, Diastase
Gelatin, Guar gum
Olive oil, Orange
Pectin, Podophyllum
Tragacanth, Turmeric
Sandalwood, Starch
Vanilla, and vasaka etc.

Chapter 23...

HERBAL DRUGS FOR HIV AND AIDS

Study of natural drugs capable of inhibiting the growth of Human Immuno-deficiency virus (HIV) and Acquired Immune Deficiency Syndrome is necessary due to its wide spread infection throughout the World.

India has more than 30000 plant-species and most of them have medicinal properties. Even as on today more than 50% marketed drug products are obtained from natural sources only.

In the light of side-effects and toxicity of synthetic chemical drugs the use of natural drugs is preferred or is advisable. Over the last two-decades notable progress has been made in research of natural drugs for the treatment of AIDS.

There is **no cure** for AIDS/HIV but many drugs are available to **control** the HIV/AIDS virus. Such treatment is known as **antiretroviral therapy** (ART).

HIV can be diagnosed by testing blood for antibodies to the virus, but it is time consuming and takes about 12 weeks, and hence not practicable for immediate use.

CD4 T Cell Count Test for Diagnosis of HIV/AIDS:

These cells (CD4-T) of white blood corpuscles are targeted and destroyed by HIV. Even if one has no symptoms of HIV, HIV infection progresses to aids when CD4 T cells go below 200 in numbers.

HIV and AIDS are the abbreviations of Human-Immuno-Deficiency Virus and Acquired Immune Deficiency Syndrome, respectively, as mentioned above.

Pandemic of HIV/AIDS is currently the most socio-economic challenge that is facing the world at large as it affects mostly the young and economically productive population.

Amongst the reputed herbal drugs which are used to reduce the severity of HIV are:

(1) *Sutherlandia frutescens*

(2) *Hypoxis hemerocallidea* (African potato)

(3) *Hypericum perforata*

(1) *Sutherlandia*

Biological Source

Sutherlandia consists of *Sutherlandia- frutescens* (synonym: Lessertia frutescens) shrubs of leguminosae family.

Geographical Source

Found in and is indigenous to South Africa i.e. Cape floristic region and Fynbos biome.

Description

It is medium sized shrub with fine grayish green leaves and red butterfly shaped flowers. The pods of the plant are quite large and balloon like with slightly reddish tint. It is mostly harvested wild and is also grown in community gardens and commercially on a small number by farmers. This plant has more than 25 recorded names.

It is a shrub of bitter taste and aromatic smell, red-orange flowers occur in spring to mid-summer.

There two herbal remedies are currently recommended by South African ministry of heath for HIV management.

(2) *Hypoxis hemerocallidea* (African Potato)

Synonym

African star grass.

Biological Source

This consists of the herb *Hypoxis hemerocallidea* belonging to family Hypoxidaceae.

Geographical Source

It is native to Southern Africa i.e. Mozambique and Zimbabwe.

Description

Hypoxis improves some symptoms secondary to prostatic hyperplasia, it also alters the activity of cytochrome P450, which may interfere with the activeness of other drugs or antiretroviral drugs.

(3) *Hypericum perforata*

Hypericum consists of dried aerial parts of the plant *Hypericum perforatum*, fam. Hypericaceae. It is perennial herb and about one metre in height. It is collected during the flowering season or shortly before flowering.

It is indigenous to England, found in Europe, Australia and India. It is found distributed in Western Himalaya upto an altitude of 1000 – 3000 metres.

The leaves of plant are green marked with black-sports. It is astringent and bitter in taste and has yellow coloured corolla (Fig. 9.6 on page 9.19).

❖❖❖

Chapter 24...

BIOLOGICALS

Although, biologicals indicate the products or preparations thereof obtained from living organisms which may include plants and animals, now-a-days its meaning is restricted to those biological products used in immunological treatment of human beings either for prophylatic or therapeutic aspects. Accordingly, the biologicals cover the valuable preparations exerting immunity like, vaccines prepared from bacteria, viruses or rickettsae; serums; toxins; antitoxins; toxoids, human blood and its products.

The present chapter mainly deals with description of individual immunological products. The description is presented with a presumption that the reader has a general idea regarding the fundamental aspects of immunology like antigens, antibodies and their reactions, humoral and cellular immunity and other related aspects.

A general account of different types of immunities is given here, which is followed by brief idei., about different vaccines and sera and other allied preparations.

TYPES OF IMMUNITIES

It is grouped under two major headings, viz,

(1) Natural or innate immunity; and

(2) Acquired immunity.

These two groups are further divided as shown below

Immunity

- Natural
  - Individual
  - Species
  - Racial
- Acquired
  - Active
    - Natural
    - Artificial
  - Passive
    - Natural
    - Artificial

(1) NATURAL IMMUNITY

The resistance to certain diseases is endowed to certain individuals, species or races by way of nature only. It should be noted that it is a congenial one and not developed during life time. A variety of biological parameters are responsible for it and as such only a single particular factor cannot be stated.

Natural immunity is retained throughout the life time of an individual.

Bacteria like skin and mucous membrane healthy body tissues; proper production of white blood cells are useful for bacterial combat; type and quality of resident bacterial flora in oral cavity, upper respiratory tract, large intestine and other part's of body. This bacterial flora has a certain role in offering is resistance towards certain invading pathogenic bacteria. The other factors that also play a significant role are gastric juice and intestinal enzymes. The former kills many types of toxic bacteria and viruses : while the latter have a role in secondary defence mechanism.

(b) Racial Immunity

Some of the races of human beings are altogether resistant to some diseases like malaria, tuberculosis, etc. The reasons for racial immunity, in general, are genetic set up, nutrition, fatigue, etc. The races with sickle cell characters have a natural resistance to malaria.

(c) Species Immunity

Owing to the specific physiological set up or in some cases natural presence of antibodies, 'human beings or animals are resistant to certain diseases. Immunity to cattle plague and hog cholera is naturally shown by human beings. The physiological reasons attributed are resistance offered by epithelial or other tissues and some other reasons, like presence of natural antibodies without the external stimulus.

(2) ACQUIRED IMMUNITY

The immunity is acquired either by introduction of specific antigens or supply of specific intibodies. By introducing the antigens of pathogenic bacteria or viruses, the human body raises the antibodies against it and acquires the active immunity. The antigens, not in all the cases, can raise the antibodies. Hence, in such cases, the antibodies raised in other individuals or animals for a particular disease are supplied by which the person attains a passive acquired immunity.

Active acquired immunity is further differentiated into naturally or artificially acquired active immunity.

(a) Naturally Acquired Active Immunity

In some individuals, even by exposure to virulent antigens, the diseased condition or' clinical infection state is not exhibited, due to increased host resistance or decreased disease causing capacity to those antigens. In such cases, the body produces *sufficient titre of antibodies* so as to acquire the ability to resist further possible affections by the same micro-organisms.

(b) Artificially Acquired Active Immunity

Vaccines and toxoids are used for this purpose. Such preparations contain the live or dead or attinuated bacteria, which when injected, raise the antibodies against those antigens from the bacteria as the dose of the antigens is also less, the diseased condition does not occur, but body is brought to a condition of resisting the future clinical infection.

(c) Passive Acquired Immunity

Human body is not responsive to all types of bacterial or viral antigens, and hence cannot raise the antibodies against the, antigens of certain diseases. In such cases, the antibodies are raised in other animals which are responsive to such diseases and injected into human body. By this way, a passive immunity is gained.

The naturally acquired passive immunity is gained by transmission of immunoglobulin G towards foetus from mother. Due to this, the young born baby may show an immunity towards some diseases like measles, mumps, diphtheria, tetanus, etc. for first six months.

In case of artificial acquired passive immunity, the preparations which are injected, offer immunity for a limited period and are used only as an immediate treatment of infections or toxicities. Such preparations are sera and antitoxins both from animals, immune globulins and hyperimmune sera from human plasma, antivenins for snake bite and spider bite.

TYPES OF BIOLOGICALS

The preparations used in active or passive acquired immunity are briefed below.

1. Vaccine

It is used for active immunity. It is a suspension of killed or attenuated bacteria and is given to induce the immunity and hence used as a prophylactic for several diseases.

2. Toxoid

It is also used for active immunity. It contains the detoxified exotoxins of bacteria, but it possesses the capacity to stimulate the formation of antitoxin and hence is used to prevent bacterial toxicities.

3. Animal Immune Serum

It is also termed as heterologous serum. It includes the following preparations.

(a) Antitoxin: It contains antibodies obtained from animal serum, like horse serum which is immunised with toxoids like tetanus or diphtheria. Antitoxins are either used for treatment or to achieve passive acquired immunity.

(b) Antiviral serum: It also contains the antibodies raised in animal, generally horse, by immunising with a specific viral vaccine. **Antirabies serum** is an example of antiviral serum.

(c) Antivenin: The venoms of animals like coral snake, rattle snake or toxic spiders like black widow spider are injected into animals, against which the antibodies are raise.

The separated serum with antibodies is then called antivenin and used against snake bite 'or spider bite to detoxify their venoms.

4. Human Immune Sera

It is also called as homologous sera.

Immune globulin : It contains antibodies from plasma of not less than 1000 health individuals. Immune globulin is used for regular maintenance of immuno-deficient persons. It is also used in measles and hepatitis A to gain passive immunity,

(b) Hyper immune serum: It is also called as **'specific immune globulin'** and is obtained from those individuals. who show high antibody raising capacity for a certain disease. The examples are Hepatitis E immune globulin, Rabies immune globulin, etc.

It is essential to maintain the standards of biologicals by taking certain precautions. The safety purity and potency of such preparations are important. The sterility and potency immunological sterility at –10°C. In other cases, the freezing should be avoided. The biologics have a limited shelf I& se of loss of potency on storage. The liquid small pox vaccine retains the potency for 90 days only while dry form of small pox vaccine can retain potency upto 18 months. Diphtheria antitoxin with excess potency retains its efficacy upto 5 years. In India the manufacture and use of all pathological preparations is governed by schedules of Drugs and Cosmetics Act. 1940.

(A) BACTERIAL VACCINES

1. ANTHRAX VACCINE ADSORBED

Biological Source

It is sterile opaque filtrate obtained during the anaerobic growth of virulent, non-encapsulated of *Bacillus anthracis*.

Preparation

The sterile filtrate obtained is adsorbed on sterile aluminium hydroxide. It is further subjected to genea-pig protection test for determination ethonium of potency. For the purpose of preservative, Benzalconium chloride (0.0025%) is used.

It is used for active immunisation against anthrax. It is also given to those persons working as veterinary health personnel, laboratory workers, those involved with works connected with hides, animal hairs, bone meal and wool.

Total 4 doses, each of 0.5 ml., given intramuscularly. The first 3 doses are separated from each other by an interval of 3 weeks. The 4th dose is given after 6 months. This is followed by 0.5 ml oeintorcing dose, per year,

2. BACILLUS CALMETTE-GUERINE VACCINE (BCG VACCINE/TUBERCULOSIS-VACCINE)

Biological Source

It is a suspension of living lyophilised cells of authentic *Bacillus Calmette - Guerin* strain of *Mycobacterium turculosis* var. bovis.

The vaccine is available as a dried product. During administration, it is reconstituted by the addition is a suitable sterile diluent, and the quantity not used is discarded.

Uses

It is used to induce active acquired immunity in case of epidemic. It is also used for those, who have negative turberculin skin tests and, repeated exposure to persistently untreated sputum positive uses of pulmonary tuberculosis.

Dose

Prophylactic, 0.1 ml intracutaneously as a single dose.

For infants upto 28 days of age, 0.05 ml.

Storage: 2-10°C

Expiry: After one year of manufacture

3. CHOLERA VACCINE

Biological Source

It is a sterile whitish suspension of a suitably killed strain of Cholera vibrio, Cholera vaccine contains equal parts of vaccines prepared from smooth strains of Inaba and Ogawa, which are 2 main serological types, along with type 0 antigen.

It contains not less than 8000 million Vibrio cholerae per dose.

Uses

It has a limited use because it gives active. acquired immunity only in 60% 'of persons and the immunity lasts for maximum 6 months.

Dose

About 0.5 ml. (4000 million organisms) by..Intramuscularly or subeutaneously, followed by 1 ml after 1-4 Weeks.

Storage: At 2 -10°C and protected from light and freezing.

Expiry: 18 months from the date of manufacture.

4. TYPHOID VACCINE

Biological Source

It is a sterile milky suspension of killed *Salmonella typhi*, containing not less than 500 million and not more than 1000 million bacteria in 1 ml. he strain should have full complement of H, 0 and Vi antigens.

Uses

It builds up acquired active immunity in 70% persons to protect from, typhoid fever. The immunity persists for 2 years only.

Dose

Primary immunisation of adult: 2 doses of 1 ml each by intramuscular or subcutaneous route with 4-6 weeks interval amongst them.

Reinforcing dose: Every 3 years, or yearly, if the epidemic is high.

Storage: Between 2-10°C and protected from light and freezing.

Expiry: 2 years from date of manufacture.

5. TYPHOID - PARATYPHOID A AND B VACCINE (TAB VACCINE)

Biological Source

It is sterile mixed suspension of Salmonella typhi, S. paratyphi A, and S. paratyphi B. One ml of this vaccine contains 1000 million S. typhi, and 500-750 million each of S. paratyphi A and S. paratyphi B.

Uses

It is used for raising acquired active immunity against typhoid and paratyphoid fevers. It takes about 1 month to acquire immunity and lasts for 2 years. This vaccine is found to be effective in 70% individuals.

Same as typhoid vaccine

Storage: Same as typhoid vaccine

Expiry: 18 months from the date of manufacture

6. VACCINE (WHOOPING COUGH VACCINE)

Biological Source

It is a sterile suspension of a formalin killed strain or strains of Bordetella, pertussis, in saline or --- appropriate solution which is isotonic with blood.

Uses

It is used for active immunization against whooping cough in infants.

Adsorbed pertussis vaccine is the pertussis vaccine adsorbed on aluminium phosphate. It is not used in individuals above 7 years of age, as the adverse reactions are increased.

Dose

(1) Primary immunisation: 0.5 ml for 3 times each with 4-8 weeks interval

(2) Reinforcing dose: 0.5 ml. one year after

Storage: Between 2 to 10° C and protected from light and freezing.

Expiry: 2 years from date of manufacture

7. PLAGUE VACCINE

Biological Source

It is a sterile, whitish suspension of suitably killed *Yersinia pestis*.

Uses

It is used for active immunisation against plague, specifically for those working in bacteriological laboratories or in those places where very bad condition of sanitation exists.

Dose

Two injections of 1 ml and after 1-3 months, 0.2 ml. Another injection of 0.2 ml is necessary after 34 months. In those areas with plague epidermic, a booster dose of 0.1 to 0.2 ml every 6 months is given.

Storage : Between 2-10°C and protected from light and freezing

8. MENINGOCOCCAL POLYSACCHARIDE VACCINE

Biological Source

It is a sterile lyophilised extract prepared from high molecular weight polysaccharide capsular antigens derived from the outer layer of *Neisseria meningitides* serogroup A and serogroup C.

The vaccine is available for each serogroup (monovalent vaccine) and with both sefogroups (bivalent vaccine).

Uses

It is used for active immunisation against *Neisseria meningitidis* in areas with disease reaching epidemic levels. Mono-or bivalent vaccines are used as per the type of infection.

Dose

0.5 ml of reconstituted vaccine

Storage : It is stored as freeze dried material at low temperatures.

9. PNEUMOCOCCAL VACCINE POLYVALENT

Biological Source

It is a sterile liquid containing mixture of purified capsular polysaccharide antigens from 23 types of *Streptococcus pneumoniae*. The capsular antigens front bacterial type are purified separately and then pooled together by which each dose of vaccine contains 25 mg of polysacharide antigen.

Uses

It is given for active i1ninUni5,16011 against infections due to 23' prevalent types of pneumococci causing prieumococcal diseases. Those individuals are more prone to pneuniococcal infectibns, who are suffering with diabetes mellitus, alcoholism, functional impairment of hepatic, renal and cardiorespiratory systems.

These diseases are also common in aged people'or those with sickle cell diseases.

Dose

0.5 ml by intramuscular or subcutaneous route

(b) VIRAL VACCINES

Preparation

Viruses are intracellular obligate parasites. They are completely dependent on host cells for their survival. Unlike bacteria, they cannot be grown on non-living cultures and hence they are difficult culture. The known methods of growing the viruses are as follows.

(1) Human diploid cell cultures: This is done with tissue cultures of human cells. The vaccine prepared are Rabies vaccine and Rubella vaccine.

(2) Embryonic egg cultures: Measles, mumps, influenza and yellow fever vaccines are prepare by this technique.

(3) Monkey cell cultures are employed for preparation of polio vaccine

(4) Skin of living calves is used for small-pox vaccine preparation.

(5) Human plasma is used as a source of surface antigen to prepare hepatitis B Vaccine.

After growing the viruses by above methods, they are separated from host cells by employing various techniques like column filtration, centrifugation, disintegration, etc. The viral vaccines exist either as whole viral vaccines or subvirion. Because of their containing live viruses they offer a sustained and boosting stimulation to the individual and produce antibodies. This is due to that continuous multiplication of viruses in human body. But some of the viral vaccines like influenza, hepatitis B. rabies vaccine, etc. are inactivated forms of viruses.

The live virus vaccines are supplied as freeze-dried products and inactivated vaccines are available as suspensions. Both of them are given by parenteral route, while **living polio vaccine is given by oral route in liquid form only**.

1. POLIO VIRUS ORAL VACCINE (SABIN VACCINE)

Biological Source

It is an aqueous suspension of suitable live attenuated strains of *Poliomyelitis* virus, types gown in suitable cell cultures. This vaccine contains any one of these types viz. type I (Mahon type 2 (MEF 1) and type 3 (Saukett), or combination of them.

Uses

It is used orally for active immunisation against poliomyelitis. It has a bifold function of formation antibodies in the blood-and also offering a local resistance in the intestinal tissues to infection virulent poliomyelitis vaccine.

Dose

Oral, 2 doses with6 - 8 weeks interval for infants starting at 6 -12 weeks age, followed by a third dose 8 -12 months later.

Storage : It should be stored at temperature of – 20ºC.

Expiry : If stored below- 20ºC. It retains potency upto 2 years and if between 0 – 4ºC, retains potency upto 3 months.

2. POLIO VIRUS VACCINE (INACTIVATED)

Biological Source

It is an aqueous suspension of suitable strain of poliomyelitis virus, types 1, 2 and 3 grown cultures of monkey kidney tissue and inactivated with a suitable method.

Uses

It is used for raising active acquired immunity against póliomyelitis. (oral polio vaccine is normal y. preferred).

Dose

(1) Primary immunisation : 1 ml, 3. times subcutaneously at an interval of 4-8 weeks each time

(2) Reinforcing dose : 1 ml, 6 -12 months after third dose.

Storage : Between 2 - 10°C and protected from light and freezing.

Expiry : About 18 months from date of manufacture

3. INFLUENZA VIRUS VACCINE

Biological Source

It is an aqueous suspension of a suitable strain or strains of influenza virus types A and B either individually or mixed, which are growing in the allantoic cavity of fertile incubated chick embryos. The viruses are inactivated so that they are non-infective, but retain their antigenic properties.

Uses

It is used for active immunisation against epidemic influenza. About 70% of persons inoculatea, achieve immunity, which is developed within 2-3 weeks, but it lasts for 3-4 months only. It is given --- medical and nursing staff and for the patients suffering with chronic, renal disease, chronic cardiac or pulmonary insufficiency.

Dose

0.5 ml single dose intramuscularly or subcutaneously

Storage : Between 2° -10°C and protected from light and freezing.

Expiry : About 18 months from the date of manufacture

4. HEPATITIS B VACCINE

Biological Source

It is a sterile white suspension of surface antigen (Australia antigen) of hepatitis B virus. It is obtained from the plasma of human carriers of hepatitis B virus.

Uses

It is used to raise active immunisation against infections by different subtypes of hepatitis B virus, especially in persons with high risk of hepatitis B infections.

Such persons are generally users of illicit drugs, homosexually active males or those coming from high hepatitis B virus endemicity.

Dose

3 injections of 1 ml each intramuscularly,'the second and third injection is liven 1 and 6 months after first dose.

5. RABIES VACCINE

Biological Source

It is a sterile lyophilised preparation of a suitably killed Wistar rabies virus strain PM -150 3-3M, which has been grown on human diploid cell cultures.

Uses

It is used for prevention of rabies in the patients to whom the rabid animals or animals suspected of being rabid have biten. It is a so used to give active immunity to persons, whose work is involved with rabid animals or handling of rabies virus, like animal handlers, researchers, veterinarians, etc.

Dose

Pre exposure : 3 doses of 1 ml each with one week's interval intramuscularly

Post exposure 5 doses of 1 ml each on each of 0, 3, 7, 14, and 28 days alongwith Rabies Globulin on day 0.

Storage : Between 2° to 10°C and protected from light and freezing.

Expiry : 6 months from the date of manufacture

6. SMALL-PDX VACCINE

Biological Source

It is the lyophilized, preparation of a suitable strain of living virus of *vaccinia* prepared from materials obtained from the lesions produced by the inoculation of *vaccinia* viru7inthe calf lymph or on chorioallantoic membranes of chick embryo.

Uses

It is given by scarification to develop active immunity against small-pox virus. Due to scarification, characteristic eruption of vaccinia is produced at the site of inoculation and hence antibodies against small-pox and cowpox are produced. The immunity lasts for several years.

Dose

2 - 3 needle punctures of vaccine, percutaneously

Storage and Expiry : It should be stored below - 25°C, so that it retains potency upto one year. After taking out from this storage, It should be stored at 4°C, so that it retains potency upto 3 months.

Note : World Health Organization has declared in May 1980, that world is free of small-pox and hence this vaccine is no longer manufactured.

7. MEASLES VIRUS VACCINE (LIVE ATTENUATED)

Biological Source

It is a bacterially sterile lyophilised preparation of an attenuated live measle virus grown in cultures of chick embryo cells.

Uses

It is given to raise active immunity against measles, which lasts for 8 years. In about 90% of children inoculated, antibodies are produced. It is given 9 months after birth as the infant already contains antibodies against measles, received from mother's blood.

Dose

Single dose subcutaneously and each dose contains not less the 1000 TCID60 (tissue culture infection (loses).

Storage : Between 2° - 10°C

Expiry : One year from the date of manufacture

8. RUBELLA VIRUS VACCINE (LIVE ATTENUATED)

Biological Source

It is a bacterially sterile lyophilised preparation of suitable live attenuated strain of rubella virus grown in human diploid cell cultures.

Uses

This vaccine is used to raise active immunisation against German measles. It raises antibodies in about 90% of children. It is given to children between the age of 12 months and puberty.

Dose

0.5 -1 ml subcutaneously

Storage and Expiry : 2ºC -10°C, retains potency for 12 months.

8. YELLOW FEVER VACCINE

Biological Source

It is a bacterially sterile, light orange freeze-dried preparation of live 17 D strain of yellow fever virus grown in chick embryos.

Uses

This vaccine is used to raise active acquired immunity. against yellow fever virus. The immunity is attained within a few days and lasts for many years.

Dose

Single dose is 0.5 ml subcutaneously and booster dose after 10 years, if necessary. Each dose contains less than 7500 mouse LD50 units.

Storage : In dark at 0°C

Expiry : It retains potency upto one year from the date of manufacture.

9. MUMPS VIRUS VACCINE (LIVE - ATTENUATED)

Biological Source

It is a bacterially sterile freeze-dried preparation of live attenuated strain of mumps virus grown in cell cultures of chick embryos.

Uses

This vaccine is used to raise active acquired immunity against mumps virus, which lasts for 6 years. About 90% of individuals respond to it.

Dose

A single dose of not less than 5000 $TCID_{50}$ subcutaneously

Storage : Between 2ºC - 10ºC

Expiry : One year from the date of manufacture

10. MUMPS AND RUBELLA VACCINE (LIVE - ATTENUATED)

Biological Source

It is a bacterially sterile freeze-dried preparation of a suitable live attenuated strain of mumps virus and a suitable live attenuated strain of rubella virus grown in cultures of chick embryo cells.

Uses

This vaccine is used to raise concomitant active immunity against mumps and. rubella (German measles).

Dose

Single dose, 0.5 ml subcutaneously. Each dose contains 5000 TCID5c) of mumps virus vaccine live and not less than 1000 TCIID,o of rubella virus vaccine live.

Storage : Between 2°-10°C.

Expiry : 1 year from the date of manufacture

11. MEASLES AND RUBELLA VIRUS VACCINE (LIVE ATTENUATED)

Biological Source

It is a bacterially sterile freeze-dried preparation of a suitable live attenuated strain of measles virus grown in cultures of chick embryo cells and a suitable live attenuated strain of rubella virus grown in cultures of duck embryo cells.

Uses

It is used for simultaneous active immunisation against measles and rubella in children above 15 months.

Dose

Single dose is 0.5 nil subcutaneously. Each dose contains not less than 1000 TCID50 of measles virus vaccine live attenuatedand 1000 TCID50 of rubella virus vaccine live attenuated.

Storage : Between 2° to 10° and to be protected from light.

Expiry : It retains the potency upto 1 year from the date of manufacture.

12. MEASLES, MUMPS AND RUBELLA VIRUS VACINE (LIVE ATTENUATED)

Biological Source

It is a bacterially sterile freeze-dried preparation of suitable live attenuated strains of measle virus, mumps virus and rubella virus grown in cultures of chick embryo or duck embryo cells.

Uses

It is used to raise the simultaneous active immunity against measles, mumps and rubella in the children above 15 months of age.

Dose

Single dose 0.5 ml subcutaneously; each dose contains not less than 1000 TC[D.,,o of measles virus vaccine live attenuated, 5000 TCID5() of mumps virus vaccine live and 1000 $TCID_{50}$ of rubella virus vaccine live attenuated.

Storage : Between 2' to 10°C protected from light

Expiry : One year from the date of manufacture

(C) RICKETTSIAL VACCINES

Like viral vaccines, they are also to be propagated on 'living media' such as monkey kidney tissue cultures or chick embryos.

TYPHUS VACCINE

Biological Source

It is a sterile suspension of killed epidemic typhus rickettsiae (*Rickettsia prowazekii*) prepared in eggs, rodent lungs, or the peritoneal cavity of gerbils.

Uses

It is used for active immunisation against flea borne and louse borne typhus. It only reduces severity of disease and does not give total protection. It is given as preventive drug for those travelling in mountaineous areas of Asia.

Dose

2 doses of 1 ml each with an interval of 7 - 10 days

Storage : It is stored between 2° to 10°C and protected from, light and freezing.

Expiry : 1 year from the elate of manufacture

TOXOIDS

Toxoids are prepared from the exotoxins of bacteria. Bacterial ell produces toxins. Those toxins which are excreted in surrounding medium are called as **exotoxins**.

For the isolation of exotoxins, after growing and killing bacteria in a medium, they are separated by bacterial filters. They are specifically obtained from highly virulent organisms. These separated toxins are further treated with formaldehyde to mitigate their toxic preparation, but to retain the antigenic character. The preparation thus, obtained is called toxoid or detoxified toxin.

The toxoids are either plain preparations or added with an adjuvant like aluminium hydroxide, aluminium phosphate or alum.' These adjuvants are insoluble materials and serve the role of keeping toxoids in tissue for prolonged periods and hence effect persisting antibody stimulation.

The toxoid, ready for use may contain single, multiple or mixed antigens. The plain toxoid has a cloudy solution form, while absorbed toxoid is in liquid suspension.

1. DIPHTHERIA, TETANUS AND PERTUSSIS VACCINE (TRIPLE VACCINE)

Biological source

It is a sterile suspension of diphtheria formal toxoid, tetanus formal toxoid and a suspension of killed Bordetella pertussis.

Uses

It is used for simultaneous active immunization against diphtheria, tetanus and whooping cough. The antigenicity of diphtheria vaccine is enhanced by the pertussis vaccine. It is used for children from 2-3 months upto 6 years of age.

Dose

Primary immunisation : 3 doses of 0.5 ml each with an interval of 4-8 weeks intramuscularly.

Reinforcing dose of 0.5 ml. after one year

Booster dose 0.5 ml for the children between 4 to 6.years of age

Storage: Between 2° -10°C to be protected from light and freezing

2. TETANUS TOXOID ADSORBED

Biological Source

It is a sterile white, slightly grey or slightly pink Suspension of purified tetanus toxoid prepared from tetanus toxin produced by the growth of *Clostridium tetani*. The toxoid is alum precipitated or adsorbed onto aluminium hydroxide or aluminium phosphate.

Uses

It is used to raise active immunisation against tetanus- which lasts for 5 years. It is inoculated in infants of 3 - 6 months age, and in the case of injury to non-immunised persons as a primary immunisation. But, generally it is given in the form of triple vaccine.

Dose

Two doses of 0.5 ml. each with an interval of 4 - 8 weeks, third reinforcing dose of 0.5 ml. after 12 months. It is given by deep subcutaneous or intramuscular route.

Storage: Between 2° to 10°C and to be e protected from light and freezing.

3. DIPHTHERIA TOXOID ADSORBED

Biological Source

It is a sterile white, slightly grey or pink suspension of purified diphtheria toxoid prepared from the toxin of *Corynebacterium diphtheriae*, adsorbed onto aluminium hydroxide.

Uses

It is used to raise active immunity against diphtheria, but it is generally given with pertussis vaccine.

Dose

Two doses of 0.5 ml each with an interval of 6 - 8 weeks, intramuscularly and third reinforcing dose of 0.5 ml after one year

Storage: It is stored between 2° and 10°C and protected from light and freezing

4. DIPHTHERIA AND TETANUS TOXOID ADSORBED

Biological Source

It is a sterile, white or slightly grey or pink or cream coloured suspension of purified diphtheria and tetanus toxoids adsorbed onto aluminium hydroxide or aluminium phosphate: These mixed toxoids for adult use are also available as alum precipitated preparation.

Uses

It is available in 2 forms, viz. paediatric and adult uses. In both the cases, it is used for active immunisation against diphtheria and tetanus either in children below 6 years of age (paediatric form) or in children above 6 years and adults (adult form). The paediatric form is given when triple vaccine is contra-indicated due to presence of pertussis component.

Dose

Two doses of 0.5 ml each with an interval of 4 - 8 weeks, followed by reinforcing dose of 0.5 mi after one year, intramuscularly. Each paediatric dosage form contains 10 - 15 Lf units of diphtheria toxoid and 5 - 10 Lf units of tetanus toxoid.

Each adult dosage form (0.5 ml) contains not more than 2 Lf units of diphtheria toxoid and 5 - 10 Lf units of tetanus toxoid.

Storage: It is stored between 2°-10°C and protected from light and freezing.

BIOLOGICALS FOR PASSIVE IMMUNIZATION

These are used to impart passive immunity either for prophylactic or therapeutic purposes. They are obtained either from human serum (homologous sera) or from animals (heterologous sera). The former includes the immune globulins obtained from the pooled plasma, free of hepatitis B antigen, of adult donors either from the general population or from hyperimmunised donors for the purpose of certain diseases. For the preparation, the pooled plasma is treated with ethyl alcohol under controlled pH and ionic strength, for fractional precipitation. By this way, the immuno-globulins which are precipitated are purified so as to contain the final product not less than 15 per cent and not more than 18 per cent of proteins. The globulin fractions are standardised which give special protection against certain diseases.

Heterologous sera are obtained from animals and include antitoxins, antiviral serum and antivenins. The specific antigens used to raise the immunity in horse are either toxin, virus or venom. They are injected into horse and several closes are given iven after certain interval in an increasing order, till the blood serum achieves a proper antibody titre. The general procedure for refining of the pooled plasma from the immunised animal is dilution with water and controlled digestion with pepsin, due to which the protein molecules are split into immunologically active fragments (immune globulins) and immunologically inactive fragments (fibrinogen and albumin). The immunologically inactive fragments are separated by fractional precipitation with ammonium sulphate by which immunoglobulins are precipitated.

Antitoxin and antiserum (antiviral serum) are available in solution form and antivenins as freeze-dried products.

Antivenins include preparations for treatment of **snake bite, scorpion bite, spider bite** and also due to poisoning by the **bite of wasp, hone bee** etc.

In case of snake venom, the deleterious fractions are neurotoxins, cytotoxins along with haemolysins, and coagulants. Elapine snakes like cobra, and mamba mainly contain neurotoxins,while the vipers, rattle snakes and adders, the venoms of which are called viperine venoms, contain cytotoxins and coagulants. The rattle snake venom exerts neurotoxic effects. It is not possible to prepare an antiserum which shall act against all types of snake venoms because of great immunological differences. For this purpose, the method derived out is to prepare an anti-snake venom serum by neutralising the venoms of the indigenous snakes. After preparing such antisera, they are/ used in combined form i.e. polyvalent antivenon serum. In Indian subcontinent, this polyvalent antivalent serum is prepared frorn.venoms of cobra, russets viper, krait and saw-scaled viper. In Great Britain, it is prepared as monovalent anti-serum from venom of common viper or adder (*Vipera berus*).

Some of scorpion venoms used to prepare their antiserum are obtained from *Adroctonus australis*, *A. crassicauda, Legurus quinquestrialus* and *Buthus occitanus*. The advantage for scorpion antivenins is that the venoms of many scorpions are antigenically similar and hence used to neutralise the venoms of a large number of species of. scorpions.

HUMAN IMMUNE SERA (HOMOLOGOUS SERA)

1. IMMUNE GLOBULIN (GAMMA GLOBULIN)

Biological Source

It is a sterile, nonpyrogenic transparent or slightly opalescent solution of globulins containing various antibodies normally present in adult human blood. It is prepared by pooling equal amounts of blood, serum or placentas from not less than 1000 donors. The human plasma of such blood should be not reactive when tested for hepatitis B surface antigen.

Uses

Immune globulin is used to induce passive immunity against measles, viral hepatitis A and also for the regular maintenance of immunodeficient persons.

Dose

It is given only by intramuscular route and the doses are varying for different conditions.

(a) 0.02 ml to 0.04 ml/kg for prophylaxis against viral hepatitis A

(b) 0.25 Ml/kg for prophylaxis against measles

(c) 0.05 ml/kg for measle modification

2. Rh_o (D) IMMUNE GLOBULIN (HUMAN)

Biological Source

It is a sterile, transparent or s' 6ntly opalescent con(.,-.itrated solution of globulins derived from human blood plasma containing antibody to erythrocyte factor Rho (D) I

Uses

The antibody present in this type of immune globulin neutralises the antigen in Rh-positive blond. This antigen sensitizes Rh - negative women and cause Rh hemolytic disease (Ervthroblastosis fatalis) of the baby. Hence, this product is used for Rh negative female, delivering Rh_o (D) positive baby. It is also used for amniocentesis and *her abdominal trauma, resulting in fetal cells entering into maternal circulation.

Dose

1 vial intramuscularly.

3. RABIES IMMUNE GLOBULIN

Biological Source

It is a sterile, nonpyrogenic, transparent or slightly opalescent solution of globulines derived from the blood, plasma of individuals immunised with rabies vaccine.

Uses

It is used for prophylaxis to rabies in patients who are severely bitten on areas like head and neck, from rabid animals or animals suspected of being rabid. It is not given to those persons who are already immunized with rabies vaccine with adequate antibody titers.

Dose

20 International units per kg of body weight intramuscularly. In some cases, half the dose is infiltrated through the wound and remaining given intramuscularly in gluteal or deltoid region.

Storage and Expiry : It is stored at 2° -10°C and protected from light and freezing, by which it retains potency upto 3 years from the date of manufacture.

4. HEPATITIS B IMMUNE GLOBULIN (HUMAN)

Biological Source

It is a sterile, non-pyrogenic solution which consists of the globulins derived from pooled blood plasma of human donors with a high titre of antibodies against hepatitis B surface antigen.

Uses

It is used for post exposure passive immunisation following the direct mucous membrane contact or parenteral exposure or oral ingestion of Hepatitis B surface antigen (HBs Ag or Australia antigen) - positive materials like blood, plasma or serum. It is also used for passive immunisation of infants born to HBs Ag positive mothers.

Dose

0.06 ml/kg of body weight intramuscularly and given within a week of exposure and second dose of same quantity given after one month.

5. PERTUSSIS IMMUNE GLOBULIN (HUMAN)

Biological Source

It is a sterile, non-pyrogenic, ogenic, transparent or slightly opalescent solution of globulins derived from the pooled blood plasma of adult human donors, who have been immunised with pertussis vaccine.

Uses

It is used in the passive prophylaxis or attenuation and treatment of pertussis.

Dose

(1) **Prophylactic:** 1:25 ml - 2.5 ml as per the age of child intramuscularly, 1 or 2 times with 1 to 2 weeks interval.

(2) **Therapeutic:** 1.2-5 to 2.5 ml intramuscularly as per the severity of infection, 1 or 2 times with 1 - 2 days interval.

6. TETANUS IMMUNE GLOBULIN (HUMAN)

Biological Source

It is a sterile, transparent, practically elear non-pyrogenic solution of globulins derived from the pooled blood plasma of adult human donors, who have been immunised with tetanus toxoid.

Uses

It is used to induce passive immunity for those persons who are immunodeficient to tetanus. It is used both for prophylactic and therapeutic purposes.

Dose

(1) **Prophylactic:** 250 units, intramuscularly and in case of severe wounds 500 units

(2) **Therapeutic:** 3000 to 6000 units intramuscularly

7. MUMPS IMMUNE GLOBULIN (HUMAN)

Biological Source

It is a sterile, non-pyrogenic solution of globulins derived from healthy adult human donor's blood, are previously hyperimmunised with mumps virus vaccine.

Uses

It is used as a prophylactic and therapeutic for mumps in both children and adults to give passive immunity.

ANIMAL IMMUNE SERA (HETEROLOGOUS SERA)

1. DIPHTHERIA ANTITOXIN

Biological Source

It is a sterile non-pyrogenic transparent or slightly opalescent solution which contains refined and concentrated antibodies mainly globulins obtained from the blood of healthy horses which are immunised against diphtheria toxin or toxoid prepared from *Corynebactoriurn diptheriace*. It is made harmless to susceptible animals.

Uses

It is used for passive immunisation in prevention and treatment of diphtheria. This antitoxin utralises the toxin produced by the C. diphthdriae locally at the site of infection, but does not reduce pathological changes induced by the toxin.

Dose

(1) **Prophylactic:** 10,000 units intramuscularly

(2) **Therapeutic:** 20,000 to 1,20,000 units intramuscularly or intravenously

Storage : It is stored at 2) to 10°C and protected from light and freezing

Expiry : 3 years from the date of manufacture

2. MIXED GAS-GANGRENE ANTITOXIN

Biological Source

It is a sterile, colourless or faintly coloured preparation prepared from mixing gas gangrene antitoxin (septicum), Gas-gangrene antitoxin (perfringens) and Gas-gangrene antitoxin (oedernatients)

Uses

It is used as a prophylactic against the development of gasgangrene, immediately after the occurrence of wounds. It is also given therapeutically to neutralize the toxaemia.

Dose

(1) **Prophylactic:** 25,000 units, intramuscularly or intravenously as a single dose

(2) **Therapeutic:** 75,000.-units intravenously

Storage : It is stored between 2° to 10°C and protected from light and freezing.

Expiry : It retains its potency upto 3 years from the date of manufacture.

TETANUS ANTITOXIN

Biological Source

It is a sterile, non-pyrogenic transparent or slightly opalescent solution containing the refined and concentrated antibodies, which are obtained from the blood of healthy horses that have been immunised against tetanus toxin or toxoid (produced by Clostridium.tetani).

Uses

It is used to give temporary passive immunity against tetanus both for preventive and therapeutic purposes.

Dose

(1) Prophylactic : Single dose,1500-5000 units intramuscularly or subcutaneously

(2) Therapeutic : 50,000 to 1 lakh units. A part of this dose is given intravenously and remaining dose is given intramuscularly

Storage : It is stored between 2° and 10°C and protected from light and freezing.

Expiry : It retains its potency upto 3 years from the date of manufacture.

4. BOTULISM ANTITOXIN (BOTULISM ANTITOXIN TRIVALENT)

Biological Source

It is a sterile, non-pyrogenic, transparent or slightly opalescent, nearly colourless solution containing refined and concentrated antitoxic antibodies, chiefly globulins obtained from the blood of healthy horses which are immunised against toxins produced by type A, type B and type E strains of Clostridium botulism

Uses

This antitoxin is used to supply passive immunity against Clostridium botulism type A, B or E both for prophylactic and therapeutic purposes. It neutralises only the circulating toxin, but has no counteracting effects on toxin already bound to receptor cells.

Dose

(1) Prophylactic : 1/5 th of one vial intramuscularly. (1 vial of this preparation contains 7500 International Units of type A strain, 5500 International units of type B strain and 8500 International units of type E. strain).

(2) Therapeutic : 1 vial intravenously diluted 1 : 10 form, which is followed by 1 vial intramuscularly. The subsequent dose is given to produce a reservoir of antitoxin in body.

5. ANTIRABIES SERUM

Biological Source

It is a sterile, non-pyrogenic, slightly opalescent or transparent solution which contains antiv substances obtained from the blood plasma of healthy horses which are immunised against rabies \N vaccine rabies.

Uses

Antirabies serum is given along with rabies vaccine, in case Rabies Immune Globulin is not available.

Dose

Prophylactic : 1000 units on post exposure. Out of this dose, 50 per cent is infiltrated throng the wound and 50 per cent given intramuscularly.

ANTIVENOMS

1. POLYVALENT CROTALINE ANTIVIENIN

Biological Source

It is a sterile, non-pyrogenic freeze dried preparation containing concentrated specific venon neutralising serum globulines which are obtained by fractionating blood plasma pooled from health) horses immunised with venoms of four species of pit vipers (crotalids) viz. *Crotalus adamanten,* (Eastern diamond rattle snake), *Cromitis itrox* (Western diamond rattle snake), *Crotalus durissu terrificus* (South American rattle-snake) and *Bothrops atrox* (South American Far de Lance).

Uses

This antivenin is use(] for treatment of poisoning (envenomation) due to the bites of these 4 pit vipers and also other pit vipers found in North, South and Central America.

Dose

Initial (lose is 20 - 150 ml, intramuscularly or intravenously. The subsequent dose may be giver) depending upon severity and extent of poisoning. Children may require double the adult dose initially.

2. BLACK WIDOW SPIDER ANTIVENIN

Biological Source

It is a sterile freeze-dried product which is prepared from the blood serum of horses immunised with the venom of black widow spider (*Latroductus amcirans*).

Uses

It is used for treatment of poisoning due to black widow spider bite.

Dose

1 vial (6000 antivenin units) intramuscularly. In case of severe poisoning, or for children 1 vial

DIAGNOSTIC SKIN ANTIGENS

1. TUBERCULIN (OLD TUBERCULIN)

Biological Source

It is a sterile heat concentrated filtrate from a fluid medium on which the tubercle bacillus (*Mycobacterium tuberculosis* or bovis) has been grown.

Uses

It is used as a diagnostic immunologic aid with multipuncture devices in testing patients, who are suspected or infected with tuberculosis. Tuberculin Js also used in epidemiological surveys to determine areas with high epidemic. It is also used in programmes for determining the priorities for other testings like chest X-ray photographs.

Dose

Diagnostic : 0.1 ml intraLutaneously (Each 0.1 ml contains 1, 5, 10 or 100 units).

Storage: It is stored between 2° and 10°C and protected from light and freezing.

2. TUBERCULIN (PURIFIED PROTEIN DERIVATIVE)

Biological Source

It is a sterile solution derived from a further puriffied protein fraction of culture ffltrates obtained from a human strain of Mycobacterium tuberculosis, which is grown on a synthetic medium free from protein.

Uses

It is used as an immunologic diagnostic acid for tuberculin sensitive individuals. It is also used in more sensitive intradermal mantoux test.

Dose

5 tuberculin units intradermally

Storage : It is stored between 2° and 10°C and protected from light and freezing.

3. COCCIDIODIN

Biological Source

It is a clear, sterile filtrate containing antigens prepared from the spherules of Coccidiodes immitis, grown in liquid synthetic medium.

Uses

It is used in diagnosis of coccidiomycosis. Also used in distinguishing coccidiomycosis from histoplasmosis, sarcoidosis, as well as, other bacterial and mycotic infections.

Dose

0.1 ml in 1 :100 dilution intradermally, on the flex or surface of the fore arm.

4. HISTOPLASMIN

Biological Source

It is a sterile standardized liquid concentrate of the soluble growth products developed by the fungus, Histoplasma capsulatum, grown in the mycelia) phase on a liquid synthetic medium.

Uses

It is used as an aid in the diagnosis of histoplasmosis.

Dose

0.1 ml intradermally into the flexor surface of the forearm. The induration reaction is measured 48 to 72 hours after injection.

5. MUMPS SKIN TEST ANTIGEN

Biological Source

It is sterile aqueous suspension containing killed mumps virus prepared from the extra-embryonic fluid of virus-infected chick embryos, concentrated and purified by centrifugation, inactivated with formaldehyde and then suspended in isotonic sodium chloride solution.

Uses

The mumps skin test is used to assess the person's previous experience (immunocompetency) with mumps virus. This antigen is not used for immunization, treatment or diagnosis of mumps.

Dose

0.1 ml intradermally on the inner surface of forearm. The readings for area of erythema formed as a positive reaction, are taken after 24 and 48 hours after injection.

6. DIPHTHERIA TOXIN FOR SCHICK TEST

Biological Source

It is a sterile filtrate containing standardised toxic products from a culture in nutrient broth of *Corynebacterium diptheriae*.

Uses

It is used in schick test for the determination of serologic immunity to diphtheria.

Dose

0.1 ml into right arm intradermally. The reaction is read after 24 to 48 hours of injection.

7. SCHICK TEST CONTROL

Biological Source

It is a sterile solution of Diphtheria toxin for schick test which has been inactivated by heat.

Uses

Due to the possibility of pseudo-reactions owing to sensitivity to diphtheria bacillus protein, schist test control is given as a control.

Dose

0.1 ml intradermally in left arm.

Chapter 25...

PLANT BASED VACCINES

Though the "Plant Genetic Engineering Technology" has progressed like anything is the last two or three decades, but nothing spectacular has been made in plant-based vaccines.

'Animal based vaccine' is very old Idea now, and several vaccines are being consumed and produced on commercial scale, but plant based vaccine is a new concept.

At present **"New Castle disease vaccine for Poultry"** is only plant-based vaccine developed by united-state Department of Agriculture USA.

It is produced in Suspension Culture Tobacco Cells on commercial scale.

Several plant based vaccines for use in humans and animals are in clinical trials only.

In conceptual sense oral plant based vaccine is ideal, because the manufacturing process is simple. No additional medical devices are needed for injection, their antigen immunogenicity and biological activities are preserved in gastro intestinal tract due to their natural bioeneapsulation in plant cell organelle.

Seeds of cereal crops (like wheat, maize, jawar) are attractive for subunit vaccine production, because vaccines produced in seeds are stable for long storage periods.

Recombinant vaccines are safer than conventional vaccines as these contain no live pathogens. Upto the end of twentieth century, vaccines were getting manufactured on commercial scale from animal and micro organisms, with the dawn of Twenty first century one more source i.e. plants, has been added to the list.

Technical terms coming across frequently while studying vaccines are defined as under:

Vaccine: A vaccine is a preparation containing living, attenuated, killed bacteria, killed rickettsia, killed viruses and are used to stimulate the production of antibodies.

Immunity: Natural resistance to infection.

Antigen: It is a harmful substance that causes the body to produce antibodies.

Chemically, antigens are proteins but exceptionally high molecular weight polysachharides are reported to be antigens.

Antibodies: These are found only in serum fraction of blood. Antibodies are also found in association with other tissues, such as lymph nodes, mucous membranes etc.

Attuniated: Weaker or free of specific infection.

Immunodeficiency: It is failure of the body to resist infection.

Virus: It is a minute organism which can cause disease.

Types of Immunity

These are two types of immunity. It can be Innate inborn immunity or it can be acquired immunity.

(a) **Active immunity:** It is acquired naturally during an infectious disease or artificially by vaccination with dead or living organisms.

(b) **Passive immunity:** It is acquired naturally when maternal antibody passes to the child via placenta or in milk or artificially by administration immune sera containing antibody obtained from animals or human beings.

Types of Vaccines

(a) **Viral vaccines:** Mumps, rubella, small pox, rabies, yellow fever, chicken pox, influenza, poliomyletis, rubeola (measales) and hepatitis.

(b) **Bacterial vaccines:** Typhoid, cholera, pleagae, pertussis (whoping cough), tubercuolsis, diphtherla and miningitis.

(c) **Reekettsial vaccines:** Reckeltsia typhi, rickettsia-prowazekii, tetanus.

❖❖❖

Chapter 26...

STUDY OF POISONOUS AND HARMFUL ANIMALS

INTRODUCTION

Scientific study of all phases of animal life is zoology. Thus from the definition one cannot understand the type and the importance of the animal under study. Depending upon the habitat, the animals can be categorised as farm-animals (Cows, horses, bulls etc.), pet-animals (Cats, dogs, etc.), pests (Rats, mice etc.). and beasts. Some of them live on the land (i.e. terrestrial), in water (either fresh water or sea water) or even in the air. They exist in great variety of forms, as quite a few of them are the largest living organisms of the world, while others are microscopic. Economical importance of animals reveals that they are very essential for the survival of human beings, while others are parasitic or pathogenic and poisonous. Sometimes the parasitic animals are not so fatal, as pathogenic animals or pathogenic micro-organisms. Poisonous animals are harmful and some of them are deadly poisonous.

Few other animals which are neither useful to human being nor disease producing but are the carriers of the diseases or disease producing organisms are known as **vectors**.

For the prevention of diseases or for the proper control over the infectious diseases it is essential to know few important vectors. Apart from their morphological characters a pharmacist must know their habit, nutritional methods, life cycle or mode of reproduction, disease or diseases caused and the methods to control their growth.

In the following table, the phylum, class, type habitat and pathogenic effects of some important disease producing organisms and vectors are mentioned. The detailed information of each of them is described thereafter while the poisonous animals have been described separately.

Table 26.1: Some disease producing organisms and vectors

Phylum	Class	Type	habitat	Pathogenic Effects
Protozoa	Sarcodina or Rhizopoda	1. *Entamoeba hystolitica*	Large intenstine	Amoebic dysentery
		2. *Entamoeba gingivalis*	Mouth	Non-pathogenic but responsible for pyorrhea.

Contd...

Phylum	Class	Type	habitat	Pathogenic Effects
Malarial - Parasites				
	Sporozoa	1. *Plasmodium vivax* (malarial)	R.B.C.	Malaria
		2. *Plasmodium malariae*	R.B.C.	Malarial fever and chill
		3. *Plasmodium falciparum*		Cerebral malaria
		4. *Giardia inestinalis*	Intestine	Giardiasis
		5. *Leishmonia donoyam*	Viscra	Kala-azar
		6. *Leishmania Tropica*	Skin	Cutaneous Leishmaniasis
		7. *Trichomonas vaginalis*	Vagina	Vaginitis and urethritis
		8. *Trypanosoma gambiense*	R.B.C.	African typanosomiasis
		9. *Trypanosoma rhodesiense*	R.B.C.	African trypanosomiasis
		10. *Trypanosoma cruzii*	R.B.C.	Trypano somiasis
	Ciliata	*Balantidium coli*	Large intestine	Sleeping-sickness Balantidial dysentery
Platyhel-minthes	Trematoda	1. *Fasciola hepatica* (*Liver-fluke*)	Liver	Vomiting persistent diarrhoea
		2. *Schistosoma haemetobium*	Blood vessels of urinary bladder	Headache, body pain, diarrhoea dysentery
		3. *Schistosoma mansoni*	Hepatic portal system	Schistosomal dysentery
		4. *Schistosoma japonicum*	Portal veins and mesentric veins	Intestinal and hepatic schisto-somasis

Contd...

Phylum	Class	Type	habitat	Pathogenic Effects
	Cestoda	1. *Taenea solium* (Pig tapeworm)	Intestine	Body pain, chronic indigestion, nervous discomfort, anaemia.
		2. *Taenea saginata* (Beef tape-worm)	Intestine	Body pain, chronic indigesion, nervous discomfort, anaemia
		3. *Diphyllobo-thrium latum* (Fish tape-worm)	Intestine	Body pain, chronic indigestion, nervous discomfort, anaemia
	Nematoda	1. *Ascaris lumbricoides* (Common round worm)	Small intestine	Colic pain, abdominal and nervous discomfort, Pneumonia.
		2. *Ancylostoma duodenale* (Hook worm)	Small intestine	Ancylostomasis anaemia
		3. *Enterobius vermicularis* (Pin worm)	Small intestine	Vaginitis, abdominal discomfort, sleeplessness.
		4. *Dracunculus mediansis* (Guinea worm)	Inside the tissues of hands and legs	Blister, abscesses and ulcers causing vomiting, nausia diarrhoea etc.
Arthropoda	Arachnida	1. *Sarcoptes scabie* (Itch mite)		Scabies in man.
		2. *Pediculus hamanus* (Louse)	Scalp	(Head-louse)
		3. *Ticks and mites*	Skin	External parasities, suck the blood of man and transfer diseases.

Contd...

Phylum	Class	Type	habitat	Pathogenic Effects
	Insecta	1. *Anopheles sp.* **(Mosquito)**	Damp and marshy places	Vector i.e. transfer various diseases to man, like malaria; filaria, yellow fever etc.
		2. *Musca domestica* **(House fly)**	Filthy and dirty places	Vector, responsible for the diseases like cholera, typhoid, diarrhoea, amoebic dysentery etc.
		3. **Cockroach**	Warm and damp places	Vector, transfer various diseases with their hairy legs.
		4. *Pulex irritas* **(Human flea)**	Human body	Plague

POISONOUS ANIMALS

The poisonous animals include the snakes, lizards, scorpion and spiders.

1. Snake:

All the snakes belong to the class-Reptilia. Most of the species of snakes are non-poisonous and few are poisonous. The poisonous snake generally live in the forest area. General body of the snake is long, slender and covered with epidermal scales of various sizes. The head is small, mouth is wide

Fig. 26.1: Cobra

with a expansible and slender bifid tongue which is tactile and olfactory too. Eyelids are fused and non-motile with no nictitating membrane. The snake replaces the outer epidermal skin by sheding the white transparent membrane, called moulting during which it suffers from temporary blindness. The snake has no ear or any hearing apparatus like ear drum or external ear etc. But it can get sensation of vibration through the epidermal scales present on the belly region.

The posionous snakes are Cobra, Krait, Viper, Pit-viper etc. It shows generally, large and broad scales extending across the belly. In cobra beside the broad belly scale, the head shows hood with a particular marking. In viper, the head is pear shaped with distinct neck and covered with small scales. In pit viper, a pit is present in between the nostril and eye. It also shows a pair of well developed fangs.

Venom of the Snake:

Snake poison is a specialised salivary secretion. It is a clear viscoid yellowish fluid, composed of modified proteins. It is acidic in nature. The venom is of two kinds
(i) haemotoxic, paralysing the circulatary system and (ii) neurotoxic, acting on central nervous system.

Symptoms of Snake Bite in Man:

In poisonous snake bite, there is a severe pain and burning sensation associated with a mark swelling of the wound. General symtoms are headachae, giddiness, nausea, vomitting, drowsiness, muscular weakness and paralysis. If the venom is haemotoxic, haemorrhage takes place and results in heart failure. In neurotoxic, the poison act on the central nervous system and paralyses the body.

Uses of Snake and its Venom:

1. Snake is a friend of farmer, it feeds on rodents i.e. rats who destroy the valuable crops and food grains.
2. Snake skin is useful in preparing the variety of fancy goods such as belts, purses, hand bags, shoes, glows etc.
3. Venom of the snake is a raw material from which, the antivenom, the only effective antidote for snake bite is prepared.
4. The venom of the snake is used as an analgesic for muscular disorders, cancer etc.

Diagnosis and Treatment:

On the vicinity of bite, fang marks are observed (generally two marks) with a burning pain to the patient.

The first aid treatment is to wipe the site of the bite and then cover with a cloth. Do not incise the bite, it does not help but often introduces infections. Apply a firm (not very tight) ligature above the bite and then immediately visit to the nearest hospital for further treatment. In hospital the patients are treated with vitamin B complex and antivenom.

2. Lizards:

Most of the lizards are not poisonous but the *Heloderma suspectum* and *H. horrilum* are poisonous. These are called as "Gila monesters". A poison apparatus is present in the lower jaw of the monester associated with the grooved teeth. The symptom of poisoning shows body paralysis.

Fig. 26.2: Wall Lizard

3. Scorpion:

It is included in the class-Arachnida with a cosmopolitan distribution. These are found in hilly regions, under stones, logs, debris etc. The body of the scorpion is long narrow and dorso-ventrally flattened with a head, thorax and abdomen. The last segment of abdomen is modified into sting with poison gland.

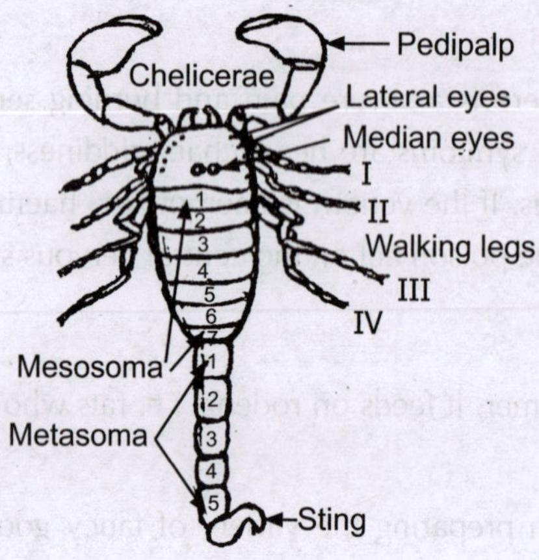

Fig. 26.3: Scorpion

Toxicity of Poison:

Scorpion does not really produce any disease in human beings but its poison may prove fatal to man.

The scorpion sting is very powerful causing uneasyness, though not very fatal. Scorpion venom is clear and colourless liquid containing a variety of toxic substances. The venom may be neurotoxic or haemolytic. It produces paralysis, nervous discomfort and pulmonary disorders. Other symptoms are headache, giddiness, nausea and vomitting, profuse perspiration, feeble pulse etc.

In case of children, it may prove fatal because of muscular cramps, profuse perspiration and vomitting. In general, the symptoms are laceration, salivation and sphincter relaxation.

Treatment:

In case of a scorpion sting in children, the sting should be thoroughly washed by a concentrated solution of potassium permanganate. In severe cases, injection of an ampoule of novocin and adrenaline are also useful when given in the vicinity of the sting.

4. Spiders:

Beside the scorpion, most of all spiders are poisonous. The venom of the spiders is injurious to the insects but few are dangerous to man, example: Latrodectus.

5. Rabid Animals:

Animals suffering from rabies like dog, fox and others. The disease is caused by very much toxic viruses present in the saliva. The bite of such animals is falal and results in madness.

❖❖❖

APPENDICES

1. GLOSSARY OF BOTANICAL TERMS

Achene	:	A simple dry one-seeded, indehiscent fruit with unfused seed coat and fruit wall, but with no special method to liberate the seeds.
Acuminate	:	With very sharp terminal tapering point.
Adventitious	:	Arising from abnormal position.
Acute	:	Tapering more broadly than acuminate.
Adventitious roots	:	Arising from stems or leaf cutting but not from primary root.
Adventitious buds	:	A bud which develops in some other place than the axils of leaves or stem apices.
Aerenchyma	:	Tissue comprising of thin walled cells with large air sacs found in different parts of plants.
Aleurone grains	:	Protein granules found in plants, quite often in seeds.
Algae	:	Simple lower plants capable of carrying out the photosynthesis and with unicellular organs or reproduction.
Androecium	:	Collective term, for the stamens of flower.
Angiosperm	:	Flowering plant.
Anther	:	Pollen bearing part of a stamen.
Anticlinal	:	Situated approximately at right angles to outer surface of plant part.
Annual	:	Plant that completes its life cycle from seed germination to seed production followed by death, in a single season.
Annual ring	:	The layer of xylem (wood) formed by one year's growth of cambium.
Annual thickening	:	Internal thickening of wall of xylem vessel so as to form rings at intervals alongwith its length.
Apical	:	Of the tip.
Apothecium	:	Cup-shaped fruit body of certain ascomycete fungi.
Aril (Arillus)	:	Succulent development of stalk or base of seeds of fungi (mycophyta).
Ascomycetes	:	Group of fungi (mycophyta).
Ascospore	:	A spore produced by sac (ascomycete).
Ascus	:	A sac-like structure within which ascospores are produced.
Auxins	:	Growth regulating substance or plant hormones.

Awn	:	Siff bristle-like appendage occurring frequently on the flowering glumes of grasses or on seeds i.e. strophanthus seed.
Axil	:	The upper angle between a twig or leaf stalk and the axis from which it grows.
Bud	:	Terminal structure consisting of meristemic tissue, covered wholly or in part by overlapping leaves.
Cambium	:	Layer or meristematic cells between xylem and phloem tissues.
Carpel	:	A female floral organ which bears and encloses ovules.
Chitin	:	Nitrogen containing polysaccharide derivative found in insects and lower plants.
Conidiospore	:	A sexual spore of certain fungi cut-off externally at apex of specialized hypha.
Conidium	:	A hypha which produces conidia.
Cork	:	Suberized tissue formed from the cork cambium on stem and root surfaces of dicot plants.
Corm	:	A short, often globose, upright, underground stem which stores food; differs from a bulb in that the latter consists chiefly of fleshy storage leaves growing from a small stem, whereas a corm is chiefly stem tissue.
Cortex	:	Tissues lying between the epidermis and endodermis of a stem or root.
Cotyledon	:	A food-digesting and food-storing part of an embryo; also known as seed-leaf.
Cryptogams	:	These are flowerless or seedless plants.
Cuticle	:	Waxy layer formed on outer walls of epidermal cells.
Cutin	:	Waxy substance containing cutinogenic acids impermeable to water; the cuticle is composed of cutin.
Drupe	:	A simple fleshy fruit in which the endocarp (inner wall of ovary) becomes hard and stony, and encloses one or two seeds.
Emergence	:	An outgrowth, consisting of epidermal and cortical tissues and lacking vascular tissues, e.g. rose prickles.
Endodermis	:	The surface layer of cell in leaves and other soft plant parts.
Ferns	:	These are extensive class of flowerless plants. Remarkable for grace and beauty of their leaves, found in shady places, walls or waterfall grown in gardens, can be multiplied by division or by spores.
Fermentation	:	A chemical reaction brought about by the action of (ferments) microorganisms, liberating gas and heat. Examples are makeup of bread, cheese and wine (alcohol).
Fibre	:	An elongated, tapering thick-walled strengthening cell occurring in various plant parts.
Glossary	:	List of words and their meaning.

Fossil : An impression or trace of a plant or plant part in the earth's crust.

Genome : Set of Chromosomes

Guard cells : Paired epidermal, chlorophyllous cell which enclose stoma.

Hilum : Scar on a seed coat, making the place of an attachment of the seed stalk to the seed.

Hydathode : Water excreting gland occurring on the edges or tips of leaves of many plants.

Hypha : A fungal filament.

Index : Alphabetical list of names or subjects referred in a book.

Inner bark : The phloem and cortex of bark.

Integument : The coat of an ovule.

Inulin : Insoluble polysaccharide or vegetable origin.

Isobilateral : A structure of which both sides get equally exposed and possesses same structure on both sides.

Kelp : Type of sea weed.

Latex : A characteristic milky fluid produced by various flowering plants, which oozes out from cut surface and rapidly coagulates on exposure to air. It contains a number of chemicals like sugars, oil, enzyme, alkaloids, etc. e.g. fig, banyan, papaya, popy capsules.

Lenticel : In woody stems and other plant parts, a pore through which exchange of gases occurs; in woody stems, lenticels occur in bark.

Lichen : Plants resembling moss are dual organism formed from symbiotic association of two plants, a fungus and an algae. Lichens are food for cattle and wild animals. They are also used in food in cosmetics, soaps and liquors. Litmus is prepared from some lichens.

Lignin : A complex aromatic compound deposited in cell walls of sclerenchyma, xylem vessels, tracheids, making them strong and rigid.

Liver-worts : Similar to mosses but without leaves.

Mericarp : One seeded portion of schizocarp.

Micropyle : The minute opening in the integument of an ovule or seeds, through which a pollen tube grows to reach the female gametophyte.

Natural fermentation: It produces concentration of alcohol in the fermenting liquid that rarely exceeds 14.0% by volume because the fermentative organisms are usually inhibited at such a high concentration of alcohol.

Nucleolus : A small particle containing RNA, in nucleus.

Outer bark : The cork and cork cambium tissue of woody stem and root.

Orchids	:	Plants with much varied form growing with flowers of many colours. Orchids are found in temperate and tropic regions. Orchids need moist and humid situation growing on rocks and upon moss-grown situation and upon branches of trees. They are terrestrial, saprophytic and epiphytic. Orchids grow on trees, exposing their roots to the air, from which they get their nourishment required. Orchids have no green leaves and are not cultivated. These are propogated by divisor of roofs or psudo-roots
Pathogen	:	Disease producing parasite.
Pericycle	:	A layer between endodermis and conducting tissue.
Perithecium	:	A globose or pear-shaped ascocarp with a small, definite pore.
Phelloderm	:	Secondary tissue formed by cork cambium.
Polyploid	:	Having more than two sets of chromosomes.
Pteridophytes	:	Includes fern and their allies.
Radicle	:	The lower portion of the hypocotyl which grows into the primary root of a seedling.
Raphides	:	Needle-like crystals of calcium oxalate occurring in bundles.
Runner	:	A stem which grows horizontally over the surface of the soil, often developing new plants at its nodes; also horizontal, surface hyphae of certain fungi (e.g. Rhizopus).
Schizocarp	:	Dry fruit formed from a syncarpous ovary that splits at maturity into its individual carpels forming one seeded fruits known as mericarps.
Sclerenchyma	:	A strengthening tissue composed of thick walled, elongated cells (fibres) or shorter cells (stone cells).
Stele	:	Vascular tissue, immediately lying internally to the endodermis which contains xylem, phloem, pericycle, and sometimes, pith and medullary rays. Detailed structure of stele may vary from plant to plant.
Suberin	:	Thick-walled mixture of oxidation and condensation products of various fatty acids. Suberin is present generally in cork cells making them impervious to water.
Thallophyta	:	These are algae, fungi, bacteria and lichens.
Tracheid	:	A type of non-perforate conducting and strengthening cell in xylem tissue, or elongated tapering form with pitted walls.
Turgidity	:	State of being plump or swollen as a result of internal water pressure.

2. PHARMACOGNOSTIC TERMS

Many pharmacognostic terms have their origin either in **Greek** or in **Latin** language, therefore etymology of some of them may be helpful to know the subject properly.

(A) GREEK PREFIXES USED WITH THEIR MEANING AND EXAMPLES

a-, without (acellular);
acro-, end (acropetal);
actino-, rayed (actinocytic);
adelphos, brother (monadelphous);
ake; pointed, Akakia (Acacia)
amphi-, both (amphibian);
ana-, up (anabolic);
andr-, of male (androecium);
anemos, wind (anemometer);
angios, a vessel (angiosperm);
anti-, opposite (antisocial);
apo-, away from (apocarpous);
bio-, life (biochemistry);
blema, covering (epiblema):
carp, fruit (mesocarp)
cata, down (catalyse);
chloro-, green (chloroplast);
chromo, colour (chromotography);
cleisto-, closed (cleistogamous);
cyto-, cell (cytotoxic)
derma, skin (Hypodermic);
di-, twice (dicotyledonus);
dynamic, energetic (tetradynamous);
endo-, within (endosperm);
epi-, on (epidural);
gamos, marriage (polygamy);
geo, earth (geotropic);
-gen, producing (exogenous);
gyn-, of female (gynaecology);
helios, sun (heliotropic);
heteros, different (heterocyclic);
histos, tissue (histology);
homos, same (homosexual);
hypo-, under (hypogynous);
logos, science (Pharmacology);
mega-, large (megawatt);
meros, part (mericarp);
meso-, middle (mesophyll);
micro-, little (microscopic);
mono-, single (monocotyledon);
morphe, form (morphography);
-oecium , house (gyanecium) ;
-oid, like (Toxoid) ;
orthos, straight (orthospermus);
peri-, around (perisperm);
-phile, loving (hydrophilous);
phobe, hating (hydrophobic);
-phore, carrying (Androphore);
phyll, leaf (chlorophyll);
phyte, plant (Phytotoxic);
plasma, anything formed (protoplasm);
pod, foot (polypodial);
poly, many (polygonal);
protos, first (protoplasmic);
pseudo, false (pseudonym);
rhiza, a root (rhizom);
sapros, rotton (saprophyte);
schizo split (schizophrenic);
scleros, hard (sclerenchymatous);
sperma, seed (endospermic);
stichos, a row (orthostichies);
syn-, together with (synchronise);
tetra, four (tetragonal);
thee, a case (thecal);
tropos, turn (trophosphere);
xero-, dry (xerophytic);
zygon, a yoke (zygomorphic);
xylon, woody tissue (xylem).

(B) LATIN PREFIXES USED WITH THEIR MEANING AND RXAMPLES

ad, to (adhere);
albus, white (albumen);
amplexus, embraced (amplexicaul);
annum : annual (perennial)
argilla, clay (argillaceous);
auricula, small ear (auricular);
axilla, armpit (axillary);
bacillum, little staff (bacillus);
bi-, twice (biannual);
bulbus, squill (bulb);
caducus, fallen (caducous);
capillus, hair (capillarity);
capsa, case (capsule);
carn-, muscular (carnivorous);
caruncula, small piece of meat (caruncle);
caulis, stem (cauline);
corolla, little crown;
corymbus, group of flowers (corymbose);
cutis, skin (cuticular);
decusso, in pairs at right angles (decussate);
dehisco, to break (dehiscent);
ex, without (exfoliate);
fid, cleft (pinnatifid);
fistula, pipe (fistulous);
flos, flower (floral);
folia, leaf (foliar);
folliculus, little sac (follicle);
fugio, to flee (fugaceous);
glaber, smooth (glabrous);
glaucus, covered with bloom (glaucous);
hasta, spear (hastate);
haustus, drawing up water (haustorium);
hispidus, hairy (hispid);
imbrex, over lapping as tiles on roof (imbricate);
impar, unequal (imparipinnate);
inter, among (intercontinental);
involucrum, ring of bracts surrounding flower (involucre);
labium, lip (bilabiate);
lignum, wood (lignified);
ligula, strap (ligulate);
loculus, little place (bilocular);
nuto, to nod (nutation);
nux, nut (nux-vomica);
ovum, egg (ovulations);
papilio, nipple-like outgrowth (papilionaceae);
par, equal (at par);
paries, wall (parietal);
pelta, shield (peltate);
persona, mask (personate);
peto, to seek (acropetal);
piper : pepper, pungent peppermint
pinna, wing (pinnate);
planus : flat, planifolia
pluma, feather (plumule);
pyxis, box (pyxidium):
racemus, bunch (racemose);
radix, rootlet (radicle);
renes, kidney (reniform);
rota, wheel (rotary);
sagitta, archer (sagittarius);
sectus, cut (pinnatisect);
serra, saw (serration);
siliqua, pod or shell;
subula, small tool for piercing (subulate);
umbella, sunshade (umbellate);
urceolus, little pitcher (urceolate);
vas, vessel (vas deferens);
verus : true - Aloe-vera
versatilis, turn (versatile);
verticillus, whirling (verticillate).

3. VARIOUS CHEMICAL RING STRUCTURES

1. Five – membered rings

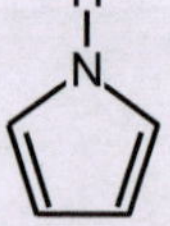

Pyrrole

Furane

Tetrahydro-furane

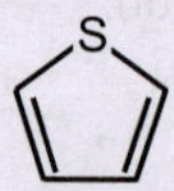

Thiophene

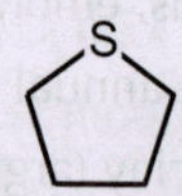

Tetrahydro-thiophene

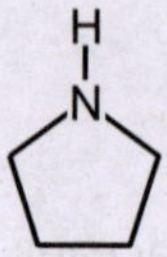

Pyrolidine

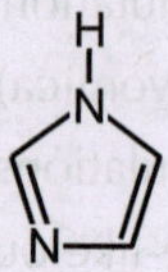

Imidazole

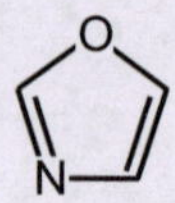

Oxazole

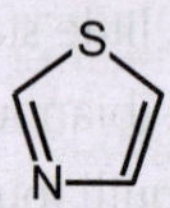

Thiazole

2. Six – membered rings

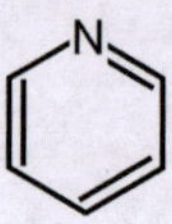

Pyridine

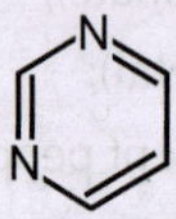

Pyrimidine

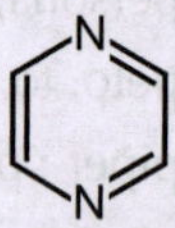

Pyrazine

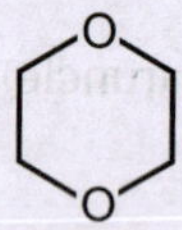

Diaoxan

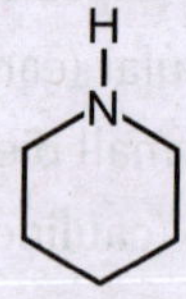

Peperidine

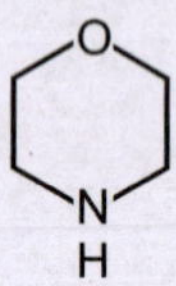

Morpholine

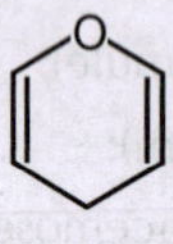

Pyrane

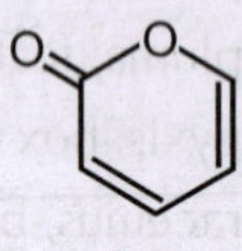

α-Pyrone

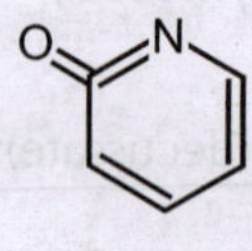

Pyridone

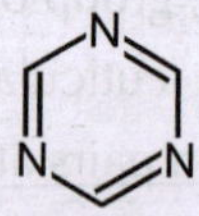

Triazine

3. Fused rings

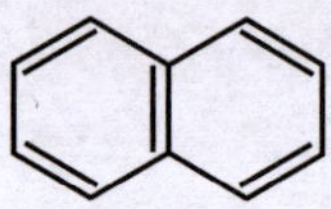

Naphthalene

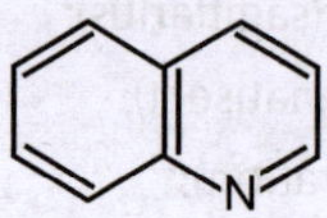

Quinoline

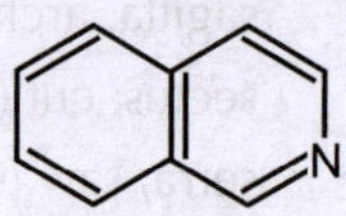

Isoquinoline

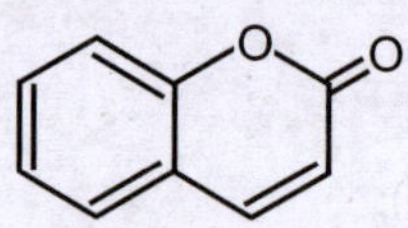

Coumarin

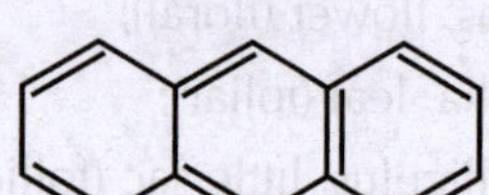

Anthracene

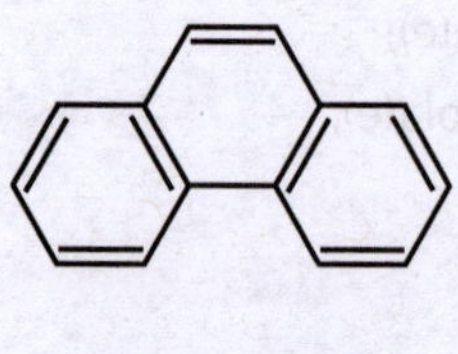

Phenanthrene

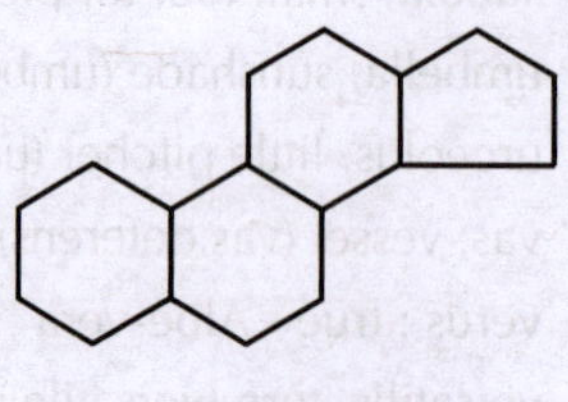

Steroidal

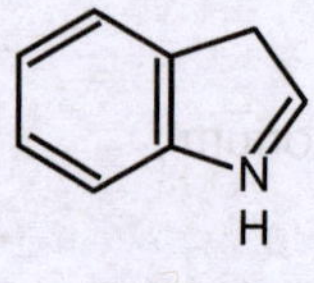

Indole

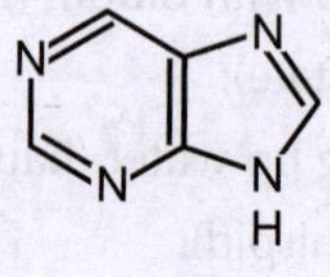

Purine

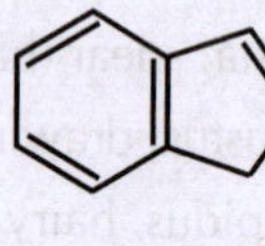

Indene

❖❖❖

4. NAMED TESTS/SOLUTIONS/REAGENTS

5. ALTERNATIVE REVISED NAMES OF HERBAL FAMILIES

International code of Botanical Nomenclature vide its article no. 18.5 and 18.6 authorises to use alternative names to following flowering plant families. Though there is controversy in using the alternative names in recent years, just to create awareness among the students they are mentioned below :

Sr. No.	Present name of plant family	Proposed alternative name
1.	Apocynaceae	Asclepiadaccae
2.	Compositae	Asteraceae
3.	Cruciferae	Brassicaceae
4.	Gramineae	Poaceae
5.	Guttiferae	Clusiaceae
6.	Labiateae	Lamiaceae
7.	Leguminosae	Fabaceae
8.	Liliaceae	Colichicaeeae
9.	Pamae	Araceae
10.	Polypodiaceae	Aspidiaceae
11.	Umbelliferae	Apiaceae
12.	Flacourtiaceae	Salicaceae

❖❖❖

6. COMMERCIAL ISOLATION OF PHYTOPHARMACEUTICALS

Sr. No.	Name of Phytopharmaceutical	Source of Isolation
1.	Amylose and Amylopectin	Starch (*Zea-maize, Oriza-sativa*)
2.	Andrographolide	Kalmegh herb (*Andrographis peniculata*)
3.	Arabinose	Gum ghati (*Anogeisus latifolia*)
4.	Atropine	Egyptian henbane herb (*Hyoscyamus muticus*)
5.	Azadiractin	Neem leaves and seeds (*Azadiracta indica*)
6.	Berberine	Daruhalad rhizomes (*Berberis aristata*)
7.	Bixin	Annato seeds (*Bixa orellana*)
8.	Boswellic acid	Sallaki oleo resin (*Boswellia serrata*)
9.	Brucine	Nux-vomica seeds (*Strychnous Nux-vomica*)
10.	Caffeine	Tea-leaves (*Thea-sinensis*)
11.	Camphor	Comphor oil (*Cinnamomum camphora*)
12.	Camptothecin	Comptotheca wood (*Camptotheca acuminata*)
13.	Capsaicin	Chillies (*Capsicum annum*)
14.	Carrageenan	Irish moss (*Chondrus cryptus*)
15.	Colchicine	Colchicum-corms (*Colchicum-autumnale*)
16.	Curcumin	Turmeric-rhizomes (*Curcuma-longa*)
17.	Dextrin	Starch (*Zea-mays, Oriza-sativa*)
18.	Digoxin	Digitalis leaves (*Digitalis-lanata*)
19.	Diosgenin	Dioscorea rhizomes (*Dioscorea deltoida*)
20.	Emetin	Ipecacunha-roots (*Cephaelis-acuminata*)
21.	Ergometrine	Ergot scherotia (*Claviceps purpurea*)
22.	Eugenol	Clove flower buds (*Eugenia caryophyllus*)
23.	Forskohlin	Coleus roots (*Coleus forskohlii*)
24.	Fructose	Inulin bulbs (*Inula helenium*)

Contd...

25.	Geraniol	Rosha oil (*Cymbopogon martini*)
26.	Glucose	Starch (*Zex-mays, Oriza-sativa*)
27.	Glycyrrhezinic acid	Liquorice roots (*Glycyrrhiza-glabra*)
28.	Guar-gum	Guar (seeds) (*Cyamopsis tetragonolobus*)
29.	Guggulosterone	Guggul oleo resin (*Commiphora weightii*)
30.	Hesperidin	Orange peels (*Citrus aurantium*)
31.	Hyoscyamine	Duboisia leaves (*Duboisia myoporoides*)
32.	Invertase	Yeast (*Saccharomyces cerevisiae*)
33.	Levodopa	Kavach seeds (*Mucuna prurence*)
34.	Menthol	Peppermint oil (*Mentha piperita*)
35.	Morphine	Opium-latex (*Papaver somniferum*)
36.	Nicotine	Tobacco leaves (*Nitotiana tobaccum*)
37.	Papain	Papita fruit (*Carica papaya*)
38.	Pectin	Lemon peels (*Citrus lemonis*)
39.	Piperine	Black-pepper fruits (*Piper nigrum*)
40.	Podophyllotoxin	Podophyllum rhizomes (*Podophyllum-hexandrum*)
41.	Psoralen	Psoralea fruits (*Psoralea corylifolia*)
42.	Quinine	Cinchona bark (*Cinchona succirubra*)
43.	Reserpine	Sarpagandha roots (*Rauwolfia serpentina*)
44.	Solasodine	Solanum berries (*Solanum - khasianum*)
45.	Stevioside	Stevia herb (*Stevia-rebaundiana*)
46.	Strychnine	Nux-vomica seeds (*Strychnous - Nux vomica*)
47.	Tannic acid	Oak galls (*Quercus infectoria*)
48.	Taxol	Taxus herb (*Taxus baccata*)
49.	Vasicine	Vasaka leaves (*Adhatoda-vasica*)
50.	Vincristine	Vinca leaves (*Catharanthus - roseus*)

7. NATURAL DRUGS AND THEIR SYNONYMS IN VARIOUS LANGUAGES

Key to Abbreviations Used

A.	Arabic	B.	Bengali;	En.	English;	Fr.	French;	G.	Gujrati;
Gr.	German;	H.	Hindi;	K.	Kannada;	M.	Malayalam;	Mr.	Marathi;
Or.	Oriyan;	P.	Punjabi;	S.	Sanskrit;	Tm.	Tamil;	T.	Telugu.
Tr.	Turkish	It.	Italian						

A. HERBAL DRUGS AND THEIR SYNONYMS

ACACIA

B. Babul
G. Gundar
H. Kikar gond
K. Gondu
M. Karuvelam
Mr. Babhul dink
S. Babbula
Tm. Krukkalamipalsi
T. Kodisapala
P. Kikar
Or. Babura
Gr. Agyptische akazie
Fr. Acacia d'egypt
It. Akazia

ACONITE

B. Katbish
G. Ativish
H. Vachnag
K. Vasannabhi
Mr. Bachnag
S. Visha
Tm. Vashanavi
T. Nabhi
P. Patees
Gr. Schulvolzbanin

AJOWAN

B. Yamani
G. Ajmo
H. Ajwain
K. Omu
M. Onva
Mr. Ova
S. Dipyaka
Tm. Omum
T. Vamu

ALFALFA

E. Lucern
H. Lasunghas
K. Vilayati-hully
Mr. Vilayati Gawat
P. Lusan
Ladakh : Hol

ALOES

B. Gherto-Kumari
G. Kumar pathu
H. Musabhar
K. Lolirasa
M. Kumari-Kathavazha
Mr. (Korphad) Kalabol
S. Ghirta Kumari
Tm. Karibolan
T. Musambaram

AMBI HALAD

B. Amada
H. Am haladi
K. Ambahalad
M. Mannayinci
Mr. Amba haladi
S. Karpura haridra
Tm. Mangai Inji
T. Manidiallum

AMLA

Ar. Amlag
B. Amla, Amlaki
En. Emblic-myrobalan
Fr. Eblique officinale
Gr. Grave myrobalane
G. Ambala, Anvala
H. Amla, Amlika, Aonla
It. Embelica
K. Amalaka, Nelli
M. Nelli
Mr. Avala
Or. Aunla
S. Amalaki
Tm. Nellikai
T. Usirikai
Tr. Amlac

ANANAS

Ar. Ananas
En. Pineapple
Fr. Ananas
Gr. Ananas, Fichtenapfel
H. Ananas
It. Ananasso ordinario
K. Kaitaccakka
M. Kazhudhchak ka
Mr. Ananas
Or. Sapuri
S. Anannas
Tm. Annasippalam
T. Annasapandu
Tr. Ananas, Festa

ANISE

B. Mauri
E. Aniseed
G. Anisa
H. Sounf
K. Sampu
P. Sounf
S. Shetpushpa
Tm. Sombu
T. Sopu

ANETHUM

B. Sewa
H. Sowa
Mr. Surva shepu
S. Pushpi
Tm. Sata Kuppi

ANNATTO

B. Latkan
En. Annato
G. Sindhuri
H. Sinduria
K. Kappumangal
Mr. Shendri
S. Raktabija
Tm. Jafra
T. Jafrachellu

ARACHIS

B. Chinerbadam
H. Mungphali
M. Nelkadala
Mr. Bhui-mug
Tm. Nila Kadalni
T. Verusenagalu

ARECA

S. Poogiphalam
H. Supari
Mr. Supari
Tm. Knmugu
Tel. Vakka
Mal. Adakka

ARJUNA

B. Arjun
G. Arjunsadra
H. Arjun
M. Venmarutti
K. Holematti
Mr. Arjun
S. Arjuna
Tm. Attumarudu
T. Tellamaddicittu
P. Arjan
Or. Arjuno

ARTEMISIA

G. Chhuvaria ajmoda
H. Kirmala
Mr. Kirmani ova
S. Chauhara
K. Murni
Gr. Strand deifuss
Fr. Armonoise
It. Assenzio

ASAFOETIDA

B. Hing
G. Hing
H. Heeng
M. Hingu
K. Kayam
Mr. Hing
S. Hinguka
Tm. Perungayam
T. Inguva
Or. Hengu
Gr. Asandisteckan
Fr. Assa-foetida
It. Assaptida

ASHOKA

B. Ashoka
G. Asopalara
H. Ashoka
K. Ashoka
Mr. Ashoka
S. Ashoka
Tm. Asogam
Or. Oshoko

ASHWAGANDHA

B. Ashwagandha
G. Asan
H. Asgand
M. Amnkkiram
K. Hiremaddinegida
Mr. Askandha
S. Ashvagandha
Tm. Ashvagandi
T. Asvtgandhis
P. Asagand

BABCHI

B. Lalakasturi
En. Psoralea
G. Bavachi
H. Babchi
M. Bavachi
K. Babchi
Mr. Bavchi
S. Karpokarishi
Tm. Kalagija
T. Kalagija
P. Babchi
Or. Bakuchi

BAEL

B. Bael
G. Bilvaphal
H. Bel
K. Bilva-patri
M. Baela-koovalom
Mr. Bel
S. Bilva
Tm. Vilvam
T. Bilvamu
P. Bela
Or. Belo
Gr. Bhelbaum
Fr. Bel indian
It. Bela Indiana

BAHAVA

En. Golden shower
G. Balla
H. Amaltas
K. Aragina
Mr. Bahava
S. Shema pusha

BEHRA

B. Behra
G. Behedo
H. Behera
M. Thannikkai
K. Shantimara
Mr. Beheda
S. Bahira
Tm. Akkam
T. Tandra
Or. Bhara
Gr. Murobala nenb
Fr. Myrobalan belleric
It. Mirobalano

BELLADONNA

B. Yebruj
H. Sag angur
Mr. Girbuti
P. Bantamaku
S. Suchi
Gr. Wolaskirche
Fr. Atrope
It. Solano insano

BENAFSHA (Wild violet)

B. Banosa
H. Banaphsha
M. Vayilettu
Mr. Banafshah
S. Banaphsha
Tm. Nilapuspha

BENZOIN

B. Luban
G. Loban
H. Lobamn
M. Sabrani
K. Lobana
Mr. Ood
S. Shriloban
Tm. Shabirani
T. Sambrani

BERBERIS

B. Aristata
G. Darhali
H. Rasaul, chitra
M. Maramanjal
Mr. Jharki-halad
S. Pita-daru
Tm. Mramanjal
T. Kasturi paspu
P. Jounsar

BETAL LEAF

B. Pan
G. Pan
H. Pan
Mr. Vidyache pan
Sk. Tembul
Tm. Vettilai
T. Tamalpaku

BHRINGRAJ

S. Bhrigaraja
En. Eclipta
B. Kesuti, Keshori
G. Bhangra
H. Mochkand
K. Garagadasoppu
M. Kyonni
Mr. Maka
Or. Kesarda
P. Bhangra
Tm. Garuga, Kaikeshi
T. Galagara

BHUI AMALA

S. Bhumi amalki
B. Bhui amala
H. Jangali amala
M. Kizhkkayinelli
Mr. Bhui-avali
Tm. Kizharalli
T. Nellavsari

BITTER ALMOND

B. Bilatibadam
G. Badam
H. Badam
Mr. Badam
S. Badama
Tm. Vadam-kottai
T. Badam vittulu

BLACK CATECHU

B. Khaer
G. Katho
H. Kathha
K. Cachu
M. Karianali
Mr. Kaat
S. Khadira
Tm. Karum-Kali
T. Chandra
P. Kher
Or. Khaira
Gr. Catechu akazie
Fr. Cachoutier
It. Acacia dicachou

BLACK PEPPER

B. Golmorich
G. Golmirch
M. Kurumuaka
Mr. Kalimiree
S. Milagu
Tm. Marichamu
T. Maricha
P. Golmirichi
Gr. Schwarzer pfeffer
Fr. Poivrenoir
It. Pepènero

BOSWELLIA

S. Sallaki
En. Indian olibanum tree
B. Salai
G. Mukul salai
H. Salai, Kundur
K. Madi, Guggula
M. Kunturukkam
Mr. Salai Sallaki guggul
Tm. Parangisambrani
T. Parangisambrani

BRAHMI

S. Jal/Neer brahmi
M. Brahmi
H. Brahmi
B. Brahmi sak
K. Niru brahmi
T. Neer Brahmi, sambrani
Mr. Brahmi
Tm. Neer brahmi

BROCCOLI

En. Wild Cabbage
B. Kopi
G. Pangobi
H. Karan Kalla
K. Karam
Tm. Gos
T. Gos
S. Dala malini

CALAMUS

S. Bhukhnashini
B. Bach
G. Gudhiloraj
H. Safed bach
K. Bajegida
M. Vekhand
Tm. Vasamghu
T. Vasa

CALOTROPIS

S. Ark, Arka
B. Acanda G. Akado
H. Ak, Madar
Mr. Mindara, AK
Tm. Vellerukku
T. Mandaramu

CAMPHORA

S. Karpoorah
En. Camphor tree
Ar. Kafur
Fr. Camphrier dechine
Gr. Japanischer Kamferbaum
G. Kapur
H. Kapur
It. Allorocanforato
M. Puram, chutakkar
Or. Karpura, Gacha
Tm. Karpurammu
T. Karppuram
Tr. Kiafur ag

CANOLA OIL

S. Kala sarshapa
B. Kali sarson
H. Kali sarson
Tm. Karappu khadigu
T. Nallaavalu
M. Karukhatuka

CANNABIS

B. Bhang
G. Ganja
H. Ganja, charas
K. Bhangi ganja
M. Kancha
Mr. Ganja
S. Bhanga
Tm. Kanja
T. Ganjayi
Gr. Echterhana
Fr. Chanurier
It. Canapa indica

CAPSICUM

B. Lal-mirachi
G. Marchan
H. Lal-mircha
K. Mensinkay
M. Mulagu
Mr. Lal mirchi
S. Mirchi-palan
Tm. Panchirmolaga
T. Mirpa-kayalu
Or. Lalilanka
Gr. Echtebeiss beere
Fr. Piment
It. Peporossa

CARAWAY

B. Jira
G. Shahjiru
H. Shahjira
K. Jiriggae
M. Shahjiira
Mr. Shahjira
S. Seeshavi
Tm. Shimai - shembu
T. Shimaisapu
P. Zira siah
Gr. Echter kummel
It. Comino tedesco

CARDAMOM

B. Elachi
G. Elaichi
H. Choti ilaychi
K. Yalakki
M. Yelam
Mr. Veldoda
S. Ela
Tm. Elakkaiz
P. Illachi
Gr. Kleine cardamomen
Fr. Cardamome petit
It. Cardamomo minore

CARTHAMUS

B. Kusum
G. Kusumbo
H. Kusum
Mr. Kardai
Sk. Agnishikha
Tm. Nila Sundurakam
T. Kushumba

CASSIA

B. Dalchini
G. Taj
H. Dalchini
M. Lavanga-patti
Mr. Dalchini
Tm. Lavangapattai
T. Dasam-chakkau

CASTOR OIL

B. Bherenda
G. Divel
H. Erand
K. Harelenne
M. Amanakku
Mr. Erand tel
S. Erandi
Tm. Amanakkuchei
T. Arnudamu
P. Rendi
Or. Jada
Gr. Echter wunderbaun
Fr. Palma christi
It. Risino

CELLERY

B. Chana, Randhuni
G. Bodiajamoda
H. Ajmoda
Mr. Badi ajmud, karap
E. Apium
S. Ajmoda
P. Bhutjata

CENTELLA

B. Tholkkari
G. Barmi
H. Braha Mandaki
T. Bokku du
Mr. Brahmi
S. Manduk parni
Tm. Bellarai

CHAULMOOGRA

B. Chalmoogra tel
H. Kava-ka tel
M. Morotti enna
Mr. Kadu Kawat
Tm. Niradimettu ennai
T. Nirodi vittulu nune

CHENOPODIUM

En. Mexican tea
B. Bathusag
G. Sheel
H. Sugandhvastuk
M. Katu
T. Parupakki
P. Bathu
S. Sugandhvastuk

CHICORY

S. Kasni
En. Chicory
B. Kasni
H. Kasni
Mr. Chicory
U. Kansi

CHIRATA

B. Chitrata
H. Chirayata
K. Nelabedu
Mr. Chiraita
S. Kirata-tkta
Tm. Nila vembu
T. Nila vem
P. Charaita

CINNAMON

B. Dalchin
G. Ceylon taj
H. Dalchini
M. Dalchini
Mr. Dalchini
S. Darushila
Tm. Lavangapatti
Gr. Zimtbaum
Fr. Cannellier deceylon

CITRONILLA

(See Lemon grass)
E. Lemon grass
G. Lilicha
H. Gandhatrina
Mr. Gavati chaha
S. Pntigandha
K. Parhati hullu

COCONUT

S. Narikelah
En. Coconut plam
Ar. Gawz el hind
B. Dab, Narikel
Fr. Cocotier
Gr. Kokospalme
G. Naliyer
H. Nariyal
It. Albero del cocco
K. Tenginamara Tengu
M. Tenna
Mr. Naral
Or. Nadia
Tm. Tenkai
T. Tenkaya
Tr. Narcilag

COFFEE-SEED

H. Coffee
G. Cofee
Mr. Coffee
B. Kafi
M. Bund

COLEUS

H. Pathar
G. Garmar
K. Maghiberu
Mr. Maeen mool

COLCHICUM

S. Suranjan
H. Hirantutiya
P. Surinijan-talkh

COLOCYNTH

B. Indrayan, makhal
G. Indrayan, makhal
H. Indrayan, makhal
K. Karandali
M. Vaikummatti, katuvelleki
Mr. Indrayan, makhal
S. Indrauanuni, Uirhala
Tm. Paedikani, Attutunnatti
T. Eti-puchcha, chittipapara
P. Kaud tumbha
Or. Kalara
Gr. Bitter zitrutte
Fr. Coloquinte
It. Coloquinda

CORIANDER

B. Dhania
G. Konphir
H. Dhania
K. Kothambri
Mr. Dhane
S. Dhanika
Tm. Kotamalli
T. Danyalu

CORN OIL

B. Bhuththe janar
G. Makai
H. Makka, Bhutta
M. Jaguny
Mr. Maka, makaibanda
S. Yanarala
Tm. Makka-chalam
T. Makka-zonnalu

COTTON

B. Kapas
G. Kapas
H. Rui
K. Hatti
M. Paruti
Mr. Kapus
S. Karpsi
Tm. Paratti
T. Patthi
P. Rui
Or. Carpaso
Gr. Kurzegeorgia baumwolle
Fr. Cotonnierd' asie

CLOVE

B. Lavang
G. Laving
H. Laung
K. Lavanga
M. Grambhu
Mr. Lavang
S. Lavangaha
Tm. Lavangapattai
T. Dasammchakkau

CUBEB

B. Sitalachini, kababchini
G. Tadamiri
H. Kababchini
M. Val - milaku
Mr. Kankol, Himsimiri
S. Sungadha - muricha
Tm. Val - milaku
T. Toka - miriyalu, chalara - miriyal
Gr. Cubeben pepffer
Fr. Poivrier cubebe

CUMMIN

B. Safed jeera, zira
G. Safed Jiraun, zero
H. Safed Jeera, zira shiajira, jira
M. Cheerakam, Jeerakam
Mr. Jeera
S. Ajaji, Jeeraka, Hrasvanga
Tm. Shimai - shombu, cheerakum
T. Jeelakara, jirana, jiraka
P. Zira safed
Gr. Echter romerkumngel
Fr. Cumin
It. Cumino

DARUHALAD

B. Daruharidra
G. Daruhaldar
H. Daruhaldi
K. Dodde Marad resin
M. Daruhalad
S. Daruharidra
Mr. Daruhalad
Tm. Magamayal

DATURA

B. Dhatura
G. Dhatoria
H. Dhatura
K. Kariummathi
M. Karrinummatham
Mr. Dhotara
S. Dhusturapattai
Tm. Vellummattai
T. Ummatha
P. Dhattur
Or. Dudura
Gr. Echter stechapfel
Fr. Stramoine
It. Stramonio

DAVANA

H. Davana
K. Davana
Mr. Davana
Tm. Davana
T. Davanamu

DHAYATI

B. Dhai
En. Downy grislea
H. Dhawai, Dhabi
K. Are
M. Dhaitti, Madhya-vasini
Mr. Dhayti
Tm. Velakhai
T. Siringi
P. Dhavy
Sk. Dhataki

DIGITALIS

En. Foxglove
H. Digitalis
S. Hrat-patri
Mr. Teel pushpi

DILL

B. Sowa
G. Suva
H. Sova
K. Sata
M. Cutakuppa
Mr. Shepu
S. Satapushpi
Tm. State kupe
T. Sompa

DIOSCOREA

S. Varahi
En. Japanese yam
Ar. Ighnam
B. Banalu, Kukuralu
Fr. Dioscoree
Gr. Yamwurze
H. Ratalu
It. Igname
Mr. Kadu-karanda
Or. Kandamula
Tm. Kokikilangu
Tr. Ignam

EPHEDRA

S. Soma
En. Ephedra
P. Asmania

FENNEL

B. Ban mauri
G. Variyali
H. Sonf
K. Bada soppa
M. Sambu
Mr. Badishep
S. Madhurika
Tm. Sohikerovirai
T. Paddejilakurra
P. Sonf
Or. Maduri
Gr. Echter fenchel
Fr. Fenovil
It. Finocchio

FENNGREEK

S. Chandrika
Tm Vendayam
T Menthi-kura
G. Methini
H. Methi
M. Methi
Mr. Methi
K. Menthe-palle

FIG

S. Anjira
B. Anjir
G. Anjeer
H. Anjir
Mr. Anjir
Tm. Simayatti
T. Anjura
P. Fagari

GALANGA

B. Kurchi vaca
En. Greater Galangal
G. Kulinjan
H. Bada, kulinjan
K. Dumparasmi
M. Kol-inji, Peraratte
Mr. Kosht-Kulinjan
Tm. Perara tthei
T. Pedda dumparashtran

GALLS

B. Bonjhau
E. Tamarize
G. Jhaunijhauda
H. Jhau
Mr. Jhau
S. Jhavu
Tm. Sirush vukka
T. Erusara
P. Jhau

GARLIC

B. Lasun
G. Lasun
H. Lasan
K. Bellulli
M. Velluli
Mr. Lasun
S. Ugragandha
Tm. Karuvelum
T. Nellatumma
P. Thom
Or. Rasun
Gr. Knoblauch
Fr. Ail
It. Aglio

GENTIAN

S. Trayanti
En. Gold thread
B. Karu, Kutki
H. Kadu, Kutki
It. Gentiana
K. Karadihanni
M. Trayamana
P. Nilkant, Nilakali
Tm. Kampantirai
T. Buroni

GINGER

B. Ada
G. Adu
H. Adrak
K. Shunti
M. Inchi
Mr. Ale
S. Adrakam
Tm. Sukka
T. Sonti
Or. Ada
Gr. Ingiver
Fr. Gingembre
It. Zenzero

GLORIOSA

S. Langali, visalya
En. Malabar glory lily
Ar. Al bahir al hindiyah
B. Bishalanguli
Fr. Superbe de Malabar
Gr. Malabarische methonika
G. Varhvardi
H. Karihari, Languli
It. Narciso superbo
K. Agnisikhe
M. Kantal, Mentonni
Mr. Indai, Watnakhi
Or. Gulia
P. Kariari, Mulim
Tm. Akkinichilam
T. Adavinabhi

GOKARNI

S. Aparajita
B. Nila aparajita
M. Aral
E. Butterfly pea
Tm. Kakkanum

GOKHRU

B. Gokhru
H. Cholagokhru
K. Negalu
M. Neringil
Mr. Lahangokharu
S. Gokshura
Tm. Nerunji
T. Palleru
P. Lotak
Or. Gokhuru
Gr. Echter burzel
Fr. Tribule terrestre
It. Tribolo basapie

GUAR GUM

G. Guar-gundar
H. Gowar Gond
K. Guvar kai
M. Guwar
Mr. Gawar dink
S. Bakuchi
Tm. Kothaveray
T. Gorchidudu

GUGGUL

B. Guggul
G. Guggul
H. Guggul
M. Guggul
Mr. Guggul
S. Guggulu
Tm. Gukkulu
Gr. Indisches bdelliumbaum
Fr. Bdellium de'Inde
It. bdellio

GUIZOTIA

B. Ram til
E. Nigar seed
H. Kalatil
G. Ram til
Mr. Karale
K. Guretil
T. Vallsulu

GULWEL

B. Gurach, Giloe
H. Gilo
M. Sittamrytu
Mr. Gulvel
S. Guduchi, Amrita
Tm. Sindil
T. Somida

GYMNEMA

B. Merasingi
G. Mardasingi
H. Gurmar
K. Sennegerasehambu
Mr. Kalikardodi
S. Meshashringi
Tm. Adigam
T. Podapatri

HENNA

S. Ragangi
En. Henna plant
Ar. Henna
B. Mendi
Fr. Henne
Gr. Agyptische
G. Medi
H. Mehendi
It. Enne, Cipro
K. Mayilanchi
M. Mailanchi
Mr. Mendhi
Or. Benjati
P. Mehndi
Tm. Marithondi
T. Goranti
Tr. Kena ag

HEMI DESMUS

B. Anantmul
H. Magrabu
M. Nanri
Mr. Anantmul
S. Ananta
Tm. Nannari
T. Gadisugandha

HIBISCUS

S. Rakta pushpi
En. Chinese hibiscus
Shoe flower
H. Gudahal, Jasut
K. Daswada
Mr. Jaswanda
B. Joba
G. Jasua
Tm. Sapattuppa
T. Dasnamn

HONEY

B. Madh
G. Madh
H. Madhu
K. Jenu Tuppa
M. Ten
Mr. Madh
S. Madhu
Tm. Ten
T. Tene

HYOSCYAMUS (Henbane)

B. Korasani ajowan
H. Khurasani ajavayan
Mr. Khorasani owa
S. Yavani, Parasikaya
Tm. Korusanai
T. Karashanivamam

INDIAN SARSAPARILLA

S. Sariva
H. Salva
G. Upalsari
B. Anantmul
U. Anantmool
T. Nannari
Tm. Nannari

INULA

S. Rasana
H. Paskarmul
K. Rasnabheda
M. Puskkaramulam
Mr. Tambadi
Tm. Puskkaramulam
T. Puskaramu

ISAPGOL

B. Eshopgol
G. Isafaghol
H. Isabghul
K. Issagolu
M. Karkalasaringi
Mr. Isabgol
S. Shigdhajiraka
Tm. Ishappukol
T. Isapagola
Gr. Sogelwegerich
Fr. Ispaghula

JAMALGOTTA

En. Purging croton
G. Nepal
H. Jamalgota
K. Kanti, Jepeda
Mr. Jamalgota
Sk. Dhanti

JASMIN

B. Chambelli
G. Chambeli
K. Ajjige
Mr. Jaee
S. Chambeli jai

JALBRAHMI

S. Manduki
En. Centella
H. Jalabrahmi
K. Nirubrahmi
M. Nirbrahmi
Mr. Mandukparni
Tm. Arakkam
T. Sugandipala
Tr. Hind Saparnasi

JATAMANSI

B. Jatamanshi
G. Jatamasi
H. Jatamanshi
M. Jetamanshi
Mr. Batacharea
S. Jatamanshi
Tm. Jalamansi
T. Jatamansi
Gr. Indischenarde
Fr. Nardindien
It. Spignardi

JUTE

En. Jute
G. Chhunachan
Mr. Chaunchan
S. Kalasaka

KALADANA

B. Kaladana
G. Kaladana
H. Nilkanal
Mr. Nilpushpi
S. Krishnabji
Tm. Sirikki
T. Jirikki
Gr. Nilwind
Fr. Etoileda matin

KALMEGH

B. Kalmegh
G. Kariytu
K. Nelaberu
H. Kiryat
Tm. Kiryat
Mr. Olikirayat
S. Kirata
Tm. Nilavembu
T. Nelavembu
En. Great chiretta

KALIJIRA

B. Mus grela
E. Black cumin
G. Kaloni giram
H. Kalanji
Mr. Kalijire
S. Upakanchika
Tm. Karunji, Shira gum
T. Nullaji lakra

KAPUR KACHARI

S. Shati
En. Spiked ginger lily
B. Kapurakachari
G. Karpurakachari
H. Kapura Kachari
K. Gandhashanti
Mr. Sontakka
P. Sheduri
Tm. Shimai Kichchilik Kishangu

KARANJA

B. Dahar kananja
H. Kuranj, kiramal
K. Karinje-kraaku
M. purgammaram
Mr. Karanj, kidumar
S. karanja, Nalatarnala
Tm. Punyamuranam
T. kamuga-chetter
P. Karanj
Or. Koranjo
Gr. Mondbohne
Fr. Arbe de' pongolote

KAVACH (Cow itch plant)

B. Alkusi
H. Kavach
M. Shoriyanam
Mr. Kach Kuhili (Kuhili)
S. Kapikachechha, Atmagupta
Tm. Punajkhali
P. Dulagondi

KOKUM BUTTER

G. Kokam
H. Kokum
K. Murgal
M. Punampuli
Mr. Ratamba, Kokamb
Tm. Murgal

KURCHI

B. Kurchi
G. Kadacchal
H. Kaureya kurachi
K. Kodamuriki
M. Kadagapala
Mr. Kudasal
S. Kutaja
Tm. Krukkalami-palsi
T. Kodisapala
P. Keor
Or. Kheraba
Gr. Iuhrohrholarrhena
Fr. Holarrhena antidysenterique

LEMON GRASS

B. Gandha benertel
E. Lemon grass
G. Giliche
H. Gandha-bina-ka-tel
Mr. Hirava chafa
Gr. Camaleshov
Fr. Citronellejonc ordorant

LEMON PEEL

B. Niboo
G. Limboonichhal
H. Nimboo-ka-chhilaka
K. Nimbe sippe
M. Cherunarangatholi
Mr. Limboo saal
S. Jambira
Tm. Elunibhaamphalam
T. Nimma-gandu

LINSEED

B. Masina
G. Alasi
H. Alasi
K. Agashi
M. Cerucannam vithu
Mr. Javas
S. Agasi bija
T. Arish vettu
Gr. Echterlein
Fr. Lincommun
It. Linousuale

LIQUORICE (Glycyrrhiza)

B. Jasthi-Medhu
G. Jethimadha
H. Mulethi/Mithi-lakadi
K. Jhthahimhuram
Mr. Jeshtamadh
S. Jashta madhu
Tm. Atimadhuram
T. Atimadhuramu
Gr. Echtes sussholz
Fr. Regulisse
It. Dolce radice

LOBELIA

B. Bantamaku
H. Nali
K. Nala
M. Kathupukyaila
Mr. Deonala
S. Bibhishna
Tm. Kattuppugaiyinai
T. Adavipogaku

LODHRA

B. Lodh
H. Lodh
Mr. Lodhra
S. Lodhra
Tm. Bhom roti
T. Loddugi

LONG PEPPER

B. Pepul
H. Pipal
M. Pippali
Mr. Lendipimpali, Pan-pimpali
S. Pipali
Tm. Thippilli
T. Pippallu

MANJISHTA

Eng. Indian madder
San. Manjista
G. Majeeth
Kan. Manjushta
M. Manchetti
H. Majeeth
Mr. Manjishta
B. Manjit
T. Tamtavatti
Tm. Manditta

MENTHA

B. Podina
H. Pudinah
M. Putiyina
Mr. Pudina
Tm. Pudina
T. Pudina
P. Bahure
Or. Odana
Gr. Baschmunze
Fr. Baumebert
It. Menta

MOMORDICA

H. Karela
B. Karela
Mr. Karle
S. Sushari
Tm. Pakal
T. Kakara
K. Hagul
M. Kaippavalli
Or. Kalara

MUSTARD

B. Raisarisha
G. Rai
H. Sarson
K. Sasive
Mr. Mohari
S. Rajica
T. Kadugu
P. Asalrai
Or. Sorosha
Gr. Sareptasent
Fr. Bredes moutarde

MYROBALAN

B. Haritaki
G. Hirdo
H. Harana
K. Alate, Harade
Mr. Hirda
S. Haritaki
Tm. Kadukkai
T. Karitaki
P. Har
Or. Haridra
Gr. Mirobalanenbaum
Fr. Myrobalan belleric
It. Mirobalano

MYRRH

B. Gandharash
G. Hirabol
H. Bol
K. Bola
Mr. Hirabol
S. Rasgandha
Tm. Vellaipalam
T. Balintrapolum
Gr. Echter myrrhenbaum
Fr. Arbreamyrrhe
It. Alberodel' mirra

NAGLI
S. Rajika
B. Marua
H. Ragi
G. Barto
K. Ragi
Mr. Nagli, Nachani
Tm. Kelveregu
T. Rajulu

NEEM
B. Nim
H. Nim
M. Vepa
Mr. Kadulimb
S. Nimba
Tm. Vepa
T. Vepa
P. Bakam
Or. Nimba
Gr. Margosa geneiner
Fr. Azadarachte
It. Azadarc

NIRGUDI
B. Nisinda
G. Nirgari
H. Nirgandi
K. Bile-Nikai
M. Indrani
Mr. Nirgudi
T. Nallavalli
S. Nirgundi
Tm. Nirkunnchi

NUTMEG
B. Jayphal
G. Jayphal
H. Jayphal
K. Jaikai
M. Jatka
Mr. Jayphal
S. Jatiphalam
Tm. Jajikaya
P. Jaiphala
Gr. Muskatnussbaum
Fr. Muscadier
It. Nocemoscata

NUX-VOMICA
B. Kuchila
G. Zeri-Kachuo
H. Kuchla
K. Kasar-kena
M. Kanjiram
Mr. Kuchala
S. Vishamushti
Tm. Yettiraman
T. Mustimanu
P. Kagophale

OPIUM
B. Poostodheri
G. Afin
H. Afim
K. Afeemu
M. Apeem
Mr. Aphoo
S. Naranga
Tm. Apin, Abhni
T. Nallamandu
P. Afim
Or. Afima
Gr. Schlafmohn
Fr. Pavot
It. Papavero

OLIVE
H. Oliu
K. Julipe
Mr. Olive
Tm. Smidun

ORANGE PEEL
B. Kamala nebu
G. Santra-ni-chhal
H. Narangi-ka-chilka
K. Musambi-sippe
M. Naranga-chi-saal
Mr. Santrya-chi-saal
S. Naranga
Tm. Kitchli

PALE CATECHU
B. Papri, khayar
H. Kath, kutha
M. Gambier
Mr. Kath
S. Khadir
Tm. Ankudu- kurna
T. Ankudu - kurra

PAPAYA
B. Papeya
H. Papaya
M. Pappayam
Mr. Papai
Tm. Pappayi
T. Boppayi
P. Randkharipuza
Or. Amrulabhanda
Gr. Papyabaum
Fr. Papayer
It. Ficodelisole

PATCHAULI
S. Ajamoda
B. Pachapat
H. Patchauli
E. Patchauli
Mr. Pachpan
Tm. Kadir pachai

PHYLLANTHUS
Eng. Phyllanthus
S. Bhumyaamlaki
H. Jangali amla
Mr. Bhuee avala
B. Bhuiamla
Tm. Kizharelli

PICRORRHIZA
B. Katki
G. Kadu
H. Kutki, kuru
M. Katu khur- obani
Mr. Kutki
S. Katuka
Tm. Kitchli
T. Katukuroni
P. Karru

PIPER BETLE

B. Tam vellitai
H. Pan
G. Pan
K. Ellabetti
M. Tambulum
Mr. Vidyache pan
Sk. Tamboli
T. Vetica

PIPLAMOOL

B. Piplaimool
G. Pippri mool
H. Piplamool
K. Hippaliberu
M. Tippali mulum
Mr. Pimpali
Tm. Kandan Tippali
T. Tippalikatte

PODOPHYLLUM

S. Giriparpat
En. Indian podophyllum
B. Papra
G. Venivel
H. Bhavan bakra
Mr. Padwel
P. Papri

PSORALEA

H. Babchi
G. Baurchi
S. Sugandha Kantak
T. Kala-ginja
M. Tea
Mar. Bauchi

PTEROCARPUS (Kino)

B. Pistal
G. Biyo
H. Bijasal
M. Veng
Mr. Asan, Dhorbenla
S. Honne
Tm. Vengai
T. Yeggi

PUNARNAVA

B. Sabunba
G. Lalsabuni
H. Lalsabunooch
K. Muchchu-joni
Mr. Pandhari-punernava
S. Punernavi
Tm. Sharunai
T. Galijeru

RAPE SEED

B. Kali sarson
H. Kali surson
S. Kala sarshopa
M. Karu-puaka-tuea
Tm. Karu ppukka dugu
T. Nalla valu

RAUWOLFIA

B. Chandra
G. Sarpagandho
H. Chhotachand
M. Sarpagandha
Mr. Sarpagandha
S. Punernavi
Tm. Govannamilpuri
T. Patala agandhi

RASNA

S. Sugandhamula kulanjan
En. Galangale, Siamese, ginger
Ar. Khawalngan
B. Kulanjan
Fr. Galanga
Gr. Galgant
G. Kulinjan
H. Kulanjan, kulinjan
It. Alpinia
K. Kumbarasme
M. Kolinji, Aratta
Mr. Kulinjan
Tm. Perarathei
T. Dumparastramu
Tr. Pedda-Dumparashtram

RHURARB

S. Gandhini
B. Bangal-arevanchini
G. Gamnir-eranchini
H. Revakinni
Mr. Ladaki-revanchini
S. Gandhini
Tm. Nettureval chini
T. Nethurewal chini
K. Revakinni
M. Variyatta

SAFED-MUSALI

G. Gholi- musali
H. Safed musli
M. Shedeveli
Mr. Safed -musali
Tm. Tanirureta

SAFFLOWER

B. Kusum, kajireh
G. Kusurubi, kasauibi kabri
H. kusum, kajireh
M. Chendurakam
Mr. Kardai, Kusumba
S. Kammalottara, Kusumba
Tm. Senderukam
T. Ayni-sikha, kusumbher

SAFFRON

B. Jafran
G. Keshar
H. Kesar
Mr. Keshar
S. Kumkuma
Tm. Kungumapu
T. Kunkumpuva

SALLAKI

S. Sallaki
B. Salai
G. Dhup
K. Sambrani
M. Muntnrukkam
Mr. Sallaki-guggul
T. Parongi-sambrani

SANDAL-WOOD

B. Chandan
G. Chandan
H. Safedchandan
K. Chandana
M. Chandnam
Mr. Chandan
Tm. Chandanam
T. Chandunamu

SAUSSUREA (Costus)

B. Kur
G. Kut
K. Koshtha
H. Kut, Pachak
M. Sepuddy
Mr. Ooplet
S. Kushtha
Tm. Koshtam
T. Koshtum
En. Kuth

SENNA LEAF

B. Sonamukhi
G. Nat-kisana
H. Sena-ka-patta
K. Nilavaka
Mr. Sonamukhi
S. Swarnamukhi
Tm. Nilavarai
T. Sunamikhi
P. Sanapati

SESAME OIL

B. Til
G. Til
H. Trillika tel
K. Achchellu
M. Ellu
Mr. Teelache tel
S. Tila
Tm. Ellu
T. Nurrullu

SHANKHAPUSHPI

B. Dankuni
G. Shankhaphuli
Mr. Shankhvel
S. Shankhapushpi

SHATAVARI

B. Shatamuli
H. Satawar
M. Shatawali
Mr. Shatavari
S. Shatmuli, Tagar
Tm. Shimaishdavari
T. Challagadda

SPEARMINT

En. Green mint
H. Pahadi Pudina
P. Pahadi Pudina
B. Pudina

SQUILL (Urginia) (Indian Squill)

B. Jangli piyaz
H. Jungli piyaz
M. Kanthanga
Mr. Jangli kanda
S. Vanapalandam
Tm. Kattuvengayam
T. Nakkavalligadda

STORAX

B. Silaras
G. Silaras
H. Silaras
M. Silaras
S. Silaras
Tm. Meri - arishippal
T. Silaras

STRAMONIUM

B. Saa-dhatura-pata
H. Dhattura
K. Ummatta
M. Ummatta
Mr. Dhotra
S. Dhuttura patra
Tm. Umma thai
T. Dhathuramu

SWEET ALMOND

S. Badam
B. Bilati badam
G. Badam
H. Badam
M. Badam
Mr. Badam
P. Badam
T. Badamane
Tm. Vedamai

TAMARIND

S. Amlika
B. Ambli
G. Amli
H. Imli
Mr. Chinch
K. Amlike
P. Imli
T. Amlika
Tm. Amilan

TAXUS

B. Sugandh, Bhirmie
H. Thuneer, Birmi, Kash
Mr. Barmi
S. Talispatra
Tm. Thalizopathari
P. Birmi, Kumaon, Thuner

TEA-LEAVES

H. Chay
P. Chai
B. Cha
Mr. Chaha
Tm. Thayilai
T. Theyaku

THEVETIA

B. Kolkaphul
G. Pila - kaner
H. Pila - kaner
M. Pachchai - alari
Mr. Pivala kanrer, sherani
S. Hapusha
Tm. Pachchchai - alari
T. Pachcha gannera

TINOSPORA

B. Gurach, Giloe
H. Gilo
M. Sittamrytu
Mr. Gulvel
S. Guduchi, Amrita
Tm. Sindil
T. Somida

TOBACCO

B. Tamak
G. Tamaku
H. Tambaku
M. Pokala
Mr. Tambakhu
S. Tamakhu
Tm. Pugal villay
T. Pogaku

TODDY

S. Taal
B. Taal
E. Ice apple
G. Tad
H. Taari
M. Pane
Mr. Taad
T. Tate-munjlu
Tm. Nugnu

TRIGONELLA

S. Methika
En. Fenugreek
Ar. Hhelbah
B. Methuka
Fr. Fenugree
Gr. Bockshom
G. Methini
H. Mutti
It. Fiena Greco
K. Menphya
M. Venidayam
Mr. Methi
Or. Methi
P. Methi
Tm. Vendayam
T. Mentulu
Tr. Cemen

TULSI

B. Tulsi
G. Tulsi
H. Tulsi
K. Vishnu-tulsi
M. Trittaru
Mr. Tulsi
S. Manjiri
Tm. Thulasi
T. Thulasi

TAMALPATRA

S. Tej patra
B. Tej patta
H. Tej patta
Mr. Tamal patra
Unani Tajpat
T. Talis patri
Tm. Talisha pattri

TURMERIC

S. Haridra, Haldi
B. Halud
G. Halada
H. Haldi
K. Arishina
M. Haladi
Mr. Halad
Tm. Manjal
T. Pasupu

URGINEA

B. Jangali pyaz
G. Iangli dun
H. Jangli piyaz
K. Kadu belulli
M. Kantena
Mr. Bhui-kanda
S. Kolakanda
Tm. Nari-vengayam
T. Advitellagadda

VAJRADANTI

S. Kuntakoranta
G. Kanthephelio
H. Katoeriya
Mr. Kate koranti
M. Kalsunda
K. Mudarangi

VALERIAN

B. Tagar
H. Tagar, Mushkbala
Mr. Kalavala
S. Tagar
K. Mandibattal
G. Tagar
O. Bada musk

VASAKA

B. Bakas
G. Adsoge
H. Arusha
K. Adusoge
M. Adalodagam
Mr. Adulsa
S. Vasa
Tm. Adatodai
T. Addasaramu

VETIVAR OIL

B. Khaskhas
G. Valo
H. Khaj, khas-bene
K. Bhanavalo
M. Ramachham
Mr. Vaala
S. Usheena Veeranam, Aurranalan
T. Kunuvaeru uetti-vellu

VIDANG

B. Vidang
G. Vavding
H. Vayavidang
K. Vayuvilanga
M. Vizalari
Mr. Vavding
Tm. Vayuvidangam
T. Vayuvidangam

VINCA

B. Nayantara
H. Sadabahar, Ratanjyot
Mr. Sadaphuli
T. Billaganneru
G. Sadaphuli
K. Billaganeru
En. Periwinkle

BAJRA

B. Bajra
G.
H. Bajra
Mr. Bajari
S. Nali
P.
Tm. Kanbu
T. Sazza

KODRA (Kodo millet)

B. Kado
G. Kodre
H. Kado adhan
K. Harada
Mr. Kadra
P. Kodra
Tm. Varagu
T. Kiraruga

NACHANI (Finger Millet)

B. Marua
S. Rhjika
H. Ragi
G. Barto
K. Ragi
Mr. Nagli
Tm. Kelveregn
T. Rajulu

CHINNA (Prose-millet)

S. China
B. China
H. China
Mr. Rai (Bhagar)
G. Rai
P. China
Tm. Vagurn
T. Vagarulu
K. Bagara

RAJAGIRA

S. Rajgiri
B. Chako
H. Chanalai
Tm. Pungi
G. Chuko
Mr. Rajgira

SAVA (Born yard mille)

B. Sanwa
S. Shyama
Mr. Bauto samul
H. Sanua
P. Sawank
Tm. Kudraivali pillu

JAWAR

S. Dirghamala
H. Jawar
B. Jawar
P. Jawar
Mr. Jawari
Tm. Cholam
T. Jannolu
M. Chavela

KANGANI (Fox-tail millet)

S. Kangavi
B. Kangani
Mr. Kangani
Tm. Tenai
T. Koralu
M. Tauna
H. Kangani
P. Kangani

KURRI (Brown top millet)

Mr. Chasura/Char

KUTKI (Little millet)

Panicum sumatrense

B. ANIMAL DRUGS AND THEIR SYNONYMS

CIVET

B. Khatase
E. Civet Cat
H. Ladona
S. Gandha marsare
Tm. Punugu puney
T. Sawad puney
Mr. Panuga majar
K. Punugin bekku

KASTURI

En. Musk
B. Kasturi
G. Kasturi
H. Kasturi
K. Kasturi
Mr. Kasturi
Tm. Kasturi
T. Kasturi
S. Mriga nabhi

SUET

B. Charbi
En. Prepared suet
G. Charbi
H. Charbi
Mr. Charbi
Tm. Mozhuppu
T. Kohu
Mrignabhi

LEECH

S. Jaluka
E. Speckled Leech
B. Jonk
G. Jalo
H. Jonk
M. Attah
Mr. Jaloo
Tm. Attei
T. Jalgalun

MYLABRIS

E. Telini fly, Mylabris beetle
H. Teleni makhi
B. Telini poka
Mr. Telini Mashi
Tm. Puis Tarini
T. Ejaloo

SHARK LIVER OIL

H. Machhi Kael
Mr. Shark Mashache
Tm. Meena aenay

SHILAJIT

B. Silajitu
H. Silajata, Ralyahudi
S. Shilaras
Tm. Pirangyum
Mr. Shilajeet

SNAKE VENUM

S. Sarpawish, Garala
H. Sap ka Jahar
G. Vadulun

BEES WAX - YELLOW

En. Yellow bess-wax
B. Mom
G. Mina
H. Mor
M. Kezhuku
K. Mena
Mr. Men
S. Madhujan, Shiktha
Tm. Mellugu
T. Mainarm

COCHINEAL

H. Beer-bouhtee
G. Kira maja
K. Kiranjee
Tm. Kiranjee
T. Kiramju

CASTOREUM

S. Gen ¨
E. Castor
H. Gondabadustan, Jundo
G. Zanda – bidastara
M. Alu-beeyum
Tm. Kasturi munai
T. Zanun, Naru

LAKH

B. Gala
E. Lac
H. Lakh
G. Lakha
M. Laksha
Mr. Lakh
S. Laksha
Tm Araku
T. Lacca, Lukkah

HONEY

B. Madh
G. Madh
H. Madhu
K. Jenu Tuppa
M. Ten
Mr. Madh
S. Madhu
Tm. Ten
T. Tane

AMBRA GRASEN

En. Ambergris
S. Amber sugandha
B. Amber
G. Ambara
Mr. Amber
Tm. Minumber

C. MINERAL DRUGS AND THEIR SYNONYMS

BORAX

S. Ras shodan
E. Sod borate
H. Tinkal
B. Sahaga
G. Tankan khar
Mr. Kankan khar
M. Ponkaram
Tm. Venkaram
T. Velligaram

CHALK

B. Karimatti
G. Chaka
H. Vilati-chuna
M. Kapur ingris
Mr. Khadu
Tm. Seemay Churnable

GOPICHANDAN

S. Shoraktri
H. Pani-soka
B. Sugandamitti
Mr. Gopichandan

ROCK SALT

H. Sendhaalon
K. Sendhara lavana
G. Sendhaluna
Tm. Indu-uppu
T. Saindhava lawan
M. Intu-upa
Mr. Sendhu lawan

KAOLIN

S. Gairika, krishna, mrittika
E. China clay kaolinite, Porcelain clay
B. Girimati
H. Chikni mitti
Mr. Multani mitti

SHILAJIT

S. Shilaras
B. Silajitu
H. Silajita, Ral yahudi
G. Shilajit
Mr. Shilajeet
Tm. Perangyum

SODIUM CHLORIDE

S. Saindhava
E. Rock salt, sea salt, Bay salt
H. Sendhalon
K. Sendhara lavana
G. Sendhluna
Tm. Indu
T. Sendhava lavana
Mr. Saindhav

SULPHUR

B. Gandhak
En. Sublimed sulphur
G. Gandhak
H. Gandhak
K. Gandhak
M. Gendagum
S. Gandhak
Tm. Gondacum
T. Gondhakakam
Mr. Gandhak

TALC

E. Soap stone
H. Singe jerahata
G. Shankh jirun
K. Veesaj
Mr. Shankh jiri
S. Abhra
Tm. Bulpam

ZINC OXIDE

En. Flowers of Zinc
H. Putty
G. Jasata bhasma

8. PATHWAY TO SCREEN PHYTOCHEMICAL NATURE OF NATURAL DRUGS

Test solution: Depending upon the type of natural drug under examination, the test solution may be a aqueous extract or alcoholic extract or extract in specific menstrum like petroleum ether,

chloroform, ethyl acetate, etc. for the tests to be performed. For testing, as far as possible clear, transparent solution should be used.

1. Acidic Compounds

(a) Alcoholic extract of the drug after treatment with sodium bicarbonate produces effervescence.

(b) Treat alcoholic extract of the drug with warm water and filter. Test the filtrate with litmus paper, it turns blue.

2. Aleurone Grains

(a) Add few drops of alcoholic iodine solution to aqueous test solution, yellowish brown to brown colour indicates presence of aleurone grains.

(b) To the sample add few drops of alcoholic trinitrophenol (picric acid), aleurone grains take yellow colour.

(c) To the sample add lml of mercuric nitrate solution, brick red colour indicate presence of aleurone grains.

3. Alkaloids

(a) **Dragendorff's reagent:** Alkaloids give reddish brown precipitate with Dragendorff's reagent (Potassium bismuth iodide solution).

(b) **Mayer's reagent:** Alkaloids give cream colour precipitate with Mayer's reagent (Potassium mercuric iodide solution).

(c) **Wagner's reagent:** Alkaloids give reddish brown precipitate with Wagner's reagent (Iodine-potassium iodide solut1*n).

(d) **Hager's reagent:** Alkaloids give yellow precipitate with Hager's reagent (Saturated solution of picric acid).

(e) **Tannic acid test:** Alkaloids give buff colour precipitate with tannic acid solution.

(f) **Picrolonic acid test:** Alkaloids give yellow colour precipitate picrolonic acid.

Apart from above precipitate forming agents, Kraunt's reagent, Marme's reagent, Gold chloride solution, phospho tungstic acid solution. Reineckate part solution are also used for detection of Alkaloids.

4. Amino Acids

(a) **Millon's test:** To the test solution add about 2ml of Millons reagent, white precipitate indicates presence of amino acids.

(b) **Ninhydrine test:** To the test solution add Ninhydrine solution, boil, violet colour indicates presence of amino acid.

5. Bitterness Value

Medicinal plant materials that have a strong bitter taste ("bitters") are employed therapeutically, mostly as appetizing agents. Their bitterness stimulates secretions in the gastrointestinal tract, especially of gastric juice. The bitter properties of plant material are determined by comparing the threshold bitter concentration of an extract of the materials with that of a dilute solution of quinine hydrochloride.

The bitterness value is expressed in units equivalent to the bitterness of a solution containing 1 g of Quinine hydrochloride R (R- Reference material) in 2000 ml.

Bitterness value calculated in units per g using the following formula:

$$\text{Bitterness value} = \frac{2000 \times c}{a \times b}$$

where,

a = The concentration of the stock test solution (ST) (mg/ml),

b = The volume of test solution ST (in ml) in the tube with the threshold bitter concentration,

c = The volume of Quinine hydrochloride R (in mg) in the tube with the threshold bitter concentration.

6. Carbohydrates (with aqueous test solution)

(a) **Molisch's test:** To the test solution add few drops of alcoholic a-naphthol, then add few drops of concentrated sulphuric acid through sides of test tube, purple to violet colour ring appears at the junction.

(b) **Barfoed's test:** 1 ml of test solution is heated with Iml of Barfoed's reagent on water bath, if red cupric oxide is formed, monosaccharide is present. Disaccharides on prolong heating (about 10 min.) may also cause reduction, owing to partial hydrolysis to monosaccharides.

(c) **Selivanoff's test (Test for ketones):** To the test solution add crystals of resorcinol and eq, volurae of concentrated hydrochloric acid and heat on a water bath, rose colour is produce (eg. Fructose, honey)

(d) **Test for pentoses:** To the test solution add equal volume of hydrochloric acid containing c small amount of phloroglucinol and heat, red colour is produced.

(e) **Osazone formation test:** Heat the test solution with solution of phenylhydrazine hydrochloride, sodium acetate and acetic acid. Examine the yellow crystals formed under microscope. These crystals are of characteristic shape for particular' sugars. (Osazones are the crystalline derivatives of sugars having particular shapes. Glucose and fructose form the same osazone-glucosazone, m.p. 205°C).

7. Cellulose

(a) To the thin section of the sample add 1-2 drops of iodinated zinc chloride solution, wait for a minute. Add 1 drop of iodine and then add 1 drop of sulphuric acid, cell walls (cellulose) stain blue or blue-violet.

(b) To the thin section of the sample add 0.1 M iodine solution and sulphuric acid, cell walls (cellulose) turns blue-violet. If cuoxam test solution is added to, cell walls swell and gradually dissolve.

Methyl cellulose: Make a solution of 0.1 gm sample in 10 ml of boiling water, allow to cool, Add 0.5 ml of 0.05 per cent solution of brilliant yellow in water. 0.05 ml of N/10 solution and sodium hydroxy if formed in colourless solution.

(c) **Thionine test:** Treat thin section of the sample with thionine solution, wash with alcohol after 15 min., cell wall turns blue.

8. Fats and Fixed Oils (Petroleum ether extract)

Fixed oils and fats can be confirmed by chemical tests for glycerine, which is produced by their hydrolysis.

(a) **Saponification test:** Add few drops of 0.5 N alcoholic potassium hydroxide to a small quantity of various extracts along with a drop of phenolphthalein separately and heat on a water bath for 1-2 hours. The formation of soap or partial neutralization of alkali indicates the presence of fixed oils and fats.

(b) Treat 5 drops of sample with Iml of 1-% copper sulphate solution, Then add 10% sodium hydroxide solution. A clear blue solution is obtained which shows glycerine is present in the sample. The cupric hydroxide formed in the reaction does not precipitate out as it is soluble in glycerine.

(c) To the 5 drops of sample add a pinch of sodium hydrogen sulphate, pungent odour emanates indicating presence of glycerine.

9. Flavonoids

(a) Shinoda test: To the test solution add few magnesium turnings and concentrated hydrochloric acid dropwise, pink scar!et, crimson red or occasionally green to blue colour appears after few minutes.

(b) Alkaline reagent test: To the test solution add few drops of sodium hydroxide solution, intense yellow colour is formed which turns to colourless on addition of few drops of dilute acid indicate presence of flavonoids.

(c) Zinc hydrochloride test: To the test solution add a mixture of zinc dust and conc. hydrochloric acid. It gives red colour after few minutes.

(d) Leucoanthocyanidine test: Leucoanthocyanidines give red colour in strong acidic media, while in weak acidic media neutral or ionised base is formed which imparts blue clolur.

10. Foaming Index

Many medicinal plant materials contain saponins that can cause a persistent foam when an aqueous decoction is shaken. The foaming ability of an aqueous decoction of plant materials and their extracts is measured in terms of a foaming index.

Foaming index is calculated by using the following formula:

$$\text{Foaming index} = \frac{1000}{a}$$

where, a = The volume in ml of the decoction used for preparing the dilution in the tube

11. Glycosides

I. General tests:

Test A: Extract 200 mg of drug with 5 ml of dilute sulphuric acid by warming on a waterbath. Filter it. Then neutralize the acid extract with 5^0% solution of sodium hydroxide. Add 0.1 ml of

Fehling's solution A and B until it becomes alkaline (test with pH paper) and heat on a water bath for 2 minutes. Note the quantity of red precipitate formed and compare with that of formed in Test B.

Test B: Extract 200 mg of the drug using 5 ml of water instead of sulphuric acid. After boiling add equal amount of water as used for sodium hydroxide in the above test. Add 0.1 ml Fehling's solution A and B until alkaline (test with pH paper) and heat on water bath for 2 minutes. Note the quantity of red precipitate formed. Compare the quantity of precipitate formed in Test B with that of formed in Test A. If the precipitate in Test A is greater than in Test B then Glycoside may be present. Since Test B represents the amount of free reducing sugar already present in the crude drug, whereas Test A represents free reducing sugar plus those related on acid hydrolysis of any glycoside in the crude drug.

12. Haemolytic Index

Many medicinal plant materials of the families Caryophyllaceae, Araliaceae, Sapindaceae, Primulaceae, and Dioscoreaceae contain saponins. The most characteristic property of saponins is their ability to cause haemolysis; when added to a suspension of blood, saponins produce changes in erythrocyte membranes, causing haemoglobin to diffuse into the surrounding medium.

The haemolytic activity of plant materials, or a preparation containing saponins, is determined by comparison with that of a reference material.

Calculate the haemolytic activity of the medicinal plant material using the following formula:

$$\text{Haemolytic index} = 1000 \times \frac{a}{b}$$

where,

1000 = The defined haemolytic activity of saponin R in relation to blood,

a = Quantity of saponin R that produces total haemolysis (g)

b = Quantity of plant material that produces total haemolysis (g)

13. Lignans

These are the low molecular weight plant products formed by oxidative coupling of p-hydroxy-phenyl propane units in which two units are linked by oxygen bridge.

14. Lignin

(a) Treat the section of drug with conc. hydrochloric acid and phlvrog{ucinol solution, pink colour is formed.

(b) Treat section of drug with saffranine solution, pink colour is formed.

(c) **Thionine test:** Treat the section of drug with thionine solution, after 15 minutes wash with alcohol, bluishviolet colour is formed.

15. Suberine (in cuticular cell wall)

In a thin section of a drug. Cells of cuticular walls are stained orange red or-red with Sudan red solution.

16. Volatile Oil

(a) To a thin section of drug add **Sudan III** solution, red colour obtained by globules indicate presence of volatile oil.

(b) To a thin section of drug add few drops of **Tincture alkana**, red colour indicates presence *of* volatile oil,

II. Chemical tests for specific glycosides:

A. Anthraquinone glycosides:

(a) Borntrager's test: Boil the test material with 1 ml of sulphuric acid in a test tube for five minutes. Filter while hot. Cool the filtrate and shake with equal volume of dichloromethane or chloroform. Separate the lower layer of dichloromethane or chloroform and shake it with half of its volume of dilute ammonia. A rose pink to red colour is produced in the ammoniacal layer.

(b) Modified Borntrager's test: Boil 200 mg of the test material with 2 ml of dilute sulphuric acid. Treat it with 2 ml of 5 % aqueous ferric chloride solution (freshly prepared) for 5 minutes, shake it with equal volume of chloroform and continue the test as above. As some plants contain anthracene aglycone in a reduced form, if ferric chloride is used during the extraction, oxidation to anthraquinones takes place, which shows response to Borntrager's test.

(c) Hydrolyzed ether extract with methanolic magnesium acetate gives pink colour in daylight and greenish orange colour under ultraviolet (UV) light.

(d) Test for Hydroxy-anthraquinones: Treat the sample with potassium hydroxide solution red colour is produced.

B. Cardiac glycosides (Cardinolides):

Libermann Burchard's test: Extract of powdered drug when trented with acctic anhydride and conc. sulphuric acid gives bluish green colour. (Test for bufodienolides).

(a) Kedde's test: Extract the drug with chloroform, evaporate to dryness. Add one drop of 90 % alcohol and 2 drops of 2 % 3,5-dinitro benzoic acid in 90 %, alcohol. Make alkaline with 20 % sodium hydroxide solution, purple colour is produced. The colour reaction with 3,5-dinitro benzoic acid depends on the presence of a, P-unsaturated lactones in the aglycone.

(b) Keller-killiani test (test for deoxy sugars): Extract the drug with chlrooform and evaporate it to dryness. Add 0.4 ml of glacial acetic acid containing trace amour* of ierr'.c chloride. Transfer to a small test tube, add carefully 0.5 ml of concentrated sulphur:c acid by the side of the test tube. Acetic acid layer shows blue colour.

(c) Raymond's test: Treat the test solution with hot methanolic alkali, violet colour is produced.

(d) Legal's test: Treat the test solution with pyridine and add alkaline sodium nitroprusside solution, blood red colour appears.

(e) Baljet's test: Treat the test solution with picric acid or sodium picrate, orange colour is formed.

(f) Libermann Burchard's test: Extract of powdered drug when treated with acidic anhydride and conc. sulphuric acid gives bluish green colour (test for bufadienolides)

C. Coumarin glycosides:

Place a small amount of sample in test tube and cover the test tube with a filter paper moistert-c with dilute sodium hydroxide solution. Place the covered test tube on water bath for several minutes. Remove the paper and expose to ultraviolet (UV) light, the paper shows green fluorescence.

D. Cyanogenetic glycosides:

(a) Place 200 mg of drug in a conical flask and moisten with few drops of water. (There should be no free liquid at the bottom of the flask as the test will not work because the hydrogen cyanide produced will dissolve in the water rather than come off as a gas to react with the paper). Moisten a piece of picric acid paper with 5 % aqueous sodium carbonate solution and suspended by means of cork in the neck of the flask. Warm gently at about 37°C. Observe the change in colour. Hydrogen cyanide is liberated from cyanogenetic glycosides by the enzyme activity and reacts with sodium picrate to form the reddish purple sodium isopurpurate.

(b) Prepare solution of Guaiacum resin in absolute alcohol and allow it to dry on paper. Treat it with copper sulphate solution. The paper turns blue due to prussic acid with HCN.

E. Saponin glycosides:

(a) Froth formation test: Place 2 ml solution of drug in water in a test tube, shake well, stable froth (foam) is formed.

(b) Haemolysis test: Add 0.2 ml of solution of saponin (prepared in 1% normal saline) to 0.2 ml of blood in normal saline and mix well. Centrifuge and note the red supernatant Compare with controltube cantaining 0.2 ml of 10 % blood in normal saline diluted with 0.2 ml of normal saline.

17. Inulin

To the test solution add the solution of a-naphthol and sulphuric acid, brownish red colour is formed.

18. Latex Cells

Latex cells when stained with iodine show dumb bell shaped starch grains.

19. Mucilage

(a) Treat the test solution with ruthenium red solution, pink colour is obtained.

(b) Treat test solution with Chinese ink, transparent, spherical dilated fragments on black background are observed.

(c) Treat the test solution with thionine solution. After 1 5minutes wash with alcohol mucilage turns violet red.

20. Naphthoquinones

(1) Juglone test: Treat 2 ml of chloroform extract and 2 ml of ethyl ether with dil ammonia solution: Pink colour indicates naphthoquinones.

(2) Dam - Karrer test: To the chloroformic plant extract add 10 % pot hydroxide solution blue colour develops.

21. Pectins

To the section Add 0.05% Toluidine blue O solution. It takes pink or reddish purple colour.

22. Pesticidal Residues

Normally, it should not be more than 1%. Pesticidal residue (in mg of pesticide per kg of plant material) can be calculated on the basis of the maximum acceptable daily intake of the pesticide for humans (ADI), as recommended WHO, and the mean daily intake (MDI) of the medicinal plant material.

$$\text{Pesticidal Residue} = \frac{ADI \times E \times 60}{MDI \times 100}$$

where,

ADI = Maximum acceptable daily intake of pesticide (mg/kg of body weight)

E = Extraction factor, which determines the transition rate of the pesticide from the plant material into the dosage form

MDI = Mean daily intake of medicinal plant product.

23. Tannins (Phenolic Compounds)

(a) Goldbeater's skin test: Add 2 % hydrochloric acid to a small piece of goldbeater's skin, **rinse** it with distilled water and place in the solution to be tested for five minutes. Then give wash of distilled water and transfer to a 1% ferrous sulphate solution. A brown or black colour on the skin indicates presence of tannins.(Goldbeater's skin is a membrane obtained from intestine of the ox and behaves similarly to an untanned hide.)

(b) Ferric chloride test: Treat the extract with ferric chloride solution, blue co!our appears if hydrolysable tannins are present and green colour appears if condensed tannins are present.

(c) Phenazone test: Add about 0.5 gm of sodium acid phosphate to 5 ml of aqueous extract. Warm it and filter. To the filtrate add 2 % phenazone solution, bulky precipitate is formed which is often coloured.

(d) Gelatin test: To the test solution add 1 % gelatin solution containing 10 % sodium chloride. Precipitate is formed.

(e) Test for catechin: Dip a matchstick in the test solution, dry it and lastly moisten with concentrated hydrochloric acid. Then warm the stick near flame. The colour of the wood changes to pink due to phloroglucinol. (Phloroglucinol is formed when catechins are treated with acids).

(f) Test for chlorogenic acid: Treat the test solution with aqueous ammonia and expose to air gradually, green colour is developed.

24. Proteins

(a) Warming test: Heat the test solution in a boiling water bath, proteins get coagulated.

(b) Test with trichloroacetic acid: To the test solution add trichloroacetic acid, precipitate is formed.

(c) Biuret test: To the test solution (2 ml) add Biuret reagent (2 ml), violet colour indicates presence of proteins.

(d) **Hydrolysis test:** Hydrolyze the test solution with hydrochloric acid or sulphuric acid. Then carry out the Ninhydrine test for amino acids.

(e) **Xanthoproteic test:** To the (5 ml) of test solution, add 1 ml of concentrated nitric acid and boil, yellow precipitate is formed. After cooling it, add 40 % sodium hydroxide solution, orange colour is formed.

25. Starch (Amylum)

To the aqueous extract add weak aqueous iodine solution, blue colour indicates presence of starch, which disappears on heating and reappears on cooling.

26. Steroids and Triterpenoids

(a) **Libermann-Burchard test:** Treat the extract with few drops of acetic anhydride, boil and cool. Then add concentrated sulphuric acid from the side of the test tube, brown ring is formed at the junction two layers and upper layer turns green which shows presence of steroids and formation of deep red colour indicates presence of **triterpenoids**.

(b) **Salkowski test:** Treat the extract with few drops of concentrated sulphuric acid, red colour at lower layer indicates presence of steroids and formation of yellow coloured lower layer indicates presence of triterpenoids.

(c) **Sulfur powder test:** Add small amount of sulfur powder to the test solution, it sinks at the bottom.

27. Swelling Index

The swelling index is the volume in ml taken up by the swelling of 1 g of plant material under specified conditions. Its determination is based on the addition of water or a swelling agent as specified in the test procedure for each individual plant material (either whole, cut or pulverized).

28. Waxes

To the test solution add alcoholic alkali solution, waxes get saponified.

9. THERAPEUTIC AGENTS

(A) DEFINITIONS WITH EXAMPLES

Abortifacient : An agent which causes an abortion. e.g. papaya, ergot.

Adaptogens : These are the biological active substances which improve physical endurance for doing work even in adverse and difficult environmental conditions. e.g. ginseng, gingko, withania.

Addiction : To form a habit or enslavement to some habit. e.g. opium, cannabis (ganja).

Adrenergic : Term applied to sympathetic nerves which liberate adrenaline and nor-adrenaline from their terminations.

Adsorbent : Substances which attract gases or dissolved substances to their surfaces as a film. i.e. kaolin, charcoal, kieselguhr.

Anaesthetic : Drugs which cause insensibility to touch or pain, may be general or local. e.g. coca, clove, veratrum.

Analeptic : Drug restoring strength after illness. e.g. lobelia, nux vomica, camphor.

Analgesic : Substance which relieves pain by acting on central nervous system. e.g. aconite, belladonna, opium, cannabis. They are of two types : (a) Narcotic analgesics : opium, Indian hemp, (b) Antipyretic analgesics : aconite.

Anodyne : An agent which relieves pain, but is milder than analgesic. e.g. belladonna, camphor.

Antacids : A substance which neutrialises excess acidity in stomach e.g. amla, bahera, myrobaium (Tripeala Churna).

Anthelmintic : Substance used to expel the worms out of intestine like thread worm, tape worm, round worm etc. They do not kill the worms essentially. e.g. santonin, quassia, vidang, malefern, chenopodium.

Anti amoebic : Which controls the infection of Entamoeba hystolitica e.g. kurchi.

Anti asthmatic : Agents used treat difficulties in breathing, to relieve cough and also for relaxation of the branchial smooth muscles. e.g. vasaka, belladonna.

Anti cancer : Recommanded in the treatment of cancer. i.e. podophyllum, vinca, taxus.

Anti coagulant : One which prevents the coagulation of blood. i.e. heparin.

Anti diabetic : Substances overcoming diabetis. e.g. stevia, bitter guard, gymnema, pterocarpus.

Anti dysentric : Agent which checks the dysentary. e.g. arjuna, amra, kino.

Anti febrile : Substance reducing fever. e.g. quinine, cinchona, chirata.

Anti fertility : Which prevents conception e.g. vidang, stevia.

Anti HIV : Given in the treatment of infection caused by HIV (Human Immuno deficiency Virus). e.g. hypericum, phyllanthus, artemisia.

Anti-inflammatory : One which reduces the inflammation. e.g. datura, linseed, ashwagandha, bahera.

Anti-leprotic : Drugs used in the treatment of leprosy. e.g. chaulmoogra oil (externally) Jalbrahmi.

Anti-malerial : Used for treatment of malerial fever. e.g. cinchona, chirata, artemisia annua.

Anti-oxytocic : One which controls the uterine contraction and prevent abortion. e.g. shatavari.

Anti-pyretic : Reducing the body temperature. e.g. guduchi, quinine.

Anti-rhumatic : One which control the rhumatism e.g. guggul, boswellia, dioscorea, colchicum.

Anti-retroviral : Drugs which are used to control (but not cure) the HIV/AIDS virus.

Anti-sialagogue : A compound that reduces or lessens the secretion of saliva. e.g. belladonna.

Anti-tumour : Used to control the malignant growth of carcinogenic tumours in the body. e.g. colchicum corm.

Anti-diaphoretic : Substance reducing the profuse sweating. e.g. belladonna, hyocyamus.

Anti-diarrhoeals : Rapid passage of faecal matter through gastrointestinal tract and frequent passage of liquid faeces. e.g. opium, catechu.

Anti-emetic : A drug that prevents vomiting e.g. lemon oil, ginger, caraway.

Anti-leucodermic : Defective skin pigmentation. e.g. manjishta, psoralea, karanja oil.

Anti-oxidant : It is an agent when present in low concentrations as compared with the oxidisable substance, significantly delays or prevents oxidization of the substrate. Oxidizable substrate encompasases almost every thing except water, found in foods, in living tissues and includes proteins, lipids carbohydrates and DNA. Flavonoids and polyphenols are potent oxidants e.g. green tea, black tea, catechin.

Anti-phlogistic : External application employed to reduce inflammation. e.g. aconite, linseed, boswellia.

Anti-protozoal : Against protozoa (e.g. unicellular micro-organisms). e.g. kurchi.

Anti-retroviral therapy : There is no cure for HIV/AIDS, but drugs which control the virus, causing HIV/AIDS are known as anti-retroviral agents. Examples: Hypericum, African potato, Sutherlandia.

Anti-septic : Substance used to check the growth or development of micro-oganisms. e.g. neem, turpentine, benzoin, myrrh.

Anti-spasmodic : Substances which relieve or control spasms (pain) of voluntary or involuntary muscles. e.g. datura, valerian, lobelia, opium, belladonna.

Anti-ulcer : Open sore in a body surface. e.g. catechu, glycyrrhiza, hyoscyamus, sandal wood.

Anti-viral : Agent used to prevent and treat the viral infestation. e.g. Guduchi (Hepatitis) Neem oil.

Aperient : Mild laxative e.g. rhubarb, isapgol, manna.

Aphrodisiac : An agent stimulating the sexual desire. e.g. ginseng, gokhru, safed musali, shilajit, cannabis.

Appetizer : Compound that stimulates the desire for food. e.g. gentian, capsicum, bitter guard.

Aromatic : Substance having fragrant agreeable odour. e.g. cardamom clove, cinnamon, nutmeg, orange peel etc.

Astringent : Precipitates proteins and causes contraction of tissues lessen body secretions e.g. tannic acid, catechu, myrobalan.

Bitter : Substances characterized by bitter taste. e.g. gentian, nux-vomica, cinchona, chirata.

Cardiotonic : Agents which increase the strength and rate of myocardial contraction and in turn increases cardiac output. e.g. digitalis, arjuna, sqill, strophanthus.

Carminative : A substance that removes gases from the G.I. tract with soothing or calming properties and relieving pain in stomach and bowels by regulating peristalsis, e.g. cardamom, peppermint, asafoetida, caraway.

Cathartic : Active purgative usually producing severe evacuation, may or may not be accompanied with pain, e.g. jalap, cotocynth, castor oil, aloe, rhubarb.

Cholinergic : Term applied to parasympathetic nerves which liberate acetylcholine at their terminations. e.g. physostigma, pilocarpus.

Condiments : Which improve the taste and flavour of sweet foods cardamon, nutmeg lavenders rose oil.

Counter-irritant : Substances producing superficial and artificial inflammation in order to exercise a good effect on the part applied. e.g. camphor, turpentine oil, eucalyptus oil, peppermint.

Cumulative : One which gets accumulated in body. e.g. digitalis, santonin.

Demulcent : Midicament that soothes or protects the mucous membrane or any other part to which it is applied. e.g. tragacanth, acacia, linseed, barley.

Deodorant : An agent which neutralises, removes or destroys foul odour. e.g. mentha oil, rose, sandal wood.

Depressant : Drugs which lower functional activities or retard the physiological action of an organ. e.g. opium, belladonna, hyoscyamus.

Detergent : One which causes cleansing. e.g. quillaia, soap nut, shikekai, hinganbet.

Diaphoretic : An agent which increases perspiration or sweating e.g. camphor, opium, ipecacunha, jaborandi, sarsaparilla.

Digestive : Drugs which assist the stomach and intestine in their normal function of promoting the digestion of foods, e.g. capsicum, papain, malt, pepsin diastase.

Diuretic : Drugs which increase the flow or secretion of urine or increase the quantity of urine, e.g. buchu, caffeine, punernava, tea, gokhru.

Edima : Accumulation of fluid.

Emetics	:	Substances which induce or produce vomitting e.g. ipcacuanha, mustard, senega, squill.
Emmenagogue	:	Substances which stimulate the menstrual flow. e.g. clove, galbanum, savin, bakyan neem.
Estrogen	:	Harmone which produce female physical and sexual characteristics.
Emollient	:	An agent that softens and smoothens the part when applied locally, e.g. honey, starch, olive oil, linseed oil, liquid paraffin, lard.
Excitant	:	An agent that stimulates the special function of the body according to the action, as motor, cerebral, etc. e.g. balsam of tolu, glycyrrhiza, senega.
Expectorant (Anti tussive)	:	Agents which stimulate and expel bronchial mucus or secretions. e.g. vasaka, tulsi, glycyrrhiza, balsam of tolu.
Febrifuge	:	An agent that lessens fever. i.e. (antipyretic) cinchona, chirata.
Flavour	:	Which removes the undesired (smell) flavour and makes pleasant. e.g. cardamom, peppermint, sandal wood, vanilla.
Galactogogue	:	An agent that increases the secretion of milk in the breast. e.g. shatavari fennel.
Gelling agents	:	One which increases the viscosity of the liquid. e.g. carageenan, gum ghati.
Gout	:	Disorder causing swelling and painful joints.
Haemostatics	:	An agent that arrests bleeding. e.g. tannic acid, kino.
Hair growth tonic	:	Which stimulates the growth of hairs. e.g. jatamansi, bhringraj, cantherides.
Hallucinogenic	:	A drug causing hallucinotions
Hallucination	:	It is a false perception occurring without any true sensory shmulus. It is a abnormal metnal condition based on misidentification of sensalon on siant.
Hepatoprotective	:	Which protects (supports) the working of liver. e.g. picrorrhiza, kalmegh.
Hydrogogue	:	Drug promoting watery evacuation of the bowel. e.g. croton oil, jalap.
Hypnotics	:	Drugs that induce sleep without cerebral excitement, e.g. hops, opium, cannabis.
Hypoglycemic	:	Which reduces the glucose in the blood serum. e.g. gymnema, bitter guard, insulin.
Hypotensive	:	One which reduces the high blood pressure. e.g. Ashwagandha.
Immuno modulators	:	Agents which modulate regulates immuno response (May be stimulative or suppressive) e.g. Ashwagandha, shatavari, tulsi, picrorrhiza, ginseng.
Insecticide	:	Causing distruction of insects. e.g. tobacco, pyrethrum, derris, sabadilla.
Irritant	:	An agent which, when used locally, produces more or less inflammation. e.g. cantharides, capsicum.
Laxative	:	Drugs that loosen the bowels, mild purgative. e.g. senna, Isapgol, karaya.
Muscle relaxant	:	Which relives/relaxes the muscles-strain. e.g. curare (skeletal), datura, opium.

Mydriatics : Remedy causing dialation of the pupil and paralysis of ciliary muscles. e.g. belladonna, datura.

Narcotic : Substance that produces stupor or sleep, e.g. Indian hemp, opium.

Nanobodies : A particle of size of one upon one hundred crore and may contain proteins, metal and mineral, naturally these are present in cells of camel and lamas.

Nervine tonic : Stimulating activities of nervous system. e.g. brahmi, asafoetida, shankhpushpi.

Nutritive : Nourishing medicines which increase vital energy. e.g. honey, malt extract, butter.

Oxytocic : An agent that causes expulsion of the contents of uterus by contracting the uterine muscles. e.g. ergot, vasicine, quinine.

Proteolytic : Causing hydrolysis of proteins. e.g. papain, ficin, pineapple.

Pungent : Irritation with a sensation of warmth. e.g. capsicum, ginger, balck-pepper, cubeba.

Purgative : An agent that causes evacuation of bowels. e.g. senna, aloe, castor oil.

(B) DISORDERS AND THEIR REMEDIES WITH NATURAL DRUGS

Acne : It is the state of the skin in which black-heads are associated with the eruption of pilose cubaecous follicles and cause red pimples. **Remedy:** Phytolaceae.

Aconite myrestica : Diseases produced by infection to skin externally. **Remedy:** Berberis, calendula enteicea.

Alopecia : This is abnormal hair loss and can be treated with Arnica flowers.

Aminorrhoea : **Remedy:** Calundula lestisticum, Ruta, gossylrium.

Anorexia : It is the condition in which the patient has no feeling to eat any thing and can be put to the right by using any one of the following remedy: Acorus, Artenuisia, Carum, Sousorea lappa herb.

Arthritis (Rhumatoid) : It is characterised by inflammation of joints and can be relieved by using **Remedy:** Gultheria, Guaicum, Apium, Cimicifuge or piper methesticum.

Asthma bronchitis : Inflammation of bronchial mucous membrane. **Remedy:** Garlic, Ephedra Lobelia.

Catarrh : Excessive mucus is the throat. **Remedy:** Garlic, commiphora.

Colaguage purgatives : These act by stimulating liver such as Rhubarbs Calomel.

Colic : Severe intestinal pain. **Remedy:** Belladonna

Colitis : Inflammation of colon. **Remedy:** Psyllum Isabgol.

Conjunctivitis : Inflammation of membrane connecting eye ball. **Remedy:** Fennel calundula.

Constipations : No emptying of bowel. **Remedy:** Berber is fistula jugalous.

Convolescence : Twisted arrangement of tissues. **Remedy:** Medicago, Swirtia, Trigonella.

Convulsescence : Folds or twists found in intestine. **Remedy:** Chordrous, medicago, swertia trigonalla, cough irritable. **Remedy:** Inula, Prunus, Pimpenella.

Cramps : Painful unvoluntaory tightening of muscles. **Remedy:** Cinchona, valerian dioscorea.

Cyst : Abnormal cavity in the body. **Remedy:** Calundulla

Cystitis : Pains of gall bladder. **Remedy:** Chondrous, Apium

Debility general : General weakness. **Remedy:** Trigonella, cinchona medicago

Dermititis : Inflammation of skin. **Remedy:** Chondrous.

Diabetis mellitus : Usual permeability of kidneys to glucose due to lack of insulin. **Remedy:** Galanga, Rhein, Sizygium.

Diabetis unspidus : There is achydratism.

Drastic purgatives : Irritate mucous membrane, such as Jalap, Colocynth.

Dyspepsic : Indigestion of food. **Remedy:** Fennel, coriander.

Dysurea : Painful urination, **Remedy:** petroselbneuon, violatricolour

Eczema : **Remedy:** Berberis, calundula, sassafras

Epilepsy : (Grand mall): Loss of conceousness. **Remedy:** Verbernia, passiflora

Flatulent : Too much gas in stomach. **Remedy:** Alpcnia, anethum carum, caraway cassia.

Gallstone : Small hard mass in gall bladder. **Remedy:** Berberis, swirtea.

Gastric ulcer : **Remedy:** Acorus, berberis, Chondrus, Liquorice.

Gingivitis : Inflammation of gums. **Remedy:** Black catechu, salvia, cominiphora.

Haemorrhoids : Swelling of veins around the anus. **Remedy:** Calundule, Giranium, Hamamais.

Hay-fever : Allergic inflammation of nasal mucor membrane. **Remedy:** Ephedra.

Headache : Hypericum, Lavandula rosemun important condition arising.

Hydroguage purgatives : Producing fluid motions such as Coroton oil, Colocynth.

Hyper chlorhydria : Excessive hydrochloric acid in the gastric juice. **Remedy:** Cinchona.

Hypochlorhydria : Decreased hydrochloric acid in the gastric juice. **Remedy:** Jetearrhiza.

Hysteria : Uncontrollable emotion. **Remedy:** Asafoltida, Rosemary, Valerian.

Impotant : Unable to achieve an errections. **Remedy:** Panax, serenoa.

Infertility : Unable to bear young. **Remedy:** Panax, Avena.

Insomnia : Disorder of sleepleness. **Remedy:** Humulus, Panax, Passiflora.

Jaundice : **Remedy:** Berberis, Taraxcum verbena.

Jaundice : Yellowing of skin due to liver disorder. **Remedy:** Berberis, tamaxcum, verbena.

Leishmaniasis : Infestation of leishmania spread by sand flies. **Remedy:** Berberis.

Leucorrhoea : Sticky whitish vaginal discharge. **Remedy:** Berberis, black-cateachu, geranium.

Mental depression : Mental unhappyness. **Remedy:** Panax, Rosemary valerian.

Migraine : Periodic severe headache. **Remedy:** Berberis, Cinchona swertia.

Mumps : Disease causing swelling of glands at the side of face. **Remedy:** Phytolacca.

Nausea : Feeding of sickness. **Remedy:** Alpenia, Nutmeg, Cinnamon.

Neuralgia : Intense pain along the nerve. **Remedy:** Capsicum, hypericum, Panax Aconite.

Peptic ulcer : Non malignant ulcers in the part of digestive system. **Remedy:** Geranium herb, glycyrrhiza.

Pertusis : Whoopiyg cough, an infections disease of children, with attach of coughing organisms causing the cough *Haemophilus purtisis*. **Remedy:** Garlic, Belladonna, Ipecac.

Pharingitis : Inflammation of pharynx. **Remedy:** Catechu, Cenchona, Cammiphora.

Psonasis : Chronic skin disease in which erythematous areas are covered by adherant scales. **Remedy:** Berberis, lappa roots, Lavendula.

Refrigerant : An agent that relieves thirst and is cooling. e.g. orange, tamarind, amla.

Rhumatic pains : Pains due to rhumatism.

Rhumatism : A disease of connective tissue with pains or stiffness or swelling of muscles and joints. **Remedy:** Centella, cimicifuga, lavendula, thuja.

Rubefacient : Agent which reddens the skin producing a local congestion, dilating the vessels and thus, resulting in the increased blood supply. e.g. oil of turpentine, capsicum, mustard, flax seed, cantharide.

Sciatica : A pain the line of distribution of sciatic nerve C buttack, back of theigh calf and foot. **Remedy:** Aconite, gaullheria, hypericum

Seabies : Paralytic skin disease caused by Itch-mite. **Remedy:** Pimpnella, phytolacea.

Sedative : Agent which lowers the functional activities, or substances which quiet the nervous system. e.g. cannabis, rauwolfia, bitter almond.

Spermicidal : Which destroyes the sperms. e.g. neem oil, gossypiol.

Stimulants : Agents which temporarily increase the functional activity, classified according to the organ upon which they act as : cardiac, bronchial, gastric, cerebral, intestinal, nervous, motor, vasomotor and respiratory. e.g. camphor, turpentine, strychnine, caffeine, ephedrine, asafoetida.

Stomachic : Substances increasing the secretion of gastric juice and also the functional activity of the stomach, e.g. dill, fennel, coriander, gentian.

Sturnutatory : Which causes sneezing, Quillaia.

Styptic : Anything that checks the haemorrhage, e.g. tannic acid, kino, pterocarpus.

Sweetening agent : Masking the undesired taste and making it sweet. e.g. stevia, honey, thaumatin.

Uterine craps : **Remedy:** Cimicifuga.

Vasoconstrictor : Causing constriction of blood vessels. e.g. ergotamine, ephedra.

Vasodilator : Causing dilatation or relaxation of blood vessels. e.g. visnaga.

Vericose veins : Disease of lower limbs affecting the important veins. Dilated veins valves of which become incompetant and blood flow may get reversed. **Remedy:** Calendula.

Vermicide : Which dislodges or kills intestinal worms. e.g. vidang, chenopodium, quassia.

Vesicant : An agent that produces blisters. e.g. mustard gas, cantharides, mylabris, marking nut.

Worts : A small hard outgrowth on the skin. **Remedy:** Podophyllum thuja.

❖❖❖

10. PLACEMENT OF NATURAL DRUGS IN PHARMACOPOEIA

(A = Drug is available in the pharmacopoeia; ■ Not available, d = deleted.)

S.N.		IP 1955	IP 66	IP 85	IP 96	IP 2007	B.P. 2010	B.P. 2007	API 2006	Remarks
		*1	*2	*3	*4	*5	*6	*7	*8	
(a)	**Vegetable Drugs**									
	A									
1.	Absorbent cotton	A	A	A	A	A	A	A	■	
2.	Acacia gum	A	A	A	A	A	A	A	■	
3.	Aconite roots	A	A	d	■	■	■	■	A	
4.	Acorus calamus	■	■	■	■	■	■	■	■	
5.	Agar	A	A	d	■	■	■	A	A	
6.	Ajowan fruits	■	A	■	■	A	A	■	■	
7.	Algin	■	A	■	A	A	■	A	■	
8.	Aloes	A	A	d	A	A	■	A	A	
9.	Amla fruits	■	■	■	■	A	A	■	A	
10.	Ammi fruits	■	■	■	■	■	■	■	■	
11.	Amra	■	■	■	■	A	A	■	■	
12.	Anise	A	■	■	■	■	■	A	■	
13.	Anant mul	A	■	■	■	A	A	■	■	
14.	Annatto seeds	■	■	■	■	■	■	■	■	
15.	Arachis oil	A	A	A	A	A	A	A	■	
16.	Arnica flowers	■	■	■	■	■	■	A	■	
17.	Arjuna bark	■	■	■	A	A	A	■	A	
18.	Arraroot starch	■	■	■	■	■	■	A	■	
19.	Artemisia	A	■	■	■	■	A	■	■	
20.	Artemisia annua	■	■	■	■	A	■	■	■	
21.	Asafoetida	A	■	d	■	■	■	■	A	
22.	Ashoka	A	A	d	■	A	■	■	A	
23.	Ashwagandha roots	A	A	d	A	A	A	■	A	
	B									
24.	Bael fruits	A	A	d	■	■	■	■	A	
25.	Bahera	■	■	■	■	A	A	■	A	
26.	Bear berry leaves	■	■	■	■	■	■	A	■	
27.	Belladonna herb	A	A	A	A	A	A	A	■	
28.	Benafsha	■	■	■	■	■	■	■	■	

contd...

		*1	*2	*3	*4	*5	*6	*7	*8	
29.	Benzoin	A	A	d	A	A	■	■	■	
30.	Bhuiamla	■	■	■	■	A	A	■	■	
31.	Bhringraj	■	■	■	■	A	A	■	A	
32.	Bitter almonds	■	■	■	■	■	■	■	■	
33.	Black catechu	A	A	d	■	■	■	■	A	
34.	Black mustard	■	■	■	■	■	■	■	■	
35.	Black mustard oil	A	A	■	■	■	■	■	■	
36.	Black pepper	■	■	■	■	A	A	■	A	
37.	Boswellia resin	■	■	■	■	A	A	■	■	
38.	Brahmi (Bacopa)	■	■	■	■	A	A	■	A	
39.	Buck wheat	■	■	■	■	■	■	A	■	
	C									
1.	Camphor	A	A	d	A	■	■	A	■	
2.	Camphor oil	■	A	■	■	■	■	■	■	
3.	Camptotheca leaves	■	■	■	■	■	■	■	■	
4.	Cannabis herb	A	A	d	■	■	■	■	■	
5.	Capsicum fruits	A	A	d	■	■	■	A	■	
6.	Capsicum oleo-resin	■	A	d	■	■	■	■	■	
7.	Caramel	A	A	d	■	A	■	■	■	
8.	Caraway	A	A	d	■	■	■	A	A	
9.	Cardamom fruit	A	A	d	■	■	■	■	A	
10.	Carnauba-wax	■	A	■	A	A	■	■	■	
11.	Carrageenan	■	■	■	■	■	■	■	■	
12.	Cascara-sagrada	■	■	■	■	■	■	■	■	
13.	Cassia-bark	A	A	d	■	■	■	■	■	
14.	Castor oil	A	A	A	A	A	A	A	A	
15.	Castor oil hydrogenated	■	■	■	A	A	■	A	■	
16.	Celery fruits	■	A	■	■	■	■	■	■	
17.	Chaulmoogra oil	A	A	d	■	■	■	■	■	
18.	Chenopodium oil	A	A	d	■	■	■	■	■	
19.	Chirata herb	A	A	d	■	■	■	■	A	
20.	Chrysarobin	A	A	d	■	■	■	■	■	
21.	Cinchona	A	A	d	■	■	■	A	■	
22.	Cinnamon	A	A	d	■	■	■	A	A	
23.	Citronella oil	■	A	■	■	■	■	■	■	
24.	Cloves	A	A	d	■	■	A	A	A	
25.	Coca-leaves	■	■	■	■	■	■	■	■	

contd...

		*1	*2	*3	*4	*5	*6	*7	*8	
26.	Cocoa	■	■	■	■	■	■	■	■	
27.	Cocoa butter	A	■	■	■	■	■	A	■	
28.	Coffee-seeds	■	■	■	■	■	■	■	■	
29.	Colchicum	A	■	■	■	■	■	■	■	
30.	Coleus forskoholii	■	■	■	■	A	A	■	■	
31.	Colocynth fruits	A	■	■	■	■	■	■	■	
32.	Colophony	A	■	■	■	■	■	■	■	
33.	Coriander	A	A	d	■	■	■	A	■	
34.	Corn oil	■	■	■	■	■	■	■	■	
35.	Cotton seed oil	■	■	■	■	■	■	A	■	
36.	Cubeb fruits	■	■	■	■	■	■	■	■	
37.	Cummin	A	■	■	■	■	■	■	A	
38.	Curare	■	■	■	■	■	■	■	■	
	D									
1.	Daruhalad	■	■	■	■	A	A	■	A	
2.	Datura herb	A	A	A	■	■	■	■	A	
3.	Davana oil	■	A	■	■	■	■	■	■	
4.	Derris roots	A	■	■	■	■	■	■	■	
5.	Dextrin	■	■	■	■	A	■	■	■	
6.	Digitalis	A	A	d	■	■	■	A	■	
7.	Dill	A	A	d	■	■	■	A	■	
8.	Dioscorea	■	■	■	■	■	■	■	■	
9.	Duboisia	■	■	■	■	■	■	■	■	
	E									
1.	Echinecea herb	■	■	■	■	■	■	■	■	
2.	Ephedra	A	A	d	■	■	■	A	■	
3.	Ergot of Rye	A	A	d	■	■	■	■	■	
4.	Eucalyptus oil	A	A	A	A	A	A	A	A	
5.	Evening primrose	■	■	■	■	■	■	■	■	
	F									
1.	Fennel fruits	A	A	d	■	A	A	■	A	
2.	Fewerfew herb	■	■	■	■	■	■	■	■	
	G									
1.	Garcinia	■	■	■	■	A	A	■	■	
2.	Garlic	■	■	■	■	■	A	A	A	
3.	Gaultheria oil	A	■	■	■	■	■	■	■	
4.	Gentian roots	■	■	■	■	■	■	A	■	

contd...

		*1	*2	*3	*4	*5	*6	*7	*8	
5.	Geranium oil	■	■	■	■	■	■	■	■	
6.	Ginger oleo resin	■	■	■	■	■	■	■	■	
7.	Ginger rhizome	A	A	d	■	A	A	A	A	
8.	Gingko leaves	■	■	■	■	■	■	■	■	
9.	Ginseng rhizomes	■	■	■	■	■	■	A	■	
10.	Ginseng rhizomes	■	■	■	■	■	■	A	■	
11.	Gloriosa	■	■	■	■	■	■	■	■	
12.	Gokhru	■	■	■	■	A	A	■	■	
13.	Gossypol	■	■	■	■	■	■	■	■	
14.	Guar gum	■	A	A	A	A	A	■	■	
15.	Guggul	■	■	■	A	A	A	■	A	
16.	Gum ghati	■	■	■	■	■	■	■	■	
17.	Gum karaya	■	■	■	■	■	■	■	■	
18.	Gymnema (Gudmar)	■	■	■	■	A	A	■	■	
	H									
1.	Hemlock									
2.	Henna	■	■	■	■	■	■	■	A	
3.	Hops	■	■	■	■	■	■	A	■	
4.	Hyoscyamus leaves	A	A	d	■	■	■	■	■	
5.	Hyperiicum herb	■	■	■	■	■	■	■	■	
	I									
1.	Inulin	■	■	■	■	■	■	A	■	
2.	Ipecacunha roots	A	A	d	■	■	■	A	■	
3.	Ipomoea roots	A	■	■	■	■	■	■	■	
4.	Isapgol	A	A	d	■	A	A	■	A	
5.	Isapgol Husk	A	A	A	■	A	■	■	■	
	J									
1.	Jaborandi leaves	■	■	■	■	■	■	■	■	
2.	Jalap roots	■	■	■	■	■	■	■	■	
3.	Jalbrahmi (Centella)	■	■	■	■	A	A	■	■	
4.	Jatamansi rhizomes	■	A	d	■	■	■	■	■	
5.	Jojoba oil	■	■	■	■	■	■	■	■	
	K									
1.	Kaladana	A	■	■	■	■	■	■	■	
2.	Kalmegh	A	A	d	■	A	A	■	■	
3.	Kapur-kachari (Shati)	■	■	■	■	A	■	■	■	
4.	Karanja oil	■	■	■	■	■	■	■	■	
5.	Kavach seeds	■	■	■	■	■	■	■	■	

contd...

		*1	*2	*3	*4	*5	*6	*7	*8	
6.	Kokam butter	A	A	d	■	■	■	■	■	
7.	Kola	■	■	■	■	■	■	■	■	
8.	Kurchi	A	A	d	■	■	■	■	A	
	L									
1.	Lavender oil	■		■	■	■	■	A	■	
2.	Lecithin	■	■	■	■	■	■	■	■	
3.	Lectins	■	■	■	■	■	■	■	■	
4.	Lemon grass oil	A	A	d	■	■	■	■	■	
5.	Lemon oil	A	A	d	■	■	■	■	■	
6.	Lemon peels	A	A	d	■	■	■	■	■	
7.	Linseed	A	A	d	■	■	■	■	■	
8.	Linseed oil	A	A	d	■	■	■	■	■	
9.	Liquid glucose	A	■	■	■	■	■	■	■	
10.	Liquorice	A	A	A	A	A	A	A	A	
11.	Lobellia herb	A	A	d	■	■	■	■	■	
12.	Locust bean gum	■	■	■		■	■	■	■	
13.	Long pepper	A	A	d	■	A	A	■	■	
	M									
1.	Maize starch	A	A	A	A	A	■	A	■	
2.	Male fern rhizomes	A	A	d	■	■	■	■	■	
3.	Malt extract	■	A	A	A	A	A	■	■	
4.	Manna	■	■	■	■	■	■	■	■	
5.	Manjishta	■	■	■	■	A	A	■	■	
6.	Mannitol	■	A	■	■	■	A	A	■	
7.	Mentha oil	■	A	■	A	A	A	■	■	
8.	Methi	■	■	■	■	■	■	■	■	
9.	Momordica fruits	■	■	■	■	■	■	■	■	
10.	Myrobalan	A	A	d	■	A	A	■	A	
11.	Myrrh	A	■	■	■	■	■	A	■	
	N									
1.	Neem herb	■	■	■	■	■	A	■	A	
2.	Neem oil	A	A	d	■	■	■	■	■	
3.	Nutmeg	A	A	d	■	■	■	A	A	
4.	Nux-vomica	A	A	d	■	■	■	■	■	
	O									
1.	Olive oil	■		■	■	■	■	A	■	
2.	Opium	A	A	A	A	A	A	A	■	

contd...

		*1	*2	*3	*4	*5	*6	*7	*8	
3.	Orange oil	A	A	■	■	■	■	A	■	
4.	Orange peels	A	A	d	■	■	■	A	■	
5.	Oubain	■	A	d	■	■	■	■	■	
	P									
1.	Pale-catechu	■	■	■	■	■	■	■	■	
2.	Palma rosa oil	■	■	■	■	■	■	■	■	
3.	Papain	A	A	A	A	A	A	■	■	
4.	Pectin	A	A	A	A	d	■	■	■	
5.	Picrorrhiza rhizomes	A	A	d	■	A	A	■	■	
6.	Peppermint oil	A	■	■	A	A	A	A	■	
7.	Peru balsam	■	■	■	■	■	■	A	■	
8.	Physostigma-seeds	■	A	■	■	■	■	A	■	
9.	Pipalamul	■	■	■	■	A	A	■	A	
10.	Podophyllum resin	A	A	d	■	■	■	A	■	
11.	Podophyllum roots	A	A	■	■	■	■	■	■	
12.	Poppy seed oil	■	■	■	■	■	■	■	■	
13.	Potato starch	A	A	A	A	A	■	A	■	
14.	Psoralea fruits	A	A	d	■	■	■	■	■	
15.	Psoralen	■	■	A	A	A	■	■		
16.	Psyllum	■	■	■		■	■	A	■	
17.	Pterocarpus herb	A	■	■	■	■	■	■	■	
18.	Punarnava herb	A	A	d	■	A	A	■	■	
19.	Pyrethrum	A	A	d	A	A	■	■	■	
	Q									
1.	Quassia	A	■	■	■	■	■	■	■	
2.	Quilliai bark	A	A	d	■	■	■	■	■	
	R									
1.	Rasna roots	■	A	d	■	■	■	■	■	
2.	Rauwolfia	A	A	d	■	A	A	■	A	
3.	Rhubarb	A	A	d	■	■	■	A	■	
4.	Rice bran oil	■	■	■	■	■	■	■	■	
5.	Rice starch	A	A	A	A	A	■	A	■	
6.	Rosemarry oil	A		■	■	■	■	A	■	
7.	Ryania roots	■	■	■	■	■	■	■	■	
	S									
1.	Sabadilla	■	■	■	■	■	■	■	■	
2.	Safed musali	■	■	■	■	■	■	■	■	

contd...

		*1	*2	*3	*4	*5	*6	*7	*8	
3.	Safflower oil	■	■	■	■	■	■	■	■	
4.	Saffron	A	■	■	■	■	■	A	A	
5.	Sandal wood oil	■	■	■	■	■	■	■	A	
6.	Santonica	A	■	■	■	■	■	■	■	
7.	Saussurea rhizomes	A	A	d	■	■	■	■	■	
8.	Senega roots	■	■	■	■	■	■	A	■	
9.	Senna leaves (Alexandrian)	A	A	A	■	■	A	A	■	
10.	Senna leaves (Indian)	A	A	A	A	A	A	■	A	
11.	Senna pods	A	A	A	A	A	A	■	■	
12.	Sesame oil	A	A	d	■	■	■	A	■	
13.	Shankhpushpi-herb	■	■	■	■	■	■	■	A	
14.	Shatavari	■	■	■	■	A	A	■	A	
15.	Shati (Hedichium)	■	■	■	■	A	■	■	■	
16.	Shellac	■	■	A	■	A	A	■	■	
17.	Silymarin seeds	■	■	■	■	■	■	■	■	
18.	Solanum khasianum	■	■	■	■	■	■	■	■	
19.	Soya bean seeds	■	■	■	■	■	■	A	■	
20.	Spearmint oil	■	■	■	■	■	■	A	■	
21.	Spirulina	■	■	■	■	■	■	■	■	
22.	Squill (European)	■	■	■	■	■	■	■	■	
23.	Starch	A	A	d	■	A	A	■	■	
24.	Stevia	■	■	■	■	■	■	■	■	
25.	Storax	A	A	d	■	A	■	■	■	
26.	Stramonium herb	A	A	d	■	■	■	A	■	
27.	Strophanthus	■	■	■	■	■	■	■	■	
	T									
1.	Tannic Acid	A	A	d	■	■	■	■	■	
2.	Tar oil	A	A	■	■	■	■	A	■	
3.	Taxus	■	■	■	■	■	■	■	■	
4.	Tea-herb	■	■	■	■	■	■	■	■	
5.	Thaumatin	■	■	■	■	■	■	■	■	
6.	Thevetia	■	■	■	■	■	■	■	■	
7.	Thyme	■		■	■	■	■	■	A	
8.	Tinospora	■	■	■	■	A	A	■	A	
9.	Tobacco leaves	■	■	■	■	■	■	■	■	
10.	Tolu balsam	A	A	d	A	A	A	A	■	
11.	Tonka beans	■	■	■	■	■	■	■	■	

contd...

		*1	*2	*3	*4	*5	*6	*7	*8	
12.	Tragacanlh	A	■	A	A	A	A	A	■	
13.	Tulsi herb	■	■	■	■	A	A	■	A	
14.	Turmeric rhizomes	A	A	d	■	A	A	A	A	
15.	Turpentine oil	A	A	d	■	■	■	A	■	
	U									
1.	Urginea	A	■	■	■	■	■	■	■	
	V									
1.	Valerian rhizomes	A	A	d	■	■	■	A	■	
2.	Vasaka leaves	A	A	d	■	A	A	A	A	
3.	Veratrum rhizomes	■	■	■	■	■	■	■	■	
4.	Vetiver oil	■	■	■	■	■	■	■	■	
5.	Vidang	A	A	d	■	■	■	■	■	
6.	Vinca-herb	■	■	■	■	■	■	■	■	
7.	Visnaga fruits	■	■	■	■	■	■	■	■	
	W									
1.	Wheat germ oil	■	■	■	■	■	■	A	■	
2.	Wheat starch	A	A	A	A	A	A	A	A	
3.	Wild cherry bark	■		■	■	■	■	■	■	
	X									
1.	Xanthan gum	■	■	■	■	■	■	A	■	
	Y									
1.	Yeast	A	■	A	d	■	■	■	■	
(b)	**Animal Drugs**									
1.	Ambergris	■	■	■	■	■	■	■	■	
2.	Bees wax-yellow	A	A	A	A	A	A	A	■	
3.	Cantherides	■	■	■	■	■	■	■	■	
4.	Casein	■	A	■	■	■	■	■	■	
5.	Castoreum	■	■	■	■	■	■	■	■	
6.	Chitin	■	■	■	■	■	■	■	■	
7.	Civet oil	■	■	■	■	■	■	■	■	
8.	Cobra-venum	A	A	A	■	■	■	■	■	
9.	Cochineal	A	■	■	■	■	■	■	■	
10.	Cod liver oil	■	■	■	■	■	■	A	■	
11.	Collagen	■	■	■	■	■	■	■	■	
12.	Gelatin	A	A	A	A	A	A	■	■	
13.	Heparin sodium	A	A	A	A	■	■	A	■	
14.	Honey	A	A	d	A	A	A	A	■	
15.	Lactose	A	A	A	A	A	A	A	■	

contd...

		*1	*2	*3	*4	*5	*6	*7	*8	
16.	Lard	■	■	■	■	■	■	■	■	
17.	Musk	■	■	■	■	■	■	■	■	
18.	Mylabris	■	■	■	■	■	■	■	■	
19.	Protamine sulphate	■	■	■	■	■	■	A	■	
20.	Protein hydrolysate	■	A	■	■	■	■	■	■	
21.	Shark-liver oil	A	A	A	d	■	A	■	■	
22.	Shellac	■	■	A	A	A	A	■	■	
23.	Silk fibres	■	■	■	■	■	■	■	■	
24.	Spermaceti	■	■	■	■	■	■	■	■	
25.	Suet fat	A	■	■	■	■	■	■	■	
26.	Wool fat	A	A	A	A	A	A	A	■	
27.	Wool fibres	■	■	■	■	■	■	■	■	
(c)	**Mineral Drugs**									
1.	Asbestos	■	■	■	■	■	■	■	■	
2.	Bentonite	A	A	A	A	A	A	A	■	
3.	Calamine	A	A	A	A	A	A	A	■	
4.	Chalk (whiting)	A	■	■	■	■	■	A	■	
5.	Fullers earth	■	A	■	■	■	■	■	■	
6.	Kaolin (China-clay)	A	A	A	A	A	A	A	■	
7.	Kieselguhr	A	A	■	■	■	■	■	■	
8.	Paraffins (liquid, hard & soft)	A	A	A	A	A	A	A	■	
9.	Shillajit	■	■		■	■	■	■	■	
10.	Talc (soap stone)	A	A	A	A	A	A	A	■	
(d)	**Enzymes - Animal origin**									
1.	Bromelain	■	■	■	■	■	■	■	■	
2.	Chymotrypsin	■	■	■	■	■	■	A	■	*
3.	Collagenase	■	■	■	■	■	■	■	■	*
4.	De-oxyribonuclease	■	■	■	■	■	■	■	■	*
5.	Diastase (Amylase)	■	■	A	A	■	■	■	■	
6.	Fibrinolysin	■	■	■	■	■	■	■	■	*
7.	Fungal-lipase	■	■	■	■	■	■	■	■	
8.	Hemecellulase	■	■	A	■	■	■	■	■	
9.	Hyaluronidase	■	A	A	■	A	A	A	■	*
10.	L-Asparginase	■	■	■	■	■	■	■	■	
11.	Muramidase	■	■	■	■	■	■	■	■	*
12.	Pancrealipase	■	■	■	■	■	■	■	■	*

contd...

		*1	*2	*3	*4	*5	*6	*7	*8	
13.	Pancreatin	A	A	A	A	A	A	A	■	*
14.	Pepsin	A	A	A	A	A	A	A	■	*
15.	Rennin	■	■	■	■	■	■	■	■	*
16.	Sutilains	■	■	■	■	■	■	■	■	
17.	Seratio-peptidase	■	■	■	■	■	■	■	■	
18.	Streptokinase	■	■	■	A	■	■	A	■	
19.	Trypsin	■	■	■	■	■	■	A	■	*
20.	Urokinase	■	■	■	A	A	A	A	■	*

*1. Indian Pharmacopoeia 1955, 1st Edition, Controller of publications, Govt. of India, Ministry of Health, New Delhi.

*2. Indian Pharmacopoeia 1966, 2nd Edition, Controller of publications, Govt. of India, Ministry of Health, New Delhi.

*3 Indian Pharmacopoeia 1985, 3rd Edition Vol. I & II, Controller of publications Govt. of India, Ministry of Health and Family Welfare, New Delhi.

*4. Indian Pharmacopoeia 1996, 4th Edition Vol. I & II, Controller of publications Govt. of India, Ministry of Health and Family Welfare, New Delhi.

*5. Indian Pharmacopoeia 2007, 5th Edition, Vol. I, II & III, Controller of publications, Govt. of India, Ministry of Health and Family Welfare, New Delhi.

*6. Indian Pharmacopoeia 2010, 6th Edition Vol. I, II & III, Govt. of India, Ministry of Health and Family Welfare, New Delhi.

*7. British Pharmacopoeia 2007, Vol. I to III, Ministry of Health and Social service for Northern Ireland UK.

*8. Ayurvedic Pharmacopoeia of India, 2001 - 2006, 1st Edition, Vol. I to IV Govt. of India. Ministry of Health and Family Welfare Department of **AYUSH**, New Delhi.

11. RETAIL PRICES OF NATURAL DRUGS DURING 2023 - 2024

ITEM	Price in ₹ per kg
ACACIA -GUM	300
ACONITE - RHIZOME	300
AGAR	450
AGAR STRIPS (SHREDS)	550
AJOWAN FRUITS	200
ALMOND (BITTER) SEED	500
ALMOND (SWEET) SEED	600
ALOE	300
AMLA - FRUITS	175
AMOMUM	240
ANISE FRUITS	500
ARJUNA BARK	060
ARTEMESIA FLOWERS	750
ASHOKA BARK	120
ASHWAGANDHA RHIZOMES	200
ASHWAGANDHA ROOT	150
BAEL FRUITS	150
BENAFSA LEAVES	200
BAVCHI FRUITS	125
BEES WAX WHITE/YELLOW	300
BEHERA FRUITS	050
BELLADONNA LEAVES	300
BELLADONNA ROOTS	350
BERBERIS ROOTS	150
BLACK PEPPER	550
BRAHMI LEAVES	250
CALAMUS RHIZOMES	200

ITEM	Price in ₹ per kg
CALUMBA ROOTS	800
CAPSICUM FRUITS	350
CARAWAY FRUITS	800
CARDAMOM FRUITS	1000
CASCARA BARK	450
CASSIA BARK	330
CASTOR SEEDS	080
CATECHU (BLACK)	500
CATECHU (PALE)	200
CELERY FRUITS	300
CHIRATA HERB	600
CHITRAK HERB	100
CINCHONA BARK	200
CINNAMON BARK	500
CLOVE FLOWER BUDS	800
COLCHICUM CORM	500
COLCHICUM SEEDS	1000
COLOCYNTH FRUITS	150
CORIANDER FRUITS	150
CUBEB	4200
CUMIN FRUITS	300
COTTON (ABSORBANT)	400
DATURA LEAVES	150
DIASTASE POWDER	2500
DIGITALIS LEAVES	800
DILL FRUITS	200
DIOSCOREA RHIZOMES	250
EPHEDRA STEMS	200
FENNEL FRUITS	250

GALANGAL RHIZOMES	350
GALLS	300
GARLIC BULB	200
GELATIN POWDER	500
GENTIAN RHIZOMES	200
GINGER RHIZOMES	250
GOKHRU FRUITS	200
GUAR GUM	250
GUDMAR LEAVES	200
GYMNEMA	250
HERDE (MYROBALLAN) FRUITS	150
HONEY	300
HYOSCYAMUS LEAVES	250
IPECACUNHA ROOTS	2000
ISAPGOL HUSK	250
ISAPGOL SEEDS	160
JALAP RHIZOMES	200
JATAMANSI RHIZOMES	250
KALADANA SEED	200
KALMEG HERB	100
KAPUR KACHARI RHIZOMES	450
KINO	150
KURCHI BARK	200
LEMON PEELS	200
LINSEED	250
LIQUORICE	250
LOBELIA HERB	300
MACE	350
MALE - FERN	500
MALKANGNI SEEDS	250
MUSTARD SEED	150
NEEM SEED	050
NUTMEG SEEDS	500
NUX VOMICA SEEDS	060

ORANGE PEELS	200
PAPAIN	1500
PECTIN POWDER	800
PTEROCARPUS	400
PICRRORHIZA RHIZOMES	300
PIPER LONGUM FRUITS	500
PODOPHYLLUM RESIN	5000
POPPY CAPSULES	2000
PUNARANAVA	100
PYRETHRUM FLOWERS	400
QUAILLIA BARK	200
QUASSIA WOOD	500
RASNA RHIZOMES	250
RAUWOLFIA ROOTS	400
RHUBARB ROOT	200
ROSEMERY	1000
SAFFRON* (per 100 gm)	7500
SANDALWOOD	500
SANTONICA FLOWERS	500
SAUSSUREA RHIZOMES	200
SENEGA ROOTS	400
SENNA LEAVES	250
SENNA PODS	250
SESAME SEEDS	200
SHANKHPUSHPI	200
SHATAVARI	200
SHILAJEET	2000
SILK FIBRES	2500
SQUILL BULBS	350
STARCH POWDER	150
STRAMONIUM HERB	300
STROPHENTHUS SEEDS	600
TALC POWDER	150
TAMARIND PULP	400

TAR	120
TOBACCO LEAVES	200
TULSI HERB	150
TURMERIC RHIZOMES	150
VALERIAN RHIZOME	250
VASAKA LEAVES	150
VIDANG SEEDS	250
VINCA HERB	350
WHITE MUSTARD FRUITS	150
WILD CHERRY BARK	400
WOOL FIBRES	2000
YEAST POWDER	600

WAXES

BEES WAX WHITE	380
BEES WAX YELLOW	360
CANDEDILLA WAX	1300
CARNUBA WAX (DARK YELLOW)	2000
CARNUBA WAX (RED)	1000
CERESINE WAX	295
MICROCRYSTALLINE WAX	1500
PARAFFIN WAX	180
SPERMACETI WAX	1000
WOOL FAT	350

RESINOIDS

ASAFOETIDA	500
BALSAM OF COPAIBA	2500
BALSAM OF PERU	2500
BALSAM OF TOLU	2500
BENZOIN	450
CALOPHONY	200
GUGGUL	600
MYRRH	700
OAK MOSS	1500
OLIBANUM	1000
SALLAKI GUGGUL	1500
SHELLAC	400
STORAX	400

FIXED OILS

ALMOND OIL	700
CHAULMOGRA OIL	700
COCO-BUTTER	250
COCONUT OIL	300
CODLIVER OIL	800
CORN OIL	300
COTTON SEED OIL	250
CASTOR OIL	500
GROUNDNUT OIL	200
JOJOBA OIL	1100
KARANJA OIL	400
KOKUM-BUTTER	350
LINSEED OIL	400
MAHUA OIL	200
MUSTARD OIL	250
NEEM SEED OIL	500
OLIVE OIL	500
PALM OIL	100
RICE BRAN OIL	200
SAFFLOWER OIL	300
SESAME OIL	250
SHARK LIVER OIL DILUTE	400
SUNFLOWER OIL	300
WHEAT GERM OIL	1500

VOLATILE OILS

Oil	Price	Oil	Price
AJOWAN OIL	1000	JASMINE OIL	2000
ALMOND OIL (BITTER)	800	JATAMANSI OIL	1000
ANETHI OIL	5000	LAVENDER OIL	2000
ANISE OIL	1400	LEMON GRASS OIL	1000
ANISEED OIL	1500	LEMON OIL	1000
BASIL OIL	600	LIME OIL	1500
BAY OIL	2800	LIME SWEET OIL	3500
BERGAMOT OIL	1600	MACE OIL	6500
BLACK PEPPER OIL	13500	MENTHA CITRATA OIL	1500
CAJEPUTI OIL	1000	METHA OIL	700
CAMPHOR	600	MENTHA PEPPERITA OIL	1500
CAMPHOR OIL	400	NAGARMOHTA OIL	5000
CARAWAY OIL	8500	NEROLI OIL	1500
CARDAMOM OIL	6500	NUTMEG OIL	1000
CASSIA OIL	900	ORANGE OIL	600
CEDARWOOD OIL	800	PALMA ROSA OIL	1200
CELERY OIL	4200	PATCHOULI OIL	3500
CHAMELI OIL	1200	PETTIGRAIN OIL	1500
CHENOPODIUM OIL	3000	PINE OIL	200
CINNAMON OIL	600	ROSE OIL	700
CITRONELLA OIL	750	ROSEMARY OIL	1800
CLOVE OIL	1000	SANDALWOOD OIL	25000
CORIANDER OIL	5500	SASSAFRAS OIL	3500
CUMIN OIL	4500	SPEARMINT OIL	1200
DILL OIL	3500	THYME OIL	1000
EUCALYPTUS OIL	400	THYMOL	2000
FENNEL OIL	2750	TULSI OIL	1600
GARLIC OIL	5000	TUNG OIL	1600
GERANIUM OIL	4000	TURPENTINE OIL	160
GINGER GRASS OIL	1000	VALERIAN OIL	12000
GINGER OIL	5500	VETIVER OIL	8100

* Prices are approximate and are subject to change depending on demand in the market.

12. SUPPLIERS OF BIO-MARKERS AND STANDARDIZED HERBAL EXTRACTS

1. Arjuna Natural Exatracts Ltd.
   Alwaye, 623 101 (Kerala).
   Tel. : 0484 – 2622644, 2622655
   Fax : 0484 – 2532404, 2622612
   E.mail : mail@arjuna natural.com
   www.arjuna natural.com

2. Shubham Pharmachem Pvt.Ltd.
   207, Laxmi plaza, laxmi industrial estate,
   New Link Road, Andheri(W),
   Mumbai-53.
   Tel. : 022-26344608, 66914797
   E.mail : shubham@shubham.co.in
   www.shubham.co.in

3. Indus Extracts
   7, Deokaran menstion
   24, vithaldas Road,
   Princess Street, Mumbai-400 002
   Tel. : 022-22014864
   www.info@indusextract.com

4. Beijing New World bio-Technology Comp. Ltd.
   715, New World-centre a Tower no. 3
   Chongwenmen Wai Street
   Beijing, China-100 062
   Tel. : 0086-10-67091198
   E.mail: nw@nwpharm.com
   www.nwpharm.com

5. Ruland chemistry (Nanjing) Co. Ltd
   Rm 1201, heping mansion, -
   No. 22 East Beijing Road,
   Nanjing China P.C210 018
   Tel : 0086-25-86897711
   E.mail : ruland@jsmail.com.cn
   www.rulandchem.com

6. Panjin Huacheng Pharmaceutical Co. Ltd.
   Dongfeng Town Dawa county,
   Panjin city Liaoning Province,
   China- 124 205
   Tel. : 0086-427-6843800, 6842218
   E.mail : pjhczy@tom.com
   www.pjhczy.com

7. Shenyang Tain Feng Bio-engineering Technology Co. Ltd.
   Room 222 No. 46-5 Wen' an street,
   Heping District shenyang,
   China- 110 003
   Tel. : 0086-24-83971058, 83971398
   E.mail : wjw@vip.163.com
   Mob : 13704019198, 13609898121
   www.taxol.com.cn

8. Sichuan Senyn Import & Export Co.Ltd.
   Weicheng Jinhe East Road, Shifang City, Sichuan province, China- 618 400
   Tel: 86-838-8108531
   Fax: 86-838-8108531
   E.mail:scsy@scsenyu.cn
   www.scsenyu.cn

9. Human Keyuan Bio-Products Co. Ltd.
   Maoping Town Chengbu County,
   Human province. China 422 503
   Tel. : 0086-739-7474658
   Fax : 0086-739-7474986
   E.maii : lzqky@vip163.com
   www.hnkeyuan.com

10. Shanghai Natural Bio-engineering Co.Ltd.
Room 1606, Shanghai Medicine Trade building,
No. 1399 Golden Bridge Road **Shanghai**,
China.
Tel. : 0086-21-58341647 Fax: 0086-21-25344207
E.mail : fuyun_123@hnkeyuan.com

11. Natural Remedies
Plot No. S B, 9`11' K.M.Stone,
Veer Sandra industrial area,
Electronic city post,
Bangalore-560 100
Tel: 91-80-40209999
Fax: 91-80-402098 17
E.mail :
phytochemistry@naturalremedy.com

12. Dr. Y. S. Bedi, Scientist
Indian Institute of Integrative Medicine
(Formerly: Regional Research Laboratory, Jammu)
Canal road, **Jammu Tawi**-180 001
Tel. : 0191-2569030, 2569111, 2569222
E.mail : ysbrrljm@gmail.com,
ysbrrljm@rediffmail.com
www.rrjammu.org

13. Sami Labs Ltd.
19/1 and 19/2 I main II phase,
Peenyz Industrial Area
Bangalore : 560058 Karnataka.
Tel. : 080-28397973, 75, 78
Fax : 080-28373035
E.mail : mail@samilabs.com

14. Amsar Private Limited
2, Hormuz Mansion,
72 B Desai Road
Mumbai-400006
Tel. : 022-23632348
Fax : 022-23673009
E-mail : amsar@amsar.com

15. Herbochem Ltd.
38 Technocrats Industrial Estate,
Balangar,
Hyderabad-500037
Tel. : 040-23077331
Fax : 040-23078059
E-mail : herbochem@rediffmail.com

16. Chemiloids
40-15-14, Brindavan colony
Venkateshwarpuram, Labbipet
Vijayawada-520010
Tel. : 0866-2476561
Fax : 066-2475278

17. Hysel India Pvt. Ltd.
1405 Chiranjiv Tower
43, Nehru place
New Delhi 110019
E-mail : hysel@del2.vsnl.net.in
Tel. L 011-26227801/02/03/04

18. Gaurao Enterprise
D-29, Kamla Nagar,
Agra 282005
E-mail : giris 640@ yahoo.com, Tel. : 0562 2883724
Fax. 0562 288 1414

❖❖❖

13. MACROSCOPIC CHARACTERS OF LEAVES

(a) Shape of the Lamina of Leaves:

Various shapes of the leaves are due to various types or shapes of lamina.

It may be one of the following:

1. **Acicular:** Needle like, Example: *Pinus*.
2. **Subulate:** With acute apex and recurved point, Example: *Ephedra sinica*.
3. **Linear:** When it is long, narrow and flat, Example: *Grasses*.
4. **Oblong:** Broad leaves with two parallel margins and abruptly tapering apex, Example: *Banana*.
5. **Lanceolate:** It look like lance or spear-shaped, Examples: *Nerium, Senna*.

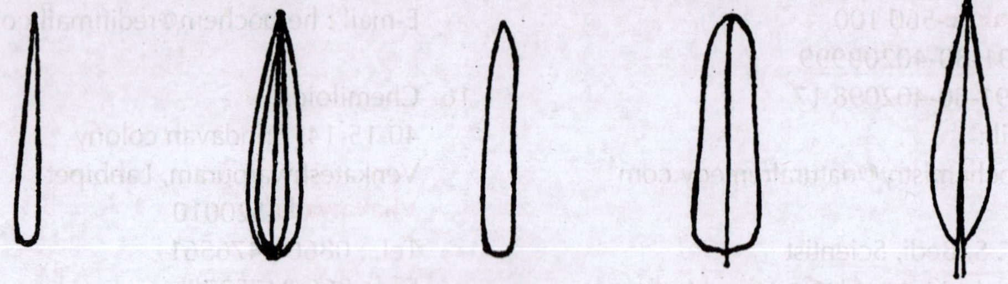

1. Acicular 2. Subulate 3. Linear 4. Oblong 5. Lanceolate

Fig. 1: Shape of Lamina of leaves

6. **Ovate:** Egg-shaped or broad base and narrow apex. Examples: *China rose, Buchu*.
7. **Obovate:** Broad apex and narrow base. Example: *Jangali-badam*.
8. **Obcordate:** Inversely heart-shaped i.e. base is narrow but apex is broad. Example: Oxalis.
9. **Spathulate:** Like spatula or spoon-shaped as in calendula and drosera.
10. **Cuneate:** Wedge-shaped as in *Pista*.

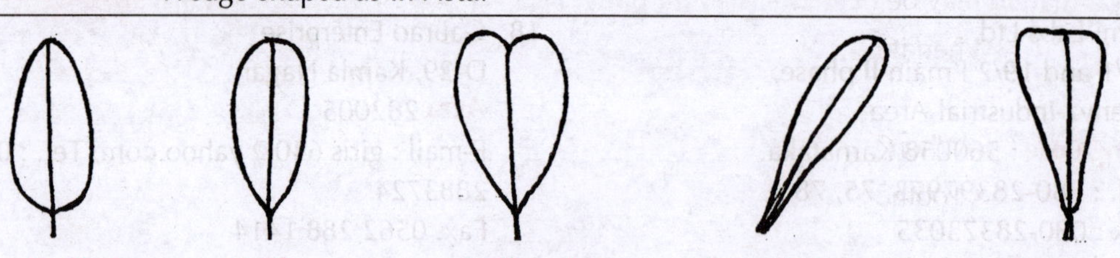

6. Ovate 7. Obovate 8. Obcordate 9. Spathulate 10. Cuneate

Fig. 2: Shape of Lamina of leaves

11. **Cordate** or heart shaped, Example: *Betel*.
12. **Sagittate:** Arrow-shaped such as in *Arum*.
13. **Hastate:** When the two lobes of sagittate-leaf are directed outwards as in *Ipomoea*.
14. **Reniform:** Kidney shaped, Example: *Jalbramhi, Indian pennywort*.
15. **Auriculate:** When the leaf has got ear-like projections at the base.

16. **Lyrate:** When it is lyre-shaped or the blade is divided into lobes with large marginal lobe, Example: *Raddish, Mustard.*

Fig. 3: Shape of lamina of leaves

17. **Runcinate:** With the lobes convex before and straight behind, pointing backward like the teeth of the double saw i.e. *Dendelion leaf.*
18. **Rotund (Orbicular):** When the blade is circular or round e.g. *lotus.*
19. **Elliptical or oval:** When the leaves are narrow at the base and apex but broad in the middle such as *guava, vinca* etc.

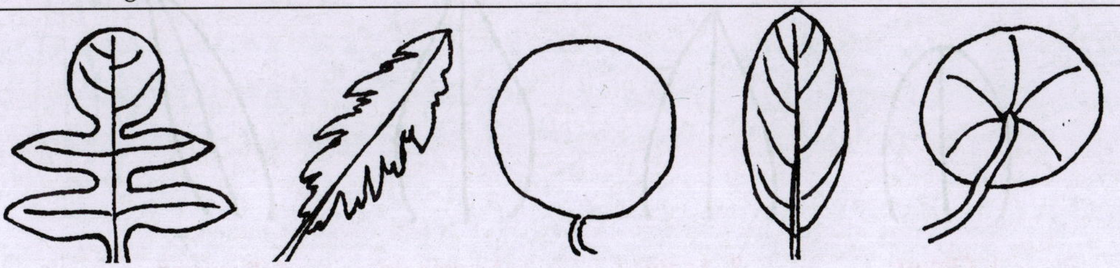

Fig. 4: Shape of Lamina of leaves

20. **Peltate:** When the lamina is shield-shaped and fixed to the stalk by the center.
21. **Falcate:** When leaves are sickle-shaped as in *Eucalyptus and Astratian acacia.*

(b) Leaf Margins:

Leaf margin may be of the following types:

(a) **Entire:** When it is even and smooth, Example: *senna, eucalyptus.*

(b) **Sinuate or wavy:** With slight undulations like *Ashok.*

(c) **Crenate:** When the teeth are round as in *digitalis.*

(d) **Dentate:** Toothed margin, teeth directing outwards such as *Margosa, Melon.*

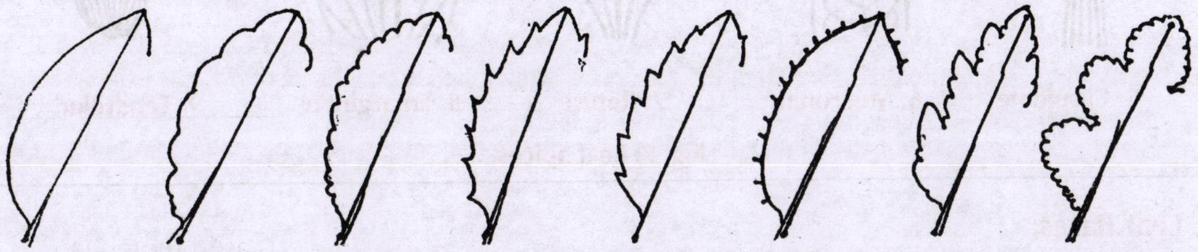

Fig. 5: Margins of leaves

(e) **Serrate:** When it is like the teeth of the saw such as *Rose, China rose* etc.

(f) **Ciliated:** It is fringed with hairs.

(g) **Spiny:** When provided with spines like *Argemone*.

(h) **Bi-serrate:** Lobed serrate margin.

(i) **Bi-crenate:** Lobed crenate margin.

(c) Leaf Apices:

The apex of the leaf may be one of the following kinds:

1. **Obtuse:** Rounded tip i.e. *Banyan*.
2. **Acute:** When it is pointed to form acute angle, but not stiff i.e. *Hibiscus*.
3. **Acuminate:** Pointed tip with much elongation, *Peepal*.
4. **Recurved:** When the apex is curved backward.

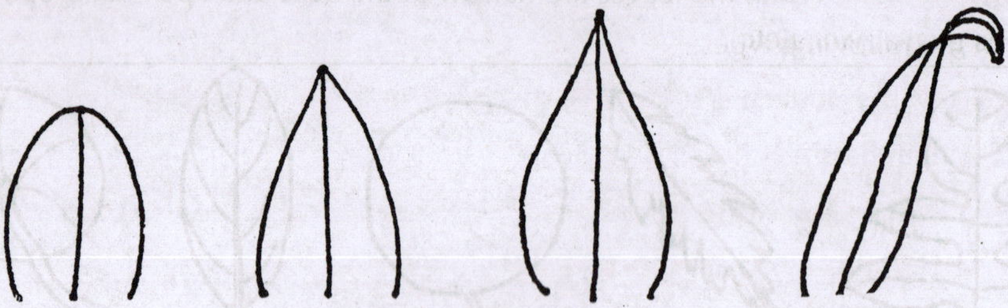

1. Obtuse 2. Acute 3. Acuminate 4. Recurred

Fig. 6: Leaf apices

5. **Cuspidate:** With spiny tip like *Date palm*.
6. **Mucronate:** Rounded apex ending abruptly in a short point i.e *vinca, ixora*.
7. **Retuse:** Broad tip with slight notch i.e. *Pistia*.
8. **Emarginate:** When tip is deeply notched as in *Bambinia*.
9. **Tendrillar:** Tip forming a tendril such as *Gloriosa - superba*.

5. Cuspidate 6. Mucronate 7. Retuse 8. Emarginate 9. Tendrillar

Fig. 7: Leaf apices

(d) Leaf Bases:

The lower extremity of the lamina of the leaf may exhibit one of the following shapes: (Fig. 5.4)

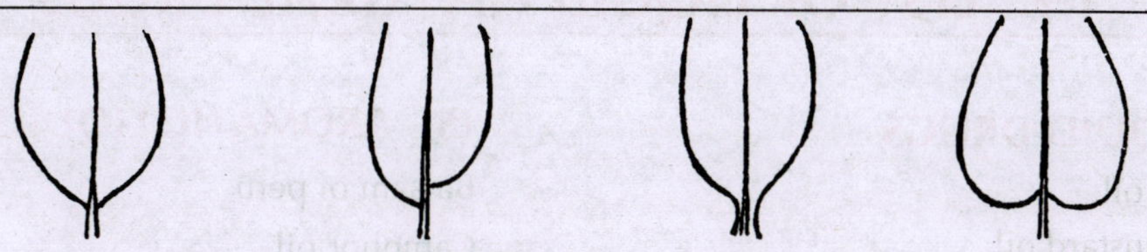

a. Symmetrical b. Asymmetrical c. Decurrent d. Cordate

Fig. 8: Leaf-bases

1. **Symmetrical:** Equal as in *vasaka*.
2. **Asymmetrical:** Unequal as in *senna* or *Datura*.
3. **Decurrent:** As in *Digitalis*.
4. **Cordate:** As in *Betel*.

(e) Leaf Surface:

It may be of the following types:

1. **Glabrous:** When surface is smooth and free of hair or any outgrowth i.e. *Vasaka, Datura*.
2. **Rough:** When harsh to touch, *Digitalis*.
3. **Hairy:** When covered with hairs.
4. **Glutinous:** When covered with sticky substance, *Tobacco*.
5. **Glaucous:** When covered with waxy coating, *Castor*.
6. **Pubescent:** Covered with straight, short hair i.e. *Senna*.

(f) Venation:

The arrangements of veins in lamina or leaf blade is known as venation. In flowering plants two types of venation exist.

1. Reticulate e.g. Henna, eucalyptus, carica.
2. Parallel e.g. Banana, fan-palm.

14. LIST OF LIQUID NATURAL DRUGS

(A) LIQUID DRUGS

Arachis oil
Black mustard oil
Capsicum oleo-resin
Castor oil
Chaulmoogra oil
*Cod liver oil
Corn oil
Cotton seed oil
Evening primrose oil
Ginger oleo resin
*Halibut liver oil
*Honey
Jojoba oil
Karanja oil
Linseed oil
Liquid glucose
Mango kernel oil
Neem oil
Olive oil
Poppy seed oil
Rice bran oil
Safflower oil
Sesame oil
*Shark-liver oil
Soyabean oil
Sunflower oil
Wheat jerm oil

* Drugs of animal origin

(B) AROMATIC LIQUIDS

Balsam of peru
Camphor oil
Chenopodium oil
Citronella oil
Civet
Davana oil
Eucalyptus oil
Gaultheria oil
Ginger oleo-resin
Guggul resin
Lavender oil
Lemon grass oil
Palma rosa oil
Peppermint oil
Pine-tar
Sandal wood oil
Turpentine oil
Vetiver oil

(C) SEMI-SOLID DRUGS

Balsam tolu
Castorium
Musk
Storax

15. ANNUAL AND PERENNIAL PLANTS

ANNUALS

1. Ajowan
2. Ashwagandha
3. Bhringraj
4. Bhuiamla
5. Black mustard
6. Black-pepper
7. Brahmi
8. Capsicum
9. Caraway
10. Centella
11. Chirata
12. Coriander
13. Cumin
14. Datura
15. Dawana
16. Dill
17. Fenugreek
18. Fennel
19. Garlic
20. Ginger
21. Gokhru
22. Guar herb
23. Haldi
24. Isapgoal
25. Jalbrahmi
26. Kalmegh
27. Linseed
28. Maize
29. Mustard
30. Pea-nut
31. Peppermint
32. Potato
33. Psoralea
34. Punarnava
35. Rice
36. Safflower
37. Senna
38. Sesame
39. Shankhpushpi
40. Spearment
41. Squill
42. Sunflower
43. Turmeric
44. Vajradanti
45. Vinca

PERENNIALS

1. Aloe
2. Annatto
3. Areca
4. Boswellia
5. Castor
6. Catechu
7. Citronetta
8. Coconut
9. Coffee
10. Ephedra
11. Guggul
12. Kapur-kachari
13. Karanj
14. Lemon
15. Lemongrass
16. Palmarosha
17. Sallaki
18. Sandal
19. Shatawari
20. Stevia
21. Tea-herb
22. Tulsi
23. Vasaka
24. Vidang

16. HERBS, SHRUBS, CLIMBERS, TREES ETC.

Herbs : These are the plants with soft stem and dies down after flowering.

Shrubs: Plants smaller than a tree and has several stems of wood coming from ground.

Trees: Tall plants which survive for long duration and have a thick central wooden trunk from which branches arise.

HERBS

Bhringraj
Bhui amla
Capsicum
Datura
Digitalis
Fennel
Galanga
Hyoscyamus
Kalmegh
Kutki
Methi

SHRUBS

Annatto
Castor
Coffee herb
Daruhalad
Ephedra
Garden rue
Glycyrrhiza
Kantakari
Karanj
Nerium
Senna
Vasaka

CLIMBERS

Black-pepper
Dioscorea
Gudmar
Gulwel
Karela
Long pepper
Manjishta
Shatavari
Tinospora
Vanilla

PROSTATE PLANTS

Bramhi
Colocynth
Gokhru
Pea-nut
Punarnava

TREES

Haritaki
Amla
Arjuna
Ashok
Behda
Cinnamon
Eucalyptus
Kurchi
Neem
Nux-vomica
Sandalwood

❖❖❖

17. ETYMOLOGY OF SCIENTIFIC NAMES OF MEDICINAL PLANTS

Sr. No.	Name of Natural Drug	Scientific (Latin) Name of Drug with family	Part of Plant in use	Origin of Word	Meaning
1.	Almond oil	*Prunus amygdalis* *Rosaceae*	Seeds	Prunus Amygdalas Amara is latin Dulsis is latin	Prunus is name of plum tree Amygdalas means Almond tree Amara means bitter variety Dulsis means sweet variety
2.	Aloe	*Aloe barbadensis* *Aloe vera* *Aloe ferox* *Aloe spicata* *Liliaceae*	Leaf-juice	Alloeh is arabic vera is from *Verus ferox* is latin spicata is latin	Alloeh meaning shining/bitter Verus means true Ferox means wild Spicata means flowers in spikes.
3.	Amala	*Emblica officinalis* Euphoribiaceae	Fresh or dried fruits	*Amlika* means Sanskrit	Sour juice of the fruit Sold in shop for service to men.
4.	Ananas	*Ananus americana* or A. *comosus* Bromeliaceae	Fruit	Ananus: Comosus:	American name Hair
5.	Arjuna	*Terminalis arjuna* Combretaceae	Bark	Terminalis Arjuna-sanscrit	At the end White coloured bark is externally white in colour.
6.	Asafoetida	*Ferula foetida* Umbellifeae	Rhizome Oleo-resin	Ferula: Foetid:	Giant funnel shaped Unpleasent
7.	Ashoka	*Saraca indica* Leguminosae	Bark	Saraca Indica	Orgzilian name the plant India
8.	Ashwagandha	*Withania somnifera* Solanaceae	Roots	Withan	Name of scientist
9.	Bael	*Aegle mormalose* Rutaceae	Fruit	Aegle: marmalose:	Goat Portugese word
10.	Bahava	*Cassia fistula* Leguminosae	Dried fruits	Cassia is from Greek kassia fruits fistula: tube like	Fruits are tube like
11.	Balanite	*Belanitus aegyptica* *Simarubaceae*	Fruit	Balanite aegyptica	Nut of Igypt

12.	Behra	*Termenalia balerica* *Combretaceae*	Fruits	*Terminus*: Ballrica:	End leaves are terminal From Aranion name Belay.
13.	Belladonna herb	*Atropa belladonna* *Solanaceae*	Aerial parts of herb	Bella italian Donna italian	Bella means beautiful Donna means a lady
14.	Benzoin	*Styrax benzoin* *Styrax tonkinensis* *Styraceae*	Oleo-resin	Styrax is Arabian Ben is arabic Zoa	Styrax means sweet smelling exudade Ben means fragrent means an exudation Tonkin is northen region of Vietnam.
15.	Betel nut	*Areca – catechu* *Palmae*	Seeds	Areca is spanish	Spanish term for betel nuts.
16.	Bhringraj	*Eclipta alba* *Asteraceae*	Entire herb	Eclipta: alba:	from ekleipo to be deficient white
17.	Black-pepper	*Piper-nigra* *piperacea*	Fruit	*Piper* *niger*	greek name of plant, black
18.	Brahmi	*Herpestis moniera* *Scrophulariaceae*	Entire plant	Herpestis: moni:	creeping Necklace
19.	Camphor	*Cinnamomun camphora* *Lauraceae*	Leaves	Cinnamon is from kayamanis Malay Comphora is arabic Kafur	Malay means sweet wood Kafur means chalk
20.	Capsicum	*Capsicum annum* *Capsicum fruitescens* *Solanaceae*	Fruits	Capsa is latin Annum is latin Fruitescen is latin	Capsa means a box, hollow box like fruit (capsicum) means annual habitate of plant Shrubby character of plant
21.	Caraway	*Carum-carvi* *Umbelliferae*	Fruits	Latin carum Carvi-Arab	caria-place where it was discovered Karawia-Plant where it was found
22.	Cardamom	*(1) Elettaria cardamomum* *(2) Kardamon – Amomoncress* Zingibereceae	Seed	Elletar is Malayalam name for Cardamomum	Amomum – A : No, momus : without impurity : pungency of seeds.
23.	Cascara sagrada	*Rhammus purshiana* *Rhamnaceae*	Bark	Cascara is spanish Purshiana is German	A name for back-thorn bark. It is name of German botanist Purshianus.

Contd...

24.	Cassia bark	*Cinnamonum cassia* *Lauraceae*	Bark	Kassica is greek	Cassia means to strip off bark
25.	Castor	*Ricinus communis* *Euphorbiaceae*	Seeds	Sominifera Ricinus Communis	Producing sleep means Tick Castor seeds resemble insect tick
26.	Castor oil	*Ricinus conmunis* *Euphorbiaceae*	Seeds	Ricinus is latin	Ricinus means tick or bug Appearance of seeds has resemblance to bug in shape and markings on seed.
27.	Catechu	*Acacia catechu* *Leguminosae*	Extract of wood	Catechu is latin	Catechu from Kachu: Name in Karnataka for exudates. Astringent extract
28.	Centella	*Centella asiatica* *Umbelliferae*	Entire plants	Centella: Asiatica:	Many Asian leaves similar to frog (Manduk-parni)
29.	Chandan	*Santalum alba* Santalaceae	Wood	Santalon: Alba:	Original greek word Santolon glandening aroma and medicinal properties, White
30.	Chaulmoogra	*Hydnocarpus heterophylla* *Hydnocarpus laurifollia* *Flacourtriaceae*	Seeds Fixed oil	Hydnocarpus: *Hydnocarpus laurifollia*	Fruit, Leaves like laurel, Leaves like laurel leaves. Laurel is glossy leaved shrub.
31.	Cinchona	*Cinchona* *Cinchona calisaya* *Cinchona succirubra* *Cinchona ledgeriana* *Rubiaceae*	Bark	In the honor of Countess of Chinchon. *Calisaya* is spanish Succus – latin Rubra – latin	The name of wife of viceroy of Peru. Name of drug in Peru Succus means dry Rubra means red colour of drug
32.	Cinnamon	*Cinnamonum zeylanicum* *Lauraceae*	Bark	Cinnamon is Arabic from Kayu manis Loureirii Zeylamicum Cassia is from Greek	Means sweet wood Name of french Botanist Loureiro means Ceylon Greek and latin historians named as a spice in the books of MOSES

Contd...

33.	Clove	*Eugenia caryophyllus Myrtaceae*	Flower buds	Caryophyllus is greek	Means nutleaf referring to nut-like flower buds
34.	Coca leaves	*Erythroxylon coca Erythroxylon truxillance Erythroxyllaceae*	Leaves	Coca is spanish name of tree Truxillance	Yielding cocain Truxillance from Truxilla coastal city in Peru
35.	Colocynth	*Citrullus colocynthus* Cucurbitaceae	Fruit	*Citrullus* from Latin	*Citrullus* from lemon family *colocynlhus*: Bitter pulped fruit.
36.	Coriander	*Coriandrum sativum Umbelliferae*	Fruits	Corys: means Sativrum:	Helmet andrum anthered. Cultivated.
37.	Cumin	*Cuminum-cyminnm* Umbelliferae	Fruit	Greek: Kyminum	For cooking
38.	*Dhayati*	*Woodfordia floribunda Lythraceae*	Fruit Flowers	*Wood fordia*: Floribuda:	Name of botanist, Getting flowers in bunches.
39.	Digitalis	Digitalis purpurea Scrophulariaceae	Leaves	Digitus is latin Purpurea is latin	Digitus meaning finger, refers to finger shaped corolla of flower Purpurea refers to purple coloured flowers of the plant.
40.	Dill	*Anethum graveolens* Umbelliferae	Fruit	*Anelhum:* *Theo*:	scented like Anise run grows rapidly
41.	Eucalyptus oil	*Eucalyptus globulus Myrtaceae*	Fresh leaves and oil	Eucalyptus is Greek	means well covered (i.e. lid like cover of bud)
42.	Fenugreek	*Trigonella foenum graecum Leguminosae*	Leaves	*Tri*: *gone*: *Faenum*: groceum :	Three, Angled Hay, Large
43.	Garlic	*Allium sativa*	Under ground bulb	Allium: *Sativa:*	Garlic Cultivated
44.	Ginger	*Zingiber officinale* Zinziberaceae	Dried rhizomes	Ginger is misnomer of Sanskrit Shrungber Officinale:	Sold in shop for service.
45.	Gokhru	*Tribulus-terrestris Zygophyllaceae*	Fruit	Tribulus Terristris Zygo	Having three sides Spreading on ground Bones of yoke

46.	Guduchi	*Tinospora cardifolia* Minispermiaeeae	Leaves	Tinospora: Cordio:	To stretch like a bow Heart shaped leaves Seeds are curved round the intruded endocarp.
47.	Indian senna	*Cassia angustifolia Leguminosae*	Leaves and fruits	Senna is arabic Acutifolia is latin Angustifolia is latin	It is native name of drug Acutifolia meaning sharply pointed leaflets Meaning narrow leaves.
48.	Ipecacunha	*Cephaelis ipecacunha Rubiaceae*	Roots	Cephaelis is greek Ipekaaguene is Portuguese	Cephaelis means head and to collect. Ipekaaguene means a creeper
49.	Isabgoal	*Plantago – Ovata Plantaginaceae*		*Plantogo*: *Ovata*:	Flat level surface Egg shaped
50.	Kaladana	*Ipomoea hederaceae Convolvulaceae*	Seeds	*Ipomoea:* *hederacea*:	Morning glory Ivy (ever green plant) like in habit
51.	Kalmegh	*Andrographis peniculata* Acanthacea	Entire herb	*Andro*: *Panicula:*	Male, A loose cluster of flowers
52.	Kantkari	*Solanum xanthocarpum Solanaceae*	Entire plant	Solanum: *Xanthum* :	Night shed yellow *Carpum* fruit.
53.	Karanj	*Pongamia glabra* Papilionacea	Bark	Pongamia from Tamil: Glabra:	Pungam Smooth hairless leaves.
54.	Kurchi	*Holerrhena antidysenterica* Apocyanaceae	Bark	*Latin Hollos* : *Arrhen*:	Whole, Greek Male, antidysenterica – against dysentery.
55.	Liquorice (Yesti)	*Glycyrrihiza glabra Glycyrrhiza glandulefera Glycyrrhiza – violaceae Leguminosae*	Rhizomes	Glycyrrhiza is greek Glabra Violacea	Glycyrrhiza means sweet root. means smooth refers to smooth pod like fruit Glandulefera means gland like swelling on roots. Violacea means violet it refers to colour of the flowers of herb.

Contd...

56.	Lodhra	*Symplocos recemosa* *Symplocaceae*	Bark	Symplocos: Racemose:	Stamens united at base Form or cluster of grapes.
57.	Manjishta	*Rubia cordifolia* *Rubiaceae*	Rhizom	Rubia: Cardiofolio	Red Heart shaped leaves
58.	Mentha	*Mentha piperitta* *Piperaceae*	Vol. oil	Mentha is greek Piperata is latin. Piper i.e. pepper	Mythical nymph metamorphosed into plant For aromatic and pungent taste
59.	Mustard	*Brassica nigra* *Brassica juneca* *Crusifareae*	Seeds	Brasic is celtic Juncea is latin Nigra is latin	Celtic means cabbages Juncea means rush Nigra means black
60.	Myrrh	*Commiphora molmol* *Burseraceae*	Oleo – gum resin	Myrr is arabic Commiphora is greek	Murr means bitter it is bitter in taste. Means gum bearing.
61.	Nagarmotha	*Cyperous rhotendus* *Cyperaceae*	Rhizom	Cyperous: *rhotendus*	Greek name round
62.	Neem	*Azadiracta indica* *Meliaceae*	Leaves and flowers	Azadiracta: Indica:	Person name means best. Indian.
63.	Nutmeg	*Myrestica fragrance* *Myristicaceae*	Seeds	Fragrance	Refering to fragrance of nutmeg.
64.	Nux Vomica	*Strychnos nux-vomica* Loganiaceae	Seed	*Strychnus* *Nux* *Vomica*	Poisonous Nut Causing vomiting
65.	Olive oil	*Olea europea* *Oleaceae*	Fruits	Olea is latin; or Olaion is greek	Olea means Olive plant Olaion means oil.
66.	Opium	*Papaver somniferum* *Papavaraceae*	Latex	Opium is Greek Sonmiferum is lain	Opion means poppy juice Sonmiferum means to produce sleep
67.	Psoralen	*Psoralea corylifolia*	Fruit	*psoralea*	Plants roughened with glandular dots
68.	Punarnava	*Boerhavia diffusa* *Nyetaginaceae*	Entire herb	Boerhauia: Diffusa:	Name of scientist Widely spreading
69.	Rhubarb	*Rheum palmatum* *Rheum webbianum* *Rheum-emodi* *Polygonaceae*	Roots and rhizomes	It is from Rha is latin latin latin	Rha is the name of volga river near which species of Rheum grow large spreading leaves. Name of Indian taxonomist. Containing emodin

70.	Sarpagandha	*Rauwolfia serpentina Apocyanaceae*	Roots	Rauwolfia Serpentina	Name scientist Pertaining to serpents
71.	Shatavari	*Asparagus-recemosus Liliaceae*	Entire plant	A-sparraso: Recemosa:	Very prickly, Having inflorescence with long axis.
72.	Soap nut	*Sapiandus trifolita* Sapondaceae	Fruit	Sapiandus Trifoliata	Yielding soap Three sided
73.	Spearmint	*Mentha spicata Labiateae*	Vol. oil	Spicata is latin	Spicata means spike Mentha is name of a plant
74.	Storax	*Liquidamber orientatlis* Hemamelidaecae	Oleo resin	Liquidus amber is arabian Orientatils	Means fluid means amber coloured pertaining to east
75.	Stramonium	*Datura stramonium Solanaceae*	Entire herb	Stramonium: Datura	Inclined to madness Name of plant
76.	Theobroma oil	*Theobroma cacao Sturculiaceae*	Seeds	Theobroma is greek	Theobroma means food of gods.
77.	Tulsi	*Ocimum sanctum* Lamiaceae	Entire herb	Ocimum Sanctum:	From greek okimon, Holy used for religious purposes.
78.	Vanilla	*Vanilla planifolia Orchidaceae*	Unripe fruits	Vanis is spanish illa is spanish Planifolia latin	means sheath like pod means small Plan means flat Folium means leaf
79.	Vasaka	*Aadhatoda vasica Acanthaceae*	Leaves	*Adhatoda:* *Vasica*:	A malbar name, Scented.
80.	Vidang	*Embelica ribes* Myrsinaceae	Seeds	ambelia: ribes:	acidic, misnomer of Rhuborb
81.	Wild cherry	*Prunus serotina Rosaceae*	Stem and bark	Prnus serotina	Classic name of plum-tree refering to late time of flowering and fruiting.

18. CONTRA-INDICATIONS OF NATURAL DRUGS

Most of herbal drugs, which are not to be utilized in diseases or disorders mentioned against each.

Sr. No.	Name of herbal Drugs	Disease or disroders
1.	Belladonna (Atropa)	Avoid in tachicardia, glaucoma and prostatic hypertophy.
2.	Indian berberry (Daruhalad)	Avoid in diarrhoea and early pregnancy.
3.	Cephalis (Ipecacunha)	General debility, exhaustion, shock and cardiac diseases.
4.	Cinchona	Pregnancy
5.	Tanacetum (Tancy)	Avoid in pregnancy due to presence of thujone.
6.	Thuja	Avoid in pregnancy, may cause abortion by reflex uterine contraction.
7.	Vinca major	Avoid in constipation.

❖❖❖

19. LIST OF POISONOUS PLANTS AND THEIR POISONS

Important poisonous plants are listed below:

Sr. No.	Name of Plant	Name of Poison	Location of Poison	Chemical type and percentage
1.	Atropa-belladonna	Hyocyamine	Leaves	Alkaloid
2.	*Calotropis gigantea* Colotropis procera	Calotropin	Latex of the plant	Protein
3.	Sapiandus saponarea Sapiandus drummondii	Saponin glycoside	Matured dried seeds	Saponin about 13.0 percent
4.	*Catharanthus roseous*	Vincrystin and vinblastin	Mainly leaves	Alkaloids
5.	*Nicotiana tabaccum*	Nicotin	Entire plant	Alkaloids
6.	*Ricinus comimunis*	Ricin	In oil cake	Ricin in protein
7.	*Erythroxylon coca*	Cocain	All parts of plant	Alkaloid
8.	*Abrus-precatorils*	Abrin	Seeds	Protein
9.	*Aconitum nepallus*	Aconitin	Roots and rhisomes	Alkaloids
10.	*Colchicum antumnale*	Colchicin	Seeds and corms	Alkaloids
11.	*Datura stramonium*	Aerial parts	Leaves	Alkaloids
12.	*Eupetorium rugosum*	Tremetol and Resin acid	Latex	
13.	*Euphorbia pulcherrima Croton tiglium*	Phorbol	Seeds	-
14.	*Jatropha curcus*	Purgative oil and lectin	Seeds	Oil
15.	*Lobelia inflata*	Lobelin	Herb	Alkaloids 0.8 – 1.15 %
16.	*Taxus-buccata*	Taxol	Dried leaves bark and roots	Terpenes 0.01%
17.	*Thevetia peruviana*	Thevetin neriifolin	Dried seeds	Glycosides
18.	*Apple Malus*	Sytvestris (APPLE) roseceae	Seeds	
18.	*Psedera quinguefolia*			
19.	*Amanita* verna (Mushrum)			

❖❖❖

20. NATURAL DRUGS BELONGING TO ONE AND THE SAME FAMILIES

(A) MONOCOTYLEDONS FAMILIES

ACANTHACEAE

1. Kalmegh
2. Vajradanti
3. Vasaka

GRAMINAE (Poaceae)

1. Andropogan sqerrosus
2. Cymbopogan citratus
3. Cymbopogan nardus
4. Cynodon dactylon
5. Sorghum vulgare
6. Saccharum officinarum
7. Vetiveria zizanoidus
8. Zea-mays

LABIATAE

1. Coleus aromaticus (Coleus)
2. Lavendula vera (Lavender)
3. Mentha pipperita (Peppermint)
4. Mentha viridis (Spearmint)
5. Ocimum sanctum (Tulsi)
6. Rosemarinus officinalis (Rosemary)
7. Thymus-vulgaris (Thyme)

LILIACEAE

1. Agave americana
2. Allium satirum (Garlic)
3. Aloe vera (Aloe)
4. Colchiam antumnale
5. Asperagus receamosus
6. Gloriosa superba

PALMAE

1. Areca catechu (Areca-nut)
2. Calomus ciliaris
3. Caryota urens (Toddy Palm)
4. Carnauba
5. Cocos-nucifera (Coconut fruit)

ZINGIBERACEAE

1. Alpenia galangs (Galanga)
2. Amomum subulatum
3. Colchicum antumnale
4. Elett-ria cardamomum
5. Hedychium spicata
6. Curcuma longa (Haldi)
7. Zingiber officinale (Ginger)

(B) DICOTYLEDO US FAMILIES

APOCYANACEAE (Asclepiadaceae)

1. Holarrhena antidysentrica (Kurchi)
2. Oleander
3. Rauwolfia serpentina (Sarpgandha)
4. Thevetia pernviana
5. Gymnema sylvestre (Gudmar)

ANACARDIACEAE

1. Anacardium occidentale
2. Gymmema sylvestre (semicarpus-anacardium
3. Hemidesmus (indicus
4. Mangifera indica (Mango)

COMPOSITAE (Astraceae)

1. Artemsia cina
2. Carthamus tinctorium
3. Chrysanthemum cinerarifolium
4. Eclipta alba
5. Helianthus tuborosus
6. Cinchorium intybus
7. Stevia reboundiana

CONVOLVULACEAE

1. Ipomoea hederaceae
2. Ipomoea purga
3. Ipomoea purpurea
4. Ipomoea turpeth
5. Ipomoea palmata

PAPAVERACEAE

1. Argemona mexicana
2. Himalayan poppy
3. Papaverum semniferum

LEGUMINOSAE

1. Acacia arabica (Babool)
2. Acacia catechu (Kaat)
3. Cajanns cajan (Toor)
4. Glycine-max
5. Ingigotena-tinctorius
6. Medicago-sativa
7. Pongamea-pinnata
8. Psoralca corylifolia
9. Pterocarpus santalum
10. Sesbenia grandiflora

RUBIACEAE

1. Cephelis Ipecacunha
2. Cinchona officinalis
3. Coffee arabica (Coffee)
4. Rubia cordifolia (Manjishta)

RUTACEAE

1. Aegle marmalase (Bael fruit)
2. Citrus lemonis (Lemon fruit)
3. Citrus aurantium (Orange fruit)
4. Citrus medica
5. Citrus senensis (Mosambi)
6. Muraya koenigi (Kadhi-patta)
7. Ruta-graveolens

SAPINDACEAE

1. Dodonea-viscosa (Jakhmi)
2. Litchi chinevsis (Lichi)
3. Sapindus trifaliatus (Ritha)

SCROPHULARIACEAE

1. Bacopa herpestis (Brahmi)
2. Digitalis purpurea (Teel-pushpa)
3. Picrorrhiza (Kutki)

SOLANACEAE

1. Atropa-belladonna (Belladonna)
2. Datura stramonium (Dhotara)
3. Nicotiana tobaccum (Tobacco-leaves)
4. Solanum tuberosum (Potato-rhizome)

STURCULACEAE

1. Cola acuminata
2. Helicteris isora (Murud sheng)
3. Theobroma-cocoa (cocoa seed)

EUPHORBIACEAE

1. Croton tiglium (Jamal-gota)
2. Emblica officinalis (Amala fruit)
3. Phyllanthus neruri (Bhui-amla)
4. Ricinus communis (Castor seed)

UMBELLIFERAE (Apiaceae)

1. Anethum graveolens (Dill)
2. Carum carvi (Caraway)
3. Apium graveolens (Celery)
4. Foeniculum capillaceum (Fennel)
5. Coriandrum sativum (Coriander)
6. Cuminum cyminum (Cumin)

MYRTACEAE

1. Rugenia caryophyllum (Clove)
2. Psidium guayava (Guava)
3. Pimenta officinalis (Allspice)
4. Melacuca leuca dendrum (Caju put)

COMBRETACEAE

1. Termenalia arguna (Arjuna)
2. Termenalia balerica (Bahera)
3. Termenalia chebuln (Myrobalan)
4. Anogeissus latifolia (Gum-ghatti)

PIPERACEAE

1. Cimicifuga-racemosa (Kawa-kawa)
2. Piper-nigrum (Black-pepper)
3. Piper longum (Long pepper)
4. Piper chebula (Cubeb)

CRUCAFARAE (Bracecaccae)

1. Brassica-nigra (Black mustered oil)
2. Brassica compestris (Rape-seed oil)

GENTIANACEAE

1. Swertia chirata (Chirata)
2. Gentiana-lutea (Gentian)
3. Canscora decussata (Shankh pushpin)

RUNANCULACEAE

1. Aconitum nepallus (Aconite)
2. Cimicifuga racemosa (Black-cohosh)
3. Nigella sativa (Kalijira)
4. Thalictrum foliolosum (Surma plant)

RHAMNACEAE

1. Rhamnus-purshiana (cascara)
2. Zizyphus jujuba (Bor)
3. Zizyphus vulgaris (Unnab)

MALVACEAE

1. Hibiscus saldariffa (Lal-ambadi)
2. Hibiscus abelmoschus (Mushkdona)
3. Hibiscus rosa sinensis (Jaswand)
4. Malvastram spicalum

LAURACEAE

1. Cassia cinnamon
2. Cinnamon zeylanicum
3. Cinnamomum camphora

21. LIST OF ORGANISED CRUDE DRUGS

1. Absorbent cotton
2. Aconite
3. Acorus
4. Ajowan
5. Alfalfa
6. Allspice
7. Ambi Halad
8. Ammi
9. Amra-bark
10. Anantmool
11. Anise
12. Annatto
13. Arica-nut
14. Arjuna
15. Arnica
16. Artemisia
17. Artemisia annea
18. Ashoka bark
19. Ashwagnadha
20. Avala
21. Bael
22. Bahava
23. Bahera
24. Bearberry
25. Belladonna
26. Benafsha
27. Betal leaf
28. Bhringraj
29. Bhui-amla
30. Bitter almond
31. Bitter orange peel
32. Black-cohosh
33. Boswellia
34. Brahmi
35. Broccoli
36. Cachineal
37. Cannabis
38. Cantharide
39. Capsicum
40. Caraway
41. Cardamom
42. Cascara bark
43. Cassia
44. Cellery
45. Chaulmoogra herb
46. Chicory
47. Chirata
48. Cinchona bark
49. Cinnamon
50. Clove
51. Coca-leaves
52. Cocoa seed
53. Coffee seed
54. Colchicum-corm
55. Colchicum-seed
56. Coleus
57. Colocynth
58. Colophony
59. Comptotheca
60. Coriander
61. Cubeb
62. Cummin
63. Daruhalad
64. Datura
65. Derris
66. Dhayati
67. Digitalis lanata
68. Digitalis purpurea
69. Dill
70. Discorea
71. Duboisia
72. Ephedra
73. Fennel
74. Fenugreek
75. Feverfew
76. Fig
77. Galls
78. Garcinia
79. Garlic
80. Gentian
81. Ginger
82. Ginkgo
83. Ginseng
84. Gloriosa
85. Gokhru
86. Gudmar
87. Guduchi
88. Guggul
89. Haldi
90. Hamamalis
91. Hemlock
92. Henna
93. Hibiscus
94. Hops
95. Hyoscyamus
96. Hypericum
97. Inula
98. Ipecacunha
99. Ipomoea
100. Jalap
101. Jalbramhi
102. Jatamansi

103. Kaladana
104. Kalmegh
105. Kapur kachari
106. Kavach
107. Kavakava
108. Kurchi bark
109. Leech
110. Lemon peel
111. Liquorice
112. Lobelia
113. Long pepper
114. Malefern
115. Manjishta
116. Milk thistle
117. Momordica
118. Mustard seed
119. Mylabris
120. Myrobalan
121. Myrrh
122. Nagarmotha
123. Nagli (Nachani)
124. Neem
125. Nirgudi
126. Nutmeg
127. Nux-vomica
128. Oleander
129. Parsley
130. Physostigma
131. Picrorriza
132. Pilocarpus
133. Pipala mool
134. Podophyllum
135. Propolis
136. Psoralea
137. Psyllum seed
138. Pterocarpus
139. Punarnava
140. Pyrethrum
141. Quassia
142. Quillaia
143. Rasberry leaves
144. Rasna
145. Rawolfia
146. Rhubarh
147. Rue
148. Ryania
149. Sabadilla
150. Safed musali
151. Saffron seed
152. Sandal wood
153. Saussurea
154. Senega
155. Senna Alexandrian
156. Senna Indian
157. Senna pod
158. Shankhpushpi
159. Shatavari
160. Silk
161. Solanum
162. Spearment
163. Spirulina
164. Squill european
165. Squill Indian
166. Stevia
167. Storax
168. Stramonium
169. Strophanthus
170. Sugarbeet
171. Sugarcane
172. Tamarind
173. Taxus
174. Tea leaves
175. Thevetia
176. Thyme
177. Tobacco
178. Tonka bean
179. Tulsi
180. Vajradanti
181. Valerian european
182. Valerian Indian
183. Vanilla pod
184. Vasaka
185. Veratrum-green
186. Veratrum-white
187. Vidang
188. Vinca
189. Visnega
190. Wild cherry bark
191. Witch-razel
192. Wool

22. LIST OF UNORGANISED NATURAL DRUGS IN THE TEXT

1. Acacia
2. Agar
3. Algin
4. Almond oil
5. Aloe
6. Asafoetida
7. Asbestos
8. Asparginase
9. Bees-wax
10. Bentonite
11. Benzoin
12. Borax
13. Bromelain
14. Buck-wheat
15. Calamin
16. Carrageenan
17. Caramel
18. Cardamom oil
19. Carnauba wax
20. Casein
21. Castor oil
22. Catechu
23. Cellulose
24. Chalk
25. Chaulmoogra oil
26. Chenopodium
27. Chitosan
28. Chymotrypsin
29. Citronella oil
30. Civet cat
31. CMC
32. Coamphor
33. Cobra venum
34. Coco butter
35. Coconut water
36. Coconut oil
37. Cod liver oil
38. Collagen
39. Colophony
40. Conola oil
41. Corn oil
42. Cotton seed oil
43. Croton oil
44. Cyclodextrin
45. Davana oil
46. Deoxy- rebonuclease
47. Dextran
48. Dextrin
49. Diastase
50. Ergot
51. Evening primrose oil
52. Fullers earth
53. Galls
54. Gelatin
55. Geranium oil
56. Gopichandan
57. Gossypol
58. Gualtheria
59. Guar gum
60. Guggul
61. Gum ghati
62. Gum karaya
63. Halibut liver oil
64. Hard paraffin
65. Hemi cellulase
66. Heparin
67. Honey
68. Hyaluronidase
69. Isapgol
70. Jojoba oil
71. Kaolin
72. Karanj oil
73. Kieselguhr
74. Kokum butter
75. Lard
76. Lavender oil
77. Lemongrass oil
78. Linseed oil
79. Liquid paraffin
80. Litmas
81. Malt extract
82. Mango-kernel oil
83. Manna
84. Muramedase
85. Musk
86. Mustard oil
87. Myrrh
88. Neem oil
89. Olive oil
90. Opium
91. Pale catechu
92. Palm oil
93. Palma rosha

94. Pancre lipase
95. Pancreatin
96. Papain
97. Patchouli oil
98. Pea-nut oil
99. Pectin
100. Penicillinase
101. Peppermint
102. Pepsin
103. Peru balsam
104. Poppy seed oil
105. Propalis
106. Rape-seed oil
107. Rennin
108. Rice bran oil
109. Rock-salt
110. Rosemary oil
111. Royal jelly
112. Sandal wood
113. Sandal wood oil
114. Satilains
115. Serratiopeptidase
116. Sesame oil
117. Shellac
118. Shilajit
119. Soft paraffin
120. Soyabean oil
121. Spearmint
122. Spermaceti
123. Spirulina
124. Starch
125. Star anise
126. Storax
127. Streptokinase
128. Suet oil
129. Sulphur
130. Safflower oil
131. Sunflower oil
132. Talcum
133. Tar
134. Tea tree oil
135. Thyme oil
136. Tolu balsam
137. Tragacanth
138. Trypsin
139. Tuna-fish liver oil
140. Turpentine oil
141. Urokinase
142. Vetiver oil
143. Wheat germ oil
144. Wool fat
145. Xanthum gum
146. Yeast

23. LIST OF ANIMAL DRUGS

1. Bees wax
2. Cantharide
3. Casein
4. Castorium
5. Chitin
6. Chitosan
7. Civet cat
8. Cobra venum
9. Cochineal
10. Cod-liver oil
11. Collagen
12. Deoxyribonuclease
13. Diastase
14. Fibrinolysin
15. Gelatin
16. Halibut liver oil
17. Heparin
18. Honey
19. Hyaluronidase
20. Lactose
21. Lard
22. Leech
23. Muranmidase
24. Musk
25. Mylabris
26. Pancreatin
27. Pancrelipase
28. Penicillinase
29. Pepsin
30. Rennin
31. Shark liver oil
32. Shellac
33. Spermaceti
34. Suet
35. Trypsin
36. Tuna fish liver oil
37. Urokinase
38. Wool fat

24. DRUGS OF MICROBIAL OCCURANCE

1. **Yeast:** Fungal micro-organisms *Saccharomyas cerevisiae* fam Saccharomycetaceae.
2. **Streptokinase :** Proteolytic enzyme from *Streptococci* group of micro organisms.
3. **Serratopeptidase:** Proteolytic enzyme from *Serrata* genus of bacteria (Bacterial enzyme)
4. **Sutilains:** Proteolytic enzyme derived from *Bacillus substilis* bacteria.
5. **Penicillinase:** Bacterial enzyme produced by many bacteria, it deactivates antibiotic penicillin.
6. **L-asparginase:** Bacterial enzyme obtained from *Escherichia coli.*
7. **Xanthum gum:** A polysaccharide produced by series of bacteria *Xanthomonas compestris.*

25. MINERAL DRUGS

1. Asbestos
2. Bentonite
3. Borax
4. Calamine
5. Chalk
6. Fullers earth
7. Gopichandan
8. Kaolin
9. Kieselguhr
10. Hard paraffin
11. Soft paraffin
12. Liquid paraffin
13. Rock salt
14. Shilajeet
15. Sulphur
16. Talcum

❖❖❖

26. COUNTRYWISE AVAILABILITY OF NATURAL DRUGS

ABYSSINIA

Ammi
Fenugreek

AFGHANISTAN

Ajowan
Manjishta
Opium
Valerian

ALGERIA

Carat
Cochineal
European squill
Fig
Geranium
Indian-squill
Linseed
Kieselguhr

AUSTRELIA

Agar
Artemisia annula
Chicory
Ephedra
Ginger
Hypericum
Honey
Leech
Jatamansi
Olive-fruits
Shatavari
Safflower
Suet
Tea-tree oil
Vinca

BELGIUM

Chiwry
Chammomile
Hyoscyamus
Inula

BAZIL

Corn oil
Cocoa
Cola seeds
Castor
Colius
Carnauba wax
Black pepper
Cinnamon
Stevia
Tonka bean

CANADA

Algin
Bearberry
Chinopodium,
Cascara-sagrada
Echinacea
Sunflower
Taxus
Ginseng
Tar
Senega
Milk-thistle
Stevia
Wild cherry bark
Winter green oil

CHINA

Camphor
Canola oil
Black-mustard
Dioscorea
Rhubarb
Pea-nut
Picrorrhiza
Jatamansi
Ginkgo
Jalbrahmi
Momordica
Henna
Maize
Seasame
Stevia
Jamalgota
Kokum butter
Soyabean
Mango kernel oil
Musk
Tuna fish liver oil
Solahum khasianum
Ginseng
Visnaga

FLORIDA

Henna
Lemon fruits
Balsam of peru
Papain
Thevetia
Wild cherry

EGYPT

Ashwagandha
Bahava
Ammi
Fig

Coriander
Indian gum
Indian senna
Locust bean gum
Safflower
Sun flower

FRANCE
Carrageenan
Chicory
Inula
Fennel
Ginkgo
Squill
Saffron
Thyme
Turpentine
Kaolin
Talcum
Bees wax

GERMANY
Aconite
Dill
Fennel
Hops
Leech
Stramonium
Cod Liver oil
Kieselguhr

GREECE
Anise
Indian squill
Linseed
Saffron
Thyme
Tragacanth
Olive oil

INDIA
Agar
Aloe
Annatto
Brahmi
Black-pepper
Chicory
Cinchona
Datura
Digitalis
Dioscorea
Fenugreek
Guar-gum
Hyoscyamus
Isapgol
Indian-senna
Mustard seeds
Momordica
Nux-vomica
Karaya gum
Kurchi
Opium
Olive tree
Lemon peel
Neem
Long-pepper
Rauwalofia
Safed musali
Seasame seed
Orange peel
Tamarind
Turmeric
Peppermint
Vasaka
Turpentine
Fenugreek

INDONESIA
Coffee
Guduchi
Jalbrahmi
Cinchona
Nutmeg
Vetiver
Stevia
Pale catechu
Shellac

IRAQ
Ajowan
Tragacanth

ISRAEL
Locust-bean gum
Olive fruit

ITALY
Bitter almond
Geranium oil
Indian squill
Linseed
Lemon-peel
Olive-fruits
Talcum
Bees wax

JAPAN
Agar
Algin
Dhayati
Buck-wheat
Chitin
Camphor
Coffee
Ginseng
Ginko

Fennel
Senega
Sesame
Stevia

MALAYSIA

Palm oil
Citronella
Ashoka bark
Long-pepper

MADAGASKAR

Civet
Clove
Dhayati
Litmus

MEXICO

Cochineal
Cocoa butter
Cannabis
Chinopodium
Dioscorea
Jojoba
Ipomoea
Saffron
Tonka-bean
Stevia
Vanilla

MOROCCOO

Bitter almond
Colocynth
Cummin
Geranium
European squill
Locust bean gum
Olive oil

MYANMAR (BRAHMADESH)

Bahava
Buck-wheat
Choulmoogra
Citronella
Vasaka
Vetiver

NEPAL

Borax
Chirata
Dioscorea
Long-pepper
Shalijit
Sulphur

NEW ZEALAND

Agar
Chicory
Hemlock
Honey
Kawa-kawa

PAKISTHAN

Artemsia
Ajowan
Colophony
Guar
Ephedra
Isapgol
Opium
Jalbramhi

PERU

Annatto
Balsam peru
Coca
Cinchona
Quillaia

PORTUGAL

Bitter almond
Lavender
Eucalyptus
Thyme

RUSSIA (USSAR)

Buck-wheat
Castor oil
Anise
Fennel
Ginseng
Hyoscyamus
Linseed
Liquorice
Opium
Suet
Musk
Wool fat
Turpentine
Sun flower

SICILY

Bitter almond
Lemon fruit
Locust bean gum
Liquorice
Orange fruit
Manna

SIKKIM

Jatamansi
Male fern
Picrorrhiza
Podophyllum
Solanum-khsiannum
Strawberry

SRI-LANKA

Acacia
Bael
Bahava
Cinchona
Citronella
Cinnamon
Cocoa
Clove
Coffee
Coleus
Gum-ghati
Guduchi
Gokhru
Jatamansi
Kalmegh
Nux-vomica
Psoralea
Vanilla
Vasaka
Coca leaves
Garcinia

SPAIN

Carrageenan
Bitter almond
Indian squill
European squill
Lemons
Ephedra
Oranges
Liquorice
Manna
Saffron
Locust bean gum
Olive oil
Anise
Spearment

SUDAN

Acacia
Alexandrian senna
Senna pod
Henna

TIBET

Artemisia
Camptotheca
Gokhru
Rhubarb
Borax

SOUTH AFRICA

Acacia
Aloe
Capsium
Celery
Coridander
Ashwagandha
Ginger
Cannabis
Gloriosa
Guggul
Lemon-grass
Myrrh
Shatavari
Vinca

YUGOSLAVIA

Colenicum
Ephedra
Gentian
Pyrethrum

UNITED KINGDOM (BRETAIN)

Aconite
Evening primrose
Belladonna
Datura
Colchicum
Rue
Fever few

USA

Agar
Alfalfa
Algin
Arnica
Black mustard
Balsam of tolu
Carrageenan
Cascara-sagrada
Echinacea
Dioscorea
Ginseng
Garlic
Ginger
Gualtheria
Eucalyptus
Jojoba
Comptotheca
Lobelia
Olive
Momordica
Milk-thistle
Senega
Hops
Sperulina
Tar
Taxas
Mustard-seed
Rauwalfia
Stramonium
Spearmint
Peppermint
Wild cherry

WEST-INDIES

Coca-leaves

Cola seeds

Tamarnd

Thevetia

IRAN (PERSIA)

Almond

Asafoetida

Ajowan

Artemisia

Fig

Galls

Opium

Liquorice

Poppy seed oil

VIETNAM

Benzoin

Jalbrahmi

UKRAIN

Sun flower oil

Wheat

HOLAND

Tonka-bean

Coriander

Calamus

MALAYA

Behra

Jamalgota

Momordica

KENYA

Annatto

Cascara sagrada

Kieselguhr

KOREA

Rhubarb

Ginseng

TURKEY (TURKASTAN)

Oilve-fruits

Galls

Fig-fruits

Storax

ZANZIBAR

Aloe

Clove

Eucalyptus

Leman-fruit

Linseed

Spermaceti

A. Mediterranean Countries are:

Greece, Tunsesia, Cyprus, Moracco, Italy, Spain, Israil, Croatia, France, Turkey, Palestine, Portugal, Lebonon.

B. Caribbeon Countries are:

West-indies, Bahama, Caico, Turki, Nicaragua, Barbidose, Jamaika, Curacao, Antigua, Barmuda, Cuba, Aruba.

* List are subject to change

C. Mesopotemea:

Iran, Issaq, Qunnet, Israil and Turky region.

27. MARATHI NAMES OF NATURAL DRUGS

1. Acacia - Babhul
2. Aconite – Bachnaag
3. Acorus – Vekhand
4. Ajowan – Owa
5. Almond - Badam
6. Aloe – Korphad
7. Anise – Saonf
8. Annatto – Shendri
9. Areea – Supari
10. Artemisia – Kirmani – owa
11. Asatoetide – Hing
12. Babchi – Bavachi
13. Beeswax – Men
14. Benzoin – Ood, Behda
15. Black cateelic – Kaat
16. Borax – Takun-khar
17. Boswellic – Salai – guggul
18. Cannabis – Ganja
19. Canola – Kali mohair
20. Capsicum – Mirchi
21. Caraway – Shahjire
22. Cardamom – Weldoda
23. Cassia – Dalchini
24. Castor – Erand
25. Celery – Mothi-ajmoda
26. Chaulmogua – Kadu-kawat
27. Cinnamon – Kalmi-Dalchini
28. Coleus - Maeen mool
29. Colocynth – Kadu - indrayan
30. Coriander – Dhane
31. Corn oil – Makyache tel
32. Cubeb – Kankol
33. Cumin – Jeera
34. Davana – Maka
35. Digitalis – Teel – pushpi
36. Dill – shepu
37. Dioscorea – Kadu karanda
38. Eclipta – Bhringraj
39. Fennel – Badishoap
40. Fig fruit – Anjir
41. Galangal – Koshte- kolingin
42. Garlic – Lasun
43. Gentian – Kadu - kutki
44. Gloriosa – Kal lawavi
45. Gymnema – Kali-kardodi
46. Hemamalis
47. Hibiscus – Jaswand
48. Hyoscyamus – Khurasani owa
49. Inula – deliha
50. Ispaghula – isapgol
51. Jalbrahmi – Manduk-parni
52. Jalbrambi – Jamalgota.
53. Kaladana – Nilpushpi
54. Kalmegh – Oli - karayat
55. Kaolin – Multani mitti
56. Kawach kaolin – Multani matti
57. Kurchi kuda
58. Leman grass – Hirva chaffa
59. Liguorice – Jeshtamadh
60. Linseed – Jawas

61. Lobelia – Deonala
62. Long pepper – Lendi pimpali
63. Momordica – Kadu – karale
64. Musk – Kasturi
65. Myrobalam - Hirda
66. Myrrh – Hirabol
67. Nagli – Nachani
68. Nnx-vomica – Kuchala
69. Opium – Aphoo
70. Pale catechu – Pandhara Kaat
71. Panchauli – Pachpan
72. Pea-nut – Bhuimug
73. Phyllauthns – Bhui-amla
74. Picrorrhizo – Kutki
75. Pterocanpus – Asan
76. Rasna – Kolinjan
77. Rauwolfic – Sarpgandha
78. Rhnbarb – Ladadi Revzehini
79. Safflower – Suryaphool
80. Saffron – Keshar
81. Sarsapurila – Anantmool
82. Senna (Alexandrian) – Sonamukhi
83. Senna (Indian) – Sonamukhi
84. Senna-pod –
85. Sesame oil – Teel Tail
86. Spearment – Jangali- pudina
87. Squll (Indian) – Bhui-kanda
88. Stranonium – Dhotra Saussuria
89. Tamarind – Chinch
90. Taxus –
91. Tealeaves – Chaha patti
92. Thevetia – Piwala-kanda
93. Tinospora – Gulwel
94. Trigonella – Methi
95. Tulsi – Tulas
96. Turmeric – Haldi
97. Vajurdanti – Kate-koranti
98. Valerian
99. Vasaka – Adulsa
100. Vegenic –
101. Vetiver – Waala
102. Vigand – Vavding
103. Vinca – Sadaphuli
104. Withanca – Ashwagandha
105. Wool fat

28. GLOSSARY OF TECHNICAL TERMS

- Alopesia – baldness from disease.
- Amenorrhoea – absence or abnormal stoppage of the menses.
- Anodyne – relieving pain.
- Antilithic – preventing the formation of stone in the kidneys.
- Aphthae – disease having whitish spots in the mouth.
- Atonic – having no tone or muscular power.
- Attenuant –an agent increasing the fluidity or thinness of the blood or other secretion.
- Biferous – producing flowers or fruit twice a year.
- Calculus – a stonelike concretion in any organ.
- Capitate – pin headed. Catarrh – cold in the head, an increased flow of mucus from the nose.
- Caudate – tailed.
- Cauline – being to or growing on stem.
- Chyluria – the presence of fat in the urine.
- Clavate – club shaped.
- Colliquative – characterized by excessive liquid discharge.
- Connate – united.
- Corjaceous – leathery.
- Corymb – a flat topped or merely convex and open flower-cluster of the inter-mediate or the centripetal order. Cymbiform – boat-shaped. Cymbiform – boat-shaped.
- Cyme – a flower cluster of centrifugal type, especially broad and flattened one.
- Cymose – bearing cymes.
- Decompoun – several times divided.
- Deltoid – shaped like an equilateral triangle.
- Didymous – growing in pairs.
- Digitate – branching into divisions like fingers.
- Dicecious – unisexual male and female in different individuals.
- Dorsal – pertaining to the back.
- Dysuria – difficult in making water.
- Fascicled – connected into a close bundle of flowers, leaves, etx.
- Filiform – thread like
- Frond – leaf of a fern.

- Gastrodynic – causing pain in the stomach.
- Glaucous – covered with a bloom.
- Glume – the chaffy two-ranked members of grass flowers.
- Hectic – habitual.
- Herpes – skin disease marked by clusters of small vesicles.
- Hispid – rough hairy.
- Keel- the two anterior and united petals of a papilionaceous morolla forming a ridge like the keel of a boat.
- Leucoderma – mottling of the skin by abnormal whiteness.
- Leucorrhoea – a whitish, viscous discharge from the vagina and uterine cavity.
- Lodicule – a small scale outside the stamens in the flower of grasses.
- Menorrhagia – excessive menstruation.
- Micturition – urination.
- Mottled – spotted in various colours.
- Nephritis – inflammation of the kidneys.
- Obcordate – inversely heart-shaped.
- Oedema – dropsical swelling.
- Palmate – shaped like the palm of a hand
- Paludine – growing in marshy places.
- Panicle – a loose flower cluster.
- Papillae – soft superficial glands.
- Peltate – a leaf attached by its lower surface to the stalk.
- Phlegm – bronchial mucus.
- Pilose – hairy
- Pistillode- organ similar to the female organ of the flower.
- Polygamous – with hermaphrodite and unisexual flowers on the same or different individuals on the same species.
- Proximal – nearest.
- Psora – itch.
- Puberulous – slightly hairy.
- Pubescent – clothed with soft hairs.
- Rain-forest – evergreen forest having over 72 inches of rainfall annually.
- Repand – with slightly uneven margin.
- Revolute – rolled backwards or down-ward.
- Rotund – round, spherical.

- Rugose – wrinkled.
- Sagittate – enlarged at the base into two acute lobes.
- Scabrous – rough to the touch.
- Scarious – thin, dry and membranous.
- Senile – pertaining to old age.
- Sinuate – with a deep wavy margin.
- Somniferous – bringing sleep.
- Soporific – an agent which induces sleep.
- Spur – a hollow and slender extension of some part of flower.
- Stellate – star-like
- Subscandent – partially climbing.
- Sudorific – inducing profuse sweating.
- Syncarpous – completely united carpels composed of two or more.
- Tomentose – softly hairy.
- Torus – that portion of the axis on which the parts of a flower are inserted.
- Tunicated – having coats.
- Urceolate – pitcher-like
- Variegated – having spots or markings of light colour.
- Vertigo – giddiness
- Vulnerary – a remedy useful in healing wounds.

27. BIBLIOGRAPHY

- **Ahluwalia, K.S.,** British Pharmaceutical Codex Plants and Their Indian Substitutes, Central Institute of Research in Indigeneous Systems of Medicine, Jamnagar, 1962.
- **Atal, C. K. and Kapur, B.M.,** Cultivation and Utilization of Medicinal Plants, R.R.L., Jammu, 1982.
- **Audus, L.J.,** Plant Growth Substances, 1, Leonard Hill, London, 1972.
- *Ayurvedic Formulary of India*, Govt. of India, New Delhi.
- *Ayurvedic Pharmacopoeia, Part I Vol. I First edition*, Ministry of Health & Family Welfare Govt. of India, New Delhi, 1986.
- **Barz, W., Reinhard, E. and Zenk, M.H.,** Plant Tissue Culture and Its Biotechnological Application, Springer, Berlin, 1977.
- *Bombay State Gazetteer (Botany) - I - Medicinal Plants*, Directorate of Govt. Printing Stationary and Publications, M.S. Bombay.
- **Brain, K. R. and Turner, T.D.,** The Practical Evaluation of Phytopharmaceuticals, Wright - Scientechnica, Bristol, 1975.
- *British herbal pharmacopoeia Keighly:* British Herbal medicine Association 1983.
- *British Pharmscopoeia,Vol. I and II,* Ministry of Health and Social Service for Northern Ireland, U.K, 1980.
- **Bruneton Jean:** *Pharmacognosy (Phytochemistry, Medicinal plants) 2nd edition,* Intercept 1999.
- *Bulletin of Medico-ethano-botanical Research* 3, Central Council for Research in Ayurveda and Siddha, Jaipur 1980.
- **Bulock, J.D.,** The Biosynthesis of Natural Products, McGraw Hill, 1965.
- *Canadian Pharmaceutical Journals*, Lloydia, Economic Botany, Planta Medica, Fitoterapia, Quarterly Journal of Crude Drug Research, Phytochemistry, Tetrahedron Letters, Journal of Pharmacy and Pharmacology, Journal of Pharmaceutical Sciences Naturwissenshaften, Plant Cell Reports, I.T.C. and UNCTAD / GATT Reports.
- **Charles J. Kokoski, Robert kokoski and Frank J. Slame,** Fluorescence of powdered vegetable Drugs under ultraviolet radiation. Journal of the American pharmaceutical association Vol. XL VII No. 10, October, 1958.
- **Chattopadhyay P.K.** Herbal Cosmetics and Auyrvedic medicines, National Institute of Indian Research Delhi.
- *Chemical Export Bulletins,* Basic Chemicals Pharmaceuticals and Cosmetics Export Promotion Council, Bombay, 1985 - 1996.
- **Chopra, R. N., Nayar, S. L. and Chopra, I.C.,** Glossary of Indian Medicinal Plants, C.S.I.R., New Delhi.
- **Clarke, E.C.G.,** Isolation and Identification of Drugs, The Pharmaceutical Press, London, 1967.
- *Database on Medical plants used in Ayurveda Vol. I (2000), Vol. II (2001), vol. III (2001)* Central Council for Research in Ayurveda and Siddha, Department of ISM and H, Ministry of Health and Family Welfare (Govt. of India) New Delhi.

- **De Mayo, P.** The Chemistry of Natural Products, 2 - 3, Interscience, New York, 1959.
- **DENSTON T.C.** A text book of Pharmacognosy – 1945.
- **Douglas Kinghorn A.,** Toxic Plants, Columbia University Press, New York, 1979.
- *Drug Index*, Passi Publications, New Delhi, 1999.
- **Dr. Weici Tan, Dr. Gerhard Cisenbrand,** *Chinese Drugs of Plant Origin – 1992,* Springer-verlay, Berline, Germany.
- **Dr. Franz-Christian Czygan, Dr. Dietrich Frohne,** *Herbal Drugs And Phytopharmaceuticals – 1994,* Medipharm Scientific Publishers, Stuttgart.
- **Dubos, R. J. and Hirsch, J. G.,** Bacterial and Mycotic Infections of Man, 4th Ed., W.B. Sounders Co., Philadelphia, 1965.
- **Duel, H.A. Jr.,** The Lipids - Their Chemistry and Biochemistry, 1 3, Interscience, New York, 1951 - 1957.
- **Dymock. W,** Vegetable materia medica of western India 1883.
- **Dyke, S.F.,** The Carbohydrates, Interscience Publishes, Inc., New York, 1960.
- **Eckey, E.W.,** Vegetable Fats and Oils, Reinhold Publishing Corporation, New York, 1954.
- **Enrique cademas & Lester Packer,** Hand book of Antioxidants, second edition 2002, Marcel Dekker inc New York.
- **Evans, E.,** Plant Diseases and Their Chemical Control, Blackwell Scientific Publications, Oxford, 1968.
- *Export Potential of Selected Medicinal Plants,* by Basic Chemicals, Pharmaceuticals and Cosmetic Export Promotion Council, Bombay, 1978 and other reports.
- **Fahn, A.,** *Plant Anatomy, 3rd Ed.,* Pergamon Press, Oxford 1981.
- **Faulkner, D. J. and Fenical W.H.,** *Marine Natural Products Chemistry (NATO conference Series 4),* Plenum Press, New York, 1977.
- **Feichter, A.,** *Advances in Biochemical Engineering* 16, Plant cell Cultures, Springer - Verlag, Berlin,1980.
- **Fitzpatrick, E.A.,** *An Introduction to Soil Science, 2nd Ed.*, Longman Group U.K Ltd., England, 1986.
- **Fletcher, W.W. and Kirwood, R.C.,** *Herbicides and Plant Growth Regulators,* Granada Publishing Ltd., London, 1982.
- **Gamborg, O.L. and Wetter, L.R.,** *Plant Tissue Culture Methods,* National Research Council of Canada, Saskatchewan, 1975.
- **George, R., Okun, R. and Cho, A.K.,** *Annual Review of Pharmacology and Toxicology*, 26 Annual Reviews Inc., Palo Alto, California, 1986.
- **Gibbs, R. Darneley,** *Chemotaxonomy of Flowering Plants, 4 Volumes,* McGill University Press, 1974.
- **Guenther, E.,** *The Essential Oils* 1 - 4, D. Van Nostrand Co., Inc., New York, 1949 - 1952.
- **Hand book of Ayurvedic medicinal plants :** L. D. Kapoor, 1990, C.R.C. Press New York, USA.
- **Harborne, J.B.** *Biochemistry of Phenolic Compounds Academic*. Press, New York, 1964.

- **Harborne, J.B.,** *Phytochemical Methods*, Chapman, and Hall, International Edition, London, 1973.
- **Heftmann, E.,** Chromatography - A Laboratory Handbook of Chromatographic and Electrophoretic Methods, 3rd Ed., Van Nostrand, Princeton, N.J., 1975.
- **Henderson, I.F. and Henderson, W.D.,** *A Dictionary of Biological Terms, 8th Ed.*, Oliver and Boyd, London, 1963.
- **Henry, T.A.,** *The Plant Alkaloids, 5th Ed.*, McGraw-Hill, New York, 1960.
- **Heywood, V.H.,** *Plant Taxonomy, 2nd Ed.*, Edward Arnold, London, 1976.
- **Hinton, H.E. and Corbet, A.S.,** *Common Insect Pests of Stored Food Products, 5th Ed.*, London, 1972.
- *Indian Council of Agricultural Research*, Ministry of Agriculture, Govt. of India, New Delhi, various bulletins, 1976 - 88.
- *Indian Journals or Publications referred.*
- Indian Drugs, Indian Journal of Pharm. Sciences, Indian Forester, Pharma-pulse Indian Journal of Experimental Biology, Medicinal and Aromatic Plants Abstracts, Ancient Science of Life, Indian Journal of Agricultural Sciences, Bulletins and Reports of Regional Research Laboratory Jammu and Jorhat, C.I.M.A.P., C.D.R.I. and I.C.A.R.
- *Indian Pharmaceutical Codex,* C.S.I.R. Publication, New Delhi, 1953.
- *Indian Pharmacopoeia Vol. I and II 3rd edition,* 1985, Ministry of Health Govt. of India, New Delhi.
- *Indian Pharmacopoeia Vol. I and II 4th edition,* 1996, Ministry of Health Govt. of India, New Delhi.
- *Indian Pharmacopoeia 1996, 4th edition, Addendum 2000, 2002, 2005 Controller of Publications Ministry of health and family welfare, Govt. of India, Delhi 110054.*
- *Indian Pharmacopoeia, 5th edition 2007, Indian Pharmacopoeia Commission, sector-23, Raj Nagar, Ghaziabad 201002.*
- *Indian Pharmacopoeia 2007 edition, Addendum 2008, Indian Pharmacopoeia Commission, sector-23, Raj Nagar, Ghaziabad 201002.*
- *Indian Species, Quarterly Journal (1982 - 1986),* The Spices Export Promotion Council, Cochin.
- *International Congress on Plant Tissue and Cell Culture (Abstracts).* for 3rd, 4th, 5th and 6th congresses).
- **James A. Duke,** *Handbook of Medicinal Herbs*, CRC press – Washington, 2nd Edition 2002.
- *Japanese Pharmacopoeia XIII*, Tokyo, Ministry of Health and Welfare 1996.
- **John S. Glasby,** Encyclopaedia of the Alkaloids, Vol. 1, Plenum Press, London, 1975.
- **Joseph M. Kadans,** Encyclopedia of Medicinal Herbs, Throsons Publishers Limited, Wellingborough, 1970.
- *Kirk - Othmer Encyclopedia of Chemical Technology, 3rd Ed.,1,* Wiley Interscience Publication, U.S.A., 1978.

- **Kokate C.K., Khandelwal, Gokhale S.B.,** *Practical Pharmacognosy 2nd edition* 1996, Nirali Prakashan, Pune.
- **Kokate, C.K.,** *Practical Pharmacognosy, 2nd Ed.*, Vallabh Prakashan. Delhi, 1988.
- **Laurence, D.R.** *Clinical Pharmacology, 4th Ed.,* English Language Book Society and Churchill, Living stone.
- **Leng albert, Foster Steven,** *Encyclopedia of Common Natural Ingredients (2nd Edition) used in Drugs, Food and Cosmatics, 1996,* John Wiley and Sons inc.
- **Liteanu, C.,** *Gradient Liquid Chromatography*, Horwood, Chichester,1974.
- **Luckner, M., Nover, L. and Boehm, H.,** *Secondary Metabolism and Cell Differentiation*, Berlin, 1977.
- **Luckner, M.,** *Secondary Metabolism in Plants and Animals*, Chapman and Hall, 1972.
- **Manitto, P.,** *The Biosynthesis of Natural Products*, Ellis Horwood, Chichester, 1981.
- **Manske, R.H.F.,** *The Alkaloids - Chemistry and Physiology*, 1-11, Academic Press, New York, 1950, 1968.
- **Martindale**, *The Extra Pharmacopoeia, 27th* Ed., Pharmaceutical Socilety of Great Britain, London, 1977.
- *Materials and Technology*, 5, Longman Group Ltd., London, 1972.
- *Medicinal Plants of India*, I, Indian Council of Medical Research, New Delhi, 1976.
- *Medicinal and Aroma plants programme in Asia (MAPPA) IDRC / SARO*, 208, Jor bagh, New Delhi – 11003.
- **Miller, L.P.,** *Phytochemistry*, 1 - 3, Van Nostrand Reinhold Co., 1973.
- **Nadkarni, A.K,** *Indian Materia Medica*, 1-2, Popular Prakashan Pvt. Ltd. Bombay 1954.
- **Natori, S.,** et al, Advances in *Natural Products Chemistry Extraction and Isolation of Biologically Active Compounds*, Wiley, New York, 1981.
- ***New all Anderson Phillipson,*** *Herbal Medicines 1996*, Pharmaceutical - press London.
- **Nickell, L.G.,** *Plant Growth Regulators* - Agricultural Uses, Springer - Verlag, Berlin, 1982.
- *Official Methods of Analysis*, Association of Official Analytical Chemists publications, Washington.
- **Pal O.C. and Jain S.K.,** *Tribal Medicines,* Naya Prakash, 206 Bidhan Sarani, Kolkatta-6, 1998.
- **Peter B. Kaufman and others** *Natural Products from Plants.* 1999. C.R.C. Press New York, USA.
- **Peach, K. and Tracey, M.V.** Modern Methods of Plant Analysis, 1-4, Narosa Publishing House, New Delhi.
- Pepper, clove ginger, coriander, Farm Bulletins, Directorate of Areca nut and Spices Development, Calicut, 1971.
- *Pharmacopoeia of India*, 1st edition, 1955, New Delhi, Ministry of Health, Govt. of India, & 2nd edition, 1966.
- *Pharmacopoeia of the People's Republich of China* (English ed.) Guanzhon Guangdong Science and Technology press – 1992.

- **Pridham, J.B. And Swain, T.,** Biosynthetic Pathways in Higher Plants, Academic Press, New York, 1965.
- **Pridham, J.B.,** *Terpenoids in Plants,* Academic Press, New York, 1967.
- Proceedings of International Symposium on Plant Cell Culture (Alfermann, A.W. and Reinhard, E.Ed.) Muenchen, 1978.
- **Purohit S.S.,** *'A laboratory Manual of Plant biotechnology'*, Agro-botanical Publishers (India) 4E - 176 J.N. Vyas Nagar, Bikaner.
- *Quality Control Methods for Medicinal Plant Material,* Reprint 2002. World Health Organisation Geneva.
- **Ram P. Rastogi, B.N. Malhotra,** *Compedium of Indian Medicinal Plants* CDRI & PID New delhi,1995.
- **Ray Noggle G. and Fritz, G.J.,** *Introductory Plant Physiology*, Prentice Hall of India Pvt. Ltd., New Delhi, 1986.
- **Reid, R.A. and Leech, R.M.,** *Biochemistry and Structure of Cell Organelles, Blackie, Glasgow* - London, 1980.
- **Reinetrt, J. and Bajaj, Y.P.S.,** *Applied and Fundamental Aspects of Plant Cell, Tissue and Organ Culture*, Berlin, 1977.
- *Remington's Pharmaceutical Sciences, 17th Ed.*, Mack Publishing Company, Pennsylvania, U.S.A.
- *Report of the Task Force on Conservation and Sustainable Use of Medicinal Plants, Planning Commission*, Govt. of India, 2000.
- **Robinson, T,** *The Biochemistry of Alkaloids*, Springer - Verlag, New York,1968.
- **Rosenthaler, L.,** *The Chemical Investigations of Plants,* G. Bell and Sons Ltd., London, 1930.
- **Ross, M.S.F. and Brain, K.R.,** *An Introduction to Phytopharmacy*, Pitaman-Medical, Kent, 1977.
- **Santapau, H.,** *Common Trees,* National Book Trust, New Delhi, 1966.
- **Santra, S. C.** *Biology of Rice-Fields Blue-green Algae*, Daya Publishing House, New Delhi.
- **Schellard, E.J.,** *Exercises in the Evaluation of Drugs and Surgical Dressings,* Pitman Medical Publishing Co. Ltd., London.
- **Schellard, E.J.,** *Practical Plant Chemistry for Pharmacy Students*, Pitman Medical London, 1957.
- **Scheuer, P.J.,** Marine Natural Products, 1 - 2, Academic Press, London, 1973-81.
- **Sim, S.K.,** Medicinal Plant Glycosides, University of Toronto Publication.
- **Sinnott, E.W.,** *Dunn, L.C. and Dobzhansky, T., Principles of Genetics*, Tata McGraw Hill Publishing Co. Ltd., New Delhi, 1985.
- **Staba, E.J.,** *Plant Tissue Culture as a Source of Biomedicinals*, CRC Press Florida, 1980.
- **Stahl, E.,** *Thin Layer Chromatography - A Laboratory Handbook, Springer - Verlag*, Berlin, 1969.
- **Street, H.E.,** *Tissue Culture and Plant Science*, Academic Press, London, 1974.
- **Stumpf, P.K. and Conn, E.E.,** *The Biochemistry of Plants: A Comprehensive Treatise,* 1-8, Academic Press, 1980-81.

- **Swain, T.,** *Chemical Plant Taxonomy*, Academic Press, London, 1963.
- **Swain, T.,** *Comparative Phytochemistry*, Academic Press, London, 1966.
- *The Ayurvedic Pharmacopoeia of India, Vol. I* Part – 1, First Edition Rep – 2001. Vol. II (1999) Vol. III (2001). Government of India Ministry of Health and Family Welfare Department of Indian Systems of Medicine and Homeopathy, New Delhi.
- *The Hindu,* Survey of Indian Agriculture, Kasturi and Sons Ltd., Chennai, 1989.
- *The Merck Index,* 10th Ed., Merck & Co., Rahway, New Jersey, U.S.A., 1983.
- *The National Formulary,* 15th Ed., United States Pharmscopoeial Convention Inc., Rockville,
- *The Wealth of India,* Raw Materials, (All volumes), Council of Scientific and Industrial Research, New Delhi.
- **Trease, G.E. and Evans, W.C.,** Pharmacognosy, 12th Ed., Bailliere Tindall, Eastbourne, U.K, 1983.
- **Trease, G.E.,** Pharmacy in History, Bailliere Tindall and Cox, London, 1964.
- **Tropics and Vickery, M.P. and Vickery, B.,** Secondary Plant Metabolism, Basingstoke, Macmillan, 1981.
- **Tyler, V.C., Brady, L.R., and Robbers, J.E.,** *Pharmacognosy,* 8th Ed, Lea and Febiger, Philadelphia, 1981.
- **Tyler, V.E. Jr. and Schwarting, A.E.,** *Experimental Pharmacognosy*, Burgess Pub. Co., Minneapolis, Minnesota.
- *United States Pharmacopoeia,* 12th Revision, United States Pharmacopoeal Convention Inc, Rockville Meddison, U.S.A.1980.
- **Upadhyay S. N.,** *Immunomodulation – 1997,* Navosa Publishing House, New Delhi.
- **Ustimenko - Bakumovsky, G.V.,** Plants Growing In Substropics, Miv. Publishers, Moscow, 1982.
- **Wallis, C.J.** Practical Biology, 2, William Heinemann Medical Books Ltd., 1958.
- **Wallis, T.E.,** Analytical Microscopy, J & A Churchill Ltd., London.
- **Wallis, T.E.,** Text Book of Pharmacognosy, 5th Ed., J & A Churchill Ltd., London, 1967.
- **Weiss, T.J.,** Food Oils and Their Uses, 2nd Ed., Ellis Horwood Ltd., England, 1983.
- **Welsh, J.R.,** Fundamentals of Plant Genetics and Breeding, Wiley, New York, 1981.
- **Whistler, R.L.,** Industrial Gums, Polysaccharides and Their Derivatives, 2nd edition, Academic Press, New York 1973.
- **Williams Sidney,** Official Methods of Analysis, 14th edition., Association of Official Analytical Chemists Inc., Arlington, Virginia, U.S.A., 1984.
- **Youngken H.W.,** *Text book of Pharmacognosy* 6th edition, Philadelphia Blackiston – 1950.

INDICES

(A) BIOLOGICAL INDEX

(B) SYNONYM INDEX

(C) SCIENTIST INDEX

(D) CHEMICAL INDEX

(E) SUBJECT INDEX

(F) ANIMAL DRUG INDEX

Note:

Note: